GENOME DAMAGE

GENOME DAMAGE

Repair and Adaptation Across Life

Edited by

VINOD K. KANNAUJIYA

Department of Botany, Mahila Mahavidyalaya, Banaras Hindu University, Varanasi, Uttar Pradesh, India

GANESH K. MAURYA

Department of Zoology, Mahila Mahavidyalaya, Banaras Hindu University, Varanasi, Uttar Pradesh, India

RAJESH P. RASTOGI

Environment Impact Assessment (EIA), Environment Division, Ministry of Environment, Forest and Climate Change, Government of India, New Delhi, India

ELSEVIER

ACADEMIC PRESS

An imprint of Elsevier

elsevier.com/books-and-journals

Academic Press is an imprint of Elsevier

125 London Wall, London EC2Y 5AS, United Kingdom
50 Hampshire Street, 5th Floor, Cambridge, MA 02139, United States

ISBN: 978-0-443-42857-9

For Information on all Academic Press publications visit our website at https://www.elsevier.com/books-and-journals

Publisher: Megan Ball
Acquisitions Editor: Michelle Fisher
Editorial Project Manager: Shreya Jain
Production Project Manager: Vijayaraj Purushothaman
Cover Designer: Raman Kumar

Working together to grow libraries in developing countries
www.elsevier.com • www.bookaid.org

Typeset by Aptara, New Delhi, India

Contents

16. Metal stress, DNA damage, and mutagenesis: Molecular mechanisms and health consequences

Megha Dave and Raktim Mukherjee

17. Apoptosis: Cell death and damage

Nimisha Joly Chennakadan, Yadhu Krishna Menon K, Chandrabose Selvaraj, Yuvaraj Sivamani, and Sumitha Elayaperumal

18. Phytochemical modulation of DNA repair: Mechanisms, epigenetics, and therapeutic potential

Prasad Andhare, Khushbu Verma, Bhargavi Sonavane, Bhavtosh Kikani, Niraj Kumar Singh, and Datta Madamwar

19. Significance of algal green chemicals in repair mechanisms

Ram Lal, Megha Jaiswal, Sukriti Chauhan, and Vinod K. Kannaujiya

20. Phyco-chemicals in DNA repair: Algal and cyanobacterial antioxidants and epigenetic modulators in cellular defense

Bhargavi Sonavane, Stuti Nareshkumar Patel, Datta Madamwar, and Niraj Kumar Singh

21. DNA damage, aging concept: Feasibility and significance

Nimisha Joly Chennakadan, Yadhu Krishna Menon K, Chandrabose Selvaraj, Yuvaraj Sivamani, Kishan Nanga Srihari Kishan Reddy, and Sumitha Elayaperumal

22. Structural and dynamic signatures of DNA damage in *Leishmania* : Mechanisms, implications, and therapeutic perspectives

Aviral Kaushik, Naureenbanu Makarani, Krupanshi Bharadava, Sanjay Kumar, and Radhey Shyam Kaushal

23. Impacts of nanoparticles on DNA damage and repair

Amol Gholap, Pankaj Khuspe, Dipali Mane, and Md Jasim Uddin

24. DNA mutagenesis and genomic instability induced by heavy metals: Biological and environmental perspectives

Naureenbanu Makarani, Krupanshi Bharadava, Aviral Kaushik, Sanjay Kumar, and Radhey Shyam Kaushal

25. Significance of plant nutraceuticals in DNA repair

Rinkal Mulani, Chandani Kaushik, and Jitendriya Panigrahi

26. DNA damage and repair: Future perspectives

Yogendra Singh Rajpurohit, Aanchal Mishra, Dhirendra Kumar Sharma, and Ishu Soni

Contributors

Moyosoreoluwa Oluwaseun Abegunde Biological Sciences Department, Redeemer's University, Ede, Nigeria

Fatima-Zahra Akensous Center of Agrobiotechnology and Bioengineering, Research Unit Labelled CNRST (CentreAgroBiotech-URL-CNRST-05), Cadi Ayyad University (UCA), Marrakesh, Morocco

Nasreen Amin Department of Botany, Mahila Mahavidyalaya, Banaras Hindu University, Varanasi, Uttar Pradesh, India

Riya Anand Department of Zoology, Mahila Mahavidyalaya, Banaras Hindu University, Varanasi, Uttar Pradesh, India

Prasad Andhare Department of Life Sciences, School of Science, GSFC University, Vadodara, Gujarat, India

Krupanshi Bharadava Biophysics & Structural Biology, Research & Development Cell, Parul University, Vadodara, Gujarat, India; Department of Life Sciences, Parul Institute of Applied Science Research & Development Cell, Parul University, Vadodara, Gujarat, India

Sukriti Chauhan Department of Zoology, Mahila Mahavidyalaya, Banaras Hindu University, Varanasi, Uttar Pradesh, India; Department of Botany, Mahila Mahavidyalaya, Banaras Hindu University, Varanasi, Uttar Pradesh, India

Nimisha Joly Chennakadan Department of Biotechnology and Bioinformatics, JSS Academy of Higher Education and Research, Mysuru, Karnataka, India

Krishna Kumar Choudhary Department of Botany, Mahila Mahavidyalaya, Banaras Hindu University, Varanasi, Uttar Pradesh, India

Rikina Choudhury Department of Botany, Mahila Mahavidyalaya, Banaras Hindu University, Varanasi, Uttar Pradesh, India

Megha Dave Department of Biotechnology, Shree P M Patel Institute of Postgraduate Studies and Research in Science, Sardar Patel University, Anand, Gujarat, India

Sumitha Elayaperumal Department of Biotechnology and Bioinformatics, JSS Academy of Higher Education and Research, Mysuru, Karnataka, India

Amol Gholap Department of Pharmaceutics, St. John Institute of Pharmacy and Research, Palghar, Maharashtra, India

Débora Brito Goulart Department of Veterinary Microbiology and Preventive Medicine, Iowa State University, Ames, IA, United States

Muhammad Bakir Hussain Bioelementology and Human Ecology, I. M. Sechenov First Moscow State Medical University, Moscow, Russia

Nizakat Hussain Institute of Biological Sciences (IBS), Khwaja Fareed University of Engineering and Information Technology, Rahim Yar Khan, Pakistan

Megha Jaiswal Department of Botany, Mahila Mahavidyalaya, Banaras Hindu University, Varanasi, Uttar Pradesh, India

Yadhu Krishna Menon K Department of Biotechnology and Bioinformatics, JSS Academy of Higher Education and Research, Mysuru, Karnataka, India

Vinod Kumar Kannaujiya Department of Botany, Mahila Mahavidyalaya, Banaras Hindu University, Varanasi, Uttar Pradesh, India

Radhey Shyam Kaushal Biophysics & Structural Biology, Research & Development Cell, Parul University, Vadodara, Gujarat, India; Department of Life Sciences, Parul Institute of Applied Science Research & Development Cell, Parul University, Vadodara, Gujarat, India

Aviral Kaushik Biophysics & Structural Biology, Research & Development Cell, Parul University, Vadodara, Gujarat, India; Department of Life Sciences, Parul Institute of Applied Science Research & Development Cell, Parul University, Vadodara, Gujarat, India

Chandani Kaushik Department of Biotechnology, Shri Alpesh N. Patel Postgraduate Institute of Science and Research, Anand, Gujarat, India

Pankaj Khuspe Department of Pharmaceutics, Shriram Shikshan Sanstha's College of Pharmacy, Solapur, Maharashtra, India

Bhavtosh Kikani Department of Biological Sciences, P. D. Patel Institute of Applied Sciences, Charotar University of Science and Technology, Anand, Gujarat, India

Swathi Kota Molecular Biology Division, Bhabha Atomic Research Centre, Mumbai, Maharashtra, India; Life Sciences, Homi Bhabha National Institute, Mumbai, Maharashtra, India

Sanjay Kumar Structural and Molecular Biology Centre for Discovery and Innovation Building, The City College of New York, NY, United States

Ankita Kumari Department of Zoology, Mahila Mahavidyalaya, Banaras Hindu University, Varanasi, Uttar Pradesh, India

Ram Lal Department of Botany, Mahila Mahavidyalaya, Banaras Hindu University, Varanasi, Uttar Pradesh, India

Datta Madamwar Department of Biological Sciences, P. D. Patel Institute of Applied Sciences, Charotar University of Science and Technology, Anand, Gujarat, India

Naureenbanu Makarani Biophysics & Structural Biology, Research & Development Cell, Parul University, Vadodara, Gujarat, India; Department of Life Sciences, Parul Institute of Applied Science Research & Development Cell, Parul University, Vadodara, Gujarat, India

Chandan Malick Department of Zoology, Kalna College, Kalna, West Bengal, India

Dipali Mane Department of Pharmacology, Shriram Shikshan Sanstha's College of Pharmacy, Solapur, Maharashtra, India

Ganesh Kumar Maurya Department of Zoology, Mahila Mahavidyalaya, Banaras Hindu University, Varanasi, Uttar Pradesh, India

Nizar El Mazouni Center of Agrobiotechnology and Bioengineering, Research Unit Labelled CNRST (CentreAgroBiotech-URL-CNRST-05), Cadi Ayyad University (UCA), Marrakesh, Morocco; College of Agriculture and Environmental Sciences, Center of Excellence in Soil and Fertilizer Research in Africa (CESFRA), AgroBio-Sciences, University Mohammed VI Polytechnic (UM6P), Ben Guerir, Morocco

Abdelilah Meddich Center of Agrobiotechnology and Bioengineering, Research Unit Labelled CNRST (CentreAgroBiotech-URL-CNRST-05), Cadi Ayyad University (UCA), Marrakesh, Morocco; African Sustainable Agriculture Research Institute (ASARI), University Mohammed VI Polytechnic (UM6P), Laayoune, Morocco

Aanchal Mishra Molecular Biology Division, Bhabha Atomic Research Centre, Mumbai, Maharashtra, India; Life Sciences, Homi Bhabha National Institute (Department of Atomic Energy, Deemed to be University), Mumbai, Maharashtra, India

Shruti Mishra Molecular Biology Division, Bhabha Atomic Research Centre, Mumbai, Maharashtra, India; Life Sciences, Homi Bhabha National Institute, Mumbai, Maharashtra, India

Arup Mistri Fish Physiology Laboratory, Department of Zoology, The University of Burdwan, Burdwan, West Bengal, India

Raktim Mukherjee Department of Biotechnology, Shree P M Patel Institute of Postgraduate Studies and Research in Science, Sardar Patel University, Anand, Gujarat, India

Rinkal Mulani Department of Biochemistry, Shri Alpesh N, Patel Postgraduate Institute of Science and Research, Anand, Gujarat, India

Temitayo Ogundimu Department of Computer and Information Sciences, Covenant University, Ota, Nigeria

Arijit Pal Department of Zoology, Raiganj Surendranath Mahavidyalaya, University of Gour Banga, Uttar Dinajpur, West Bengal, India

Jitendriya Panigrahi Department of Biotechnology, Shri Alpesh N, Patel Postgraduate Institute of Science and Research, Anand, Gujarat, India

Stuti Nareshkumar Patel Małopolska Centre of Biotechnology, Jagiellonian University, Krakow, Poland

Yogendra Singh Rajpurohit Molecular Biology Division, Bhabha Atomic Research Centre, Mumbai, Maharashtra, India; Life Sciences, Homi Bhabha National Institute (Department of Atomic Energy, Deemed to be University), Mumbai, Maharashtra, India

Kishan Nanga Srihari Kishan Reddy Department of Biotechnology and Bioinformatics, JSS Academy of Higher Education and Research, Mysuru, Karnataka, India

Sahar Rostami Department of Pathobiology and Population Medicine, Mississippi State University, Starkville, MS, United States

Sourav Sardar Fish Physiology Laboratory, Department of Zoology, The University of Burdwan, Burdwan, West Bengal, India

Chandrabose Selvaraj Department of Marine Biotechnology, CSRDD Lab, Bioinformatics Division, AMET University (Deemed to be University), Kanathur, Tamil Nadu, India

Dhirendra Kumar Sharma Molecular Biology Division, Bhabha Atomic Research Centre, Mumbai, Maharashtra, India; Life Sciences, Homi Bhabha National Institute (Department of Atomic Energy, Deemed to be University), Mumbai, Maharashtra, India

Nandini Shreya Department of Biotechnology, School of Earth, Biological and Environmental Sciences, Central University of South Bihar, Gaya, Bihar, India

Ajeet Singh Department of Zoology, Mahila Mahavidyalaya, Banaras Hindu University, Varanasi, Uttar Pradesh, India

Karuna Singh Department of Zoology, Mahila Mahavidyalaya, Banaras Hindu University, Varanasi, Uttar Pradesh, India

Niraj Kumar Singh Department of Science and Technology, Gujarat Biotechnology Research Centre (GBRC), Government of Gujarat, Gandhinagar, Gujarat, India; School of Forensic Sciences, National Forensic Sciences University (NFSU), Nagpur, Maharashtra, India

Sindhuja Singh Molecular Biology Division, Bhabha Atomic Research Centre, Mumbai, Maharashtra, India; Life Sciences, Homi Bhabha National Institute, Mumbai, Maharashtra, India

Yuvaraj Sivamani Crescent School of Pharmacy, B.S. Abdur Rahman Crescent Institute of Science & Technology, Chennai, Tamil Nadu, India

Bhargavi Sonavane Department of Biological Sciences, P. D. Patel Institute of Applied Sciences, Charotar University of Science and Technology, Anand, Gujarat, India

Ishu Soni Molecular Biology Division, Bhabha Atomic Research Centre, Mumbai, Maharashtra, India; Life Sciences, Homi Bhabha National Institute (Department of Atomic Energy, Deemed to be University), Mumbai, Maharashtra, India

Himani Tewari Molecular Biology Division, Bhabha Atomic Research Centre, Mumbai, Maharashtra, India; Life Sciences, Homi Bhabha National Institute, Mumbai, Maharashtra, India

Md Jasim Uddin Department of Pharmaceutical Technology, Faculty of Pharmacy, Universiti Malaya, Kuala Lumpur, Malaysia

Khushbu Verma Department of Life Sciences, School of Science, GSFC University, Vadodara, Gujarat, India

About the editors

Dr. Vinod K. Kannaujiya is an Assistant Professor in the Botany Section, MMV, Banaras Hindu University, Varanasi, India. He received MSc and PhD degrees in Botany from the Centre of Advanced Study in Botany, Banaras Hindu University, Varanasi, India. He has been working on the ultraviolet stress response of cyanobacteria, specifically focusing on phycobiliproteins, photoprotective metabolites, and extracellular polymers. He has served as the Principal Investigator of high-impact research projects funded by UGC, CRG-SERB, IoE-BHU, and ANRF, India. He is life member fellow of Indian Crystallographic Society, Indian Science Congress, Indian Photobiology Society, Association of Microbiologists of India (AMI), and Society of Biological Chemists (SBC), alongside his active membership in the American Society for Microbiology (ASM). Dr. Kannaujiya has published a number of research papers and review articles in high-impact journals of international repute. He has also published book chapters and edited/authored three books in Elsevier/Springer.

Dr. Ganesh K. Maurya is an Assistant Professor in the Zoology Section, MMV, Banaras Hindu University, Varanasi, India. He received MSc degree in Zoology from the Centre of Advanced Study in Zoology, Banaras Hindu University, Varanasi, India, and PhD in Life Science from Bhabha Atomic Research Centre, HBNI, Mumbai, India. He has been working on genome maintenance under oxidative stress, including Gamma radiation, in the polyextremophile *Deinococcus radiodurans*. Besides this, he is also working on antibiotic resistance and phage therapy against pathogenic bacteria. He has received gold medals in MSc and PhD. He is a recipient of a prestigious "Young Scientist Award from the National Academy of Sciences, India." He is a recipient of UGC-Startup, SRG-SERB, and IoE project grants from various funding agencies. He is life member and fellow of the Society of Biological Chemists (India). He has published more than 25 original research papers, reviews, and book chapters. He was a Visiting Professor at the School of Biological Sciences, Ulsan National Institute of Science and Technology (UNIST), South Korea.

Dr. Rajesh P. Rastogi is currently working as a Scientist/Joint Director at the Ministry of Environment, Forest and Climate Change, New Delhi, India. He completed his PhD on UV-induced DNA Damage and Repair at Banaras Hindu University, Varanasi, India. Dr. Rastogi is the recipient of some national and international awards and fellowships. He was a Visiting Scientist at Friedrich Alexander University, Nuremberg, Germany, and served as a Visiting Professor of Biochemistry at Chulalongkorn University, Thailand. His major research areas are freshwater or marine algae and cyanobacterial ecology and biotechnology, plant physiology, photochemistry, photobiology, molecular microbiology, and stress biology. Dr. Rastogi has published a number of research papers and review articles in high-impact journals of international repute. He has also published several book chapters and five books from renowned international publishers (Elsevier/Springer/Nova). He is serving as an editorial board member of some national and international journals and is a life member of several scientific organizations.

CHAPTER

1

Genome research: Basic concept and recent advances

Débora Brito Goulart[a], *Muhammad Bakir Hussain*[b], *Moyosoreoluwa Oluwaseun Abegunde*[c], *Temitayo Ogundimu*[d], *Sahar Rostami*[e] *and Nizakat Hussain*[f]

[a]Department of Veterinary Microbiology and Preventive Medicine, Iowa State University, Ames, IA, United States [b]Bioelementology and Human Ecology, I.M Sechenov First Moscow State Medical University, Moscow, Russia [c]Biological Sciences Department, Redeemer's University, Ede, Nigeria [d]Department of Computer and Information Sciences, Covenant University, Ota, Nigeria [e]Department of Pathobiology and Population Medicine, Mississippi State University, Starkville, MS, United States [f]Institute of Biological Sciences (IBS), Khwaja Fareed University of Engineering and Information Technology, Rahim Yar Khan, Pakistan

1.1 Introduction

The genome represents "life" on a grand scale. Genome research deals with the organization, manipulation, and application of genetic material. This includes genome mapping, the building of genetic libraries, genome engineering, and recombinant deoxyribonucleic acid (DNA). The nomenclature used in this field is broad and can also apply to biotechnology in general. The study of genome research begins with an understanding of the genome itself. Picture an ancient universal language written not with ink, but with molecules. If one imagines nature as a language, the genome would be the most complex code ever created-a master script existing in every living being. The genome is the entire genetic material, DNA in most organisms, and ribonucleic acid (RNA) in some viruses, present in an organism. It encompasses coding regions (e.g., a set of genes) and noncoding sequences that might affect gene expression, replication, and cellular functions. The genome contains the information necessary for an organism to grow, develop, and reproduce (Venter et al., 2001).

Genome Damage: Repair and Adaptation Across Life.
DOI: https://doi.org/10.1016/B978-0-443-42857-9.00004-9

The mammalian genome is about 2 m long, and even though it fits perfectly within a nucleus that is only a few micrometers wide, the entire genome can fit in the spherical nucleus at the microscopic level. This compaction is achieved through supercoiling, which guarantees that chromatin, made of DNA and proteins, is closely packed and configured into a complex spatial structure. This architecture assures proper gene expression, DNA replication, and other cellular events (Magnitov and de Wit, 2024). The functional regions of the genome are referred to as genes. Genes are the basic units that direct and control the genetic traits of every organism. Genes are made of DNA sequences that code for proteins or RNA, which in turn serve as tools to control various traits. Genes have two key components: coding regions, which function like instruction manuals for producing proteins and RNA, and noncoding regions, which act like the control panels. These control panels have promoters and enhancers that decide how and when the instructions are used (Mansouri and Parker, 2025).

The size and complexity of genomes vary greatly among living organisms, as well as among different types of genomes within a single organism. Members of different groups, or even members of the same group, may have genomes that differ in size and complexity due to varying amounts of repetitive elements, such as transposable elements and satellite DNA. These repetitive elements may also contribute to variations in genome size within closely related species. Moreover, species may differ in the total number of genes or DNA segment copies, a variation classified as "copy number variation" (Kawaguchi et al., 2023). Intraspecies genome size differences can arise from environmental exposures and population dynamics, which can also affect genomic structure (Chapelle and Silvestre, 2022). The association between the complexity of biological life forms and genome size is not straightforward. An organism with a smaller genome can be more complex than one with a larger genome, which proves there is no simple biological paradigm connecting size and complexity hierarchies. A larger genome is often associated with a higher frequency of repetitive elements. These might serve as a springboard for evolutionary innovation or as a source of "raw material." Contrarily, small genomes tend to experience stronger purifying selection, creating an efficiently streamlined genetic structure (Simões et al., 2021).

This chapter will cover important aspects of genome research, including: the types of genomes (e.g., nuclear, mitochondrial, chloroplast, and plasmid); the central dogma of molecular biology (e.g., the roles of transcription and translation in gene expression, genetic variation and mutations, types of mutations); tools and technologies in genome research (e.g., DNA sequencing technologies, genome assembly and annotation, clustered regularly interspaced palindromic repeats [clustered regularly interspaced short palindromic repeats (CRISPR)/Cas9], bioinformatics); recent advances in genome research (e.g., high-throughput sequencing and big data, single-cell genomics, epigenomics, metagenomics, synthetic biology and genome design, pangenomics); applications of genome research across living organisms (e.g., humans, animals, plants, microbes); ethical, legal, and social implications of genome research (e.g., privacy concerns in genomic data, ethical issues in genome editing, regulatory frameworks for genomics technologies); and future directions in genome research. By the end of this chapter, the reader should have a broad understanding of the importance of genome research in improving the diagnosis, prevention, and treatment of diseases.

1.2 Types of genomes

Genomes vary across different forms of life. The nuclear, mitochondrial, chloroplast, and plasmid genomes are the commonly found types. The nuclear genome is the most extensive and complex of all genomes; it encodes the majority of an organism's genetic material. Eukaryotic cells have their nuclear genome organized into linear chromosomes within the nucleus and contain both coding (exons) and noncoding (introns) DNA. Typically, each cell contains two copies of nuclear DNA (Lim, 2024). One of the recent discoveries about the nuclear genome is that it is dynamic, with the aid of long-read sequencing technologies, which have enabled the characterization of structural variations and repetitive elements that were previously undetected yet are important for evolution and disease (Smith, 2020). Moreover, epigenetic modifications, such as DNA methylation and histone acetylation, might influence pooled gene expression without altering the original nucleotide sequence (Cedar and Bergman, 2009).

Because it encodes important subunits of the respiratory chain responsible for oxidative phosphorylation and thus vital for energy conversion, the mitochondrial genome (mtDNA) is frequently referred to as the powerhouse of the cell. It is located in the mitochondrial matrix and is surrounded by a double-layered membrane. Interestingly, mtDNA copy number can reach up to 100,000 copies per cell (Rackham and Filipovska, 2022). The human mtDNA has 16,569 base pairs (bp) and is a double-stranded circular DNA molecule that carries 37 genes essential for oxidative phosphorylation (OXPHOS). These genes include two ribosomal RNAs, 22 transfer RNAs required for mitochondrial translation, and 13 protein-coding genes that make up subunits of the OXPHOS (Kopinski et al., 2021).

Chloroplasts enable plants to carry out photosynthesis. Within these chloroplasts resides self-regulating genetic material known as chloroplast DNA (cpDNA). Usually composed of a circular molecule approximately 150 kb in size, cpDNA contains between 120 and 130 genes that produce 50–200 proteins. The cpDNA's structure includes a large single-copy area and two inverted repeat regions (A and B) (Liu et al., 2024). Highly conserved in both content and structure, the plant chloroplast genome provided insightful information for evolutionary research and genetic improvement initiatives (Yu et al., 2024). Maternally inherited, cpDNA has multiple copies per cell and remains haploid, with no recombination. Because it contains extensive genetic information, cpDNA enables the creation of large databases that are essential for comparative research (Li et al., 2021a).

Plasmids are small, circular pieces of DNA found in bacteria and some eukaryotes. They carry genes that allow bacteria to resist antibiotics, cause disease, or perform various metabolic functions, so they can respond quickly to environmental challenges. Currently, scientists are using CRISPR technology to track these plasmids and combat antimicrobial resistance (AMR) (Moitra et al., 2024).

1.3 Central dogma of molecular biology

The processes of transcription, translation, and DNA replication are defined by the fundamental tenet of molecular biology (Crick, 1970). All genetic information is stored in a four-nucleotide genetic alphabet (e.g., guanine [G], cytosine [C], adenine [A], and thymine [T]) and is deciphered in groups of nucleotide triplets, or codons, according to the genetic code. Here, we discuss the factors related to transcription and translation in gene expression, mutations, and genetic variability.

1.3.1 Role of transcription and translation in gene expression

One of the principles in molecular biology is the model of transcription and translation commonly referred to as the central dogma. Gene expression is a complex process that turns genetic information encoded in DNA into proteins essential for cellular function and, ultimately, survival. Gene expression consists of two main processes: transcription, in which messenger RNA (mRNA) is synthesized from DNA, and translation, in which amino acids are assembled into proteins using the mRNA as a template. Although transcription and translation are separate processes, both are tightly regulated and coordinated to allow efficient and accurate gene expression. Transcription refers to the process by which RNA polymerase synthesizes mRNA from a template strand of DNA. In prokaryotes, transcription occurs in the cytoplasm, whereas in eukaryotes, it occurs in the nucleus. In contrast, translation is the process by which ribosomes synthesize proteins from mRNA templates. In prokaryotes, translation and transcription are often coupled, with ribosomes initiating translation of mRNA before transcription is complete. This is referred to as transcription–translation coupling. Transcription and translation are crucial processes that occur in living cells, enabling them to respond to external environmental changes through gene expression optimization and protein synthesis regulation. Malfunctions in transcription and translation are associated with various diseases, including cancer and neurodevelopmental disorders (Smith, 2020). Both processes convert gene-encoded information into proteins that have important roles in cell structure, function, and regulation.

1.3.2 Genetic variation and mutations

Genetic variation in living organisms is defined as the differences among individuals of the same species that may be expressed genetically or morphologically. Variation is a key factor in biological diversity and may determine ecological interactions and evolutionary trajectories. It is also essential in medical practice, as it influences the understanding and treatment of genetic diseases using tools such as personalized medicine (Zhang et al., 2025b). Genetic variation may be caused by mutations, which are random changes in DNA, or by recombination, which creates new allele combinations (Johnston, 2024). Gene flow due to migration and epigenetic modifications caused by chemical changes can also drive genetic variation (Chapelle and Silvestre, 2022).

Current research in molecular biology has focused on understanding mutation rates, which are key to medical genetics and conservation biology. Genetic mutations and variations are the primary causes of diversity between species. Mutations, which provide genetic variation, are generally noted to be heritable changes in genetic material that do not involve recombination. These heritable changes, which can involve single-nucleotide changes or alterations to entire chromosomes, can be influenced by several factors, including genome size, population size, and environmental factors (Johnston, 2001). Animal mitochondrial genomes often have more mutations than those of plants. Mutations can have a range of effects on organisms, including neutral, beneficial, or negative outcomes. However, most mutations are neutral (Franjić, 2023).

1.3.3 Types of genetic mutations

There are different types of genetic mutations based on their size and structure, functional impact, and origin. Genetic mutations based on size and structure are classified as point mutations, insertions and deletions (indels), and large-scale mutations. Point mutations, which occur basically as a single-nucleotide change, occur in various forms. Examples of point mutations include silent mutations, in which the nucleotide change does not alter the amino acid sequence (Yamasaki et al., 2025). A missense mutation is another type of point mutation, in which a nucleotide change leads to the substitution of one amino acid for another. For instance sickle cell anemia disease is caused by a mutation in which the codon GAG is changed to GTG, resulting in the replacement of acid glutamic acid with valine (Tsukahara et al., 2024). A nonsense mutation, on the other hand, creates a premature stop codon, which truncates the protein. Cystic fibrosis is an example of a disease caused by a nonsense mutation (Titoli et al., 2025). Insertions and deletions are caused by the insertion or deletion of nucleotides. These are divided into two major types. Frameshift mutations are insertions or deletions that do not occur in multiples of 3 and are capable of shifting the reading frame; Huntington's disease is a classic example (Purohit et al., 2025). In-frame mutations are insertions or deletions that occur in multiples of three and do not cause frameshifts (Tien et al., 2024). Large-scale point mutations may include duplications, where a DNA segment is copied multiple times; deletions, involving the loss of a DNA segment; inversions, where a DNA segment is reversed; and translocations, which occur when DNA segments move between nonhomologous chromosomes (Mendiratta et al., 2021).

Regarding genetic mutations based on functional impact, a mutation can lead to a loss of function when protein activity is reduced or eliminated, as in the case of p53 mutations in cancer. Alternatively, it may result in a gain of protein function, such as in RAS oncogene mutations (Zhang et al., 2025a). There are also neutral mutations, which generally do not affect the organism (Monteiro and Aguiar, 2021). Regarding genetic mutations based on origin, when they occur in the germline, they affect reproductive cells and can be transmitted to the next generation. However, when they occur in somatic cells, they are not inherited. They could also occur spontaneously or be induced by environmental factors (Johnston, 2001). Transcription, translation, and genetic mutations are molecular processes that shape life's blueprint. Variations and errors in DNA not only lead to evolution but also cause disease. Interestingly, even a single altered nucleotide can lead to significant changes, revealing the dynamic and fragile nature of gene expression.

1.4 Tools and technologies in genome research

Genome research technologies offer significant advantages, enabling rapid, accurate, and large-scale analysis of genetic material. In this session, we discuss genome research technologies such as sequencing methods, genome assembly and annotation, CRISPR analysis, and computational tools.

1.4.1 DNA sequencing technologies

DNA sequencing has developed through three major "generations," defined by methodological innovations and throughput, read length, and core application. Together, these technologies, including Sanger sequencing, next-generation sequencing (NGS), and third-generation sequencing (TGS), have transformed genomics, facilitating larger, more accurate, and cheaper DNA analyses (Satam et al., 2023; Warburton and Sebra, 2023). Started in 1977 the Sanger (chain-termination) sequencing method is the conservative technique used for precise and error-free, but low-throughput applications. It uses dideoxynucleotide triphosphates (ddNTPs) to terminate DNA sequence synthesis at specific bases, followed by capillary electrophoresis to separate the resulting pieces (Crossley et al., 2020). Typically, read lengths range from 600 to 1000 bp, with an accuracy of more than 99.99%. Thus Sanger sequencing is ideal for the validation of specific mutations, confirmation of clinical mutations, and analysis of plasmids or forensic analyses (Akintunde et al., 2024). The most significant disadvantages of this approach are its high cost per base and a total capacity of only a few kilobases per run, making it unrealistic for large-scale or whole-genome projects (Akintunde et al., 2024).

Since the mid-2000s the democratization of whole-genome, transcriptome, and epigenome analyses has been driven by the massive parallelization of DNA reads of short sizes (e.g., 100–300 bp) enabled by NGS platforms. Illumina's sequencing-by-synthesis is the most widely used technology, routinely generating more than 1 billion paired-end reads (2 × 150 bp) per run, with Q30 error rates higher than 85% (Satam et al., 2023). However, other chemistries exist, catering to niche applications. The NGS workflow consists of DNA fragmentation, adapter ligation, clonal amplification, and cyclic detection of nucleotide incorporations. Downstream bioinformatics pipelines include alignment, variant calling, and annotation (Zhong et al., 2021). Despite limitations with structural variant resolution and repetitive regions without complex assembly approaches, the speed and cost reduction of NGS (approximately U.S.$5 per gigabase [gb]) support large-scale activities in population genomics, rare variant discovery, and clinical diagnostics (Warburton and Sebra, 2023).

Platforms based on TGS have the capability of generating read lengths ranging from tens to more than 100 kilobases (kb) by sequencing single DNA molecules directly, thus overcoming short-read constraints. Pacific Biosciences single-molecule real-time (SMRT) sequencing can produce high-fidelity ("HiFi") circular consensus reads (~ 15–20 kb, >99.9% accuracy) using sequencing templates that are continuously closed in zero-mode waveguides (Espinosa et al., 2024). Oxford Nanopore Technologies (ONTs) instruments sequence native DNA or RNA molecules through protein nanopores, which read ionic current disruptions in real time and simultaneously detect base modifications (e.g., 5-methylcytosine) (Lin et al., 2021). Although raw error rates remain high (PacBio~1%; ONT~5%–15%), ongoing chemistry and base-calling improvements are narrowing the gap (Scarano et al., 2024). TGS offers effective *de novo* assembly, haplotype phasing, and structural-variant analysis that allow the process completion without any gap and perform real-time pathogen surveillance in the field. However, per-sample costs and the need for specialized analysis may limit its adoption compared to NGS (Cuber et al., 2023; Warburton and Sebra, 2023).

The decision to use Sanger, NGS, or TGS mainly relies on the objectives of a project, financial resources, sample size, and the required resolution. Sanger sequence is necessary for validating small-scale samples; NGS is useful for population-level detection and cost-efficiency, whereas TGS remains a solution for resolving complex genomic regions. Convergent approaches that integrate both short- and long-read datasets are becoming increasingly popular, since they associate the high accuracy of NGS with the long-range contiguity of TGS to improve genome assembly and variant discovery (Murigneux et al., 2020). Table 1.1 shows the main sequencing methods used in genome research.

1.4.2 Genome assembly and annotation

Assembly and annotation of a genome are the processes that transform the raw sequence data into a biologically meaningful reference, which is the foundation for subsequent research in comparative and functional genomics (Del Angel et al., 2018; Jung et al., 2020).

1.4.2.1 Genome assembly

Genome assembly reconstructs contiguous sequences (e.g., contigs) from millions of sequencing reads and further orders them into scaffolds that approximate chromosomes (Del Angel et al., 2018). Two principal strategies are widely employed (Table 1.2).

TABLE 1.1 Overview of sequencing methods.

Feature	Sanger sequencing	Next-generation sequencing (NGS)	Third-generation sequencing (TGS)
Principle	Chain-termination/ddNTP	Massively parallel sequencing	Single-molecule long-read sequencing
Read length	800–1000 bp	100–300 bp (short-read)	>10 kb (TGS); up to 100 kb+ (ONT)
Throughput	Low (one/few sequences)	High (millions to billions)	Moderate to high (gigabase scale)
Accuracy	Very high (99.99%)	High (now >99% for Illumina)	Improving, high per run with consensus (PacBio HiFi >99%)
Applications	Targeted sequencing, validation, microbial ID, clinical diagnostics	Whole genome/exome, RNA-seq, metagenomics, clinical diagnostics	Structural variant detection, de novo assembly, epigenetics, and direct RNA sequencing
Limitations	Low throughput, expensive for large-scale, and limited scalability	Short-read length, assembly/structural variant challenges	Higher raw error rates[a], cost/efficiency for very high throughput[a]

ddNTP, dideoxynucleotide triphosphate; *ONT*, Oxford Nanopore Technology; *RNA*, ribonucleic acid.
[a] Error rates and throughput vary with platform and chemistry; improvements are ongoing (Crossley et al., 2020; Athanasopoulou et al., 2021; Satam et al., 2023).

TABLE 1.2 Characteristics of different assembly types.

Assembly type	Use case	Advantages	Limitations
De novo	Novel genomes	No reference bias, detects structural variants	Computationally intensive
Reference-guided	Re-sequencing	Faster, less resource-heavy	Lack novel regions
Hybrid	Complex genomes	Balances accuracy and completeness	Higher cost

- *De novo* assembly: performed when no reference genome is present, uses either the overlap-layout-consensus or de Bruijn graph algorithms. The algorithms are utilized by assemblers, including SPAdes for short reads and Canu for long reads. These approaches are used to generate contigs directly from read overlaps (Wajid and Serpedin, 2016).
- Reference-guided assembly: aligns reads to a closely related genome, facilitating scaffolding and gap filling, and is widely used in resequencing projects (Ghurye and Pop, 2019).

Combining highly accurate Illumina short reads with longer but lower consistency reads from platforms such as PacBio or Oxford Nanopore results in hybrid assemblies that better resolve repeats and structural variants (Ghurye and Pop, 2019). Preprocessing steps, such as quality

TABLE 1.3 Challenges associated with genome assembly.

Challenges	Description	Impact on assembly	Mitigation strategies
Repetitive sequences	Highly similar repeats longer than the read length cause ambiguous assembly	Fragmentation, misassembly, and collapsed repeats	Long reads, paired-end reads, and scaffolding data
Sequencing errors	Errors in reads create false overlaps or break true overlaps	Erroneous contigs, fragmented assembly	Error correction, hybrid assembly
Polymorphism/heterozygosity	Allelic variation misinterpreted as errors or distinct loci	Haplotypes collapse, misassembly	Haplotype-aware assemblers, phased assembly
Genome size/complexity	Large, complex genomes require high-quality DNA and computational power	Incomplete assemblies, resource-intensive	Improved DNA extraction, high-performance computing
Fragmented assemblies	Short reads and repeats cause many small contigs and gaps	Incomplete gene models, missing regions	Integrative scaffolding, long-read sequencing
Bioinformatics limits	Algorithms cannot fully resolve all complexities	Lower assembly quality, need for manual curation	Development of new algorithms, multidata integration

filtering and error correction (e.g., Trimmomatic), are used to increase the read accuracy (El-Metwally et al., 2013). During the graph construction process, read compatibility is modeled based on overlaps or k-mer connectivity, and scaffolding methods (e.g., paired-end links, 10x Genomics barcodes, or Hi-C chromatin contacts) are applied to infer long-range genomic structure (Ghurye and Pop, 2019).

Obstacles are encountered during genome assembly (Table 1.3). One crucial problem is the presence of repetitive genomic elements, which comprise more than 60%–80% of eukaryotic genomes. This repetitiveness leads to confusion in graph structures, which induce gaps or misassemblies when not covered by ultra-long reads (Claros et al., 2012; Uengwetwanit et al., 2021). Also, long-read platforms with high error rates give rise to false overlaps, which require further polishing steps to be eliminated (Aury and Istace, 2021). Moreover, the presence of heterozygosity in diploid or polyploid genomes is an additional challenge in the assembly process, since divergent alleles may be collapsed or misassigned without proper phasing (Collins, 2018). Finally, the assembly of large genomes is challenging when computational resources are limited. The process generally requires the integration of optical maps and genetic linkage data to reach chromosome-scale contiguity.

1.4.2.2 Genome annotation

Annotation is the process that identifies genes and other functional elements in assembled sequences, transforming contigs into maps of coding and noncoding features (Li et al., 2021b). The principal steps are as follows: (1) repeat masking, a method that excludes transposable

TABLE 1.4 Tools for annotating genes and functional elements.

Tool/platform	Primary function	Organism focus	Key features
GeneMark	Gene prediction	Prokaryotes, eukaryotes	Hidden Markov model (HMM)-based, integrates with pipelines
Augustus	Ab initio gene prediction	Eukaryotes	Species-specific training
Prodigal	Prokaryotic gene prediction	Prokaryotes	Fast, accurate
RNAmmer	rRNA gene prediction	All	HMM-based, rapid
GenSAS	Structural and functional annotation	Eukaryotes, prokaryotes	Pipeline with visualization
NCBI Genome Annotation Pipeline	Automated gene and functional annotation	Eukaryotes	Integrates multiple data types
RASTtk/VIGOR4	Bacterial and viral annotation	Bacteria, viruses	Subsystem-based metabolic annotation
MAKER/MAKER-P	Integrated annotation pipeline	Plants, animals	Combines ab initio and evidence-based methods
Artemis	Visualization and manual curation	All	Genome browser with editing capabilities
Geneious	Sequence analysis and annotation	All	Automated annotation, ORF prediction

NCBI, National Center for Biotechnology Information.

elements to prevent false gene predictions, (2) evidence alignment, the process of mapping RNA-seq reads, expressed sequence tags, and protein homologs to define exon-intron boundaries, (3) gene prediction, the strategy of consolidating ab initio tools (e.g., AUGUSTUS) with homology-based methods to unveil the coding sequences (Brůna et al., 2021), (4) functional assignment, which labels predicted genes with Gene Ontology terms, protein domains via InterPro, and pathway associations using the UniProt and eggNOG databases (Li et al., 2021b).

Several tools have been created to aid in the annotation of genes and functional elements (Table 1.4). Automated pipelines such as BRAKER2 and BRAKER3 integrate RNA-seq and protein evidence to refine gene models iteratively, achieving high sensitivity even in complex eukaryotic genomes (Brůna et al., 2021). In prokaryotes, the National Center for Biotechnology Information's (NCBI) PGAP and tools such as Bakta leverage curated protein family models for rapid, high-confidence annotation (Schwengers et al., 2021). Despite automation, alternative splicing, noncoding RNAs, and lineage-specific gene models often require manual curation to resolve discrepancies. Modern integrative pipelines generate standardized outputs (e.g., GFF3 and JSON) and leverage multiomics data, driving more accurate and comprehensive genome annotations. Together, advances in long-read sequencing, hybrid assembly strategies, and evidence-rich annotation pipelines continue to improve the quality and utility of genome references across all domains of life.

1.4.3 CRISPR-Cas9 and genome editing

Genome editing technologies have changed the way scientists conduct biological research by making the DNA modification across a variety of organisms more precise, effective, and cost-effective. Previous methods, such as zinc-finger nucleases and transcription activator-like effector nucleases, have established a groundwork. However, CRISPR/Cas systems have become the predominant tool, mainly because of their programmability via RNA–DNA base pairing, simplicity of design, and scalability for multiplex editing (Hille and Charpentier, 2016; Manghwar et al., 2020).

The CRISPR-Cas9 system consists of two highly important components: a Cas9 endonuclease, which breaks the target DNA by introducing a double-strand break (DSB), and a single-guide RNA (sgRNA), which directs Cas9 to a 20-nucleotide genomic locus immediately upstream of an "NGG" protospacer adjacent motif (PAM) (Allen et al., 2021). After PAM detection and RNA–DNA hybridization, Cas9's HNH and RuvC nuclease domains each cleave one DNA strand through a two-metal (Mg^{2+})-dependent mechanism, generating a site-specific DSB (Casalino et al., 2020). The cell's endogenous repair pathways then resolve the break either by nonhomologous end joining (NHEJ), which often yields small insertions or deletions like indels that disrupt gene function, or by homology-directed repair (HDR), when a donor template is provided, allowing precise insertions or corrections (Hille and Charpentier, 2016; Asmamaw and Zawdie, 2021). The CRISPR innovation landscape has been enriched by new tools. Two notable recent advances are base editors and prime editors. Base editors are Cas9 mutants fused with deaminase enzymes, which enable the conversion of single nucleotides without inducing DSBs, greatly reducing the risk of unwanted insertions or deletions (Nadakuduti and Enciso-Rodríguez, 2021). Prime editing employs a reverse transcriptase-Cas9 nickase complex, the latter of which writes new DNA sequences directly at the target site. This method can introduce insertions, deletions, and all 12 possible point mutations with minimal off-target effects (Bosch et al., 2021). Furthermore, engineered Cas variants that alter PAM recognition specificity or enhance editing precision can further broaden the scope of target sites and minimize off-target effects (Aschenbrenner et al., 2020).

Currently, CRISPR-Cas9 is regarded as the most impactful genome editing technology in biosciences, allowing life science innovations that are transforming existing infrastructure and future development trajectories (Wang and Doudna, 2023). Nevertheless, the construction and synthesis of CRISPR components remain a primary hurdle, mostly in agricultural and therapeutic contexts. Whereas plasmids or viral vectors have been extensively utilized, ribonucleoprotein (RNP) complexes, which contain preassembled Cas9 protein and single-guided RNA (sgRNA), exert a transient, nonpermanent mode of action that lessens the rate of gene silencing—nuclease expression, off-target effects, and gaps of faulty cuts (Zhang et al., 2021a). Recent advances in nanotechnology have produced lipid nanoparticles, exosomes, and polymer-based carriers that improve cellular uptake, tissue targeting, and immune compatibility (Duan et al., 2021; Ahmar et al., 2021).

In plants, advanced allele replacements and targeted knockouts have led to the development of crops with increased stress tolerance, enhanced yields, and better nutritional value, such as salinity-resistant soybeans developed through the multiplexed editing of *AITR* genes (Liu et al., 2021; Wang et al., 2021). In animal models, CRISPR greatly accelerates the generation of disease-specific knockout mice and enables agricultural enhancement in aquaculture, cattle, and poultry (Khwatenge and Nahashon, 2021). Microbial strains optimized for industrial purposes, such as biofuel, enzymes, and pharmaceuticals, enhance production through cryptic biosynthetic

TABLE 1.5 CRISPR-Cas9 applications across domains.

Domain	Applications	Outcomes
Plants	Trait improvement, disease resistance, stress tolerance, and yield enhancement	Salinity-tolerant soybean, disease-resistant wheat (Wang et al., 2021)
Animals	Disease models, therapeutic correction, agricultural and biopharma improvement	Knockout models, disease gene correction, universal CAR-T-cells (Li et al., 2020)
Microorganisms	Secondary metabolite production, industrial strain optimization, gene function studies	Enhanced antibiotic production, novel bioactive compounds (Jiang et al., 2021)
Human Medicine	Somatic gene therapy, ex vivo correction, diagnostic tools, functional genomics	Sickle cell therapy, immunotherapy advances (Wang and Doudna, 2023)

pathways, particularly in fungi (Jiang et al., 2021). In clinical practice, CRISPR is advancing ex vivo therapies, including the correction of sickle cell disease in hematopoietic stem cells. Moreover, CRISPR is entering early in vivo trials for monogenic disorders, albeit with ongoing concerns over delivery, immunogenicity, and ethical governance (Ayanoğlu et al., 2020; Wang and Doudna, 2023). Table 1.5 shows the CRISPR-Cas9 applications among humans, animals, plants, and microorganisms.

As CRISPR technologies continue to evolve, along with advances in guide RNA (gRNA) design informed by machine learning (ML), improvements in delivery platforms, and rigorous safety evaluations, will greatly assist in maximizing the potential of genome editing. The ongoing development of strong ethics frameworks and regulatory oversight will be necessary to safeguard responsible use and public trust.

1.4.4 Bioinformatics in genome research

Bioinformatics has become inextricably associated with genome studies, ranging from bacterial genomes to human-directed therapeutic whole-exome or whole-genome sequencing (WGS). It transforms streams of sequence data into usable biological information through computational pipelines, databases, and visualization tools. Because high-throughput sequencing creates terabytes of new data daily, efficient and reliable algorithms that warrant reproducibility, particularly for read mapping, functional interpretation, and all intermediate steps, are paramount (Sayers et al., 2022; Deshpande et al., 2023). At its core, a bioinformatics workflow consists of sequence alignment, which is performed using software such as BWA and Bowtie (Madeira et al., 2024). This step identifies conserved regions by mapping short reads to a reference genome, forming the foundation for downstream analyses. Variant callers, such as GATK, FreeBayes, and vclib, have advanced variant detection by using filtering and normalization methods to detect single-nucleotide polymorphisms (SNPs), insertions and deletions (e.g., indels), and structural variants in both population and clinical genomics. Genomic assembly tools such as SPAdes and JCVI establish long contiguous sequences from short, often fragmented, reads. These technologies use graph-based genome assembly algorithms with quality metrics to address repetitive elements, sequencing errors, and incomplete coverage (Tang et al., 2024). Annotation pipelines, including BRAKER, MAKER, and PGAP from NCBI, use a combination of *ab initio* gene prediction, homology-based evidence, and transcript evidence to generate catalogs of coding regions, regulatory motifs, and noncoding RNAs (Cotto et al., 2023).

Primary resources such as GenBank, the European Nucleotide Archive, and the DNA Data Bank of Japan operate synchronized global nucleotide repositories that store more than 25 trillion bps of data (Sayers et al., 2022). Ensembl augments these data with deep annotations for vertebrates and model organisms and offers the variant effect predictor application to predict mutation impact (Harrison et al., 2024). The NCBI aggregates gene-centric (e.g., Gene), variation (e.g., dbSNP, Clinvar), and phenotypic (e.g., dbGaP) databases under a single set of APIs, while genome browsers (e.g., GDV, UCSC) enable the exploration of assemblies, annotations, and multiomics tracks in an interactive environment (Rangwala et al., 2021). Web-based, zero-code platforms such as OmicShare, Sangerbox, and Galaxy, employ drag-and-drop workflows, interactive visualizations, and built-in statistical tools to connect users without programming backgrounds to the growing volume of bioinformatics data (Shen et al., 2022; Chen et al., 2024; Mu et al., 2024). These environments contain hundreds of modules, from read preprocessing to pathway enrichment, which significantly decrease the need for hypothesis testing and promote reproducibility.

Recent developments utilize deep learning architectures, such as transformers and graph neural networks, to enhance feature extraction for various tasks, including enhancer discovery, pathogenic variant impact interpretation, and pan-genome profiling (Choi and Lee, 2023). Tools, such as Augur, allow real-time phylogenetic analysis of pathogen outbreaks, whereas R-based ecosystems, such as Bioconductor, support statistical modeling and visualization of expression data in a comprehensive manner (Huddleston et al., 2021; Giorgi et al., 2022). Furthermore, spatial genomics methods, such as in situ genome sequencing (IGS) and Hi-C, leverage cytometry-based sequencing analysis to map chromatin architecture at submicrometer resolution, leading to in situ discovery of three-dimensional genome structures (Payne et al., 2021; Wen and Tang, 2022). Visualization frameworks like Gosling.js and NGenomeSyn help with the representation of synteny and variant landscapes in a customizable, scalable, web-native format, while interactive tree-view packages such as ggtreeExtra synthesize phylogenies with metadata layers (Lyi et al., 2021; He et al., 2023).

Genome science has evolved within an ecosystem built on the integration of emerging molecular biology applications, high-throughput and NGS techniques (e.g., nanopore, prime editing), new sample preparation, automated workflows, and scalable data storage and management infrastructure. New bioinformatics pipelines, for sequence alignment, variant calling, assembly, annotating, and data visualization, operate either via web applications or machine-learned supervised algorithms, ensuring continual data stewardship and dynamic interpretation of large genomics datasets. Thus the ecosystem not only provides efficiency across stages of discovery but also enables application and translational efforts in genetics, evolution, and medicine. It continues to grow through innovations such as deep learning and interactive visualization, contributing to the accuracy, accessibility, and impact of genomics for scientists and clinicians.

1.5 Recent advances in genome research

In recent years, genome research has experienced remarkable advancements across several domains. This session delves into many pivotal areas.

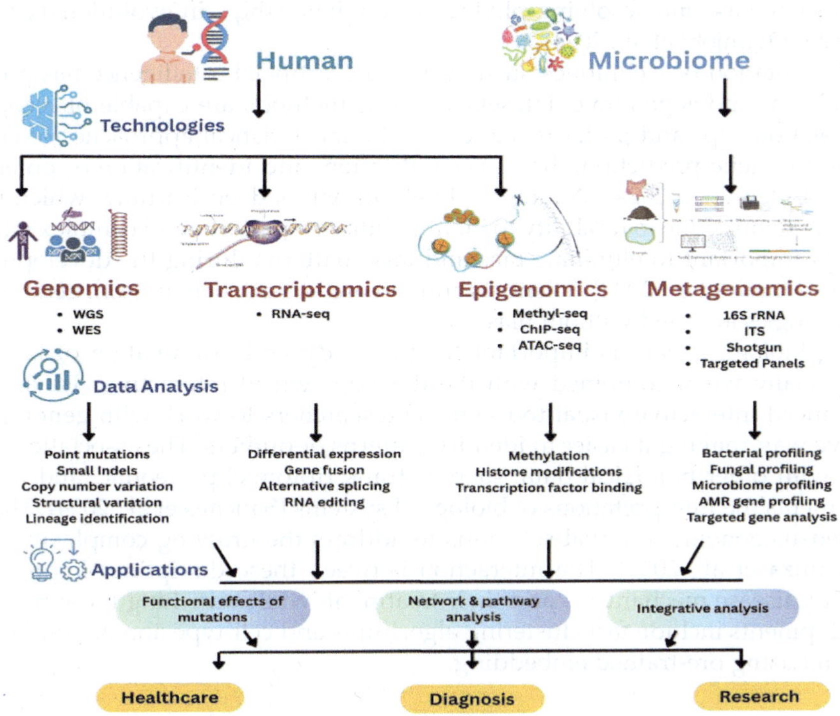

FIGURE 1.1 A summary of NGS technologies and their applications with different techniques.

1.5.1 High-throughput sequencing and big data

NGS technologies are changing the landscape of genomics by delivering unprecedented capability for high-throughput, rapid DNA sequencing. This technology has impacted genomics research, as researchers can now access genomic information more quickly and with greater precision than in previous decades (Satam et al., 2023). NGS has widened the horizons of biological and medical research through applications such as WGS, exome sequencing, RNA sequencing, and metagenomics (Horgan et al., 2024). As illustrated in Fig. 1.1, the scope of NGS technologies includes diverse techniques that are important for genome elucidation and clinical diagnosis.

The data produced by NGS technologies is extensive and poses considerable challenges related to storage, processing, and analysis (Satam et al., 2023). NGS data are referred to as large-scale genomic datasets, which require significant computational infrastructure and bioinformatics tools. Approaches for large-scale datasets processing and analysis usually include the development of efficient methods and algorithms for data compression, alignment, variant detection, and annotation (Abdi et al., 2024). NGS data analysis is increasingly approached with high-performance computing clusters, resource management systems, cloud computing, and other computational mechanisms. Research in statistical and ML methods, as well as analytic methods for large genomic datasets, is particularly important in the context of such analysis. Identifying genes linked to certain diseases, predicting physiological responses to pharmaceutical agents, and resolving phylogenetic relationships in evolutionary biology is a prime example (Ogunjobi et al., 2024).

The implementation of techniques such as ML and artificial intelligence has greatly eased the work with enormous genomic datasets, as these methods are capable of recognizing sophisticated relationships and patterns that conventional statistical approaches often miss. ML can be applied to gene prediction, disease classification, and identification of potential drugs (Adeyanju and Ogunjobi, 2024). Another subfield of ML is deep learning, which has gained popularity in genomics due to its ability to identify intricate patterns and complex relationships. It is extremely important to eliminate bias and false patterns during the development of ML algorithms (D'Amour et al., 2022). Scientists and researchers can benefit from genomic analyses if these technologies are used without bias.

Integrating biological data is important for the study and examination of large genomic datasets, especially when combined with the development of interactive visualization techniques. Advanced interactive visual tools enable researchers to work with genomic data in a more intuitive way, making it easier to identify patterns or outliers. The association of genomic information with other biological data, such as transcriptomic, proteomic, and metabolomic data, supplements the interpretations of biological systems (Sanches et al., 2024). The construction of consensus genomic interval sets aims to address the growing complexity of genomic datasets (Rymuza et al., 2024). The interaction between these disciplines improves our understanding of disease mechanisms and their treatment, while enriching these research fields. Recent developments include fast clustering algorithms and cell-type annotation for single-cell ATAC-seq data using pre-trained embedding.

1.5.2 Single-cell genomics

Transformative breakthroughs in cell biology have stemmed from single-cell genomics. Exploring cellular heterogeneity, rare cell identification, and active cellular processes has become possible due to advanced single-cell genomic technologies such as scRNA-seq, ATAC-seq, and single-cell proteomics (Chappell et al., 2018). In addition, single-cell genomics techniques enable multiparametric measurements of gene expression, chromatin accessibility, and protein quantification (Dimitriu et al., 2022). Improved microfluidics and better combinatorial indexing have improved the precision and efficiency of single-cell genomics, allowing researchers to examine single cells with precision. Furthermore, progress in the computational aspects of forward and reverse genomics has enabled rapid data analysis for WGS.

Single-cell genomics applications facilitate the understanding of cellular diversity across various biological systems (Griffiths et al., 2018). Along with cell type and subtype identification, single-cell transcriptomics enables the mapping of developmental trajectories and the disentanglement of complex molecular mechanisms that drive cell fate decisions for advanced transformations. Cell characterization becomes imperative in complex tissues and organs, since cellular heterogeneity can affect tissue function and disease mechanisms. Computational models have been devised to integrate single-cell and spatial transcriptomic datasets, enhancing the characterization of cellular constituents and their interactions within tissues (Wang et al., 2023).

Insights regarding multiple diseases have broadened through the application of single-cell genomics. For example, in cancer research, single-cell expression profiling has been employed to classify different types of tumor cells based on their heterogeneity, identify cancer stem cells, and study the mechanisms behind drug resistance (Zhang et al., 2021b). Furthermore, single-cell

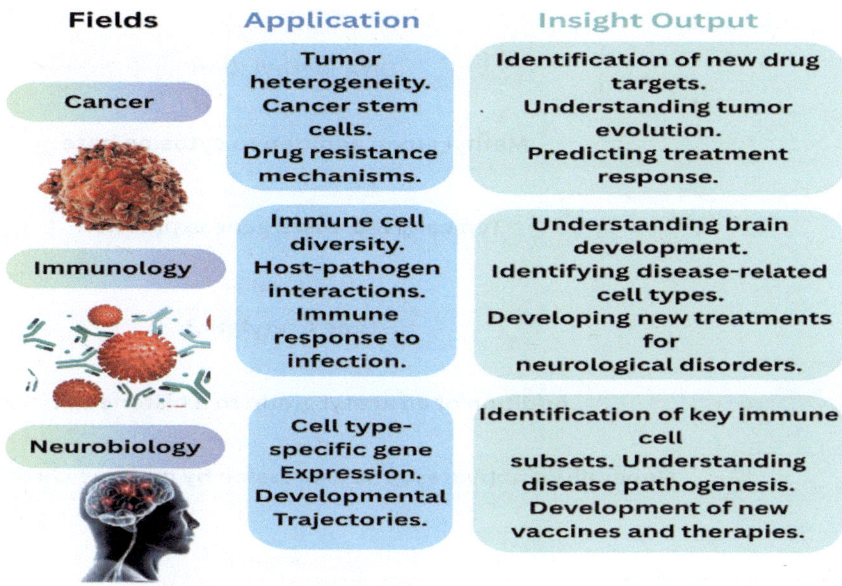

FIGURE 1.2 Single-cell genomics application.

sequencing enables defining more complex interactions between tumor cells and the immune system, thereby enhancing our understanding of immunological responses and immune escape mechanisms in immunotherapy. Single-cell genomics has also been utilized to study host–pathogen interactions, identify the most susceptible cell types to infection, viral replication, and the host immune response (Gioacchino et al., 2024). Finally, single-cell RNA analysis has been successfully applied to determine organ-specific cell types most vulnerable to severe acute respiratory syndrome coronavirus 2 (SARS-CoV-2).

Single-cell genomics analyzes the onset of autoimmune and other complex diseases. By examining expression profiles, researchers can delineate the molecular pathways that are disrupted and identify the cell types most affected by the disease in the relevant tissue sample. Considering new emerging research, single-cell genomics may also simplify drug design and development by clarifying therapeutic targets and biomarkers, enhancing treatment strategies. For example, researchers proposed a bioinformatics framework for single-cell sequencing analysis in amyotrophic lateral sclerosis, which contributes to explaining the complex and multifaceted phenomena revealed through such analysis (Liu et al., 2023). Moreover, single-cell genomics can work alongside proteomics and metabolomics to better understand cellular functions and disease mechanisms. The application of single-cell genomics in disciplines such as oncology, immunology, and neurobiology is summarized in Fig. 1.2.

1.5.3 Epigenomics

Epigenomics is the study of epigenetic modifications that are heritable and do not involve changes to the DNA sequence (Kumar et al., 2020). DNA methylation modification, histone structure alteration, and regulation by noncoding RNAs play important roles in the regulation

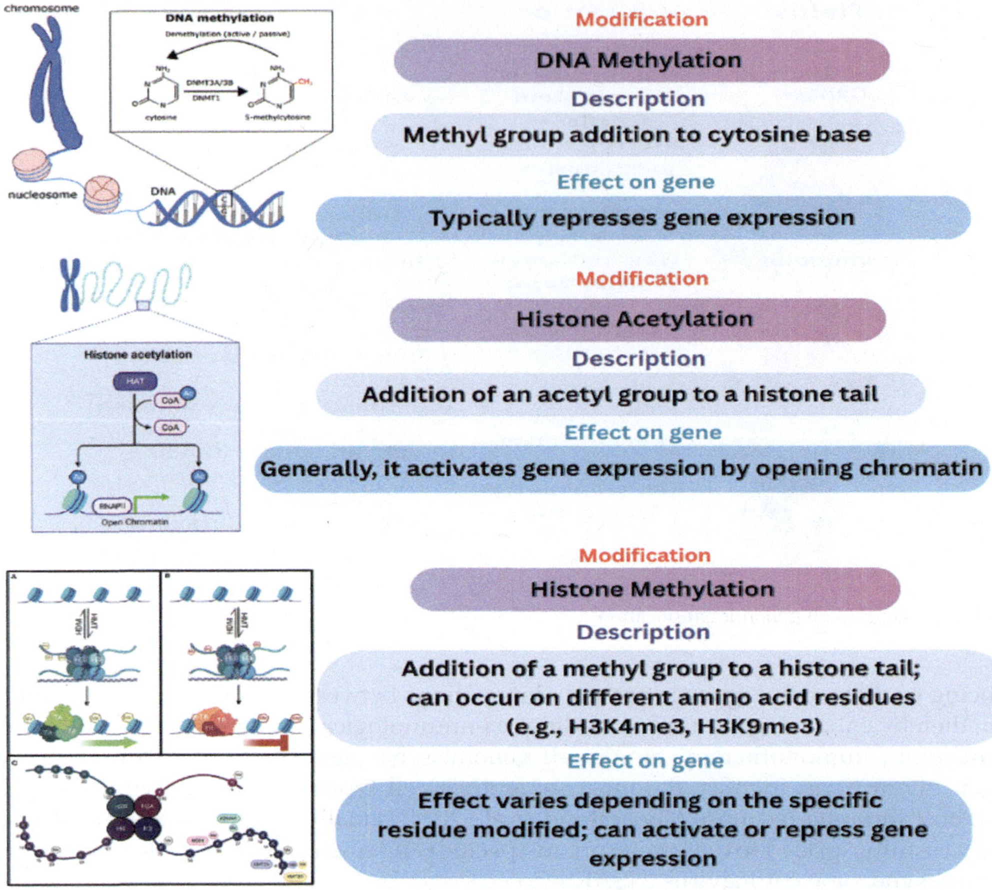

FIGURE 1.3 Epigenomic modification and its effects.

of genes, chromatin architecture, and the maintenance of epigenetic marks. The development of epigenomic research has furthered our knowledge of epigenetic modifications, their role in gene regulation, and their implications for heredity. High-throughput technologies now allow the mapping of DNA methylation and histone modifications across the whole genome. These techniques enable the exploration of various epigenetic marks throughout the organism, as illustrated in Fig. 1.3. DNA methylation tends to inhibit transcription, while histone acetylation promotes it by loosening the chromatin structure.

The term "DNA methylation" refers to the addition of a methyl group to one of the bases in the DNA molecule, usually cytosine. The patterns of DNA methylation, which play a role in gene regulation and variation, are established and maintained by a family of proteins called DNA methyltransferases (DNMTs) (Liu et al., 2025). DNA methylation represents functionality in embryonic development, genomic imprinting, and silencing transposons. Abnormalities of DNA methylation patterns have been identified in different types of cancers, neurological disorders, such as Alzheimer's disease, autoimmune issues, and even environmental effects on gene expression. These patterns regulate gene functionality, and disruptions in DNA methylation can contribute to health-related issues (Khodadadi et al., 2021).

Histone modification is the chemical marking of histone proteins, the most abundant protein components of chromatin, and includes acetylation, methylation, phosphorylation, and ubiquitination (Bae et al., 2025). Histone modifications are regulated by specific enzymes, including histone demethylases (HDMs), histone acetyltransferases (HATs), histone methyltransferases (HMTs), and histone deacetylases (HDACs) (Yao et al., 2024). These enzymes dynamically regulate gene expression by modifying histones, which in turn modulate cellular function, development, and disease.

Epigenetics is particularly crucial for regulating an organism's biological processes, as it controls gene expression by altering the structure and accessibility of chromatin. DNA acetylation and methylation, along with histone markers, change the accessibility of transcription factors and other proteins to the DNA, thereby influencing gene expression (Singh and Paramanik, 2025). Usually, DNA methylation within a gene's promoter region silences its expression, whereas histone acetylation has been associated with increased gene expression. Changes in chromatin methylation and acetylation are key factors that support and regulate the long-term control of gene expression.

Epigenetic inheritance refers to the transmission of gene expression modifications without changes to the DNA sequence itself. The transmission of epigenetic modifications across generations in diverse organisms such as plants, invertebrates, and mammals is known as transgenerational epigenetic inheritance (Elder et al., 2025). These epigenetic changes can be stably passed on and have a widespread impact (Banushi et al., 2025). Studies suggest that epigenetic inheritance is crucial in several biological phenomena, including development, behavior, and adaptation.

1.5.4 Metagenomics

Metagenomics is the analysis of genetic material obtained directly from environmental samples, such as soil, water, or the human gut, with the ability to detect the entire microbial population in a sample. Metagenomics enables the study of genetic diversity and functional capacity of microorganisms in an ecosystem without the need for isolation, cultivation, or identification of individual species (Curto et al., 2025). Moreover, metagenomic analyses provide more details into microbial life and its interactions within ecosystems, addressing ecology and evolution. Recent advances in metagenomics include improved DNA extraction, the development of high-throughput sequencing platforms, and the utilization of bioinformatics for pattern recognition in complex metagenomic data (Lema et al., 2023).

Metagenomics has numerous applications in ecology, enabling the study of the structure, dynamics, functions, and communities of microbes in previously inaccessible environments. Scientists estimate and identify community changes using environmental DNA (eDNA) from soil, water, and air. These studies provide knowledge of species composition, demographic structure, and functional capabilities (Adamska et al., 2017). Such information is essential for assessing the impact of environmental changes, such as pollution, climate change, and habitat destruction, on microbial biodiversity. Metagenomics also facilitates the study of microbial diversity in extreme environments, including the deep sea, hydrothermal vents, polar ice, and hypersaline lakes. The application of viral metagenomics is also valuable for identifying emerging and reemerging tick-borne viruses (Manuel and Snyder, 2024). Fig. 1.4 presents the

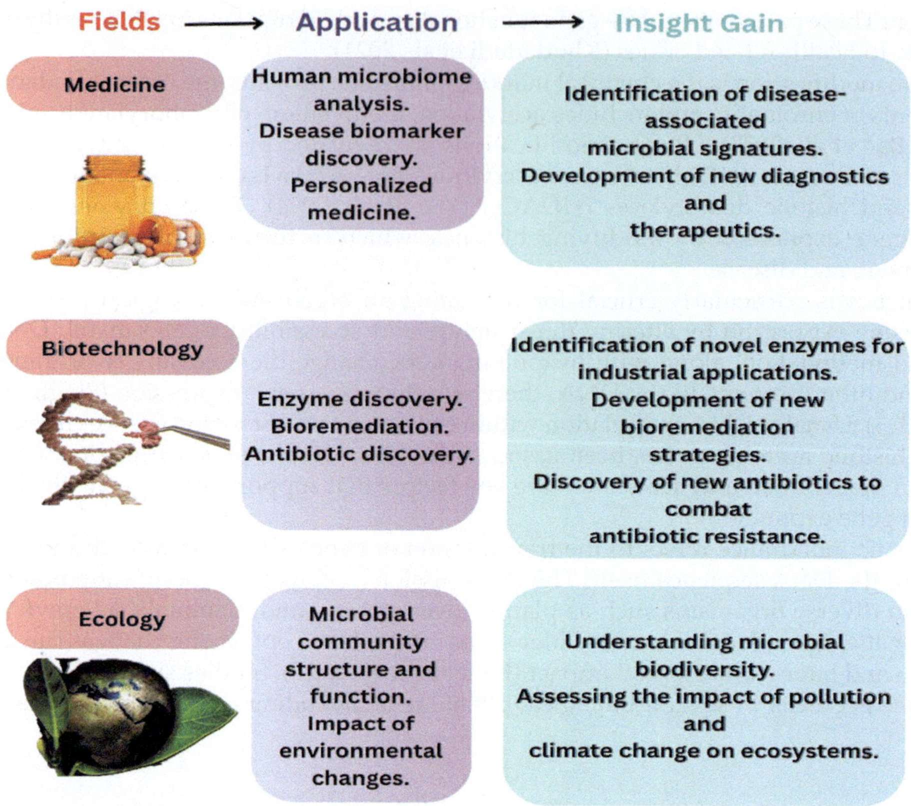

FIGURE 1.4 Application of metagenomics in different fields.

use of metagenomics in diverse research fields.

In medicine, the use of metagenomics is specifically referred to as human metagenomics, which seeks to retrieve data about the human microbiome, a unit within biology that denotes the collective microorganisms that either colonize or reside symbiotically in the human body (Kim et al., 2024). The human microbiome plays an important role in several health-related processes such as digestion, immune response, and even mental health. Metagenomic research has demonstrated that variations in the structure and function of the human microbiome are associated with various diseases, such as obesity, diabetes, inflammatory bowel disease, cancer, and others. Metagenomics can also be used for the discovery of novel biomarkers and the development of personalized medicine based on an individual's microbiome (Athanasopoulou et al., 2023).

Metagenomics has numerous applications in biotechnology. It can be applied to discover novel enzymes, bioproducts, and antibiotic research (Coughlan et al., 2015). By searching metagenomic libraries, scientists can identify genes that code for enzymes with potential applications in diverse industries, including biofuels, bioremediation, and even food production. For example, researchers used metagenomics in the search for new antibiotics to combat resistant bacteria (Berglund et al., 2019). While the scope of metagenomics begins with genomics, it also extends to biotechnology that utilizes microorganisms to remediate environmental pollution (Chettri et al., 2024).

1.5.5 Synthetic biology and genome design

1.5.5.1 Creating synthetic genomes

Synthetic biology represents a paradigm shift in the biological sciences, moving beyond the mere observation and analysis of biological systems toward the rational design and construction of entirely novel systems or extensive re-engineering of existing ones. Synthesizing whole genomes from scratch is a key tenet of this field, which involves the chemical synthesis of DNA sequences—either codon-optimized or designed for specific functions—and their subsequent assembly into functional chromosomes and genomes within host cells. Some of the early landmark achievements include the synthesis of the poliovirus genome and the controversial development of the first synthetic bacterial cell, *Mycoplasma mycoides* JCVI-syn1.0 (Gibson et al., 2010). Further attempts in genome minimization led to JCVI-syn3.0, a bacterial cell with the smallest known genome capable of autonomous replication, providing valuable information about the minimal gene set required for life (Sandberg et al., 2023). More recently, the field has directed its attention to the challenge of eukaryotic genome synthesis. The Synthetic Yeast Genome Project (Sc2.0) aims to build a fully synthetic and designer *Saccharomyces cerevisiae* genome, incorporating features such as codon reassignment, inducible evolution systems (e.g., SCRaMbLE), and the removal of unstable elements such as transposons (Dymond and Boeke, 2012; Jones, 2018). Progress has been made, with several chromosomes and chromosome arms synthesized and functionally tested (Jiang et al., 2023; Zhao et al., 2023). These endeavors have driven advancements in DNA synthesis technologies, hierarchical genome assembly plans most often based on homologous recombination in yeast, and advanced genome-wide debugging strategies to recognize and rectify design flaws or lethal mutations introduced during the construction process (Ostrov et al., 2020).

1.5.5.2 Applications in medicine, agriculture, and industry

The design principles and technologies underlying the development of synthetic biology and genome engineering actively shape innovations in various sectors. In medicine, an increase in the use of synthetic biology in vaccine development has been impressive. The rapid design, manufacture, and clinical efficacy of mRNA vaccines, especially in combating SARS-CoV-2, vividly explain synthetic mRNA technology's capability to lead to faster responses than have been observed before (Kitney et al., 2021; Tan et al., 2021; Kwon et al., 2022). Synthetic genomics set the stage for the establishment of viral genomes, accelerating reverse genetics studies, and promoting vaccine design and production (Dormitzer et al., 2013; Venter et al., 2022). Apart from infectious diseases, synthetic biology has been the engine of mammalian cell and microbial engineering to treat several diseases. Developing more advanced cell-based therapies, such as CAR-T cells for cancer and engineered microbes for metabolic diseases, illustrates the transformative potential of this technology (Katz et al., 2018). In addition, rapid, low-cost disease detection is becoming possible through novel synthetic gene circuits and CRISPR-based technology (Tan et al., 2021). Similarly, in agriculture, biotechnology is important for providing tools and scientific breakthroughs that can address nutritional deficiencies and improve crop production (Li et al., 2022; Edwards et al., 2024). Scientists are exploring methods to transfer omega-3 fatty acids and other animal-derived compounds directly into plant oils (Han et al., 2022; Edwards et al., 2024). Synthetic biology also enables the development of plant protection approaches and drives the creation of alternative food sources based on consumer preferences (Katz et al., 2018; Clarke and Kitney, 2020).

Through synthetic biology, industrial biotechnology achieves remarkable results because it enables sustainable methods to produce chemicals, materials, biofuels, and enzymes via optimized microbial fermentation (Katz et al., 2018; Clarke and Kitney, 2020). Engineered microbial platforms, such as *Escherichia coli* and yeast, have been programmed to generate essential molecules, including artemisinic acid, 1, 3-propanediol, 1, 4-butanediol, and isoprenoids (Meadows et al., 2016; Katz et al., 2018). The industrial shift from fossil fuels to bio-based manufacturing receives rapid acceleration through synthetic biology platforms, which implement design-build-test-learn cycles for on-demand production (Clarke and Kitney, 2020; Kitney et al., 2021).

1.5.6 Pan-genomics

1.5.6.1 Studying the entire gene pool of a species

Scientists who began sequencing large genomes quickly learned that a singular reference genome does not adequately represent the full genetic material of any species (Bayer et al., 2020; Sherman and Salzberg, 2020). Microbial genomics led to the recognition that a reference genome could not capture the complete genetic diversity of a species, which led to the creation of the pan-genome concept in 2005 (Medini et al., 2005). The pan-genome analysis framework enables researchers to understand genetic diversity because it goes beyond reference-based read mapping, which depends on a single linear sequence (Liu et al., 2020).

Scientific studies define the pan-genome through its two essential parts: the core genome, shared by all tested subjects, and the variable genome, which contains sequences specific to individual groups or even individual organisms (Medini et al., 2005; Bayer et al., 2020; Liu et al., 2020). Researchers build the pan-genome by evaluating genetic material from various members of the same species. This comparative method is necessary because many species possess extensive genetic diversity, which includes structural variants (SVs), presence/absence variations (PAVs), copy number variations (CNVs), and novel sequences that cannot be identified through single-reference analysis (Bayer et al., 2020; Liu et al., 2020; Sherman and Salzberg, 2020). Human research has revealed that a complete human pan-genome could include tens to hundreds of megabases of sequence not present in current reference assemblies (Sherman and Salzberg, 2020). Scientists need sophisticated computational frameworks such as graph-based genome representations to properly handle the intricate combination of core and variable genetic sequences (Liu et al., 2020).

1.5.6.2 Insights into genetic diversity and adaptation

The ability to conduct pan-genomic analyses allows for a deeper understanding of variations present in a species, beyond those captured using single-reference phylogenetic approaches and the inherent SNPs of the references (Bayer et al., 2020; Sherman and Salzberg, 2020; Li et al., 2023). The greatest level of diversity is generally found in the dispensable, or variable, genome component, which includes PAVs, CNVs, and other SVs (Liu et al., 2020; Tong et al., 2022). For example, the dispensable genome component may be involved in the adaptation to specific niches, resistance, responses to environmental stress (biotic and abiotic), and secondary metabolism. In contrast, the core genome usually contains essential housekeeping genes (Qin et al., 2021; Kang et al., 2023). In addition to the structural variations themselves, their derived versus ancestral states, and the mechanism by which they arise, such as transposable elements, NHEJ, or horizontal gene transfer in microbes, offer significant insights regarding the evolutionary processes affecting a specific species (Medini et al., 2005; Bayer et al., 2020; Qin et al., 2021; Li et al., 2023).

Pan-genomes enable researchers to explore ways of understanding how genetic diversity drives adaptation and artificial selection. For instance, specific SVs have been associated with local adaptation to alpine environments in *Arabidopsis thaliana*, adaptation and domestication traits in rice, immune response traits in pigs, and economically important or ecologically adaptive traits in silkworms (Qin et al., 2021; Tong et al., 2022; Li et al., 2023; Kang et al., 2023). Genes with variable contributions to fitness evolve more rapidly than core genes, demonstrating different selective pressures and the continuous generation of adaptive potential. Therefore pan-genomes provide a concise and powerful framework for linking genomic variation directly to evolution, adaptation, domestication, and breeding initiatives.

In summary, the introduction of NGS technologies is revolutionizing the paradigms of biomedical sciences through single-cell genomic, epigenomic, and metagenomic techniques. Combining advanced computational tools with cloud computing enables detailed analysis of genes and environmental factors. Standardized and affordable genome studies will accelerate the translation of research into personalized medicine. This will also advance ecological conservation, global health, and precision medicine in the postgenomic era.

1.6 Applications of genome research across living organisms

1.6.1 Applications of genome research in humans

This holistic approach to cancer therapeutics, intrinsic to pharmaceutical and clinical chemistry, has changed the deeply rooted practices in medicine (Chin et al., 2011; Feero and Guttmacher, 2014). Personalized medicine aims to target interventions to an individual's unique genome, thereby moving away from generic treatment approaches (Feero and Guttmacher, 2014). Pharmacogenomics is the major pillar of this approach, studying genetic variability in relation to differential drug response. CYP2D6, a drug-metabolizing enzyme involved in the metabolism of approximately 20% of clinically used drugs, represents the challenges and advancements found in pharmacogenomics (Kehinde et al., 2023).

CYP2D6 is highly polymorphic, with a multitude of star alleles, including SVs and hybrid forms, that determine metabolizer status (e.g., poor, intermediate, normal, or ultrarapid). These variations significantly affect the efficacy and toxicity of widely prescribed drugs, including opioids and antidepressants (Chen et al., 2021; Kehinde et al., 2023). Accurate genotyping of all important complex variants, coupled with consideration of population-specific allele frequencies, remains a major challenge to clinical implementation, requiring sophisticated bioinformatics tools (Chen et al., 2021). The promotion of pharmacoequity requires addressing the underrepresentation of diverse populations in pharmacogenomic studies, as allele frequencies and even allele function can vary greatly by ancestry (Kehinde et al., 2023). Furthermore, gut microbiome metabolism contributes to drug efficacy, adding another level of complexity to personalized drug response (Heinken et al., 2021).

Genomics revolutionized our understanding of genetic diseases, yet protocols for testing single-gene Mendelian diseases (e.g., hereditary cancer syndromes associated with BRCA1/2) are still being refined (Feero and Guttmacher, 2014; Bharucha et al., 2020). Meanwhile, genome-wide association studies (GWAS) and the increasing use of polygenic risk scores (PRSs) are assisting in estimating susceptibility to complex diseases (Feero and Guttmacher, 2014; Kachuri et al., 2024). While PRS accounts for genetic susceptibility, their predictive accuracy and across-population transferability remain important issues (Martin et al., 2019; Kachuri et al., 2024).

Whole-exome sequencing and WGS are also being used to identify previously undiagnosed or rare conditions (Feero and Guttmacher, 2014). Cancer genomics has fundamentally transformed oncology, shifting diagnosis paradigms from morphological features to genetics (Chin et al., 2011). Large-scale projects have cataloged somatic mutations, distinguishing pathogenic drivers from incidental variants (Chin et al., 2011). Identified driver mutations in EGFR, BRAF, and ALK, among others, are prime candidates for targeted therapies and must be considered in conjunction with the biology of each tumor (Feero and Guttmacher, 2014).

1.6.2 Applications of genome research in animals

Animal science researchers rely heavily on genome research, as it provides essential strategies to safeguard endangered species and enhance domestic animal breeding programs (Ryder, 2004; Steiner et al., 2013). The management of endangered species through conservation genomics results from the analysis of genome-wide information (Ryder, 2004; Hohenlohe et al., 2021). WGS and reduced-representation methods, such as RADseq, enable accurate evaluations of genetic diversity and other factors, even when studying nonmodel organisms or using noninvasive samples (Steiner et al., 2013; Hohenlohe et al., 2021; Schmidt et al., 2024). These data support the creation of appropriate conservation units (e.g., evolutionarily significant units and management units) for management purposes (Hohenlohe et al., 2021).

Genomic research methods excel at determining inbreeding measurements by analyzing runs of homozygosity and detecting harmful mutations that influence genetic load (Hohenlohe et al., 2021; Schmidt et al., 2024). The genetic foundation of adaptation becomes clearer through genotype-environment association and GWAS, which assess population responses to environmental modifications and support the development of genetic rescue and assisted migration approaches (Steiner et al., 2013; Hohenlohe et al., 2021; Schmidt et al., 2024). Researchers can leverage high-resolution genomic technologies by integrating these data with traditional pedigree analyses, producing a powerful tool for validating and incorporating essential demographic and life-history information (Galla et al., 2022).

Through genomic tools, livestock producers achieve improved animal health and productivity. GWAS have revealed genetic patterns underlying economically important traits such as growth speed rate, milk production, and disease resistance (Steiner et al., 2013). The combination of marker-assisted selection and genomic selection enables faster genetic improvement than traditional breeding methods (Ryder, 2004; Hohenlohe et al., 2021). The understanding of disease resistance in genomics enables the development of animals with greater resilience, leading to cost reductions and enhanced animal welfare.

1.6.3 Applications of genome research in plants

The study of genomes in plants has completely transformed plant science by enhancing agricultural development and ensuring food production for the future (Bevan et al., 2017).

In modern breeding programs, GWAS and genomic selection (GS) represent core methods for crop development (Wang et al., 2018; Mir et al., 2023). Through GWAS, scientists can detect genetic markers associated with beneficial agronomic characteristics, including resistance against Fusarium head blight and nonbiological stressors (Mir et al., 2023).

GS uses whole-genome marker data to evaluate an individual's genetic value, known as genomic estimated breeding values, through which breeders select candidates without extensive phenotyping and achieve faster breeding (Wang et al., 2018; Bartholomé et al., 2022). The optimization of GS requires researchers to properly select markers, construct appropriate training populations, use models such as GBLUP or Bayesian approaches, and incorporate multienvironmental testing data to detect genotype-environment interactions (Wang et al., 2018; Bartholomé et al., 2022).

The CRISPR-Cas genome editing system represents a groundbreaking advancement, offering exceptional accuracy for trait modification in crops. Through this technology, scientists can perform precise gene knockouts to enhance fragrance or provide disease resistance, along with base editing for nucleotide modifications, gene insertions, and complex chromosomal modifications. These tools surpass traditional breeding methods through their capability to directly insert or modify genetic features, achieve efficient quantitative trait locus (QTL) stacking, and disrupt undesirable genetic linkages (Gao, 2021).

The study of plant evolution through genomics reveals the central role of whole-genome duplication and polyploidy in shaping the genetic structure of crop plants (Bevan et al., 2017). Understanding plant domestication has identified specific genes that evolved during agricultural adaptation and simultaneously reveals genetic bottlenecks (Bevan et al., 2017; Gao, 2021). Introducing novel genetic traits to develop stress-resistant crops requires precise knowledge of the genetic diversity present in wild crop relatives, as revealed through genome sequencing (Bevan et al., 2017). Genome editing also enables the rapid domestication of wild species and the improvement of orphan crops, contributing to the development of climate-resilient and nutritious food sources (Gao, 2021).

1.6.4 Applications of genome research in microbes

The study of microbial genomics offers both fundamental science and practical solutions that benefit a broad range of industrial biotechnology sectors (Rubin, 2008; Liu et al., 2021). The application of genomic science in sustainable biofuel development results in faster utilization of lignocellulosic feedstock through microbial enzyme identification and biomass deconstruction (Rubin, 2008). Scientists obtain extensive enzymatic libraries through genome sequencing of bacteria, fungi, and termite hindgut microbial consortia, which include cellulases and hemicellulases essential for transforming plant polymers into fermentable sugars (Rubin, 2008). Genomic findings also enhance metabolic engineering approaches, allowing the creation of efficient microbial chassis systems (e.g., *E. coli* and yeasts) that convert both hexose and pentose sugars into ethanol and butanol (Rubin, 2008).

Bioremediation strongly relies on genomics and metagenomics to discover and analyze microbial pathways able to break down persistent pollutants, including PAHs, organohalides, and other xenobiotics (Liu et al., 2021; Oyewusi et al., 2021; Yang et al., 2022). Knowledge of essential degradative enzymes, such as dehalogenases, dioxygenases, peroxidases, and cytochrome P450s, combined with their genetic structure, regulatory mechanisms, and microbial associations, allows the optimization of environmental cleanup approaches (Liu et al., 2021; Oyewusi et al., 2021; Yang et al., 2022).

Environmental metagenomics is rapidly expanding, particularly in wastewater surveillance, as a powerful means of evaluating both pollutant degradation and AMR dispersal (Oyewusi et al., 2021; Wheeler et al., 2023). WGS has transformed microbial epidemiology and deepened our understanding of pathogenicity (Hill et al., 2021; Castillo-Ramírez, 2022). Genomic epidemiology provides ultra-high resolution for tracking both viral and bacterial transmission of pathogens across micro (e.g., hospital outbreaks) and macro (e.g., global pandemics) spatiotemporal scales (Leekitcharoenphon et al., 2021).

Pan-genomic methods, which compare variation in core genome and accessory genome, provide better resolution among closely related isolates in outbreak investigations, especially given the rapid evolution of accessory gene content (Castillo-Ramírez, 2022). Genomics is critical in AMR prevention and control. Using WGS, researchers can accurately identify AMR genes, predict resistance phenotypes, and determine the contribution of mobile genetic elements (MGEs)—such as plasmids, transposons, and integrons—to AMR dissemination within and between bacterial species and environments (Leekitcharoenphon et al., 2021; Wheeler et al., 2023). Monitoring specific AMR genes or plasmids enables targeted surveillance of resistance and, when combined with isolate sequencing and metagenomics, supports resistance tracking in health and environmental contexts under a One Health perspective.

1.7 Ethical, legal, and social implications (ELSI) of genome research

Genome research has transformed medicine, agriculture, and almost every field of human endeavor. However, it raises important concerns related to privacy, ethics, and legal frameworks. These issues need to be addressed carefully to ensure future advances in genome research. The protection of genetic data remains vital to preventing unfair treatment and discrimination (Hudson et al., 2008; Gymrek et al., 2013). The implementation of cloud computing systems demands enhanced mechanisms for data security (Shabani et al., 2015). The CRISPR-Cas9 genome editing method enables genetic disease correction, yet it raises significant ethical issues regarding human embryo editing, equitable access, and unforeseen risks (Doudna and Charpentier, 2014; Lanphier et al., 2015). Legal frameworks such as the Genetic Information Nondiscrimination Act (GINA) and General Data Protection Regulation (GDPR) were established to safeguard genetic privacy through responsible research procedures (Hudson et al., 2008). However, existing governance remains incomplete, necessitating ongoing adjustments and worldwide collaboration (Araki and Ishii, 2014). This session emphasizes the need to balance scientific progress while safeguarding human dignity and societal values.

1.7.1 Privacy concerns in genomic data

Genome research produces large amounts of personal genetic information, making data protection significantly important. Genomic data contains sensitive details about a person's health, ancestry, and potential disease risks. If this information is misused, it can lead to discrimination or emotional harm (Gymrek et al., 2013). Many organizations store genetic data for research purposes. However, not all follow strong security protocols. There are risks that hackers could access and steal this private information. Even if names are removed from datasets, researchers can sometimes trace the data back to individuals, a process known as "re-identification."

Insurance companies and employers might misuse genomic information. For example, a company could refuse to hire someone based on their genetic risk of developing an illness.

Therefore strong privacy laws such as GINA are necessary to protect individuals (Hudson et al., 2008). Another issue is consent. Individuals must clearly understand how their genetic data will be used before agreeing to share it. Sometimes, researchers use stored data for new studies without informing the original donors, raising ethical questions regarding respect and autonomy. Emerging technologies, such as cloud computing, raise additional concerns, as storing genomic data increases the risk of cyberattacks. For example, the responsible use of existing security methods, such as encryption and controlled access, must be further developed and implemented to protect against unauthorized use (Shabani et al., 2016). While there are great benefits to genomic research, its privacy risks cannot be ignored. Researchers, policymakers, and industry stakeholders must collaborate toward developing safe and ethical frameworks for storing genetic data. Public trust is fundamental to the success of future genome studies.

1.7.2 Ethical issues in genome editing

Genome editing technologies, particularly CRISPR Cas9, have opened the door to new possibilities in science. Scientists can now change DNA sequences to correct genetic disorders. However, these technologies raise several ethical concerns (Doudna and Charpentier, 2014). Editing human embryos is one major problem. Some scientists argue that if genetic issues can be corrected before birth, many disorders could be prevented. However, others claim that it is ethically problematic to make permanent changes to the genome, changes that will be transmitted to future generations. Additionally, there is concern about "designer babies," in which parents might choose characteristics such as intelligence or appearance, leading to greater social inequality (Lanphier et al., 2015).

Another concern is fairness. Gene editing could be expensive and therefore reserved only for the wealthy, widening the wealth gap and creating new forms of discrimination. Safety also presents ethical concerns. Unintended consequences from gene editing can occur, and even undetectable errors could lead to irreversible harm to an individual's health. Scientists must act with extreme caution, and treatments must be rigorously tested before being implemented in humans (Sciences et al., 2017).

Concerns also exist over editing plants and animals. Genes in crops can be modified to improve food production, but this can negatively impact ecosystems. Genetically modified animals could have similar effects on natural populations. Additionally, views on genome editing are influenced by religious and cultural beliefs. It is crucial to develop laws that respect diverse values and opinions toward genetic technologies. Ultimately, genome editing is promising but must be used responsibly. Open discussions among scientists, ethicists, and the public are required. To ensure genome editing is safe and benefits all members of society, strict guidelines and regulations should be developed.

1.7.3 Regulatory frameworks for genomic technologies

Genome research and technologies are governed by regulations to ensure they are used safely and ethically. Laws regarding the control and use of genetic information and technologies have been developed worldwide. For instance GINA protects individuals from discrimination based on their genetic data (Hudson et al., 2008). It prohibits employers and insurance companies from employing genetic information unfairly. Gene therapies are also regulated by the Food and Drug Administration (FDA) to guarantee they are safe and effective (Takefman and Bryan, 2012).

In Europe, the GDPR sets strict rules for handling personal genetic data (Phillips, 2018). Before collecting and using genomic information, researchers must have clear and informed consent. The GDPR also grants individuals the right to access their data and the "right to be forgotten," which is the right to request that their data be deleted. In most countries, human genome editing is tightly controlled. Editing of embryos for reproduction is typically banned or strictly limited. The World Health Organization (WHO) and the United Nations Educational, Scientific, and Cultural Organization have appealed for global cooperation (Loizides et al., 2021). The WHO has recommended the implementation of international guidelines for managing genome editing, especially in humans. However, these laws have not fully addressed all ethical gaps. In some regions, regulations are weak or outdated. Since science occurs rapidly, it is important to regularly update laws to promote research, protect human rights, public health, and social equity.

1.8 Conclusion

The central dogma of molecular biology encompasses the passage of the genetic flow from DNA to RNA to protein, and this has been established as the foundation of molecular genetics. The central dogma has been the basis for genome research and the development of new molecular biology tools, including CRISPR-Cas9, NGS, and bioinformatics platforms. The rate of knowledge generation resulting from these technologies and their application to genomic structures and functions has increased rapidly. Genome applications can be found in every organism system to improve crop yield and resilience regarding plant systems, disease modeling in animals, and personalized medicine in humans, allowing for individual therapies based on genomic information. Advances in genome science challenge health equity and create important ELSI. Issues range from germline editing to the use of GINA in the U.S. to the implications of the GDPR in the European Union. As genome research expands, it is important to establish and respect scientists' responsibility to society, to strive for equitable distribution of the benefits of science and innovations, and to prioritize ethical standards.

References

Abdi, G., Tarighat, M.A., Jain, M., Tendulkar, R., Tendulkar, M., Barwant, M., 2024. Revolutionizing genomics: exploring the potential of next-generation sequencing. In: Singh, V., Kumar, A. (Eds.), Advances in Bioinformatics. Springer, Singapore.

Adamska, A., Domenichini, A., Falasca, M., 2017. Pancreatic ductal adenocarcinoma: current and evolving therapies. Int. J. Mol. Sci. 18 (7), 1338.

Adeyanju, S.A., Ogunjobi, T.T., 2024. Machine learning in genomics: applications in whole genome sequencing, whole exome sequencing, single-cell genomics, and spatial transcriptomics. Medinformatics doi:10.47852/bonviewMEDIN42024120.

Ahmar, S., Mahmood, T., Fiaz, S., Mora-Poblete, F., Shafique, M.S., Chattha, M.S., et al., 2021. Advantage of nanotechnology-based genome editing system and its application in crop improvement. Front. Plant Sci. 12, 663849.

Akintunde, O., Tucker, T., Carabetta, V.J., 2024. The evolution of next-generation sequencing technologies. In: Carabetta, V., Akintunde, O. (Eds.), High Throughput Gene Screening: Methods and Protocols. Springer, New York, NY.

Allen, D., Rosenberg, M., Hendel, A., 2021. Using synthetically engineered guide RNAs to enhance CRISPR genome editing systems in mammalian cells. Front Genome Ed. 2, 617910.

Araki, M., Ishii, T., 2014. International regulatory landscape and integration of corrective genome editing into in vitro fertilization. Reproductive Biology and Endocrinology. 12, 1–12.

Aschenbrenner, S., Kallenberger, S.M., Hoffmann, M.D., Huck, A., Eils, R., Niopek, D., 2020. Coupling Cas9 to artificial inhibitory domains enhances CRISPR-Cas9 target specificity. Sci. Adv. 6 (6), eaay0187.

Asmamaw, M., Zawdie, B., 2021. Mechanism and applications of CRISPR/Cas-9-mediated genome editing. Biol. Targets Therapy 15, 353–361.

Athanasopoulou, K., Adamopoulos, P.G., Scorilas, A., 2023. Unveiling the human gastrointestinal tract microbiome: the past, present, and future of metagenomics. Biomedicines. 11 (3), 827.

Athanasopoulou, K., Boti, M.A., Adamopoulos, P.G., Skourou, P.C., Scorilas, A., 2021. Third-generation sequencing: the spearhead towards the radical transformation of modern genomics. Life. 12 (1), 30.

Aury, J.-M., Istace, B., 2021. Hapo-G, haplotype-aware polishing of genome assemblies with accurate reads. NAR Genom. Bioinform. 3 (2), lqab034.

Ayanoğlu, F.B., Elçin, A.E., Elçin, Y.M., 2020. Bioethical issues in genome editing by CRISPR-Cas9 technology. Turkish J. Biol. 44 (2), 110–120.

Bae, W., Ra, E.A., Lee, M.H., 2025. Epigenetic regulation of reprogramming and pluripotency: insights from histone modifications and their implications for cancer stem cell therapies. Front. Cell Dev. Biol. 13, 1559183.

Banushi, B., Collova, J., Milroy, H., 2025. Epigenetic echoes: bridging nature, nurture, and healing across generations. Int. J. Mol. Sci. 26 (7), 3075.

Bartholomé, J., Prakash, P.T., Cobb, J.N., 2022. Genomic prediction: progress and perspectives for rice improvement. Methods Mol. Biol. 2467, 569–617.

Bayer, P.E., Golicz, A.A., Scheben, A., Batley, J., Edwards, D., 2020. Plant pan-genomes are the new reference. Nat. Plants. 6 (8), 914–920.

Berglund, F., Österlund, T., Boulund, F., Marathe, N.P., Larsson, D.J., Kristiansson, E., 2019. Identification and reconstruction of novel antibiotic resistance genes from metagenomes. Microbiome. 7, 1–14.

Bevan, M.W., Uauy, C., Wulff, B.B., Zhou, J., Krasileva, K., Clark, M.D., 2017. Genomic innovation for crop improvement. Nature. 543 (7645), 346–354.

Bharucha, P.P., Chiu, K.E., François, F.M., Scott, J.L., Khorjekar, G.R., Tirada, N.P., 2020. Genetic testing and screening recommendations for patients with hereditary breast cancer. Radiographics. 40 (4), 913–936.

Bosch, J.A., Birchak, G., Perrimon, N., 2021. Precise genome engineering in Drosophila using prime editing. Proc. Natl. Acad. Sci. 118 (1), e2021996118.

Brůna, T., Hoff, K.J., Lomsadze, A., Stanke, M., Borodovsky, M., 2021. BRAKER2: automatic eukaryotic genome annotation with GeneMark-EP+ and AUGUSTUS supported by a protein database. NAR Genom. Bioinform. 3 (1), lqaa108.

Casalino, L., Nierzwicki, Ł., Jinek, M., Palermo, G., 2020. Catalytic mechanism of non-target DNA cleavage in CRISPR-Cas9 revealed by ab initio molecular dynamics. ACS Catal. 10 (22), 13596–13605.

Castillo-Ramírez, S., 2022. Beyond microbial core genomic epidemiology: towards pan genomic epidemiology. Lancet Microbe 3 (4), e244–e245.

Cedar, H., Bergman, Y., 2009. Linking DNA methylation and histone modification: patterns and paradigms. Nat. Rev. Genet. 10 (5), 295–304.

Chapelle, V., Silvestre, F., 2022. Population epigenetics: the extent of DNA methylation variation in wild animal populations. Epigenomes 6 (4), 31.

Chappell, L., Russell, A.J., Voet, T., 2018. Single-cell (multi) omics technologies. Annu. Rev. Genomics Hum. Genet. 19 (1), 15–41.

Chen, D., Xu, L., Xing, H., Shen, W., Song, Z., Li, H., et al., 2024. Sangerbox 2: enhanced functionalities and update for a comprehensive clinical bioinformatics data analysis platform. Imeta 3 (5), e238.

Chen, X., Shen, F., Gonzaludo, N., Malhotra, A., Rogert, C., Taft, R.J., et al., 2021. Cyrius: accurate CYP2D6 genotyping using whole-genome sequencing data. Pharmacogenom. J. 21 (2), 251–261.

Chettri, D., Verma, A.K., Chirania, M., Verma, A.K., 2024. Metagenomic approaches in bioremediation of environmental pollutants. Environ. Pollut. 363, 125297.

Chin, L., Andersen, J.N., Futreal, P.A., 2011. Cancer genomics: from discovery science to personalized medicine. Nat. Med. 17 (3), 297–303.

Choi, S.R., Lee, M., 2023. Transformer architecture and attention mechanisms in genome data analysis: a comprehensive review. Biology 12 (7), 1033.

Clarke, L., Kitney, R., 2020. Developing synthetic biology for industrial biotechnology applications. Biochem. Soc. Trans. 48 (1), 113–122.

Claros, M.G., Bautista, R., Guerrero-Fernández, D., Benzerki, H., Seoane, P., Fernández-Pozo, N., 2012. Why assembling plant genome sequences is so challenging. Biology 1 (2), 439–459.

Collins, A., 2018. The challenge of genome sequence assembly. Open Bioinform. J. 11 (1), 231–239.

Cotto, K.C., Feng, Y.-Y., Ramu, A., Richters, M., Freshour, S.L., Skidmore, Z.L., et al., 2023. Integrated analysis of genomic and transcriptomic data for the discovery of splice-associated variants in cancer. Nat. Commun. 14 (1), 1589.

Coughlan, L.M., Cotter, P.D., Hill, C., Alvarez-Ordóñez, A., 2015. Biotechnological applications of functional metagenomics in the food and pharmaceutical industries. Front. Microbiol. 6, 672.

Crick, F., 1970. Central dogma of molecular biology. Nature 227 (5258), 561–563.

Crossley, B.M., Bai, J., Glaser, A., Maes, R., Porter, E., Killian, M.L., et al., 2020. Guidelines for Sanger sequencing and molecular assay monitoring. J. Vet. Diagn. Invest. 32 (6), 767–775.

Cuber, P., Chooneea, D., Geeves, C., Salatino, S., Creedy, T.J., Griffin, C., et al., 2023. Comparing the accuracy and efficiency of third generation sequencing technologies, Oxford Nanopore Technologies, and Pacific Biosciences, for DNA barcode sequencing applications. Ecolog. Genet. Genom. 28, 100181.

Curto, M., Veríssimo, A., Riccioni, G., Santos, C.D., Ribeiro, F., Jentoft, S., et al., 2025. Improving whole biodiversity monitoring and discovery with environmental DNA metagenomics. Mol. Ecol. Resour. 25, e14105.

D'amour, A., Heller, K., Moldovan, D., Adlam, B., Alipanahi, B., Beutel, A., et al., 2022. Underspecification presents challenges for credibility in modern machine learning. J. Mach. Learn Res. 23 (226), 1–61.

Del Angel, V.D., Hjerde, E., Sterck, L., Capella-Gutierrez, S., Notredame, C., Pettersson, O.V., et al., 2018. Ten steps to get started in Genome Assembly and Annotation. F1000Research 7, ELIXIR–148.

Deshpande, D., Chhugani, K., Chang, Y., Karlsberg, A., Loeffler, C., Zhang, J., et al., 2023. RNA-seq data science: from raw data to effective interpretation. Front. Genet. 14, 997383.

Dimitriu, M.A., Lazar-Contes, I., Roszkowski, M., Mansuy, I.M., 2022. Single-cell multiomics techniques: from conception to applications. Front. Cell Dev. Biol. 10, 854317.

Dormitzer, P.R., Suphaphiphat, P., Gibson, D.G., Wentworth, D.E., Stockwell, T.B., Algire, M.A., et al., 2013. Synthetic generation of influenza vaccine viruses for rapid response to pandemics. Sci. Transl. Med. 5 (185), 185ra68.

Doudna, J.A., Charpentier, E., 2014. The new frontier of genome engineering with CRISPR-Cas9. Science 346 (6213), 1258096.

Duan, L., Ouyang, K., Xu, X., Xu, L., Wen, C., Zhou, X., et al., 2021. Nanoparticle delivery of CRISPR/Cas9 for genome editing. Front. Genet. 12, 673286.

Dymond, J., Boeke, J., 2012. The Saccharomyces cerevisiae SCRaMbLE system and genome minimization. Bioeng. Bugs. 3 (3), 170–173.

Edwards, R.A., Ng, X.Y., Tucker, M.R., Mortimer, J.C., 2024. Plant synthetic biology as a tool to help eliminate hidden hunger. Curr. Opin. Biotechnol. 88, 103168.

El-Metwally, S., Hamza, T., Zakaria, M., Helmy, M., 2013. Next-generation sequence assembly: four stages of data processing and computational challenges. PLoS Comput. Biol. 9 (12), e1003345.

Elder, E., Lemieux, A., Legault, L.-M., Caron, M., Bertrand-Lehouillier, V., Dupas, T., et al., 2025. Rescuing DNMT1 fails to fully reverse the molecular and functional repercussions of its loss in mouse embryonic stem cells. Nucleic Acids Res. 53 (4), gkaf130.

Espinosa, E., Bautista, R., Larrosa, R., Plata, O., 2024. Advancements in long-read genome sequencing technologies and algorithms. Genomics. 116 (3), 110842.

Feero, W.G., Guttmacher, A.E., 2014. Genomics, personalized medicine, and pediatrics. Acad. Pediatr. 14 (1), 14–22.

Franjić, S., 2023. A few words about genetic variation. Int. J. Immunol. Res. 5 (2), 5.

Galla, S.J., Brown, L., Couch-Lewis, Y., Cubrinovska, I., Eason, D., Gooley, R.M., et al., 2022. The relevance of pedigrees in the conservation genomics era. Mol. Ecol. 31, 41–54.

Gao, C., 2021. Genome engineering for crop improvement and future agriculture. Cell 184 (6), 1621–1635.

Ghurye, J., Pop, M., 2019. Modern technologies and algorithms for scaffolding assembled genomes. PLoS Comput. Biol. 15 (6), e1006994.

Gibson, D.G., Glass, J.I., Lartigue, C., Noskov, V.N., Chuang, R.-Y., Algire, M.A., et al., 2010. Creation of a bacterial cell controlled by a chemically synthesized genome. Science. 329 (5987), 52–56.

Gioacchino, E., Vandelannoote, K., Ruberto, A.A., Popovici, J., Cantaert, T., 2024. Unraveling the intricacies of host-pathogen interaction through single-cell genomics. Microbes Infect. 26 (7), 105313.

Giorgi, F.M., Ceraolo, C., Mercatelli, D., 2022. The R language: an engine for bioinformatics and data science. Life 12 (5), 648.

Griffiths, J.A., Scialdone, A., Marioni, J.C., 2018. Using single-cell genomics to understand developmental processes and cell fate decisions. Mol. Syst. Biol. 14 (4), e8046.

Gymrek, M., Mcguire, A.L., Golan, D., Halperin, E., Erlich, Y., 2013. Identifying personal genomes by surname inference. Science 339 (6117), 321–324.

Han, L., Silvestre, S., Sayanova, O., Haslam, R.P., Napier, J.A., 2022. Using field evaluation and systematic iteration to rationalize the accumulation of omega-3 long-chain polyunsaturated fatty acids in transgenic Camelina sativa. Plant Biotechnol. J. 20 (9), 1833–1852.

Harrison, P.W., Amode, M.R., Austine-Orimoloye, O., Azov, A.G., Barba, M., Barnes, I., et al., 2024. Ensembl 2024. Nucleic Acids Res. 52 (D1), D891–D899.

He, W., Yang, J., Jing, Y., Xu, L., Yu, K., Fang, X., 2023. NGenomeSyn: an easy-to-use and flexible tool for publication-ready visualization of syntenic relationships across multiple genomes. Bioinformatics 39 (3), btad121.

Heinken, A., Basile, A., Hertel, J., Thinnes, C., Thiele, I., 2021. Genome-scale metabolic modeling of the human microbiome in the era of personalized medicine. Annu. Rev. Microbiol. 75 (1), 199–222.

Hill, V., Ruis, C., Bajaj, S., Pybus, O.G., Kraemer, M.U., 2021. Progress and challenges in virus genomic epidemiology. Trends Parasitol. 37 (12), 1038–1049.

Hille, F., Charpentier, E., 2016. CRISPR-Cas: biology, mechanisms and relevance. Phil. Trans. R. Soc. B: Biol. Sci. 371, 20150496.

Hohenlohe, P.A., Funk, W.C., Rajora, O.P., 2021. Population genomics for wildlife conservation and management. Mol. Ecol. 30 (1), 62–82.

Horgan, D., Pesapane, F., Van Der Buckle, M., De Maria, R., Dube, F., Singh, J., et al., 2024. From theory to practice: implementing next-generation sequencing and public health genomics in healthcare systems. Crit. Rev. Oncol. Hematol. 201, 104433.

Huddleston, J., Hadfield, J., Sibley, T.R., Lee, J., Fay, K., Ilcisin, M., et al., 2021. Augur: a bioinformatics toolkit for phylogenetic analyses of human pathogens. J. Open Source Softw. 6 (57), 2906.

Hudson, K.L., Holohan, M., Collins, F.S., 2008. Keeping pace with the times—the Genetic Information Nondiscrimination Act of 2008. N. Engl. J. Med. 358 (25), 2661–2663.

Jiang, C., Lv, G., Tu, Y., Cheng, X., Duan, Y., Zeng, B., et al., 2021. Applications of CRISPR/Cas9 in the synthesis of secondary metabolites in filamentous fungi. Front. Microbiol. 12, 638096.

Jiang, S., Luo, Z., Wu, J., Yu, K., Zhao, S., Cai, Z., et al., 2023. Building a eukaryotic chromosome arm by de novo design and synthesis. Nat. Commun. 14 (1), 7886.

Johnston, M.O., 2001. Mutations and new variation: overview. eLS. Wiley, New York, NY doi:10.1038/npg.els.0004165.

Johnston, S.E., 2024. Understanding the genetic basis of variation in meiotic recombination: past, present, and future. Mol. Biol. Evol. 41 (7), msae112.

Jones, S., 2018. SCRaMbLE does the yeast genome shuffle. Nat. Biotechnol. 36 (6), 503.

Jung, H., Ventura, T., Chung, J.S., Kim, W.-J., Nam, B.-H., Kong, H.J., et al., 2020. Twelve quick steps for genome assembly and annotation in the classroom. PLoS Comput. Biol. 16 (11), e1008325.

Kachuri, L., Chatterjee, N., Hirbo, J., Schaid, D.J., Martin, I., Kullo, I.J., et al., 2024. Principles and methods for transferring polygenic risk scores across global populations. Nat. Rev. Genet. 25 (1), 8–25.

Kang, M., Wu, H., Liu, H., Liu, W., Zhu, M., Han, Y., et al., 2023. The pan-genome and local adaptation of Arabidopsis thaliana. Nat. Commun. 14 (1), 6259.

Katz, L., Chen, Y.Y., Gonzalez, R., Peterson, T.C., Zhao, H., Baltz, R.H., 2018. Synthetic biology advances and applications in the biotechnology industry: a perspective. J. Ind. Microbiol. Biotechnol. 45 (7), 449–461.

Kawaguchi, Y.W., Tsuchikane, Y., Tanaka, K., Taji, T., Suzuki, Y., Toyoda, A., et al., 2023. Extensive copy number variation explains genome size variation in the unicellular zygnematophycean alga, Closterium peracerosum–strigosum–littorale complex. Genome Biol. Evolut. 15 (8), evad115.

Kehinde, O., Ramsey, L.B., Gaedigk, A., Oni-Orisan, A., 2023. Advancing CYP2D6 pharmacogenetics through a pharmacoequity lens. Clin. Pharmacol. Ther. 114 (1), 69–76.

Khodadadi, E., Fahmideh, L., Khodadadi, E., Dao, S., Yousefi, M., Taghizadeh, S., et al., 2021. Current advances in DNA methylation analysis methods. Biomed. Res. Int. 2021 (1), 8827516.

Khwatenge, C.N., Nahashon, S.N., 2021. Recent advances in the application of CRISPR/Cas9 gene editing system in poultry species. Front. Genet. 12, 627714.

Kim, N., Ma, J., Kim, W., Kim, J., Belenky, P., Lee, I., 2024. Genome-resolved metagenomics: a game changer for microbiome medicine. Exp. Mol. Med. 56 (7), 1501–1512.

Kitney, R.I., Bell, J., Philp, J., 2021. Build a sustainable vaccines industry with synthetic biology. Trends Biotechnol. 39 (9), 866–874.

Kopinski, P.K., Singh, L.N., Zhang, S., Lott, M.T., Wallace, D.C., 2021. Mitochondrial DNA variation and cancer. Nat. Rev. Cancer 21 (7), 431–445.

Kumar, H., Chaudhary, A., Singh, A., Sukhija, N., Panwar, A., Saravanan, K., et al., 2020. A review on epigenetics: manifestations, modifications, methods & challenges. J. Entomol. Zool. Stud. 8 (4), 01–06.

Kwon, S., Kwon, M., Im, S., Lee, K., Lee, H., 2022. mRNA vaccines: the most recent clinical applications of synthetic mRNA. Arch. Pharmacal Res. 45 (4), 245–262.

Lanphier, E., Urnov, F., Haecker, S.E., Werner, M., Smolenski, J., 2015. Don't edit the human germ line. Nature 519 (7544), 410–411.

Leekitcharoenphon, P., Johansson, M.H.K., Munk, P., Malorny, B., Skarżyńska, M., Wadepohl, K., et al., 2021. Genomic evolution of antimicrobial resistance in Escherichia coli. Sci. Rep. 11 (1), 15108.

Lema, N.K., Gemeda, M.T., Woldesemayat, A.A., 2023. Recent advances in metagenomic approaches, applications, and challenges. Curr. Microbiol. 80 (11), 347.

Li, C., Mei, H., Hu, Y., 2020. Applications and explorations of CRISPR/Cas9 in CAR T-cell therapy. Brief. Funct. Genom. 19 (3), 175–182.

Li, F., Xie, X., Huang, R., Tian, E., Li, C., Chao, Z., 2021a. Chloroplast genome sequencing based on genome skimming for identification of Eriobotryae Folium. BMC Biotech. 21, 1–17.

Li, J., Scarano, A., Gonzalez, N.M., D'orso, F., Yue, Y., Nemeth, K., et al., 2022. Biofortified tomatoes provide a new route to vitamin D sufficiency. Nat. Plants. 8 (6), 611–616.

Li, W., O'neill, K.R., Haft, D.H., Dicuccio, M., Chetvernin, V., Badretdin, A., et al., 2021b. RefSeq: expanding the prokaryotic genome annotation pipeline reach with protein family model curation. Nucleic Acids Res. 49 (D1), D1020–D1028.

Li, Z., Liu, X., Wang, C., Li, Z., Jiang, B., Zhang, R., et al., 2023. The pig pangenome provides insights into the roles of coding structural variations in genetic diversity and adaptation. Genome Res. 33 (10), 1833–1847.

Lim, K., 2024. Mitochondrial genome editing: strategies, challenges, and applications. BMB Reports 57 (1), 19.

Lin, B., Hui, J., Mao, H., 2021. Nanopore technology and its applications in gene sequencing. Biosensors 11 (7), 214.

Liu, H., Guan, L., Deng, M., Bolund, L., Kristiansen, K., Zhang, J., et al., 2023. Integrative genetic and single cell RNA sequencing analysis provides new clues to the amyotrophic lateral sclerosis neurodegeneration. Front. Neurosci. 17, 1116087.

Liu, Q., Yang, F., Zhang, J., Liu, H., Rahman, S., Islam, S., et al., 2021. Application of CRISPR/Cas9 in crop quality improvement. Int. J. Mol. Sci. 22 (8), 4206.

Liu, Y., Du, H., Li, P., Shen, Y., Peng, H., Liu, S., et al., 2020. Pan-genome of wild and cultivated soybeans. Cell 182 (1), 162–176.

Liu, Y., Wang, X.-Q., Zhang, P., Haghparast, A., He, W.-B., Zhang, J.-J., 2025. Research progress of DNA methylation on the regulation of substance use disorders and the mechanisms. Front. Cell. Neurosci. 19, 1566001.

Liu, Y., Zhan, J., Li, J., Lian, M., Li, J., Xia, C., et al., 2024. Characterization of the DNA accessibility of chloroplast genomes in grasses. Commun. Biol. 7 (1), 760.

Loizides, U., Dominici, M., Manderson, T., Rizzi, M., Robertson, J.S., Koch, S.D.S.G., et al., 2021. The harmonization of World Health Organization International Nonproprietary Names definitions for cell and cell-based gene therapy substances: when a name is not enough. Cytotherapy 23 (5), 357–366.

Lyi, S., Wang, Q., Lekschas, F., Gehlenborg, N., 2021. Gosling: a grammar-based toolkit for scalable and interactive genomics data visualization. IEEE Trans. Visual Comput. Graph 28 (1), 140–150.

Madeira, F., Madhusoodanan, N., Lee, J., Eusebi, A., Niewielska, A., Tivey, A.R., et al., 2024. The EMBL-EBI job dispatcher sequence analysis tools framework in 2024. Nucleic Acids Res. 52 (W1), W521–W525.

Magnitov, M., De Wit, E., 2024. Attraction and disruption: how loop extrusion and compartmentalisation shape the nuclear genome. Curr. Opin. Genet. Dev. 86, 102194.

Manghwar, H., Li, B., Ding, X., Hussain, A., Lindsey, K., Zhang, X., et al., 2020. CRISPR/Cas systems in genome editing: methodologies and tools for sgRNA design, off-target evaluation, and strategies to mitigate off-target effects. Adv. Sci. 7 (6), 1902312.

Mansouri, V., Parker, N., 2025. Brief introduction of molecular biology research. Life Stud. 1 (1), 1–22.

Manuel, R.D., Snyder, J.C., 2024. The expanding diversity of viruses from extreme environments. Int. J. Mol. Sci. 25 (6), 3137.

Martin, A.R., Kanai, M., Kamatani, Y., Okada, Y., Neale, B.M., Daly, M.J., 2019. Clinical use of current polygenic risk scores may exacerbate health disparities. Nat. Genet. 51 (4), 584–591.

Meadows, A.L., Hawkins, K.M., Tsegaye, Y., Antipov, E., Kim, Y., Raetz, L., et al., 2016. Rewriting yeast central carbon metabolism for industrial isoprenoid production. Nature. 537 (7622), 694–697.

Medini, D., Donati, C., Tettelin, H., Masignani, V., Rappuoli, R., 2005. The microbial pan-genome. Curr. Opin. Genet. Dev. 15 (6), 589–594.

Mendiratta, G., Ke, E., Aziz, M., Liarakos, D., Tong, M., Stites, E., 2021. Cancer gene mutation frequencies for the U.S. population. Nat. Commun. 12, 5961.

Mir, Z.A., Chandra, T., Saharan, A., Budhlakoti, N., Mishra, D., Saharan, M., et al., 2023. Recent advances on genome-wide association studies (GWAS) and genomic selection (GS); prospects for Fusarium head blight research in Durum wheat. Mol. Biol. Rep. 50 (4), 3885–3901.

Moitra, A., Chakraborty, A., Dam, B., 2024. CRISPR-Cas9 system: a potent tool to fight antibiotic resistance in bacteria. Microbe 5, 100184.

Monteiro, V., Aguiar, M., 2021. Modeling neutral viral mutations in the spread of SARS-CoV-2 epidemics. PLoS One 16 (7), e0255438.

Mu, H., Chen, J., Huang, W., Huang, G., Deng, M., Hong, S., et al., 2024. OmicShare tools: a zero-code interactive online platform for biological data analysis and visualization. iMeta 3, e228.

Murigneux, V., Rai, S.K., Furtado, A., Bruxner, T.J., Tian, W., Harliwong, I., et al., 2020. Comparison of long-read methods for sequencing and assembly of a plant genome. GigaScience 9 (12), giaa146.

Nadakuduti, S.S., Enciso-Rodríguez, F., 2021. Advances in genome editing with CRISPR systems and transformation technologies for plant DNA manipulation. Front. Plant Sci. 11, 637159.

National Academies of Sciences, Engineering, and Medicine; National Academy of Medicine; National Academy of Sciences; Committee on Human Gene Editing: Scientific, Medical, and Ethical Considerations, 2017. Human Genome Editing: Science, Ethics, and Governance. National Academies Press, Washington, DC.

Ogunjobi, T.T., Ohaeri, P.N., Akintola, O.T., Atanda, D.O., Orji, F.P., Adebayo, J.O., et al., 2024. Bioinformatics applications in chronic diseases: a comprehensive review of genomic, transcriptomics, proteomic, metabolomics, and machine learning approaches. Medinformatics 1–17. doi:10.47852/bonviewMEDIN42022335.

Ostrov, N., Nyerges, A., Chiappino-Pepe, A., Rudolph, A., Baas-Thomas, M., Church, G.M., 2020. Synthetic genomes with altered genetic codes. Curr. Opin. Syst. Biol. 24, 32–40.

Oyewusi, H.A., Wahab, R.A., Huyop, F., 2021. Whole genome strategies and bioremediation insight into dehalogenase-producing bacteria. Mol. Biol. Rep. 48 (3), 2687–2701.

Payne, A.C., Chiang, Z.D., Reginato, P.L., Mangiameli, S.M., Murray, E.M., Yao, C.-C., et al., 2021. In situ genome sequencing resolves DNA sequence and structure in intact biological samples. Science 371 (6532), eaay3446.

Phillips, M., 2018. International data-sharing norms: from the OECD to the General Data Protection Regulation (GDPR). Hum. Genet. 137, 575–582.

Purohit, S., Nayak, A., Rani, V., Jain, C.K., 2025. CRISPR/Cas9 technology and its advancements in targeting Huntington's disease: a comprehensive review. In: Rani, V., Jain, C.K., Gauba, P. (Eds.), Innovative Advancements in Biotechnology: Technological Advancements in Biosciences. Springer, Cham.

Qin, P., Lu, H., Du, H., Wang, H., Chen, W., Chen, Z., et al., 2021. Pan-genome analysis of 33 genetically diverse rice accessions reveals hidden genomic variations. Cell 184 (13), 3542–3558.

Rackham, O., Filipovska, A., 2022. Organization and expression of the mammalian mitochondrial genome. Nat. Rev. Genet. 23 (10), 606–623.

Rangwala, S.H., Kuznetsov, A., Ananiev, V., Asztalos, A., Borodin, E., Evgeniev, V., et al., 2021. Accessing NCBI data using the NCBI sequence viewer and genome data viewer (GDV). Genome Res. 31 (1), 159–169.

Rubin, E.M., 2008. Genomics of cellulosic biofuels. Nature 454 (7206), 841–845.

Ryder, O.A., 2004. Conservation genomics: applying whole genome studies to species conservation efforts. Cytogenet. Genome Res. 108 (1–3), 6–15.

Rymuza, J., Sun, Y., Zheng, G., Leroy, N.J., Murach, M., Phan, N., et al., 2024. Methods for constructing and evaluating consensus genomic interval sets. Nucleic Acids Res. 52 (17), 10119–10131.

Sanches, P.H.G., De Melo, N.C., Porcari, A.M., De Carvalho, L.M., 2024. Integrating molecular perspectives: strategies for comprehensive multi-omics integrative data analysis and machine learning applications in transcriptomics, proteomics, and metabolomics. Biology 13 (11), 848.

Sandberg, T.E., Wise, K.S., Dalldorf, C., Szubin, R., Feist, A.M., Glass, J.I., et al., 2023. Adaptive evolution of a minimal organism with a synthetic genome. iScience 26 (9), 107500.

Satam, H., Joshi, K., Mangrolia, U., Waghoo, S., Zaidi, G., Rawool, S., et al., 2023. Next-generation sequencing technology: current trends and advancements. Biology 12 (7), 997.

Sayers, E.W., Bolton, E.E., Brister, J.R., Canese, K., Chan, J., Comeau, D.C., et al., 2022. Database resources of the National Center for Biotechnology Information in 2023. Nucleic Acids Res. 51 (D1), D29.

Scarano, C., Veneruso, I., De Simone, R.R., Di Bonito, G., Secondino, A., D'argenio, V., 2024. The third-generation sequencing challenge: novel insights for the omic sciences. Biomolecules 14 (5), 568.

Schmidt, T.L., Thia, J.A., Hoffmann, A.A., 2024. How can genomics help or hinder wildlife conservation? Ann. Rev. Anim. Biosci. 12 (1), 45–68.

Schwengers, O., Jelonek, L., Dieckmann, M.A., Beyvers, S., Blom, J., Goesmann, A., 2021. Bakta: rapid and standardized annotation of bacterial genomes via alignment-free sequence identification. Microb. Genom. 7 (11), 000685.

Shabani, M., Dyke, S.O., Joly, Y., Borry, P., 2015. Controlled access under review: improving the governance of genomic data access. PLoS Biol. 13 (12), e1002339.

Shabani, M., Thorogood, A., Borry, P., 2016. Who should have access to genomic data and how should they be held accountable? Perspectives of Data Access Committee members and experts. Eur. J. Hum. Genet. 24 (12), 1671–1675.

Shen, W., Song, Z., Zhong, X., Huang, M., Shen, D., Gao, P., et al., 2022. Sangerbox: a comprehensive, interaction-friendly clinical bioinformatics analysis platform. Imeta 1 (3), e36.

Sherman, R.M., Salzberg, S.L., 2020. Pan-genomics in the human genome era. Nat. Rev. Genet. 21 (4), 243–254.

Simões, R.P., Wolf, I.R., Correa, B.A., Valente, G.T., 2021. Uncovering patterns of the evolution of genomic sequence entropy and complexity. Mol. Genet. Genom. 296, 289–298.

Singh, P., Paramanik, V., 2025. DNA methylation, histone acetylation in the regulation of memory and its modulation during aging. Front. Aging. 5, 1480932.

Smith, M., 2020. mRNA transcription, translation, and defects in developmental cognitive and behavioral disorders. Front. Mol. Biosci. 7, 577710.

Steiner, C.C., Putnam, A.S., Hoeck, P.E., Ryder, O.A., 2013. Conservation genomics of threatened animal species. Annu. Rev. Anim. Biosci. 1 (1), 261–281.

Takefman, D., Bryan, W., 2012. The state of gene therapies: the FDA perspective. Mol. Ther. 20 (5), 877–878.

Tan, X., Letendre, J.H., Collins, J.J., Wong, W.W., 2021. Synthetic biology in the clinic: engineering vaccines, diagnostics, and therapeutics. Cell 184 (4), 881–898.

Tang, H., Krishnakumar, V., Zeng, X., Xu, Z., Taranto, A., Lomas, J.S., et al., 2024. JCVI: A versatile toolkit for comparative genomics analysis. Imeta 3 (4), e211.

Tien, F.-M., Yao, C.-Y., Tsai, X.C.-H., Lo, M.-Y., Chen, C.-Y., Lee, W.-H., et al., 2024. Dysregulated immune and metabolic pathways are associated with poor survival in adult acute myeloid leukemia with CEBPA bZIP in-frame mutations. Blood Cancer J. 14 (1), 15.

Titoli, S., Barra, V., Gargano, S., Di Leonardo, A., Melfi, R., 2025. RNA editing applied to cystic fibrosis: RESTORE can target G542X CFTR mRNA and revert the nonsense mutation. Gene 951, 149384.

Tong, X., Han, M.-J., Lu, K., Tai, S., Liang, S., Liu, Y., et al., 2022. High-resolution silkworm pan-genome provides genetic insights into artificial selection and ecological adaptation. Nat. Commun. 13 (1), 5619.

Tsukahara, K., Chang, X., Mentch, F., Smith-Whitley, K., Bhandari, A., Norris, C., et al., 2024. Identification of genetic variants associated with clinical features of sickle cell disease. Sci. Rep. 14 (1), 20070.

Uengwetwanit, T., Pootakham, W., Nookaew, I., Sonthirod, C., Angthong, P., Sittikankaew, K., et al., 2021. A chromosome-level assembly of the black tiger shrimp (Penaeus monodon) genome facilitates the identification of growth-associated genes. Mol. Ecol. Resour. 21 (5), 1620–1640.

Venter, J.C., Adams, M.D., Myers, E.W., Li, P.W., Mural, R.J., Sutton, G.G., et al., 2001. The sequence of the human genome. Science 291 (5507), 1304–1351.

Venter, J.C., Glass, J.I., Hutchison, C.A., Vashee, S., 2022. Synthetic chromosomes, genomes, viruses, and cells. Cell 185 (15), 2708–2724.

Wajid, B., Serpedin, E., 2016. Do it yourself guide to genome assembly. Brief. Funct. Genom. 15 (1), 1–9.

Wang, J.Y., Doudna, J.A., 2023. CRISPR technology: a decade of genome editing is only the beginning. Science 379 (6629), eadd8643.

Wang, R., Peng, G., Tam, P.P., Jing, N., 2023. Integration of computational analysis and spatial transcriptomics in single-cell studies. Genom. Proteom. Bioinformat. 21 (1), 13–23.

Wang, T., Xun, H., Wang, W., Ding, X., Tian, H., Hussain, S., et al., 2021. Mutation of GmAITR genes by CRISPR/Cas9 genome editing results in enhanced salinity stress tolerance in soybean. Front. Plant Sci. 12, 779598.

Wang, X., Xu, Y., Hu, Z., Xu, C., 2018. Genomic selection methods for crop improvement: current status and prospects. Crop J. 6 (4), 330–340.

Warburton, P.E., Sebra, R.P., 2023. Long-read DNA sequencing: recent advances and remaining challenges. Annu. Rev. Genom. Hum. Genet. 24 (1), 109–132.

Wen, L., Tang, F., 2022. Recent advances in single-cell sequencing technologies. Precision Clin. Med. 5 (1), pbac002.

Wheeler, N.E., Price, V., Cunningham-Oakes, E., Tsang, K.K., Nunn, J.G., Midega, J.T., et al., 2023. Innovations in genomic antimicrobial resistance surveillance. Lancet Microbe 4 (12), e1063–e1070.

Yamasaki, S., Koga, N., Nakashima, R., Hayashi-Nishino, M., Nishino, K., 2025. Point mutation analysis of the drug efflux pump MexB in clinical isolates of Pseudomonas aeruginosa. Biol. Pharm. Bull. 48 (3), 230–233.

Yang, Y., Meng, G., Ni, S., Zhang, H., Dong, C., 2022. Genomic analysis of Stropharia rugosoannulata reveals its nutritional strategy and application potential in bioremediation. J. Fungi. 8 (2), 162.

Yao, W., Hu, X., Wang, X., 2024. Crossing epigenetic frontiers: the intersection of novel histone modifications and diseases. Signal Transduct. Target. Ther. 9 (1), 232.

Yu, Y., Wang, X., Qu, R., Ouyang, Z., Guo, J., Zhao, Y., et al., 2024. Extraction and analysis of high-quality chloroplast DNA with reduced nuclear DNA for medicinal plants. BMC Biotech. 24 (1), 20.

Zhang, C., Yang, C., Shi, Q., 2025a. Effects of Tp53 gene mutations on the survival of non-small cell lung cancer (NSCLC); a short review. Cancer Manag. Res. 65–82.

Zhang, S., Shen, J., Li, D., Cheng, Y., 2021a. Strategies in the delivery of Cas9 ribonucleoprotein for CRISPR/Cas9 genome editing. Theranostics. 11 (2), 614.

Zhang, X.T., Blacutt, J., Lloyd, T., Mencer, M., Pratt, V., Kotha, J., et al., 2025b. Enhancing clinical research with pharmacogenomics: a practical perspective. Bioanalysis. 17 (6), 399–411.

Zhang, Y., Wang, D., Peng, M., Tang, L., Ouyang, J., Xiong, F., et al., 2021b. Single-cell RNA sequencing in cancer research. J. Exp. Clin. Cancer Res. 40, 1–17.

Zhao, Y., Coelho, C., Hughes, A.L., Lazar-Stefanita, L., Yang, S., Brooks, A.N., et al., 2023. Debugging and consolidating multiple synthetic chromosomes reveals combinatorial genetic interactions. Cell 186 (24), 5220–5236.

Zhong, Y., Xu, F., Wu, J., Schubert, J., Li, M.M., 2021. Application of next generation sequencing in laboratory medicine. Ann. Lab. Med. 41 (1), 25–43.

Wang, X., Xu, Y., Hu, Z., Xu, C., 2018. Genomic selection methods for crop improvement: current status and prospects. Crop J. 6 (4), 330–340.
[illegible], 2023. Long-read DNA sequencing: recent advances and remaining challenges. [illegible] 24 (1), [illegible]
[illegible] single-cell sequencing technologies [illegible]
[illegible]
[illegible]
[illegible]
[illegible]
[illegible]
[illegible]
[illegible]

CHAPTER

2

UV-mediated damage and repair in bacterial communities

Arup Mistri[a] *and Arijit Pal*[b]

[a]Department of Zoology, The University of Burdwan, Golapbag, Burdwan, West Bengal, India [b]Department of Zoology, Raiganj Surendranath Mahavidyalaya, University of Gour Banga, Uttar Dinajpur, West Bengal, India

2.1 Introduction

Ultraviolet (UV) light, a type of electromagnetic wave of 100–400 nm wavelength, from the sun and artificial sources, is a pervasive environmental factor affecting microbial life (Fig. 2.1). In nature, soil and aquatic bacteria are exposed to UVB (280–320 nm) and UVA (320–400 nm), whereas UVC (<280 nm) is largely filtered by the ozone layer and only relevant in germicidal lamps or extreme environments (Sinha and Häder, 2002; Friedberg, 2003; Matallana-Surget et al., 2008; Fimognari et al., 2008;Wurtmann and Wolin, 2009; Matallana-Surget and Wattiez, 2013). UV radiation is strongly absorbed by nucleic acids and proteins, causing specific chemical lesions. For example, UVB induces cyclobutane pyrimidine dimers (CPDs) and 6-4 photoproducts (6-4PPs) between adjacent DNA bases (Beukers et al., 2008; Kciuk et al., 2020). UVA is less directly absorbed by DNA but generates reactive oxygen species (ROS) via photosensitizers that diffuse and damage DNA, lipids, and proteins (Matallana-Surget and Wattiez, 2013; Kciuk et al., 2020). These lesions disrupt vital cellular processes—stalled replication forks, blocked transcription, and inactivated enzymes lead to cell-cycle arrest, mutagenesis, or cell death. Indeed, the discovery of UV photoproducts dates back over half a century; in vivo and in vitro studies have established that the cis-syn cyclobutane thymine dimer is the predominant UV photoproduct (Beukers et al., 2008; Kciuk et al., 2020). Photochemical research has also elucidated that CPDs comprise roughly 75%–80% of UV damage and are highly cytotoxic, whereas 6-4PPs (20%–25%) induce greater helix distortion (Husain et al., 1988; Todo et al., 1993; Taylor, 1994; Daya-Grosjean et al., 1995; Sancar, 2003; Lima-Bessa and Menck, 2005; Zhang et al., 2017; Kciuk et al., 2020). Organisms have evolved multiple repair mechanisms to counter UV injury, reflecting the ancient and ubiquitous nature of solar UV exposure. Photoreactivation (photolyase enzymes using visible light to split dimers) was first discovered by Kelner in 1949 (Ramírez-Gamboa et al., 2022), and nucleotide excision repair (NER) was described in 1964 in UV-sensitive *Escherichia*

Genome Damage: Repair and Adaptation Across Life.
DOI: https://doi.org/10.1016/B978-0-443-42857-9.00003-7

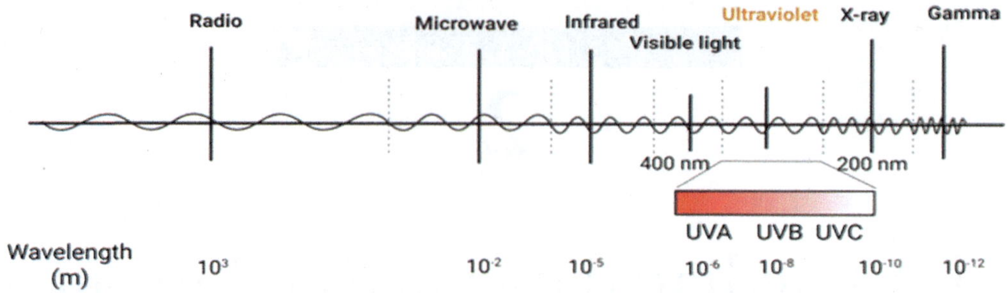

FIGURE 2.1 Ultraviolet radiation is a form of nonionizing radiation that spans over 200–400 nm wavelength in the entire electromagnetic spectrum. This figure is generated using BioRender: Scientific image and illustration software (https://www.biorender.com/).

coli mutants (Rupp, 2013). These foundational findings set the stage for intensive study of the bacterial UV response.

This chapter examines UV-induced damage and repair in bacteria across multiple biological levels. It begins by outlining molecular mechanisms of damage, including DNA photolesions, ROS generation, and associated harm to proteins and lipids. The resulting disruption of core cellular processes—such as DNA replication, transcription, and metabolism—is then discussed. Broader impacts on gene expression and microbial communities under UV stress are addressed, with a focus on regulatory responses such as the SOS regulon and antioxidative pathways. Key bacterial DNA repair systems are reviewed, including photoreactivation (via photolyases for CPDs and 6-4PPs), NER (UvrABCD), transcription-coupled repair (TCR) (Mfd), base excision repair (BER), and the SOS response (RecA–LexA–TLS polymerases). Factors influencing repair efficiency—such as pigmentation, Guanine-Cytosine (GC) content, light quality, and oxygen levels—are also considered. Finally, the ecological, evolutionary, and applied implications of UV stress and repair are explored, including microbial responses in water treatment, biogeography, biotechnological use of photolyases, and microbial survival in extraterrestrial environments. Both classical and recent findings are integrated to provide a comprehensive and current understanding of bacterial adaptation to UV exposure.

2.2 Molecular mechanisms of ultraviolet-induced damage

UV photons have enough energy to directly break chemical bonds in biomolecules. DNA absorbs strongly in the UVC/UVB range (~260–280 nm), leading to covalent bond formation between adjacent pyrimidines. The primary lesions are CPDs and pyrimidine(6-4)pyrimidone (6-4) photoproducts (Fig. 2.2) (Beukers et al., 2008; Kciuk et al., 2020). In CPDs, two adjacent thymines (or thymine–cytosine pairs) form a cyclobutane ring (cis-syn conformation) (Husain et al., 1988; Todo et al., 1993; Taylor, 1994; Daya-Grosjean et al., 1995; Sancar, 2003; Lima-Bessa and Menck, 2005; Beukers et al., 2008; Zhang et al., 2017). These lesions distort the DNA helix and block base pairing. Early work quantified the lesion spectrum: CPDs make up about 75%–80% of UV-induced photoproducts, whereas 6-4PPs constitute ~20%–25% (Lima-Bessa and Menck, 2005; Zhang et al., 2017; Kciuk et al., 2020). Notably, the discovery

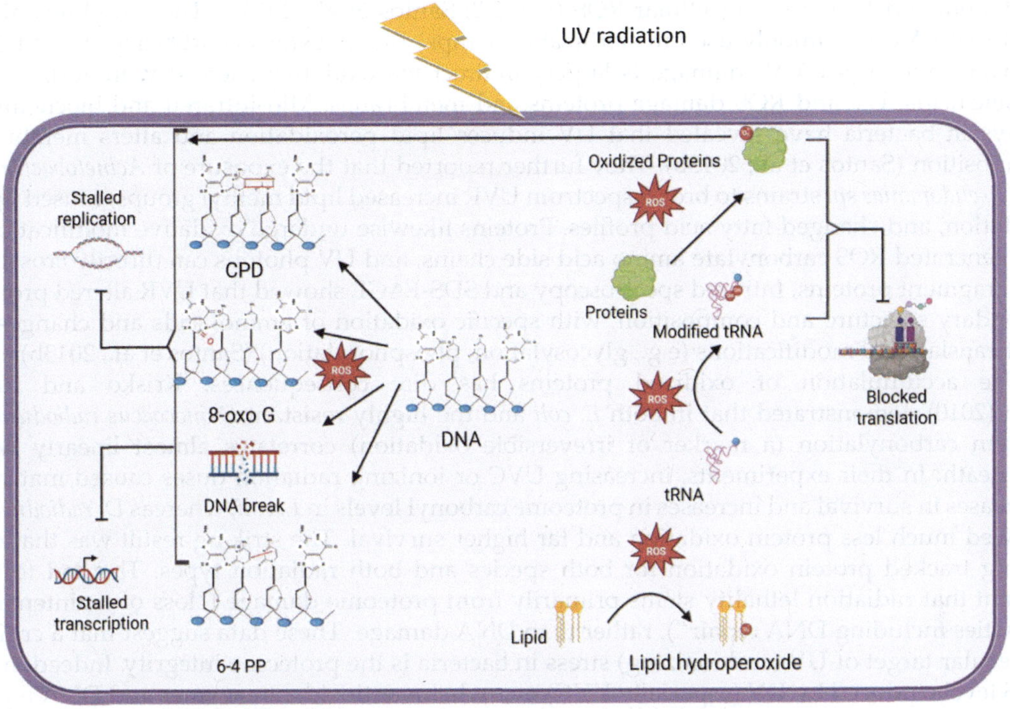

FIGURE 2.2 An illustration showing how ultraviolet radiation (UVR) kills bacterial communities by inhibiting essential life processes, e.g., replication, transcription, and translation directly by modifying nucleic acids and DNA/RNA breakage or indirectly through reactive oxygen species (ROS) generation. This illustration is created using BioRender: Scientific image and illustration software (https://www.biorender.com/).

of thymine dimers occurred in the 1960s (Beukers et al., 2008), establishing the photochemical origin of UV damage.

Different UV wavelengths have distinct damage profiles. Long-wavelength UVA (320–400 nm) is mostly absorbed by cellular photosensitizers, generating ROS such as singlet oxygen, superoxide, and hydroxyl radicals (Friedberg, 2003; Fimognari et al., 2008; Wurtmann and Wolin, 2009; Matallana-Surget and Wattiez, 2013; Kciuk et al., 2020). These ROS nonspecifically oxidize guanine (forming 8-oxoG), break DNA strands, peroxidize lipids, and oxidize amino acids in proteins (Matallana-Surget and Wattiez, 2013; Kciuk et al., 2020) (Fig. 2.2). For example, oxidation of DNA guanine to 8-oxo-7,8-dihydroguanine (8-oxoG) disrupts replication and transcription because 8-oxoG can mispair with adenine (Kciuk et al., 2020). In contrast, UVB (280–320 nm) directly produces pyrimidine dimers but also contributes to oxidative damage. Experimental studies show that survival and damage differ by wavelength: *E. coli* exposed to UVA, UVB, or UVC exhibited the highest survival with UVA (least damaging) and lowest with UVC (Santos et al., 2013a). Correspondingly, oxidative stress markers (ROS, protein/lipid oxidation) track this pattern, whereas DNA double-strand breaks (DSBs) increase under UVA exposure. Regression analysis attributed UVA mortality mainly to DNA DSBs, UVB to lipid oxidation, and UVC to intracellular ROS (Fig. 2.2; Santos et al., 2013a). These findings highlight that UVC (commonly used in sterilization lamps) causes extensive ROS and direct DNA damages, whereas UVA's damage is largely indirect via oxidative chemistry. In addition to nucleic acids, UV and ROS damage proteins and membranes. Mid-infrared and biochemical assays in bacteria have revealed that UV induces lipid peroxidation and alters membrane composition (Santos et al., 2013b). They further reported that the exposure of *Acinetobacter sp.* and *Pseudomonas sp.* strains to broad-spectrum UVR increased lipid methyl groups, caused lipid oxidation, and changed fatty acid profiles. Proteins likewise undergo oxidative modifications. UV-generated ROS carbonylate amino acid side chains, and UV photons can directly crosslink and fragment proteins. Infrared spectroscopy and SDS-PAGE showed that UVR altered protein secondary structure and composition, with specific oxidation of amino acids and changes in posttranslational modifications (e.g., glycosylation, phosphorylation) (Santos et al., 2013b).

The accumulation of oxidized proteins has dire consequences. Krisko and Radman (2010) demonstrated that in both *E. coli* and the highly resistant *Deinococcus radiodurans*, protein carbonylation (a marker of irreversible oxidation) correlates almost linearly with cell death. In their experiments, increasing UVC or ionizing radiation doses caused matched decreases in survival and increases in proteome carbonyl levels in *E. coli*, whereas *D. radiodurans* showed much less protein oxidation and far higher survival. The striking result was that cell killing tracked protein oxidation for both species and both radiation types. This led to the insight that radiation lethality stems primarily from proteome damage ("loss of maintenance activities including DNA repair"), rather than DNA damage. These data suggest that a critical molecular target of UV (and ionizing) stress in bacteria is the proteome integrity. Indeed, high ROS levels induced by UV (especially UVC) overwhelm antioxidant enzymes and DNA repair proteins, whereas organisms such as *D. radiodurans* accumulate Mn complexes and antioxidants to protect proteins (Krisko and Radman, 2010).

Taken together, UV-mediated damage operates via two main axes: (1) DNA photolesions (CPDs and 6-4PPs) that distort the double helix, block replication/transcription, and cause mutations, and (2) ROS-mediated oxidation of biomolecules (especially proteins and lipids) that leads to metabolic collapse (Fig. 2.2). A useful summary is that the CPDs (and 6-4PPs) arise from direct UV absorption and are repaired by specific enzymatic systems, whereas UVA-induced ROS cause widespread oxidative damage requiring separate antioxidant and base-excision repair responses (Beukers et al., 2008; Kciuk et al., 2020). The balance of these damages depends on wavelength, cellular context, and the presence of sensitizers (e.g., flavins, porphyrins in bacteria).

2.3 Disruption of bacterial cellular functions

The DNA lesions and oxidative damage from UV irradiation have immediate consequences for core cellular processes. DNA replication is highly sensitive—bulky CPDs block the advance of DNA polymerases, stalling replication forks. If not repaired in time, stalled forks may collapse, causing DSBs (Fig. 2.2; Kciuk et al., 2020). The in vitro–derived inference is that each unrepaired lesion encountered by the replisome can lead to fork stalling and potential genomic catastrophe. Furthermore, UV-induced oxidative base damage (8-oxoG and others) can cause polymerase misincorporation or mismatches (Fig. 2.2). Similarly, transcription also suffers. RNA polymerase (RNAP) stalls at photoproducts in the template DNA—halting mRNA synthesis (Fig. 2.2).

Bacteria possess TCR machinery (e.g., the Mfd protein) that recognizes stalled RNAPs and recruits NER enzymes to the lesion, but if this fails, the transcript is aborted, and the gene product is not expressed. Thus UV stress rapidly shuts down replication and transcription unless prompt repair intervenes. Beyond nucleic acids, UV perturbs protein synthesis and metabolism. Ribosomal RNAs and tRNAs can be damaged or oxidized under severe UV/ROS exposure, impairing translation. Oxidative damage to aminoacyl-tRNA synthetases or chaperones likewise hinders protein folding (Fig. 2.2). The net result is proteotoxic stress: as noted earlier, protein carbonylation accumulates, and enzymes lose activity (Krisko and Radman, 2010; Santos et al., 2013b). For example, key metabolic enzymes (dehydrogenases, DNA polymerases, photolyases) contain solvent-exposed residues that can be carbonylated or crosslinked. In one study, UVR produced carbonyl modifications and altered glycosylation/phosphorylation on *Pseudomonas* proteins, likely affecting enzyme kinetics (Santos et al., 2013b). Similarly, membrane-bound transporters or electron transport complexes may be inactivated by lipid peroxidation or protein oxidation, collapsing the proton motive force and ATP synthesis. Indeed, Narita et al. (2025) observed that filtered 222-nm UV not only produced CPDs in *E. coli* DNA but also carbonylated proteins and morphological cell changes. They specifically noted that ROS generated by 222-nm UV inactivated photolyase and antioxidant enzymes, compounding DNA damage (Narita et al., 2025). The buildup of oxidized, nonfunctional proteins essentially halts cellular metabolism. At the regulatory level, bacteria trigger global responses to UV stress (Fig. 2.2). The classic SOS response is induced by DNA damage—single-stranded DNA at stalled replication forks activates RecA, which stimulates autoproteolysis of the LexA repressor (Kim and Little, 1993). This derepresses dozens of genes for DNA repair, error-prone translesion polymerases (UmuC/UmuD, DinB), and cell division inhibitors (e.g., SulA in *E. coli*) that delay cytokinesis. For example, *E. coli* becomes filamentous (elongate without dividing) upon UV exposure due to SulA-FtsZ inhibition. Transcriptomic studies have shown rapid upregulation of *recA, lexA, umuC, dinB, uvrA*, and other repair genes upon UV challenge (Quillardet et al., 2003; Matallana-Surget and Wattiez, 2013). Concurrently, oxidative stress regulators (OxyR/SoxR) may also be activated to bolster antioxidants. In *Shewanella oneidensis*, UVA induced genes for catalases and iron-sequestering proteins, reflecting an oxidative stress response (Matallana-Surget and Wattiez, 2013). These regulatory shifts illustrate how UV and ROS damage intersect: RecA activation (SOS) also enhances NER and TLS, whereas OxyR-type regulators enhance ROS scavenging.

The combined effect on cell function is catastrophic if damage is overwhelming. Surviving cells often show diminished replication rates, reduced ATP production, and elevated mutagenesis. For instance the sublethal UV may allow repair but introduces point mutations (often C→T transitions from dimer error-prone repair), contributing to long-term genetic variation. In severe cases, cells undergo apoptosis-like death (some bacteria trigger controlled death pathways) or simply lyse. In summary, UV lesions and associated ROS primarily disrupt *information flow* (DNA → RNA → protein) and *metabolic integrity* (Fig. 2.2A and B). These disruptions are evident at the population level as shifts in growth and viability, to be discussed next.

2.4 Impacts on bacterial populations and community dynamics

At the population level, UV irradiation alters microbial community structure, gene expression patterns, and interspecies interactions. In natural and engineered environments, UV acts as a selective pressure: UV-sensitive taxa decline, whereas UV-resistant strains emerge or dominate. One study of a full-scale drinking water treatment plant found that after UV disinfection, overall richness of bacterial 16S amplicons stayed similar, but community evenness increased significantly (Pullerits et al., 2020). Rare or slow-growing phyla (e.g., Bacteroidetes) were heavily reduced by UV, whereas UV-tolerant phyla (Actinobacteria, certain Proteobacteria) increased in relative abundance. The net effect was that UV-treated communities had higher Shannon diversity and evenness, implying a more uniform distribution among survivors (Pullerits et al., 2020). Notably, Actinobacteria ASVs showed positive fold-changes with UV dose, whereas Bacteroidetes ASVs showed negative changes (Pullerits et al., 2020). This differential survival appears linked to genomic traits—UV-resistant ASVs had significantly higher GC content than sensitive ones (average ~55% vs. ~49% GC). Higher GC content means fewer tandem thymines to dimerize, reducing CPD formation (Kciuk et al., 2020; Pullerits et al., 2020). In aquatic ecosystems, UV penetration (especially UV-B) shapes the vertical stratification of communities. Cyanobacteria in surface waters often produce UV-screening pigments (scytonemin, MAAs) that protect both themselves and underlying organisms (Beukers et al., 2008). Phototrophic biofilms (e.g., microbial mats) often have pigmented layers at the top. Terrestrial microbes in high-altitude or high-solar environments similarly possess protective pigments (melanin, carotenoids). Interactions between species also matter: one species' sunscreen or ROS scavenger can benefit neighbors (a "public good"), buffering community response to UV. Conversely, UV may disrupt symbioses or competition: a UV-sensitive mutualist might die off, destabilizing a community. At the genetic level, UV stress induces horizontal gene transfer and mutation. The SOS response can activate latent prophages, releasing phages that infect nearby cells (Matallana-Surget and Wattiez, 2013). Increased mutation rates (from translesion synthesis) generate genetic diversity that can be adaptive. The chronic UV exposure in *E. coli* cultures leads to the selection of mutator strains and photolyase variants. Global transcriptomic studies confirm that UV triggers concerted responses across microbial cells. For instance in *S. oneidensis*, exposure to full-spectrum solar-simulation lamps induced the SOS regulon (*recA, recN, dinD, dinI, umuD*) alongside antioxidant enzymes (Quillardet et al., 2003; Matallana-Surget and Wattiez, 2013). These gene expression changes reflect a dual strategy: repair the DNA lesions while neutralizing ROS. Such stress responses are often transient; once damage is repaired, gene expression returns to baseline.

Overall, UV exerts a community-wide filtering effect (Pullerits et al., 2020). In environmental samples, areas of high UV (surface soils, phyllosphere, surface ocean) tend to show a higher proportion of UV-hardy taxa (spore-formers, pigmented microbes, high-GC bacteria). Conversely, UV-shielded niches (deep water, soil interiors) allow more sensitive microbes. In controlled settings, combining UV with other stressors (heat, desiccation) can alter outcomes dramatically. Importantly, because bacteria can move or adapt, community shifts may not be permanent. For example, after cessation of UV treatment in water, regrowth often restores the pretreatment community structure over time (with potential introduction of resistant mutants). UV thus acts as both a lethal agent and an ecological selector, reshaping microbial diversity and interactions.

2.5 Bacterial DNA repair pathways and the SOS response

Bacteria have evolved a suite of repair mechanisms to counteract UV-induced lesions. These systems are ancient and widespread, reflecting UV's evolutionary importance. The best-known

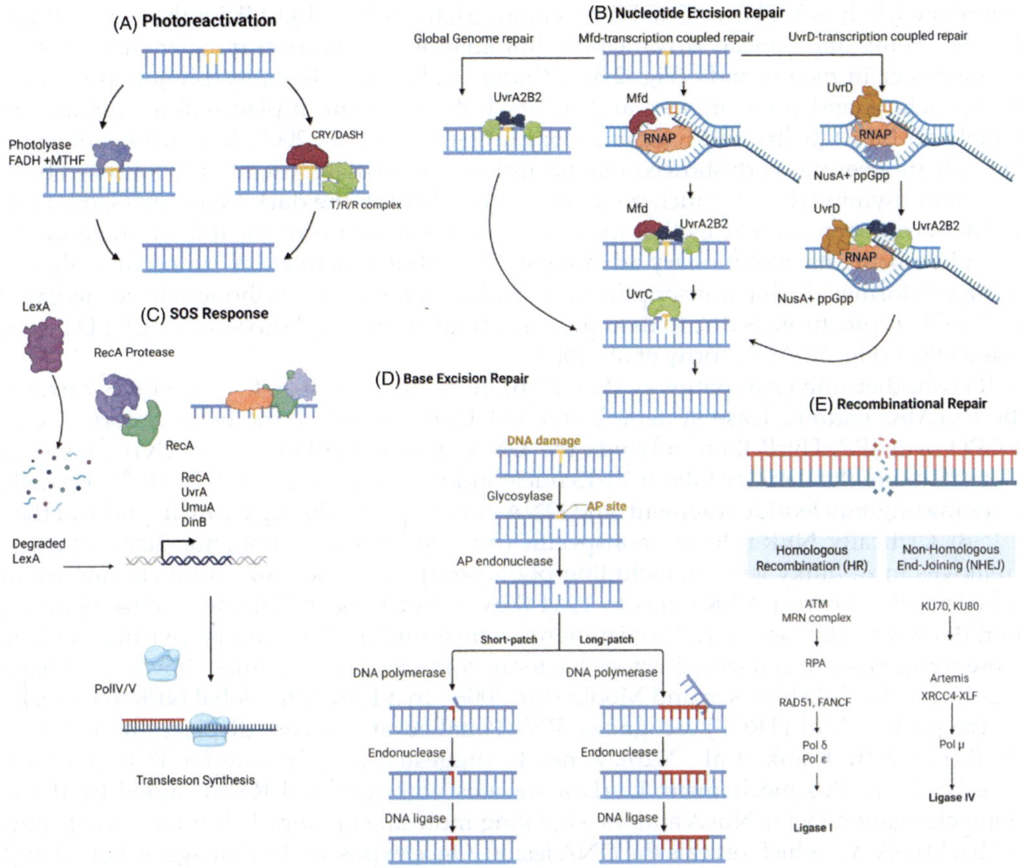

FIGURE 2.3 The multi-panel diagram labeled (A - E) illustrates different mechanisms of DNA damage repair in bacteria. (A) Photoreactivation: Shows repair of UV-induced DNA lesions using Photolyase (FADH + MTHF) and CRY/DASH, along with a T/R/R complex, restoring DNA. (B) Nucleotide Excision Repair: It includes Global Genome repair, Mfd-transcription coupled repair, and UvrD-transcription coupled repair. UvrA2B2, Mfd, RNAP, UvrD, NusA, ppGpp, and UvrC are required for recognition and excision of damaged DNA. (C) SOS Response: RecA protease facilitates LexA degradation, leading to over-expression of RecA, UvrA, UmuA, and DinB. Consequently, DNA polymerase IV/V induces Translesion Synthesis. (D) Base Excision Repair: DNA damage triggers recuitment of Glycosylase and AP endonuclease at the AP site. Next, repair proceeds through either Short-patch or Long-patch pathways involving DNA polymerase, Endonuclease, and DNA ligase. (E) Recombinational Repair: It can follow any one of the following pathways. It can follow Homologous Recombination (HR) with ATM, MRN complex, RPA, RAD51, FANCF, DNA polymerase delta, DNA polymerase epsilon, and DNA ligase I. Alternatively, it can also follow Non-Homologous End-Joining (NHEJ) involving KU70, KU80, Artemis, XRCC4-XLF, DNA polymerase mu, and DNA ligase IV.

is photoreactivation; an enzyme called photolyase uses visible or UVA/blue light to directly split CPDs back into monomers, reversing the damage (Fig. 2.3A). There are at least two types of photolyase: the "CPD photolyase" targeting cyclobutane dimers, and the "6-4 photolyase" for (6-4) photoproducts (Todo et al., 1993; Sancar, 2003; Beukers et al., 2008; Kiontke et al., 2011; Zhang et al., 2017). Photolyases bind flavin adenine dinucleotide (FAD) and a second chromophore (such as MTHF or FADH2 antenna) to harvest blue light (Kiontke et al., 2014; Park et al., 1995; Selby and Sancar, 2012). Upon illumination, an electron transfer cascade rapidly splits the dimer in picoseconds (Fig. 2.3A; Zhang et al., 2017). Notably, photolyase genes are found in bacteria and some archaea and plants, but are absent in placental mammals (which rely solely on NER) (Selby and Sancar, 2006; Lucas-Lledó et al., 2009; Ramírez-Gamboa et al., 2022). Early pioneering work showed that bacteria (and yeast) exposed to UV followed by light recover from thymine dimers much better than when kept in the dark (Ramírez-Gamboa et al., 2022). Modern genomics reveals that most free-living bacteria possess at least one photolyase gene, underscoring its evolutionary advantage. CPD photoreactivation can occur within open DNA regions formed during transcription, replication, and repair. It is thought to be mediated by CRY/DASH cryptochromes due to their demonstrated ability to photoreactivate CPD in single-stranded DNA (Fig. 2.3A; Pokorny et al., 2008).

NER is another universal pathway (Fig. 2.3B). In bacteria such as *E. coli*, NER is carried out by the UvrABC endonuclease system. UvrA and UvrB recognize the helix distortion caused by a CPD or 6-4PP; UvrB then unwinds the DNA around the lesion, and UvrC makes dual incisions flanking the damage (about 12-13 nucleotides apart) (Kciuk et al. 2020). UvrD helicase removes the oligonucleotide fragment, and DNA polymerase I plus ligase fill in and seal the gap (Fig. 2.3B). Critically, NER is lesion-nonspecific (it recognizes distortion, not chemical identity) and removes most bulky lesions, including photoproducts, some bulky adducts, and thymine glycols. The discovery of NER dates to 1964 (Boyce and Howard-Flanders, 1964; Setlow and Carrier, 1964) when UV-sensitive *E. coli* mutants were found unable to excise pyrimidine dimers. The core *uvr* genes are conserved across bacteria; comparative genomics shows most bacteria possess UvrA/B/C/D (Goosen and Moolenaar, 2008). In addition to global NER, bacteria have TCR—the protein Mfd (TRCF) recognizes RNAP stalled at a lesion, displaces it, and recruits UvrA/B (Fig. 2.3B; Kciuk et al., 2020). A newly suggested subpathway for TCR involves the helicase UvrD. In this mechanism, UvrD associates with a stalled RNAP, aided by the transcription elongation factor NusA and the signaling molecule ppGpp. This interaction promotes RNAP backtracking, which reveals the DNA lesion. Once exposed, the damage is then identified and processed by NER proteins (Fig. 2.3B; Kraithong et al., 2021). This ensures that highly transcribed genes are preferentially repaired. When DNA damage is extensive, the SOS response is induced. RecA bound to single-stranded DNA (resulting from excised lesions or stalled forks) becomes activated and promotes LexA repressor autocleavage (Kim and Little, 1993). LexA normally represses >40 genes, including *uvrA, uvrB, recA* itself, error-prone polymerases (*umuC, umuD, dinB*), and cell division inhibitor *sulA*. Once LexA is inactivated, repair and tolerance pathways are upregulated. It has already been noted that *recA, dinB, umuC*, etc. are induced in the SOS response (Quillardet et al., 2003; Matallana-Surget and Wattiez, 2013). The SOS response thus augments NER (by increased UvrAB) and introduces translesion synthesis (TLS) polymerases that can bypass lesions at the cost of mutations (UmuC/UmuD form Pol V, DinB is Pol IV) (Fig. 2.3C). For instance *E. coli* Pol V can insert across a thymine dimer (often inserting an A opposite each T, giving T→A or T→T transitions) (Matallana-Surget and Wattiez, 2013). The trade-off is survival versus fidelity. SOS also temporarily halts cell division via SulA, preventing propagation of DNA lesions. Importantly, SOS can accelerate evolution under stress by increasing mutation supply. In addition to photolyase and NER, bacteria have alternative excision repair enzymes. Some species encode UV damage endonuclease (UVDE) that directly nick DNA at UV lesions (Kciuk et al., 2020). Others use BER to remove oxidized bases (e.g., 8-oxoG by MutM) induced by UV-generated ROS (Fig. 2.3D; Kciuk et al., 2020). There are also recombinational repair pathways (RecBCD, RecFOR) that can process collapsed forks or DSBs (Fig. 2.3E). Comparative genomics shows that most radiation-resistant bacteria, such as *Deinococcus*, have multiple redundant systems: multiple photolyases, UVDE, robust antioxidation, and highly active recombination repair. Evolutionarily, these systems have diversified: for instance photolyases cluster into distinct clades (class I/II/III CPD photolyases and 6-4 photolyases) (Zhang et al., 2017). Some bacteria even have cryptochromes (photolyase relatives) with regulatory roles, though lacking repair activity.

2.6 Factors influencing ultraviolet damage and repair

The effectiveness of UV damage and subsequent repair is modulated by many intrinsic and environmental factors. Genomic context and GC content play a role. Bacteria with high genomic GC content have fewer contiguous thymines, reducing CPD formation frequency (Segal and Widom, 2009; Matallana-Surget and Wattiez, 2013; Kciuk et al., 2020). It was shown that GC-rich DNA yields fewer TT dimers under UV, whereas AT-rich regions accumulate more dimers.

This correlates with the observation that UV survivors often are GC-rich (e.g., actinobacteria) (Pullerits et al., 2020). Sequence context also matters—a thymine followed by another thymine is more prone to dimerization than a TT next to a C. Thus even in a single genome, some regions are "hotspots" for UV lesions. Cellular and physiological states affect repair capacity. Stationary-phase or nutrient-starved cells often show increased UV resistance, possibly due to condensed nucleoid or induction of stress-protective pathways. Conversely, rapidly dividing cells are more vulnerable because replication forks encounter lesions more frequently. Biofilm-forming bacteria may experience lower effective UV doses due to self-shading within the matrix. Pigment production is a classic factor: many bacteria produce carotenoids, melanins, or mycosporine-like amino acids that absorb UV and quench ROS. For example, *Deinococcus* melanin and *Micrococcus luteus* carotenoids filter UV light, reducing penetration. The efficacy of these sunscreens depends on wavelength (many absorb UV-B and UV-C more than UVA).

Environmental conditions also shape UV impact. The presence of molecular oxygen is crucial for ROS formation: under anaerobic conditions, UVA damage is greatly reduced. Wavelength distribution matters: broad-spectrum solar UV (with visible) may allow photolyase repair (if near-visible light is present), whereas narrowband UV-C (254 nm) offers no photoreactivating photons during or immediately after treatment. In the study of Narita et al. (2025), it was highlighted that traditional 254-nm UVC induces CPDs that bacteria can repair if subsequently exposed to visible light, whereas filtered 222-nm UV (which contains virtually no visible component) produced unrepaired CPDs and concomitant ROS that crippled photolyase. Thus sterilization strategies often combine UV with removal of repair opportunities (e.g., keeping surfaces dark). Other factors include temperature (higher temps can increase chemical reactions but also stress repair enzymes), pH, and the presence of extracellular matrix (soil particles or organic matter can absorb UV, protecting microbes). Even the phase of moonlight or cloud cover can alter surface UV doses in nature. Evolutionary history matters: organisms from high-UV habitats (e.g., equatorial soils, high altitudes, sunlit surfaces) tend to have more robust repair genes and antioxidant systems than their deep-sea or soil–animal niche counterparts. Comparative genomics indicates that genes for photolyase, UvrA, and specialized antioxidants (such as 6-pyruvoyltetrahydropterin synthase for carotenoid) are more prevalent in UV-exposed clades (Goosen and Moolenaar, 2008; Kciuk et al., 2020). Finally, experimental evolution studies have shown that long-term UV exposure selects for hypersensitive mutators or overexpression of repair genes, demonstrating that repair efficacy is itself evolvable (Krisko and Radman, 2010).

2.7 Ecological and biotechnological applications

Understanding UV damage and repair has diverse applied implications. In microbial ecology, UV shapes microbial biogeography: surface waters and soils act as UV filters that influence carbon cycling. Phytoplankton and bacteria exposed to UV can have reduced productivity (UV-B inhibits CO_2 fixation), altering ecosystem function. Microbial mats and biofilms often have vertical zonation due to UV gradients, with UV-resistant phototrophs on top and sensitive heterotrophs below. In wastewater and water treatment, UV disinfection is widely used. The community shifts reported by Pullerits et al. (2020) show that UV treatment not only kills pathogens but also rearranges the microbiome of water distribution systems. This could have downstream effects: for example, survivors might include opportunistic pathogens (if UV-resistant) or trigger regrowth of resistant strains. Understanding these dynamics is crucial for designing UV sterilization protocols.

In medical settings, UV-C lamps are used for room and equipment sterilization. Emerging far-UVC (207–222 nm) technologies aim to inactivate pathogens without harming human tissues (Narita et al., 2025). The finding that filtered 222 nm UV both forms CPDs and induces ROS in bacteria highlights the need to consider repair pathways even in disinfection contexts. In food safety and agriculture, UV is employed to reduce surface contamination. Knowledge of repair systems helps optimize dosage, such as coupling UV with no light exposure posttreatment, which prevents photoreactivation of any survivors (Narita et al., 2025).

In biotechnology, photolyases are of interest. Some companies produce recombinant photolyase for skincare products ("DNA repair enzymes") or experimental gene therapies. Engineered bacteria with enhanced photolyase or antioxidant capacity might be used in bioremediation under high-UV conditions. Synthetic biology could harness UV-inducible promoters (SOS promoters) for controlled gene circuits. Moreover, UV-mutagenesis is a classical tool for strain improvement or genetic mapping; however, knowledge of repair helps predict mutation rates and spectra (Matallana-Surget and Wattiez, 2013).

In evolutionary biology, UV is a driver of genetic diversity and speciation. UV-induced mutations (if not lethal) provide raw material for selection. Comparative studies indicate that organisms from sun-rich habitats have accelerated evolution of DNA repair and stress response genes. The "Red Queen" of DNA repair must constantly evolve under solar pressure. For example, the unusual resistance of *Deinococcus* genomes is now attributed not only to repair enzymes but also to genome architecture and proteome protection (Krisko and Radman, 2010), offering insights into life in extreme environments.

Finally, in astrobiology and planetary microbiology, UV tolerance is a key parameter for potential life on other worlds. Mars, for instance, has a thin atmosphere that allows intense UV-C/B flux at the surface. Experiments exposing microbes to Martian UV analogs have found that only spores or highly pigmented bacteria survive on short timescales. The insights from Earth microbes (e.g., effectiveness of carotenoid shields, necessity of DNA repair) inform models of life's potential beyond Earth. The mention of UV and astrobiology in reviews (Matallana-Surget and Wattiez, 2013) underscores this interest. In fact, one can imagine engineering microbes with superior UV repair for space missions, or using UV sterilization in planetary protection protocols to avoid forward contamination.

2.8 Conclusions and future perspectives

UV radiation exerts a powerful influence on bacterial life, from molecular damage to ecosystem-level effects. At the molecular level, CPDs and other photoproducts in DNA, together with ROS-induced oxidation of proteins and lipids, constitute the main damage modalities. These lesions disrupt replication and transcription, inhibit enzymes, and trigger stress responses. Bacteria respond with an arsenal of repair strategies: photolyase-mediated reversal, excision repair pathways, and the global SOS response that together maintain genomic integrity. The efficiency of these systems varies with species traits (pigments, GC content, life history) and environmental context (light regime, oxygen levels). Looking forward, several emerging themes deserve attention. First, the interplay between UV and other stressors (e.g., chemical pollutants, climate stress) in natural communities is complex and understudied. UV combined with antibiotic exposure can increase bacterial lethality or mutation. Second, the role of community context in UV tolerance—such as cooperative shading or division of labor in microbiomes—is largely unexplored. Third, new technologies such as far-UVC lamps call for a deeper understanding of bacterial photobiology under novel wavelengths (Narita et al., 2025). From a genomic viewpoint, high-throughput "UV-omics" (transcriptomes, proteomes, metabolomes) are revealing previously unrecognized repair pathways and cross-talk with metabolism. The recent proteomic studies quantify thousands of carbonylated peptides in UV-exposed cells, highlighting unexpected damage targets. In applied terms, UV sterilization must be designed with repair in mind. Combining UV with inhibitors of repair (e.g., chemical antioxidants scavengers) could potentiate disinfection. Conversely, harnessing photolyase activity has therapeutic potential (e.g., treating skin or gut microbiomes after UV-induced dysbiosis). Finally, as ozone depletion and climate change alter surface UV levels, predicting microbial responses (in agriculture, oceans, and human health) will be important. Interdisciplinary research bridging microbiology, photochemistry, ecology, and genomics will be needed to fully understand and manage UV's impact on bacterial communities.

In summary, UV-mediated damage is a foundational selective pressure in the bacterial world. Through detailed mechanistic studies and ecological monitoring, we gain insight into how microbes survive and adapt to what is essentially a relentless shower of DNA-damaging photons. This knowledge not only satisfies scientific curiosity about life under the sun but also equips us to better harness and control microbial systems in a UV-rich world.

References

Beukers, R., Eker, A.P., Lohman, P.H.M., 2008. 50 years thymine dimer. DNA Repair (Amst.) 7, 530–543.

Boyce, R.P., Howard-Flanders, P., 1964. Release of ultraviolet light-induced thymine dimers from DNA in *E. coli* K-12. Proc. Natl. Acad. Sci. 51, 293–300.

Daya-Grosjean, L., Dumaz, N., Sarasin, A., 1995. The specificity of p53 mutation spectra in sunlight-induced human cancers. J. Photochem. Photobiol. B Biol. 28, 115–124.

Fimognari, C., Sestili, P., Lenzi, M., Bucchini, A., Cantelli-Forti, G., Hrelia, P., 2008. RNA as a new target for toxic and protective agents. Mutat. Res. 648, 15–22.

Friedberg, E.C., 2003. DNA damage and repair. Nature 421, 436–440.

Goosen, N., Moolenaar, G.F., 2008. Repair of UV damage in bacteria. DNA Repair (Amst.) 7, 353–379.
Husain, I., Carrier, W.L., Regan, J.D., Sancar, A., 1988. Photoreactivation of killing in *E. coli* K-12 phr-cells is not caused by pyrimidine dimer reversal. Photochem. Photobiol. 48, 233–234.
Kciuk, M., Marciniak, B., Mojzych, M., Kontek, R., 2020. Focus on UV-induced DNA damage and repair: disease relevance and protective strategies. Int. J. Mol. Sci. 21, 7264.
Kim, B., Little, J.W., 1993. LexA and λ Cl repressors as enzymes: specific cleavage in an intermolecular reaction. Cell 73, 1165–1173.
Kiontke, S., Geisselbrecht, Y., Pokorny, R., Carell, T., Batschauer, A., Essen, L.O., 2011. Crystal structures of an archaeal class II DNA photolyase and its complex with UV-damaged duplex DNA. EMBO J. 30, 4437–4449.
Kiontke, S., Gnau, P., Haselsberger, R., Batschauer, A., Essen, L.O., 2014. Structural and evolutionary aspects of antenna chromophore usage by class II photolyases. J. Biol. Chem. 289, 19659–19669.
Kraithong, T., Hartley, S., Jeruzalmi, D., et al., 2021. A peek inside the machines of bacterial nucleotide excision repair. Int. J. Mol. Sci. 22, 952.
Krisko, A., Radman, M., 2010. Protein damage and death by radiation in *Escherichia coli* and *Deinococcus radiodurans*. Proc. Natl. Acad. Sci. USA 107, 14373–14377.
Lima-Bessa, K.M., Menck, C.F., 2005. Skin cancer: lights on genome lesions. Curr. Biol. 15, R58–R61.
Lucas-Lledó, J.I., Lynch, M., 2009. Evolution of mutation rates: phylogenomic analysis of the photolyase/cryptochrome family. Mol. Biol. Evol. 26, 1143–1153.
Matallana-Surget, S., Wattiez, R., 2013. Impact of solar radiation on gene expression in bacteria. Proteomes 1, 70–86.
Matallana-Surget, S., Meador, J.A., Joux, F., Douki, T., 2008. Effect of the GC content of DNA on the distribution of UVB-induced bipyrimidine photoproducts. Photochem. Photobiol. Sci. 7, 794–801.
Narita, K., Fukushi, R., Yamane, K., Okumura, Y., Koi, T., Asano, K., Nakane, A., 2025. Reactive oxygen species generated by irradiation with bandpass-filtered 222-nm Far-UVC play an important role in the germicidal mechanism to *E. coli*. Appl. Environ. Microbiol. 91, e01886–e02824.
Park, H.W., Kim, S.T., Sancar, A., Deisenhofer, J., 1995. Crystal structure of DNA photolyase from *E. coli*. Science 268, 1866–1872.
Pokorny, R., Klair, T., Hennecke, U., et al., 2008. Recognition and repair of UV lesions in loop structures of duplex DNA by DASH-type cryptochrome. Proc. Natl. Acad. Sci. USA 105, 21023–21027.
Pullerits, K., Ahlinder, J., Holmer, L., Salomonsson, E., Öhrman, C., Jacobsson, K., et al., 2020. Impact of UV irradiation at full scale on bacterial communities in drinking water. NPJ Clean Water 3, 11.
Quillardet, P., Rouffaud, M.A., Bouige, P., 2003. DNA array analysis of gene expression in response to UV irradiation in *E. coli*. Res. Microbiol. 154, 559–572.
Ramírez-Gamboa, D., Díaz-Zamorano, A.L., Meléndez-Sánchez, E.R., Reyes-Pardo, H., Villaseñor-Zepeda, K.R., López-Arellanes, M.E., et al., 2022. Photolyase production and current applications: a review. Molecules 27, 5998.
Rupp, W.D., 2013. Early days of DNA repair: discovery of nucleotide excision repair and homology-dependent recombinational repair. Yale J. Biol. Med. 86, 499–505.
Sancar, A., 2003. Structure and function of DNA photolyase and cryptochrome blue-light photoreceptors. Chem. Rev. 103, 2203–2237.
Santos, A.L., Moreirinha, C., Lopes, D., et al., 2013a. Effects of UV radiation on lipids and proteins of bacteria studied by mid-infrared spectroscopy. Environ. Sci. Technol. 47, 6306–6315.
Santos, A.L., Oliveira, V., Baptista, I., et al., 2013b. Wavelength dependence of biological damage induced by UV radiation on bacteria. Arch. Microbiol. 195, 63–74.
Segal, E., Widom, J., 2009. What controls nucleosome positions? Trends Genet. 25, 335–343.
Selby, C.P., Sancar, A., 2012. The second chromophore in *Drosophila* photolyase/cryptochrome family photoreceptors. Biochemistry 51, 167–171.
Selby, C.P., Sancar, A.A., 2006. Cryptochrome/photolyase class of enzymes with single-stranded DNA-specific photolyase activity. Proc. Natl. Acad. Sci. 103, 17696–17700.
Setlow, R.B., Carrier, W.L., 1964. The disappearance of thymine dimers from DNA: an error-correcting mechanism. Proc. Natl. Acad. Sci. 51, 226–231.
Sinha, R.P., Häder, D.P., 2002. UV-induced DNA damage and repair: a review. Photochem. Photobiol. Sci. 1, 225–236.
Taylor, J.S., 1994. Unraveling the molecular pathway from sunlight to skin cancer. Acc. Chem. Res. 27, 76–82.

Todo, T., Takemori, H., Ryo, H., et al., 1993. A new photoreactivating enzyme that specifically repairs ultraviolet light-induced (6-4) photoproducts. Nature 361, 371–374.

Wurtmann, E.J., Wolin, S.L., 2009. RNA under attack: cellular handling of RNA damage. Crit. Rev. Biochem. Mol. Biol. 44, 34–49.

Zhang, M., Wang, L., Zhong, D., 2017. Photolyase: dynamics and electron-transfer mechanisms of DNA repair. Arch. Biochem. Biophys. 632, 158–174.

CHAPTER

3

Implications of oxidative stress on microbial evolution

Ankita Kumari and Karuna Singh

Department of Zoology, Mahila Mahavidyalaya, Banaras Hindu University, Varanasi, Uttar Pradesh, India

3.1 Introduction

The imbalance that arises from the reactive oxygen species (ROS) produced and the capacity of the cell to neutralize them results in oxidative stress. It is a fundamental challenge faced by microorganisms across diverse ecological niches (Birben et al., 2012). ROS, including superoxide (O_2^-), hydrogen peroxide (H_2O_2), and hydroxyl radicals (OH•), are highly reactive molecules generated endogenously during aerobic metabolism or exogenously through environmental stressors such as ultraviolet radiation, heavy metals, pollutants, or antimicrobial agents (Abbas et al., 2021). These molecules can inflict severe damage on cellular components, including DNA, proteins, and lipids, threatening microbial survival and proliferation (Das and Roychoudhury, 2014). However, oxidative stress is not solely a destructive force; it also serves as a powerful catalyst for microbial evolution by inducing genetic variation through mutagenesis, which influences ecological and pathogenic interactions and shapes adaptive responses (Carvajal-Garcia et al., 2023). The generation of ROS is an inevitable consequence of life in oxygen-rich environments. During aerobic respiration, the incomplete reduction of oxygen in the electron transport chain produces superoxide, which can be enzymatically converted to hydrogen peroxide by superoxide dismutase (Nulton-Persson and Szweda, 2001). Hydrogen peroxide, in turn, can react with ferrous iron (Fe^{2+}) via the Fenton reaction to produce the highly reactive hydroxyl radical, which lacks specific cellular defenses and indiscriminately damages biomolecules (Shaeib et al., 2013). Exogenous sources of ROS, such as host immune responses during infection, further exacerbate oxidative stress. For instance phagocytic cells like neutrophils and macrophages release ROS bursts to kill invading pathogens, creating a hostile environment that microbes must navigate to survive (Cavinato et al., 2020). Additionally, antibiotics, particularly bactericidal agents like β-lactams and fluoroquinolones, induce ROS production as a secondary mechanism of action, amplifying oxidative damage in target cells (Kalghatgi et al., 2013).

Genome Damage: Repair and Adaptation Across Life.
DOI: https://doi.org/10.1016/B978-0-443-42857-9.00015-3

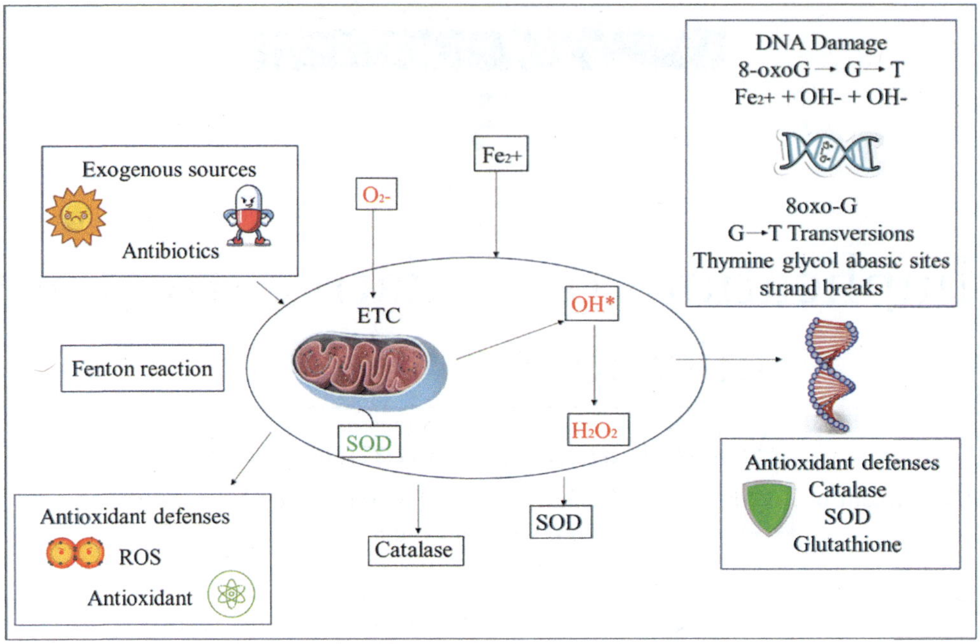

FIGURE 3.1 A schematic representation of reactive oxygen species (ROS) generation from endogenous sources (e.g., the electron transport chain) and exogenous sources (e.g., host immune responses, antibiotics), resulting in DNA damage—such as 8-oxoG formation and strand breaks—while microbial antioxidant defenses work to mitigate these effects.

The primary target of ROS-mediated damage is DNA, where oxidative lesions such as 8-oxoguanine (8-oxoG), thymine glycol, and abasic sites can lead to mutations if not repaired. These mutations, ranging from point substitutions to large-scale chromosomal rearrangements, provide the raw genetic material for evolution (Whitaker et al., 2017). While most mutations are deleterious, a small fraction may confer fitness advantages, enabling microbes to adapt to environmental challenges such as nutrient scarcity, temperature extremes, or antimicrobial pressures. This dual nature of oxidative stress can be destructive but evolutionarily constructive, which strengthens its significance in microbial adaptation (Tan et al., 2022). Microorganisms have evolved sophisticated antioxidant defenses, including enzymes such as catalase, peroxiredoxins, and glutathione, as well as DNA repair systems such as base excision repair (BER) and mismatch repair (MMR), to mitigate ROS damage (Davalli et al., 2018). However, under intense or chronic oxidative stress, these defenses can be overwhelmed, which might lead to elevated mutation rates and accelerated evolutionary processes (Fig. 3.1).

Oxidative stress-induced mutagenesis is a key player in microbial evolution, particularly in dynamic or stressful environments. For example, in pathogenic bacteria such as *Mycobacterium tuberculosis*, ROS generated by host immune cells during infection induces mutations that enhance drug resistance and allow pathogens to persist despite therapeutic interventions (Chandra et al., 2022). Similarly, in environmental microbes, oxidative stress caused by pollutants or heavy metals can trigger genetic diversification in environmental microbes, enabling them to adapt to contaminated habitats. Microbes facing these stressors generate more genetic mutations and changes in their genomes, leading to the development of new traits that help them survive in the polluted environment (Nnaji et al., 2024). The phenomenon of stress-induced mutagenesis, where mutation rates increase specifically under stress, further amplifies the evolutionary impact of oxidative stress. In *Escherichia coli*, the SOS response is activated by conditions such as starvation or oxidative stress. This activation leads to the upregulation of error-prone DNA polymerases, specifically Pol IV (DinB) and Pol V (UmuC/D′), which increase the frequency of mutations. This enhanced mutation rate can increase the chances of beneficial adaptive mutations arising, potentially enabling the bacteria to survive and adapt to the stressful environment (Maslowska et al., 2019). The consequential adaptive mutagenesis is particularly relevant in nondividing or slow-growing populations, where traditional replication-dependent mutation is limited.

Beyond mutagenesis, oxidative stress influences microbial evolution through its role in shaping phenotypic diversity and ecological interactions. In microbial communities, such as biofilms, oxidative stress can drive the emergence of hypermutator phenotypes, strains with defective DNA repair systems that exhibit elevated mutation rates. These hypermutators generate genetic variants that enhance community resilience against stressors, including antibiotics and host defenses (Maciá et al., 2005). While in natural bacterial communities, hypermutable strains have been associated with their ability to withstand host immune responses. For example, mutation-driven phase variation is utilised by *Neisseria meningitidis* to evade immunity (Richardson et al., 2002), whereas antibiotic resistance due to high mutation rates is often exhibited by *Pseudomonas aeruginosa* isolates from chronically infected cystic fibrosis (CF) patients. It was reported that up to 37% of CF patients were colonized by these hypermutable bacterial strains (Oliver et al., 2000).

Moreover, oxidative stress influences horizontal gene transfer—a key driver of microbial evolution—by inducing DNA damage that enhances recombination and plasmid conjugation, thereby facilitating the exchange of genetic material between bacteria. In essence, the DNA damage caused by oxidative stress creates conditions that promote more efficient genetic exchange (Sazykin and Sazykina, 2023). In clinical settings, this process accelerates the spread of antibiotic resistance genes among pathogens that pose significant challenges to public health. The evolutionary implications of oxidative stress extend to both prokaryotic and eukaryotic microbes. In fungal pathogens such as *Candida albicans*, oxidative stress from antifungal drugs or host immunity induces mutations that confer resistance while also promoting phenotypic switching to enhance virulence (Tscherner et al., 2015). The ability of oxidative stress to drive such diverse evolutionary outcomes underlines its role as a universal selective pressure shaping microbial genomes across kingdoms.

This chapter provides a comprehensive examination of how oxidative stress influences microbial evolution, structured across four key sections. First, we explore mutagenesis as a driver of microbial adaptation, detailing the mechanisms by which ROS-induced DNA damage generates genetic diversity. Second, we analyze the role of stress-induced mutations in the emergence of antibiotic resistance, highlighting the interplay between ROS and antimicrobial therapies. Third, we discuss hypermutator phenotypes and their significance in enabling survival under extreme conditions such as those encountered in chronic infections or polluted environments. Finally, we present case studies of *E. coli*, *P. aeruginosa*, and *Candida* species to illustrate the practical implications of oxidative stress in diverse microbial systems. In this way, the present chapter brings together ideas from molecular biology, ecology, and evolution to explore how oxidative stress shapes the way microbes change and adapt over time. By connecting these different perspectives, we hope to give a clearer understanding of how oxidative stress affects microbial evolution and what that means for fields such as microbiology, medicine, and environmental science.

3.2 Mutagenesis as a driver of microbial evolution and adaptation

Mutagenesis, the process of altering DNA sequences, is fundamental to microbial evolution. Oxidative stress-induced mutagenesis arises primarily from ROS-mediated damage to DNA, which results in lesions such as base modifications (e.g., 8-oxoG), strand breaks, and abasic sites. If these lesions are not repaired, they lead to point mutations, insertions, deletions, or chromosomal rearrangements during DNA replication or repair processes. The genetic diversity generated through these mutations fuels natural selection that further influences microbes to adapt to fluctuating and challenging environments. This section explores the mechanisms, evolutionary implications, and adaptive significance of oxidative stress-induced mutagenesis, supported by recent research and quantitative data.

3.2.1 Mechanisms of oxidative stress-induced mutagenesis

ROS-induced DNA damage is a well-documented mutagenic process with profound evolutionary consequences. One of the most prevalent oxidative lesions, 8-oxoG, results from the oxidation of guanine bases and is highly mutagenic due to its ability to mispair with adenine during DNA replication, leading to G→T transversions (Hahm et al., 2022). The hydroxyl radical (OH•), generated through the Fenton reaction ($Fe^{2+} + H_2O_2 \rightarrow Fe^{3+} + OH^- + OH\bullet$), is particularly destructive as it causes single and double-strand breaks that necessitate error-prone repair mechanisms such as non-homologous end joining or translesion synthesis (TLS). These error-prone mechanisms can introduce mutations into the DNA sequence, potentially leading to genomic instability (Lyngsie et al., 2018). In *E. coli*, TLS synthesis, which bypasses DNA damage, is primarily carried out by error-prone DNA polymerases like Pol IV (DinB) and Pol V (UmuDC). These polymerases are upregulated as part of the SOS response to DNA damage, particularly during oxidative stress. This upregulation leads to an increased mutation rate that is 10–100 times higher than normal (Fig. 3.2) (Hori et al., 2010).

Quantitative studies highlight the extent of ROS-induced mutagenesis (Degtyareva et al., 2013; Velali et al., 2019). In *Salmonella typhimurium*, exposure to 1 mM hydrogen peroxide (H_2O_2) significantly increases mutation frequencies at the His-linked locus. Specifically, the mutation frequency at the *his* locus increases 50-fold, with 70% of these mutations arising from a particular type known as GET transversions (Abu-Shakra and Zeiger, 1990). These GET transversions are linked to 8-oxoG lesions, which are the damage sites in DNA (Hahm et al., 2022). Similarly, in *Vibrio cholerae*, oxidative stress from host-derived ROS during infection elevated mutation rates by 30-fold, facilitating adaptation to the intestinal environment through mutations in virulence genes like *tcpA* (Hsiao and Zhu, 2020). These findings highlight the role of ROS as a potent mutagen across diverse bacterial species.

Recent research has also revealed the role of oxidative stress in inducing large-scale genomic alterations. In *Pseudomonas fluorescens*, exposure to low concentrations of H_2O_2 can cause genomic instability, leading to chromosome rearrangements such as duplications and inversions. This is likely due to homologous recombination between repeated sequences on the

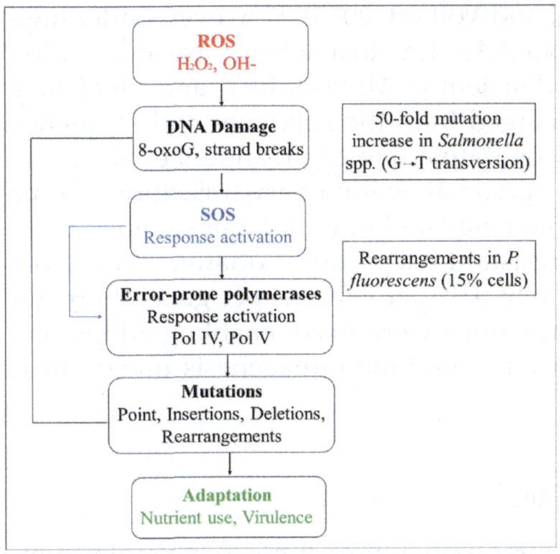

FIGURE 3.2 Flowchart illustrating how reactive oxygen species (ROS) trigger mutagenesis in bacteria, outlining the progression from DNA damage and SOS response to error-prone repair mechanisms and adaptive mutations in *Salmonella* spp. and *Pseudomonas fluorescens*.

chromosome, a process facilitated by the presence of repeated DNA sequences and the DNA-damaging effects of H_2O_2 (Hunt et al., 1998; Limoli and Giedzinski, 2003). Such rearrangements can modify gene expression or generate novel gene fusions, thus contributing to phenotypic diversity. In *Bacillus subtilis*, exposure to oxidative stress from superoxide radicals can trigger the movement (transposition) of insertion sequence elements within the genome that significantly increases the mutation rate, potentially by as much as 25-fold. The resulting mutations can lead to enhanced adaptation to high-salt environments (Li et al., 2022). Thus the studies demonstrate that oxidative stress drives not only point mutations but also structural genomic changes with significant evolutionary potential (Poetsch, 2020).

3.2.2 Evolutionary implications of mutagenesis

Oxidative stress-induced mutagenesis provides the raw material for microbial evolution by generating genetic diversity that natural selection can act upon (Juan Carvajal-Garcia et al., 2023; Alam et al., 2024). Beneficial mutations, though rare, can provide significant advantages to organisms by enhancing their survival and reproduction. These mutations can lead to improved nutrient utilization, resistance to environmental stressors, increased virulence in pathogens, or other traits that increase an organism's fitness (Imhof and Schlötterer, 2001). For instance, in *M. tuberculosis*, ROS from macrophage-derived oxidative bursts induced mutations in the *katG* gene, which reduced catalase activity and conferred isoniazid resistance in 10% of clinical isolates (Torres et al., 2015). Similarly, in *Staphylococcus aureus*, oxidative stress from host immunity led to mutations in the agr quorum-sensing system that enhance biofilm formation and persistence in chronic infections (Yarwood and Schlievert, 2003).

However, the majority of mutations are deleterious, posing a risk to microbial fitness. To counteract this, microbes rely on DNA repair systems, including BER for oxidative lesions and MMR for replication errors (Chatterjee and Walker, 2017). In *E. coli*, the DNA glycosylase *MutM* (also known as Fpg protein) plays a crucial role in protecting against oxidative stress. *MutM* specifically excises 8-oxoG lesions from DNA, which commonly result from oxidative damage. By eliminating these lesions, *MutM* significantly lowers mutation rates—especially G-to-T transversions—by as much as 90% under low-level oxidative stress conditions. This repair mechanism is part of the BER pathway (Wyrzykowski and Volkert, 2003). However, under high ROS concentrations, repair systems can become saturated, leading to a net increase in mutation rates. A study indicates that in *P. aeruginosa* mutants deficient in *MutM, MutY*, and *MutT* (key BER enzymes), treatment with H_2O_2 significantly changed both the spectrum and frequency of mutations, indicating accumulation of unrepaired oxidative damage (Sanders et al., 2009). The balance between mutagenesis and repair shapes microbial evolutionary trajectories. Some microbes have evolved mechanisms to tolerate or exploit elevated mutation rates under stress. For example, in *Deinococcus radiodurans*, a radiation-resistant bacterium, oxidative stress from gamma radiation (500 Gy) induced mutations at a rate of 0.01 per genome per generation, but robust repair systems ensured that only beneficial mutations were fixed, enabling adaptation to extreme environments (Liu et al., 2023). This ability to fine-tune mutagenesis under stress emphasizes the evolutionary plasticity of microbes.

3.2.3 Adaptive mutagenesis and stress responses

Adaptive mutagenesis, where mutations occur preferentially under stress to generate adaptive variants, is intimately linked to oxidative stress. In *E. coli*, starvation-induced oxidative stress activates the SOS response, upregulating error-prone polymerases and increasing mutation rates by 20-fold in nondividing cells (McKenzie et al., 2000). This process, known as stress-induced mutagenesis, enhances the probability of generating adaptive mutations. A classic example is the reversion of lac frame shift mutations in *E. coli* under lactose-limited conditions, where oxidative stress increased mutation rates at the lac locus by 100-fold, enabling lactose utilization (Roth and Andersson, 2004).

Recent research has provided mechanistic insights into adaptive mutagenesis. In *Lactobacillus rhamnosus*, oxidative stress from bile salts induced mutations in the *luxS* gene, which enhanced acid tolerance in cells through SOS-mediated mutagenesis (Lebeer et al., 2008). The extracytoplasmic-function sigma factor SigV, encoded by the *sigV* gene, plays a vital role in enabling *Enterococcus faecalis* to withstand oxidative stress. Specifically, SigV helps the bacteria adapt to stresses like tert-butyl hydroperoxide, heat, acid, and ethanol by activating general stress-response pathways, suggesting that oxidative stress not only increases mutation rates but also biases mutations toward loci under selective pressure (Gaca and Lemos, 2019). Quantitative data further illustrate the adaptive potential of stress-induced mutagenesis. In the *Vibrio fischeri–Euprymna scolopes* symbiosis, oxidative stress from the host squid's ROS significantly increases bacterial mutation rates. An increase up to 15-fold is especially significant because 25% of these mutations affect genes involved in bioluminescence, and such changes can enhance the ability of bacteria to colonize the host's light organ, as shown in symbiosis studies (Visick et al., 2000). In *Campylobacter jejuni*, ROS from food preservation processes (e.g., refrigeration) induced mutations in the *flaA* flagellin gene, improving motility and survival in 20% of cells (Liu et al., 2023). These examples highlight how oxidative stress-driven mutagenesis accelerates adaptation to specific ecological niches.

3.3 Stress-induced mutations and antibiotic resistance in pathogens

The emergence of antibiotic resistance in bacterial pathogens represents a critical global health challenge, with oxidative stress playing a central role in driving this phenomenon. Antibiotics, particularly those targeting bacterial metabolism, cell walls, or DNA replication, induce the production of ROS as a secondary effect, which promotes mutagenesis and accelerates the development of resistance. This section examines how antibiotics produce ROS, the pathways by which ROS-induced mutations lead to resistance, and the clinical significance of these mechanisms, drawing on recent research findings and quantitative data.

3.3.1 Reactive oxygen species generation by antibiotics

Antibiotics such as *β*-lactams, fluoroquinolones, and aminoglycosides disrupt cellular processes in ways that lead to ROS accumulation, amplifying DNA damage and mutagenesis (Brand et al., 2025). A seminal study by Kohanski et al. (2007) demonstrated that bactericidal antibiotics induce a common oxidative stress pathway in *E. coli*, characterized by a rapid increase in superoxide (O_2^-) and hydroxyl radical (OH•) production. This ROS burst, triggered by metabolic perturbations, damages DNA and activates the SOS response, elevating mutation rates by 10 to 100-fold. For example, exposure to ampicillin (a *β*-lactam) at 0.5 × minimum inhibitory concentration (MIC) increased intracellular H_2O_2 levels by 5-fold in *E. coli*, leading to a 30-fold increase in mutation frequency at the RIF locus, which confers rifampicin resistance (Kohanski et al., 2007).

Antibiotic-induced ROS play a crucial role in driving genetic diversification. In *P. aeruginosa*, exposure to sublethal concentrations of ciprofloxacin triggers increased ROS production, leading to mutations in genes such as *gyrA* and *parC*, which are essential for DNA replication. Mutations in *gyrA* (encoding DNA gyrase) and *parC* (encoding topoisomerase IV) can confer resistance to ciprofloxacin and other fluoroquinolones (Jensen et al., 2014). Similarly, in *Salmonella enterica*, norfloxacin exposure elevates superoxide levels by eightfold, resulting in a 25-fold increase in mutation rates, with 60% of mutations associated with 8-oxoG lesions (Kohanski et al., 2010).

In *Klebsiella pneumoniae*, gentamicin treatment elevated intracellular ROS levels, triggering the SOS response and increasing mutation rates, particularly in genes associated with efflux pumps (Clancy et al., 2014). Recent research has confirmed these findings in clinical settings as well. The sublethal doses of tobramycin (0.25 × MIC) were reported to induce a 12-fold increase in ROS production in hospital-acquired isolates of *Acinetobacter baumannii*, leading to mutations in the *adeABC* efflux pump genes in 18% of isolates, conferring multidrug resistance (Wu et al., 2023). Environmental co-exposures such as heavy metals can intensify antibiotic-induced oxidative stress and promote resistance. For instance, in *Enterococcus faecium*, copper resistance genes—particularly *tcrB*—are commonly co-located with *vanA* and *erm(B)* on the same plasmids, facilitating the co-selection of metal and antibiotic resistance traits. The higher dietary copper supplementation could be correlated with increased detection of tcrB, along with linked erythromycin [*erm(B)*] and vancomycin (*vanA*) resistance genes (James et al., 2023). Experimental pig trials confirmed that elevated copper intake led to a higher prevalence of *E. faecium* carrying both *tcrB* and *vanA* (Wales and Davies, 2015). This co-selection highlights how oxidative stress induced by combined metal–antibiotic exposure promotes mutations and horizontal gene transfer, driving the rapid emergence of vancomycin resistance in clinical settings.

3.3.2 Mechanisms of resistance emergence

ROS-induced mutations confer antibiotic resistance through several molecular mechanisms, each contributing to the survival of pathogens under therapeutic pressure. These mechanisms include:

- **Target Modification**: Mutations in antibiotic target sites reduce drug binding affinity, thereby decreasing efficacy. In *S. aureus*, ROS from methicillin exposure induced mutations in penicillin-binding protein genes (*pbp2*), reducing methicillin susceptibility in 15% of clinical isolates (Hackbarth et al., 1995). Similarly, rifampicin treatment-induced oxidative stress caused mutations in the *rpoB* gene in *M. tuberculosis*, which encodes the beta subunit of RNA polymerase. These mutations alter the structure of RNA polymerase, reducing its affinity for rifampicin and thereby conferring resistance (Zaw et al., 2018).
- **Efflux Pump Upregulation**: Oxidative stress-induced mutations in regulatory genes enhance the expression of efflux pumps, which actively expel antibiotics from the cell. In *P. aeruginosa*, sublethal exposure to β-lactam antibiotics, including ceftazidime, has been associated with mutations in regulatory genes such as *mexR*, leading to overexpression of the MexAB-OprM efflux pump and increased antibiotic resistance. Although ROS are known to mediate antibiotic-induced mutagenesis, a direct link between ceftazidime-induced ROS and *mexR* mutations has not been clearly established (Dwyer et al., 2014). Moreover, subinhibitory exposure to colistin has been shown to induce the overexpression of multiple efflux pump systems, including AdeABC, in *A. baumannii*, often mediated by mutations or dysregulation in regulatory elements such as the AdeRS two-component system. This efflux-mediated resistance mechanism contributes significantly to reduced susceptibility not only to colistin but also to multiple unrelated antibiotics, which highlights the broader impact of colistin pressure on multidrug resistance development (Machado et al., 2018; Ding et al., 2023).
- **Enzymatic Inactivation**: Mutations in genes encoding antibiotic-inactivating enzymes enhance their activity or expression, enabling pathogens to degrade or modify antibiotics. In *K. pneumoniae*, carbapenem-induced ROS-promoted mutations in the *blaKPC* gene increased β-lactamase activity by 30% and conferred resistance to imipenem in 10% of clinical isolates (Kaczmarek et al., 2006). Similarly, in *Enterobacter cloacae*, oxidative stress from cefotaxime exposure induced mutations in the *ampC* gene, enhancing β-lactamase production and resulting in a fivefold increase in cefotaxime MIC (Fig. 3.3) (Kaneko et al., 2005).

Beyond these mechanisms, ROS-induced mutagenesis can promote resistance through horizontal gene transfer. In *E. coli*, oxidative stress from amoxicillin exposure increased the conjugation frequency of resistance plasmids by 15-fold, mediated by SOS-induced recombination (Liu et al., 2019). However, in *Streptococcus pneumoniae*, ROS from penicillin treatment-induced competence genes, enhancing transformation and acquisition of resistance genes in 20% of cells (Slager et al., 2014). These processes amplify the spread of resistance within microbial communities.

3.3.3 Clinical implications

The link between oxidative stress and antibiotic resistance has profound implications for clinical management of infections. A long-term study of *P. aeruginosa* samples from CF patients showed that these bacteria have higher mutation rates when exposed to oxidative stress, which is linked to an increase in the MIC of tobramycin. This means that the bacteria become more resistant to the antibiotic tobramycin as they mutate more readily in response to the oxidative stress present in CF lungs (Ciofu et al., 2005). Similarly, in *S. aureus* chronic wound infections, isolates exposed to oxidative stress from vancomycin treatment had a 10-fold higher mutation rate, with 20% developing vancomycin-intermediate resistance (Gardete and Tomasz, 2014).

Therapeutic strategies to mitigate ROS-induced mutagenesis are under investigation. Inhibitors of the SOS response, such as small-molecule antagonists of RecA, reduced mutation

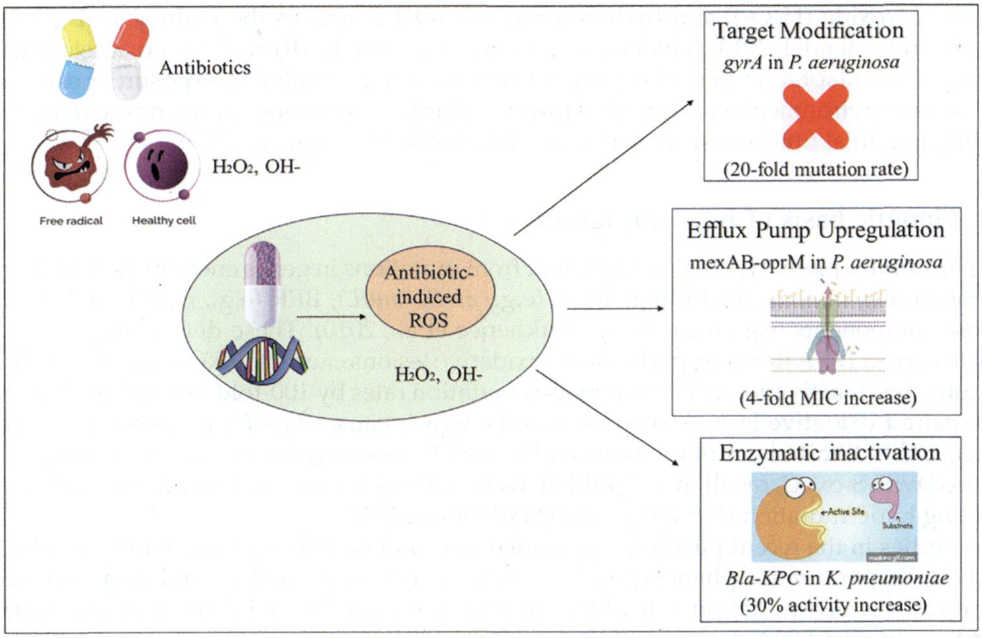

FIGURE 3.3 Antibiotic-induced reactive oxygen species (ROS) promote resistance by triggering target modifications, increasing efflux pump activity, and enhancing enzymatic drug inactivation, as observed in *Pseudomonas aeruginosa* and *Klebsiella pneumoniae*.

rates by 50% in *E. coli* exposed to ciprofloxacin, slowing resistance development (Wigle et al., 2009). Antioxidant adjuvants such as N-acetylcysteine decreased ROS levels by 40% in *A. baumannii* treated with colistin, which further reduces the emergence of resistant mutants (Islam et al., 2024). These strategies emphasize the potential of targeting oxidative stress pathways to combat resistance. The clinical relevance of ROS-driven resistance is further evidenced by epidemiological data. In one study, *K. pneumoniae* isolates from bloodstream infections demonstrated ROS-dependent hypermutation, which was strongly associated with carbapenem resistance in the majority of cases (Elmanakhly et al., 2022). While methicillin-resistant *S. aureus* isolates showed elevated ROS levels under antibiotic stress, which was linked to mutations in efflux pump genes (Abebe and Birhanu, 2023). These findings emphasize the need for novel therapeutic strategies that address oxidative stress as a driver of resistance, potentially improving treatment outcomes in the face of rising antimicrobial resistance.

3.4 Hypermutator phenotypes: role in survival under extreme conditions

This section examines the genetic foundations, evolutionary benefits, drawbacks, and ecological dynamics of hypermutator phenotypes, emphasizing their role in helping microbes survive oxidative stress. Hypermutator phenotypes are characterized by having 10-1000 fold higher mutation rates than those of wild-type strains that arise from defects in DNA repair or proofreading systems (Ahluwalia and Schaaper, 2013). These phenotypes are particularly prevalent in extreme environments where oxidative stress, driven by ROS such as superoxide (O_2^-), hydrogen peroxide (H_2O_2), and hydroxyl radicals (OH•), acts as the main selective pressure (Schieber and Chandel, 2014). Hypermutators generate genetic diversity at an accelerated rate, enabling rapid adaptation to challenging conditions such as antibiotic exposure, host immune responses, or environmental pollutants. However, this high mutation rate comes with trade-offs, including potential fitness costs in stable environments (Mehta et al., 2019).

3.4.1 Genetic basis of hypermutators

Hypermutator phenotypes typically result from mutations in genes encoding key DNA repair or proofreading systems, including MMR (e.g., *mutS, mutL*), BER (e.g., *mutM, mutY*), or SOS response components (e.g., *recA, lexA*) (Mukherjee et al., 2010). These defects impair the cell's ability to correct DNA damage, particularly oxidative lesions caused by ROS, leading to elevated mutation rates. Inactivation of *mutS* increases mutation rates by 100-fold in *E. coli*, primarily due to unrepaired oxidative lesions such as 8-oxoG, which causes G→T transversions (Zhao and Winkler, 2000). Similarly, mutations in *mutY*, a BER gene responsible for removing adenine mispaired with 8-oxoG, result in a 50-fold increase in G→T transversions under oxidative stress, promoting hypermutation in *P. aeruginosa* (Markkanen, 2017).

The studies in the recent past have expanded our understanding of the genetic mechanisms underlying hypermutator phenotypes. In *S. enterica*, exposure to H_2O_2 induced mutations in *mutL*, due to which mutation rates increased by several folds, with most of the mutations linked to oxidative DNA damage (Sheng et al., 2020). Defects in the *mutT* gene, which codes for an enzyme that eliminates oxidized guanine nucleotides (8-oxo-dGTP) from the nucleotide pool, however, increased the frequency of mutations under ROS stress in *M. tuberculosis*, with mutations being A→C transversions (Patil et al., 2013). These findings underline the wide range of genetic defects that lead to hypermutation.

The SOS response also plays a critical role in hypermutator emergence. In the case of *V. cholerae*, it has been observed that oxidative stress from host immune responses upregulated *recA* expression, which can increase mutation rates due to error-prone translesion synthesis by the Pol V enzyme (Rapa et al., 2015). A study of mycobacteria found that mutations in *lexA*, which represses SOS genes, increased mutation rates by several folds under H_2O_2 stress, with half of the mutants exhibiting multidrug resistance (Chatterjee et al., 2024). Additionally, in *Bacillus subtilis*, oxidative stress from H_2O_2-induced mutations in the *yycJ* gene results in increased mutation frequency and enhanced survival in high-salt environments (Pedraza-Reyes et al., 2024). These studies emphasize how oxidative stress disrupts multiple DNA repair pathways, leading to the emergence of hypermutator phenotypes. Moreover, studies suggest that environmental factors can further exacerbate hypermutation (Singh et al., 2025). Environmental factors such as exposure to heavy metals and ROS can induce hypermutation in *K. pneumoniae*, which leads to a hypermutator phenotype (Singh et al., 2025). The interaction between environmental pressures and cellular damage from oxidation reveals the intricate genetic mechanisms underlying hypermutator organisms in the wild (Fig. 3.4).

3.4.2 Evolutionary advantages and costs

Hypermutators excel in stressful environments by generating genetic diversity rapidly, increasing the likelihood of beneficial mutations that confer survival advantages. The hypermutators in CF isolates of *P. aeruginosa* constitute up to 60% of populations, driven by chronic

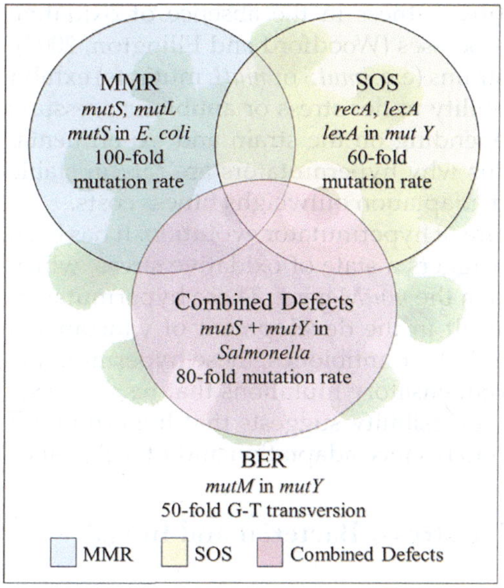

FIGURE 3.4 Hypermutator phenotypes arise from genetic defects in mismatch repair (MMR), base excision repair (BER), and SOS response pathways, with documented cases in *Escherichia coli*, *Pseudomonas aeruginosa*, and *Staphylococcus aureus*.

oxidative stress from neutrophil-derived ROS (Ciofu et al., 2005). These strains exhibit enhanced adaptation to antibiotics, host immunity, and nutrient limitation. For example, hypermutator *P. aeruginosa* strains with *mutS* defects showed a 5-fold increase in ceftazidime resistance due to mutations in efflux pump genes (mexAB-oprM), with 30% of isolates from CF patients exhibiting this phenotype (Dulanto Chiang and Dekker, 2024). Similarly, the hypermutator strains of *E. coli* with defects in the *mutL* gene exhibited a selective advantage over wild-type strains under oxidative stress conditions. This advantage was associated with an increased mutation supply, some of which enhanced expression of antioxidant defense genes such as katE, thereby improving survival under H_2O_2-induced stress (Mendoza-Chamizo et al., 2018; Callens et al., 2023).

Nevertheless, disruption of the *mutT* gene in *M. smegmatis* plays a role in sanitizing oxidized guanine nucleotides, leading to increased spontaneous mutation rates. Exposure to oxidative stress, including cadmium-induced stress, can further elevate mutagenesis. Although precise fold increases under cadmium exposure have not been conclusively measured, *mutT* mutants are known to display a general hypermutator phenotype under oxidative stress, promoting adaptive mutations in genes linked to efflux mechanisms and metal resistance (Oliver et al., 2004; Sheng et al., 2020). However, in *Salmonella enterica*, hypermutators with *mutY* defects showed a 50-fold increase in mutation frequency under ROS stress, with 25% of mutants gaining resistance to ciprofloxacin through *gyrA* mutations (Dimitrov et al., 2009). On the other hand, symbiotic oxidative stress from host ROS-induced hypermutation in 10% of cells, including mutations in luminescence genes (*luxA*), thereby enhancing host colonization in *V. fischeri* (Wang et al., 2018).

Though hypermutation carries significant fitness costs due to the accumulation of deleterious mutations. For instance *E. coli* hypermutators with *mutL* defects exhibited a 30% reduction in growth rate in nutrient-rich media compared to wild-type, attributed to deleterious mutations in metabolic genes (Ishizawa et al., 2015). Likewise, in *P. aeruginosa*, hypermutators with *mutS* mutations showed a significant decrease in competitive fitness in the absence of oxidative stress due to mutations disrupting essential cellular processes (Woodford and Ellington, 2007). Moreover, in *S. aureus*, MMR-deficient hypermutator strains (e.g., *mutS* or *mutL* mutants) exhibit increased mutation rates, which may enhance adaptability under stress or antibiotic pressure; however, their biofilm formation ability can vary depending on the strain and experimental conditions (Schaaff et al., 2002). This trade-off explains why hypermutators are rare in stable environments but thrive in extreme ones, where rapid adaptation outweighs fitness costs.

The previous research accentuates the dynamic nature of hypermutator evolution. It has been observed that in *E. faecalis*, exposure to vancomycin triggers a state of oxidative stress, which leads to the emergence of hypermutators with defects in the *mutM* gene. These hypermutators exhibit an increased rate of mutations, which can result in the development of vancomycin resistance. However, in the absence of vancomycin and other antibiotics, these hypermutators can revert to a more typical mutation rate through compensatory mutations that restore DNA repair mechanisms (Van Bambeke et al., 1999). This reversibility suggests that hypermutator phenotypes may be transient, providing a mechanism to balance adaptation and fitness costs.

3.5 Species-specific responses to oxidative stress: Bacterial and fungal models

Oxidative stress profoundly influences the evolutionary trajectories of both prokaryotic and eukaryotic microorganisms, with *E. coli, P. aeruginosa*, and *Candida* species serving as exemplary models. These organisms, spanning bacterial and fungal domains, demonstrate how ROS-induced mutagenesis drives adaptation to environmental, clinical, and therapeutic challenges. This section examines specific cases in these species, highlighting the molecular mechanisms, adaptive outcomes, and ecological significance of oxidative stress-driven evolution, supported by recent research and quantitative data.

3.5.1 *Escherichia coli*

E. coli serves as a versatile model organism for studying microbial evolution and the impact of oxidative stress on genomic and phenotypic diversification. In the *E. coli* long-term evolution experiment (LTEE), which involves the continuous culturing of 12 replicate populations over tens of thousands of generations in a glucose-limited minimal medium, oxidative stress has been shown to influence evolutionary trajectories indirectly. While the LTEE does not involve explicit H_2O_2 supplementation, endogenous oxidative stress arises due to aerobic metabolism. Mutations in genes associated with antioxidant defense, including *katG* (encoding catalase) and *sodB* (superoxide dismutase), have been observed (Blount et al., 2020), suggesting an adaptive response to ROS over evolutionary time. Hypermutator phenotypes caused by defects in DNA repair genes such as *mutS*, *mutL*, or *mutT* have independently emerged in several LTEE populations. These hypermutators exhibit mutation rates approximately 100-fold higher than wild-type strains and have been associated with rapid adaptation to the experimental environment through accumulation of beneficial mutations in metabolic and regulatory genes, such as *pykF* (pyruvate kinase) and *nadR* (NAD biosynthesis regulator), rather than *glpK* (Jayaraman, 2009; Good et al., 2017). Although the precise fitness advantage varies across populations, some hypermutators demonstrated increased relative fitness in the glucose-limited medium, underscoring the role of stress-induced mutagenesis in adaptive evolution.

Antibiotic resistance in *E. coli* is strongly linked to ROS-mediated mutagenesis. Exposure to sublethal concentrations of ciprofloxacin (0.05 μg/mL) resulted in a fivefold elevation of intracellular ROS levels, inducing mutations in *gyrA* (encoding DNA gyrase) and *marR* (a repressor of efflux pumps). These genetic alterations conferred a fourfold increase in ciprofloxacin MIC in 25% of the mutants (Jørgensen et al., 2013). Clinical isolates from urinary tract infections (UTIs) further illustrate this phenomenon. In clinical UTI *E. coli* isolates from 80 patients, hypermutator variants were detected in only 5% of patients. Within those affected individuals, hypermutator subpopulations comprised 4.2%–62.5% of isolates and were associated with defects in the GO system (including *mutY* or *mutM*) and/or MMR genes like *mutL* (Couce et al., 2016). Additionally, among six clinical *E. coli* isolates, 50% carried constitutive activating mutations in *soxR*, significantly higher than the 20% previously reported. These mutations led to continuous SoxRS-mediated oxidative stress-response activation, which in turn correlated with multidrug resistance (Koutsolioutsou et al., 2005; Rozwadowski and Gawel, 2022).

Oxidative stress contributes significantly to the adaptive response of *E. coli* under environmental pressures such as heavy metal exposure. The *CusRS* two-component regulatory system, consisting of the sensor kinase *CusS* and the response regulator *CusR*, governs the expression of the CusCFBA efflux pump, which plays a critical role in copper and silver detoxification. Upon exposure to copper-induced oxidative stress, activation of the CusCFBA operon enhances efflux of toxic ions, thereby improving bacterial survival. Disruptions in *CusR* or genes within its associated operon have been demonstrated to increase the susceptibility of *E. coli* to copper toxicity (Yamamoto and Ishihama, 2005; Fridman et al., 2014). Additionally, *E. coli* populations exposed to UV-induced ROS developed mutations in the *uvrB* DNA repair gene, increasing mutation rates by 40-fold and conferring resistance to UV radiation in 15% of cells (Ghosh and Orman, 2024). Under oxidative stress, *E. coli* biofilm formation is enhanced via redox-sensitive regulatory circuits (e.g., *OxyR, RpoS*) that upregulate CsgD-controlled matrix production. This cascade culminates with activation of *AdrA* and increased c-di-GMP levels, which stimulate *BcsA*-mediated cellulose synthase activity, thereby improving adhesion and biofilm robustness (Hufnagel et al., 2018). These findings highlight the diverse adaptive strategies employed by *E. coli* in response to oxidative stress across clinical and environmental contexts.

3.5.2 *Pseudomonas aeruginosa*

P. aeruginosa is an opportunistic pathogen that adapts rapidly in oxidative stress-rich environments such as CF lungs. Chronic infection exposes it to repeated ROS bursts generated by neutrophils, which select for hypermutable strains (Nguyen et al., 2017). Mutations in DNA MMR genes, *mutS* or *mutL,* disrupt proofreading and can elevate mutation rates by over 20-fold. In CF lung isolates, hypermutators are found in approximately 27%–60% of patients; most studies report around 40%–60%, with one multicohort study finding 57% of patients had at least one hypermutator isolate (Hall et al., 2022). These hypermutators facilitate rapid adaptation to both host defenses and antibiotic pressure. For example, mutations in the *mexZ* repressor or in related regulators lead to derepression of the MexXY-OprM efflux pump, promoting resistance to aminoglycosides like tobramycin (Alguel et al., 2010). One study showed isolates with *mexZ* mutations had 2 to 16-fold increased tobramycin MICs, with inactivated *mexZ* variants often observed in strains with elevated MICs (Rees et al., 2019). Similarly, mutations in *lasR*, a key quorum-sensing transcriptional regulator, are frequently observed in roughly 40% of CF isolates and are associated with metabolic shifts and immune evasion (Cruz et al., 2020). Biofilm formation is a crucial persistence strategy in *P. aeruginosa*, offering significant protection against oxidative stress. Catalase activity, particularly through the *katA* gene, plays a central role in detoxifying hydrogen peroxide within biofilms. A study demonstrated that *katA* mutants exhibit a significant reduction in viability under continuous H_2O_2 exposure, confirming its critical role in oxidative stress defense within the biofilm matrix (Elkins et al., 1999). Hypermutator strains, such as those with *mutT* mutations, are known to increase mutation rates and confer antibiotic resistance (Chopra et al., 2003). There is currently no direct evidence linking *mutT*-associated hypermutation with enhanced *katA* expression or improved biofilm tolerance to hydrogen peroxide.

In a collection of CF *P. aeruginosa* isolates, hypermutator strains—characterized by defects in DNA MMR genes—are significantly more likely to exhibit the mucoid, alginate-overproducing phenotype. Although one specific study did not examine exactly 80 isolates, others have found hypermutability to be closely associated with mucoidy (due to mutations primarily in *mucA*, leading to overactivation of algU, also known as algT) (Hall et al., 2022). Mutations in *mucA* (and occasionally in *algU/mucB/mucD*) derepress AlgU, thereby activating the alginate biosynthesis operon (algD–algA) (Malhotra et al., 2019). Additionally, alginate-overproducing biofilms exhibit enhanced structural integrity and persistence in hostile environments, including under ROS exposure and antibiotic challenge (Hentzer et al., 2001).

In biofilm-based chronic infections, *mutS*-deficient hypermutators (defective in MMR) do not specifically show enhanced survival under macrophage ROS, but they display significantly greater phenotypic diversification and adaptability in structured biofilms, giving them a competitive edge over nonmutators (Luján et al., 2011). Across both environmental and clinical niches, oxidative stress, whether from host immune responses or external sources, promotes biofilm formation through upregulation of polysaccharide matrix components (Psl, Pel, and alginate) that protect them against ROS (Chen et al., 2024). In wound environments, *P. aeruginosa* expresses a metallothionein (PmtA) that confers antioxidant defense and resistance to macrophage phagocytosis (Forman and Torres, 2002). These findings highlight how oxidative stress drives both mutagenesis and biofilm adaptation, shaping *P. aeruginosa* evolution in diverse clinical settings.

3.5.3 *Candida* species

Candida species, notably *C. albicans* and *C. glabrata*, are fungal pathogens where oxidative stress drives antifungal resistance and evolutionary adaptation. In *C. albicans*, ROS from antifungals like amphotericin B (2 μg/mL) induce mutations in *ERG11*, encoding lanosterol demethylase, conferring azole resistance in clinical isolates by reducing drug binding affinity (Flowers et al., 2015). Over 50% of clinical *C. glabrata* isolates harbor loss-of-function mutations in *MSH2*, a key MMR gene, conferring a mutator phenotype that accelerates the emergence of antifungal resistance both *in vitro* and in animal models (Healey et al., 2016). These mutator strains frequently (approximately 71%) acquire gain-of-function mutations in *PDR1*, leading to overexpression of efflux pumps (CDR1/CDR2) and high-level fluconazole resistance (Zhang et al., 2024). Oxidative stress from host immune responses plays a pivotal role in

Mutagenesis	Hypermutators	Resistance
katG mutations (LTEE, 30% ROS detoxification)	*mutS* defects (40-60% CS isolates)	*FKS1/FKS2* mutations (3-fold ciprofloxacin MIC)
mutS defects (40-60% CF isolates)	*mutS* defects (40-60% CF isolates)	*FKS1/FKS2* defects (3-fold fosfomycin MIC)
gyrA mutations (4-fold ciprofloxacin)	mexAB-oprM (4-fold tobramycin MIC)	*FKS1/FKS2* mutations (3-fold caspofungin MIC)

FIGURE 3.5 Comparative illustration showing oxidative stress-driven mutagenesis, hypermutator emergence, and antimicrobial resistance across *Escherichia coli*, *Pseudomonas aeruginosa*, and *Candida* species. Bacterial and fungal cells exhibit distinct genetic adaptations under oxidative pressure, including mutations in *katG*, *mutS*, and *FKS1/2*, contributing to evolutionary fitness and drug resistance.

shaping *Candida* evolution and antifungal resistance. In *C. albicans*, macrophage phagocytosis elevates the internal production of ROS, which damages mitochondrial DNA and promotes the emergence of respiratory-deficient "petite" mutants. Microevolution experiments estimate that ROS exposure increases petite mutant frequency by approximately 10%–15%, and these mutants exhibit a two- to fourfold elevation in amphotericin B MIC, indicating enhanced drug tolerance (Cheng et al., 2007). Similarly, in *C. glabrata*, ROS produced by neutrophils during infection contributes to mutations in the *FKS1* and *FKS2* genes, which encode the β-1,3-glucan synthase targeted by echinocandins. Genomic surveys revealed that roughly 10%–15% of clinical isolates harbor *FKS* hotspot mutations resulting from ROS-related stress, and these genetic alterations correspond to a striking threefold increase in caspofungin MIC, often reaching clinical resistance thresholds (Pham et al., 2014). Moreover, environmental and clinical stressors further drive adaptations in *Candida* species. Fluconazole exposure shapes *C. albicans* biofilm adaptation but not primarily through SOS-like mutation bursts. Instead, fluconazole at clinically relevant concentrations triggers enhanced expression of drug-efflux pumps (CDR1, MDR1) and alters sterol biosynthesis pathways, helping biofilm cells tolerate azole stress. While *BCR1* remains a central biofilm regulator (Mathé and Van Dijck, 2013). In *Candida auris*, exposure to the azole antifungal voriconazole has been associated with increased production of ROS, which can promote mutagenesis. Recent studies have identified mutations in the *TAC1B* gene, a

TABLE 3.1 Summary of oxidative stress-mediated genetic adaptations in key microorganisms.

Microorganism	Oxidative stress source	Affected genes/pathways	Resulting mutation/adaptation	Implication	References
Escherichia coli	Antibiotics, host ROS, UV	*mutS*, *mutY*, *soxR*, *uvrB*, *katG*	Hypermutator phenotype, antioxidant gene mutations, and SOS activation	Antibiotic resistance, redox tolerance	Koutsolioutsou et al. (2005), Couce et al. (2016)
Pseudomonas aeruginosa	Neutrophil-derived ROS, antibiotics	*mutS*, *mexZ*, *mucA*, *katA*	Elevated mutation rate, efflux pump derepression, mucoidy, biofilm resilience	Multidrug resistance in CF patients	Elkins et al. (1999), Ciofu et al. (2005), Hall et al. (2022)
Candida albicans	Amphotericin B, macrophage ROS	*ERG11*, *MDR1*	Azole resistance via mutations and efflux activation	Drug resistance and survival in the host	Cheng et al. (2007), Flowers et al. (2015)
Candida glabrata	Neutrophil ROS, fluconazole	*MSH2*, *FKS1/2*, *PDR1*	Hypermutation, echinocandin resistance	Multidrug resistance emergence	Pham et al. (2014), Healey et al. (2016)
Salmonella typhimurium	Hydrogen (H_2O_2) exposure	His locus (G→T transversions)	8-oxoG lesions, point mutations	Stress adaptation	Abu-Shakra and Zeiger (1990), Hahm et al. (2022)
Vibrio cholerae	Host ROS in the intestine	*tcpA*, *recA*	Virulence regulation, SOS response	Colonization and immune evasion	Rapa et al. (2015), Hsiao and Zhu (2020)

(*continued on next page*)

Mycobacterium tuberculosis	Macrophage ROS, isoniazid	*katG, mutT, lexA*	Isoniazid resistance, mutagenesis	Treatment failure and persistence	Torres et al. (2015), Chatterjee et al. (2024)
Staphylococcus aureus	Host immune ROS, vancomycin	*vraS, graR, sodA, katA*	Enhanced oxidative stress resistance, cell wall modifications	VISA phenotype, immune evasion	Clements et al. (1999), Jousselin et al. (2013, 2019)
Helicobacter pylori	Gastric epithelial ROS, host immunity	*ahpC, sodB, katA*	Peroxidase deficiency, SOD mutations	Gastric colonization defects, stress sensitivity	Olczak et al. (2002)
Aspergillus fumigatus	ROS generated by macrophages and antifungals	*oxrA, catA, YAP1*	Oxidative stress resistance, transcriptional reprogramming	Pathogenesis and antifungal tolerance	Lamarre et al. (2008), Qiao et al. (2021)
Aspergillus flavus	Environmental oxidative stress	Secondary metabolite clusters, *catA*	Aflatoxin production modulation, stress tolerance	Food safety and mycotoxin contamination	Fountain et al. (2014, 2016)
Acinetobacter baumannii	Host ROS, antibiotics	*oxyR, katE, bfmR*	Catalase overexpression, biofilm formation	Multidrug resistance, hospital persistence	Jacobs et al. (2014), Gebhardt et al. (2015)
Klebsiella pneumoniae	Neutrophil ROS, carbapenems	*oxyR, rpoS, blaKPC*	Stress-response activation, β-lactamase expression	Carbapenem resistance, nosocomial infections	Tan et al. (2020)

(*continued on next page*)

TABLE 3.1 Summary of oxidative stress-mediated genetic adaptations in key microorganisms—cont'd

Microorganism	Oxidative stress source	Affected genes/pathways	Resulting mutation/adaptation	Implication	References
Enterococcus faecalis	Host ROS, antibiotics	*sodA, katA, vanA*	Superoxide dismutase variants, vancomycin resistance	VRE emergence, biofilm persistence	Verneuil et al. (2004), Kaplan et al. (2012, 2013)
Streptococcus pneumoniae	Macrophage ROS, β-lactams	*sodA, spxB, pbp*	Pyruvate oxidase mutations, PBP alterations	Penicillin resistance, immune evasion	Regev-Yochay et al. (2007), Martner et al. (2009)
Neisseria gonorrhoeae	Neutrophil ROS, antibiotics	*mtrR, penA, recA*	Efflux pump derepression, SOS response	Multidrug resistance, antigenic variation	Shafer et al. (2001), Stohl et al. (2013)
Cryptococcus neoformans	ROS generated by macrophages and antifungals	*laccase, SOD1, CAT1*	Antioxidant enzymes, melanin production	CNS invasion, antifungal tolerance	Missall et al. (2004), Zaragoza et al. (2009), Kumari et al. (2021)
Burkholderia cepacia	CF lung ROS, antibiotics	*katA, oxyR, mexR*	Catalase overproduction, efflux activation	CF lung colonization, antibiotic resistance	Bernier and Sakol (2005), Cunha et al. (2004, 2010)

transcriptional regulator of efflux pump genes, including *CDR1* as well. These mutations lead to overexpression of *CDR1*, contributing to azole resistance. For example, *TAC1B* gain-of-function mutations such as A640V and F214S were shown to increase *CDR1* expression by up to two- to fourfold, resulting in elevated fluconazole MICs and resistance phenotypes in a significant proportion of clinical isolates (Rhodes et al., 2018; Rybak et al., 2020).

Additionally, in *C. glabrata*, oxidative stress triggered by H_2O_2 exposure activates the transcription factor *YAP1*, a central regulator of the oxidative stress response. Although specific mutational induction in *YAP1* under 1 mM H_2O_2 has not been experimentally validated, *YAP1* activation contributes to the transcription of antioxidant genes, enhancing resistance to oxidative stress. Functional deletion of *YAP1* significantly reduces survival under oxidative stress, indicating its central role in oxidative stress tolerance (Cuéllar-Cruz et al., 2008; Roetzer et al., 2011). These findings highlight the critical role of oxidative stress in driving antifungal resistance and adaptation in *Candida* species across diverse clinical contexts (Fig. 3.5). The key oxidative stress–mediated genetic adaptations across major microbial systems are summarised in Table 3.1.

3.6 Conclusion

Oxidative stress stands as a powerful evolutionary force that intricately remodels microbial genomes through a spectrum of genetic and phenotypic alterations. This chapter explores how ROS, despite being inherently cytotoxic, paradoxically enhances microbial resilience by triggering mutagenesis, accelerating antibiotic resistance, and fostering hypermutator phenotypes in both bacteria and fungi. From the activation of SOS pathways to the emergence of mutator strains under immune and environmental pressure, oxidative stress reshapes microbial communities with far-reaching clinical and ecological consequences. The dynamic interplay between oxidative damage, DNA repair fidelity, and horizontal gene transfer reveals an evolutionary tightrope where survival depends on the balance between genetic innovation and stability. As our understanding deepens, targeting redox-mediated evolutionary pathways through antioxidant adjuvants or stress-response inhibitors may offer transformative approaches to controlling microbial adaptation and resistance in the Anthropocene era. Ultimately, oxidative stress is not merely a threat to microbial life but a sculptor of its evolutionary destiny.

References

Abbas, K., Alam, M., Kamal, S., 2021. Heavy metals contamination in water bodies and its impact on fish health and fish nutritional value: a review. Int. J. Fauna Biol. Stud. 8, 43–49. https://doi.org/10.22271/23940522.2021.v8.i3a.833.

Abebe, A.A., Birhanu, A.G., 2023. Methicillin resistant *Staphylococcus aureus*: molecular mechanisms underlying drug resistance development and novel strategies to combat. Infect. Drug Resist. 16, 7641–7662. https://doi.org/10.2147/idr.S428103.

Abu-Shakra, A., Zeiger, E., 1990. Effects of *Salmonella* genotypes and testing protocols on H_2O_2-induced mutation. Mutagenesis 5 (5), 469–473. https://doi.org/10.1093/mutage/5.5.469.

Ahluwalia, D., Schaaper, R.M., 2013. Hypermutability and error catastrophe due to defects in ribonucleotide reductase. Proc. Natl. Acad. Sci. USA 110 (46), 18596–18601. doi:10.1073/pnas.1310849110.

Alam, M., Abbas, K., Zehra, Z., Kamil, F., 2024. Genetic advancement, global trade dynamics, persistent challenges and future prospects in ornamental fish culture. Asian J. Res. Zool. 7 (1), 32–46. https://doi.org/10.1108/QROM-09-2019-1825.

Alguel, Y., Lu, D., Quade, N., Sauter, S., Zhang, X., 2010. Crystal structure of MexZ, a key repressor responsible for antibiotic resistance in *Pseudomonas aeruginosa*. J. Struct. Biol. 172 (3), 305–310. https://doi.org/10.1016/j.jsb.2010.07.012.

Bernier, S.P., Sokol, P.A., 2005. Use of suppression-subtractive hybridization to identify genes in the *Burkholderia cepacia* complex that are unique to *Burkholderia cenocepacia*. J. Bacteriol. 187 (15), 5278–5291. doi:10.1128/JB.187.15.5278-5291.2005.

Birben, E., Sahiner, U.M., Sackesen, C., Erzurum, S., Kalayci, O., 2012. Oxidative stress and antioxidant defense. World Allergy Organ J. 5 (1), 9–19. https://doi.org/10.1097/WOX.0b013e3182439613.

Blount, Z.D., Maddamsetti, R., Grant, N.A., Ahmed, S.T., Jagdish, T., Baxter, J.A., et al., 2020. Genomic and phenotypic evolution of Escherichia coli in a novel citrate-only resource environment. eLife 9, e55414. https://doi.org/10.7554/eLife.55414.

Brand, C., Newton-Foot, M., Grobbelaar, M., Whitelaw, A., 2025. Antibiotic-induced stress responses in Gram-negative bacteria and their role in antibiotic resistance. J. Antimicrob. Chemother. 80 (5), 1165–1184. https://doi.org/10.1093/jac/dkaf068.

Callens, M., Rose, C.J., Finnegan, M., Gatchitch, F., Simon, L., Hamet, J., et al., 2023. Hypermutator emergence in experimental *Escherichia coli* populations is stress-type dependent. Evol. Lett. 7 (4), 252–261. https://doi.org/10.1093/evlett/qrad019.

Carvajal-Garcia, J., Samadpour, A.N., Viera, H., A., J., Merrikh, H., 2023. Oxidative stress drives mutagenesis through transcription-coupled repair in bacteria. Proc. Natl. Acad. Sci. USA 120 (27), e2300761120. https://doi.org/10.1073/pnas.2300761120.

Carvajal-Garcia, J., Samadpour, A.N., Viera, H., A., J., Merrikh, H., 2023. Oxidative stress drives mutagenesis through transcription-coupled repair in bacteria. Proc. Natl Acad. Sci. 120 (27), e2300761120. https://doi.org/10.1073/pnas.2300761120.

Cavinato, L., Genise, E., Luly, F.R., Di Domenico, E.G., Del Porto, P., Ascenzioni, F., 2020. Escaping the phagocytic oxidative burst: the role of *SODB* in the survival of *Pseudomonas aeruginosa* within macrophages. Front. Microbiol. 11, 326. doi:10.3389/fmicb.2020.00326.

Chandra, P., Grigsby, S.J., Philips, J.A., 2022. Immune evasion and provocation by *Mycobacterium tuberculosis*. Nat. Rev. Microbiol. 20 (12), 750–766. https://doi.org/10.1038/s41579-022-00763-4.

Chatterjee, C., Mohan, G.R., Chinnasamy, H.V., Biswas, B., Sundaram, V., Srivastava, A., et al., 2024. Anti-mutagenic agent targeting *LexA* to combat antimicrobial resistance in mycobacteria. J. Biol. Chem. 300 (9), 107650. https://doi.org/10.1016/j.jbc.2024.107650.

Chatterjee, N., Walker, G.C., 2017. Mechanisms of DNA damage, repair, and mutagenesis. Environ. Mol. Mutagen. 58 (5), 235–263. https://doi.org/10.1002/em.22087.

Chen, Y.T., Lohia, G.K., Chen, S., Riquelme, S.A., 2024. Immunometabolic regulation of bacterial infection, biofilms, and antibiotic susceptibility. J. Innate Immun. 16 (1), 143–158. https://doi.org/10.1159/000536649.

Cheng, S., Clancy, C.J., Nguyen, K.T., Clapp, W., Nguyen, M.H., 2007. A *Candida albicans* petite mutant strain with uncoupled oxidative phosphorylation overexpresses *MDR1* and has diminished susceptibility to fluconazole and voriconazole. Antimicrob. Agents Chemother. 51 (5), 1855–1858. https://doi.org/10.1128/aac.00182-07.

Chopra, I., O'Neill, A.J., Miller, K., 2003. The role of mutators in the emergence of antibiotic-resistant bacteria. Drug Resist. Updates 6 (3), 137–145. ttps://doi.org/10.1016/S1368-7646(03)00041-4.

Ciofu, O., Riis, B., Pressler, T., Poulsen, H.E., Høiby, N., 2005. Occurrence of hypermutable *Pseudomonas aeruginosa* in cystic fibrosis patients is associated with the oxidative stress caused by chronic lung inflammation. Antimicrob. Agents Chemother. 49 (6), 2276–2282. https://doi.org/10.1128/aac.49.6.2276-2282.2005.

Clancy, C.J., Hao, B., Shields, R.K., Chen, L., Perlin, D.S., Kreiswirth, B.N., et al., 2014. Doripenem, gentamicin, and colistin, alone and in combinations, against gentamicin-susceptible, KPC-producing *Klebsiella pneumoniae* strains with various *ompK36* genotypes. Antimicrob. Agents Chemother. 58 (6), 3521–3525. https://doi.org/10.1128/aac.01949-13.

Clements, M.O., Watson, S.P., Foster, S.J., 1999. Characterization of the major superoxide dismutase of *Staphylococcus aureus* and its role in starvation survival, stress resistance, and pathogenicity. J. Bacteriol. 181 (13), 3898–3908. doi:10.1128/JB.181.13.3898-3903.1999.

Couce, A., Alonso-Rodriguez, N., Costas, C., Oliver, A., Blázquez, J., 2016. Intrapopulation variability in mutator prevalence among urinary tract infection isolates of *Escherichia coli*. Clin. Microbiol. Infect. 22 (6), 566.e561–566.e567. https://doi.org/10.1016/j.cmi.2016.03.008.

Cruz, R.L., Asfahl, K.L., Van den Bossche, S., Coenye, T., Crabbé, A., Dandekar, A.A., 2020. RhlR-regulated acyl-homoserine lactone quorum sensing in a cystic fibrosis isolate of *Pseudomonas aeruginosa*. mBio 11 (2). https://doi.org/10.1128/mBio.00532-20.

Cuéllar-Cruz, M., Briones-Martin-del-Campo, M., Cañas-Villamar, I., Montalvo-Arredondo, J., Riego-Ruiz, L., Castaño, I., et al., 2008. High resistance to oxidative stress in the fungal pathogen *Candida glabrata* is mediated by a single catalase, Cta1p, and is controlled by the transcription factors YAP1p, Skn7p, Msn2p, and Msn4p. Euk. Cell 7 (5), 814–825. https://doi.org/10.1128/ec.00011-08.

Cunha, M.V., Sousa, S.A., Leitao, J.H., Moreira, L.M., Videira, P.A., Sá-Correia, I., 2004. Studies on the involvement of the exopolysaccharide produced by cystic fibrosis-associated isolates of the *Burkholderia cepacia* complex in biofilm formation and in persistence of respiratory infections. JCM 42 (7), 3052–3058. https://doi.org/10.1128/jcm.42.7.3052-3058.2004.

Das, K., Roychoudhury, A., 2014. Reactive oxygen species (ROS) and response of antioxidants as ROS-scavengers during environmental stress in plants. Front. Environ. Sci. 2, 53. https://doi.org/10.3389/fenvs.2014.00053.

Davalli, P., Marverti, G., Lauriola, A., D'Arca, D., 2018. Targeting oxidatively induced DNA damage response in cancer: opportunities for novel cancer therapies. Oxid. Med. Cell Longev. 2018, 2389523. https://doi.org/10.1155/2018/2389523.

Degtyareva, N.P., Heyburn, L., Sterling, J., Resnick, M.A., Gordenin, D.A., Doetsch, P.W., 2013. Oxidative stress-induced mutagenesis in single-strand DNA occurs primarily at cytosines and is DNA polymerase zeta-dependent only for adenines and guanines. Nucleic Acids Res. 41 (19), 8995–9005. https://doi.org/10.1093/nar/gkt671.

Dimitrov, T., Dashti, A.A., Albaksami, O., Udo, E.E., Jadaon, M.M., Albert, M.J., 2009. Ciprofloxacin-resistant *Salmonella enterica* serovar typhi from Kuwait with novel mutations in *gyrA* and *parC* genes. J. Clin. Microbiol. 47 (1), 208–211. https://doi.org/10.1128/jcm.01161-08.

Ding, Y., Hao, J., Xiao, W., Ye, C., Xiao, X., Jian, C., et al., 2023. Role of efflux pumps, their inhibitors, and regulators in colistin resistance. Front. Microbiol 14, 1207441. https://doi.org/10.3389/fmicb.2023.1207441.

Dulanto Chiang, A., Dekker, J.P., 2024. Efflux pump-mediated resistance to new beta lactam antibiotics in multidrug-resistant gram-negative bacteria. Commun. Med. 4 (1), 170. https://doi.org/10.1038/s43856-024-00591-y.

Dwyer, D.J., Belenky, P.A., Yang, J.H., MacDonald, I.C., Martell, J.D., Takahashi, N., et al., 2014. Antibiotics induce redox-related physiological alterations as part of their lethality. Proc. Natl Acad. Sci. 111 (20), E2100–E2109. doi:10.1073/pnas.1401876111.

Elkins, J.G., Hassett, D.J., Stewart, P.S., Schweizer, H.P., McDermott, T.R., 1999. Protective role of catalase in *Pseudomonas aeruginosa* biofilm resistance to hydrogen peroxide. Appl. Environ. Microbiol. 65 (10), 4594–4600. https://doi.org/10.1128/aem.65.10.4594-4600.1999.

Elmanakhly, A.R., Bendary, M.M., Safwat, N.A., Awad, E.A.-E., Alhomrani, M., Alamri, A.S., et al., 2022. Carbapenem-resistant *Klebsiella pneumoniae*: diversity, virulence, and antimicrobial resistance. Infect. Drug Resist 5, 6177–6187. https://doi.org/10.2147/IDR.S387742.

Flowers, S.A., Colón, B., Whaley, S.G., Schuler, M.A., Rogers, P.D., 2015. Contribution of clinically derived mutations in *ERG11* to azole resistance in *Candida albicans*. Antimicrob. Agents Chemother. 59 (1), 450–460. https://doi.org/10.1128/aac.03470-14.

Forman, H., Torres, M., 2002. Reactive oxygen species and cell signaling: respiratory burst in macrophage signaling. Am. J. Respir. Crit. Care Med. 166 (12 Pt 2), S4–S8. https://doi.org/10.1164/rccm.2206007.

Fountain, J.C., Scully, B.T., Ni, X., Kemerait, R.C., Lee, R.D., Chen, Z.Y., et al., 2014. Environmental influences on maize-*Aspergillus flavus* interactions and aflatoxin production. Front. Microbiol. 5, 40. https://doi.org/10.3389/fmicb.2014.00040.

Fridman, O., Goldberg, A., Ronin, I., Shoresh, N., Balaban, N.Q., 2014. Optimization of lag time underlies antibiotic tolerance in evolved bacterial populations. Nature 513 (7518), 418–421. https://doi.org/10.1038/nature13469.

Gaca, A.O., Lemos, J.A., 2019. Adaptation to adversity: the intermingling of stress tolerance and pathogenesis in enterococci. Microbiol. Mol. Biol. Rev. 83 (3). https://doi.org/10.1128/mmbr.00008-19.

Gardete, S., Tomasz, A., 2014. Mechanisms of vancomycin resistance in *Staphylococcus aureus*. J. Clin. Invest. 124 (7), 2836–2840. https://doi.org/10.1172/jci68834.

Gebhardt, M.J., Gallagher, L.A., Jacobson, R.K., Usacheva, E.A., Peterson, L.R., Zurawski, D.V., et al., 2015. Joint transcriptional control of virulence and resistance to antibiotic and environmental stress in *Acinetobacter baumannii*. mBio 6 (6), 10–128. doi:10.1128/mBio.01660-15.

Ghosh, S., Orman, M.A., 2024. UV-induced DNA repair mechanisms and their effects on mutagenesis and culturability in *Escherichia coli*. Biorxiv. https://doi.org/10.1101/2024.11.14.623584.

Good, B.H., McDonald, M.J., Barrick, J.E., Lenski, R.E., Desai, M.M., 2017. The dynamics of molecular evolution over 60,000 generations. Nature 551 (7678), 45–50. https://doi.org/10.1038/nature24287.

Hackbarth, C.J., Kocagoz, T., Kocagoz, S., Chambers, H.F., 1995. Point mutations in *Staphylococcus aureus* PBP 2 gene affect penicillin-binding kinetics and are associated with resistance. Antimicrob. Agents Chemother. 39 (1), 103–106. https://doi.org/10.1128/aac.39.1.103.

Hahm, J.Y., Park, J., Jang, E.S., Chi, S.W., 2022. 8-Oxoguanine: from oxidative damage to epigenetic and epitranscriptional modification. Exp. Mol. Med. 54 (10), 1626–1642. https://doi.org/10.1038/s12276-022-00822-z.

Hall, K.M., Pursell, Z.F., Morici, L.A., 2022. The role of the *Pseudomonas aeruginosa* hypermutator phenotype on the shift from acute to chronic virulence during respiratory infection. Front. Cell. Infect. Microbiol. 12, 943346. https://doi.org/10.3389/fcimb.2022.943346.

Healey, K.R., Zhao, Y., Perez, W.B., Lockhart, S.R., Sobel, J.D., Farmakiotis, D., et al., 2016. Prevalent mutator genotype identified in fungal pathogen *Candida glabrata* promotes multi-drug resistance. Nat. Commun. 7, 11128. https://doi.org/10.1038/ncomms11128.

Hentzer, M., Teitzel, G.M., Balzer, G.J., Heydorn, A., Molin, S., Givskov, M., et al., 2001. Alginate overproduction affects *Pseudomonas aeruginosa* biofilm structure and function. J. Bacteriol. 183 (18), 5395–5401. https://doi.org/10.1128/jb.183.18.5395-5401.2001.

Hori, M., Yonekura, S.-I., Nohmi, T., Gruz, P., Sugiyama, H., Yonei, S., et al., 2010. Error-prone translesion DNA synthesis by *Escherichia coli* DNA polymerase IV (DinB) on templates containing 1,2-dihydro-2-oxoadenine. J. Nucleic Acids 2010 (1), 807579. https://doi.org/10.4061/2010/807579.

Hsiao, A., Zhu, J., 2020. Pathogenicity and virulence regulation of *Vibrio cholerae* at the interface of host-gut microbiome interactions. Virulence 11 (1), 1582–1599. https://doi.org/10.1080/21505594.2020.1845039.

Hufnagel, D.A., Price, J.E., Stephenson, R.E., Kelley, J., Benoit, M.F., Chapman, M.R., 2018. Thiol starvation induces redox-mediated dysregulation of *Escherichia coli* biofilm components. J. Bacteriol. 200 (1). https://doi.org/10.1128/jb.00389-17.

Hunt, C.R., Sim, J.E., Sullivan, S.J., Featherstone, T., Golden, W., Von Kapp-Herr, C., et al., 1998. Genomic instability and catalase gene amplification induced by chronic exposure to oxidative stress. Cancer Res. 58 (17), 3986–3992.

Imhof, M., Schlötterer, C., 2001. Fitness effects of advantageous mutations in evolving *Escherichia coli* populations. Proc. Natl. Acad. Sci. 98 (3), 1113–1117. doi:10.1073/pnas.98.3.1113.

Ishizawa, Y., Ying, B.-W., Tsuru, S., Yomo, T., 2015. Nutrient-dependent growth defects and mutability of mutators in *Escherichia coli*. Genes Cells 20 (1), 68–76. https://doi.org/10.1111/gtc.12199.

Islam, M.M., Jung, D.E., Shin, W.S., Oh, M.H., 2024. Colistin resistance mechanism and management strategies of colistin-resistant *Acinetobacter baumannii* infections. Pathogens 13 (12), 1049. https://www.mdpi.com/2076-0817/13/12/1049.

Jacobs, A.C., Thompson, M.G., Black, C.C., Kessler, J.L., Clark, L.P., McQueary, C.N., et al., 2014. AB5075, a highly virulent isolate of *Acinetobacter baumannii*, as a model strain for the evaluation of pathogenesis and antimicrobial treatments. mBio 5 (3), 10–128. doi:10.1128/mBio.01076-14.

James, C., James, S.J., Onarinde, B.A., Dixon, R.A., Williams, N., 2023. A critical review of AMR risks arising as a consequence of using biocides and certain metals in food animal production. Antibiotics 12 (11), 1569. https://www.mdpi.com/2079-6382/12/11/1569.

Jayaraman, R., 2009. Mutators and hypermutability in bacteria: the *Escherichia coli* paradigm. J. Genet. 88 (3), 379–391.

Jensen, P.Ø., Briales, A., Brochmann, R.P., Wang, H., Kragh, K.N., Kolpen, M., et al., 2014. Formation of hydroxyl radicals contributes to the bactericidal activity of ciprofloxacin against *Pseudomonas aeruginosa* biofilms. Pathog. Dis. 70 (3), 440–443. https://doi.org/10.1111/2049-632x.12120.

Jørgensen, K.M., Wassermann, T., Jensen, P., Hengzuang, W., Molin, S., Høiby, N., et al., 2013. Sublethal ciprofloxacin treatment leads to rapid development of high-level ciprofloxacin resistance during long-term experimental evolution of *Pseudomonas aeruginosa*. Antimicrob. Agents Chemother. 57 (9), 4215–4221. https://doi.org/10.1128/aac.00493-13.

Jousselin, A., Kelley, W.L., Barras, C., Lew, D.P., Renzoni, A., 2013. The *Staphylococcus aureus* thiol/oxidative stress global regulator Spx controls *trfA*, a gene implicated in cell wall antibiotic resistance. Antimicrob. Agents Chemother. 57 (7), 3283–3292. doi:10.1128/AAC.00220-13.

Kaczmarek, F.M., Dib-Hajj, F., Shang, W., Gootz, T.D., 2006. High-level carbapenem resistance in a *Klebsiella pneumoniae* clinical isolate is due to the combination of bla (ACT-1) beta-lactamase production, porin OmpK35/36 insertional inactivation, and down-regulation of the phosphate transport porin phoe. Antimicrob. Agents Chemother. 50 (10), 3396–3406. https://doi.org/10.1128/aac.00285-06.

Kalghatgi, S., Spina, C.S., Costello, J.C., Liesa, M., Morones-Ramirez, J.R., Slomovic, S., et al., 2013. Bactericidal antibiotics induce mitochondrial dysfunction and oxidative damage in mammalian cells. Sci. Transl. Med. 5 (192), 192ra185. https://doi.org/10.1126/scitranslmed.3006055.

Kaneko, K., Okamoto, R., Nakano, R., Kawakami, S., Inoue, M., 2005. Gene mutations responsible for overexpression of *AmpC* beta-lactamase in some clinical isolates of *Enterobacter cloacae*. J. Clin. Microbiol. 43 (6), 2955–2958. https://doi.org/10.1128/jcm.43.6.2955-2958.2005.

Kaplan, J.B., Izano, E.A., Gopal, P., Karwacki, M.T., Kim, S., Bose, J.L., et al., 2012. Low levels of β-lactam antibiotics induce extracellular DNA release and biofilm formation in *Staphylococcus aureus*. mBio 3 (4), 10–128. doi:10.1128/mBio.00198-12.

Kohanski, M.A., DePristo, M.A., Collins, J.J., 2010. Sublethal antibiotic treatment leads to multidrug resistance via radical-induced mutagenesis. Mol. Cell 37 (3), 311–320. https://doi.org/10.1016/j.molcel.2010.01.003.

Kohanski, M.A., Dwyer, D.J., Hayete, B., Lawrence, C.A., Collins, J.J., 2007. A common mechanism of cellular death induced by bactericidal antibiotics. Cell 130 (5), 797–810. https://doi.org/10.1016/j.cell.2007.06.049.

Koutsolioutsou, A., Peña-Llopis, S., Demple, B., 2005. Constitutive *soxR* mutations contribute to multiple-antibiotic resistance in clinical *Escherichia coli* isolates. Antimicrob. Agents Chemother. 49 (7), 2746–2752. https://doi.org/10.1128/aac.49.7.2746-2752.2005.

Kumari, K., Uttam, G., Singh, K., Kumari, A., Tirkey, N.N., 2021. *In-silico* studies on virulence factors of *Cryptococcus* species: phylogenetic analysis and B-cell epitope prediction. Biointerf. Res. Appl. Chem. 11 (6), 14775–14793. https://doi.org/10.33263/BRIAC116.1477514793.

Lamarre, C., Sokol, S., Debeaupuis, J.P., Henry, C., Lacroix, C., Glaser, P., et al., 2008. Transcriptomic analysis of the exit from dormancy of *Aspergillus fumigatus* conidia. BMC Genom. 9, 1–5. doi:10.1186/1471-2164-9-417.

Lebeer, S., Claes, I.J., Verhoeven, T.L., Shen, C., Lambrichts, I., Ceuppens, J.L., et al., 2008. Impact of *luxS* and suppressor mutations on the gastrointestinal transit of *Lactobacillus rhamnosus* GG. Appl. Environ. Microbiol. 74 (15), 4711–4718. https://doi.org/10.1128/aem.00133-08.

Li, X., Liang, X., Li, W., Yao, A., Liu, W., Wang, Y., et al., 2022. Isolation and functional analysis of *mbcbf2*, a *Malus baccata* (L.) Borkh CBF transcription factor gene, with functions in tolerance to cold and salt stress in transgenic *Arabidopsis thaliana*. Int. J. Mol. Sci. 23 (17). https://doi.org/10.3390/ijms23179827.

Limoli, C.L., Giedzinski, E., 2003. Induction of chromosomal instability by chronic oxidative stress. Neoplasia 5 (4), 339–346. https://doi.org/10.1016/s1476-5586(03)80027-1.

Liu, F., Li, N., Zhang, Y., 2023. The radioresistant and survival mechanisms of *Deinococcus radiodurans*. Radiat. Med. Prot. 04 (02), 70–79. doi:10.1016/j.radmp.2023.03.001.

Liu, G., Bogaj, K., Bortolaia, V., Olsen, J.E., Thomsen, L.E., 2019. Antibiotic-induced, increased conjugative transfer is common to diverse naturally occurring ESBL plasmids in *Escherichia coli*. Front. Microbiol. 10, 2019. https://doi.org/10.3389/fmicb.2019.02119.

Liu, X., Yao, H., Zhao, X., Ge, C., 2023. Biofilm formation and control of foodborne pathogenic bacteria. Molecules 28 (6). https://doi.org/10.3390/molecules28062432.

Luján, A.M., Maciá, M.D., Yang, L., Molin, S., Oliver, A., Smania, A.M., 2011. Evolution and adaptation in *Pseudomonas aeruginosa* biofilms driven by mismatch repair system-deficient mutators. PLoS One 6 (11), e27842. https://doi.org/10.1371/journal.pone.0027842.

Lyngsie, G., Krumina, L., Tunlid, A., Persson, P., 2018. Generation of hydroxyl radicals from reactions between a dimethoxyhydroquinone and iron oxide nanoparticles. Sci. Rep. 8 (1), 10834. https://doi.org/10.1038/s41598-018-29075-5.

Machado, D., Antunes, J., Simões, A., Perdigão, J., Couto, I., McCusker, M., et al., 2018. Contribution of efflux to colistin heteroresistance in a multidrug resistant *Acinetobacter baumannii* clinical isolate. J. Med. Microbiol. 67 (6), 740–749. https://doi.org/10.1099/jmm.0.000741.

Maciá, M.D., Blanquer, D., Togores, B., Sauleda, J., Pérez, J.L., Oliver, A., 2005. Hypermutation is a key factor in development of multiple-antimicrobial resistance in *Pseudomonas aeruginosa* strains causing chronic lung infections. Antimicrob. Agents Chemother. 49 (8), 3382–3386. https://doi.org/10.1128/aac.49.8.3382-3386.2005.

Malhotra, S., Hayes, D., J.R., Wozniak, D.J., 2019. Cystic Fibrosis and *Pseudomonas aeruginosa*: the host-microbe interface. Clin. Microbiol. Rev. 32 (3). https://doi.org/10.1128/cmr.00138-18.

Markkanen, E., 2017. Not breathing is not an option: how to deal with oxidative DNA damage. DNA Repair (Amst.) 59, 82–105. https://doi.org/10.1016/j.dnarep.2017.09.007.

Martner, A., Dahlgren, C., Paton, J.C., Wold, A.E., 2009. Pneumolysin released during *Streptococcus pneumoniae* autolysis is a potent activator of intracellular oxygen radical production in neutrophils. Infect. Immun. 76 (9), 4079–4087. doi:10.1128/IAI.01747-07.

Maslowska, K.H., Makiela-Dzbenska, K., Fijalkowska, I.J., 2019. The SOS system: a complex and tightly regulated response to DNA damage. Environ. Mol. Mutagen. 60 (4), 368–384. https://doi.org/10.1002/em.22267.

Mathé, L., Van Dijck, P., 2013. Recent insights into *Candida albicans* biofilm resistance mechanisms. Curr. Genet. 59 (4), 251–264. https://doi.org/10.1007/s00294-013-0400-3.

McKenzie, G.J., Harris, R.S., Lee, P.L., Rosenberg, S.M., 2000. The SOS response regulates adaptive mutation. Proc. Natl Acad. Sci. 97 (12), 6646–6651. doi:10.1073/pnas.120161797.

Mehta, H.H., Prater, A.G., Beabout, K., Elworth, R.A.L., Karavis, M., Gibbons, H.S., et al., 2019. The essential role of hypermutation in rapid adaptation to antibiotic stress. Antimicrob. Agents Chemother. 63 (7). https://doi.org/10.1128/aac.00744-19.

Mendoza-Chamizo, B., Løbner-Olesen, A., Charbon, G., 2018. Coping with reactive oxygen species to ensure genome stability in *Escherichia coli*. Genes 9 (11), 565. https://www.mdpi.com/2073-4425/9/11/565.

Missall, T.A., Lodge, J.K., McEwen, J.E., 2004. Mechanisms of resistance to oxidative and nitrosative stress: implications for fungal survival in mammalian hosts. Eukaryot. Cell. (4) 835–846. doi:10.1128/EC.3.4.835-846.2004.

Mukherjee, S., Ridgeway, A.D., Lamb, D.J., 2010. DNA mismatch repair and infertility. Curr. Opin. Urol. 20 (6), 525–532. https://doi.org/10.1097/MOU.0b013e32833f1c21.

Nguyen, G.T., Green, E.R., Mecsas, J., 2017. Neutrophils to the ROScue: mechanisms of NADPH oxidase activation and bacterial resistance. Front. Cell. Infect. Microbiol. 7, 373. https://doi.org/10.3389/fcimb.2017.00373.

Nnaji, N.D., Anyanwu, C.U., Miri, T., Onyeaka, H., 2024. Mechanisms of heavy metal tolerance in bacteria: a review. Sustainability 16 (24), 11124. https://www.mdpi.com/2071-1050/16/24/11124.

Nulton-Persson, A.C., Szweda, L.I., 2001. Modulation of mitochondrial function by hydrogen peroxide. J. Biol. Chem. 276 (26), 23357–23361. https://doi.org/10.1074/jbc.M100320200.

Olczak, A.A., Olson, J.W., Maier, R.J., 2002. Oxidative-stress resistance mutants of *Helicobacter pylori*. J. Bacteriol. 184 (12), 3186–3193. https://doi.org/10.1128/jb.184.12.3186-3193.2002.

Oliver, A., Cantón, R., Campo, P., Baquero, F., Blázquez, J., 2000. High frequency of hypermutable *Pseudomonas aeruginosa* in cystic fibrosis lung infection. Science 288 (5469), 1251–1254. https://doi.org/10.1126/science.288.5469.1251.

Oliver, A., Levin, B.R., Juan, C., Baquero, F., Blázquez, J., 2004. Hypermutation and the preexistence of antibiotic-resistant *Pseudomonas aeruginosa* mutants: implications for susceptibility testing and treatment of chronic infections. Antimicrob. Agents Chemother. 48 (11), 4226–4233. https://doi.org/10.1128/aac.48.11.4226-4233.2004.

Patil, A.G., Sang, P.B., Govindan, A., Varshney, U., 2013. *Mycobacterium tuberculosis* MutT1 (Rv2985) and ADPRase (Rv1700) proteins constitute a two-stage mechanism of 8-oxo-dGTP and 8-oxo-GTP detoxification and adenosine to cytidine mutation avoidance. J. Biol. Chem. 288 (16), 11252–11262. https://doi.org/10.1074/jbc.M112.442566.

Pedraza-Reyes, M., Abundiz-Yañez, K., Rangel-Mendoza, A., Martínez, L.E., Barajas-Ornelas, R.C., Cuéllar-Cruz, M., et al., 2024. *Bacillus subtilis* stress-associated mutagenesis and developmental DNA repair. Microbiol. Mol. Biol. Rev. 88 (2), e0015823. https://doi.org/10.1128/mmbr.00158-23.

Pham, C.D., Iqbal, N., Bolden, C.B., Kuykendall, R.J., Harrison, L.H., Farley, M.M., et al., 2014. Role of *FKS* Mutations in *Candida glabrata*: MIC values, echinocandin resistance, and multidrug resistance. Antimicrob. Agents Chemother. 58 (8), 4690–4696. https://doi.org/10.1128/aac.03255-14.

Poetsch, A.R., 2020. The genomics of oxidative DNA damage, repair, and resulting mutagenesis. Comput. Struct. Biotechnol. J. 18, 207–219. https://doi.org/10.1016/j.csbj.2019.12.013.

Qiao, J., Kontoyiannis, D.P., Wan, Z., Li, R., Liu, W., 2021. Antifungal activity of statins against *Aspergillus* species. Med Mycol J 45 (7), 589–593. https://doi.org/10.3389/fmicb.2014.00040.

Rapa, R.A., Islam, A., Monahan, L.G., Mutreja, A., Thomson, N., Charles, I.G., et al., 2015. A genomic island integrated into *recA* of *Vibrio cholerae* contains a divergent *recA* and provides multi-pathway protection from DNA damage. Environ. Microbiol. 17 (4), 1090–1102. https://doi.org/10.1111/1462-2920.12512.

Rees, V.E., Deveson Lucas, D.S., López-Causapé, C., Huang, Y., Kotsimbos, T., Bulitta, J.B., et al., 2019. Characterization of hypermutator *Pseudomonas aeruginosa* isolates from patients with cystic fibrosis in Australia. Antimicrob. Agents Chemother. 63 (4). https://doi.org/10.1128/aac.02538-18.

Regev-Yochay, G., Trzcinski, K., Thompson, C.M., Malley, R., Lipsitch, M., 2007. Interference between *Streptococcus pneumoniae* and *Staphylococcus aureus*: in vitro hydrogen peroxide-mediated killing by *Streptococcus pneumoniae*. J. Bacteriol. 188 (13), 4996–5001. doi:10.1128/JB.00317-06.

Rhodes, J., Abdolrasouli, A., Farrer, R.A., Cuomo, C.A., Aanensen, D.M., Armstrong-James, D., et al., 2018. Genomic epidemiology of the UK outbreak of the emerging human fungal pathogen *Candida auris*. Emerg. Microbes Infect. 7 (1), 43. https://doi.org/10.1038/s41426-018-0045-x.

Richardson, A.R., Yu, Z., Popovic, T., Stojiljkovic, I., 2002. Mutator clones of *Neisseria meningitidis* in epidemic serogroup A disease. Proc. Natl. Acad. Sci. USA 99 (9), 6103–6107. https://doi.org/10.1073/pnas.092568699.

Roetzer, A., Klopf, E., Gratz, N., Marcet-Houben, M., Hiller, E., Rupp, S., et al., 2011. Regulation of *Candida glabrata* oxidative stress resistance is adapted to host environment. FEBS Lett. 585 (2), 319–327. https://doi.org/10.1016/j.febslet.2010.12.006.

Roth, J.R., Andersson, D.I., 2004. Adaptive mutation: how growth under selection stimulates Lac (+) reversion by increasing target copy number. J. Bacteriol. 186 (15), 4855–4860. https://doi.org/10.1128/jb.186.15.4855-4860.2004.

Rozwadowski, M., Gawel, D., 2022. Molecular factors and mechanisms driving multidrug resistance in uropathogenic *Escherichia coli*- an update. Genes (Basel) 13 (8). https://doi.org/10.3390/genes13081397.

Rybak, J.M., Muñoz, J.F., Barker, K.S., Parker, J.E., Esquivel, B.D., Berkow, E.L., et al., 2020. Mutations in *TAC1B*: a novel genetic determinant of clinical fluconazole resistance in *Candida auris*. mBio 11 (3). https://doi.org/10.1128/mBio.00365-20.

Sanders, L.H., Sudhakaran, J., Sutton, M.D., 2009. The GO system prevents ROS-induced mutagenesis and killing in *Pseudomonas aeruginosa*. FEMS Microbiol. Lett. 294 (1), 89–96. https://doi.org/10.1111/j.1574-6968.2009.01550.x.

Sazykin, I.S., Sazykina, M.A., 2023. The role of oxidative stress in genome destabilization and adaptive evolution of bacteria. Gene 857, 147170. https://doi.org/10.1016/j.gene.2023.147170.

Schaaff, F., Reipert, A., Bierbaum, G., 2002. An elevated mutation frequency favors development of vancomycin resistance in *Staphylococcus aureus*. Antimicrob. Agents Chemother. 46 (11), 3540–3548. https://doi.org/10.1128/aac.46.11.3540-3548.2002.

Schieber, M., Chandel, N.S., 2014. ROS function in redox signaling and oxidative stress. Curr. Biol. 24 (10), R453–R462. https://doi.org/10.1016/j.cub.2014.03.034.

Shaeib, F., Banerjee, J., Maitra, D., Diamond, M.P., Abu-Soud, H.M., 2013. Impact of hydrogen peroxide-driven Fenton reaction on mouse oocyte quality. Free Radic. Biol. Med. 58, 154–159. https://doi.org/10.1016/j.freeradbiomed.2012.12.007.

Shafer, W.M., Qu, X.D., Waring, A.J., Lehrer, R.I., 2001. Modulation of *Neisseria gonorrhoeae* susceptibility to vertebrate antibacterial peptides due to a member of the resistance/nodulation/division efflux pump family. Proc. Natl Acad. Sci. 95 (4), 1829–1833. doi:10.1073/pnas.95.4.1829.

Sheng, H., Huang, J., Han, Z., Liu, M., Lü, Z., Zhang, Q., et al., 2020. Genes and proteomes associated with increased mutation frequency and multidrug resistance of naturally occurring mismatch repair-deficient *Salmonella* hypermutators. Front. Microbiol. 11, 770. https://doi.org/10.3389/fmicb.2020.00770.

Singh, R.P., Kapoor, A., Sinha, A., Ma, Y., Shankar, M., 2025. Virulence factors of *Klebsiella pneumoniae*: insights into canonical and emerging mechanisms driving pathogenicity and drug resistance. Microbe 7, 100289. https://doi.org/10.1016/j.microb.2025.100289.

Slager, J., Kjos, M., Attaiech, L., Veening, J.-W., 2014. Antibiotic-induced replication stress triggers bacterial competence by increasing gene dosage near the origin. Cell 157 (2), 395–406. https://doi.org/10.1016/j.cell.2014.01.068.

Stohl, E.A., Criss, A.K., Seifert, H.S., 2005. The transcriptome response of *Neisseria gonorrhoeae* to hydrogen peroxide reveals genes with previously uncharacterized roles in oxidative damage protection. Mol. Microbiol. 58 (2), 520–532. doi:10.1111/j.1365-2958.2005.04839.x.

Tan, D., Zhang, Y., Cheng, M., Le, S., Gu, J., Bao, J., et al., 2019. Characterization of *Klebsiella pneumoniae* ST11 isolates and their interactions with lytic phages. Viruses 11 (11), 1080. doi:10.3390/v11111080.

Tan, Y.-S., Zhang, R.-K., Liu, Z.-H., Li, B.-Z., Yuan, Y.-J., 2022. Microbial adaptation to enhance stress tolerance. Front. Microbiol. 13, 888746. https://doi.org/10.3389/fmicb.2022.888746.

Torres, J.N., Paul, L.V., Rodwell, T.C., Victor, T.C., Amallraja, A.M., Elghraoui, A., et al., 2015. Novel *katG* mutations causing isoniazid resistance in clinical *M. tuberculosis* isolates. Emerg. Microbes. Infect., 4 (7), e42. https://doi.org/10.1038/emi.2015.42.

Tscherner, M., Zwolanek, F., Jenull, S., Sedlazeck, F.J., Petryshyn, A., Frohner, I.E., et al., 2015. The *Candida albicans* histone acetyltransferase Hat1 regulates stress resistance and virulence via distinct chromatin assembly pathways. PLoS Pathog. 11 (10), e1005218. https://doi.org/10.1371/journal.ppat.1005218.

Van Bambeke, F., Chauvel, M., Reynolds, P.E., Fraimow, H.S., Courvalin, P., 1999. Vancomycin-dependent *Enterococcus faecalis* clinical isolates and revertant mutants. Antimicrob. Agents Chemother. 43 (1), 41–47. https://doi.org/10.1128/aac.43.1.41.

Velali, E., Pantazaki, A., Besis, A., Choli-Papadopoulou, T., Samara, C., 2019. Oxidative stress, DNA damage, and mutagenicity induced by the extractable organic matter of airborne particulates on bacterial models. Regul. Toxicol. Pharmacol. 104, 59–73. https://doi.org/10.1016/j.yrtph.2019.03.004.

Verneuil, N., Mazé, A., Sanguinetti, M., Laplace, J.M., Benachour, A., Auffray, Y., et al., 2006. Implication of (Mn) superoxide dismutase of *Enterococcus faecalis* in oxidative stress responses and survival inside macrophages. Microbiology 152 (9), 2579–2589. doi:10.1099/mic.0.28922-0.

Visick, K.L., Foster, J., Doino, J., McFall-Ngai, M., Ruby, E.G., 2000. *Vibrio fischeri* lux genes play an important role in colonization and development of the host light organ. J. Bacteriol. 182 (16), 4578–4586. doi:10.1128/jb.182.16.4578-4586.2000.

Wales, A.D., Davies, R.H., 2015. Co-selection of resistance to antibiotics, biocides and heavy metals, and its relevance to foodborne pathogens. Antibiotics (Basel) 4 (4), 567–604. https://doi.org/10.3390/antibiotics4040567.

Wang, H., Xing, X., Wang, J., Pang, B., Liu, M., Larios-Valencia, J., et al., 2018. Hypermutation-induced *in vivo* oxidative stress resistance enhances *Vibrio cholerae* host adaptation. PLoS Pathog. 14 (10), e1007413.

Whitaker, A.M., Schaich, M.A., Smith, M.R., Flynn, T.S., Freudenthal, B.D., 2017. Base excision repair of oxidative DNA damage: from mechanism to disease. Front. Biosci. (Landmark Ed) 22 (9), 1493–1522. https://doi.org/10.2741/4555.

Wigle, T.J., Sexton, J.Z., Gromova, A.V., Hadimani, M.B., Hughes, M.A., Smith, G.R., et al., 2009. Inhibitors of *RecA* activity discovered by high-throughput screening: cell-permeable small molecules attenuate the SOS response in *Escherichia coli*. J. Biomol. Screen. 14 (9), 1092–1101. https://doi.org/10.1177/1087057109342126.

Woodford, N., Ellington, M.J., 2007. The emergence of antibiotic resistance by mutation. Clin. Microbiol. Infect. 13 (1), 5–18. https://doi.org/10.1111/j.1469-0691.2006.01492.x.

Wu, H.-J., Xiao, Z.-G., Lv, X.-J., Huang, H.-T., Liao, C., Hui, C.-Y., et al., 2023. Drug-resistant *Acinetobacter baumannii*: From molecular mechanisms to potential therapeutics. Exp Ther Med 25 (5), 209. https://doi.org/10.3892/etm.2023.11908.

Wyrzykowski, J., Volkert, M.R., 2003. The *Escherichia coli* methyl-directed mismatch repair system repairs base pairs containing oxidative lesions. J. Bacteriol. 185 (5), 1701–1704. https://doi.org/10.1128/jb.185.5.1701-1704.2003.

Yamamoto, K., Ishihama, A., 2005. Transcriptional response of *Escherichia coli* to external copper. Mol. Microbiol. 56 (1), 215–227. https://doi.org/10.1111/j.1365-2958.2005.04532.x.

Yarwood, J.M., Schlievert, P.M., 2003. Quorum sensing in *Staphylococcus infections*. J. Clin. Invest. 112 (11), 1620–1625. https://doi.org/10.1172/JCI20442.

Zaragoza, O., Rodrigues, M.L., De Jesus, M., Frases, S., Dadachova, E., Casadevall, A., 2009. The capsule of the fungal pathogen *Cryptococcus neoformans*. Adv. Appl. Microbiol. 68, 133–216. https://doi.org/10.1080/21505594.2018.1431087.

Zaw, M.T., Emran, N.A., Lin, Z., 2018. Mutations inside rifampicin-resistance determining region of *rpoB* gene associated with rifampicin-resistance in *Mycobacterium tuberculosis*. J. Infect. Public Health 11 (5), 605–610. https://doi.org/10.1016/j.jiph.2018.04.005.

Zhang, G., Chen, Y., Chen, J., Yao, D., 2024. Association of multilocus sequence typing, *MSH2* gene mutations, and antifungal resistance in *Candida glabrata*: implications for clinical outcomes in Chinese hospitals. Ann. Clin. Microbiol. Antimicrob. 23 (1), 100. https://doi.org/10.1186/s12941-024-00758-4.

Zhao, J., Winkler, M.E., 2000. Reduction of GC ->TA transversion mutation by overexpression of *MutS* in *Escherichia coli* K-12. J. Bacteriol. 182 (17), 5025–5028. https://doi.org/10.1128/jb.182.17.5025-5028.2000.

CHAPTER

4

DNA photolyase: Structure, mode of action, and therapeutic applications

Ajeet Singh[a], *Sukriti Chauhan*[a,b], *Ganesh Kumar Maurya*[a] *and Vinod Kumar Kannaujiya*[b]

[a] **Department of Zoology, Mahila Mahavidyalaya, Banaras Hindu University, Varanasi, Uttar Pradesh, India** [b] **Department of Botany, Mahila Mahavidyalaya, Banaras Hindu University, Varanasi, Uttar Pradesh, India**

4.1 Introduction

Genome maintenance and its proper functioning are essential for better working and survival of all life forms on Earth. Since DNA is the fundamental constituent of the genome of most animals, except for a few RNA-containing plant viruses, its stability and integrity are of primary concern for the efficient functioning and existence of life. There are multiple types and sources of DNA damage (Ciccia and Elledge, 2010). Any irreparable damage to DNA results in mutations, which can lead to diseases. The sun is the source of various types of radiation, including ultraviolet radiation (UVR), which interacts with DNA, resulting in DNA lesions. Depletion of the ozone layer by ozone-depleting substances has increased the chances of UVR reaching all life forms directly, interacting with their DNA, and resulting in damaged DNA. Based on the range of wavelengths, UVR is generally differentiated into three forms i.e., UV-A with the wavelength of 400–320 nm also known as black light, not absorbed by ozone layer, UV-B with the wavelength of 320–280 nm known as medium wave, partially absorbed by the ozone layer and last UV-C with the wavelength of 280–100 nm also known as short wave (Shetta and Areaf, 2009), completely absorbed by the ozone layer and atmosphere. The shorter wavelength of UVR has a higher energy level, so UV-C has a much more damaging effect due to the higher level of energy in comparison to the other two UV-A and UV-B radiations (Katerova and Todorova, 2009). Exposure of UVR on human skin for a longer duration can be a factor for causing mutation

Genome Damage: Repair and Adaptation Across Life.
DOI: https://doi.org/10.1016/B978-0-443-42857-9.00012-8

in DNA, photoaging inside the skin, immune system inhibition, and skin cells related cancer such as melanoma, basal cell, and squamous cell carcinoma (Shetta and Areaf, 2009). Skin is the largest organ of humans, so skin cells directly come under the threat of UV-induced DNA damage, leading to the formation of the most prevalent DNA lesions. A DNA lesion is a change in DNA produced by a break in the sugar-phosphate backbone or some changes in the nucleotide bases. If removal of damaged cells is not done by apoptosis, the mutagenic property of DNA lesions is expressed as the inactivation of tumor suppression genes and activation of proto-oncogenes, which further leads to the initiation of skin cancer (Katerova and Todorova, 2009). UVR is also involved in promoting skin cancer via the induction of inflammation, oxidative stress, and DNA damage through the formation of 8-oxoguanine and cyclobutane pyrimidine dimers (CPDs) (Cadet and Douki, 2018). In nearly 90% of human nonmelanoma skin cancer, UV exposure is the only responsible factor (Rahman et al., 2021). So, there is a need for such compounds or biological molecules that have the potential to overcome these problems and provide protection against the harmful effects of UV exposure to different organisms. The initial phase of the Earth's environment lacked an oxygen and ozone layer in the atmosphere; this led to the direct exposure of UV irradiation on the Earth's surface (Cockell, 1998). Therefore the initial form of life must possess the efficient competency of DNA repair against the threat of UV-induced DNA lesions, including CPD and a pyrimidine–pyrimidone (6-4) photoproducts (6-4PP) formation (Beukers et al., 2008; Rastogi et al., 2020). As a cure against UV-induced DNA lesions, photolyase (PL)-mediated direct damage reversal using the external blue-light energy was one of the available choices for the primordial organisms (Sancar, 2003; Vechtomova et al., 2020). PLs have a wide distribution throughout all kingdoms of life, which provides the indication of very early evolution of it as an ancient biological molecule (Lucas-Lledo and Lynch, 2009). PLs are enzymes belonging to the cryptochrome/photolyase protein family (CPF) that have been identified as a potential biological molecule for repairing single-stranded DNA (ssDNA) as well as double-stranded DNA (dsDNA) damage by directly reversing UVR-induced DNA damages of CPD and 6-4PPs using near UV or blue light as an energy source (Sancar, 2003; Rastogi et al., 2010). The other function of PLs is working as a circadian photoreceptor in circadian rhythms, and they also have involvement in transcriptional regulation. PL shows much more variation and diversity in function, with a wide distribution in life forms. *Escherichia coli*, an enteric bacterium that is not under the threat of UV light, has PL; whereas soil-inhabiting bacteria, *Bacillus subtilis*, which is under continuous threat of UVR, do not have PL. It is absent in humans and other mammals, but at the same time present in some marsupial mammals like the opossum (Ozer et al., 1995). In this chapter, UV-induced DNA damage, evolution and diversity, structure, mode of action, and therapeutic application of DNA PL have been described.

4.2 Different types of DNA damage and sources

Defective DNA is formed either due to modification in DNA composition leading to DNA lesions via multiple sources, such as mutation, endogenous and exogenous DNA damage, errors in DNA replication, instead of having sophisticated proof-reading activity, all these need to be prevented to ensure perfect transmission of intact and stable genetic information into all living cells. Multiple DNA-damaging agents and sources can be broadly classified into two types: exogenous sources of DNA damage and endogenous sources of DNA damage. These sources of DNA damage can be listed as shown in Table 4.1.

TABLE 4.1 Different types of DNA lesions are broadly classified under the two broad categories of endogenous and exogenous sources, respectively.

Sr. No.	Types of DNA lesions	Cause	Specific lesion site	Estimated lesions/cell/day
Endogenous sources of DNA lesions				
1.	Depurination	Spontaneous hydrolysis	AP site	~10,000
2.	Cytosine deamination	Spontaneous hydrolysis	Base transition (e.g., C→T)	100–500
3.	S-adenosylmethionine (SAM)-induced	Alkylation (methylation)	3-methyladenine (3meA)	600
4.	SAM-induced	Alkylation (methylation)	7-methylguanine (7meG)	4000
5.	SAM-induced	Alkylation (methylation)	O6-methylguanine (O6meG)	10–30
6.	Reactive oxygen species from metabolism	Oxidative stress	8-oxoguanine (8oxoG)	400–1500
Exogenous sources of DNA lesions				
Sr. No.	**Types of DNA lesions**	**Cause**	**Specific lesion site**	**Estimated lesions/cell/day**
1.	CPDs including minor contribution of 6-4 Photoproducts (6-4PPs)	UV-B, UV-C radiation	Covalent bonds between adjacent thymine or cytosine bases	~10^5 CPDs per epidermal cell
2.	6-4 Photoproducts (6-4PPs)	UV radiation	Cross-links between adjacent pyrimidine bases	~12,500 (Eightfold less in comparison to cyclobutane pyrimidine dimer (CPD) lesions)

(*continued on next page*)

TABLE 4.1 Different types of DNA lesions are broadly classified under the two broad categories of endogenous and exogenous sources, respectively—cont'd

Endogenous sources of DNA lesions				
Sr. No.	**Types of DNA lesions**	**Cause**	**Specific lesion site**	**Estimated lesions/cell/day**
3.	Single-strand breaks (SSBs)	X-rays, gamma rays, beta particles	Sugar-phosphate backbone (one strand)	~1000 per Gy
4.	Double-strand breaks (DSBs)	Ionizing radiation (e.g., gamma rays, α-particles)	Both DNA strands have closely spaced breaks	~25–40 per Gy
5.	Cigarette smoke	Polycyclic aromatic hydrocarbons, tobacco-specific nitrosamines (TSNAs), aromatic amines, etc.	Aromatic DNA adducts	~45–1029
6.	Aflatoxin-DNA adducts	Aflatoxin B1 (from mold-contaminated food)	N7 position of guanine	10–100
7.	DNA intercalation	Doxorubicin, ethidium bromide	Between stacked base pairs	Dose-dependent

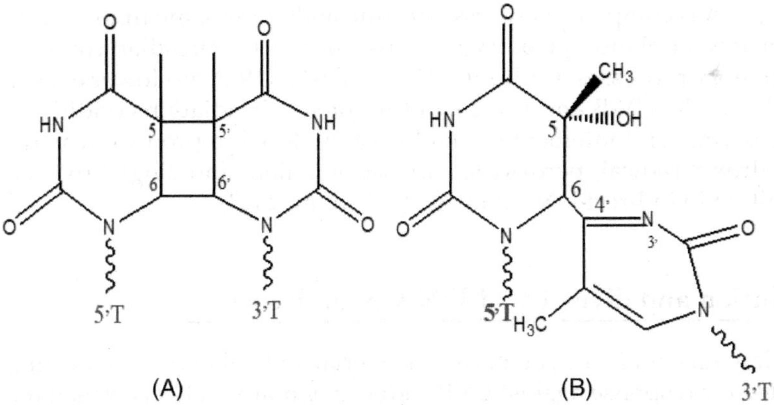

FIGURE 4.1 Two common UVR-induced DNA lesions: (A) Cyclobutane pyrimidine dimers and (B) pyrimidine–pyrimidone (6-4) photoproducts. *UVR*, Ultraviolet radiation.

UVR as an exogenous source of radiation can produce up to 10^5 DNA lesions per cell per day and hence can play a role in the generation of skin cancer (Adeyemi, 2023). So if the DNA gets damaged, it needs to be repaired, and for this purpose, there are generally two types of DNA repair mechanisms universally seen in all forms of life as evolutionary aspects, which are ssDNA damage repair and dsDNA damage repair. The ssDNA damage repair is further classified into nucleotide excision repair, base excision repair, mismatch repair, and direct reversal repair, whereas dsDNA damage repair is further classified into homologous recombination and nonhomologous end joining repairs (Kavakli et al., 2019). Direct reversal repair which is responsible for repairing ssDNA damage is a particular type of repair mechanism where no DNA synthesis or template is involved and also it cannot be used in repairing all type of DNA damage, it depends on a special type of enzyme in the case of UV-induced DNA damage such as PL which causes direct reversal of DNA damage utilizing blue light, O6-alkylguanine-DNA alkyltransferases (AGTs) enzyme mediated direct damage reversal of O-alkylated DNA damage and AlkB family dioxygenases enzyme mediated direct damage reversal of N-alkylated base adducts (Yi and He, 2013).

4.3 Ultraviolet-induced DNA damage

The UV-induced DNA damage leads to the formation of two dimeric photoproducts between the bases of adjacent pyrimidines, including CPDs and 6-4PPs, with their chemical structure shown in Fig. 4.1. Two adjacent thymine pyrimidine bases of the same DNA strand are attached covalently to form a cyclobutane ring, CPD, as a major photoproduct responsible for approximately 70%–90% of total UV-induced DNA damage, and 6-4PPs as a minor photoproduct responsible for remaining 10%–30% of damaging DNA, however variations in these ratios are also possible depending on the UV wavelength, DNA sequence and the type of organism included in the study (Masuma et al., 2013).

Continuous formation due to a lack of PL activity leads to the accumulation of these photoproducts, which further disrupts the normal functioning of cells by interfering with replication and transcription of DNA (Katerova and Todorova, 2009). Photons from UV-B show interaction with DNA, resulting in direct DNA damage through the formation of DNA lesions in the form of CPD and 6-4PPs. Whereas UV-A is comparatively less harmful, and at the same time, its low dose is needed to initiate signaling in photoregulatory proteins, such as for circadian rhythms and many other different regulatory processes (Cadet and Douki, 2018). UV-A has lower energy in comparison to the other two UVR (UV-B and UV-C), but in intense irradiation conditions, it can penetrate deep into the dermis and interact with biological material to produce reactive oxygen species, including hydroxyl radical, peroxides, superoxide anion, and singlet oxygen (Tyrrell and Reeve, 2006) resulting in indirect DNA damage (Adeyemi, 2023).

4.4 Evolution and diversity of DNA photolyases

PLs are found in all three domains of life, i.e., archaea, eubacteria, and eukaryotes, reflecting their early evolution as a protective response against UVR-induced genotoxic effects. When the environment of the Earth becomes aerobic due to the abundance of photosynthetic bacteria and plants producing oxygen, this results in the formation of the ozone layer. This ozone layer, further working as a shield, decreases the UV exposure to the Earth's surface (Cockell, 1998). This decreased amount of UV exposure also reduces the pressure for evolution of DNA repair machinery involving DNA PL, especially for the UV-induced DNA damage. Thereafter, the importance of PLs might have been lowered to some extent, allowing some living organisms to evolve to their functionally diverse homologs of PLs, called cryptochromes (Chaves et al., 2011). PLs are reported to be ancient DNA repair enzymes, which helped in the survival of ancient organisms during sunlight exposure, and later these PLs became parent molecules from which cryptochromes, as a special class of blue-light photoreceptor, evolved (Müller and Carell, 2009). Cryptochrome exhibits sequence similarity with PL. The role of PL has also been reported in enhancing excision repair in *E. coli,* which gets further culmination during evolution, where in placental mammals, primary photoreactivating activity is completely replaced by dominating its secondary function, performing excision repair, which was earlier seen in *E. coli* (Yamamoto et al., 1983; Sancar et al., 1984). Phylogenetic analysis using amino acid sequences of hundreds of CPF indicates more classes of CPF than what we expect so far (Kavakli et al., 2019), but broadly PLs can be classified into two types based on their involvement in particular substrate, these are CPD PL and 6-4PP PL (6-4 PL), both types of PLs use 350–450 nm of near-UV light or blue light as an energy source for repairing UVR-induced damages (Sancar, 2003; Beukers et al., 2008).

The evolutionary connections for individual members of this CPF family were always discussed controversially, but now CPD PL has been widely accepted as a common ancestor (Müller and Carell, 2009). CPD PL are further subdivided into class I, II, III, and ssDNA PL (Zhang et al., 2017). Based on amino acid sequence similarity, microbial PLs are generally included in CPD PL class I, and animal and plant PLs are included in CPD PL class II (Yasui et al., 1994). DASH-cryptochrome (DASH = *Drosophila, Arabidopsis, Synechocystis, Homo*), described by Brudler et al. (2003) as a new class of cryptochrome found across diverse taxa, including animals, plants, and bacteria, taking part in photorepair of ssDNA, shows its broad phylogenetic distribution and evolutionarily conserved function. But through the recent data, it is reported that the DASH cryptochromes are ssDNA-specific CPD PL (Selby and Sancar, 2006; Pokorny et al., 2008).

Inside the 6-4 PL, a new class of PL is included containing an iron–sulfur cluster [4Fe–4S] named as iron–sulfur bacterial cryptochromes and PLs (FeS-BCPs). FeS-BCPs are found in some prokaryotes containing a unique antenna cofactor 6,7-dimethyl 8-ribityl-lumazine (DMRL)

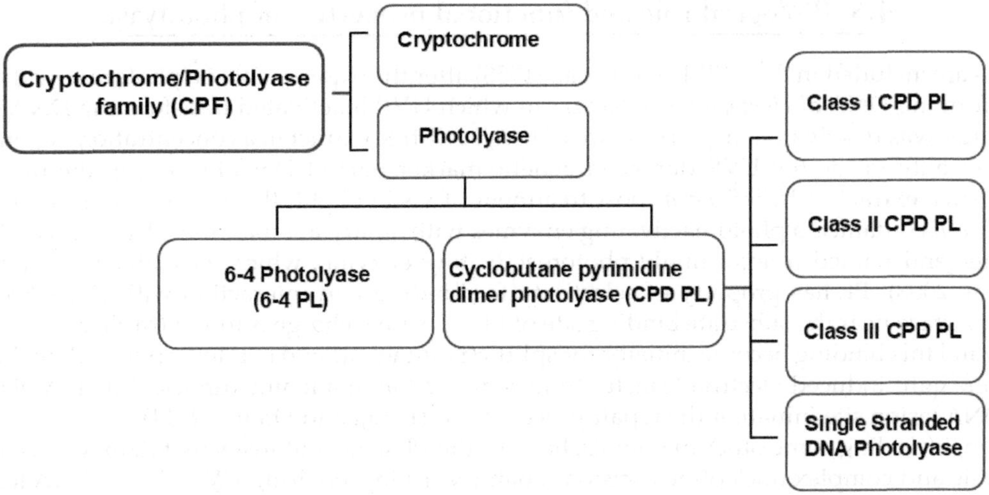

FIGURE 4.2 Flow diagram of classification of cryptochrome/photolyase family (CPF) protein.

TABLE 4.2 Summary of types of photolyases and their related examples.

Sr. No.	Types of photolyases	Examples
1.	6-4 photolyase	FeS-BCPs protein, PhrB gene encoded photolyase of *Agrobacterium tumefaciens* and other prokaryotic 6-4 photolyase, etc.
2.	Class I CPD photolyase	*Escherichia coli, Synechocystis* species
3.	Class II CPD photolyase	*Drosophila melanogaster, Arabidopsis thaliana*
4.	Class III CPD photolyase	The PhrA gene encoded photolyase of *Agrobacterium tumefaciens* and *Caulobacter crescentus*
5.	Single-stranded DNA photolyase	DASH cryptochromes

CPD, Cyclobutane pyrimidine dimer.

having an iron–sulfur cluster at the C-terminal domain (Geisselbrecht et al., 2012). Phylogenetic analysis also supports that prokaryotic 6-4 PLs having an iron–sulfur cluster were the initial ancestor for the other classes of CPF (Zhang et al., 2013). A soil bacterium, *Agrobacterium tumefaciens*, has two genes responsible for encoding PL-related proteins A and B (PhrA and PhrB, respectively) (Oberpichler et al., 2011). PhrA belongs to the class III CPD PL, whereas PhrB is included in the new class of FeS-BCPs inside the 6-4 PLs. Mutant study also revealed that both PhrA and PhrB are needed for photoreactivation and neither of them is redundant (Zhang et al., 2013). In short, PLs can be classified as follows, as indicated in Fig. 4.2, with their related examples shown in Table 4.2.

4.5 Biological role and functional properties of photolyase

PLs are included in the CPF. Rupert et al. (1958) after the experiments on photoreactivation in transforming DNA of *Hemophilus influenzae* in which UVR inactivated transforming DNA of *H. influenzae* was reactivated in the presence of *E. coli* cell-free extract in a concentration dependent manner only inside the UVR damaged genetic marker part of DNA but not in the presence of cell-free extract of *H. influenzae* post treatment of visible light, this experimentation reveals that *E. coli* contained a photoreactivating enzymes with ability to repair the UVR-induced DNA damage and named it accordingly photoreactivating enzyme, which was later named as PL (Sancar, 2008). PL has property of selectivity for binding and interaction with DNA because at the surface near the substrate binding site of PLs, positive charge is found (Müller and Carell, 2009) and this binding become initiator for splitting mutagenic and cytotoxic pyrimidine dimers via blue light induced electron transfer from reduced flavin adenine dinucleotide ($FADH^-$) to the DNA lesion site initiating the repair process for it (Faraji and Dreuw, 2014).

Man, including some other mammals, has lost the PL gene, but instead of it, they have more versatile and complex nucleotide excision repair (NER) for repairing UV-induced DNA lesions (Lima-Bessa et al., 2008). But in case of NER-deficient cells, both CPD PL and 6-4PL play a preventive role against UVR-induced apoptosis, whereas in case of NER proficient cells, only CPD PL prevents UV-induced apoptosis (Lima-Bessa et al., 2008).

It is also proposed that in many organisms, PLs are found deep inside the tissue where light is inaccessible, and due to the tight packaging of DNA and some other minor changes, the excision repair pathway also becomes unable to recognize CPD lesion (Guintini et al., 2015). In these cases damaged DNA sites are removed together with the PLs and original structures restored (Vechtomova et al., 2021) and other possible functional property of DNA PLs include where PLs bind with drug damaged site of DNA such as to the cisplatin-dGpG diadduct that is not recognized by DNA excision repair enzyme and this type of damage is also removed along with PLs leading to promoting cell survival and removing damaged DNA (Sancar, 2003).

4.6 Structural features of DNA photolyase

DNA PLs are included in a class of structurally related flavoprotein family called CPF found in all kingdoms of life. This CPF is further subdivided into two major groups: PL and cryptochrome (CRY), in which PLs are involved in repairing UV-induced DNA damage, whereas cryptochromes (CRYs) are responsible for utilizing blue light to regulate a variety of developmental as well as adaptive responses from lower prokaryotes to humans.

4.6.1 Structural similarity and differences of photolyases and cryptochromes

Protein architectures of CRYs show similarity with PLs and also possess the flavin adenine dinucleotide (FAD) cofactor, as shown in Fig. 4.3. An additional signal-transmitting domain is also found attached either to the N-termini or C-termini of CRYs (Müller and Carell, 2009). Ahmad and Cashmore first discovered the CRYs in 1993 in *Arabidopsis thaliana*, which represent similarity with PLs. So, the receptors are homologs of DNA PLs, especially named cryptochrome. DNA-PL sequence shows 25%–40% homology with amino acid sequences of CRY, although

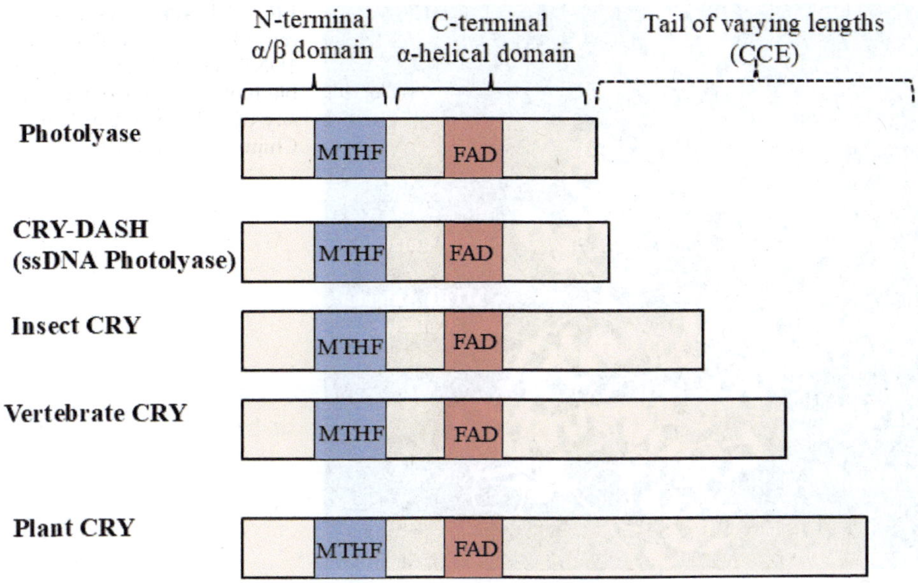

FIGURE 4.3 Comparative domain structure of DNA photolyases and cryptochromes from different source organisms: Domain structure of CRY has similarity with photolyase, with FAD as an essential cofactor and an optional antenna cofactor not present in all, but present in some other CRY, found associated with photolyase ancestors containing some property of DNA repair. At the same time, photolyases lack a varying length of the C-terminal domain as a strikingly differentiating fact. *CRY*, cryptochrome; *FAD*, flavin adenine dinucleotide.

in higher plants and animals, it has lost its active site ability to bind with DNA; therefore it cannot be involved in repairing the UV-induced DNA damage (Katerova and Todorova, 2009). CRY has two domains, the first one is an N-terminal domain, also known as the PHR domain (PL homologous region), exhibiting similarity with PL and consisting of nearly 500 residues, and a second C-terminal domain, also known as the CCE domain (CRY C-terminal extension) consists of a varying length of amino acid sequence (Wang and Lin, 2025). The PHR domain is a highly conserved chromophore-binding domain that binds noncovalently with FAD, and it is also reported that evolutionarily it is derived from DNA PL (Cashmore, 2003; Sancar, 2003). The structural similarity of both PL and cryptochrome is also revealed in the other form, where the N-terminal α/β subdomain contains a binding pocket for secondary antenna cofactor such as 5,10-methenyl-tetrahydrofolate (MTHF), and similarly, the C-terminal helical subdomain contains a binding pocket for FAD cofactor (Brautigam et al., 2004; Huang et al., 2006). CRY PHR from different source organisms like plants, insects, and vertebrates have been co-purified with chromophores but not from the mammals (Ozturk et al., 2009; Czarna et al., 2013; Levy et al., 2013; Zoltowski et al., 2019). FAD is situated inside the active center of all CPF. The C-terminal domain has an additional tail, which plays an essential role in protein localization as well as interacting with other regulatory proteins, and also regulates gene expression after binding of the protein to it. The presence of extensions either at the C-terminal or at the N-terminal of cryptochrome, involved in the signal transduction step, is the key point of structural difference between PLs and cryptochromes (Müller and Carell, 2009).

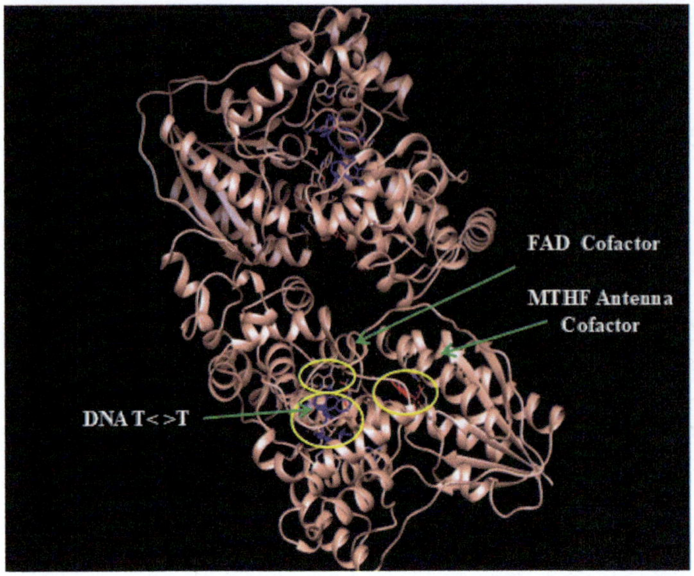

FIGURE 4.4 Crystal structure of *E. coli* photolyase. (PDB accession code: 1DNP, created by using an extensible molecular modeling system University of California, San Francisco (UCSF) Chimera version 1.19.

PLs are monomeric proteins consisting of 450–550 amino acids (AA) having two chromophores: first one is FAD needed as a cofactorfor catalyzing DNA repair and second chromophore consist of natural antenna pigment with any one of a pteridine architecture such as 5,10-methenyltetrahydrofolate (MTHF), flavin mononucleotide (FMN) or FAD, 8-hydroxy-7,8-didemethyl-5-deazariboflavin (8-HDF) and last one 6,7-dimethyl-8-ribityllumazine (DMRL) containing lumazine work as a photoantenna (Ramírez et al., 2021) and take the energy of blue-light spectrum photons and this energy of photon is further delivered to FAD for the sake of increasing enzyme repair efficiency because FAD does not directly absorb the blue-light spectrum (Kavakli et al., 2019). As of today, the crystal structure of PLs has been studied from different source organisms. But the first crystal structure study was done on the CPD PL crystal structure from *E. coli* in 1995 (Park et al., 1995) and then of *Anacystis nidulans* in 1997 (Tamada et al., 1997). *E. coli* PLs, as shown in Fig. 4.4, are more frequently studied, so understanding the structural features of this PL provides a better understanding of all other PLs present in different organisms. The study about the crystal structure of PL reveals that the enzyme has a positively charged surface containing a binding pocket for accommodating the redox cofactor. $FADH^-$ noncovalently gets accommodated inside this pocket with such an orientation to recruit DNA lesion and the second light-harvesting cofactor.

E. coli DNA PL utilizes 350–450 nm blue light as an energy source for repairing CPD DNA lesions as one of the major types of UVR-induced lesions (Lo et al., 2005), and there are approximately 10-20 molecules of active DNA PL found per cell of *E. coli* (Harm et al., 1968). *E. coli* PLs are made up of a single 471 AA long polypeptide chain having two well-defined domains: an N-terminal α/β domain consisting of 1–131 residues and a C-terminal α-helical domain consisting of 204–471 residues, both domains are connected through a long interdomain loop consisting of 132–203 residues wrapped around the α/β domain as shown in Fig. 4.5 (Sancar et al., 1984; Saxena et al., 2004).

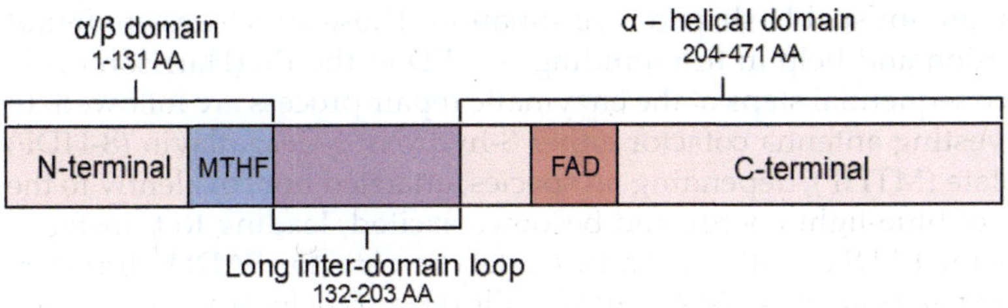

FIGURE 4.5 Schematic domain structure of *E. coli* photolyase.

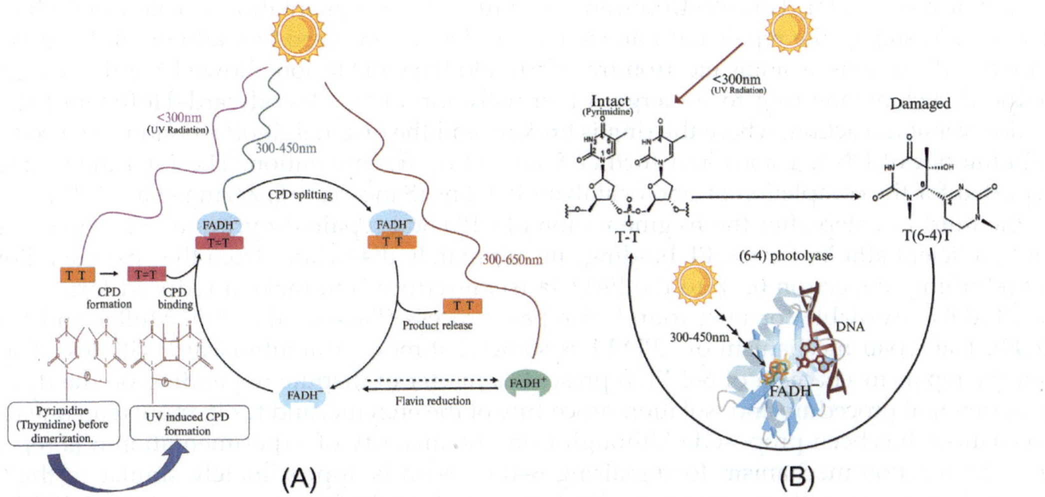

FIGURE 4.6 Mode of action of DNA photolyase: (A) CPD photolyase in repairing CPD DNA lesion, in which intact pyrimidine (T-T) is indicated with *orange* color, and a UV-induced pyrimidine dimer is indicated with *light red* color shade. (B) 6-4 photolyase for repair of 6-4 PP photolesion. *CPD*, Cyclobutane pyrimidine dimer; *UV*, ultraviolet.

4.7 Mode of action

The flavin cofactor present in PLs can exist in three different redox states: one-electron reduced state known as semiquinone, which is also considered a radical due to an unpaired electron, two-electron reduced state known as hydroquinone, and the third oxidized state. FAD cofactor in PLs is found in a two-electron reduced state ($FADH^-$) working as the electron-donating entity (Zhang et al., 2017). After the disclosure of the first crystal structure of CPD PL complex with a DNA duplex containing CPD lesion in 2004, it opened up a breakthrough for decoding the molecular mechanism of PL-mediated DNA lesion repair (Mees et al., 2004). The mode of action of DNA PLs is considered one of the classic examples of photocatalyzed enzymatic reaction in which one crucial stage depends on light as a sole cause of photoreactivation (Sancar, 2003, 2016), as shown in Fig. 4.6. When the site of pyrimidine dimer lesions comes into contact with PL, its detection becomes easy because of the flipping out of the pyrimidine dimer, and this outward flipped region gets recruited into the active site cavity of PL, having catalytic cofactor $FADH^-$, leading to the formation of the enzyme–substrate complex. This binding site is present on the surface of the helical domain in the form of a pocket. FAD is found attached to the bottom of this pocket in an unusual U-shaped conformation. These structures are found complementary to the CPD lesion and help in firm binding of CPD to the PL (Harrison et al., 2005). After the detection, three sequential steps of the enzymatic repair process are followed. In the initial first step, light-harvesting antenna cofactor either 8-hydroxy-5-deazaflavin (8-HDF) or 5,10-methenyltetrahydrofolate (MTHF), depending on species, attached noncovalently to the enzyme, absorbs the photons of blue-light energy and becomes excited, leading to transfer of electrons to the catalytic cofactor $FADH^-$. Ultimately, this catalytic cofactor $FADH^-$ transfers the electrons to the DNA lesion site (Kao et al., 2008a, 2008b). Electron transfer from the excited $FADH^-$ to the DNA lesion site, either to the CPD or to the 6-4PP lesion, is a key step for the PL repair mechanism. This electron transfer induces the fragmentation reaction in both cases. The second step of the repair mechanism shows electron-induced breakdown of the lesion. In case of CPD lesions, a single electron transferred to the CPD lesion allowed the single electron reduced cyclobutane ring to undergo a thermally forbidden (Woodward–Hoffmann) [2 + 2] cycloreversion reaction, where the ring is broken, and the original double bond is restored after flipping out of DNA lesions and further followed by fragmentation. The total time duration reported for the completion of photocatalysis is 1.2 ns (Sancar, 2016; Zhang et al., 2017). Finally in the third last step after the fragmentation of CPD and repaired pyrimidine restored, which lack sufficient affinity for the PL binding site, ultimately dissociates from the enzyme followed by returning of electron back to the FADH• (semi-reduced) to make a reduced form of FAD, i.e., $FADH^-$ available for next round of repair process (Park et al., 1995; Müller and Carell, 2009). The repair mechanism of CPD PL is somewhat more straightforward with less dispute, but the repair mechanism of 6-4 PL represents a matter of dispute depending on the different experimental procedure and isolation procedure of the enzyme, and for it, more than one repair mechanism has been proposed. Although from the majority of experimentation it is reported that the reaction mechanism for repairing 6-4PP lesion is approximately similar to the CPD repair mechanism with some minute differences such as the incorporation of the 6-4PP lesion into the active enzyme complex as an initial light independent step, an intermediate 6-iminium ion is formed via proton transfer induced by highly conserved two histidines present among the 6-4PP which spontaneously rearrange to form an oxetane cycle (Maul et al., 2008), further breaking of this cycle become more complicated in comparison to breaking of CPD resulting in the less quantum yield of photoreparation of 6-4PL (0.3) in comparison to CPD PL (0.7–1) (Todo et al., 1993; Kim et al., 1994).

4.8 Therapeutic applications

PLs are therapeutically used naturally from the beginning, when the environment was too harsh for supporting life under the direct UVR exposure as a DNA repair enzyme. The attention on the therapeutic application of PLs became a more valuable and research-oriented biological molecule of interest after the in vitro experiment of Rupert et al. on photoreactivation in ultraviolet-inactivated *H. influenzae* transforming DNA. PL has therapeutic applications in various skin cells related to cancer, such as melanoma, basal cell, and squamous cell carcinoma, and actinic keratosis (AK). PL extracted from the algae *A. nidulans* was applied topically with sunscreens, showing a significant reduction in the formation of CPDs and UVR-induced apoptosis in human skin (Berardesca et al., 2012). During the chemotherapy and radiotherapy for treating cancerous cells, at the same time, healthy cells also become damaged and affected

(Vechtomova et al., 2021). By the use of PLs, the DNA-damaging effects of UVR can also be controlled to some extent during chemotherapy and radiotherapy. Topical use of PLs containing sunscreen formula shows 93% reduction in CPD formation before UVR exposure over a week; comparatively, traditional sunscreens resulted in 62% reduction (Emanuele, 2011). Also, the use of the enzyme topically on human skin before UVR exposure prevents the premature senescence of cells (Emanuele et al., 2013). The effectiveness of the therapeutic potential of PLs is enhanced when these are combined with UV filters and encapsulated into liposomes (Nascimento et al., 2024). This combined enzyme therapy provides better photoprotection and repair against UVR-induced damage (Stege et al., 2000).

4.9 Future perspectives

The repair mechanism of PLs exclusively depends on the excitation of $FADH^-$ by blue light; at the same time, living tissues are also less permeable to blue light (Terai et al., 2020). This inevitably leads to the need for a secondary light-harvesting chromophore that has the ability to provide excitation energy to the ground-state $FADH^-$ for efficient DNA repair. So there is a need for exploration of some chemical modification in PLs using a fluorophore as an alternative to light-harvesting chromophore for very stable and efficient DNA repair without any side effects. There is also a need for Antarctic organisms' exploration for novel and robust PLs that are in continuous exposure to a gradual increase of UVR intensity due to the continuous ozone layer depletion, as well as efficient encapsulation and drug delivery methods, which have also opened a new area of exploration (Ramírez et al., 2021).

4.10 Conclusions

PLs are the enzymes responsible for UVR-induced DNA damage repair via photoreactivation, and work as a safeguard against skin cancer, photoaging, and AK. What makes it so important is not only the event of evolution of life form and their genome maintenance against the damaging effects of UVR, when there was no shield of ozone layer but modern days also needs PL with efficient DNA repair mechanism through the involvement of chemical modification, genetic engineering, nanobiotechnology for its stable and accurate delivery at the site of DNA lesions with broad spectrum therapeutic applications including cosmetic products, sunscreens, and also in agricultural field to ensure survival of plants, animals and all other life forms against more additive UV-induced DNA damage due to pollution and ozone layer depletion day by day.

References

Adeyemi, R.O., 2023. Transcription and DNA repair collide after UV exposure. Proc. Natl. Acad. Sci. 120 (16), e2303201120.

Berardesca, E., Bertona, M., Altabas, K., Altabas, V., Emanuele, E., 2012. Reduced ultraviolet-induced DNA damage and apoptosis in human skin with topical application of a photolyase-containing DNA repair enzyme cream: clues to skin cancer prevention. Mol. Med. Reports 5 (2), 570–574.

Beukers, R., Eker, A.P., Lohman, P.H., 2008. 50 years thymine dimer. DNA Repair (Amst.) 7 (3), 530–543.

Brautigam, C.A., Smith, B.S., Ma, Z., Palnitkar, M., Tomchick, D.R., Machius, M., et al., 2004. Structure of the photolyase-like domain of cryptochrome 1 from *Arabidopsis thaliana*. Proc. Natl. Acad. Sci. 101 (33), 12142–12147.

Brudler, R., Hitomi, K., Daiyasu, H., Toh, H., Kucho, K.I., Ishiura, M., et al., 2003. Identification of a new cryptochrome class: structure, function, and evolution. Mol. Cell 11 (1), 59–67.

Cadet, J., Douki, T., 2018. Formation of UV-induced DNA damage contributing to skin cancer development. Photochem. Photobiol. Sci. 17 (12), 1816–1841.

Cashmore, A.R., 2003. Cryptochromes: enabling plants and animals to determine circadian time. Cell 114 (5), 537–543.

Chaves, I., Pokorny, R., Byrdin, M., Hoang, N., Ritz, T., Brettel, K., et al., 2011. The cryptochromes: blue light photoreceptors in plants and animals. Annu. Rev. Plant Biol. 62 (1), 335–364.

Ciccia, A., Elledge, S.J., 2010. The DNA damage response: making it safe to play with knives. Mol. Cell 40 (2), 179–204.

Cockell, C.S., 1998. Biological effects of high ultraviolet radiation on early Earth—a theoretical evaluation. J. Theor. Biol. 193 (4), 717–729.

Czarna, A., Berndt, A., Singh, H.R., Grudziecki, A., Ladurner, A.G., Timinszky, G., et al., 2013. Structures of Drosophila cryptochrome and mouse cryptochrome1 provide insight into circadian function. Cell 153 (6), 1394–1405.

de Lima-Bessa, K.M., Armelini, M.G., Chiganças, V., Jacysyn, J.F., Amarante-Mendes, G.P., Sarasin, A., et al., 2008. CPDs and 6-4PPs play different roles in UV-induced cell death in normal and NER-deficient human cells. DNA Repair (Amst.) 7 (2), 303–312.

Emanuele, E., Altabas, V., Altabas, K., Berardesca, E., 2013. Topical application of preparations containing DNA repair enzymes prevents ultraviolet-induced telomere shortening and c-FOS proto-oncogene hyperexpression in human skin: an experimental pilot study. J. Drugs Dermatol. 12 (9), 1017–1021.

Faraji, S., Dreuw, A., 2014. Physicochemical mechanism of light-driven DNA repair by (6-4) photolyases. Annu. Rev. Phys. Chem. 65 (1), 275–292.

Geisselbrecht, Y., Frühwirth, S., Schroeder, C., Pierik, A.J., Klug, G., Essen, L.O., 2012. CryB from *Rhodobacter sphaeroides*: a unique class of cryptochromes with new cofactors. EMBO Rep. 13 (3), 223–229.

Guintini, L., Charton, R., Peyresaubes, F., Thoma, F., Conconi, A., 2015. Nucleosome positioning, nucleotide excision repair, and photoreactivation in *Saccharomyces cerevisiae*. DNA Repair (Amst.) 36, 98–104.

Harm, W., Harm, H., Rupert, C.S., 1968. Analysis of photoenzymatic repair of UV lesions in DNA by single light flashes: II. In vivo studies with *Escherichia coli* cells and bacteriophage. Mutat. Res. 6 (3), 371–385.

Harrison, C.B., O'Neil, L.L., Wiest, O., 2005. Computational studies of DNA photolyase. J. Phys. Chem. A 109 (32), 7001–7012.

Huang, Y., Baxter, R., Smith, B.S., Partch, C.L., Colbert, C.L., Deisenhofer, J., 2006. Crystal structure of cryptochrome 3 from *Arabidopsis thaliana* and its implications for photolyase activity. Proc. Natl. Acad. Sci. 103 (47), 17701–17706.

Kao, Y.T., Saxena, C., He, T.F., Guo, L., Wang, L., Sancar, A., et al., 2008a. Ultrafast dynamics of flavins in five redox states. J. Am. Chem. Soc. 130 (39), 13132–13139.

Kao, Y.T., Tan, C., Song, S.H., Öztürk, N., Li, J., Wang, L., et al., 2008b. Ultrafast dynamics and anionic active states of the flavin cofactor in cryptochrome and photolyase. J. Am. Chem. Soc. 130 (24), 7695–7701.

Katerova, Z.I., Todorova, D., 2009. Endogenous polyamines lessen membrane damage in pea plants provoked by enhanced ultraviolet-C radiation. Plant Growth Regul. 57 (2), 145–152.

Kavakli, I.H., Ozturk, N., Gul, S., 2019. DNA repair by photolyases. In: Donev, R (Ed.). In: Advances in Protein Chemistry and Structural Biology, 115. Academic Press, pp. 1–19. doi:10.1016/bs.apcsb.2018.10.003.

Kim, S.T., Malhotra, K., Smith, C.A., Taylor, J.S., Sancar, A., 1994. Characterization of (6-4) photoproduct DNA photolyase. J. Biol. Chem. 269 (11), 8535–8540.

Levy, C., Zoltowski, B.D., Jones, A.R., Vaidya, A.T., Top, D., Widom, J., et al., 2013. Updated structure of Drosophila cryptochrome. Nature 495 (7441), E3–E4.

Lo, H.L., Nakajima, S., Ma, L., Walter, B., Yasui, A., Ethell, D.W., Owen, L.B., 2005. Differential biologic effects of CPD and 6-4PP UV-induced DNA damage on the induction of apoptosis and cell-cycle arrest. BMC Cancer 5 (135), 1–9. doi:10.1186/1471-2407-5-135.

Lucas-Lledó, J.I., Lynch, M., 2009. Evolution of mutation rates: phylogenomic analysis of the photolyase/cryptochrome family. Mol. Biol. Evol. 26 (5), 1143–1153.

Masuma, R., Kashima, S., Kurasaki, M., Okuno, T., 2013. Effects of UV wavelength on cell damage caused by UV irradiation in PC12 cells. J. Photochem. Photobiol. B 125, 202–208.

Maul, M.J., Barends, T.R., Glas, A.F., Cryle, M.J., Domratcheva, T., Schneider, S., Carell, T., et al., 2008. Crystal structure and mechanism of a DNA (6-4) photolyase. Angew. Chem. Int. Ed. 47 (52), 10076–10080.

Mees, A., Klar, T., Gnau, P., Hennecke, U., Eker, A.P., Carell, T., Essen, L.O., 2004. Crystal structure of a photolyase bound to a CPD-like DNA lesion after in situ repair. Science 306 (5702), 1789–1793.

Müller, M., Carell, T., 2009. Structural biology of DNA photolyases and cryptochromes. Curr. Opin. Struct. Biol. 19 (3), 277–285.

Nascimento, N.S., Torres-Obreque, K.M., Oliveira, C.A., Rabelo, J., Baby, A.R., Long, P.F., et al., 2024. Enzymes for dermatological use. Exp. Dermatol. 33 (1), e15008.

Oberpichler, I., Pierik, A.J., Wesslowski, J., Pokorny, R., Rosen, R., Vugman, M., et al., 2011. A photolyase-like protein from Agrobacterium tumefaciens with an iron-sulfur cluster. PLoS One 6 (10), e26775.

Ozer, Z., Reardon, J.T., Hsu, D.S., Malhotra, K., Sancar, A., 1995. The other function of DNA photolyase is: stimulation of excision repair of chemical damage to DNA. Biochemistry 34 (49), 15886–15889.

Ozturk, N., Selby, C.P., Song, S.H., Ye, R., Tan, C., Kao, Y.T., et al., 2009. Comparative photochemistry of animal type 1 and type 4 cryptochromes. Biochemistry 48 (36), 8585–8593.

Park, H.W., Kim, S.T., Sancar, A., Deisenhofer, J., 1995. Crystal structure of DNA photolyase from *Escherichia coli*. Science 268 (5219), 1866–1872.

Pokorny, R., Klar, T., Hennecke, U., Carell, T., Batschauer, A., Essen, L.O., 2008. Recognition and repair of UV lesions in loop structures of duplex DNA by DASH-type cryptochrome. Proc. Natl. Acad. Sci. 105 (52), 21023–21027.

Rahman, H., Kumar, D., Liu, T., Okwundu, N., Lum, D., Florell, S.R., et al., 2021. Aspirin protects melanocytes and keratinocytes against UVB-induced DNA damage in vivo. J. Invest. Dermatol. 141 (1), 132–141.

Ramírez, N., Serey, M., Illanes, A., Piumetti, M., Ottone, C., 2021. Immobilization strategies of photolyases: challenges and perspectives for DNA repairing application. J. Photochem. Photobiol. B 215, 112113.

Rastogi, R.P., Kumar, A., Tyagi, M.B., Sinha, R.P., 2010. Molecular mechanisms of ultraviolet radiation-induced DNA damage and repair. J. Nucleic Acids 2010 (1), 592980.

Rastogi, R.P., Madamwar, D., Nakamoto, H., Incharoensakdi, A., 2020. Resilience and self-regulation processes of microalgae under UV radiation stress. J. Photochem. Photobiol. C 43, 100322.

Rupert, C.S., Goodgal, S.H., Herriott, R.M., 1958. Photoreactivation in vitro of ultraviolet-inactivated *Hemophilus influenzae* transforming factor. J. Gen. Physiol. 41 (3), 451.

Sancar, A., 2016. Mechanisms of DNA repair by photolyase and excision nuclease (nobel lecture). Angew. Chem. Int. Ed. (30) 55.

Sancar, A., 2008. Structure and function of photolyase and in vivo enzymology: 50th anniversary. J. Biol. Chem. 283 (47), 32153–32157.

Sancar, A., 2003. Structure and function of DNA photolyase and cryptochrome blue-light photoreceptors. Chem. Rev. 103 (6), 2203–2238.

Sancar, G.B., Smith, F.W., Lorence, M.C., Rupert, C.S., Sancar, A., 1984. Sequences of the *Escherichia coli* photolyase gene and protein. J. Biol. Chem. 259 (9), 6033–6038.

Saxena, C., Sancar, A., Zhong, D., 2004. Femtosecond dynamics of DNA photolyase: energy transfer of antenna initiation and electron transfer of cofactor reduction. J. Phys. Chem. B 108 (46), 18026–18033.

Selby, C.P., Sancar, A., 2006. A cryptochrome/photolyase class of enzymes with single-stranded DNA-specific photolyase activity. Proc. Natl. Acad. Sci. 103 (47), 17696–17700.

Shetta, N., Aref, I., 2009. Impact of ultraviolet-C radiation on seed germination and chlorophyll concentration of some woody trees grown in Saudi Arabia. In: J. Agri. Food Environ. Sci, 8, pp. 1–21.

Stege, H., Roza, L., Vink, A.A., Grewe, M., Ruzicka, T., Grether-Beck, S., et al., 2000. Enzyme plus light therapy to repair DNA damage in ultraviolet-B-irradiated human skin. Proc. Natl. Acad. Sci. 97 (4), 1790–1795.

Tamada, T., Kitadokoro, K., Higuchi, Y., Inaka, K., Yasui, A., de Ruiter, P.E., et al., 1997. Crystal structure of DNA photolyase from *Anacystis nidulans*. Nat. Struct. Biol. 4 (11), 887–891.

Terai, Y., Sato, R., Matsumura, R., Iwai, S., Yamamoto, J., 2020. Enhanced DNA repair by DNA photolyase bearing an artificial light-harvesting chromophore. Nucleic Acids Res. 48 (18), 10076–10086.

Todo, T., Takemori, H., Ryo, H., Lhara, M., Matsunaga, T., Nikaido, O., et al., 1993. A new photoreactivating enzyme that specifically repairs ultraviolet light-induced (6-4) photoproducts. Nature 361 (6410), 371–374.

Tyrrell, R.M., Reeve, V.E., 2006. Potential protection of skin by acute UVA irradiation—From cellular to animal models. Prog. Biophys. Mol. Biol. 92 (1), 86–91.

Vechtomova, Y.L., Telegina, T.A., Buglak, A.A., Kritsky, M.S., 2021. UV radiation in DNA damage and repair involving DNA-photolyases and cryptochromes. Biomedicines 9 (11), 1564.

Vechtomova, Y.L., Telegina, T.A., Kritsky, M.S., 2020. Evolution of proteins of the DNA photolyase/cryptochrome family. Biochemistry (Moscow) 85 (Suppl 1), 131–153.

Wang, X., Lin, C., 2025. The two action mechanisms of plant cryptochromes. Trends Plant Sci. 30 (7), 775–791.

Yamamoto, K., Satake, M., Shinagawa, H., Fujiwara, Y., 1983. Amelioration of the ultraviolet sensitivity of an *Escherichia coli* recA mutant in the dark by photoreactivating enzyme. Mol. Gen. Genet. 190 (3), 511–515.
Yasui, A., Eker, A.P., Yasuhira, S., Yajima, H., Kobayashi, T., Takao, M., et al., 1994. A new class of DNA photolyases is present in various organisms, including aplacental mammals. EMBO J. 13 (24), 6143–6151.
Yi, C., He, C., 2013. DNA repair by reversal of DNA damage. Cold Spring Harb. Perspect. Biol. 5 (1), a012575.
Zhang, F., Scheerer, P., Oberpichler, I., Lamparter, T., Krauß, N., 2013. Crystal structure of a prokaryotic (6-4) photolyase with an Fe-S cluster and a 6, 7-dimethyl-8-ribityllumazine antenna chromophore. Proc. Natl. Acad. Sci. 110 (18), 7217–7222.
Zhang, M., Wang, L., Zhong, D., 2017. Photolyase: dynamics and electron-transfer mechanisms of DNA repair. Arch. Biochem. Biophys. 632, 158–174.
Zoltowski, B.D., Chelliah, Y., Wickramaratne, A., Jarocha, L., Karki, N., Xu, W., et al., 2019. Chemical and structural analysis of a photoactive vertebrate cryptochrome from pigeon. Proc. Natl. Acad. Sci. 116 (39), 19449–19457.

CHAPTER

5

DNA repair in extremophiles

Yogendra Singh Rajpurohit[a,b], *Dhirendra Kumar Sharma*[a,b] *and Ishu Soni*[a,b]

[a]Molecular Biology Division, Bhabha Atomic Research Centre, Mumbai, Maharashtra, India [b]Life Sciences, Homi Bhabha National Institute (Department of Atomic Energy, Deemed to be University), Mumbai, Maharashtra, India

5.1 Introduction

Extremophiles are organisms that live in extremely hostile environments, such as those with extreme temperatures and pressure, high acidity, elevated salinity, intense drought, and radiation-rich conditions (Atomi and Reeve, 2019; Coker, 2019). For instance some extremophiles can survive at temperatures ranging from −12°C, such as *Psychromoas* species (Auman et al., 2006; Nogi et al., 2007; Auman et al., 2010), to as high as 129°C, such as *Geogemma barossii* (Kashefi and Lovley, 2003; Cowen, 2004), pH values to near zero (Johnson et al., 2012), as seen in *Picrophilus torridus*, to as high as 13, which includes species such as *Plectonema nostocorum* and *Hydrogenophaga* sp. (Horikoshi, 1999; Moran-Reyna and Coker, 2014; Mamo, 2020), saline conditions that exceed saturation for NaCl and KCl, as demonstrated by *Haloferax volcanii* (Mullakhanbhai and Larsen, 1975; Kessel et al., 1988; Perez-Arnaiz et al., 2020), and pressures that go beyond 100 MPa, such as *Shewanella benthica* (Kato et al., 1998). Additionally, some extremophiles can withstand intense ultraviolet radiation greater than 100 J/m^2 and gamma radiation exceeding 12 kGy, as seen in *Halobacterium* sp. NRC-1 and *Deinococcus radiodurans* (Cox and Battista, 2005; Slade and Radman, 2011; Kawaguchi et al., 2020). These remarkable adaptations in diverse niches highlight the resilience of life.

In extreme environments, extremophiles are subjected to external threats that vary with environmental conditions, including chemical cross-linkers, ultraviolet light, ionizing radiation, heat-induced DNA denaturation, and external reactive oxygen species (ROS). These threats induces undesired mutations that negatively impact cellular and organismal fitness. For example, halophilic archaea in UV-exposed salt plains and hyperthermophiles at DNA-denaturing temperatures experience increased rates of deamination, depurination, and oxidation. Additionally, errors during DNA replication, such as mismatches or the incorporation of ribonucleotide monophosphates (rNMPs), along with ROS generated by metabolism, further jeopardize DNA

Genome Damage: Repair and Adaptation Across Life.
DOI: https://doi.org/10.1016/B978-0-443-42857-9.00022-0

integrity. Despite these challenges, extremophiles maintain their genomic stability comparable to that of mesophilic prokaryotes such as *Escherichia coli*. Interestingly, no unique DNA repair pathways have been identified in archaea or extremophilic bacteria such as *D. radiodurans*. Instead, these organisms depend on well-established repair mechanisms to preserve their genome integrity in hostile habitats, revealing how life can persist in harsh conditions. In this chapter, we will discuss the DNA repair adaptations that organisms in different extreme environments have evolved to survive. We will also emphasize important scientific findings related to the DNA repair abilities of these organisms as model systems.

5.2 DNA repair in acidophiles

Acidophiles are adapted to live in acidic environments. Their cell membranes are highly impermeable to protons, aided by unique lipid structures, such as tetraether lipids in archaea, which resist acid hydrolysis (Amaro et al., 1991; Albers et al., 2001; Konings et al., 2002). Additionally, some organisms, such as *Acidithiobacillus ferrooxidans*, have smaller membrane channel pores to limit proton entry (Baker-Austin and Dopson, 2007), or some organisms, such as *Bacillus* and *Thermoplasma sp.*, use active pumping of protons to balance cellular pH (Michels and Bakker, 1985; Coker, 2019). Acidophiles do not carry any specific genetic signature for acid resistance but have smaller genomes compared to their neutrophile counterparts, with unclear evolutionary benefit. A study focusing on the oxidative stress response and UV tolerance of two acidophilic iron-oxidizing bacteria: *Leptospirillum ferriphilum* and *A. ferrooxidans*, suggested that *L. ferriphilum has* higher ROS mitigation utilizing superoxide dismutase and thioredoxin over *A. ferrooxidans* (Farias et al., 2021). Additionally, the DNA damage caused by UV radiation, primarily in the form of cyclobutane pyrimidine dimers (CPDs), was significantly lower in *L. ferriphilum* than in *A. ferrooxidans* across all recovery periods analyzed. This means that *L. ferriphilum* possesses effective DNA protection and repair mechanisms that enhance its resilience to UV stress (Farias et al., 2021). Class I photolyase gene found in *L. ferriphilum*, which is supposed to play an important role in the DNA damage that occurs due to UV irradiation. The absence of this gene in *A. ferrooxidans* explains the reason for its enhanced sensitivity to UV radiation due to the lack of a photoreactivation mechanism for effective repair of DNA lesions caused by UV (Farias et al., 2021). Acidophiles are crucial in the biochemical process of bioleaching and help recover metal from sulfidic ores by oxidizing ferrous ions (Rohwerder et al., 2003; Vera et al., 2013).

5.3 DNA repair in halophiles

Halophiles, organisms that grow in saline environments and have evolved two primary strategies to survive in high-salinity conditions: the "salt-in" and "salt-out" approaches (Yin et al., 2015). Salt-in organisms maintain similar ion concentrations inside and outside their cells, with acidic proteins dominating their proteome to mitigate the dehydrating effects of high salinity (Brininger et al., 2018). These acidic proteins form protective "water cages" that prevent aggregation and precipitation (Reed et al., 2013; Kumar et al., 2018). This adaptation involves significant evolutionary changes, confining these organisms to stable saline environments, and is exclusive to prokaryotes. In contrast, salt-out organisms actively regulate internal salt concentrations by accumulating organic osmolytes (da Costa et al., 1998; Roberts, 2005). Though more energy-intensive, this strategy allows them to tolerate a broader range of salinity levels, demonstrating greater environmental flexibility.

Halophile *Halobacterium* NRC-1 survives in environments with intense solar radiation and harbors numerous DNA repair mechanisms to combat damage caused by UV exposure (Dassarma et al., 2001; Boubriak et al., 2008). Photoreactivation, a direct DNA repair process mediated by photolyases, is particularly active in *Halobacterium*, with two photolyase homologs potentially addressing UV-induced lesions such as CPDs and 6-4 photoproducts (Ng et al., 2000; McCready and Marcello, 2003; Baliga et al., 2004). The organism also employs excision repair pathways: base excision repair is facilitated by glycosylases such as Ogg, AlkA, *MutY*, and Nth, alongside XthA for subsequent steps (Grasso and Tell, 2014). The Rad2-family flap endonuclease is critical in *Halobacterium* sp. NRC-1, potentially serving as an alternative flap endonuclease employed by archaea for long-patch repair resolution (Berquist et al., 2007). Nucleotide excision repair involves both bacterial-type (UvrA, UvrB, UvrC, and UvrD) and eukaryotic-type (Rad2, Rad3, Rad25, ERCC4) homologs, suggesting the active utilization of excision repair (Ng et al., 2000; Dassarma et al., 2001; Crowley et al., 2006). Mismatch repair genes *MutS*1 and *MutL* are also present in these archaea, correct replication errors, and suggest active strand discrimination, potentially through methylation or restriction endonuclease pathways (Ng et al., 2000; Busch and DiRuggiero, 2010; Perez-Arnaiz et al., 2020). Homologous recombination and recombinational repair are likely carried out by RadA1 and RadA2 recombinases, akin to RecA/Rad51, supported by distant homologs of Mre11 and Holliday junction resolvase (Patoli et al., 2017). Besides this, the Mre11-Rad50 complex has been shown to promote genome compaction as part of the DNA damage response, which accelerates cell recovery by assisting DNA repair proteins in locating their targets and facilitating the search for intact DNA sequences during homologous recombination (Delmas et al., 2013). RadA mutants of *H. volcanii* are radiation sensitive (Woods and Dyall-Smith, 1997; Hogrel et al., 2020), whereas RadA paralogue (RadB) functions as a recombination mediator for proficient recombination repair and cell survival against DNA-damaging agents (Wardell et al., 2017). DNA replication, essential for repair, involves four DNA polymerase homologs, whereas a homolog of UmuC/DinB allows damage tolerance through translesion synthesis (Perez-Arnaiz et al., 2020). DNA ligation is mediated by an archaeal-type ATP-dependent ligase (Zhao et al., 2006; Perez-Arnaiz et al., 2020) or NAD-dependent ligase (Poidevin and MacNeill, 2006). Collectively, the homologs of key bacterial and eukaryotic repair proteins underscore a highly efficient DNA repair system in halophile archaea, enabling them to tolerate extreme saline environments with robust genomic stability mechanisms. Additionally, *H. salinarum*, *Haloferax mediterranei*, and *H. volcanii* have all been shown to be polyploid, a feature that contributes to DNA repair through homologous recombination (Zerulla and Soppa, 2014; Ludt and Soppa, 2019).

5.4 DNA repair in radiophiles

Radiophiles are organisms that can withstand high levels of radiation, specifically ionizing (gamma) and UV radiation (Coker, 2019). *D. radiodurans* is the best-known example of a radiophile and is renowned for its exceptional resistance to extreme conditions, particularly ionizing radiation, UV radiation, and desiccation (Battista, 1997; Cox and Battista, 2005; Blasius et al., 2008; Daly, 2009). The key resistance mechanisms of *D. radiodurans*include proficient repair of

DNA double-strand breaks (DSBs) (Minton, 1994; Cox and Battista, 2005; Misra et al., 2006; Zahradka et al., 2006; Slade et al., 2009; Bentchikou et al., 2010; Bouthier de la Tour et al., 2011; Slade and Radman, 2011; Satoh et al., 2012; Uranga et al., 2017; Jin et al., 2019; Rajpurohit et al., 2021; Ujaoney et al., 2021; Sharma et al., 2024a), enzymes protection from oxidation (Daly et al., 2007; Tian et al., 2009; Daly et al., 2010; Krisko and Radman, 2010; Daly, 2012; Culotta and Daly, 2013), cellular antioxidants protection (Rajpurohit et al., 2008; Ji, 2010; Shrivastava et al., 2010; Slade and Radman, 2011; Misra et al., 2012; Culotta and Daly, 2013; Khairnar et al., 2013; Misra et al., 2013; Rajpurohit et al., 2013; Liu et al., 2014; Zhao et al., 2014; Jeong et al., 2016; Bihani et al., 2018; Lim et al., 2019; Santos et al., 2019; Adamec et al., 2020; Gao et al., 2020; Qi et al., 2020; Dai et al., 2021; Cai et al., 2023; Kuzucu, 2023; Yu et al., 2023), novel DNA repair proteins (Makarova et al., 2001; Earl et al., 2002; Harris et al., 2004; Narumi et al., 2004; Wang et al., 2008; Kota et al., 2010; Sugiman-Marangos and Junop, 2010; Das and Misra, 2012; Selvam et al., 2013; Munteanu et al., 2015; Kota et al., 2016; Banneville et al., 2022), cellular signaling constitute serine/threonine kinase and cell division regulation (Rajpurohit et al., 2008; Kamble et al., 2010; Rajpurohit and Misra, 2010; Desai et al., 2011; Rajpurohit and Misra, 2013a, 2013b; Rajpurohit et al., 2016; Maurya et al., 2018; Sharma et al., 2020; Hou et al., 2022; Rajpurohit et al., 2022; Sharma et al., 2022a; Chaudhary et al., 2023a; Chaudhary et al., 2023b; Misra and Rajpurohit, 2024), and a compact nucleoid structure (Bouthier de la Tour et al., 2009; Toueille et al., 2012; Banneville et al., 2022; Hou et al., 2022). All working together to ensure cell survival after exposure to extremely high doses of gamma rays, prolonged desiccation, and other DNA-damaging agents.

5.4.1 Unique double-strand break repair of *Deinococcus radiodurans*

The DNA DSB repair machinery of *D. radiodurans* is of special interest, which enables it to repair a highly fragmented genome with high accuracy. The bacterium uses a complex system consisting of two main, interdependent routes for repairing DSBs: single-strand annealing (SSA) and extended synthesis-dependent strand annealing (ESDSA) (Fig. 5.1; Minton, 1994; Minton and Daly, 1995; Misra et al., 2006; Zahradka et al., 2006; Slade et al., 2009; Ithurbide et al., 2015; Sugiman-Marangos et al., 2016). The SSA pathway is a crucial RecA-independent process for repairing DSBs in *D. radiodurans* (Ithurbide et al., 2015; Sugiman-Marangos et al., 2016). This pathway is necessary during the initial phases of recovery from radiation-induced DNA damage, particularly when the genome is heavily fragmented. The process includes protection of DNA ends by proteins such as DdrA and DdrB, inhibiting nuclease-mediated degradation, and promoting DdrB-mediated DNA strand annealing based on microhomology with reasonable accuracy (Harris et al., 2004; Tanaka et al., 2004; Xu et al., 2010; Bouthier de la Tour et al., 2011; Selvam et al., 2013). ESDSA is a novel DNA repair mechanism utilized by *D. radiodurans* to restore its genome after it has been broken by radiation or desiccation (Zahradka et al., 2006; Slade et al., 2009). The process is defined by massive DNA synthesis, in contrast to conventional synthesis-dependent strand annealing, which utilizes limited DNA synthesis. ESDSA employs overlapping homology chromosomal fragments as templates and primers for massive complementary single-strand synthesis. The mechanism is triggered by RecJ-supported exonucleolytic resection of DNA fragments, resulting in single-stranded 3′ overhangs (Zahradka et al., 2006; Misra et al., 2006). These overhangs are used as primers to initiate massive DNA synthesis on overlapping fragments. The priming of DNA repair synthesis is a RecA- and RadA-mediated process, which is critical for strand invasion (Zahradka et al., 2006; Misra et al., 2006). The RecFOR

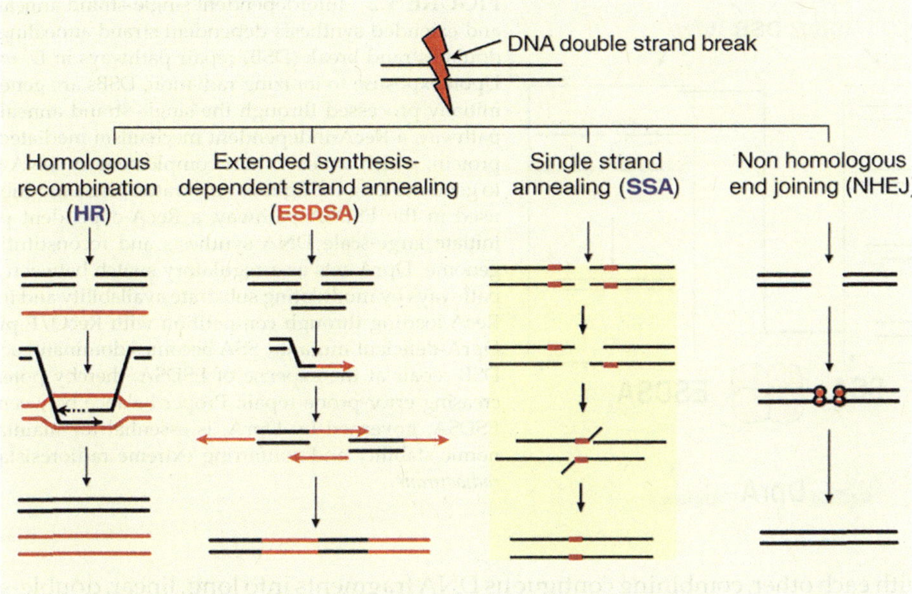

FIGURE 5.1 Four major pathways for repairing double-strand breaks (DSBs) in eukaryotes and bacteria: homologous recombination (HR), extended synthesis-dependent strand annealing (ESDSA), single-strand annealing (SSA), and nonhomologous end joining (NHEJ). HR is an error-free mechanism that uses a homologous sequence, typically from the sister chromatid, to accurately repair the break through strand invasion and DNA synthesis. ESDSA discovered in *D. radiodurans* and involves strand invasion followed by extensive DNA synthesis (*red arrow*), with the newly synthesized strands annealing to complementary regions without crossover to form patch-work repair pattern of old and newly synthesized DNA arranged in tandem. SSA repairs DSBs between homologous repeat sequences by minor resecting DNA ends, aligning homologous regions (*green*), and deleting the intervening sequence. In contrast, NHEJ directly ligates the broken DNA ends without requiring homology, making it a faster but error-prone repair process.

pathway helps the loading of RecA onto ssDNA, as *D. radiodurans* lacks RecBC homologs of *E. coli* (Bentchikou et al., 2010; Satoh et al., 2012). Although RecA is vital for the process as a whole, RadA plays a minor role in priming when RecA is available. Furthermore, distinctive aspects of *D. radiodurans* RecA, including promotion of inverse strand exchange, interaction with PprA, and regulation via phosphorylation, increase its activity in ESDSA repair (Kim and Cox, 2002; Rajan and Bell, 2004; Repar et al., 2010; Rajpurohit et al., 2016; Sharma et al., 2020; Rajpurohit et al., 2021, 2022).

During ESDSA, extensive DNA synthesis is triggered by DNA polymerase III (Pol III) and DNA polymerase I (Pol I) and is needed for efficient elongation of the single strands (Zahradka et al., 2006; ; Slade et al., 2009), and their absence leads to the degradation of DNA fragments and rapid cell death. The speed and coordination of the ESDSA are impressive; the process is completed in few hours after irradiation and relies on a high genome copy number and approx. five times more DNA synthesis than unirradiated wild-type cells (Slade et al., 2009). The elongated single strands overhangs, which can be as long as 20–30 kb, enable the more accurate annealing of complementary strands. The complementary single-stranded extensions

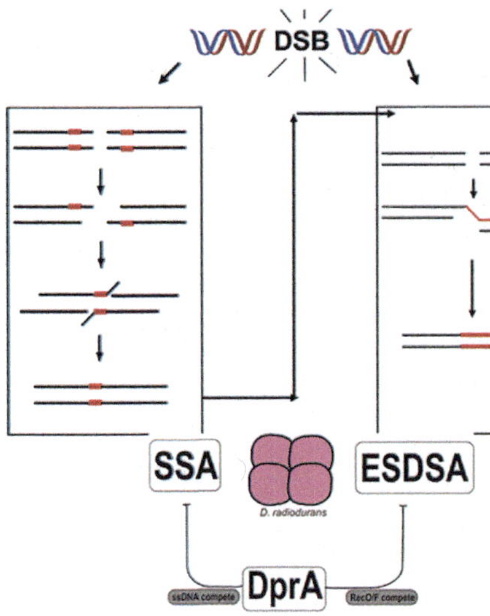

FIGURE 5.2 Interdependent single-strand annealing (SSA) and extended synthesis-dependent strand annealing (ESDSA) double-strand break (DSB) repair pathways in *D. radiodurans*. Upon exposure to ionizing radiation, DSBs are generated and initially processed through the single-strand annealing (SSA) pathway, a RecA-independent mechanism mediated by DdrB protein, which anneals short complementary DNA fragments to generate long contiguous substrates. These are subsequently used in the ESDSA pathway, a RecA-dependent process, to initiate large-scale DNA synthesis and reconstitution of the genome. DprA acts as a regulatory switch between these two pathways by modulating substrate availability and influencing RecA loading through competition with RecO/F proteins. In DprA-deficient mutants, SSA becomes dominant, accelerating DSB repair at the expense of ESDSA, thereby potentially increasing error-prone repair. Proper balance between SSA and ESDSA, governed by DprA, is essential for maintaining genomic stability and conferring extreme radioresistance in *D. radiodurans*.

anneal with each other, combining contiguous DNA fragments into long, linear, double-stranded intermediates. The reassembly of newly synthesized fragments is aided by a third "old-DNA bridging fragment," creating a patchwork of old and new DNA blocks (Fig. 5.1). The long linear intermediates generated by ESDSA are not yet finished circular chromosomes and need to go through RecA-dependent crossovers to become mature circular chromosomes. The DNA ligases, both NAD type (*ligA*) and ATP type (*ligB*), are crucial in both SSA and ESDSA repair (Narumi et al., 2004; Blasius et al., 2007; Kota et al., 2010; Kota et al., 2014).

5.4.2 The interdependence of single-strand annealing and extended synthesis-dependent strand annealing repair in *Deinococcus radiodurans*

SSA and ESDSA are interdependent DNA DSB repair pathways in *D. radiodurans*, especially after ionizing radiation exposure (Fig. 5.2; Xu et al., 2010; Bouthier de la Tour et al., 2011; Ithurbide et al., 2015). SSA, through the action of the DdrB protein, is an initial, RecA-independent process, and it brings together small DNA pieces into large DNA fragments by annealing single-stranded DNA, which is important since high doses of gamma radiation induce a lot of small fragments of the genome (Bouthier de la Tour et al., 2011; Ithurbide et al., 2015). Then the ESDSA pathway, a RecA-dependent process, utilizes these assembled fragments as primers and templates for large-scale DNA synthesis, generating long linear intermediates that then mature into circular chromosomes through homologous recombination (Fig. 5.2; Zahradka et al., 2006; Misra et al., 2006; Slade et al., 2009). These processes can compete for substrates, and SSA is capable of compensating for the loss of homologous recombination. Without RecA, the SSA pathway is a principal pathway of DNA repair (Bouthier de la Tour et al., 2011; Ithurbide et al., 2015). Furthermore, SSA is used to rescue stalled replication forks, especially in the absence of RecA or RecO proteins (Cox et al., 2000; Ithurbide et al., 2015). Therefore, coordinated action of SSA and ESDSA is an important factor contributing to the extreme radioresistance of *D. radiodurans* (Bouthier de la Tour et al., 2011; Ithurbide et al., 2015).

The natural transformation-specific protein DprA regulates balancing the utilization of these two pathways (Sharma et al., 2022b; Sharma et al., 2023; Sharma et al., 2024b; Sharma et al., 2024a). A DprA deletion leads to an accelerated DSB repair, mainly through the SSA pathway, suggesting that DprA might inhibit ESDSA (Sharma et al., 2024a; Sharma et al., 2024b). DprA might also interfere with the loading of RecA onto single-stranded DNA by the RecFOR complex, slowing down the ESDSA repair and indirectly increasing the SSA repair preference (Sharma et al., 2024). However, the function of DprA is multifaceted, as it also appears to suppress SSA repair, maybe due to interference with the access of DdrA/DdrB to ssDNA or through the generation of extra DSBs (Sharma et al., 2024a, 2024b). Thus, DprA facilitates control over the decision between SSA and ESDSA pathways, such that its presence favors more balanced usage of both pathways, whereas its loss brings about a more rapid, but potentially more error-prone, repair through SSA (Fig. 5.2; Sharma et al., 2024a; Sharma et al., 2024b). Therefore, a proper balance between the SSA and ESDSA pathways is necessary to ensure genomic stability in *D. radiodurans*.

5.4.3 Novel proteins, regulatory and other protective mechanisms, and roles in radioresistance of *Deinococcus radiodurans*

D. radiodurans possesses several novel proteins involved in its extreme radioresistance, a majority of which are involved in DNA repair, protection, and stress response (White et al., 1999; Makarova et al., 2001; Tanaka et al., 2004; Selvam et al., 2013; Agapov and Kulbachinskiy, 2015; Ujaoney et al., 2017; Gao et al., 2020). Transcriptome analysis of cells recovering from ionizing radiation or desiccation revealed several highly upregulated genes, such as *ddrA, ddrB, pprA, ddrC,* and *ddrD, which* were found to be specific to the *Deinococcaceae* family (Karlin and Mrazek, 2001; Makarova et al., 2001; Liu et al., 2003; Tanaka et al., 2004; Xu et al., 2010; Selvam et al., 2013; Bouthier de la Tour et al., 2017; Wang et al., 2019; Xiong et al., 2022; Szabla et al., 2024). The DdrA, DdrB, and PprA proteins are DNA-binding proteins that play key roles in radioresistance and have been extensively studied both in vivo and in vitro (Harris et al., 2004; Narumi et al., 2004; Tanaka et al., 2004; Ohba et al., 2005; Kota and Misra, 2006; Murakami et al., 2006; Blasius et al., 2007; Bauermeister et al., 2009; Xu et al., 2010; Bouthier de la Tour et al., 2011; Devigne et al., 2013; Selvam et al., 2013; Adachi et al., 2014; Kota et al., 2014; Devigne et al., 2016; Kota et al., 2016; Kumar Maurya et al., 2019; Rajpurohit et al., 2021). DdrB facilitates an SSA repair, which is critical for assembling small chromosomal fragments generated by high radiation exposure (Bouthier de la Tour et al., 2011; Ithurbide et al., 2015; Sugiman-Marangos et al., 2016). DdrA preferentially binds to 3′ single-stranded DNA ends in vitro and is thought to be part of an end-protection system (Harris et al., 2004; Gutsche et al., 2008; Selvam et al., 2013). PprA protects dsDNA from nuclease degradation, enhances DNA ligase activity (Narumi et al., 2004; Blasius et al., 2007), and has been recently suggested to participate in chromosome segregation following DNA repair (Devigne et al., 2013; Kota et al., 2014; Devigne et al., 2016). DdrC, yet another novel protein with a unique function in circularization and compacting the DNA, and able to bind single-strand nicks and DSBs (Banneville et al., 2022; Szabla et al., 2024), whereas DdrD binds with ssDNA and dsDNA with 5′ ends and may restore plasmid transformation in SSA-defective Δ*ddrB* mutant (de la Tour et al., 2021).

In addition to the presence of novel proteins, unique structural and functional properties of certain repair proteins (RecA, RecD, RecQ, RecN, SbcCD, DNA glycosylases, UvrABC, UvrD,

MutS1, MutL, and SSB), proteins involved in DNA ends processing, genome partitioning, histidine kinases, and Nudix hydrolases, may also contribute to the bacterium's radioresistance (Evans and Moseley, 1983; Moseley and Evans, 1983; Minton, 1994; Makarova et al., 2001; Xu et al., 2001; Kim et al., 2002; Mennecier et al., 2004; Fisher et al., 2004; Buchko et al., 2008; Wang et al., 2008; Chen et al., 2009; Reyes et al., 2010; Kamble and Misra, 2010; Desai et al., 2011; Charaka and Misra, 2012; Satoh et al., 2012; Stelter et al., 2013; Lockhart and DeVeaux, 2013; Charaka et al., 2013; Ngo et al., 2013; Maurya et al., 2016; Maurya et al., 2019; Khairnar et al., 2019; Kumar Maurya et al., 2019; Fan and Su, 2021; Chaudhary et al., 2021; Maurya et al., 2021; Maurya and Misra, 2021; Kota et al., 2021; Mishra et al., 2022; Seck et al., 2022; Seck et al., 2023). Furthermore, *D. radiodurans* contains several proteins with large hydrophilic regions, akin to desiccation-responsive proteins found in plants and animals, and proteins with disordered regions enhance stability under low-water conditions during desiccation (Krisko et al., 2010; Awile et al., 2010; Rajpurohit and Misra, 2013a; Liu et al., 2021). The bacterium also expresses several oxidative stress protective and salvage enzymes, metabolites, and carotenoids at high levels. These components are induced after irradiation and play a critical role in its radiation resistance (Bruce, 1964; Markillie et al., 1999; Khairnar et al., 2003; Misra et al., 2004; Daly et al., 2004; Kobayashi et al., 2006; Khairnar et al., 2007; Daly et al., 2007; Tian et al., 2007; Tian et al., 2009; Shrivastava et al., 2010; Daly et al., 2010; Shashidhar et al., 2010; Chen et al., 2010; Slade and Radman, 2011; Misra et al., 2012; Khairnar et al., 2013; Bihani et al., 2018; Gaidamakova et al., 2022; Sadowska-Bartosz and Bartosz, 2023). Moreover, *D. radiodurans* genome polyploidy and the compact nucleoid structure have been implicated in the radioresistance (Levin-Zaidman et al., 2003; Eltsov and Dubochet, 2005; Cox and Battista, 2005; Blasius et al., 2008). However, their exact contributions are not fully understood. Polyploidy provides a substrate for recombinational repair, and the ring-like nucleoid may provide a scaffold for DNA repair. There is conflicting evidence regarding their importance in radioresistance, and, likely, these are not the primary factors in its remarkable resistance to radiation (Zimmerman and Battista, 2005; Gao et al., 2007).

In the context of gene expression regulation under stress, *D. radiodurans* does not have functionally active SOS transcriptional regulation (Narumi et al., 2001; Bonacossa de Almeida et al., 2002; Sheng et al., 2004). Rather, the radiation/desiccation response (RDR) mechanism, which is regulated by IrrE and DdrO, is necessary for controlling the transcription of DNA repair genes (Bauermeister et al., 2009; Agapov and Kulbachinskiy, 2015; Devigne et al., 2015; Blanchard et al., 2017; Lim et al., 2019; Blanchard and de Groot, 2021; Narasimha and Basu, 2021). Contrary to the SOS regulation in other bacteria, the RDR mechanism is SOS-independent and is triggered by oxidative stress (Blanchard et al., 2017; Blanchard and de Groot, 2021). The metallopeptidase IrrE cleaves and inactivates the transcriptional repressor DdrO, which is bound to the radiation/desiccation response motif (RDRM) present in the promoter regions of RDR genes (Fig. 5.3; Ludanyi et al., 2014; Devigne et al., 2015; de Groot et al., 2019). The activity of IrrE is controlled by the availability of zinc ions released during oxidative stress. The elevated intracellular concentration of zinc activates IrrE, leading to DdrO cleavage. Cleavage of DdrO results in the de-repression of RDR genes, including *recA* and other DNA repair genes, for the repair of damage caused by radiation or desiccation (Ohba et al., 2005; Devigne et al., 2015; Agapov and Kulbachinskiy, 2015; Anaganti et al., 2017; Wang et al., 2019; Blanchard and de Groot, 2021). DdrO is critical for cell viability, and a depletion of DdrO can cause cell death (Devigne et al., 2015). DdrO exists in a dynamic balance between monomers and dimers, and the IrrE cleaves DdrO monomers. DdrO can re-accumulate when the stress is removed, and the zinc signal is gone. The IrrE/DdrO system is conserved across many *Deinococcus* species and is also

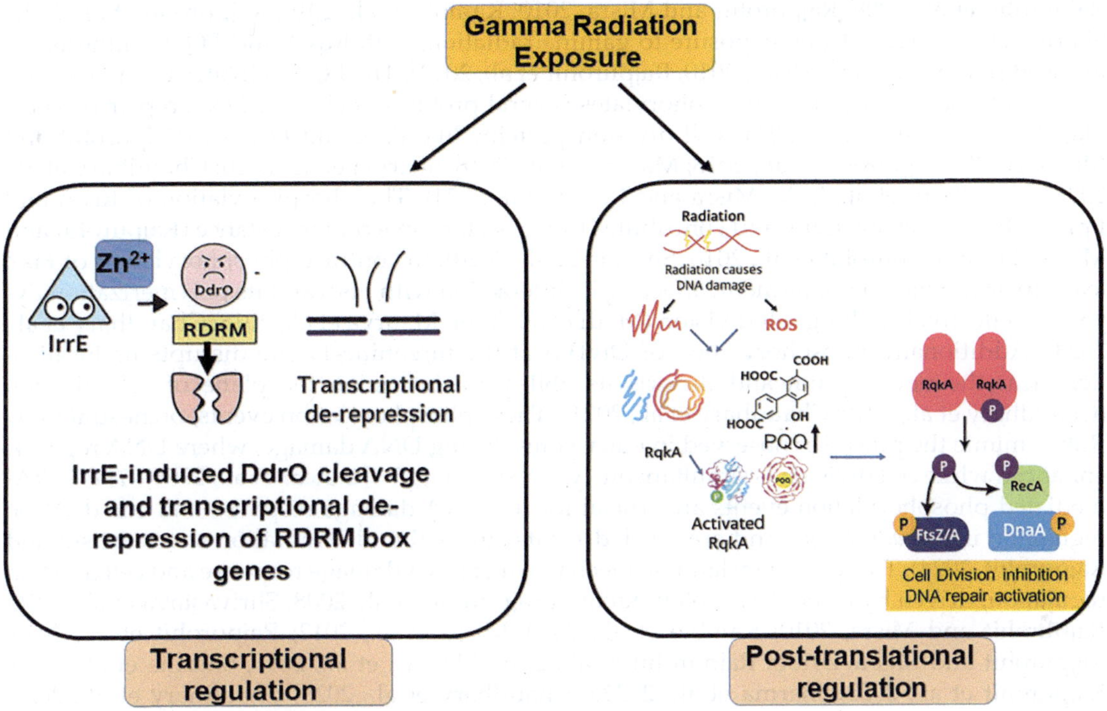

FIGURE 5.3 Schematic representation of transcriptional and posttranscriptional regulation of DNA repair and cell division in the radioresistant bacterium *D. radiodurans*. Gamma radiation exposure leads to the generation of reactive oxygen species (ROS), which triggers the release of Zn^{2+} ions that activate the protease IrrE. Activated IrrE cleaves the DdrO repressor, thereby de-repressing the radiation/desiccation response motif (RDRM) regulon involved in DNA repair and oxidative stress response. Posttranscriptional regulation is further mediated by the kinase RqkA and the synthesis of pyrroloquinoline quinone (PQQ), which activates RqkA's autokinase activity. RqkA then phosphorylates key proteins such as RecA and PprA, enhancing their DNA repair functions, as well as FtsZ and FtsA, central proteins in cell division. Phosphorylation of FtsZ inhibits its function, thereby reducing cell division activity. These phosphorylation events occur in a temporally regulated manner during postirradiation recovery, collectively contributing to efficient DNA repair while transiently suppressing cell division to maintain genomic stability.

found in other bacteria, indicating a widespread stress response mechanism (Blanchard et al., 2017; Blanchard and de Groot, 2021). Furthermore, it has been demonstrated that stabilizing guanine quadruplex (G4) DNA structures using N-methyl mesoporphyrin inhibits postirradiation recovery, highlighting the crucial role of G4-mediated regulation in the transcriptional response of *D. radiodurans*. G4 DNA motifs are located in the promoter regions of numerous DNA repair genes, and their stabilization suppresses the expression of these genes following irradiation, while leaving the expression of housekeeping genes associated with other cellular processes unaffected (Beaume et al., 2013; Kota et al., 2015; Mishra et al., 2019; Kumari and Raghavan, 2021).

Besides these, in *D. radiodurans, the* post-translational regulation of the DNA repair and cell division in response to DNA damage is mediated by the eukaryotic-type Ser/Thr protein kinase RqkA, which is activated by pyrroloquinoline quinone (PQQ) (Fig. 5.3; Khairnar et al., 2003;

Rajpurohit et al., 2008; Rajpurohit and Misra, 2010; Kamble et al., 2010; Rajpurohit et al., 2022; Sharma et al., 2022a). Upon exposure to gamma radiation, both RqkA and PQQ synthesis are induced (Rajpurohit and Misra, 2010; Rajpurohit et al., 2013). The PQQ activates the autokinase activity of RqkA, which then phosphorylates several proteins, including DNA repair proteins like RecA and PprA, as well as cell division proteins like FtsZ and DivIVA (Rajpurohit and Misra, 2013b; Rajpurohit et al., 2016; Maurya et al., 2018; Sharma et al., 2020; Chaudhary et al., 2023; Chaudhary et al., 2023; Misra and Rajpurohit, 2024). The phosphorylation of RecA and PprA enhances their activity and contributes to the bacterium's radioresistance (Rajpurohit and Misra, 2013b; Rajpurohit et al., 2016; Sharma et al., 2020). In contrast, phosphorylation of FtsZ leads to its functional impairment, affecting its interaction with FtsA and its polymerization dynamics, effectively halting cell division (Modi et al., 2014; Maurya et al., 2018; Chaudhary et al., 2023). Additionally, phosphorylation of DivIVA at the threonine-19 site disrupts its localization near the new septum and distorts its ability to determine the plane of cell division (Chaudhary et al., 2019; Chaudhary et al., 2021). These phosphorylation events, orchestrated by RqkA, mimic the processes observed in eukaryotes during DNA damage, where DNA repair is enhanced while cell division is simultaneously arrested (Fig. 5.3; Sancar et al., 2004). The RqkA-mediated phosphorylation events are crucial for the DNA damage response and cell division regulation in *D. radiodurans* and are needed for its survival under radiation stress (Misra and Rajpurohit, 2024). This post-translational mechanism of DNA damage response and cell division regulation, driven by a Ser/Thr protein kinase (Rajpurohit et al., 2008; Shrivastava et al., 2010; Rajpurohit and Misra, 2010; Kamble et al., 2010; Misra et al., 2012; Rajpurohit et al., 2013; Rajpurohit and Misra, 2013b; Rajpurohit et al., 2016; Maurya et al., 2018; Sharma et al., 2020; Rajpurohit et al., 2022; Sharma et al., 2022a; Chaudhary et al., 2023; Chaudhary et al., 2023; Misra and Rajpurohit, 2024), is distinct from the canonical SOS response observed in *E. coli* (Radman, 1975), or RDR regulation of *D. radiodurans*. This mechanism highlights an example synthesis independent approach to DNA damage response and cell division regulation in bacteria particularly under intense stress where new synthesis-dependent approch may not feasible, suggesting that similar STPK-based mechanisms might exist in other bacteria that have that survive in extreme environments or hostile conditions.

5.5 Adaptations and DNA repair in psychrophiles and thermophiles

Psychrophiles, which are best grown in extreme cold conditions, have evolved several mechanisms to counter the challenges presented by low temperatures, such as repair of DNA damage (De Maayer et al., 2014; Collins and Margesin, 2019; Choudhary et al., 2024). Although not discussed specifically as a major psychrophile stressor, general principles of DNA repair and protection apply. In order to be protected from the action of temperature, psychrophiles make a number of physiological adjustments, such as control of fluidity of their membranes, for example, increasing the proportion of branched-chain or unsaturated fatty acids, or reducing the length of the fatty-acyl chains, or both (Jin et al., 2022; Dhaulaniya et al., 2019). Psychrophiles also produce temperature-related chaperones, which assist in protein refolding, and antifreeze molecules, which inhibit ice crystal formation (De Maayer et al., 2014; Collins and Margesin, 2019; Choudhary et al., 2024). They also utilize compatible solutes as cryoprotectors for reducing the freezing point of cytoplasm and stabilizing membranes (Kawahara, 2002; Casanueva et al., 2010; Choudhary et al., 2024; Lopes et al., 2024). These mechanisms, although not specific to DNA repair, aid the survival and functioning of psychrophiles at cold temperatures.

In the Antarctic bacterium *Pseudomonas syringae*, the activity of the RecBCD pathway is critical for low-temperature growth as well as for DNA repair (Regha et al., 2005; Satapathy et al., 2008; Pavankumar et al., 2010). As opposed to *E. coli*, all three subunits of the RecBCD enzyme (RecB, RecC, and RecD) play a vital role in *P. syringae*, and their absence results in inhibition of growth at low temperatures and sensitivity to UV and mitomycin C (Pavankumar et al., 2010). Deletion mutants of any one of these subunits have linear chromosomal DNA and smaller DNA fragments that accumulate, reflecting defective DNA repair (Pavankumar et al., 2010). Both RecB and RecD ATP-dependent helicase activity is vital for DNA repair, but the nuclease activity of RecB is not essential (Pavankumar et al., 2010). Also, it seems that the DNA repair process involves cooperation between the RecBCD and RecFOR pathways and RecJ activity (Pavankumar et al., 2010). These results show the significance of the RecBCD pathway and its unique requirements for psychrophilic bacteria under conditions of low-temperature adaptation and repair of DNA damage.

Thermophiles, which survive in high-temperature environments, have developed unique strategies to protect their DNA and maintain cellular function. Unlike psychrophiles, which experience problems with cold denaturation, thermophiles need to reverse the denaturing effects of heat on macromolecules (Berezovsky and Shakhnovich, 2005; Luke et al., 2007). Hyperthermophilic archaea (HA) DNA is stabilized by the introduction of positive supertwists, introduced by the reverse gyrase enzyme (D'Amaro et al., 2007; Jamroze et al., 2014). Supertwists stabilize DNA at elevated temperatures. Further stabilization of the DNA double helix is provided by an enhanced content of guanine–cytosine base pairs in certain regions of the DNA, such as stem-loops (Siliakus et al., 2017). Archaeal thermophiles employ histones, like those of eukaryotes, to condense DNA and elevate its melting temperature, which also serves to avoid denaturation (Siliakus et al., 2017). Another key characteristic of HA is that their DNA polymerases stall at dU (uracil) in the template strand, a phenomenon not observed in bacteria or eukaryotes (Fogg et al., 2002; Shuttleworth et al., 2004; Kim et al., 2008; Wardle et al., 2008; Connolly, 2009). While the primary rationale for this dU-induced stalling is thought to be the prevention of G: C to A: T transition mutations, this mechanism is more complex, as it hinders base excision repair by sequestering dU within the polymerase's binding pocket, necessitating replication fork reversal or disassembly for repair (Asami et al., 2006; Richardson et al., 2013; Grasso and Tell, 2014). The functional implications of dU accumulation are not fully understood; however, it is proposed that dU may serve as a "biochemical time stamp" marking newly synthesized DNA, potentially regulating replication (Grogan, 2015). HA genomes contain multiple uracil DNA glycosylases (UDGs), and while these are expected to perform BER, deletion mutants of some UDGs exhibit impaired growth, suggesting additional roles beyond BER (She et al., 2009; Fujikane et al., 2010; Zhang et al., 2013). This suggests a more intricate system involving dU than simple BER, and the roles of these enzymes may be more complex than initially assumed (Sakai et al., 2008).

In contrast to other organisms, HA also exhibits an apparent lack of functionality of conventional nucleotide excision repair (NER) (Dorazi et al., 2007; Romano et al., 2007). While they possess homologs of eukaryotic NER proteins (XP proteins), deletions of these genes generally do not affect cell survival after exposure to DNA-damaging agents (Ogrunc et al., 1998; Makarova et al., 2002). This lack of conventional NER, along with the absence of UvrABC homologs (found in bacteria and mesophilic archaea), points to an alternative method of bulky lesion removal (Dorazi et al., 2007; Romano et al., 2007). One proposed mechanism for this is the use of replication fork breakage to target and remove bulky DNA lesions (Grogan, 2000; Grogan, 2015; Ishino and Narumi, 2015). In this scenario, when replication is blocked by a lesion, endonucleases like Hef/Xpf or Fen1/Xpg cleave the DNA at the base of the stalled fork, resulting in a double-stranded DNA end with the lesion attached and processed through homologous recombination (HR) pathways (Grogan, 2015; Ishino and Narumi, 2015). In addition, HA also lacks MutSL homologs (Fitz-Gibbon et al., 2002; Grogan, 2015). These proteins are critical for mismatch repair (MMR) in bacteria (Li, 2008), but HA maintains accurate replication despite their absence (Grogan et al., 2001; Mao and Grogan, 2017). This indicates the possibility of an alternative, not fully understood, mechanism for mismatch repair or an exceptionally accurate replication machinery. The analysis of genetic recombination in *Sulfolobus acidocaldarius suggested* an intriguing hypothesis involving replication fork reversal, where the reversed fork exposes newly synthesized daughter strands that could allow for the removal of mismatches (Grogan, 2015). Another significant and unique aspect of HA is the essentiality of HR proteins (Seitz et al., 1998; Seitz and Kowalczykowski, 2000; Graham and Haseltine, 2013). Unlike in bacteria and eukaryotes, where HR can be eliminated with loss-of-function mutations, the *radA* gene, along with *mre11, rad50, herA,* and *nurA*, is essential for viability in HA (Spies et al., 2000; Constantinesco et al., 2004; Quaiser et al., 2008; Zhang et al., 2013; Huang et al., 2015). These proteins are implicated in DSBs processing during HR, and the essentiality of these proteins may be directly linked to the fork-breakage mechanism proposed for DNA lesion removal (Grogan, 2015; Ishino and Narumi, 2015). Thus, the proposed strategies for HA genome maintenance appear to be an integration of DNA repair, recombination, and replication, which is a marked departure from how these processes are thought to be handled in other forms of life.

References

Adachi, M., Hirayama, H., Shimizu, R., Satoh, K., Narumi, I., Kuroki, R., 2014. Interaction of double-stranded DNA with polymerized PprA protein from Deinococcus radiodurans. Protein Sci. 23, 1349–1358.

Adamec, F., Farci, D., Bina, D., Litvin, R., Khan, T., Fuciman, M., et al., 2020. Photophysics of deinoxanthin, the keto-carotenoid bound to the main S-layer unit of Deinococcus radiodurans. Photochem. Photobiol. Sci. 19, 495–503.

Agapov, A.A., Kulbachinskiy, A.V., 2015. Mechanisms of stress resistance and gene regulation in the radioresistant bacterium Deinococcus radiodurans. Biochemistry 80, 1201–1216.

Albers, S.V., Van de Vossenberg, J.L., Driessen, A.J., Konings, W.N., 2001. Bioenergetics and solute uptake under extreme conditions. Extremophiles 5, 285–294.

Amaro, A.M., Chamorro, D., Seeger, M., Arredondo, R., Peirano, I., Jerez, C.A., 1991. Effect of external pH perturbations on in vivo protein synthesis by the acidophilic bacterium Thiobacillus ferrooxidans. J. Bacteriol. 173, 910–915.

Anaganti, N., Basu, B., Mukhopadhyaya, R., Apte, S.K., 2017. Proximity of the radiation desiccation response motif to the core promoter is essential for basal repression as well as gamma radiation-induced gyrB gene expression in Deinococcus radiodurans. Gene 615, 8–17.

Asami, Y., Murakami, M., Shimizu, M., Pisani, F.M., Hayata, I., Nohmi, T., 2006. Visualization of the interaction between archaeal DNA polymerase and uracil-containing DNA by atomic force microscopy. Genes Cells 11, 3–11.

Atomi, H., Reeve, J., 2019. Microbe profile: thermococcus kodakarensis: the model hyperthermophilic archaeon. Microbiology (Reading) 165, 1166–1168.

Auman, A.J., Breezee, J.L., Gosink, J.J., Kampfer, P., Staley, J.T., 2006. Psychromonas ingrahamii sp. nov., a novel gas vacuolate, psychrophilic bacterium isolated from Arctic polar sea ice. Int. J. Syst. Evol. Microbiol. 56, 1001–1007.

Auman, A.J., Breezee, J.L., Gosink, J.J., Schumann, P., Barnes, C.R., Kampfer, P., et al., 2010. Psychromonas boydii sp. nov., a gas-vacuolate, psychrophilic bacterium isolated from an Arctic sea-ice core. Int. J. Syst. Evol. Microbiol. 60, 84–92.

Awile, O., Krisko, A., Sbalzarini, I.F., Zagrovic, B., 2010. Intrinsically disordered regions may lower the hydration free energy in proteins: a case study of nudix hydrolase in the bacterium Deinococcus radiodurans. PLoS Comput. Biol. 6, e1000854.

Baker-Austin, C., Dopson, M., 2007. Life in acid: pH homeostasis in acidophiles. Trends Microbiol. 15, 165–171.

Baliga, N.S., Bjork, S.J., Bonneau, R., Pan, M., Iloanusi, C., Kottemann, M.C., et al., 2004. Systems level insights into the stress response to UV radiation in the halophilic archaeon Halobacterium NRC-1. Genome Res. 14, 1025–1035.

Banneville, A.S., Bouthier de la Tour, C., De Bonis, S., Hognon, C., Colletier, J.P., Teulon, J.M., et al., 2022. Structural and functional characterization of DdrC, a novel DNA damage-induced nucleoid associated protein involved in DNA compaction. Nucleic Acids. Res. 50, 7680–7696.

Battista, J.R., 1997. Against all odds: the survival strategies of Deinococcus radiodurans. Annu. Rev. Microbiol. 51, 203–224.

Bauermeister, A., Bentchikou, E., Moeller, R., Rettberg, P., 2009. Roles of PprA, IrrE, and RecA in the resistance of Deinococcus radiodurans to germicidal and environmentally relevant UV radiation. Arch. Microbiol. 191, 913–918.

Beaume, N., Pathak, R., Yadav, V.K., Kota, S., Misra, H.S., Gautam, H.K., et al., 2013. Genome-wide study predicts promoter-G4 DNA motifs regulate selective functions in bacteria: radioresistance of D. radiodurans involves G4 DNA-mediated regulation. Nucleic Acids. Res. 41, 76–89.

Bentchikou, E., Servant, P., Coste, G., Sommer, S., 2010. A major role of the RecFOR pathway in DNA double-strand-break repair through ESDSA in Deinococcus radiodurans. PLoS Genet. 6, e1000774.

Berezovsky, I.N., Shakhnovich, E.I., 2005. Physics and evolution of thermophilic adaptation. Proc. Natl. Acad. Sci. USA. 102, 12742–12747.

Berquist, B.R., DasSarma, P., DasSarma, S., 2007. Essential and non-essential DNA replication genes in the model halophilic Archaeon, Halobacterium sp. NRC-1. BMC Genet. 8, 31.

Bihani, S.C., Panicker, L., Rajpurohit, Y.S., Misra, H.S., Kumar, V., 2018. drFrnE represents a hitherto unknown class of eubacterial cytoplasmic disulfide oxido-reductases. Antioxid. Redox. Signal. 28, 296–310.

Blanchard, L., de Groot, A., 2021. Coexistence of SOS-dependent and SOS-independent regulation of DNA repair genes in radiation-resistant Deinococcus bacteria. Cells 10, 924.

Blanchard, L., Guerin, P., Roche, D., Cruveiller, S., Pignol, D., Vallenet, D., et al., 2017. Conservation and diversity of the IrrE/DdrO-controlled radiation response in radiation-resistant Deinococcus bacteria. Microbiologyopen 6, e477.

Blasius, M., Buob, R., Shevelev, I.V., Hubscher, U., 2007. Enzymes involved in DNA ligation and end-healing in the radioresistant bacterium Deinococcus radiodurans. BMC Mol. Biol. 8, 69.

Blasius, M., Sommer, S., Hubscher, U., 2008. Deinococcus radiodurans: what belongs to the survival kit? Crit. Rev. Biochem. Mol. Biol. 43, 221–238.

Bonacossa de Almeida, C., Coste, G., Sommer, S., Bailone, A., 2002. Quantification of RecA protein in Deinococcus radiodurans reveals involvement of RecA, but not LexA, in its regulation. Mol. Genet. Genom. 268, 28–41.

Boubriak, I., Ng, W.L., DasSarma, P., DasSarma, S., Crowley, D.J., McCready, S.J., 2008. Transcriptional responses to biologically relevant doses of UV-B radiation in the model archaeon, Halobacterium sp. NRC-1. Saline Syst. 4, 13.

Bouthier de la Tour, C., Boisnard, S., Norais, C., Toueille, M., Bentchikou, E., Vannier, F., et al., 2011. The deinococcal DdrB protein is involved in an early step of DNA double strand break repair and in plasmid transformation through its single-strand annealing activity. DNA Repair (Amst.) 10, 1223–1231.

Bouthier de la Tour, C., Mathieu, M., Meyer, L., Dupaigne, P., Passot, F., Servant, P., et al., 2017. In vivo and in vitro characterization of DdrC, a DNA damage response protein in Deinococcus radiodurans bacterium. PLoS One 12, e0177751.

Bouthier de la Tour, C., Toueille, M., Jolivet, E., Nguyen, H.H., Servant, P., Vannier, F., et al., 2009. The Deinococcus radiodurans SMC protein is dispensable for cell viability yet plays a role in DNA folding. Extremophiles 13, 827–837.

Brininger, C., Spradlin, S., Cobani, L., Evilia, C., 2018. The more adaptive to change, the more likely you are to survive: protein adaptation in extremophiles. Semin. Cell Dev. Biol. 84, 158–169.

Bruce, A.K., 1964. Extraction of the radioresistant factor of Micrococcus radiodurans. Radiat. Res. 22, 155–164.

Buchko, G.W., Litvinova, O., Robinson, H., Yakunin, A.F., Kennedy, M.A., 2008. Functional and structural characterization of DR_0079 from Deinococcus radiodurans, a novel Nudix hydrolase with a preference for cytosine (deoxy)ribonucleoside 5′-Di- and triphosphates. Biochemistry 47, 6571–6582.

Busch, C.R., DiRuggiero, J., 2010. MutS and MutL are dispensable for maintenance of the genomic mutation rate in the halophilic archaeon Halobacterium salinarum NRC-1. PLoS One 5, e9045.

Cai, J., Zhang, M., Chen, Z., Zhao, Y., Xu, H., Tian, B., et al., 2023. MoaE is involved in response to oxidative stress in Deinococcus radiodurans. Int. J. Mol. Sci. 24, 2441.

Casanueva, A., Tuffin, M., Cary, C., Cowan, D.A., 2010. Molecular adaptations to psychrophily: the impact of 'omic' technologies. Trends Microbiol. 18, 374–381.

Charaka, V.K., Mehta, K.P., Misra, H.S., 2013. ParA encoded on chromosome II of Deinococcus radiodurans binds to nucleoid and inhibits cell division in Escherichia coli. J. Biosci. 38, 487–497.

Charaka, V.K., Misra, H.S., 2012. Functional characterization of the role of the chromosome I partitioning system in genome segregation in Deinococcus radiodurans. J. Bacteriol. 194, 5739–5748.

Chaudhary, R., Gupta, A., Kota, S., Misra, H.S., 2019. N-terminal domain of DivIVA contributes to its dimerization and interaction with genome segregation proteins in a radioresistant bacterium Deinococcus radiodurans. Int. J. Biol. Macromol. 128, 12–21.

Chaudhary, R., Kota, S., Misra, H.S., 2021. DivIVA regulates its expression and the orientation of new septum growth in Deinococcus radiodurans. J. Bacteriol. 203, e0016321.

Chaudhary, R., Kota, S., Misra, H.S., 2023a. DivIVA phosphorylation affects its dynamics and cell cycle in radioresistant Deinococcus radiodurans. Microbiol. Spectr. 11, e0314122.

Chaudhary, R., Mishra, S., Maurya, G.K., Rajpurohit, Y.S., Misra, H.S., 2023b. FtsZ phosphorylation brings about growth arrest upon DNA damage in Deinococcus radiodurans. FASEB Bioadv. 5, 27–42.

Chen, H., Huang, L., Hua, X., Yin, L., Hu, Y., Wang, C., et al., 2009. Pleiotropic effects of RecQ in Deinococcus radiodurans. Genomics 94, 333–340.

Chen, H., Wu, R., Xu, G., Fang, X., Qiu, X., Guo, H., et al., 2010. DR2539 is a novel DtxR-like regulator of Mn/Fe ion homeostasis and antioxidant enzyme in Deinococcus radiodurans. Biochem. Biophys. Res. Commun. 396, 413–418.

Choudhary, P., Bhatt, S., Chatterjee, S., 2024. From freezing to functioning: cellular strategies of cold-adapted bacteria for surviving in extreme environments. Arch. Microbiol. 206, 329.

Coker, J.A., 2019. Recent advances in understanding extremophiles. F1000Res 8 F1000 Faculty Rev-1917.

Collins, T., Margesin, R., 2019. Psychrophilic lifestyles: mechanisms of adaptation and biotechnological tools. Appl. Microbiol. Biotechnol. 103, 2857–2871.

Connolly, B.A., 2009. Recognition of deaminated bases by archaeal family-B DNA polymerases. Biochem. Soc. Trans. 37, 65–68.

Constantinesco, F., Forterre, P., Koonin, E.V., Aravind, L., Elie, C., 2004. A bipolar DNA helicase gene, herA, clusters with rad50, mre11 and nurA genes in thermophilic archaea. Nucleic Acids. Res. 32, 1439–1447.

Cowen, D.A., 2004. The upper temperature of life–where do we draw the line? Trends Microbiol. 12, 58–60.

Cox, M.M., Battista, J.R., 2005. Deinococcus radiodurans - the consummate survivor. Nat. Rev. Micro. 3, 882–892.

Cox, M.M., Goodman, M.F., Kreuzer, K.N., Sherratt, D.J., Sandler, S.J., Marians, K.J., 2000. The importance of repairing stalled replication forks. Nature 404, 37–41.

Crowley, D.J., Boubriak, I., Berquist, B.R., Clark, M., Richard, E., Sullivan, L., et al., 2006. The uvrA, uvrB and uvrC genes are required for repair of ultraviolet light induced DNA photoproducts in Halobacterium sp. NRC-1. Saline Syst. 2, 11.

Culotta, V.C., Daly, M.J., 2013. Manganese complexes: diverse metabolic routes to oxidative stress resistance in prokaryotes and yeast. Antioxid. Redox. Signal. 19, 933–944.

D'Amaro, A., Rossi, M., Ciaramella, M., 2007. Reverse gyrase: an unusual DNA manipulator of hyperthermophilic organisms. Ital. J. Biochem. 56, 103–109.

da Costa, M.S., Santos, H., Galinski, E.A., 1998. An overview of the role and diversity of compatible solutes in bacteria and archaea. Adv. Biochem. Eng. Biotechnol. 61, 117–153.

Dai, S., Xie, Z., Wang, B., Yu, N., Zhao, J., Zhou, Y., et al., 2021. Dynamic polyphosphate metabolism coordinating with manganese ions defends against oxidative stress in the extreme bacterium Deinococcus radiodurans. Appl. Environ. Microb. 87 e02785-20.

Daly, M.J., 2009. A new perspective on radiation resistance based on Deinococcus radiodurans. Nat. Rev. Micro. 7, 237–245.

Daly, M.J., 2012. Death by protein damage in irradiated cells. DNA Repair (Amst.) 11, 12–21.

Daly, M.J., Gaidamakova, E.K., Matrosova, V.Y., Kiang, J.G., Fukumoto, R., Lee, D.Y., et al., 2010. Small-molecule antioxidant proteome-shields in Deinococcus radiodurans. PLoS One 5, e12570.

Daly, M.J., Gaidamakova, E.K., Matrosova, V.Y., Vasilenko, A., Zhai, M., Leapman, R.D., et al., 2007. Protein oxidation implicated as the primary determinant of bacterial radioresistance. PLoS Biol. 5, e92.

Daly, M.J., Gaidamakova, E.K., Matrosova, V.Y., Vasilenko, A., Zhai, M., Venkateswaran, A., et al., 2004. Accumulation of Mn(II) in Deinococcus radiodurans facilitates gamma-radiation resistance. Science 306, 1025–1028.

Das, A.D., Misra, H.S., 2012. DR2417, a hypothetical protein characterized as a novel beta-CASP family nuclease in radiation resistant bacterium, Deinococcus radiodurans. Biochim. Biophys. Acta 1820, 1052–1061.

Dassarma, S., Kennedy, S.P., Berquist, B., Victor Ng, W., Baliga, N.S., Spudich, J.L., et al., 2001. Genomic perspective on the photobiology of Halobacterium species NRC-1, a phototrophic, phototactic, and UV-tolerant haloarchaeon. Photosynth. Res. 70, 3–17.

de Groot, A., Siponen, M.I., Magerand, R., Eugenie, N., Martin-Arevalillo, R., Doloy, J., et al., 2019. Crystal structure of the transcriptional repressor DdrO: insight into the metalloprotease/repressor-controlled radiation response in Deinococcus. Nucleic Acids. Res. 47, 11403–11417.

de la Tour, C.B., Mathieu, M., Servant, P., Coste, G., Norais, C., Confalonieri, F., 2021. Characterization of the DdrD protein from the extremely radioresistant bacterium Deinococcus radiodurans. Extremophiles 25, 343–355.

De Maayer, P., Anderson, D., Cary, C., Cowan, D.A., 2014. Some like it cold: understanding the survival strategies of psychrophiles. EMBO Rep. 15, 508–517.

Delmas, S., Duggin, I.G., Allers, T., 2013. DNA damage induces nucleoid compaction via the Mre11-Rad50 complex in the archaeon Haloferax volcanii. Mol. Microbiol. 87, 168–179.

Desai, S.S., Rajpurohit, Y.S., Misra, H.S., Deobagkar, D.N., 2011. Characterization of the role of the RadS/RadR two-component system in the radiation resistance of Deinococcus radiodurans. Microbiology (Reading) 157, 2974–2982.

Devigne, A., Guerin, P., Lisboa, J., Quevillon-Cheruel, S., Armengaud, J., Sommer, S., et al., 2016. PprA protein is involved in chromosome segregation via its physical and functional interaction with DNA gyrase in irradiated Deinococcus radiodurans bacteria. mSphere 1 e00036-15.

Devigne, A., Ithurbide, S., Bouthier de la Tour, C., Passot, F., Mathieu, M., Sommer, S., et al., 2015. DdrO is an essential protein that regulates the radiation desiccation response and the apoptotic-like cell death in the radioresistant Deinococcus radiodurans bacterium. Mol. Microbiol. 96, 1069–1084.

Devigne, A., Mersaoui, S., Bouthier-de-la-Tour, C., Sommer, S., Servant, P., 2013. The PprA protein is required for accurate cell division of gamma-irradiated Deinococcus radiodurans bacteria. DNA Repair (Amst.) 12, 265–272.

Dhaulaniya, A.S., Balan, B., Kumar, M., Agrawal, P.K., Singh, D.K., 2019. Cold survival strategies for bacteria, recent advancement and potential industrial applications. Arch. Microbiol. 201, 1–16.

Dorazi, R., Gotz, D., Munro, S., Bernander, R., White, M.F., 2007. Equal rates of repair of DNA photoproducts in transcribed and non-transcribed strands in Sulfolobus solfataricus. Mol. Microbiol. 63, 521–529.

Earl, A.M., Mohundro, M.M., Mian, I.S., Battista, J.R., 2002. The IrrE protein of Deinococcus radiodurans R1 is a novel regulator of recA expression. J. Bacteriol. 184, 6216–6224.

Eltsov, M., Dubochet, J., 2005. Fine structure of the Deinococcus radiodurans nucleoid revealed by cryoelectron microscopy of vitreous sections. J. Bacteriol. 187, 8047–8054.

Evans, D.M., Moseley, B.E., 1983. Roles of the uvsC, uvsD, uvsE, and mtcA genes in the two pyrimidine dimer excision repair pathways of Deinococcus radiodurans. J. Bacteriol. 156, 576–583.

Fan, H.F., Su, S., 2021. The regulation mechanism of the C-terminus of RecA proteins during DNA strand-exchange process. Biophys. J. 120, 3166–3179.

Farias, R., Norambuena, J., Ferrer, A., Camejo, P., Zapata, C., Chavez, R., et al., 2021. Redox stress response and UV tolerance in the acidophilic iron-oxidizing bacteria Leptospirillum ferriphilum and Acidithiobacillus ferrooxidans. Res. Microbiol. 172, 103833.

Fisher, D.I., Cartwright, J.L., Harashima, H., Kamiya, H., McLennan, A.G., 2004. Characterization of a nudix hydrolase from Deinococcus radiodurans with a marked specificity for (deoxy)ribonucleoside 5′-diphosphates. BMC Biochem. 5, 7.

Fitz-Gibbon, S.T., Ladner, H., Kim, U.J., Stetter, K.O., Simon, M.I., Miller, J.H., 2002. Genome sequence of the hyperthermophilic crenarchaeon Pyrobaculum aerophilum. Proc. Natl. Acad. Sci. USA. 99, 984–989.

Fogg, M.J., Pearl, L.H., Connolly, B.A., 2002. Structural basis for uracil recognition by archaeal family B DNA polymerases. Nat. Struct. Biol. 9, 922–927.

Fujikane, R., Ishino, S., Ishino, Y., Forterre, P., 2010. Genetic analysis of DNA repair in the hyperthermophilic archaeon, Thermococcus kodakaraensis. Genes Genet. Syst. 85, 243–257.

Gaidamakova, E.K., Sharma, A., Matrosova, V.Y., Grichenko, O., Volpe, R.P., Tkavc, R., et al., 2022. Small-molecule Mn antioxidants in Caenorhabditis elegans and Deinococcus radiodurans supplant MnSOD enzymes during aging and irradiation. mBio 13, e0339421.

Gao, G., Lu, H., Yin, L., Hua, Y., 2007. Ring-like nucleoid does not play a key role in radioresistance of Deinococcus radiodurans. Sci. China C Life Sci. 50, 525–529.

Gao, L., Zhou, Z., Chen, X., Zhang, W., Lin, M., Chen, M., 2020. Comparative proteomics analysis reveals new features of the oxidative stress response in the polyextremophilic Bacterium Deinococcus radiodurans. Microorganisms 8 451.

Graham, W.J., Haseltine, C.A., 2013. A recombinase paralog from the hyperthermophilic crenarchaeon Sulfolobus solfataricus enhances SsoRadA ssDNA binding and strand displacement. Gene 515, 128–139.

Grasso, S., Tell, G., 2014. Base excision repair in Archaea: back to the future in DNA repair. DNA Repair (Amst.) 21, 148–157.

Grogan, D.W., 2000. The question of DNA repair in hyperthermophilic archaea. Trends Microbiol. 8, 180–185.

Grogan, D.W., 2015. Understanding DNA repair in Hyperthermophilic Archaea: persistent gaps and other reasons to focus on the fork. Archaea 2015, 942605.

Grogan, D.W., Carver, G.T., Drake, J.W., 2001. Genetic fidelity under harsh conditions: analysis of spontaneous mutation in the thermoacidophilic archaeon Sulfolobus acidocaldarius. Proc. Natl. Acad. Sci. USA. 98, 7928–7933.

Gutsche, I., Vujicic-Zagar, A., Siebert, X., Servant, P., Vannier, F., Castaing, B., et al., 2008. Complex oligomeric structure of a truncated form of DdrA: a protein required for the extreme radiotolerance of Deinococcus. Biochim. Biophys. Acta 1784, 1050–1058.

Harris, D.R., Tanaka, M., Saveliev, S.V., Jolivet, E., Earl, A.M., Cox, M.M., et al., 2004. Preserving genome integrity: the DdrA protein of Deinococcus radiodurans R1. PLoS Biol. 2, e304.

Hogrel, G., Lu, Y., Alexandre, N., Bosse, A., Dulermo, R., Ishino, S., et al., 2020. Role of RadA and DNA polymerases in recombination-associated DNA synthesis in Hyperthermophilic archaea. Biomolecules 10 1045.

Horikoshi, K., 1999. Alkaliphiles: some applications of their products for biotechnology. Microbiol. Mol. Biol. Rev. 63, 735–750.

Hou, J., Dai, J., Chen, Z., Wang, Y., Cao, J., Hu, J., et al., 2022. Phosphorylation regulation of a histone-like HU protein from Deinococcus radiodurans. Protein Pept. Lett. 29, 891–899.

Huang, Q., Liu, L., Liu, J., Ni, J., She, Q., Shen, Y., 2015. Efficient 5′-3′ DNA end resection by HerA and NurA is essential for cell viability in the crenarchaeon Sulfolobus islandicus. BMC Mol. Biol. 16, 2.

Ishino, Y., Narumi, I., 2015. DNA repair in hyperthermophilic and hyperradioresistant microorganisms. Curr. Opin. Microbiol. 25, 103–112.

Ithurbide, S., Bentchikou, E., Coste, G., Bost, B., Servant, P., Sommer, S., 2015. Single strand annealing plays a major role in RECA-independent recombination between repeated sequences in the radioresistant Deinococcus radiodurans Bacterium. PLoS Genet. 11, e1005636.

Jamroze, A., Perugino, G., Valenti, A., Rashid, N., Rossi, M., Akhtar, M., et al., 2014. The reverse gyrase from Pyrobaculum calidifontis, a novel extremely thermophilic DNA topoisomerase endowed with DNA unwinding and annealing activities. J. Biol. Chem. 289, 3231–3243.

Jeong, S.W., Jung, J.H., Kim, M.K., Seo, H.S., Lim, H.M., Lim, S., 2016. The three catalases in Deinococcus radiodurans: only two show catalase activity. Biochem. Biophys. Res. Commun. 469, 443–448.

Ji, H.F., 2010. Insight into the strong antioxidant activity of deinoxanthin, a unique carotenoid in Deinococcus radiodurans. Int. J. Mol. Sci. 11, 4506–4510.

Jin, M., Xiao, A., Zhu, L., Zhang, Z., Huang, H., Jiang, L., 2019. The diversity and commonalities of the radiation-resistance mechanisms of Deinococcus and its up-to-date applications. AMB Expr. 9, 138.

Jin, S., Wang, Y., Zhao, X., 2022. Cold-adaptive mechanism of psychrophilic bacteria in food and its application. Microb. Pathog. 169, 105652.

Johnson, D.B., Kanao, T., Hedrich, S., 2012. Redox transformations of iron at extremely low pH: fundamental and applied aspects. Front. Microbiol. 3, 96.

Kamble, V.A., Misra, H.S., 2010. The SbcCD complex of Deinococcus radiodurans contributes to radioresistance and DNA strand break repair in vivo and exhibits Mre11-Rad50 type activity in vitro. DNA Repair (Amst.) 9, 488–494.

Kamble, V.A., Rajpurohit, Y.S., Srivastava, A.K., Misra, H.S., 2010. Increased synthesis of signaling molecules coincides with reversible inhibition of nucleolytic activity during postirradiation recovery of Deinococcus radiodurans. FEMS Microbiol. Lett. 303, 18–25.

Karlin, S., Mrazek, J., 2001. Predicted highly expressed and putative alien genes of Deinococcus radiodurans and implications for resistance to ionizing radiation damage. Proc. Natl. Acad. Sci. USA. 98, 5240–5245.

Kashefi, K., Lovley, D.R., 2003. Extending the upper temperature limit for life. Science 301, 934.

Kato, C., Li, L., Nogi, Y., Nakamura, Y., Tamaoka, J., Horikoshi, K., 1998. Extremely barophilic bacteria isolated from the Mariana Trench, challenger deep, at a depth of 11,000 meters. Appl. Environ. Microb. 64, 1510–1513.

Kawaguchi, Y., Shibuya, M., Kinoshita, I., Yatabe, J., Narumi, I., Shibata, H., et al., 2020. DNA damage and survival time course of deinococcal cell pellets during 3 years of exposure to outer space. Front. Microbiol. 11, 2050.

Kawahara, H., 2002. The structures and functions of ice crystal-controlling proteins from bacteria. J. Biosci. Bioeng. 94, 492–496.

Kessel, M., Buhle, Jr., E.L., Cohen, S., Aebi, U., 1988. The cell wall structure of a magnesium-dependent halobacterium, Halobacterium volcanii CD-2, from the Dead Sea. J. Ultrastruct. Mol. Struct. Res. 100, 94–106.

Khairnar, N.P., Joe, M.H., Misra, H.S., Lim, S.Y., Kim, D.H., 2013. FrnE, a cadmium-inducible protein in Deinococcus radiodurans, is characterized as a disulfide isomerase chaperone in vitro and for its role in oxidative stress tolerance in vivo. J. Bacteriol. 195, 2880–2886.

Khairnar, N.P., Kamble, V.A., Mangoli, S.H., Apte, S.K., Misra, H.S., 2007. Involvement of a periplasmic protein kinase in DNA strand break repair and homologous recombination in Escherichia coli. Mol. Microbiol. 65, 294–304.

Khairnar, N.P., Maurya, G.K., Pandey, N., Das, A., Misra, H.S., 2019. DrRecQ regulates guanine quadruplex DNA structure dynamics and its impact on radioresistance in Deinococcus radiodurans. Mol. Microbiol. 112, 854–865.

Khairnar, N.P., Misra, H.S., Apte, S.K., 2003. Pyrroloquinoline-quinone synthesized in Escherichia coli by pyrroloquinoline-quinone synthase of Deinococcus radiodurans plays a role beyond mineral phosphate solubilization. Biochem. Biophys. Res. Commun. 312, 303–308.

Kim, J.I., Cox, M.M., 2002. The RecA proteins of Deinococcus radiodurans and Escherichia coli promote DNA strand exchange via inverse pathways. Proc. Natl. Acad. Sci. USA. 99, 7917–7921.

Kim, J.I., Sharma, A.K., Abbott, S.N., Wood, E.A., Dwyer, D.W., Jambura, A., et al., 2002. RecA protein from the extremely radioresistant bacterium Deinococcus radiodurans: expression, purification, and characterization. J. Bacteriol. 184, 1649–1660.

Kim, Y.J., Cha, S.S., Lee, H.S., Ryu, Y.G., Bae, S.S., Cho, Y., et al., 2008. Sensing domain and extension rate of a family B-type DNA polymerase determine the stalling at a deaminated base. J. Microbiol. Biotechnol. 18, 1377–1385.

Kobayashi, I., Tamura, T., Sghaier, H., Narumi, I., Yamaguchi, S., Umeda, K., et al., 2006. Characterization of monofunctional catalase KatA from radioresistant bacterium Deinococcus radiodurans. J. Biosci. Bioeng. 101, 315–321.

Konings, W.N., Albers, S.V., Koning, S., Driessen, A.J., 2002. The cell membrane plays a crucial role in survival of bacteria and archaea in extreme environments. Antonie Van Leeuwenhoek 81, 61–72.

Kota, S., Charaka, V.K., Misra, H.S., 2014. PprA, a pleiotropic protein for radioresistance, works through DNA gyrase and shows cellular dynamics during postirradiation recovery in Deinococcus radiodurans. J. Genet. 93, 349–354.

Kota, S., Chaudhary, R., Mishra, S., Misra, H.S., 2021. Topoisomerase IB interacts with genome segregation proteins and is involved in multipartite genome maintenance in Deinococcus radiodurans. Microbiol. Res. 242, 126609.

Kota, S., Dhamodharan, V., Pradeepkumar, P.I., Misra, H.S., 2015. G-quadruplex forming structural motifs in the genome of Deinococcus radiodurans and their regulatory roles in promoter functions. Appl. Microbiol. Biotechnol. 99, 9761–9769.

Kota, S., Kamble, V.A., Rajpurohit, Y.S., Misra, H.S., 2010. ATP-type DNA ligase requires other proteins for its activity in vitro and its operon components for radiation resistance in Deinococcus radiodurans in vivo. Biochem. Cell. Biol. 88, 783–790.

Kota, S., Misra, H.S., 2006. PprA: A protein implicated in radioresistance of Deinococcus radiodurans stimulates catalase activity in Escherichia coli. Appl. Microbiol. Biotechnol. 72, 790–796.

Kota, S., Rajpurohit, Y.S., Charaka, V.K., Satoh, K., Narumi, I., Misra, H.S., 2016. DNA Gyrase of Deinococcus radiodurans is characterized as type II bacterial topoisomerase and its activity is differentially regulated by PprA in vitro. Extremophiles 20, 195–205.

Krisko, A., Radman, M., 2010. Protein damage and death by radiation in Escherichia coli and Deinococcus radiodurans. Proc. Natl. Acad. Sci. USA. 107, 14373–14377.

Krisko, A., Smole, Z., Debret, G., Nikolic, N., Radman, M., 2010. Unstructured hydrophilic sequences in prokaryotic proteomes correlate with dehydration tolerance and host association. J. Mol. Biol. 402, 775–782.

Kumar, A., Alam, A., Tripathi, D., Rani, M., Khatoon, H., Pandey, S., et al., 2018. Protein adaptations in extremophiles: an insight into extremophilic connection of mycobacterial proteome. Semin. Cell Dev. Biol. 84, 147–157.

Kumar Maurya, G., Kota, S., Kumar, N.N., Tewari, R., Misra, H.S., 2019. ParA proteins of secondary genome elements cross-talk and regulate radioresistance through genome copy number reduction in Deinococcus radiodurans. Biochem. J. 476, 909–930.

Kumari, N., Raghavan, S.C., 2021. G-quadruplex DNA structures and their relevance in radioprotection. Biochim. Biophys. Acta Gen. Subj. 1865, 129857.

Kuzucu, M., 2023. Extremophilic solutions: the role of deinoxanthin in counteracting uv-induced skin harm. Curr. Iss. Mol. Biol. 45, 8372–8394.

Levin-Zaidman, S., Englander, J., Shimoni, E., Sharma, A.K., Minton, K.W., Minsky, A., 2003. Ringlike structure of the Deinococcus radiodurans genome: a key to radioresistance? Science 299, 254–256.

Li, G.M., 2008. Mechanisms and functions of DNA mismatch repair. Cell Res. 18, 85–98.

Lim, S., Jung, J.H., Blanchard, L., de Groot, A., 2019. Conservation and diversity of radiation and oxidative stress resistance mechanisms in Deinococcus species. FEMS Microbiol. Rev. 43, 19–52.

Liu, C., Wang, L., Li, T., Lin, L., Dai, S., Tian, B., et al., 2014. A PerR-like protein involved in response to oxidative stress in the extreme bacterium Deinococcus radiodurans. Biochem. Biophys. Res. Commun. 450, 575–580.

Liu, Y., Zhang, C., Wang, Z., Lin, M., Wang, J., Wu, M., 2021. Pleiotropic roles of late embryogenesis abundant proteins of Deinococcus radiodurans against oxidation and desiccation. Comput. Struct. Biotechnol. J. 19, 3407–3415.

Liu, Y., Zhou, J., Omelchenko, M.V., Beliaev, A.S., Venkateswaran, A., Stair, J., et al., 2003. Transcriptome dynamics of Deinococcus radiodurans recovering from ionizing radiation. Proc. Natl. Acad. Sci. USA. 100, 4191–4196.

Lockhart, J.S., DeVeaux, L.C., 2013. The essential role of the Deinococcus radiodurans ssb gene in cell survival and radiation tolerance. PLoS One 8, e71651.

Lopes, J.C., Kinasz, C.T., Luiz, A.M.C., Kreusch, M.G., Duarte, R.T.D., 2024. Frost fighters: unveiling the potential of microbial antifreeze proteins in biotech innovation. J. Appl. Microbiol. 135 (6), lxae140.

Ludanyi, M., Blanchard, L., Dulermo, R., Brandelet, G., Bellanger, L., Pignol, D., et al., 2014. Radiation response in Deinococcus deserti: IrrE is a metalloprotease that cleaves repressor protein DdrO. Mol. Microbiol. 94, 434–449.

Ludt, K., Soppa, J., 2019. Polyploidy in halophilic archaea: regulation, evolutionary advantages, and gene conversion. Biochem. Soc. Trans. 47, 933–944.

Luke, K.A., Higgins, C.L., Wittung-Stafshede, P., 2007. Thermodynamic stability and folding of proteins from hyperthermophilic organisms. FEBS J. 274, 4023–4033.

Makarova, K.S., Aravind, L., Grishin, N.V., Rogozin, I.B., Koonin, E.V., 2002. A DNA repair system specific for thermophilic Archaea and bacteria predicted by genomic context analysis. Nucleic Acids. Res. 30, 482–496.

Makarova, K.S., Aravind, L., Wolf, Y.I., Tatusov, R.L., Minton, K.W., Koonin, E.V., et al., 2001. Genome of the extremely radiation-resistant bacterium Deinococcus radiodurans viewed from the perspective of comparative genomics. Microbiol. Mol. Biol. Rev. 65, 44–79.

Mamo, G., 2020. Challenges and adaptations of life in alkaline habitats. Adv. Biochem. Eng. Biotechnol. 172, 85–133.

Mao, D., Grogan, D.W., 2017. How a genetically stable extremophile evolves: modes of genome diversification in the Archaeon Sulfolobus acidocaldarius. J. Bacteriol. 199, e00177-17.

Markillie, L.M., Varnum, S.M., Hradecky, P., Wong, K.K., 1999. Targeted mutagenesis by duplication insertion in the radioresistant bacterium Deinococcus radiodurans: radiation sensitivities of catalase (katA) and superoxide dismutase (sodA) mutants. J. Bacteriol. 181, 666–669.

Maurya, G.K., Chaudhary, R., Pandey, N., Misra, H.S., 2021. Molecular insights into replication initiation in a multipartite genome harboring bacterium Deinococcus radiodurans. J. Biol. Chem. 296, 100451.

Maurya, G.K., Kota, S., Misra, H.S., 2019. Characterisation of ParB encoded on multipartite genome in Deinococcus radiodurans and their roles in radioresistance. Microbiol. Res. 223-225, 22–32.

Maurya, G.K., Misra, H.S., 2021. Characterization of ori and parS-like functions in secondary genome replicons in Deinococcus radiodurans. Life Sci. Alliance 4 e202000856.

Maurya, G.K., Modi, K., Banerjee, M., Chaudhary, R., Rajpurohit, Y.S., Misra, H.S., 2018. Phosphorylation of FtsZ and FtsA by a DNA damage-responsive Ser/Thr protein kinase affects their functional interactions in Deinococcus radiodurans. mSphere 3 e00325-18.

Maurya, G.K., Modi, K., Misra, H.S., 2016. Divisome and segrosome components of Deinococcus radiodurans interact through cell division regulatory proteins. Microbiology (Reading) 162, 1321–1334.

McCready, S., Marcello, L., 2003. Repair of UV damage in Halobacterium salinarum. Biochem. Soc. Trans. 31, 694–698.

Mennecier, S., Coste, G., Servant, P., Bailone, A., Sommer, S., 2004. Mismatch repair ensures fidelity of replication and recombination in the radioresistant organism Deinococcus radiodurans. Mol. Genet. Genomics. 272, 460–469.

Michels, M., Bakker, E.P., 1985. Generation of a large, protonophore-sensitive proton motive force and pH difference in the acidophilic bacteria Thermoplasma acidophilum and Bacillus acidocaldarius. J. Bacteriol. 161, 231–237.

Minton, K.W., 1994. DNA repair in the extremely radioresistant bacterium Deinococcus radiodurans. Mol. Microbiol. 13, 9–15.

Minton, K.W., Daly, M.J., 1995. A model for repair of radiation-induced DNA double-strand breaks in the extreme radiophile Deinococcus radiodurans. Bioessays 17, 457–464.

Mishra, S., Chaudhary, R., Singh, S., Kota, S., Misra, H.S., 2019. Guanine quadruplex DNA regulates gamma radiation response of genome functions in the radioresistant bacterium Deinococcus radiodurans. J. Bacteriol. 201, e00154-19.

Mishra, S., Misra, H.S., Kota, S., 2022. FtsK, a DNA motor protein, coordinates the genome segregation and early cell division processes in Deinococcus radiodurans. mBio 13, e0174222.

Misra, H.S., Khairnar, N.P., Barik, A., Indira Priyadarsini, K., Mohan, H., Apte, S.K., 2004. Pyrroloquinoline-quinone: a reactive oxygen species scavenger in bacteria. FEBS Lett. 578, 26–30.

Misra, H.S., Khairnar, N.P., Kota, S., Shrivastava, S., Joshi, V.P., Apte, S.K., 2006. An exonuclease I-sensitive DNA repair pathway in Deinococcus radiodurans: a major determinant of radiation resistance. Mol. Microbiol. 59, 1308–1316.

Misra, H.S., Rajpurohit, Y.S., 2024. DNA damage response and cell cycle regulation in bacteria: a twist around the paradigm. Front. Microbiol. 15, 1389074.

Misra, H.S., Rajpurohit, Y.S., Khairnar, N.P., 2012. Pyrroloquinoline-quinone and its versatile roles in biological processes. J. Biosci. 37, 313–325.

Misra, H.S., Rajpurohit, Y.S., Kota, S., 2013. Physiological and molecular basis of extreme radioresistance in Deinococcus radiodurans. Curr. Sci. 104, 194–205.

Modi, K.M., Tewari, R., Misra, H.S., 2014. FtsZDr, a tubulin homologue in radioresistant bacterium Deinococcus radiodurans is characterized as a GTPase exhibiting polymerization/depolymerization dynamics in vitro and FtsZ ring formation in vivo. Int. J. Biochem. Cell Biol. 50, 38–46.

Moran-Reyna, A., Coker, J.A., 2014. The effects of extremes of pH on the growth and transcriptomic profiles of three haloarchaea. F1000Res 3, 168.

Moseley, B.E., Evans, D.M., 1983. Isolation and properties of strains of Micrococcus (Deinococcus) radiodurans unable to excise ultraviolet light-induced pyrimidine dimers from DNA: evidence for two excision pathways. J. Gen. Microbiol. 129, 2437–2445.

Mullakhanbhai, M.F., Larsen, H., 1975. Halobacterium volcanii spec. nov., a Dead Sea halobacterium with a moderate salt requirement. Arch. Microbiol. 104, 207–214.

Munteanu, A., Uivarosi, V., Andries, A., 2015. Recent progress in understanding the molecular mechanisms of radioresistance in Deinococcus bacteria. Extremophiles 19, 707–719.

Murakami, M., Narumi, I., Satoh, K., Furukawa, A., Hayata, I., 2006. Analysis of interaction between DNA and Deinococcus radiodurans PprA protein by atomic force microscopy. Biochim. Biophys. Acta 1764, 20–23.

Narasimha, A., Basu, B., 2021. New insights into the activation of radiation desiccation response regulon in Deinococcus radiodurans. J. Biosci. 46, 10.

Narumi, I., Satoh, K., Cui, S., Funayama, T., Kitayama, S., Watanabe, H., 2004. PprA: a novel protein from Deinococcus radiodurans that stimulates DNA ligation. Mol. Microbiol. 54, 278–285.

Narumi, I., Satoh, K., Kikuchi, M., Funayama, T., Yanagisawa, T., Kobayashi, Y., et al., 2001. The LexA protein from Deinococcus radiodurans is not involved in RecA induction following gamma irradiation. J. Bacteriol. 183, 6951–6956.

Ng, W.V., Kennedy, S.P., Mahairas, G.G., Berquist, B., Pan, M., Shukla, H.D., et al., 2000. Genome sequence of Halobacterium species NRC-1. Proc. Natl. Acad. Sci. USA. 97, 12176–12181.

Ngo, K.V., Molzberger, E.T., Chitteni-Pattu, S., Cox, M.M., 2013. Regulation of Deinococcus radiodurans RecA protein function via modulation of active and inactive nucleoprotein filament states. J. Biol. Chem. 288, 21351–21366.

Nogi, Y., Hosoya, S., Kato, C., Horikoshi, K., 2007. Psychromonas hadalis sp. nov., a novel piezophilic bacterium isolated from the bottom of the Japan Trench. Int. J. Syst. Evol. Microbiol. 57, 1360–1364.

Ogrunc, M., Becker, D.F., Ragsdale, S.W., Sancar, A., 1998. Nucleotide excision repair in the third kingdom. J. Bacteriol. 180, 5796–5798.

Ohba, H., Satoh, K., Yanagisawa, T., Narumi, I., 2005. The radiation responsive promoter of the Deinococcus radiodurans pprA gene. Gene 363, 133–141.

Patoli, B.B., Winter, J.A., Patoli, A.A., Delahay, R.M., Bunting, K.A., 2017. Co-expression and purification of the RadA recombinase with the RadB paralog from Haloferax volcanii yields heteromeric ring-like structures. Microbiology (Reading) 163, 1802–1811.

Pavankumar, T.L., Sinha, A.K., Ray, M.K., 2010. All three subunits of RecBCD enzyme are essential for DNA repair and low-temperature growth in the Antarctic Pseudomonas syringae Lz4W. PLoS One 5, e9412.

Perez-Arnaiz, P., Dattani, A., Smith, V., Allers, T., 2020. Haloferax volcanii-a model archaeon for studying DNA replication and repair. Open Biol. 10, 200293.

Poidevin, L., MacNeill, S.A., 2006. Biochemical characterisation of LigN, an NAD+-dependent DNA ligase from the halophilic euryarchaeon Haloferax volcanii that displays maximal in vitro activity at high salt concentrations. BMC Mol. Biol. 7, 44.

Qi, H.Z., Wang, W.Z., He, J.Y., Ma, Y., Xiao, F.Z., He, S.Y., 2020. Antioxidative system of Deinococcus radiodurans. Res. Microbiol. 171, 45–54.

Quaiser, A., Constantinesco, F., White, M.F., Forterre, P., Elie, C., 2008. The Mre11 protein interacts with both Rad50 and the HerA bipolar helicase and is recruited to DNA following gamma irradiation in the archaeon Sulfolobus acidocaldarius. BMC Mol. Biol. 9, 25.

Radman, M., 1975. SOS repair hypothesis: phenomenology of an inducible DNA repair which is accompanied by mutagenesis. Basic Life. Sci. 5A, 355–367.

Rajan, R., Bell, C.E., 2004. Crystal structure of RecA from Deinococcus radiodurans: insights into the structural basis of extreme radioresistance. J. Mol. Biol. 344, 951–963.

Rajpurohit, Y.S., Bihani, S.C., Waldor, M.K., Misra, H.S., 2016. Phosphorylation of Deinococcus radiodurans RecA regulates its activity and may contribute to radioresistance. J. Biol. Chem. 291, 16672–16685.

Rajpurohit, Y.S., Desai, S.S., Misra, H.S., 2013. Pyrroloquinoline quinone and a quinoprotein kinase support gamma-radiation resistance in Deinococcus radiodurans and regulate gene expression. J. Basic Microbiol. 53, 518–531.

Rajpurohit, Y.S., Gopalakrishnan, R., Misra, H.S., 2008. Involvement of a protein kinase activity inducer in DNA double strand break repair and radioresistance of Deinococcus radiodurans. J. Bacteriol. 190, 3948–3954.

Rajpurohit, Y.S., Misra, H.S., 2010. Characterization of a DNA damage-inducible membrane protein kinase from Deinococcus radiodurans and its role in bacterial radioresistance and DNA strand break repair. Mol. Microbiol. 77, 1470–1482.

Rajpurohit, Y.S., Misra, H.S., 2013a. DR1769, a protein with N-terminal beta propeller repeats and a low-complexity hydrophilic tail, plays a role in desiccation tolerance of Deinococcus radiodurans. J. Bacteriol. 195, 3888–3896.

Rajpurohit, Y.S., Misra, H.S., 2013b. Structure-function study of deinococcal serine/threonine protein kinase implicates its kinase activity and DNA repair protein phosphorylation roles in radioresistance of Deinococcus radiodurans. Int. J. Biochem. Cell Biol. 45, 2541–2552.

Rajpurohit, Y.S., Sharma, D.K., Misra, H.S., 2021. PprA protein inhibits DNA strand exchange and ATP hydrolysis of Deinococcus RecA and regulates the recombination in gamma-irradiated cells. Front. Cell Dev. Biol. 9, 636178.

Rajpurohit, Y.S., Sharma, D.K., Misra, H.S., 2022. Involvement of serine/threonine protein kinases in DNA damage response and cell division in bacteria. Res. Microbiol. 173, 103883.

Reed, C.J., Lewis, H., Trejo, E., Winston, V., Evilia, C., 2013. Protein adaptations in archaeal extremophiles. Archaea 2013, 373275.

Regha, K., Satapathy, A.K., Ray, M.K., 2005. RecD plays an essential function during growth at low temperature in the antarctic bacterium Pseudomonas syringae Lz4W. Genetics 170, 1473–1484.

Repar, J., Cvjetan, S., Slade, D., Radman, M., Zahradka, D., Zahradka, K., 2010. RecA protein assures fidelity of DNA repair and genome stability in Deinococcus radiodurans. DNA Repair (Amst.) 9, 1151–1161.

Reyes, E.D., Patidar, P.L., Uranga, L.A., Bortoletto, A.S., Lusetti, S.L., 2010. RecN is a cohesin-like protein that stimulates intermolecular DNA interactions in vitro. J. Biol. Chem. 285, 16521–16529.

Richardson, T.T., Gilroy, L., Ishino, Y., Connolly, B.A., Henneke, G., 2013. Novel inhibition of archaeal family-D DNA polymerase by uracil. Nucleic Acids. Res. 41, 4207–4218.

Roberts, M.F., 2005. Organic compatible solutes of halotolerant and halophilic microorganisms. Saline Syst. 1, 5.

Rohwerder, T., Gehrke, T., Kinzler, K., Sand, W., 2003. Bioleaching review part A: progress in bioleaching: fundamentals and mechanisms of bacterial metal sulfide oxidation. Appl. Microbiol. Biotechnol. 63, 239–248.

Romano, V., Napoli, A., Salerno, V., Valenti, A., Rossi, M., Ciaramella, M., 2007. Lack of strand-specific repair of UV-induced DNA lesions in three genes of the archaeon Sulfolobus solfataricus. J. Mol. Biol. 365, 921–929.

Sadowska-Bartosz, I., Bartosz, G., 2023. Antioxidant defense of Deinococcus radiodurans: how does it contribute to extreme radiation resistance? Int. J. Radiat. Biol. 99, 1803–1829.

Sakai, T., Tokishita, S., Mochizuki, K., Motomiya, A., Yamagata, H., Ohta, T., 2008. Mutagenesis of uracil-DNA glycosylase deficient mutants of the extremely thermophilic eubacterium Thermus thermophilus. DNA Repair (Amst.) 7, 663–669.

Sancar, A., Lindsey-Boltz, L.A., Unsal-Kacmaz, K., Linn, S., 2004. Molecular mechanisms of mammalian DNA repair and the DNA damage checkpoints. Annu. Rev. Biochem. 73, 39–85.

Santos, S.P., Yang, Y., Rosa, M.T.G., Rodrigues, M.A.A., De La Tour, C.B., Sommer, S., et al., 2019. The interplay between Mn and Fe in Deinococcus radiodurans triggers cellular protection during paraquat-induced oxidative stress. Sci. Rep. 9, 17217.

Satapathy, A.K., Pavankumar, T.L., Bhattacharjya, S., Sankaranarayanan, R., Ray, M.K., 2008. ATPase activity of RecD is essential for growth of the Antarctic Pseudomonas syringae Lz4W at low temperature. FEBS J. 275, 1835–1851.

Satoh, K., Kikuchi, M., Ishaque, A.M., Ohba, H., Yamada, M., Tejima, K., et al., 2012. The role of Deinococcus radiodurans RecFOR proteins in homologous recombination. DNA Repair (Amst.) 11, 410–418.

Seck, A., De Bonis, S., Saint-Pierre, C., Gasparutto, D., Ravanat, J.L., Timmins, J., 2022. In vitro reconstitution of an efficient nucleotide excision repair system using mesophilic enzymes from Deinococcus radiodurans. Commun. Biol. 5, 127.

Seck, A., De Bonis, S., Stelter, M., Okvist, M., Senarisoy, M., Hayek, M.R., et al., 2023. Structural and functional insights into the activation of the dual incision activity of UvrC, a key player in bacterial NER. Nucleic Acids. Res. 51, 2931–2949.

Seitz, E.M., Brockman, J.P., Sandler, S.J., Clark, A.J., Kowalczykowski, S.C., 1998. RadA protein is an archaeal RecA protein homolog that catalyzes DNA strand exchange. Genes Dev. 12, 1248–1253.

Seitz, E.M., Kowalczykowski, S.C., 2000. The DNA binding and pairing preferences of the archaeal RadA protein demonstrate a universal characteristic of DNA strand exchange proteins. Mol. Microbiol. 37, 555–560.

Selvam, K., Duncan, J.R., Tanaka, M., Battista, J.R., 2013. DdrA, DdrD, and PprA: components of UV and mitomycin C resistance in Deinococcus radiodurans R1. PLoS One 8, e69007.

Sharma, D.K., Bihani, S.C., Siddiqui, M.Q., Misra, H.S., Rajpurohit, Y.S., 2022a. WD40 domain of RqkA regulates its kinase activity and role in extraordinary radioresistance of D. radiodurans. J. Biomol. Struct. Dyn. 40, 1246–1259.

Sharma, D.K., Misra, H.S., Bihani, S.C., Rajpurohit, Y.S., 2023. Biochemical properties and roles of DprA protein in bacterial natural transformation, virulence, and pilin variation. J. Bacteriol. 205, e0046522.

Sharma, D.K., Misra, H.S., Soni, I., Rajpurohit, Y.S., 2022b. Characterization of DNA processing protein A (DprA) of the radiation-resistant bacterium Deinococcus radiodurans. Microbiol. Spectr. 10, e0347022.

Sharma, D.K., Siddiqui, M.Q., Gadewal, N., Choudhary, R.K., Varma, A.K., Misra, H.S., et al., 2020. Phosphorylation of deinococcal RecA affects its structural and functional dynamics implicated for its roles in radioresistance of Deinococcus radiodurans. J. Biomol. Struct. Dyn. 38, 114–123.

Sharma, D.K., Soni, I., Misra, H.S., Rajpurohit, Y.S., 2024a. Natural transformation-specific DprA coordinate DNA double-strand break repair pathways in heavily irradiated D. radiodurans. Appl. Environ. Microb. 90, e0194823.

Sharma, D.K., Soni, I., Rajpurohit, Y.S., 2024b. Surviving the storm: exploring the role of natural transformation in nutrition and DNA repair of stressed Deinococcus radiodurans. Appl. Environ. Microb. 91, e0137124.

Shashidhar, R., Kumar, S.A., Misra, H.S., Bandekar, J.R., 2010. Evaluation of the role of enzymatic and nonenzymatic antioxidant systems in the radiation resistance of Deinococcus. Can. J. Microbiol. 56, 195–201.

She, Q., Zhang, C., Deng, L., Peng, N., Chen, Z., Liang, Y.X., 2009. Genetic analyses in the hyperthermophilic archaeon Sulfolobus islandicus. Biochem. Soc. Trans. 37, 92–96.

Sheng, D., Zheng, Z., Tian, B., Shen, B., Hua, Y., 2004. LexA analog (dra0074) is a regulatory protein that is irrelevant to recA induction. J. Biochem. 136, 787–793.

Shrivastava, M., Rajpurohit, Y.S., Misra, H.S., D'Souza, S.F., 2010. Survival of phosphate-solubilizing bacteria against DNA damaging agents. Can. J. Microbiol. 56, 822–830.

Shuttleworth, G., Fogg, M.J., Kurpiewski, M.R., Jen-Jacobson, L., Connolly, B.A., 2004. Recognition of the pro-mutagenic base uracil by family B DNA polymerases from archaea. J. Mol. Biol. 337, 621–634.

Siliakus, M.F., van der Oost, J., Kengen, S.W.M., 2017. Adaptations of archaeal and bacterial membranes to variations in temperature, pH and pressure. Extremophiles 21, 651–670.

Slade, D., Lindner, A.B., Paul, G., Radman, M., 2009. Recombination and replication in DNA repair of heavily irradiated Deinococcus radiodurans. Cell 136, 1044–1055.

Slade, D., Radman, M., 2011. Oxidative stress resistance in Deinococcus radiodurans. Microbiol. Mol. Biol. Rev. 75, 133–1391.

Spies, M., Kil, Y., Masui, R., Kato, R., Kujo, C., Ohshima, T., et al., 2000. The RadA protein from a hyperthermophilic archaeon Pyrobaculum islandicum is a DNA-dependent ATPase that exhibits two disparate catalytic modes, with a transition temperature at 75 degrees C. Eur. J. Biochem. 267, 1125–1137.

Stelter, M., Acajjaoui, S., McSweeney, S., Timmins, J., 2013. Structural and mechanistic insight into DNA unwinding by Deinococcus radiodurans UvrD. PLoS One 8, e77364.

Sugiman-Marangos, S., Junop, M.S., 2010. The structure of DdrB from Deinococcus: a new fold for single-stranded DNA binding proteins. Nucleic Acids. Res. 38, 3432–3440.

Sugiman-Marangos, S.N., Weiss, Y.M., Junop, M.S., 2016. Mechanism for accurate, protein-assisted DNA annealing by Deinococcus radiodurans DdrB. Proc. Natl. Acad. Sci. USA. 113, 4308–4313.

Szabla, R., Li, M., Warner, V., Song, Y., Junop, M., 2024. DdrC, a unique DNA repair factor from D. Radiodurans, senses and stabilizes DNA breaks through a novel lesion-recognition mechanism. Nucleic Acids. Res. 52, 9282–9302.

Tanaka, M., Earl, A.M., Howell, H.A., Park, M.J., Eisen, J.A., Peterson, S.N., et al., 2004. Analysis of Deinococcus radiodurans's transcriptional response to ionizing radiation and desiccation reveals novel proteins that contribute to extreme radioresistance. Genetics 168, 21–33.

Tian, B., Sun, Z., Shen, S., Wang, H., Jiao, J., Wang, L., et al., 2009. Effects of carotenoids from Deinococcus radiodurans on protein oxidation. Lett. Appl. Microbiol. 49, 689–694.

Tian, B., Xu, Z., Sun, Z., Lin, J., Hua, Y., 2007. Evaluation of the antioxidant effects of carotenoids from Deinococcus radiodurans through targeted mutagenesis, chemiluminescence, and DNA damage analyses. Biochim. Biophys. Acta 1770, 902–911.

Toueille, M., Mirabella, B., Guerin, P., Bouthier de la Tour, C., Boisnard, S., Nguyen, H.H., et al., 2012. A comparative proteomic approach to better define Deinococcus nucleoid specificities. J. Proteomics. 75, 2588–2600.

Ujaoney, A.K., Padwal, M.K., Basu, B., 2017. Proteome dynamics during post-desiccation recovery reveal convergence of desiccation and gamma radiation stress response pathways in Deinococcus radiodurans. Biochim. Biophys. Acta Proteins Proteom. 1865, 1215–1226.

Ujaoney, A.K., Padwal, M.K., Basu, B., 2021. An in vivo interaction network of DNA-repair proteins: a snapshot at double strand break repair in Deinococcus radiodurans. J. Proteome Res. 20, 3242–3255.

Uranga, L.A., Reyes, E.D., Patidar, P.L., Redman, L.N., Lusetti, S.L., 2017. The cohesin-like RecN protein stimulates RecA-mediated recombinational repair of DNA double-strand breaks. Nat. Commun. 8, 15282.

Vera, M., Schippers, A., Sand, W., 2013. Progress in bioleaching: fundamentals and mechanisms of bacterial metal sulfide oxidation–part A. Appl. Microbiol. Biotechnol. 97, 7529–7541.

Wang, L., Xu, G., Chen, H., Zhao, Y., Xu, N., Tian, B., et al., 2008. DrRRA: a novel response regulator essential for the extreme radioresistance of Deinococcus radiodurans. Mol. Microbiol. 67, 1211–1222.

Wang, W., Ma, Y., He, J., Qi, H., Xiao, F., He, S., 2019. Gene regulation for the extreme resistance to ionizing radiation of Deinococcus radiodurans. Gene 715, 144008.

Wardell, K., Haldenby, S., Jones, N., Liddell, S., Ngo, G.H.P., Allers, T., 2017. RadB acts in homologous recombination in the archaeon Haloferax volcanii, consistent with a role as recombination mediator. DNA Repair (Amst.) 55, 7–16.

Wardle, J., Burgers, P.M., Cann, I.K., Darley, K., Heslop, P., Johansson, E., et al., 2008. Uracil recognition by replicative DNA polymerases is limited to the archaea, not occurring with bacteria and eukarya. Nucleic Acids. Res. 36, 705–711.

White, O., Eisen, J.A., Heidelberg, J.F., Hickey, E.K., Peterson, J.D., Dodson, R.J., et al., 1999. Genome sequence of the radioresistant bacterium Deinococcus radiodurans R1. Science 286, 1571–1577.

Woods, W.G., Dyall-Smith, M.L., 1997. Construction and analysis of a recombination-deficient (radA) mutant of Haloferax volcanii. Mol. Microbiol. 23, 791–797.

Xiong, Y., Wei, L., Xin, S., Min, R., Liu, F., Li, N., et al., 2022. Comprehensive temporal protein dynamics during postirradiation recovery in Deinococcus radiodurans. Oxid. Med. Cell Longev. 2022, 1622829.

Xu, G., Lu, H., Wang, L., Chen, H., Xu, Z., Hu, Y., et al., 2010. DdrB stimulates single-stranded DNA annealing and facilitates RecA-independent DNA repair in Deinococcus radiodurans. DNA Repair (Amst.) 9, 805–812.

Xu, W., Shen, J., Dunn, C.A., Desai, S., Bessman, M.J., 2001. The Nudix hydrolases of Deinococcus radiodurans. Mol. Microbiol. 39, 286–290.

Yin, J., Chen, J.C., Wu, Q., Chen, G.Q., 2015. Halophiles, coming stars for industrial biotechnology. Biotechnol. Adv. 33, 1433–1442.

Yu, S., Kim, S., Kim, J., Kim, J.W., Kim, S.Y., Yeom, B., et al., 2023. Highly water-dispersed and stable deinoxanthin nanocapsule for effective antioxidant and anti-inflammatory activity. Int. J. Nanomed. 18, 4555–4565.

Zahradka, K., Slade, D., Bailone, A., Sommer, S., Averbeck, D., Petranovic, M., et al., 2006. Reassembly of shattered chromosomes in Deinococcus radiodurans. Nature 443, 569–573.

Zerulla, K., Soppa, J., 2014. Polyploidy in haloarchaea: advantages for growth and survival. Front. Microbiol. 5, 274.

Zhang, C., Tian, B., Li, S., Ao, X., Dalgaard, K., Gokce, S., et al., 2013. Genetic manipulation in Sulfolobus islandicus and functional analysis of DNA repair genes. Biochem. Soc. Trans. 41, 405–410.

Zhao, A., Gray, F.C., MacNeill, S.A., 2006. ATP- and NAD+-dependent DNA ligases share an essential function in the halophilic archaeon Haloferax volcanii. Mol. Microbiol. 59, 743–752.

Zhao, Z., Zhou, Z., Li, L., Xian, X., Ke, X., Chen, M., et al., 2014. A copper-responsive gene cluster is required for copper homeostasis and contributes to oxidative resistance in Deinococcus radiodurans R1. Mol. Biosyst. 10, 2607–2616.

Zimmerman, J.M., Battista, J.R., 2005. A ring-like nucleoid is not necessary for radioresistance in the Deinococcaceae. BMC Microbiol. 5, 17.

CHAPTER

6

Recent developments in ultraviolet-induced deoxyribonucleic acid damage and repair in higher plants

Fatima-Zahra Akensous[a], *Nizar El Mazouni*[a,b] and *Abdelilah Meddich*[a,c]

[a]Center of Agrobiotechnology and Bioengineering, Research Unit Labelled CNRST (CentreAgroBiotech-URL-CNRST-05), Cadi Ayyad University (UCA), Marrakesh, Morocco [b]College of Agriculture and Environmental Sciences, Center of Excellence in Soil and Fertilizer Research in Africa (CESFRA), AgroBioSciences, University Mohammed VI Polytechnic (UM6P), Ben Guerir, Morocco [c]African Sustainable Agriculture Research Institute (ASARI), University Mohammed VI Polytechnic (UM6P), Laayoune, Morocco

6.1 Introduction

Climate crisis' current and projected trends in the context of the 21st century indicate that climatic manifestations like global warming are exacerbating and will continue to intensify by the end of the current century (Shukla et al., 2019). Probably, 2024 marked the warmest year ever since 1850 (Olonscheck et al., 2021). Moreover, the last decade constitutes the warmest ever recorded in more than 100 observed years (Bricknell et al., 2021). Ever since the occurrence of ozone (O_3) layer damage around the late 1970s, its recovery has been impeded due to climatic factors like the interchange between the O_3 layer and the increased greenhouse gas (GHG) emissions. Consequently, the ultraviolet (UV) radiation has been increased to an unprecedented level, alarmingly warming Earth's surface (Bernhard et al., 2020). Such consequences have been negatively impacting the atmospheric O_3 layer, a critical player in balancing out temperatures that reach Earth's surface. The O_3 layer depletion losses have been mainly driven by the anthropogenic factor due to the release of detrimental chemicals (e.g., chlorine)

Genome Damage: Repair and Adaptation Across Life.
DOI: https://doi.org/10.1016/B978-0-443-42857-9.00018-9

(Jin et al., 2025). Along with global warming-increasing GHG concentrations (e.g., carbon dioxide [CO_2], chlorofluorocarbons [CFCs]), the O_3 layer depletion heavily contributes to the exacerbation of the GHG negative impact (Chipperfield et al., 2020). Being sessile, the kingdom of plants is constantly subject to environmental conditions that can be drastic like abiotic stressors. The exposure to UV-type B (UV-B), for instance, can have detrimental effects on plants that can be lethal due to deoxyribonucleic acid (DNA) damage (Chen et al., 2022). The UV-disturbed genomic functions of plants lead to detrimental consequences on the structure and function levels owing to *a highly unstable* DNA (Vanhaelewyn et al., 2020). Solar radiation is vital for living organisms. As far as plants are concerned, blue and red lights are particularly crucial to ensure the optimal functioning and yielding of the photosynthetic apparatus, thus an optimal plant productivity (Liu and van Iersel, 2021). Nevertheless, the exposure to UV radiation, most especially when excessive, can easily and significantly affect the *genomic machinery* of plants, bringing about DNA lesions, such as cyclobutane–pyrimidine dimers (CPDs) and 6-4 photoproducts (6-4 PPs) (Strzałka et al., 2020). Nevertheless, plants have learned to adapt to UV-induced DNA damage through effective mechanistic damage repair systems such as photoreactivation, *programmed* excisions, and homologous recombination (Verma et al., 2020). Therefore, the present book section delves into recent advances in UV-induced DNA damage consequences and repairing mechanisms in higher plants. Furthermore, the book section underscores remarkable challenging bottlenecks and futuristic research avenues of high interest in this regard.

6.2 Greenhouse gas emission effects and ultraviolet-induced deoxyribonucleic acid damage in higher plants

GHG emissions have witnessed unprecedented upsurging trends in the context of the 21st century, essentially driven by the anthropogenic factor (e.g., fossil fuel combustion, industrial rapid growth, intensive agriculture, deforestation) (Gakaev et al., 2024). GHG gases, such as CO_2, methane (CH_4), and nitrous oxide (N_2O), constitute the predominant GHGs that largely contribute to the climate crisis (Gu et al., 2022; Filonchyk et al., 2024). Such gases have experienced increasing atmospheric concentrations, leading to major shifts in the environmental conditions under which higher plants thrive and evolve (Li et al., 2024a). Elevated rates of CO_2, one of the major GHGs, have complex and often contradictory effects on plant physiology (Table 6.1). On the one hand, increased CO_2 can enhance photosynthetic efficiency in C_3 plants, plants that rely on the Calvin cycle pathway for C fixation, by increasing the carboxylation rate of ribulose-1,5-bisphosphate carboxylase/oxygenase (RuBisCO) and reducing photorespiration, potentially stimulating biomass accumulation (Wang et al., 2020). On the other hand, the fertilization effect of CO_2 is constrained by nutrient availability, particularly nitrogen (N), and may be *offset* by adverse alterations in temperature and precipitation patterns (Leung et al., 2022). The observed elevated temperature trends, another direct consequence of exacerbating GHG emissions, have been leading to frequent and prolonged episodes of heat and drought stresses with significant implications for plant development, flowering periods, reproductive *success*, and yielding (Qian et al., 2025). Elevated temperatures can destabilize protein structures, impair enzymatic activity, and accelerate the production of reactive oxygen species (ROS), leading to oxidative stress and cellular damage (Devireddy et al., 2021). Parallelly, altered precipitation

TABLE 6.1 Greenhouse gas emissions-ultraviolet nexus' effects on higher plants.

Main objective	Higher plant	Irradiation type	Notable effects	Reference
Investigation of UV-caused mutations	*Arabidopsis thaliana* (*supF*-transgenic plants)	UV-C	Mutation frequency increase, with G:C-to-A:T transitions dominating in the spectrum of mutation and distributed across single, tandem, and multiple base substitutions due to induced cyclobutane-type pyrimidine dimers and 6-4 pyrimidine	Nakamura et al. (2021)
Investigation of aquatic plant function to remove N and GHGs	*Iris pseudacorus*	-	-Increase in total N removal rate in wetland -Abundance of NH_3 oxidation bacteria (e.g., *Nitrosospira*, *Nitrosomonas*) and *amoA* gene maintenance -Plants acting as C sink and covering N_2O and CH_4 emissions	Gu et al. (2022)
Investigation of plant kaempferol and quercetin biosynthesis under drought and/or UV exposure	*Oryza sativa* (wild and transgenic)	UV-B	-Overaccumulation of flavonoids -Transcriptional regulation of dehydrin and UV-B resistance genes	Jan et al. (2022)
Investigation of plant discrete responses to excessive UV-B radiation	*Triticum aestivum*, *Triticum monococcum*	UV-B	-Pronounced plant H_2O_2 levels -Antioxidant defense machinery's gene upregulation	Kirova et al. (2024)

amoA, ammonia-oxidizing archaea gene; *A:T*, adenine:thymine; *C*, carbon; *CH_4*, methane; *GHGs*, greenhouse gases; *G:C*, guanine:cytosine; *H_2O_2*, hydrogen peroxide; *N*, nitrogen; *N_2O, nitrous oxide*; *NH_3*, ammonia; *supF*, suppressor factor F gene; *UV*, ultraviolet.

regimes, including sporadic and irregular rainfall patterns, have been disrupting plant-water relations (Feldman et al., 2024). Water stress, particularly drought stress, severely affects vital physiological parameters, such as plant turgor pressure and stomatal conductance, triggering systemic stress signaling pathways, subsequently leading to DNA damage (Ehonen et al., 2020). Moreover, N_2O emissions from intensive N fertilization have been linked to alterations in soil microbial communities and plant nutrient dynamics, further compounding stress conditions (Lazcano et al., 2021). CH_4, though less directly impactful on plant metabolism, significantly contributes to global warming and indirectly exacerbates abiotic stress (Kabange et al., 2022). The overaccumulation of ROS under stressful conditions like the exposure to UV radiation conducts to DNA damage in higher plants (Nakamura et al., 2021; Jan et al., 2022; Kirova et al., 2024; Lingwan, 2024). As a consequence, single-strand DNA breaks and double-strand DNA breaks (DSBs) are formed; the damage that occurs in DNA can lead to base loss and the formation of miscoding/noncoding lesions, besides sugar-phosphate backbone breakage (Wolter et al., 2021). The escalation of such DNA damage occurrences can bring about detrimental mutations, adding to genomic instability, which can prove to be lethal for plants (Szurman-Zubrzycka et al., 2023).

6.3 Ultraviolet radiation effects on higher plants

6.3.1 Notable morphophysiological and biochemical effects

The stratospheric O_3 layer serves as a protective *shield* that filters out the most harmful wavelengths of solar radiation, such as UV-B (Table 6.2). However, anthropogenic emissions of CFCs and other O_3-depleting substances have led to persistent O_3-*thinning*, particularly at high latitudes and during spring seasons (Western et al., 2023). This depletion has resulted in increased surface-level exposure of higher plants to UV-B radiation, with critical implications for their physiological and developmental processes (Shi and Liu, 2021). Upon *perforating* the plant *canopy*, UV-B radiation is primarily absorbed by epidermal cells and leaf tissues, where it initiates a cascade of morphophotobiological and biochemical responses (Yoon et al., 2021). One of the most immediate effects is the inhibition of plant cell division and elongation, especially in meristematic zones, leading to reduced leaf expansion and stunted shoot and root growth (Sheridan et al., 2022). At the molecular level, UV-B disrupts hormonal balance (e.g., auxin signaling, gibberellin signaling), thereby altering growth patterns and morphogenesis (Miao et al., 2021). Photosynthesis, the central energy-supplying phenomenon in plants, is particularly vulnerable to UV-B-induced damage (Piccini et al., 2020). The photosynthetic protein systems, particularly photosystem II, are subject to structural and functional disruptions under significant UV exposure. Chlorophyll degradation, impaired electron transport chains, and reduced RuBisCO activity collectively lead to decreased photosynthetic efficiency and lowered biomass accumulation (Piccini et al., 2021). Moreover, UV-B engenders excessive production of ROS, such as hydroxyl radicals, which cause oxidative damage to vital biomolecules like lipids, proteins, and nucleic acids (Kumari et al., 2022). DNA molecules within plant cells constitute direct targets of UV-B photons, leading to the formation of CPDs and 6-4 PPs (Strzałka et al., 2020). These lesions, if unrepaired, can result in mutations, cell-cycle abruption, and apoptosis, among others, thereby compromising vigor and reproduction in higher plants (Asik et al., 2025).

TABLE 6.2 Ultraviolet-induced effects on higher plants.

Main objective	Higher plant	Irradiation type	Notable effects	Reference
Evaluation of the antioxidant, pro-oxidant, and photoprotective impact of olive mill wastewater	*Olea europaea* (olive mill wastewater)	UV-A	Efficient antioxidant activity	Lecci et al. (2021)
Investigation of flavonoids and linked to the expression of genes in wheat under UV-B radiation	*Triticum* sp. (blue-grained wheat)	UV-B	Overaccumulation of metabolites and flavonoids decrease, but flavones and flavanols increase	Li et al. (2024a)
Study of photoreceptors' roles in sustaining photosynthetic capacity in *Arabidopsis* subject to UV-A1, UV-B, and blue light	*Arabidopsis* sp.	UV-A, UV-B	Protection by photoreceptors against photodamage under UV radiation and epidermal flavonols' overconcentration that compromises photoprotection	Morales et al. (2025)
Revealing of a conserved mechanism throughout plant lineages distinguished by a reciprocal antagonistic relationship between UV response and P adaptation	*Oryza sativa*, *Glycine max*, *Triticum aestivum*, and *Zea mays*	UV	P availability regulates plant UV tolerance and the expression of UV radiation response genes in a phosphate starvation response protein fashion, inducing UV tolerance	Ren et al. (2025)

P, phosphorus; *UV*, ultraviolet.

6.3.2 Notable oxidative and molecular effects

The excessive exposure of higher plants to UV radiation as a *hormetic* stress engenders the overaccumulation of secondary metabolites (e.g., flavone, flavonoids, nicotinamide) (del Socorro Sánchez Correa et al., 2023). Higher plants can readjust their primary metabolism as well to deal with the overaccumulation of ROS (Table 6.2; Shi and Liu, 2021; Li et al., 2024b; Morales et al., 2025). Plant responsive overproduction of coumarins under UV-B exposure was registered as well, with fluctuations in, for instance, protein production (e.g., proteins associated with photorespiration), as reported in the book chapter by del Socorro Sánchez Correa et al. (2023), indicating ROS modulation facets by higher plants. The overexposure to UV radiation induces gene expression alterations in higher plants as well (Ren et al., 2025). It was shown that UV-B radiation could prompt molecular pathways linked to cell growth and morphology (e.g., modification of transcription factor expression) (Tao et al., 2022). *UV-B light-insensitive* mutant plants, associated with hypocotyl growth linked to wild-type plants (Strzałka et al., 2020). A basic leucine zipper transcription factor responsible for morphogenesis in plants, *LONG HYPOCOTYL5,* was observed to be regulated under *mild* UV-B exposure (Zhou et al., 2022). The transcription factor, which is light-activated, is also crucial for phytochrome and cryptochrome regulation networking, implying that it may not be directly responsive to UV-B (Tavridou et al., 2020), as reported in the book chapter by del Socorro Sánchez Correa et al. (2023).

6.4 Ultraviolet radiation-induced deoxyribonucleic acid impairment in higher plants

6.4.1 Cyclobutane-pyrimidine dimers

Among the most deleterious consequences of UV radiation on higher plants is the direct impairment of the DNA molecule (Table 6.3). The most prevalent and mutagenic of these lesions are CPDs (Banaś et al., 2020). CPDs are formed when two adjacent pyrimidine bases, typically thymine or cytosine, become covalently bonded through the formation of a cyclobutane ring (Lu et al., 2021). This photochemical reaction occurs when UV-B photons are absorbed by DNA, leading to an excited electronic state that facilitates dimerization. In higher plants, CPD formation is dose-dependent and accumulates with prolonged or intense UV-B exposure (Asik et al., 2025). If unrepaired, CPDs can lead to point mutations, deletions, or chromosomal rearrangements, ultimately affecting gene expression, genome stability, and heritable traits (Strzałka et al., 2020). The extent of CPD accumulation and its physiological impact vary with leaf age, tissue type, and developmental stage, reflecting differential exposure and repair capacity across plant organs (Qian et al., 2021). In response to CPD-induced DNA damage, higher plants have evolved specialized repair mechanisms. Photoreactivation represents a capital mechanism, a light-dependent pathway mediated by photolyase enzymes that specifically recognize and monomerize CPDs using blue light/UV-A radiation as an energy source (Banaś et al., 2020). Additionally, nucleotide excision repair operates in a light-independent manner, involving a multistep process of lesion recognition, excision, repair synthesis, and ligation (Strzałka et al., 2020).

TABLE 6.3 Ultraviolet-induced deoxyribonucleic acid damage in higher plants.

Main objective	Higher plant	Irradiation type	Notable effects	Reference
Distinguishing the effects of varied UV-B intensities on *Arabidopsis thaliana* than those of UV-A	*Arabidopsis thaliana*	UV-A, UV-B	-Elongation response, mediated by UV-A, is phenotypically distinct from classical shade avoidance -Inhibitory effects exertion by UV-B on the length of the petiole, blade, and leaf area	Cunningham et al. (2024)
Examination of the impact of UV-A within the 350–400 nm range (UV-A1 range)	*Arabidopsis thaliana*	UV-A	UV-A affecting plant shoot growth (possibly linked to DNA damage)	Giza et al. (2025)
Identification and characterization of genomic structural variations in plant somatic cells subject to UV-B and UV-C	*Arabidopsis thaliana*	UV-B, UV-C	Genomic regions localized in heterochromatin are prone to forming genomic structural variations in comparison with those localized in euchromatin	Sall et al. (2025)
Investigation of whether prolonged UV exposure would enhance epigenetic diversity more than it would for genetic diversity	*Arabidopsis thaliana*	UV-C	Continuous exposure to UV-C enhances (epi)genomic diversity, with epigenetic alterations occurring at the level of several stress-responsive pathways	Virgen et al. (2025)

UV, ultraviolet.

6.4.2 6-4 photoproducts

UV-induced DNA damage in higher plants was evidenced in studies, especially carried out on *Arabidopsis* (Table 6.3). In addition to CPDs, UV-B radiation induces another class of photolesions in the DNA of higher plants known as 6-4 PPs. Although less abundant than CPDs, 6-4 PPs are structurally more distorting and biologically more disruptive, posing a significant threat to genome stability (Strzałka et al., 2020; Sall et al., 2025). 6-4 PPs are formed when a covalent bond is established between the C6 position of a 5′-end pyrimidine (typically thymine or cytosine) and the C4 position of an adjacent 3′-end pyrimidine. In contrast to CPDs, 6-4 PPs are recognized more rapidly and repaired more efficiently by plant DNA repair systems, reflecting their heightened cytotoxicity (Asik et al., 2025). Nonetheless, their presence activates a robust DNA damage response (DDR), including cell-cycle checkpoints and stress signaling pathways, aimed at mitigating the deleterious effects on cellular homeostasis and development (Asik et al., 2025; Virgen et al., 2025). Like CPDs, 6-4 PPs can be reversed through photoreactivation, a light-dependent repair mechanism mediated by a specific class of photolyase enzymes known as 6-4 photolyases (Zgłobicki et al., 2024). These enzymes absorb blue light/UV-A radiation and use the energy to cleave the covalent bond linking the pyrimidine bases, thereby restoring native DNA structure (Banaś et al., 2020; Cunningham et al., 2024; Giza et al., 2025).

6.5 Deoxyribonucleic acid damage repair in higher plants

6.5.1 Photolyase-assisted photoreactivation and cell-cycle control

Higher plants, constantly exposed to UV radiation, have developed intricate and efficient mechanisms to repair DNA lesions, ensuring the reestablishment of genomic integrity (Table 6.4). Among such important mechanisms, photoreactivation represents one of the most direct and evolutionarily conserved repair pathways (Mmbando, 2024). This light-dependent process is specifically mediated by photolyase enzymes, which target and reverse UV-induced DNA photoproducts, principally CPDs and 6-4 PPs (Douki and Cadet, 2023; Zgłobicki et al., 2024). Photolyases are flavoproteins that utilize visible light, predominantly in the blue light/UV-A radiation spectrum (320-480 nm), to catalyze the monomerization of UV-induced DNA lesions (Mmbando et al., 2021). Upon photon absorption through their chromophore cofactors, typically flavin adenine dinucleotide, and either methenyltetrahydrofolate or 8-hydroxy-5-deazaflavin, photolyases are activated, transferring an electron to break the abnormal covalent bonds linking adjacent pyrimidines (Ramírez-Gamboa et al., 2022). In higher plants, genes encoding photolyase proteins are differentially expressed in response to UV-B exposure, circadian rhythms, developmental stages, and tissue type (Yadav et al., 2020; Otake et al., 2024). For instance, CPD photolyase is more abundantly expressed in photosynthetically active tissues, such as leaves, where exposure to UV radiation is most intense (Mmbando et al., 2023). Cell-cycle control is a central component of the cellular response to DNA damage in higher plants, functioning as a critical regulatory mechanism that ensures genomic integrity under environmental stress conditions, including UV radiation exposure (Gentric et al., 2021). When DNA lesions like CPDs and 6-4 PPs are detected, plants activate sophisticated *surveillance* systems, known as DNA damage checkpoints, that transiently block cell-cycle progression to allow sufficient time for repair processes to operate before DNA replication or cell division proceeds (Mmbando et al., 2020; Strzałka et al., 2020). In plants, the intricate relationship between DNA repair mechanisms and

TABLE 6.4 Ultraviolet-induced deoxyribonucleic acid damage repair in higher plants.

Main objective	Higher plant	Irradiation type	Notable effects	Reference
Investigation of UV-B radiation resistance and the activity of cyclobutane–pyrimidine dimer (CPD) photolyase in different cultivars of rice	*Oryza sativa*	UV-B	-Variation in the resistance to UV-B correlated with the total activity of CPD photolyase -Observation of low enzyme activity engendered by polymorphisms and decreased cellular CPD photolyase contents within super-hypersensitive cultivars to UV-B	Mmbando et al. (2020)
Characterization of AtRAD16 and AtRAD16b in UV tolerance	*Arabidopsis thaliana*	UV	Enhanced UV tolerance	Alrayes et al. (2023)
Analysis of the genome-wide excision-repairing mediation of 6-4 PPs photoproducts	*Arabidopsis thaliana*	UV	-Repairing of 6-4 PPs through global repair -Open chromatin genome regions are swiftly repaired in comparison with heterochromatin regions -Important repairs at the level of promoter, start site, and end site of transcription in genes	Kaya et al. (2024)

(*continued on next page*)

TABLE 6.4 Ultraviolet-induced deoxyribonucleic acid damage repair in higher plants—cont'd

Main objective	Higher plant	Irradiation type	Notable effects	Reference
Elucidation of the anti-UV impact of Rad5 ortholog and two Rad16 paralogs in plants lacking Rad7 homolog and assessment of their anti-UV effects' dependency on the NER-dependent dark reactivation/photorepair-dependent photoreactivation	*Beauveria bassiana*	UV-B	Photoreactivation dependency of anti-UV effects of Rad5, Rad16A, and Rad16B and mechanistic distinction from homologs	Luo et al. (2024)
Identification of sequences localized at the level of the N-terminal region	*Oryza sativa*	UV-B	OsPHR transport/localization to chloroplasts is being orchestrated by a vesicle transport system under serine-phosphorylation control	Otake et al. (2024)
Study of whether and the extent of Rad2 and Rad14 activities in the photoreactivation and dark reactivation processes, and their anti-UV-B part in in vivo photoreactivation	*Beauveria bassiana*	UV-B	Important photoreactivation activity of Rad2/Rad14 is due to a close link to an important protein network shaped by photolyase modulators and anti-UV proteins via protein-protein interactions	Yu et al. (2024)
Analysis of the activity and localization of GFP-tagged AtUVR3, the 6-4 photoproduct-specific photolyase	*Arabidopsis thaliana*, *Nicotiana benthamiana*, *Nicotiana tabacum*	UV	-Positive correlation between AtUVR3 levels and 6-4 photoproduct removal rate -Photoreactivation is the main repairing pathway in chloroplasts and mitochondria	Zgłobicki et al. (2024)

AtRAD, *Arabidopsis thaliana* DNA repair protein; *AtUVR3*, UV repair defective 3 protein; *GFP*, green fluorescent protein; *N*, nitrogen; *NER*, nucleotide excision repair; *OsPHR*, *Oryza sativa* photolyase; *6-4 PPs*, 6-4 pyrimidine–pyrimidone/photoproducts; *Rad*, anti-UV radiation protein family; *UV*, ultraviolet.

cell-cycle control is vital for maintaining genomic integrity in response to various stressors. Unlike animals that utilize p53 for DNA damage signaling, plants rely on the suppressor of gamma response 1 (SOG1) transcription factor, a major component of the plant DDR pathway. SOG1 activation is mediated through phosphorylation by ataxia-telangiectasia mutated (ATM) and ATM/rad3-related kinases, which respond to DSBs in DNA (Meschichi et al., 2022; Luo et al., 2024; Yu et al., 2024; Takahashi et al., 2025). This signaling cascade ensures that plants can enact cell-cycle checkpoints that facilitate timely DNA repair before cell division resumes, thus preventing the propagation of genetic errors. DNA DSBs can induce endoreduplication rather than apoptosis, a unique adaptation allowing for continued cellular growth despite damage (Szurman-Zubrzycka et al., 2023). This programmed response showcases how specific cell-cycle transitions can be modulated to prioritize repair and genomic stability over cell division (Gentric et al., 2021). Furthermore, during instances of DNA damage, plants exhibit a halt in cell-cycle progression, allowing the necessary time for efficient repair processes to occur (Pedroza-Garcia et al., 2022).

6.5.2 Base excision, apoptosis, and homologous recombination

Since the past two decades, research on base excision repair (BER) has led to insightful advances, most especially with work conducted on *Arabidopsis thaliana* (Table 6.4; Hindi et al., 2021; Alrayes et al., 2023; Kaya et al., 2024). When it comes to DNA damage repair, BER constitutes a crucial DNA repair mechanistic process in organisms like higher plants, enabling the removal of damaged bases, brought about by stressful factors like excessive UV exposure (Asik et al., 2025). The process of BER is *commenced* by DNA glycosylases that proceed by excising the damaged base through hydrolytic cleavage of the N-glycosidic bond located between C1′ of 2′-deoxyribose and the N-atom at the aimed base, completed by supplementary proteins removing the *residual* sugar-phosphate moiety, filling up the consequent gap, and performing ligation (Roldán-Arjona et al., 2019; Grin et al., 2023). As a responsive mechanism to UV-induced DNA damage, the DDR pathway is stimulated in higher plants (e.g., SOG1, a transcription factor implicated in DDR regulation). Generally, the DDR signaling pathway prompts transcriptional and epigenetic modulators that instrument cell-cycle *cessation* and DNA repair mechanisms (Grin et al., 2023). The cessation of the cell cycle allows time for DNA damage repair before replication. If DNA damage repair culminates in success, reactivation of the cell cycle occurs. If DNA-repairing mechanistic processes are unsuccessful, however, bringing about DNA lesions' accumulation, plant cells begin apoptosis (Szurman-Zubrzycka et al., 2023). Higher plants are equipped with many DNA damage repair and recombination pathways directed toward removing/tolerating damaged DNA lesions and sustaining the overall genome integrity (Grin et al., 2023). When chromosomal ends in vegetative cells of higher plants break due to stressful factors like excessive UV radiation, they often *heal* through DNA ends' ligation into unrelated sequences/microhomologies-based sequences as a DNA-strand-break repairing mechanism (Ray and Langer, 2002).

6.6 General conclusions and research avenues

Being sessile organisms, more than ever, higher plants are subject to exacerbating, intensifying, and frequent and prolonged episodes of stressful factors in the context of the 21st century.

At the level of the higher plant genome, deoxyribonucleic acid represents a highly unstable biomolecule, subject to that damage following exposure to highly intensified environmental conditions. Solar radiation constitutes a vital component of paramount importance for higher plants' processes, beginning with the photosynthetic apparatus, powered by photosynthetically active radiation occurring between 400 and 700 nm. Nevertheless, solar ultraviolet radiation, namely ultraviolet A, ultraviolet B, and ultraviolet C, found between 100 and 400 nm, particularly ultraviolet A and ultraviolet B, represents detrimental factors for higher plants, causing damage at the molecular level, especially under excessive exposure. Ultraviolet-induced deoxyribonucleic acid damage in higher plants, such as cyclobutane–pyrimidine dimers and 6-4 photoproducts, when accumulating in somatic cells, can engender lethal consequences like mutations, menacing genomic integrity. In response, higher plants deploy mechanistic strategies such as photoreactivation, cell-cycle control, base excision, apoptosis, and homologous recombination, which help reverse, encounter, and tolerate damaged deoxyribonucleic acid lesions and by-products (e.g., reactive oxygen species as *signaling* and resulting chemicals). The last few years have seen a surging interest in exploring and investigating the mechanisms by which higher plants deal with ultraviolet radiation exposure, with regard to the deployed mechanisms. Besides, researchers have been taking interest in the development of stress-tolerant cultivars under an ever-intensifying climate crisis. Besides, further studies are required to decipher the exact intricate orchestrated molecular bases governing higher plant responsive mechanisms. Answering questions like why a mechanistic process is preferred over another and in which way they act synergistically and complementarily, in addition to the dynamics in (hetero/eu)chromatins, can prove to be of paramount importance in this regard.

References

Alrayes, L., Stout, J., Schroeder, D., 2023. Arabidopsis RAD16 homologues are involved in UV tolerance and growth. Genes 14 (8), 1552. https://doi.org/10.3390/genes14081552.

Asik, E., Kashif, S.Z., Oztas, O., 2025. Plant tolerance mechanisms to DNA-damaging UV stress. J. Exp. Bot. 1–24. https://doi.org/10.1093/jxb/eraf272.

Banaś, A.K., Zgłobicki, P., Kowalska, E., Bażant, A., Dziga, D., Strzałka, W., 2020. All you need is light. Photorepair of UV-induced pyrimidine dimers. Genes 11 (11), 1304. https://doi.org/10.3390/genes11111304.

Bernhard, G.H., Neale, R.E., Barnes, P.W., Neale, P.J., Zepp, R.G., Wilson, S.R., et al., 2020. Environmental effects of stratospheric ozone depletion, UV radiation and interactions with climate change: UNEP Environmental Effects Assessment Panel, update 2019. Photochem. Photobiol. Sci. 19 (5), 542–584. https://doi.org/10.1039/d0pp90011g.

Bricknell, I.R., Birkel, S.D., Brawley, S.H., Van Kirk, T., Hamlin, H.J., Capistrant-Fossa, K., et al., 2021. Resilience of cold water aquaculture: a review of likely scenarios as climate changes in the Gulf of Maine. Rev. Aquacult. 13 (1), 460–503. https://doi.org/10.1111/raq.12483.

Chen, Z., Dong, Y., Huang, X., 2022. Plant responses to UV-B radiation: signaling, acclimation and stress tolerance. Stress Biol. 2 (1), 51. https://doi.org/10.1007/s44154-022-00076-9.

Chipperfield, M.P., Hossaini, R., Montzka, S.A., Reimann, S., Sherry, D., Tegtmeier, S., 2020. Renewed and emerging concerns over the production and emission of ozone-depleting substances. Nature Rev. Earth Environ. 1 (5), 251–263. https://doi.org/10.1038/s43017-020-0048-8.

Cunningham, N., Crestani, G., Csepregi, K., Coughlan, N.E., Jansen, M.A.K., 2024. Exploring the complexities of plant UV responses; distinct effects of UV-A and UV-B wavelengths on Arabidopsis rosette morphology. Photochem. Photobiol. Sci. 23 (7), 1251–1264. https://doi.org/10.1007/s43630-024-00591-w.

del Socorro Sánchez Correa, M., el Rocío Reyero Saavedra, M., Antonio Estrella Parra, E., Nolasco Ontiveros, E., del Carmen Benítez Flores, J., Gerardo Ortiz Montiel, J., et al., 2023. Ultraviolet radiation and its effects on plants. Abiotic Stress in Plants - Adaptations to Climate Change. IntechOpen, United Kingdom. https://doi.org/10.5772/intechopen.109474.

Devireddy, A.R., Tschaplinski, T.J., Tuskan, G.A., Muchero, W., Chen, J.-G., 2021. Role of reactive oxygen species and hormones in plant responses to temperature changes. Int. J. Mol. Sci. 22 (16), 8843. https://doi.org/10.3390/ijms22168843.

Douki, T., Cadet, J., 2023. DNA damage. In: Encyclopedia of Astrobiology. Springer, Berlin, pp. 827–832.

Ehonen, S., Hölttä, T., Kangasjärvi, J., 2020. Systemic signaling in the regulation of stomatal conductance. Plant Physiol. 182 (4), 1829–1832. https://doi.org/10.1104/pp.19.01543.

Feldman, A.F., Feng, X., Felton, A.J., Konings, A.G., Knapp, A.K., Biederman, J.A., et al., 2024. Plant responses to changing rainfall frequency and intensity. Nature Rev. Earth Environ. 5 (4), 276–294. https://doi.org/10.1038/s43017-024-00534-0.

Filonchyk, M., Peterson, M.P., Zhang, L., Hurynovich, V., He, Y., 2024. Greenhouse gases emissions and global climate change: examining the influence of CO2, CH4, and N2O. Sci. Total Environ. 935, 173359. https://doi.org/10.1016/j.scitotenv.2024.173359.

Gakaev, R., Gatsaeva, L., Eskiev, M., 2024. Impacts of greenhouse gas-induced climate change: risks, vulnerabilities, and adaptation strategies. E3S Web of Conferences 537, 03009. https://doi.org/10.1051/e3sconf/202453703009.

Gentric, N., Genschik, P., Noir, S., 2021. Connections between the cell cycle and the DNA damage response in plants. Int. J. Mol. Sci. 22 (17), 9558. https://doi.org/10.3390/ijms22179558.

Giza, A., Hermanowicz, P., Ważny, R., Domka, A., Rozpądek, P., Łabuz, J., 2025. Effect of UV-A on endophyte colonisation of Arabidopsis thaliana. PLoS One 20 (5 May), 1–26. https://doi.org/10.1371/journal.pone.0323576.

Grin, I.R., Petrova, D.V., Endutkin, A.V., Ma, C., Yu, B., Li, H., et al., 2023. Base excision DNA repair in plants: Arabidopsis and beyond. Int. J. Mol. Sci. 24 (19), 14746. https://doi.org/10.3390/ijms241914746.

Gu, X., Chen, D., Wu, F., Tang, L., He, S., Zhou, W., 2022. Function of aquatic plants on nitrogen removal and greenhouse gas emission in enhanced denitrification constructed wetlands: Iris pseudacorus for example. J. Cleaner Prod. 330 (July 2021), 129842. https://doi.org/10.1016/j.jclepro.2021.129842.

Hindi, N.N., Elsakrmy, N., Ramotar, D., 2021. The base excision repair process: comparison between higher and lower eukaryotes. Cell. Mol. Life Sci. 78 (24), 7943–7965. https://doi.org/10.1007/s00018-021-03990-9.

Jan, R., Khan, M.A., Asaf, S., Lubna, W., M., P., J., R., et al., 2022. Drought and UV radiation stress tolerance in rice is improved by overaccumulation of non-enzymatic antioxidant flavonoids. Antioxidants 11 (5), 1–20. https://doi.org/10.3390/antiox11050917.

Jin, W., Yan, Y., Qiu, X., Peng, L., Li, Z., Tang, Y., 2025. Characterizing full-phase chlorine species emissions from domestic coal combustion in China: implications for significant impacts on air pollution and ozone-layer depletion. Environ. Pollut. 372, 126043. https://doi.org/10.1016/j.envpol.2025.126043.

Kabange, N.R., Mun, B.G., Lee, S.M., Kwon, Y., Lee, D., Lee, G.M., et al., 2022. Nitric oxide: a core signaling molecule under elevated GHGs (CO2, CH4, N2O, O3)-mediated abiotic stress in plants. Front. Plant Sci. 13 (November), 1–14. https://doi.org/10.3389/fpls.2022.994149.

Kaya, S., Erdogan, D.E., Sancar, A., Adebali, O., Oztas, O., 2024. Global repair is the primary nucleotide excision repair subpathway for the removal of pyrimidine-pyrimidone (6-4) damage from the Arabidopsis genome. Sci. Rep. 14 (1), 1–9. https://doi.org/10.1038/s41598-024-53472-8.

Kirova, E., Moskova, I., Manova, V., Koycheva, Y., Tsekova, Z., Borisova, D., et al., 2024. Exogenous cytokinin 4PU-30 modulates the response of wheat and einkorn seedlings to ultraviolet B radiation. Plants 13 (10), 1401. https://doi.org/10.3390/plants13101401.

Kumari, P., Thakur, R., Singh, N., Rastogi, A., Yadav, S., 2022. Relationships of oxidative stress and ultraviolet-B radiation in plants. In: UV-B Radiation and Crop Growth. Plant Life and Environment Dynamics. Springer, Singapore, pp. 277–291.

Lazcano, C., Zhu-Barker, X., Decock, C., 2021. Effects of organic fertilizers on the soil microorganisms responsible for N2O emissions: a review. Microorganisms 9 (5), 983. https://doi.org/10.3390/microorganisms9050983.

Lecci, R.M., D'Antuono, I., Cardinali, A., Garbetta, A., Linsalata, V., Logrieco, A.F., et al., 2021. Antioxidant and pro-oxidant capacities as mechanisms of photoprotection of olive polyphenols on UVA-damaged human keratinocytes. Molecules 26 (8), 2153. https://doi.org/10.3390/molecules26082153.

Leung, F., Sitch, S., Tai, A.P.K., Wiltshire, A.J., Gornall, J.L., Folberth, G.A., et al., 2022. CO_2 fertilization of crops offsets yield losses due to future surface ozone damage and climate change. Environ. Res. Lett. 17 (7), 074007. https://doi.org/10.1088/1748-9326/ac7246.

Li, K., Du, L., Qin, C., Bolan, N., Wang, H., Wang, H., 2024a. Microplastic pollution as an environmental risk exacerbating the greenhouse effect and climate change: a review. Carbon Research 3 (1), 9. https://doi.org/10.1007/s44246-023-00097-7.

Li, L., Jiang, G., Li, H., Liu, J., Zhang, P., Wang, Q., et al., 2024b. UV-B induced flavonoid accumulation and related gene expression in blue- grained wheat at different periods of time. Front. Plant Sci. 15 (December), 1–17. https://doi.org/10.3389/fpls.2024.1520543.

Lingwan, M., 2024. Damage and repair: how Poaceae plants fix DNA damaged by UV-B radiation. Plant Physiol. 195 (1), 265–267. https://doi.org/10.1093/plphys/kiae094.

Liu, J., van Iersel, M.W., 2021. Photosynthetic Physiology of blue, green, and red light: light intensity effects and underlying mechanisms. Front. Plant Sci. 12, 1–14. https://doi.org/10.3389/fpls.2021.619987.

Lu, C., Gutierrez-Bayona, N.E., Taylor, J.-S., 2021. The effect of flanking bases on direct and triplet sensitized cyclobutane pyrimidine dimer formation in DNA depends on the dipyrimidine, wavelength and the photosensitizer. Nucleic Acids Res. 49 (8), 4266–4280. https://doi.org/10.1093/nar/gkab214.

Luo, X.-C., Yu, L., Xu, S.-Y., Ying, S.-H., Feng, M.-G., 2024. Photoreactivation activities of Rad5, Rad16A and Rad16B help Beauveria bassiana to recover from solar ultraviolet damage. J. Fungi 10 (6), 420. https://doi.org/10.3390/jof10060420.

Meschichi, A., Zhao, L., Reeck, S., White, C., Da Ines, O., Sicard, A., et al., 2022. The plant-specific DDR factor SOG1 increases chromatin mobility in response to DNA damage. EMBO Rep. 23 (12). https://doi.org/10.15252/embr.202254736.

Miao, T., Li, D., Huang, Z., Huang, Y., Li, S., Wang, Y., 2021. Gibberellin regulates UV-B-induced hypocotyl growth inhibition in Arabidopsis thaliana. Plant. Signal. Behav. 16 (11). https://doi.org/10.1080/15592324.2021.1966587.

Mmbando, G.S., 2024. Variation in ultraviolet-B (UV-B)-induced DNA damage repair mechanisms in plants and humans: an avenue for developing protection against skin photoaging. Int. J. Radiat. Biol. 100 (11), 1505–1516. https://doi.org/10.1080/09553002.2024.2398081.

Mmbando, G.S., Ando, S., Takahashi, H., Hidema, J., 2023. High ultraviolet-B sensitivity due to lower CPD photolyase activity is needed for biotic stress response to the rice blast fungus, Magnaporthe oryzae. Photochem. Photobiol. Sci. 22 (6), 1309–1321. https://doi.org/10.1007/s43630-023-00379-4.

Mmbando, G.S., Teranishi, M., Hidema, J., 2020. Very high sensitivity of African rice to artificial ultraviolet-B radiation caused by genotype and quantity of cyclobutane pyrimidine dimer photolyase. Sci. Rep. 10 (1), 1–14. https://doi.org/10.1038/s41598-020-59720-x.

Mmbando, G.S., Teranishi, M., Hidema, J., 2021. Transgenic rice Oryza glaberrima with higher CPD photolyase activity alleviates UVB-caused growth inhibition. GM Crops Food 12 (1), 435–448. https://doi.org/10.1080/21645698.2021.1977068.

Morales, L.O., Shapiguzov, A., Rai, N., Aphalo, P.J., Brosché, M., 2025. Protection of photosynthesis by UVR8 and cryptochromes in Arabidopsis under blue and UV radiation. Plant Cell Environ. 48 (8), 6321–6335. https://doi.org/10.1111/pce.15608.

Nakamura, M., Nunoshiba, T., Hiratsu, K., 2021. Detection and analysis of UV-induced mutations in the chromosomal DNA of Arabidopsis. Biochem. Biophys. Res. Commun. 554, 89–93. https://doi.org/10.1016/j.bbrc.2021.03.087.

Olonscheck, D., Schurer, A.P., Lücke, L., Hegerl, G.C., 2021. Large-scale emergence of regional changes in year-to-year temperature variability by the end of the 21st century. Nat. Commun. 12 (1), 7237. https://doi.org/10.1038/s41467-021-27515-x.

Otake, M., Teranishi, M., Komatsu, C., Hara, M., Yoshiyama, K.O., Hidema, J., 2024. Poaceae plants transfer cyclobutane pyrimidine dimer photolyase to chloroplasts for ultraviolet-B resistance. Plant Physiol. 195 (1), 326–342. https://doi.org/10.1093/plphys/kiae060.

Pedroza-Garcia, J.A., Xiang, Y., De Veylder, L., 2022. Cell cycle checkpoint control in response to DNA damage by environmental stresses. Plant J. 109 (3), 490–507. https://doi.org/10.1111/tpj.15567.

Piccini, C., Cai, G., Dias, M.C., Araújo, M., Parri, S., Romi, M., et al., 2021. Olive varieties under UV-B Stress show distinct responses in terms of antioxidant machinery and isoform/activity of RubisCO. Int. J. Mol. Sci. 22 (20), 11214. https://doi.org/10.3390/ijms222011214.

Piccini, C., Cai, G., Dias, M.C., Romi, M., Longo, R., Cantini, C., 2020. UV-B radiation affects photosynthesis-related processes of two Italian Olea europaea (L.) varieties differently. Plants 9 (12), 1712. https://doi.org/10.3390/plants9121712.

Qian, D., Wang, M., Niu, Y., Yang, Y., Xiang, Y., 2025. Sexual reproduction in plants under high temperature and drought stress. Cell Rep. 44 (3), 115390. https://doi.org/10.1016/j.celrep.2025.115390.

Qian, M., Rosenqvist, E., Prinsen, E., Pescheck, F., Flygare, A.-M., Kalbina, I., et al., 2021. Downsizing in plants—UV light induces pronounced morphological changes in the absence of stress. Plant Physiol. 187 (1), 378–395. https://doi.org/10.1093/plphys/kiab262.

Ramírez-Gamboa, D., Díaz-Zamorano, A.L., Meléndez-Sánchez, E.R., Reyes-Pardo, H., Villaseñor-Zepeda, K.R., López-Arellanes, M.E., et al., 2022. Photolyase production and current applications: a review. Molecules 27 (18), 5998. https://doi.org/10.3390/molecules27185998.

Ray, A., Langer, M., 2002. Homologous recombination: ends as the means. Trends Plant Sci. 7 (10), 435–440. https://doi.org/10.1016/S1360-1385(02)02327-0.

Ren, J., Li, T., Guo, M., Zhang, Q., Ren, S., Wang, L., et al., 2025. A PHR-dependent reciprocal antagonistic interplay between UV response and P-deficiency adaptation in plants. Plant Commun. 6 (1), 101140. https://doi.org/10.1016/j.xplc.2024.101140.

Roldán-Arjona, T., Ariza, R.R., Córdoba-Cañero, D., 2019. DNA base excision repair in plants: An unfolding story with familiar and novel characters. Front. Plant Sci. 10, 1–18. https://doi.org/10.3389/fpls.2019.01055.

Sall, S.O., Alioua, A., Staerck, S., Graindorge, S., Pellicioli, M., Schuler, J., et al., 2025. Characterization of radiations-induced genomic structural variations in Arabidopsis thaliana. Plant J. 121 (1), 1–19. https://doi.org/10.1111/tpj.17180.

Sheridan, M.L., Simonelli, L., Giustozzi, M., Casati, P., 2022. Ultraviolet-B radiation represses primary root elongation by inhibiting cell proliferation in the meristematic zone of Arabidopsis seedlings. Front. Plant Sci. 13, 1–17. https://doi.org/10.3389/fpls.2022.829336.

Shi, C., Liu, H., 2021. How plants protect themselves from ultraviolet-B radiation stress. Plant Physiol. 187 (3), 1096–1103. https://doi.org/10.1093/plphys/kiab245.

Shukla, P.R., Skea, J., Buendia, C., E., M.-D., V., P.ö, H., O., et al. (2019). *IPCC, 2019: Climate Change and Land: an IPCC special report on climate change, desertification, land degradation, sustainable land management, food security, and greenhouse gas fluxes in terrestrial ecosystems*. United Nations Climate Change, Germany. https://www.ipcc.ch/srccl/

Strzałka, W., Zgłobicki, P., Kowalska, E., Bażant, A., Dziga, D., Banaś, A.K., 2020. The dark side of UV-induced DNA lesion repair. Genes 11 (12), 1450. https://doi.org/10.3390/genes11121450.

Szurman-Zubrzycka, M., Jędrzejek, P., Szarejko, I., 2023. How do plants cope with DNA damage? A concise review on the DDR pathway in plants. Int. J. Mol. Sci. 24 (3), 2404. https://doi.org/10.3390/ijms24032404.

Takahashi, N., Ogita, N., Koike, T., Nishimura, K., Inagaki, S., Zhang, Y., et al., 2025. The transcriptional repressors IAA5 and IAA29 participate in DNA damage-induced stem cell death in Arabidopsis roots. Plant Physiol. 198 (3). https://doi.org/10.1093/plphys/kiaf303.

Tao, M., Zhu, W., Han, H., Liu, S., Liu, A., Li, S., et al., 2022. Mitochondrial proteomic analysis reveals the regulation of energy metabolism and reactive oxygen species production in Clematis terniflora DC. leaves under high-level UV-B radiation followed by dark treatment. J. Proteomics 254, 104410. https://doi.org/10.1016/j.jprot.2021.104410.

Tavridou, E., Pireyre, M., Ulm, R., 2020. Degradation of the transcription factors PIF4 and PIF5 under UV-B promotes UVR8-mediated inhibition of hypocotyl growth in Arabidopsis. Plant J. 101 (3), 507–517. https://doi.org/10.1111/tpj.14556.

Vanhaelewyn, L., Van Der Straeten, D., De Coninck, B., Vandenbussche, F., 2020. Ultraviolet radiation from a plant perspective: The plant-microorganism context. Front. Plant Sci. 11, 1–18. https://doi.org/10.3389/fpls.2020.597642.

Verma, P., Tandon, R., Yadav, G., Gaur, V., 2020. Structural aspects of DNA repair and recombination in crop improvement. Front. Genet. 11, 1–30. https://doi.org/10.3389/fgene.2020.574549.

Virgen, A.L., Yadav, N.S., Byeon, B., Ilnytskyy, Y., Kovalchuk, I., 2025. Genetic and epigenetic changes in Arabidopsis thaliana exposed to ultraviolet-C radiation stress for 25 generations. Life 15 (3), 1–26. https://doi.org/10.3390/life15030502.

Wang, F., Gao, J., Yong, J.W.H., Wang, Q., Ma, J., He, X., 2020. Higher atmospheric CO2 levels favor C3 plants over C4 plants in utilizing ammonium as a nitrogen source. Front. Plant Sci. 11, 1–16. https://doi.org/10.3389/fpls.2020.537443.

Western, L.M., Vollmer, M.K., Krummel, P.B., Adcock, K.E., Crotwell, M., Fraser, P.J., et al., 2023. Global increase of ozone-depleting chlorofluorocarbons from 2010 to 2020. Nat. Geosci. 16 (4), 309–313. https://doi.org/10.1038/s41561-023-01147-w.

Wolter, F., Schindele, P., Beying, N., Scheben, A., Puchta, H., 2021. Different DNA repair pathways are involved in single-strand break-induced genomic changes in plants. Plant Cell 33 (11), 3454–3469. https://doi.org/10.1093/plcell/koab204.

Yadav, A., Singh, D., Lingwan, M., Yadukrishnan, P., Masakapalli, S.K., Datta, S., 2020. Light signaling and UV-B-mediated plant growth regulation. J. Integr. Plant Biol. 62 (9), 1270–1292. https://doi.org/10.1111/jipb.12932.

Yoon, H.I., Kim, H.Y., Kim, J., Oh, M.-M., Son, J.E., 2021. Quantitative analysis of UV-B radiation interception in 3D plant structures and intraindividual distribution of phenolic contents. Int. J. Mol. Sci. 22 (5), 2701. https://doi.org/10.3390/ijms22052701.

Yu, L., Xu, S.Y., Luo, X.C., Ying, S.H., Feng, M.G., 2024. High photoreactivation activities of Rad2 and Rad14 in recovering insecticidal Beauveria bassiana from solar UV damage. J. Photochem. Photobiol. B 251 (January), 112849. https://doi.org/10.1016/j.jphotobiol.2024.112849.

Zgłobicki, P., Hermanowicz, P., Kłodawska, K., Bażant, A., Łabuz, J., Grzyb, J., et al., 2024. The photoreactivation of 6-4 photoproducts in chloroplast and nuclear DNA depends on the amount of the Arabidopsis UV repair defective 3 protein. BMC Plant Biol. 24 (1), 1–17. https://doi.org/10.1186/s12870-024-05439-0.

Zhou, J., Meng, J., Zhang, S., Chi, R., Wang, C., Wang, D., et al., 2022. The UV-B-induced transcription factor HY5 regulated anthocyanin biosynthesis in Zanthoxylum bungeanum. Int. J. Mol. Sci. 23 (5), 2651. https://doi.org/10.3390/ijms23052651.

CHAPTER

7

Molecular dynamics of ultraviolet-B induced DNA lesions and repair mechanisms in higher plants

Rikina Choudhury and Krishna Kumar Choudhary

Department of Botany, Mahila Mahavidyalaya, Banaras Hindu University, Varanasi, Uttar Pradesh, India

7.1 Introduction

Solar radiation reaching Earth consists of infrared (IR: >700 nm), visible light (400–700 nm), and ultraviolet (UV: 100–400 nm) spectra, each essential for sustaining life by providing light, heat, and energy (Shen and Yin, 2022). However, these radiations, particularly UV, can harm living organisms by damaging cellular components, disrupting biological functions, and inducing DNA lesions (Correa et al., 2023). Genotoxic effects of UV radiation, mainly UV-A and ultraviolet-B (UV-B), have been extensively studied. UV exposure often causes dipyrimidine lesions leading to helix distortion and transcriptional dysregulation (Schuch et al., 2017). According to Huang and Zhou (2021), organisms mitigate such damage via DNA repair mechanisms, including direct light-dependent photoreactivation and light-independent nucleotide excision repair (NER), mismatch repair (MMR), homologous recombination (HR), BER, and nonhomologous end joining (NHEJ). These pathways involve precise lesion recognition, influenced by chromatin structure and epigenomic landscapes. Advances in genome-wide mapping and chromatin profiling have revealed that DNA damage formation and repair efficiency are closely linked to genome and epigenome organization (Kciuk et al., 2020; Huang and Zhou, 2021).

Research has been going on for the last century about the UV effect on plants; however, interest in UV-B intensified after ozone depletion was identified in the 1980s (Bais et al., 2015). Stratospheric ozone blocks most UV radiation, especially UV-C and part of UV-B, while UV-A and some UV-B still reach the Earth's surface. CO_2, along with other greenhouse gases (GHGs), drives global warming and also weakens the ozone layer (Singh and Choudhary, 2025c). Recent

Genome Damage: Repair and Adaptation Across Life.
DOI: https://doi.org/10.1016/B978-0-443-42857-9.00024-4

data collected during lockdowns of COVID-19 temporarily reduced emissions of GHGs, with nitrogen oxides (NOx) declining by nearly 15% between February and June 2020 (Miyazaki et al., 2021). In China, concentrations of CO, SO_2, and volatile organic compounds (VOCs) fell by 21%–26%, 16%–26%, and 37%–57%, while ozone (O_3) increased by about 20.5% during the same period (Li et al., 2020). Findings from 22 Indian cities indicated 10% decline in CO levels with no variation in SO_2, while O_3 increased by 17% (Sharma et al., 2020). Overall, 40% reduction in GHG emissions was observed during March 2020 as compared to the same period of 2019 in India specifically (Liaqat et al., 2024). Recent data reflected that UV-B is still a concern for plants and humans (Neale et al., 2021). Without the Montreal Protocol, erythemal UV irradiance below 50° latitude would have increased by 10%–15% between 1996 and 2020 (Bernhard et al., 2023). In March 2020, the ultraviolet index (UVI) of the Canadian Arctic was recorded 70% increment during 2005–2019, while springtime UVI at the South Pole exceeded by 100%, and southern latitudes above 50° rose by 25% annually (Bernhard et al., 2023). Despite the Montreal Protocol, UV-B influx due to ozone depletion broke records in the Arctic region during 2020 (Neale et al., 2021). Hence, UV-B did not decline, although ozone recovery has begun as reported by Bais et al. (2015). Although the practice of using ozone-degrading chemicals has declined, algorithms suggested that UV-B will rise over time and climate change is going to intensify this effect adversely (Barnes et al., 2022; Bera et al., 2022) even if all nations fully implement the Montreal Protocol. As O_3 declines due to excess chlorofluorocarbons (CFCs), harmful UV-B wavelengths pass through the ozone hole (Mmbando and Ngongolo, 2024). Banning CFCs resulted in ozone recovery in Antarctica, but global ozone levels continue to fall (UN, 2021). Hence, 1% reduction in the ozone layer leads to a 2%–3% rise in UV radiation reaching the lower atmosphere (Liaqat et al., 2024). Consequently, that UV-B radiation influx leads to damage to DNA and causes skin cancer in humans as well as harm plant cells by reducing their growth and productivity. Projections of UV-B exposure remain uncertain and worrisome because of the interaction of GHGs add complexity to atmospheric processes. Understanding plant responses to UV-B is crucial, as it is a constant part of sunlight. However, limited information is available for the plant's DNA repair system and related components, as DNA damage response (DDR) is better characterized in mammals than in plants because plants lack the p53 (guardian of the cell) gene (Ma et al., 2016; Szurman-Zubrzycka et al., 2023; Mmbando, 2024a).

Plants are affected both directly and indirectly by the influx of UV-B. Aside from direct absorption by photosynthetic components, UV-B disrupts photosynthesis mainly by degrading D1 and D2 proteins in photosystem II (Jumrani and Joshi-Paneri, 2023). It can also damage and cause photomorphogenic modification of PSI, but the effect is less compared to PS II, and this directs ROS generation and loss of cellular integrity (Piccini et al., 2020; Tevini, 2023). Strong links exist between ROS and gene expression, although UV-B-triggered ROS signaling is not fully explored.

Several cellular components absorb UV-B radiation, including nucleic acids and proteins. UV-B photon absorption excites electrons, thereby causing molecular changes like reversible protein conformations or DNA lesion formation, etc. (Gao et al., 2023; Laad et al., 2023). Unlike UV-C or gamma rays, UV-B lacks energy to eject electrons but can still damage DNA by forming photodimers that block transcription and replication, leading to mutations if remained unrepair (De et al., 2022). Impact of UV-B absorption on RNA and proteins remains poorly understood; however, RNA transcription of chl a/b associated proteins, along with Dicer expression are mostly affected (Andrade-Lima et al., 2015; Sun et al., 2020; Aktary et al., 2024). A unique protein, i.e., UV resistance locus8 (UVR8) found to perceive UV-B radiation and activate downstream gene expression for flavonoid production, which helps shield inner tissues by absorbing UV-B in plants. UV-B induces flavonoid biosynthesis through CHS expression, aided by the UVR8 photoreceptor (Cuadra et al., 2020). High dose absorption by flavonoids through UVR8 dependent pathway is found to provide protection but does not facilitate signaling pathway activation (Fang et al., 2022; Santin et al., 2023; Neugart et al., 2024). UV-B activates mitogen-activated protein kinase (MAPK) signaling, including UVR8-independent pathways that indicate the presence of multiple UV-B detection systems in plants (Chen et al., 2022).

UV-B direct absorption by cellular components triggers responses either through immediate effects or secondary consequences. These include reduced height, biomass, and hypocotyl elongation, along with altered cotyledon and leaf development (Bera et al., 2022). UV-B induces genes in the phenylpropanoid pathway, which ultimately promote flavonoid and anthocyanin accumulation to shield internal tissues from radiation damage (Singh et al., 2023). Mutants lacking UV-B screening compounds show stunted growth and reduced photosynthesis. Though UV-B can damage DNA, generate ROS, and impair photosynthesis, such effects are often exaggerated in lab settings. Natural conditions show minimal changes in photosynthesis despite growth reduction due to changes in cell division (Tripathi et al., 2017; Özel et al., 2021). Early studies linked UV-B responses to DNA as a potential photoreceptor, especially under more damaging and shorter wavelengths (280–300 nm), though this idea remains debated. Many studies excluded shorter UV-B wavelengths, which may bypass actual plant responses against UV-B radiation. Long-wavelength UV-B (above 300 nm) is now associated with photomorphogenic effects like growth inhibition and flavonoid production (Lee et al., 2022). UV-B effects also vary with fluence, where low doses trigger development and high doses cause stress. Overall, plants perceive UV-B through multiple mechanisms, with UV-B acting as both a damaging agent and developmental signal (Mishra et al., 2021; Cunningham et al., 2024). DNA repair systems respond to UV-B by removing photodimers and ROS-induced damage via photoreactivation, NER, or HR. Unrepaired lesions activate Ataxia telangiectasia mutated (ATM)/ ATM and Rad3 related (ATR) pathways, mainly by blocking replication. UV-B upregulates DNA repair genes, especially those for HR, while NER genes show low expression (Asik et al., 2025). This chapter outlines UV-induced DNA lesions, explores their interplay with epigenetic modifications, and examines the molecular features of lesion recognition proteins across repair pathways.

7.2 Effects of ultraviolet-B radiation on plants

7.2.1 Ultraviolet-B impact on plant morphology, yield, and cellular ultrastructure

Plants exhibit wide inter and intraspecific variation in growth, biomass production, and physiological responses to UV-B. Some species remain unaffected or even show improved growth, while most of the crops, including maize, wheat, and rice, etc. are vulnerable. UV-B commonly reduces plant size, alters morphology, and disrupts development. A study revealed that exposure to UV-B for 3 weeks on different rice varieties significantly reduced shoot and root growth-associated traits, with leaf area showing the highest decline (54% in indica, 48% in japonica), and root traits were less affected in japonica as compared to indica (Mathur et al., 2024). Frequent UV-B effects include thicker leaves, curling, and changes in shape, with reduced leaf area being the most consistent response. This decline results from UV-B induced suppressed cell expansion and division, though the extent differs among species. In peas, stem shortening is frequently observed and is associated with reduced gibberellin accumulation in young tissues, as UV-B exposure suppresses the expression of GA biosynthetic genes (GA20-oxidase and GA3-oxidase) while accelerating DELLA protein degradation (Roro et al., 2017). UV-B may influence shoot growth through effects on both cell division and elongation (Dotto and Casati, 2017). Evidence also shows enhanced axillary branching and tillering, altering biomass allocation toward roots. Flowering is often delayed, as observed in maize, mung bean, and *Arabidopsis*, but in common bean, though the molecular basis remains unclear. UV-B reduces inflorescence elongation, increases branching, and can modify fruit quality by altering secondary metabolites. UV-B irradiation also reduces stem growth, biomass, and floral primordia while advancing fruiting in well-watered plants. UV-B alters leaf anatomy and morphology, leading to structural changes that reduce the efficiency of photosynthesis (Wijewardana et al., 2016). UV-B exposure resulted in altered leaf thickness, weight, and cuticular wax deposition as well as trichome development, which further modify the light penetration level and plant morphology (Mmbando, 2024b). Overall, UV-B treatment promoted phenolic accumulation for stress defense, and this mechanistic resource partitioning redirects the plant's overall mechanism to stress management rather than maintaining total biomass and yield. Overall, high UV-B exposure consistently lowers relative growth rate, nitrogen use efficiency, biomass, and dry matter yield (Kumari et al., 2023).

Structural damage at the cellular level is one of the earliest indicators of UV-B stress in plants. The nuclear membrane often shows signs of disruption, which compromises the stability of genetic material and impairs cellular regulation. Chloroplasts, being highly sensitive organelles, undergo noticeable swelling, and in severe cases, their outer wall ruptures. Thylakoid membranes within the chloroplasts also show dilation followed by a breakdown of their organized structure. As the double membrane surrounding the chloroplast disintegrates, large starch granules begin to accumulate due to impaired photosynthetic and metabolic activity (Dwivedi et al., 2023). Beyond chloroplasts, the endoplasmic reticulum displays significant changes, most notably the enlargement of cisternae. Mitochondria and plastids also suffer structural damage, leading to reduced energy production and metabolic imbalance. In parallel, the plasmalemma and tonoplast frequently exhibit vesiculation, which destabilizes membrane integrity and disrupts cellular compartmentalization. Prolonged UV-B exposure not only modifies the shape and size of individual cells but also distorts their internal architecture. These changes highlight the vulnerability of plant cells to environmental stressors and underscore the central role of organelle stability in sustaining normal physiological processes (Takshak and Agrawal, 2019).

7.2.2 Ultraviolet-B impact on plant physiology

UV-B radiation disrupts plant cellular integrity and impairs physiological processes. Dicotyledons are generally more vulnerable than monocotyledons (Mavrič Čermelj et al., 2021). DNA lesions and replication errors arise under stress, while cortical microtubule breakdown limits cell division (Gustavsson, 2023). UV-B-induced amino acid degradation leads to enzyme and protein inactivation. Lipid peroxidation by UV-B damages cell membranes by generating free oxygen radicals (Singh and Choudhary, 2025a). These effects contribute to altered metabolism and growth inhibition (Sun et al., 2022). The preliminary target of UV-B radiation is photosynthetic machinery as it reduces key photosynthetic pigment content, particularly chlorophyll a (Singh and Choudhary, 2025b). Carotenoids act as protective pigments and are often increased under UV-B (Kurinjimalar et al., 2019). Additionally, UV-B modifies stomatal behavior and reduces CO_2 uptake for photosynthesis. UV-A promotes the opening of stomata while UV-B is responsible for their closure (Brestic et al., 2023), thereby sharply hampering the internal cellular processes of the plant. Moreover, UV-B reduces the water conductance that leads to decreased assimilation rates (Cechin et al., 2018; Reyes et al., 2020). It also downregulates photosynthetic genes and damages key proteins, including D1 and D2 of PSII as well as reaction centers and quinone acceptors (Yadav et al., 2020). Secondary metabolism also shifts under UV-B stress. Flavonoids increase as protective compounds, while the role of anthocyanins remains disputed. UV-B also upregulated genes associated with phenylpropanoid metabolism, such as PAL and CHS, and Mn-SOD expression, which ultimately enhances flavonoid production (Sharma et al., 2019; Tian et al., 2024). Rodríguez-Calzada et al. (2019) reported significant enhancement of flavonoid content in epidermis (particularly chlorogenic acid and apigenin derivatives) and, in combination with drought, found to induce a synergistic increase in luteolin glycosides. Auxin metabolism is negatively affected in plants under UV-B radiation, which results in reduced stem elongation, Rubisco activity, CO_2 fixation, and oxygen evolution (Kataria et al., 2013). UV-B triggers signaling pathways that activate ROS, phytohormones such as salicylic acid, gibberellins, and jasmonic acid, as well as several growth-regulating factors (GRFs), which enhance resistance to DNA damage by regulating defense gene expression (Mmbando et al., 2023).

7.2.3 Ultraviolet-B impact on photosynthesis

Depletion of stratospheric ozone affects plant growth, particularly by targeting the photosynthetic machinery. Although UV-B comprises ~0.5% of solar radiation, it holds disproportionately high energy, capable of damaging vital macromolecules, including DNA, proteins, and pigments. Crop studies confirm reduced chlorophyll levels in wheat, cotton, barley, sorghum, and amaranth under UV-B stress (Klem et al., 2015). Incidence of UV-B on the plant surface downregulates the gene expression and blocks the production of essential rubisco and Chl a/b binding proteins (Lhcb). Additionally, nuclear-encoded genes including Rubisco small subunit (rbcs), ATP synthase gamma subunit (atpc), and chlorophyll a/b binding protein of the PSII antenna (cab) are highly sensitive to UV-B radiation (Kataria et al., 2014; Ozel et al., 2021). In contrast, chloroplast-encoded genes such as rubisco large subunit (rbcL), ATP synthase Beta and Epsilon subunits (atpB and atpE), cytochrome b (petB), subunit IV of the cytochrome b/f complex (petD), and D1 and D2 proteins (psbA3) show relatively lower sensitivity to UV-B radiation (Kataria et al., 2014). The photosynthetic apparatus, especially PSII, is primarily affected by UV-B stress more than PSI. Critical sites of damage include the oxygen-evolving complex (OEC), manganese cluster, and reaction center proteins D1 and D2, all essential for water-splitting and electron transport (Singh and Choudhary, 2022). Damage to chlorophyll biosynthesis, carotenoids, and Rubisco activity further diminishes carbon fixation and ATP generation, thereby compromising plant productivity due to UV-B-induced injury (Kataria et al., 2014; Xiao et al., 2023). This multitargeted impairment leads to substantial declines in photosynthetic efficiency and crop yield.

7.2.4 Ultraviolet-B impact on oxidative status

Chloroplasts are major sources of ROS under UV-B stress, with photosystem I and II producing superoxide and singlet oxygen (Li and Kim, 2022). Excess ROS arises from electron transport chain overload, peroxidases, oxidases, photosensitizers, and peroxisomes, leading to membrane damage and oxidative stress. Hydroxyl radicals near the Mn cluster and Rubisco inhibition further intensify redox imbalance. Plants mitigate this damage through UV-absorbing phenolics, DNA and protein repair, and antioxidant defenses. Key enzymatic antioxidants include superoxide dismutase (SOD), peroxidases, catalase, glutathione reductase, ascorbate peroxidase, dehydroascorbate reductase, and monodehydroascorbate reductase, supported by ascorbate, glutathione, carotenoids, and α-tocopherol (Tan et al., 2023). Ascorbate plays a central role in detoxification, signaling, and recovery, maintained through recycling enzymes. Effective regulation of these systems confers UV-B tolerance. UV-B induced generation of oxidative stress on two native Antarctic vascular plant species, i.e., *Colobanthus quitensis* and *Deschampsia Antarctica*, was evident within their natural habitat. These plants were found to have DNA damage under field conditions compared to individuals grown in controlled, less stressful laboratory environments. Using Comet assay coupled with specific enzymatic treatments, the researchers distinguished between different types of DNA damage: Oxidized purines (via FPG enzyme), oxidized pyrimidines (via Endo-III), and UV-B-induced pyrimidine dimers (via T4 endonuclease V) (Muruzabal et al., 2021; Zheng et al., 2023). ROS accumulation intensifies oxidative stress and disrupts electron transport, which is generally detected by single-cell gel electrophoresis (SCGE) assay (Jha, 2023). Overall, high UV-B exposure oxidizes proteins, phospholipids, and nucleic acids, leading to severe cellular damage in plants by the generated ROS (Piccini et al., 2020). In contrast, low UV-B acts as a signaling cue that helps regulate redox homeostasis (Ashokkumar et al., 2024)

7.3 Comet assay to measure DNA damage

The comet assay, or SCGE, was particularly adopted for detecting DNA damage at the individual cell level by tracking the denatured DNA migration through an agarose gel under an electric field. DNA fragments are visualized using ethidium bromide (EtBr), silver staining, or fluorescent dyes. Disrupted DNA having strand breaks migrates further toward the anode than intact DNA, forming a celestial comet-like pattern (Jha, 2023). The pH level used in DNA unwinding influences the type of damage detected by the comet assay. It is called the alkaline comet assay at pH 13, which identifies SSB, DSB, alkali-labile sites, and incomplete repair, though without distinction (Bivehed et al., 2023). It can also reveal DNA–DNA or DNA protein cross-links, shown by reduced DNA migration compared with controls. A study using *Allium cepa* root cells demonstrated that salicylic acid induces DNA damage in a dose-dependent manner, evaluated by alkaline comet assay (Tyutereva et al., 2024). Recovery analysis further showed time-dependent decreases in comet parameters, where such damage decreases over time, indicating repair within about 48 hours.

7.4 Ultraviolet-B signaling in plants

7.4.1 UV resistance locus8-mediated perception of ultraviolet-B and photomorphogenic signaling

UV-B radiation constitutes a small fraction of solar radiation, but it exerts profound effects on plant physiology and development. Moderate doses of UV-B act as regulatory signals promoting photomorphogenic responses, including suppression of hypocotyl elongation, expansion of cotyledons, and enhanced biosynthesis of flavonoids, which function as natural UV screens (Rai et al., 2021). However, high levels of UV-B impose significant stress by inducing DNA lesions (e.g., cyclobutane pyrimidine dimers (CPDs), disrupting photosynthetic components, and generating ROS. Central to UV-B perception in plants is the UVR8 photoreceptor. Mutant UVR8-1 was first detected due to its higher susceptibility to UV-B in comparison with the wild-type (Jenkins, 2014). These mutants fail to induce CHS due to UV-B and show elevated levels of PR1 and PR5, which are proteins involved in pathogen defense. UVR8 is mostly found in the cytoplasm, but a small portion is already present in the nucleus. Still, gene expression regulated by UVR8 only occurs after UV-B exposure. UVR8 protein levels do not change following UV-B treatment and remain stable in both dark and light-grown plants. Before UV-B exposure, the cells contain the UVR8 dimer and a basic leucine zipper (bZIP) transcription factor elongated hypocotyl 5 (HY5). HY5 acts as a molecular switch that remains "OFF" in darkness (degraded) and "ON" in light (stabilized). UVR8 monomerize upon the exposure to UV-B and subsequently it starts interacting with constitutively photomorphogenic 1 (COP1) which is a multifunctional E3 ubiquitin ligase which subsequently leads to the stabilization of HY5 activity during UV-B influx (Hu et al., 2024; Liu et al., 2024) and degrade proteins those are harmful to UV protection and promotes de-etiolation (photomorphogenesis) as well as UV-B stress tolerance. Suppressor of phyA-105 1–4 (COP1–SPA) are co-factors of COP1, together with the cullin4–damaged DNA-binding protein 1 (CUL4–DDB1) ubiquitination–degradation complex, suppress HY5 in dark, keeping HY5 protein levels very low and promote skotomorphogenesis (etiolation) (Takeda, 2021). Plants perceive UV-B radiation, which leads to the accumulation of HY5, which then activates a suite of genes responsible for photoprotection, including those involved in flavonoid biosynthesis and antioxidative defense (Chen et al., 2022). This cascade activates several TF, such as myeloblastosis (MYB) and bZIP, which regulate genes responsible for UV protection, repair, and stress response (Song et al., 2023). Furthermore, UVR8 also integrates with other regulatory proteins such as WRKY36, BES1, BIM1, and MYB13, illustrating the multilayered transcriptional control of plant responses to UV-B (Table 7.1). This UVR8-COP1-HY5 axis is tightly regulated by negative feedback components, including RUP1, RUP2, and BBX22/24 (Table 7.1) and interacts with broader light signaling networks involving cryptochromes and phytochromes (Yang et al., 2018; Liang et al., 2018; Yang et al., 2020).

Plant responses to environmental signals are modulated via downstream signaling elements that influence gene expression and growth. COP1 acts as a modulator of photomorphogenesis together with de-etiolated 1 protein (DET1) and ultimately targets the degradation of functional proteins. COP1 and DET1 function as negative regulators of photomorphogenesis because their mutants display light-grown traits even in darkness. In UV-B signaling, COP1 functions differently by degrading HY5 in the absence of light but becoming inactive in light, allowing HY5 to activate gene expression. UVR8-COP1 interaction facilitates UV-B photomorphogenesis by inducing HY5, which then activates downstream genes (Han et al., 2020). Repressor of UV-B photomorphogenesis1/2 (RUP1/2) controls this UVR8 pathway by limiting its activity. These proteins share similarity with COP1 and contain WD40 repeats. Their expression is triggered by UV-B and depends on the UVR8-COP1-HY5 pathway. Different types of light can also induce RUP1/2, suggesting broader light response roles (Chen et al., 2022). CHS expression increases significantly in rup2 mutants after UV-B but is nearly absent in overexpression lines. The rup1rup2 double mutant shows stronger inhibition of hypocotyl growth under UV-B but adapts better to UV-B exposure. RUP 1/2 likely helps UVR8 revert to its inactive form after

TABLE 7.1 Genes associated with UV-B induced DNA damage and the repair system of plants.

Gene	Function	Target site	Impact on DDR	References
BBX22	Promotes antioxidant accumulation and DNA repair	DNA repair-related genes	Enhances DNA repair and reduces UV-B-induced DNA damage	Job et al. (2025)
MED17	Mediator subunit regulating transcription under light/darkness	DNA damage responses and light-responsive genes	Modulates DNA repair signaling under UV-B	Giustozzi et al. (2025)
E2Fd/DEL2	Cell cycle transcription factor	Cell cycle and DNA repair genes	Activates DDR genes after UV-B exposure	Simonelli et al. (2025)
E2Ff/DEL3	DDR transcription factor (shoot-specific)	Shoot DDR genes	Enhances UV-B-induced DNA repair in shoots	Simonelli et al. (2025).
UVR8	UV-B photoreceptor	HY5 and DNA repair pathway genes	Initiates UV-B signaling for DNA repair	Asik et al., (2025)
HY5	Transcription factor downstream of UVR8	Light and DNA repair genes	Induces photorepair gene expression under UV-B	Hu et al. (2023)
COP1	E3 ubiquitin ligase	HY5, photolyase regulators	Mediates HY5 stability affecting DNA repair	Chen et al. (2022)
PHR1 (Photolyase)	Directly repairs UV-induced CPDs	Cyclobutane pyrimidine dimers	Removes UV-B induced DNA lesions	Zgłobicki et al. (2024)
DDB1	DNA damage-binding protein	DNA damage recognition sites	Enhances nucleotide excision repair	Chen et al. (2022)
XPC	Damage recognition in NER	UV-induced bulky DNA lesions	Detects UV-B DNA lesions to trigger repair	Paul et al. (2021)

(*continued on next page*)

RAD1/10	Endonuclease in NER	UV-damaged DNA strands	Cleaves DNA at damage sites for excision repair	Yu et al. (2022)
RAD51	Homologous recombination protein	Double-strand breaks	Promotes DNA repair under UV-B stress	Bonilla et al. (2020)
ATM	DNA damage checkpoint kinase	DDR signaling cascades	Activates repair gene transcription under UV-B	Wagh et al. (2020)
ATR	Stress-responsive kinase	Replication stress & DNA lesions	Coordinates repair of UV-B-induced DNA lesions	Asik et al. (2025)
SOG1	Master regulator of DDR	Multiple DNA repair genes	Activates transcription of DNA repair pathways after UV-B	Sakamoto et al. (2021)

activation by binding to its C27 domain, even without UV-B (Wang et al., 2023). DET1 and COP1 also control DNA repair by regulating the expression of phosphate starvation response proteins1 (PHR1) and UVR3 through HY5/HYH degradation in darkness (Wang et al, 2023; Ali et al., 2025; Ren et al., 2025) (Fig. 7.1). DET1 is needed for NER by interacting with DNA damage sensors damaged DNA-binding protein 2 (DDB2) which detect DNA distortion or stalled transcription (Desgarnier et al., 2018). Two of these proteins associate with the CUL4-DDB1 complex, which binds DET1 during regular developmental progression and activates UV-B resistance. DDB2 degradation by this complex requires ATR, linking its control to DNA damage checkpoints. DNA damage-associated signaling and repair are critical for the development of plant and it is regulated by key photomorphogenesis components COP1 and DET1 (Castells et al., 2011).

7.4.1.1 *Mechanism of ultraviolet-B perception by UV resistance locus8*

Recent studies have revealed that UVR8 functions as a UV-B photoreceptor specifically under in vitro conditions. Initial studies showed its similarity to the human gene regulator of chromatin condensation (RCC1), which is a guanine nucleotide exchange factor for the G-protein Ran (Binkert et al., 2016). UVR8 forms a self-interacting dimer that splits into monomers upon UV-B exposure. Certain tryptophan residues are essential to maintain the dimer structure and serve as the chromophore for UV-B absorption. These tryptophan pyramids are formed between monomers and are stabilized by surrounding charged aromatic residues which further form salt bridges. UV-B light breaks these bridges, causing monomerization. The monomer is the active form, and it gets associated with COP1 to regulate genome expression. UVR8 consists of a β-propeller domain important for its interaction with COP1 after UV-B exposure, while its C-terminal region is essential for UV-B-specific signaling. A central cluster of tryptophans was required for both dimerization and binding of COP1; however, mutation studies showed that disrupting dimerization does not always block COP1 interaction. One mutant, $UVR8^{W285A}$, bound COP1 without forming dimers, while $UVR8^{W285F}$ formed dimers but did not respond to UV-B or interact with COP1. Although $UVR8^{W285A}$ was expected to show constant activation of UVR8-regulated genes like HY5 or CHS, plants with this mutation resembled UVR8 mutants and lacked typical UV-B responses (Cuypers et al., 2016). Structural analysis showed that $UVR8^{W285A}$ closely resembled the wild-type dimer, explaining its inactivity. However, later studies found that this mutant did induce constant photomorphogenic responses to UV-B. These included steady HY5 and CHS expression, possibly due to UVR8 binding to chromatin even without UV-B. It remains unclear whether this binding involves the dimer or monomer.

7.4.2 Systemic and UV resistance locus8-independent ultraviolet-B protective strategies

In addition to the UVR8-dependent signaling, plants possess complementary UV-B defense systems that operate through UVR8-independent pathways (Chen et al., 2022). Numerous pieces of evidence show that some UV-B induced physiological changes in plants occur without the involvement of UVR8, suggesting the presence of multiple perception mechanisms. These include MAPK cascades, particularly involving MAP kinase phosphatase 1 (MKP1), MPK3, and MPK6, which collectively help to modulate the acute UV-B response and are crucial for shoot protection (Zhang and Zhang, 2022; Chen et al., 2024; Xin et al., 2024). It remains unclear whether MAPK pathways are triggered by UV-induced DNA damage, formation of ROS, or

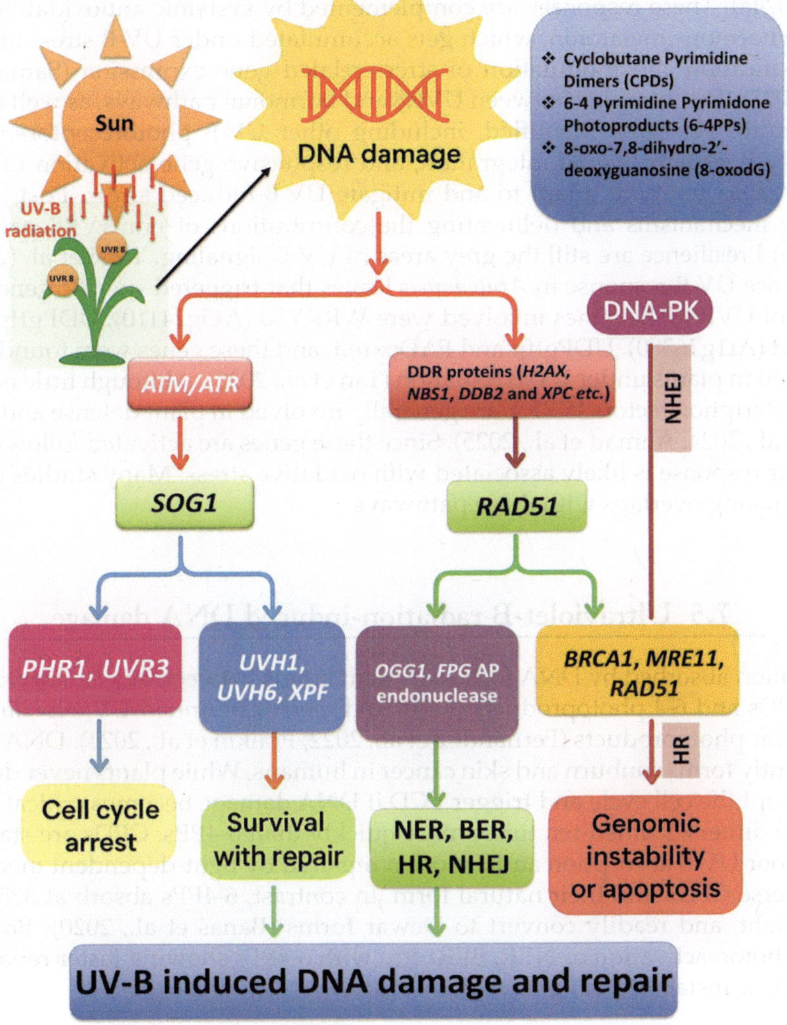

FIGURE 7.1 Schematic illustration of UV-B-induced DNA damage and repair. In darkness, photoreactivation genes PHR1 and UVR3 remain at basal expression due to promotion by HY5 and HYH and repression by DET1 and COP1. Under light, HY5 and HYH enhance PHR1 expression, while nuclear-localized DET1 continues to act as a strong inhibitor. Dual role for DET1 in controlling expression of light-responsive and DNA repair genes and the contribution of photomorphogenic regulators in the control of light-dependent DNA repair. Similarly, RAD51 repairs DSBs primarily via HR, while OGG1 and FPG repair oxidized bases via base excision repair (BER). Additionally, DNA protein kinases (DNA-PK) do quick error-prone repair through nonhomologous end joining (NHEJ), whereas RAD51 is central to HR, mediating strand invasion and exchange. BRCA1 promotes HR by antagonizing DNA-PK and facilitating end resection through its cooperation with the MRN complex (MRE11-RAD50-NBS1). MRE11, as part of MRN, initiates DNA end resection, creating single-stranded DNA that RAD51 loads onto, with BRCA1's assistance.

other signals. ATM and ATR kinases are traditionally associated with DDRs and play pivotal roles in signaling damage and coordinating repair, especially in roots (Martens et al., 2020; Mmbando, 2024a). These responses are complemented by systemic antioxidative mechanisms involving the hormone melatonin, which gets accumulated under UV-B stress and contributes to ROS detoxification and modulation of stress-related gene expression (Samanta and Roychoudhury, 2023). The interplay between UVR8 and hormonal pathways, as well as the possible role of others are yet to be identified, including other UV-B photoreceptors. This intricate network of UV-B sensing, signal integration, and responsive gene activation underscores the plant's remarkable ability to adapt to and mitigate UV-B-induced stress. Post-transcriptional UV-B sensing mechanisms and delineating the contributions of nonUVR8 pathways to UV acclimation and resilience are still the grey areas of UV-B signaling. Tossi et al. (2019) reported that high-fluence UV-B response in *Arabidopsis* leaves that triggered specific gene expression is independent of UVR8. The genes involved were WRKY30 (At5g24110), UDPgtfp (At1g05680), and FADoxred (At1g26380). UDPgtfp and FADoxred, and these genes were found to respond to H_2O_2 generated in plants under UV-B radiation (Tan et al., 2023). Although little is known about WRKY30, transcription factors WRKY are generally involved in plant defense and senescence of leaves (Cui et al., 2024; Ahmad et al., 2025). Since these genes are activated following high UV-B exposure, their response is likely associated with oxidative stress. Many studies have explored how UV-B signaling overlaps with these pathways.

7.5 Ultraviolet-B radiation-induced DNA damage

UV-B radiation absorbed by DNA forms covalent bonds between adjacent pyrimidine bases, producing CPDs and 6-4 photoproducts (6-4PPs) dimers. Continued UV exposure converts 6-4PPs into Dewar photoproducts (Fernández et al., 2022; Fraikin et al., 2024). DNA absorbs UV-B and subsequently forms sunburn and skin cancer in humans. While plants never develop cancer, UV-B can disrupt the cell cycle and trigger PCD if DNA damage becomes critical. CPDs are the most common dimers, formed ten times more quickly than 6-4PPs. CPDs are stable in natural light due to poor UV-B absorption and are often repaired by light-dependent mechanisms, and UV-C can reverse CPDs into their natural form. In contrast, 6-4PPs absorb at 325 nm, degrade faster in sunlight, and readily convert to Dewar forms (Banaś et al., 2020). Both dimers are removed by photoreactivation or NER, BER, etc., with 6-4PPs showing faster repair in the dark, likely due to their instability in light.

7.5.1 DNA damage responses

Plants utilize molecular mechanisms to detect, signal, and repair DNA damage caused by environmental and endogenous stresses. The DNA lesions, ranging from SSB and base modifications to DSBs, are caused by several genotoxic agents such as UV-B radiation, ultimately leading to oxidative stress. Due to their immobile nature, plants are especially vulnerable to such damage and rely on the DDR pathway to secure chromosomal integrity (Szurman-Zubrzycka et al., 2023). The DDR process includes a series of events starting from damage sensing by key proteins like ATM and ATR kinases, followed by the activation of signaling networks that halt the cell cycle and promote damage repair (Nisa et al., 2019). A pivotal regulator in

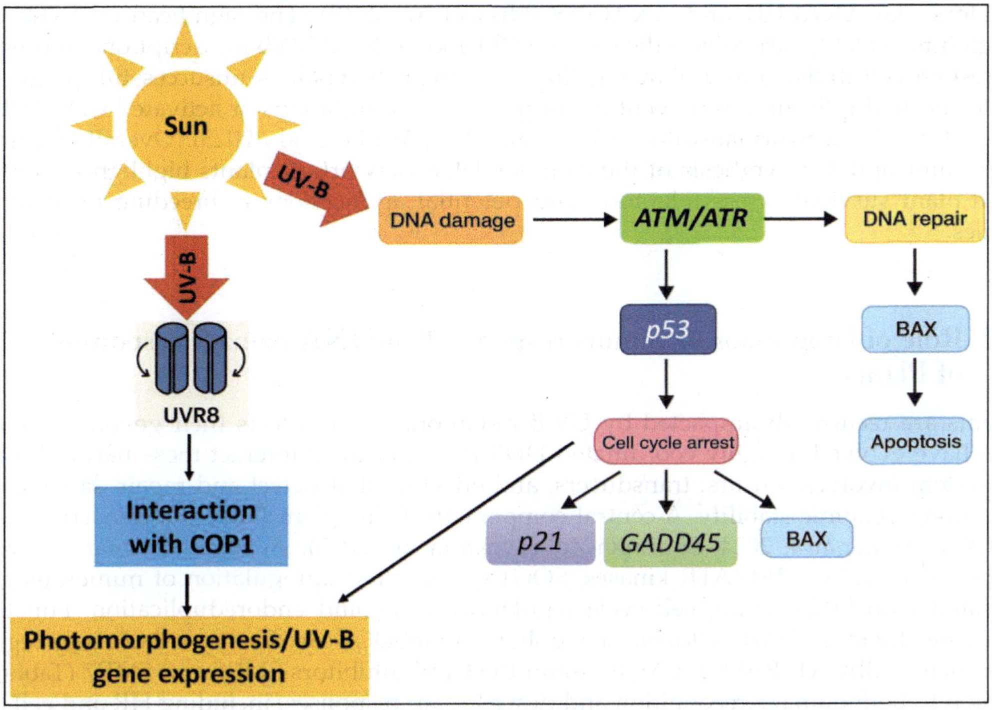

FIGURE 7.2 Diagram showing UV-B perception and downstream cellular responses. UV-B is detected by the UVR8 photoreceptor, which monomerizes and binds with COP1 to activate HY5/HYH-regulated genes. At the same time, DNA absorbs UV-B and forms photodimers (DNA damage). These lesions are repaired by photoreactivation or nucleotide excision repair, though repair is often incomplete. Unrepaired photodimers trigger cell cycle arrest through Ataxia Telangiectasia Mutated/Ataxia Telangiectasia and Rad3-related (ATM/ATR) signaling by activating the p53 gene. As well as apoptosis is induced by BAX (Bcl-2–associated X protein) in case the DNA repair system failed. Together, these pathways regulate photomorphogenesis and UV-B induced gene expression.

plant DDR is the suppressor of gamma response 1 (SOG1), a NAC transcription factor that is the functional analog of p53 in animals. SOG1 governs the transcription of genes responsible for DNA repair, endoreduplication, cell cycle arrest, and even programmed cell death (by differential expression of BAX and GADD, etc., genes) (Fig. 7.2) when damage is irreparable (Zaragoza et al., 2024; Wada et al., 2025). Additionally, a unique mediator complex, MED17, plays a regulatory role in plant responses to environmental signals, including UV-B radiation. Giustozzi et al. (2025) reported that MED17 of *Arabidopsis thaliana* is linked to the DDR pathway and aids in mediating transcriptional reprogramming when plants are under UV-B (Table 7.1). Mutants lacking MED17 display higher sensitivity to UV-B due to impaired repair mechanisms and altered expression of stress-responsive genes.

There are specific repair mechanisms engaged in the removal of heterogeneous DNA defects. For SSBs, plants utilize BER, NER, and MMR, while DDBs are resolved through HR or NHEJ (Szurman-Zubrzycka et al., 2023). *Arabidopsis thaliana* serves as a model system with many identified homologs of DNA repair genes conserved in eukaryotes, yet also featuring unique plant-specific adaptations such as specialized BER enzymes and plant-exclusive transcriptional regulators like ANAC044 and ANAC085 (Nisa et al., 2019). The significance of cell cycle checkpoints in DDR, particularly the roles of WEE1 kinase and MYB transcriptional repressors, in arresting cell division and allowing time for repair. If repair is unsuccessful, plants may initiate endoreduplication to prevent the propagation of mutations or activate PCD (Pedroza-Garcia et al., 2022; Szurman-Zubrzycka et al., 2023; Wada et al., 2025). Overall, a comprehensive and updated synthesis of the complex DDR network in plants highlights its crucial role in plant survival, stress tolerance, and potential applications in breeding resilient crop varieties.

7.5.2 Role of suppressor of gamma response 1 in DNA damage responses of Plants

Plants are recurrently impacted by UV-B radiation, which affects their genomic integrity. Plants have evolved a highly coordinated DDR pathway to counteract these harmful effects. This system involves sensors, transducers, and effectors that detect and repair damage, thus maintaining genomic stability. A central component of this plant DDR is a transcription factor SOG1 (Mahapatra, 2024; Szurman-Zubrzycka et al., 2025). Activated primarily through phosphorylation by ATM/ATR kinases, SOG1 governs the upregulation of numerous genes associated with DNA repair, cell cycle regulation, PCD, and endoreduplication. Functional studies, mostly in *Arabidopsis thaliana*, have shown that SOG1 can directly or indirectly regulate genes such as BRCA1, RAD51, CYCB1;1 and the CDK inhibitors SMR5 and SMR7 (Table 7.1), linking it to both damage recognition and downstream responses, including HR and cell cycle arrest (Ogita et al., 2018; Takahashi et al., 2019; Gentric et al., 2021).

The structural and biochemical attributes of SOG1 highlight its unique NAC domain and five C-terminal serine-glutamine (SQ) motifs essential for ATM-mediated phosphorylation (Sakamoto et al., 2021; Pagano et al., 2022; Szurman-Zubrzycka et al., 2023). These posttranslational modifications are critical for SOG1's ability to bind DNA and activate transcription. Moreover, SOG1 is involved in regulating the accumulation of transcriptional repressors such as Rep-MYBs via activation of other NAC family proteins like ANAC044 and ANAC085, ensuring tight control over the G2/M checkpoint (Takahashi et al., 2019; Carneiro et al., 2021; Shimotohno et al., 2021; de Luxán-Hernández et al., 2023). Further studies to unravel the evolutionary origins and functional diversity of SOG1 in other plant species, which could have profound implications for understanding plant resilience and genome stability in the face of climatic fluctuations. This knowledge can ultimately inform crop improvement strategies aimed at enhancing stress tolerance.

DNA damage detection is essential for plant resistance to environmental stresses like UV-B radiation. A conserved network of proteins identifies DNA damage and initiates signaling to halt the cell cycle, limiting mutation spread. This system is activated by protein kinases ATM and ATR in response to double or single-stranded breaks. UV-B exposure causes photodimer accumulation and stalled replication, activating DDRs. In plants, SOG1 mediates UV-B induced responses downstream of ATM and ATR (Fig. 7.3). Studies using *Arabidopsis* mutants showed that DNA-absorbed UV-B light inhibits hypocotyl growth via DNA damage rather than UVR8-dependent pathways. Mutants like xpf-3 and uvr1-1 were more sensitive to UV-B at lower fluences, suggesting photodimers drive this photomorphogenic response (Dotto and Casati, 2017). The xpf sog1-1 double mutant lacked hypersensitivity, confirming the involvement of SOG1 (Biever and Gardner, 2016). UV-B exposure also triggered cell cycle arrest, which is seen through

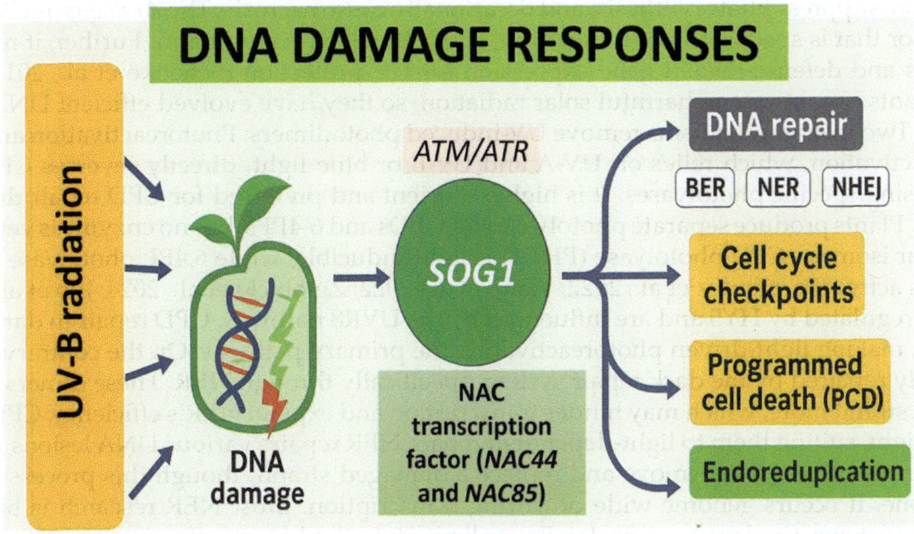

FIGURE 7.3 Overview of UV-B induced DNA damage responses signaling in plants. In response to DNA damage, ATM and/or ATR kinases are activated. ATM is activated upon DSBs, whereas ATR is activated upon SSBs and replication stress. They phosphorylate and activate SOG1 (both direct and indirect regulation through NAC transcription factor, like NAC44 and NAC85), which regulates genes responsible for cell cycle stoppage and DNA repair.

CYCB1;1-GUS accumulation, and acted independently of UVR8, which mainly regulates CHS induction. UVR8 mutants retained growth inhibition but lost CHS induction (Biever et al., 2014). These findings reveal a photomorphogenic pathway activated by photodimers independent of UVR8. UV-B activates signaling pathways, causing growth arrest or PCD (Fig. 7.1). UV-B can also induce endoreduplication instead of arrest, aiding stress tolerance. In Arabidopsis, this process depends on UVR8 and is evident in mutants like uvi4, which show reduced sensitivity due to extra endoreduplication cycles (Carneiro et al., 2021).

7.6 Damage repair

7.6.1 Light repair and dark repair

Plants employ two protective strategies against UV damage. One is light-dependent repair pathway called photoreactivation (PR), which is mostly recruited in low frequency damages as the main enzyme photolyase (activated by blue light ranging from 300 to 600 nm) directly targets DNA lesions (Ramírez-Gamboa et al., 2022; Zgłobicki et al., 2024). Plants protect themselves by accumulating UV-absorbing molecules (flavonoids and anthocyanins) and reflecting radiation through epicuticular waxes and cuticular protective features. And the other one is direct repair (DR), which is the simplest repair mechanism as it is accompanied by a single enzyme methylguaninemethyltransferase (MGMT) (Mostofa et al., 2018; Strzałka et al., 2020; Asik et al., 2025). UV-protective sunscreen production in *Arabidopsis* is transcriptionally repressed by AtMyB4, and mutants lacking this repressor display higher UV-B tolerance than wild types, as observed by Cathie Martin's team (John Innes Center). UV-B exposure in plants can trigger ROS generation, suppress photosynthesis, and occasionally cause necrosis. The *Arabidopsis* UVR8 acts as a sensor that is specific to UV-B radiation, and it binds with chromatin. Further, it regulates the stress and defense-related gene expression for UV protection (Schenke et al., 2011; Saini, 2019). Plants cannot escape harmful solar radiation, so they have evolved efficient DNA repair systems. Two main mechanisms remove UV-induced photodimers: Photoreactivation and NER. Photoreactivation, which relies on UV-A and UV-B or blue light, directly reverses CPDs and 6-4PPs using specific photolyases. It is highly efficient and preferred for CPD repair due to its accuracy. Plants produce separate photolyases for CPDs and 6-4PPs, but no enzyme is yet known for Dewar isomers. CPD photolyase (PHR1) is light-inducible, while 6,4PP photolyase (UVR3) is always active (Fernández et al., 2022; Wang et al., 2022; Zgłobicki et al., 2024; Job et al., 2025). Both are regulated by HY5 and are influenced by the UVR8 pathway. CPD repair in darkness is minimal, making light-driven photoreactivation the primary pathway. On the contrary, 6-4PPs are mostly repaired by the dark repair system, specifically through NER. These dimers disrupt the DNA strand more, which may hinder transcription and explain NER's efficiency. CPDs form more in light, suiting them to light-dependent repair. NER repairs various DNA lesions without light, using enzymes that remove and replace a damaged strand, though this process is more error-prone. It occurs genome-wide or during transcription. Most NER research is based on nonplant systems, so plant-specific details are limited. Both photoreactivation and NER protect plants from UV-B. Mutants lacking these enzymes are more UV-sensitive, often showing growth defects. Since NER proteins also act at stalled replication or transcription sites, their mutation causes broader stress sensitivity. HR may also remove CPDs but not 6-4PPs. HR activity increases with CPD levels and depends on light, though not on photolyase (Stein, 2017; Hung et al., 2020; Szurman-Zubrzycka et al., 2023). Since CPDs are more frequent, they are likely the main HR targets. A lack of HR data for 6-4PPs does not rule out its involvement. HR appears to act when other repair systems fail, as shown in mutants lacking NER components like CENTRIN2 (Raina et al., 2021).

Plants utilize a special mechanism in the absence of light to remove DNA damage caused by UV-B radiation, where photoreactivation is ineffective. UV-B incidence develops many types of DNA lesions, including CPDs, 6-4PPs, and oxidative modifications such as 8-oxo-dG (induced by OGG1). While photoreactivation via photolyases efficiently repairs CPDs and 6-4PPs under light conditions, its efficiency diminishes in low light or dark conditions, prompting reliance on alternate repair systems collectively referred to as dark repair. This includes NER, BER, MMR, HR, and NHEJ. The article elaborates on how these mechanisms contribute to genome stability by repairing lesions that photolyases cannot address, emphasizing the importance of NER for the repair of bulky UV-B induced lesions and BER in addressing oxidative damage.

Plants share homologs of DNA repair genes known from yeast and mammals, although certain components, like XPA in the NER pathway, are missing in plants, raising questions about plant-specific alternatives. The functional relevance of plant homologs such as AtDDB2, AtUVH, and AtCSA in NER and AtOGG1 and AtFPG1 in BER (Fig. 7.1) is substantiated by mutant studies showing hypersensitivity to UV-B and subsequently impaired repair efficiency (Strzałka et al., 2020). Moreover, dark repair pathways extend beyond the nucleus, with components of BER and HR identified in mitochondria and chloroplasts, indicating the universal need for genome integrity across organelles. Additionally, DNA damage tolerance pathways like translesion synthesis are utilized to bypass replication-blocking lesions. Overall, this comprehensive synthesis underscores the complexity and redundancy of plant DNA repair systems, revealing how plants maintain genome integrity under persistent UV stress, even in the absence of light, a critical insight in the context of increasing environmental challenges and potential agricultural applications (Khatib et al., 2024).

7.7 Conclusion and future perspective

Recent studies have greatly improved our knowledge regarding the responses of plants under UV-B radiation by identifying the UVR8 photoreceptor and its signaling pathways. Although UVR8 is essential for sensing UV-B, it does not explain the entire UV-B responses in plants. Plant responses to UV-B are diversified, suggesting the presence of multiple perception mechanisms. While this idea is partly accepted, past classifications of UV-B responses have restricted how we interpret them, often labelling them as either damage-related or photomorphogenic. It is unlikely that plants rely on a single UV-B photoreceptor, especially when they possess multiple or overlapping receptors for other light types. As discussed in this chapter, UV-B absorption triggers various processes and responses, including photomorphogenesis, that may involve mechanisms beyond the UVR8 pathway.

Earlier studies suggest that photomorphogenic responses (like inhibited hypocotyl growth in dark-grown *Arabidopsis* seedlings) can be triggered by UV-B and influence DNA damage in UVR8 independent manner. Hence, plants have multiple UV-B sensing pathways. Early hypotheses on UV-B sensing in plants were based on this similarity. In seedlings emerging from the soil, low flavonoid levels and light exposure leave them unprotected from UV-B, which may directly damage DNA and form photodimers. Pathways involving ATM/ ATR, or SOG1, are triggered, leading to growth inhibition through halting the cell cycle if the DNA repair system detects these photodimers. UVR8 helps later in development by promoting flavonoid production and UV-B protection, but the direct DNA-sensing pathway remains relevant as UV-B still reaches inner tissues. These two UV-B sensing mechanisms work together to shape plant growth. While plant DNA repair is better understood, many biochemical steps remain unclear. Enzymes in NER also serve other repair roles and might be key to initiating DNA damage signaling. Although they function broadly, their UV-B-specific activity is evident. Core proteins for DNA damage recognition known from yeast and humans are Rad23B, RPA, XPC, XPA, TFIIH, and CENTRIN2, which are mostly present in plants except XPA. Their roles in plants need more research. Clarifying how ATM and ATR contribute to UV-B signaling is essential, especially since persistent photodimers mainly block replication, implicating ATR. Understanding its role may reveal how UV-B damage leads to downstream effects like cell cycle arrest. UV-B may also reduce auxin transport, which affects both hypocotyl growth and the cell cycle. Measuring auxin movement could show whether flavonoids or other factors interfere with this process. Exploring UV-B in association with other environmental factors will improve our knowledge regarding plants' responses to varying intensities of UV-B.

References

Ahmad, Z., Ramakrishnan, M., Wang, C., Rehman, S., Shahzad, A., Wei, Q., 2025. Unravelling the role of WRKY transcription factors in leaf senescence: genetic and molecular insights. J. Adv. Res. 74, 191–206.

Aktary, Z., Petit, V., Berlin, I., Raymond, J., Berger, F., Charoenchon, N., Sage, E., Bertrand, J., Larue, L., 2024. UVB radiation suppresses Dicer expression through β-catenin. J. Cell Sci. 137 (22), jcs261978.

Ali, A., Santoro, P., Mori, J., Ferrante, A., Cocetta, G., 2025. Intermittent UV-B irradiation optimizes secondary metabolite production and growth in red rubin basil. Theor. Exp. Plant Physiol. 37 (1), 1–15.

Andrade-Lima, L.C., Veloso, A., Paulsen, M.T., Menck, C.F., Ljungman, M., 2015. DNA repair and recovery of RNA synthesis following exposure to ultraviolet light are delayed in long genes. Nucleic. Acids. Res. 43 (5), 2744–2756.

Ashokkumar, V., Thirugnanasambantham, K., Palanisamy, S., 2024. Unveiling UVB resilience in *Oryza sativa* L.: integrative analysis of physiological, molecular and microRNA responses. Plant Stress 12, 100495.

Asik, E., Kashif, S.Z., Oztas, O., 2025. Plant tolerance mechanisms to DNA-damaging UV stress. J. Exp. Bot. 77 (1), 27–50 eraf272.

Bais, A.F., McKenzie, R.L., Bernhard, G., Aucamp, P.J., Ilyas, M., Madronich, S., Tourpali, K., 2015. Ozone depletion and climate change: impacts on UV radiation. Photochem. Photobiol. Sci. 14 (1), 19–52.

Banaś, A.K., Zgłobicki, P., Kowalska, E., Dziga, D., Strzałka, W., 2020. All you need is light. Photorepair of UV-induced pyrimidine dimers. Genes 11 (11), 1304.

Barnes, P.W., Robson, T.M., Neale, P.J., Williamson, C.E., Zepp, R.G., Madronich, S., Wilson, S.R., Andrady, A.L., Heikkilä, A.M., Bernhard, G.H., Bais, A.F., 2022. Environmental effects of stratospheric ozone depletion, UV radiation, and interactions with climate change: UNEP environmental effects assessment panel, Update 2021. Photochem. Photobiol. Sci. 21 (3), 275–301.

Bera, K., Ball, K., Ghosh, S., Sadhukhan, S., Dutta, P., 2022. UV radiation: plant responses and an in-depth mechanism of sustainability under climatic extremities. https://www.researchsquare.com/article/rs-2219238/v1.

Bernhard, G.H., Bais, A.F., Aucamp, P.J., Klekociuk, A.R., Liley, J.B., McKenzie, R.L., 2023. Stratospheric ozone, UV radiation, and climate interactions. Photochem. Photobiol. Sci. 22 (5), 937–989.

Biever, J.J., Gardner, G., 2016. The relationship between multiple UV-B perception mechanisms and DNA repair pathways in plants. Environ. Exp. Bot. 124, 89–99.

Biever, J.J., Brinkman, D., Gardner, G., 2014. UV-B inhibition of hypocotyl growth in etiolated *Arabidopsis thaliana* seedlings is a consequence of cell cycle arrest initiated by photodimer accumulation. J. Exp. Bot. 65 (11), 2949–2961.

Binkert, M., Crocco, C.D., Ekundayo, B., Lau, K., Raffelberg, S., Tilbrook, K., Yin, R., Chappuis, R., Schalch, T., Ulm, R., 2016. Revisiting chromatin binding of the Arabidopsis UV-B photoreceptor UVR8. BMC Plant Biol. 16 (1), 42.

Bivehed, E., Hellman, B., Fan, Y., Haglöf, J., Buratovic, S., 2023. DNA integrity under alkaline conditions: an investigation of factors affecting the comet assay. Mutat. Res. Genet. Toxicol. Environ. Mutagen. 891, 503680.

Bonilla, B., Hengel, S.R., Grundy, M.K., Bernstein, K.A., 2020. RAD51 gene family structure and function. Annu. Rev. Genet. 54 (1), 25–46.

Brestic, M., Zivcak, M., Vysoka, D.M., Barboricova, M., Gasparovic, K., Yang, X., Kataria, S., 2023. Acclimation of photosynthetic apparatus to UV-B radiation. In: S. Kataria and M. Brestic (Eds.). UV-B Radiation and Crop Growth. Singapore: Springer Nature Singapore, pp. 223–260. doi:10.1007/978-981-19-3620-3_11.

Carneiro, A.K., Montessoro, P.D.F., Fusaro, A.F., Araújo, B.G., Hemerly, A.S., 2021. Plant CDKs—driving the cell cycle through climate change. Plants 10 (9), 1804.

Castells, E., Molinier, J., Benvenuto, G., Bourbousse, C., Zabulon, G., Zalc, A., Cazzaniga, S., Genschik, P., Barneche, F., Bowler, C., 2011. The conserved factor DE-ETIOLATED 1 cooperates with CUL4–DDB1DDB2 to maintain genome integrity upon UV stress. EMBO J. 30 (6), 1162–1172.

Chen, X., Zhang, R., Xin, Y., Yin, Y., Jia, X., Tian, G., Zhan, Y., Zeng, F., 2024. Multi-omics analysis reveals the perception and defense of *Betula platyphylla* Suk. to UltravioletB. Ind. Crops Prod. 214, 118565.

Cechin, I., Gonzalez, G.C., Corniani, N., Fumis, T.D.F., 2018. The sensitivity of sunflower (Helianthus annuus L.) plants to UV-B radiation is altered by nitrogen status. Cienc. Rural 48 (2), e20170369. doi:10.1590/0103-8478cr20170369.

Chen, Z., Dong, Y., Huang, X., 2022. Plant responses to UV-B radiation: signaling, acclimation and stress tolerance. Stress Biology 2 (1), 51.

Correa, M.D.S.S., Saavedra, M.E.R.R., Parra, E.A.E., Ontiveros, E.N., Flores, J.D.C.B., Montiel, J.G.O., Contreras, J.E.C., Urrutia, E.L., Acevedo, J.G.Á., Nopala, G.E.J., González, A.M.E., 2023. Ultraviolet radiation and its effects on plants. Abiotic Stress in Plants-Adaptations to Climate Change. IntechOpen doi:10.5772/intechopen.109474.

Cuadra, P., Guajardo, J., Carrasco-Orellana, C., Stappung, Y., Fajardo, V., Herrera, R., 2020. Differential expression after UV-B radiation and characterization of chalcone synthase from the Patagonian hairgrass Deschampsia antarctica. Phytochemistry 169, 112179.

Cui, X., Fan, X., Xu, S., Wang, S., Niu, F., Zhao, P., Yang, B., Liu, W., Guo, X., Jiang, Y.Q., 2024. WRKY47 transcription factor modulates leaf senescence through regulating PCD-associated genes in *Arabidopsis*. Plant Physiol. Biochem. 213, 108805.

Cunningham, N., Crestani, G., Csepregi, K., Coughlan, N.E., Jansen, M.A., 2024. Exploring the complexities of plant UV responses; distinct effects of UV-A and UV-B wavelengths on *Arabidopsis* rosette morphology. Photochem. Photobiol. Sci. 23 (7), 1251–1264.

Cuypers, A., Hendrix, S., Amaral dos Reis, R., De Smet, S., Deckers, J., Gielen, H., Jozefczak, M., Loix, C., Vercampt, H., Vangronsveld, J., Keunen, E., 2016. Hydrogen peroxide, signaling in disguise during metal phytotoxicity. Front. Plant Sci. 7, 470.

de Luxán-Hernández, C., Lohmann, J., Tranque, E., Chumova, J., Binarova, P., Salinas, J., Weingartner, M., 2023. MDF is a conserved splicing factor and modulates cell division and stress response in *Arabidopsis*. Life Sci. Allian. 6 (1), e202201507.

De, S., Jose, J., Pal, A., Roy Choudhury, S., Roy, S., 2022. Exposure to low UV-B dose induces DNA double-strand breaks mediated onset of endoreduplication in *Vigna radiata* (L.) R. Wilczek seedlings. Plant Cell Physiol. 63 (4), 463–483.

Desgarnier, M.C.D., Rochette, P.J., 2018. Enhancement of UVB-induced DNA damage repair after a chronic low-dose UVB pre-stimulation. DNA Repair (Amst.) 63, 56–62.

Dotto, M., Casati, P., 2017. Developmental reprogramming by UV-B radiation in plants. Plant Sci. 264, 96–101.

Dwivedi, S., Khan, E., Ahmad, I.Z., 2023. MicroRNA-mediated regulation of UV radiation stress response. In: Plant MicroRNAs and Stress Response. CRC Press, pp. 144–166. doi:10.1201/9781003322214-8.

Fang, F., Lin, L., Zhang, Q., Lu, M., Skvortsova, M.Y., Podolec, R., Zhang, Q., Pi, J., Zhang, C., Ulm, R., Yin, R., 2022. Mechanisms of UV-B light-induced photoreceptor UVR8 nuclear localization dynamics. New Phytol. 236 (5), 1824–1837.

Fernández, M.B., Latorre, L., Correa-Aragunde, N., Cassia, R., 2022. A putative bifunctional CPD/(6-4) photolyase from the cyanobacteria Synechococcus sp. PCC 7335 is encoded by a UV-B inducible operon: new insights into the evolution of photolyases. Front. Microbiol. 13, 981788.

Fraikin, G.Y., Belenikina, N.S., Rubin, A.B., 2024. Photochemical processes of cell DNA damage by UV radiation of various wavelengths: biological consequences. Mol. Biol. 58 (1), 1–16.

Gao, M., Kwaria, D., Norikane, Y., Yue, Y., 2023. Visible-light-switchable azobenzenes: molecular design, supramolecular systems, and applications. Nat. Sci. 3 (1), e220020.

Gentric, N., Genschik, P., Noir, S., 2021. Connections between the cell cycle and the DNA damage response in plants. Int. J. Mol. Sci. 22 (17), 9558.

Giustozzi, M., Freytes, S.N., Ferreyra, M.L.F., Cerdán, P., Casati, P., 2025. Mediator subunit 17 regulates light and darkness responses in Arabidopsis plants. Plant Sci. 350, 112285. doi:10.1016/j.plantsci.2024.112285.

Gustavsson, M.A., 2023. Investigating the effects of UV-B and temperature on survival and stomatal physiology in *Arabidopsis thaliana* seedlings. University of Bristol.

Han, X., Huang, X., Deng, X.W., 2020. The photomorphogenic central repressor COP1: conservation and functional diversification during evolution. Plant Commun. 1 (3), 100044.

Hu, S., Chen, Y., Qian, C., Ren, H., Liang, X., Tao, W., Chen, Y., Wang, J., Dong, Y., Han, J., Ouyang, X., 2024. Nuclear accumulation of rice UV-B photoreceptors is UV-B-and OsCOP1-independent for UV-B responses. Nat. Commun. 15 (1), 6396.

Hu, Y., Rosado, D., Lindbäck, L.N., Micko, J., Pedmale, U.V., 2023. Cryptochromes and UBP12/13 deubiquitinases antagonistically regulate DNA damage response in *Arabidopsis*. Biorxiv doi:10.1101/2023.01.15.524001.

Huang, R., Zhou, P.K., 2021. DNA damage repair: historical perspectives, mechanistic pathways and clinical translation for targeted cancer therapy. Signal Transduct. Target. Ther. 6 (1), 254.

Hung, K.F., Sidorova, J.M., Nghiem, P., Kawasumi, M., 2020. The 6-4 photoproduct is the trigger of UV-induced replication blockage and ATR activation. Proc. Natl. Acad. Sci. 117 (23), 12806–12816.

Jenkins, G.I., 2014. Structure and function of the UV-B photoreceptor UVR8. Curr. Opin. Struct. Biol. 29, 52–57.

Jha, A.N., 2023. Measuring DNA modifications with the comet assay: a compendium of protocols. Nat. Protoc. doi:10.1038/s41596-022-00754-y. www.nature.com/nprot.

Job, N., Dwivedi, S., Lingwan, M., Datta, S., 2025. BBX22 enhances the accumulation of antioxidants to inhibit DNA damage and promotes DNA repair under high UV-B. Physiol. Plant. 177 (1), e70038.

Jumrani, K., Joshi-Paneri, J., 2023. Major Influence on Photosynthetic Apparatus Under UV-B Exposure. In: UV-B Radiation and Crop Growth. Springer Nature Singapore, Singapore, pp. 55–70.

Kataria, S., Jajoo, A., Guruprasad, K.N., 2014. Impact of increasing Ultraviolet-B (UV-B) radiation on photosynthetic processes. J. Photochem. Photobiol. B 137, 55–66.

Kciuk, M., Marciniak, B., Mojzych, M., Kontek, R., 2020. Focus on UV-induced DNA damage and repair—disease relevance and protective strategies. Int. J. Mol. Sci. 21 (19), 7264.

Khatib, J.B., Nicolae, C.M., Moldovan, G.L., 2024. Role of translesion DNA synthesis in the metabolism of replication-associated nascent strand gaps. J. Mol. Biol. 436 (1), 168275.

Klem, K., Holub, P., Štroch, M., Nezval, J., Špunda, V., Tříska, J., Jansen, M.A., Robson, T.M., Urban, O., 2015. Ultraviolet and photosynthetically active radiation can both induce photoprotective capacity allowing barley to overcome high radiation stress. Plant Physiol. Biochem. 93, 74–83. doi:10.1016/j.plaphy.2015.01.001.

Kumari, P., Thakur, R., Singh, N., Rastogi, A., Yadav, S., 2023. Relationships of oxidative stress and ultraviolet-B radiation in plants. In: UV-B Radiation and Crop Growth. Springer Nature Singapore, Singapore, pp. 277–291.

Kurinjimalar, C., Kavitha, G., Thevanathan, R., Kulandaivelu, G., Rengasamy, R., 2019. Impact of ultraviolet-B radiation on growth and biochemical composition of *Botryococcus braunii* Kutz. Curr. Sci. 116 (1), 89–95. https://www.jstor.org/stable/27137794.

Laad, P., Patel, P., Guruprasad, K.N., 2023. UVR8 signaling, mechanism, and integration with other pathways. In: UV-B Radiation and Crop Growth. Springer Nature Singapore, Singapore, pp. 193–221.

Lee, J.H., Tanaka, S., Goto, E., 2022. Growth and biosynthesis of phenolic compounds of canola (*Brassica napus* L.) to different ultraviolet (UV)-B wavelengths in a plant factory with artificial light. Plants 11 (13), 1732.

Li, L., Li, Q., Huang, L., Wang, Q., Zhu, A., Xu, J., Liu, Z., Li, H., Shi, L., Li, R., Azari, M., 2020. Air quality changes during the COVID-19 lockdown over the yangtze river delta region: an insight into the impact of human activity pattern changes on air pollution variation. Sci. Total Environ. 732, 139282.

Li, M., Kim, C., 2022. Chloroplast ROS and stress signaling. Plant Communications 3 (1), 100264.

Liang, T., Mei, S., Shi, C., Yang, Y., Peng, Y., Ma, L., Wang, F., Li, X., Huang, X., Yin, Y., Liu, H., 2018. UVR8 interacts with BES1 and BIM1 to regulate transcription and photomorphogenesis in *Arabidopsis*. Dev. Cell 44 (4), 512–523.

Liaqat, W., Altaf, M.T., Barutçular, C., Nawaz, H., Ullah, I., Basit, A., Mohamed, H.I., 2024. Ultraviolet-B radiation in relation to agriculture in the context of climate change: a review. Cereal Res. Commun. 52 (1), 1–24.

Liu, W., Giuriani, G., Havlikova, A., Li, D., Lamont, D.J., Neugart, S., Velanis, C.N., Petersen, J., Hoecker, U., Christie, J.M., Jenkins, G.I., 2024. Phosphorylation of *Arabidopsis* UVR8 photoreceptor modulates protein interactions and responses to UV-B radiation. Nat. Commun. 15 (1), 1221.

Ma, H., Song, T., Wang, T., Wang, S., 2016. Influence of human p53 on plant development. PLoS One 11 (9), e0162840.

Mahapatra, K., 2024. Unveiling the structure and interactions of SOG1, a NAC domain transcription factor: an in-silico perspective. J. Genetic. Engineering. Biotechnol. 22 (1), 100333.

Martens, M., Horres, R., Wendeler, E., Reiss, B., 2020. The importance of ATM and ATR in physcomitrella patens DNA damage repair, development, and gene targeting. Genes 11 (7), 752.

Mathur, S., Bheemanahalli, R., Jumaa, S.H., Kakar, N., Reddy, V.R., Gao, W., Reddy, K.R., 2024. Impact of ultraviolet-B radiation on early-season morpho-physiological traits of indica and japonica rice genotypes. Front. Plant Sci. 15, 1369397.

Mavrič Čermelj, A., Golob, A., Vogel-Mikuš, K., Germ, M., 2021. Silicon mitigates negative impacts of drought and UV-b radiation in plants. Plants 11 (1), 91.

Mishra, P., Prasad, S.M., 2021. Low dose UV-B radiation induced mild oxidative stress impact on physiological and nutritional competence of Spirulina (Arthrospira) species. Plant Stress 2, 100039.

Miyazaki, K., Bowman, K., Sekiya, T., Takigawa, M., Neu, J.L., Sudo, K., Osterman, G., Eskes, H., 2021. Global tropospheric ozone responses to reduced NO x emissions linked to the COVID-19 worldwide lockdowns. Sci. Adv. 7 (24), eabf7460.

Mmbando, G.S., Ngongolo, K., 2024. Environmental & health impacts of ultraviolet radiation: current trends and mitigation strategies. Discover Sustainability 5 (1), 436.

Mmbando, G.S., 2024a. The recent possible strategies for breeding ultraviolet-B-resistant crops. Heliyon 10 (6), e27806.

Mmbando, G.S., 2024b. Variation in ultraviolet-B (UV-B)-induced DNA damage repair mechanisms in plants and humans: an avenue for developing protection against skin photoaging. Int. J. Radiat. Biol. 100 (11), 1505–1516.

Mmbando, G.S., Ando, S., Takahashi, H., Hidema, J., 2023. High ultraviolet-B sensitivity due to lower CPD photolyase activity is needed for biotic stress response to the rice blast fungus, Magnaporthe oryzae. Photochem. Photobiol. Sci. 22 (6), 1309–1321.

Mostofa, A.G.M., Punganuru, S.R., Madala, H.R., Srivenugopal, K.S., 2018. S-phase specific downregulation of human O6-methylguanine DNA methyltransferase (MGMT) and its serendipitous interactions with PCNA and p21cip1 proteins in glioma cells. Neoplasia 20 (4), 305–323.

Muruzabal, D., Collins, A., Azqueta, A., 2021. The enzyme-modified comet assay: past, present and future. Food Chem. Toxicol. 147, 111865.

Neale, R.E., Barnes, P.W., Robson, T.M., Neale, P.J., Williamson, C.E., Zepp, R.G., Wilson, S.R., Madronich, S., Andrady, A.L., Heikkilä, A.M., Bernhard, G.H., 2021. Environmental effects of stratospheric ozone depletion, UV radiation, and interactions with climate change: UNEP environmental effects assessment panel, Update 2020. Photochem. Photobiol. Sci. 20 (1), 1–67.

Neugart, S., Steininger, V., Fernandes, C., Martínez-Abaigar, J., Núñez-Olivera, E., Schreiner, M., Strid, Å., Viczián, A., Albert, A., Badenes-Pérez, F.R., Castagna, A., 2024. A synchronized, large-scale field experiment using *Arabidopsis thaliana* reveals the significance of the UV-B photoreceptor UVR8 under natural conditions. Plant, Cell Environ. 47 (10), 4031–4047.

Nisa, M.U., Huang, Y., Benhamed, M., Raynaud, C., 2019. The plant DNA damage response: signaling pathways leading to growth inhibition and putative role in response to stress conditions. Front. Plant Sci. 10, 653.

Ogita, N., Okushima, Y., Tokizawa, M., Yamamoto, Y.Y., Tanaka, M., Seki, M., Makita, Y., Matsui, M., Okamoto-Yoshiyama, K., Sakamoto, T., Kurata, T., 2018. Identifying the target genes of SUPPRESSOR OF GAMMA RESPONSE 1, a master transcription factor controlling DNA damage response in *Arabidopsis*. Plant J. 94 (3), 439–453.

Ozel, H.B., Abo Aisha, A.E.S., Cetin, M., Sevik, H., Zeren Cetin, I., 2021. The effects of increased exposure time to UV-B radiation on germination and seedling development of Anatolian black pine seeds. Environ. Monit. Assess. 193 (7), 388.

Pagano, A., Gualtieri, C., Mutti, G., Raveane, A., Sincinelli, F., Semino, O., Balestrazzi, A., Macovei, A., 2022. Identification and characterization of SOG1 (Suppressor of Gamma Response 1) homologues in plants using data mining resources and gene expression profiling. Genes 13 (4), 667.

Paul, D., Mu, H., Tavakoli, A., Dai, Q., Chakraborty, S., He, C., Ansari, A., Broyde, S., Min, J.H., 2021. Impact of DNA sequences on DNA 'opening'by the Rad4/XPC nucleotide excision repair complex. DNA Repair (Amst.) 107, 103194.

Pedroza-Garcia, J.A., Xiang, Y., De Veylder, L., 2022. Cell cycle checkpoint control in response to DNA damage by environmental stresses. Plant J. 109 (3), 490–507.

Piccini, C., Cai, G., Dias, M.C., Romi, M., Longo, R., Cantini, C., 2020. UV-B radiation affects photosynthesis-related processes of two Italian Olea europaea (L.) varieties differently. Plants 9 (12), 1712.

Rai, N., Morales, L.O., Aphalo, P.J., 2021. Perception of solar UV radiation by plants: photoreceptors and mechanisms. Plant Physiol. 186 (3), 1382–1396.

Raina, A., Sahu, P.K., Laskar, R.A., Rajora, N., Sao, R., Khan, S., Ganai, R.A., 2021. Mechanisms of genome maintenance in plants: playing it safe with breaks and bumps. Front. Genet. 12, 675686.

Ramírez-Gamboa, D., Díaz-Zamorano, A.L., Meléndez-Sánchez, E.R., Reyes-Pardo, H., Villaseñor-Zepeda, K.R., López-Arellanes, M.E., Sosa-Hernández, J.E., Coronado-Apodaca, K.G., Gámez-Méndez, A., Afewerki, S., Iqbal, H.M., 2022. Photolyase production and current applications: a review. Molecules 27 (18), 5998.

Ren, J., Li, T., Guo, M., Zhang, Q., Ren, S., Wang, L., Wu, Q., Niu, S., Yi, K., Ruan, W., 2025. A PHR-dependent reciprocal antagonistic interplay between UV response and P-deficiency adaptation in plants. Plant Communications 6 (1), 101140.

Reyes, T.H., Pompeiano, A., Ranieri, A., Volterrani, M., Guglielminetti, L., Scartazza, A., 2020. Photosynthetic performance of five cool-season turfgrasses under UV-B exposure. Plant Physiol. Biochem. 151, 181–187. doi:10.1016/j.plaphy.2020.03.025.

Rodríguez-Calzada, T., Qian, M., Strid, A., Neugart, S., Schreiner, M., Torres-Pacheco, I., Guevara-González,, R.G., 2019. Effect of UV-B radiation on morphology, phenolic compound production, gene expression, and subsequent drought stress responses in chili pepper (*Capsicum annuum* L.). Plant Physiol. Biochem. 134, 94–102. doi:10.1016/j.plaphy.2018.06.025.

Roro, A.G., Dukker, S.A., Melby, T.I., Solhaug, K.A., Torre, S., Olsen, J.E., 2017. UV-B-induced inhibition of stem elongation and leaf expansion in pea depends on modulation of gibberellin metabolism and intact gibberellin signalling. J. Plant Growth Regul. 36 (3), 680–690.

Saini, P., 2019. Protein-protein interaction network study of shoot stem cell niche derived transcription factors revealed the contrasting roles of ELONGATED HYPOCOTYL5 and DEWAX in UV-B stress in *Arabidopsis thaliana*. IISERM.

Sakamoto, A.N., Sakamoto, T., Yokota, Y., Teranishi, M., Yoshiyama, K.O., Kimura, S., 2021. SOG1, a plant-specific master regulator of DNA damage responses, originated from nonvascular land plants. Plant Direct 5 (12), e370.

Samanta, S., Roychoudhury, A., 2023. Crosstalk of melatonin with major phytohormones and growth regulators in mediating abiotic stress tolerance in plants. S. Afr. J. Bot. 163, 201–216.

Santin, M., Simoní, S., Vangelisti, A., Giordani, T., Cavallini, A., Mannucci, A., Ranieri, A., Castagna, A., 2023. Transcriptomic analysis on the Peel of UV-B-exposed peach fruit reveals an Upregulation of phenolic-and UVR8-related pathways. Plants 12 (9), 1818.

Schenke, D., Boettcher, C., Scheel, D., 2011. Crosstalk between abiotic ultraviolet-B stress and biotic (flg22) stress signalling in *Arabidopsis* prevents flavonol accumulation in favor of pathogen defence compound production. Plant, Cell Environ. 34 (11), 1849–1864.

Schuch, A.P., Moreno, N.C., Schuch, N.J., Menck, C.F.M., Garcia, C.C.M., 2017. Sunlight damage to cellular DNA: focus on oxidatively generated lesions. Free Radical Biol. Med. 107, 110–124.

Sharma, A., Shahzad, B., Rehman, A., Bhardwaj, R., Landi, M., Zheng, B., 2019. Response of phenylpropanoid pathway and the role of polyphenols in plants under abiotic stress. Molecules 24 (13), 2452.

Sharma, S., Zhang, M., Gao, J., Zhang, H., Kota, S.H., 2020. Effect of restricted emissions during COVID-19 on air quality in India. Sci. Total Environ. 728, 138878.

Shen, L., Yin, X., 2022. Solar spectral management for natural photosynthesis: from photonics designs to potential applications. Nano Convergence 9 (1), 36.

Shimotohno, A., Aki, S.S., Takahashi, N., Umeda, M., 2021. Regulation of the plant cell cycle in response to hormones and the environment. Annu. Rev. Plant Biol. 72 (1), 273–296.

Simonelli, L., Gómez, M.S., Casati, P., 2025. Similar but different: unveiling the roles of atypical E2Fd/DEL2 and E2Ff/DEL3 transcription factors during the DNA damage response after UV-B exposure in Arabidopsis. Plant J. 123 (1), e70348.

Singh, P., Choudhary, K.K., 2022. Potential impacts of ultraviolet-B radiation on crop plants and its consequences. In: Molecular Response and Genetic Engineering for Stress in Plants, Volume 1: Abiotic stress. IOP Publishing, Bristol, UK, pp. 1–29.

Singh, P., Choudhary, K.K., 2025a. UV-B orchestration of growth, yield and grain quality traits highlights modifications of source-to-sink relationship in pearl millet cultivars. Physiol. Plant. 177 (2), e70141.

Singh, P., Choudhary, K.K., 2025b. Mechanistic insights on physiological, biochemical, and metabolite profiling of pearl millet cultivars focusing bioactive compounds under elevated UV-B radiation. Plant Physiol. Biochem. 222, 109682.

Singh, P., Choudhary, K.K., 2025c. Research hotspots and global trends in plant responses under UV-B radiation. Discover Plants 2 (1), 142.

Singh, P., Singh, A., Choudhary, K.K., 2023. Revisiting the role of phenylpropanoids in plant defense against UV-B stress. Plant Stress 7, 100143.

Song, Y., Ma, B., Guo, Q., Zhou, L., Zhou, X., Ming, Z., You, H., Zhang, C., 2023. MYB pathways that regulate UV-B-induced anthocyanin biosynthesis in blueberry (Vaccinium corymbosum). Front. Plant Sci. 14, 1125382.

Stein, A.I., 2017. The Role of Homologous Recombination in the Maintenance and Repair of the Mitochondrial Genome. University of Rochester, Rochester, New York.

Strzałka, W., Zgłobicki, P., Kowalska, E., Dziga, D., Banaś, A.K., 2020. The dark side of UV-induced DNA lesion repair. Genes 11 (12), 1450.

Sun, Q., Liu, M., Cao, K., Xu, H., Zhou, X., 2022. UV-B irradiation to amino acids and carbohydrate metabolism in *Rhododendron chrysanthum* leaves by coupling deep transcriptome and metabolome analysis. Plants 11 (20), 2730.

Sun, Y., Zhang, X., Zhang, L., Huang, Y., Yang, Z., Montagnes, D., 2020. UVB radiation suppresses antigrazer morphological defense in Scenedesmus obliquus by inhibiting algal growth and carbohydrate-regulated gene expression. Environ. Sci. Technol. 54 (7), 4495–4503.

Szurman-Zubrzycka, M., Jędrzejek, P., Szarejko, I., 2023. How do plants cope with DNA damage? A concise review on the DDR pathway in plants. Int. J. Mol. Sci. 24 (3), 2404.

Szurman-Zubrzycka, M., Kocjan, A., Spałek, E., Gajecka, M., Jędrzejek, P., Nawrot, M., Szarejko, I., Kwasniewska, J., 2025. To divide or not to divide? NAC8 (SOG1) as a key regulator of DNA damage response in barley (Hordeum vulgare L.). DNA Repair (Amst.) 146, 103810.

Takahashi, N., Ogita, N., Takahashi, T., Taniguchi, S., Tanaka, M., Seki, M., Umeda, M., 2019. A regulatory module controlling stress-induced cell cycle arrest in *Arabidopsis*. eLife 8, e43944.

Takeda, J., 2021. Molecular mechanisms of UVR8-mediated photomorphogenesis derived from revaluation of action spectra. Photochem. Photobiol. 97 (5), 903–910.

Takshak, S., Agrawal, S.B., 2019. Defense potential of secondary metabolites in medicinal plants under UV-B stress. J. Photochem. Photobiol. B 193, 51–88.

Tan, Y., Duan, Y., Chi, Q., Wang, R., Yin, Y., Cui, D., Li, S., Wang, A., Ma, R., Li, B., Jiao, Z., 2023. The role of reactive oxygen species in plant response to radiation. Int. J. Mol. Sci. 24 (4), 3346.

Tevini, M., 2023. UV-B effects on plants. In: Agrawal, S.B., Agrawal, M. (Eds.), Environmental Pollution and Plant Responses. Routledge/CRC Press, Boca Raton, FL, USA, pp. 83–97.

Tian, X., Hu, M., Yang, J., Yin, Y., Fang, W., 2024. Ultraviolet-B radiation stimulates flavonoid biosynthesis and antioxidant systems in buckwheat sprouts. Foods 13 (22), 3650.

Tossi, V.E., Regalado, J.J., Iannicelli, J., Laino, L.E., Burrieza, H.P., Escandón, A.S., Pitta-Álvarez, S.I., 2019. Beyond arabidopsis: differential UV-B response mediated by UVR8 in diverse species. Front. Plant Sci. 10, 780.

Tripathi, D.K., Singh, S., Singh, V.P., Prasad, S.M., Dubey, N.K., Chauhan, D.K., 2017. Silicon nanoparticles more effectively alleviated UV-B stress than silicon in wheat (Triticum aestivum) seedlings. Plant Physiol. Biochem. 110, 70–81.

Tyutereva, E.V., Strizhenok, A.D., Kiseleva, E.I., Voitsekhovskaja, O.V., 2024. Comet assay: multifaceted options for studies of plant stress response. Horticulturae 10 (2), 174.

United Nation (UN). 2021. Rebuilding the ozone layer: how the world came together for the ultimate repair job. https://www.unep.org/news-and-stories/story/rebuilding-ozone-layer-how-world-came-together-ultimate-repair-job.

Wada, T., Sakamoto, A.N., Umeda, M., Takahashi, N., 2025. Suppresor of gamma response 1 promotes early onset of endoreplication upon DNA double-strand breaks by inducing CCS52A1 expression in *Arabidopsis* roots. J. Plant Res. 38 (4), 679–693.

Wagh, V., Joshi, P., Jariyal, H., Chauhan, N., 2020. ATM and ATR checkpoint kinase pathways: a concise review. Advances in Human Biology 10 (2), 51–59.

Wang, L., Wang, Y., Chang, H., Ren, H., Wu, X., Wen, J., Guan, Z., Ma, L., Qiu, L., Yan, J., Zhang, D., 2023. RUP2 facilitates UVR8 redimerization via two interfaces. Plant Commun. 4 (1), 100428.

Wang, Z., Li, Z., Lei, Y., Liu, Y., Feng, Y., Chen, D., Ma, S., Xiao, Z., Hu, M., Deng, J., Wang, Y., 2022. Recombinant photolyase-thymine alleviated UVB-induced photodamage in mice by repairing CPD photoproducts and ameliorating oxidative stress. Antioxidants 11 (12), 2312.

Wijewardana, C., Henry, W.B., Gao, W., Reddy, K.R., et al., 2016. Interactive effects on CO_2, drought, and ultraviolet-B radiation on maize growth and development. J. Photochem. Photobiol. B: Biol. 160, 198–209. doi:10.1016/j.jphotobiol.2016.04.004.

Xiao, S., Li, D., Tang, Z., Wei, H., Zhang, Y., Yang, J., Zhao, C., Liu, Y., Wang, W., 2023. Supplementary UV-B radiation effects on photosynthetic characteristics and important secondary metabolites in *Eucommia ulmoides* leaves. Int. J. Mol. Sci. 24 (9), 8168.

Xin, J., Li, C., Liu, X., Shi, X., Sun, Y., Shang, J.X., 2024. Emerging functions of protein tyrosine phosphatases in plants. Int. J. Mol. Sci. 25 (22), 12050.

Yadav, A., Singh, D., Lingwan, M., Yadukrishnan, P., Masakapalli, S.K., Datta, S., 2020. Light signaling and UV-B-mediated plant growth regulation. J. Integr. Plant Biol. 62 (9), 1270–1292. doi:10.1111/jipb.12932.

Yang, Y., Liang, T., Zhang, L., Shao, K., Gu, X., Shang, R., Shi, N., Li, X., Zhang, P., Liu, H., 2018. UVR8 interacts with WRKY36 to regulate HY5 transcription and hypocotyl elongation in *Arabidopsis*. Nat. Plants 4 (2), 98–107.

Yang, Y., Zhang, L., Chen, P., Liang, T., Li, X., Liu, H., 2020. UV-B photoreceptor UVR8 interacts with MYB73/MYB77 to regulate auxin responses and lateral root development. EMBO J. 39 (2), e101928.

Yu, L., Xu, S.Y., Luo, X.C., Ying, S.H., Feng, M.G., 2022. Rad1 and Rad10 tied to photolyase regulators protect insecticidal fungal cells from solar UV damage by photoreactivation. J. Fungi 8 (11), 1124.

Zaragoza, J.Z., Klap, K., Heidstra, R., Zhou, W., Scheres, B., 2024. The dual role of the retinoblastoma-related protein in the DNA damage response is coordinated by the interaction with LXCXE-containing proteins. Plant J. 118 (4), 1194–1206.

Zgłobicki, P., Hermanowicz, P., Kłodawska, K., Łabuz, J., Grzyb, J., Dutka, M., Kowalska, E., Jawor, J., Leja, K., Banaś, A.K., 2024. The photoreactivation of 6– 4 photoproducts in chloroplast and nuclear DNA depends on the amount of the *Arabidopsis* UV repair defective 3 protein. BMC Plant Biol. 24 (1), 723.

Zhang, M., Zhang, S., 2022. Mitogen-activated protein kinase cascades in plant signaling. J. Integr. Plant Biol. 64 (2), 301–341.

Zheng, C., Collins, A., Brunborg, G., van Schooten, F.J., Nordengen, A.L., Shaposhnikov, S., Godschalk, R., 2023. Assay conditions for estimating differences in base excision repair activity with Fpg-modified comet assay. Cell Biol. Toxicol. 39 (6), 2775–2786.

CHAPTER

8

Ionizing radiation-induced mammalian cell damage and repairs

Shruti Mishra[a,b], *Himani Tewari*[a,b], *Sindhuja Singh*[a,b] *and Swathi Kota*[a,b]

[a]Molecular Biology Division, Bhabha Atomic Research Centre, Mumbai, Maharashtra, India [b]Life Sciences, Homi Bhabha National Institute, Mumbai, Maharashtra, India

8.1 Introduction

Ionizing radiation (IR) is a form of high-energy radiation that can displace tightly bound electrons from atoms, leading to the creation of ions. This characteristic allows IR to engage with biological molecules and disturb cellular structures, particularly DNA. IR can originate from natural sources, such as cosmic rays, solar radiation, and radioactive materials present in the earth's crust, as well as from human activities such as medical imaging (e.g., X-rays, CT scans), cancer treatment, and specific industrial uses. Because of its high energy, IR can penetrate bodily tissues and cause a range of molecular alterations (Hall, 1989).

Mammalian cells are continuously exposed to various forms of IR. The capacity of IR to induce diverse DNA lesions, ranging from subtle base modifications to highly disruptive double-strand breaks (DSBs), presents a fundamental challenge to genomic integrity. In response to these genotoxic threats, mammalian cells have evolved remarkably sophisticated and interconnected defense systems collectively known as the DNA damage response (DDR). The DDR functions as a surveillance and repair system, detecting DNA damage, arresting the cell cycle to prevent replication or division of damaged DNA, and activating specific repair pathways. If the damage is irreparable, the DDR can direct cells to undergo programmed cell death (apoptosis) or enter a permanent state of growth arrest (senescence), thereby preventing the propagation of mutations.

Understanding how IR damages cells and how the DDR mitigates its effects is critical for several reasons. It deepens our knowledge of fundamental cellular processes and helps explain the molecular basis of cancer and various genetic disorders linked to defective DNA repair.

Genome Damage: Repair and Adaptation Across Life.
DOI: https://doi.org/10.1016/B978-0-443-42857-9.00013-X

Moreover, it helps in the development of novel therapeutic approaches, including radiotherapy enhancement, targeted DDR inhibitors, and protective measures against radiation exposure. Continued research in this area is essential for improving public health and expanding the therapeutic applications of radiation in medicine.

8.2 Effects of ionizing radiation on DNA

Gamma rays, X-rays, high-energy UV (UVC), alpha particles, beta particles, and neutrons are the different kinds of IR. The standard international unit for measuring absorbed radiation dose is the **gray (Gy),** which is defined as the absorbed dose of 1 J of radiation energy per kilogram of tissue. When IR interacts with biological tissues, it can inflict DNA damage through two primary mechanisms:

- **Direct Damage:** The radiation directly strikes and alters the DNA molecule itself, breaking phosphodiester bonds between nucleotides or disrupting the DNA's structural integrity. This direct interaction can lead to immediate and substantial alterations to the DNA helix (Borrego et al., 2015).
- **Indirect Damage:** IR primarily interacts with water molecules, especially relevant given that cells are composed of nearly 80% water. This process, known as water radiolysis, generates highly reactive oxygen species (ROS) such as hydroxyl radicals (•OH), superoxide radicals ($O_2^{\bullet -}$), hydrogen peroxide (H_2O_2), and reactive nitrogen species. These ROS are potent oxidizing agents that can subsequently attack DNA, leading to a wide range of oxidative damage (Maynard et al., 2009).

The biological impact of the same absorbed dose can vary depending on the type of radiation and the characteristics of the exposed tissue. The unit **sievert (Sv)** is used to represent the **equivalent dose,** which accounts for these differences. On average, individuals are exposed to about **2.4 millisieverts (mSv)** of IR annually from natural sources. Cells with high rates of proliferation, such as bone marrow, skin, and digestive tract epithelial cells, are more radiosensitive than cells that are quiescent and divide slowly, such as those that make up the nervous system. The impact of IR on cellular DNA is not uniform; rather, it is highly dependent on factors like the proximity of radiation tracks to the nucleus, intrinsic properties of the radiation itself (e.g., alpha, beta, gamma rays), duration or intensity of exposure, its dose, energy, and linear energy transfer, i.e., LET (Fig. 8.1). LET refers to the amount of energy transferred by IR to the material through which it passes per unit distance.

- **High LET radiation** (e.g., alpha particles, heavy ions) deposits a large amount of energy over a short distance, leading to a high density of ionization events. This dense ionization often causes more direct and complex DNA damage, resulting in clustered DNA damage (CDD). CDDs are multiple DNA damages (such as DSBs, base modifications, and abasic sites) occurring in proximity on the DNA strand (Mavragani et al., 2019).
- **Low LET radiation** (e.g., X-rays, gamma rays, beta particles, high-energy UV) deposits energy less densely and over a broader area compared to high LET radiation. Thus they have sparse ionization tracks. A significant proportion of the damage caused by low LET radiation is indirect, mediated through the radiolysis of water. This typically results in a

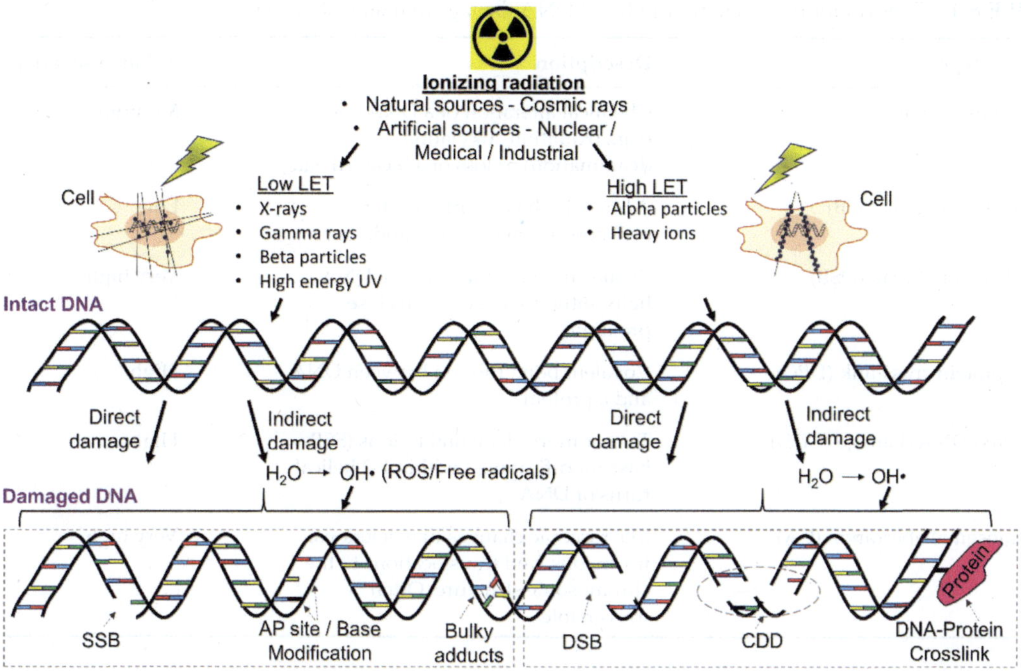

FIGURE 8.1 DNA damages caused by different types of ionizing radiation (IR). IR arises from two primary sources: natural and artificial. It is classified into low linear energy transfer (LET) radiation and high LET radiation depending on the ionization events along their track. IR can cause damage either directly by interacting with the DNA backbone or indirectly through water radiolysis, generating reactive oxygen species and free radicals. Low LET radiation predominantly causes single-strand breaks (SSBs), abasic (AP) sites, and base modifications. In contrast, high LET radiation induces more complex and clustered lesions, such as double-strand breaks (DSBs), bulky adducts, clustered DNA damage (CDD), and DNA–protein cross-links (DPCs).

greater proportion of indirect damage via ROS and simpler, more isolated lesions (Smith et al., 2023).

8.2.1 Key types of DNA damage induced by ionizing radiation

IR causes a wide range of DNA alterations, such as sugar damage, base damage, single-strand breaks (SSBs), DSBs, DNA–DNA and DNA–protein cross-links (DPCs) (Table 8.1). As depicted in Fig. 8.1, different types of IR cause separate DNA damage.

- **Base Modifications:** IR can lead to chemical alterations to individual DNA bases (adenine, thymine, cytosine, guanine), such as oxidation, deamination, and formation of abasic (apurinic or apyrimidinic) sites. These often result from indirect damage by ROS. $OH^{\bullet}$ acts preferentially on C8 of purines (adenine and guanine) in DNA to generate C8-OH adduct, which further undergoes either one-electron oxidation to form 8-oxo-7,8-dihydroguanine (8-oxoGua) and 8-oxo-7,8-dihydroadenine (8-oxoAde), respectively, or one-electron reduction to generate formamidopyrimidine (Fapy)-Gua/Ade (Cadet and Wagner, 2013). 8-oxoG is

TABLE 8.1 Types of ionizing radiation-induced DNA damage in mammalian cells.

Damage type	Description	Relative severity
Base modification	Chemical alteration of a DNA base (e.g., oxidation, alkylation, deamination) or loss of a base (AP site).	Medium
Single-strand break (SSB)	A break in the phosphodiester backbone of one DNA strand.	Low
Double-strand break (DSB)	Breaks in both strands of the DNA helix, either opposite or in close proximity.	Very high
DNA–protein cross-link (DPC)	Covalent bond formed between DNA and a protein.	High
Clustered DNA damage (CDD)	Two or more elemental lesions (SSBs, base modifications) within 1-2 helical turns of DNA.	High
Chromosomal aberration (CA)	Macroscopic changes like deletions, inversions, and translocations to the chromosome structure. Often irreversible.	Very high

the most commonly formed oxidative lesion that can mispair with adenine during DNA replication, leading to G-C to T-A transversions (point mutations). If not repaired, these changes can lead to heritable mutations (David et al., 2007).

- **Single-Strand Breaks (SSBs):** A discontinuity or break in one of the two phosphodiester backbones of the DNA double helix results in SSBs. This can be caused directly by radiation or indirectly by ROS. While often considered less severe, unrepaired or misrepaired SSBs can be converted into DSBs during DNA replication (Cannan and David, 2016).
- **Double-Strand Breaks (DSBs):** Simultaneous breaks in both strands of the DNA helix, either at the same nucleotide position or in proximity (within 10–20 base pairs). While DSBs are less frequent than base damage and SSBs, they are considered the primary cause of cell death from radiation exposure. They can be caused directly by high-energy radiation (e.g., gamma rays, X-rays). The ratio of SSB to DSB can vary depending on factors such as radiation type and dose. If not accurately repaired, DSBs can lead to chromosomal rearrangements, loss of genetic material, cell cycle arrest, cell death, or the initiation of oncogenic transformation (Mahaney et al., 2009).
- **DNA Cross-links:** Covalent bonds formed between two strands of DNA (interstrand cross-links) or within the same strand (intrastrand cross-links). IR can induce chemical reactions that create these cross-links. DNA cross-linking prevents the DNA strands from unwinding, thereby blocking essential cellular processes such as replication and transcription (Dextraze et al., 2010). Cross-links can also be formed between DNA and proteins (DPCs). In the presence of oxygen, DNA damages such as SSBs and DSBs are encouraged when cells are exposed to radiation. On the other hand, DPCs are enhanced when oxygen is not available, indicating their significance in hypoxic cells, like those found in tumor cells. DPC is

caused by the DNA and protein radicals generated by hydroxyl radicals (i.e., indirect effect of radiation). Furthermore, guanine radical cations produced by the direct effect can also create DPCs (Nakano et al., 2017).

- **Clustered DNA Damage (CDD):** This refers to multiple lesions (e.g., base damage, SSBs, DSBs) occurring in proximity (within one or two helical turns) on the DNA molecule. CDD is a hallmark of high LET radiation, such as alpha particles or heavy ions. These lesions could be either simple or complex. Simple clustered DNA damage includes those base damages that occur in two neighboring bases or near DSB ends. Complex clustered DNA damages are those that occur in three or more nearby bases or near DSB ends. The high density of lesions in CDD significantly increases the difficulty of repair and is highly mutagenic and genotoxic, posing a substantial challenge to genomic stability (Nickoloff et al., 2020; Zhang et al., 2024).
- **Chromosomal aberrations (CA):** IR causes a variety of chromosomal aberrations consisting of genome rearrangements produced by unjoining or misrejoining of DSB ends (Ballarini and Carante, 2016). This leads to the formation of dicentric, centric ring, and acentric chromosomes, creating micronuclei (MN), and eventually causing cell death. The evaluation of CA is considered one of the first tools in radiation biodosimetry (Durante and Formenti, 2018).

8.2.2 DNA-damage signaling and repair pathways

Cells have evolved sophisticated DDR mechanisms to detect, signal, and repair DNA lesions, thereby maintaining genomic integrity and preventing a range of diseases, including cancer, neurodegeneration, and age-related disorders. The DDR involves a coordinated network of signaling cascades that not only recognize the type and extent of damage but also initiate cell cycle checkpoints, temporarily halting cell division to allow time for the appropriate DNA repair machinery to act (Chatterjee and Walker, 2017) (Fig. 8.2). This continuous surveillance and repair system is essential because cells are constantly exposed to endogenous (such as ROS from metabolism or replication errors) and exogenous (including IR, UV light, and chemical agents) threats. In addition to maintaining genome integrity, DDR networks coordinate additional cellular reactions such as apoptosis and senescence, in cases where damage is irreparable. When these systems are compromised, through inherited mutations or environmental stressors, genomic instability occurs, which is a hallmark of cancer and other degenerative conditions, as depicted in Fig. 8.2. At least four major DNA repair pathways are active throughout different stages of the cell cycle post-IR exposure: base excision repair (BER), nucleotide excision repair (NER), homologous recombination (HR), and nonhomologous end joining (NHEJ) (Chen et al., 2024). Each pathway consists of different proteins that are specialized to address distinct types of DNA lesions, contributing to the overall resilience of the genome (Table 8.2).

8.2.2.1 Base excision repair

BER is the primary DNA repair pathway responsible for correcting small, non-helix-distorting base lesions. These lesions commonly arise from endogenous processes such as oxidative damage, alkylation, deamination, or spontaneous hydrolysis (Robertson et al., 2009).

The BER pathway is initiated by DNA glycosylases, a family of enzymes (at least 11 distinct mammalian types, Table 8.3) that recognize and excise specific damaged bases, creating an apurinic/apyrimidinic (AP) site (Gohil et al., 2023). To efficiently scan the vast genome for rare lesions, glycosylases employ facilitated diffusion mechanisms, combining one-dimensional (1D) diffusion along the DNA (sliding) with three-dimensional (3D) diffusion in

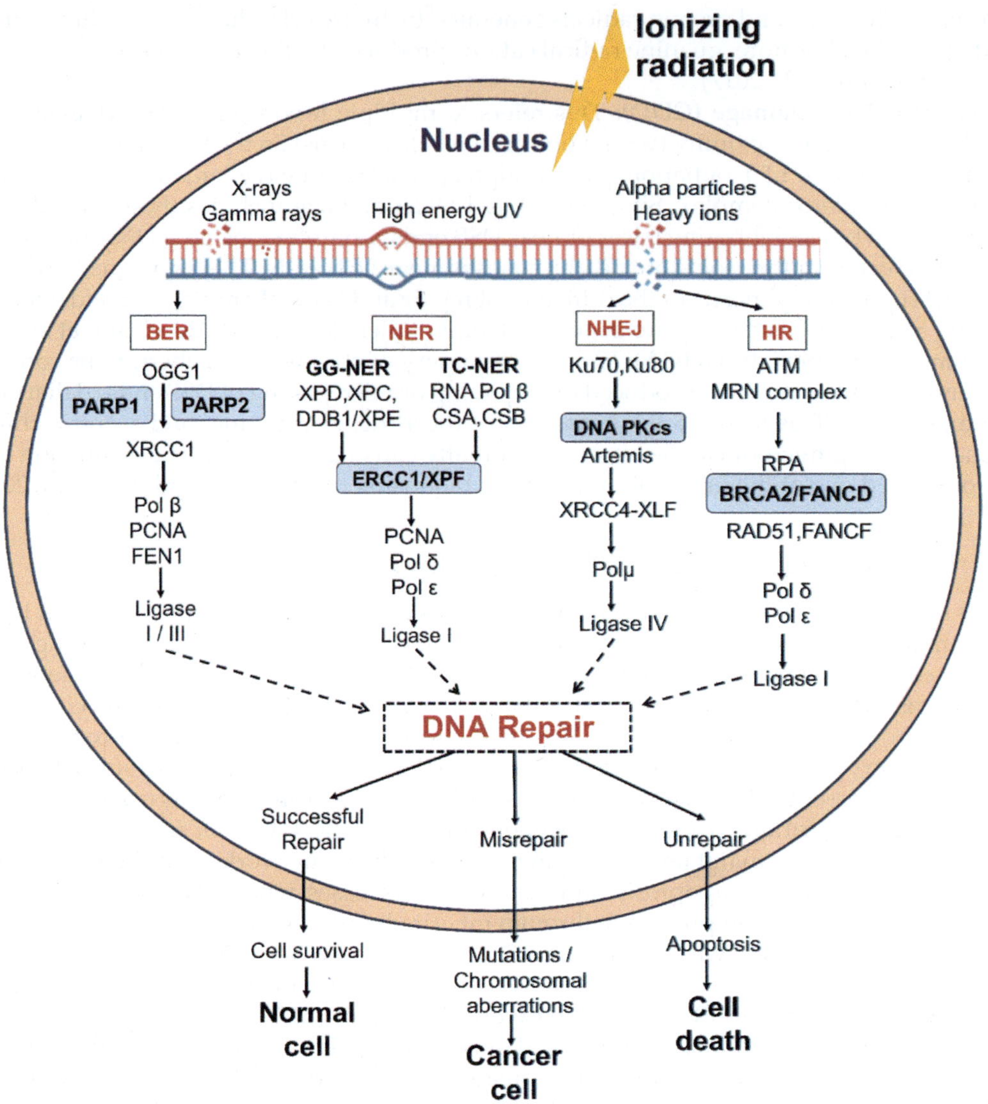

FIGURE 8.2 Schematic overview of the cellular DNA damage response (DDR) to ionizing radiation. Upon exposure to ionizing radiation (IR), cells activate a complex DDR network that coordinates lesion recognition, signal transduction, and repair. IR can induce a spectrum of DNA lesions, including single-strand breaks (SSBs), double-strand breaks (DSBs), abasic sites, and cross-links. These are detected by damage sensors, which activate key transducers such as ataxia telangiectasia mutated kinase (ATM), ATM-related kinase (ATR), and DNA-dependent protein kinase catalytic subunit (DNA-PKcs). The downstream effectors regulate cell cycle checkpoints to halt cell division and focus on DNA repair pathways, and, if necessary, apoptosis. Major repair pathways include base excision repair (BER) for small base lesions, nucleotide excision repair (NER) for adducts formed by high-energy UV, such as UVC, nonhomologous end joining (NHEJ) for quick DSB repair, and homologous recombination (HR) for high-fidelity DSB repair.

TABLE 8.2 List of key proteins and major DNA repair pathways in mammalian cells.

Pathway	Primary damage type(s) repaired	Key characteristics	Representative key proteins
Nonhomologous end joining (NHEJ)	Double-strand breaks (DSBs)	Error-prone, template-independent, active throughout cell cycle, rapid	Ku70/Ku80, DNA-PKcs, XRCC4, DNA Ligase IV, XLF, PAXX, Artemis, Pol μ/λ
Homologous recombination (HR)	Double-strand breaks (DSBs), cross-links	High-fidelity, template-dependent (sister chromatid), active in S/G2 phases	ATM, ATR, BRCA1, BRCA2, RAD51, RPA, RAD52
Base excision repair (BER)	Small base lesions (e.g., oxidized, alkylated, deaminated bases, AP sites), single-strand breaks (SSBs)	High-fidelity, two subpathways (short-patch, long-patch)	DNA glycosylases, APE1, Pol β, Pol δ/ε, XRCC1, PARP1, FEN1, PCNA, DNA ligase I/III
Nucleotide excision repair (NER)	Bulky lesions (e.g., UV photoproducts)	High-fidelity, "cut-and-paste," two subpathways (GG-NER, TC-NER)	XPC, UV-DDB, TFIIH (XPB, XPD), XPA, RPA, XPG, ERCC1-XPF, DNA Pol $\delta/\varepsilon/\kappa$, DNA ligases

solution (hopping) (Lee and Lee, 2025). Subsequently, an AP endonuclease, primarily APE1 (Apurinic/Apyrimidinic Endonuclease 1) in humans, cleaves the DNA backbone at this AP site, generating SSBs with a 3′-hydroxyl and 5′-deoxyribose phosphate (dRP) termini. APE1 is essential for cell survival and also interacts with DNA Polymerase β (Pol β), an interaction crucial for the subsequent gap filling (Krokan and Magnar 2013).

BER proceeds via two main subpathways:

- **Short-Patch BER (SP-BER):** A single nucleotide at the damaged site is usually repaired by this pathway. DNA Polymerase β (Pol β) performs a vital role after APE1 creates an SSB at the damaged site. Pol β has two separate enzymatic activities: AP lyase activity (removing the 5′-dRP group) and DNA synthesis (adding a single nucleotide to the 3′-hydroxyl end). This synchronized action leads to a smooth DNA-damage repair without accumulating intermediate lesions. As a scaffold protein, X-ray repair cross-complementing protein 1 (XRCC1) coordinates interactions between Pol β and DNA Ligase III (LIG3). Poly (ADP-ribose) polymerase (PARP1) is crucial in detecting SSBs, bringing XRCC1 and LIG3 close to the damage site, and assembling the repair factors. LIG3, in conjunction with XRCC1, then works to finish the SP-BER by sealing the nick in the DNA backbone.
- **Long-Patch BER (LP-BER):** This pathway repairs a longer stretch of 2–12 nucleotides. It is typically initiated when the 5′-dRP terminus generated by APE1 resists removal by Pol β, necessitating the recruitment of additional proteins. DNA Polymerase δ/ε (Pol δ/ε) is critical for LP-BER, synthesizing a DNA stretch beyond the initial nucleotide gap and displacing part of the damaged strand to form a flap. Flap endonuclease 1 (FEN1) then cleaves this 5′-flap structure, ensuring efficient removal of the displaced DNA. Proliferating cell nuclear antigen (PCNA) serves as a processivity factor for Pol δ/ε, forming a clamp around the DNA

TABLE 8.3 DNA glycosylases and recognized DNA damage in base excision repair.

DNA glycosylase (human/mammalian)	Enzyme family/class	Type of DNA damage recognized
Uracil-DNA glycosylase (UNG)	Uracil-DNA glycosylase (UDG)	Uracil (U) in DNA (e.g., from cytosine deamination or dUMP misincorporation), 5-fluorouracil (5-FU)
Single-strand-specific monofunctional uracil-DNA glycosylase 1 (SMUG1)	Uracil-DNA Glycosylase (UDG)	Uracil (U), 5-hydroxymethyluracil (5-hmU) in single- and double-stranded DNA
Thymine DNA glycosylase (TDG)	Helix–hairpin–helix (HhH) glycosylase	G: T mismatches (from 5-methylcytosine deamination), G: U mismatches
Methyl-CpG-binding domain protein 4 (MBD4)	HhH glycosylase	T: G mismatches and U: G mismatches in CpG contexts
8-oxoguanine DNA glycosylase 1 (OGG1)	HhH glycosylase	8-oxoguanine (8-oxoG) paired with cytosine (C), Formamidopyrimidine (FapyG)
MutY homolog DNA glycosylase (MUTYH)	MutY homolog	Adenine (A) mispaired with 8-oxoguanine (8-oxoG), A opposite C or G
Methylpurine DNA glycosylase (MPG), also known as alkyladenine DNA glycosylase (AAG)	Monofunctional alkylpurine DNA glycosylase	3-methyladenine (3-meA), 7-methylguanine (7-meG), 3-methylguanine (3-meG), Hypoxanthine (Hx), ε-adenine (εA)
Endonuclease III-like 1 (NTHL1)	Endonuclease III-like (NEIL) glycosylase	Thymine glycol (Tg), 5-hydroxycytosine (5-hC), 5-hydroxyuracil (5-hU), formamidopyrimidine (FapyG)
Endonuclease VIII-like glycosylase 1 (NEIL1)	Endonuclease VIII-like (NEIL) glycosylase	Oxidized pyrimidines and purines like thymine glycol (Tg), formamidopyrimidine (FapyG/FapyA), 8-oxoguanine (8-oxoG), dihydrouracil (DHU), spiroiminodihydantoin (Sp), guanidinohydantoin (Gh) in single- and double-stranded DNA
Endonuclease VIII-like glycosylase 2 (NEIL2)	Endonuclease VIII-like (NEIL) glycosylase	Similar range of oxidative lesions to NEIL1, often with a preference for single-stranded DNA
Endonuclease VIII-like glycosylase 3 (NEIL3)	Endonuclease VIII-like (NEIL) glycosylase	Formamidopyrimidines (FapyG, FapyA), spiroiminodihydantoin (Sp), guanidinohydantoin (Gh) in single-stranded DNA

to enhance synthesis efficiency and boost FEN1 activity by maintaining the DNA substrate in an open flap conformation. Finally, DNA Ligase I (LIG1) seals the remaining nick after FEN1 cleavage, completing the long-patch repair.

The BER pathway exemplifies the cell's strategies for efficiency and adaptability in DNA repair. The existence of two distinct subpathways allows for tailored repair based on lesion complexity, whereas mechanisms such as facilitated diffusion optimize the search for rare lesions across a vast genome.

8.2.2.2 Nucleotide excision repair

NER is a highly versatile DNA repair pathway responsible for removing a wide variety of bulky DNA lesions that distort the DNA double helix (Schärer, 2013). These lesions include UV-induced photoproducts (e.g., cyclobutane pyrimidine dimers [CPDs], 6-4 photoproducts), adducts from environmental carcinogens (e.g., benzo[*a*]pyrene), oxidative endogenous lesions, and certain chemotherapy drugs (e.g., carboplatin, cisplatin) (Zachayus et al., 2025). Deficiencies in NER are linked to xeroderma pigmentosum, an inherited disorder characterized by extreme sensitivity to sunlight and a significantly increased risk of skin cancer due to the inability to repair UV-induced lesions.

The NER process involves a multistep "cut-and-paste" mechanism: lesion recognition, excision of a short oligonucleotide (24-32 nucleotides) containing the damage, synthesis of a new DNA strand using the undamaged strand as a template, and final ligation. NER is initiated by two subpathways:

- **Global Genome NER (GG-NER):** This pathway operates on the entire genome and is initiated by a specific factor XPC-RAD23B, sometimes functioning along with UV-damaged DNA-binding protein (UV-DDB). XPC-RAD23B is the initial damage recognition factor in GG-NER, important for the assembly of other core NER factors. Instead of detecting specific chemical structures of the lesions, Xeroderma Pigmentosum complementation group C (XPC) protein recognizes DNA damage by sensing thermodynamic destabilization of the DNA double helix. This provides an advantage to XPC in binding to a variety of bulky DNA lesions. XPC-RAD23B does not interact with one type of bulky DNA lesion—cyclobutane pyrimidine dimers (CPDs) with high specificity, so UV-DDB (DDB1/DDB2) is crucial for recognizing them. In addition to having a strong affinity for UV-damaged DNA, DDB2 is a component of the CUL4-ROC1 ubiquitin ligase complex, which ubiquitinates XPC and histones to enable NER within chromatin.
- **Transcription-Coupled NER (TC-NER):** This pathway works toward the accelerated repairing of the lesions in the transcribed strand of active genes. During transcription, if RNA polymerase recognizes a lesion, it gets stalled there. This initiates the process of recruiting other TC-NER specific factors, CSA (Cockayne syndrome A protein) , CSB (Cockayne syndrome B protein), and XAB2 (XPA-binding protein 2).

Both pathways require the action of the core NER factors to bring about the excision process. Following initial damage site recognition, TFIIH is recruited, which is a multi-subunit complex with roles in both transcription and NER. TFIIH contains two helicase subunits, XPB and XPD, which open up the DNA around the lesion. XPB possesses an ATPase activity, which is required for TFIIH loading, while XPD has helicase and ATPase activities, which are required to separate the dsDNA. As DNA lesions do not fit through the ssDNA tunnel of XPD, its helicase activity gets blocked, allowing XPD to act as a candidate for damage verification. After TFIIH, other factors like XPA, replication protein A (RPA), and XPG are recruited independently, whereas XPC-RAD23B often leaves the site. Xeroderma Pigmentosum group A (XPA) is a central component, interacting with multiple factors (TFIIH, RPA, ERCC1-XPF) and ensuring all NER factors are correctly positioned for incision. RPA wraps the nondamaged DNA strand, helping to correctly position the endonucleases ERCC1-XPF and XPG at the damaged DNA strand. ERCC1-XPF performs the initial 5′ incision, producing a free 3′ hydroxyl group for the repair synthesis, while XPG performs the 3′ incision later in the reaction. After excision, the lesion-containing oligonucleotide is released, and the gap is filled by DNA polymerases (Pol δ, Pol ε, Pol κ) and sealed by DNA ligases (DNA ligase IIIα and XRCC1 in quiescent cells, DNA ligase I in replicating cells). NER also operates on DNA within chromatin, requiring NER proteins to enlist chromatin-modifying enzymes like CAF-1 (chromatin assembly factor) and chromatin-remodeling complexes (e.g., ALC1, SWI/SNF, INO80) to access lesions and restore chromatin structure post-repair. UV-induced lesions also potently induce a DNA-damage-signaling response. A long stretch of ssDNA covered with RPA is a principal signal for DDR activation, which can be generated by the EXO1 nuclease processing incomplete NER repair synthesis products.

The NER pathway demonstrates the cell's capacity for adaptive and context-dependent repair. Its ability to handle a wide range of structurally diverse bulky lesions, coupled with its intricate coordination with chromatin remodeling and DNA-damage signaling, showcases a highly sophisticated and flexible repair system. The differentiation between global genome repair (GG-NER) and transcription-coupled repair (TC-NER) further highlights this adaptability, allowing the cell to prioritize repair in actively transcribed genes, thereby safeguarding essential cellular functions. This multi-layered approach ensures comprehensive genome maintenance under various genotoxic stresses.

8.2.2.3 Nonhomologous end joining

NHEJ is a crucial and highly efficient pathway for repairing DNA DSBs in mammalian cells. This pathway of DSB repair is termed "nonhomologous" because it directly ligates the broken DNA ends without requiring a homologous DNA template. This allows NHEJ to operate on a wide variety of DNA end configurations, making it a versatile repair strategy (Davis and Chen, 2013). NHEJ is not restricted to specific phases of the cell cycle and is active throughout the cell cycle (Brandsma and Gent, 2012).

The general mechanism of NHEJ involves a series of sequential steps. First, DNA end recognition and assembly of the NHEJ complex occur. The Ku70/Ku80 heterodimer is the primary sensor of DSBs in NHEJ, forming a ring structure that detects and encircles the broken DNA ends (Davis and Chen, 2013). This heterodimer then recruits the catalytic subunit of DNA-dependent protein kinase, DNA-PKcs (Zhang et al., 2024). DNA-PKcs is a key enzyme in NHEJ; upon recruitment by Ku, it undergoes autophosphorylation, leading to its activation. Activated DNA-PKcs phosphorylates numerous downstream proteins, signaling damage and initiating repair. Second, bridging and stabilization of DNA ends are achieved, often with the recruitment of X-ray repair cross-complementing protein 4 (XRCC4)-like factor (XLF) and PAXX (XRCC4 and XLF paralog), which interact with the Ku70/Ku80 heterodimer to stabilize the NHEJ complexes on DNA, facilitating efficient repair. Third, DNA end processing may be necessary if the broken ends contain damaged nucleotides or overhangs. Artemis, when complexed with DNA-PKcs, is phosphorylated and converted from a nucleic acid exonuclease to an endonuclease, enabling the processing of such complex DNA ends. DNA polymerases μ and λ are also recruited by Artemis and interact with Ku to promote end processing and DNA hairpin opening, restoring DNA integrity. Finally, ligation of the ends completes the repair. XRCC4 physically links polynucleotide kinase with DNA Ligase IV (LIG4) to form a complex essential for this final ligation step of DSBs (Pannunzio et al., 2018).

NHEJ's rapid, template-independent nature makes it the primary defense against lethal DSBs, preventing catastrophic chromosomal loss. However, the absence of a homologous template means that precise restoration of the original sequence is not guaranteed. Indeed, NHEJ is known to be an error-prone pathway, often leading to small deletions or insertions at the repair site. This highlights a fundamental evolutionary compromise in maintaining genome integrity, where rapid repair of life-threatening lesions takes precedence over absolute sequence accuracy, contributing to mutations and genomic instability if misrepaired events occur.

8.2.2.4 Homologous recombination

HR is a high-fidelity DNA repair pathway that precisely repairs DSBs, DPCs, and stalled replication forks. Interestingly, DPCs caused by radiation heal more slowly than DSBs. In order to repair both DSBs and radiation-induced DPCs, HR is essential. The resultant DNA–peptide cross-links were circumvented by translesion synthesis in a recently reported unique process of DPC repair mediated by a DPC protease. Its accuracy is achieved by using a homologous DNA sequence, typically the sister chromatid, as a template to guide repair synthesis (Prakash et al., 2015). HR is primarily active during the S and G2 phases of the cell cycle, when a sister chromatid is available to serve as the template. The defining step of HR is homologous strand exchange, directed by the RAD51 protein (Rossi et al., 2021). The process begins with end resection, where nucleases process the broken DNA ends to create single-stranded DNA (ssDNA) overhangs. These ssDNA regions are then coated by RPA, a single-stranded DNA-binding protein that protects the DNA and prepares it for RAD51 loading. Subsequently, RAD51 binds to the ssDNA to form a nucleoprotein filament, which then promotes strand invasion into a homologous duplex (typically the sister chromatid) to initiate repair synthesis. The 3′ end of the broken strand primes DNA synthesis from the homologous template, ensuring precise restoration of the original sequence (Prakash et al., 2015).

Several key proteins are involved in the HR pathway, such as ataxia telangiectasia mutated kinase (ATM), a pivotal protein kinase and a master regulator of the DDR. ATM is activated by DSBs and phosphorylates numerous HR proteins. Cells deficient in ATM exhibit extreme radio sensitivity, underscoring its critical role (Tseng et al., 2024). ATM-related kinase (ATR) is another key protein kinase that primarily responds to replication stress and DNA damage. ATR regulates cell cycle checkpoint control and promotes HR by phosphorylating HR substrates, such as BRCA1 and Partner and Localizer of BRCA2 (PALB2), and by inhibiting cyclin-dependent kinases (Brown et al., 2014). Breast Cancer Type 1 and 2 Susceptibility Proteins (BRCA1 and BRCA2) are critical tumor-suppressor proteins and key players in the HR pathway (Rosen, 2013). BRCA1 is a multifunctional protein that, in response to DNA damage, is phosphorylated by ATM/ATR and forms complexes that recruit other repair factors, including BRCA2. BRCA2 is essential for HR, specifically controlling RAD51 by mediating its delivery to single-stranded DNA. Mutations in BRCA1/2 are frequently associated with familial breast and ovarian cancers, underscoring HR's importance in maintaining genomic stability. While RAD52 is critical for recombination in yeast, its role in mammalian cells has been historically overlooked due to the lack of a strong DNA repair or recombination phenotype in knockouts. However, recent work indicates it is crucial for backup DNA repair pathways in HR-deficient cancer cells, promoting

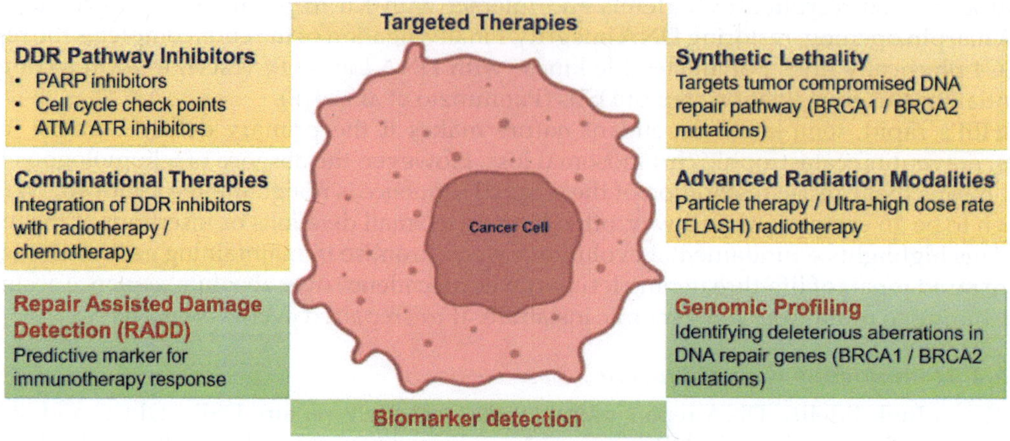

FIGURE 8.3 Targeting DNA damage response network components for cancer diagnostic tools and therapeutics. The **DNA damage response** (DDR) network provides various ways for the development of targeted therapies and diagnostic instruments in oncology. Inhibitors of the DDR pathway, such as poly (ADP-ribose) polymerase (PARP) and ATM-related kinase/ataxia telangiectasia mutated kinase (ATR/ATM) inhibitors, can make cancer cells more sensitive to chemotherapy and radiation therapy. Using these in combination with other treatments makes them work better. Synthetic lethality strategies take advantage of problems with tumor DNA repair pathways, such as breast cancer type 1 and 2 (BRCA1/2) mutations. Advanced radiation therapies such as particle therapy and FLASH radiotherapy target cancer cells with high precision while causing less damage to healthy tissues. Repair-assisted damage detection (RADD) acts as a predictive biomarker for immunotherapy response, and genomic profiling helps find mutations in DDR genes that can be used for personalized cancer treatment.

DNA pairing (D-loop formation), single-strand DNA and DNA: RNA annealing, and inverse strand exchange (Rossi et al., 2021).

HR's reliance on a homologous template ensures error-free repair, making it crucial for maintaining genomic stability, especially during DNA replication. The severe consequences of HR pathway defects, exemplified by individuals with BRCA1/2 mutations, directly link the integrity of this molecular repair pathway to cancer susceptibility. This highlights HR as a critical tumor-suppressor mechanism, and its vulnerabilities are actively exploited in cancer therapies, such as the synthetic lethality observed with PARP inhibitors in HR-deficient tumors (Toulany, 2019).

8.3 Targeted therapies and biomarkers for early detection of cancer

Recent breakthroughs in our understanding of DDR pathways activated by IR have opened up exciting new avenues in cancer biology. By elucidating the complex cellular pathways involved in detecting and repairing DNA damage, researchers are now able to translate this knowledge into targeted therapeutic strategies and diagnostic tools, offering new possibilities for early detection, personalized treatment, and improved patient outcomes (Kelley et al., 2024). Some of the ways are mentioned here and represented in Fig. 8.3.

- **DDR Pathway Inhibitors:** One of the most promising strategies in targeted cancer therapy involves the development of small-molecule inhibitors that disrupt key components of the

DDR network. By selectively targeting proteins critical for DNA repair and checkpoint control, these agents can sensitize tumor cells to DNA-damaging treatments or induce lethal DNA-damage accumulation on their own. For instance inhibitors targeting kinases such as ATR, DNA-PKcs, WEE1, and enzymes such as PARP1 are showing promise (Cheng et al., 2022).

- **Synthetic Lethality:** Tumor cells often exhibit defects in specific DNA repair mechanisms, such as HR or mismatch repair (MMR), which make them particularly vulnerable to the inhibition of compensatory pathways. This powerful concept is at the forefront of DDR-targeted therapies. This has been effectively exploited in the use of **PARP inhibitors** to treat cancers with BRCA1 or BRCA2 mutations. By selectively targeting the tumor's compromised DNA repair capacity, these therapies lead to an unsustainable level of genomic instability and cancer cell death. Healthy cells, with intact alternative repair pathways, remain largely unaffected, thereby widening the therapeutic window. Ongoing research is actively focused on identifying additional gene pairs and pathway dependencies that can be exploited across various tumor types (Toulany, 2019). Table 8.4 summarizes the DDR inhibitors that are currently active in clinical trial studies for treating various types of cancers.
- **Combination Therapies:** The integration of DDR inhibitors with conventional cancer treatments such as **radiotherapy** and **chemotherapy** represents a promising strategy to enhance therapeutic efficacy and overcome treatment resistance. Many of the standard cancer therapies exert their effects by inducing DNA damage, either directly, as in the case of IR, or indirectly through DNA-damaging chemotherapeutic agents such as platinum-based drugs, topoisomerase inhibitors, and alkylating agents. By sensitizing cancer cells to DNA damage, these inhibitors can amplify the effectiveness of conventional treatments, potentially overcoming tumor radioresistance and improving patient outcomes (Li et al., 2021)
- **Advanced Radiation Modalities:** Beyond inhibitors, a deeper understanding of DDR informs the development of new radiation techniques. Particle therapies, such as proton beam therapy, offer superior dose distribution, minimizing damage to healthy tissues. Ultra-high dose rate fast-acting linear accelerator driven stereotactic high dose rate radiotherapy (FLASH) radiotherapy is another promising development, aiming to maximize tumor killing while sparing healthy tissue (Melia and Jason, 2023).

The insights into DDR pathways are also revolutionizing cancer diagnostics by enabling the development of sophisticated biomarkers:

- **Genomic Profiling:** Advances in genomic profiling and molecular biology have significantly enhanced the identification of oncogenes and tumor-suppressor genes, which are crucial for discovering cancer biomarkers. Identifying deleterious aberrations in DNA repair genes, such as BRCA1/2 mutations in HRR, can directly be used for treatment, as these mutations are actionable targets for therapies such as PARP inhibitors (Passaro et al., 2024).
- **DNA-Damage Measurements:** Emerging tools, such as the repair-assisted damage detection assay, can measure a broad spectrum of DNA lesions within tumors. These measurements have shown potential for monitoring treatment response, guiding targeted therapies, and even indicating the susceptibility of a tumor to immunotherapy. For example, higher levels of oxidative DNA lesions have been correlated with decreased recurrence-free survival in ovarian cancer, and specific DNA-damage types can influence the tumor microenvironment and immune responses (Dave et al., 2025).

TABLE 8.4 List of inhibitors that target DDR network proteins and are currently active in clinical trials for treating different cancers.

Pathway	Target	Drug name	Type of cancer
NHEJ	DNA-PKcs	XRD-0394	Squamous cell carcinoma and triple-negative breast cancer
		AZD7648	Advanced malignancy, Adult soft tissue sarcoma
		M9831	Advanced solid tumor
		M3814	Glioblastoma, gliosarcoma, ovarian cancer
		CC-115	Prostate cancer, advanced malignancy, glioblastoma
		BR101801	Colorectal cancer, breast cancer, lung cancer
		PEPOSERTIB	Locally advanced rectal cancer, head and neck squamous cell carcinoma, synovial sarcoma, and metastatic melanoma
BER	PARP	E7016	Melanoma
		NIRAPARIB	Ovarian cancer, breast cancer
		OLAPARIB	Lung cancer, carcinoma of the esophagus, head and neck cancer, breast cancer, ovarian cancer, gastric cancer, pancreatic cancer
		RUCAPARIB	Prostate cancer, ovarian carcinoma
		TALAZOPARIB	Prostate cancer, breast cancer
		AMXI-500I	Advanced malignant neoplasms, including breast and ovarian cancers
		AZD9574	Solid tumors and brain tumors
		BASROPARIB	Ovarian cancer, breast cancer, and prostate cancer
		CK-102	Breast cancer and ovarian cancer
		CVL218	Ovarian cancer, breast cancer, and prostate cancer
		E7016	Melanoma and advanced solid tumors
		FUZULAPARIB	Breast cancer and ovarian cancer
		HS-10502	Breast cancer and ovarian cancer
		PAMIPARIB	Ovarian cancer, breast cancer, and prostate cancer
		RP12146	Locally advanced or metastatic solid tumors
		SARUPARIB	Ovarian cancer, breast cancer and prostate cancer
		SIMMIPARIB	Metastatic castration-resistant prostate cancer (mCRPC), breast cancer, and ovarian cancer
		SENAPARIB	Ovarian cancer

(*continued on next page*)

TABLE 8.4 List of inhibitors that target DDR network proteins and are currently active in clinical trials for treating different cancers—cont'd

Pathway	Target	Drug name	Type of cancer
		STENOPARIB	Ovarian cancer
		TALAZOPARIB	Breast cancer
		TQB3823	Ovarian cancer, breast cancer, and prostate cancer
		TSL1502	Ovarian cancer, breast cancer, and pancreatic cancer
		VENADAPARIB	Gastric cancer and ovarian cancer
		SC10914	Breast cancer and ovarian cancer
		HTMC0435	Breast cancer and ovarian cancer
		HWH340	Small cell lung cancer (SCLC), BRCA-deficient breast cancer, and BRCA-deficient pancreatic cancer
		NESUPARIB	Breast cancer and ovarian cancer
		NMS-293	Small-cell lung cancer and BRCA wild-type ovarian carcinoma
		VELIPARIB	Advanced solid malignancy, epithelial ovarian cancer, fallopian cancer, breast cancer, pancreatic cancer, nonsmall cell lung cancer (NSCLC)
HR	ATM	AZD1390	Brain cancer, glioblastoma, glioma, NSCLC, and adult soft tissue sarcoma
		AZD0156	Solid tumor
	ATR	AZD6738	Refractory cancer, gastric adenocarcinoma, malignant melanoma
		M6620	Esophageal adenocarcinoma, Solid tumor, and squamous cell carcinoma
		BAY-1895344	Solid tumor, ovarian cancer, non-Hodgkin's lymphoma
		M1774	Endometrial carcinoma, ovarian carcinoma, solid tumor, Merkel cell carcinoma, refractory prostate carcinoma
		ART0380	ATM null tumors, platinum-resistant ovarian cancer, and solid tumors
		M4344	Ovarian cancer, breast cancer, and solid tumor
		ATG-018	Advanced solid tumors and hematologic malignancies
		ATRN-119	Human castration-resistant prostate cancer and human colon cancer

(*continued on next page*)

TABLE 8.4 List of inhibitors that target DDR network proteins and are currently active in clinical trials for treating different cancers—cont'd

Pathway	Target	Drug name	Type of cancer
		BERZOSERTIB	ATM-deficient advanced colorectal cancer
		CAMONSERTIB	Advanced solid tumors
		DN020198	Ovarian cancer, breast cancer, lung cancer, and pancreatic cancer
		ELIMUSERTIB	Breast cancer, ovarian cancer, and urothelial carcinoma
		HRS2398	Breast cancer and ovarian cancer
		IMP9064	Advanced solid tumors
		SC0245	Breast cancer
		TUVUSERTIB	NSCLC, gastric cancer, and high-grade serous ovarian cancer

Recent advancements in understanding the DDR have profoundly affected the fields of molecular oncology and translational medicine. The discovery of novel cellular recovery mechanisms reveals new vulnerabilities unique to cancer cells. These insights not only deepen our knowledge of tumor biology but also lay the foundation for the development of targeted treatments that are both more precise and less toxic. Continued investment in fundamental DDR research, coupled with robust translational and clinical development, will be essential to realize the full potential of these advances, ultimately leading to more effective, personalized, and durable solutions for cancer.

References

Ballarini, F., Carante, M.P., 2016. Chromosome aberrations and cell death by ionizing radiation: evolution of a biophysical model. Radiat. Phys. Chem. 128, 18–25.

Borrego-Soto, G., Ortiz-López, R., Rojas-Martínez, A., 2015. Ionizing radiation-induced DNA injury and damage detection in patients with breast cancer. Genet. Mol. Biol. 38, 420–432.

Brandsma, I., van Gent, D.C., 2012. Pathway choice in DNA double strand break repair: observations of a balancing act. Genome Integr. 3, 1–10.

Brown, A.D., Sager, B.W., Gorthi, A., Tonapi, S.S., Brown, E.J., Bishop, A.J., 2014. ATR suppresses endogenous DNA damage and allows completion of homologous recombination repair. PloS One 9 (3), e91222.

Cadet, J., Wagner, J.R., 2013. DNA base damage by reactive oxygen species, oxidizing agents, and UV radiation. Cold Spring Harb. Perspect. Biol. 5 (2), a012559.

Cannan, W.J., Pederson, D.S., 2016. Mechanisms and consequences of double-strand DNA break formation in chromatin. J. Cell. Physiol. 231 (1), 3–14.

Chatterjee, N., Walker, G.C., 2017. Mechanisms of DNA damage, repair, and mutagenesis. Environ. Mol. Mutagen. 58 (5), 235–263.

Chen, J., Potlapalli, R., Quan, H., Chen, L., Xie, Y., Pouriyeh, S., et al., 2024. Exploring DNA damage and repair mechanisms: a review with computational insights. BioTech 13 (1), 3.

Cheng, B., Pan, W., Xing, Y., Xiao, Y., Chen, J., Xu, Z., 2022. Recent advances in DDR (DNA damage response) inhibitors for cancer therapy. Eur. J. Med. Chem. 230, 114109.

Dave, D., Arend, R.C., Hedlich-Dwyer, J., Dal Zotto, V., Gassman, N., 2025. DNA damage measurements as a potential biomarker for immunotherapy response in ovarian cancer. J. Clin. Oncol. 43, e17600.

David, S.S., O'Shea, V.L., Kundu, S., 2007. Base-excision repair of oxidative DNA damage. Nature 447 (7147), 941–950.

Davis, A.J., Chen, D.J., 2013. DNA double strand break repair via non-homologous end-joining. Transl. Cancer Res. 2 (3), 130.

Dextraze, M.E., Gantchev, T., Girouard, S., Hunting, D., 2010. DNA interstrand cross-links induced by ionizing radiation: an unsung lesion. Mutat. Res. 704 (1-3), 101–107.

Durante, M., Formenti, S.C., 2018. Radiation-induced chromosomal aberrations and immunotherapy: micronuclei, cytosolic DNA, and interferon-production pathway. Front. Oncol. 8, 192.

Gohil, D., Sarker, A.H., Roy, R., 2023. Base excision repair: mechanisms and impact in biology, disease, and medicine. Int. J. Mol. Sci. 24 (18), 14186.

Hall, E.J., 1989. Radiation and life. Bull. N. Y. Acad. Med. 65 (4), 430.

Kelley, M.R., Logsdon, D., Fishel, M.L., 2014. Targeting DNA repair pathways for cancer treatment: what's new? Future Oncol. 10 (7), 1215–1237.

Krokan, H.E., Bjørås, M., 2013. Base excision repair. Cold Spring Harb. Perspect. Biol. 5 (4), a012583.

Lee, D., Lee, G., 2025. Single-molecule studies of repair proteins in base excision repair. BMB Rep 58 (1), 17.

Li, L.Y., Guan, Y.D., Chen, X.S., Yang, J.M., Cheng, Y., 2021. DNA repair pathways in cancer therapy and resistance. Front. Pharmacol. 11, 629266.

Mahaney, B.L., Meek, K., Lees-Miller, S.P., 2009. Repair of ionizing radiation-induced DNA double-strand breaks by non-homologous end-joining. Biochem. J. 417 (3), 639–650.

Mavragani, I.V., Nikitaki, Z., Kalospyros, S.A., Georgakilas, A.G., 2019. Ionizing radiation and complex DNA damage: from prediction to detection challenges and biological significance. Cancers 11 (11), 1789.

Maynard, S., Schurman, S.H., Harboe, C., de Souza-Pinto, N.C., Bohr, V.A., 2009. Base excision repair of oxidative DNA damage and association with cancer and aging. Carcinogenesis 30 (1), 2–10.

Melia, E., Parsons, J.L., 2023. DNA damage and repair dependencies of ionising radiation modalities. Biosci. Rep. 43 (10), BSR20222586.

Nakano, T., Xu, X., Salem, A.M., Shoulkamy, M.I., Ide, H., 2017. Radiation-induced DNA–protein cross-links: mechanisms and biological significance. Free Radic. Biol. Med. 107, 136–145.

Nickoloff, J.A., Sharma, N., Taylor, L., 2020. Clustered DNA double-strand breaks: biological effects and relevance to cancer radiotherapy. Genes 11 (1), 99.

Pannunzio, N.R., Watanabe, G., Lieber, M.R., 2018. Nonhomologous DNA end-joining for repair of DNA double-strand breaks. J. Biol. Chem. 293 (27), 10512–10523.

Passaro, A., Al Bakir, M., Hamilton, E.G., Diehn, M., André, F., Roy-Chowdhuri, S., et al., 2024. Cancer biomarkers: emerging trends and clinical implications for personalized treatment. Cell 187 (7), 1617–1635.

Prakash, R., Zhang, Y., Feng, W., Jasin, M., 2015. Homologous recombination and human health: the roles of BRCA1, BRCA2, and associated proteins. Cold Spring Harb. Perspect. Biol. 7 (4), a016600.

Robertson, A.B., Klungland, A., Rognes, T., Leiros, I., 2009. DNA repair in mammalian cells: base excision repair: the long and short of it. Cell. Mol. Life Sci. 66, 981–993.

Rosen, E.M., 2013. BRCA1 in the DNA damage response and at telomeres. Front. Genet. 4, 85.

Rossi, M.J., DiDomenico, S.F., Patel, M., Mazin, A.V., 2021. RAD52: paradigm of synthetic lethality and new developments. Front. Genet. 12, 780293.

Schärer, O.D., 2013. Nucleotide excision repair in eukaryotes. Cold Spring Harb. Perspect. Biol. 5 (10), a012609.

Smith, J., Jones, A.B., Williams, C.D., 2023. The cellular response to ionizing radiation-induced DNA damage in proton beam therapy. Int. J. Part. Ther. 15 (2), 123–145.

Toulany, M., 2019. Targeting DNA double-strand break repair pathways to improve radiotherapy response. Genes 10 (1), 25.

Tseng, C.C., Ku, M.H., Wu, Y.C., Huang, W.L., Wu, W.M., Pai, C.H., et al., 2024. Therapeutic options targeting the ataxia-telangiectasia mutated (ATM)-mediated DNA damage response, macropinocytosis, and adaptive immunity in ovarian cancer. Anticancer Res. 44 (4), 1353–1364.

Zachayus, A., Loup-Forest, J., Cura, V., Poterszman, A., 2025. Nucleotide excision repair: insights into canonical and emerging functions of the transcription/DNA repair factor TFIIH. Genes 16 (2), 231.

Zhang, C., Liu, J., Wu, J., Ranjan, K., Cui, X., Wang, X., et al., 2024. Key molecular DNA damage responses of human cells to radiation. Front. Cell Dev. Biol. 12, 1422520.

CHAPTER

9

Ultraviolet-induced DNA damage and repair in mammals

Sukriti Chauhan[a,b], *Ajeet Singh*[b], *Ram Lal*[a], *Vinod K. Kannaujiya*[a] *and Ganesh K. Maurya*[b]

[a]Department of Botany, Mahila Mahavidyalaya, Banaras Hindu University, Varanasi, India [b]Department of Zoology, Mahila Mahavidyalaya, Banaras Hindu University, Varanasi, India

9.1 Introduction

Ultraviolet (UV) rays, a natural component of solar radiation, have beneficial roles such as aiding in vitamin D synthesis in humans and photosynthesis in plants, but they also have detrimental effects on animals and human health due to their mutagenic and carcinogenic properties. It was initially discovered by German physicists in 1801. UV radiation covers the wavelength range of 100–400 nm. It comprises UV-C (<280 nm), UV-B (280–315 nm), and UV-A (315–400 nm) radiation (Tang et al., 2024). The UV-A and UV-B bands have been further delineated into narrowband UV-B (311–313 nm), UV-A2 (320–340 nm), and UV-A1 (340–400 nm) (Ahmad, 2017). Among all the radiations, UV-C, a shorter-wavelength radiation, is ecologically insignificant, as it is absorbed by the oxygen and ozone layer in the Earth's atmosphere. While longer wavelength radiations, UV-A and UV-B, can have a deleterious effect on the life forms of the Earth, even though the majority of the UV-B is absorbed by the ozone layer (Utrillas et al., 2016; Schmalwieser et al., 2019; Zhou et al., 2019; Bernhard et al., 2023; Tang et al., 2024). The intensity and biological impact of UV exposure depend on several factors, such as the sun's angle in the sky, geographical latitude, cloud cover, altitude, ozone layer thickness, and ground reflection. There are some artificial sources of UV radiation also, including arc lamps, fluorescent lamps, TL-01 UV lamps, black-light ILTs, UV light-emitting diodes, shortwave UV lamps, UV lasers, and narrowband UV band C lamps, but all individuals are unavoidably exposed to solar UV radiation due to ozone layer depletion because of chlorofluorocarbon (CFC) emissions that compromise the protective shield (Lickley et al., 2020; Tang et al., 2024). As a result, exposure to solar UV radiation reaching Earth's surface is on the rise.

DNA is the primary target of UV radiation in both prokaryotes and higher animals, including placental mammals like humans. DNA damage is associated with both cytotoxic and genotoxic

Genome Damage: Repair and Adaptation Across Life.
DOI: https://doi.org/10.1016/B978-0-443-42857-9.00021-9

effects. The genotoxic effects of solar UV radiation may have prohibited terrestrial life for 2 or 3 billion years before the ozone layer formed (Tornaletti and Pfeifer, 1996; Sinha and Häder, 2002). Short-term exposure is beneficial for health, but long-term or excessive exposure can have a detrimental impact on the organisms due to the minimal quantity of UV-B absorbed by cellular DNA, which can induce the formation of photolesions, for instance, cyclobutane pyrimidine dimers (CPDs) and 6-4 photoproducts (6-4PPs), primarily responsible for direct DNA damage. UV-A wavelength is less effective at inducing DNA damage due to its nonabsorption by DNA, but penetrates deeply into the tissues and generates free radicals, including hydroxyl and oxygen radicals, and consequently damages DNA, proteins, and lipids via indirect photosensitizing reactions initiated by absorption of UVA by unidentified chromophores (Ichihashi et al., 2003; Hiraku et al., 2007; Zhou et al., 2019). The exposure to UV-A and residual UV-B elevates the likelihood of enduring effects, including photoaging, photoimmunosuppression, and photocarcinogenesis. UV-A significantly affects human and mammalian skin health by inducing pigmentation, providing a degree of protection from solar exposure; however, it also contributes to skin aging and wrinkles, which may be cosmetically unpleasant or potentially carcinogenic. UV-A radiation contributes to photobiological consequences, such as nonmelanoma skin cancer, cataracts, and immunosuppression, despite its lesser efficacy than UV-B radiation (Schmalwieser et al., 2019). Whereas changes that occur due to UV-B radiation are limited to the epidermis, although it also infiltrates the upper region of the dermis. The adverse impact of UV radiation exposure primarily encompasses skin-related issues, including sunburn, photodermatoses, hyperpigmentation, photoaging, and precancerous lesions and skin malignancies (Debacq-Chainiaux et al., 2012; Zhang et al., 2014).

In response to the harmful effects of UV radiation, cells have evolved different DNA repair pathways, such as nucleotide excision repair (NER), base excision repair (BER), and double-strand break repair (DSBR), to minimize DNA damage (McKinnon et al., 2009). Mutations in DNA repair genes that encode proteins essential for these pathways are strongly linked to neurological disorders and cancer predisposition. Disorders such as xeroderma pigmentosum (XP), ataxia-telangiectasia (AT), cockayne syndrome (CS), fanconi anemia (FA), huntington's disease (HD), and Nijmegen breakage syndrome (NBS) arise from defects in one or more DNA repair processes (McKinnon et al., 2009; Aykan et al., 2022). This chapter summarizes the UV-induced DNA damage and associated repair mechanisms, as well as diseases associated with defects in the repair mechanism.

9.2 Types of DNA damage induced by ultraviolet radiation

UV radiation, particularly UV-B (280–315 nm) and UV-A (315–400 nm), induces several types of DNA damage by being directly absorbed by the DNA molecule and disrupting its chemical structure. The most common categories of carcinogenic DNA lesions induced directly by UV-B radiation are CPDs and 6-4PPs (Fig. 9.1A), and the less frequent Dewar isomers. These lesions are the primary cause of distorting the DNA helix, interfering with base pairing, and can block DNA replication and transcription if unrepaired (Kciuk et al., 2020). Less energetic yet 20 times more powerful UVA (315–400 nm) radiation also induces the production of CPDs. Additionally, UV-A indirectly damage DNA by generating reactive oxygen species (ROS), which results in oxidative base modification such as 8–oxo–7,8–dihydroguanine (8-oxoGua) and single-strand breaks (SSBs) (Cadet and Douki, 2018). The DNA bases, due to their aromatic ring structure,

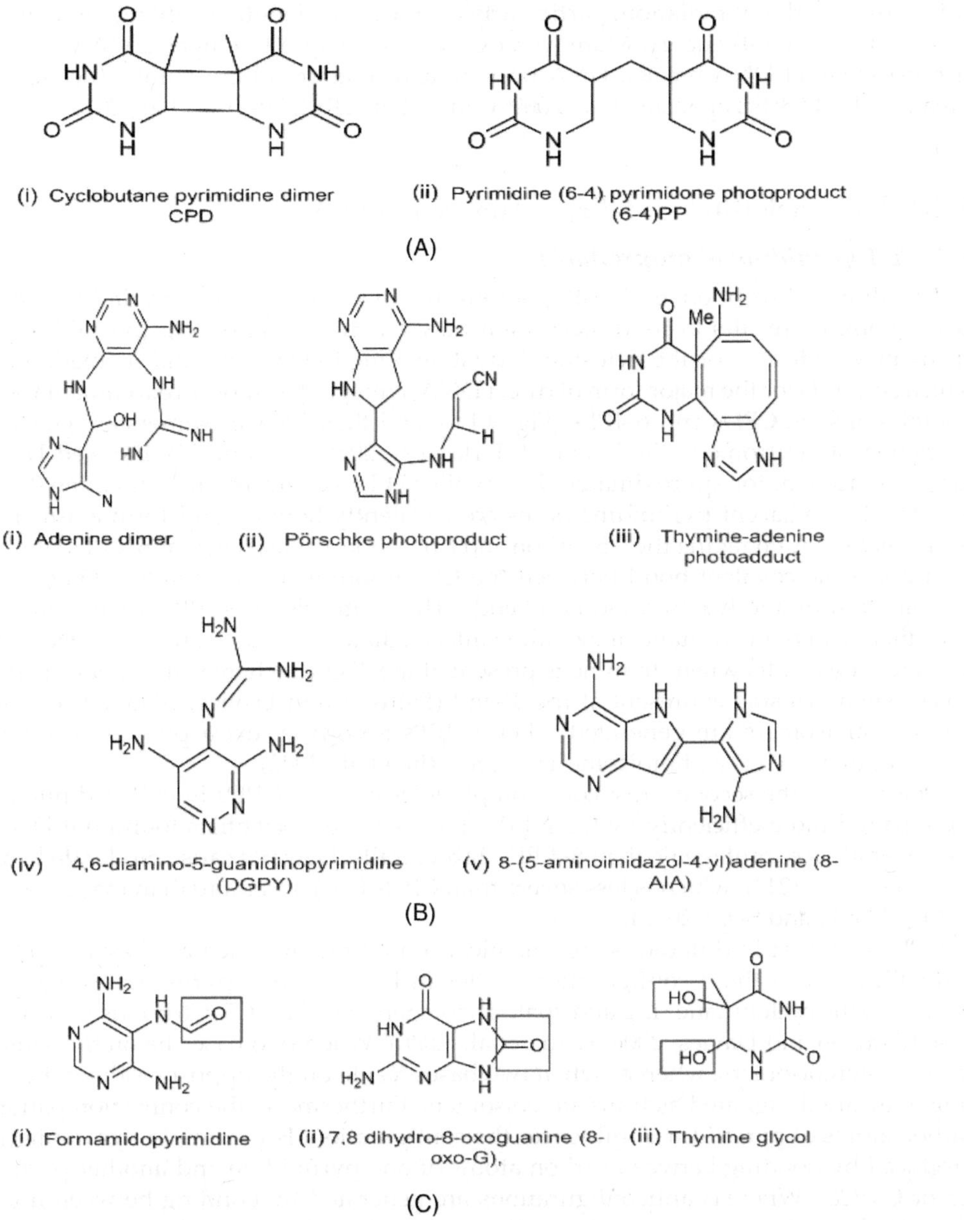

FIGURE 9.1 (A) Ultraviolet (UV)-C induced the most toxic and mutagenic photoproducts. (1) Represents cyclobutane pyrimidine dimers (CPDs) that arise between two adjacent thymines (pyrimidines), constituting 75% damage, and (2) thymine-cytosine dimers that comprise 25% of damage. (B) Structure of Photoproducts of Purine: (1) Adenine dimer, (2) pörschke photoproduct, (3) thymine-adenine photoadduct, (4) 4,6-diamino-5-guanidinopyrimidine (DGPY), and (5) 8-(5-aminoimidazol-4-yl) adenine (8-AIA). (C) UV radiation generates reactive oxygen species (ROS) that indirectly damages purines and pyrimidines by oxidation and produce mutagenic lesions such as (1) formamidopyrimidine derivative of adenine (Fapy-A), (2) 7,8-dihydro-8-oxoguanine (8-oxo-G), and (3) thymine glycol. Alterations in the structure of bases by ROS are highlighted with the blue boxes. Created with Chemdraw.

efficiently absorb UV radiation, particularly of short wavelength, enabling them to capture and trap UV-C and UV-B energy. Moreover, excessive exposure to longer UV-A wavelengths may also cause small DNA breaks, DNA-protein, and interstrand cross-links (Perrier et al., 2006; Crean et al., 2008; Sczepanski et al., 2009; Cadet et al., 2012; Silerme et al., 2014).

9.2.1 Ultraviolet-induced direct damage in DNA

9.2.1.1 Pyrimidine photoproducts

The direct absorption of UV-B photons by the photoreactive pyrimidine bases of the DNA molecule results in its dimerization between two neighboring pyrimidines, especially at thymine, cytosine, or the cytosine derivative 5-methylcytosine and its oxidation products, which accounts for the major form of direct DNA damage. The most predominant UV-B-induced photolesions are CPDs and 6-4PPs (Fig. 9.1A), and their Dewar isomers are generated by the formation of electronic excited states (Markovitsi, 2016). Among UV-induced DNA damage, the CPDs account for approximately 75% while 6-4 PPs comprise the remaining 25%. The CPD results when adjacent pyrimidine bases are covalently bonded and form a 4-membered ring conformation by engaging the C5 carbon and C6 carbon atoms, whereas 6-4 PPs synthesis occurs by a noncyclic covalent bond between the C6 carbon of one pyrimidine (5′end) and the C4 carbon atom of another pyrimidine (3′end). The synthesis of 6-4PPs occurs via spontaneous rearrangement of the oxetane or azetidine intermediates in double-stranded DNA. The oxetane intermediate results when thymine is present at the 3′ end, whereas the azetidine intermediate arises when cytosine is present at the 3′-end (Pattison and Davies, 2006; Rastogi et al., 2010). The Dewar isomers are generated when 6-4PPs absorb an extra photon at around 320 nm wavelength, which is a photoisomerized form (Jin et al., 2022).

Dewar's are the second most common photolesion after CPDs in cultured human cells and are repaired more efficiently by the NER repair pathway. Scientists found out Dewar isomers are generally less mutagenic than 6-4 PPs. Structurally, Dewar's cause moderate bending in the DNA molecule (21°), which is less severe than CPDs (44°) and more than that caused by 6-4PPs (<10°) (Douki and Sage, 2016).

CPDs can exist in different stereochemical configurations, such as cis-syn CPDs and trans-syn CPDs, where these configurations differ in how the two pyrimidine rings are oriented relative to the cyclobutane ring and spatial arrangement of the C5–C6 bonds in each pyrimidine base (Pattison and Davies, 2006; Kciuk et al., 2020). When two bases lie on the same side, they form cis stereoisomers, whereas when two bases reside on the opposite side of the cyclobutane ring, they are designated as trans stereoisomers. Furthermore, the connection pattern between carbon atoms of pyrimidines influences the configuration. For example, syn configurations are produced by bonding between carbon atoms of one pyrimidine and another pyrimidine (C5–C5 or C6–C6). Whereas anticonfigurations are generated by bonding between nonequivalent atoms (C5–C6) between pyrimidines. Experimental studies have demonstrated that cis-syn-configured CPD lesions are the primary photoproducts produced far more frequently than trans-syn-configured CPDs (Carell and Epple, 1998; Kciuk et al., 2020). The trans-syn isomers occurred more frequently in single-stranded DNA due to enhanced flexibility of the DNA backbone (Ravanat et al., 2001; Douki et al., 2003). In contrast to CPDs, 6-4PPs are also present in two stereoisomers postirradiation. Although 6-4PPs cause more significant distortions in the DNA helix structure compared to CPDs. But according to De Lima-Bessa, CPDs are considered

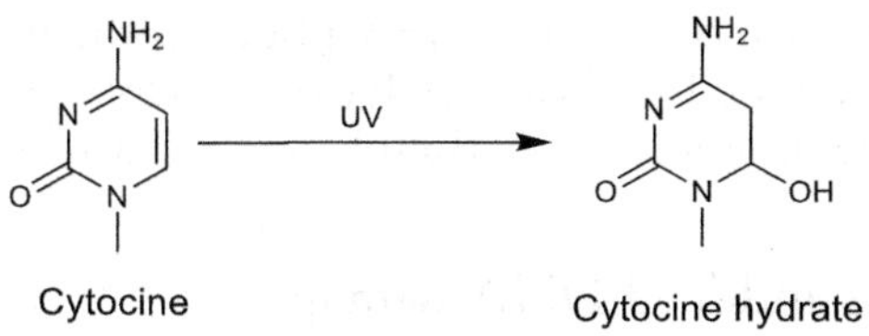

FIGURE 9.2 Emergence of cytosine photohydrate (6-hydroxy-5,6-dihydrocytosine) from cytosine through a photohydration reaction. Created with Chemdraw.

primary cytotoxic lesions accountable for cell apoptosis after UV exposure due to fast repair of 6-4PPs by NER (De Lima-Bessa et al., 2008). Interestingly, some research revealed that 6-4PPs are responsible for cellular mortality due to the inability of cells to bypass these cellular lesions. On the other hand, some studies found that 6-4PPs may not play an important role in mutagenesis (Cadet and Douki, 2018).

Furthermore, UV-B-induced CPDs in the DNA helix occur mostly at 5′TT and 5′TC sequences, followed by 5′CC and 5′CT dinucleotide sequences, with frequencies of 53%, 34%, 8%, and 5%, respectively (Jin et al., 2021). The UV-B-induced mutations primarily exhibit C > T alterations at 5′TC and 5′CC dipyrimidine sequences, but such changes hardly arise at 5′TT locations. The negligible mutagenicity at 5′TT is due to the identification of specialized DNA polymerase eta (POLH) that accurately avoids these photolesions by incorporating two adenine bases across the lesions, generally known as "A rule" (Pfeifer, 2020). The most common explanation for the UV-B-induced C>T alteration is the deamination of cytosines (or 5-methylcytosines) within CPDs (Jiang and Taylor, 1993; Pfeifer, 2020), which produces uracil bases (or thymines) within the CPDs (Douki, 2013). Such dimers are replicated through the insertion of adenines opposite the deaminated bases by DNA polymerase eta (POLH), ultimately resulting in a C > T mutation at the deaminated CPD of dipyrimidines (Tu et al., 1998; Lee and Pfeifer, 2003). Moreover, cytosine bases within CpG sequences can undergo methylation to transform into 5-methylcytosine, which may subsequently be deaminated to form thymine bases (Kciuk et al., 2020).

In addition to major pyrimidine lesions, UV radiation can induce minor photoproducts, such as cytosine photohydrates (e.g., 6-hydroxy-5,6-dihydrocytosine), formed through the addition of water to cytosine's C5–C6 double bond (photohydration reaction) (Fig. 9.2) (Wierzchowski and Shugar, 1961). The cytosine hydrates and their stereoisomers are very unstable and short-lived. Due to rapid degradation of these photoproducts, little information is known regarding their formation in UV-exposed DNA; hence, they represent only minor photoproducts (Mitchell et al., 1991; Su et al., 2010).

9.2.1.2 Purine photoproducts

Although UV-B radiation primarily forms dipyrimidine photoproducts, UV-induced variations in DNA purine bases have also been acknowledged as biologically significant. Purine photoproducts involve at least one adenine residue that reacts with nearby adenine or thymine through photocycloaddition upon UV-B exposure. Although the frequency of adenine-thymine (A-T) is minimal, i.e., 1 in 100,000 bases in native DNA, it has been demonstrated to be mutagenic (Bose and Davies, 1984). Adenine dimerization occurs via a cycloaddition between the N7–C8 bond of one adenine and the C6–C5 bond of the adjacent adenine, leading to an unstable azetidine intermediate. This intermediate may then result in the formation of two main adenine photoproducts: the adenine dimer (A=A) and the Pörschke photoproducts. Both the photoproducts can be further converted into identifiable compounds like 4,6-diamino-5-guanidinopyrimidine (DGPY) and 8-(5-aminoimidazol-4-yl) adenine (8-AIA), detectable after acid hydrolysis of UV-exposed DNA (Fig. 9.1B). Studies indicate that A=A photoproducts are significantly reduced when complexation of UV-irradiated poly(dA)-poly(dT) in DNA is present, suggesting that base pairing suppresses adenine photoreactivity (Kciuk et al., 2020).

9.2.2 Indirect damage—Reactive oxygen species-induced DNA damage

The direct effect of UV rays causes CPDs and 6–4 photoproducts lesions, which are well-known contributors to mutagenesis and carcinogenesis. A significant portion of cellular damage and other DNA damage induced by UV-A occurs through indirect pathways. Unlike UV-B, which acts directly on DNA bases, UV-A radiation penetrates deeper into the cells and tissues and is involved in ROS generation indirectly through the photosensitization process, directly affecting cellular components. UV-A indirectly excites chromophores like pterins, folates, flavins, melanins, 6-formylindolo[3,2-*b*]carbazole, porphyrins, and aromatic amino acids like tyrosine, tryptophan, and cytosine (double-bonded cysteine), as well as biopolymers such as proteins and nucleic acids (Park et al., 2015; Buglak et al., 2016; Baptista et al., 2021; Fu et al., 2021). Upon excitation, these chromophores transfer an electron to molecular oxygen, leading to the formation of ROS (De Jager et al., 2017), which includes a range of species such as O^{2-}, singlet oxygen, hydroxyl radicals, and hydrogen peroxide (OH•) (Kimeswenger et al., 2016). Among them, hydroxyl radicals (OH•) produced from Haber-Weiss reactions possess a very short half-life (10^{-9} s) and have significant DNA-damaging capacity due to their high reactivity (Juan et al., 2021). The primary mechanism by which DNA is compromised by hydroxyl radical stress involves the generation of oxidized nitrogenous bases (pyrimidines and purines) and an SSB ratio of 1:1:3 through an oxidative mechanism. This process starts with the abstraction of a proton by radicals from any site on the deoxyribose, resulting in the production of several products (Vechtomova et al., 2021). Over 100 types of oxidatively formed base lesions and 2 sugar modifications have been identified, indicating the extensive range of potential oxidative damage (Cadet et al., 2010). For purines, oxidation often yields formamidopyrimidine derivatives, 8-oxo derivatives (7,8-dihydro-8-oxoguanine), and 2-hydroxyadenine, while pyrimidine oxidation results in hydrates and glycol derivatives (Fig. 9.1C). Guanine is particularly susceptible to oxidation and alters the stability of the genome. Furthermore, 8-oxo-7,8-dihydroguanine, the most frequently occurring DNA lesion associated with skin cancer, possesses mutagenic potential and has the ability to block transcription and replication processes (Cadet et al., 2015). In addition to base damage, one of the consequences of ROS production is the introduction of SSBs by directly targeting and damaging the DNA sugar-phosphate backbone. If occurring on both strands, SSBs can turn into more deadly double-strand breaks (DSBs). DSBs can also occur during replication fork arrest of unrepaired photoproducts. Moreover, it has been found that ROS-induced DNA strand breaks are linked to topoisomerase malfunction. The replication stresses caused by UV radiation can block topoisomerase I and II, resulting in the generation of cleavage complexes that can be turned into DSBs (Strumberg et al., 2000; Banáth and Olive, 2003; Ohnishi et al., 2009). Additionally, ROS can also affect other components of cells, such as lipids and proteins. ROS can cause lipid peroxidation by forming products such as malondialdehyde (MDA) and 4-hydroxynonenal (4-HNE) that can form DNA adducts with bases like adenine, cytosine, and guanine, further leading to genomic instability and mutagenesis (Bauer et al., 2015; Chatterjee and Walker, 2017; Poetsch, 2020). ROS can cause protein damage, which is essential for cellular functions, by oxidizing amino acid residues, cleaving peptide bonds, and aggregating proteins (Juan et al., 2021). At low ROS levels, cells may accumulate mutations; at moderate levels, they may enter senescence, and at high levels, apoptosis or necrosis occurs. Importantly, this oxidative damage has been associated not only with skin aging and immune suppression but also with the initiation and progression of melanoma and nonmelanoma skin cancers (Wei et al., 2024).

9.3 DNA repair mechanisms in mammals

Over the past three decades, significant advances have been made in our comprehension of many DNA repair mechanisms in both prokaryotes and eukaryotes. Majorly four types of repair pathways exist, namely, direct reversal of damage (DR); NER, which encompasses global genome repair (GG-NER) and transcription-coupled repair (TC-NER); BER; and finally, recombination repair, including homologous repair and nonhomologous end joining (NHEJ). Bacteria use the DNA photolyase enzyme for the direct reversal of DNA damage (CPDs and 6-4PPs), but in humans and most mammals, there is no DNA photolyase enzyme, and the only system responsible for removing the UV-induced damage is the NER pathway (Vechtomova et al., 2021). This chapter mainly focuses on the DNA repair pathways that occur in mammalian cells. If these repair pathways do not work properly, this will lead to mutagenesis and cancer progression, ultimately leading to cell death (Sancar, 2016).

9.3.1 Nucleotide excision repair

Living organisms throughout the evolutionary spectrum protect their genetic material from the continuous exposure of terrestrial UV rays and byproducts of cellular metabolism. DNA modifications such as UV photoproducts, bulky DNA adducts, and bases damaged by ROS, which alter the structure of the DNA helix and interrupt RNA polymerase progressions, are specifically addressed by one of two subpathways of NER: GG-NER or transcription-coupled (TC-NER) (Spivak, 2015; Kciuk et al., 2020). Each NER subpathway can be classified into five principal molecular events: (1) Detection of damage, (2) verification of damage, (3) incision of sugar-phosphate backbone of DNA around the damage to excise a damage-bearing oligonucleotide, (4) repairing of the nucleotide patch by using the undamaged complementary strand, and (5) ligation of the remaining nick to finalize the process as shown in Fig. 9.3 (Tiwari and Wilson, 2019). The former repair pathway GG-NER is activated by precise identification of a DNA structure-distorting lesion within the genome, while the second process TC-NERoccurs during transcription, when RNA polymerase II (RNAPII) stalls at the damage site within a gene that is actively transcribed. Both pathways harness common repair proteins but only differ in the initial step of DNA damage recognition. Research demonstrates that GG-NER needs hours to eliminate substrate lesions, but TC-NER acts in a rapid manner, generally in minutes, to resume transcription (Rapin, 2013). Both repair mechanisms start with the remodeling of chromatin to allow the CPD damage site to be accessible for the repair proteins (Adam and Polo, 2012; Czaja et al., 2012).

Deficiencies in this mechanism result in XP disease. People afflicted by this condition exhibit maximum sensitivity to sunlight and are prone to the development of skin malignancies. Genetic screening of these individuals identified mutations in primary genes named XPA-XPG, which are important to the NER function (Sancar, 2016).

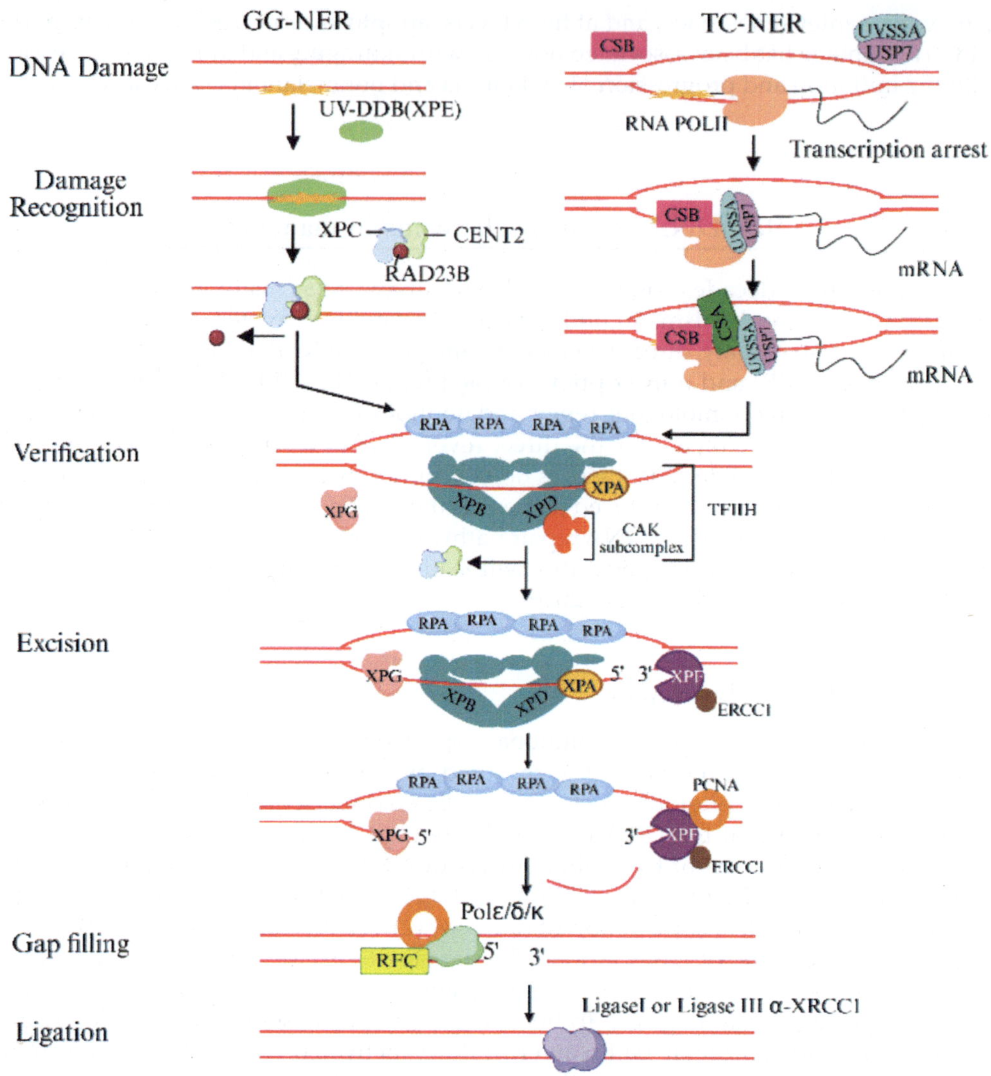

FIGURE 9.3 Global genome NER and transcription-coupled NER repair mechanisms. Created with Biorender.com.

9.3.1.1 Global genome-nucleotide excision repair

In GG-NER, about 20–30 proteins work consecutively, as shown in Fig. 9.3. The process starts with the XPC protein (damage recognition protein) functioning together with UV excision repair protein RAD23 homolog B (RAD23B) and centrin2 (CENT2) to identify UV-induced damage in the genome (Chatterjee and Walker, 2017). Sometimes, the UV radiation-DNA damage binding protein (UV-DDB) complex, acting upstream of the XPC-RAD23B complex comprising DDB1 (XPE-binding protein) and the DDB2 protein, detects small disturbances caused by CPDs, which are further detected by XPC. The assembled complex scans the DNA for helix instability. XPC is associated with the intact strand opposite to the photodamaged one (Scharer, 2013; Lee et al., 2014; Kciuk et al., 2020). Moreover, the XPC-RAD23B complex is not only an essential recognition factor for GG-NER but also important for the assembly of downstream NER components and ensuring the effective progression of the repair response (Scharer, 2013). XPC acts as a platform for recruiting the transcription factor II H (TFIIH) complex at the 5′ lesion site. TFIIH consists of 10 protein subunits, including ATP-dependent helicases XPB (also called ERCC3) (3′-5′ helicase) and XPD (called ERCC2) (5′-3′ helicase), which employ them to move along both DNA strands to unwind the DNA toward the damage site (Coin et al., 2007; Kciuk et al., 2020). XPD encounters a bulky lesion on the DNA helix, and both XPD and XPB become halted and recruit XPA (which enhances the stalling effect), replication protein A (RPA), and XPG, which together form the preincision complex (Li et al., 2015). RPA binds with the intact strand arising from the action of XPB and XPD helicases and renders it accessible to the lesion-cleaving endonucleases XPF/ERCC1, which incise the lesion-containing strand on the 5′ side of the bubble, and XPG, which cleaves the same strand on the 3′ side of the lesion (Araújo and Kuraoka, 2019). As a result, a DNA fragment of 24–32 nucleotides encompassing the dimerized pyrimidine remains bound to TFIIH until it is excised and subsequently digested by nucleases (Tsodikov et al., 2007; Staresincic et al., 2009; Kemp et al., 2012). In the next step, the resultant gap is further filled by DNA polymerases $\delta/\varepsilon/\kappa$ along with the proliferating cell nuclear antigen (PCNA), replication factor C (RFC), and replication protein A (RPA). In replicating cells and nonreplicating cells, the patch is filled by pol ε, pol δ, and κ, respectively (Ogi et al., 2010; Lehmann, 2011). Finally, broken phosphodiester bonds are ligated by the DNA ligase I in replicating cells, while Ligase IIIα works with X-ray cross-complementing protein 1 (XRCC1) in quiescent cells (Lehmann, 2011; Spivak, 2015). This NER pathway is the primary DNA repair mechanism that addresses the UV-induced photolesions in mammalian cells.

9.3.1.2 *Transcription-coupled-nucleotide excision repair*

Transcription-coupled nucleotide excision repair (TC-NER) is a specialized repair mechanism associated with actively transcribed genes. This pathway is of prime importance, as DNA damage-induced transcriptional blockage can disrupt gene expression, ultimately resulting in impairing cell growth and viability (Vermeulen and Fousteri, 2013). In this pathway, lesions are repaired fast with accuracy on the transcribed strand of active genes (Pfeifer, 2020). The process starts when DNA lesions are encountered by RNAPII; it leads to its blockage at the damage site, triggering the initiation of TC-NER (Vermeulen and Fousteri, 2013). After stalling, recruitment of two main proteins occurs, named CSA (ERCC8), which is mutated in Cockayne syndrome type A, and CSB (ERCC6), mutated in Cockayne syndrome type B (Fig. 9.3) (Vermeulen and Fousteri, 2013; Pfeifer, 2020). CSB, an ATP-driven remodeler that wraps DNA around itself and loosens the RNAPII–DNA interface. CSB, a master coordinator in TC-NER, appoints CSA complex (Spivak, 2016). CSA, being the essential component of the pathway, recruits other TC-NER components. CSA, along with DDB1, cullin 4A, and E3 ubiquitin ligase RBX1, makes an E3-ubiquitin ligase complex that is inactivated at the damaged site by the constitutive photomorphogenesis 9 (COP9) signalosome (CSN), which deneddylates Cullin (Fousteri and Mullenders, 2008). CSB, in conjunction with CSA, promotes RNAPII backtracking, resulting in exposing damaged sites to repair proteins, including XPA-binding protein 2 (XAB2), high mobility group nucleosome-binding domain-containing protein 1 (HMGN1), ubiquitin-specific processing protease 7 (USP7), p300, SPT16, and transcription factor IIS (TFIIS)

(Fousteri and Mullenders, 2008; Fei and Chen, 2012). XAB2 binds with the XPA protein and RNAPII, acting as a scaffold for further recruitment of additional TC-NER components. HMGN1 associates with RNAPII in a CSA/CSB-dependent manner, which loosens nucleosomes by competing with histone H1 (Fousteri and Mullenders, 2008). TFIIS cleaves the nascent transcript to remove mismatched nucleotides by backtracking of RNAPII. Upon transcriptional arrest, USP7 and UVSSA interact strongly with RNAPII (Nakazawa et al., 2012; Schwertman et al., 2012; Zhang et al., 2012). USP7 stabilizes CSB by deubiquitylating it. This action counteracts CSA-mediated breakdown of CSB, which under normal conditions is important for reactivation of transcription (Brooks, 2013). UV-stimulated scaffold protein A (UVSSA) and ubiquitin-specific processing protease 7 (USP7) collaboratively not only protect CSB but also promote the CSA and CSB-mediated ubiquitylation of stalled phosphorylating RNAPII, assist in remodeling, and allow USP7-mediated deubiquitylation, which helps in conversion to nonphosphorylated RNAPII for the initiation of transcription (Schwertman et al., 2013; Spivak and Ganesan, 2014). After recognition, the rest of the pathway is similar to GG-NER (Scharer, 2013; Spivak, 2015).

9.3.2 Base excision repair

BER is the primary mechanism for fixing small-scale DNA damage such as oxidized DNA bases (e.g., 8-oxoG), alkylated nucleotides, and deaminated bases. It involves five major steps. (1) base excision of the damaged base by the specific DNA glycosylase, resulting in the formation of an apurinic/apyrimidinic (AP) site; (2) incision at the AP site by AP-endonuclease or AP-lyase; (3) processing of ends; (4) gap-filling; (5) ligation (Fortini, 2003). The repair starts with DNA glycosylases traversing the genome, identifying and removing the damaged bases by removing the *N*-glycosidic bond that connects the base to the deoxyribose sugar, resulting in the formation of an AP site (Krokan and Bjørås, 2013; Quiñones and Demple, 2016). Mammals possess approximately 11 varying damage-specific DNA glycosylases with overlapping but distinct substrates (Krokan and Bjørås, 2013), such as uracil deoxyribonucleic acid glycosylase (UNG), single-strand selective monofunctional uracil glycosylase (SMUG1), thymine DNA glycosylase, methylpurine glycosylase, *Escherichia coli* Nei endonuclease VIII-like 1/2/3 (NEIL1/2/3), *E. coli* nth endonuclease III-like 1 (NTHL1), *E. coli* MutY homolog (MUTYH), methyl binding domain 4 protein and 8-oxoguanine DNA glycosylase (OGG1) (Tiwari and Wilson, 2019). Based on enzymatic function, these glycosylases are classified as monofunctional (possessing only glycosylase activity, e.g., UNG, TDG, and MPG) or bifunctional (exhibiting both glycosylase and AP-lyase activity, e.g., OGG1 and NEIL1–3). Bifunctional glycosylases directly cleave the phosphodiester bonds at the 3′ side of the AP site via β-elimination or β,δ-elimination, resulting in a 3′-phospho-α,β-unsaturated aldehyde (PUA) and phosphate groups on both ends (Tudek, 2007). SSBs generated from bifunctional DNA glycosylases need to be processed by 3′SSB repair enzymes such as polynucleotide kinase phosphatase (PNKP), tyrosyl-DNA phosphodiesterase 1 (TDP1), and aprataxin (APTX) before gap-filling and ligation, as shown in Fig. 9.4. PNKP eliminates 3′ phosphates, resulting in a 3′ OH group, and phosphorylates 5′ hydroxyl termini of DNA molecules; TDP-1 eliminates covalent adducts via hydrolysis of a 3′-phosphodiester bond, and APTX, which excises adenosine monophosphate (AMP) residues that remain on DNA due to the abortive function of DNA Ligase IV during nonhomologous end joining (NHEJ). After eliminating the damaged base, poly ADP-ribose polymerase-1 (PARP-1), a nick sensor protein, is recruited to the damaged sites to enhance the effectiveness of the repair process. The main

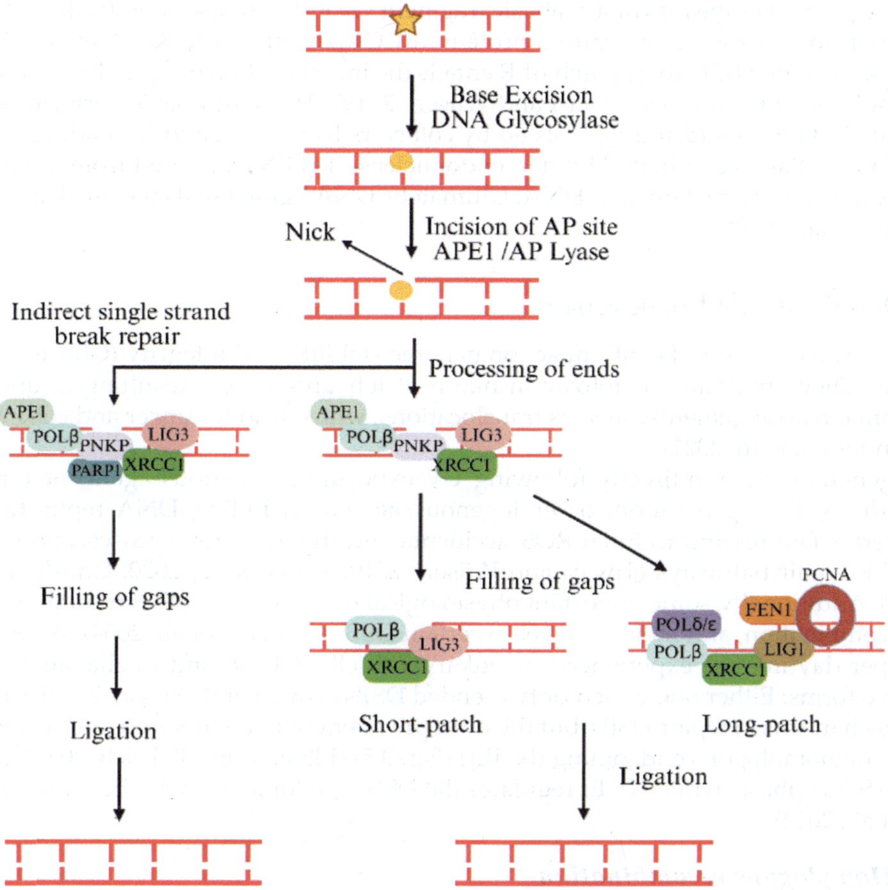

FIGURE 9.4 The base excision-based short-patch and long-patch repair mechanisms. Created with Biorender.com.

AP-endonuclease in mammals is APEI, that turn 3′ α,β-unsaturated aldehydes into hydroxyl groups at 3′ ends ready to be filled by DNA Polβ and ligated by LIG3α (Kciuk et al., 2020).

Monofunctional glycosylases rely on mammalian APEI to cut the phosphodiester bond at the 5′ end of the abasic (AP site), generating a SSB that carries a hydroxyl group at the 3′ site and deoxyribose phosphate (5′dRP) at the 5′ end (Friedberg and Meira, 2006; Li and Wilson, 2014). 3′ ends containing phosphoglycolates or phosphates undergo further processing and preparation for DNA synthesis. APEI possesses 3′ phosphodiesterase activity and effectively removes 3′ phosphoglycolate groups, although it exhibits limited 3′ phosphatase activity (Kelley and Parsons, 2001). Phosphate groups at the 3′ hydroxyls are likely eliminated by polynucleotide kinase PNKP (Wiederhold et al., 2004; Tudek, 2007).

Following incision by AP endonuclease/AP-lyase and end processing, BER may advance through two alternative pathways: the "short-patch" (or single-nucleotide) or "long-patch" repair pathway, represented in Fig. 9.4. Short-patch BER, apparently the preferred mechanism, typically includes the replacement of a single-nucleotide by the gap-filling DNA polymerase β.

Polymerase β removes the residual 5′ abasic fragment via its 5′ dRPase activity, then DNA ligase 3α (LIG3α) in complex with the scaffold protein XRCC1 seals the nick (Kciuk et al., 2020).

Unlike short-patch BER, long-patch BER entails the incorporation of up to 13 nucleotides and strand-displacement synthesis (Tiwari and Wilson, 2019). The synthesis is performed by high-processivity polymerase (δ and ε), assisted by cofactors like PCNA and its loading factor RFC. The resulting 5′ flap is eliminated by flap endonuclease 1 (FEN1), created from the interaction of DNA polymerase with damaged DNA. Ultimately, DNA ligase I seals the break in long-patch BER (Kciuk et al., 2020).

9.3.3 Double-strand break repair

DSBs are causing a significant impact on genome stability and integrity (Ghosh and Raghavan, 2021). They are mainly cytotoxic in nature if left unresolved, resulting in apoptosis or chromosomal rearrangements, such as translocations, which lead to cancer and cell cycle arrest (Ghosh and Raghavan, 2021).

They generally arise indirectly following UV exposure (exogenous agent) or through the photoproducts, ROS generation, or endogenous agents, including DNA replication errors, topoisomerase failures, intracellular ROS, accidental nicking by endonucleases, or development during BER repair pathways (Tiwari and Wilson, 2019; Kciuk et al., 2020; Ghosh and Raghavan, 2021). Additionally, some important physiological processes like V(D)J recombination, class switch recombination, and meiosis develop certain DSBs (Couëdel et al., 2004). Approximately 10 DSBs per day may be experienced by eukaryotic cells (Ghosh and Raghavan, 2021). They exist in two forms: Either one-ended or two-ended DSBs (Tiwari and Wilson, 2019). Cells utilize several mechanisms to repair DSBs, but the most prevalent ones are homologous recombination (HR) and nonhomologous end joining (NHEJ) (Fig. 9.5) (Elbakry and Löbrich, 2021). HR occurs in the late S/G2 phase, while NHEJ regulates the DSBs predominantly in the G1/early S phase (Aykan et al., 2022).

9.3.3.1 Homologous recombination

HR is the most efficient (high-fidelity) repair mechanism that employs sister chromatids (homologous copies) as templates to address detected DSBs (Fig. 9.5). This high-fidelity repair mechanism is particularly suited to resolve one-ended DSBs generated from replication fork collapse (Ait Saada et al., 2018; Wright et al., 2018; Tiwari and Wilson, 2019). The process commences with the detection and binding of the MRN complex, consisting of meiotic recombination 11 homologue A (Mre11)–RAD50–Nijmegen breakage syndrome 1 (NBS1), to DSBs, resulting in recruitment of ataxia-telangiectasia mutated kinase (ATM kinase) (Kumar et al., 2016). ATM kinase is a pivotal protein kinase that identifies the DSBs and activates the G1, S, and G2/M checkpoints by phosphorylating several checkpoint control protein such as structural maintenance of chromosomes 1 (SMC1), checkpoint kinase 2 (CHK2), breast cancer 1 (BRCA1), p53 and NBS1 (Clouaire et al., 2017; Tiwari and Wilson, 2019). HR is generally categorized into four key stages: DNA end resection, strand invasion, repair synthesis, and strand dissociation.

After DSB identification, the resection phase is initiated when C-terminal binding protein 1 (CtBP1), with its nuclease activity in conjunction with the Mre11/Rad50/Nbs1 (MRN) complex, facilitates 5′ DNA end resection to generate single-stranded 3′ overhangs (Aykan et al., 2022). This step is regulated by CtBP-interacting protein (CtIP), which works alongside breast cancer 1 (BRCA1), while other nucleases like Werner syndrome ATP-dependent helicase

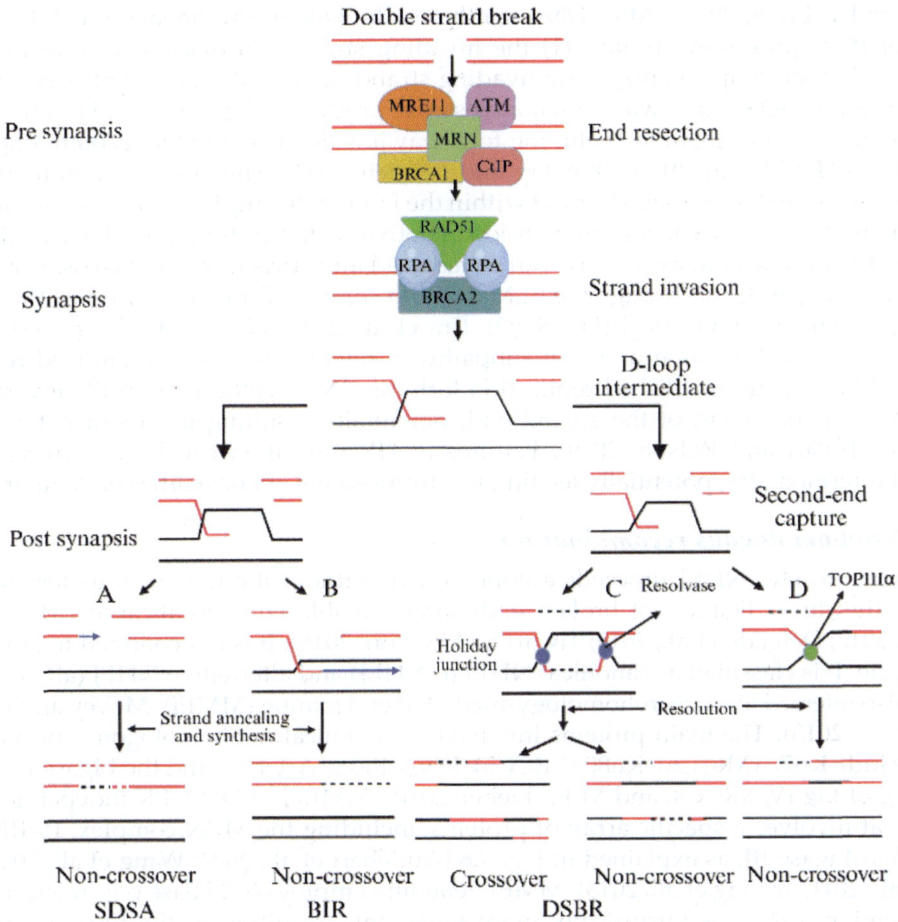

FIGURE 9.5 Homologous recombination-mediated DSB repair. Created with Biorender.com. *DSB*, Double-strand break

(WRN), Bloom syndrome protein (BLM), DNA replication ATP-dependent helicase/nuclease DNA2 (DNA2), and exonuclease 1 (EXO1), also participate in this process (Kciuk et al., 2020). The resulting ssDNA is coated by replication protein A (RPA) to stabilize the strand, which is later replaced by RAD51 recombinase, with the assistance of breast cancer 2 (BRCA2), to facilitate the presynapsis phase of HR (Jasin and Rothstein, 2013). RAD51 forms a nucleoprotein filament and initiates a homology search. In the synapsis phase, RAD51 promotes strand invasion into the homologous duplex, resulting in the development of a displacement loop (D-loop) where homology is established between the invading and the template strand (Kurtoğlu and Tekedereli, 2015; Aykan et al., 2022). During the postsynaptic phase, RAD51 in a calcium-dependent manner enhances the ATPase activity of RAD54 to facilitate the interweaving of the 3′ invader strand with the corresponding template strand, thereby creating a primer-template junction conducive to DNA synthesis

(Elbakry and Löbrich, 2021). After DNA synthesis, the D-loop intermediate may be resolved via one of three processing routes: (1) the invading strand can disengage from the D-loop and reanneal to its complementary noninvading strand, representing the synthesis-dependent strand annealing (SDSA) pathway with noncrossover products. (Fig. 9.5A); (2) The other process involves joint structure creation, including four-way junctions among the recombining strands referred to as Holliday junctions (Elbakry and Löbrich, 2021). This results when noninvading DNA end anneals to the displaced strand within the D-loop during a second-end capture phase, or potentially through the concurrent invasion of both resected ends into the donor, followed by extension. The double holiday junction can be resolved into crossover or noncrossover products by resolvases (Fig. 9.5C) or through the BLM-TOPIIIα complex [Bloom syndrome DNA helicase (BLM)-topoisomerase (TOPIIIα)] (Fig. 9.5D). (Ira et al., 2003; Wu and Hickson, 2003) (3) HR can also follow the third noncanonical subpathway referred to as break-induced replication (BIR), identified by an established replication fork for DNA synthesis from the invading DSB end, without involvement of the second-end, potentially resulting in loss of heterozygosity (Fig. 9.5B) (Tiwari and Wilson, 2019). Failures in HR resolution can lead to intricate DNA structural intermediates, potentially leading to chromosomal abnormalities or cellular demise.

9.3.3.2 Nonhomologous recombination

In contrast to HR, NHEJ proceeds autonomously without the homologous templates and promotes the direct ligation of broken ends after suitable end modifications (Fishel, 2016; Liu et al., 2017; Delgado et al., 2018; Tiwari and Wilson, 2019). It is not confined to any stage of the cell cycle. It is classified as canonical NHEJ (c-NHEJ) and alternative NHEJ (alt-NHEJ). Alt-NHEJ is also referred to as microhomology-mediated end joining (MMEJ) (McVey and Lee, 2008; Decottignies, 2013). The main proteins involved in canonical nonhomologous end joining (c-NHEJ) include Ku70 (XRCC6), Ku80 (XRCC5), DNA-PKcs, Artemis, and the Ligase IV complex (consisting of Lig IV, XRCC4, and XLF) (Lieber, 2010). MMEJ is a DNA-PK-independent repair method that involves a specific array of proteins, including the MRN complex, PARP-1, CtIP, XRCC1, and Ligase III, as explained in Fig. 9.6 (Audebert et al., 2004; Wang et al., 2006; Della-Maria et al., 2011; Truong et al., 2013). MMEJ generally employs 5–25 base pair microhomology for end joining. MMEJ is intrinsically more mutagenic, resulting in the loss of nucleotides between microhomology areas (Yu and McVey, 2010). Like other DNA repair mechanisms, NHEJ is followed by multiple steps, including identification, end bridging, processing, DNA synthesis, and ligation.

The classical pathway begins with recognition and binding to broken DNA ends by a crucial factor named Ku heterodimers (Ku70/Ku80), which prevents excessive end resection. Ku, in conjunction with DNA damage-sensing protein 53BP1 and its effectors, prevents DNA end resection, which is a crucial step for NHEJ repair (Bunting et al., 2010; Escribano-Díaz et al., 2013; Mirman and Lange, 2020). Structurally, the toroid-shaped Ku heterodimer acts as a platform to recruit additional proteins.

This includes DNA-PKcs, which forms synaptic complexes to hold DNA ends together (Tiwari and Wilson, 2019). DNA-PKcs interacts with Artemis, a nuclease belonging to the metallo-β-lactamase family. The autophosphorylated DNA-PKcs activates Artemis by phosphorylating its C-terminal autoinhibitory region of the Artemis nuclease. Upon phosphorylation, Artemis acquires both 5′ and 3′ endonuclease activity (Ma et al., 2005; Goodarzi et al., 2006; Pawelczak and Turchi, 2010; Ghosh and Raghavan, 2021). This processing activity generates ligatable ends by processing them. In addition to this, other proteins are also contributing

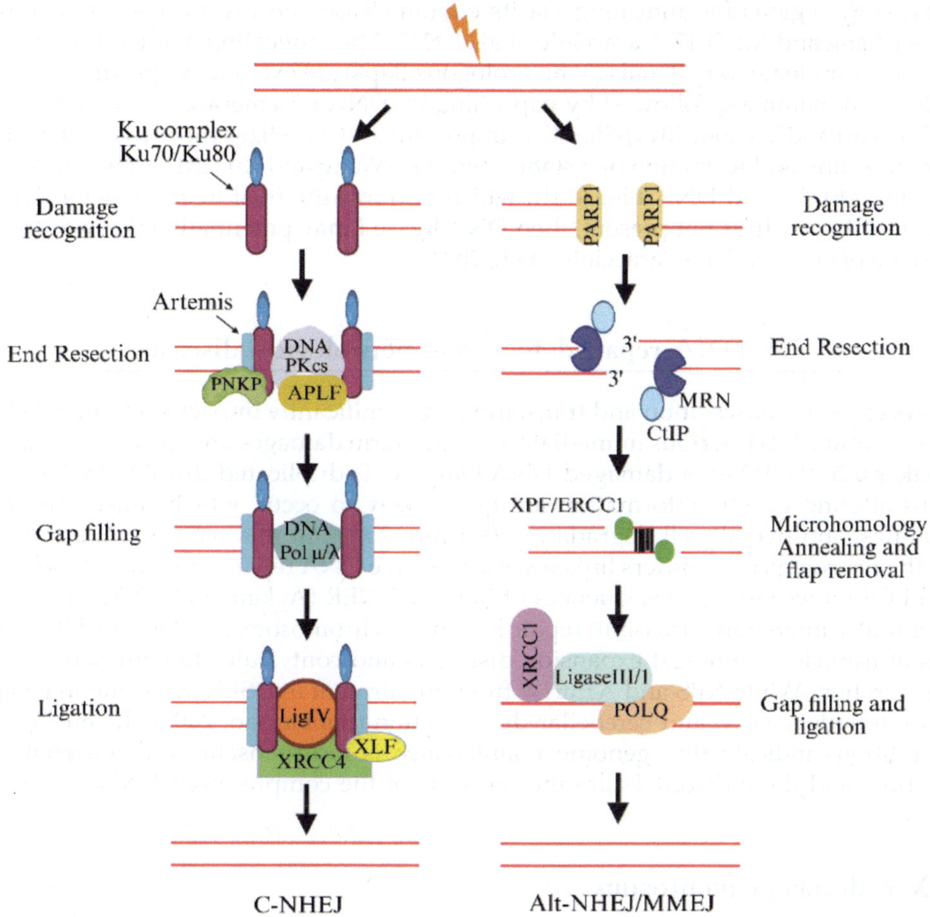

FIGURE 9.6 Nonhomologous recombination pathway. This pathway is further categorized into the classical and alternate-NHEJ repair mechanisms of DSB repair. Created with Biorender.com.

to end processing, such as PNKP, PNKP-like factor (APLF), TDP1, APTX, and tyrosyl-DNA phosphodiesterase 2 (TDP2) (Gómez-Herreros et al., 2013; Heo et al., 2015; Chang et al., 2017; Ghosh and Raghavan, 2021). After processing, DNA polymerase λ (Pol. λ) and polymerase μ (Pol. μ) fill the gap. Finally, ends are ligated by the DNA Ligase IV–XRCC4–XLF complex (Menon and Povirk, 2016; Tiwari and Wilson, 2019). XRCC4 increases the end-joining activity by augmenting the adenylation of the Ligase IV enzyme (Grawunder et al., 1997). The filamentous complex XRCC4-XLF stabilizes the broken ends by connecting them (Ropars et al., 2011; Brouwer et al., 2016; Tiwari and Wilson, 2019).

When classical NHEJ is impaired, cells use alternate pathways that are more error-prone, known as MMEJ or Alt-NHEJ. Unlike the classical pathway, MMEJ operated independently of KU or the XRCC4–Ligase IV complex (Mateos-Gomez et al., 2015). Instead, it starts with DSB identification by PARP1 and initiates end resection by the MRN complex and CtIP, producing microhomology regions for annealing via its endonuclease and exonuclease functions at the repair site (Wang and Xu, 2017; Caracciolo et al., 2021). After annealing of microhomologous sequences, noncomplementary 3′ tails (3′ heterologous flaps) are excised by specific endonuclease XPF/ERCC1 in mammals, followed by gap-filling by DNA polymerase Q (POLQ) (Wang and Xu, 2017; Tiwari and Wilson, 2019). POLQ is important as it stabilizes the 3′ overhangs, facilitating their annealing and formation of a stable complex (Wang and Xu, 2017). The MRN complex then recruits XRCC1 and DNA Ligase III, which perform the final ligation steps. If the most efficient DNA Ligase III is not present, then DNA ligase I may potentially take over to catalyze the last stage of end ligation (Caracciolo et al., 2021).

9.4 DNA repair defects and neurological diseases

The processes of transcription and translation are significantly impacted by the DNA lesions (Giglia-Mari et al., 2011). Serious immediate and long-term damages are caused by gene damage (Hoeijmakers, 2001). When a damaged DNA template is duplicated during DNA replication, mutations altering genetic information are quite likely to occur, which cause cancer, innate abnormalities, and overall cell degradation (Karran, 2000). In this section, we are only highlighting the neurological disorders linked with defective DNA repair pathways. Disorders such as XP and CS represent the consequences of impaired NER (Aykan et al., 2022). FA shows the impact of faulty interstrand crosslink repair leading to chromosomal instability. HD represents the class of trinucleotide repeat expansion disorders and contributes to neuronal dysfunction and degeneration. While NBS and AT arise from impairment in DSBR, resulting in progressive ataxia, immunodeficiency, and cerebellar degeneration (Subba Rao, 2007). Taken together, all these conditions indicate that genome maintenance is indispensable for neuronal survival and function, and these disorders are the outcome of the compromised DNA damage repair pathways.

9.4.1 Xeroderma pigmentosum

XP is an uncommon autosomal recessive DNA repair condition that is characterized by light sensitivity and UV radiation-induced malignancies of the skin and mucosal membranes (Aykan et al., 2022). This condition is caused by mutations that impair the NER pathway, which removes DNA lesions caused by UV rays. Defects in DNA polymerase η and seven proteins of the NER pathway were discovered in XP cells (DiGiovanna and Kraemer, 2012). Those suffering from XP are extremely sensitive to sunlight and often acquire deadly skin cancer and neurological complications (Aykan et al., 2022). DNA repair is crucial for preventing cancer in the general population, as evidenced by clinical research conducted at the National Institutes of Health that revealed a nearly 10,000-fold increase in skin cancer in XP patients under the age of 20. About 25% of XP patients experience increasing neurological degeneration and neuronal loss over time. This is most likely due to DNA damage brought on by oxidative metabolism, which destroys the nervous system's nondividing cells (DiGiovanna and Kraemer, 2012).

9.4.2 Cockayne syndrome

Mutations in the CSB gene are the cause of Cockayne syndrome, in which the TC-NER pathway, which repairs DNA damage, is impaired or faulty in CS patients (Karikkineth et al., 2017;

Aykan et al., 2022). Recent research, however, indicates that flaws in mitochondrial processes and base excision DNA repair might potentially be important. As a result, molecular treatments in CS and, by extension, aging may become possible. It is a hereditary autosomal recessive condition. The disorder's most obvious symptoms include early-onset aging, impaired brain development due to several organ degeneration, neurological abnormality, hypersensitivity to sunlight, and ambulatory and feeding difficulties, leading to death by 12 years of age on average. In addition to two complementation groups (CSA and CSB) and many phenotypes (Karran, 2000; Hoeijmakers, 2001; Giglia-Mari et al., 2011), CS shares similarities with XP. It has been categorized as progeria, and a lot of its symptoms are similar to those of accelerated aging. Therefore, CS research offers a chance to learn more about the fundamental processes of aging (Karikkineth et al., 2017; Aykan et al., 2022).

9.4.3 Fanconi anemia

Numerous symptoms are seen in FA, an autosomal recessive hereditary condition (Auerbach, 2009). Bone marrow failure and susceptibility to tumor development are two examples. Skin cancers and pigmentation disorders (hyperpigmentation, hypopigmentation, and café-au-lait spots) are the main signs of dermatology. It is brought on by mutations in genes that fix interstrand cross-links, frequently in conjunction with HR genes like RAD51, BRCA1, and BRCA2 (Pollard and Gatti, 2009; Soulier, 2011). Patients experience congenital anomalies, bone marrow failure, and an increased risk of leukemia and other malignancies.

9.4.4 Huntington's disease

HD is a progressive neurological illness that is autosomal-dominant. Its distinctive phenotype includes incoordination, behavioral issues, cognitive decline, and chorea and dystonia.

While the illness can appear at any point between infancy and senescence, symptoms usually begin in middle life after afflicted persons have kids (Bates et al., 2015). Huntington's disease-related huntingtin is a mutant protein in which an enlarged CAG repeat leads to a variable-length polyglutamine strand at the N-terminus. Evidence points to a hazardous increase in function from this tail (Martin, 1999; Sieradzan and Mann, 2001). The base/base mismatches arising during replication are fixed by the mismatch repair system; Oxidative DNA damage results in numerous repeats, which causes HD (Jeppesen et al., 2011).

9.4.5 Ataxia-telangiectasia

AT is caused by mutations in the ATM gene, which encodes a protein kinase crucial for DNA damage signaling. Patients suffering from AT show neurodegeneration, weakened immunity, and motor delays (Aykan et al., 2022). Children with progressive cerebellar ataxia, which starts in early infancy, are at a very high risk of developing AT. Deletion of ATM does not disrupt neuronal development; nevertheless, it blocks the ATM-mediated response to DNA damage (Subba Rao, 2007).

9.4.6 Nijmegen breakage syndrome

NBS patients frequently exhibit mental and developmental delays, microcephaly (small head size), unique facial traits, compromised immunity, and an increased risk of cancer (Aykan et al., 2022). Their cells, particularly fibroblasts and lymphocytes, are extremely vulnerable to gamma and X-ray radiation, which increases DNA damage and causes aberrant cell death (Burgt et al., 1996). Mutations in the NBS1 gene, which codes for the Nibrin protein, are the cause of the condition (Pollard and Gatti, 2009). The MRN complex, which is formed by Nibrin, Mre11, and Rad50, is essential for identifying and fixing DNA DSBs. As a result of this abnormality, NBS shares several clinical characteristics with AT.

9.5 Conclusion and future perspectives

UV light is the most predominant DNA-damaging genotoxic agent that damages DNA in different ways in mammalian cells. It generates a variety of DNA lesions, such as CPD and 6-4 photoproducts, and ROS that damage the bases and cause breaks in the DNA strand. To survive such damage, cells have developed different repair mechanisms that protect them from the effects of damage. Organisms are protected from diseases because of the efficient repair mechanism that maintains genomic stability. NER is the main repair mechanism that cells employ to remove DNA photoproducts. While damage induced by ROS, as well as SSBs, is repaired by BER, DSBs are mended via HR and nonhomologous recombination pathways (Tiwari and Wilson, 2019). Genetic impairments in these repair pathways will lead to severe disorders such as XP and CS in the case of NER, neurological abnormalities, premature aging, and cancer predisposition (Aykan et al., 2022). People suffering from XP exhibit high sensitivity to sunlight and a predisposition to cancer and premature aging. Furthermore, extreme exposure to sunlight harbors UV rays, which are directly linked to the development of malignant melanoma (MMs), squamous cell carcinoma (SCC), and basal cell carcinoma (BCC). Thus, knowledge related to UV-induced DNA damage and repair defects in mammalian cells is important for understanding how the stability of the genome is maintained under external stress.

Despite significant research, many areas of UV-induced DNA damage and repair need to be elucidated. Highlighting the molecular basis of disorders connected with defective DNA damage repair can provide detailed insights into cellular pathways and their interaction, and the role of individual repair proteins in response to UV-induced damage. Different strategies have been adopted to protect from UV radiation. Currently, one such strategy is the development of sunscreen formulation, a first line of defense against UV, but its true effectiveness needs to be re-evaluated. The photochemical interaction between formulation ingredients should be studied in detail to design more effective and stable formulations (Kciuk et al., 2020). Additionally, topical solutions of DNA repair enzymes, such as DNA photolyases, have been used, which prove to be effective and provide protection against UV-induced skin cancer. Moreover, topical application of liposome-encapsulated photolyases shows positive results in repairing CPD lesions in skin. But the major limitation of its use is variable transfer efficiency and its timely application. Further studies could serve as promising therapeutic targets, as they may benefit the patients suffering from diseases like XP who are more prone to UV damage (Jans et al., 2005).

References

Adam, S., Polo, S.E., 2012. Chromatin dynamics during nucleotide excision repair: histones on the move. Int. J. Mol. Sci. 13, 11895–11911.

Ahmad, S.I., Christensen, L., Baron, E., 2017. History of UV lamps, types, and their applications. In: Ahmad, S. (Ed.). In: Ultraviolet Light in Human Health, Diseases, and environment. Adv. Exp. Med. Biol., 996. Springer, Cham, pp. 3–11.

Ait Saada, A., Lambert, S.A., Carr, A.M., 2018. Preserving replication fork integrity and competence via the homologous recombination pathway. DNA Repair (Amst.) 71, 135–147.

Araújo, S.J., Kuraoka, I., 2019. Nucleotide excision repair genes shaping embryonic development. Open Biol. 9, 190166.

Audebert, M., Salles, B., Calsou, P., 2004. Involvement of poly (ADP-ribose) polymerase-1 and XRCC1/DNA ligase III in an alternative route for DNA double-strand breaks rejoining. J. Biol. Chem. 279, 55117–55126.

Auerbach, A.D., 2009. Fanconi anemia and its diagnosis. Mutat. Res. Mol. Mech. Mutagen. 668, 4–10.

Aykan, K., Altuntaş, İ., Erbaş, O., 2022. DNA repair mechanisms: DNA repair defects and related diseases. D. J. Med. Sci. 8 (3), 130–140.

Banáth, J.P., Olive, P.L., 2003. Expression of phosphorylated histone H2AX as a surrogate of cell killing by drugs that create DNA double-strand breaks. Cancer Res. 63, 4347–4350.

Baptista, M.S., Cadet, J., Greer, A., Thomas, A.H., 2021. Photosensitization reactions of biomolecules: definition, targets, and mechanisms. Photochem. Photobiol. 97, 1456–1483.

Bates, G.P., Dorsey, R., Gusella, J.F., Hayden, M.R., Kay, C., Leavitt, B.R., Nance, M., Ross, C.A., Scahill, R.I., Wetzel, R., Wild, E.J., Tabrizi, S.J., 2015. Huntington disease. Nat. Rev. Dis. Primer. 1, 15005.

Bauer, N.C., Corbett, A.H., Doetsch, P.W., 2015. The current state of eukaryotic DNA base damage and repair. Nucleic Acids Res. 43, 10083–10101.

Bernhard, G.H., Bais, A.F., Aucamp, P.J., Klekociuk, A.R., Liley, J.B., McKenzie, R.L., 2023. Stratospheric ozone, UV radiation, and climate interactions. Photochem. Photobiol. Sci. 22, 937–989.

Bose, S.N., Davies, R.J.H., 1984. The photoreactivity of T-A sequences in oligodeoxyribonucleotides and DNA. Nucleic Acids Res. 12, 7903–7914.

Brooks, P.J., 2013. Blinded by the UV light: how the focus on transcription-coupled NER has distracted from understanding the mechanisms of Cockayne syndrome neurologic disease. DNA Repair (Amst.) 12, 656–671.

Brouwer, I., Sitters, G., Candelli, A., Heerema, S.J., Heller, I., Melo De, A.J., Zhang, H., Normanno, D., Modesti, M., Peterman, E.J.G., Wuite, G.J.L., 2016. Sliding sleeves of XRCC4–XLF bridge DNA and connect fragments of broken DNA. Nature 535, 566–569.

Buglak, A.A., Telegina, T.A., Kritsky, M.S., 2016. A quantitative structure—property relationship (QSPR) study of singlet oxygen generation by pteridines. Photochem. Photobiol. Sci. 15, 801–811.

Bunting, S.F., Callén, E., Wong, N., Chen, H.-T., Polato, F., Gunn, A., Bothmer, A., Feldhahn, N., Fernandez-Capetillo, O., Cao, L., Xu, X., Deng, C.-X., Finkel, T., Nussenzweig, M., Stark, J.M., Nussenzweig, A., 2010. 53BP1 inhibits homologous recombination in *Brca1*-deficient cells by blocking resection of DNA breaks. Cell 141, 243–254.

Burgt, I.van der, Chrzanowska, K.H., Smeets, D., Weemaes, C., 1996. Nijmegen breakage syndrome. J. Med. Genet. 33, 153–156.

Cadet, J., Douki, T., 2018. Formation of UV-induced DNA damage contributing to skin cancer development. Photochem. Photobiol. Sci. 17, 1816–1841.

Cadet, J., Douki, T., Ravanat, J., 2015. Oxidatively Generated Damage to cellular DNA by UVB and UVA radiation. Photochem. Photobiol. 91, 140–155.

Cadet, J., Douki, T., Ravanat, J.-L., 2010. Oxidatively generated base damage to cellular DNA. Free Radic. Biol. Med. 49, 9–21.

Cadet, J., Ravanat, J.-L., TavernaPorro, M., Menoni, H., Angelov, D., 2012. Oxidatively generated complex DNA damage: tandem and clustered lesions. Cancer Lett. 327, 5–15.

Caracciolo, D., Riillo, C., Di Martino, M.T., Tagliaferri, P., Tassone, P., 2021. Alternative non-homologous end-joining: error-prone DNA repair as cancer's achilles' heel. Cancers 13, 1392.

Carell, T., Epple, R., 1998. Repair of UV Light induced DNA lesions: a comparative study with model compounds. Eur. J. Org. Chem. 1998, 1245–1258.

Chang, H.H.Y., Pannunzio, N.R., Adachi, N., Lieber, M.R., 2017. Non-homologous DNA end joining and alternative pathways to double-strand break repair. Nat. Rev. Mol. Cell Biol. 18, 495–506.

Chatterjee, N., Walker, G.C., 2017. Mechanisms of DNA damage, repair, and mutagenesis. Environ. Mol. Mutagen. 58, 235–263.

Clouaire, T., Marnef, A., Legube, G., 2017. Taming tricky DSBs: ATM on duty. DNA Repair (Amst.) 56, 84–91.

Coin, F., Oksenych, V., Egly, J.M., 2007. Distinct roles for the XPB/p52 and XPD/p44 subcomplexes of TFIIH in damaged DNA opening during nucleotide excision repair. Mol. Cell 26, 245–256.

Couëdel, C., Mills, K.D., Barchi, M., Shen, L., Olshen, A., Johnson, R.D., Nussenzweig, A., Essers, J., Kanaar, R., Li, G.C., Alt, F.W., Jasin, M., 2004. Collaboration of homologous recombination and nonhomologous end-joining factors for the survival and integrity of mice and cells. Genes Dev. 18, 1293–1304.

Crean, C., Uvaydov, Y., Geacintov, N.E., Shafirovich, V., 2008. Oxidation of single-stranded oligonucleotides by carbonate radical anions: generating intrastrand cross-links between guanine and thymine bases separated by cytosines. Nucleic Acids Res. 36, 742–755.

Debacq-Chainiaux, F., Leduc, C., Verbeke, A., Toussaint, O., 2012. UV, stress, and aging. Dermato-endocrinol. 4, 236–240.

Decottignies, A., 2013. Alternative end-joining mechanisms: a historical perspective. Front. Genet. 4, 48.

Czaja, W., Mao, P., Smerdon, M.J., 2012. The emerging roles of ATP-dependent chromatin remodeling enzymes in nucleotide excision repair. Int. J. Mol. Sci. 13, 11954–11973.

De Jager, T.L., Cockrell, A.E., Du Plessis, S.S., 2017. Ultraviolet light induced generation of reactive oxygen species. In: Ahmad, S.I. (Ed.). In: Ultraviolet light in human health, diseases and environment, Adv. Exp. Med. Biol., 996. Springer, Cham, pp. 15–23.

Delgado, J.L., Hsieh, C.-M., Chan, N.-L., Hiasa, H., 2018. Topoisomerases as anticancer targets. Biochem. J. 475, 373–398.

De Lima-Bessa, K.M., Armelini, M.G., Chiganças, V., Jacysyn, J.F., Amarante-Mendes, G.P., Sarasin, A., Menck, C.F.M., 2008. CPDs and 6-4PPs play different roles in UV-induced cell death in normal and NER-deficient human cells. DNA Repair (Amst.) 7, 303–312.

Della-Maria, J., Zhou, Y., Tsai, M.-S., Kuhnlein, J., Carney, J.P., Paull, T.T., Tomkinson, A.E., 2011. Human Mre11/human Rad50/Nbs1 and DNA ligase IIIα/XRCC1 protein complexes act together in an alternative nonhomologous end joining pathway. J. Biol. Chem. 286, 33845–33853.

DiGiovanna, J.J., Kraemer, K.H., 2012. Shining a light on xeroderma pigmentosum. J. Invest. Dermatol. 132, 785–796.

Douki, T., 2013. The variety of UV-induced pyrimidine dimeric photoproducts in DNA as shown by chromatographic quantification methods. Photochem. Photobiol. Sci. 12, 1286–1302.

Douki, T., Reynaud-Angelin, A., Cadet, J., Sage, E., 2003. Bipyrimidine photoproducts rather than oxidative lesions are the main type of DNA damage involved in the genotoxic effect of solar UVA radiation. Biochemistry 42, 9221–9226.

Douki, T., Sage, E., 2016. Dewar valence isomers, the third type of environmentally relevant DNA photoproducts induced by solar radiation. Photochem. Photobiol. Sci. 15, 24–30.

Elbakry, A., Löbrich, M., 2021. Homologous recombination subpathways: a tangle to resolve. Front. Genet. 12, 723847.

Escribano-Díaz, C., Orthwein, A., Fradet-Turcotte, A., Xing, M., Young, J.T.F., Tkáč, J., Cook, M.A., Rosebrock, A.P., Munro, M., Canny, M.D., Xu, D., Durocher, D., 2013. A cell cycle-dependent regulatory circuit composed of 53BP1-RIF1 and BRCA1-CtIP controls DNA repair pathway choice. Mol. Cell 49, 872–883.

Fei, J., Chen, J., 2012. KIAA1530 protein is recruited by Cockayne syndrome complementation group protein A (CSA) to participate in transcription-coupled repair (TCR). J. Biol. Chem. 287, 35118–35126.

Fishel, R., Lee, J.B., 2016. Mismatch repair. In: Hanaoka, F., Sugasaw, K. (Eds.), DNA replication, recombination, and repair. Springer, Tokyo, pp. 305–339.

Fortini, P., Pascucci, B., Parlanti, E., D'errico, M., Simonelli, V., Dogliotti, E., 2003. The base excision repair: mechanisms and its relevance for cancer susceptibility. Biochimie. 85, 1053–1071.

Fousteri, M., Mullenders, L.H., 2008. Transcription-coupled nucleotide excision repair in mammalian cells: molecular mechanisms and biological effects. Cell Res. 18, 73–84.

Friedberg, E.C., Meira, L.B., 2006. Database of mouse strains carrying targeted mutations in genes affecting biological responses to DNA damage version 7. DNA Repair (Amst.) 5, 189–209.

Fu, S., Xue, S., Chen, J., Shang, S., Xiao, H., Zang, Y., Tang, X., 2021. Effects of different short-term UV-B radiation intensities on metabolic characteristics of *Porphyra haitanensis*. Int. J. Mol. Sci. 22, 2180.

Ghosh, D., Raghavan, S.C., 2021. Nonhomologous end joining: new accessory factors fine tune the machinery. Trends Genet. 37, 582–599.

Giglia-Mari, G., Zotter, A., Vermeulen, W., 2011. DNA damage response. Cold Spring Harb. Perspect. Biol. 3, a000745.

Gómez-Herreros, F., Romero-Granados, R., Zeng, Z., Álvarez-Quilón, A., Quintero, C., Ju, L., Umans, L., Vermeire, L., Huylebroeck, D., Caldecott, K.W., Cortés-Ledesma, F., 2013. TDP2–dependent non-homologous end-joining protects against topoisomerase II–induced DNA breaks and genome instability in cells and in vivo. PLoS Genet. 9, e1003226.

Goodarzi, A.A., Yu, Y., Riballo, E., Douglas, P., Walker, S.A., Ye, R., Härer, C., Marchetti, C., Morrice, N., Jeggo, P.A., Lees-Miller, S.P., 2006. DNA-PK autophosphorylation facilitates Artemis endonuclease activity. EMBO J. 25, 3880–3889.

Grawunder, U., Wilm, M., Wu, X., Kulesza, P., Wilson, T.E., Mann, M., Lieber, M.R., 1997. Activity of DNA ligase IV stimulated by complex formation with XRCC4 protein in mammalian cells. Nature 388, 492–495.

Heo, J., Li, J., Summerlin, M., Hays, A., Katyal, S., McKinnon, P.J., Nitiss, K.C., Nitiss, J.L., Hanakahi, L.A., 2015. TDP1 promotes assembly of non-homologous end-joining protein complexes on DNA. DNA Repair (Amst.) 30, 28–37.

Hiraku, Y., Ito, K., Hirakawa, K., Kawanishi, S., 2007. Photosensitized DNA damage and its protection via a novel mechanism. Photochem. Photobiol. 83, 205–212.

Hoeijmakers, J.H., 2001. DNA repair mechanisms. Maturitas. 38, 17–22.

Ichihashi, M., Ueda, M., Budiyanto, A., Bito, T., Oka, M., Fukunaga, M., Tsuru, K., Horikawa, T., 2003. UV-induced skin damage. Toxicol. 189, 21–39.

Ira, G., Malkova, A., Liberi, G., Foiani, M., Haber, J.E., 2003. Srs2 and Sgs1–Top3 suppress crossovers during double-strand break repair in Yeast. Cell 115, 401–411.

Jans, J., Schul, W., Sert, Y.-G., Rijksen, Y., Rebel, H., Eker, A.P.M., Nakajima, S., Van Steeg, H., De Gruijl, F.R., Yasui, A., Hoeijmakers, J.H.J., Van Der Horst, G.T.J., 2005. Powerful skin cancer protection by a CPD-photolyase transgene. Curr. Biol. 15, 105–115.

Jasin, M., Rothstein, R., 2013. Repair of strand breaks by homologous recombination. Cold Spring Harb. Perspect. Biol. 5, a012740.

Jeppesen, D.K., Bohr, V.A., Stevnsner, T., 2011. DNA repair deficiency in neurodegeneration. Prog. Neurobiol. 94, 166–200.

Jiang, N., Taylor, J.S., 1993. In vivo evidence that UV-induced C T mutations at dipyrimidine sites could result from the replicative bypass of cis-syn cyclobutane dimers or their deamination products. Biochemistry 32, 472–481.

Jin, S.-G., Padron, F., Pfeifer, G.P., 2022. UVA radiation, DNA damage, and melanoma. ACS Omega 7, 32936–32948.

Jin, S.-G., Pettinga, D., Johnson, J., Li, P., Pfeifer, G.P., 2021. The major mechanism of melanoma mutations is based on deamination of cytosine in pyrimidine dimers, as determined by circle damage sequencing. Sci. Adv. 7, eabi6508.

Juan, C.A., Pérez De Lastra, J.M., Plou Gasca, F.J., Pérez-Lebeña, E., 2021. The chemistry of reactive oxygen species (ROS) revisited: outlining their role in biological macromolecules (DNA, lipids, and proteins) and induced pathologies. Int. J. Mol. Sci. 22, 4642.

Karikkineth, A.C., Scheibye-Knudsen, M., Fivenson, E., Croteau, D.L., Bohr, V.A., 2017. Cockayne syndrome: clinical features, model systems and pathways. Ageing Res. Rev. 33, 3–17.

Karran, P., 2000. DNA double-strand break repair in mammalian cells. Curr. Opin. Genet. Dev. 10, 144–150.

Kciuk, M., Marciniak, B., Mojzych, M., Kontek, R., 2020. Focus on UV-induced DNA damage and repair—disease relevance and protective strategies. Int. J. Mol. Sci. 21, 7264.

Kelley, M.R., Parsons, S.H., 2001. Redox regulation of the DNA repair function of the human AP endonuclease Ape1/ref-1. Antioxid. Redox Signal 3, 671–683.

Kemp, M.G., Reardon, J.T., Lindsey-Boltz, L.A., Sancar, A., 2012. Mechanism of release and fate of excised oligonucleotides during nucleotide excision repair. J. Biol. Chem. 287, 22889–22899.

Kimeswenger, S., Schwarz, A., Födinger, D., Müller, S., Pehamberger, H., Schwarz, T., Jantschitsch, C., 2016. Infrared A radiation promotes survival of human melanocytes carrying ultraviolet radiation-induced DNA damage. Exp. Dermatol. 25, 447–452.

Krokan, H.E., Bjørås, M., 2013. Base excision repair. Cold Spring Harb. Perspect. Biol. 5, a012583.

Kumar, A., Purohit, S., Sharma, N.K., 2016. Aberrant DNA double-strand break repair threads in breast carcinoma: orchestrating genomic insult survival. J. Cancer Prev. 21, 227–234.

Kurtoğlu, E.L., Tekedereli, İ., 2015. DNA onarim mekanizmalarI. Balıkesir Sağlık Bilim. Derg. 4, 169–177.

Lee, D.-H., Pfeifer, G.P., 2003. Deamination of 5-methylcytosines within cyclobutane pyrimidine dimers is an important component of UVB mutagenesis. J. Biol. Chem. 278, 10314–10321.

Lee, Y.-C., Cai, Y., Mu, H., Broyde, S., Amin, S., Chen, X., Min, J.-H., Geacintov, N.E., 2014. The relationships between XPC binding to conformationally diverse DNA adducts and their excision by the human NER system: is there a correlation? DNA Repair (Amst.) 19, 55–63.

Lehmann, A.R., 2011. DNA polymerases and repair synthesis in NER in human cells. DNA Repair (Amst.) 10, 730–733.

Li, C.-L., Golebiowski, F.M., Onishi, Y., Samara, N.L., Sugasawa, K., Yang, W., 2015. Tripartite DNA lesion recognition and verification by XPC, TFIIH, and XPA in nucleotide excision repair. Mol. Cell 59, 1025–1034.

Li, M., Wilson, D.M., 2014. Human apurinic/apyrimidinic endonuclease 1. Antioxid. Redox Signal. 20, 678–707.

Lickley, M., Solomon, S., Fletcher, S., Velders, G.J.M., Daniel, J., Rigby, M., Montzka, S.A., Kuijpers, L.J.M., Stone, K., 2020. Quantifying contributions of chlorofluorocarbon banks to emissions and impacts on the ozone layer and climate. Nat. Commun. 11, 1380.

Lieber, M.R., 2010. The mechanism of double-strand DNA break repair by the nonhomologous DNA end-joining pathway. Annu. Rev. Biochem. 79, 181–211.

Liu, D., Keijzers, G., Rasmussen, L.J., 2017. DNA mismatch repair and its many roles in eukaryotic cells. Mutat. Res. Rev. Mutat. Res. 773, 174–187.

Ma, Y., Schwarz, K., Lieber, M.R., 2005. The Artemis: DNA-PKcs endonuclease cleaves DNA loops, flaps, and gaps. DNA Repair (Amst.) 4, 845–851.

Markovitsi, D., 2016. UV-induced DNA Damage: the role of electronic excited states. Photochem. Photobiol. 92, 45–51.

Martin, J.B., 1999. Molecular basis of the neurodegenerative disorders. N. Engl. J. Med. 340, 1970–1980.

Mateos-Gomez, P.A., Gong, F., Nair, N., Miller, K.M., Lazzerini-Denchi, E., Sfeir, A., 2015. Mammalian polymerase θ promotes alternative NHEJ and suppresses recombination. Nature 518, 254–257.

McKinnon, P.J., 2009. DNA repair deficiency and neurological disease. Nat. Rev. Neurosci. 10, 100–112.

McVey, M., Lee, S.E., 2008. MMEJ repair of double-strand breaks (director's cut): deleted sequences and alternative endings. Trends Genet. 24, 529–538.

Menon, V., Povirk, L.F., 2016. End-processing nucleases and phosphodiesterases: an elite supporting cast for the nonhomologous end joining pathway of DNA double-strand break repair. DNA Repair (Amst.) 43, 57–68.

Mirman, Z., de Lange, T., 2020. 53BP1: a DSB escort. Genes Dev. 34, 7–23.

Mitchell, D.L., Jen, J., Cleaver, J.E., 1991. Relative induction of cyclobutane dimers and cytosine photohydrates in DNA irradiated in vitro and in vivo with ultraviolet-C and ultraviolet-B light. Photochem. Photobiol. 54, 741–746.

Nakazawa, Y., Sasaki, K., Mitsutake, N., Matsuse, M., Shimada, M., Nardo, T., Takahashi, Y., Ohyama, K., Ito, K., Mishima, H., Nomura, M., Kinoshita, A., Ono, S., Takenaka, K., Masuyama, R., Kudo, T., Slor, H., Utani, A., Tateishi, S., Yamashita, S., Stefanini, M., Lehmann, A.R., Yoshiura, K., Ogi, T., 2012. Mutations in UVSSA cause UV-sensitive syndrome and impair RNA polymerase IIo processing in transcription-coupled nucleotide-excision repair. Nat. Genet. 44, 586–592.

Ogi, T., Limsirichaikul, S., Overmeer, R.M., Volker, M., Takenaka, K., Cloney, R., Nakazawa, Y., Niimi, A., Miki, Y., Jaspers, N.G., Mullenders, L.H.F., Yamashita, S., Fousteri, M.I., Lehmann, A.R., 2010. Three DNA polymerases, recruited by different mechanisms, carry out NER repair synthesis in human cells. Mol. Cell 37, 714–727.

Ohnishi, T., Mori, E., Takahashi, A., 2009. DNA double-strand breaks: their production, recognition, and repair in eukaryotes. Mutat. Res. Mol. Mech. Mutagen. 669, 8–12.

Park, S.L., Justiniano, R., Williams, J.D., Cabello, C.M., Qiao, S., Wondrak, G.T., 2015. The tryptophan-derived endogenous aryl hydrocarbon receptor ligand 6-formylindolo[3,2-b]carbazole is a nanomolar UVA photosensitizer in epidermal keratinocytes. J. Invest. Dermatol. 135, 1649–1658.

Pattison, D.I., Davies, M.J., 2006. Action of ultraviolet light on cellular structures. In: Bignold, L.P. (Ed.). In: Cancer: Cell Structures, Carcinogens and Genomic Instability, Experientia Suppl., 96. Springer, Birkhäuser Basel, Switzerland, pp. 131–157.

Pawelczak, K.S., Turchi, J.J., 2010. Purification and characterization of exonuclease-free Artemis: implications for DNA-PK-dependent processing of DNA termini in NHEJ-catalyzed DSB repair. DNA Repair (Amst.) 9, 670–677.

Perrier, S., Hau, J., Gasparutto, D., Cadet, J., Favier, A., Ravanat, J.-L., 2006. Characterization of lysine–guanine cross-links upon one-electron oxidation of a guanine-containing oligonucleotide in the presence of a trilysine peptide. J. Am. Chem. Soc. 128, 5703–5710.

Pfeifer, G.P., 2020. Mechanisms of UV-induced mutations and skin cancer. Genome Instab. Dis. 1, 99–113.

Poetsch, A.R., 2020. The genomics of oxidative DNA damage, repair, and resulting mutagenesis. Comput. Struct. Biotechnol. J. 18, 207–219.

Pollard, J.M., Gatti, R.A., 2009. Clinical radiation sensitivity with DNA repair disorders: an overview. Int. J. Radiat. Oncol. 74, 1323–1331.

Quiñones, J.L., Demple, B., 2016. When DNA repair goes wrong: BER-generated DNA-protein crosslinks to oxidative lesions. DNA Repair (Amst.) 44, 103–109.

Rapin, I., 2013. Chapter 167- disorders of nucleotide excision repair. In: Handbook of Clinical Neurology. Part III, 113. Elsevier, Amsterdam, Netherland, pp. 1637–1650.

Rastogi, R.P., Richa, Kumar, A., Tyagi, M.B., Sinha, R.P., 2010. Molecular mechanisms of ultraviolet radiation-induced DNA damage and repair. J. Nucleic Acids 2010, 592980.

Ravanat, J.-L., Douki, T., Cadet, J., 2001. Direct and indirect effects of UV radiation on DNA and its components. J. Photochem. Photobiol. B 63, 88–102.

Ropars, V., Drevet, P., Legrand, P., Baconnais, S., Amram, J., Faure, G., Márquez, J.A., Piétrement, O., Guerois, R., Callebaut, I., Le Cam, E., Revy, P., De Villartay, J.-P., Charbonnier, J.-B., 2011. Structural characterization of filaments formed by human Xrcc4–Cernunnos/XLF complex involved in nonhomologous DNA end-joining. Proc. Natl. Acad. Sci. 108, 12663–12668.

Sancar, A., 2016. Mechanisms of DNA repair by photolyase and excision nuclease (Nobel Lecture). Angew. Chem. Int. Ed. 55, 8502–8527.

Scharer, O.D., 2013. Nucleotide excision repair in eukaryotes. Cold Spring Harb. Perspect. Biol. 5, a012609.

Schmalwieser, A.W., Klotz, B., Schwarzmann, M., Baumgartner, D.J., Schreder, J., Schauberger, G., Blumthaler, M., 2019. The Austrian UVA-Network. Photochem. Photobiol. 95, 1258–1266.

Schwertman, P., Lagarou, A., Dekkers, D.H.W., Raams, A., Van Der Hoek, A.C., Laffeber, C., Hoeijmakers, J.H.J., Demmers, J.A.A., Fousteri, M., Vermeulen, W., Marteijn, J.A., 2012. UV-sensitive syndrome protein UVSSA recruits USP7 to regulate transcription-coupled repair. Nat. Genet. 44, 598–602.

Schwertman, P., Vermeulen, W., Marteijn, J.A., 2013. UVSSA and USP7, a new couple in transcription-coupled DNA repair. Chromosoma 122, 275–284.

Sczepanski, J.T., Jacobs, A.C., Majumdar, A., Greenberg, M.M., 2009. Scope and mechanism of interstrand cross-link formation by the C4′-oxidized abasic site. J. Am. Chem. Soc. 131, 11132–11139.

Sieradzan, K.A., Mann, D.M.A., 2001. The selective vulnerability of nerve cells in Huntington's disease. Neuropathol. Appl. Neurobiol. 27, 1–21.

Silerme, S., Bobyk, L., Taverna-Porro, M., Cuier, C., Saint-Pierre, C., Ravanat, J.-L., 2014. DNA-polyamine cross-links generated upon one electron oxidation of DNA. Chem. Res. Toxicol. 27, 1011–1018.

Sinha, R.P., Häder, D.-P., 2002. UV-induced DNA damage and repair: a review. Photochem. Photobiol. Sci. 1, 225–236.

Soulier, J., 2011. Fanconi anemia. Hematology Am. Soc. Hematol. Educ. Program 2011, 492–497.

Spivak, G., 2015. Nucleotide excision repair in humans. DNA Repair (Amst.) 36, 13–18.

Spivak, G., 2016. Transcription-coupled repair: an update. Arch. Toxicol. 90, 2583–2594.

Spivak, G., Ganesan, A.K., 2014. The complex choreography of transcription-coupled repair. DNA Repair (Amst.) 19, 64–70.

Staresincic, L., Fagbemi, A.F., Enzlin, J.H., Gourdin, A.M., Wijgers, N., Dunand-Sauthier, I., Giglia-Mari, G., Clarkson, S.G., Vermeulen, W., Schärer, O.D., 2009. Coordination of dual incision and repair synthesis in human nucleotide excision repair. EMBO J. 28, 1111–1120.

Strumberg, D., Pilon, A.A., Smith, M., Hickey, R., Malkas, L., Pommier, Y., 2000. Conversion of Topoisomerase I cleavage complexes on the leading strand of ribosomal DNA into 5′-phosphorylated DNA double-strand breaks by replication runoff. Mol. Cell. Biol. 20, 3977–3987.

Su, D.G., Taylor, J.-S.A., Gross, M.L., 2010. A new photoproduct of 5-Methylcytosine and adenine characterized by high-performance liquid chromatography and mass spectrometry. Chem. Res. Toxicol. 23, 474–479.

Subba Rao, K., 2007. Mechanisms of disease: DNA repair defects and neurological disease. Nat. Clin. Pract. Neurol. 3, 162–172.

Tang, X., Yang, T., Yu, D., Xiong, H., Zhang, S., 2024. Current insights and future perspectives of ultraviolet radiation (UV) exposure: friends and foes to the skin and beyond the skin. Environ. Int. 185, 108535.

Tiwari, V., Wilson, D.M., 2019. DNA damage and associated DNA repair defects in disease and premature aging. Am. J. Hum. Genet. 105, 237–257.

Tornaletti, S., Pfeifer, G.P., 1996. UV damage and repair mechanisms in mammalian cells. Bioessays 18, 221–228.

Truong, L.N., Li, Y., Shi, L.Z., Hwang, P.Y.-H., He, J., Wang, H., Razavian, N., Berns, M.W., Wu, X., 2013. Microhomology-mediated end joining and homologous recombination share the initial end resection step to repair DNA double-strand breaks in mammalian cells. Proc. Natl. Acad. Sci. 110, 7720–7725.

Tsodikov, O.V., Ivanov, D., Orelli, B., Staresincic, L., Shoshani, I., Oberman, R., Schärer, O.D., Wagner, G., Ellenberger, T., 2007. Structural basis for the recruitment of ERCC1-XPF to nucleotide excision repair complexes by XPA. EMBO J. 26, 4768–4776.

Tu, Y., Dammann, R., Pfeifer, G.P., 1998. Sequence and time-dependent deamination of cytosine bases in UVB-induced cyclobutane pyrimidine dimers in vivo. J. Mol. Biol. 284, 297–311.

Tudek, B., 2007. Base excision repair modulation as a risk factor for human cancers. Mol. Aspects Med. 28, 258–275.

Utrillas, M.P., Marín, M.J., Esteve, A.R., Salazar, G., Suarez, H., Castillo, J., Martínez-Lozano, J.A., 2016. UVER and UV index at high altitude in Northwestern Argentina. J. Photochem. Photobiol. B. 163, 290–295.

Vechtomova, Y.L., Telegina, T.A., Buglak, A.A., Kritsky, M.S., 2021. UV radiation in DNA damage and repair involving DNA-photolyases and cryptochromes. Biomedicines 9, 1564.

Vermeulen, W., Fousteri, M., 2013. Mammalian transcription-coupled excision repair. Cold Spring Harb. Perspect. Biol. 5, a012625.

Wang, H., Xu, X., 2017. Microhomology-mediated end joining: new players join the team. Cell Biosci. 7, 1–6.

Wang, M., Wu, W., Wu, W., Rosidi, B., Zhang, L., Wang, H., Iliakis, G., 2006. PARP-1 and Ku compete for repair of DNA double-strand breaks by distinct NHEJ pathways. Nucleic Acids Res. 34, 6170–6182.

Wei, M., He, X., Liu, N., Deng, H., 2024. Role of reactive oxygen species in ultraviolet-induced photodamage of the skin. Cell Div. 19, 1.

Wiederhold, L., Leppard, J.B., Kedar, P., Karimi-Busheri, F., Rasouli-Nia, A., Weinfeld, M., Tomkinson, A.E., Izumi, T., Prasad, R., Wilson, S.H., Mitra, S., Hazra, T.K., 2004. AP endonuclease-independent DNA base excision repair in human cells. Mol. Cell. 15, 209–220.

Wierzchowski, K.L., Shugar, D., 1961. Photochemistry of cytosine nucleosides and nucleotides. II. Biochim. Biophys. Acta 8, 219–234.

Wright, W.D., Shah, S.S., Heyer, W.-D., 2018. Homologous recombination and the repair of DNA double-strand breaks. J. Biol. Chem. 293, 10524–10535.

Wu, L., Hickson, I.D., 2003. The Bloom's syndrome helicase suppresses crossing over during homologous recombination. Nature 426, 870–874.

Yu, A.M., McVey, M., 2010. Synthesis-dependent microhomology-mediated end joining accounts for multiple types of repair junctions. Nucleic Acids Res. 38, 5706–5717.

Zhang, J., Yin, Z., Ma, L., Yin, Z.-Q., Hu, Y., Xu, Y., Wu, D., Permatasari, F., Luo, D., Zhou, B., 2014. The protective effect of baicalin against UVB irradiation-induced photoaging: an in vitro and in vivo study. PLoS One 9, e99703.

Zhang, X., Horibata, K., Saijo, M., Ishigami, C., Ukai, A., Kanno, S., Tahara, H., Neilan, E.G., Honma, M., Nohmi, T., Yasui, A., Tanaka, K., 2012. Mutations in UVSSA cause UV-sensitive syndrome and destabilize ERCC6 in transcription-coupled DNA repair. Nat. Genet. 44, 593–597.

Zhou, Y., Meng, X., Belle, J.H., Zhang, H., Kennedy, C., Al-Hamdan, M.Z., Wang, J., Liu, Y., 2019. Compilation and spatio-temporal analysis of publicly available total solar and UV irradiance data in the contiguous United States. Environ. Pollut. 253, 130–140.

CHAPTER

10

Oxidative DNA damage: Repair mechanisms and associated disorders in humans

Amol Gholap[a], *Pankaj Khuspe*[b], *Dipali Mane*[c] *and Md Jasim Uddin*[d]

[a]Department of Pharmaceutics, St. John Institute of Pharmacy and Research, Palghar, Maharashtra, India [b]Department of Pharmaceutics, Shriram Shikshan Sanstha's College of Pharmacy, Solapur, Maharashtra, India [c]Department of Pharmacology, Shriram Shikshan Sanstha's College of Pharmacy, Solapur, Maharashtra, India [d]Department of Pharmaceutical Technology, Faculty of Pharmacy, Universiti Malaya, Kuala Lumpur, Malaysia

10.1 Introduction

10.1.1 Overview of DNA damage in human cells

The human genome is under numerous threats on a daily basis, as many physical, chemical, and biological agents lay siege on its stability. The DNA damage is structural and refers to a change in nucleotide sequence or, chemically, a change in the DNA bases, sugars, or phosphodiester backbone (Dabin et al., 2023). The origins of such damage may be environmental insults, metabolic by-products, replication errors, or enzymatic reactions. There are tens of thousands of lesions per day in a typical mammalian cell, and they include single- and double-stranded breaks, base modifications, cross-links, and mismatches. Unresolved lesions have the potential of impairing fidelity of genetic information, interference with rearrangement and replication, and the induction of genomic instability–one of the main characteristics of carcinogenesis and diverse degenerative diseases (Schumacher et al., 2021). To counteract these insults, cells have developed a complex interconnection of DNA damage response (DDR) pathways that include, but are not limited to, base excision repair (BER), NER, mismatch repair (MMR), and double-strand break repair. Oxidative lesions are considered one of the most widespread and biologically impactful types of DNA damage, which are produced by reactive oxygen species

Genome Damage: Repair and Adaptation Across Life.
DOI: https://doi.org/10.1016/B978-0-443-42857-9.00014-1

(ROS) formed during the process of aerobic metabolism. Unchecked accumulation of oxidative lesions also has the potential to increase the rate of aging, induce mutagenesis, and perhaps lead to the development of complex pathologies, highlighting the importance of effective repair and detection mechanisms (Yousefzadeh et al., 2021).

10.1.2 Importance of oxidative DNA damage

Oxidative DNA damage represents a subset of damages that are formed by ROS, such as the superoxide anion ($O_2^{\bullet -}$), hydroxyl radical ($^{\bullet}OH$), and hydrogen peroxide (H_2O_2) on nucleobases or deoxyribose units. These reactions may lead to base damage, fragmentation of sugar molecules, breaks of one strand of the DNA (SSBs), or even clustered lesions in DNA (Azzouz and Palaniyar, 2024). Among the best studied oxidative lesions is 8-oxo-7, 8-dihydroguanine (8-oxoG), which is the product of guanine oxidation, a very high mutation-prone product because this molecule can mispair with adenine, leading to G:C to T:A transversions during the replication process (Pravin et al., 2024).

The biological importance of oxidative DNA damage is due to its commonality, its endurance of long-term oxidative stress, and its ability to interfere with cellular homeostasis. Though physiological amounts of ROS play significant roles in signal transduction and defense, they cause pathological oxidative stress when the Alzheimer's disease (AD) capacity has been defeated by an excess of ROS production (Zhang et al., 2024). Oxidative damage to DNA may not be repaired by cell survival mechanisms and continues to accumulate in the genome, to trigger cell survival decisions, whether through apoptosis, cellular senescence, or oncogenesis, as presented in Fig. 10.1. Thus, the balance between ROS production and the repair system is crucial to the genomic stability, longevity of an organism, and healthy aging and freedom from cancer, neurodegeneration, and cardiovascular diseases (Wang et al., 2025).

10.1.3 Sources of oxidative stress (endogenous vs. exogenous)

Oxidative stress is the result of insufficient antioxidant defense against the production of pro-oxidants. The endogenous sources are mainly associated with cell metabolism, with the main source of ROS being the mitochondrial electron transport chain that produces ROS during the oxidative phosphorylation process. The formation of ROS is the product of other endogenous enzymatic systems that include: NADPH oxidases, xanthine oxidase, cytochrome P450 enzymes, and peroxisomal oxidases (Krishnamurthy et al., 2024). Through the oxidative burst, immune cells, especially neutrophils and macrophages, purposely generate ROS to attack invaders, but this happens at the expense of the host as DNA is unintentionally oxidized. Also, cellular production of ROS can be enhanced by the metal-catalyzed Fe Fenton and Haber-Weiss reactions (Wang et al., 2024).

Such exogenous factors include lifestyle and environmental factors. Water radiolysis caused by ionizing radiations, gamma-rays and X-rays releases $^{\bullet}OH$ radicals which directly target DNA. ROS can also be generated through a photosensitization reaction under the influence of ultraviolet (UV) radiation. Redox cycling is triggered in tissues by chemical pollutants, including polycyclic aromatic hydrocarbons, heavy metals (e.g., cadmium, arsenic), and pesticides (Younus et al., 2024). Oxidative stress is further enhanced by cigarette smoke, which is an excellent source of oxidants as well as dietary ingredients having carcinogens or oxidative lipids. Some drugs or chemotherapeutic agents, which are useful in treatment, can trigger ROS as side effects. Interactions between endogenous metabolic generation of ROS and constant exposure

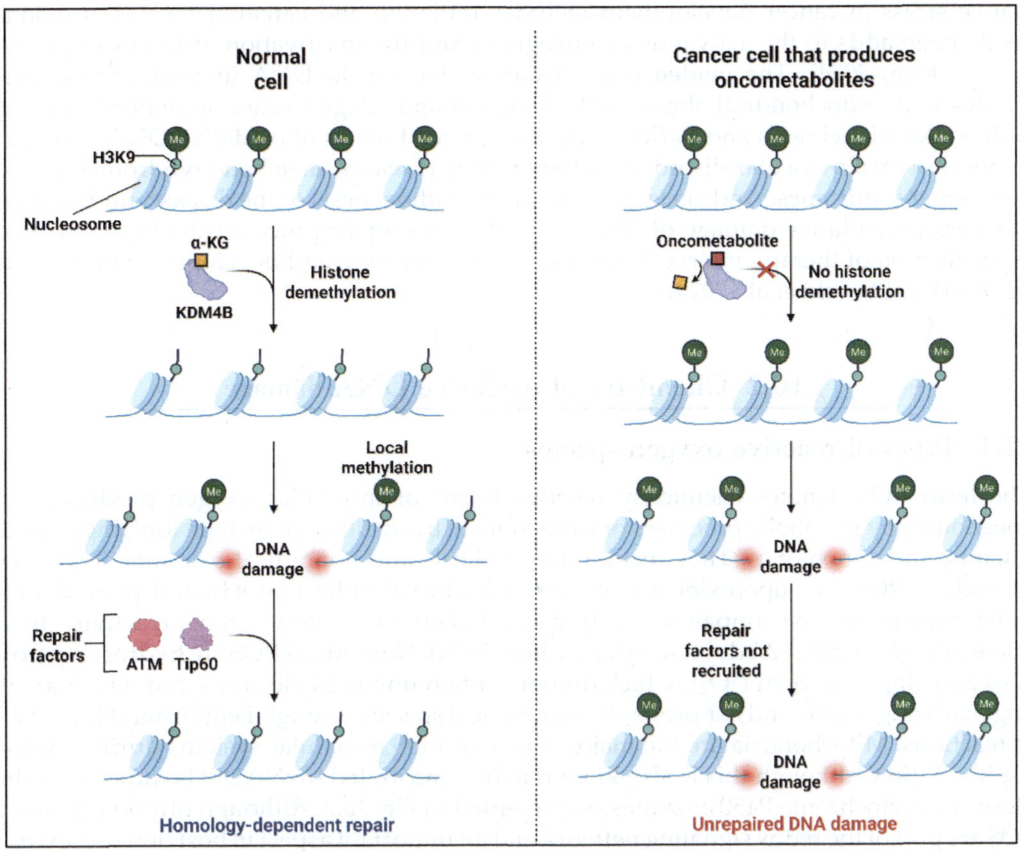

FIGURE 10.1 Mechanistic comparison of DNA damage repair in normal cells and cancer cells that generate oncometabolites. In un-stimulated cells (left), H3K9 demethylation is regulated by KDM4B in an α-KG-dependent manner, which promotes local demethylation and permits the recruitment of repair factors (ATM and Tip60) to sites of DNA damage. This enhances repair and stability of the genome, which is dependent on homology. In comparison, oncometabolites accumulate to suppress histone demethylation in cancer cells (right), causing permanent methylation marks. DNA damage is therefore achieved, it is unable to stimulate repair factors, leading to inefficient repair, and accumulation of repair-looking DNA damage (created using BioRender.com).

to external pro-oxidants subject human DNA to an oxidative stress load. Depending on the type of ROS, its location, and the length of exposure, resulting damage varies and depends on the antioxidant and DNA repair capacity of the affected cell (Di Carlo and Sorrentino, 2024; Yang et al., 2024).

10.1.4 Significance for human health

Oxidative damage to DNA has significant health consequences because it is highly linked to mutagenesis, aging, and a myriad of pathologies. Unresolved oxidative damages may be generated through modification of coding regions, regulatory components, and epigenetic signatures, cross-modifying the expression of genes and protein functionality. The role of chronic oxidative stress in cancer development includes initiation and carcinogenesis, but oxidative DNA damage adds to the activation of oncogenes and the inactivation of tumor suppressors (Lim and Kim, 2024). The evidence of oxidative damage to DNA in postmitotic neurons associates with mitochondrial abnormalities and synaptic degeneration in neurodegenerative disorders like Alzheimer's and Parkinson's. Exaggerated levels of oxidative DNA damage are also found in cardiovascular disorders, inflammatory diseases, metabolic syndromes, and premature aging syndromes. And since it is core in the pathogenesis of the disease, understanding the process of oxidative damage of DNA as well as the repair process is highly important in the specification of therapy interventions, diagnostic accuracy, and development of preventive measures (Wróblewski et al., 2024).

10.2 Chemistry of oxidative DNA damage

10.2.1 Types of reactive oxygen species

The term ROS denotes chemically reactive forms of molecular oxygen produced either endogenously by metabolic processes or exogenously by insults that include ionizing radiation, pollutants, and xenobiotics. They include free radicals and nonradical oxidants. Examples of some radical ROS are superoxide anion ($O_2{\bullet}^-$), hydroxyl radical (•OH), and peroxyl radical (ROO•), which all have unpaired electrons and therefore a very strong reactivity toward biomolecules (Mostafavi Abdolmaleky and Zhou, 2024). Nonradical ROS are hydrogen peroxide (H_2O_2) and singlet oxygen ($^1O_2$), which do not contain unpaired electrons, but which are very strong oxidizing agents and/or progenitors of radical species through Fenton and Haber Weiss-type reactions. Mitochondria are the major source of ROS in cellular systems during oxidative phosphorylation, although this is also known to be contributed by NADPH oxidases, xanthine oxidase, and cytochrome P450 enzymes, as presented in Fig. 10.2. Although physiological levels of ROS are part of the redox signaling network and an important aspect in host defense, excessive production of ROS or failure to neutralize the ROS with antioxidant systems leads to a redox shift (cellular oxidative state) that exposes the genomic DNA to chemical alterations in DNA structure and fidelity replication that render the DNA vulnerable to deterioration of its quality. These factors in particular are critical to the DNA damage spectrum and severity, the chemical diversity, half-life, and diffusibility of individual species of ROS (Zhou et al., 2024).

10.2.2 Mechanisms of reactive oxygen species-induced DNA oxidation

Oxidative DNA damage through ROS can be carried out in a direct and an indirect fashion, which depends on the speciation and the availability of the target DNA. Extremely reactive •OH, produced in situ by transition metal-catalyzed Fenton chemistry, has near diffusion-controlled reaction rates, and reacts with nucleobases, as well as with the deoxyribose backbone, in an indiscriminate manner (Xu et al., 2024). This is one-electron oxidation or abstraction of the hydrogen atom, leading to the formation of the base radicals or sugar-centered radicals. Superoxide anion, although not as reactive to DNA per se, serves as a precursor to other, more potent oxidants through disproportionation to form H_2O_2, which are contributors to the formation of hydroxyl radicals. The singlet oxygen also affects guanine more preferably, and the endoperoxides formed can undergo rearrangement to form mutagenic compounds such as 8-oxo-7,8-dihydroguanine (8-oxoG). Also, indirectly, electrophilic products of lipid peroxidation,

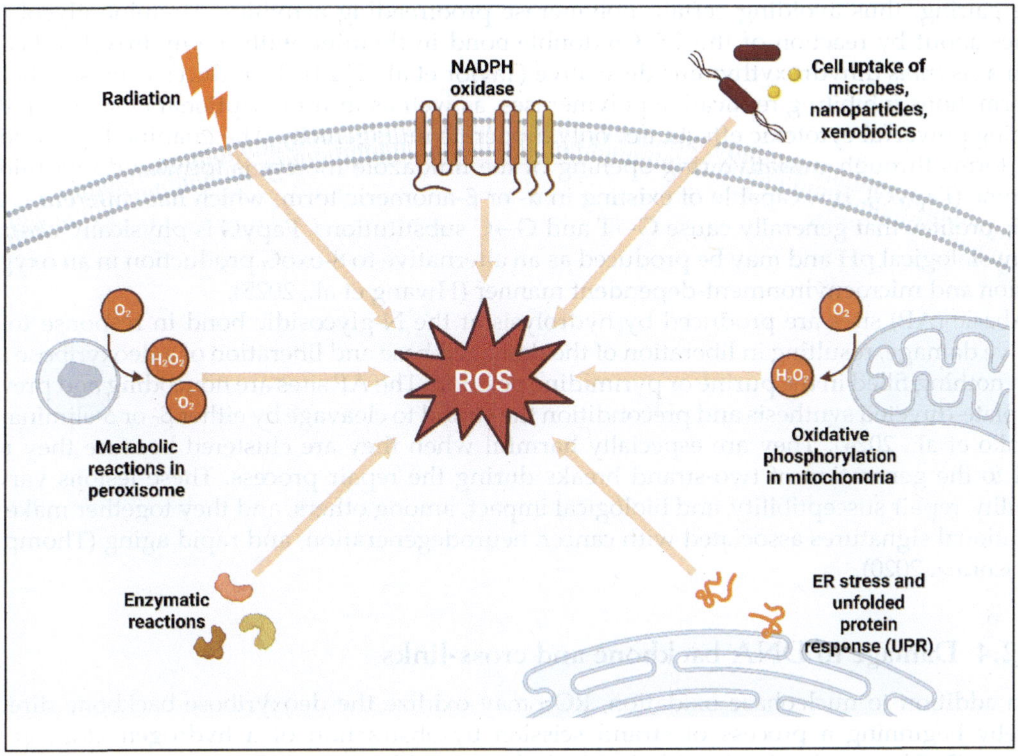

FIGURE 10.2 Diagnostic representation of the key cellular sources and pathways to the production of reactive oxygen species (ROS). The formation of ROS occurs in various pathways, such as by radiation, stimulation of NADPH oxidase, ingestion of microbes, nanoparticles, or xenobiotics. Intracellular ROS can be further promoted by metabolic responses in the peroxisome, enzymatic responses, and oxidative phosphorylation in the mitochondria. Additionally, endoplasmic reticulum (ER) stress and the unfolded protein response (UPR) act as significant contributors to ROS formation. The interaction of these pathways underlines the complex genesis of ROS in cells and their central position in oxidative stress, signal transduction, and pathophysiological states (created using BioRender.com).

such as malondialdehyde and 4-hydroxynonenal, have the capacity to generate exocyclic DNA adducts as a secondary initiation point in the genotoxicity of ROS. The oxidative reactions are also site-specific and are localized to metal ion DNA complexes, causing either site-specific strand cleavage or lesion clusters (An et al., 2024). Lesion distribution and persistence depend on the complex interplay between the chemistry of ROS, the conformation of the DNA and chromatin context, and may have implications on mutagenesis, apoptosis, or senescence should repair processes be excluded or inaccurate because overwhelmed (Liu et al., 2024).

10.2.3 Common oxidative modification

The most common oxidative base lesion is 8-oxoG, which is mostly generated by the attack of a hydroxyl radical at position 8 of guanine. In its mutant, modified electronic structure, it biases bonding to adenine so that replication yields G:C→T:A transversions, the clinical earmark of oxidative mutagenesis. Structurally, 8-oxoG exists in a syn-conformation that allows Hoogsteen base-pairing, thus avoiding certain polymerase proofreading activities. Thymine glycol (Tg) comes about by reaction of the C5–C6 double bond in thymine with the hydroxyl radical to form a *cis trans* dihydroxythymine derivative (Taylor et al., 2024). Tg induces extensive helical deformations inhibiting replicative polymerases, as well as in transcriptional elongation, thus causing powerful cytotoxic effects, but only moderate mutagenicity. The guanine-type product that forms through oxidative ring opening of the imidazole moiety is formamidopyrimidine guanine (FapyG). It is capable of existing in α- or β-anomeric form, which has different mutagenic profiles that generally cause G→T and G→C substitutions. FapyG is physically unstable at physiological pH and may be produced as an alternative to 8-oxoG production in an oxygen-tension and microenvironment-dependent manner (Hwang et al., 2025).

Abasic (AP) sites are produced by hydrolysis at the N-glycosidic bond in response to oxidative damage, resulting in liberation of the damaged base and liberation of a deoxyribose unit with nothing filled in the purine or pyrimidine position. The AP sites are noncoding and prevent template-directed synthesis and precondition the strand to cleavage by either β- or δ-elimination (Minko et al., 2024). They are especially harmful when they are clustered because they may lead to the generation of two-strand breaks during the repair process. These lesions vary in stability, repair susceptibility, and biological impact, among others, and they together make the mutational signatures associated with cancer, neurodegeneration, and rapid aging (Thompson and Cortez, 2020).

10.2.4 Damage to DNA backbone and cross-links

In addition to nucleobase oxidation, ROS may oxidize the deoxyribose backbone directly, thereby beginning a process of strand scission by abstraction of a hydrogen atom at C1, C4, or C5. The result of C4 oxidation includes unstable intermediates that give rise to 3-phosphoglycolate termini, thereby slowing ligation and requiring specialized enzymes to process the end (Wojtaszek and Williams, 2024). The ROS may also enhance the formation of intra- and interstrand cross-links via covalent bridging of DNA bases or DNA and protein bridges. These cross-links generally occur during secondary reactions of aldehydic products of lipid peroxidation or ROS-activated transition metals that trap DNA in conformations that stall replication and transcriptional machinery. Compared to simple picnic cross-links, cross-linked DNA represents an unusual challenge to repair machineries, and repair often involves cooperation of the NER, homologous recombination (HR), and specialized translesion synthesis polymerases. Any ongoing backbone breakages or cross-links are very cytotoxic and may induce chromosomal instability irreparably (Oswalt and Eichman, 2024).

10.3 Detection and quantification of oxidative DNA lesions

10.3.1 Biochemical assays

Other techniques, such as biochemical assay analyses, show high levels of sensitivity, specificity, and reproducibility, and therefore form the most accurate methods to detect and quantify oxidative DNA lesions. The densely packed HPLC with electrochemical detector (HPLC-ECD) system is especially well adapted to the quantitation of redox-active base adducts, including 8-oxoguanine (8-oxoG), with detection limits of the femtomoles. The identification of a wide range of oxidative lesions is possible with gas chromatography with isotope-dilution-mass spectrometry (GC-MS) after derivatization, providing the choice and discrimination of isomeric products (Zheng et al., 2025). Nevertheless, the use of GC-MS can lead to artificial oxidation of the samples during the preparation process, the plan of control measures is stringent. Cellular and tissue lesion profiling have both become highly compatible with liquid chromatography-tandem mass spectrometry (LC-MS/MS) in terms of selectivity, minimal sample manipulation, and compatibility with all types of lesions, including oxidized purines, pyrimidines, and abasic sites (Shybyray et al., 2025). With the recent introduction of ultra-high-performance LC-MS/MS (UHPLC-MS/MS), throughput and signal-to-noise have been increased dramatically, and allowed quantifying lesions multiplexed from small biological material. Taken together, such chromatographic-mass spectrometric methods enable quantitative biomarker analysis of oxidative stress in normal and disease processes in terms of mechanistic investigations and translational biomonitoring (Yu et al., 2024).

10.3.2 Immunoassays for lesion-specific detection

Monoclonal or polyclonal antibodies against selected oxidative adducts of DNA and particularly of 8-oxoG, thymine glycol, and ethanol adducts are used in immunoassays. Semi-quantitative high-throughput screening in biofluids or extracted DNA can be performed because ELISA formats can be used, and spatial localization of lesions can be performed within the tissue architecture using immunohistochemistry (IHC). Narrow quantitative specificity aside, immunoassays are valuable in screening, histopathological correlation, and rapid screening. Due to recent progress in antibody engineering, there is increased specificity (Amemiya et al., 2025).

10.3.3 Comet assay for oxidative strand breaks

Sensitive methods that can detect DNA strand breaks, as well as alkali-labile sites, at a single-cell level include the alkaline single-cell gel electrophoresis (comet) assay. During the measurement of oxidative damage, the addition of enzymes that repair only the lesions of interest, formamidopyrimidine DNA glycosylase (Fpg) or endonuclease III, allows oxidized purines and pyrimidines of interest to be preferentially converted into strand breaks that can be measured (Nordengen et al., 2024). The migration of DNA, measured in terms of tail length/tail moment, is an indicator of the oxidative lesion burden. The comet assay is flexible in measuring DNA integrity in a variety of cell types and is sensitive to nontoxic levels of damage, which is attractive in environmental studies of genotoxicity and clinical biomonitoring (Tyutereva et al., 2024).

10.3.4 Fluorescent probes and live-cell imaging

The use of fluorescent probes makes it possible to visualize oxidative DNA damage, in real-time, in living cells, therefore, creating dynamic information regarding lesion formation and repair rates. To detect lesions with high specificity, small-molecule probes, e.g., 8-oxoG-binding fluorophores or genetically encoded reporters based on lesion-recognizing protein domains fused to fluorescent proteins have been developed (Ohno et al., 2024). Such tools, combined with confocal or super-resolution microscopy in many cases, allow spatial mapping of damage with subcellular resolution. In perceptual aspects, live-cell imaging has shown that lesions are detected and repair factors recruited very fast to the chromatin domains, providing more insight into the signaling pathways that are involved in damage via oxidation. The key issues are not resolved yet and lie in the realization of single-cell discrimination of probes, reduction of phototoxicity, and detection of multiple lesions in complicated biological systems (Ramos Cáceres et al., 2024).

10.4 DNA repair pathways for oxidative damage

10.4.1 Base excision repair pathway

10.4.1.1 Initiation by DNA glycosylases

Lesion-specific DNA glycosylases induce the BER pathway by cleaving naturally damaged bases, without the need to cut the DNA backbone. These enzymes work by using a so-called base-flipping mechanism, pushing a catalytic nucleotide into the DNA helix and thrusting the damaged nucleotide into the active site (Hua and Sweasy, 2024). Of all the oxidative lesions, one of the major mutational products is 8-oxoguanine (8-oxoG), which is principally corrected by 8-oxoguanine DNA glycosylase 1 (OGG1), which excises 8-oxoG in a cytosine pairing with G: C to A: T transversion blockage. In a way, MUTYH acts as a mutagenesis gatekeeper because it acts downstream of OGG1 and removes the wrongly inserted adenines opposite 8-oxoG. This action prevents noncorrected oxidative mispairing events from being skipped during replication. The NTH1 (endonuclease III-like 1) mainly can clear the oxidized pyrimidines, including thymine glycol, which is produced due to the oxidation of the cytosine or thymine by ROS. The way it is structurally arranged in domain position facilitates single-stranded as well as double-stranded specificity in DNA (Le Meur et al., 2024). A major DNA repair mechanism is presented in Fig. 10.3.

The NEIL glycosylases (NEIL1-3) are less substrate-specific and act on hydantoin oligomers, formamidopyrimidines (FapyG, FapyA), and oxidized bases bound within crowded DNA conformations, such as replication and transcription hotspots. The example of NEIL1, involved in replication-coupled repair, and NEIL3, which is biased toward non-B-DNA structures and has a major role in highly oxidatively labile regions of the genome, including telomeres, is provided (Zhong et al., 2025). All these glycosylases combine to be able to act in a lesion-type-specific and temporally coordinated fashion. Excessive substrate specificity allows redundancy, which is a fail-safe mechanism, and expression and recruitment are tightly regulated by redox state, cell cycle stage, and interaction with repair scaffolding proteins such as XRCC1. Some of these, including OGG1 and MUTYH, are associated with cancer predisposition syndromes, such as MUTYH-associated polyposis, showing the importance of these genes in ensuring genomic integrity (Shahi and Kidane, 2025; Vodicka et al., 2025).

10.4.1.2 AP endonuclease activity

After removal of the base by glycosylase, an AP site is created, and this is potentially harmful to strand breakage and pause in transcription unless it is repaired. The most common mammalian AP endonuclease, named AP endonuclease 1 (APE1), catalyzes hydrolytic scission of the phosphodiester bond on the 5 side of the AP site to produce a single-strand break with a 3 and a 5 end and a 5 termini of deoxyribose phosphate (5-dRP) (Jordano-Raya et al., 2025). APE1 has both endonuclease and 3-phosphodiesterase functions; it has been shown to be capable of dealing with the blocked DNA ends caused by ROS damage or otherwise due

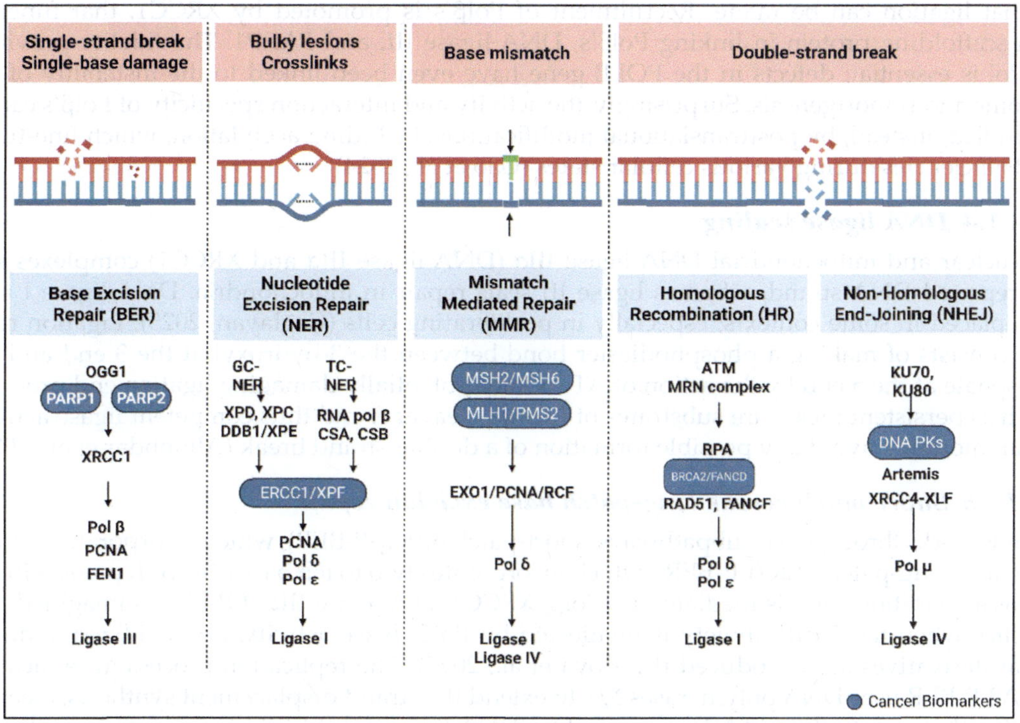

FIGURE 10.3 Overview of major DNA damage types and their corresponding repair pathways. Base excision repair (BER) mends single-strand breaks and single-base damage, with participation of enzymes including OGG1, PARP1/2, XRCC1, and DNA polymerase β. Nucleotide excision repair (NER) provides repair of bulky lesions and cross-links, mediated by XPC, XPD, XPA, CSA/CSB, and ERCC1/XP. Mismatch repair (MMR), which is a system that interacts with MSH2/MSH6 and MLH1/PMS2 complexes, corrects base mismatches. Repairs of double-strand breaks proceed by homologous recombination (HR), which is coordinated by ATM, BRCA2, FANCD, and RAD51, or by nonhomologous end joining (NHEJ), coordinated by KU70/KU80, DNA-PKcs, Artemis, and XRCC4. Biomarkers of cancer are suggested (created using BioRender.com).

to abortive topoisomerase reactions. In addition to its enzymatic activity, APE1 transcription factor redox regulator of AP-1, p53, and NF-κB, thus linking DNA repair and stress-response pathways. Insufficiency or failure of APE1 causes the AP sites to accumulate and sensitivity to oxidative stress induction, thereby contributing to neurodegeneration, cancer liability, and inflammatory diseases. The central nature of APE1 makes it of interest as a target in oncology, and therapeutic agents that inhibit it are being sought to enhance the efficacy of DNA-damaging chemotherapeutic agents (Liu et al., 2024).

10.4.1.3 DNA polymerase β-mediated gap filling

During short-patch BER, the one-nucleotide gap that occurs following APE1 cleavage is again filled mainly through the enzyme DNA polymerase β (Polβ). Pol β also has two different enzyme activities: a polymerase domain, which can add the proper nucleotide in the opposite direction of the undamaged base, and a lyase domain, which removes the remaining 5 - dRP group (Lebedeva et al., 2024). This coupling allows fast recovery of the DNA backbone chemistry so that ligation can be made. Recruitment of Polβ's is promoted by XRCC1, that function as a scaffolding protein in linking Polβ's, DNA ligase III, and PARP1. The fidelity provided by Pol is essential; defects in the POLB gene have even been linked to the instability of the genome and tumorigenesis. Surprisingly, the activity and interaction specificity of Polβ's can be controlled, instead, by posttranslational modifications, including acetylation, which fine-tunes BER efficiencies during various cellular states (Gao et al., 2024).

10.4.1.4 DNA ligase sealing

Nuclear and mitochondrial DNA ligase IIIα (DNA ligase IIIα and XRCC1) complexes seal the repaired DNA strand, whereas ligase III seals repair in mitochondria. DNA ligase I may be replaced in some contexts, especially in proliferating cells (Çağlayan, 2025). Ligation reaction consists of making a phosphodiester bond between the 3-hydroxyl at the 3 end and the phosphate at the 5 end by the action of ATP. Other potentially damaging ligation end products (such as persistent nicks) are substrates of further cleavage, and thus competent ligase activity is paramount to avoid any possible formation of a double-strand break (Almohdar et al., 2024).

10.4.1.5 Short-patch versus long-patch base excision repair

BER works through two subpathways: short-patch BER (SP-BER), which incorporates a single base, and long-patch BER (LP-BER), which incorporates two to ten bases. SP-BER predominates in basal conditions and is mediated by Polβ, XRCC1, and ligase IIIα. LP-BER is imagined into practice when the 5 dRP moiety is unaffected by Polβ -lyase activity, e.g., oxidized/reduced sugar derivatives are introduced (Burdova et al., 2025). The replication process in replicating DNA LP-BER uses DNA polymerases δ/ε to extend the strand displacement synthesis, creating a 5 flap that is cut off by FEN1 and sealed by DNA ligase I. LP-BER plays a role in particular importance in replicating cells and when repairing clustered oxidative lesions in avoiding replication fork collapse (Kutuzov et al., 2025).

10.4.1.6 Base excision repair regulation by posttranslational modifications

BER proteins are dynamically regulated by posttranslational modifications (PTMs), which allow fine-tuning following oxidative stress and the cell cycle. Key proteins are regulated in their activity, stability, and interaction by phosphorylation, acetylation, SUMOylation, and ubiquitination. By way of example, phosphorylation of OGG1 stimulates its chromatin binding property, but acetylation of Polβ can promote its catalytic activity at the expense of Polβ interaction with XRCC1 (Chen et al., 2025). The structure of APE1 is determined by acetylation and redox-sensitive conformational switching, but it has effects on DNA repair and transcriptional regulation. XRCC1 is SUMOylated, and this controls the kinetics of XRCC1 recruitment to sites of DNA damage. This type of control by PTM will guarantee that BER will be turned on at a high rate during oxidative stress but inhibited when in an unfavorable situation, hence the avoidance of undesired DNA modification.

10.4.1.7 Base excision repair in nuclear versus mitochondrial DNA repair

BER is functional both in nuclear and in mitochondrial genomes; however, with different features. BER chromatin remodeling factors are spatially synchronized with BER in the nucleus, allowing a loosely organized DNA to be reached. There are various isoforms of DNA repair proteins, which include OGG1 and ligase IIIα, which are directed to the nucleus by localization signals (Xiao et al., 2025). BER is the most important pathway of DNA repair in mitochondria, which functions to counteract the extreme oxidative stress present in the organelle, as well as the absence of NER. Mitochondrial BER depends on a fraction of the nuclear-encoded proteins that is imported into the organelle, targeting OGG1, NEIL1, Polγ (rather than Polβ), and ligase III. Histones are not present, and the condensation of the mitochondrial DNA provides kinetic benefits specific to nitrogen bonding as well as increasing vulnerability to repair defects, which have been associated with aging and neurodegenerative diseases (Amoah et al., 2025).

10.4.2 Nucleotide excision repair

10.4.2.1 Role in repairing bulky oxidative adducts

Nucleotide excision repair (NER) is a highly adaptable pathway of repair that has specialized in the detection and excision of structurally damaging lesions involving modification of bases or alteration of the DNA core structure; it recognizes and removes bulky oxidative adducts that disrupt the DNA helix. Whereas BER plays a dominant role in repairing minor oxidative damage, NER is required in cases where oxidative damage causes a substantial base alteration, crosslink, or cyclopurine lesion, usually 8,5′-cyclo-2′-deoxyadenosine, whose resulting helical distortion severely hinders transcription or replication in both directions (Ma et al., 2024). NER takes place in a series of events wherein the damaged-sensing complexes (XPC-RAD23B in global repair or RNA polymerase II stalling in transcription-coupled repair) recognize a lesion, local unwinding by the TFIIH helicase occurs, and the XPF-ERCC1 (5) XPG (3) endonucleases then make dual incisions, leaving a gap that is filled by the DNA polymerases δ / ε and the gap that is then sealed by the ligases. The flexibility of the NER in carving out heterogeneous oxidative adducts stands out as the reason it is essential to genome maintenance under circumstances of exceptionally high quantities of ROS. The physiological relevance of NER is associated with diseases or disorders such as xeroderma pigmentosum (XP) and Cockayne's syndrome associated with impairments (Quiñones et al., 2024).

10.4.2.2 Global genome nucleotide excision repair versus transcription-coupled nucleotide excision repair

NER occurs in two subpathways, namely Global genome NER (Gg-NER) and Transcription-Coupled NER, which identify lesions differently yet share a common excision mechanism. With damage sensors such as the XPC-RAD23B or DDB2, GG-NER sweeps the whole genome, looking at helix-distorting lesions, and can then get the preemptive repair done in both transcribed and nontranscribed sequences. The pathway plays a very important role in the prevention of mutagenesis in the quiescent chromatin (Gong et al., 2024). Instead, TC-NER is scrupulous, and it only focuses on lesions that hamper the progress of elongating RNA polymerase II in actively transcribed genes. In this case, CSA and CSB proteins assist in the removal of stalled polymerase and replace it with repair factors and the removal of the lesion in an attempt to reestablish transcriptional fidelity. Whereas GG-NER is used to implement widespread genomic monitoring, TC-NER focuses on transcriptional rescue that reduces the resultant months-long arrest of RNA polymers, which becomes cytotoxic. GG-NER defects are mostly expressed as phenotypes predisposed to cancer (e.g., XP), whereas deficiencies in TC-NER result in neurodegenerative disorders of transcription (e.g., Cockayne syndrome) since each system has a unique role in the repair of reactive oxygen-induced DNA damages (Pani and Nudler, 2025).

10.4.3 Mismatch repair

10.4.3.1 Recognition of mispairs from oxidative damage

Base modifications, e.g., 8-oxoguanine (8-oxoG), can often arise as a result of oxidative stress and are subsequently replicated opposite adenine, forming a transition mutation in the case of the 8-oxoG example, unless these base-modified products can be recognized and repaired before replication. These fine-scale structural peculiarities are identified by the MMR system by probing helical distortion and altered hydrogen-bonding in the DNA duplex. Sophisticated recognition complexes tease out oxidatively induced mispairs as compared to canonical base pairs when there is no gross structural injury. Such high-fidelity detecting depends on modes of conformational proofreading, which incorporates base geometry appraisal as well as kinetic discrimination of DNA scanning. Inhibition of mutagenic lesion repair by preventing lesion repair is related to causing fixation of genomic stability and reducing the mutational burden caused by repeated exposure to chronic oxidative challenges (Xu et al., 2025).

10.4.3.2 Role of MutSα and MutLα

The MSH2and MSH6 heterodimericMutα a forms the chief sensor of oxidative damage-induced base mispairings due to 8-oxoG: A and other noncanonical base pairs in particular. When a lesion is identified, the MutS is exposed to a conformational rearrangement (ATP-dependent), leading to the recruitment of MutLα-a heterodimer comprising MLH1 and PMS2, which plays the role of matchmaker between lesion detection and the excision machinery (Erie and Weninger, 2024). MutLα organizes endonucleolytic cleavage close to the mispaired location, allowing strand removal by exonucleases, and the next resynthesis using the intact strand as a template. This coordinated MutSα–MutLα cascade removes not only replication-linked oxidative errors but also connects with BER, also increasing efficiency of lesion processing in the face of oxidative stress conditions (Kadyrova et al., 2024).

10.4.4 Homologous recombination and nonhomologous end joining

10.4.4.1 Oxidative double-strand breaks repair

One of the most cytotoxic ROS-induced lesions is oxidative DNA double-strand breaks (DSBs) that are frequently indirectly induced as a result of oxidized base damage and collapse of the stalled replication fork. In eukaryotes, two major repair pathways are organized to eliminate a wide range of damage, including HR and nonhomologous end joining (NHEJ). HR is an unmutated process and necessitates a sister chromatid as a basis, and the activity is mostly limited to S and G2 phases, starting with the resection of DNA ends by a complex (MRE11 RAD50 NBS1) and by CtIP in association with further stages of resection by the complex and then invasion of the strands by RAD51 (Li et al., 2024). On the other hand, NHEJ occurs during the cell cycle and entails the recognition of the KU70/80 heterodimer, the activation of DNA-PKcs, and the ligation working through the XRCC4-Ligase IV complex, but with greater mutagenicity. Pathway choice and efficiency are affected by oxidative alterations on repair proteins, chromatin environment, and ATP-dependent remodeling. Oxidative stress dysregulates these mechanisms and forms the basis of genomic instabilities that form the pathological basis of a variety of degenerative and neoplastic pathologies (Yang and Lan, 2025).

10.4.4.2 Interplay with oxidative stress response

Oxidative DSB repair is closely connected with the general exposure to oxidative stress events via redox-responsive signaling circuits. DSB sensing and activation of ATM and ATR kinases not only activate HR or NHEJ but also stimulate transcriptional programs in collaboration with p53, NRF2, and FOXO that are involved in the resolution of redox imbalance. ROS are conservative and damaging, while they also act as signaling molecules promoting PTMs (phosphorylation, ubiquitination, and PARylation) of components of HR/NHEJ to enhance repair dynamics (Iqbal et al., 2024). The redox environment required to facilitate enzymatic repair protein activity is provided by antioxidant pathways (such as the glutathione and thioredoxin systems), which favorably affect the reduction potential of repair enzymes, especially those containing the oxidizable cysteine cluster. The interactions of DNA repair with mitochondrial dynamics also coordinate other processes, including the coordination of ROS generation and fidelity of repair. Disruptions of this interaction aggravate the phenomenon of mutation accumulation, accelerate the process of senescence, or trigger tumor development; therefore, the strategies of therapeutic intervention should be aimed not only at the repair of DNA but also at the regulation of oxidative stress as a way to increase genome stability in the pathological state (Bae et al., 2024).

10.5 Regulation of oxidative DNA repair

10.5.1 Cell cycle control of repair activity

The cell cycle controls the efficiency and selection of the oxidative DNA repair pathways largely to maintain genomic integrity and also to avoid the replication-linked mutagenesis. The oxidative lesions that are mostly addressed by BER, including 8-oxoG, are actively repaired during the G1 phase, when the lesions can be corrected prior to synthesis. Certain BER enzymes, such as OGG1 and APE1, are also phase-regulated and have PTMs involved in the regulation of their activity (Pellarin et al., 2025). In the S phase, repair mechanisms associated with replication are activated at high speed to avoid insertion of damaged bases, and this is facilitated by recruitment of repair proteins, which augment replication activities through proliferating cell nuclear antigen. During G2/M, repair efficiency decreases, in part because chromatin condensation and checkpoint signaling cause a defocus toward repair and a focus toward chromosome segregation. This temporal regulation is mediated by cyclin-dependent kinases by direct phosphorylation of the repair factors in connection to the DNA checkpoint and processing of oxidative lesions. Such coordination means that repair is maximally faithful, but that repair activities do not cause damaging disruption to replication or mitotic events (Khamidullina et al., 2024).

10.5.2 Cross-talk with antioxidant defense systems (NRF2 pathway)

Nuclear factor erythroid 2-related factor 2 (NRF2) pathway is the master regulator of the responses to oxidative stress, which in turn directly affects the oxidative DNA repair capacity. In the absence of stimulus or under basal conditions, NRF2 is kept in the cytoplasm by KEAP1 and is ubiquitinated and degraded. Oxidative stress interferes with the process of KEAP1-mediated repression and allows translocation of the NRF2 into the nucleus and interaction with antioxidant response elements (AREs). In addition to increasing the expression of traditional antioxidant enzymes (e.g., superoxide dismutase, glutathione peroxidase), NRF2 also induces the expression of genes encoding proteins involved in DNA repair (e.g., OGG1, XRCC1, APE1), increasing the BER capacity (Manavi et al., 2025). This synergistic activity lowers the ROS load as well as enhances the effectiveness of eliminating lesions. More so, NRF2 signaling overlaps with mitochondrial biogenesis and redox homeostasis, conceiving secondary oxidative damage. New evidence indicates that sustained NRF2 activation, which is protective of acute and oxidative stress, has a predisposing role in promoting cell survival of DNA-damaged cells in oncogenesis, where balanced regulation is a necessity. Therefore, the NRF2-DNA repair axis is a life-saving adaptive mechanism between oxygen sensing and genomic repair mechanisms (Shan et al., 2024).

10.5.3 p53-Mediated regulation

p53 is a core transcription factor that mediates between the detection of oxidative DNA damage and apoptotic pathway activation or triggering repair responses. p53 stabilizes in response to oxidative stress through ATM/ATR-dependent phosphorylation, triggering gene transcription of enzymes in BER pathways (e.g., POL9, APE1) and reinforcing antioxidants like sestrins. p53 additionally regulates the NER mediators of bulky adducts. In parallel, p53 inhibits progression of the cell cycle via p21, to give the cell the time to resolve the lesion. In extreme oxidative genotoxicity, the p53 proteins select pro-apoptotic transcripts (e.g., BAX, PUMA) over repair in order to exclude dissemination of severely damaged genomes. It is a selective regulatory switch, which guarantees repair fidelity as well as tumor suppression (Osbourne and Thayer, 2025).

10.5.4 Role of chromatin remodeling in repair accessibility

Nucleosomal locations of repair-impeding histone-DNA interactions tend to be the sites of many oxidative DNA lesions. Chromatin remodeling complexes (e.g., SWI/SNF, INO80, and ISWI) reposition or dislodge the nucleosomes at the sites of damage, opening them to damage repair and other repair machineries. Such repair factors are recruited by histone PTMs, such as H3K9 acetylation and H2A ubiquitination. Chromatin relaxation to recognize the lesion is also available by PARP1. On the other hand, heterochromatin areas undergo more remodeling events, which slows down the repair times. Therefore, dynamic chromatin restructuring plays a key role in the efficiency of oxidative DNA repair, controlling the rate at which lesions are detected as well as the postrepair genome-wide completion of repair (Downs and Gasser, 2024).

10.6 Oxidative DNA damage in mitochondria

10.6.1 Mitochondrial vulnerability to reactive oxygen species

The most effective location of ROS production is mitochondria as a result of electrons leaking out of the oxidative phosphorylation (OXPHOS) systems, namely complexes I and III. MtDNA is an excellent target of oxidative damage because it is in proximity to the inner mitochondrial membrane. The inner mitochondrial membrane is a site at which ROS are generated. In contrast, as opposed to nuclear DNA, mtDNA lacks; therefore, protective histones and is arranged in nucleoids with little structural protection (Song et al., 2024). Further, even putatively available mitochondrial antioxidant defenses (e.g., MnSOD, peroxiredoxins) are comparatively weak as in nuclear systems. The elevated transcriptional activity of the mtDNA also adds more exposure to ROS-driven DNA damage lesions in the form of 8-oxoguanine, single-strand breaks, and abasic sites. Since every mitochondrion has many copies of mtDNA inside it, oxidative damage may multiply clonally during the process of mitochondrial duplication. As a result, ROS sensitivity of the mitochondrion is the key factor determining the cellular oxidative burden, bioenergetic loss, and predisposition to degenerative disease (Lin et al., 2025).

10.6.2 mtDNA repair pathways (base excision repair predominance)

Compared to nuclear DNA, the repair processes in the mitochondria are rather scarce, and BER is the most dominant process of removing oxidative lesions. Lesion-specific DNA glycosylases that include OGG1 (8-oxoguanine DNA glycosylase), NTH1 (endonuclease III-like protein 1) initiate BER in mitochondria by binding and hydrolyzing the damaged-base linkage (Chaudhari et al., 2025; Hatvate and Omri, 2025; Jadhav et al., 2025; Prajapat et al., 2025; Sagar et al., 2025; Uddin et al., 2025b). To complete these processes, APE1 is used to enzymatically convert the DNA backbone into a cut at an abasic site. The gap introduced by the excision is repaired by DNA polymerase γ (Pol γ), the only mitochondrial replicase, which inserts the correct nucleotide to fill the gap, and a nick is sealed by DNA ligase III to reestablish DNA integrity (Urrutia et al., 2024). BER is necessary to keep mtDNA stable, as opposed to the nuclear compartment, which contains strong NER and MMR mechanisms. Evidence is also emerging that there is a role of small levels of HR and microhomology-mediated end joining in resolving a double-strand break, but this remains less well-studied. An important interaction between mitochondrial BER and mitochondrial dynamics is that fusion allows dilution of the lesions between several copies of the mitochondrial genome, and fission can segregate heavily damaged mitochondria so they can enter the process of mitophagy. Mutation in any of the BER processes, which may occur either through genetic mutation or through oxidative overload, increases mutation rates, diminishes the capacity of functional respiratory chain activity, and predisposes cell fates of apoptosis or senescence, highlighting its pivotal role in maintaining genomics of mitochondria (Bhattacharjee and Das, 2025).

10.6.3 Impact of defective mtDNA repair on aging and disease

Lack of repair in mtDNA contributes to the accumulation of oxidative lesions that get replicated, bypassed, or mispaired, therefore causing point mutations, deletions, and depletion of the entire mitochondrial genome. These defects aggravate further to inhibit OXPHOS functioning, which expounds the production of ROS in a positive feedback cycle of oxidative damage (Quiñones et al., 2024; Xiao et al., 2025). This translates to a reduction in mitochondrial bioenergetics, disturbed metabolic signaling, and apoptotic vulnerability, which are close to the mitochondrial free radical theory of aging in old age. Causally, poor repair of mtDNA leads to a variety of illnesses such as mitochondrial encephalomyopathies, cardiomyopathies, and neurodegenerative conditions, including Parkinson's and Alzheimer's diseases, where the oxidative dysfunction of the mitochondrion is a characteristic feature. Tumorigenesis by inducing metabolic reprogramming is also mediated by somatic mutations of mtDNA that occur during repair deficits. Consequently, preserving the efficiency of mitochondrial DNA repair, especially BER fidelity, is a key not only to the cell lifespan but also to therapeutic intervention to reduce the occurrence of age-related functional impairment and oxidative-stress-triggered pathogenesis (Hua and Sweasy, 2024; Jordano-Raya et al., 2025; Shahi and Kidane, 2025; Vodicka et al., 2025).

10.7 Associated human disorders

10.7.1 Cancer

10.7.1.1 Role of oxidative mutations in carcinogenesis

One of the key causes of genomic instability, which is one of the hallmarks of cancer, is oxidative DNA lesions caused by ROS. Among these lesions, 8-oxoguanine (8-oxoG) is both a highly mutagenic lesion, as it is a frequent substrate of misincorporation by replication polymerases; it yields G:C→T:A transversions via cross-recognition by the replicating polymerase as an A. The build-up of such mutations in proto-oncogenes (e.g., RAS), tumor suppressor genes (e.g., TP53), and genes regulating DNA repair affects the critical pathways in the cell, i.e., proliferation, apoptosis, and fidelity of DNA repair. Oxidative stress can be so persistent as with chronic inflammation, metabolic blowout, and carcinogenic exposure to the environment, and overwhelm BER capacity, fixing a mutation during replication cycles (Lebedeva et al., 2024; Vodicka et al., 2025).

In addition, oxidative DNA damage mediated via ROS enhances epigenetic perturbation, such as the differences in methylation, failed chromatin remodeling, which may result in the repression of tumor suppressor loci or trigger the activity of oncogenic promoters. Because of its being close to the ROS production sites and its limited repair mechanisms, mtDNA is especially vulnerable, resulting in mitochondrial dysfunction, metabolism reprogramming, and subsequent increased ROS production, thus forming a carcinogenic self-amplifying loop. Due to the effects of oxidative mutagenesis, impaired DNA repair, and pro-oncogenic signal transduction pathways, oxidative stress can be described as the trigger and accelerator of carcinogenesis. More contemporary research points toward the prospects of oxidative damage-specific DNA repair proteins as both the biomarkers of cancer risk and as the target of cancer treatment in redox-based tumors (Di Carlo and Sorrentino, 2024; Yang et al., 2024; Younus et al., 2024).

10.7.1.2 MUTYH-associated polyposis

MUTYH-associated polyposis (MAP) is an autosomal recessive heritable disorder of colorectal carcinoma related to biallelic germline mutations in the MUTYH gene, a fundamental BER synthetic glycosylase that evacuates an incorrect incorporation of adenine in front of the oxidized compound 8-oxoG. Inactivation of MUTYH allows fixation of G:C→T:A transversions, which are of exceptionally short wavelength and specific to the APC tumor suppressor gene and lead to adenomatous polyp formation and malignant progression. Clinically, MAP exhibits numerous colorectal adenomas that are often more than 10-100 polyps in number, and their lifetime risk of colorectal cancer is substantially increased. Early genetic screening and monitoring would be important because MAP is highly responsive to prospective colonoscopic treatment and preventive surgical procedures (Pani and Nudler, 2025).

10.7.1.3 *OGG1 mutations and cancer risk*

The major BER enzyme to recognize and remove 8-oxoG lesions is 8-oxoguanine DNA glycosylase 1 (OGG1). The OGG1 germline or somatic mutations impair substrate recognition and efficiency of the enzyme, leading to the accumulation of oncogenes and tumor suppressor mutagenic transversions. Epidemiological studies found OGG1 polymorphisms such as Ser326Cys to associate with increased cancer risk involving the lungs, the stomach, and hepatocellular carcinomas, especially when people are subjected to high levels of oxidative stressors through smoking habits, diets, or inflammation. OGG1 loss-of-function increases mutation load, collaborates with seq. Other BER defects and the IND agent interact with each other to promote tumorigenesis. OMG1 polymorphisms may be excellent markers of overall cancer tendencies and precision prevention interventions (Zhong et al., 2024).

10.7.2 Neurodegenerative diseases

10.7.2.1 *Alzheimer's disease*

AD is a progressive neurodegenerative disease that is associated with cognitive decline and memory dysfunction, along with synaptic dysfunction, and oxidative DNA damage is considered one of the main factors of the pathogenesis. Mitochondrial dysfunction, amyloid-β (Aβ) oligomers, and neuroinflammation produce ROS that cause lesions in the genomes of neurons in the form of 8-oxoguanine (8-oxoG) and single-strand breaks (SSB). Deficit in BER, most notably a decreased activity of 8-oxoguanine DNA glycosylase 1 (OGG1) and APE1, accumulates lesions that cause neurotoxicity and increases the rate of neuronal death. Tau hyperphosphorylation and synaptic loss are also enhanced by oxidative DNA damage, and this neurodegeneration leads to a vicious cycle. Current research has demonstrated the interaction between defective DNA repair and epigenetic dysregulation in relation to AD development. Reduction in oxidative stress, repair and protection of DNA, and protection at the mitochondrial level—therapeutic strategies that include polyphenolic antioxidants, NAD+ enhancing agents, and BER-promoting agents—are also under consideration as prospective disease-modifying therapies (Buccellato et al., 2021).

10.7.2.2 *Parkinson's disease*

Parkinson's disease (PD) is a motor disorder that is primarily caused by progressive injury of dopaminergic nerve cells in the substantianigra pars compacta, and oxidative injury, causing DNA damage, is revealed as a major contributor to the mechanistic processes. The high levels of ROS, which are produced as a result of the abnormal functioning of mitochondrial complex I, dopamine-mediated auto-oxidation, and neuroinflammation, lead to DNA damage comprising 8-oxoG, DSB, and DNA-protein cross-links. Such high metabolic requirements and low regenerative potential of the nigral neurons make them especially sensitive to breakdowns in repair. Malfunctioning BER, specifically decreased these protein levels of OGG1 and XRCC1, impairs the process of removing the lesions, which favors the development of genomic instability and apoptotic signaling. Besides, oxidative vulnerability is worsened by mutations in the DNA repair-related gene PARK7 (DJ-1). This ongoing DNA damage triggers p53-mediated cell death pathways, which adds to the loss of neurons. These early studies indicate that neuroprotection could be achieved by pharmacologically improving DNA repair-system enzymes, and by mitochondrial stabilization with a specific series of antioxidants, providing opportunities to make interventions disease-modifying in PD (Dong-Chen et al., 2023).

10.7.3 Cardiovascular and metabolic disorders

10.7.3.1 Oxidative DNA damage in atherosclerosis

The notion of atherosclerosis being an oxidative stress-mediated vascular pathology whose etiology involves the production of ROS that cause prolonged oxidative stress-related lesions, particularly 8-oxo-7,8-dihydroguanine (8-oxoG) on endothelial and smooth muscle cells, is gaining traction. These lesions lead to the activation of DDR pathways, triggering the p53-dependent cell cycle arrest, cellular senescence, and the pro-inflammatory transcriptional programs to promote plaque formation. An increase in ROS then acts to enhance the damage to mtDNA, setting into motion a recursive oxidative process. Defect in BER, especially a deficient OGG1 action, has been associated with rapid lesion progression. These genomic insults alone destabilize already present atherosclerotic plaques, but they also serve to increase the thrombogenic nature of the disease, a phenomenon which is highlighted as a valid therapeutic objective when treating cardiovascular disease on the basis of oxidative DNA damage alone (Rahimi et al., 2025).

10.7.3.2 Oxidative DNA damage in diabetes

Chronic hyperglycemia in diabetes mellitus increases intracellular ROS by autoxidation of glucose, protein glycation, and mitochondrial overload that results in widespread nuclear DNA and mtDNA oxidation. The build-up of 8-oxoG and strand breaks destabilizes the genome of pancreatic 8-oxoG and break-up and hampers insulin production/secreting abilities. DNA damage with hyperglycemia initiates the enzyme known as poly(ADP-ribose) polymerase-1 (PARP-1), which poisons higher NAD+ and ATP, compounding cellular dysfunction. Maladaptations in BER enzymes like MUTYH and NEIL1 promote oxidative mutagenesis, which develops microvascular and macrovascular complications. Oxidative DNA damage in the vascular endothelium perpetuates inflammation, fibrosis, and atherogenesis and is credited with the increased cardiovascular risk status exhibited by diabetic patients (Spoto et al., 2025).

10.7.4 Premature aging syndromes

Rare, genetically determinant, premature aging disorders, Werner syndrome (WS) and Cockayne syndrome (CS), are genetically characterized by improper DNA repair. The pathology of WS, which is a result of WRN helicase/exonuclease mutations, includes defects in replication, recombinational repair, and telomere maintenance and causes genomic instability, rapid onset of cellular aging and premature aging diseases. On the other hand, CS is caused by mutations in ERCC6 (CSB) or ERCC8 (CSA) and causes failure to perform transcription-coupled NER (TC-NER), and reconstitution is impaired in oxidative DNA damage. WS is characterized by cataracts, atherosclerosis, and diabetes in clinical practice, and CS is characterized by neurodegeneration, growth failure, and photosensitivity. The two syndromes emphasize the role of oxidative damage of DNA repair in aging and in the maintenance of tissues (Onisiforou et al., 2024).

10.8 Experimental and therapeutic approaches

10.8.1 Pharmacological antioxidants

Pharmacological antioxidants can form one of the main intervention routes to reduce oxidative injury to DNA through direct elimination of ROS or indirectly, through reserves of the host defense. Examples of classical agents are vitamin E (α-tocopherol), ascorbate (vitamin C), and carotenoids that intercept free radicals before they engage nucleic acids. Synthetic products, including edaravone and tempol, have been developed to be more stable ROS traps with greater tissue penetration. Some antioxidants, such as N-acetylcysteine (NAC), in addition to direct scavenging, also allow the synthesis of glutathione to strengthen redox buffering within the cell. Mitochondria-specific antioxidants (e.g., MitoQ, SkQ1) can provide targeted protection at one of the main sites of ROS production, which eliminates DNA damage of the mitochondrial genome that spreads nuclear genomic instability. In clinical use, there are limitations of varying pharmacokinetics, bioavailability, and even the paradoxical danger of disrupting physiologic ROS signaling. Recent interests are in combinatorial antioxidant therapy combined with targeted delivery systems (liposomes, polymeric nanoparticles) that will increase tissue selectivity, reduce dose-exposure to the systemic circulation, and maximize therapeutic effects (Meulmeester et al., 2022).

10.8.2 Gene therapy for defective repair genes

Gene therapy provides a curative solution to the diseases caused by mutation of oxidative DNA repair genes, especially involved in the BER pathway, including OGG1, MUTYH, and XRCC1. Stable transgenic expression in affected tissues can be performed with viral vectors, most notably adeno-associated viruses, to restore enzymatic activity and thus prevent the deposition of lesions. Besides viruses, efforts have been made in nonviral delivery of lipid nanoparticles or CRISPR/Cas9-dependent gene correction, which have been tested in preclinical models of inherited DNA repair syndromes, including MAP. The most important question that faces us yet is the superior, efficient expression that is specific and devoid of off-target effects and insertional mutagenesis. The promise of safety is being refined by improvements in tissue-specific promoters, epigenetic control, and transient delivery of mRNA. Currently experimental, gene therapy has the potential to become a key modality to address rare variants of genetic repair deficient, and perhaps to increase repair capacity in other high-risk groups (Kabziński and Majsterek, 2022).

10.8.3 Small molecules enhancing base excision repair activity

Selective small molecules that can boost the BER activity are a new pharmacological niche. The function of these agents is to up-regulate repair enzyme expression, stabilize repair protein complexes, or enhance enzyme kinetics. Examples include resveratrol and the polyphenolic derivatives, which enhance the transcription of BER-related genes by activation of such signaling pathways (e.g., SIRT1, Nrf2). There are also other modulators, such as methoxyamine, that interfere with AP endonuclease, which tilts the balance toward accurate repair rather than mutagenic bypass. Using high-throughput screening, synthetic modulators have been detected that can support OGG1 catalytic turnover or enhance the affinity of DNA glycosylases to oxidized guanine lesions. The therapeutic reasoning is that it will reduce age-related deterioration in repair rates, and immune and neural cells in the high-turnover tissues will be shielded against the accumulation of oxidative mutagenesis. The pharmacodynamic variability will have to be overcome, and repair enhancement must not have a bias toward the survival of genomically unstable cells to translate success into clinical practice (Hua and Sweasy, 2024; Lebedeva et al., 2024; Shahi and Kidane, 2025).

10.8.4 Lifestyle and dietary modulation of oxidative stress

Lifestyle and nutritional modulation have a significant adjuvant role in the reduction of oxidative DNA damage. Polyphenol-rich diets and omega-3 fatty acids, as well as microelements like selenium and zinc, accumulate antioxidant defense and recovery mechanisms. Mitochondrial efficiency can be enhanced by caloric control and exercise, which leads to a decrease in ROS leakage. On the other hand, smoking, consumption of too much alcohol, and high-fat/sugar diets increase oxidative stress and override repair mechanisms. Plant-based dietary patterns and intermittent fasting show emerging evidence of their value in reducing the occurrence of systemic redox imbalance as an adjunct to pharmacological and genetic interventions (Azzouz and Palaniyar, 2024; Pravin et al., 2024; Yang et al., 2024).

10.9 Role of nanotechnology in DNA damage and repair

The repair of DNA damage and its subsequent repair are basic biological processes that affect carcinogenesis, aging, and therapeutic outcomes in oncology (Chaudhari et al., 2025; Hatvate and Omri, 2025; Jadhav et al., 2025). Conventional approaches to the study or modulation of DNA repair depend on biochemical assays and small molecules, which are not always specific and sensitively detecting (Gholap et al., 2025; Gholap Amol et al., 2026a, 2026b, 2026c, 2026d; Pardeshi et al., 2026; Phade et al., 2025). Nanotechnology has become a paradigm shift technology; it provides specific means to investigate the presence of DNA lesions, understand repair kinetics, and even regulate the repair pathway to increase the efficacy of therapy. Researchers are redefining the approaches to measuring and manipulating DNA integrity in biomedical assays by utilizing nanoscale materials like quantum dots, metallic nanoparticles, liposomes, and polymeric carriers. One of the most direct applications of nanotechnology lies in the detection of DNA damage at high sensitivity and resolution (Dadarkar and Dubey, 2025; Khairnar et al., 2025; Prajapat et al., 2025; Sagar et al., 2025; Uddin et al., 2025b). DNA probes have been conjugated to gold nanoparticles (AuNPs) or quantum dots (QDs) to detect strand breaks or base modifications, or the establishment of adducts. Guanine nanoparticles (AuNP) colorimetric and QD-based fluorescence assays are able to identify femtomolar concentrations of broken DNA (Desai and Vora, 2025; Gholap and Omri, 2025; Khuspe et al., 2025a, 2025b, 2025c). As an example, oligonucleotide-conjugated AuNP probes would be aggregate or dispersed, respectively, in the presence of mismatched or damaged sequences, which is measurable as a change of color. These nanoprobes can be used to monitor radiation, oxidative stress, or chemotherapeutic damage of DNA in real-time (Hatvate et al., 2025; Khan et al., 2025; Pardeshi et al., 2024b; Uddin et al., 2025a).

DNA repair systems, such as BER, NER, HR, and NHEJ, are very dynamic. Other traditional tests, like comet tests or immunofluorescence, lack throughput and time resolution. Nanotechnology-based biosensors overcome these challenges. As an example, nanowire-based field-effect transistors are capable of real-time detection of repair protein binding to broken DNA substrates. In a parallel fashion, magnetic nanoparticles functionalized with lesion-specific oligonucleotides have the capacity to isolate repair intermediates, allowing mechanistic studies of enzymatic repair kinetics. These methods have enlarged our knowledge of repair enzyme-binding and lesion-excision recognition, as well as cancer cell-manipulation of repair efficiency to evade therapy (Faiyazuddin et al., 2024; Kumar et al., 2024; Pardeshi et al., 2024a; Sawant et al., 2024a, 2024b).

Another highly significant application of nanotechnology is the targeted and controlled delivery of DNA-damaging agents or repair inhibitors (Anand, 2011; Borude, 2011; Gholap, 2012; Magar Tushar V., 2011a; Rajurkar, 2012a, 2012b). Conventional chemotherapeutics such as cisplatin induce cross-links and strand breaks, but systemic toxicity limits their efficacy. Liposomes, dendrimers, or polymeric nanoparticles can be encapsulated, which increases tumor-specific accumulation through the enhanced permeability and retention effect and decreases off-target effects. Moreover, DNA repair inhibitors (e.g., PARP inhibitors) can be coloaded into the nanocarrier and serve as a complementary means of increasing the DNA damage and rendering tumors sensitive to radiotherapy or chemotherapy. The approach takes advantage of the principle of synthetic lethality, especially in tumors containing deficiencies of BRCA1/2 (Mesut et al., 2024; Srivastava and Ibrahim, 2024; Pardeshi et al., 2024b; Patil et al., 2024; Rojekar et al., 2024a; Rojekar et al., 2024b; Shaikh et al., 2024; Vardikar et al., 2025).

Radiation therapy relies on DNA DSBs for cytotoxicity. Metallic nanoparticles, particularly those of gold or hafnium oxide, are radiosensitizers that enhance local radiation dose by secondary emission of electrons. This causes better localization of DNA damage in tumor tissues, but not in normal cells. Certain nanoparticles also have their engineered capability to release ROS on irradiation, which further enhances the potential to scissile DNA strands. These bifunctional nanomaterials generate and detect DNA lesions to form theranostic platforms toward personalized radiotherapy (Bhowmik et al., 2023; Hatvate et al., 2024; Khuspe et al., 2024; Said et al., 2023; Pardeshi et al., 2023; Pardeshi and Giram, 2023; S et al., 2023; Uddin et al., 2024). At the other extreme, nanotechnology can be used to guard against the destruction of DNA or to facilitate a faster rate of repair. Nanoparticles that carry antioxidants like cerium oxide nanostructures are capable of eliminating ROS and alleviating oxidative damage to DNA (Dattatray, 2012a, 2012b; Dighe, 2012a, 2012b; Puri, 2018; Rajurkar, 2012c; Singh, Tribhuvan et al., 2016; Tupe, 2012). Also, gene delivery through nanoparticles is capable of repairing damaged DNA repair genes. Indicatively, liposomal vectors with functional copies of p53 or BRCA1 can rescue tumor suppressor programs in cells lacking products, and enhance genomic stability. These protective measures have the prospects of reducing genotoxicity in normal tissues that are either irradiated or subjected to environmental mutagens (Borude and Mahajan, 2011; Gupta et al., 2024; Hatvate et al., 2023b; Kapare et al., 2024; Khedkar, 2012a, 2012b; Pardeshi et al., 2024a; Parit et al., 2024; Rojekar et al., 2024a; Sawant et al., 2023).

In spite of these developments, there are problems in the translation of nanoscale knowledge to clinical practice. Biocompatibility, clearance, and off-target interactions should be strictly considered. In addition, repairing the DNA is a complex process that requires multimodal nanoplatforms with the ability to sense, deliver, and regulate at the same time. New directional approaches such as CRISPR-based nanocarriers, DNA origami nanostructures, and nanoparticle design with the help of artificial intelligence are expected to continue to transform this area (Bhagat, 2012; Borude and Mahajan, 2011; Bothiraja et al., 2014; Hatvate et al., 2023a; Sayyad et al., 2023). Finally, nanotechnology is not only a peephole into DNA damage and repair, but an effective tool to play with these two processes to achieve therapeutic advantage (Magar, 2011b; Rajurkar, 2012).

10.10 Future perspectives

Next-generation sequencing, in addition to a lesion-specific enrichment approach, is revolutionizing the single-base resolution map of oxidative damage to DNA throughout the genome. Single-nucleotide precision in the distribution of damage patterns can be identified by high-throughput lesion mapping techniques like Excision Repair-sequencing (XR-seq) and Damage-seq. The methods enable correlating lesion hotspots with transcription, chromatin state, and mutational signatures in disease settings. Combining these datasets with epigenomic and transcriptomic data has the potential to explain the contextually variable susceptibility to oxidative stress. Nonetheless, there is a problem of better lesion specificity, quantitation accuracy, and interrelation with clinical biospecimens to implement these techniques into the use of precision medicine (Lim and Kim, 2024; Wróblewski et al., 2024).

DNA repair enzymes designed and nanomaterial-aided delivery systems are one of the frontiers of therapeutic modulation of oxidative DNA damage. In novel versions of artificial enzymes (metalloprotein mimetics or catalytic nucleic acids), lesion-selective excision or base restoration can be engineered with increased specificity toward a substrate. At the same time, nanocarriers such as lipid polymer hybrids, QDs, and DNA origami scaffolds may synergize with dedicated enzymes or gene editors to localize them within a cell at specific sites under threat of damage. Such technologies have the prospect of reducing oxidative genotoxicity in cancer and degenerative diseases. The key translational challenges are the biocompatibility, immune tolerance, and nuclear activity, and these issues require multidisciplinary optimization (Yousefzadeh et al., 2021; Khamidullina et al., 2024; Osbourne and Thayer, 2025; Zhong et al., 2025).

10.11 Conclusions

Oxidative DNA damage, usually caused by oxidative radiations, of which ROS plays a central part, in DNA damage is a significant risk to genomic stability and a leading cause of mutagenesis, cellular aging, and cancer. Central lesions, 8-oxoG, abasic sites, and SSB are effectively addressed by repair mechanisms like BER, NER, and MMR. Failure to properly execute or adequate levels of such mechanisms are closely linked to human diseases, such as neurodegenerative disorders, cancer, and premature aging syndromes. Progress in molecular biology has led to an understanding of repair enzyme structure, specificities, and signal networks, which has opened the possibility of new treatments of oxidative genome instability. Translationally, knowledge of oxidative DNA repair is a virtual necessity in disease prevention, diagnostics, and precision medicine. Single-nucleotide polymorphisms in DNA glycosylases or repair policy mediate personal predisposition to disorders of oxidative stress, which can be used as potential markers of risk. Altering repair capacity (by increasing endogenous repair, or targeting repair-deficient tumor cells with repair enzymes) is of great promise. There is, however, a problem of finding the balance between repair fidelity and controlled cytotoxicity when used in therapy against malignancies. Profiling involving omics assessment, CRISPR-based gene regulation, and redox-based therapy will become the key approaches of future research countering oxidative genome damage and related clinical consequences.

References

Almohdar, D., Gulkis, M., Ortiz, A., Tang, Q., Sobol, R.W., Çağlayan, M., 2024. Impact of polβ/XRCC1 interaction variants on the efficiency of nick sealing by DNA ligase IIIα in the base excision repair pathway. J. Mol. Biol. 436, 168410.

Amemiya, K., Hirotsu, Y., Iimuro, Y., Tajiri, R., Oyama, T., Obi, S., et al., 2025. Decoding genomic diversity to guide tumor lesion-specific treatment of multifocal hepatocellular carcinoma. Cancer Med. 14, e70814.

Amoah, D.P., Hussein, S.K., Johnson, J.L., LaPointe, P., 2025. Ordered hydrolysis in the Hsp90 chaperone is regulated by Aha1 and a conserved post-translational modification. Protein Sci. 34, e5255.

An, X., Yu, W., Liu, J., Tang, D., Yang, L., Chen, X., 2024. Oxidative cell death in cancer: mechanisms and therapeutic opportunities. Cell Death. Dis. 15, 556.

Anand, M., 2011. Smart crystals technology: a review. Pharmacologyonline 1, 238–243.

Azzouz, D., Palaniyar, N., 2024. How do ROS Induce NETosis? Oxidative DNA damage, DNA repair, and chromatin decondensation. Biomolecules 14, 1307.

Bae, T., Hallis, S.P., Kwak, M.-K., 2024. Hypoxia, oxidative stress, and the interplay of HIFs and NRF2 signaling in cancer. Exp. Mol. Med. 56, 501–514.

Bhagat, V., 2012. Gold nanorods: a promising drug delivery vehicle. Inven. Rapid NDDS 1, 1–4.

Bhattacharjee, S., Das, B.B., 2025. Mitochondrial topoisomerases, nucleoid architecture and mtDNA repair in human disease. J. Cell Sci. 138, jcs263638.

Bhowmik, D.D., Deshmukh, A.Y., Hatvate, N.T., 2023. Quintessential impact of dendrimer bioconjugates in targeted drug delivery. In: Verma, S.K., Das, A.K. (Eds.), Bioconjugated Materials Part 2 Applications in Drug Delivery, Vaccine Formulations and Important Conjugates for Cancer Therapy, Comprehensive Analytical Chemistry. Elsevier, Netherlands, pp. 257–302.

Borude, S., 2011. Aquasomes: a potential drug delivery carrier. Pharmacologyonline 1, 230–237.

Borude, S.S., Mahajan, A.M., 2011. Recent updates on Q fever: diagnosis and treatment. Pharmacologyonline 1, 254–264.

Borude, S.S., Mahajan, A.M., 2011. POEMS syndrome: a review. Pharmacologyonline 1, 265–274.

Bothiraja, C., Gholap, K.S., Pawar, A.P., 2014. Investigation of ethyl cellulose microsponge gel for topical delivery of eberconazole nitrate for fungal therapy. Ther. Deliv. 5, 781–794.

Buccellato, F.R., D'Anca, M., Fenoglio, C., Scarpini, E., Galimberti, D., 2021. Role of oxidative damage in Alzheimer's disease and neurodegeneration: from pathogenic mechanisms to biomarker discovery. Antioxidants 10, 1353.

Burdova, K., Hailstone, R., Hanzlikova, H., Caldecott, K.W., 2025. FEN1 is critical for rapid single-strand break repair in G1 phase. Nucleic. Acids. Res. 53, gkaf710.

Çağlayan, M., 2025. Repair pathway coordination from gap filling by polβ and subsequent nick sealing by LIG1 or LIG3α governs BER efficiency at the downstream steps. DNA Repair (Amst.) 148, 103826.

Chaudhari, Y.S., Chaudhari, M.Y., Alam, M.I., Khalid, M., Webster, T.J., Sundaram, G.S., et al., 2025. Surface engineering of nano magnesium alloys for orthopedic implants: a systematic review of strategies to mitigate corrosion and promote bone regeneration. Front. Bioeng. Biotechnol. 13, 1617585.

Chen, Y., Ding, K., Zheng, S., Gao, S., Xu, X., Wu, H., et al., 2025. Post-translational modifications in DNA damage repair: mechanisms underlying temozolomide resistance in glioblastoma. Oncogene 44, 1781–1792.

Dabin, J., Mori, M., Polo, S.E., 2023. The DNA damage response in the chromatin context: a coordinated process. Curr. Opin. Cell Biol. 82, 102176.

Dadarkar, Y.P., Dubey, A.S., 2025. Bioremediation of pharmaceuticals with cow dung and earthworms. In: Biomedical Waste Management. Apple Academic Press, Palm Bay, FL, pp. 239–261.

Dattatray, T., 2012a. Artificial viral envelopes: emerging vehicle for gene delivery. PharmaTutor 1, 1–3.

Dattatray, T., 2012b. Potential applications of botanical extract. Inven. Rapid Ethnopharmacol 1, 1–4.

Desai, H., Vora, L.K., 2025. Impact of parenteral microemulsion on healthcare therapeutics. Pharmaceutical Microemulsions for Parenteral Delivery: From Bench to Bedside, first ed. Apple Academic Press, New York.

Di Carlo, E., Sorrentino, C., 2024. Oxidative stress and age-related tumors. Antioxidants 13, 1109.

Dighe, S.B., 2012a. Role of sebum in human health & acne. Inven. Rapid Cosmeceuticals 1, 1–3.

Dighe, S.B., 2012b. Dandruff: pathogenesis and treatment. Inven. Rapid Cosmeceuticals 1, 1–4.

Dong-Chen, X., Yong, C., Yang, X., Chen-Yu, S., Li-Hua, P., 2023. Signaling pathways in Parkinson's disease: molecular mechanisms and therapeutic interventions. Signal Transduct. Target. Ther. 8, 73.

Downs, J.A., Gasser, S.M., 2024. Chromatin remodeling and spatial concerns in DNA double-strand break repair. Curr. Opin. Cell Biol. 90, 102405.

Erie, D.A., Weninger, K.R., 2024. Combining single-molecule and structural studies reveals protein and DNA conformations and assemblies that govern DNA mismatch repair. Curr. Opin. Struct. Biol. 89, 102917.

Faiyazuddin, M., Maqsood, S., Abbas, S., Chaudhari, Y.S., Sharmila, S., Webster, T.J., 2024. Prospects and ethics of ChatGPT in biomedical research and medical practice: an emphasis on nanotechnology. J. Biomed. Nanotechnol. 20, 1661–1678.

Gao, S., Oden, P.N., Ryan, B.J., Yang, H., Freudenthal, B.D., Greenberg, M.M., 2024. Biochemical and structural characterization of Fapy• dG replication by human DNA polymerase β. Nucleic. Acids. Res. 52, 5392–5405.

Gholap, A.D., 2012. Translational nanomedicine: status assessment and opportunities. Int. J. Res. Pharm. Biomed. Sci. 4, 1454–1463.

Gholap, A.D., Khuspe, P.R., Mane, D.V., Galave, V., Mali, A., Mohite, P., et al., 2025. Challenges and future directions biomaterial-based additive manufacturing in tissue engineering and regeneration. In: Biomaterial-Based Additive Manufacturing in Tissue Engineering and Regeneration. Biomaterials, Bioengineering and Sustainability. Springer Nature,, Singapore,, pp. 491–529.

Gholap, A.D., Omri, A., 2025. Advances in artificial intelligence-envisioned technologies for protein and nucleic acid research. Drug Discov. Today 30, 104362.

Gong, W., Holmberg, H., Lu, C., Huang, M., Li, S., 2024. Interplay of the Tfb1 pleckstrin homology domain with Rad2 and Rad4 in transcription coupled and global genomic nucleotide excision repair. Nucleic. Acids. Res. 52, 6333–6346.

Gupta, J., Kamandar, P., Bhowmik, D.D., Rojekar, S., Faiyazuddin, M., Hatvate, N.T., et al., 2024. Harnessing nanovaccines for effective immunization–a special concern on COVID-19: facts, fidelity, and future prospective. ACS Biomater. Sci. Eng. 10, 271–297.

Hatvate, N., Omri, A., 2025. Reprogramming therapeutics: synthetic biological constructs at the frontier of precision medicine. Discov. Med. 37, 1718.

Hatvate, N.T., Bhowmik, D., Sawant, S.D., Londhe, K.M., Rojekar, S., Faiyazuddin, M., 2025. Engineered artificial microRNA in plant: an overview. In: MicroRNA Advances and Application in Plant Biology. Elsevier, Netherlands, pp. 311–341.

Hatvate, N.T., Khuspe, P.R., Mandhare, T.A., Kashid, P., Gaikwad, V.D., 2023a. Proteomics in oncology: retrospect and prospects. In: S., K. (Ed.), Novel Technologies in Biosystems, Biomedical & Drug Delivery. Springer Nature,, Singapore, pp. 243–269.

Hatvate, N.T., Khuspe, P.R., Mandhare, T.A., Kashid, P., Vora, L.K., 2024. Decoding of nanoneutraceuticals and their delivery. In: Mohanta, Y.K. (Ed.), Sustainable Green Nanotechnology. CRC Press, New York, pp. 284–309.

Hatvate, N.T., Sawant, S.D., Khemka, P.N., Vora, L.K., 2023b. Unlocking the potential of nanobiohybrids to combat environmental pollution. In: Lens, P. (Ed.), Nanobiohybrids for Advanced Wastewater Treatment and Energy Recovery. IWA Publishing, United Kingdom, pp. 169–190.

Hua, A.B., Sweasy, J.B., 2024. Functional roles and cancer variants of the bifunctional glycosylase. Environ. Mol. Mutagen. 65, 40–56.

Hwang, Y., Kang, S.-J., Kang, J., Choi, J., Kim, S.-J., Jang, S., 2025. DNA repair and disease: insights from the human DNA glycosylase NEIL family. Exp. Mol. Med. 57, 524–532.

Iqbal, M.J., Kabeer, A., Abbas, Z., Siddiqui, H.A., Calina, D., Sharifi-Rad, J., et al., 2024. Interplay of oxidative stress, cellular communication and signaling pathways in cancer. Cell Commun. Signal. 22, 7.

Jadhav, K., Shevalkar, G., Sugandhi, V.V., Pai, R., Parikh, K., Prajapati, M.K., et al., 2025. Exploring protein aggregation in biological products: from mechanistic understanding to practical solutions. AAPS PharmSciTech. 26, 189.

Jordano-Raya, M., Schrader, C.E., Ariza, R.R., Roldán-Arjona, T., Córdoba-Cañero, D., 2025. Divergent evolution of opposite base specificity and single-stranded DNA activity in animal and plant AP endonucleases. Nucleic. Acids. Res. 53, gkae1297.

Kabziński, J., Majsterek, I., 2022. Association of base excision repair pathway genes OGG1, XRCC1 and MUTYH polymorphisms and the level of 8-oxo-guanine with increased risk of colorectal cancer occurrence. Int. J. Occup. Med. Environ. Health 35, 625–633.

Kadyrova, L.Y., Mieczkowski, P.A., Kadyrov, F.A., 2024. MutLα suppresses error-prone DNA mismatch repair and preferentially protects noncoding DNA from mutations. J. Biol. Chem. 300, 107406.

Kapare, H.S., Pagar, S., Kamandar, P., Bhowmik, D., Vishwakarma, N., Raikwar, S., et al., 2024. Exploring modified chitosan-based gene delivery technologies for therapeutic advancements. Int. J. Biol. Macromol. 260, 129581.

Khairnar, R.A., Sawant, S.D., Parab, T.D., 2025. Microbial strategies for the biodegradation of pharmaceutical pollutants. In: Biomedical Waste Management. Apple Academic Press, New York, pp. 297–317.

Khamidullina, A.I., Abramenko, Y.E., Bruter, A.V., Tatarskiy, V.V., 2024. Key proteins of replication stress response and cell cycle control as cancer therapy targets. Int. J. Mol. Sci. 25, 1263.

Khan, A.R., Grewal, N.S., Jun, Z., Khalid, M., Zhang, H.-J., 2025. Advances in smart hybrid scaffolds: a strategic approach for regenerative clinical applications. Eng. Regen. 6, 85–110.

Khedkar Ajay, T.D., 2012a. Pharmacosomes: an emerging vesicular system. Inven. Rapid NDDS 1, 1–3.

Khedkar Ajay, T.D., 2012b. Cubosomes: emerging drug delivery system. Inven. Rapid NDDS 1, 1–3.

Khuspe, P.R., Bharati, D.K., Pardeshi, S.R., Ahmad, M.D., Hossain, A.S., Prajapati, B.G., et al., 2025a. Computational Neuropharmacology in Psychiatry. Wiley, New York,, pp. 207–244.

Khuspe, P.R., Faiyazuddin, M., Uddin, M.J., Bhowmik, D.D., Said, R.P., Sonawane, K.S., et al., 2025b. Lipid-based vaccines against viruses including COVID-19. In: Lipids in Pulmonary Drug Delivery. Elsevier, Netherlands, pp. 409–433.

Khuspe, P.R., Faiyazuddin, M., Uddin, M.J., Gupta, J., Kamandar, P., Bhowmik, D.D., et al., 2025c. Current scenario and challenges of lipid-based nanomedicines for pulmonary drug delivery. In: Lipids in Pulmonary Drug Delivery. Elsevier, Netherlands, pp. 21–47.

Gholap Amol, D., Khuspe, P.R., Mane, D.V., 2026a. Artificial intelligence driven discovery of anticancer phytochemicals: From plants to lead compounds. In: Emerging Trends in Phytotherapy of Cancer. Elsevier, Amsterdam, Netherlands, pp. 441–460.

Gholap Amol, D., Khuspe, P.R., Mane, D.V., 2026b. Emerging machine learning techniques for the development and analysis of anticancer drugs from plants. In: Emerging Trends in Phytotherapy of Cancer. Elsevier, Amsterdam, Netherlands, pp. 461–493.

Khuspe, P.R., Pardeshi, S.R., Uddin, M.J., Das, U., Hatvate, N.T., Rojekar, S., et al., 2024. Achieving optimal health with host-directed therapies (HDTs) in infectious diseases—a new horizon. Adv. Ther., 2400169.

Gholap Amol, D., Khuspe, P.R., Rebello, N.L., Jain, M.N., Desai, H.R., Giram, P., 2026c. Photon-avalanching nanoparticles for biomedical applications. In: Upconversion Nanoparticles for Biomedical Applications. Elsevier, Amsterdam, Netherlands, pp. 185–225.

Gholap Amol, D., Khuspe, P.R., Rebello, N.L., Jain, M.N., Desai, H.R., Giram, P., 2026d. Upconversion nanoparticles for drug and gene delivery. In: Upconversion Nanoparticles for Biomedical Applications. Elsevier, Amsterdam, Netherlands, pp. 517–575.

Krishnamurthy, H.K., Pereira, M., Rajavelu, I., Jayaraman, V., Krishna, K., Wang, T., et al., 2024. Oxidative stress: fundamentals and advances in quantification techniques. Front. Chem. 12, 1470458.

Kumar, D., Kumar, S., Kumari, R., Tanwar, R., Kumar, V., Khalid, M., et al., 2024. A comprehensive review on the biomass-mediated synthesis of silver nanoparticles: opportunities and challenges. J. Environ. Chem. Eng., 115133.

Kutuzov, M., Sayfullina, D., Belousova, E., Lavrik, O., 2025. HPF1 regulates Pol β efficiency in nucleosomes via the modulation of total poly(ADP-Ribose) synthesis. Int. J. Mol. Sci. 26, 1794.

Le Meur, R.A., Pecen, T.J., Le Meur, K.V., Nagel, Z.D., Chazin, W.J., 2024. Molecular basis and functional consequences of the interaction between the base excision repair DNA glycosylase NEIL1 and RPA. J. Biol. Chem. 300, 107579.

Lebedeva, N.A., Anarbaev, R.O., Maltseva, E.A., Sukhanova, M.V., Rechkunova, N.I., Lavrik, O.I., 2024. DNA repair protein XRCC1 stimulates activity of DNA polymerase λ under conditions of microphase separation. Int. J. Mol. Sci. 25, 6927.

Li, X., Yang, C., Wu, H., Chen, H., Gao, X., Zhou, S., et al., 2024. DSB-induced oxidative stress: uncovering crosstalk between DNA damage response and cellular metabolism. DNA Repair (Amst.) 141, 103730.

Lim, E.Y., Kim, G.-D., 2024. Particulate matter-induced emerging health effects associated with oxidative stress and inflammation. Antioxidants. 13, 1256.

Lin, Y.-H., Lin, K.-J., Shao, I.-H., 2025. The critical role of mitochondrial dysfunction in ADT-induced neurotoxicity. Prostate Cancer Prostatic Dis 28, 1013–1014.

Liu, L., Li, S., Zhu, W., Bao, Q., Liang, Y., Zhao, T., et al., 2024. Study on the mechanism of ROS-induced oxidative stress injury and the broad-spectrum antimicrobial performance of nickel ion-doped V6O13 powder. Sci. Rep. 14, 22374.

Liu, Z., Yan, B., Liu, H., Liu, X., Xiao, X., Ming, Z., 2024. Enhancing APE1 detection through apurinic/apyrimidinic site inhibition of DNA polymerase: an innovative, highly sensitive approach. Chem. Commun. 60, 4695–4698.

Ma, J., Qi, R., Harcourt, E.M., Chen, Y.-T., Barbosa, G.M., Peng, Z., et al., 2024. 3, N 4-Etheno-5-methylcytosine blocks TET1-3 oxidation but is repaired by ALKBH2, 3 and FTO. Nucleic. Acids. Res. 52, 12378–12389.

Magar Tushar, V., 2011a. Advances in the field of nanooncology: an overview. Inven. Impact NDDS. 4, 1–3.

Magar Tushar, V., 2011b. Elastic vesicles as drugs carriers through the skin. Inven. Rapid NDDS. 1, 1–2.

Manavi, M.A., Mohammad Jafari, R., Shafaroodi, H., Dehpour, A.R., 2025. The Keap1/Nrf2/ARE/HO-1 axis in epilepsy: crosstalk between oxidative stress and neuroinflammation. Int. Immunopharmacol. 153, 114304.

Mesut, B., Al-Mohaya, M., Yeşilkaya, E., Das, U., Akhtar, M.S., Sah, R., et al., 2024. Demystifying the potential of lipid-based nanocarriers in targeting brain malignancies. Naunyn. Schmiedebergs. Arch. Pharmacol. 397, 9243–9279.

Meulmeester, F.L., Luo, J., Martens, L.G., Mills, K., van Heemst, D., Noordam, R., 2022. Antioxidant supplementation in oxidative stress-related diseases: what have we learned from studies on alpha-tocopherol? Antioxidants 11, 2322.

Minko, I.G., Moellmer, S.A., Luzadder, M.M., Tomar, R., Stone, M.P., McCullough, A.K., et al., 2024. Interaction of mitoxantrone with abasic sites - DNA strand cleavage and inhibition of apurinic/apyrimidinic endonuclease 1, APE1. DNA Repair (Amst.) 133, 103606.

Mostafavi Abdolmaleky, H., Zhou, J.-R., 2024. Gut microbiota dysbiosis, oxidative stress, inflammation, and epigenetic alterations in metabolic diseases. Antioxidants 13, 985.

Nordengen, A.L., Kværner, A.S., Krutto, A., Alavi, D.T., Henriksen, H.B., Henriksen, C., et al., 2024. DNA base oxidation in relation to TNM stages and chemotherapy treatment in colorectal cancer patients 2–9 months post-surgery. Free Radic. Biol. Med. 212, 174–185.

Ohno, H., Sasaki, E., Yamada, S., Hanaoka, K., 2024. Recent advances in Si-rhodamine-based fluorescent probes for live-cell imaging. Org. Biomol. Chem. 22, 3099–3108.

Onisiforou, A., Zanos, P., Georgiou, P., 2024. Molecular signatures of premature aging in major depression and substance use disorders. Sci. Data. 11, 698.

Osbourne, R., Thayer, K.M., 2025. Structural and mechanistic diversity in p53-mediated regulation of organismal longevity across taxonomical orders. PLoS Comput. Biol. 21, e1012382.

Oswalt, L.E., Eichman, B.F., 2024. NEIL3: a unique DNA glycosylase involved in interstrand DNA crosslink repair. DNA Repair (Amst.) 139, 103680.

Pani, B., Nudler, E., 2025. Transcription-coupled repair: protecting genome across generations. Curr. Opin. Genet. Dev. 94, 102385.

Pardeshi, Sagar, More, M., Togre, N., Rebello, N., Giram, P., 2023. Dendrimers based antibacterial and antiviral materials. In: Antibacterial and Antiviral Functional Materials. ACS Publications, USA, pp. 139–169.

Pardeshi, S., Giram, P., 2023. Preparation, antibacterial and antiviral activity measurements and detection methods. In: Deshmukh, K. (Ed.), Antibacterial and Antiviral Functional Materials. ACS Publications, USA, pp. 33–64.

Pardeshi, S., Kapare, H., More, A., Rebello, N., Giram, P., 2024a. Coating approaches of functionalized magnetic nanoparticles for theranostic applications. In: Functionalized Magnetic Nanoparticles for Theranostic Applications. Wiley, New York, pp. 131–171.

Pardeshi, S., Pillai, D., Giram, P., Pardeshi, C., Naik, J., 2024b. Nanocarriers for ocular drug delivery. In: Nanocarrier Drug Delivery Systems. De Gruyter, Germany, pp. 469–500.

Pardeshi, S.R., Hatvate, N.T., Dhorkule, N., Sayyad, S.F., Faiyazuddin, M., Khalid, M., 2024a. Environmental implications and nanotechnological advances in octocrylene-enriched sunscreen formulations: a comprehensive review. Chemosphere 358, 142235.

Pardeshi, S.R., Hatvate, N.T., Gharat, K.D., Naik, J.B., Omri, A., 2024b. Advances in dorzolamide hydrochloride delivery: harnessing nanotechnology for enhanced ocular drug delivery in glaucoma management. Discov. Nano. 19, 199.

Pardeshi, S.R., Sen, N.K., Khandagale, P.D., Khadka, I.M., Pardeshi, P.R., Datta, D., et al., 2026. Prospects and applications of nanoemulsions in cosmeceuticals. In: Nanotechnology in Cosmeceuticals. Elsevier, Netherlands, pp. 369–386.

Parit, S., Manchare, A., Mundhe, P., Hatvate, N., Rojekar, S., Patravale, V., 2024. Antibody-drug conjugates: a promising breakthrough in cancer therapy. Int. J. Pharm. 659, 124211.

Patil, M., Dhumal, D., Rojekar, S., Jain, H., Pingale, P., Kumar Prajapati, M., 2024. Nanocarriers for antimicrobial therapy. In: Nanocarrier Drug Delivery Systems. De Gruyter, Germany, pp. 561–602.

Pellarin, I., Dall'Acqua, A., Favero, A., Segatto, I., Rossi, V., Crestan, N., et al., 2025. Cyclin-dependent protein kinases and cell cycle regulation in biology and disease. Signal Transduct. Target. Ther. 10, 11.

Phade, S.A., Mane, D.V., Hatvate, N.T., Puri, A.V., Rojekar, S., 2025. Advanced smart biomaterials and constructs for hard tissue engineering and regeneration through additive manufacturing. In: Biomaterial-Based Additive Manufacturing in Tissue Engineering and Regeneration. Springer Nature, Singapore, pp. 131–166.

Prajapat, M., Shinde, S., Padhiyar, D., Butani, S., Shah, S., Rajput, A., 2025. 3D printing technology in microneedles: an emerging era in transdermal drug delivery. Hybrid Adv. 10, 100447.

Pravin, B., Nanaware, V., Ashwini, B., Wondmie, G.F., Jardan, Y.A.Bin, Bourhia, M., 2024. Assessing the antioxidant properties of Naringin and Rutin and investigating their oxidative DNA damage effects in breast cancer. Sci. Rep. 14, 15314.

Puri, A.V., 2018. Comparative evaluation of papaya and banana starch as disintegrant in diclofenac sodium tablet formulation. Int. J. Pharm. Biol. Sci. 8, 512–522.

Quiñones, J.L., Tang, M., Fang, Q., Sobol, R.W., Demple, B., 2024. C-terminal residues of DNA polymerase β and E3 ligase required for ubiquitin-linked proteolysis of oxidative DNA-protein crosslinks. DNA Repair (Amst.) 143, 103756.

Rahimi, M., Soufi, F.G., Koochakkhani, S., Rahnama Inchehsablagh, B., Azarbad, A., Mahmoudi, M., et al., 2025. The association between subclinical atherosclerosis serum markers and oxidative <scp>DNA</scp>damage in normoglycemic normotolerant offspring of diabetic parents. J. Diabet. 17, e70133.

Rajurkar, V.G., 2012a. Potential applications of capillary electromigration methods for the improvement of separation efficiency. Inven. Rapid Pharm Anal. Qual. Assur. 1, 1–3.

Rajurkar, V.G., 2012b. Recent technologies in nanomedicine for treatment of cancer. Inven. Rapid Pharm. Tech. 4, 1–3.

Rajurkar, V.G., 2012c. Potential applications of superparmagnetic nanoparticles technology. Inven. Rapid: NDDS 1, 1–3.

Rajurkar, V.G., Dighe, S., 2012. Phonophoresis: ultrasound enhanced transdermal drug delivery. Inven. Rapid: NDDS. 1, 1–3.

Ramos Cáceres, E., Kemperman, L., Bonger, K.M., 2024. Environment-sensitive turn-on fluorescent probe enables live cell imaging of myeloperoxidase activity during NETosis. Commun. Chem. 7, 262.

Rojekar, Satish, Togre, N., Bhoj, P., Haeck, C., Hatvate, N., Singh, N., et al., 2024a. Current status of mannose receptor-targeted drug delivery for improved anti-HIV therapy. J. Control. Release 372, 494–521.

Rojekar, S., Kapare, H.S., Vishwakarma, N., Raikwar, S., Garkal, A., Mehta, T.A., et al., 2024b. Chitosan scaffolds: expanding horizons in biomedical applications. Carbohydr. Polym. 323, 121394.

Rojekar, S., Parit, S., Manchare, A., Nangare, S.N., Hatvate, N., Sugandhi, V.V., et al., 2024c. Revolutionizing eye care: exploring the potential of microneedle drug delivery. Pharmaceutics 16, 1398.

S, J., Kamandar, P.A., Banchhod, G.V, Hatvate, N.T., 2023. Antibody-drug conjugates for cancer therapy: an up-to-date review on the chemistry and pharmacology. In: Verma, S.K., Das, A.K. (Eds.), Bioconjugated Materials Part 2 Applications in Drug Delivery, Vaccine Formulations and Important Conjugates for Cancer Therapy, Comprehensive Analytical Chemistry. Elsevier, Netherlands, pp. 105–190.

Sagar, Giram, P., Nangare, S., Sodha, S., Kapare, H., More, M., et al., 2025. Breathing new life nanomedicines for pulmonary drug delivery: targeting approaches, experimental models, and regulatory aspects. J. Basic Appl. Sci. 14, 65.

Said, P.R., Pawar, R.D., Ambore, G.S., Hatvate, N.T., 2023. Importance of carbohydrate-drug conjugates in vaccine development: a detailed review. In: Verma, S.K., Das, A.K. (Eds.), Bioconjugated Materials Part 2 Applications in Drug Delivery, Vaccine Formulations and Important Conjugates for Cancer Therapy, Comprehensive Analytical Chemistry. Elsevier, Netherlands, pp. 191–256.

Sawant, S.D., Pardeshi, S.R., Khuspe, P.R., Sayyad, S.F., Chavan, M., Hatvate, N.T., et al., 2024a. Omics tools for determining microbial composition: a systematic review. Microbiome Eng. 1, 35–50.

Sawant, S.D., Pardeshi, S.R., Khuspe, P.R., Sayyad, S.F., Chavan, M., Hatvate, N.T., et al., 2024b. The role of inulin in altering the animal microbiome: a comprehensive review. Microbiome Eng 1, 79–102.

Sawant, S.D., Sayyad, S.F., Hatvate, N.T., Chavan, M., Rojekar, S., Faiyazuddin, M., 2023. Food biopackaging for human benefits: status and perspectives. In: Ahmed, S. (Ed.), Biobased Packaging Materials. Springer Nature, Singapore, pp. 1–35.

Sayyad, S.F., Hatvate, N.T., Dhumal, V.V., Pardeshi, S.R., Chavda, V.P., Vora, L.K., 2023. Drug delivery strategies for avobenzone: a case study of photostabilization. Pharmaceutics 15, 1008.

Schumacher, B., Pothof, J., Vijg, J., Hoeijmakers, J.H.J., 2021. The central role of DNA damage in the ageing process. Nature 592, 695–703.

Shahi, A., Kidane, D., 2025. Decoding mitochondrial DNA damage and repair associated with H. pylori infection. Front. Cell. Infect. Microbiol. 14, 1529441.

Shaikh, M.S., Faiyazuddin, M., Khan, M.S., Pathan, S.K., Syed, I.J., Akhtar, M.S., et al., 2024. Chikungunya virus vaccine: a decade of progress solving epidemiological dilemma, emerging concepts, and immunological interventions. Front. Microbiol. 15, 1413250.

Shan, C., Wang, Yuan, Wang, Yin, 2024. The Crosstalk between autophagy and Nrf2 signaling in cancer: from biology to clinical applications. Int. J. Biol. Sci. 20, 6181–6206.

Shybyray, Y., Liu, L., Kaiyrkulova, A., Huang, X., Jenis, J., Li, J., et al., 2025. LC–MS/MS-guided discovery of sinapic acid derivatives from Lepidium sativum seeds and their anti-inflammatory activity. J. Agric. Food Chem. 73, 16393–16407.

Singh, Tribhuvan, Kumar, V.Ravi, Sanap, G., Anjaiah, B., Roja, K., 2016. In vitro antipyrial activity of Psidium guajava leaf extract. Eur. J. Pharm. Med. Res. 3, 312–315.

Song, J., Xiao, L., Zhang, Z., Wang, Y., Kouis, P., Rasmussen, L.J., et al., 2024. Effects of reactive oxygen species and mitochondrial dysfunction on reproductive aging. Front. Cell Dev. Biol. 12.

Spoto, B., Politi, C., Pizzini, P., Parlongo, R.M., Testa, A., Mobrici, M., et al., 2025. 8-hydroxy-2′-deoxyguanosine, a biomarker of oxidative DNA injury, in diabetic kidney disease. Nutr. Metab. Cardiovasc. Dis. 35, 103722.

, 2024. Manipulating the rhizosphere microbiome for plant health. In: Srivastava, N., Ibrahim, S.A. (Eds.), Microbiome Engineering. CRC Press, New York, p. 19.

Taylor, K.E., Miller, L.G., Contreras, L.M., 2024. RNA-binding proteins that preferentially interact with 8-oxoG-modified RNAs: our current understanding. Biochem. Soc. Trans. 52, 111–122.

Thompson, P.S., Cortez, D., 2020. New insights into abasic site repair and tolerance. DNA Repair (Amst). 90, 102866.

Tupe, D., 2012. The threat of photoaging: clinical conditions. Evaluation and Treatment. Inven. Rapid Cosmeceuti. 1, 1–3.

Tyutereva, E.V., Strizhenok, A.D., Kiseleva, E.I., Voitsekhovskaja, O.V., 2024. Comet assay: multifaceted options for studies of plant stress response. Horticulturae 10, 174.

Uddin, M.J., Faiyazuddin, M., Omri, A., Gowri, S., Khalid, M., 2024. Advances in artificial intelligence for drug delivery and development: a comprehensive review. Comput. Biol. Med. 178, 108702.

Uddin, M.J., Faiyazuddin, M., Saeed, S.M., Sadiq, S.B., Koly, I.A.C., 2025a. Lipid-based drug delivery for systemic effect through pulmonary route of administration. In: Lipids in Pulmonary Drug Delivery. Elsevier, Netherlands, pp. 255–299.

Uddin, M.J., Zulfiqar, I., Sadiq, S.B., Lucky, S., Lubna, K.A., 2025b. Natural biomaterials used in wound healing and diabetic foot. In: Md. Faiyazuddin, M.S. (Ed.), Design and Processing of Green Materials. Springer, Germany, pp. 329–368.

Urrutia, K., Chen, Y.H., Tang, J., Hung, T.I., Zhang, G., Xu, W., et al., 2024. DNA sequence and lesion-dependent mitochondrial transcription factor A (TFAM)-DNA-binding modulates DNA repair activities and products. Nucleic. Acids. Res. 52, 14093–14111.

Vardikar, A., Das, U., Mandal, S., Mandlik, S., Bari, D., Pardeshi, C., et al., 2025. Surface engineered multimodal magnetic nanoparticles for neurodegenerative diseases. In: Targeted Therapy for the Central Nervous System. Elsevier, Netherlands, pp. 121–153.

Vodicka, P., Vodenkova, S., Danesova, N., Vodickova, L., Zobalova, R., Tomasova, K., et al., 2025. Mitochondrial DNA damage, repair, and replacement in cancer. Trends Cancer 11, 62–73.

Wang, L., Wei, C., Jing, J., Shao, M., Wang, Z., Wen, B., et al., 2025. The effects of polyphenols on doxorubicin-induced nephrotoxicity by modulating inflammatory cytokines, apoptosis, oxidative stress, and oxidative <scp>DNA</scp>damage. Phyther. Res. 39, 2147–2164.

Wang, X., Zhu, H., Chen, B., Zhang, Y., Kok, A., van Knegsel, A., et al., 2024. Effects of endogenous DHA milk and exogenous DHA milk on oxidative stress and cognition in SAMP8 mice. Biomed. Pharmacother. 174, 116467.

Wojtaszek, J.L., Williams, R.S., 2024. From the TOP: formation, recognition and resolution of topoisomerase DNA protein crosslinks. DNA Repair (Amst.) 142, 103751.

Wróblewski, M., Wróblewska, W., Sobiesiak, M., 2024. The role of selected elements in oxidative stress protection: key to healthy fertility and reproduction. Int. J. Mol. Sci. 25, 9409.

Xiao, S., Yu, Y., Liao, M., Song, D., Xu, X., Tian, L., et al., 2025. Post-translational modification of p62: roles and regulations in autophagy. Cells 14, 1016.

Xu, L., Ma, S., Qu, M., Li, N., Sun, X., Wang, T., et al., 2024. Parthanatos initiated by ROS-induced DNA damage is involved in intestinal epithelial injury during necrotizing enterocolitis. Cell Death Discov 10, 345.

Xu, X., Pang, Y., Fan, X., 2025. Mitochondria in oxidative stress, inflammation and aging: from mechanisms to therapeutic advances. Signal Transduct. Target. Ther. 10, 190.

Yang, H., Lan, L., 2025. Transcription-coupled repair protects genome stability upon oxidative stress-derived strand breaks. FEBS Lett. 599, 168–176.

Yang, J., Luo, J., Tian, X., Zhao, Y., Li, Y., Wu, X., 2024. Progress in understanding oxidative stress, aging, and aging-related diseases. Antioxidants 13, 394.

Younus, A.M., Yamanaka, T., Shimada, M., 2024. The protective effects of antioxidants against endogenous and exogenous oxidative stress on bull sperm. Vitr. Cell. Dev. Biol. - Anim. 60, 969–982.

Yousefzadeh, M., Henpita, C., Vyas, R., Soto-Palma, C., Robbins, P., Niedernhofer, L., 2021. DNA damage—how and why we age? eLife 10, e62852.

Yu, H., Wu, L., Xuan, D., Peng, Q., Qu, W., Zhou, Y., 2024. Development and validation of a GC–MS/MS method for the determination of iodoacetic acid in biological samples. Anal. Bioanal. Chem. 416, 3185–3194.

Zhang, K., Li, J., Dong, W., Huang, Q., Wang, X., Deng, K., et al., 2024. Luteolin alleviates cadmium-induced kidney injury by inhibiting oxidative DNA damage and repairing autophagic flux blockade in chickens. Antioxidants 13, 525.

Zheng, J., Pan, W., Liu, M., Yu, Y., Zhao, Y., Su, K., et al., 2025. Antioxidant profiling and quality assessment of Lithocarpus polystachyus sweet tea using LC-ECD and LC-MS/MS. Sci. Rep. 15, 13163.

Zhong, R., Zhang, W., Qu, X., Xiang, Y., Ji, M., 2025. Molecular duality of OGG1: from genomic guardian to redox-sensitive modulator in diseases. Antioxidants 14, 980.

Zhong, Y., Zhang, X., Feng, R., Fan, Y., Zhang, Z., Zhang, Q., et al., 2024. OGG1: an emerging multifunctional therapeutic target for the treatment of diseases caused by oxidative DNA damage. Med. Res. Rev. 44, 2825–2848.

Zhou, S., Guo, C., Dai, Y., Pan, X., Luo, X., Qin, P., et al., 2024. Association between polycyclic aromatic hydrocarbon exposure and liver function: the mediating roles of inflammation and oxidative stress. Environ. Pollut. 342, 123068.

CHAPTER

11

Detection of DNA damage and repair: A technological development

Nasreen Amin, Ram Lal and Vinod K. Kannaujiya

Department of Botany, Mahila Mahavidyalaya, Banaras Hindu University, Varanasi, Uttar Pradesh, India

11.1 Introduction

The integrity of genomic DNA is constantly affected by a variety of endogenous and exogenous agents. Several factors, such as reactive oxygen species (ROS), replication errors, ultraviolet radiation, and environmental mutagens, generate a wide range of DNA lesions, including single-strand breaks, double-strand breaks, base modifications, and cross-links (Lindahl, 1993; Jackson and Bartek, 2009). These unrepaired damages compromise genomic integrity by disruption of normal cellular processes, chromosomal instability, and cell death. To counteract these threats, cells have evolved highly conserved DNA damage response (DDR) pathways that detect lesions, activate repair machinery, and coordinate cell cycle checkpoints (Ciccia and Elledge, 2010). The key repair mechanisms, including base excision repair (BER), nucleotide excision repair (NER), mismatch repair (MR), homologous recombination (HR), and nonhomologous end joining (NHEJ), play a core role in rectifying specific types of DNA damage in cells (Hoeijmakers, 2009).

The failure or dysfunction of DNA repair systems is strongly associated with cancer development, neurodegenerative disorders, and premature aging syndromes. For example, mutations in BRCA1/2 impair HR and predispose individuals to breast and ovarian cancers (Venkitaraman, 2014), while defects in NER proteins cause xeroderma pigmentosum, a disorder marked by extreme UV sensitivity on human skin (Cleaver et al., 2009). Importantly, DNA repair machinery also serves as a therapeutic target in oncology, PARP inhibitors, and synthetic lethality in BRCA-deficient tumors to selectively induce cancer cell death (Lord and Ashworth, 2017). Therefore, regular maintenance of genomic stability through efficient DNA repair is essential for survival and cellular health. The outline of this chapter not only provides insights into disease but also the detection mechanism for developing targeted therapies.

Genome Damage: Repair and Adaptation Across Life.
DOI: https://doi.org/10.1016/B978-0-443-42857-9.00020-7

11.2 Molecular approaches for evaluating DNA damage and repair

Advances in molecular biology have provided several sensitive methods to evaluate DNA damage and repair at the nucleotide level. These techniques are particularly valuable for detecting low-frequency lesions and characterizing repair efficiency across different genomic regions.

11.2.1 Polymerase chain reaction-based approaches

Polymerase chain reaction (PCR) and its derivatives remain among the most widely applied techniques for assessing DNA integrity. DNA lesions, such as strand breaks, oxidative modifications, or bulky adducts, can block the progression of DNA polymerase, leading to reduced amplification efficiency. Quantitative PCR (qPCR) extends this principle by allowing precise measurement of lesion frequency and amplification yield, which inversely correlates with the extent of damage (Ayala-Torres et al., 2000).

11.2.2 Long amplicon polymerase chain reaction and terminal transferase-dependent polymerase chain reaction

Long amplicon polymerase chain reaction (LMPCR) is especially useful for assessing rare DNA lesions or double-strand breaks in specific genomic regions. Amplifying long DNA fragments (>10 kb) allows researchers to detect interruptions caused by lesions that block polymerase activity (Bohr et al., 1987). Similarly, transferase-dependent PCR (TDPCR) incorporates terminal deoxynucleotidyl transferase to detect DNA breaks with single-nucleotide resolution, making it useful for mapping repair sites (Pfeifer et al., 1991).

11.2.3 Interference polymerase chain reaction and short interspersed nuclear element-polymerase chain reaction

The interference PCR (I-PCR) utilizes lesion-specific endonucleases or repair enzymes to introduce nicks at damaged sites before amplification for enhancing lesion detection sensitivity (Paz-Elizur et al., 2007). Short interspersed nuclear element (SINE)-PCR targets repetitive genomic regions and enables high-throughput monitoring of DNA repair capacity in human populations exposed to genotoxic agents (Collins et al., 2008).

11.2.4 DNA repair proteins as biomarkers

In addition to PCR-based lesion detection, molecular assays often focus on DNA repair proteins as biomarkers of damage and repair activity. Phosphorylation of histone H2AX (γH2AX) is one of the earliest cellular responses to double-strand breaks, which is quantified by immunofluorescence or flow cytometry (Rogakou et al., 1998). Similarly, proteins such as Ku70/Ku80 (NHEJ) and XRCC1 (BER) serve as indicators of specific repair pathways (Jeggo and Löbrich, 2007). Collectively, these molecular approaches provide powerful tools for mechanistic studies of DNA damage and repair, and offer both qualitative and quantitative insights into genomic stability.

11.3 Fluorescence-based strategies for DNA damage and repair analysis

Fluorescence-based assays are widely employed in DNA damage due to their high sensitivity, visualization capability, and ability to provide both qualitative and quantitative data. These methods exploit fluorescent probes based on labeling techniques to detect DNA strand breaks, fragmentation, or apoptotic signals at the single-cell level.

11.3.1 Comet assay (single-cell gel electrophoresis)

This assay is one of the most versatile and widely applied methods for assessing DNA strand breaks at the single-cell level by embedding cells in agarose, followed by lysis and electrophoresis. The damaged DNA fragments migrate away from the nucleus and appear as a comet-like tail after staining with ethidium bromide or SYBR Green. The tail length and intensity correlate with the extent of DNA damage (Olive and Banáth, 2006). Variants of the assay, including alkaline comet (for single- and double-strand breaks), neutral comet (for double-strand breaks), and enzyme-modified comet (Fpg, EndoIII, or hOGG1 to reveal oxidative damage), enhance its specificity (Azqueta and Collins, 2013) (Tables 11.1, 11.2 and Figs. 11.1, 11.2).

11.3.2 Halo assay

The halo assay is a simpler alternative for evaluating DNA integrity. It measures chromatin dispersion, where damaged DNA diffuses out of the nuclear matrix to form a halo under fluorescent stain. The size of the halo correlates with the number of DNA strand breaks (Fernández et al., 2005). This method has been frequently used in reproductive biology to assess sperm DNA fragmentation.

11.3.3 Terminal deoxynucleotidyl (transferase dUTP nick-end labeling) assay

The transferase dUTP nick-end labeling (TUNEL) assay specifically detects DNA fragmentation associated with apoptosis. Terminal deoxynucleotidyl transferase incorporates fluorescently labeled dUTP at the 3′-OH ends of DNA breaks, allowing visualization under fluorescence microscopy and quantification by flow cytometry. However, TUNEL cannot always distinguish between apoptotic and necrotic DNA fragmentation without additional markers (Gavrieli et al., 1992) (Table 11.1, Figs. 11.2, 11.3).

11.3.4 DNA breakage detection–fluorescence in situ hybridization

DNA breakage detection–fluorescence in situ hybridization (DBD-FISH) combines the sensitivity of FISH with DNA break detection. Cells embedded in agarose are denatured under alkaline conditions, exposing single-stranded DNA at damage sites. Fluorescently labeled probes are then hybridized to specific genomic sequences, enabling the detection of DNA damage in defined chromosomal regions (Ribas et al., 1999). This technique provides valuable spatial information that is not available in standard comet or TUNEL assays.

TABLE 11.1 Common methods for the detection of DNA damage *in vivo* condition.

Methods	Principle	Type of DNA damage detected	Model/specimen used	Advantages	Limitations	References
Comet assay (single-cell gel electrophoresis)	Migration of DNA fragments under an electric field from individual cells	Single- and double-strand breaks (DSBs), alkali-labile sites	Whole organism tissues (e.g., liver, blood cells)	Sensitive, quantitative, low sample requirement	Requires fresh cells, limited to accessible tissues	Ventura et al. (2013)
γ-H2AX foci assay	Immunodetection of phosphorylated histone H2AX at DSBsites	Double-strand breaks	Animal tissues, biopsies	High specificity for DSBs, visual localization	Requires microscopy and antibodies	Rogakou et al. (1999)
Micronucleus assay	Detection of chromosomal fragments or whole chromosomes in the cytoplasm	Chromosomal breaks and missegregation	Bone marrow, blood erythrocytes	Simple, in vivo genotoxicity biomarker	Indirect measure of DNA damage	Fenech (2000)
TUNEL assay	Labeling of free 3′-OH ends of fragmented DNA	Apoptotic DNA fragmentation	Tissue sections, embryos	Detects programmed cell death and DNA fragmentation	May overestimate apoptosis	Gavrieli et al. (1992)
DNA adduct detection (e.g., 32P-postlabeling, LC-MS/MS)	Chemical labeling or mass detection of DNA adducts	Covalent modifications (e.g., UV, ROS, carcinogens)	Animal organs, blood, and DNA	Sensitive and specific to the type of adduct	Complex analysis, costly instrumentation	Phillips and Arlt (2007)

LC-MS, liquid chromatography–mass spectrometry; *ROS*, reactive oxygen species; *TUNEL*, transferase dUTP nick-end labeling; *UV*, ultraviolet.

TABLE 11.2 Comparison between *in vivo* and *in vitro* DNA damage detection methods.

Criteria	In vivo detection	In vitro detection	Remarks
Experimental system	Conducted in whole organisms (e.g., animal models, tissues, or plants)	Performed on isolated or cultured cells, plasmids, or cell-free extracts	In vitro models are more controlled, while in vivo models represent physiological complexity
Biological relevance	Reflects systemic biological responses, including metabolism, repair, and immune activity	Represents direct cellular or molecular effects under controlled conditions	In vivo data are more predictive of organismal effects
Common techniques	Micronucleus test, comet assay, γ-H2AX assay, TUNEL assay, DNA adduct analysis	Alkaline/neutral comet assay, DNA fragmentation assay, 8-oxo-dG quantification, reporter gene assays	Several techniques (like the Comet assay) can be adapted for both conditions
Type of damage detected	DNA strand breaks, chromosomal aberrations, adducts, and apoptosis	Strand breaks, oxidative lesions, base modifications, mutations	Both systems detect complementary types of damage
Sensitivity and specificity	High biological variability; specificity depends on tissue and exposure	High sensitivity and reproducibility under standardized protocols	In vitro systems allow finer control of variables
Time and cost	Time-consuming; requires animal care and ethical clearance	Faster, cost-effective, and ethically simpler	In vitro methods are suitable for preliminary screening
Ethical considerations	Requires approval from animal ethics committees	Minimal ethical issues	Reduction in animal use encourages in vitro preference
Data interpretation	Complex due to systemic interactions	Straightforward, focused on molecular events	Integration of both gives mechanistic and physiological insights
Applications	Biomonitoring, toxicology, risk assessment, and environmental genotoxicity	Mechanistic studies, mutagenicity screening, and drug testing	Combining both improves translational relevance
Limitations	Difficult to isolate direct DNA effects; inter-tissue variability	May not mimic metabolic activation or repair accurately	Complementary use strengthens conclusions

TUNEL, transferase dUTP nick-end labeling.

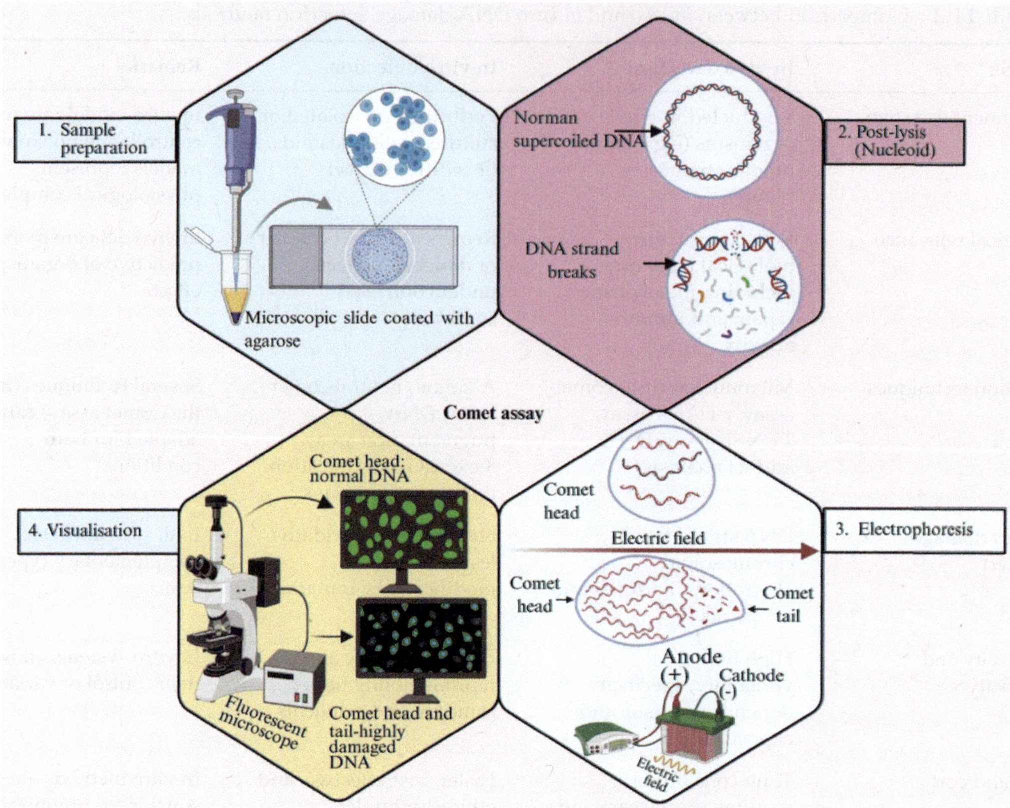

FIGURE 11.1 Schematic illustration of the comet assay showing the migration of fragmented DNA from the nucleus under an electric field, where increased tail length and intensity indicate higher levels of DNA damage. *This figure has been drawn using BioRender (https://BioRender.com).*

11.4 Flow cytometry-based approaches

Flow cytometry can be used to quantify DNA fragmentation by fluorescent labeling. One common strategy employs Annexin V (a protein that binds phosphatidylserine residues externalized in apoptosis) in combination with propidium iodide staining to discriminate between viable, apoptotic, and necrotic cells (Vermes et al., 1995). Additionally, γH2AX immunofluorescence combined with flow cytometry allows quantification of double-strand breaks in large populations of cells.

11.5 Chemiluminescence-based strategies for DNA damage and repair

Chemiluminescence-based assays rely on antibody–antigen interactions coupled with enzyme-linked detection systems, providing highly specific and quantitative means of assessing DNA lesions and repair proteins. These assays are advantageous for their sensitivity, ability to detect rare lesions, and applicability in both basic and clinical studies.

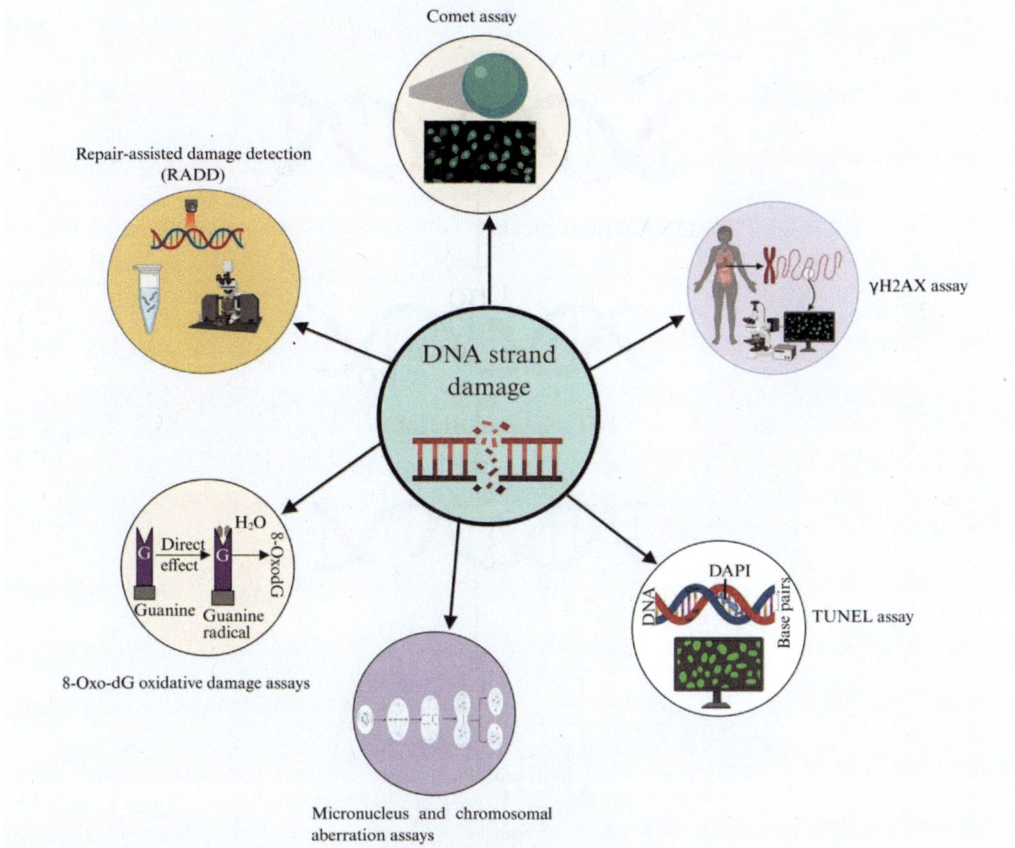

FIGURE 11.2 Diagrammatic representation of detection of DNA damage assay. *This figure has been drawn using BioRender (https://BioRender.com).*

11.5.1 Enzyme-linked immunosorbent assay

Enzyme-linked immunosorbent assay (ELISA) is widely employed to detect specific DNA adducts or repair proteins in biological samples. Antibodies raised against modified nucleotides, such as 8-oxoguanine or thymine dimers, enable sensitive quantification of oxidative or UV-induced damage (Shen et al., 2012). Variants such as competitive ELISA allow the measurement of DNA repair enzyme activity by monitoring substrate recognition (Fig. 11.4).

11.5.2 Immunohistochemistry

Immunohistochemistry (IHC) combines antigen–antibody recognition with chemiluminescent or chromogenic detection to visualize DNA damage markers in tissue sections. Commonly, targeted proteins include phosphorylated H2AX (γH2AX), ATM, and PARP, which are activated in response to DNA double-strand breaks (Bartkova et al., 2005). This technique provides spatial context, allowing researchers to study DNA repair in specific cell populations within heterogeneous tissues.

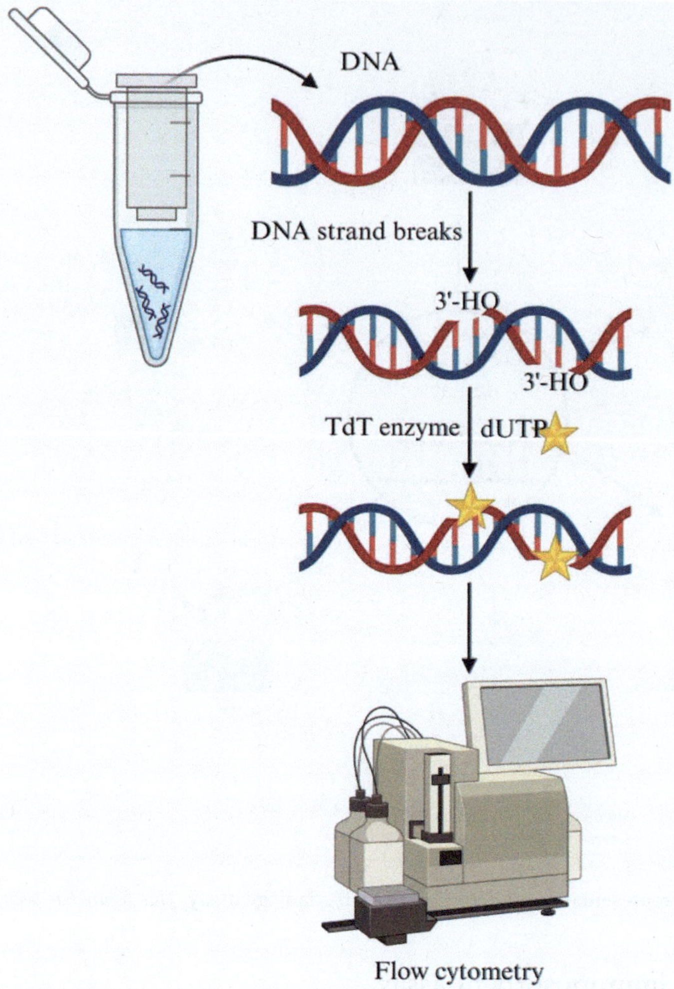

FIGURE 11.3 Illustration of the TUNEL assay showing terminal deoxynucleotidyl transferase-mediated labeling of free 3′-OH DNA ends generated during DNA damage. *TUNEL,* Transferase dUTP nick-end labeling. *This figure has been drawn using BioRender (https://BioRender.com).*

11.5.3 Immunological slot-blot assay

The slot-blot assay is a high-throughput immunoassay for DNA adducts. DNA is immobilized on a membrane and probed with lesion-specific antibodies, followed by chemiluminescent detection. This approach has been extensively applied in biomonitoring studies to assess exposure to genotoxic carcinogens such as benzo pyrene in human (Kawanishi et al., 2001). Its advantages include rapid processing and the ability to analyze multiple samples simultaneously.

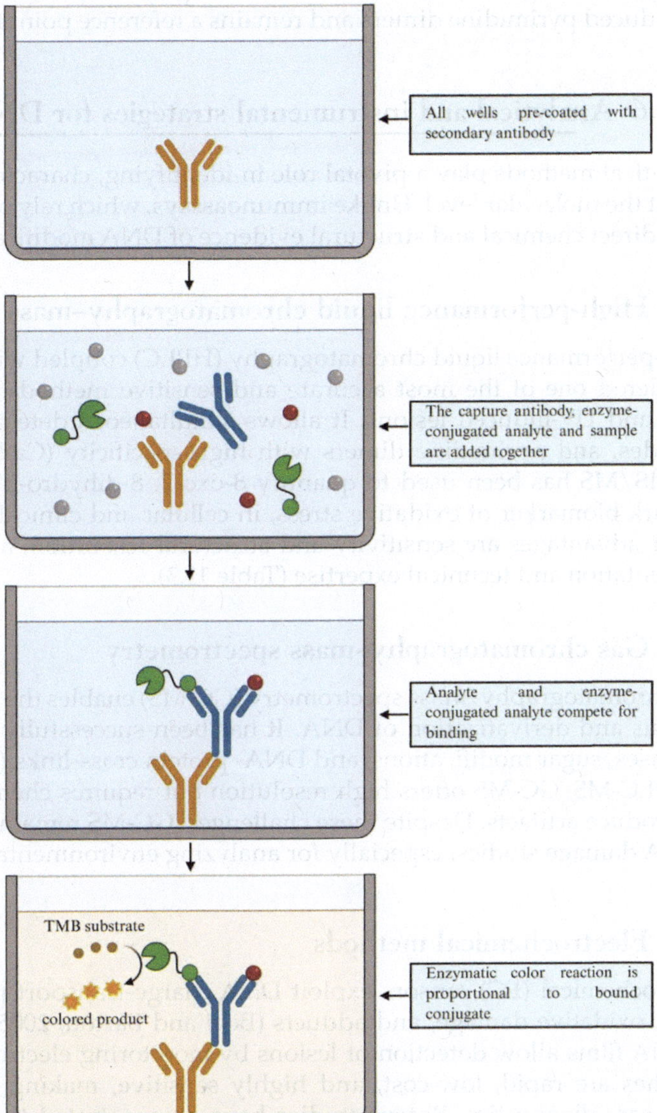

FIGURE 11.4 Conceptual representation of DNA damage assessment using a direct ELISA format based on antigen–antibody interactions. *ELISA*, Enzyme-linked immunosorbent assay. *This figure has been drawn using BioRender (https://BioRender.com).*

11.5.4 Radioimmunoassay

Radioimmunoassay (RIA) employs radiolabeled antigens to compete with DNA lesions or repair proteins for antibody binding. Although highly sensitive, the method has been largely replaced by ELISA and chemiluminescence-based systems due to concerns over radioactive handling (Friedberg et al., 2006). Nevertheless, it played a pivotal role in the early quantification of UV-induced pyrimidine dimers and remains a reference point for lesion detection sensitivity.

11.6 Analytical and instrumental strategies for DNA damage and repair

Analytical methods play a pivotal role in identifying, characterizing, and quantifying DNA lesions at the molecular level. Unlike immunoassays, which rely on antibodies, these techniques provide direct chemical and structural evidence of DNA modifications.

11.6.1 High-performance liquid chromatography–mass spectrometry

High-performance liquid chromatography (HPLC) coupled with tandem mass spectrometry is considered one of the most accurate and sensitive methods for detecting oxidative DNA damage and UV-induced lesions. It allows simultaneous detection of modified nucleobases, nucleosides, and pyrimidine dimers with high specificity (Cadet et al., 2010). For example, HPLC-MS/MS has been used to quantify 8-oxo-7, 8-dihydro-2′-deoxyguanosine (8-oxo-dG), a hallmark biomarker of oxidative stress, in cellular and clinical samples (Cooke et al., 2006). Its major advantages are sensitivity and structural resolution, although it requires expensive instrumentation and technical expertise (Table 11.3).

11.6.2 Gas chromatography–mass spectrometry

Gas chromatography–mass spectrometry (GC-MS) enables the detection of DNA lesions after hydrolysis and derivatization of DNA. It has been successfully applied to measure oxidized nucleobases, sugar modifications, and DNA- protein cross-links (Dizdaroglu, 2012). Compared with HPLC-MS, GC-MS offers high resolution but requires chemical derivatization steps that may introduce artifacts. Despite these challenges, GC-MS remains a gold standard in quantitative DNA damage studies, especially for analyzing environmental genotoxins.

11.6.3 Electrochemical methods

Electrochemical (EC) sensors exploit DNA charge-transport properties to detect base mismatches, oxidative damage, and adducts (Boal and Barton, 2005). Modified electrodes coated with DNA films allow detection of lesions by monitoring electron transfer disruptions. These approaches are rapid, low-cost, and highly sensitive, making them attractive for potential point-of-care diagnostics. Recent studies have demonstrated their utility in detecting single-base modifications and oxidative lesions at femtomole levels (Pereira-Rodrigues et al., 2014). However, their main limitation lies in distinguishing complex DNA lesions in heterogeneous samples.

11.6.4 Nuclear magnetic resonance spectroscopy and other biophysical methods

Although less commonly used than MS, nuclear magnetic resonance (NMR) provides detailed structural insights into DNA lesions and their interactions with repair enzymes

TABLE 11.3 Common methods for the detection of DNA damage in vitro conditions.

Methods	Principle	Type of DNA damage detected	Cell type/system	Advantages	Limitations	References
Alkaline comet assay	Electrophoretic migration of fragmented DNA in alkaline conditions	SSBs, DSBs, alkali-labile sites	Cultured cells, lymphocytes	Rapid, sensitive, quantitative	Requires standardized protocol	Collins (2004)
Neutral comet assay	Neutral pH electrophoresis for intact double-strand break detection	DSBs	Mammalian cell lines, bacteria	Specific to DSBs	Less sensitive than the alkaline version	Olive and Banáth (2006)
8-oxo-dG quantification (HPLC-ECD or LC-MS)	Measurement of oxidized guanine base	Oxidative DNA lesions	Cell extracts, plasmids	Specific to oxidative stress	Expensive instrumentation	Poulsen et al. (2014)
DNA fragmentation assay (Agarose gel electrophoresis)	Visualization of fragmented DNA as a "ladder"	Apoptotic DNA cleavage	Cultured mammalian cells	Simple, visual	Qualitative, low sensitivity	Facchinetti et al. (1991)
Reporter gene mutation assays (e.g., HPRT, TK)	Mutation frequency in a target reporter gene	Point mutations, deletions	Cell lines (CHO, L5178Y)	Functional assessment of mutagenicity	Time-consuming	Moore et al. (2006)
Comet-FISH	Combines Comet assay with fluorescent in situ hybridization	Site-specific DNA strand breaks	Mammalian cells	Identifies gene-specific damage	Requires fluorescent probes	Glei et al. (2009)

DSBs, double-strand breaks; *FISH*, fluorescence in situ hybridization; *HPLC*, high-performance liquid chromatography; *SSBs*, single-strand breaks; HPRT, hypoxanthine phosphoribosyltransferase; TK, thymidine kinase; CHO, Chinese hamster ovary.

(Leroy, 2019). Circular dichroism spectroscopy and Fourier-transform infrared (FTIR) spectroscopy have also been applied to characterize conformational alterations in DNA upon damage. These biophysical tools complement MS-based approaches by elucidating the mechanistic effects of lesions at the atomic level.

11.7 Conclusion and future direction

DNA damage and repair mechanisms are central to maintaining genomic stability and preventing disease onset. As highlighted throughout this chapter, a wide range of experimental strategies, such as PCR-based approaches, fluorescence- and chemiluminescence-based methods, advanced analytical technologies, such as HPLC-MS, GC-MS, and EC biosensors, enable the detection, quantification, and characterization of DNA lesions. Each method offers unique advantages: molecular techniques provide gene-specific insights, fluorescence assays allow single-cell resolution, chemiluminescence approaches deliver high sensitivity, and analytical tools provide structural and quantitative precision. The clinical relevance of these approaches is profound. DNA repair proteins such as Ku, XRCC1, and γH2AX serve as biomarkers for tumor prognosis and therapy response, while assays like comet, TUNEL, and DBD-FISH are now routinely used in research and diagnostics to monitor genotoxicity. Importantly, therapeutic strategies often exploit DNA repair deficiencies in cancer cells. For example, the use of PARP inhibitors in BRCA-mutated cancers demonstrates that understanding repair pathways can guide precision medicine. Looking forward, the integration of multi-omics technologies, high-throughput sequencing, and biosensor-based platforms promises to expand our ability to study DNA damage and repair in real time. Emerging approaches, such as CRISPR-based DNA damage mapping and nanopore sequencing, offer unprecedented opportunities to understand repair pathway dynamics in health and disease. Moreover, combining traditional methods with machine learning and systems biology could allow predictive modeling of DNA repair efficiency and treatment outcomes.

In summary, continued advances in DNA damage detection and repair analysis will not only deepen our mechanistic understanding of genomic stability but will also provide critical tools for cancer diagnosis, therapy optimization, and biomonitoring of environmental genotoxins. The future lies in integrating sensitive molecular assays with cutting-edge analytical platforms to translate benchside findings into clinically actionable biomarkers and therapeutic targets.

Acknowledgments

Nasreen Amin is thankful to University Grants Commission (UGC), Banaras Hindu University (BHU), for granting the fellowship. Ram Lal is grateful to the Council of Scientific and Industrial Research (CSIR), Ministry of Science and Technology, Government of India for the award of a Junior Research Fellowship (JRF) (File No: 09/ 0013(21769)/2025-EMR-I). Inclusivity Research Grant (IRG) Scheme, Anusandhan National Research Foundation (ANRF), New Delhi (ANRF/IRG/2024/000209/LS), India, to Vinod K. Kannaujiya is highly acknowledged.

References

Ayala-Torres, S., Chen, Y., Svoboda, T., Rosenblatt, J., Van Houten, B., 2000. Analysis of gene-specific DNA damage and repair using quantitative polymerase chain reaction. Methods 22 (2), 135–147.

Azqueta, A., Collins, A.R., 2013. The essential comet assay: a comprehensive guide to measuring DNA damage and repair. Arch. Toxicol. 87 (6), 949–968.
Bartkova, J., Hořejší, Z., Koed, K., Krämer, A., Tort, F., Zieger, K., et al., 2005. DNA damage response as a candidate anti-cancer barrier in early human tumorigenesis. Nature 434 (7035), 864–870.
Boal, A.K., Barton, J.K., 2005. Electrochemical detection of lesions in DNA. Bioconjugate Chem. 16 (2), 312–321.
Bohr, V.A., Smith, C.A., Okumoto, D.S., Hanawalt, P.C., 1985. DNA repair in an active gene: removal of pyrimidine dimers from the DHFR gene of CHO cells is much more efficient than in the genome overall. Cell 40 (2), 359–369.
Cadet, J., Douki, T., Ravanat, J.L., 2010. Oxidatively generated base damage to cellular DNA. Free Radical Biol. Med. 49 (1), 9–21.
Ciccia, A., Elledge, S.J., 2010. The DNA damage response: making it safe to play with knives. Mol. Cell 40 (2), 179–204.
Cleaver, J.E., Lam, E.T., Revet, I., 2009. Disorders of nucleotide excision repair: the genetic and molecular basis of heterogeneity. Nat. Rev. Genet. 10 (11), 756–768.
Cooke, M.S., Evans, M.D., Dizdaroglu, M., Lunec, J., 2003. Oxidative DNA damage: mechanisms, mutation, and disease. FASEB J. 17 (10), 1195–1214.
Collins, A.R., 2004. The comet assay for DNA damage and repair: principles, applications, and limitations. Mol. Biotechnol. 26 (3), 249–261.
Collins, A.R., Azqueta, A., Langie, S.A., 2012. Effects of micronutrients on DNA repair. Eur. J. Nutr. 51 (3), 261–279.
Dizdaroglu, M., 2015. Oxidatively induced DNA damage and its repair in cancer. Mutat. Res. - Rev. Mutat. Res. 763, 212–245.
Facchinetti, A., Tessarollo, L., Mazzocchi, M., Kingston, R., Collavo, D., Biasi, G., 1991. An improved method for the detection of DNA fragmentation. J. Immunol. Methods 136 (1), 125–131.
Fenech, M., 2000. The in vitro micronucleus technique. Mutat. Res. - Fundam. Mol. Mech. Mutagen. 455 (1-2), 81–95.
Fernández, J.L., Muriel, L., Rivero, M.T., Goyanes, V., Vazquez, R., Alvarez, J.G., 2003. The sperm chromatin dispersion test: a simple method for the determination of sperm DNA fragmentation. J. Androl. 24 (1), 59–66.
Friedberg, E.C., Walker, G.C., Siede, W., Wood, R.D., 2005. DNA Repair and Mutagenesis. American Society for Microbiology Press, Washington, DC.
Gavrieli, Y., Sherman, Y., Ben-Sasson, S.A., 1992. Identification of programmed cell death in situ via specific labeling of nuclear DNA fragmentation. J. Cell Biol. 119 (3), 493–501.
Glei, M., Hovhannisyan, G., Pool-Zobel, B.L., 2009. Use of Comet-FISH in the study of DNA damage and repair. Mutat. Res. - Rev. Mutat. Res. 681 (1), 33–43.
Hoeijmakers, J.H., 2009. DNA damage, aging, and cancer. N. Engl. J. Med. 361 (15), 1475–1485.
Jackson, S.P., Bartek, J., 2009. The DNA-damage response in human biology and disease. Nature 461 (7267), 1071–1078.
Jeggo, P.A., Lobrich, M., 2007. DNA double-strand breaks: their cellular and clinical impact? Oncogene 26 (56), 7717–7720.
Kawanishi, S., Hiraku, Y., Oikawa, S., 2001. Mechanism of guanine-specific DNA damage by oxidative stress and its role in carcinogenesis and aging. Mutat. Res. - Rev. Mutat. Res. 488 (1), 65–76.
Leroy, J.L., 2019. Structural studies of DNA damage by NMR. Chem. Rev. 119 (10), 5889–5911.
Lindahl, T., 1993. Instability and decay of the primary structure of DNA. Nature 362 (6422), 709–715.
Lord, C.J., Ashworth, A., 2017. PARP inhibitors: synthetic lethality in the clinic. Science 355 (6330), 1152–1158.
Moore, I., Samalova, M., Kurup, S., 2006. Transactivated and chemically inducible gene expression in plants. Plant J. 45 (4), 651–683.
Olive, P.L., Banáth, J.P., 2006. The comet assay: a method to measure DNA damage in individual cells. Nat. Protoc. 1 (1), 23–29.
Paz-Elizur, T., Krupsky, M., Blumenstein, S., Elinger, D., Schechtman, E., Livneh, Z., 2003. DNA repair activity for oxidative damage and risk of lung cancer. J. Natl. Cancer Inst. 95 (17), 1312–1319.
Pereira-Rodrigues, N., Sussulini, A., Arruda, M.AZ., 2014. Electrochemical DNA biosensors in cancer diagnostics and prognosis. Clin. Chim. Acta 430, 1–11.
Pfeifer, G.P., Steigerwald, S.D., Mueller, P.R., Wold, B., Riggs, A.D., 1989. Genomic sequencing and methylation analysis by ligation mediated PCR. Science 246 (4931), 810–813.
Phillips, D.H., Arlt, V.M., 2007. The 32P-postlabeling assay for DNA adducts. Nat. Protoc. 2 (11), 2772–2781.
Poulsen, H.E., Nadal, L.L., Broedbaek, K., Nielsen, P.E., Weimann, A., 2014. Detection and interpretation of 8-oxodG and 8-oxoGua in urine, plasma and cerebrospinal fluid. Biochim. Biophys. Acta 1840 (2), 801–808.

Ribas, G., Frenz, C.M, Marcos, R., Creus, A., 1999. DNA damage detection by the comet assay and FISH in human lymphocytes. Mutagenesis 14 (6), 637–643.

Rogakou, E.P., Pilch, D.R., Orr, A.H., Ivanova, V.S., Bonner, W.M., 1998. DNA double-stranded breaks induce histone H2AX phosphorylation on serine 139. J. Biol. Chem. 273 (10), 5858–5868.

Rogakou, E.P., Boon, C., Redon, C., Bonner, W.M., 1999. Megabase chromatin domains involved in DNA double-strand breaks in vivo. J. Cell Biol. 146 (5), 905–916.

Shen, J., Deininger, P., Hunt, J.D., Zhao, H., 2007. 8-Hydroxy-2′-deoxyguanosine (8-OH-dG) as a potential survival biomarker in patients with nonsmall-cell lung cancer. Cancer 109 (3), 574–580.

Venkitaraman, A.R., 2014. Cancer suppression by the chromosome custodians, BRCA1 and BRCA2. Science 343 (6178), 1470–1475.

Ventura, L., Giovannini, A., Savio, M., Donà, M., Macovei, A., Buttafava, A., Carbonera, D., Balestrazzi, A., et al., 2013. Single cell gel electrophoresis (Comet) assay with plants: research on DNA repair and ecogenotoxicity testing. Chemosphere 92 (1), 1–9.

Vermes, I., Haanen, C., Steffens-Nakken, H., Reutellingsperger, C., 1995. A novel assay for apoptosis flow cytometric detection of phosphatidylserine expression on early apoptotic cells using fluorescein labelled annexin V. J. Immunol. Methods 184 (1), 39–51.

CHAPTER

12

Perfluorinated compounds in fish: Ecotoxicological effects, DNA damage and implications for aquatic and human health

Arup Mistri and Sourav Sardar

Fish Physiology Laboratory, Department of Zoology, The University of Burdwan, Burdwan, West Bengal, India

12.1 Introduction

Perfluorinated compounds (PFCs), a subset of the broader class of per- and polyfluoroalkyl substances (PFAS), are synthetic organofluorine chemicals characterized by fully fluorinated carbon backbones (Stahl et al., 2011). Among the various types of PFASs, perfluorooctane sulfonate (PFOS) and perfluorooctanoic acid (PFOA) are the most widely used chemicals in industrial and commercial sectors for over 50 years in applications such as fabrics resistant to stains, popcorn bags, nonstick cookware coatings, paper plates, and firefighting foams (Buck et al., 2011; Gluge et al., 2020; Wang et al., 2021; Kurwadkar et al., 2022). Carbon–fluorine (C-F) bonds are among the strongest known in chemistry, giving these compounds exceptional stability and high resistance to chemical and biological degradation (Rericha et al., 2023). Due to the wide range of applications of PFASs in industrial and commercial sectors, they have been detected in surface waters, groundwater, soil, air, biota, and even human blood and breast milk (Buck et al., 2011; Wang et al., 2021). Due to their persistence and bioaccumulative potential, PFASs are increasingly considered emerging persistent organic pollutants. Fish integrate PFC contamination through the food chain and bioaccumulate these compounds, acting as sentinels of polluted environments (Buck et al., 2011; Gluge et al., 2020; Rericha et al., 2023). Additionally, fish represent a significant source of human exposure through diet in many regions (Lee et al., 2020; Soudani et al., 2024). In laboratory studies, fish models (such as zebrafish) have been extensively used to elucidate PFC toxicokinetics, developmental toxicity, and molecular mechanisms (Han et al., 2012; Vogs et al., 2019; Menger et al., 2020; Gaballah et al., 2020;

Genome Damage: Repair and Adaptation Across Life.
DOI: https://doi.org/10.1016/B978-0-443-42857-9.00025-6

Satbhai et al., 2022; Rericha et al., 2023). Numerous field surveys also document global PFC residues in wild marine and freshwater fish (Lee et al., 2020; Soudani et al., 2024). Importantly, PFC exposure in fish has been linked to diverse adverse outcomes: altered growth and survival, endocrine disruption, hepatic and gill damage, reproductive impairment, immunotoxicity, and genotoxic effects (Shi et al., 2008; Du et al., 2009; Hu, 2009; Huang et al., 2010; Hagenaars et al., 2011; Kim et al., 2011; Wang et al., 2011; Zheng et al., 2012; Xia et al., 2013a, 2013b; Li et al., 2015; Zhao et al., 2016; Annunziato et al., 2020). Several recent reviews have begun synthesizing PFC hazards in fish and other aquatic organisms (Lee et al., 2020; Ma et al., 2023; Rericha et al., 2023). Lee et al. (2020) classified PFC toxicity in fish into major categories, including metabolic disturbance, reproductive disruption, oxidative stress, developmental toxicity, and thyroid disruption. Ma et al. (2023) similarly summarized that PFCs interact with nuclear receptors, induce reactive oxygen species, and trigger molecular cascades leading to diverse toxic endpoints. Rericha et al. (2023) highlighted advances using zebrafish as an integrative model, showing structure-activity trends and the value of high-throughput approaches for PFAS hazard assessment.

This study provides a comprehensive global overview of PFC ecotoxicology in fish, with a special emphasis on DNA damage and its implications for both fish and human health. Drawing on research from the early 2000s to recent studies published up to 2025, this study covers freshwater and marine species, examining environmental occurrence, bioaccumulation, and both acute and chronic toxic effects at the developmental, physiological, cellular, and molecular levels, and focuses on genotoxic impacts on fish DNA, while also emphasizing human health. Thus the study would be helpful for researchers and policymakers concerned with PFC risks in aquatic ecosystems and fish health.

12.2 Occurrence and bioaccumulation of perfluorinated compounds in aquatic environments

PFCs enter aquatic systems from industrial effluents, wastewater, atmospheric deposition, and firefighting runoff (Stahl et al. 2011; Lee et al., 2020). Freshwater bodies and coastal marine areas worldwide show measurable PFC levels because of their high water solubility (especially for ionic PFAS) and resistance to degradation. Fish accumulate PFCs primarily through diet (consuming contaminated prey), gill uptake from water, and maternal transfer to eggs. PFAS are known to bioconcentrate and biomagnify: higher trophic-level fish often show greater tissue burdens (Lee et al., 2020; Rericha et al., 2023).

Bioaccumulation depends on chain length and functional group. Generally, long-chain PFAS (e.g., PFOS, PFNA, PFHxS) bioaccumulate more strongly than short-chain analogs. Rericha et al. (2023) note that in zebrafish, uptake rates increase with alkyl chain length and for sulfonate headgroups (PFSAs) compared to carboxylate groups (PFCAs). Several studies have measured PFAS in fish muscle, liver, and plasma. PFOS often dominates in biota. This includes analysis of wild freshwater fish from Swiss lakes, which detected PFOS in 100% of samples, frequently at concentrations above European safety limits (2 μg/kg) (Soudani et al., 2024). In the Swiss study, an optimized extraction and LC-MS/MS analysis of six lake fish species (including *Cyprinus carpio, Oncorhynchus mykiss, and Salmo trutta*) found average PFOS levels of up to ~20 μg/kg, with PFHxS also being abundant. These PFAS burdens exceeded the tolerable weekly intake thresholds for human consumers in many samples.

PFAS distribution in fish tissues can vary. PFOS is generally known to partition into blood, liver, and kidneys due to its binding to proteins (e.g., albumin). In contrast, PFOA is more evenly distributed but still bioaccumulates in the kidney and liver (Rericha et al., 2023). Recent studies using zebrafish and other models have found that adult fish exposed to PFAS for long periods accumulate these compounds in organs such as the liver, muscle, and ovaries, with evidence of maternal transfer to offspring (Rericha et al., 2023). Sharpe et al. (2010) and Wen et al. (2017, 2019) demonstrated that adult zebrafish retain PFAS in their liver and muscle, and females pass some of this burden to their eggs. Biotransformation of precursors also occurs: e.g., 6:2 fluorotelomer alcohols (6:2 FTOH) are metabolized to perfluoroalkyl acids that then accumulated in zebrafish with sex-specific differences (Shi et al., 2019). Therefore fish living in contaminated waters are routinely exposed to PFAS, often at levels that bioaccumulate. Tissue surveys worldwide (freshwater and marine species) consistently find PFOS and related PFAS in fish, reflecting historical and ongoing pollution. These contamination patterns warrant concern about potential toxic effects on fish health and, ultimately, on food webs, including human consumers (Lee et al., 2020; Soudani et al., 2024).

12.3 Toxicokinetics and accumulation patterns in fish

The toxicokinetics of PFCs in fish—absorption, distribution, metabolism, and elimination—influence their toxicity. Fish absorb PFCs predominantly from water via gills and skin, and from diet (ingesting contaminated prey) (Buck et al., 2011; Wang et al., 2021; Rericha et al., 2023). Studies in zebrafish and carp demonstrate higher uptake rates for longer carbon chains and sulfonate groups. Wen et al. (2017, 2019) exposed adult zebrafish to mixtures of perfluoroalkyl acids and found that uptake rates increased with chain length and were higher for PFSAs (such as PFOS) than for analogous PFCAs. Conversely, elimination (depuration) rates are generally slower for longer-chain PFAS, leading to higher bioconcentration factors (BCF). Rericha et al. (2023) note that PFAS uptake and elimination result in bioaccumulation trends similar to those of other persistent organic compounds, with high-molecular-weight and strongly binding compounds (such as PFOS and PFNA) accumulating more in fish tissues. Few fish can metabolize PFCAs and PFSAs; they are mainly excreted unchanged. However, precursors (fluorotelomer or phosphate-ester compounds) can be biotransformed into PFCAs inside organisms. For example, 6:2 FTOH is converted to perfluorohexanoic acid (PFHxA) and other acids by fish metabolism. Rericha et al. (2023) note that 6:2 fluorotelomer alkyl phosphate diesters (6:2 FTAB) were metabolized in zebrafish to 6:2 fluorotelomer alkyl acid (6:2 FTAA), which then bioaccumulated in a sex-dependent manner. These metabolic processes can complicate exposure assessment, as PFCAs may arise from precursors in the diet or the environment.

Tissue distribution in adult fish is species- and compound-dependent. Commonly, PFOS tends to accumulate in the liver, blood plasma, and kidneys due to its affinity for proteins (Hu, 2009; Shi and Zhou, 2010; Zhao et al., 2016; Sant et al., 2021; Ma et al., 2023). Muscle tissue often has lower PFAS concentrations compared to the liver, but still reflects the overall burden. Studies in wild carp and eel reported PFOS levels to be highest in the liver, with albumin-rich plasma concentrations also elevated (Hoff et al., 2003, 2005; Hagenaars et al., 2008; Arukwe and Mortensen, 2011). Some experiments have mapped PFC distributions: PFAS often concentrate in organs rich in blood flow and proteins (liver > gill > muscle), whereas reproductive organs can contain transferred compounds that affect eggs and gonads (Rericha et al., 2023). Importantly,

PFAS exhibit maternal transfer in oviparous fish. Female fish exposed to PFAS can deposit these compounds in eggs, thereby exposing developing embryos (Du et al., 2016). They also found that adult zebrafish exposed to PFOS pass a substantial load to their eggs, leading to developmental toxicity in embryos. Overall, toxicokinetics studies underscore that fish do not readily eliminate PFCs. Instead, they accumulate over time, especially in long-chain forms. This bioaccumulation underlies chronic toxicity risks: even low environmental levels can lead to significant body burdens that disturb biological systems.

12.4 Acute and chronic toxicity: Survival, growth, and development

12.4.1 Lethal and sublethal effects

Many laboratory studies have quantified the lethal toxicity (LC50) of PFCs to fish. Zebrafish embryos and larvae are a standard model: PFOS has moderate acute toxicity, with reported LC50 values in the range of tens of mg/L over 48-120 h. Zebrafish embryo 96-h LC50s for PFOS range roughly 10–70 mg/L (Shi et al., 2008; Du et al., 2009;Hu, 2009; Huang et al., 2010; Hagenaars et al., 2011; Kim et al., 2011; Wang et al., 2011; Zheng et al., 2012; Xia et al., 2013a, 2013b; Li et al., 2015; Zhao et al., 2016; Annunziato et al., 2020). PFOA is generally less acutely toxic than PFOS; one review notes a 96-h LC50 for zebrafish embryos of 371–662 mg/L for PFOA (versus much lower values for PFOS) (Ma et al., 2023). In adult fish, acute LC50 values are also typically in the mg/L range (e.g., PFOS 96-h LC50 ~22.6 mg/L in red sea bream *Pagrosomus major*, and as low as 4.7 mg/L in fathead minnow (*Pimephales promelas*) (Ma et al., 2023). These values classify PFOS as a moderately toxic contaminant, though environmental concentrations are much lower (ng–μg/L) in practice, implying that acute lethality is uncommon in the field.

Sublethal chronic effects occur at much lower concentrations. Rericha et al. (2023) reported that PFOS at ~5 μM (~2 mg/L) causes substantial mortality in zebrafish embryos by 5 days post-fertilization, and morphological malformations occur at even lower exposures. Many studies have found that exposure in the low-μM or sub-μM range (≪1 mg/L) can impair growth and development. Du et al. (2016) exposed zebrafish embryos to 0.4–1.6 mg/L PFOS and observed significant developmental toxicity, including delayed hatching and edema. Chronic growth inhibition has been reported by several researchers. PFOS exposure reduced body length and weight in juvenile and adult fish (e.g., zebrafish) (Du et al., 2009; Xia et al., 2013a, 2013b; Chen et al., 2014). In a multigenerational exposure study of fathead minnows, Suski et al. (2020) observed that the growth of the F1 generation was reduced because their parents had been exposed to PFOS (Ankley et al., 2005). In summary, high doses of PFCs cause rapid death, and deleterious responses have been observed even at environmentally relevant PFC concentrations. Chronic exposure to PFCs deteriorates health, reduces energy storage, and minimizes growth, especially at vulnerable life stages, even when survival is unaffected.

12.4.2 Developmental effects

Recent studies have shown that zebrafish exposed to PFAS can induce developmental abnormalities (Du et al., 2009; Zhao et al., 2016), such as pericardial edema (Shi et al., 2008; Zheng et al., 2012; Zhao et al., 2016), yolk sac, fin deformities (Xia et al., 2013a, 2013b), curvature of the tail (Zhao et al., 2016), and abnormal organ development (Dong et al., 2021). Findings from histological analysis depict impairment in gill arches (Hu et al., 2009) and liver and digestive tissue in embryos (Chen et al., 2014; Cui et al., 2016). Cui et al. (2016) found that PFOS can disrupt the microvilli structure of the intestine and liver in zebrafish, indicating impairment in nutrient processing during digestion.

Altered physiology and behavior can often cause malformations in fish, such as embryos exposed to PFOS, which show hyper- and hypoactivity in light and dark assays, reflecting neurodevelopmental impacts (Zheng et al., 2012; Xia et al., 2013a, 2013b). According to Du et al. (2009), zebrafish offspring of the F0 generation treated with PFOS showed developmental anomalies that may be due to epigenetic consequences and maternal transfer. Cui et al. (2016) also noted that PFOS exposure in F0 generation can affect metabolic alterations in glucose and lipid dysregulation in F1 generations, which suggests heritable metabolic changes. These studies highlight that PFCs are developmental toxicants in fish. While specific malformations vary among experiments, consistent themes are disrupted organ development (especially heart, liver, and endocrine tissues), delayed hatching, and behavioral changes in larvae. Given the ecological importance of early survival and recruitment, such developmental toxicity could have population-level consequences in contaminated waters.

12.4.3 Ecotoxicological impacts: DNA damage, oxidative stress, and genotoxicity

The ecotoxicological impacts of PFCs on fish are diverse, encompassing DNA damage, oxidative stress, and genotoxicity. These effects are critical indicators of cellular and genetic integrity disruption, potentially leading to broader health consequences for aquatic species, like fish.

12.4.3.1 Genotoxic and DNA-related effects in fish

A critical concern is the genotoxicity of PFCs in fish. Several studies have demonstrated DNA damage in fish tissues following PFC exposure, using assays such as the comet assay (single-cell gel electrophoresis), the micronucleus test, and mutation frequency assays. Du et al. (2016) showed that adult zebrafish exposed to PFOS (0.4–1.6 mg/L) showed significant DNA strand breaks in peripheral blood cells, assessed by the comet assay, and increased micronuclei in erythrocytes. Zinc-oxide nanoparticle co-exposure can cause DNA damage, suggesting interactive toxicity. Similarly, after being exposed to PFOA, hepatocytes of Nile tilapia showed increased DNA damage (comet tail intensity), a decrease in cell viability, and activated caspases (Abdel-Gawad et al., 2016). Hoff et al. (2003) reported that carp from contaminated waterways showed abnormalities in damaged hepatocyte membrane integrity and DNA metabolism. Chen et al. (2016) measured the mutagenicity of PFOS using a transgenic medaka (*Oryzias latipes*) mutation assay. Fish exposed to low concentrations of PFOS (6.7–87.6 μg/L) for 30 days showed a twofold increase in mutation frequency in the liver at the highest dose. This may be due to the frameshift events causing mutations. This experiment provides direct evidence that even sublethal PFOS can elevate mutation rates in fish liver DNA, which is a strong genotoxic signal. At the gene expression level, PFOS exposure has been shown to induce mutagenic changes. In zebrafish embryos, PFOS increased the mutation rate in the tumor suppressor gene p53 (Chen et al., 2009) and altered transcription of hundreds of other genes (including apoptosis and repair pathways). While not all PFCs exhibit equal genotoxicity (PFOS is often more so than PFOA), the weight of evidence suggests that PFAS can damage fish DNA. Likely mechanisms include ROS-mediated oxidative DNA damage and interference with DNA repair (both indicated by upregulation of DNA repair and stress genes in exposed (Du et al., 2016; Chen et al., 2016). Kim et al. (2020) found global DNA hypomethylation in zebrafish exposed to Perfluoroalkyl phosphinic acids (PFPiAs) PFOS. Although data are still sparse, these epigenetic effects could contribute to the transgenerational toxicities observed. PFAS-exposed fish show genotoxic effects, including DNA strand breaks, point mutations, chromosomal aberrations, and potentially, epigenetic changes (Du et al., 2016; Chen et al., 2016). This finding highlights long-term risks of PFAS pollution in fish populations, heritable defects, and leads to cancer.

12.4.3.2 *Oxidative stress-induced DNA damage in fish*

PFASs are known to induce oxidative stress, which can lead to DNA damage (Erinc et al., 2021). PFAS can bind with the grooves of DNA duplex through ionic, van der Waals forces, and hydrophobic interactions, i.e., PFOS interacts hydrophobically with the homolateral bases of DNA via alkyl groups running parallel to the phosphate backbone. These hydrophobic interactions cause the contiguous base pairs to draw closer, altering DNA's conformation (Zhang et al., 2009). The damage to fish DNA is likely due to the production of reactive oxygen species (ROS), such as hydroxyl radical (OH-), superoxide anion (O2-), and hydrogen peroxide (H_2O_2), which can cause strand breakage and other cellular damage (Akcha et al., 2003; Peña-Llopis et al., 2003; Banudevi et al., 2006). High concentrations of ROS can disrupt the redox equilibrium, leading to oxidative stress that damages various macromolecules, including DNA, lipids, and proteins (Fig. 12.1; Liguori et al., 2018; Pacchini et al., 2025). A study on Japanese medaka (*Oryzias latipes*) exposed to PFDoA at concentrations of 0.1, 0.5, and 2.5 mg/L for 28 days revealed pronounced nuclear abnormalities in erythrocytes, including micronuclei, fragmented apoptotic nuclei, lobed nuclei, and bean-shaped cells (Ayanda et al., 2018). The frequency of erythrocyte abnormalities peaked on day 14, whereas DNA damage (measured as percent tail length) reached its maximum on day 7, followed by a gradual decline. These findings indicate that subchronic PFDoA exposure induces significant DNA damage in medaka, accompanied by concurrent increases in multiple erythrocyte nuclear abnormalities (Ayanda et al., 2018). Liu et al. (2007) observed that DNA laddering, a hallmark of apoptosis, was observed after 48 h of exposure to PFOS or PFOA at concentrations of 15 and 30 mg/L. This indicates that these chemicals induce apoptosis in a dose-dependent manner and also states that PFOS and PFOA activate caspases, specifically caspase-3, -8, and -9, which are key enzymes in the apoptotic pathway. This activation ultimately leads to apoptosis.

Oxidative stress enzymes have the ability to interact with DNA, which can further cause alterations in DNA methylation, as further supported by Kim et al. (2020). He observed that in zebrafish, 144 h of exposure to 8:8 PFPiA caused a reduction of methylation in cytosine, whereas in control groups, 4.034% were methylated. In contrast, in the study of Ding et al. (2015), in sea urchins, PFOS exposure not only caused an increase in SOD activity but also increased DNA methylation polymorphism and methylation rate, whereas 8:8 PFPiA showed a decrease in methylation of DNA in zebrafish (Ding et al., 2015; Kim et al., 2020). When PFOA is exposed to zebrafish, it disrupts the mitochondrial electron transport chain and causes ROS production, which further causes modifications in nitrogenous bases and interferes with replication fidelity, contributing to genotoxicity (Fig. 12.1; Hagenaars et al., 2013). Araujo et al. (2026) state that strand breaks of DNA may depend on exposure duration and concentration, as specific changes, such as methylation of promoter regions, are more sensitive to alterations than morphological

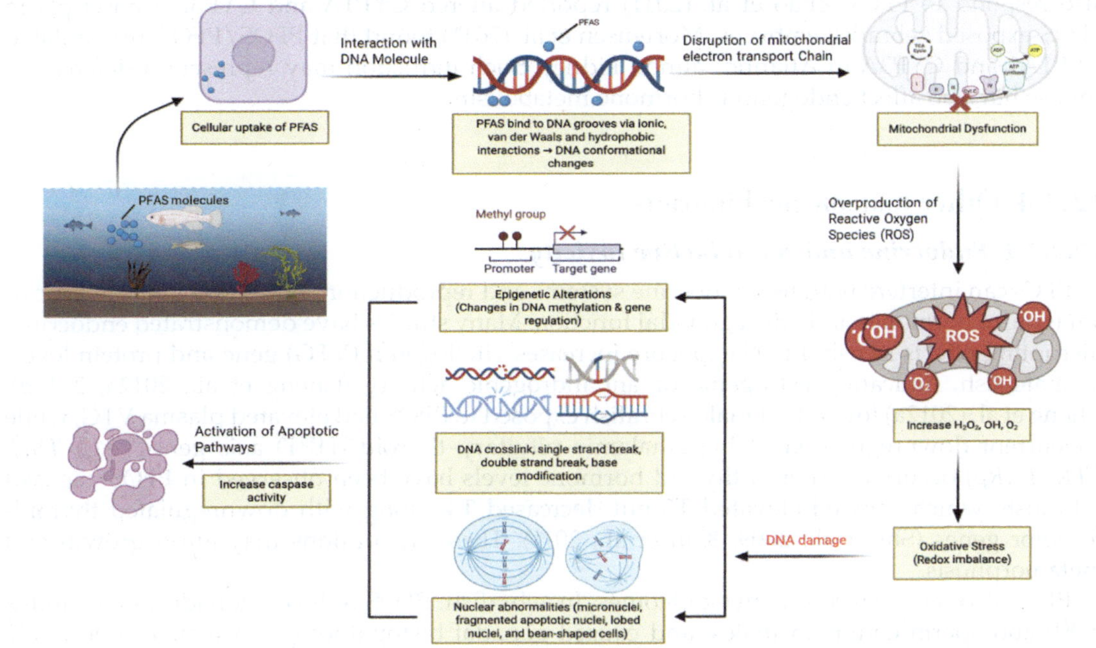

FIGURE 12.1 Overview of the biochemical cascade through which PFAS exposure induces oxidative stress and contributes to DNA damage in aquatic organisms (illustrated using BioRender software). *PFAS*, Perfluoroalkyl substance.

changes (Bouwmeester et al., 2016). This suggests that epigenetic changes might occur before the visible physiological changes or visible damage (Araujo et al., 2026). The mechanism by which PFAS-induced oxidative stress contributes to DNA damage is summarized in Fig. 12.1.

12.4.3.3 Oxidative stress and metabolic disruption

A unifying mechanism in PFC toxicity appears to be oxidative stress. Many fish studies report that PFAS exposure elevates ROS and lipid peroxidation, while perturbing antioxidant defenses (Sant et al., 2021). In zebrafish, PFOS significantly increased ROS production and malondialdehyde (MDA, a lipid-peroxidation marker), accompanied by altered activities of detoxification enzymes (SOD, CAT, GPx) (Hu, 2009; Shi and Zhou, 2010; Yao, 2018).Shi and Zhou (2010) found that in zebrafish larvae, PFOS can downregulate master oxidative stress regulators (i.e., Nrf2 signaling) and can induce antioxidant enzyme genes. Similar oxidative responses occur with PFOA: Liu et al. (2007) observed that PFOA increased lipid peroxidation and ROS in tilapia hepatocytes.

Oxidative stress is linked to metabolic effects. PFAS alter lipid and carbohydrate metabolism in fish, often via nuclear receptor pathways (e.g., PPARs) and peroxisomal proliferation. Cheng et al. (2016) showed that chronic PFOS reduced serum LDL/VLDL in male zebrafish and induced fatty liver (hepatic steatosis). In the same study, key genes of fatty acid β-oxidation and transport (e.g., *acox1, acadm, fabp10a*) were up- or downregulated in the liver, reflecting disrupted lipid handling. Enzymes involved in xenobiotic metabolism (cytochrome P450s, etc.)

also respond to PFCs. Zhao et al. (2011) reported altered CYP1A and CYP3A transcripts in PFOS-exposed zebrafish embryos. Mortensen et al. (2011) found that PFOS/PFOA upregulated CYP1A1 and CYP3A in Atlantic salmon kidney. Such induction may represent a defense response but also affect endogenous hormone metabolism.

12.4.4 Other toxicological impacts

12.4.4.1 Endocrine and reproductive toxicity

PFCs can interfere with fish endocrine systems and reproduction. PFOS and PFOA mimic or antagonize hormones and affect gonadal function. Many studies have demonstrated endocrine-disrupting effects in fish. PFOS exposure increases vitellogenin (VTG) gene and protein levels in male fish, indicating estrogenic or antiandrogenic activity (Cheng et al., 2012a, 2012b). Cheng et al. (2012a) found that male zebrafish exposed to PFOS had elevated plasma VTG, while concurrent downregulation of hypothalamic–pituitary–thyroid (HPT) axis genes (e.g., *TSH, TTR, THRβ*) occurred. Altered thyroid hormone levels have been observed in PFOS-exposed zebrafish, which showed elevated T3 but decreased T4, along with downregulated thyroid-receptor genes (Shi et al., 2009; Kim et al., 2011). These disruptions may affect growth and metamorphosis.

PFCs also impair reproductive outcomes. In zebrafish, PFOS reduced gonadosomatic index (GSI) and sperm counts in males, and caused gonadal histopathology in females (Du et al., 2009; Chen et al., 2014). In fathead minnows, PFOS exposure decreased GSI in males, disrupted ovarian development in females, and lowered fecundity [eggs produced (Ankley et al., 2005; Suski et al., 2020)]. Male fathead minnows showed increased testosterone levels after PFOS, whereas PFOA had milder effects on hormone titers. Nile tilapia hepatocytes in culture produced VTG when exposed to PFOS/PFOA, indicating estrogenic gene induction (Liu et al., 2007; Han et al., 2012). Conversely, some PFCs (or mixtures) have antiestrogenic actions, reducing exogenous estradiol-induced VTG (Liu et al., 2007).

Behavioral and developmental outcomes related to the endocrine system have also been documented. PFCs disrupt endocrine regulation in fish, as evidenced by alterations in reproductive hormones, impaired gonadal development, and disturbances in vitellogenesis (Shi et al., 2009; Sant et al., 2021; Ma et al., 2023).

12.4.4.2 Immune and neurotoxic effects

PFCs can alter immune functions in fish. In Nile tilapia (*Oreochromis niloticus*), exposure to PFAS (i.e., PFOS and PFOA) can lead to shifts in leukocyte populations. Han et al. (2012) reported that increasing levels of granulocytes, macrophages, and B-cells after exposure to either of these compounds. Very similar reports were observed by Wang et al. (2020) that PFOS can induce immune-related genes (i.e., cytochrome c oxidase, c-type lectin) and heat shock proteins, HSP70 in *Cynoglossus semilaevis* (tongue sole), which suggests immunotoxic-related stress response. In the head kidney of crucian carp (Carassius auratus), an increase in lymphocyte apoptosis was found after exposure to PFOA, causing immune impairment and reducing immune competence (Wang et al., 2015). Sublethal doses of PFOS can cause hyperactivity in larvae under light and dark conditions (Zheng et al., 2012), and they can also disrupt the expression of neuronal genes (Zhao et al., 2016), which demonstrates that PFCs also cause neurobehavioral toxicity, which further causes locomotion and learning in zebrafish. Exposure to PFOS reduces acetylcholinesterase (AChE) activity in the brains of zebrafish, which indicates cholinergic toxicity. Locomotor experiments revealed that PFAS can alter swimming patterns, i.e., higher swimming speeds and disrupted behavioral rhythms (Wang et al., 2011; Chen et al., 2012). Taken together, PFCs can trigger immune dysregulation and neurofunctional changes in fish. Although these effects are less well-characterized than metabolic/endocrine disruptions, evidence of altered immune cell profiles, antioxidant gene expression, and neurobehavior suggests broad systemic toxicity.

12.4.3 Comparative toxicity and emerging PFCs

Most toxicology research has focused on "legacy" PFAS: long-chain PFOA and PFOS. However, global phase-outs of these chemicals have led to increased use of alternatives (GenX, PFBS, etc.). Comparative studies suggest that chain length and functional group correlate with toxicity. Rericha et al. (2023) demonstrated that among the tested PFAS in zebrafish, long-chain PFCs generally elicited a stronger response in fish morphology and behavior than other shorter-chain PFCs. Gaballah et al. (2020) found that PFNA (C9) caused zebrafish malformations, whereas shorter PFCAs (C5-C8) had minimal developmental effects. Similar patterns are observed in toxicity assays: PFOS is consistently more toxic than PFOA on an equimolar basis (Ma et al., 2023). GenX (hexafluoropropylene oxide dimer acid) and F-53B (a chlorinated PFOS alternative used in plating) are newer PFAS of concern. Previous studies indicate these compounds can also harm fish. Gaballah et al. (2020) showed that exposure to GenX can cause gene expression changes and developmental delays in zebrafish embryos. Ahrens and Bundschuh (2014) reviewed emerging PFAS toxicity and found that many follow similar modes of action (PPAR activation, oxidative stress), albeit often at higher dose thresholds. Although substitution with shorter-chain or alternative PFAS may reduce bioaccumulation, initial evidence suggests these chemicals are not benign. They can still perturb fish physiology, especially at chronic exposures. As one review notes, the rapid introduction of new PFAS outpaces toxicological evaluation (Lee et al., 2020), making fish studies essential for hazard identification.

12.4.4 Global and species-specific perspectives

PFC ecotoxicology in fish has been studied globally, encompassing tropical to temperate and marine to freshwater ecosystems. Field surveys in North America (e.g., Great Lakes, Chesapeake Bay), Europe (e.g., Baltic Sea, Danube), and Asia (e.g., Yangtze River, Indian lakes) report PFOS, PFOA, and other PFAS in diverse fish species. Kannan et al. (2005) documented PFOS in Great Lakes fish as early evidence of bioaccumulation. Recent work by Valsecchi et al. (2017) derives water-quality criteria based on observed PFAS in fish from European rivers. Species differences are notable. Fatty acid metabolism and habitat drive variability: carnivorous predatory fish tend to have higher PFAS levels than planktivores. Marine mammals rank highest in coastal systems, but marine fish (e.g., tuna, cod) also accumulate PFAS. Wu et al. (2019) compared a predatory and herbivorous carp from Lake Chaohu (China) and found higher PFAS in the predator. In saltwater, studies of fish like red sea bream and salmon have reported tissue PFAS and linked them to hepatic effects (Krøvel et al., 2008). In freshwater, the Swiss lake survey found predatory perch (*Perca fluviatilis*) had the highest PFOS levels among six species tested (Soudani et al., 2024). An important aspect is human exposure via fish consumption. Risk assessments (e.g., Domingo and Nadal, 2019) estimate that PFAS intake from fish can be significant, especially in regions where fish is a dietary staple. The Swiss study computed that even modest fish intake could exceed tolerable weekly intakes of PFAS for many consumers (Soudani et al., 2024). Thus protecting fish not only conserves wildlife but also safeguards food safety.

12.5 Health risks and future concern

The widespread presence of PFCs in fish and their demonstrated ecotoxicological effects pose significant health risks not only to aquatic ecosystems but also to human consumers (Lu et al., 2011; Zhao et al., 2011; Zhang et al., 2011a, 2011b; Squadrone et al., 2014; Melake et al., 2022; Takdastan et al., 2023). Human exposure to PFCs through fish consumption is a recognized pathway (Zhang et al., 2011a, 2011b; Zhao et al., 2011; Squadrone et al., 2014). Squadrone et al. (2014) in their study in Italy have identified fish consumption as a source of human exposure to PFCs. Studies aim to quantify the potential for adverse effects from such exposure, including human consumption patterns and concentrations in fish (Zhao et al., 2011; Takdastan et al., 2023; Li et al., 2022, 2024). Cooking procedures, such as boiling, baking, and frying, have been proven to reduce PFOA and PFOS levels in fish; however, the extent of reduction varies (Sungur et al., 2019).

12.5.1 Health risks to humans via fish consumption

PFCs in fish present a significant human-health concern because dietary intake of contaminated fish is a major and well-documented pathway through which people are exposed to these persistent environmental pollutants (Stahl et al., 2011; Christensen et al., 2016).

12.5.2 Exposure pathways and bioaccumulation in humans

Humans can be exposed to PFCs through various pathways (Fig. 12.2). These include diet, water consumption, consumer or daily-use products, and indoor air (Lau, 2012). Although the ingestion of contaminated fish, particularly those at higher trophic levels, can be a major source of exposure due to biomagnification (Stahl et al., 2011; Christensen et al., 2016; Niaz et al., 2024). Previous studies have demonstrated that fish consumption can result in elevated levels of PFCs in human serum (Christensen et al., 2016; Wu et al., 2017). Christensen et al. (2016) found in their study that the consumption habits of older male anglers in Wisconsin can influence their levels of other persistent contaminants, including PFCs. Once ingested, PFCs are efficiently absorbed and remain in the human body for prolonged periods due to their long half-lives, resulting in their accumulation in various tissues, particularly the liver and serum (Lau, 2012; Wu et al., 2017).

12.5.3 Human-health effects

The direct link between PFC-contaminated fish consumption and specific human-health outcomes is complex and requires further epidemiological research. Extensive toxicological and

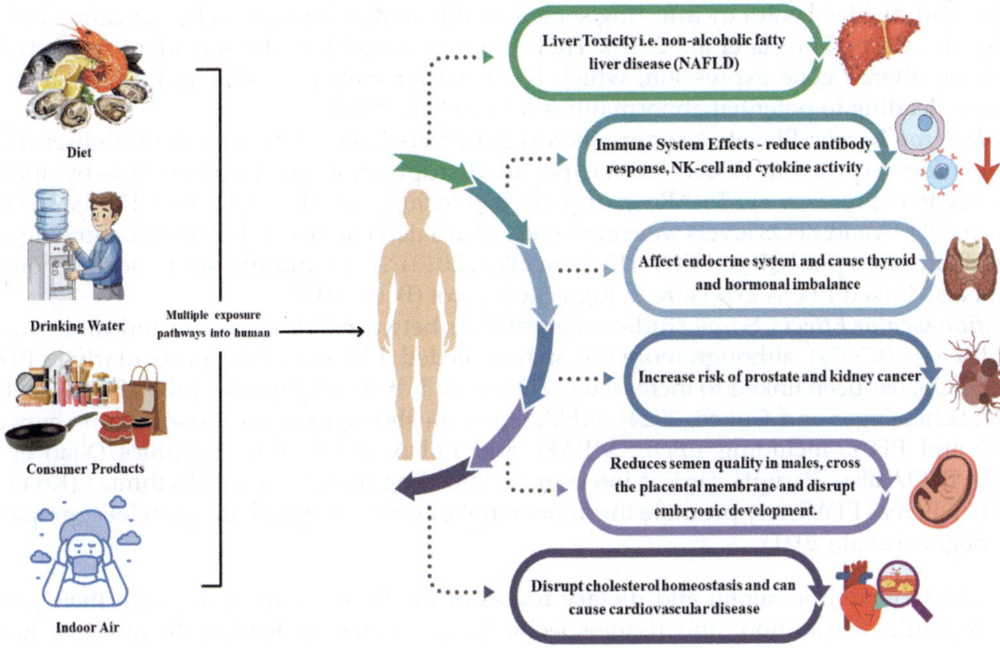

FIGURE 12.2 PFAS exposure in humans and their bioaccumulation across tissues, illustrating the resulting endocrine, metabolic, immunological, reproductive, and carcinogenic health effects (illustrated using BioRender software). *PFAS*, Perfluoroalkyl substance.

epidemiological studies have associated PFC exposure with a range of adverse health effects in humans (Fig. 12.2). These include:

- ***Liver Toxicity:*** PFCs are known hepatotoxicants. Exposure can lead to liver damage, increased liver enzyme levels, and potentially contribute to conditions such as nonalcoholic fatty liver disease (NAFLD) (Niaz et al., 2024).
- ***Immune System Effects:*** PFCs can suppress the immune system, leading to reduced antibody responses to vaccines and increased susceptibility to infections (Brieger et al., 2011; Zhang et al., 2022). PFOS and PFOA have been shown to decrease natural killer-cell activity and reduce proinflammatory cytokine release in human leukocytes (Brieger et al., 2011). Chronic inflammation and immunosuppression are key characteristics of PFAS exposure (Zhang et al., 2023).
- ***Endocrine Disruption:*** Similar to their effects in fish, PFCs can disrupt human endocrine systems, affecting thyroid function and reproductive hormones, and potentially contributing to developmental issues (Stahl et al., 2011).
- ***Cancer Risk:*** Some PFCs have been associated with increased risks of certain cancers, including kidney and testicular (Vieira et al., 2013; Kahn et al., 2020; Imir et al., 2021; Seyyedsalehi and Boffetta, 2023; Wei et al., 2023; Zheng et al., 2024). Prostate cancer progression has also been linked to PFC exposure (Wei et al., 2023).
- ***Developmental and Reproductive Effects:*** PFCs can cross the placental barrier, expose the fetus, and potentially disrupt reproductive development. Evidence suggests they impair

male reproductive health by affecting semen quality and hormone levels, indicating antiandrogenic effects (Foresta et al., 2018). Their developmental toxicity is evident from studies showing altered gene expression, which is crucial for embryogenesis in human stem cell models, leading to potential abnormalities (Zhao et al., 2024).

- ***Cholesterol Levels:*** Elevated serum cholesterol levels have been consistently observed in populations exposed to PFCs. PFOA exposure disrupts cholesterol homeostasis by altering liver gene expression via PPARα pathways (Schlezinger et al., 2020). In a Danish cohort, plasma PFOA and PFOS levels were positively associated with total cholesterol, showing a 4.4 mg/dL increase per IQR of PFOA (Eriksen et al., 2013). A community study near a chemical plant also linked PFOA exposure to high cholesterol (Betts, 2014).
- ***Cardiovascular Effects:*** Some studies suggest links between PFC exposure and cardiovascular diseases (CVDs), although more research is needed. PFC exposure, particularly to PFOA and PFOS, has been linked to increased CVD risk and dyslipidemia (Abdullah Soheimi et al., 2021; Schlezinger and Gokce, 2024). NHANES data showed positive associations between CVD and PFCs, including PFDE, MPAH, and PFUA, across demographics (Xiao et al., 2025). PFOA also disrupts cardiac ion channels, leading to potential arrhythmias (Ko et al., 2024), whereas PFAS may promote thromboembolic events by enhancing platelet aggregation (Meneguzzi et al., 2021).

The primary environmental and dietary routes of PFAS exposure, along with their persistence, systemic distribution, and tissue-specific bioaccumulation leading to multiple health impacts—including endocrine disruption, immunotoxicity, metabolic disorders, reproductive abnormalities, carcinogenic potential, and organ-specific toxic effects—are summarized in Fig. 12.2.

The health risks posed by PFCs are further exacerbated by their long persistence in the human body, leading to chronic exposure and potential accumulation of adverse effects over a lifetime. This highlights the importance of monitoring PFC levels in consumed fish and implementing measures to reduce environmental contamination. Future research should focus on a more comprehensive understanding of the long-term, sublethal effects of PFCs on fish populations, including reproductive impairments, developmental abnormalities, and immune system dysregulation. Investigating the interplay between PFCs and other environmental stressors could provide a more realistic assessment of their impacts. Furthermore, developing and implementing effective strategies for PFC remediation in aquatic environments and stricter regulations on their production and use are crucial for mitigating their pervasive threat to aquatic biodiversity and human health. The development of novel procedures for predicting chronic toxicities and ecological risks of PFCs in aquatic environments is an ongoing area of research (Li et al., 2022).

12.6 Conclusions and future direction

PFCs present a clear ecotoxicological threat to fish. Decades of research document that PFCs are ubiquitous in aquatic environments and bioaccumulate in fish, disrupting physiology at multiple levels. This review has emphasized global studies and a range of fish species, combining laboratory and field data. The collected literature indicates that PFC toxicity in fish is multifactorial, involving oxidative, endocrine, immune, and genotoxic pathways. Given the structural diversity of PFAS, each compound may target different systems, but PFOS/PFOA have reliably emerged as reference toxicants. Emerging PFAS substitutes (e.g., GenX, PFBS) require similar scrutiny, as initial data suggest they also harm fish, albeit sometimes at higher concentrations.

Future research should prioritize understudied species (e.g., tropical marine fish), long-term multigenerational studies, and mechanistic insights at the molecular level (e.g., transcriptomics, epigenetics). Monitoring efforts must continue to track PFAS in fish, especially new and short-chain PFAS, to inform risk assessments. Moreover, to advance the understanding and management of PFAS contamination, several key research and policy areas must be prioritized. These include generating comprehensive toxicity data for all emerging PFAS, investigating mixture effects, and elucidating biotransformation pathways and the toxicity of metabolites. The development of sensitive molecular biomarkers and refined bioaccumulation models that consider protein binding and environmental factors is essential. Standardized monitoring protocols and localized exposure assessments are needed to improve risk evaluation, particularly for vulnerable populations. Additionally, efforts must focus on identifying, reducing, and remediating PFAS sources, supported by interdisciplinary collaboration to address this complex environmental challenge. Ultimately, addressing PFC pollution is critical not only for fish conservation but also for broader ecological and public health.

Acknowledgment

AM acknowledges the support and assistance from the Anusandhan National Research Foundation (ANRF), India (File Number: TAR/2022/000560). We also express our appreciation to the Head, Department of Zoology, The University of Burdwan, for providing essential research infrastructure facilities.

References

Abdel-Gawad, F.K., Khalil, W.K.B., El-Kady, A.A., Waly, A.I., Abdel-Wahhab, M.A., 2016. Carboxymethyl chitosan modulates the genotoxic risk and oxidative stress of perfluorooctanoic acid in Nile tilapia *(Oreochromis niloticus)*. J. Saudi Soc. Agric. Sci. 15, 57–66.

Abdullah Soheimi, S.S., Rahman, A.A., Latip, N.A., Ibrahim, E., Kadir, S.H.S.A., 2021. Understanding the Impact of perfluorinated compounds on cardiovascular diseases and their risk factors: a meta-analysis study. Int. J. Environ. Res. Public Health 18, 8345.

Ahrens, L., Bundschuh, M., 2014. Fate and effects of poly- and perfluoroalkyl substances in the aquatic environment: a review. Environ. Toxicol. Chem. 33, 1921–1929.

Akcha, F., Hubert, F.V., Pfhol-Leszkowicz, A., 2003. Potential value of the comet assay and DNA adduct measurement in dab (*Limandalimanda*) for assessment of in situ exposure to genotoxic compounds. Mutat. Res. 534, 21–32.

Ankley, G.T., Kuehl, D.W., Kahl, M.D., Jensen, K.M., Linnum, A., Leino, R.L., 2005. Reproductive and developmental toxicity and bioconcentration of perfluorooctanesulfonate in a partial life-cycle test with the fathead minnow (*PimephalesPromelas*). Environ. Toxicol. Chem. 24, 2316–2324.

Annunziato, K.M., Doherty, J., Lee, J., Clark, J.M., Liang, W., Clark, C.W., 2020. Chemical characterization of a legacy aqueous film-forming foam sample and developmental toxicity in zebrafish (*Danio rerio*). Environ. Health Perspect. 128 (9), 97006.

Araujo, T.B.G., Junior, S.F.S., de Farias Araujo, G., Medeiros, R.J., Soares, L.L., Ikehara, B.R.M., et al., 2025. Multi-endpoint evaluation of PFOA toxicity in *Danio rerio*: oxidative, genotoxic, and metabolomic responses. J. Haz. Mat. 501, 140704.

Arukwe, A., Mortensen, A.S., 2011. Lipid peroxidation and oxidative stress responses of salmon fed a diet containing perfluorooctane sulfonic- or perfluorooctane carboxylic acids. Comp. Biochem. Physiol. Part C. Toxicol. Pharmacol. 154, 288–295.

Ayanda, I.O., Yang, M., Yu, Z., Zha, J., 2018. Cytotoxic and genotoxic effects of perfluorododecanoic acid (PFDoA) in Japanese medaka. Knowl. Manag. Aquat. Ecosyst. 419, 9.

Banudevi, S., Krishnamoorthy, G., Venkataraman, P., Vignesh, C., Aruldhas, M.M., Arunakaran, J., 2006. Role of α-tocopherol on antioxidant status in liver, lung and kidney of PCB exposed male albino rats. Food Chem. Toxicol. 44, 2040–2046.

Betts, K.S., 2014. PFOA and high cholesterol: basis for the finding of a probable link. Environ. Health Perspect. 122 (12), A338.

Bouwmeester, M.C., Ruiter, S., Lommelaars, T., Sippel, J., Hodemaekers, H.M., van den Brandhof, E.J., et al., 2016. Zebrafish embryos as a screen for DNA methylation modifications after compound exposure. Toxicol. Appl. Pharmacol. 291, 84–96.

Brieger, A., Bienefeld, N., Hasan, R., Goerlich, R., Haase, H., 2011. Impact of perfluorooctanesulfonate and perfluorooctanoic acid on human peripheral leukocytes. Toxicol. Vitro 25, 960–968.

Buck, R.C., Franklin, J., Berger, U., Conder, J.M., Cousins, I.T., de Voogt, P., et al., 2011. Perfluoroalkyl and polyfluoroalkyl substances in the environment: terminology, classification, and origins. Integr. Environ. Assess. Manag. 7, 513–541.

Chen, J., Das, S.R., La Du, J., Corvi, M.M., Bai, C., Chen, Y., 2012. Chronic PFOS exposures induce life stage-specific behavioral deficits in adult zebrafish and produce malformation and behavioral deficits in F1 offspring. Environ. Toxicol. Chem. 32, 201–206.

Chen, J., Tanguay, R.L., Tal, T.L., Gai, Z., Ma, X., Bai, C., 2014. Early life perfluorooctanesulphonic acid (PFOS) exposure impairs zebrafish organogenesis. Aquat. Toxicol. 150, 124–132.

Chen, W.F., Zhao, Q.S., Yin, D.Q., 2009. Treatment with PFOS causes point mutations in zebrafish embryos genome. Environ. Chem. 28, 215–219.

Chen, Y., Hu, W., Huang, C., Hua, S., Wei, Q., Bai, C., 2016. Subchronicperfluorooctanesulfonate (PFOS) exposure induces elevated mutant frequency in an *in vivo* λ transgenic medaka mutation assay. Sci. Rep. 6, 38466.

Cheng, J., Lv, S., Nie, S., Liu, J., Tong, S., Kang, N., et al., 2016. Chronic perfluorooctane sulfonate (PFOS) exposure induces hepatic steatosis in zebrafish. Aquat. Toxicol. 176, 45–52.

Cheng, Y., Cui, Y., Dang, Z.C., Xie, W.P., Li, H.S., Yin, H.H., 2012a. Effects of perfluorooctane sulfonate (PFOS) exposure on vitellogenin mRNA level in zebrafish (*BrachyDanio rerio*). Environ. Sci. 33, 1865–1870.

Cheng, Y., Cui, Y., He, P., Dang, Z., Xie, W., Yin, H., 2012b. Effect of perfluorooctane sulfonate (PFOS) on vitellogenin levels in blood plasma and tissue homogenate of zebrafish (*BrachyDanio rerio*). Asian J. Ecotoxicol. 7, 65–70.

Christensen, K.Y., Raymond, M., Thompson, B.A., Anderson, H.A., 2016. Perfluoroalkyl substances in older male anglers in Wisconsin. Environ. Int. 91, 312–318.

Cui, Y., Lv, S., Liu, J., Noe, S., Chen, J., Dong, Q., 2016. Chronic perfluorooctanesulfonic acid exposure disrupts lipid metabolism in zebrafish. Hum. Exp. Toxicol. 36, 207–217.

Ding, G., Wang, L., Zhang, J., Wei, Y., Wei, L., Li, Y., et al., 2015. Toxicity and DNA methylation changes induced by perfluorooctane sulfonate (PFOS) in sea urchin *Glyptocidariscrenularis*. Chemosphere 128, 225–230.

Domingo, J.L., Nadal, M., 2019. Human exposure to per- and polyfluoroalkyl substances (PFAS) through drinking water: a review of the recent scientific literature. Environ. Res. 177, 108648.

Dong, G., Zhang, R., Huang, H., Lu, C., Xia, Y., Wang, X., 2021. Exploration of the developmental toxicity of TCS and PFOS to zebrafish embryos by whole-genome gene expression analyses. Environ. Sci. Pollut. Res. 28, 56032–56042.

Du, J., Wang, S., You, H., Jiang, R., Zhuang, C., Zhang, X., 2016. Developmental toxicity and DNA damage to zebrafish induced by perfluorooctane sulfonate in the presence of ZnO nanoparticles. Environ. Toxicol. 31, 360–371.

Du, Y., Shi, X., Liu, C., Yu, K., Zhou, B., 2009. Chronic effects of water-borne PFOS exposure on growth, survival and hepatotoxicity in zebrafish: a partial life-cycle test. Chemosphere 74, 723–729.

Eriksen, K.T., Raaschou-Nielsen, O., McLaughlin, J.K., Lipworth, L., Tjønneland, A., Overvad, K., et al., 2013. Association between plasma PFOA and PFOS levels and total cholesterol in a middle-aged Danish population. PLoS One 8 (2), e56969.

Erinc, A., Davis, M.B., Padmanabhan, V., Langen, E., Goodrich, J.M., 2021. Considering environmental exposures to per-and polyfluoroalkyl substances (PFAS) as risk factors for hypertensive disorders of pregnancy. Environ. Res. 197, 111113.

Foresta, C., Tescari, S., Di Nisio, A., 2018. Impact of perfluorochemicals on human health and reproduction: a male's perspective. J. Endocrinol. Investig. 41, 639–645.

Gaballah, S., Swank, A., Sobus, J.R., Howey, X.M., Schmid, J., Catron, T., 2020. Evaluation of developmental toxicity, developmental neurotoxicity, and tissue dose in zebrafish exposed to GenX and other PFAS. Environ. Health Perspect. 128, 047005.

Gluge, J., Scheringer, M., Cousins, I.T., DeWitt, J.C., Goldenman, G., Herzke, D., et al., 2020. An overview of the uses of per- and polyfluoroalkyl substances (PFAS). Environ. Sci. Process. Impacts 22, 2345–2373.

Hagenaars, A., Knapen, D., Meyer, I.J., van der Ven, K., Hoff, P., De Coen, W., 2008. Toxicity evaluation of perfluorooctane sulfonate (PFOS) in the liver of common carp (*Cyprinus carpio*). Aquat. Toxicol. 88, 155–163.

Hagenaars, A., Vergauwen, L., Benoot, D., Laukens, K., Knapen, D., 2013. Mechanistic toxicity study of perfluorooctanoic acid in zebrafish suggests mitochondrial dysfunction to play a key role in PFOA toxicity. Chemosphere 91, 844–856.

Hagenaars, A., Vergauwen, L., De Coen, W., Knapen, D., 2011. Structure activity relationship assessment of four perfluorinated chemicals using a prolonged zebrafish early life stage test. Chemosphere 82, 764–772.

Han, Z., Liu, Y., Wu, D., Zhu, Z., Lü, C., 2012. Immunotoxicity and hepatotoxicity of PFOS and PFOA in tilapia (*Oreochromis niloticus*). Chin. J. Geochem. 31, 424–430.

Hoff, P.T., Van Campenhout, K., Van de Vijver, K., Covaci, A., Bervoets, L., Moens, L., 2005. Perfluorooctane sulfonic acid and organohalogen pollutants in liver of three freshwater fish species in flanders (Belgium): relationships with biochemical and organismal effects. Environ. Pollut. 137, 324–333.

Hoff, P.T., Van Dongen, W., Esmans, E.L., Blust, R., De Coen, W.M., 2003. Evaluation of the toxicological effects of perfluorooctane sulfonic acid in the common carp (*Cyprinus carpio*). Aquat. Toxicol. 62, 349–359.

Hu, Q., Zhou, Z., Zhou, Q., Zhang, J., Liang, Y., 2009. Acute effects of perfluorooctane sulfonate on microstructure of the gill of zebrafish (*Danio rerio*). Asian J. Ecotoxicol. 4, 530–536.

Huang, H., Huang, C., Wang, L., Ye, X., Bai, C., Simonich, M.T., 2010. Toxicity, uptake kinetics and behavior assessment in zebrafish embryos following exposure to perfluorooctanesulphonicacid (PFOS). Aquat. Toxicol. 98, 139–147.

Imir, O.B., Kaminsky, A.Z., Zuo, Q.-Y., Liu, Y.-J., Singh, R., Spinella, M.J., et al., 2021. Per- and polyfluoroalkyl substance exposure combined with high-fat diet supports prostate cancer progression. Nutrients 13, 3902.

Kahn, L.G., Philippat, C., Nakayama, S.F., Slama, R., Trasande, L., 2020. Endocrine-disrupting chemicals: implications for human health. Lancet. Diabetes Endocrinol. 8 (8), 703–718.

Kannan, K., Tao, L., Sinclair, E., Pastva, S.D., Jude, D.J., Giesy, J.P., 2005. Perfluorinated compounds in aquatic organisms at various trophic levels in a Great Lakes food chain. Arch. Environ. Contam. Toxicol. 48, 559–566.

Kim, S., Ji, K., Lee, S., Lee, J., Kim, J., Kim, S., 2011. Perfluorooctane sulfonic acid exposure increases cadmium toxicity in early life stage of zebrafish, *Danio rerio*. Environ. Toxicol. Chem. 30, 870–877.

Kim, S., Stroski, K.M., Killeen, G., Smitherman, C., Simcik, M.F., Brooks, B.W., 2020. 8:8 Perfluoroalkyl phosphinic acid affects neurobehavioral development, thyroid disruption, and DNA methylation in developing zebrafish. Sci. Total Environ. 736, 139600.

Ko, M.Y., Chon, S., Park, H., Min, E., Kim, Y.H., Cha, S.-W., et al., 2024. Perfluorooctanoic acid induces cardiac dysfunction in human induced pluripotent stem cell-derived cardiomyocytes. Ecotoxicol. Environ. Saf. 273, 116170.

Krøvel, A.V., Søfteland, L., Torstensen, B., Olsvik, P.A., 2008. Transcriptional effects of PFOS in isolated hepatocytes from Atlantic salmon *Salmo salar* L. Comp. Biochem. Physiol. Part C. Toxicol. Pharmacol. 148, 14–22.

Kurwadkar, S., Dane, J., Kanel, S.R., Nadagouda, M.N., Cawdrey, R.W., Ambade, B., et al., 2022. Per- and polyfluoroalkyl substances in water and wastewater: a critical review of their global occurrence and distribution. Sci. Total Environ. 809, 151003.

Lau, C., 2012. Perfluorinated compounds. In: Luch, A. (Ed.). In: Molecular, Clinical and Environmental Toxicology., 101. Springer, Basel, pp. 47–86 Environmental Toxicology.

Lee, J.W., Choi, K., Park, K., Seong, C., Yu, S.D., Kim, P., 2020. Adverse effects of perfluoroalkyl acids on fish and other aquatic organisms: a review. Sci. Total Environ. 707, 135334.

Li, J., Sun, J., Chao, L., Chen, J., Huang, L., Kang, B., 2024. Exposure, spatial distribution, and health risks of perfluoroalkyl acids in commercial fish species in the Beibu Gulf. Mar. Pollut. Bull. 209, 117101.

Li, Q., Wang, P., Wang, C., Hu, B., Wang, X., 2022. A novel procedure for predicting chronic toxicities and ecological risks of perfluorinated compounds in aquatic environment. Environ. Res. 215, 114132.

Li, Y., Han, Z., Zheng, X., Ma, Z., Liu, H., Giesy, J.P., 2015. Comparison of waterborne and in ovonanoinjection exposures to assess effects of PFOS on zebrafish embryos. Environ. Sci. Pollut. Res. 22, 2303–2310.

Liguori, I., Russo, G., Curcio, F., Bulli, G., Aran, L., Della-Morte, D., et al., 2018. Oxidative stress, aging, and diseases. Clin. Interv. Aging. 13, 757–772.

Liu, C., Du, Y., Zhou, B., 2007. Evaluation of estrogenic activities and mechanism of action of perfluorinated chemicals determined by vitellogenin induction in primary cultured tilapia hepatocytes. Aquat. Toxicol. 85, 267–277.

Liu, C., Yu, K., Shi, X., Wang, J., Lam, P.K., Wu, R.S., et al., 2007. Induction of oxidative stress and apoptosis by PFOS and PFOA in primary cultured hepatocytes of freshwater tilapia (*Oreochromis niloticus*). Aquat. Toxicol 82, 135–143.

Lu, G.-H., Yang, Y.L., Taniyasu, S., Yeung, L.W.Y., Pan, J., Zhou, B., et al., 2011. Potential exposure of perfluorinated compounds to Chinese in Shenyang and Yangtze River Delta areas. Environ. Chem. 8, 407–418.

Ma, T., Wu, P., Wang, L., Li, Q., Li, X., Luo, Y., 2023. Toxicity of per- and polyfluoroalkyl substances to aquatic vertebrates. Front. Environ. Sci. 11, 1101100.

Melake, B.A., Bervoets, L., Nkuba, B., Groffen, T., 2022. Distribution of perfluoroalkyl substances (PFASs) in water, sediment, and fish tissue, and the potential human health risks due to fish consumption in Lake Hawassa. Ethiopia. Environ. Res. 204, 112033.

Meneguzzi, A., Fava, C., Castelli, M., Minuz, P., 2021. Exposure to perfluoroalkyl chemicals and cardiovascular disease: experimental and epidemiological evidence. Front. Endocrinol. 12, 706352.

Menger, F., Pohl, J., Ahrens, L., Carlsson, G., Örn, S., 2020. Behavioural effects and bioconcentration of per- and polyfluoroalkyl substances (PFASs) in zebrafish (*Danio rerio*) embryos. Chemosphere 245, 125573.

Mortensen, A.S., Letcher, R.J., Cangialosi, M.V., Chu, S., Arukwe, A., 2011. Tissuebioaccumulation patterns, xenobiotic biotransformation and steroid hormone levels inAtlantic salmon (*Salmo salar*) fed a diet containing perfluoroactane sulfonic orperfluorooctane carboxylic acids. Chemosphere 83, 1035–1044.

Niaz, W., Zhang, D., Ahmad, Z., Shen, N., Haider, W., Ali, I., et al., 2024. Efficient removal of perfluorooctanoic acid (PFOA) and perfluorooctane sulfonic acid (PFOS) from aqueous solution using modified biochar: preparation, performance, and mechanistic insights. J. Environ. Chem. Eng. 12, 114894.

Pacchini, S., Vanzan, G., Piva, E., Schumann, S., Cortese, M., Drago, L., et al., 2025. Chronic PFAS exposure induces the activation of selenium-dependent glutathione peroxidases and catalase as antioxidant defences in the European chub (*Squalius cephalus*) (Linnaeus, 1758) kidney. Aquat. Toxicol. 287, 107524.

Peña-Llopis, S., Ferrando, M.D., Peña, J.B., 2003. Fish tolerance to organophosphate-induced oxidative stress is dependent on the glutathione metabolism and enhanced by N-acetylcysteine. Aquat. Toxicol. 65, 337–360.

Rericha, Y., Simonich, M.T., Truong, L., Tanguay, R.L., 2023. Review of the zebrafish as a model to investigate per- and polyfluoroalkyl substance toxicity. Toxicol. Sci. 194, 138–152.

Sant, K.E., Annunziato, K., Conlin, S., Teicher, G., Chen, P., Venezia, O., 2021. Developmental exposures to perfluorooctanesulfonic acid (PFOS) impact embryonic nutrition, pancreatic morphology, and adiposity in the zebrafish, *Danio rerio*. Environ. Pollut. 275, 116644.

Satbhai, K., Vogs, C., Crago, J., 2022. Comparative toxicokinetics and toxicity of PFOA and its replacement GenX in the early stages of zebrafish. Chemosphere 308, 1361.

Schlezinger, J., Gokce, N., 2024. Perfluoroalkyl/polyfluoroalkyl substances: links to cardiovascular disease risk. Circ. Res. 134, 1136–1159.

Schlezinger, J.J., Puckett, H., Oliver, J., Nielsen, G., Heiger-Bernays, W., Webster, T.F., 2020. Perfluorooctanoic acid activates multiple nuclear receptor pathways and skews expression of genes regulating cholesterol homeostasis in liver of humanized PPARα mice fed an American diet. Toxicol. Appl. Pharmacol. 405, 115204.

Seyyedsalehi, M.S., Boffetta, P., 2023. Per- and poly-fluoroalkyl substances (PFAS) exposure and risk of kidney, liver, and testicular cancers: a systematic review and meta-analysis. Med. Lav. 114 (5), e2023040.

Sharpe, R.L., Benskin, J.P., Laarman, A.H., MacLeod, S.L., Martin, J.W., Wong, C.S., et al., 2010. Perfluorooctane sulfonate toxicity, isomer-specific accumulation, and maternal transfer in zebrafish (*Danio rerio*) and rainbow trout (*Oncorhynchus mykiss*). Environ. Toxicol. Chem. 29, 1957–1966.

Shi, G., Cui, Q., Zhang, H., Cui, R., Guo, Y., Dai, J., 2019. Accumulation, biotransformation, and endocrine disruption effects of fluorotelomer surfactant mixtures on zebrafish. Chem. Res. Toxicol. 32, 1432–1440.

Shi, X., Du, Y., Lam, P.K.S., Wu, R.S.S., Zhou, B., 2008. Developmental toxicity and alteration of gene expression in zebrafish embryos exposed to PFOS. Toxicol. Appl. Pharmacol. 230, 23–32.

Shi, X., Liu, C., Wu, G., Zhou, B., 2009. Waterborne exposure to PFOS causes disruption of the hypothalamus-pituitary-thyroid axis in zebrafish larvae. Chemosphere 77, 1010–1018.

Shi, X., Zhou, B., 2010. The role of Nrf2 and MAPK pathways in PFOS-induced oxidative stress in zebrafish embryos. Toxicol. Sci. 115, 391–400.

Soudani, M., Hegg, L., Rime, C., Coquoz, C., Grosjean, D.B., Danza, F., et al., 2024. Determination of per- and polyfluoroalkyl substances (PFAS) in six different fish species from Swiss lakes. Anal. Bioanal. Chem. 416, 6377–6386.

Squadrone, S., Ciccotelli, V., Favaro, L., Scanzio, T., Prearo, M., Abete, M.C., 2014. Fish consumption as a source of human exposure to perfluorinated alkyl substances in Italy: analysis of two edible fish from Lake Maggiore. Chemosphere 114, 181–186.

Stahl, T., Mattern, D., Brunn, H., 2011. Toxicology of perfluorinated compounds. Environ. Sci. Eur. 23, 38.

Sungur, Ş., Kanan, E., Köroğlu, M., 2019. A comparison of levels of perfluorooctanoic acid (PFOA) and perfluorooctane sulfonic acid (PFOS) in raw and cooked fish. Toxin Rev 40, 59–64.

Suski, J.G., Salice, C.J., Chanov, M.K., Ayers, J., Rewarts, J., Field, J., 2020. Sensitivity and accumulation of perfluorooctanesulfonate (PFOS) and perfluorohexanesulfonic acid (PFHxS) in Fathead Minnows (*Pimephalespromelas*) exposed over critical life-stages of reproduction and development. Environ. Toxicol. Chem. 40, 811–819.

Takdastan, A., Babaei, A.A., Jorfi, S., Ahmadi, M., Tahmasebi Birgani, Y., Jamshidi, B., 2023. Perfluorooctanoic acid (PFOA) and perfluorooctane sulfonate (PFOS) in water and edible fish species of Karun River, Ahvaz, Iran: spatial distribution, human health, and ecological risk assessment. Int. J. Environ. Health Res. 34, 803–814.

Valsecchi, S., Conti, D., Crebelli, R., Polesello, S., Rusconi, M., Mazzoni, M., et al., 2017. Deriving environmental quality standards for perfluorooctanoic acid (PFOA) and related short chain perfluorinated alkyl acids. J. Hazard. Mater. 323, 84–98.

Vieira, V.M., Hoffman, K., Shin, H.M., Weinberg, J.M., Webster, T.F., Fletcher, T., 2013. Perfluorooctanoic acid exposure and cancer outcomes in a contaminated community: a geographic analysis. Environ. Health Perspect. 121 (3), 318.

Vogs, C., Johanson, G., Naslund, M., Wulff, S., Sjodin, M., Hellstrandh, M., et al., 2019. Toxicokinetics of perfluorinated alkyl acids influences their toxic potency in the zebrafish embryo (*Danio rerio*). Environ. Sci. Technol. 53, 3898–3907.

Wang, Z., Buser, A.M., Cousins, I.T., Demattio, S., Drost, W., Johansson, O., et al., 2021. A new OECD definition for per- and polyfluoroalkyl substances. Environ. Sci. Technol. 55, 15575–15578.

Wang, M., Chen, J., Lin, K., Chen, Y., Hu, W., Tanguay, R.L., 2011. Chronic zebrafish PFOS exposure alters sex ratio and maternal related effects in F1 offspring. Environ. Toxicol. Chem. 30, 2073–2080.

Wang, M.M., Chen, H.H., Zha, J.M., Wang, Z.J., 2015. Ecotoxicity of alternatives of typical perfluorooctane sulfonate (PFOS) to four native aquatic organisms. Asian J. Ecotoxicol. 10, 230–235.

Wang, X.Y., Dai, Y.Y., Jia, L., Wang, Q.S., Chen, C.X., 2020. Effects of perfluorooctane sulfonate (PFOS) on immune-related gene expression in half-smooth tongue sole (*Cynoglossussemilaevis*). Asian J. Ecotoxicol. 15, 71–80.

Wei, Y., He, H., Han, T., Wang, B., Ji, P., Wu, X., et al., 2023. Environmental explanation of prostate cancer progression based on the comprehensive analysis of perfluorinated compounds. Ecotoxicol. Environ. Saf. 263, 115267.

Wen, W., Xia, X., Hu, D., Zhou, D., Wang, H., Zhai, Y., et al., 2017. Long-chain perfluoroalkyl acids (PFAAs) affect the bioconcentration and tissue distribution of short-chain PFAAs in zebrafish (*Danio rerio*). Environ. Sci. Technol. 51, 12358–12368.

Wen, W., Xia, X.H., Zhou, D., Wang, H.T., Zhai, Y.W., Lin, H., et al., 2019. Bioconcentration and tissue distribution of shorter and longer chain perfluoroalkyl acids (PFAAs) in zebrafish (*Danio rerio*): effects of perfluorinated carbon chain length and zebrafish protein content. Environ. Pollut. 249, 277–285.

Wu, J.Y., Liu, W.X., He, W., Xu, F.L., 2019. Comparisons of tissue distributions and health risks of perfluoroalkyl acids (PFAAs) in two fish species with different trophic levels from Lake Chaohu. China. Ecotoxicol. Environ. Saf. 185, 109666.

Wu, M., Sun, R., Wang, M., Liang, H., Ma, S., Han, T., et al., 2017. Analysis of perfluorinated compounds in human serum from the general population in Shanghai by liquid chromatography-tandem mass spectrometry (LC-MS/MS). Chemosphere 168, 100–105.

Xia, J.G.J.G., Niu, C.J., Sun, L.Y., 2013a. Ecotoxicological effects of exposure to PFOS on embryo and larva of zebrafish *Danio rerio*. Acta Ecol. Sin. 33, 7408–7416.

Xia, J.J., Fu, S., Cao, Z., Peng, J., Peng, J., Dai, T., 2013b. Ecotoxicological effects of waterborne PFOS exposure on swimming performance and energy expenditure in juvenile goldfish (*Carassius auratus*). J. Environ. Sci. 25, 1672–1679.

Xiao, W., Jian, G., Ma, F., Li, H., Yang, X., Zhang, H., et al., 2025. Exposure to specific polyfluoroalkyl chemicals is associated with cardiovascular disease in US adults: a population-based study. Front. Cardiovasc. Med. 11, 1487956.

Yao, D., Wu, H., Jiang, M., 2018. Effects of perfluorooctane sulfonate on the liver metabolism of *Danio rerio*. Asian J. Ecotoxicol. 13, 97–106.

Zhang, L., Li, Y., Chen, T., Xia, W., Zhou, Y., Wan, Y., 2011a. Abnormal development of motor neurons in perfluorooctane sulphonate exposed zebrafish embryos. Ecotoxicology 20, 643–652.

Zhang, T., Sun, H., Lin, Y., Wang, L., Zhang, X., Liu, Y., et al., 2011b. Perfluorinated compounds in human blood, water, edible freshwater fish, and seafood in China: daily intake and regional differences in human exposures. J. Agric. Food Chem. 59, 11168–11176.

Zhang, X., Chen, L., Fei, X.C., Ma, Y.S., Gao, H.W., 2009. Binding of PFOS to serum albumin and DNA: insight into the molecular toxicity of perfluorochemicals. BMC Mol. Bio. 10, 16.

Zhang, X., Xue, L., Deji, Z., Wang, X., Liu, P., Lu, J., et al., 2022. Effects of exposure to per-and polyfluoroalkyl substances on vaccine antibodies: a systematic review and meta-analysis based on epidemiological studies. Environ. Pollut. 306, 119442.

Zhang, X., Zhao, L., Ducatman, A., Deng, C., Von Stackelberg, K.E., Danford, C.J., et al., 2023. Association of per-and polyfluoroalkyl substance exposure with fatty liver disease risk in US adults. JHEP Rep. 5, 100694.

Zhao, M., Yin, N., Yang, R., Li, S., Zhang, S., Faiola, F., 2024. Assessment and Comparison of Early Developmental Toxicity of Six Per- and Polyfluoroalkyl Substances with Human Embryonic Stem Cell Models. Environ. Sci. Technol. 58, 8215–8227.

Zhao, X., Ren, X., Duan, X., Luo, Z., Zhu, R., 2016. On the effect of PFOS on developmental toxicity and oxidative stress in zebrafish embryos. J. Tangshan Univ. 29, 12–16.

Zhao, Y.G., Wan, H.T., Law, A.Y.S., Wei, X., Huang, Y.Q., Giesy, J.P., et al., 2011. Risk assessment for human consumption of perfluorinated compound-contaminated freshwater and marine fish from Hong Kong and Xiamen. Chemosphere 85, 277–283.

Zheng, J., Liu, S., Yang, J., Zheng, S., Sun, B., 2024. Per-and polyfluoroalkyl substances (PFAS) and cancer: detection methodologies, epidemiological insights, potential carcinogenic mechanisms, and future perspectives. Sci. Total Environ. 953, 176158.

Zheng, X.M., Liu, H.L., Shi, W., Wei, S., Giesy, J.P., Yu, H.X., 2012. Effects of perfluorinated compounds on development of zebrafish embryos. Environ. Sci. Pollut. Res. 19, 2498–2505.

CHAPTER

13

Effects of climate change on nucleic acid composition

Chandan Malick
Department of Zoology, Kalna College, Kalna, West Bengal, India

13.1 Introduction

Investigating the impact of climate change on nucleic acid composition across all domains of life has emerged as a crucial area of research due to its profound implications for organismal adaptation and ecosystem stability. Over recent decades, research has shed light on how environmental factors, such as temperature fluctuations and ocean acidification, affect molecular structures and functions, with early studies highlighting thermal adaptation in microbial genomes (Goncearenco et al., 2014; Huete-Stauffer et al., 2016). Recent research has expanded this understanding to include epigenetic alterations and transcriptome responses in plants, animals, and marine microbes (Metzger and Schulte, 2017; Wan et al., 2023; Jin et al., 2024). The swift advancement of global warming and its impact on genetic and epigenetic mechanisms underscore the essentiality of this research, evidenced by changes in gene expression, DNA methylation patterns, and nucleic acid stability that affect biodiversity and evolutionary trajectories (McGaughran et al., 2021; Skinner, 2022). Notwithstanding progress, significant knowledge gaps persist regarding the integrative impacts of climate change on nucleic acid chemistry and functionality across several taxa. Thermal adaptation has been linked to differences in codon use and GC content in bacteria and archaea (Goncearenco et al., 2014; Liu et al., 2020). The extent to which these molecular modifications provide functional resistance or vulnerability is still debated (Huete-Stauffer et al., 2016; Thomas et al., 2022). The role of epigenetic regulation, including DNA methylation and small RNA pathways, in promoting rapid phenotypic plasticity and transgenerational inheritance in response to climate stress is still insufficiently understood (Metzger and Schulte, 2017; Skinner, 2022; Sammarco et al., 2023). Conflicting evidence indicates that epigenetic alterations may either facilitate adaptability or lead to maladaptive responses, highlighting the complexity of these systems (Houri-Zeevi et al., 2019; Brennan et al., 2024). The consequences of these shortcomings are substantial, including limiting prediction models of species survival and ecosystem responses to ongoing environmental change (Dellaert and Putnam, 2023). This chapter utilizes a conceptual framework that combines nucleic

Genome Damage: Repair and Adaptation Across Life.
DOI: https://doi.org/10.1016/B978-0-443-42857-9.00010-4

acid structural dynamics, epigenetic regulation, and evolutionary adaptation to elucidate the impact of climate-induced environmental stresses on nucleic acid composition and function. Essential concepts include the thermodynamic sensitivity of RNA and DNA structures (Thomas et al., 2022; Rissone et al., 2024), the interplay between genetic and epigenetic modifications, and the molecular basis of phenotypic plasticity and stress memory (Srikant and Drost, 2021). This method facilitates a comprehensive understanding of the molecular underpinnings of climate resilience and vulnerability across several life domains.

13.2 Temperature effect on nucleic acid structure and stability

13.2.1 Molecular dynamics and thermal stability

Temperature is the most direct and well-researched climatic variable that alters nucleic acid content. Experimental and computational studies have consistently demonstrated that increased temperatures enhance DNA flexibility, leading to a more compact conformation of double-stranded DNA chains. Molecular dynamics simulations indicate that the persistence length of DNA diminishes linearly with rising temperature. Standard values range from −0.21 to −0.28 nm/K (Chen and Sari 2023; Xie 2024).

The thermodynamic stability of nucleic acids is primarily determined by the balance between enthalpy and entropy (Olivares and Capote 2022). As temperature increases, the entropic input to free energy gains significance, rendering unfolded or partially melted states more probable. This relationship is particularly pertinent for RNA molecules due to their more intricate secondary and tertiary structures compared to DNA (Vieregg, 2015). Investigations employing genomic and folding models have elucidated the sequence and compositional determinants influencing temperature adaptation, such as codon bias and GC content, which impact nucleic acid stability in Archaea and bacteria (Table 13.1; Goncearenco et al., 2014; Liu et al., 2020; Dohnalová et al., 2024). Experimental techniques, including calorimetric force spectroscopy, have revealed widespread RNA phase transitions at low temperatures, clarifying the molecular characteristics of RNA folding dynamics under heat stress (Rissone et al., 2024). The correlation between GC content and thermal adaptation remains contentious, since several studies reveal conflicting patterns across taxa (Liu et al., 2020). Many studies focus on prokaryotes or model organisms, limiting their relevance to eukaryotes. Furthermore, the complexities of in vivo nucleic acid dynamics in response to fluctuating environmental factors are insufficiently depicted by in vitro or computational models (Goncearenco et al., 2014; Rissone et al., 2024).

13.2.2 RNA dynamics dependent on temperature

RNA molecules demonstrate considerable temperature sensitivity due to their single-stranded structure and complex folding patterns. Investigations using RNA thermometers in bacteria have shown that minor temperature variations (5°C–10°C) can elicit substantial structural alterations that affect gene expression. In *Escherichia coli*, cold shock proteins increase in prevalence at 10°C due to the temperature's impact on RNA structure, which diminishes translation efficiency (Punetha et al., 2014).

As the temperature increases, the half-life of mRNA molecules significantly decreases. The median mRNA half-life in budding yeast diminishes from 3.4 min at 20°C to 1.5 min at 42°C.

TABLE 13.1 Thermodynamic influence on nucleic acid structural stability.

Macromolecule type	Bending persistence length (nm)	Stretch modulus (pN)	Twist rigidity (nm)	Thermal dependence slope (nm/K)	Amino acid bias predictor	Source
dsDNA	50	1490 ± 12	109 ± 1	−0.21 to −0.28	IVYWREL	Dohnalová et al. (2024)
dsRNA	60 to 81	657 ± 4	123 ± 1	−0.41 ± 0.02	IVYWREL	
DNA–RNA hybrid	78 ± 0	790 ± 2	123 ± 1	−0.42 ± 0.06	Not in source	

The temperature dependence of RNA stability has considerable implications for gene expression regulation under climate stress conditions across different organisms (Table 13.2; Neupert et al., 2008; Loh et al., 2018; Pelechano et al., 2022).

13.2.3 Impacts on DNA repair mechanisms

Variations in temperature significantly influence DNA repair pathways, potentially compromising genomic integrity in a fluctuating climate. Hyperthermia (temperatures over 40°C) has been shown to hinder many DNA repair pathways, including base excision repair, nucleotide excision repair, and homologous recombination (Jiricny, 2015; Kanaar, 2018). Temperature significantly influences the functionality of DNA repair proteins, particularly DNA polymerases and ligases. Frigid temperatures may hinder DNA repairs, whereas excessive heat can alter the configuration of repair enzymes. This situation restricts organisms from efficiently preserving their DNA within a narrow temperature range, rendering them susceptible to fluctuations in temperature due to meteorological conditions (Dillon, 2014).

13.3 Ocean acidification and nucleic acid biochemistry

13.3.1 Effects on marine organisms

Ocean acidification, resulting from increased atmospheric CO_2 absorption, poses a significant threat to marine ecosystems (Doney et al., 2009). The chemical composition of saltwater directly influences the cellular environment of marine organisms, potentially altering the stability and functionality of nucleic acids.

Corals, mollusks, and echinoderms are examples of calcifying creatures that are particularly susceptible to acidification (Table 13.3; Taylor et al., 2011; Strader et al., 2020). These organisms demonstrate reduced survival, calcification rates, and growth when exposed to projected future ocean pH levels (Riebesell et al., 2000; Kennedy, 2020). The direct impacts on nucleic acids are not well researched; however, the metabolic strain of maintaining cellular pH equilibrium likely influences DNA and RNA metabolism (Philk, 2008).

13.3.2 Cellular pH and nucleic acid activity

The stability of nucleic acid structures depends on pH, as acidic conditions often facilitate the protonation of nucleotide bases and may alter base-pairing interactions (Speck and Ames, 2015). Marine bacteria must regulate internal pH while acclimating to more acidic exterior environments. This combined stress may impact DNA replication, transcription, and translation.

13.4 Ultraviolet radiation and DNA impairment

13.4.1 Increased ultraviolet exposure

Climate change, particularly the degradation of the ozone layer, has resulted in increased ultraviolet-B (UV-B) radiation reaching the Earth's surface (Bais et al., 2022). Ultraviolet exposure

TABLE 13.2 RNA dynamics dependent on temperature across different organisms.

Organism	Regulated gene	Gene function	RNAT type/structure	Switch temperature (°C)	Mechanism of action	Half-life change (inferred)	Source
Listeria monocytogenes	*prfA*	Transcriptional regulator; activates major virulence factors	Single stem-loop structure	37	Occludes the Ribosome Binding Site (RBS) at low temperatures; melts at 37°C to permit ribosome binding	Not in source	Loh et al. (2018); Neupert et al. (2008)
Vibrio cholerae	*toxT*	Transcriptional regulator; activates toxin-coregulated pilus and cholera toxin	FourU element; single stem-loop	37	Pairs with AGAG of the Shine-Dalgarno sequence; melts at 37°C to allow ribosome binding	Not in source	
Yersinia pseudo-tuberculosis	*lcrF*	Transcriptional regulator; activates T3SS and virulence factors	Two hairpin structures; second contains FourU element	37	Occludes Shine-Dalgarno region in the intergenic region; impairs translation at 25°C	A decrease in mRNA half-life is observed at higher temperatures.	
Salmonella enterica	*agsA*	Small heat shock protein; aggregation-suppressing A	FourU element	45	Base pairs with Shine-Dalgarno (SD) sequence at low temperatures; unpaired above 45°C	Not in source	
Neisseria meningitidis	*cssA*	UDP-N-acetylglucosamine 2-epimerase; capsule biosynthesis	Single stem-loop; 8-bp tandem repeats	37	Occludes RBS at nasopharyngeal temperatures (32°C–34°C); RBS is accessible at 37°C	Not in source	

(*continued on next page*)

TABLE 13.2 RNA dynamics dependent on temperature across different organisms—cont'd

Organism	Regulated gene	Gene function	RNAT type/structure	Switch temperature (°C)	Mechanism of action	Half-life change (inferred)	Source
Leptospira interrogans	*ligA/ligB*	Surface-exposed lipoprotein: adhesion and immune evasion	Two stem-loops; the second loop occludes RBS	37	Occludes RBS at lower temperatures; temperature-dependent translational control	Not in source	
Pseudomonas aeruginosa	*rhlR*	Transcriptional regulator; quorum-sensing master regulator	ROSE element	37	Inhibits translation at 30°C, leading to premature transcription termination of the operon	Not in source	
Escherichia coli	*rpoH*	Sigma-32; heat shock master regulator	Complex secondary structure (>200 nt)	42	Secondary structure between regions A and B represses translation at low temperatures	Not in source	
Salmonella enterica	*htrA*	Protease; protein quality control and stress response	FourU element; single hairpin	37	Occludes RBS at low temperatures; cooperative unfolding at the common transition temperature	Not in source	
Escherichia coli	U6 (Synthetic)	GFP or lacZ reporter	Single stem-loop structure; large loop	37	Masks the ribosome binding site at low temperatures (17°C–30°C); melts efficiently at 37°C	No significant temperature-dependent change in RNA accumulation	

TABLE 13.3 Effect of ocean acidification and nucleic acid biochemistry across different organisms.

Species/taxon	Life-history stage	Stressor (e.g., pH, pCO2, temperature)	Measured phenotype (e.g., growth, respiration, calcification)	Gene expression response	Key molecular pathways (inferred)	Metabolic/energetic effect (inferred)	Source
Strongylocentrotus purpuratus	Larva (pluteus)	*pCO2* 1264–1271 μatm	Growth, Respiration, Feeding rate	Upregulation of metabolic genes and *Na+/K+*-ATPase in feeding larvae	Ion transport (ATP-driven), Tricarboxylic acid (TCA) cycle, cellular stress response	Metabolic depression/delay; reduced scope for growth (39%–45% available energy vs 78%–80% in control)	Strader et al. (2020); Tayloret al. (2011)
Crassostrea hongkongensis	Larva (pediveliger)	pH 7.4	Metamorphosis rate, substrate selection	Differential DNA methylation of 130 genes (exon regions)	Cytoskeletal and signal transduction, oxidative stress, metabolic processes	Higher metamorphosis success but poor habitat selection ability; epigenetic adaptive plasticity	
Acropora millepora	Juvenile	pCO2 750–1000 ppm	Calcification, acclimation	Upregulation of heat shock proteins (HSPs) and antiapoptotic genes	Cellular stress response (CSR), skeletal organic matrix protein synthesis	Trade-off favoring tissue biomass and survival over skeletal growth	
Mytilus edulis	Adult	pCO2 1120–2400 μatm	Respiration (oxygen consumption), calcification	Downregulation of 2 subunits of ATP-synthase in mantle tissue	ATP production, oxidative phosphorylation	Increased metabolic costs for homeostasis despite localized gene downregulation	

(*continued on next page*)

TABLE 13.3 Effect of ocean acidification and nucleic acid biochemistry across different organisms—cont'd

Species/taxon	Life-history stage	Stressor (e.g., pH, pCO2, temperature)	Measured phenotype (e.g., growth, respiration, calcification)	Gene expression response	Key molecular pathways (inferred)	Metabolic/energetic effect (inferred)	Source
Heliconoides inflatus	Adult	pCO2 617–720 μatm	Calcification, nervous system function	Upregulation of all annotated calcification genes (matrix proteins, mucins)	Biomineralization, neural ion transport (GABAergic, glutamatergic)	Metabolic strain from increased energetic investment in shell maintenance	
Amphiurafiliformis	Adult	pH 7.0–7.3	Resting metabolic rate, arm regeneration	Downregulation of lactate dehydrogenase and glucose 6-phosphate dehydrogenase	Metabolic pathways, anaerobic metabolism	Metabolic depression: reduced capacity for regeneration and growth	
Coccolithuspelagicus ssp. *braarudii*	Unicellular alga (adult stage)	*pHo* 6.5–7.0	Calcification rate, intracellular pH homeostasis	Expression of HVCN1 (*H+* channel)	Ion transport (voltage-gated *H+* efflux), pH homeostasis	Maintenance of calcification via rapid *H+* removal to prevent cytosolic acidosis	

directly harms DNA by forming cyclobutane pyrimidine dimers (CPDs) and 6,4-photoproducts (Setlow, 2010; Cadet and Douki, 2018).

Between 50°N and 50°S latitude, DNA-damaging UV radiation is projected to diminish until 2050, followed by an increase due to climate change impacts on atmospheric ozone and cloud cover. This pattern of alteration induces UV-related genomic stress to vary throughout time and space (Bais et al., 2022).

13.4.2 DNA repair and ultraviolet adaptation

Organisms, including humans, have evolved sophisticated DNA repair mechanisms to address damage inflicted by ultraviolet radiation (Table 13.4; Kciuk et al., 2020). However, when temperature and UV exposure are combined, it creates complex scenarios in which repair effectiveness may be diminished. Low temperatures can impede DNA repair, whereas high temperatures can alter the conformation of repair proteins (Jian et al., 2022).

Research on amphibian larvae demonstrates that different species display unique responses to DNA damage caused by ultraviolet radiation, with temperature significantly influencing the effectiveness of damage repair. Certain species have reduced DNA damage at lower temperatures, coupled with impaired repair rates, highlighting the complex interactions among many climate stressors (Jian et al., 2022).

13.5 Epigenetic modification and climate response

13.5.1 Patterns of DNA methylation

DNA methylation is a crucial epigenetic mechanism that organisms employ to respond to environmental stressors. Factors associated with climate change, including temperature, dryness, and light intensity, significantly influence global DNA methylation patterns (Mohammad et al., 2023).

Investigations into mountain grasses (*Festuca* species) reveal that climatic stresses, including aridity, heightened temperatures, and augmented isolation, raise global DNA methylation levels. The magnitude of these modifications depends on species ploidy levels and climatic origin, suggesting that epigenetic responses are evolutionarily tailored to local climatic conditions (Kusal et al., 2017; Pál et al., 2022).

13.5.2 Temperature-sensitive methylation

Research on human populations demonstrates that environmental temperature affects DNA methylation patterns. Alterations in the methylation of specific genes crucial for cardiovascular function and stress response are associated with low temperatures. These epigenetic modifications may facilitate organismal adaptation to climatic variations and could be transmitted between generations (Breton et al., 2020; Yang and Schwartz, 2020).

13.5.3 Epigenetic regulation mediated by RNA

Noncoding RNAs play a crucial role in memory and adaptation to climatic stressors in organisms. MicroRNAs (miRNAs), small interfering RNAs (siRNAs), and long noncoding RNAs (lncRNAs) contribute to the epigenetic regulation of genes in humans as well as other

TABLE 13.4 Ultraviolet radiation and DNA impairment across different organisms.

Species	Radiation wavelength (nm)	DNA/RNA damage type	Repair mechanism	Environmental factors (e.g., temperature)	Mutation effect	Genomic stress level (inferred)	Source
Homo sapiens (humans)	UV-A (315–400 nm) and UV-B (280–315 nm)	CPDs, 6-4PPs, ROS-induced strand breaks, and bulky adducts	Nucleotide excision repair (NER), base excision repair (BER)	High temperatures can alter the conformation of repair proteins)	Basal cell carcinoma, squamous cell carcinoma, malignant melanoma, and aging	Critical (lack of photolyase makes humans solely reliant on NER/BER)	Kciuk et al. (2020)
Limnodynastesperonii (tadpoles)	UV-B (280–315 nm)	Cyclobutane pyrimidine dimers (CPDs), 6,4-photoproducts (6-4PPs)	Photolyase	20°C and 30°C (repair twice as fast at 30°C)	Block DNA replication, mutations, apoptosis, and developmental deformities	High (significant accumulation of damage at low temperatures)	
Escherichia coli (bacteria)	UV-C (< 280 nm), UV-B (280-315 nm)	Dipyrimidine DNA lesions, CPDs	UvrABC system (NER), photolyase, translesion synthesis (TLS)	Not in source	Inhibition of replication, SOS response activation	High (if SOS error-prone repair is activated)	
Schizosaccharomyces pombe (yeast)	UV (generic)	CPDs, 6-4PPs	UVDE-mediated repair (UVER), mitochondria repair	Not in source	Lethal hits (1–2 unrepaired dimers lethal in mutants)	High (rapid accumulation of nicks during rapid repair)	

organisms that respond to stress (Table 13.5; Li and Niu, 2023; Sánchez-Quintero and Cubillos, 2024).

Heat stress enhances miRNA-mediated gene silencing without elevating miRNA expression levels. This indicates that temperature affects the effectiveness of RNA interference mechanisms, providing a means for rapid transcriptional adjustment to thermal stress (Wang et al., 2014).

13.6 Adaptations of extremophiles and evolutionary perspectives

13.6.1 Archaeal heat shock responses

Hyperthermophilic archaea provide valuable insights into the adaptations of nucleic acids to extreme heat environments. These species possess unique systems for heat shock proteins that significantly differ from those of bacteria and eukaryotes (Macario et al., 1999).

Hyperthermophilic archaea lack some typical heat shock proteins (Hsp70, Hsp40, and GrpE), yet possess unique chaperonins known as thermosomes. These changes facilitate survival at temperatures exceeding 100°C, demonstrating the evolutionary potential for nucleic acid stability under extreme thermal stress (Oshima, 2010; Wang et al., 2022).

13.6.2 DNA repair in extremophiles

Microorganisms in Antarctica have evolved superior DNA repair mechanisms to withstand various stressors, including UV radiation, desiccation, and freeze-thaw cycles. Metagenomic investigations reveal the presence of numerous distinct DNA repair enzymes that function effectively at subzero temperatures. ATP-dependent ligases are involved in repair actions during the stationary phase (Bueno et al., 2023).

13.7 Responses of bacteria and fungi to climatic stressors

13.7.1 Repair of bacterial DNA under thermal stress

Bacteria frequently replicate genes and enhance their DNA repair mechanisms in response to temperature stress. Investigations on *Escherichia coli* acclimatized to heightened temperatures (41.5°C) reveal chromosomal duplications that include genes associated with protein repair and survival in the stationary phase (Koboldt, 2021).

The DNA repair mechanisms in bacteria are highly robust. Certain bacteria can repair double-strand fractures at temperatures below freezing (−15°C) (Imashimizu et al., 2014). This illustrates the need to maintain genomic integrity under varying climatic conditions (Table 13.6; Lee, 2022).

13.7.2 Fungal thermal tolerance

Fungi exhibit varied responses to thermal stress, and proteome analyses have revealed that heat shock proteins, metabolic enzymes, and RNA-binding proteins undergo intricate alterations (Kim et al., 2020). Alterations in fungal proteomes due to temperature fluctuations indicate that stress responses are significantly regulated posttranscriptionally (Vilcinskas, 2020).

TABLE 13.5 Epigenetic modification and climate response.

Organism	Stressor	Epigenetic mechanism	Key proteins/factors	Target genes/sequences	Observed biological outcome	Inheritance type	Source
Mammalian cells (HeLa)	Mild hyperthermia (40°C for 12 h)	miRNA-mediated gene silencing	Endogenous microRNAs (miRNAs), AGO2, DICER	*COL4A2, DICER1, LAMC1, DNMT1, GAPDH*	Enhanced gene silencing activity; downregulation of targeted endogenous genes to inhibit expression of heat-denatured proteins	Not in source	Skinner (2018), Xu et al. (2020)
Humans	Preconception cold exposure	DNA methylation	Not in source	Sperm DNA	Altered brown fat characteristics and metabolic phenotypes; improved adaptation to overnutrition and hypothermia in offspring	Paternally transmitted; intergenerational inheritance	
Humans	Ambient temperature/cold	DNA methylation	Not in source	*ZKSCAN4, ICAM-1, CRAT, F3, TLR-2, iNOS*	Short-term cold affects global DNA methylation; methylation on *ZKSCAN4* and *ICAM-1* mediates high blood pressure and cardiovascular risk	Not in source	

(*continued on next page*)

Dactylorhizamajalis / *D. traunsteineri*	Water availability and temperature	DNA (cytosine) methylation	MET1, CMT3, DRM methyltransferases	Epiloci/restriction sites (5′-CCGG)	Species-specific epigenetic patterns; stable epigenetic divergence responsible for persistent ecological differences and adaptation	Stable inheritance across generations; meiotic persistence
Festuca ovina (wild grass)	Water stress	Transcriptional regulation and gene expression modulation	Transcription factors	117 *F. ovina* homologs of drought-responsive plant genes	Nonlinear transcriptomic changes and linear declines in physiological measurements; natural drought tolerance	Not in source

TABLE 13.6 Responses of bacteria and fungi to climatic stressors.

Species	Climatic stressor	Biological response mechanism	Genetic or molecular components	Impact on survival or virulence	Ecosystem or host context	Source
Listeria monocytogenes	Host temperature (37°C)	Thermosensing via RNA structure melting	*prfA* RNA thermometer (RNAT)	Triggers translation of the master regulator for virulence genes	Human host context	Samtani et al. (2022); Abu Bakar et al. (2020); Dieser et al.(2013); Fang and St Leger (2010); Hessenauer et al. (2025)
Aspergillus fumigatus	Heat stress (supra-optimal temperatures)	Increased production of heat-shock proteins and chaperones to minimize protein misfolding	Hsf1, HSP70, HSP88	Enhances thermotolerant characteristics responsible for pathogenicity	Human host tissues	
Penicillium marneffei	High temperature (37°C)	Phase transition (dimorphism) and upregulation of virulence factors	Hsp30, Hsp70, RanA, HSP60	Facilitates cell-host adherence and transition to pathogenic yeast form	Human host tissues	
Psychrobacter arcticus 273-4	Permafrost/low temperature (−15°C)	Active DNA repair (double-strand break repair)	RecA, homologous recombination proteins	Maintains genome integrity during dormancy, extending microbial longevity in frozen matrices	Siberian permafrost	

(*continued on next page*)

Metarhizium anisopliae	Cold stress / Chilling	RNA binding and destabilization of RNA secondary structure to facilitate translation	CRP1 (cold shock protein), CRP2 (glycine-rich protein)	Increased conidial germination rates and tolerance to freezing; CRP1 increases virulence to caterpillars	Entomopathogenic (insect host context)
Bacillus subtilis	Heat shock	Activation of heat shock proteins and alternative sigma factors	HrcA, GroE, sigma B (σB)	Protects vegetative cells and coordinates sporulation for survival	Soil/general environment
Synechocystis	Cold stress (drop from 32°C to 22°C)	Membrane remodeling to regain fluidity	Acyl-lipid desaturases (*desA*, *desB*, *desD*), Hik33	Maintains membrane integrity and metabolic function in cold conditions	Aquatic environments
Ophiostoma novo-ulmi	Humidity and precipitation	Genomic adaptation and transport protein remodeling	Checkpoint kinase 1, EHBP1 protein	Affects the spread and colonization in forest ecosystems	Forests (Dutch elm disease)
Mrakiapsychrophila	Temperature fluctuation (4°C vs 20°C)	Energy metabolism modification and unfolded protein response (UPR)	DNAJ, MFS transporters	Adaptation of obligate psychrophiles to sub-optimal temperatures	Cold environments

Pathogenic fungi demonstrate heightened virulence at increasing temperatures, alongside structural alterations in nucleic acid-binding proteins that may enhance adaptability to host environments (Brown and Denning, 2013). These alterations highlight the ability of climate change to influence pathogen–host interactions through nucleic acid-mediated mechanisms (Table 13.6).

13.8 Stability of RNA and memory of stress

13.8.1 RNA turnover depends on temperature

RNA stability is highly contingent upon temperature across diverse organisms. Elevated temperatures accelerate RNA degradation and reduce the half-life of biological transcripts. Conversely, low temperatures can enhance RNA stability by reducing molecular mobility and strengthening weak interactions. Studies on plants indicate that RNA stability mechanisms play a crucial role in the formation of stress memory. Alternative splicing of pre-mRNAs, regulated by temperature-sensitive RNA-binding proteins, contributes to adaptive responses and transgenerational stress inheritance (Li et al., 2022; Chan and Lowe, 2023).

13.8.2 Ribozyme function and climate

Ribozymes, RNA molecules capable of catalysis, serve as effective models for elucidating the impact of temperature on RNA functionality. Investigations into thermostable ribozymes reveal that mutations that improve thermal stability are often context-dependent and may require multiple coordinated modifications (Miyamoto and Clark, 2001).

Synthetic ribozyme systems show that RNA structures that change with temperature can work as molecular thermometers, controlling gene expression as temperature changes (Saleski and Joyce, 1991; Martick and Scott, 2004). These findings suggest potential applications in biotechnology and synthetic biology for climate change adaptation.

13.9 Mitochondrial DNA and climatic adaptation

13.9.1 Variations at the population level

Human mitochondrial DNA (mtDNA) exhibits specific markers of climate adaptation, with populations in colder environments showing diminished mitochondrial diversity and specific mutations associated with temperature adaptation. Two significant alterations in the ND3 and ATP6 genes are intricately associated with environmental temperature and may affect mitochondrial energy utilization efficiency (Balloux et al., 2009; Ruiz-Pesini, 2004).

13.9.2 Temperature-triggered mitochondrial DNA damage

The ambient temperature directly influences the integrity of mitochondrial DNA in the human population. For instance, elderly individuals subjected to elevated temperatures during the summer exhibit a greater number of mtDNA lesions (Coskun et al., 2017). The correlation between temperature fluctuations and mitochondrial damage indicates that climate change may accelerate cellular aging (Table 13.7; Wallace, 2003).

TABLE 13.7 Mitochondrial DNA and climatic adaptation.

Organism or population	Geographic region	Climatic conditions	mtDNA variable (e.g., Content, Haplogroup, SNP)	Biological/ functional adaptation effect	Health or aging impact	Adaptation Type (Inferred)	Source
Human (Arctic populations)	Arctic	Extreme cold environments	SNP 8701A (ATP6) and 10398A (ND3)	Altered mitochondrial pH and calcium dynamics are associated with metabolic thermogenesis via uncoupled respiration	Linked to longevity and modulated risk for neurodegenerative and metabolic disorders	Long-term evolutionary selection	Balloux et al. (2009); Wallace (2007); Cheng et al. (2013); Friedrich et al. (2022); Peng et al. (2017); Haak et al. (2009)
Human (Haplogroup J)	Europe	Temperate to cold climates	Haplogroup J (polymorphisms at *Qi* and *Qo* sites of Complex III)	Reduced coupling efficiency (one-third decrease); lower ATP production; higher metabolic heat generation	Slowed aging process (enriched in centenarians); protective against Parkinson's; increased risk for LHON penetrance	Long-term evolutionary selection	
Human (Northern populations)	Northern China	Colder northern climates	Higher mtDNA copy number (content)	Increased expression of energy metabolism enzymes; accelerated thermogenesis	Not in source	Long-term evolutionary selection	
Human (Siberian and Central Asian lineages)	Siberia and Central Asia	Arctic and Subarctic cold	Sequence variations in the ATP6 gene	Increased uncoupling phosphorylation and metabolic heat production	Not in source	Long-term evolutionary selection	

(continued on next page)

TABLE 13.7 Mitochondrial DNA and climatic adaptation—cont'd

Organism or population	Geographic region	Climatic conditions	mtDNA variable (e.g., Content, Haplogroup, SNP)	Biological/ functional adaptation effect	Health or aging impact	Adaptation Type (Inferred)	Source
Human (Tibetans)	High altitude Tibet	High altitude hypoxia and cold climate	*dN/dS* ratios >1 in ATP8, CYTB, and ATP6	Protein components in the mitochondrial respiration chain are modified for adaptive selection	Not in source	Long-term evolutionary selection	
Human (Southern populations)	Southern China	Warmer southern climates	Lower mtDNA copy number (content)	Lower expression of energy metabolism enzymes; reduced thermogenesis	Not in source	Long-term evolutionary selection	
Human (Older Adults)	Phoenix, AZ, USA	Extreme heat and humidity (over 32∘C)	DNA methylation patterns (epigenetic markers)	Alteration in mitochondrial gene expression	Accelerated epigenetic aging (up to 14 months additional biological aging); increased risk for heart disease and mortality	Short-term physiological response	
Human (Older Adults)	Boston, MA, USA	Elevated air temperature (Summer)	mtDNA lesions	Altered mitochondrial structural integrity	Increased pathophysiological responses; potential acceleration of cellular aging	Short-term physiological response	
Fischer 344 rats (Old)	Laboratory setting	Physiological heat stress (41°C)	Loss of cytochrome c; 4HNE-modified proteins	Blunted mitochondrial stress response (Hsp60/Hsp10); structural morphological damage (swelling, loss of cristae)	Increased vulnerability to oxidative damage; impaired repair mechanisms; potential stimulation of apoptotic processes	Short-term physiological response	

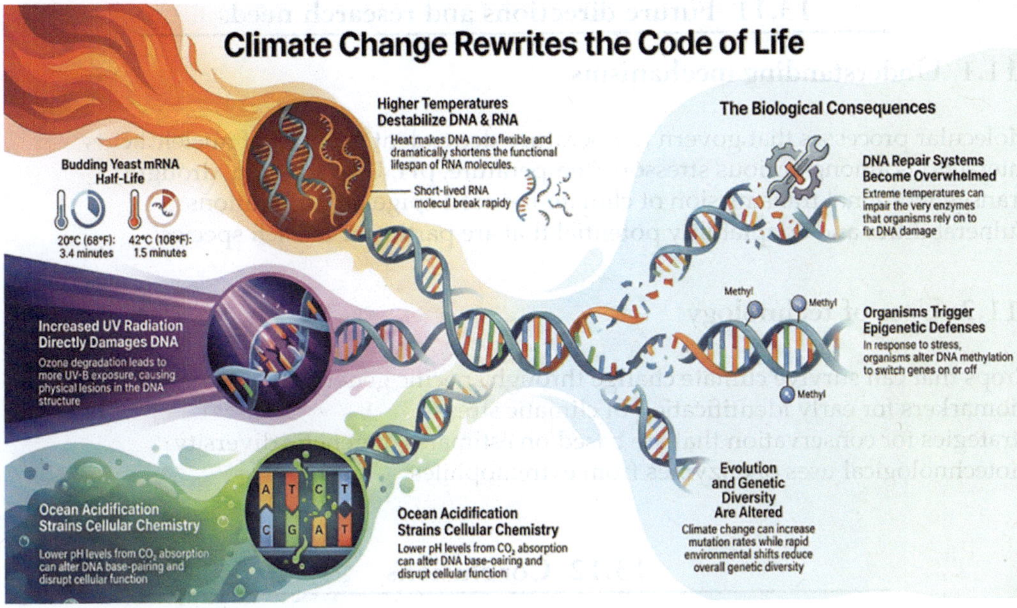

FIGURE 13.1 Effects of climate change on nucleic acid composition (google notebook lm-https://notebooklm.google.com).

13.10 Consequences for genetic variation and evolution

13.10.1 Alterations in distribution and genetic implications

Alterations in the distribution of flora and fauna resulting from climate change have intricate effects on genetic diversity. Mathematical models indicate that slow climate change preserves genetic diversity by spatial mixing of genetic elements, whereas sudden changes markedly reduce diversity (Excoffier and Ray, 2008).

Tree species may be particularly vulnerable to climate-induced loss of genetic diversity due to their prolonged generation times. Ecological niche modeling and genetic simulations suggest that climate change will lead to reduced genetic connectivity and diminished adaptive capacity in several forest species (De Lafontaine et al., 2019; Moreno-Villena and Alvarez-Ponce, 2021).

13.10.2 Mutation rates and natural selection

Rising temperatures are anticipated to intensify natural selection by amplifying the impact of DNA sequence variation on protein stability. Experimental studies indicate that elevated temperatures enhance mutation rates in plants, particularly indel mutations (insertions/deletions) (Fig. 13.1; Arnold, 2022).

The relationship between temperature and mutation rates suggests that climate change may accelerate evolutionary processes, potentially leading to increased genetic load and reduced population fitness (Arnold, 2022).

13.11 Future directions and research needs

13.11.1 Understanding mechanisms

- Molecular processes that govern temperature-dependent changes of nucleic acids
- Interactions among various stressors (temperature, pH, UV radiation, drought)
- Transgenerational transmission of climate-induced epigenetic alterations
- Vulnerabilities and adaptability potential that are particular to each species

13.11.2 Uses of technology

- Crops that can survive climate change through specific genetic changes
- Biomarkers for early identification of climatic stress
- Strategies for conservation that are based on estimates of genetic diversity
- Biotechnological uses of enzymes from extremophiles

13.12 Conclusions

Climate change significantly and intricately influences the composition of nucleic acids across all domains of life. These effects operate via multiple routes, including direct thermal impacts on molecular stability, altered DNA repair efficacy, epigenetic modifications, and changes in RNA-mediated gene regulation. The impacts encompass alterations in nucleotide structure at the molecular level and variations in genetic diversity and evolutionary potential at the population level.

The evidence analyzed herein demonstrates that nucleic acids serve as both targets and mediators of the effects of climate change. Certain organisms exhibit remarkable adaptability due to their evolved mechanisms for enhancing DNA repair and modulating gene expression in response to temperature fluctuations. Nonetheless, numerous species may lack the capacity to adapt swiftly enough to cope with climatic change. Understanding the relationships between nucleic acids and climate is important when forecasting ecosystem responses to climate change, developing conservation strategies, and potentially engineering species that can withstand climatic fluctuations. The complexity of these linkages makes it imperative to use interdisciplinary approaches that combine molecular biology, ecology, evolution, and climate studies to address one of the most pressing issues of our era. This chapter analyzes findings that provide a foundation for future research on climate-nucleic acid interactions and highlight the essential function of genetic pathways in determining species survival under changing environmental conditions. As climate change accelerates, comprehending and protecting the molecular foundations of life is essential for preserving biodiversity and ecological functionality in our evolving environment.

References

Abu Bakar, N., Karsani, S.A., Alias, S.A., 2020. Fungal survival under temperature stress: a proteomic perspective. PeerJ. 8, e10423. https://doi.org/10.7717/peerj.10423.

Arnold, J., 2022. Climate change and genetic diversity. Number Analyt.. https://www.numberanalytics.com/blog/climate-change-genetic-diversity.

Bais, A.F., et al., 2022. Ozone depletion, UV radiation, and climate change: impacts on humans and the environment. Atmos. Chem. Phys. 22, 12827–12892. https://doi.org/10.5194/acp-22-12827-2022.

Balloux, F., Handley, L.J., Jombart, T., Liu, H., Manica, A., 2009. Climate shaped the worldwide distribution of human mitochondrial DNA sequence variation. Proc. R. Soc. B. Biol. Sci. 276 (1672), 3447–3455. https://doi.org/10.1098/rspb.2009.0752.

Brennan, R.S., deMayo, J.A., Finiguerra, M., Baumann, H., Dam, H.G., Pespeni, M.H., 2024. Epigenetic and evolutionary mechanisms uniquely contribute to rescue from global change. Biorxiv. https://doi.org/10.1101/2024.03.20.585843.

Breton, C.V., et al., 2020. Environmental temperature and DNA methylation. Epigenetics 15 (1–2), 56–67. https://doi.org/10.1080/15592294.2019.1706592.

Brown, G.D., Denning, D.W., 2013. The impact of temperature on fungal pathogenicity. Virulence 4 (3), 359–370. https://doi.org/10.4161/viru.23641.

Bueno, E., Jensen, O.N., Florin-Christensen, J., 2023. DNA repair capacity of Antarctic microorganisms. Front. Microbiol. 14, 1156817. https://doi.org/10.3389/fmicb.2023.1156817.

Cadet, J., Douki, T., 2018. Formation of UV-induced DNA damage in cells and its repair. Photochem. Photobiol. 94 (3), 291–302. https://doi.org/10.1111/php.12804.

Chan, P.P., Lowe, T.M., 2023. Temperature affects RNA degradation in living cells. J. Biol. Chem. 174 (9), 733–744. https://doi.org/10.1016/j.jbc.2023.100498.

Chen, G., Sari, R., 2023. Temperature dependence of DNA elasticity: an all-atom simulation study. J. Chem. Phys. 158 (9), 094902. https://doi.org/10.1063/5.0138373.

Cheng, Y.T., Liu, J., Yang, L.Q., Sun, C., Kong, Q.P., 2013. Mitochondrial DNA content contributes to climate adaptation using Chinese populations as a model. PLoS One 8 (11), e79536. https://doi.org/10.1371/journal.pone.0079536.

Coskun, P.E., et al., 2017. Environmental stress and mitochondrial DNA lesions in humans. Mutat. Res. 778, 24–30. https://doi.org/10.1016/j.mrfmmm.2016.12.002.

De Lafontaine, G., et al., 2019. Climate change and genetic diversity in trees. Mol. Ecol. Resour. 19 (2), 220–234. https://doi.org/10.1111/1755-0998.12973.

Dellaert, Z., Putnam, H.M., 2023. Reconciling the variability in the biological response of marine invertebrates to climate change. J. Exp. Biol. 226 (17), jeb245834. https://doi.org/10.1242/jeb.245834.

Dieser, M., Battista, J.R., Christner, B.C., 2013. DNA double-strand break repair at −15°C. Appl. Environ. Microbiol. 79 (24), 7662–7668. https://doi.org/10.1128/AEM.02845-13.

Dillon, L., 2014. DNA repair in bacteria and archaea under fluctuating temperatures. Open Biology 4, 140008. https://doi.org/10.1098/rsob.140008.

Dohnalová, H., Matoušková, E., Lankaš, F., 2024. Temperature-dependent elasticity of DNA, RNA, and hybrid double helices. Biophys. J. 123 (5), 572–583. https://doi.org/10.1016/j.bpj.2024.01.032.

Doney, S.C., Fabry, V.J., Feely, R.A., Kleypas, J.A., 2009. Ocean acidification: the other CO_2 problem. Ann. Rev. Marine Sci. 1, 169–192. https://doi.org/10.1146/annurev.marine.010908.163834.

Excoffier, L., Ray, N., 2008. Innovation, dispersal and the genetic structure of populations. Nat. Rev. Genet. 9 (3), 277–293. https://doi.org/10.1038/nrg2338.

Fang, W., St Leger, R.J., 2010. RNA binding proteins mediate the ability of a fungus to adapt to the cold. Environ. Microbiol. 12 (3), 810–820. https://doi.org/10.1111/j.1462-2920.2009.02127.x.

Friedrich, V.K., Rubel, M.A., Schurr, T.G., 2022. Mitochondrial genetic variation in human bioenergetics, adaptation, and adult disease. Am. J. Hum. Biol. 34 (2), e23629. https://doi.org/10.1002/ajhb.23629.

Goncearenco, A., Ma, B.G., Berezovsky, I.N., 2014. Molecular mechanisms of adaptation emerging from the physics and evolution of nucleic acids and proteins. Nucleic Acids Res. 42 (5), 2879–2892. https://doi.org/10.1093/nar/gkt1336.

Haak, J.L., Buettner, G.R., Spitz, D.R., Kregel, K.C., 2009. Aging augments mitochondrial susceptibility to heat stress. Am. J. Physiol. Regul. Integr. Comp. Physiol. 296 (3), R812–R820. https://doi.org/10.1152/ajpregu.90708.2008.

Hessenauer, P., et al., 2025. Genomic exploration of climate-driven evolution and evolutionary convergence in forest pathogens. Genome Biol. Evolut. 17 (5), evaf069. https://doi.org/10.1093/gbe/evaf069.

Houri-Zeevi, L., Teichman, G., Gingold, H., Rechavi, O., 2019. Stress resets transgenerational small RNA inheritance. BioRxiv. https://doi.org/10.1101/669051.

Huete-Stauffer, T.M., et al., 2016. Experimental warming decreases the average size and nucleic acid content of marine bacterial communities. Front. Microbiol. 7, 730. https://doi.org/10.3389/fmicb.2016.00730.

Imashimizu, M., Oshima, T., Nishimura, A., 2014. DNA repair at subzero temperatures in bacterial systems. Appl. Environ. Microbiol. 80 (6), 1937–1943. https://journals.asm.org/doi/full/10.1128/aem.02845-13.

Jian, Z., et al., 2022. DNA repair of UV-induced damage in amphibian larvae across temperature gradients. Biol. Lett. 18, 20220358. https://doi.org/10.1098/rsbl.2022.0358.

Jin, Q., et al., 2024. Epigenetic regulation for heat stress adaptation in plants: new horizons for crop improvement under climate change. Agronomy 14 (9), 2105. https://doi.org/10.3390/agronomy14092105.

Jiricny, J., 2015. Heat-induced DNA damage and repair. Mutat. Res./DNA Rep. 764, 45–62. https://doi.org/10.1016/j.mrfmmm.2014.11.003.

Kanaar, R., 2018. Hyperthermia and DNA Damage Repair Pathways. Springer, Germany. https://repub.eur.nl/pub/108211.

Kciuk, M., Marciniak, B., Mojzych, M., Kontek, R., 2020. Focus on UV-induced DNA damage and repair—disease relevance and protective strategies. Int. J. Mol. Sci. 21 (19), 7264. https://doi.org/10.3390/ijms21197264.

Kennedy, J., 2020. How does ocean acidification affect marine life? Natural History Museum. https://www.nhm.ac.uk/discover/quick-questions/how-does-ocean-acidification-affect-marine-life.html.

Kim, D., et al., 2020. Fungal proteomes and adaptation to thermal stress. Front. Microbiol. 11, 774. https://doi.org/10.3389/fmicb.2020.00774.

Koboldt, D.C., 2021. Genetic analysis of the adaptation to temperature stress: a role for gene duplications. The Society for Integrative and Comparative Biology, Herndon, VA. https://sicb.org/abstracts/genetic-analysis-of-the-adaptation-to-temperature-stress-a-role-for-gene-duplications/.

Kusal, K., et al., 2017. Climatic adaptation and DNA methylation in *Festuca* grasses. Turk. J. Agric. For. 41 (5), 445–455.

Lee, M., 2022. Bacterial DNA repair mechanisms. Number Analyt. https://www.numberanalytics.com/blog/bacterial-dna-repair-mechanisms.

Li, Y., Dubbury, S.J., Baum, D.A., 2022. RNA structure adjustment in response to stress in plant cells. RNA Biol. 19, 936–949. https://doi.org/10.1080/15476286.2022.2074271.

Li, Y., Niu, S., 2023. Stress-responsive non-coding RNAs and epigenetic inheritance in plants. RNA Biol. 20 (1), 1–12. https://doi.org/10.1080/15476286.2022.2146756.

Liu, Q., et al., 2020. Phenotypic divergence of thermotolerance related to intrinsic DNA flexibility in glacier-inhabiting *Cryobacterium*. Environ. Microbiol. 22 (4), 1409–1420. https://doi.org/10.1111/1462-2920.14957.

Loh, E., et al., 2018. RNA thermometers in bacterial pathogens. Microbiol. Spectr. 6 (2). https://doi.org/10.1128/microbiolspec.RWR-0012-2017.

Macario, A.J., et al., 1999. Chaperonins in Archaea: evolution and unique features. Microbiol. Mol. Biol. Rev. 63 (4), 923–967.

Martick, M., Scott, W.G., 2004. Influence of temperature on RNA-catalyzed reactions. RNA 10 (12), 1916–1924. https://doi.org/10.1261/rna.7190804.

McGaughran, A., Laver, R.J., Fraser, C.I., 2021. Evolutionary responses to warming. Trends Ecol. Evol. 36 (7), 591–600. https://doi.org/10.1016/j.tree.2021.02.014.

Metzger, D.C.H., Schulte, P.M., 2017. Persistent and plastic effects of temperature on DNA methylation. Proc. R. Soc. B. 284, 20171667. https://doi.org/10.1098/rspb.2017.1667.

Miyamoto, M.M., Clark, A.G., 2001. Context-Dependent Mutational Effects on Ribozyme Activity. University of California, Berkeley, CA.

Mohammad, A., Arefeen, T., Roy, S., 2023. Epigenetic responses to climate change: mechanisms and implications. Curr. Genet. 69 (2), 1–14. https://doi.org/10.1007/s00294-022-01258-3.

Moreno-Villena, J.J., Alvarez-Ponce, D., 2021. Modeling genetic connectivity under climate change in trees. Front. Plant Sci. 12, 671336. https://doi.org/10.3389/fpls.2021.671336.

Neupert, J., Karcher, D., Bock, R., 2008. Design of simple synthetic RNA thermometers. Nucleic Acids Res. 36 (19), e124. https://doi.org/10.1093/nar/gkn545.

Olivares, E.C., Capote, T., 2022. Environmental control of nucleic acid structure. Int. J. Mol. Sci. 23 (15), 8577. https://doi.org/10.3390/ijms23158577.

Oshima, T., 2010. Molecular adaptation to extreme environments in hyperthermophilic Archaea. Encyclopedia of Life Support Systems VI, 1–14. https://www.eolss.net/sample-chapters/c03/E6-73-02-06.pdf.

Pál, J.C., et al., 2022. Climatic stress influences DNA methylation in alpine grasses. Sci. Rep. 12, 12125. https://doi.org/10.1038/s41598-022-12125-4.

Pelechano, V., Wei, W., Steinmetz, L.M., 2022. Genome-wide decay rates of mRNAs in budding yeast. Nucleic Acids Res. 50 (2), 1092–1110. https://doi.org/10.1093/nar/gkab1235.

Peng, C., Sanchez-Guerra, M., Wilson, A., Mehta, A.J., Zhong, J., Zanobetti, A., et al., 2017. Short-term effects of air temperature and mitochondrial DNA lesions within an older population. Environ. Int. 103, 23–29. https://doi.org/10.1016/j.envint.2017.03.017.

Philk, B., 2008. Impact of ocean acidification on marine nucleic acids. ICES J. Mar. Sci. 65 (3), 414–419.

Punetha, S., Meyer, M.M., Narberhaus, F., 2014. Regulation of gene expression by RNA thermometers in vivo. Front. Microbiol. 5, 171. https://doi.org/10.3389/fmicb.2014.00171.

Riebesell, U., et al., 2000. Ocean acidification and calcifying marine biota. ICES J. Mar. Sci. 57, 548–558.

Rissone, P., et al., 2024. Universal cold RNA phase transitions. BioRxiv. https://doi.org/10.1101/2024.03.22.586224.

Ruiz-Pesini, E., et al., 2004. Effects of environmental adaptation on human mitochondrial DNA variation. Proc. Nat. Acad. Sci. U.S.A. 101 (17), 283–289.

Saleski, T.E., Joyce, G.F., 1991. Directed evolution of a ribozyme with high thermal tolerance. Proc. Nat. Acad. Sci. U.S.A. 88 (13), 5919–5923.

Sammarco, I., et al., 2023. DNA methylation in the wild mediates adaptation in wild strawberry. New Phytol. https://doi.org/10.1111/nph.19464.

Samtani, H., Unni, G., Khurana, P., 2022. Microbial mechanisms of heat sensing. Indian J. Microbiol. 62 (2), 175–186. https://doi.org/10.1007/s12088-022-01009-w.

Sánchez-Quintero, M.J., Cubillos, F.A., 2024. Noncoding RNAs and plant adaptive responses to climate stress. Front. Plant Sci. 15, 1387575. https://doi.org/10.3389/fpls.2024.1387575.

Setlow, R.B., 2010. The effects of ultraviolet radiation on nucleic acids. J. Photochem. Photobiol. B. 99, 111–117.

Skinner, M.K., 2018. Preconception cold-induced epigenetic inheritance. Nat. Med. 24 (9), 1308–1309. https://doi.org/10.1038/s41591-018-0187-3.

Skinner, M.K., 2022. Environmental epigenetics and climate change. Current Zool. 68 (1), 9–17. https://doi.org/10.1093/cz/zoab063.

Speck, C., Ames, B., 2015. The effect of cellular pH on nucleic acid conformation and stability. Prog. Mol. Biol. Transl. Sci. 131, 239–277.

Srikant, T., Drost, H.G., 2021. How stress facilitates phenotypic innovation through epigenetic diversity. Front. Plant Sci. 11, 606800. https://doi.org/10.3389/fpls.2020.606800.

Strader, M.E., Wong, J.M., Hofmann, G.E., 2020. Ocean acidification promotes broad transcriptomic responses. Front. Zool. 17, 7. https://doi.org/10.1186/s12983-020-0350-9.

Taylor, A.R., et al., 2011. A voltage-gated H^+ channel underlying pH homeostasis in coccolithophores. PLoS Biol. 9 (6), e1001085. https://doi.org/10.1371/journal.pbio.1001085.

Thomas, S.E., Balcerowicz, M., Chung, B.Y.W., 2022. RNA structure-mediated thermoregulation in plants. Front. Plant Sci. 13, 938570. https://doi.org/10.3389/fpls.2022.938570.

Vieregg, J., 2015. Thermodynamics of Nucleic Acids. University of Chicago, Chicago, IL.

Vilcinskas, A., 2020. Heat shock proteins in fungi: regulation and applications. PeerJ. 8, e10423.

Wallace, D.C., 2003. A mitochondrial paradigm for complex diseases and aging. Proc. Nat. Acad. Sci. U.S.A. 100 (7), 2177–2183.

Wallace, D.C., 2007. Why do we still have a maternally inherited mitochondrial DNA? Insights from evolutionary medicine. Annu. Rev. Biochem. 76, 781–821. https://doi.org/10.1146/annurev.biochem.76.081205.1509.

Wan, J., et al., 2023. DNA methylation and gene transcription drive adaptation of a marine diatom. J. Exp. Bot.. https://doi.org/10.1093/jxb/erad150.

Wang, S., et al., 2022. Heat shock response in hyperthermophilic archaea. Front. Microbiol. 13, 9133787.

Wang, Y., et al., 2014. Heat stress enhances miRNA-mediated gene silencing in plants. PLoS One 9 (7), e103130.

Xie, Z., Cui, Z., Sun, Y., 2024. Temperature-dependence of DNA elasticity. Nucleic Acids Res. 52 (7), 1–14. https://doi.org/10.1093/nar/gkae123.
Xu, R., et al., 2020. Environmental temperature and human epigenetic modifications. Environ. Pollut. 259, 113840. https://doi.org/10.1016/j.envpol.2019.113840.
Yang, I.V., Schwartz, D.A., 2020. Environmental Temperature and Human Epigenetic Modifications. Monash University, Australia.

CHAPTER

14

RNA damage and repair mechanisms

Nandini Shreya[a], *Riya Anand*[b] *and*
Ganesh Kumar Maurya[b]

[a]Department of Biotechnology, School of Earth, Biological and Environmental Sciences, Central University of South Bihar, Gaya, Bihar, India [b]Department of Zoology, Mahila Mahavidyalaya, Banaras Hindu University, Varanasi, Uttar Pradesh, India

14.1 Introduction

Ribonucleic acid (RNA) is a complex compound of high molecular weight that plays a vital role in cellular protein synthesis and is found in the majority of living organisms and viruses (Crick, 1970). RNA is composed of nucleotides, which consist of ribose sugar, nitrogenous bases, and phosphate groups. The ribose sugar, with carbon numbered from 1′ to 5′, connects the base at the 1′ position and the phosphate group between the 3′ position of one ribose and the 5′ position of the next ribose. The ribose sugar is a cylindrical structure having a reactive hydroxyl (–OH) group attached to the second carbon, making RNA prone to hydrolysis. The nitrogenous bases include adenine, guanine, uracil, and cytosine (Jung and Gutell, 2004). RNA typically exists as a single-stranded molecule, but in some special RNA viruses, it exists in a double-stranded form.

14.1.1 Different forms of cellular ribonucleic acid

RNA serves as a messenger, regulator, catalyst, and structural component within the cell.

14.1.1.1 Coding ribonucleic acid

- **Messenger RNA (mRNA)** serves as a template for translation, carrying genetic information from DNA to ribosomes, where proteins are synthesized. It acts as an important intermediate molecule in the gene expression process (Cooper and Adams, 2023).

Genome Damage: Repair and Adaptation Across Life.
DOI: https://doi.org/10.1016/B978-0-443-42857-9.00016-5

14.1.1.2 *Noncoding RNAs*

Noncoding RNAs (ncRNAs) do not code for protein but play critical roles in regulating gene expression in both plants and animals (Esteller, 2011). Earlier, they were thought to be useless, but research has shown that they are deeply involved in many vital processes such as switching genes on and off, DNA organization, helping in stress responses, and playing roles in cancer-like diseases (Anastasiadou et al., 2018). ncRNAs are broadly categorized into housekeeping ncRNAs (like rRNA and tRNA, which are essential for protein synthesis and basic cellular functions) and regulatory ncRNAs (including miRNA, siRNA, lncRNA, and circRNA) (Palazzo and Lee, 2015).

- **Transfer RNA (tRNA)** serves as an essential adaptor in the process of translation of mRNA to proteins. It has a conserved overall L-shaped 3-D structure, including an anticodon loop that pairs with the corresponding codon on the mRNA, while the 3′ CCA tail attaches to a specific site for amino acid. tRNA acts not just as a translation tool, but it is also involved in gene regulation, stress response, evolution, and disease mechanisms(Berg and Brandl, 2021).
- **Ribosomal RNA (rRNA)** is a vital part of the ribosome, where the proteins are made. It not only provides the framework for protein assembly but also gives the ribosomes their shape. In bacteria, the rRNA aids in binding the mRNA by initiating translation by matching with a particular sequence. For precise protein synthesis, it also guarantees that the right tRNA pairs with the mRNA codon. The primary function of the big subunit's 23S rRNA is peptidyl transferase activity, which combines amino acids to make proteins. During protein construction, rRNA also facilitates the attachment of crucial aid proteins known as elongation factors to the ribosome (Liljas and Moore, 2012).
- **Regulatory ncRNAs** modulate gene expression at various levels. ncRNAs interact with proteins to guide or inhibit their functions, with mRNAs to influence protein production, and with DNA to modify gene accessibility (Palazzo and Lee, 2015). Some regulatory ncRNAs, such as lncRNA and circRNA, act as "sponges" to absorb miRNAs, forming intricate networks that finely regulate gene activity. Because they are present in all living organisms and perform a wide range of functions, ncRNAs are now considered fundamental to life and hold great potential for advancing disease diagnosis and therapy (Hansen et al., 2013).

14.1.2 Vulnerability of RNA to damage

The preservation of the genetic information encoded in DNA and RNA is fundamental to ensuring that proteins retain their proper function in the parent and daughter cells. Both DNA and RNA face constant threats from damaging factors that can potentially induce modifications and disrupt their encoded instructions. Many research on nucleic acid-damaging agents has focused on DNA and metabolic pathways; recent advances in RNA analysis are shifting attention toward RNA as an important and previously overlooked target (Simms and Knight, 2023). RNA, especially messenger RNA (mRNA), is more vulnerable to the damaging agents than DNA due to its single-stranded structure, localization in the cytoplasm, lack of protective base pairing, and chromatin organization. Unlike DNA, which is confined to the relatively low-oxidation nuclear compartment, RNA is largely cytoplasmic and subject to more oxidative conditions. As a result, RNA is at greater risk of damage from damaging agents such as reactive oxygen species (ROS) released by mitochondria, alkylating agents, crosslinking agents, and environmental toxins (Yan and Zaher, 2019).

14.1.3 Types of RNA damage and cellular consequences

A variety of exogenous and endogenous agents can damage RNA (Wurtmann and Wolin, 2009), resulting in diverse types of lesions:

14.1.3.1 Oxidative damage from reactive oxygen species

Leads to common modifications like 8-oxo-guanine (8-oxoG), abasic sites, and altered bases (Wurtmann and Wolin, 2009).

14.1.3.2 Alkylating agents

Introduce lesions like N1-methyladenosine (m1A) and O6-methylguanosine (m6G) (Yan and Zaher, 2019).

14.1.3.3 Crosslinking agents and environmental toxins

Agents like formaldehyde and acetaldehyde, along with environmental toxins and chemotherapeutic agents, further contribute to diverse RNA lesions that impair RNA stability and function (Cordes et al., 2025).

The function of RNA is critically affected by damage from both endogenous sources, like ROS, and exogenous agents such as ultraviolet (UV) radiation and alkylating agents, ultimately compromising cellular homeostasis (Wurtmann and Wolin, 2009; Yan and Zaher, 2019). Lesions such as 8-oxo-7,8-dihydroguanosine (8-oxoG), caused by oxidative stress, can disrupt RNA structure, resulting in translation defects like ribosome stalling, amino acid misincorporation, and decreased efficiency of protein synthesis (Yan and Zaher, 2019; Thomas et al., 2020). Disrupted translation interferes with proteostasis, potentially leading to the buildup of misfolded or incomplete proteins that may activate stress pathways and induce apoptosis (Kong and Lin, 2010; Yan and Zaher, 2019). Moreover, lesions in noncoding RNAs like rRNA and tRNA are crucial as they compromise their functional integrity, aggravating the overall disruption of cellular processes (Yan and Zaher, 2019; Olatunji and Liu, 2025). To counteract such damage, cells utilize surveillance pathways, including no-go decay (NGD) and ribosome-associated quality control (RQC), to eliminate faulty RNAs and resolve stalled ribosomes (Thomas et al., 2020; Olatunji and Liu, 2025). Long-lasting RNA damage is a contributing factor in ageing and disease, including cancer and neurodegenerative conditions, as it impairs gene regulation and compromises genome integrity (Kong and Lin, 2010; Li et al., 2020).

14.2 Sources of RNA damage

14.2.1 Endogenous sources

A variety of endogenous factors that are created inside a system, such as a cell, an organism, or a model, such as reactive oxygen species (ROS), alkylating agents, are responsible for RNA damage (Yan and Zaher, 2019).

14.2.1.1 Reactive oxygen species

Cytosolic and membrane-bound enzymes produce highly reactive superoxide anion, which is the key component of ROS under normal physiological conditions. In mitochondria, in the electron transport chain, high levels of ROS are produced in side reactions between molecular oxygen (O_2) and reduced nicotinamide adenine dinucleotide (NADH) and FAD ($FADH_2$). This makes mitochondria and cellular respiration the major endogenous source of ROS production. In recent studies on human liver cells, it has been reported that other organelles, such as peroxisomes, microsomes, and the endoplasmic reticulum, are also capable of producing high levels of ROS (Yan and Zaher, 2019).

14.2.1.2 *Alkylating agents*

Just like oxidizing agents, alkylating chemicals are spread widely, endogenously; they are found as a result of cellular metabolism. S- adenosylmethionine (SAM), nitrosated amines, and bile acids are some of the endogenous alkylating agents. SAM stands out because of its millimolar-range cellular concentration, which is comparatively high and compensates for its poor nonenzymatic reactivity with biomolecules. SAM forms several adducts in vitro through its reaction with nucleic acids. Unlike SAM, bile acids and nitrosated amines are produced via enzymatic side reactions (Yan and Zaher, 2019).

14.2.1.3 *Metal ions*

Metal ions help in the folding of RNA molecules, neutralizing the negative charge present on the phosphodiester backbone, which allows the nucleic acid to adopt its globular structure (Forconi and Herschlag, 2009).

14.2.2 Exogenous

14.2.2.1 *Alkylating agents*

Halogenated hydrocarbons are the most abundant environmental alkylating agents. Chloromethane, bromomethane, and iodomethane are some of the examples; these compounds readily react with nucleic acids, but their contribution to RNA damage is unknown (Yan and Zaher, 2019).

14.2.2.2 *Chemotherapeutic agents/drugs*

Cyclophosphamide, streptozotocin, and temozolomide are some of the well-known alkylating agents used in the treatment of cancers. They all modify RNA along with DNA, and the ability of these medications to change RNA and its function is largely what makes them effective in treating cancer (Feyzi et al., 2007; Yan and Zaher, 2019).

14.2.2.3 *Ultraviolet radiation*

Several types of damage, such as photochemical modification, crosslinking, and oxidative damage, are caused by UV irradiation. The majority of research on UV damage to RNA has been done in vitro, while some studies indicate that damage may also occur in vivo under physiological circumstances. Furthermore, as the photochemistry of nucleic acids mostly takes place at the nucleobase, the thorough research on UV damage to DNA helps speculate about the potential consequences of UV damage on RNA (Wurtmann and Wolin, 2009).

14.3 Types of RNA damage

14.3.1 Oxidation of bases

The term oxidative stress can be described as an imbalance favoring pro-oxidants, such as ROS, reactive nitrogen species (RNS), reactive lipid species, and free radicals over the cell's antioxidant defenses (Sies, 2015). Oxidative damage to RNA occurs as a result of exposure to ionizing radiations, various chemical agents, and as byproducts of normal metabolic processes (Liguori et al., 2018). These reactive agents can induce a wide range of chemical modifications in RNA, including base oxidation, ribose damage, strand breaks, and alterations that may affect RNA stability, structure, and function (Simms et al., 2014).

14.3.1.1 Through reactive oxygen species

A wide variety of free radicals, such as superoxide($O_2^{\cdot}$), hydroxyl radical ($OH^{\cdot}$), nitric oxide (NO), etc., are generated during normal metabolism and by exogenous sources. $OH^{\cdot}$ are extremely reactive and cause the majority of oxidative damage in biological macromolecules. They are unable to spread from their sites of creation and can readily alter RNA when they are created close to it, resulting in the most diverse type of RNA damage. It has been determined that $OH^{\cdot}$ can cause more than 20 distinct kinds of base damage, which include 8-hydroxyguanine, 8-hydroxyadenine, 2,6-diamino-4-hydroxy-5-formamidopyrimidine (FapyGua) from guanine, 4,6-diamino-5-formamidopyrimidine from adenine, and cytosine glycol. 8-hydroxyguanosine (8-OHG) is identified as the most common oxidized base in RNA out of all the base damage. Initially, Guanine and the extremely reactive $OH^{\cdot}$ combine, leading to the generation of a C8-OH adduct radical. After this, 8-OHG is produced by the loss of an electron and a proton (Feyzi et al., 2007; Kong and Lin, 2010; Yan and Zaher, 2019).

14.3.1.2 Through alkylation

A class of chemicals that transfer alkyl groups to nucleophilic sites on nucleic acids, including both DNA and RNA, is known as alkylating agents (Beranek, 1990). ROS are generated by certain alkylating agents, which can lead to oxidative base modification. Traditionally mutagenic effects have been seen in DNA, but recent evidences show that alkylating agents contribute to oxidative damage in RNA, including oxidative demethylation and methylation-based base alterations (Li et al., 2006). Alkylating compounds often use nucleophilic substitution processes to change N or O atoms at different locations in DNA and RNA. They can be subdivided into S_N1 agents (e.g., methyl nitrosourea, MNU), which modify both oxygen (O) and nitrogen (N) atoms, and S_N2 agents (e.g., methyl methane sulfonate, MMS), which mainly affect nitrogen atoms. Damage sites in RNA include reactive sites, e.g., N1 in purines and N3 in pyrimidines. In the case of DNA presence of such lesions results in replication blocks and, if not repaired, can eventually lead to cell death. Unlike DNA, the effects of RNA methylation in living cells are not yet fully understood (Feyzi et al., 2007; Yan and Zaher, 2019).

14.3.1.3 Ultraviolet-induced

UV radiation is also considered a source of oxidative RNA damage. Many pathways of excitation and oxidation have been extensively studied for DNA damage. Direct absorption of UVB light by a nucleobase can result in the formation of radical anion species, which can subsequently react further to produce oxidized nucleotides. Additionally, UVA can cause oxidative damage, which can be mediated by the production of ROS or photosensitizer molecules. The most common result of UVB and UVA-induced oxidation of DNA is the formation of (8-oxodG), while photooxidation of adenine, cytosine, and thymine bases is also present in DNA from irradiated cells (Warner and Wei, 1997).

Crucially, measurements of UV-induced oxidative damage to RNA in cells show that a large portion of the chemistry outlined for photooxidation of DNA also applies to RNA (Li et al., 2006).

Human skin fibroblasts have been used in studies to demonstrate that UV damages RNA oxidatively. According to an HPLC-EC measurement of total RNA, fibroblasts exposed to UVA at sublethal dosages had noticeably higher amounts of 8-oxoG in their RNA. As UVA irradiation intensity rose, 8-oxoG rose in a dose-dependent manner (Feyzi et al., 2007; Wurtmann and Wolin, 2009).

Coincidentally, the RNA damage outweighed the DNA damage in other oxidative stress situations. Significantly, studies supported the physiological significance of UVA-induced oxidative damage to RNA by using UVA at dosages similar to the minimal erythema dose (MED).

14.3.2 Strand break

14.3.2.1 Through reactive oxygen species

ROS also induces strand cleavage and highly reactive •OH are primarily responsible for this. These radicals are extremely short-lived and act at the site of their formation, where they can attack the ribose sugar and phosphate backbone of RNA. This leads to structural modifications that destabilize the molecule. Onewell-characterized modification is the formation of 8-hydroxyguanosine (8-OHG), which arises when •OH reacts with guanine, as discussed earlier (Seixas et al., 2022). Because of the presence of a 2′-hydroxyl group on its ribose sugar, RNA's susceptibility to hydrolytic cleavage is increased as the 2′-hydroxyl group promotes nucleophilic assault on the nearby phosphate. Strand breaks can be produced by backbone scission, which is the outcome of these chemical processes. The functional results, such as translation slowing and ribosome malfunction, are consistent with fragmented or cleaved RNA templates, even if they are not shown as breaks in this investigation. Furthermore, it was discovered that RNA is more vulnerable to oxidative stress than proteins, lipids, or DNA, which supports the idea that it is fragile in environments with a lot of ROS (Wurtmann and Wolin, 2009).

14.3.2.2 Metal ions induced

The complex 3D structure of RNA is stabilized by metal ions, especially by divalent cations like Mg^{+2} which plays a crucial role in neutralizing the negative charge present on the phosphodiester backbone, allowing it form a stable globular structure, Furthermore, site-specific metal ions may play a role in nucleic acid enzyme catalysis; notably, several RNA enzymes (ribozymes) employ carefully placed Mg^{2+} ions to aid in catalysis. $Mg^{+2}$and other metal ions can promote nucleic acid cleavage when added to a solution in micro- to millimolar concentrations (Zagorowska et al., 1998; Forconi and Herschlag, 2009).

Metal ions promote the hydrolysis of the phosphodiester bond in RNA. Hydrated metal ions cleave RNA acting as Bronsted bases, proton is abstracted from the 2′-OH group of the ribose, which results in the formation of nucleophile (2′-O^-), which attacks the adjacent phosphorus atom, leading to the break in the RNA backbone. The "in-line" geometry (the attacking nucleophile and the leaving group are positioned about 180°) greatly facilitates the reaction; thus, the metal ion-induced RNA cleavage occurs more readily in loops, bulges, and unpaired regions, which make these regions hotspots for damage. Alternative mechanisms occur in ribozymes for ion-induced RNA cleavage (Zagorowska et al., 1998; Forconi and Herschlag, 2009).

14.3.3 Crosslinking

14.3.3.1 Ultraviolet light-induced

Evidence from in vitro experiments indicates that UV can also create long-range covalent crosslinks involving RNA. This was initially proposed by the experimental investigation of ribosome structure using UV irradiation, which revealed both RNA-RNA and RNA-protein crosslinks (Noah, 2000). UVC and UVB can both cause crosslinks. The distance and angle between nucleotides, as well as their capacity to temporarily modify their conformation to permit crosslinking within the timescale of photoexcitation, are all connected to the frequency of crosslinking between nucleotides. In the case of RNA-protein crosslinking, the geometry and photo-reactivity of the nucleotides determine the likelihood of crosslinking (Noah, 2000). Similar to RNA-RNA crosslinks, RNA-protein crosslinking has a lengthy experimental history and has been used to investigate ribonucleoprotein (RNP) complexes, such as the ribosome. These studies shed light on the different kinds of crosslinks that could form. For instance, it has been observed that double-stranded areas typically do not experience crosslinking (Noah, 2000). As a result, RNA structure may make some cellular RNPs more or less vulnerable to UV crosslinking damage. Importantly, the experimental creation of RNA-protein crosslinks using UV light reflects the inherent photo-reactivity of RNPs. When Escherichia coli is exposed to appropriate dosages of broad-spectrum UVA radiation (320–405 nm), RNA-RNA crosslinking takes place within the tRNA under physiological conditions (Wurtmann and Wolin, 2009).

Cytidine and the naturally altered pseudo-uridine nucleotide form the crosslinks that have been observed. tRNAs with this crosslink had reduced rates of aminoacylation and resulted in ineffective amino acid incorporation in translation assays, according to in vitro research.

Although the RNAs involved have not been identified, RNA-protein crosslinks have been reported in insect eggs exposed to UVB radiation. In the maize study, ecologically relevant UVB concentrations caused covalent crosslinks to form between protein and RNA within the ribosome. In the 3D structure of the ribosome, the crosslinks that were seen happened where RNA and protein are adjacent to one another. Furthermore, the cell may experience functional effects as a result of this crosslinking damage (Welp et al., 2024). As the UVB exposure increases, the amounts of crosslinked products accumulated and the overall cellular level of protein synthesis decline. Although there are other potential UV-induced mechanisms for translation suppression, this correlation raises the possibility that protein-RNA crosslinks may have played a role in the loss of ribosome function (Wurtmann and Wolin, 2009).

14.3.3.2 Through oxidation

Oxidation also results in the crosslinking of rRNA to ribosomal protein. In yeast cells treated with H_2O_2, 85% of the ribosomal proteins were found to be oxidized, and HPLC-MS/MS identified numerous sites of ribosomal protein nucleotide crosslinking. Oxidative stress often results in the production of protein-rRNA crosslinks. These crosslinks are mediated by free radicals rather than Schiff base synthesis, which is typical of protein-protein interactions. The discovered structures point to a common oxidative route since they mimic those observed in protein-DNA crosslinking. Studies indicate that covalent protein-RNA complexes may interfere with protein expression, while the precise mechanism is yet unclear. Fast-growing microorganisms might be able to recover from such damage, but mammalian cell-particularly those in slow-dividing tissues like the brain in elderly people, may be more severely affected, which could lead to illness and compromised cellular function (Mirzaei and Regnier, 2006).

14.3.3.3 *Aldehyde crosslinkers*

Aldehydes have been used to crosslink biological samples since 1800s; formaldehyde was the first to be used to preserve tissues. Aldehydes for crosslinking are introduced externally and used by scientists for sample preservation experiments and RNA/RNA interaction mapping. Formaldehyde works by forming chemical bonds (adducts and crosslinks) with RNA and DNA bases, especially adenine, guanine, and cytosine. This protects them from breaking down. The bonds formed are reversible, allowing the stored RNA to be analyzed later. Formaldehydes can also be produced naturally as a byproduct of cellular metabolism, such as demethylation of DNA or protein, oxidative stress, or lipid peroxidation. When this happens, formaldehyde can accidentally react with nucleic acids (RNA/DNA) and proteins to form crosslinks (Velema and Lu, 2023).

14.3.4 Abasic RNA

Apurinic/apyrimidinic (AP or abasic) sites are nucleotides in DNA or RNA that have missing bases. These are created when the glycosidic connection between the nitrogenous base and the (deoxy)ribose sugar is broken (Kim and Wilson, 2012).

14.3.4.1 *Through hydrolysis*

Less is known about the properties and effects of abasic sites in RNA. Abasic sites are produced in RNA when the glycosidic bond that connects the ribose sugar to the nitrogenous base is hydrolyzed. A nitrogen atom is protonated at the start of the process, leaving the base positively charged. This usually happens to purines instead of pyrimidines because the glycosyl bonds in pyrimidines are 20 times more stable than those in purines. To facilitate the relocation of the electrons forming the glycosidic bond and ultimately lead to cleavage, purines can delocalize charge around their dual-ring structure. Purine base loss may also result via hydrolysis from base methylation, free radical activity, and RNA N-glycosylase enzyme activity (Table 14.1) (Prashar et al., 2023). In the following sections, we go over a few of these formation techniques in brief (Prashar et al., 2023).

14.3.4.2 *Through oxidation*

Oxidative stress from ROS and RNS can also induce abasic site formation in RNA. It was studied that γ-radiation, Fenton reaction, and peroxynitrite were used to oxidize RNA in vitro, which led to the increased formation of abasic sites. A highly sensitive Aldehyde reactive probe (ARP) assay was done, which helped in the detection of abasic sites. These probes especially targeted the aldehyde groups on the sugar of depurinated RNA. The study also showed that

TABLE 14.1 Types of hydrolysis leading to the formation of an abasic site in RNA.

Type of hydrolysis	Mechanism	Trigger	Result
Spontaneous hydrolysis	The N-Glycosidic bond is cleaved under acidic conditions	Protonation at N-7 of purines	The oxocarbenium ion is formed on the pentose sugar and charge on the water molecule is resolved in the presence of a water molecule
Base methylation-induced hydrolysis	The addition of methyl group destabilizes the glycosidic bond	Methylation by alkylating agents (e.g., dimethyl sulphate) or methyltransferases	Electron rearrangement and bond cleavage are caused by a localized positive charge created by methylation, followed by resolution from the water nucleophilic attack
Free radical/oxidation-induced hydrolysis	Hydrogen is abstracted from ribose by ROS, leading to the formation of unstable intermediates	Reactive oxygen species (e.g., superoxide, hydroxyl radicals from metabolism or stress)	Base is released from ribose to give the Cl′ alcohol intermediate, and water is added to form carbonyl (ribonolactone)
RNA N-Glycosylase enzyme-mediated hydrolysis	Bases are removed by an enzyme (glycosylase)	Enzyme activity: protonation of purine by arginine; stabilization by other amino acids	An oxocarbenium ion is formed, which is resolved by water; an abasic site is left (e.g., in rRNA)

the abasic RNA increased under oxidative stress in cells, which links it to its biological relevance (Wurtmann and Wolin, 2009).

14.3.5 Chlorinating and nitrating damage

Multiple intracellular and cellular sources of chlorine and nitrogen atoms are present, which can cause damage to RNA. Radicals like hypochlorous acid (HOCl), nitric oxide ($NO^\cdot$), and peroxynitrate ($ONOO^\cdot$) are produced by phagocytic cells, which are known to fight infections. Nitric oxides are used by cells as a signaling molecule. Nevertheless, these radical species have the ability to cause biological molecules, including RNA, to be chlorinated, nitrated, and oxidized (Wurtmann and Wolin, 2009).

14.4 RNA surveillance and quality control pathways: Molecular strategies for transcriptome integrity

Cells rely on accurate gene expression and regulation to maintain viability and adapt to external signals. Cells often encounter transcription errors, misprocessing, and environmental damage, which pose risks to RNA quality. RNA surveillance systems oversee both coding and noncoding RNA, identifying and removing defective or misprocessed RNA molecules. Such surveillance systems are essential for maintaining transcriptomic integrity and preventing the expression of aberrant proteins that could be detrimental to the cell (Shoemaker and Green, 2012; Wolin and Maquat, 2019).

14.4.1 Nuclear surveillance

Nuclear RNA surveillance acts primarily through kinetic competition with the normal RNA processing pathway. Nascent RNAs that fail to undergo correct splicing, capping, polyadenylation, or ribonucleoprotein (RNP) assembly are retained in the nucleus and degraded. Aberrant or misprocessed RNAs are recognized and degraded to prevent the accumulation of nonfunctional, immunogenic, or potentially harmful molecules. This process is governed by a network of RNA-binding proteins, decay machineries, and cis-regulatory RNA features, forming a surveillance system critical for maintaining transcriptome fidelity, developmental regulation, and genome integrity (Wolin and Maquat, 2019).

The RNA exosome and its adaptors: Eukaryotic nuclear RNA degradation is primarily driven by the RNA exosome, a conserved multiprotein complex with both 3′–5′ exoribonuclease and endoribonuclease functions (Wolin and Maquat, 2019). The exosome is recruited to its RNA targets via different adaptor complexes. Two exosome adaptor complexes have been identified in the nucleoplasm of mammalian cells are Nuclear Exosome Targeting (NEXT) complex and the poly(A) Tail Exosome Targeting (PAXT) connection (Soles et al., 2025).

- The NEXT complex, consisting of the MTR4 helicase, RBM7 RNA-binding protein, and the scaffold protein ZCCHC8, targets nonpolyadenylated nuclear RNAs for degradation by the exosome. RBM7 binds to U-rich motifs located upstream, while ZCCHC8 promotes dimerization of the complex and anchors RBM7 to MTR4, positioning RNA for capture. MTR4 sequesters the RNA 3′ end and translocates it toward the exosome. The C-terminal domain of ZCCHC8 modulates RNA path and blocks exosome docking surfaces, necessitating remodeling for handoff to the exosome. This cooperative assembly facilitates surveillance and turnover of pervasive transcripts, promoter upstream RNAs, and defective nuclear RNAs, protecting transcriptome integrity (Puno and Lima, 2022).
- In contrast, the PAXT connection, which includes MTR4 helicase, ZFC3H1 scaffold, and PABPN1, recognizes polyadenylated nuclear RNAs. These RNAs often arise from intronic polyadenylation (IPA) or other premature cleavage events. Recognition of both the 5′ splice site and the poly(A) junction serves as a signal for the cooperative recruitment of the PAXT complex by U1 snRNP and CPA factors. By unwinding RNA, MTR4 channels it to the nuclear exosome for 3′ to 5′ degradation via the nuclear exosome, thereby avoiding the accumulation of potentially harmful transcripts (Puno and Lima, 2022).

14.4.2 Cytoplasmic surveillance

In the cytoplasm, translation surveillance pathways monitor the fidelity of protein synthesis. Nonsense-mediated mRNA decay (NMD) serves as a quality control mechanism by degrading mRNAs with premature stop codons. NGD acts to resolve stalled ribosomes during translation elongation, whereas nonstop decay (NSD) targets mRNAs missing a stop codon for degradation.

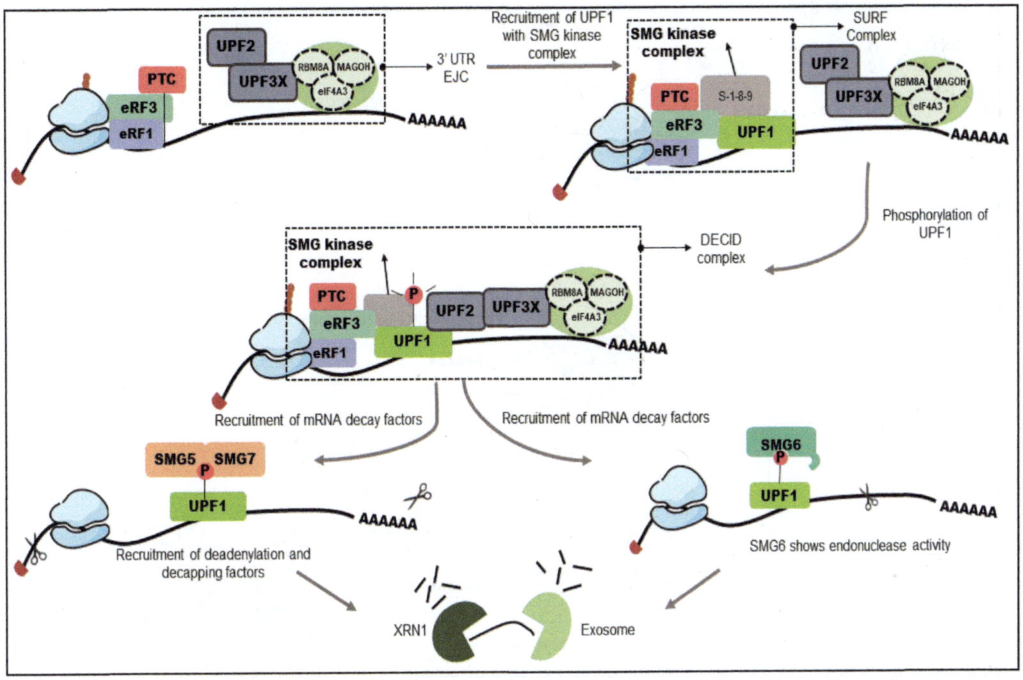

FIGURE 14.1 Nonsense-mediated decay pathway. Surveillance mechanism ensuring the quality of transcripts by targeting mRNA containing premature stop codons (PTCs). (Illustration was prepared using BioRender software).

Ribosome stalling serves as the central signal for initiating both the nonstop and no-go mRNA decay pathways (Glover et al., 2020).

- *NMD*: As a major RNA quality control mechanism in eukaryotes, the NMD pathway (Fig. 14.1) safeguards cells by identifying and removing mRNAs with premature termination codons (PTCs), thus preventing the accumulation of faulty protein products (Nicholson and Mühlemann, 2010). NMD is initiated in the nucleus, where mRNAs are scanned for premature stop codons, and the process continues in the cytoplasm for degradation (Wilkinson and Shyu, 2002). Two main mechanisms of target discrimination in NMD are EJC-dependent and EJC-independent (or long 3′ UTR-mediated) pathways, with EJC-dependent NMD generally being more efficient (Kurosaki et al., 2019). Faulty transcripts are identified by NMD through the occurrence of a PTC situated more than 50–55 nucleotides upstream of the exon junction complex (EJC) during the initial round of translation (Nagy and Maquat, 1998). NMD pathway's core factors include the helicase UPF1 associated with the UPF2 and UPF3X (UPF3B) (Chamieh et al., 2008). The SMG kinase complex (SMG1, SMG8, and SMG9) phosphorylates UPF1, facilitating the decay of defective mRNAs (Gupta and Li, 2018). While the SMG5, SM6, and SMG7 execute the endonucleolytic or exonucleolytic activities to degrade the faulty transcript. In the EJC-dependent pathway, the SURF complex (Smg-1–Upf1-release factor) interacts with the downstream EJC-bound UPF3X and UPF2, forming the DECID (decay-inducing) complex, which activates SMG1-mediated phosphorylation

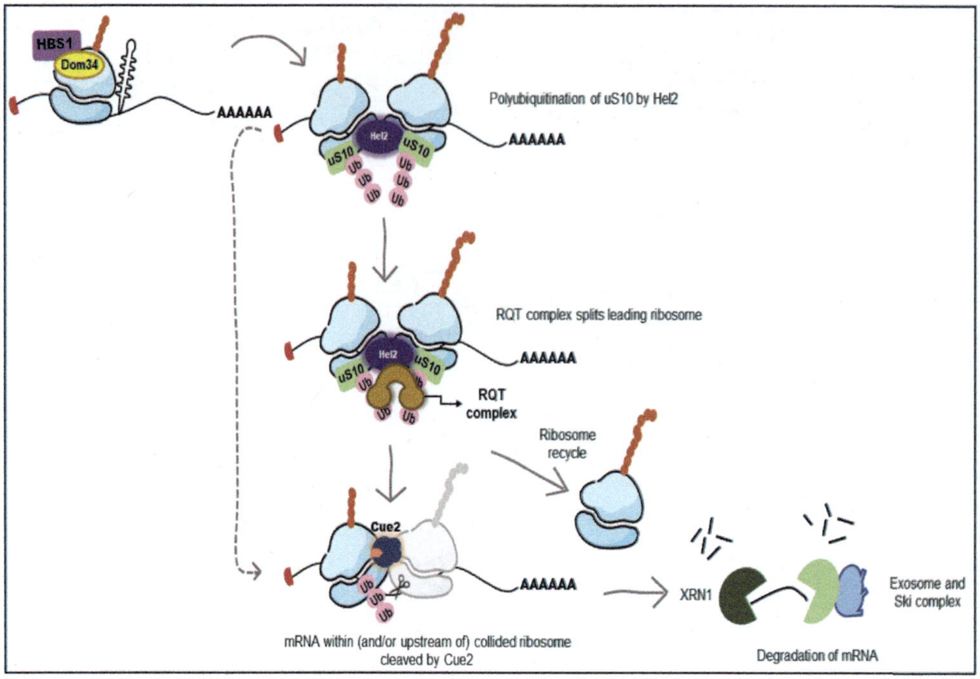

FIGURE 14.2 No-go mRNA surveillance pathway. The mRNA quality control mechanism leads to degradation of stalled mRNAs during translational elongation. (Illustration was prepared using BioRender software)

of UPF1. Once phosphorylated, UPF1 recruits SMG6 to cleave near the PTC or SMG5–7 to recruit deadenylation and decapping factors, with degradation completed by exosome, which manifests both 3′ to 5′ exoribonucleolytic and endoribonucleolytic activities; and 5′ to 3′ exoribonuclease XRN1 (Kurosaki et al., 2019; Wolin and Maquat, 2019).

- *NGD*: In eukaryotic cells, the NGD surveillance pathway (Fig. 14.2) degrades mRNAs that cause ribosome stalling during elongation, typically due to stable secondary structures, enzymatic cleavage sites, chemically altered bases, rare codon usage, or polylysine/arginine-encoding regions (Simms et al., 2017; Navickas et al., 2020). In NGD, the core event is ribosome collision, where a stalled ribosome causes the following ribosomes to stack, forming a disome structure. The Dom34 (Pelota in mammals)–Hbs1–GTP complex senses the stalled ribosomes and recruits the E3 ligase Hel2 (ZNF598 in mammals) for ubiquitination of ribosomal components such as uS10 (or eS10 in mammals) in yeast cells (Simms et al., 2017; Ikeuchi et al., 2019). Ubiquitination serves as a signal for RQT complex (Rqt4, Slh1, and Cue3) to split the leading ribosome and/ or Cue2 (N4BP2 in human) to endonucleolytic cleavage of the mRNA, about 8 nucleotides upstream of the stalling point, within the ribosomal exit tunnel of the disome (Navickas et al., 2020; Powers et al., 2020; Müller et al., 2024). Exonucleolytic degradation follows cleavage, with the exosome and Ski complex degrading 5′ fragments in the 3′ to 5′ direction, and Xrn1 breaking down 3′ fragments in the 5′ to 3′ direction (Guydosh et al., 2017; Ikeuchi et al., 2019; Navickas et al., 2020). While degradation proceeds, the Dom34-Hbs1 complex rescues stalled ribosomes by dissociating them into subunits, supporting ribosome recycling and safeguarding against depletion (Guydosh et al., 2017; Powers et al., 2020).

- *NSD*: A related surveillance pathway to the NGD triggered when transcripts lacking an in-frame stop codon are degraded through the NSD pathway, thereby avoiding ribosome stalling at the 3′ end and the formation of aberrant nonstop peptides (NSPs). Ski7 contributes to mRNA surveillance in eukaryotic cells by acting like eRF3, the translation termination factor, and directing the Ski-exosome complex to degrade mRNA from the 3′ to 5′ end. NSPs are concurrently targeted for proteasomal degradation through ubiquitination by the E3 ligase Ltn1 (Listerin) (Hashimoto et al., 2017). To maintain protein homeostasis, NSD degrades aberrant mRNAs and their toxic peptide products resulting from errors such as inappropriate polyadenylation, mutations, or RNA cleavage (Simms et al., 2017).

14.5 Repair mechanisms

14.5.1 Oxidative demethylases AlkB and hABH3 reverse methylation in RNA

AlkB enzymes are present in many plants and viruses and are conserved from bacteria to humans. They are primarily known for their role in DNA and RNA repair. In *E. coli,* AlkB has long been known for alkylation damage repair. Eight human homologues (hABH1-8) have been identified, of which hABH2 and hABH3 have been demonstrated to be functional homologues of AlkB. hABH3 can repair methylated RNA, whereas hABH2 works mainly on DNA.

These two enzymes are located in different parts of the cells. hABH2 is localized strictly to the nucleus and nucleoli, indicating a role in the repair of DNA close to the replication fork. The other one, i.e., hABH3, is found both in the nucleus and the cytoplasm, and it targets other RNAs. AlkB would immediately reverse alkylation damage by hydroxylating the methyl group on damaged DNA bases. AlkB enzymatically demethylates the DNA bases adenine and cytosine through oxidative demethylation, in which the methyl group is changed into a hydroxymethyl group that subsequently exits the base as formaldehyde (Fig. 14.3). This prediction was independently verified by the Seeberg and Sedgwick labs(Mirzaei and Regnier, 2006; Wurtmann and Wolin, 2009).

14.5.2 ATP-dependent RNA repair

Bacteriophage T4 polynucleotide kinase-phosphatase (PNkp) was the first end healing enzyme discovered in 1965–1966 by the laboratories of Charles Richardson (99) and Jerard Hurwitz. Hurwitz et al. also discovered T4 RNA ligase (now known as Rnl1) as the first RNA-specific end-joining enzyme in 1972, after the discovery and purification of DNA ligases in 1967 were made possible by the availability of T4 Pnkp as a 5′-end-labeling reagent. ATP-dependent RNA repair pathway is one of the most well-characterized RNA repair pathways, which involves three nucleotidyl transfer steps (Fig. 14.4).

Step 1: ATP and ligase combine to generate a covalent ligase-(lysyl-Nz)–adenosine monophosphate (AMP) intermediate, which releases phosphorus (PPi).
Step 2: An RNA-adenylate intermediate (AppRNA) is created when AMP is moved from the ligase–adenylate to the 5′-PO4 RNA end.

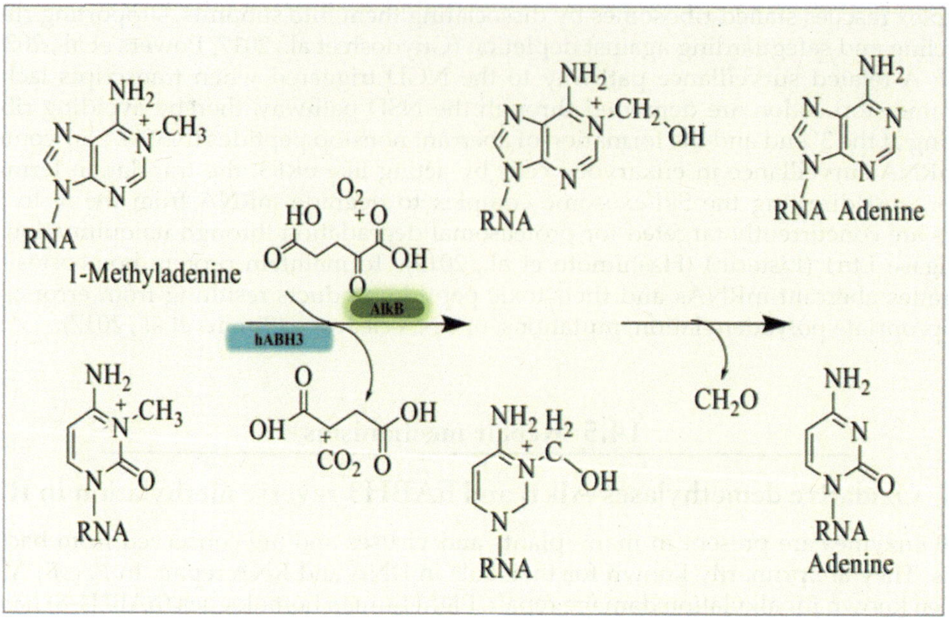

FIGURE 14.3 AlkB and hABH3 reverse methylation damage in RNA. The co-substrate 2OG is converted into succinate by the oxidative demethylases AlkB and hABH3, which also hydroxylate the methyl group in 1-meA and 3-meC. Both molecular oxygen and ferrous ions are necessary for the reaction. The hydroxymethyl group spontaneously releases as formaldehyde, returning the base to its natural state. A schematic diagram has been created by using ChemDraw (Version: 15.1.0.144).

Step 3: Ligase catalyzes an RNA 3′-OH assault on the RNA-adenylate to release AMP and seal the two ends with a phosphodiester bond.

The ligase process requires a divalent cation cofactor at each step. With the exception of step 2, where they create AppDNA intermediates, all ATP-dependent RNA ligases follow the same three-step mechanism as ATP-dependent DNA ligases. A superfamily of covalent nucleotidyl transferases, including ATP-dependent RNA ligases, DNA ligases, and GTP-dependent mRNA capping enzymes, functions through ligase-(lysyl-Nz)–NMP intermediates. In order for bacteriophage T4 to avoid host defense mechanisms, especially when it comes to undoing damage caused by tRNA-targeting ribotoxins like PrrC, this repair pathway is essential (Shuman, 2023).

14.5.3 A bacterial RNA repair system that confers immunity to recurrent RNA damage

This repair system is characterized as a defense mechanism against ribotoxins, which are enzymes that cleave RNA in response to cellular stress or competition. Many bacteria have developed a unique RNA repair system. This is composed of three types of proteins that are separately encoded: (1) a T4 end healing enzyme, Pnkp1, which consists of an N-terminal

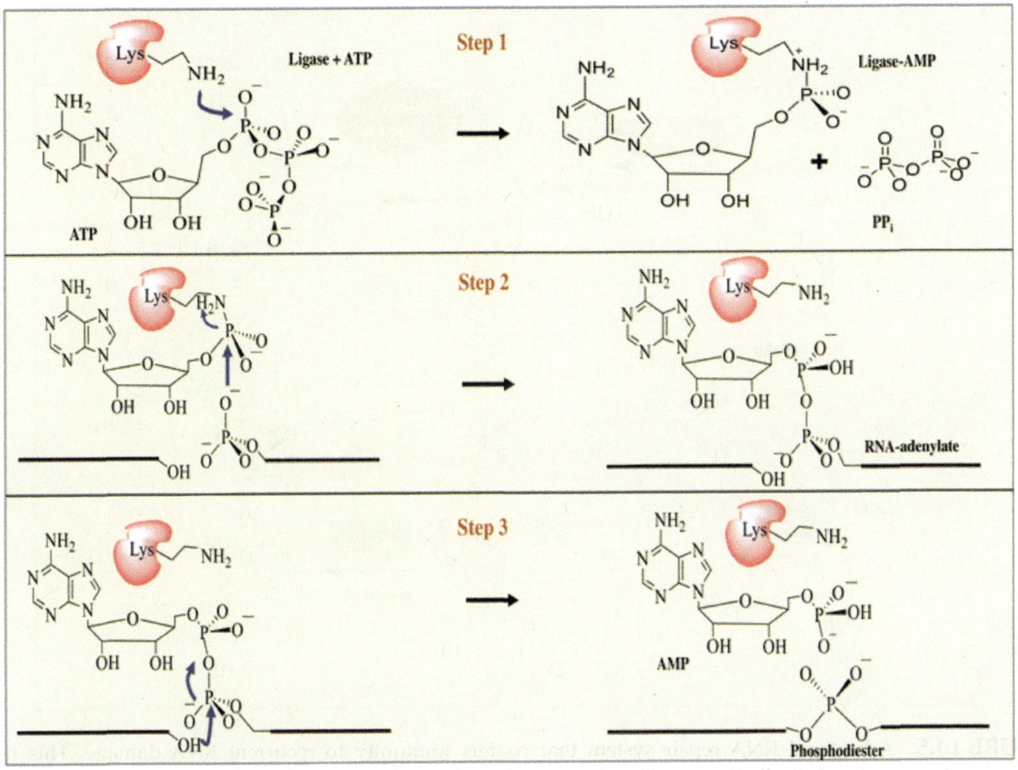

FIGURE 14.4 Steps of RNA ligation by T4 RNA ligase 1. (1) Using ATP, the ligase binds AMP to itself (at a lysine residue) and releases pyrophosphate. (2) The enzyme forms an RNA-adenylate by moving AMP from itself to the RNA's 5′-phosphate end. (3) The free 3′-OH of another RNA attacks the activated 5′-end, forming a phosphodiester bond that seals the RNA ends together, and AMP is released. A schematic diagram has been created by using ChemDraw (Version: 15.1.0.144).

polynucleotide kinase domain and a C-terminal DxDxT family 3′-phosphatase domain; (2) ATP-dependant ligase (Rn1); and (3) a Hen1 2′-O-methyltransferase. These enzymes function in concert to repair and seal the damaged RNA ends while simultaneously shielding the repaired regions from additional harm. They are frequently encoded together in an operon-like gene order.

Steps involved (Fig. 14.5).

Pnkp's end healing: The 5′-OH end of Pnkp is changed into the ligatable form 5′-phosphate by the addition of a phosphate group by the kinase domain. A 3′-OH, which is also necessary for ligation, is produced when the phosphatase domain extracts the cyclic phosphate from the 3′ end.

Hen1 methylation: A methyl group from S-adenosylmethionine (AdoMet) is transferred by Hen1 to the 2′-OH of the RNA's 3′-terminal ribose prior to ligation. By acting as a protective mark, this 2′-O-methylation prevents ribotoxins from further cleaving the newly formed RNA junction.

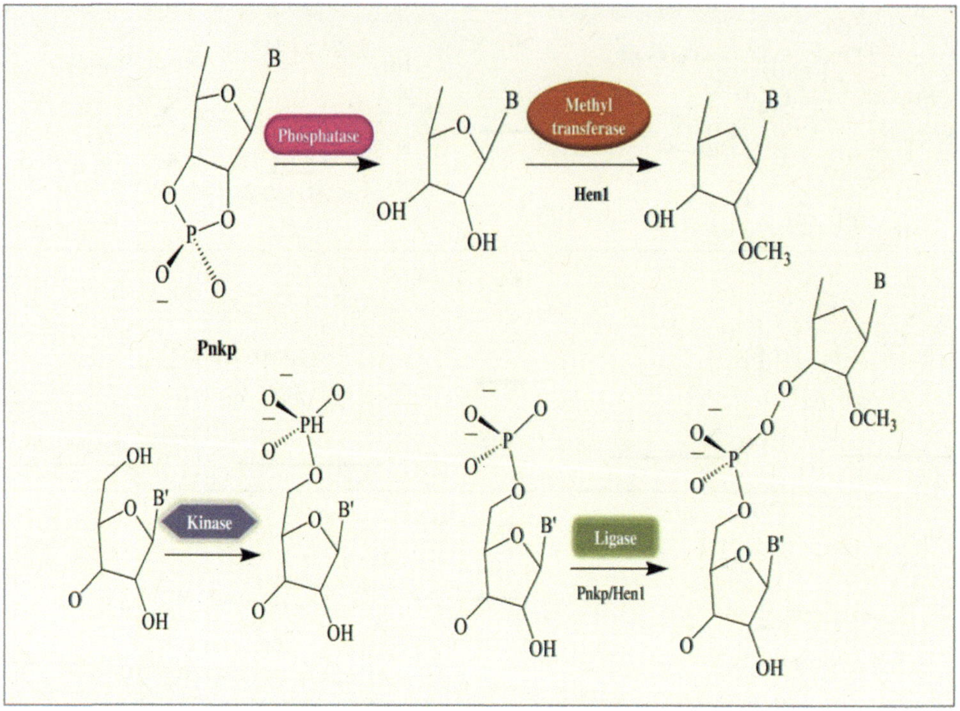

FIGURE 14.5 A bacterial RNA repair system that confers immunity to recurrent RNA damage. This repair mechanism heals and seals the 2′,3′>p and 5′-OH ends by installing a 2′-O-methyl mark at the repair junction, which stops a trans-esterifying nuclease from cleaving them repeatedly. A schematic diagram has been created by using ChemDraw (Version: 15.1.0.144).

Ligase activity-assisted end sealing: The repair is completed when the ligase domain of Pnkp, which is activated by the N-terminal domain of Hen1, forms a phosphodiester bond with the healed 5′-PO_4 and 3′-OH ends (Shuman, 2023).

14.6 Implications of RNA damage in health and disease

RNA damage represents a molecular dysfunction with profound implications for cellular function and disease pathogenesis. Unlike DNA, RNA is predominantly single-stranded, lacks protective histones, and resides in close proximity to mitochondrial ROS sources, rendering it exceptionally vulnerable to oxidative, alkylative, and UV-induced damage (Yan and Zaher, 2019; Li et al., 2020). With RNA constituting 80%–90% of cellular nucleic acids and playing central roles in translation and gene regulation, its chemical integrity is crucial for proteostasis and cellular homeostasis (Liu et al., 2020). This section examines the consequences of RNA damage in neurodegenerative diseases and cancer, highlighting shared mechanisms and disease-specific implications.

14.6.1 Neurodegenerative diseases: RNA damage as a central pathogenic mechanism

14.6.1.1 *Alzheimer's disease*

a) *Evidence of widespread RNA oxidation*: Postmortem AD brains exhibit significantly elevated 8-hydroxyguanosine (8-OHG) in vulnerable hippocampal and cortical neurons, with immunoelectron microscopy confirming ribosomal localization (Nunomura et al., 2006; Castellani et al., 2008). Oxidative damage affects multiple RNA species:
- *rRNA*: Comprises >80% of cellular RNA and shows 13-fold higher oxidation than tRNA due to iron-binding capacity (Nunomura et al., 2006).
- *mRNA*: 30%–70% of poly(A)+ mRNAs are oxidized in AD frontal cortices versus 2% in controls, targeting transcripts like p21ras, MAPK kinase 1, and Cu/Zn SOD—proteins implicated in AD pathogenesis (Nunomura et al., 2006).

b) *Early pathogenic event*: RNA oxidation precedes significant amyloid plaque formation and tau tangle formation, appearing in mild cognitive impairment (MCI) stages and correlating with mitochondrial dysfunction (Nunomura et al., 2006; Castellani et al., 2008). This positions RNA damage upstream of protein aggregation cascades.

c) *Functional consequences*: Ribosomes loaded with oxidized mRNA exhibit translational stalling and produce truncated, misfolded proteins that may seed amyloidogenic aggregates (Yan and Zaher, 2019; Xu et al., 2021). Oxidized rRNA impairs ribosomal assembly and fidelity, exacerbating proteostatic collapse (Liu et al., 2020).

14.6.1.2 *Parkinson's disease*

a) *Selective vulnerability*: Substantia nigra dopaminergic neurons show cytoplasmic 8-OHG accumulation, correlating with mitochondrial complex I defects and elevated iron levels that catalyze ROS via Fenton reactions (Nunomura et al., 2006; Liu et al., 2020).

b) *tRNA fragmentation*: Oxidative stress cleaves tRNA at the anticodon loop, generating tRNA-derived stress-induced fragments (tiRNAs) of 30–45 nucleotides (Liu et al., 2020). These fragments impair global translation while paradoxically upregulating α-synuclein expression—A key pathogenic protein in PD (Liu et al., 2020).

c) *Diagnostic biomarkers*: Elevated 8-OHG in cerebrospinal fluid (CSF) and serum of PD patients correlates with disease progression, offering potential for early detection (Nunomura et al., 2006).

14.6.1.3 *Shared mechanisms in neurodegeneration*

a) *Mitochondrial dysfunction*: As the primary source of neuronal ROS, dysfunctional mitochondria generate OH˙ via Fenton chemistry ($Fe^{2+} + H_2O_2 \rightarrow Fe^{3+} + \bullet OH + OH^-$). The brain consumes 20%–25% of bodily oxygen yet has only 10%–20% of hepatic catalase activity, creating an oxidative milieu (Liu et al., 2020).

b) *Impaired RNA quality control*: Neurons exhibit deficient expression of RNA repair enzymes like hABH3 (AlkB homolog 3), which demethylates damaged bases. Additionally, surveillance pathways like NMD become overwhelmed by oxidized transcripts (Li et al., 2020; Xu et al., 2021).

c) *Vicious cycle of proteotoxicity*: Damaged RNA generates misfolded proteins that further increase oxidative stress, creating a self-propagating cycle that drives neuronal loss (Castellani et al., 2008).

14.6.2 Cancer: RNA damage as a therapeutic target and resistance mechanism

14.6.2.1 Chemotherapy-induced RNA damage

a) *5-Fluorouracil (5-FU)*: This antimetabolite exemplifies Class 1a RNA-damaging drugs:
   - Metabolized to fluorouridine triphosphate (FUTP), incorporated into RNA at 3000–15,000× higher rates than its DNA metabolite (FdUTP) (Simms and Knight, 2023).
   - Incorporation disrupts RNA processing and function, altering splicing, ribosomal biogenesis, and microRNA maturation. This contributes significantly to cytotoxicity independently of thymidylate synthase inhibition (Simms and Knight, 2023).
   - Resistance mechanisms involve upregulation of RNA-binding proteins (e.g., HuR) that recognize fluorinated RNA and enhance stress adaptation (Simms and Knight, 2023).
   - Other RNA-Damaging Agents: Cisplatin forms RNA adducts, while doxorubicin oxidizes RNA more efficiently than DNA in rat liver models (Xu et al., 2021; Simms and Knight, 2023).

14.6.2.2 RNA damage in carcinogenesis

a) *Chronic inflammation and ROS*: Persistent inflammation generates RNS/ROS that modify RNA bases, potentially creating mutagenic templates during viral reverse transcription or endogenous retrotransposition (Li et al., 2020).

b) *Impaired surveillance in tumors*: Downregulation of RNA quality control factors like hPNPase (polynucleotide phosphorylase) allows accumulation of damaged RNA, potentially accelerating mutagenesis and genomic instability (Li et al., 2020).

14.7 Conclusion and future perspectives

RNA not only acts as a messenger but also plays a key role in regulation, catalysis, and structure inside the cell. However, because of its single-stranded form and 2′-hydroxyl group, it is more prone to damage from internal sources like ROS, alkylating metabolites, and ions, external sources like drugs and UV light, making it more brittle than DNA. These damages disrupt normal RNA processing like splicing, translation, and ribosome assembly and appear as oxidized bases, alkylated nucleotides, abasic sites, crosslinks, or strand breaks. According to studies, damaged RNA has been linked to major issues such as translational mistakes, the synthesis of defective proteins, etc. Its fragility is highlighted by the fact that RNA is frequently more susceptible to oxidative and UV stress than DNA.

RNA surveillance is a complex system essential for preserving the integrity of the transcriptome. Considering subcellular location, RNA type, and temporal dynamics offers a comprehensive view of RNA quality control. Extensive research has been conducted, yet significant mechanistic gaps remain, especially in organelle biology and long noncoding RNA regulation—highlighting promising directions for future research.

RNA repair systems are present to cope with the damage. Methyl damage can be removed directly by enzymes such as AlkB and its human counterparts, whereas damaged RNA ends are sealed by ATP-dependent ligases and kinases. Special RNA repair processes (Pnkp, Hen1, and ligases) in bacteria guard against recurring ribotoxin damage. RNA surveillance and repair are essential for the survival and health of the cell. Diseases like cancer, neurodegeneration, and ageing problems are caused by malfunctions in these systems. RNA damage transcends its historical status as a molecular bystander to emerge as a central pathogenic mechanism in neurodegeneration and a deliberate therapeutic target in cancer. The shared vulnerability of neurons and cancer cells to RNA damage underscores the importance of cell type-specific responses and repair capacity. Future therapeutics must balance the protective removal of damaged RNA in neurodegeneration with the strategic induction of RNA damage in cancer, all while navigating delivery challenges. Understanding the molecular grammar of RNA damage responses from quality control pathways to repair enzymes like AlkB homologs will unlock precisely targeted interventions across these devastating diseases. These insights shed light on fundamental RNA biology and provide fresh approaches to therapeutic interventions focusing on RNA stability mechanisms and repair in human illness.

References

Anastasiadou, E., Jacob, L.S., Slack, F.J., 2018. Non-coding RNA networks in cancer. Nat. Rev. Cancer. 18 (1), 5–18. https://doi.org/10.1038/nrc.2017.99.

Beranek, D.T., 1990. Distribution of methyl and ethyl adducts following alkylation with monofunctional alkylating agents. Mutat. Res. Fundam. Mol. Mech. Mutagen. 231 (1), 11–30.

Berg, M.D., Brandl, C.J., 2021. Transfer RNAs: diversity in form and function. RNA biol. 18 (3), 316–339. https://doi.org/10.1080/15476286.2020.1809197.

Castellani, R.J., Nunomura, A., Rolston, R.K., Moreira, P.I., Takeda, A., Perry, G., Smith, M.A., 2008. Sublethal RNA oxidation as a mechanism for neurodegenerative disease. Int. J. Mol. Sci. 9 (5), 789–806.

Chamieh, H., Ballut, L., Bonneau, F., Le Hir, H., 2008. NMD factors UPF2 and UPF3 bridge UPF1 to the exon junction complex and stimulate its RNA helicase activity. Nat. Struct. Mol. Biol. 15 (1), 85–93.

Cooper, G., Adams, K., 2023. *The cell: a molecular approach*, 9th Oxford University Press.

Cordes, J., Zhao, S., Engel, C.M., Stingele, J., 2025. Cellular responses to RNA damage. Cell 188 (4), 885–900. https://doi.org/10.1016/j.cell.2025.01.005.

Crick, F., 1970. Central dogma of molecular biology. Nature 227 (5258), 561–563. https://doi.org/10.1038/227561a0. In press.

Esteller, M., 2011. Non-coding RNAs in human disease. Nat. Rev. Genet. 12 (12), 861–874. https://doi.org/10.1038/nrg3074.

Feyzi, E., Sundheim, O., Westbye, M.P., Aas, P.A., Vagbo, C.B., Otterlei, M., Slupphaug, G., Krokan, H.E., 2007. RNA base damage and repair. Curr. Pharm. Biotechnol. 8 (6), 326–331.

Forconi, M., Herschlag, D., 2009. Metal ion-based RNA cleavage as a structural probe. Methods Enzymol. 468, 91–106.

Glover, M.L., Burroughs, A.M., Monem, P.C., Egelhofer, T.A., Pule, M.N., Aravind, L., Arribere, J.A., 2020. NONU-1 encodes a conserved endonuclease required for mRNA translation surveillance. Cell Rep. 30 (13), 4321–4331.

Gupta, P., Li, Y.R., 2018. Upf proteins: highly conserved factors involved in nonsense mRNA mediated decay. Mol. Biol. Rep. 45 (1), 39–55.

Guydosh, N.R., Kimmig, P., Walter, P., Green, R., 2017. Regulated Ire1-dependent mRNA decay requires no-go mRNA degradation to maintain endoplasmic reticulum homeostasis in S. pombe. eLife 6, e29216.

Hansen, T.B., Jensen, T.I., Clausen, B.H., Bramsen, J.B., Finsen, B., Damgaard, C.K., Kjems, J., 2013. Natural RNA circles function as efficient microRNA sponges. Nature 495 (7441), 384–388. https://doi.org/10.1038/nature11993.

Hashimoto, Y., Takahashi, M., Sakota, E., Nakamura, Y., 2017. Nonstop-mRNA decay machinery is involved in the clearance of mRNA 5′-fragments produced by RNAi and NMD in Drosophila melanogaster cells. Biochem. Biophys. Res. Commun. 484 (1), 1–7.

Ikeuchi, K., Izawa, T., Inada, T., 2019. Recent progress on the molecular mechanism of quality controls induced by ribosome stalling. Front. Genet. 9, 743.
Jung, C., Gutell, R.R., 2004. Diversity of base-pair conformations and their occurrence in rRNA structure and RNA structural motifs. J. Mol. Biol. 344 (5), 1225–1249. https://doi.org/10.1016/j.jmb.2004.09.072.
Kim, Y.J., Wilson III, D.M., 2012. Overview of base excision repair biochemistry. Curr. Mol. Pharmacol. 5 (1), 3–13.
Kong, Q., Lin, C.L.G., 2010. Oxidative damage to RNA: mechanisms, consequences, and diseases. Cell. Mol. Life Sci. 67 (11), 1817–1829.
Kurosaki, T., Popp, M.W., Maquat, L.E., 2019. Quality and quantity control of gene expression by nonsense-mediated mRNA decay. Nat. Rev. Mol. Cell Biol. 20 (7), 406–420.
Li, Z., Chen, X., Liu, Z., Ye, W., Li, L., Qian, L., Ding, H., Li, P., Aung, L.H.H., 2020. Recent advances: molecular mechanism of RNA oxidation and its role in various diseases. Front Mol. Biosci. 7, 184.
Li, Z., Wu, J., DeLeo, C.J., 2006. RNA damage and surveillance under oxidative stress. IUBMB Life 58 (10), 581–588.
Liguori, I., Russo, G., Curcio, F., Bulli, G., Aran, L., Della-Morte, D., Gargiulo, G., Testa, G., Cacciatore, F., Bonaduce, D., Abete, P., 2018. Oxidative stress, aging, and diseases. Clin. Interv. Aging 757–772.
Liljas, A., Moore, P., 2012. Ribosomes-structure and function. Curr. Opin. Struct. Biol. 22 (6), 730–732. https://doi.org/10.1016/j.sbi.2012.10.007.
Liu, Z., Chen, X., Li, Z., Ye, W., Ding, H., Li, P., Aung, L.H.H., 2020. Role of RNA oxidation in neurodegenerative diseases. Int. J. Mol. Sci. 21 (14), 5022.
Mirzaei, H., Regnier, F., 2006. Protein–RNA cross-linking in the ribosomes of yeast under oxidative stress. J. Proteome Res. 5 (12), 3249–3259. https://doi.org/10.1021/pr060337l.
Müller, M., Tittel, L.S., Petfalski, E., Iyer, K.V., Pastore, S., Butto, T., Winz, M.L., 2024. Yeast elongation factor homolog new1 protects a subset of mRNAs from degradation by no-go decay. Biorxiv doi:10.1101/2024.11.26.625534, 2024-11.
Nagy, E., Maquat, L.E., 1998. A rule for termination-codon position within intron-containing genes: when nonsense affects RNA abundance. Trends Biochem. Sci 23 (6), 198–199.
Navickas, A., Chamois, S., Saint-Fort, R., Henri, J., Torchet, C., Benard, L., 2020. No-Go Decay mRNA cleavage in the ribosome exit tunnel produces 5′-OH ends phosphorylated by Trl1. Nat. Commun. 11 (1), 122.
Nicholson, P., Mühlemann, O., 2010. Cutting the nonsense: the degradation of PTC-containing mRNAs. Biochem. Soc. Trans. 38 (6), 1615–1620.
Noah, J.W., Shapkina, T., Wollenzien, P., 2000. UV-induced crosslinks in the 16S rRNAs of Escherichia coli, Bacillus subtilis and Thermus aquaticus and their implications for ribosome structure and photochemistry. Nucleic Acids Res. 28 (19), 3785–3792.
Nunomura, A., Honda, K., Takeda, A., Hirai, K., Zhu, X., Smith, M.A., Perry, G., 2006. Oxidative damage to RNA in neurodegenerative diseases. Biomed. Res. Int. 2006 (1), 082323.
Olatunji, M., Liu, Y., 2025. RNA damage and its implications in genome stability. DNA Repair (Amst.) 147, 103821.
Palazzo, A.F., Lee, E.S., 2015. Non-coding RNA: what is functional and what is junk? Front. Genet. 6, 2. https://doi.org/10.3389/fgene.2015.00002.
Powers, K.T., Szeto, J.Y.A., Schaffitzel, C., 2020. New insights into no-go, non-stop and nonsense-mediated mRNA decay complexes. Curr. Opin. Struct. Biol. 65, 110–118.
Prashar, T., De La Selle, F., Hudak, K.A., 2023. Abasic RNA: its formation and potential role in cellular stress response. RNA Biology 20 (1), 348–358.
Puno, M.R., Lima, C.D., 2022. Structural basis for RNA surveillance by the human nuclear exosome targeting (NEXT) complex. Cell 185 (12), 2132–2147.
Seixas, A.F., Quendera, A.P., Sousa, J.P., Silva, A.F., Arraiano, C.M., Andrade, J.M., 2022. Bacterial response to oxidative stress and RNA oxidation. Front. Genet. 12, 821535.
Shoemaker, C.J., Green, R., 2012. Translation drives mRNA quality control. Nat. Struct. Mol. Biol. 19 (6), 594–601.
Shuman, S., 2023. RNA repair: hiding in plain sight. Annu. Rev. Genet. 57 (1), 461–489.
Sies, H., 2015. Oxidative stress: a concept in redox biology and medicine. Redox. Biol. 4, 180–183.
Simms, C.L., Hudson, B.H., Mosior, J.W., Rangwala, A.S., Zaher, H.S., 2014. An active role for the ribosome in determining the fate of oxidized mRNA. Cell Rep. 9 (4), 1256–1264.
Simms, C.L., Yan, L.L., Zaher, H.S., 2017. Ribosome collision is critical for quality control during no-go decay. Mol. Cell 68 (2), 361–373.
Simms, N., Knight, J.R., 2023. RNA damage: the forgotten target of clinical compounds. Front. RNA Res. 1, 1248236.

Soles, L.V., Liu, L., Zou, X., Yoon, Y., Li, S., Tian, L., Valdez, M., Yu, A.M., Yin, H., Li, W., Ding, F., 2025. A nuclear RNA degradation code is recognized by PAXT for eukaryotic transcriptome surveillance. Mol. Cell 85 (8), 1575–1588.

Thomas, E.N., Kim, K.Q., McHugh, E.P., Marcinkiewicz, T., Zaher, H.S., 2020. Alkylative damage of mRNA leads to ribosome stalling and rescue by trans translation in bacteria. eLife 9, e61984.

Velema, W.A., Lu, Z., 2023. Chemical RNA cross-linking: mechanisms, computational analysis, and biological applications. Jacs Au 3 (2), 316–332.

Wamer, W.G., Wei, R.R., 1997. In vitro photooxidation of nucleic acids by ultraviolet A radiation. Photochem. Photobiol. 65 (3), 560–563.

Welp, L.M., Wulf, A., Chernev, A., Horokhovskyi, Y., Moshkovskii, S., Dybkov, O., Neumann, P., Pašen, M., Siraj, A., Raabe, M., Göthert, H., 2024. Chemical crosslinking extends and complements UV crosslinking in analysis of RNA/DNA nucleic acid–protein interaction sites by mass spectrometry. Biorxiv doi:10.1101/2024.08.29.610268, 2024-08.

Wilkinson, M.F., Shyu, A.B., 2002. RNA surveillance by nuclear scanning? Nat. Cell Biol. 4 (6), E144–E147.

Wolin, S.L., Maquat, L.E., 2019. Cellular RNA surveillance in health and disease. Science 366 (6467), 822–827.

Wurtmann, E.J., Wolin, S.L., 2009. RNA under attack: cellular handling of RNA damage. Crit. Rev. Biochem. Mol. Biol. 44 (1), 34–49.

Xu, Z., Huang, J., Gao, M., Guo, G., Zeng, S., Chen, X., Wang, X., Gong, Z., Yan, Y., 2021. Current perspectives on the clinical implications of oxidative RNA damage in aging research: challenges and opportunities. Geroscience 43 (2), 487–505.

Yan, L.L., Zaher, H.S., 2019. How do cells cope with RNA damage and its consequences? J. Biol. Chem. 294 (41), 15158–15171.

Zagórowska, I., Kuusela, S., Lönnberg, H., 1998. Metal ion-dependent hydrolysis of RNA phosphodiester bonds within hairpin loops. A comparative kinetic study on chimeric ribo/2′-O-methylribo oligonucleotides. Nucleic Acids Res. 26 (14), 3392–3396.

CHAPTER

15

DNA damage and cancer: Advances, risk assessment, and opportunities

Amol Gholap[a], *Pankaj Khuspe*[b], *Dipali Mane*[c] *and Md Jasim Uddin*[d]

[a]Department of Pharmaceutics, St. John Institute of Pharmacy and Research, Palghar, Maharashtra, India [b]Department of Pharmaceutics, Shriram Shikshan Sanstha's College of Pharmacy, Solapur, Maharashtra, India [c]Department of Pharmacology, Shriram Shikshan Sanstha's College of Pharmacy, Solapur, Maharashtra, India [d]Department of Pharmaceutical Technology, Faculty of Pharmacy, Universiti Malaya, Kuala Lumpur, Malaysia

15.1 Introduction

15.1.1 Background on DNA damage and cancer

DNA damage is any modification in the structure or chemistry of the DNA molecule that negatively affects its integrity and has a possible effect of derailing the proper transfer of genetic information. This may cause such changes as base modifications, single- and double-strand breaks (SSBs and DSBs), crosslinks, and helix-distorting lesions (Groelly et al., 2023). When such lesions are not repaired or are repaired erroneously, the expression of genes may change, proteins may malfunction, or mutations may occur. In a basic sense, the disease of cancer is a genomic instability disease, which is described by unchecked mobility of the cells, escape from apoptosis, and progressive garnering of an invasive capability. Mutation of key regulatory genes, e.g., proto-oncogenes, tumor suppressor genes, and DNA repair genes, may be the initial stages of developing tumors. The promotion of tumor and malignant transformation is realized with the progressive accumulation of genetic and epigenetic changes (Huang and Zhou, 2021).

The first direct conceptual connection between DNA damage and carcinogenesis was made early in the 20th century during investigations of the effect of exposure to ionizing radiation and select chemicals that had been found to increase the incidence of cancer. The mutagenic

Genome Damage: Repair and Adaptation Across Life.
DOI: https://doi.org/10.1016/B978-0-443-42857-9.00011-6

properties of ultraviolet (UV) light, as well as the relationship between adduct formation in DNA and tumorigenesis, were both shown in landmark discoveries (Moon et al., 2023). The two findings cemented DNA damage as a main etiological factor for cancer. Subsequent progress in molecular biology has now shown that unresolved DNA damage, replication stress, and defective repair modules all play direct roles in the mutational burden and clonal evolution of cancer cells. In the modern world, it is still critical to learn more about the relationship between DNA damage, repair mechanisms, and the oncogenic pathways, and focus on these factors as the cornerstones of cancer risk assessment and targeted therapeutic interventions development (Jurkovicova et al., 2022).

15.1.2 Sources of DNA damage

Insults caused by endogenous and exogenous sources constantly challenge the integrity of the DNA, exposing it to significant lesions per cell per day. Normal cellular metabolism is the main source of endogenous sources (Zhu et al., 2024). Reactive oxygen species (ROS) that occur during the process of oxidative phosphorylation can oxidize bases (e.g., 8-oxoguanine) or cause SSBs and DSBs. Misincorporations of bases (or small insertions/deletions) may also arise because of replication errors that survive despite proofreading and mismatch repair. Due to spontaneous hydrolysis, there is loss of purine bases (depurination) or loss of cytosine (deamination), which changes base-pairing and makes mutation possible (Ma et al., 2023). Physical, chemical, and biological agents are exogenous. The presence of UV radiation generates cyclobutane pyrimidine dimers (CPDs) and 6-4 photoproducts, which alter the DNA helix. Direct ionization and free radical formations attribute strand breaks and oxidative base damages induced by ionizing radiations (Liao et al., 2022; Vaddavalli and Schumacher, 2022). Bulky DNA adducts formed by chemical carcinogens, e.g., polycyclic aromatic hydrocarbons, alkylating agents, and aflatoxins, have been shown to hinder replication and transcription. Other viruses (such as human papillomavirus (HPV) and Epstein-Barr virus (EBV)) cause DNA damage indirectly via the expression of oncogenes and inflammation, or directly as a result of viral insertion into the host genome. The mutation spectrum in human cancers has its basis in the interplay of these deleterious forces coupled with incomplete repair. Questions regarding the comparative role of endogenous and exogenous DNA damage in certain types of cancer, in particular, are still major epidemiological and molecular research topics as presented in Fig. 15.1 (Curtin, 2023; Yang et al., 2024).

15.1.3 Significance of DNA damage response in cancer biology

An elaborate cellular network that participates in the DDR involves lesion recognition, signal transduction, and repair. It has been referred to as a genomic protector: it keeps things stable by regulating and ensuring consistent repair of the genome with cell-cycle checkpoints, transcriptional programs, and apoptosis when the damage is irreparable. Integration of the damage signals is achieved through key mediators, which include ATM, ATR, CHK1, CHK2, and p53, which prevent propagation of mutations (Li et al., 2024). DDR contributes to cancer in two ways. It inhibits cancer by repairing lesions prior to cell division or destroying greatly damaged cells, as a tumor suppressor. In one example, functional p53 arrests the cell cycle and causes apoptosis in cells with serious genomic damage. On the other hand, DDR defects, either inherited (e.g., BRCA1/2 mutation in breast and ovarian cancer) or acquired, predispose

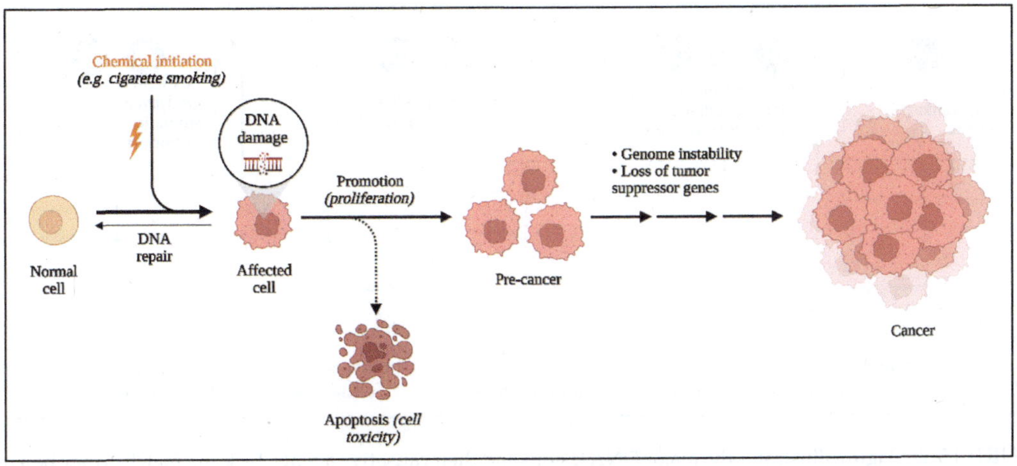

FIGURE 15.1 Schematic representation of the multistage process of chemical carcinogenesis. Normal cells become exposed to chemical initiators (e.g., cigarette smoke), and these expose DNA damage that can be repaired through intrinsic DNA repair systems. Unrepaired, cells containing the lesions in their DNA become affected, and either they undergo apoptosis (cell toxicity) or aberrant proliferation. Promotion stage results in the growth of precancerous cell populations, which is marked by cell cycle deregulation. Progressive accumulation of genomic instability and loss of tumor suppressor genes further drives malignant transformation. Finally, the uncontrolled proliferation leads to the development of cancer, and it is the balance between repair, cell death, and oncogenic progression of carcinogenesis. (Created by using BioRender.com).

genomic instability and allow tumor evolution, heterogeneity, and resistance to therapy (Drew et al., 2025). Ironically, impairment of DDR in tumor cells causes treatment susceptibilities. The agents (e.g., PARP inhibitors) take advantage of synthetic lethality to treat human cancers with homologous recombination (HR) defects and provide precision therapy approaches. Nevertheless, there may be drug resistance conferred by adaptive responses and secondary mutations, which create clinical problems. Therefore, the deconstruction of the DDR pathways provides prognostic value and translational value to individualized treatment of cancer (Huang and Zhou, 2021).

15.2 Types of DNA damage and their relevance to carcinogenesis

15.2.1 Single-strand DNA damage

Single-strand DNA damage includes base modifications, single-strand breakage (SSBs), and spontaneous base loss [depyrimidination/depurination]. Replacements of the bases, which also include oxidative damages (e.g., 8-oxoguanine), alkylation, or deamination, are a product of either endogenous products of metabolism or environmental factors. When the phosphodiester backbone is interrupted in one strand of DNA, e.g., as part of the base excision repair process or because of reactive oxygen species (ROS) attack, SSBs are produced as presented in Fig. 15.2 (Chatterjee and Walker, 2017). Abasic (AP) sites are produced as a consequence of the hydrolytic degradation of the N-glycosidic bond by a process known as depurination. AP

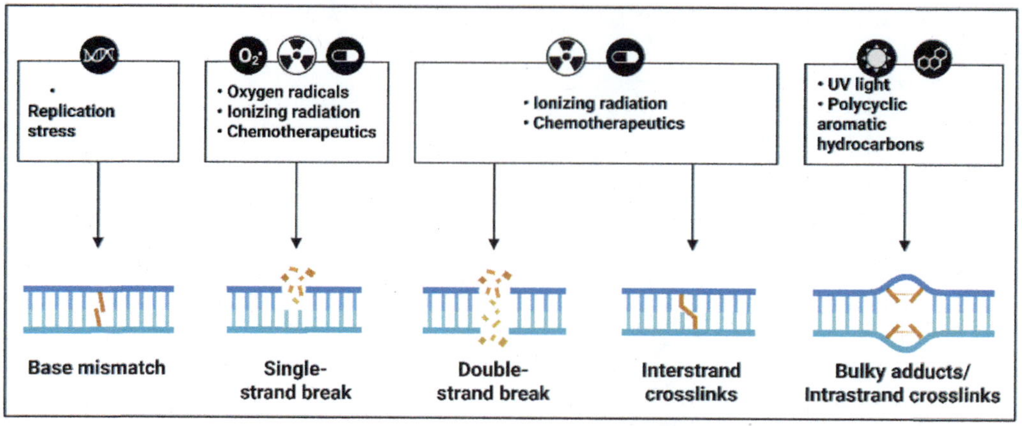

FIGURE 15.2 Figure illustrates the major DNA lesions and their causative agents. Base mismatches may be a result of replication stress during the process of DNA synthesis. Single-strand breaks are caused by oxygen radicals, ionizing radiation, and chemotherapeutic agents, whereas more serious damage from ionizing radiation and chemotherapy causes double-strand breaks. The ionizing radiation and some chemotherapeutics cause interstrand crosslinks that covalently bind two DNA strands and prevent replication. Bulky adducts or intrastrand crosslinks occurring due to UV light or polycyclic aromatic hydrocarbons lead to distortion of the DNA helix and disrupt transcription and replication. Taken together, these lesions impair genome stability and, depending on whether repaired or not, initiate DNA repair pathways, apoptosis, or carcinogenesis. (Created by using BioRender.com).

sites left unrepaired are mutagenic. These are lesions that can replicate inaccurately, giving rise to point mutations and/or frameshifts. Mutagenic capacity of single-strand break caused by a single-strand damage involves triggering either errors in the replication process or halting transcription, followed by activation of proto-oncogenes or loss of propriety in tumor-suppressor genes. Whereas most SSBs and base alterations are corrected with high efficiencies by cellular repair pathways, including base excision repair (BER) and mismatch repair (MMR), unresolved lesions lead to further genomic instability, which represents a molecular substrate of carcinogenic transformation (Andrés et al., 2023).

15.2.2 Double-strand breaks

DSBs constitute some of the most cytotoxic lesions of DNA, in which a combination of breaks in two chains of a double-stranded helix occurs simultaneously. They may originate due to the collapse of the replication fork in response to replication stress, ionizing radiation, topoisomerase II poison, or mechanical stress on chromatin (Huang and Zhou, 2020). The effect of improperly repaired or unrepaired DSBs relates to chromosomal translocations, deletion, and amplification, these occurrences being common in cancerous conditions, like leukemias and lymphomas. The mispairing of the ends of DNA can result in the activation of oncogenes (e.g., BCR-ABL fusion in chronic myeloid leukemia) or the silencing of a tumor suppressor. Pathways DSB repair nonhomologous end joining (NHEJ) and homologous recombination (HR) have different fidelity, whereby the bkrotein NHEJ is highly error-prone. HR deficiency (BRCA1/2 mutation) not only predisposes people to breast and ovarian cancer but also results in phenotypes in which PARPi could be used. The permanence of DSBs provokes ATM/ATR-responsive checkpoints, which in cancer is frequently distorted to enable blemished cells to spread (Morimoto et al., 2019).

15.2.3 DNA crosslinks and bulky adducts

Interstrand crosslinks (ICLs) and intrastrand crosslinks are crosslinkages between DNA strands that connect strands covalently, thus blocking replication and transcription to a significant extent. Chemical carcinogens (e.g., benzo[a]pyrene diol epoxide) or chemotherapeutic agents e.g., cisplatin form bulky adducts that impede the helical structure and halt the advancement of polymerases (Li et al., 2021). Such lesions induce complicated repair pathways that include nucleotide excision repair pathway (NER), translesion synthesis (TLS), and the Fanconi anemia pathway. Failure to repair correctly or successfully will result in mutagenesis or a gross chromosomal rearrangement, which contributes to carcinogenesis. Although they are therapeutically deployed in oncology, crosslinking agents provide an example of the role that persistent adducts can play in the induction of tumor formation (Jones et al., 2025).

15.2.4 Replication stress and fragile sites

Replication stress arises when the impediments to DNA synthesis include stalled proteins, secondary DNA structures, or incomplete repair of lesions. The unstable replication fork can distort into DSB until supported by the ATR-dependent mechanisms. The regions that contain fragile sites can be linked to genomic loci having cancer-related genes, and they are prone to breakage, especially in circumstances with replication stress. Tumors have frequent deletions and translocation and amplifications at these sites. Replication stress can be amplified by overexpression of oncogenes (e.g., MYC and RAS), phase-locked back to proliferative signaling to form a positive feedback loop. Targeting replication stress response pathways as an anticancer intervention is becoming an interesting concept (Saxena and Zou, 2022).

15.2.5 Mitochondrial DNA damage and cancer

Because of its location next to the electron transport chain and low repair capability, Mitochondrial DNA (mtDNA) is extraordinarily prone to oxidative damage. Normally, in the tumor, there are characteristic mutations of the mtDNA of genes that synthesize the mitochondrial components of oxidative phosphorylation and thus change energy metabolism to glycolysis (Warburg effect) (Xia et al., 2025). These mutations can also lead to an increase in ROS production, which increases nuclear DNA damage and potentiates a pro-oncogenic microenvironment. Mutations in the mitochondrial genome are therefore at the root of not only metabolic reprogramming but also of sustaining a mutagenic intracellular environment (Kopinski et al., 2021).

15.3 DNA damage response pathways in cancer

15.3.1 Base excision repair

The BER is the major mechanism to correct minor lesions or damage that does not affect the helical shape of the DNA, like oxidative and alkyl adduct-induced DNA damage that, when not fixed, can cause mutagenesis and oncogenesis. Examples of reactive oxidative products include 8-oxoguanine (8-oxoG); this is detected and repaired by an enzyme, the bifunctional glycosylase OGG1, which removes the damaged base and cleaves the DNA backbone by incision of the AP site. Breaking of the phosphodiester bond is then achieved by apurinic/apyrimidinic endonuclease 1 (APE1), and a single-strand break is made to have a 3-terminus of a hydroxyl (3 -hydroxyl) (Allkanjari and Baldock, 2021). DNA polymerase β (Pol 8) is then used to insert the rightful nucleotide and excise the 5-deoxyribose phosphate residue, and finally, ligation by DNA ligase III in combination with XRCC1. Recent discoveries show that BER is dynamically regulated by posttranslational modifications and protein-protein interactions, allowing the rapid response of cancer cells to genotoxic stress. Malfunction of OGG1 or APE1 has been associated with tumor growth, chemotherapy resistance, and aberrant redox signaling, which makes them candidates for prognosis biomarkers and targets of the therapy. The development of small-molecule modulators of BER enzymes provides hopeful directions in precision oncology, especially that of synthetic lethality (Gohil et al., 2023).

15.3.2 Nucleotide excision repair

NER is a highly conservative DNA repair pathway that plays an essential role in the elimination of helix-distorting lesions that include bulky lesions like UV-induced cyclobutane pyrimidine dimers (CPDs), 6-4 oxygen photo-product, and large adducts of genotoxic agents. It was done through the recognition of the damage, cutting both sides of the lesion, removal of a short oligonucleotide, gap filling by polymerases, and subsequent ligation. Two subpathways, namely, global genomic NER (GG-NER) and transcription-coupled NER (TC-NER), exist to preserve genomic integrity both in transcribed and nontranscribed areas (Pasqui et al., 2022). Mutations in NER genes lead to the rare autosomal recessive xeroderma pigmentosum (XP) phenotype that is manifested in profound UV hypersensitivity and hyperacceleration of cutaneous photoaging with an over 1000-fold increased risk of skin cancer caused by the cumulative bypass of mutagenic lesions. Disruption of lesion recognition, excision, or repair synthesis in XP complementation groups (XPA-XPG) directly correlates to carcinogenesis, which is directly correlated to NER defectiveness. New data is also emerging that NER modulation is relevant to the outcome of cancer therapy, with tumors that have a defective NER being more sensitive to platinum-based chemotherapeutics, and that these findings present interesting opportunities in precision oncology as presented in Fig. 15.3 (Szatkowska and Zdrada-Nowak, 2025).

15.3.3 Mismatch repair

Mismatch repair (MMR) is a postreplicative surveillance process that resolves base mismatches and smaller insertion-deletion loops created during the replication of the genome. These MMR proteins form the core, which includes MutS [MutS alpha (MutS alpha) and MutL (MutL alpha) (MLH1-PMS2)], and they identify and remove errors in DNA segments, allowing high-fidelity polymerases to resynthesize them. It is faithful, represses mutagenesis, and maintains genomic stability. A deficiency in MMR causes microsatellite instability (MSI), a mutational pattern consisting of alterations in the length of short tandem repeats (Jiang et al., 2021). Clinically, Lynch syndrome (hereditary nonpolyposis colorectal cancer) is characterized by loss of MMR, which occurs both as an autosomal dominant trait due to germline mutations in several MMR genes, most of which involve MLH1, MSH2, MSH6, and PMS2. The Lynch tumors have increased the speed of acquiring oncogenic mutations, which build up in colorectal, endometrial, and other cancer types. More recent developments involve the use of high sensitivity MSI detection algorithms together with immunotherapy, since in MMR-deficient MSI tumors, there will be increased neoantigens, which are sensitive to immune checkpoint blockade. Nevertheless, the timely screening of MMR defects is still essential in terms of risk stratification and specific disease prevention of cancer (Arai et al., 2021).

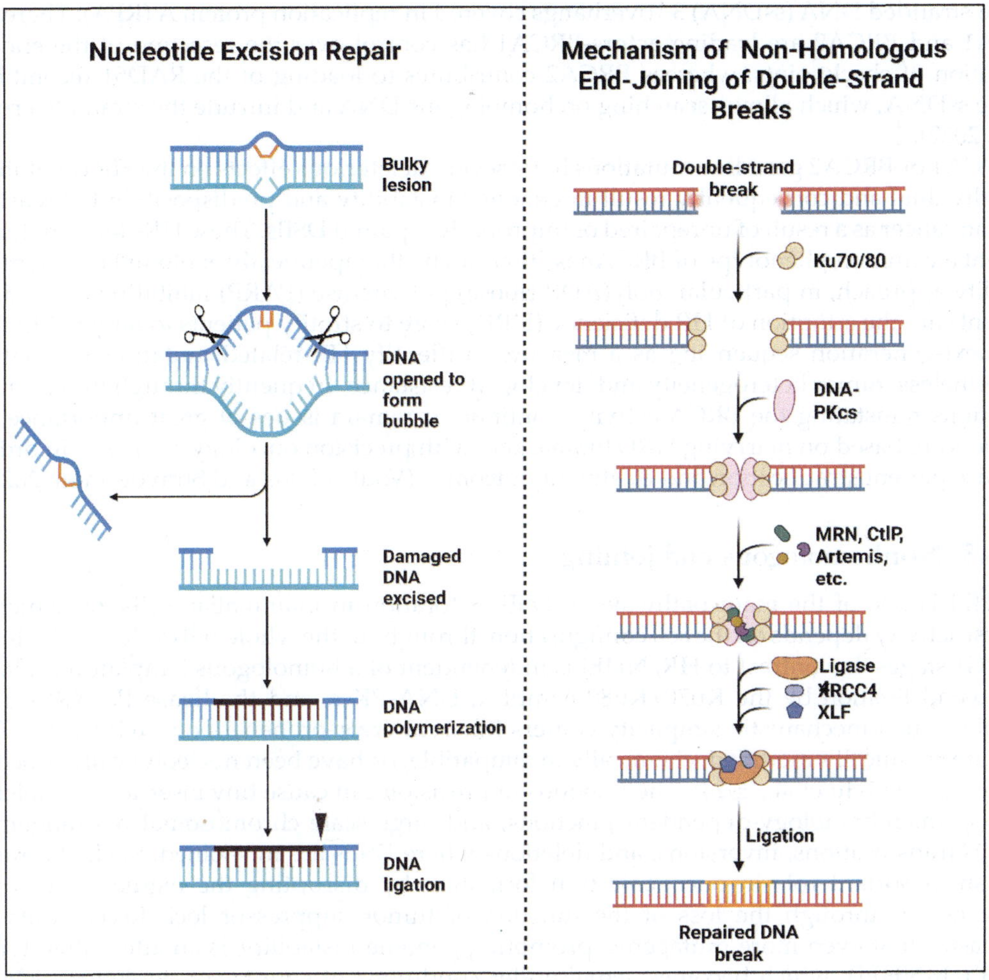

FIGURE 15.3 Shown in a schematic diagram, different DNA repair pathways are depicted. *Left*: Nucleotide excision repair (NER): The NER is able to repair large lesions like UV-induced adducts. It is performed by local unwinding of the DNA into a bubble, excising the damaged strand, filling the gap through polymerization of DNA, and a final ligation to repair the integrity. The nonhomologous end joining (NHEJ) mechanism of repairing double-strand breaks is displayed on the right. Ku70/80 proteins identify the ends of the DNA, which are then recruited by DNA-PKcs and end-processing factors (MRN complex, CtIP, and Artemis). Later, DNA ligase IV, XRCC4, and XLF ligation recover the DNA duplex to maintain genomic stability. (Created by using BioRender.com).

15.3.4 Homologous recombination

DNA repair mechanisms, HR, are a high-fidelity repair process in which a sister chromatid is used as a template to maintain error-free repair of genetic information during repair of a DNA double-strand break (DSB). Rigorous control occurs during the S and G2 phases of the cell cycle, at which time sister chromatids are present. In the HR initiation, resection of the ends of the DNA is carried out by the MRE11-RAD50-NBS1 (MRN) complex and CtIP, resulting in single-stranded DNA (ssDNA) 3 ′ overhangs covered in replication protein A (RPA). Thereafter, BRCA1 and BRCA2 are leading actors-BRCA1 has control over the resection of the end and activation of checkpoints, whereas BRCA2 contributes to loading of the RAD51 recombinase on the ssDNA, which allows searching on homologous DNA and intrude the strand (Cerniglia et al., 2022).

BRCA1 or BRCA2 germline mutations have severe deleterious effects on the HR capability of an individual and consequently result in genomic instability and predisposition to breast and ovarian cancer as a result of unrepaired or improperly repaired DSBs. These HR-deficient tumors present a common phenotype of BRCAness that can be therapeutically exploited by a synthetic lethality approach, in particular, poly(ADP-ribose) polymerase (PARP) inhibition. New developments are the adoption of HR deficiency (HRD) score to stratify patients to targeted therapy, and next-generation sequencing as a measure to identify HR-related mutational signatures. Nevertheless, tumor heterogeneity and developed resistance, frequently through the secondary mutations reinstating the BRCA activity, continue to remain issues of great importance. The frontiers are based on marrying HRD biomarkers with precision oncology strategies in order to enhance patient stratification and treatment outcomes (Voutsadakis and Stravodimou, 2023).

15.3.5 Nonhomologous end joining

NHEJ is one of the main pathways of DSB restoration in mammalian cells, on which the biggest activity depends on the cell configuration, throughout the whole cell cycle, and including G0 / G1 stages. In contrast to HR, NHEJ is independent of a homologous template, employing end-to-end ligation by the Ku70/Ku80 complex, DNA-PKcs, and the ligase IV/XRCC4-XLF complex. Such mechanistic simplicity confers inherent error-proneness on NHEJ when DNA ends are chemically modified, chemically incompatible, or have been nucleolytically processed before ligation (Hu et al., 2024). The previous imprecision can cause tiny insertions or deletions (indels), microhomology-dependent junctions, and large-scale chromosomal rearrangements such as translocations, inversions, and deletions where DSBs are mis-joined. Such aberrations play an important role in oncogenic transformation by disrupting the regulation of proto-oncogenes or through the loss of the function of tumor suppressor loci. Recent data also emphasize that even more mutagenic, promoting genome instability, is an alternative type of NHEJ (alt-NHEJ), which becomes active under conditions of core factor depletion. Selective manipulation of NHEJ fidelity holds promise as a therapeutic approach to reduce mutagenesis during genome editing and radiotherapy and to take advantage of NHEJ addiction in cancer cells as a source of synthetic lethality (Gehrke et al., 2022).

15.3.6 DNA damage tolerance and translesion synthesis

Translesion synthesis (TLS) and other DNA damage tolerance pathways allow replication machineries to overcome the roadblock caused by unrepaired lesions in DNA that would otherwise stagnate compromised replication forks. The TLS is carried out by special low-fidelity DNA polymerases (e.g., Pol -, Pol -, Pol -, and Rev1) that can distort templates in their active sites and enable replication of bulky adducts, oxidation damage, or UV-induced photoproducts. Even though critical to keep replication forging ongoing and avoiding fork collapse, the TLS tool comes with a high probability of causing mutagenesis associated with low proofreading activity and imprecise incorporation of nucleotides. Recent data reveal that altered TLS activity is a hallmark of oncogenesis, chemoresistance, and genomic instability and that it is both a survival and treatment target (Anand et al., 2023).

NHEJ is a major mechanism of repairing DSB in mammalian cells and occurs widely during the cell cycle, both during G0/G1. In contrast to HR, which uses a homologous template, NHEJ uses direct ligation of the DNA ends by Ku70/Ku80 complex, DNA-PKcs, ligase IV-XRCC4-XLF complex; this does not require a homologous template. Such a mechanistic want makes NHEJ intrinsically error-prone, particularly in the case that DNA ends are chemically altered, not compatible, or subjected to nucleolytic processing before ligation (Khatib et al., 2024). This imprecision may include small insertions/deletions (indels), microhomology-mediated fusions, and, in mis-joined long distances between DSBs, also macro-scale chromosomal rearrangement in forms including translocations, inversion, and deletion. These mutations also play a role in the oncogenic transformation because they deregulate proto-oncogenes or shut down a tumor suppressor locus. Recent data emphasize the point that in conditions of a lack of core factors, the alternative of NHEJ (alt-NHEJ) appears, which turns out to be even more mutagenic, aggravating genome instability. Altering NHEJ fidelity specifically is a potential therapeutic avenue to limit mutagenesis during genome editing and radiotherapy, as well as to leverage the NHEJ addiction of cancer cells in synthetic lethality applications (Nayak et al., 2021).

15.4 Genomic instability as a hallmark of cancer

15.4.1 Types of genomic instability

Genomic stability implies enhanced susceptibility of the genome to gain modification, which involves nucleotide change, structural rearrangement, and aneuploidy, and thus loss of genomic integrity. It is one of the characteristic features of cancer, serving as the malignant transformation driver and contributor. Genomic instability has broadly been categorized into two types of genomic instabilities, namely, MSI and chromosomal instability (CIN). MSI can be the result of impaired DNA MMR systems, thereby causing variance in the length of short tandem repeats randomly structured throughout the genome. This leads to frameshift mutations in coding areas of tumor suppressor genes, oncogenes, or apoptotic genes, which assist in oncogenesis, especially in colorectal, gastric, and endometrial cancer (Hanahan, 2022).

CIN refers, on the other hand, to extensive genomic changes, e.g., gains, losses, and translocations of whole chromosomes or large pieces of chromosomes. CIN can occur through the loss of mitotic checkpoint quality control, defective repair of DNA double-strand breaks, or dysfunctional telomeres, and further aneuploidy and karyotypic heterogeneity. In contrast to MSI, CIN, being the major contributor to tumor development, is based on altered gene dosage, reshuffling of the structural genome, and stimulation of oncogenic pathways. MSI and CIN promote tumor heterogeneity, drug resistance, and recurrence of the disease, which is why both aspects are important in precision oncology (Clarke and Mostoslavsky, 2022).

15.4.2 Mechanistic link between DNA damage response defects and cancer

DNA damage response (DDR) coordinates lesion sensing, signaling, and repair in order to preserve genome integrity. The protective network of DDR components breaks down due to the presence of defects and establishes a mutagenic environment promoting carcinogenesis.

Mutation accumulation is one of the major processes. Defective repair mechanisms, e.g., HR, nonhomologous end joining (NHEJ), or MMR, may allow surviving or misrepair lesions to occur, giving rise to base substitutions, insertions/deletions, and structural rearrangements. These changes take place over repeated cell divisions and interrupt regulatory networks of genes, permitting malignant transitions (Malamos et al., 2024).

A second key connection is the activation of an oncogene and the loss of tumor suppressor. DDR flaws may foster proto-oncogenes activation by amplifying or gain-of-function mutations, and tumor suppressor inactivation by deletion, nonsense mutations, or promoter hypermethylation. These include, by way of example, inactivation of the p53 pathway by TP53 mutation or MDM2 amplification to negate cell cycle arrest and apoptosis. Equally, ATM or BRCA1/2 loss reduces checkpoint signaling and HR repair and induces increased genomic instability. Synergistically, these molecular perturbations lead to unregulated proliferation, survival of genotoxic stress, and clonal evolvement. Moreover, mutated DDR increases the resistance to replication stress tolerance and the development of resistant apoptosis, making it a central hub in cancer pathogenesis (Kwok et al., 2024).

15.4.3 Examples of DNA damage response gene mutations in human cancers

Many mutations of the DDR gene contain powerful oncogenic associations. The most common HR repair defects are the BRCA1 and BRCA2 mutations that are characteristic of hereditary breast and ovarian cancers and make cells susceptible to the accumulation of double-strand breaks. MLH1 and MSH2 mutations, the characteristics of Lynch syndrome, give MSI and increase the risk of colorectal and endometrial cancers (Wu et al., 2025). The ATM mutations defect checkpoint activation and are common in lymphoid neoplasms. TP53, which is mutated in more than 50% of human cancers, breaks down genome surveillance and apoptotic guardianship. Taken together, these genetic alterations do not just cause tumorigenesis but also determine therapeutic sensitivities, which form the basis of synthetic lethality-based therapies, such as PARP inhibition (Zhao et al., 2024).

15.5 Advances in DNA damage detection and analysis

15.5.1 High-throughput genomics and mutational signatures

The latest development in next-generation sequencing (NGS) has transformed the sensitivity to a new level of detecting and characterizing DNA damage in cancer. Whole-genome sequencing (WGS) and whole-exome sequencing (WES) allow identification of base substitutions, small deletions/insertions, and large-scale structural variants caused by malfunctions in the DNA repair process. These comprehensive data have spurred the process of mutational signature analysis, in which certain patterns of nucleotide change represent characteristic processes of genotoxic exposure, e.g., UV light-induced cyclobutane dimers or COS cells APOBEC cytosine deamination (Muyas et al., 2024). Several dozen well-characterized mutational signatures have now been catalogued in the catalogue of somatic mutations in cancer (COSMIC) database, an extremely useful structure through which to begin finding links between genomic change, DNA damage-causative agents, and repair defects. This type of analysis can be important in precision oncology by enabling clinicians to surmise carcinogen exposure histories, hereditary repair syndromes, and therapy susceptibilities (e.g., sensitivity to PARP inhibitors in homologous recombination-defective tumors). Moreover, complementary transcriptomic and epigenomic data improve the functional context of interpretation and allow correlating mutational burden with gene expression variations and chromatin accessibility, as well as immune evasion products. When coupled with machine learning, high-throughput information of genomic profiling is increasingly being applied to predict tumor evolution and inform therapeutic courses of action in individually treated patients (Mertz et al., 2025).

15.5.2 Single-cell genomics and spatial profiling

Single-cell sequencing methods have become invaluable tools to define the discrepancy of DNA damage and repair landscapes within the same tumor. These methods define subclonal landscapes by profiling individual cells on a mutational and transcriptomic level, illuminate rare DNA damage-related variants, and characterize populations of cells refractory to therapy (Gulati et al., 2025). Spatial transcriptomics also combines genomic analysis with the tissue architecture, which allows for locating DNA repair defects in certain tumor microenvironment niches. This spatial resolution is especially important to describe hypoxia, inflammation, and metabolic stress gradients-parameters that determine the level of DNA damage accumulation and the rate of repair. Single-cell genomics combined with spatial profiling, therefore, allows viewing tumor evolution in a multidimensional context, helping to realize more accurate biomarker discovery and targeted intervention strategies. The methods can be used to improve the prediction of aggressive subclones in their early stages of growth, adaptive therapies, and localized treatment delivery (Shiao et al., 2024).

15.5.3 CRISPR/Cas9 functional screens

Genome-wide loss-of-function and gain-of-function screens of CRISPR/Cas9 have become imperative in understanding the effects of functional consequences of perturbation of the DNA damage repair genes. These screens determine synthetic lethal interactions by inactivating or activating genes against genes throughout the genome one by one and then testing the outcome, allowing them to identify specific progenitor dependencies that exist between genetic lesions (synthetic lethality) (Tafrishi et al., 2024). As a specific example, vulnerability to PARP inhibition is generated by loss of BRCA1/2, a synthesis now prominently used in cancer treatment. CRISPR-based methods also permit the characterization of novel DDR-regulators, which may be important in terms of novel repair pathways as well as novel drug targets. Combining CRISPR screens with DNA damage reagents or high-content imaging allows direct mapping of gene x drug combinations to directly reveal context-specific gene dependencies. Target prioritization is fine-tuned by following an integration of CRISPR screening data with genomic and proteomic datasets, which increases the pace of translating basic DDR biology into clinical practice (Zhou et al., 2024).

15.5.4 Molecular imaging and biosensors

New technologies in the field of molecular imaging and biosensors give the possibility to track and measure the DNA damage in real time in organisms and living cells. Dynamic monitoring of the formation and repair kinetics of double-strand breaks is possible with fluorescence resonance energy transfer (FRET)-based probes, immunofluorescence of γ-H2AX, and bioluminescent reporters (Gest et al., 2024). Noninvasive imaging of the tumor DDR can be achieved in vivo using imaging methods that include positron emission tomography (PET) using DNA damage-targeted probes, which can be used to enable a treatment response to be measured. Nanoparticle-based biosensors that are developed provide high sensitivity in the detection of oxidative lesions on DNA and alkylation products directly in physiologic systems. Such tools fill the niche between molecular tests and functional imaging to zoom into the genomic instability at spatiotemporal resolution in both preclinical and clinical applications (Batarchuk et al., 2024).

15.5.5 Emerging liquid biopsy biomarkers

Minimally invasive cancer detection of DNA damage-related genome changes has been in the spotlight because of circulating tumor DNA (ctDNA) in liquid biopsy strategies. Fragmentomics, mutational signature analysis, and methylation profiling of ctDNA would facilitate the identification of the active processes of DNA breaking and repair deficiencies without sampling any tumor tissues (Rayamajhi et al., 2024). Biomarkers, to identify the therapies and trace disease progression, various damage-associated patterns, relative strand-specific fragmentation, and the prevalence of oxidative lesions, could play a role. This type of assay comes in especially useful in longitudinal follow-up interventions, where the changes in tumor evolution can be dynamically assessed under therapeutic pressure. It has a synergistic effect in combination with other circulating biomarkers, such as exosomal RNA and protein signature, that enhance the specificity and applicability of findings in the clinical setting (Ou et al., 2025).

15.6 Risk assessment models in DNA damage–driven cancer

15.6.1 Genetic risk factors

There is a major role of genetic predisposition in regulating the predisposition of individual susceptibilities to malignancies based on the DNA damage. Defects in DDR genes, which include germline mutations in BRCA1, BRCA2, ATM, TP53, and MLH1, are firmly linked to failing genomic maintenance, which confers high mutational loads and an increase in incidence rates of cancer across organ systems. Such variants tend to have autosomal dominant inheritance, variable penetrance, and require specific genetic screening where the risk of developing the findings is high (Zhang et al., 2022). Moreover, the recent developments in genome-wide association studies (GWAS) have made possible the calculation of polygenic risk scores (PRS), which include the multiplicative impact of many genes of low penetrance on the pathways of DDR. PRS are used to categorize people into various groups of risks to refine surveillance and preventive efforts. Notably, gene-gene and gene-environment interactions have to be considered during risk modeling because the penetrance of pathogenic variants depends significantly on lifestyle, hormonal, and metabolic factors. The incorporation of germline variant profiling into clinical risk models is becoming more of a reality with NGS and large-scale biobanking, providing the opportunity to identify potential disease at an early stage and to design therapeutically useful intervention strategies (Hayes et al., 2025).

15.6.2 Epigenetic modifiers

The effect of epigenetic control of DDR gene expression and subsequent risk of cancer is considerable. Genes CpG islands in promoter regions, aberrant CpG island DNA methylation leading to transcriptional silencing of a key regulatory gene (e.g., MGMT, MLH1, and BRCA1), can be mimicked by the absence of functional proteins caused by loss-of-function mutations (Jaafari, 2024). In a similar manner, dysregulated histone modifications (hypoacetylation of histone H4 or abnormal methylation of histone H3 at sites of lysines) change chromatin accessibility and inhibit the recruitment of repair complexes with damaged sites. These alterations tend to be environmentally caused and transmissible to cell divisions, hence leading to chronic genomic instability. The biomarkers referred to as epigenetic can be measured through a liquid biopsy or high-throughput methylome profiling and have the opportunity of becoming a noninvasive measure of cancer risks. Application of epigenetic signatures in predictive models improves risk assessment, especially nonsignatures, and creates potential for epigenetic chemopreventive models (Ferrer-Diaz et al., 2024).

15.6.3 Environmental and lifestyle risk factors

The development of cancer associated with DNA damage is very important and dependent on environmental factors and lifestyles. Polycyclic aromatic hydrocarbons and nitrosamines in tobacco smoke cause bulky DNA adducts, overload the repair, while interacting as synergists with deficient DDR genotypes. Oxidative stress and nucleotide pool balance are modulated according to dietary factors, such as high intakes of processed meats (containing large amounts of N-nitroso compounds) or low intakes of antioxidants. Chronic exposure to ionizing radiation, asbestos, or industrial solvents, occupational hazards, produces cumulative DNA lesions, which catalyze malignant transformation in a predisposed person. In addition, crosslink repair is compromised because alcohol metabolism produces a genotoxic metabolite, acetaldehyde. New quantitative risk assessment models include exposure, biomarkers of DNA damage (e.g., 2-5A-H2AX foci), and genetic susceptibility diagnoses to improve both personal and population-only predictions of cancer risk (Tang et al., 2022).

15.6.4 Integrated multiomics approaches

The development of high-throughput technologies has supported multiomics-type interventions in holistic cancer risk assessment in the DNA damage capacity. Genomics gives us variant-level knowledge about DDR gene mutations, whereas epigenomics reports methylation and chromatin status changes that regulate repair capacity. Transcriptomics shows DDR pathway dysregulatIon in response to environmental stress, and proteomics measures posttranslational modification necessary to assemble repair complexes. The new field of exposomics, which involves a systematic assessment of lifetime environmental exposures, introduces an exogenous level that can causally be associated with genomic or epigenomic signatures: mutagenic exposures (Sarvari et al., 2022). On the one hand, the combined multiomics patterns resulting in complex interactions of many genetic, epigenetic, and environmental factors can be identified via an integrated multiomics model using machine learning algorithms, which gives the possibility to perform high-resolution risk stratification. These models would specifically be useful in identifying early (preclinical) changes that precede tumorigenesis. Nevertheless, problems still exist in normalizing data across nonhomogenous platforms, privacy, and the translation of findings to driving clinical tools. Dynamic and longitudinal multiomics surveillance is also likely to become the future of cancer risk prediction, whereby preventive activities can be undertaken on a personalized basis, and monitoring can be adjusted accordingly (Ovejero and Moreaux, 2021).

15.7 Translational opportunities

15.7.1 Synthetic lethality in cancer therapy

Synthetic lethality takes advantage of the idea that simultaneous impairment of two different pathways involved in DNA repair can cause tumor-selective cell death but not normal cells. The example, which is paradigm-shifting, is the implementation of poly(ADP-ribose) polymerase (PARP) inhibitors in tumors with harmful BRCA1/2 mutations. High-fidelity homologous recombination repair (HRR) of DNA double-strand breaks depends on BRCA proteins. Without them, cells heavily rely on PARP-mediated BER in order to survive (Previtali et al., 2024). PARP1/2 inhibition pharmacologically sequesters the enzyme on DNA, whereby the single-strand breaks accumulated are replicated to form lethal double-strand breaks. Clinically validated drugs like olaparib, niraparib, and rucaparib have shown meaningful benefit in ovarian cancers, breast cancers, pancreatic cancers, and prostate cancers with germline or somatic BRCA events. There are, however, mechanisms of resistance, such as re-establishment of HRR through additional BRCA mutations or increasing drug efflux transporters, that are important issues. Work in this area seeks to broaden the field of synthetic lethality that covers BRCA deficiencies to tumors other than homologous recombination deficiency (HRD), ATM loss, or replication fork protection defects, which will thereby extend the spectrum of PARP inhibitors in the same field of oncology (Muvaffak and Coleman, 2024).

15.7.2 DDR inhibitors beyond poly(ADP-ribose) polymerase

Several kinases compose the DDR network and coordinate the activation of the checkpoints, repair, recombination, and extraction of the replication stress. In addition to PARP inhibitors, there are practical methods of tumor-specific cytotoxicity by DDR-regulators of the upstream. Ataxia telangiectasia and Rad3-related kinase (ATR) plays a crucial role in stabilization of stalled replication forks and activation of S/G 2 mediated by CHK1. Inhibitors of ATR (e.g., ceralasertib and elimusertib) can cause a replication catastrophe in cancers that have a high replication stress or lack ATM (Tang et al., 2024). In malignancies that have functional-deficient p53, ATM kinase, which participates mainly in signaling and double-strand breaks, is a new target. Inhibition of CHK1 interferes with the protection of replication forks and cell cycle arrest at checkpoints and predisposes cells to the effects of genotoxic agents. The G2/M checkpoint kinase, WEE1, has an inhibitory effect on the activity of CDK1; by inhibiting WEE1 (e.g., adavosertib), it is possible to trigger mitotic entry of damaged cells prematurely, leading to mitotic catastrophe. To increase the therapeutic index, combinatorial biomarker-based selection of patients will be used, in which the selection of tumors with TP53, ATM, or BRCA mutations will be involved. Nevertheless, achieving the right balance between on-target tumoricidal activity and myelosuppression plus GI toxicity, that is, doing so without dose-limiting toxicity, has been a translational challenge.

To maximize their clinical utility, these DDR inhibitors must be optimized in dosing and scheduling, combined with other DNA-damaging agents, and improved predictive biomarkers are essential (Qian et al., 2024).

15.7.3 DDR inhibitors with immunotherapy and radiotherapy

DDR inhibitors in combination with other treatments have the potential to utilize mechanistic synergy to address intrinsic and acquired resistance. Radiotherapy causes DNA breaks by breaking the strand, which enhances the cytotoxic effect of DDR perturbation, especially by interfering with the repair machinery, e.g., HRR or nonhomologous end joining. PARP inhibitors, ATR, and ATM inhibitors allow more radiosensitization by crippling damage repair and enforcing long-standing genomic damage. At the same time, the immunotherapy in this context can be expanded by DDR inhibition, which stimulates the cGAS-stimulator of interferon generation (STING) signaling that leads to the activation of the type I interferon response that enhances tumor immunogenicity and PD-L1 expression (Yang et al., 2025). PARP or ATR inhibition can potentiate the efficacy of PD-1/PD-L1 checkpoint inhibitors in HRD or mismatch repair-deficient tumors, both in preclinical and early clinical studies in patients. Furthermore, WEE1 inhibition has the potential to increase the generation of mutations and present neoantigens, increasing immune-mediated clearance further. Sound timing is essential- stepwise or rationed use can help avoid cumulative [toxicities] like hematologic suppression. Immune evasion, resistance, and heterogeneity of the tumor demand the presence of biomarker-based methods, which consist of genomic profiling and immune landscape evaluation. Triple combinations of DDR inhibitors, immune checkpoint blockade, and radiotherapy could be used as future strategies to obtain durable responses in refractory oncologies with a manageable level of toxicity (Jurkovicova et al., 2022).

15.7.4 Exploiting tumor-specific DNA repair defects

The idea of synthetic lethality is being used more in precision oncology to target tumor weakness in the repair of the DNA damage (DDR) pathway. A large fraction of malignancies have germline or somatic variants in selected DDR genes, like BRCA1/2, ATM, ATR, PALB2, and mismatch repair (MMR) proteins, that produce vulnerabilities to therapeutic competition. In such cases, BRCA-mutated tumors have defective HRR that makes them exceptionally responsive to poly(ADP-ribose) polymerase (PARP) blockade, thereby further degrading single-strand break repair, resulting in replication-coupled double-oligomeric DNA breaks. High-throughput genomic profiling allows accurate stratification of patients to DDR mutation signatures, and the targeted application of such agents as PARP, ATR, and DNA-PK inhibitors may then proceed in a biomarker-driven manner. Furthermore, being associated with imperfect repair, mutational signatures can predict medications (e.g., COSMIC signatures of HRD or MMR deficiency). A combination of multiomic data with functional assay-based repair-deficient tumor identification, i.e., RAD51 foci formation tests, can enhance the sensitivity of the process. Such a precision system also maximizes efficacy of therapy, reducing the off-target toxicity through targeted killing of the malignant cells without unwanted damage to healthy body tissues (Clarke and Mostoslavsky, 2022; Curtin, 2023).

15.7.5 Overcoming resistance to DDR-targeted therapies

Adaptive resistance of DDR-targeted agents is a serious clinical problem despite early responses. These are mechanisms of drug toleration, such as restoring HRR by secondary mutations that restore BRCA loss-of-function, augmenting drug efflux pumps, reprogramming of replication stress pathways, and enhanced utilization of alternative repair pathways, including microhomology-mediated end joining. The therapeutic escape is also mediated by epigenetic reversal of promoter methylation in DDR genes, and escape via the resulting initiation of cell cycle checkpoints (Morimoto et al., 2019). Next-generation, more potent, more selective, and efflux-resistant inhibitors are in development to counteract these adaptive changes. Dual-targeting approaches, experimental strategies employing the combination of PARP inhibitors with ATR, CHK1, or WEE1 inhibitors to activate synthetic lethality by targeting multiple repair pathways to minimize the risk of functional evasion, are also under investigation to accomplish. Also, combinations of immunotherapy are getting attention, using the high neoantigen load caused by genomic instability to further make the tumor immunogenic. More sophisticated organoid and xenograft-derived models of patient material, as well as the monitoring of circulating tumor DNA (ctDNA), are also facilitating early detection of resistance development so that therapeutic adjustments can be made as a last resort. Finally, the combination of dynamic resistance monitoring and rational drug development will pose potential in maintaining the clinical impact of DDR-induced anticancer therapy (Qian et al., 2024; Tang et al., 2024).

15.8 Immuno-oncology and DNA damage

15.8.1 DNA damage–induced neoantigens

The landmark feature of malignant trans differentiation is genomic instability due to continuing DNA damage, and a central determinant of tumor immunogenicity. High tumor mutational burden (TMB), usually caused by defective DNA repair like mismatch repair (MMR) defect, BER deficiency, or loss of HR, produces a panorama of nonsynonymous mutations (Chaudhari et al., 2025; Gholap et al., 2025; Jadhav et al., 2025; Vardikar et al., 2025; Khuspe et al., 2026a, 2026b; Khuspe and Mane, 2026a, 2026b; Pardeshi et al., 2026). The mutations provide tumor-specific neoantigens with no existence in the regular proteome and can be considered as something that is inappropriate to the host immune system as a foreign body. Expression of neoantigens through major histocompatibility complex (MHC) class I/II molecules leads to the activation of cytotoxic T lymphocytes (CTL), which have the ability to kill cancer cells (Zhang et al., 2022; Hayes et al., 2025). The discovery of the beneficial outcome of high TMB under the treatment of immune checkpoint inhibitors (ICIs) in patients is related to the fact that it enhances the probability of productive immune recognition. Yet, the development of a tumor may result in immune editing, which describes the process of preferential killing of highly immunogenic clones and the emergence of immune-evasive ones. This tight interplay demonstrates the paradoxical nature of DNA damage as it increases immunogenicity but can selectively exert pressure that leads to immune escape, which needs the application of combinatorial approaches to maintain antitumor immunity in DNA damage-rich cancers (Wang et al., 2024).

15.8.2 DNA damage response defects and immune checkpoint therapy response

The defects of DDR pathways significantly affect the interactions between the tumor and the immune system and the responses in immunotherapies with ICIs. Mutations of genes, including BRCA1/2, ATM, and POLE, which are loss-of-function changes, increase genomic instability itself and also arouse innate immune sensing due to the excess of cytosolic DNA fragments. This increases the type I interferon (IFN) signaling, therefore facilitating immune infiltration at the tumor microenvironment (TME) (Morimoto et al., 2019; Moon et al., 2023). Tumors with DDR defects, especially MMR-deficient or HR-defendant (HRD), are very sensitive to PD-1 / PD-L1 and CTLA-4 blockade, and this is due to increased antigenicity and an inflamed TME. In addition, checkpoint blockade can be explained by the fact that DDR-deficient tumors usually have a high expression of PD-L1 as an adaptation response when faced with activated immunity. Although these benefits are presented, resistance to therapy may develop via reconstitution of DNA repair resources or exhaustion of effector immune cells, and requires reasonably implemented combinations like PARP debilitators and ICIs. Recent evidence warrants the use of DDR status as a predictive biomarker of responsiveness to immunotherapy, and trials are ongoing to determine precise patient stratification to treatment based on DDR gene signatures (Pan et al., 2024).

15.8.3 Stimulator of interferon generation pathway activation by cytosolic DNA

Not only might excessive DNA damage and/or inhibited repair lead to the release of microgram amounts of double-stranded DNA (dsDNA) into the cytoplasm, but the presence of these messengers can be sensed by the cyclic GMP AMP synthase (cGAS) STING pathway (Guo et al., 2024). The cleavage of this pathway initiates phosphorylation of TBK1 and IRF3, which, downstream, catalyze strong production of type I IFN and the release of pro-inflammatory cytokines. It will increase dendritic cell maturation, antigen presentation, and T cell priming as effector cells, thus achieving a connection between genotoxic stress and antitumor immunity. STING pharmacologic activation is also being pursued to synergize with ICIs and DNA-damaging agents that have a clear benefit of turning immunologically cold tumors into hot tumors (Dvorkin et al., 2024).

15.9 Role of nanotechnology in DNA damage and cancer

Nanotechnology has become a revolutionary instrument in the field of oncology, providing, on the one hand, unprecedented possibilities in the detection and treatment of cancer, and, on the other hand, significant issues related to the safety of genomes (Chaudhari et al., 2025; Dadarkar and Dubey, 2025; Hatvate and Omri, 2025; Khairnar et al., 2025; Vardikar et al., 2025). Due to their distinctive physicochemical characteristics such as size, surface area, charge and tunable surface chemistry nanomaterials directly interact with biological macromolecules, such as DNA, and this dual outcome presents the potential of both therapeutic advantage and toxicity (Desai and Vora, 2025; Khan et al., 2025; Prajapat et al., 2025; Sagar et al., 2025; Uddin et al., 2025b).

Nanotechnology as a cancer treatment has offered precision delivery platforms, which increase the concentration of drugs in the tumors and reduce the toxicity of the system. Many nanoparticles of polymeric materials, liposomes, dendrimers, and inorganic nanostructures (gold, silica, and quantum dots) have been engineered to enhance the pharmacokinetics and stability of drugs. Notably, carriers based on nanoscale facilitate co-delivery of chemotherapeutics and gene-editing molecules, which provides the opportunity to induce malignant cells with DNA damage selectively. An example is that doxorubicin-loaded nanoparticles increase intercalation into tumor DNA and reduce cardiotoxicity, whereas delivery of CRISPR-Cas9 by lipid or polymeric nanoparticles allows specific genome editing (Rojekar et al., 2024b; Gholap and Omri, 2025; Hatvate et al., 2025; Jadhav et al., 2025; Khuspe et al., 2025a, 2025b, 2025c; Uddin et al., 2025a). In addition to delivery, nanomaterials per se can be radiosensitizers or photosensitizers. High atomic number causes gold nanoparticles to increase the local radiation-induced DNA double-strand breaks in cancer cells. Likewise, photosensitizer-conjugated nanoparticles in the form of upconversion nanoparticles produce ROS to selectively cause localized DNA damage in tumors. These nanosystems take advantage of regulated DNA damage as a therapeutic approach in the transformation of the genotoxic properties of nanomaterials into a therapeutic benefit (Kapare et al., 2024; Mesut et al., 2024; Rojekar et al., 2024a; Sawant et al., 2024a, 2024b; Shaikh et al., 2024).

Although therapeutic uses are encouraging, an increasing body of work indicates that engineered nanomaterials are also capable of causing unintended DNA damage. They are small enough to penetrate the nucleus, and surface reactivity can drive the production of ROS, which leads to oxidative damage, including 8-oxoguanine, single- and double-strand breakages, and chromosomal aberrations. Metal oxide nanoparticles such as TiO_2, ZnO, and carbon-based nanomaterials (graphene oxide, carbon nanotubes) have been involved in the instability of DNA through oxidative stress pathways. Notably, genotoxic effects can be dependent upon particle size, shape, charge, aggregation, and functionalization (Hatvate et al., 2024; Kumar et al., 2024; Pardeshi et al., 2024a; Pardeshi et al., 2024b, 2024a; Parit et al., 2024; Uddin et al., 2024).

Nanotechnology risk assessment, therefore, needs to be multidisciplinary, involving nanotoxicology, genomics, and systems biology. Conventional assays, such as the comet assay, micronucleus test, and γ-H 2AX foci assays, have been modified to identify DNA lesions induced by nanoparticles. Additionally, omics platforms with high-throughput can now show mechanistic details of pathways between nanoparticle exposure and DNA repair, apoptosis, and mutagenesis. The problem faced by regulators is the definition of exposure levels and safety profiles because the traditional toxicology paradigm might not adequately address nanoscale-specific interactions (Pardeshi et al., 2023; Pardeshi and Giram, 2023; Sawant et al., 2023; Faiyazuddin et al., 2024; Gupta et al., 2024; Khuspe et al., 2024; Srivastava et al., 2024; Pardeshi et al., 2024b; Patil et al., 2024; Rojekar et al., 2024).

Nonetheless, there are ample prospects of exploring the potential of nanotechnology in order to control the repair of DNA damage in cancer. To begin with, genotoxic stress can be specifically induced in tumor cells and normal tissues spared by engineered nanocarriers, which facilitates targeted chemotherapy and radiosensitization. Second, a nanoparticle-based delivery of DNA repair inhibitors (e.g., PARP inhibitors) is synergistic with DNA-damaging therapies, especially in BRCA-mutated cancers. Third, nanodevices that can detect DNA adducts and repair biomarkers in a manner that is ultrasensitive can transform the future of early detection of the risk of cancer and monitoring of therapies (Singh et al., 2016; Puri, 2018; Bhowmik et al., 2023; Hatvate et al., 2023b, 2023a; Said et al., 2023; Gupta et al., 2023; Sayyad et al., 2023).

An additional prospect would be the design of safe-by-design nano materials. Through control over surface coating, charge, and degradability, investigators can reduce off-target interactions with the DNA and retain therapeutic activity. A combination of artificial intelligence and machine learning with nanotoxicology data can further predict genotoxic potential, screen materials faster, and inform regulatory frameworks (Bhagat, 2012; Dattatray, 2012a; Dighe, 2012a; Khedkar Ajay, 2012a; Rajurkar, 2012a, 2012b; 2012d; Dattatray, 2012; Bothiraja et al., 2014). Nanotechnology is at the border of hope and danger in the face of DNA damage and cancer. As nanoparticle-based platforms have enhanced cancer therapy through targeted DNA damage, radiosensitization, and gene-editing, the same characteristics create concern over unwanted genomic toxicity (Borude, 2011; Borude and Mahajan, 2011; Magar Tushar V., 2011a, 2011b). A balanced solution based on mechanistic knowledge, stringent risk evaluation, and safe-by-design solutions will dictate whether nanotechnology will achieve the full potential of oncology related to nanotechnology. Following further developments, the sphere provides the chance to change the state of cancer treatment and provide the genomic safety of patients and the environment (Anand, 2011; Borude and Mahajan, 2011; Dattatray, 2012b; Dighe, 2012b; Gholap, 2012; Khedkar Ajay, 2012b; Rajurkar, 2012c; Rajurkar, 2012a, 2012b).

15.10 Challenges and limitations

The mutual linkage between DNA damage and oncogenesis is inherently multifactorial in nature, comprising multilocal interactions between genetic, epigenetic, and environmental determinants. Although it is well established that the development of certain lesions, specifically, double-strand breaks or oxidative lesions, is mutagenic, postlesion events in terms of transformation to malignancy depend on DNA repair and cell cycle checkpoint properties, as well as the surrounding microenvironment. Besides, cancer is usually a result of repeated and chronologically distinct events, making the root cause difficult to attribute to one event of DNA damage. Epigenomic reprogramming, chromatin remodeling, and nonmutagenic changes through transcriptional stress also endow the mechanistic connection. There is a further layer of complexity in the inter-patient heterogeneity in genetic background and exposure history, which makes it difficult to discern driver mutation from incidental so-called passenger mutations. These confounders constrain the predictive accuracy of damage biomarkers and cause difficulties in developing universal causality models even with the development of high-throughput genomic and single-cell sequencing technology (Gehrke et al., 2022; Pasqui et al., 2022; Muyas et al., 2024; Szatkowska and Zdrada-Nowak, 2025).

There is a gap that needs to be fulfilled in DNA damage detection platforms, especially regarding sensitivity, specificity, and temporal resolution, despite high technological advancement. Techniques like comet assay, γ-H2AX foci, and mass spectrometry-based adduct profiling, in most cases, fail to distinguish between repairable transient lesions and those that are not repaired to generate carcinogenesis. Mapping unusual heterogeneous tumor tissues is not ideal, and single-cell resolutions are limited by their costs and throughput factors. The majority of the assays offer a snapshot and not dynamic and time-resolved observations, which hinder longitudinal measurements of accumulation and repair kinetics of DNA damage. This limits the applicability of laboratory findings to strong clinical diagnostics (Stinson and Loparo, 2021; Clarke and Mostoslavsky, 2022; Clarke and Mostoslavsky, 2022; Anand et al., 2023; Xia et al., 2025).

15.11 Future directions

Due to the involvement of artificial intelligence (AI) in the field of oncology, the forecasting of the pattern of DNA damage and optimization of the treatment plan are changing. Dense models of machine learning (ML) can learn the mutational signatures, chromosomal instability signature, and deficits in repair pathways with high accuracy, trained on multiomics data. The models allow early prediction of genomic instability and preselection of therapy, based on predicting the sensitivity of the tumor to DNA-damaging agents (e.g., in response to the platinum-based chemotherapeutics or PARP inhibitors) (Huang and Zhou, 2020; Curtin, 2023). There are also AI-based platforms that can simulate cell response to different treatment regimens and thus reduce trial and error. Moreover, dynamic longitudinal patient data contribute to the real time analysis of longitudinal patient data that enables adaptation of the line of treatment, which allows more precise treatment and less toxic side effects. The promise of immensely personalized cancer care is the inclusion of AI-based prediction systems into the clinical workflow to fill the all-too-common gap between genomic data production and clinically actionable decision-making, to speed the journey of evidence-based insights to clinical utility (Vora et al., 2023; Gholap et al., 2024).

15.12 Conclusion

Genomic integrity is shown to be earth-shatteringly important in the context of the complex relationship between oncogenesis and DNA damage. The cause of mutational events that activate oncogenes or inactivate tumor suppressor pathways and lead to malignant transformation is persistence or unrepaired lesions, especially lesions caused by oxidative stress, replication errors, or environmental genotoxins. High-throughput technologies to conduct high-throughput sequencing, single-cell genomics, and multiplex damage-mapping tools were taking steps forward, facilitating an improved way of very minute resolution with the ability to detect, quantify, and characterize the pattern of DNA damage, and as a result, have advanced the population to further stratification of risk and the development of resulting specific interventions. Most recent advancements in precision oncology exploit biomarkers of genomic instability/DNA repair defects, including HRR gene alterations, to inform selective treatment options, a success story of which is PARP inhibitors. At the same time, functional genomics via CRISPR continues to discover new regulating mechanisms of DDR, providing new avenues of targeted therapy. Despite these developments, major hurdles remain in the translation aspects, such as the heterogeneity of tumor types; the nonpredictive nature of the problem; adaptation of resistance mechanisms; and the ability to use predictive models based on the genetic, epigenetic, and environmental determinants of the susceptibility to damage. Multiomic and AI-based predictive analytics are rapidly becoming part of the risk assessment programs in order to anticipate the multifaceted nature of DNA damage, repair capacity, and cancer development. There is potential in this paradigm shift towards earlier and more precise diagnosis, risk profiling of individual patients, and the design of combinatorial combinations that take advantage of synthetic tools at a minimum dose of toxicity. The future is the current possibilities of harmonizing the emerging molecular diagnostics with the next-generation of therapeutics that specifically target the vulnerabilities in DDR, and preventive approaches that limit exposure to DNA-damaging agents. Conclusively, enhanced knowledge of the mechanistic interaction within the pathway related to DNA damage will continue to play a central role in suppressing the occurrence of cancer, the specificity of therapy, and patient prognosis.

References

Allkanjari, K., Baldock, R.A., 2021. Beyond base excision repair: an evolving picture of mitochondrial DNA repair. Biosci. Rep. 41, BSR20211320.

Anand, J., Chiou, L., Sciandra, C., Zhang, X., Hong, J., Wu, D., et al., 2023. Roles of trans-lesion synthesis (TLS) DNA polymerases in tumorigenesis and cancer therapy. NAR Cancer 5, zcad005.

Anand, M., 2011. Smart crystals technology: a review. Pharmacologyonline 1, 238–243.

Andrés, C.M.C., de la Lastra, J.M.P., Juan, C.A., Plou, F.J., Pérez-Lebeña, E., 2023. Chemical insights into oxidative and nitrative modifications of DNA. Int. J. Mol. Sci. 24, 15240.

Arai, H., Elliott, A., Xiu, J., Wang, J., Battaglin, F., Kawanishi, N., et al., 2021. The landscape of alterations in DNA damage response pathways in colorectal cancer. Clin. Cancer Res. 27, 3234–3242.

Batarchuk, V., Shepelytskyi, Y., Grynko, V., Kovacs, A.H., Hodgson, A., Rodriguez, K., et al., 2024. Hyperpolarized xenon-129 chemical exchange saturation transfer (HyperCEST) molecular imaging: achievements and future challenges. Int. J. Mol. Sci. 25, 1939.

Bhagat, V., 2012. Gold nanorods: a promising drug delivery vehicle. Inven. Rapid NDDS 1, 1–4.

Bhowmik, D.D., Deshmukh, A.Y., Hatvate, N.T., 2023. Quintessential impact of dendrimer bioconjugates in targeted drug delivery. In: Verma, S.K., Das, A.K. (Eds.), Bioconjugated materials Part 2 applications in drug delivery, vaccine formulations and important conjugates for cancer therapy, comprehensive analytical chemistry. Elsevier, Amsterdam, Netherlands, pp. 257–302.

Borude, S., 2011. Aquasomes: a potential drug delivery carrier. Pharmacologyonline 1, 230–237.

Borude, S.S., Mahajan, A.M., 2011. Recent updates on Q fever: diagnosis and treatment. Pharmacologyonline 1, 254–264.

Borude Santosh, S., Mahajan, A.M., 2011. POEMS syndrome: a review. Pharmacologyonline 1, 265–274.

Bothiraja, C., Gholap, K.S., Pawar, A.P., 2014. Investigation of ethyl cellulose microsponge gel for topical delivery of eberconazole nitrate for fungal therapy. Ther. Deliv. 5, 781–794.

Cerniglia, M., Xiu, J., Grothey, A., Pishvaian, M.J., Baca, Y., Hwang, J.J., et al., 2022. Association of homologous recombination–DNA damage response gene mutations with immune biomarkers in gastroesophageal cancers. Mol. Cancer Ther. 21, 227–236.

Chatterjee, N., Walker, G.C., 2017. Mechanisms of DNA damage, repair, and mutagenesis. Environ. Mol. Mutagen. 58, 235–263.

Chaudhari, Y.S., Chaudhari, M.Y., Alam, M.I., Khalid, M., Webster, T.J., Sundaram, G.S., et al., 2025. Surface engineering of nano magnesium alloys for orthopedic implants: a systematic review of strategies to mitigate corrosion and promote bone regeneration. Front. Bioeng. Biotechnol. 13, 1617585.

Clarke, T.L., Mostoslavsky, R., 2022. DNA repair as a shared hallmark in cancer and ageing. Mol. Oncol. 16, 3352–3379.

Curtin, N.J., 2023. Targeting the DNA damage response for cancer therapy. Biochem. Soc. Trans. 51, 207–221.

Dadarkar, Y.P., Dubey, A.S., 2025. Bioremediation of pharmaceuticals with cow dung and earthworms. Biomedical Waste Management. Apple Academic Press, pp. 239–261.

Dattatray, T., 2012. The threat of photoaging: clinical conditions, evaluation, and treatment. Inven. Rapid Cosmeceuticals 1, 1–3.

Dattatray, T., 2012a. Potential applications of botanical extract. Inven. Rapid Ethnopharmacol. 1, 1–4.

Dattatray, T., 2012b. Artificial viral envelopes: emerging vehicle for gene delivery. Pharmatutor-ART-1439 1, 1–3.

Desai, H., Vora, L.K., 2025. Impact of parenteral microemulsion on healthcare therapeutics. Pharm. Microemulsions Parenter. Deliv. 1–20.

Dighe, S.B., 2012a. Dandruff: pathogenesis and treatment. Inven. Rapid Cosmeceuticals 1, 1–4.

Dighe, S.B., 2012b. Role of sebum in human health & acne. Inven. Rapid Cosmeceuticals 1, 1–3.

Drew, Y., Zenke, F.T., Curtin, N.J., 2025. DNA damage response inhibitors in cancer therapy: lessons from the past, current status, and future implications. Nat. Rev. Drug Discov. 24, 19–39.

Dvorkin, S., Cambier, S., Volkman, H.E., Stetson, D.B., 2024. New frontiers in the cGAS-STING intracellular DNA-sensing pathway. Immunity 57, 718–730.

Faiyazuddin, M., Maqsood, S., Abbas, S., Chaudhari, Y.S., Sharmila, S., Webster, T.J., 2024. Prospects and ethics of ChatGPT in biomedical research and medical practice: an emphasis on nanotechnology. J. Biomed. Nanotechnol. 20, 1661–1678.

Ferrer-Diaz, A.I., Sinha, G., Petryna, A., Gonzalez-Bermejo, R., Kenfack, Y., Adetayo, O., et al., 2024. Revealing role of epigenetic modifiers and DNA oxidation in cell-autonomous regulation of Cancer stem cells. Cell Commun. Signal. 22, 119.

Gehrke, F., Schindele, A., Puchta, H., 2022. Nonhomologous end joining as key to CRISPR/Cas-mediated plant chromosome engineering. Plant Physiol. 188, 1769–1779.

Gest, A.M.M., Sahan, A.Z., Zhong, Y., Lin, W., Mehta, S., Zhang, J., 2024. Molecular spies in action: genetically encoded fluorescent biosensors light up cellular signals. Chem. Rev. 124, 12573–12660.

Gholap, A.D., 2012. Translational nanomedicine: status assessment and opportunities. Int. J. Res. Pharm. Biomed. Sci. 4, 1454–1463.

Gholap, A.D., Khuspe, P.R., Mane, D.V., Galave, V., Mali, A., Mohite, P., et al., 2025. Challenges and future directions biomaterial-based additive manufacturing in tissue engineering and regeneration. Biomaterial-based additive manufacturing in tissue engineering and regeneration. biomaterials, bioengineering and sustainability. Springer Nature, Singapore, pp. 491–529.

Gholap, A.D., Omri, A., 2025. Advances in artificial intelligence-envisioned technologies for protein and nucleic acid research. Drug Discov. Today 30, 104362.

Gholap, A.D., Uddin, M.J., Faiyazuddin, M., Omri, A., Gowri, S., Khalid, M., 2024. Advances in artificial intelligence for drug delivery and development: a comprehensive review. Comput. Biol. Med. 178, 108702.

Gohil, D., Sarker, A.H., Roy, R., 2023. Base excision repair: mechanisms and impact in biology, disease, and medicine. Int. J. Mol. Sci. 24, 14186.

Groelly, F.J., Fawkes, M., Dagg, R.A., Blackford, A.N., Tarsounas, M., 2023. Targeting DNA damage response pathways in cancer. Nat. Rev. Cancer 23, 78–94.

Gulati, G.S., D'Silva, J.P., Liu, Y., Wang, L., Newman, A.M., 2025. Profiling cell identity and tissue architecture with single-cell and spatial transcriptomics. Nat. Rev. Mol. Cell Biol. 26, 11–31.

Guo, X., Yang, L., Wang, J., Wu, Y., Li, Y., Du, L., et al., 2024. The cytosolic DNA-sensing cGAS STING pathway in neurodegenerative diseases. CNS Neurosci. Ther. 30 (3), e14671.

Gupta, J., Kamandar, P., Bhowmik, D.D., Rojekar, S., Faiyazuddin, M., Hatvate, N.T., et al., 2024. Harnessing nanovaccines for effective immunization–a special concern on COVID-19: facts, fidelity, and future prospective. ACS Biomater. Sci. Eng. 10, 271–297.

Gupta, J.S., Kamandar, P.A., Banchhod, G.V, Hatvate, N.T., 2023. Antibody-drug conjugates for cancer therapy: an up-to-date review on the chemistry and pharmacology. In: Verma, S.K., Das, A.K. (Eds.), Bioconjugated materials part 2 applications in drug delivery, vaccine formulations and important conjugates for cancer therapy, comprehensive analytical chemistry. Elsevier, Amsterdam, Netherlands, pp. 105–190.

Hanahan, D., 2022. Hallmarks of cancer: new dimensions. Cancer Discov. 12, 31–46.

Hatvate, N., Omri, A., 2025. Reprogramming therapeutics: synthetic biological constructs at the frontier of precision medicine. Discov. Med. 37, 1718.

Hatvate, N.T., Bhowmik, D., Sawant, S.D., Londhe, K.M., Rojekar, S., Faiyazuddin, M., 2025. Engineered artificial MicroRNA in plant: an overview. MicroRNA advances and application in plant biology. Elsevier, pp. 311–341.

Hatvate, N.T., Khuspe, P.R., Mandhare, T.A., Kashid, P., Gaikwad, V.D., 2023a. Proteomics in oncology: retrospect and prospects. In: S., K. (Ed.), Novel technologies in biosystems, biomedical & drug delivery. Springer Nature, Singapore, pp. 243–269.

Hatvate, N.T., Khuspe, P.R., Mandhare, T.A., Kashid, P., Vora, L.K., 2024a. Decoding of nanoneutraceuticals and their delivery. In: Mohanta, Y.K. (Ed.), Sustainable green nanotechnology. CRC Press, Florida, New York, pp. 284–309.

Hatvate, N.T., Sawant, S.D., Khemka, P.N., Vora, L.K., 2023b. Unlocking the potential of nanobiohybrids to combat environmental pollution. In: Lens, P. (Ed.), Nanobiohybrids for advanced wastewater treatment and energy recovery. IWA Publishing, UK., pp. 169–190.

Hayes, M.N., Cohen-Gogo, S., Kee, L., Xiong, X., Weiss, A., Layeghifard, M., et al., 2025. DNA damage response deficiency enhances neuroblastoma progression and sensitivity to combination PARP and ATR inhibition. Cell Rep. 44, 115537.

Hu, Q., Espejo Valle-Inclán, J., Dahiya, R., Guyer, A., Mazzagatti, A., Maurais, E.G., et al., 2024. Non-homologous end joining shapes the genomic rearrangement landscape of chromothripsis from mitotic errors. Nat. Commun. 15, 5611.

Huang, R., Zhou, P.-K., 2021. DNA damage repair: historical perspectives, mechanistic pathways and clinical translation for targeted cancer therapy. Signal Transduct. Target. Ther. 6, 254.

Huang, R.-X., Zhou, P.-K., 2020. DNA damage response signaling pathways and targets for radiotherapy sensitization in cancer. Signal Transduct. Target. Ther. 5, 60.

Jaafari, A., 2024. Potential epigenetic modifiers targeting the alteration of methylation in colorectal cancer. Gene Expr. 23, 139–152.

Jadhav, K., Shevalkar, G., Sugandhi, V.V., Pai, R., Parikh, K., Prajapati, M.K., et al., 2025. Exploring protein aggregation in biological products: from mechanistic understanding to practical solutions. AAPS PharmSciTech. 26, 189.

Jiang, M., Jia, K., Wang, L., Li, W., Chen, B., Liu, Y., et al., 2021. Alterations of DNA damage response pathway: biomarker and therapeutic strategy for cancer immunotherapy. Acta Pharm. Sin. B 11, 2983–2994.

Jones, K., Dye, E., Gilkes, N., Xia, D., Jiang, S., Li, W., 2025. Mechanisms of DNA repair and mutagenesis induced by acetaldehyde, acrolein, aristolochic acids, and vinyl chloride. J. Environ. Sci. Heal. Part C 1–38.

Jurkovicova, D., Neophytou, C.M., Gašparović, A.Č., Gonçalves, A.C., 2022. DNA damage response in cancer therapy and resistance: challenges and opportunities. Int. J. Mol. Sci. 23, 14672.

Kapare, H.S., Pagar, S., Kamandar, P., Bhowmik, D., Vishwakarma, N., Raikwar, S., et al., 2024. Exploring modified chitosan-based gene delivery technologies for therapeutic advancements. Int. J. Biol. Macromol. 260, 129581.

Khairnar, R.A., Sawant, S.D., Parab, T.D., 2025. Microbial strategies for the biodegradation of pharmaceutical pollutants. Biomedical Waste Management. Apple Academic Press, pp. 297–317.

Khan, Ahsan Riaz, Grewal, N.S., Jun, Z., Khalid, M., Zhang, H.-J., 2025. Advances in smart hybrid scaffolds: a strategic approach for regenerative clinical applications. Eng. Regen. 6, 85–110.

Khatib, J.B., Nicolae, C.M., Moldovan, G.-L., 2024. Role of translesion DNA synthesis in the metabolism of replication-associated nascent strand gaps. J. Mol. Biol. 436, 168275.

Khedkar Ajay, T.D., 2012a. Pharmacosomes: an emerging vesicular system. Inven. Rapid NDDS 1, 1–3.

Khedkar Ajay, T.D., 2012b. Cubosomes: emerging drug delivery system. Inven. Rapid NDDS 1, 1–3.

Khuspe, P.R., Bharati, D.K., Pardeshi, S.R., Ahmad, M.D., Hossain, A.S., Prajapati, B.G., et al., 2025a. Computational neuropharmacology in psychiatry. Computational Neuropharmacology: Fundamentals and Clinical Aspects. Wiley, pp. 207–244.

Khuspe, P.R., Faiyazuddin, M., Uddin, M.J., Bhowmik, D.D., Said, R.P., Sonawane, K.S., et al., 2025b. Lipid-based vaccines against viruses including COVID-19. Lipids in pulmonary drug delivery. Elsevier, pp. 409–433.

Khuspe, P.R., Faiyazuddin, M., Uddin, M.J., Gupta, J., Kamandar, P., Bhowmik, D.D., et al., 2025c. Current scenario and challenges of lipid-based nanomedicines for pulmonary drug delivery. Lipids in pulmonary drug delivery. Elsevier, pp. 21–47.

Khuspe, P.R., Mane, D.V., 2026a. Emerging machine learning techniques for the development and analysis of anticancer drugs from plants. Emerging trends in phytotherapy of cancer. Elsevier, pp. 461–493.

Khuspe, P.R., Mane, D.V., 2026b. Artificial intelligence driven discovery of anticancer phytochemicals: from plants to lead compounds. Emerging trends in phytotherapy of cancer. Elsevier, pp. 441–460.

Khuspe, P.R., Pardeshi, S.R., Uddin, M.J., Das, U., Hatvate, N.T., Rojekar, S., et al., 2024. Achieving optimal health with host-directed therapies (HDTs) in infectious diseases—a new horizon. Adv. Ther., 2400169.

Khuspe, P.R., Rebello, N.L., Jain, M.N., Desai, H.R., Giram, P., 2026a. Photon-avalanching nanoparticles for biomedical applications. Upconversion Nanoparticles for Biomedical Applications. Elsevier, pp. 185–225.

Khuspe, P.R., Rebello, N.L., Jain, M.N., Desai, H.R., Giram, P., 2026b. Upconversion nanoparticles for drug and gene delivery. Upconversion Nanoparticles for Biomedical Applications. Elsevier, pp. 517–575.

Kopinski, P.K., Singh, L.N., Zhang, S., Lott, M.T., Wallace, D.C., 2021. Mitochondrial DNA variation and cancer. Nat. Rev. Cancer 21, 431–445.

Kumar, D., Kumar, S., Kumari, R., Tanwar, R., Kumar, V., Khalid, M., et al., 2024. A comprehensive review on the biomass-mediated synthesis of silver nanoparticles: opportunities and challenges. J. Environ. Chem. Eng., 115133.

Kwok, M., Agathanggelou, A., Stankovic, T., 2024. DNA damage response defects in hematologic malignancies: mechanistic insights and therapeutic strategies. Blood 143, 2123–2144.

Li, J., Jia, Z., Dong, L., Cao, H., Huang, Y., Xu, H., et al., 2024. DNA damage response in breast cancer and its significant role in guiding novel precise therapies. Biomark. Res. 12, 111.

Li, Q., Dudás, K., Tick, G., Haracska, L., 2021. Coordinated cut and bypass: replication of interstrand crosslink-containing DNA. Front. Cell Dev. Biol. 9, 699966.

Liao, M., Chu, W., Sun, X., Zheng, W., Gao, S., Li, D., et al., 2022. Reduction of H3K27cr modification during DNA damage in colon cancer. Front. Oncol. 12, 924061.

Ma, F., Ma, Y., Liu, K., Gao, J., Li, S., Sun, X., et al., 2023. Resveratrol induces DNA damage-mediated cancer cell senescence through the DLC1–DYRK1A–EGFR axis. Food Funct 14, 1484–1497.

Magar Tushar, V., Musmade, D.S., 2011b. Elastic vesicles as drugs carriers through the skin. Inven. Rapid NDDS 1, 1–2.

Magar Tushar, V., Musmade, D.S., 2011a. Advances in the field of nanooncology: an overview. Inven. Impact NDDS 4, 1–3.

Malamos, P., Papanikolaou, C., Gavriatopoulou, M., Dimopoulos, M.A., Terpos, E., Souliotis, V.L., 2024. The interplay between the DNA damage response (DDR) Network and the mitogen-activated protein kinase (MAPK) signaling pathway in multiple myeloma. Int. J. Mol. Sci. 25, 6991.

Mertz, T.M., Kockler, Z.W., Coxon, M., Cordero, C., Raval, A.K., Brown, A.J., et al., 2025. Defining APOBEC-induced mutation signatures and modifying activities in yeast. Methods in enzymology. Academic Press, Elsevier, Amsterdam, Netherlands, pp. 115–161.

Mesut, B., Al-Mohaya, M., Yeşilkaya, E., Das, U., Akhtar, M.S., Sah, R., et al., 2024. Demystifying the potential of lipid-based nanocarriers in targeting brain malignancies. Naunyn. Schmiedebergs. Arch. Pharmacol. 397, 9243–9279.

Moon, J., Kitty, I., Renata, K., Qin, S., Zhao, F., Kim, W., 2023. DNA damage and its role in cancer therapeutics. Int. J. Mol. Sci. 24, 4741.

Morimoto, S., Tsuda, M., Bunch, H., Sasanuma, H., Austin, C., Takeda, S., 2019. Type II DNA topoisomerases cause spontaneous double-strand breaks in genomic DNA. Genes (Basel) 10, 868.

Muvaffak, A., Coleman, K.G., 2024. PARP inhibitor synthetic lethality in ATM biallelic mutant cancer cell lines is associated with BRCA1/2 and RAD51 downregulation. Front. Oncol. 14, 1380633.

Muyas, F., Sauer, C.M., Valle-Inclán, J.E., Li, R., Rahbari, R., Mitchell, T.J., et al., 2024. De novo detection of somatic mutations in high-throughput single-cell profiling data sets. Nat. Biotechnol. 42, 758–767.

Nayak, S., Calvo, J.A., Cantor, S.B., 2021. Targeting translesion synthesis (TLS) to expose replication gaps, a unique cancer vulnerability. Expert Opin. Ther. Targets 25, 27–36.

Ou, X., Chen, P., Liu, B.-F., 2025. Liquid biopsy on microfluidics: from existing endogenous to emerging exogenous biomarkers analysis. Anal. Chem. 97, 8625–8640.

Ovejero, S., Moreaux, J., 2021. Multi-omics tumor profiling technologies to develop precision medicine in multiple myeloma. Explor. Target. Anti-tumor Ther. 2 (1), 65.

Pan, D., Wang, Q., Shen, A., Qi, Z., Zheng, C., Hu, B., 2024. When damage responses meet tumor immunity: from mechanism to therapeutic opportunity. Int. J. Cancer 155, 384–399.

Pardeshi, S., Giram, P., 2023. Preparation, antibacterial and antiviral activity measurements and detection methods. In: Deshmukh, K. (Ed.), Antibacterial and Antiviral Functional Materials. ACS Publications, USA, pp. 33–64.

Pardeshi, S., Giram, P., Nangare, S., Sodha, S., Kapare, H., More, M., et al., 2025. Breathing new life nanomedicines for pulmonary drug delivery: targeting approaches, experimental models, and regulatory aspects. Basic Appl. Sci. 14, 65.

Pardeshi, S., Kapare, H., More, A., Rebello, N., Giram, P., 2024a. Coating approaches of functionalized magnetic nanoparticles for theranostic applications. Functionalized magnetic nanoparticles for theranostic applications. Wiley, New Jersey, U.S.A, pp. 131–171.

Pardeshi, S., More, M., Togre, N., Rebello, N., Giram, P., 2023. Dendrimerzs based antibacterial and antiviral materials. Antibacterial and Antiviral Functional Materials. ACS Publications, USA, pp. 139–169.

Pardeshi, S., Pillai, D., Giram, P., Pardeshi, C., Naik, J., 2024b. Nanocarriers for ocular drug delivery. Nanocarrier drug delivery systems. De Gruyter, Germany, pp. 469–500.

Pardeshi, S.R., Hatvate, N.T., Dhorkule, N., Sayyad, S.F., Faiyazuddin, M., Khalid, M., 2024a. Environmental implications and nanotechnological advances in octocrylene-enriched sunscreen formulations: a comprehensive review. Chemosphere 358, 142235.

Pardeshi, S.R., Hatvate, N.T., Gharat, K.D., Naik, J.B., Omri, A., 2024b. Advances in dorzolamide hydrochloride delivery: harnessing nanotechnology for enhanced ocular drug delivery in glaucoma management. Discov. Nano 19, 199.

Pardeshi, S.R., Sen, N.K., Khandagale, P.D., Khadka, I.M., Pardeshi, P.R., Datta, D., et al., 2026. Prospects and applications of nanoemulsions in cosmeceuticals. Nanotechnology in cosmeceuticals. Elsevier, pp. 369–386.

Parit, S., Manchare, A., Mundhe, P., Hatvate, N., Rojekar, S., Patravale, V., 2024. Antibody-drug conjugates: a promising breakthrough in cancer therapy. Int. J. Pharm. 659, 124211.

Pasqui, A., Boddi, A., Campanacci, D.A., Scoccianti, G., Bernini, A., Grasso, D., et al., 2022. Alteration of the nucleotide excision repair (NER) pathway in soft tissue sarcoma. Int. J. Mol. Sci. 23, 8360.

Patil, M., Dhumal, D., Rojekar, S., Jain, H., Pingale, P., Kumar Prajapati, M., 2024. Nanocarriers for antimicrobial therapy. Nanocarrier drug delivery systems. De Gruyter, Germany, pp. 561–602.

Prajapat, M., Shinde, S., Padhiyar, D., Butani, S., Shah, S., Rajput, A., 2025. 3D printing technology in microneedles: an emerging era in transdermal drug delivery. Hybrid Adv. 10, 100447.

Previtali, V., Bagnolini, G., Ciamarone, A., Ferrandi, G., Rinaldi, F., Myers, S.H., et al., 2024. New horizons of synthetic lethality in cancer: current development and future perspectives. J. Med. Chem. 67, 11488–11521.

Puri, A.V., 2018. Comparative evaluation of papaya and banana starch as disintegrant in diclofenac sodium tablet formulation. Int. J. Pharm. Biol. Sci. 8, 512–522.

Qian, J., Liao, G., Chen, M., Peng, R.-W., Yan, X., Du, J., et al., 2024. Advancing cancer therapy: new frontiers in targeting DNA damage response. Front. Pharmacol. 15, 1474337.

Rajurkar, V.G., 2012a. Recent technologies in nanomedicine for treatment of cancer. Inven. Rapid Pharm Tech 4, 1–3.

Rajurkar, V.G., 2012b. Potential applications of capillary electromigration methods for the improvement of separation efficiency. Inven. Rapid Pharm Anal. Qual. Assur. 1, 1–3.

Rajurkar, V.G., 2012c. Potential applications of superparmagnetic nanoparticles technology. Inven. Rapid NDDS 1, 1–3.

Rajurkar, V.G., Dighe, S.B., 2012d. Phonophoresis: ultrasound enhanced transdermal drug delivery. Inven. Rapid NDDS 1, 1–3.

Rayamajhi, S., Sipes, J., Tetlow, A.L., Saha, S., Bansal, A., Godwin, A.K., 2024. Extracellular vesicles as liquid biopsy biomarkers across the cancer journey: from early detection to recurrence. Clin. Chem. 70, 206–219.

Rojekar, S., Kapare, H.S., Vishwakarma, N., Raikwar, S., Garkal, A., Mehta, T.A., et al., 2024a. Chitosan scaffolds: expanding horizons in biomedical applications. Carbohydr. Polym. 323, 121394.

Rojekar, S., Parit, S., Manchare, A., Nangare, S.N., Hatvate, N., Sugandhi, V.V., et al., 2024b. Revolutionizing eye care: exploring the potential of microneedle drug delivery. Pharmaceutics 16, 1398.

Rojekar, S., Togre, N., Bhoj, P., Haeck, C., Hatvate, N., Singh, N., et al., 2024. Current status of mannose receptor-targeted drug delivery for improved anti-HIV therapy. J. Control. Release 372, 494–521.

Said, R.P., Pawar, R.D., Ambore, G.S., Hatvate, N.T., 2023. Importance of carbohydrate-drug conjugates in vaccine development: a detailed review. In: Verma, S.K., Das, A.K. (Eds.), Bioconjugated materials part 2 applications in drug delivery, vaccine formulations and important conjugates for cancer therapy, comprehensive analytical chemistry. Elsevier, Amsterdam, Netherlands, pp. 191–256.

Sarvari, P., Sarvari, P., Ramírez-Díaz, I., Mahjoubi, F., Rubio, K., 2022. Advances of epigenetic biomarkers and epigenome editing for early diagnosis in breast cancer. Int. J. Mol. Sci. 23, 9521.

Sawant, S.D., Pardeshi, S.R., Khuspe, P.R., Sayyad, S.F., Chavan, M., Hatvate, N.T., et al., 2024a. Omics tools for determining microbial composition: a systematic review. Microbiome Eng. 1, 35–50.

Sawant, S.D., Pardeshi, S.R., Khuspe, P.R., Sayyad, S.F., Chavan, M., Hatvate, N.T., et al., 2024b. The role of inulin in altering the animal microbiome: a comprehensive review. Microbiome Eng 1, 79–102.

Sawant, S.D., Sayyad, S.F., Hatvate, N.T., Chavan, M., Rojekar, S., Faiyazuddin, M., 2023. Food biopackaging for human benefits: status and perspectives. In: Ahmed, S. (Ed.), Biobased Packaging Materials. Springer Nature, Singapore, pp. 1–35.

Saxena, S., Zou, L., 2022. Hallmarks of DNA replication stress. Mol. Cell 82, 2298–2314.

Sayyad, S.F., Hatvate, N.T., Dhumal, V.V., Pardeshi, S.R., Chavda, V.P., Vora, L.K., 2023. Drug delivery strategies for avobenzone: a case study of photostabilization. Pharmaceutics 15, 1008.

Shaikh, M.S., Faiyazuddin, M., Khan, M.S., Pathan, S.K., Syed, I.J., Akhtar, M.S., et al., 2024. Chikungunya virus vaccine: a decade of progress solving epidemiological dilemma, emerging concepts, and immunological interventions. Front. Microbiol. 15, 1413250.

Shiao, S.L., Gouin, K.H., Ing, N., Ho, A., Basho, R., Shah, A., et al., 2024. Single-cell and spatial profiling identify three response trajectories to pembrolizumab and radiation therapy in triple negative breast cancer. Cancer Cell 42, 70–84.e8.

Singh, T., Ravi Kumar, V., Sanap, G., Anjaiah, B., Roja, K., 2016. N vitro antipyrial activity of psidium guajava leaf extract. Eur. J. Pharm. Med. Res. 3, 312–315.

Srivastava, N., Ibrahim, S.A., Nasr, M.H.A., 2024. Manipulating the rhizosphere microbiome for plant health, Microbiome Engineering. CRC Press, Florida, p. 19.

Stinson, B.M., Loparo, J.J., 2021. Repair of DNA double-strand breaks by the nonhomologous end joining pathway. Annu. Rev. Biochem. 90, 137–164.

Szatkowska, M., Zdrada-Nowak, J., 2025. Genetic polymorphisms in base excision repair (BER) and nucleotide excision repair (NER) pathways as potential biomarkers for gynecological cancers: a comprehensive literature review. Cancers (Basel) 17, 2170.

Tafrishi, A., Trivedi, V., Xing, Z., Li, M., Mewalal, R., Cutler, S.R., et al., 2024. Functional genomic screening in Komagataella phaffii enabled by high-activity CRISPR-Cas9 library. Metab. Eng. 85, 73–83.

Tang, M., Lee, H.-W., Weng, M., Wang, H.-T., Hu, Y., Chen, L.-C., et al., 2022. DNA damage, DNA repair, and carcinogenicity: Tobacco smoke versus electronic cigarette aerosol. Mutat. Res. Mutat. Res. 789, 108409.

Tang, Q., Wang, X., Wang, H., Zhong, L., Zou, D., 2024. Advances in ATM, ATR, WEE1, and CHK1/2 inhibitors in the treatment of PARP inhibitor-resistant ovarian cancer. Cancer Biol. Med. 20, 915–921.

Uddin, M.J., Faiyazuddin, M., Omri, A., Gowri, S., Khalid, M., 2024. Advances in artificial intelligence for drug delivery and development: a comprehensive review. Comput. Biol. Med. 178, 108702.

Uddin, M.J., Faiyazuddin, M., Saeed, S.M., Sadiq, S.B., Koly, I.A.C., 2025a. Lipid-based drug delivery for systemic effect through pulmonary route of administration. Lipids in pulmonary drug delivery. Elsevier, pp. 255–299.

Uddin, M.J., Zulfiqar, I., Sadiq, S.B., Lucky, S., Lubna, K.A., 2025b. Natural biomaterials used in wound healing and diabetic foot. In: Md.Faiyazuddin, M.S. (Ed.), Design and processing of green materials. Springer, pp. 329–368.

Vaddavalli, P.L., Schumacher, B., 2022. The p53 network: cellular and systemic DNA damage responses in cancer and aging. Trends Genet. 38, 598–612.

Vardikar, A., Das, U., Mandal, S., Mandlik, S., Bari, D., Pardeshi, C., et al., 2025. Surface engineered multimodal magnetic nanoparticles for neurodegenerative diseases. Targeted therapy for the central nervous system. Elsevier, USA, pp. 121–153.

Vora, L.K., Jetha, K., Thakur, R.R.S., Solanki, H.K., Chavda, V.P., 2023. Artificial intelligence in pharmaceutical technology and drug delivery design. Pharmaceutics 15, 1916.

Voutsadakis, I.A., Stravodimou, A., 2023. Homologous recombination defects and mutations in DNA damage response (DDR) genes besides BRCA1 and BRCA2 as breast cancer biomarkers for PARP inhibitors and other DDR targeting therapies. Anticancer Res. 43, 967–981.

Wang, J., Wu, C., Wang, Y., Shen, Y., Wu, K., Shi, Y., et al., 2024. Nano-enabled regulation of DNA damage in tumor cells to enhance neoantigen-based pancreatic cancer immunotherapy. Biomaterials 311, 122710.

Wu, T., Xu, Y., Song, Q., Liao, X., Yao, W., Wang, G., et al., 2025. DNA damage and repair (DDR) gene mutation profiles in driver gene wild-type advanced non-small cell lung cancer and the predictive role of response to platinum-based chemotherapy. Transl. Lung Cancer Res. 14, 1089–1103.

Xia, L., Yan, X., Zhang, H., 2025. Mitochondrial DNA-activated cGAS-STING pathway in cancer: mechanisms and therapeutic implications. Biochim. Biophys. Acta - Rev. Cancer 1880, 189249.

Yang, H., Lin, P., Zhang, B., Li, F., Ling, D., 2024. A Nucleophilicity-engineered DNA ligation blockade nanoradiosensitizer induces irreversible DNA damage to overcome cancer radioresistance. Adv. Mater. 36 (45), 2410031.

Yang, L., Wei, W., Yuan, X., Guo, E., Peng, P., Wang, J., et al., 2025. Targeting DNA damage repair to enhance antitumor immunity in radiotherapy: mechanisms and opportunities. Int. J. Mol. Sci. 26, 3743.

Zhang, D., Xu, X., Wei, Y., Chen, X., Li, G., Lu, Z., et al., 2022. Prognostic role of DNA Damage response genes mutations and their association with the sensitivity of olaparib in prostate cancer patients. Cancer Control 29, 10732748221129451.

Zhao, B., Li, J., Sinha, S., Qin, Z., Kou, S.H., Xiao, F., et al., 2024. Pathogenic variants in human DNA damage repair genes mostly arose in recent human history. BMC Cancer 24, 415.

Zhou, H., Ye, P., Xiong, W., Duan, X., Jing, S., He, Y., et al., 2024. Genome-scale CRISPR-Cas9 screening in stem cells: theories, applications, and challenges. Stem Cell Res. Ther. 15, 218.

Zhu, H., Cui, M., Tang, Q., Zhao, H., Zhang, P., Zeng, S., et al., 2024. Photoactivated full-API nanodrug (FAND): harnessing transition metal complexes and MTH1 inhibitor for enhanced DNA damage in cancer cells. Biomater. Sci. 12, 3154–3162.

CHAPTER

16

Metal stress, DNA damage, and mutagenesis: Molecular mechanisms and health consequences

Megha Dave and Raktim Mukherjee

Department of Biotechnology, Shree P M Patel Institute of Postgraduate Studies and Research in Science, Sardar Patel University, Anand, Gujarat, India

16.1 Introduction and scope of metal genotoxicity

Environmental exposure to toxic metals has emerged as one of the most persistent, complex, and underestimated challenges to genomic integrity and human health. Although aging is a complex process with many contributing factors, the role of environmental influences, such as exposure to heavy metals, has not been studied in much detail (Mesnage, 2025). For many years, scientists have known that metals like cadmium (Cd), lead (Pb), hexavalent chromium [Cr(VI)], arsenic (As), mercury (Hg), and nickel (Ni) affect living systems. However, recent research has greatly expanded insight into their connections with genome activity, mutation processes, and the likelihood of disease development. Balali-Mood et al. (2021) highlighted that As, Cr(VI), and Cd uniquely disrupt genomic stability by inducing free radical-mediated DNA damage and impairing DNA repair pathways, thereby contributing to their carcinogenic potential. This chapter aims to reposition the narrative around metal genotoxicity—not merely as a toxicological issue, but as a dynamic, systems-level disturbance of the genome-environment interface.

16.1.1 From industrial legacy to modern exposure landscapes

Traditionally, the study of metal toxicity has been linked to well-known occupational exposures, uncontrolled discharge of industrial effluents, and episodes of acute poisoning. Classic examples include the tragic outbreak of Minamata disease in Japan, which was the result of

Genome Damage: Repair and Adaptation Across Life.
DOI: https://doi.org/10.1016/B978-0-443-42857-9.00008-6

chronic Hg pollution, and the large-scale As contamination of groundwater that has severely affected populations in Bangladesh and the neighboring state of West Bengal. These incidents not only highlighted the immediate health risks posed by heavy metals but also brought global attention to their long-term environmental and public health consequences (Shaji et al., 2021). Early toxicology focused largely on organ-level effects: kidney damage from Cd, neurotoxicity from Pb, or carcinogenicity from Cr(VI). However, with the advent of genome-centric science, it has become increasingly clear that the genome itself—its structure, stability, expression, and repair—is a primary and early target of chronic metal exposure (Rehman et al., 2025). According to Kotnala et al. (2025), heavy metals pose significant public health risks due to their persistence, bioaccumulation, and widespread environmental release. Toxicity depends on multiple variables, including exposure route, concentration, and co-exposure to other substances. Environmental contamination by heavy metals affects water, plants, and human systems, with serious ecotoxicological consequences. There is an urgent need for novel monitoring and assessment techniques.

Today, the exposure matrix has become vastly more complex. Metals are no longer confined to clearly identifiable pollution sources. They are ubiquitously embedded in urban air particulates, electronic waste, household paints, cosmetics, contaminated soils, fertilizers, and even food and water systems (Shetty et al., 2025). This shift has broadened the affected demographic from high-risk workers to vulnerable general populations—including pregnant women, children, and individuals with genetic predispositions to repair defects. According to Balali-Mood et al. (2021), heavy metals such as Hg, Pb, Cr(VI), Cd, and As harm cells through common pathways. These include overproduction of reactive oxygen species (ROS), weakening the antioxidant defense system, and blocking important enzymes. In addition, some metals act in specific ways—for example, Pb blocks the enzymes δ-aminolevulinic acid dehydratase and ferrochelatase, which are needed for heme synthesis. Furthermore, Cr(VI), Cd, and As have the added effect of damaging DNA and interfering with DNA repair, which increases their potential to cause cancer.

16.1.2 Genome as a frontline sensor of metal stress

Unlike traditional toxicants, metals do not need to be metabolized into active forms to exert their biological influence. Their high affinity for nucleic acids, proteins, and sulfhydryl groups makes them inherently reactive and able to bypass many conventional detoxification mechanisms. At the genomic level, this translates into a variety of direct and indirect threats: DNA strand breaks, base mispairing, covalent adduct formation, epigenetic reprogramming, and disruption of DNA repair machinery (Guo et al., 2018; Sharma et al., 2025).

Metal-induced genotoxicity is unique because it can cause two types of harm. On one hand, it produces direct or "hard" damage, such as mutations, chromosomal changes, and DNA crosslinks. On the other hand, it also brings about indirect or "soft" changes, like altering epigenetic marks and disturbing gene expression patterns. Together, these effects make heavy metals powerful disruptors of the genome, as they can trigger both permanent mutations and long-lasting changes in how genes function, even without changing the DNA sequence itself (Wu et al., 2025).

16.1.3 A systems view of genotoxic risk

A key novelty in the modern framing of metal genotoxicity is the appreciation of its systemic and context-dependent nature. Unlike isolated chemical insults, metals often act in a combinatorial and cumulative fashion, interacting with other environmental stressors, genetic susceptibilities, nutritional deficiencies, and developmental windows of vulnerability (Sharma et al., 2025). The outcome is shaped by dose, duration, developmental timing, and individual genome repair capacity.

Inside the cell, Cr(VI) is converted into Cr(III), which attaches tightly to DNA and proteins, forming crosslinks, while also producing ROS. Cd causes oxidative stress indirectly by disturbing the normal function of antioxidant enzymes. Pb, on the other hand, can wrongly insert itself into the DNA structure and block important DNA repair enzymes such as APE1 and POLβ (Zhou et al., 2021). As is now recognized for its ability to induce both mutagenic DNA lesions and heritable epigenetic alterations (Rehman et al., 2025).

By adopting a systems biology approach, metal genotoxicity is now understood as a network-level perturbation involving cross-talk between the genome, epigenome, transcriptome, and metabolome. Multiomics studies have demonstrated that chronic exposure to metal ions can rewire cellular signaling pathways involving inflammation, oxidative stress, DNA repair suppression, and apoptosis (Thévenod and Lee, 2013; Wei et al., 2025).

16.1.4 Public health implications in the genomic era

The genomic consequences of metal exposure are not only of academic interest—they have urgent public health implications. Chronic exposure to As, Cd, and Pb has been linked to neurodevelopmental disorders, infertility, cardiovascular diseases, and cancer, many of which originate from early, subclinical genomic insults (Albores-Garcia et al., 2021; Ding et al., 2023). Additionally, evidence now supports a "hit-and-hide" mechanism, where DNA damage induced by metals remains latent until triggered by replication stress or aging.

Molecular biomarkers such as γ-H2AX, 8-oxo-dG, and circulating miRNAs are being used to detect DNA damage even in asymptomatic individuals. Liquid biopsies based on cfDNA and exosomal RNA have shown promise in early detection of metal-induced genome damage, especially in high-exposure populations (Li et al., 2023). These modern protocols are reshaping the landscape of environmental health surveillance.

Research shows that metal exposure can cause epigenetic changes (epimutations) that are passed on to the next generation, especially through germ cells. Studies in mice and humans have found that when mothers are exposed to As or Pb during pregnancy, their children often show altered DNA methylation patterns, which may increase the risk of metabolic problems and learning or memory disorders (Fragou et al., 2011; Martinez-Zamudio and Ha, 2011; Albores-Garcia et al., 2021).

16.1.5 Beyond damage: Adaptation, resilience, and opportunity

While the deleterious effects of heavy metals are well-documented, recent research also explores adaptive cellular responses. Some low-dose exposures may upregulate antioxidant and DNA repair pathways through stress sensors such as NRF2 or p53. This "hormesis-like" phenomenon has opened up new questions about thresholds, resilience, and therapeutic windowing (Koyama et al., 2024). Understanding the balance between damage and adaptation is crucial—not just for regulatory toxicology purview, but also for guiding therapeutic interventions. Chelation therapies, epigenetic modifiers, and even genome editing are under consideration for mitigating the long-term effects of metal genotoxicity.

16.1.6 Defining the scope of this chapter

This chapter aims to synthesize the mechanistic, molecular, and translational advances in the field of metal genotoxicity. It covers:

- Molecular mechanisms of DNA damage induced by toxic metals
- Epigenetic reprogramming and chromatin remodeling under metal stress
- Mutation signatures and biomarker innovations
- Integration of omics platforms and biosensors in exposure detection
- Translational insights into risk mitigation, diagnostics, and public health

In highlighting recent trends, this chapter aims to support future directions in biomonitoring, mechanistic research, and precision environmental health.

16.2 Mechanisms of DNA damage induced by heavy metals

Most heavy metals are pervasive environmental contaminants that pose significant threats to genomic stability (Balali-Mood et al., 2021). Their genotoxic potential arises from a constellation of direct and indirect mechanisms that culminate in various forms of DNA damage (Bona et al., 2023). These mechanisms include the generation of reactive oxygen (ROS) and nitrogen species (RNS), disruption of redox homeostasis, formation of DNA adducts, base mismatches, strand breaks, and crosslinking events (Fig. 16.1). Recent advancements in molecular docking and quantum chemical modeling have shed new light on the structural and energetic aspects of metal-DNA interactions, providing mechanistic clarity at the atomic level.

16.2.1 Metal-induced oxidative stress and reactive oxygen species/reactive nitrogen species generation

Heavy metals do not directly produce free radicals, but they can catalyze redox reactions or deplete antioxidant defenses, indirectly generating ROS. For example, Fe^{2+} and Cu^{+} lead to Fenton and Haber–Weiss reactions, transforming hydrogen peroxide into hydroxyl radicals (•OH), one of the most reactive and damaging ROS (Hussain et al., 2025; Jomova et al., 2025). Although Cd^{2+} and Pb^{2+} are redox-inactive, they displace essential redox-active metal ions (like Zn^{2+} in superoxide dismutase) and inhibit antioxidant enzymes such as catalase and glutathione peroxidase, promoting intracellular ROS accumulation (Bovio et al., 2021). Some metals also increase the production of RNS. They do this by activating an enzyme called inducible nitric oxide synthase (iNOS), which leads to the formation of harmful molecules like peroxynitrite ($ONOO^{-}$) (Abd Elnabi et al., 2023). These RNS can nitrate guanine residues or disrupt DNA–protein crosslinks, amplifying genotoxicity (Andrés et al., 2023). When cells have too much ROS or RNS, they can damage DNA in several ways. This includes creating 8-oxo-2′-deoxyguanosine (8-oxo-dG), breaking single or double strands of DNA, and forming spots where DNA bases are missing (AP sites). If these DNA damages are not repaired, they can lead to mutations. (Jomova et al., 2025).

Double-strand breaks (DSBs) and SSBs are severe consequences of ROS-mediated DNA oxidation. Cr(VI), upon cellular uptake, is reduced to Cr(III) through intermediate valence states [Cr(V) and Cr(IV)], generating abundant ROS during the process. Cr(III) forms stable complexes with DNA and histones, leading to chromatin condensation, DNA rigidity, and DSBs

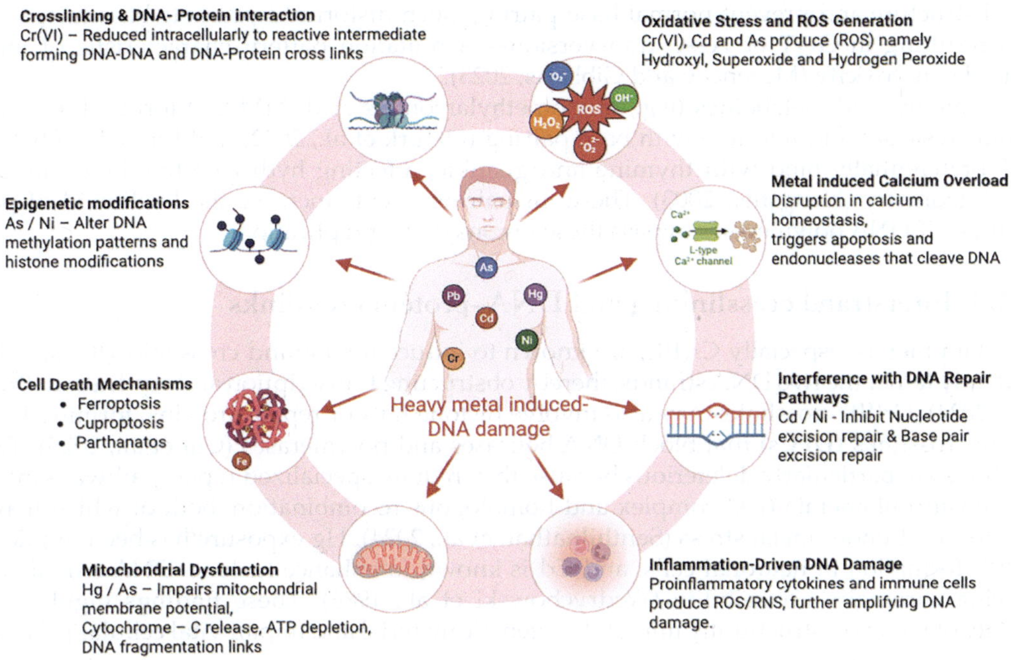

FIGURE 16.1 Schematic representation of multiple mechanisms by which heavy metals induce DNA damage. Heavy metals exert genotoxic effects through diverse and interconnected pathways. Major mechanisms include (1) oxidative stress and excessive ROS generation leading to base modifications and strand breaks; (2) inhibition of DNA repair pathways such as nucleotide excision repair (NER) and base excision repair (BER); (3) epigenetic alterations including DNA methylation changes and histone modifications; (4) disruption of essential metal ion homeostasis, impairing zinc-finger DNA repair proteins; (5) formation of DNA–DNA and DNA–protein crosslinks; (6) interference with cell cycle checkpoints (e.g., p53, ATM/ATR); (7) mitochondrial dysfunction resulting in ROS leakage and DNA fragmentation; (8) calcium overload–mediated activation of endonucleases; (9) activation of regulated cell death pathways such as ferroptosis, cuproptosis, and parthanatos; and (10) inflammation-driven DNA damage via cytokine-induced ROS/RNS production. Together, these mechanisms contribute to mutation accumulation, genomic instability, and disease progression.2.2 DNA Strand Breaks and Chromatin Destabilization (Created with BioRender.com.).

(Zhang et al., 2023a). As, via trivalent arsenite [As(III)], has been shown to destabilize chromatin by depleting poly(ADP-ribose) polymerase (PARP) activity, thus inhibiting DNA repair and prolonging the persistence of breaks (Nemati-Mansour et al., 2024).

In live-cell models and human lymphocytes, exposure to $CdCl_2$ and $Pb(NO_3)_2$ has been associated with increased γ-H2AX foci—a biomarker of DSBs—demonstrating direct DNA fragmentation (Yıldız et al., 2024). Moreover, exposure to Ni^{2+} and Cr(VI) in lung epithelial cells induces replication stress, stalling replication forks, and leading to fork collapse, which is a well-known pathway for DSB generation (Zhao et al., 2025).

16.2.2 DNA adduct formation and base mispairing

Many heavy metals directly form **DNA adducts** by binding to nucleophilic centers on DNA bases or the phosphate backbone (Sharma et al., 2025). Cr(III), the reduced product of Cr(VI), forms stable complexes with the N7 position of guanine and adenine, which can distort the helical structure and prevent normal base pairing. Such distortions promote base mispairing, often resulting in G→T or A→C transversions—a mutation pattern characteristic of metal-induced genotoxicity (Mezencev and Gibbons, 2023).

As's methylated metabolites (e.g., monomethylarsonous acid, MMA^{+3}) form adducts with guanine residues and interfere with base pairing (Ozturk et al., 2022). Additionally, Pb^{2+} and Hg^{2+} preferentially bind with thymine and guanine, affecting hydrogen bonding and DNA helical geometry (Hengstler, 2003). These mispairing events increase the likelihood of point mutations if DNA polymerases bypass these lesions during replication.

16.2.3 Interstrand crosslinking and DNA–protein crosslinks

Certain metals, especially Cr(III), are known to induce interstrand crosslinks (ICLs), which covalently tether the two DNA strands, thereby obstructing transcription and replication (Brown et al., 2024). Cr(III) crosslinks may also involve histone tails or repair proteins, forming DNA–protein crosslinks **(DPCs)** that block DNA helicases and polymerases (Cui et al., 2025). These structures are particularly deleterious because they require specialized repair pathways involving the Fanconi anemia (FA) complex and homologous recombination, both of which may be compromised under metal stress (Senthilnathan et al., 2024). Hg exposure has been implicated in DPC formation in neuronal cells, and Cd is known to enhance histone–DNA crosslinking, reducing chromatin accessibility (Wedrychowski et al., 1986). These findings highlight the capacity of metals to structurally immobilize genetic material and impair vital cellular processes.

16.2.4 New insights from molecular docking and quantum chemical modeling

Recent computational toxicology approaches, including molecular docking, density functional theory (DFT), and quantum chemical simulations, have illuminated the binding preferences and energetics of metal-DNA interactions (Kar and Leszczynski, 2019). In silico studies reveal that Cd^{2+}, Ni^{2+}, and Cr(III) preferentially bind to the minor groove of DNA, particularly at guanine-rich sequences, owing to the high electron density and coordination geometry favorable to metal ions (Mukherjee et al., 2025).

Docking studies using software such as autodock vina and GOLD demonstrate that Cr(III) and Cd^{2+} exhibit high affinity for DNA sequences at telomeric and CpG-rich regions, potentially linking metal exposure to telomere instability and epigenetic silencing. Quantum chemical analyses further reveal that the association of Cr(III) and guanine alters the HOMO-LUMO gap of the base, reducing its redox potential and increasing susceptibility to electron-transfer-induced damage (Islam et al., 2024).

Such findings are supported by molecular dynamics simulations showing DNA conformational changes—such as unwinding, bending, and groove widening—upon metal-binding. These insights not only validate empirical observations but also pave the way for predictive toxicology models based on structural bioinformatics and AI-assisted docking algorithms (Lu et al., 2019).

16.2.5 Synergistic and combined metal toxicity

In real-world exposures, metals often occur in mixtures rather than isolation (Kong et al., 2025). Studies have shown synergistic genotoxic effects of metal mixtures, such as Cd + Pb or As + Ni, which amplify oxidative insult and compound DNA damage (Kwon et al., 2023).

For instance, simultaneous exposure to Cr(VI) and Ni(II) in human airway lining cells led to more persistent DSBs and higher mutation frequencies than individual exposures (O'Hara et al., 2006; Honda et al., 2015). This synergism may arise from overlapping ROS generation pathways, shared DNA-binding sites, or cumulative inhibition of repair proteins.

Furthermore, research employing high-content screening and network toxicology has mapped shared target proteins and genetic pathways affected by multiple metals, such as p53, ATM, ATR, and NF-κB (Taboureau et al., 2020). These systems-level approaches reinforce the need to study metal-induced DNA damage in the context of integrated exposomics (Hui et al., 2025) (Table 16.1).

Heavy metals induce DNA damage via a complex interplay of oxidative mechanisms, direct chemical interactions, and chromatin-level disruptions. While ROS/RNS-mediated strand breaks remain a central theme, metal-induced adducts, crosslinks, and repair inhibition add layers of mutagenic risk. New computer-based models and advanced lab techniques are helping scientists better understand how metals affect the body. These tools could make it possible to detect problems early and develop targeted ways to reduce the harmful effects in people exposed to metals.

16.3 Epigenetic rewiring and chromatin disruption under metal exposure

Environmental exposure to toxicants, particularly heavy metals, is increasingly recognized as a significant factor influencing epigenetic processes. By altering mechanisms such as DNA methylation, histone modification, and noncoding RNA regulation, these metals can disrupt normal gene expression patterns. Such epigenetic disturbances are now understood to play a crucial role in the onset and progression of a wide range of chronic diseases (Sobral et al., 2025). These include cancer, neurodevelopmental disorders, reproductive dysfunction, and systemic immune alterations (Parida and Patel, 2023; Queiroz et al., 2024). A common thread across various environmental exposures is the induction of DNA damage, either directly or via oxidative stress pathways, which not only compromise genome integrity but also provoke chromatin remodeling and epigenetic dysregulation (Agu and Aja, 2025). The dynamic response to DNA damage necessitates alterations in chromatin architecture, including histone modifications and ATP-dependent chromatin remodeling, to facilitate repair. While these changes are typically transient, persistent exposure to toxicants or chronic inflammation can lead to lasting epigenetic alterations, potentially altering gene expression in a heritable manner (Wu et al., 2025).

The International Agency for Research on Cancer (IARC) has classified several toxic metals, including As, Cd, Cr(VI), and Ni, as environmental carcinogens due to their established cancer-causing properties. These metals do not always damage DNA directly; instead, they can change how genes are switched on or off through epigenetic mechanisms. The main ways they do this include abnormal patterns of DNA methylation, changes in histone modifications (which control how tightly DNA is packed), and disruption of noncoding RNAs that normally help regulate gene expression (Kozak, 2025; Wang et al., 2025b). These epigenetic disruptions can silence tumor suppressor genes, which normally act as safeguards against uncontrolled cell growth. At the same time, they may activate oncogenes, leading to increased cell proliferation and survival. Together, these changes play a critical role in the initiation, development, and progression of neoplastic diseases (Beyersmann, 2002; Zhu and Costa, 2020). Real-life exposure scenarios—from occupational settings to contaminated food and water—highlight the urgency of understanding these mechanisms to guide risk mitigation strategies and novel therapeutic development (Dutta et al., 2024).

TABLE 16.1 Summary of DNA damage mechanisms induced by select heavy metals.

S. no.	Metal	Major mechanism of DNA damage	Biological/cellular context	References
1	Cadmium (Cd)	Indirect DNA damage via ROS generation; inhibition of DNA repair enzymes (NER and BER); interference with mismatch repair	Human lung cells, kidney tissues, and embryonic models	Waisberg et al. (2003), Ghosh and N (2018), Yang et al. (2025), Liu et al. (2025)
2	Nickel (Ni)	Histone modification and chromatin condensation; inhibition of DNA repair proteins (XPA and PARP); induction of oxidative DNA lesions	Respiratory epithelium, occupational exposure models	Salnikow and Zhitkovich (2008), Costa (2022), Cameron et al. (2011), Mo et al. (2024)
3	Chromium (Cr, especially Cr(VI))	DNA adducts and crosslinks; strand breaks; ROS-mediated oxidative lesions (8-oxo-dG); impaired BER	Human airway epithelial cells, embryonic heart, and lung	Wise et al. (2002), O'Hara et al. (2006), Salnikow and Zhitkovich (2008), Flora and Pachauri (2010), Czarnek et al. (2025)
4	Lead (Pb)	Indirect DNA damage via oxidative stress; interference with DNA repair enzymes; chromosomal aberrations	Blood leukocytes, neuronal cells, and chronic exposure	Winder and Bonin (1993), Meng et al. (2021), Dey et al. (2024), Jomova et al. (2025)
5	Mercury (Hg)	Binding to thiol groups of DNA repair proteins; ROS overproduction; mitochondrial DNA damage	Neural tissues and kidney cells	De Flora et al. (1994), Crespo-López et al. (2011), Wu et al. (2025)
6	Arsenic (As)	ROS-mediated oxidative DNA damage; inhibition of poly(ADP-ribose) polymerase (PARP); epigenetic alterations (DNA methylation)	Human skin keratinocytes, liver, and embryonic cells	Hei and Filipic (2004), Liu et al. (2005), Ozturk et al. (2022), Khalid and Azmat (2025)

Cr(VI) has been studied extensively because of its wide industrial use and its harmful effects on reproduction. Animal studies, especially in mice, have shown that exposure to Cr(VI) during lactation can damage the ovaries. This damage appears as *follicular atresia* (loss of ovarian follicles) and hormonal imbalance. At the molecular level, Cr(VI) changes the way genes are regulated through epigenetic mechanisms. It increases the activity of DNA methyltransferases (DNMT3a/3b), which add methyl groups to DNA. It also raises repressive histone marks (H3K27me3 and H3K9me3), while reducing activating histone marks (H3K9ac and H3K27ac). Together, these changes silence genes needed for follicle growth and cell survival, while turning on genes that trigger cell death (apoptosis). These findings highlight that epigenetic regulation plays a major role in how Cr(VI) causes ovarian toxicity and infertility (Dutta et al., 2024).

The implications of epigenetic toxicity are not confined to humans. Similar molecular responses are observed in plant systems exposed to metal stress, offering a broader ecological context. Epigenetic pathways, including DNA methylation and small RNA regulation, modulate plant tolerance against heavy metal stress, thereby contributing to environmental resilience and agricultural sustainability. Advances in sequencing technologies and functional genomics have enabled the identification of epigenetic markers of metal tolerance in plants, fostering the development of crop varieties better suited to contaminated environments. These findings emphasize the growing importance of epigenetics in understanding plant responses to stress and point to the potential application of this knowledge for improving agricultural practices worldwide (Iqbal et al., 2024).

As, a widely studied toxic metal, it can cause cancer through multiple epigenetic changes. It can change DNA methylation patterns, both across the whole genome and at specific genes. As also affects histone modifications, including acetylation and methylation, and alters the levels of noncoding RNAs such as miRNAs and lncRNAs. In addition, it can interfere with RNA methylation and alternative splicing, adding extra layers of gene regulation. Together, these changes create an environment that promotes tumor growth and may provide potential biomarkers for early cancer detection and therapy (Islam et al., 2024).

Recent findings have also elucidated how Cr(VI) disrupts chromatin architecture through DNA adduct formation, nucleosomal disorganization, and interference with transcription factor binding—particularly in euchromatin regions rich in regulatory elements. These chromatin perturbations result in genomic instability, marked by aneuploidy and chromosomal aberrations, which are hallmarks of cancer. Long-term, low-dose exposure to Cr(VI) further exacerbates these effects by silencing tumor suppressor genes and diminishing the cellular capacity to respond to other environmental insults, illustrating the compounding nature of epigenetic toxicity (VonHandorf et al., 2021).

Beyond metals in ionic form, nanomaterials such as gold nanoparticles (AuNPs) have emerged as a new frontier in environmental toxicology. AuNPs can induce primary DNA damage, oxidative stress, and epigenetic modifications, including altered gene-specific methylation and miRNA expression. The physicochemical characteristics of the nanoparticles, including their size, shape, and surface chemistry, influence these effects. The variability in genotoxic outcomes and challenges in standardized toxicity testing highlight the urgent need for rigorous, adapted assessment protocols for nanomaterial safety (Ramos-Pan et al., 2024).

Epidemiological investigations provide compelling evidence linking long-term residence in contaminated environments, such as mining regions, to transcriptomic and immunologic alterations in exposed populations. For instance, Stepanyan et al. (2025) observed sex-specific differences in peripheral blood mononuclear cells (PBMCs) of residents exposed to

As, thallium, Ni, and other nonessential metals. While females exhibited robust immune activation, males showed dysregulation in ion transport and epigenetic gene regulation. The study also documented shifts in immune cell populations, including reduced CD4+ memory cells and increased NK cells, emphasizing the systemic impacts of prolonged metal exposure.

Maternal exposure to hazardous metals during pregnancy is particularly concerning in light of the developmental origins of health and disease (DOHaD) hypothesis. Epigenetic modifications established in utero—such as DNA methylation and histone changes—can have lasting effects that persist into adulthood and potentially across generations. For example, maternal exposure to Cd, Pb, or Hg is associated with altered epigenetic regulation in the fetus, which may predispose offspring to chronic conditions later in life. The transgenerational inheritance of these changes complicates risk assessment and underlines the importance of preconception and prenatal environmental surveillance. The mechanistic underpinnings of metal-induced epigenetic toxicity are further exemplified by Fragou et al. (2011), who detailed specific alterations such as Ni-induced histone methylation, Pb-associated DNA hypomethylation, and the dual methylation effects of Cr(VI). These alterations can dysregulate critical pathways that regulate cell division, maintain DNA integrity, and control programmed cell death, serving as potential biomarkers of exposure and disease prognosis.

Heavy metal exposure also impairs the function of stem cells by disrupting epigenetic machinery. Cd, Ni, Hg, and Pb have been shown to alter DNA methylation, decrease hydroxymethylation, and modify histone marks in stem cells. These disruptions compromise stem cell pluripotency and differentiation, potentially impairing tissue regeneration and increasing susceptibility to degenerative diseases. Moreover, alterations in enzymes such as DNMTs and TETs underscore the intimate link between metal exposure and the loss of stem cell identity and repair capacity (Kim et al., 2024).

Among sensitive groups like children, exposure to toxic metals from the environment has been linked to problems in brain development, including conditions such as autism spectrum disorder (ASD). A study by La-Ane et al. (2025) in Makassar, Indonesia, identified elevated hair Hg levels and maternal behaviors (e.g., use of skin-lightening creams) as significant predictors of ASD risk. Interestingly, lower Pb levels correlated with reduced ASD risk, suggesting that exposure windows, metal interactions, and genetic predisposition all contribute to the multifactorial etiology of neurodevelopmental disorders.

Chromatin immunoprecipitation combined with sequencing (ChIP-seq) has become an important technique for examining epigenetic modifications across the entire genome that are triggered by heavy metal exposure (Zou et al., 2023). This technique enables the mapping of DNA–protein interactions, histone modifications, and transcription factor binding events, providing critical insights into how toxic metals alter chromatin landscapes and gene regulation. In recent years, ChIP-seq has been increasingly applied to study the impact of metals such as As, Cd, and Cr(VI) on histone modification patterns and chromatin accessibility in both in vitro and in vivo systems (Raza et al., 2024).

For instance, ChIP-seq analyses in Cd-exposed cell lines revealed global enrichment of repressive histone marks such as H3K27me3 and H3K9me3 at promoter regions of tumor suppressor genes, contributing to their silencing (Chandrasekaran et al., 2025). Conversely, decreased levels of activating histone marks like H3K4me3 and H3K27ac have been observed at loci involved in DNA repair and apoptosis pathways, suggesting an epigenetic mechanism through which Cd promotes genomic instability and cell survival. Similar studies in As-exposed models demonstrated redistribution of H3K9ac and loss of CTCF binding, indicating compromised insulator activity and disrupted chromatin boundaries (Ma et al., 2016).

When cells are exposed to Cr(VI), studies using ChIP-seq show that important transcription factors like **AP-1** and **CTCF** do not bind to DNA in the normal way. These changes happen in euchromatin (the active parts of DNA). As a result, the normal communication between enhancers and promoters is disturbed, which leads to changes in how genes are turned on or off (VonHandorf et al., 2021). Cr(VI) can change the position of nucleosomes (the basic DNA–protein units) near important gene control regions. This shifting makes it harder for certain genes to stay active. As a result, genes that normally control the cell cycle and help the immune system keep watch can become permanently switched off (Ge et al., 2022).

Overall, ChIP-seq provides a high-resolution framework to decipher the epigenomic remodeling induced by heavy metals. It allows for the identification of specific chromatin marks and transcriptional regulators targeted during exposure, thereby advancing our conceptual knowledge of the mechanistic basis of metal-induced detrimental effects and facilitating new avenues of epigenetic biomarkers for exposure and intervention.

Together, these studies demonstrate that epigenetic changes caused by metal exposure are central to the long-term health effects of environmental toxicants. Such changes—ranging from DNA methylation and histone modifications to alterations in noncoding RNAs and chromatin structure—can disrupt gene expression and cellular balance across multiple biological systems. Beyond explaining the underlying mechanisms, these findings point toward new opportunities for developing diagnostic biomarkers and therapeutic strategies. In the face of growing environmental challenges, linking molecular epigenetic insights with epidemiological monitoring and policy measures will be essential for protecting human health and promoting sustainable environmental management.

16.4 Epigenetic mechanisms of heavy metal-induced DNA damage

Heavy metals not only directly harm DNA but also disrupt the epigenome, which governs gene expression, DNA repair mechanisms, and genomic stability (Kurita et al., 2025). These epigenetic changes include DNA methylation, histone modifications, noncoding RNA regulation, and chromatin remodeling. Importantly, these changes are not just side effects but active mechanisms that influence how metals promote mutations and disease (Fasani et al., 2023).

16.4.1 DNA methylation alterations

DNA methylation (adding a methyl group to cytosine in CpG islands) controls gene expression and chromatin structure. Metals such as Ni, Cd, Cr(VI), and As alter DNA methylation patterns (Freydenzon et al., 2022).

- *Global hypomethylation*: Cd exposure in zebrafish and human bronchial epithelial cells (HBECs) results in the depletion of DNA methylation, causing chromosomal instability and activation of oncogenes (Green et al., 2018).
- *Promoter hypermethylation*: In contrast, tumor suppressor genes like *p16*, *MLH1*, and *BRCA1* become hypermethylated after As or Ni exposure, switching off important DNA repair and apoptosis pathways.

High-throughput profiling (WGBS, RRBS) confirms that metals produce distinct methylation "signatures." For example, Ni induces hypermethylation of the *ZIC1* promoter in BEAS-2B cells, leading to its silencing and promoting cancer-like changes (Guo et al., 2018; Bozack et al., 2021).

16.4.2 Histone posttranslational modifications

Histones are proteins around which DNA is tightly wrapped, helping to organize and compact the genetic material inside the nucleus. These proteins are not just structural; they can undergo chemical modifications such as methylation, acetylation, or phosphorylation. Such modifications change how tightly DNA is packaged, thereby regulating whether specific genes are accessible for transcription or remain silenced. Metals disrupt these modifications by affecting histone-modifying enzymes (Zhao et al., 2025).

- *Histone acetylation*: Cd and Ni reduce H3K9ac and H4K16ac (normally linked to active transcription). This occurs through reduced HAT activity and increased HDAC activity.
- *Histone methylation*: As and Ni increase repressive marks (H3K9me3 and H3K27me3) at DNA repair and cell cycle genes, silencing them.

For example, in As-treated keratinocytes, H3K27me3 was enriched at *ATM* and *XRCC5*, suppressing DNA repair (Zhang et al., 2023b). Interestingly, some of these changes are reversible, opening the door for epigenetic therapies.

16.4.3 Chromatin remodeling and 3D genome structure

Metals also affect **chromatin remodeling complexes** and higher-order DNA folding.

- Ni blocks the ATPase activity of the SWI/SNF complex, causing chromatin condensation and repression of DNA repair genes (Li et al., 2023).
- In Cr(VI)-treated fibroblasts, Hi-C studies showed altered TADs (topologically associated domains) and enhancer–promoter interactions, silencing protective genes and activating inflammatory ones (Otsuka et al., 2024).

Thus, metals can disrupt genome architecture at both local and global levels.

16.4.4 Noncoding RNAs (miRNAs and lncRNAs)

Noncoding RNAs fine-tune gene expression and are strongly influenced by metal exposure.

- *miRNAs*: For instance, miR-21 is upregulated by Cd and Cr in lung cells, suppressing *PTEN* and *PDCD4* to promote survival. Similarly, miR-155 and miR-210 rise in As-exposed liver cells, enhancing inflammatory and hypoxic responses.
- *lncRNAs*: Molecules like MALAT1, HOTAIR, and NEAT1 scaffold chromatin modifiers. MALAT1, for example, recruits EZH2 (part of PRC2) to DNA repair genes, increasing H3K27me3 and silencing them in Cd-exposed cells (Li et al., 2021; Maliyam et al., 2024).

RNA-seq consistently shows altered noncoding RNAs targeting *TP53*, DNA repair genes, and chromatin remodelers, suggesting they actively drive metal-induced mutagenesis (Pal and Firdous, 2024).

16.4.5 Integrative multiomic insights

Modern multiomic studies reveal how metals simultaneously alter DNA methylation, histone marks, and transcription.

- Cr(VI)-exposed HepG2 cells showed DNA repair gene hypermethylation, reduced H3K9ac, and corresponding transcriptional downregulation, impairing base excision repair (Hu et al., 2018).
- Systems biology models show Ni acts through a regulatory hub involving HDAC1, DNMT1, and EZH2, silencing checkpoint genes essential for DNA damage response (Zhou et al., 2021).

These integrative approaches confirm that heavy metals do not act through a single pathway but reprogram the epigenome at multiple levels.

In short, heavy metals change the epigenome in many ways. They modify patterns of DNA methylation and chemical tags on histones, along with altering chromatin structure and the activity of noncoding RNAs. Together, these changes weaken DNA repair, silence tumor suppressors, activate oncogenes, and destabilize the genome, making epigenetic regulation a central mechanism of heavy metal-induced genotoxicity.

16.5 Mutation signatures and biomarker advances in metal genotoxicity

Metal-induced DNA damage has emerged as a pivotal driver of genomic instability, giving rise to diverse mutational spectra and molecular alterations that are increasingly recognized as distinct "mutation signatures." These signatures represent the cumulative outcome of oxidative stress, DNA adduct formation, crosslinking, and impaired repair pathways triggered by metals such as Cd, Cr(VI), As, and Ni (Hartwig, 1995; Järup, 2003; Valko et al., 2016). Importantly, progress in high-throughput genomics and algorithmic toxicology has enabled the delineation of these mutation patterns with greater precision, linking specific metals to characteristic molecular fingerprints of genotoxicity (Kumar et al., 2025).

16.5.1 Mutation spectrum: Single-nucleotide polymorphisms, copy number variations, and microsatellite instability

Heavy metals like Cd, As, Cr(VI), and Ni disrupt DNA replication fidelity and repair pathways, leading to characteristic mutational patterns:

- *Single-nucleotide polymorphisms (SNPs)*: Metal exposure frequently results in point mutations due to oxidative base lesions, mispairing, or faulty repair. Studies have revealed C > A and G > T transversions, often linked to 8-oxoguanine formation, as dominant mutations in As- or Cr(VI)-exposed cells (Halasova et al., 2012).
- *Copy number variations (CNVs)*: Repetitive metal-induced damage and impaired repair often result in large-scale deletions or amplifications. For example, Cd can create CNV hotspots in genes that control the cell cycle and programmed cell death (apoptosis) in mouse embryonic fibroblasts (Kaur et al., 2021).
- *Microsatellite instability (MSI)*: Inhibition of mismatch repair (MMR) proteins like MLH1 and MSH2 by metals such as Ni and As leads to MSI, a phenomenon commonly associated

with tumorigenesis. Recent genomic analyses of Ni-exposed HBECs confirmed MSI at loci controlling DNA damage response genes (Zhang et al., 2023c).

Whole genome and exome sequencing have been instrumental in decoding these mutation signatures. Computational deconvolution tools (e.g., MutationalPatterns and SigProfiler) now allow comparison of metal-induced mutation spectra with known COSMIC signatures, helping trace environmental exposures to cancer mutagenesis.

16.5.2 Biomarkers of metal-induced genotoxicity

A suite of molecular biomarkers has emerged to detect, quantify, and monitor DNA damage due to metal exposure, often even at early or subtoxic stages.

- *γ-H2AX*: The phosphorylated form of histone H2AX, known as γ-H2AX, is widely recognized as a rapid and sensitive marker of DNA DSBs. When DNA damage occurs, γ-H2AX forms distinct foci at the break sites, making it useful for detecting and monitoring DNA injury. Elevated γ-H2AX foci have been observed in PBMCs and cultured keratinocytes exposed to toxic metals such as As and Cr(VI), indicating that these metals can induce significant DNA damage. This highlights the role of γ-H2AX as a valuable biomarker for assessing genotoxic stress caused by environmental metal exposure (Zhou et al., 2021). Quantification using immunofluorescence or flow cytometry has enabled early detection of genotoxic insult.
- *8-oxo-2′-deoxyguanosine (8-oxoG)*: A hallmark of genotoxic insult, 8-oxoG is often elevated in the urine or serum of individuals chronically exposed to Cd or Ni. ELISA and LC-MS/MS-based quantification of 8-oxoG now serve as standard tools in metal toxicity assessment (Wang et al., 2023).
- *Circulating cell-free DNA (cfDNA)*: cfDNA originates from dying cells, either through apoptosis or necrosis. Because it contains the genetic fingerprints of cellular damage, researchers are exploring its role in monitoring environmental exposures. Fragmentation patterns, methylation states, and copy number in cfDNA from As-exposed populations have shown potential as noninvasive biomarkers (Maji et al., 2020).
- *MicroRNAs (miRNAs)*: Metal-altered miRNA profiles offer both mechanistic insight and biomarker utility. For instance, studies on Cd and Cr(VI) exposure models have shown increased expression of miR-21 and miR-155, along with reduced levels of miR-34a, changes that are linked to defective DNA repair and higher rates of mutagenesis. These miRNAs are detectable in biofluids and have entered clinical testing pipelines.

16.5.3 Advances in biosensors and detection platforms

Recent technological strides have led to the development of rapid, sensitive, and often field-deployable biosensor systems to detect metal-induced DNA damage:

- *Electrochemical biosensors*: Sensors using gold nanoparticles, graphene oxide, or carbon nanotube-modified electrodes can detect DNA adducts, oxidative lesions (e.g., 8-oxoG), or damage induced protein expression (like γ-H2AX) with high sensitivity. Some systems now integrate smartphone readouts and microfluidic devices for real-time analysis (Sha et al., 2025; Wangmo et al., 2025).

- *DNA aptamer-based sensors*: Engineered aptamers specific to 8-oxoG or Cr(VI)-DNA adducts have been developed for point-of-care detection. These tools hold promise for high-throughput biomonitoring in occupational and environmental settings (Sha et al., 2025).
- *Quantum dot and fluorescence sensors*: Quantum dot-labeled DNA strands or DNA repair proteins (e.g., OGG1 and APE1) are used in fluorescence-based detection platforms to track real-time interaction with damaged DNA (Biranje et al., 2021).

These sensor technologies, though still under refinement, are making genotoxic screening more accessible and scalable.

16.5.4 Omics and liquid biopsy platforms

- *Multiomic profiling*: Integration of genomics, epigenomics, transcriptomics, and proteomics has provided a holistic view of cellular response to metals. For instance, transcriptomic analysis in Ni-exposed human lung cells revealed upregulation of pro-mutagenic genes and suppression of mismatch repair pathways, validating observed mutation spectra (Yuan et al., 2022).
- *Liquid biopsy approaches*: Liquid biopsies, which analyze cfDNA, miRNAs, and exosomes from blood or urine, are gaining traction for early genotoxic surveillance. For populations exposed to As-contaminated water, longitudinal analysis of cfDNA and miRNA profiles has shown promise in predicting cancer risk and guiding intervention strategies (Hsu and Wu, 2022).
- *Single-cell and spatial omics*: Advanced techniques such as single-cell RNA sequencing (scRNA-seq) and ATAC sequencing (ATAC-seq) have been applied to study how individual cells respond differently to metal-induced DNA damage and repair. These approaches reveal cell-to-cell variation that would otherwise be hidden in bulk analyses, providing deeper insights into the mechanisms of toxicity. Spatial transcriptomics in Cr(VI)-exposed mouse liver tissue identified mutation-prone zones with suppressed DNA repair gene expression, suggesting location-specific vulnerability (Wu et al., 2024).

16.5.5 Clinical and translational implications

The identification of reliable biomarkers and mutation signatures has far-reaching implications:

- *Early diagnosis*: γ-H2AX and 8-oxoG are now being proposed as early diagnostic markers for populations at risk of metal-related carcinogenesis (Cheng et al., 2024).
- *Exposure biomonitoring*: miRNA signatures and cfDNA fragmentation indices are being incorporated into environmental health surveillance frameworks, particularly in industrial zones (Rehman et al., 2025).
- *Risk stratification*: The integration of genomic and epigenomic damage profiles is increasingly recognized as a powerful approach to **risk stratification in populations exposed to toxic metals**. By analyzing mutation signatures, DNA methylation patterns, and chromatin remodeling events, researchers can classify individuals into **high-, medium-, or low-risk categories** based on their susceptibility to metal-induced genotoxicity (Ren et al., 2011; Kucab et al., 2019). This stratification provides a framework for personalized preventive strategies, allowing early intervention in high-risk groups through dietary supplementation, lifestyle modification, or enhanced medical surveillance. Importantly, such approaches move beyond generalized

exposure assessment, offering a precision toxicology model that parallels the advances made in precision oncology (Hou et al., 2012).

- *Therapeutic targeting*: In parallel, therapeutic targeting of molecular regulators altered by metal stress is emerging as a promising strategy. Among these, microRNAs (miRNAs) have garnered considerable attention due to their role in regulating DNA repair, apoptosis, and oxidative stress pathways. For instance, miR-34a, often suppressed in metal-induced toxicity models, has been implicated in the dysregulation of DNA repair machinery. Preclinical research indicates that introducing miR-34a mimics can enhance DNA repair mechanisms, potentially mitigating the genotoxic and cancer-promoting effects of metals like Cd and As (Chang et al., 2007; Taghizadeh et al., 2019). This area of research highlights the promise of RNA-based therapies in reducing the long-term effects of metal exposure, linking mechanistic understanding with potential clinical applications.

Heavy metal exposure disrupts biological systems through complex, multi-tiered mechanisms involving gene regulation, epigenetic modifications, protein alterations, and metabolic dysfunction. Employing multi-omics approaches illuminates these mechanisms and helps identify sensitive biomarkers early in exposure.

16.5.5.1 Multiomics integration

- *Transcriptomics and toxicogenomics*: Gene expression profiling in organisms exposed to Cd or Cr(VI) has revealed activation of stress-response genes (e.g., *mt2, hsp70*) and detoxification enzymes like glutathione-S-transferase and catalase.
- Proteomics extends this analysis to protein abundance and posttranslational modifications, capturing disruptions in metal-binding proteins (metallothionein), redox enzymes, and stress signaling cascades.
- Metabolomics provides a snapshot of metabolic disruptions—highlighting oxidative stress via glutathione imbalance, TCA cycle derangement, and altered energy metabolism in heavy metal-exposed samples.
- Integration across these layers enhances the detection of early-stage biomarkers and offers mechanistic insight into metal-induced mutagenesis.

16.5.5.2 Novel genotoxicity markers

- *γ-H2AX*: This phosphorylated histone (Ser139) is among the most sensitive markers for detecting DNA DSBs following exposure to genotoxic substances. It demonstrates ~91% predictivity in clastogenic and aneugenic assays.
- *8-oxoG (8-oxo-7,8-dihydroguanine)*: A key oxidative DNA lesion that accumulates following ROS generation—common during metal exposure.
- *Cell-free DNA (cfDNA)*: Acts as an indicator of cellular injury / apoptosis and rises during early toxicological stress, reflecting systemic damage.
- *5-hmC (5-hydroxymethylcytosine)*: Altered in response to epigenetic reprogramming induced by heavy metals; serves as a marker of disrupted demethylation dynamics and genomic instability.

Together, these markers provide a comprehensive assessment of DNA damage and potential mutagenesis beyond simple cytotoxic endpoints.

16.5.5.3 Biosensor platforms

- **Whole-cell bioreporters (WCBs)**, such as *E. coli* sensing through *recA::egfp* or *zntA::mCherry*, detect genotoxic stress and bioavailable metal ions simultaneously, providing real-time functional insights.
- **Nanomaterial-based sensors**, including DNA-network and hairpin-based platforms, offer ultra-sensitive detection of specific metal ions (e.g., UO_2^{2+}), achieving detection thresholds at the femtomolar level.
- **Electrochemical biosensors**, such as MWCNT–β-cyclodextrin electrodes, allow field-deployable and sensitive detection of lead and other metals at subppb concentrations.

These platforms are increasingly geared toward multiplex detection—linking metal exposure to genotoxic events, making them powerful tools for environmental and biological monitoring.

Thus, by combining **multiomics profiling, genotoxicity markers**, and **advanced biosensor technologies**, we gain a multidimensional understanding of heavy metal-induced mutagenesis. This integrative strategy empowers early detection, mechanistic understanding, and tailored intervention—strengthening the scientific foundation for the book chapter on *Metal Stress and Mutagenesis*.

16.6 Metal stress and DNA mutagenesis: Disease associations and translational implications

An increasing corpus of evidence highlights that heavy metals are not isolated toxicants but systemic disruptors that affect multiple organs and biological pathways. Importantly, their toxic effects are context-dependent, shaped by cumulative exposures, genetic vulnerabilities, and developmental timing. This is evident across cancer biology, neurodegeneration, reproductive health, and even transgenerational inheritance.

16.6.1 Heavy metals and cancer risk

Emerging studies are shedding light on how metals influence cancer progression and prognosis. For example, Liu et al. (2024) reported that urinary supra-threshold levels of As, Cd, Hg, and Pb were significantly associated with worse overall survival in patients with advanced ovarian cancer (HGSOC) in China. Their findings not only confirmed dose–response relationships but also highlighted the role of heavy metal mixtures, with lead exerting the strongest influence on outcomes. This work emphasizes how environmental pollutants can directly impact cancer prognosis and survival.

Complementing this, Kaur et al. (2024) provided a comprehensive review of As, underscoring its systemic and neurotoxic effects. They highlighted inorganic As—particularly in its trivalent form—as especially harmful, capable of inducing oxidative stress, DNA damage, and epigenetic alterations. Such findings underscore how a single element like As contributes to both cancer risk and neurological decline, exemplifying the dual toxicity profile of many metals.

Other metals have also been implicated in cardiotoxic and carcinogenic pathways. Rajpoot and Sharma (2025) demonstrated that lead nitrate exposure in mice disrupted lipid metabolism and induced inflammatory cascades. However, they found that Allium sativum essential oil

(ASEO) provided cardioprotection, reducing inflammation and preserving myocardial architecture. These results highlight a promising avenue for therapeutic interventions in metal-induced cardiovascular toxicity.

The role of metals extends into lung cancer biology as well. Liu et al. (2025) identified microplastic–metal complexes (m-MPs) in lung adenocarcinoma tissues, showing strong correlations with aluminum (Al) and calcium (Ca) levels. Their study revealed how m-MPs activate oncogenic pathways, worsen survival outcomes, and modulate gene expression, offering new biomarkers for exposure assessment and prognosis. Taken together, these cancer-related findings underscore the diverse mechanisms—ranging from oxidative stress to signaling pathway disruption—through which metals can exacerbate malignancies.

16.6.2 Mechanistic insights: Oxidative stress and cell death

Mechanistically, oxidative stress is a unifying pathway of metal toxicity. To this end, Wang et al. (2023) highlighted the role of oxidative stress in ferroptosis and cuproptosis, two programmed cell death pathways driven by iron and copper dysregulation, respectively. Their review stressed that metals not only trigger oxidative stress but also reshape redox and ion homeostasis, tipping cells toward death or malignant transformation. These insights deepen our understanding of how metals exploit fundamental cellular processes to drive disease.

Hg toxicity provides a clear example. Queiroz et al. (2024) showed that inorganic Hg exacerbates hypercholesterolemia and accelerates atherosclerosis in mice through mitochondrial dysfunction and oxidative stress. This demonstrates how a systemic redox imbalance induced by metals translates into organ-specific pathologies like cardiovascular disease.

Cd, too, illustrates multigenerational effects. Wang et al. (2025a) showed that Cd exposure in *Drosophila melanogaster* induced reproductive toxicity and reduced fecundity, with adverse outcomes persisting for multiple generations via epigenetic mechanisms. Such findings underscore how even temporary exposures can leave enduring biological imprints.

16.6.3 Epigenetics, aging, and reproduction

The epigenetic dimension of heavy metal toxicity is increasingly recognized. Ryoo et al. (2025) showed that exposure to Pb and Cd speeds up epigenetic aging in older adults, with DNA methylation patterns indicating accumulated biological stress. Similarly, Zhang et al. (2024) showed that Cd exposure impairs testicular function in rats but that recovery occurs after cessation, suggesting that some reproductive toxicities may be reversible.

In females, As and Hg impair oocyte viability and folliculogenesis (Wang et al., 2025c), while transgenerational studies link Cd exposure to imprinting errors at genes such as *H19* and *IGF2* (Nava-Rivera et al., 2021). These insights highlight reproductive health as a critical frontier in understanding and mitigating heavy metal exposure.

16.6.4 Environmental mixtures and human risk

Epidemiological studies further highlight the risks posed by metal mixtures. Yu et al. (2025) found associations between iron (Fe) and Ni exposures and thyroid cancer, particularly in women, while Njale et al. (2024) reported that the urinary burden of aluminum, As, Ni, Cd, and Pb were significantly higher in breast cancer patients. Similarly, Fernández-Martínez et al. (2024)

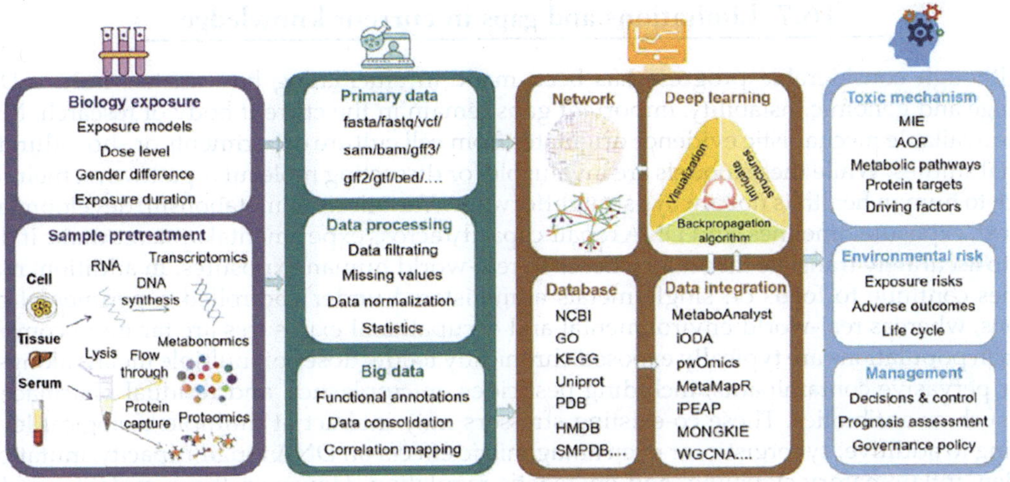

FIGURE 16.2 Schematic representation of the workflow for multiomics analysis. The workflow integrates genomics, transcriptomics, proteomics, metabolomics, and other omics data to provide a comprehensive understanding of biological systems. Key steps include sample collection, data acquisition, preprocessing, normalization, integration, and downstream bioinformatics and statistical analyses to identify molecular interactions, biomarkers, and pathway-level insights. *Courtesy Tools and Applications in Environmental Toxicology by Shi et al. (2024) with permission from Elsevier.*

showed that copper–zinc imbalance strongly predicted breast cancer risk, emphasizing how essential and toxic metals interact in shaping disease outcomes.

Fish and aquatic ecosystems serve as another critical interface between environmental metals and disease outcomes. Afifi et al. (2024) documented the combined effects of metal bioaccumulation and parasitic infections in farmed fish, raising concerns for food safety. Maraschi et al. (2024) added that rare earth elements (REEs) accumulate in tilapia tissues, damaging DNA and lipids, and highlighting new classes of underexplored environmental risks.

16.6.5 From biomarkers to policy

Modern toxicology is increasingly integrating multiomics approaches to capture the complexity of metal stress (Fig. 16.2). Shi et al. (2024) demonstrated how transcriptomics, proteomics, metabolomics, and lipidomics together provide a holistic view of metal-induced molecular alterations. Biomarker-based studies, such as Buonaurio et al. (2021), reveal that urinary oxidative stress markers like 8-oxoGuo can sensitively discriminate between metal-exposed workers and controls, even at low exposure levels.

Policy frameworks such as the EU REACH regulations, IARC monographs, and EPA/ATSDR guidelines are vital but inconsistently enforced worldwide (Briffa et al., 2020). Advances in biomarker science—ranging from γ-H2AX to circulating cfDNA—promise earlier detection and better risk stratification (Alnasser, 2025; Shilnikova et al., 2025). Alongside these, therapeutic interventions such as chelation therapy, antioxidant supplementation, and emerging epigenetic modulators are under active investigation (Ahamed and Siddiqui, 2007; Flora and Pachauri, 2010).

16.7 Limitations and gaps in current knowledge

Although considerable progress has been made in elucidating how metals induce DNA damage and genomic instability, important gaps remain in the current body of research. Much of the available mechanistic evidence originates from cell culture experiments or short-duration animal studies. While these models are invaluable for dissecting molecular pathways, their relevance to human health is not always straightforward. Variations in metabolism, developmental context, exposure timelines, and DNA repair capacity across experimental systems make it difficult to accurately translate laboratory doses to real-world human exposures. In addition, many studies continue to focus on single metals administered under controlled experimental conditions, whereas real-world environmental and occupational exposures are far more complex. Human populations are typically exposed chronically to low doses of multiple metals alongside other pervasive contaminants, including pesticides, microplastics, and residual pharmaceuticals such as antibiotics. These co-existing stressors may interact at multiple biological levels, leading to additive, synergistic, or even antagonistic effects on DNA repair capacity, mutational burden, inflammatory signaling, and epigenetic regulation. However, the cumulative and interactive consequences of such mixed-exposure scenarios remain inadequately characterized. Addressing this complexity represents a critical unmet need in environmental mutagenesis and genome stability research and will be essential for improving risk assessment and public health relevance.

At the same time, the adoption of high-resolution genomic and epigenomic techniques, including ChIP-seq, ATAC-seq, and related approaches, has provided deeper insight into metal-induced changes in chromatin organization and transcriptional control. Despite their power, these methods introduce their own challenges. Signals generated from heterogeneous tissues often reflect averaged responses, potentially obscuring cell-type–specific effects of metal exposure. Furthermore, differences in experimental protocols, antibody performance, sequencing strategies, and computational analyses have contributed to variability across studies, limiting reproducibility and comparability. The lack of universally accepted standards for data generation and interpretation further complicates the integration of findings across laboratories. As a result, conclusions drawn from these datasets should be interpreted with care and, where possible, supported by complementary functional assays to establish direct causal relationships between metal exposure, chromatin dynamics, and mutagenic processes.

16.8 Conclusion

In summary, heavy metals exert their deleterious effects through multilayered mechanisms, spanning oxidative stress, DNA damage, epigenetic reprogramming, and immune modulation. Their impacts manifest in diverse diseases—from cancer and neurodegeneration to reproductive dysfunction and accelerated aging—while transgenerational and ecological effects further compound the burden. Addressing these challenges requires not only mechanistic insights from molecular and organoid models but also robust public health policies, biomarker-driven surveillance, and therapeutic innovations. As research continues to evolve, integrating these perspectives into coherent strategies remains essential for safeguarding human health against the growing global threat of heavy metal exposure.

AI usage statement

Artificial intelligence–based language assistance was used to improve clarity and readability of the text; the authors are solely responsible for the scientific content, interpretations, and conclusions.

References

Abd Elnabi, M.K., Elkaliny, N.E., Elyazied, M.M., Azab, S.H., Elkhalifa, S.A., Elmasry, S., et al., 2023. Toxicity of heavy metals and recent advances in their removal: a review. Toxics 11, 580. https://doi.org/10.3390/toxics11070580.

Afifi, M.A.M., Radwan, M., Abbas, M.M.M., Hwihy, H.M., Alabssawy, A.N., Khalaf-Allah, H.M.M., 2024. Threat of heavy metal pollutants and parasites to freshwater fish with special reference to their risk of cancer to humans in Egypt. Aquaculture 587, 740833. https://doi.org/10.1016/j.aquaculture.2024.740833.

Agu, P.C., Aja, P.M., 2025. Generational insights into testicular toxicity: unravelling typical causes, diagnostics, and nutraceutical strategies. Toxicol. Environ. Chem. 107, 1339–1390. https://doi.org/10.1080/02772248.2025.2527656.

Ahamed, M., Siddiqui, M.K.J., 2007. Environmental lead toxicity and nutritional factors. Clin. Nutr. 26, 400–408. https://doi.org/10.1016/j.clnu.2007.03.010.

Albores-Garcia, D., McGlothan, J.L., Guilarte, T.R., 2021. Early-life lead exposure and neurodevelopmental disorders. Curr. Opin. Toxicol. 26, 22–27. https://doi.org/10.1016/j.cotox.2021.03.007.

Alnasser, S.M., 2025. Revisiting the approaches to DNA damage detection in genetic toxicology: insights and regulatory implications. BioData Min. 18, 33. https://doi.org/10.1186/s13040-025-00447-8.

Andrés, C.M.C., Lastra, J.M.P.D.L., Juan, C.A., Plou, F.J., Pérez-Lebeña, E., 2023. Chemical insights into oxidative and nitrative modifications of DNA. Int. J. Mol. Sci. 24, 15240. https://doi.org/10.3390/ijms242015240.

Balali-Mood, M., Naseri, K., Tahergorabi, Z., Khazdair, M.R., Sadeghi, M., 2021. Toxic mechanisms of five heavy metals: mercury, lead, chromium, cadmium, and Arsenic. Front. Pharmacol. 12, 643972. https://doi.org/10.3389/fphar.2021.643972.

Beyersmann, D., 2002. Effects of carcinogenic metals on gene expression. Toxicol. Lett. 127, 63–68. https://doi.org/10.1016/S0378-4274(01)00484-2.

Biranje, A., Azmi, N., Tiwari, A., Chaskar, A., 2021. Quantum dots based fluorescent probe for the selective detection of heavy metal ions. J. Fluoresc. 31, 1241–1250. https://doi.org/10.1007/s10895-021-02755-8.

Bona, D., Lucian, M., Feretti, D., Silvestri, S., Zerbini, I., Merzari, F., et al., 2023. Phytotoxicity and genotoxicity of agro-industrial digested sludge hydrochar: the role of heavy metals. Sci. Total Environ. 871, 162138. https://doi.org/10.1016/j.scitotenv.2023.162138.

Bovio, F., Sciandrone, B., Urani, C., Fusi, P., Forcella, M., Regonesi, M.E., 2021. Superoxide dismutase 1 (SOD1) and cadmium: a three models approach to the comprehension of its neurotoxic effects. Neurotoxicology 84, 125–135. https://doi.org/10.1016/j.neuro.2021.03.007.

Bozack, A.K., Rifas-Shiman, S.L., Coull, B.A., Baccarelli, A.A., Wright, R.O., Amarasiriwardena, C., et al., 2021. Prenatal metal exposure, cord blood DNA methylation and persistence in childhood: an epigenome-wide association study of 12 metals. Clin. Epigenetics 13, 208. https://doi.org/10.1186/s13148-021-01198-z.

Briffa, J., Sinagra, E., Blundell, R., 2020. Heavy metal pollution in the environment and their toxicological effects on humans. Heliyon 6, e04691. https://doi.org/10.1016/j.heliyon.2020.e04691.

Brown, S., Marchi, S., Thomas, C.S., Hale, A.R., Lockart, M., Bowman, M.K., et al., 2024. Forming a chromium-based interstrand DNA crosslink: implications for carcinogenicity. J. Inorg. Biochem. 251, 112439. https://doi.org/10.1016/j.jinorgbio.2023.112439.

Buonaurio, F., Astolfi, M.L., Pigini, D., Tranfo, G., Canepari, S., Pietroiusti, A., et al., 2021. Oxidative stress biomarkers in urine of metal carpentry workers can be diagnostic for occupational exposure to low level of welding fumes from associated metals. Cancers 13, 3167. https://doi.org/10.3390/cancers13133167.

Cameron, K.S., Buchner, V., Tchounwou, P.B., 2011. Exploring the molecular mechanisms of nickel-induced genotoxicity and carcinogenicity: A literature review. Reviews on Environmental Health 26 (2), 81–92. https://doi.org/10.1515/reveh.2011.012.

Chandrasekaran, B., Tyagi, B., Tyagi, A., Shukla, V., Schatz, M., Devanarayanan, T.N., et al., 2025. Mechanistic role of metal-responsive transcription factor-1 (MTF-1) in cadmium-induced prostate carcinogenesis. Int. J. Biol. Sci. 21, 3614–3630. https://doi.org/10.7150/ijbs.110174.

Chang, T.-C., Wentzel, E.A., Kent, O.A., Ramachandran, K., Mullendore, M., Lee, K.H., et al., 2007. Transactivation of miR-34a by p53 broadly influences gene expression and promotes apoptosis. Mol. Cell 26, 745–752. https://doi.org/10.1016/j.molcel.2007.05.010.

Cheng, C., Wang, H., Zhao, J., Wang, Y., Zhao, G., Zhang, Y., et al., 2024. Advances in the application of metal oxide nanozymes in tumor detection and treatment. Colloids Surf. B Biointerfaces 235, 113767. https://doi.org/10.1016/j.colsurfb.2024.113767.

Costa, P.M., 2022. Current aspects of DNA damage and repair in ecotoxicology: a mini-review. Ecotoxicology 31, 1–11. https://doi.org/10.1007/s10646-021-02487-2.

Crespo-López, M.E., Macêdo, G.L., Arrifano, G.P.F., Pinheiro, M.D.C.N., Do Nascimento, J.L.M., Herculano, A.M., 2011. Genotoxicity of mercury: contributing for the analysis of Amazonian populations. Environ. Int. 37, 136–141. https://doi.org/10.1016/j.envint.2010.08.009.

Cui, L., Zhang, Yu, Zhang, Youru, Chu, J., Kong, Q., Zhou, Y., Zhang, H., 2025. Effect of 4-n-nonylphenol and cadmium co-exposure on liver apoptosis and lipid metabolism in zebrafish (Danio rerio). Ecotoxicol. Environ. Saf. 295, 118165. https://doi.org/10.1016/j.ecoenv.2025.118165.

Czarnek, K., Tatarczak-Michalewska, M., Blicharska, E., Siwicki, A.K., Maciejewski, R., 2025. Genotoxic effects of chromium(III) and cobalt(II) and their mixtures on the selected cell lines. Int. J. Mol. Sci. 26, 5056. https://doi.org/10.3390/ijms26115056.

De Flora, S., Bennicelli, C., Bagnasco, M., 1994. Genotoxicity of mercury compounds. a review. Mutat. Res. Genet. Toxicol. 317, 57–79. https://doi.org/10.1016/0165-1110(94)90012-4.

Dey, K.K., Kamila, S., Das, T., Chattopadhyay, A., 2024. Lead induced genotoxicity and hepatotoxicity in zebrafish (Danio rerio) at environmentally relevant concentration: Nrf2-Keap1 regulated stress response and expression of biomarker genes. Environ. Toxicol. Pharmacol. 107, 104396. https://doi.org/10.1016/j.etap.2024.104396.

Ding, M., Shi, S., Qie, S., Li, J., Xi, X., 2023. Association between heavy metals exposure (cadmium, lead, arsenic, mercury) and child autistic disorder: a systematic review and meta-analysis. Front. Pediatr. 11, 1169733. https://doi.org/10.3389/fped.2023.1169733.

Dutta, S., Sivakumar, K.K., Erwin, J.W., Stanley, J.A., Arosh, J.A., Taylor, R.J., et al., 2024. Alteration of epigenetic methyl and acetyl marks by postnatal chromium(VI) exposure causes apoptotic changes in the ovary of the F1 offspring. Reprod. Toxicol. 123, 108492. https://doi.org/10.1016/j.reprotox.2023.108492.

Fasani, E., Giannelli, G., Varotto, S., Visioli, G., Bellin, D., Furini, A., et al., 2023. Epigenetic control of plant response to heavy metals. Plants 12, 3195. https://doi.org/10.3390/plants12183195.

Fernández-Martínez, N.F., Rodríguez-Barranco, M., Huerta, J.M., Gil, F., Olmedo, P., et al., 2024. Breast cancer risk for the joint exposure to metals and metalloids in women: results from the EPIC-Spain cohort. Sci. Total Environ. 912, 168816. https://doi.org/10.1016/j.scitotenv.2023.168816.

Flora, S.J.S., Pachauri, V., 2010. Chelation in metal intoxication. Int. J. Environ. Res. Public. Health 7, 2745–2788. https://doi.org/10.3390/ijerph7072745.

Fragou, D., Fragou, A., Kouidou, S., Njau, S., Kovatsi, L., 2011. Epigenetic mechanisms in metal toxicity. Toxicol. Mech. Methods 21, 343–352. https://doi.org/10.3109/15376516.2011.557878.

Freydenzon, A., Nabais, M.F., Lin, T., Williams, K.L., Wallace, L., Henders, A.K., et al., 2022. Association between DNA methylation variability and self-reported exposure to heavy metals. Sci. Rep. 12, 10582. https://doi.org/10.1038/s41598-022-13892-w.

Ge, X., He, J., Wang, L., Zhao, L., Wang, Y., Wu, G., et al., 2022. Epigenetic alterations of CXCL5 in Cr(VI)-induced carcinogenesis. Sci. Total Environ. 838, 155713. https://doi.org/10.1016/j.scitotenv.2022.155713.

Ghosh, K., N, I., 2018. Cadmium treatment induces echinocytosis, DNA damage, inflammation, and apoptosis in cardiac tissue of albino Wistar rats. Environ. Toxicol. Pharmacol. 59, 43–52. https://doi.org/10.1016/j.etap.2018.02.009.

Green, A.J., Hoyo, C., Mattingly, C.J., Luo, Y., Tzeng, J.-Y., Murphy, S.K., et al., 2018. Cadmium exposure increases the risk of juvenile obesity: a human and zebrafish comparative study. Int. J. Obes. 42, 1285–1295. https://doi.org/10.1038/s41366-018-0036-y.

Guo, X., Chen, X., Wang, J., Liu, Z., Gaile, D., Wu, H., et al., 2018. Multi-generational impacts of arsenic exposure on genome-wide DNA methylation and the implications for arsenic-induced skin lesions. Environ. Int. 119, 250–263. https://doi.org/10.1016/j.envint.2018.06.024.

Halasova, E., Matakova, T., Musak, L., Polakova, V., Letkova, L., Dobrota, D., et al., 2012. Evaluating chromosomal damage in workers exposed to hexavalent chromium and the modulating role of polymorphisms of DNA repair genes. Int. Arch. Occup. Environ. Health 85, 473–481. https://doi.org/10.1007/s00420-011-0684-x.

Hartwig, A., 1995. Current aspects in metal genotoxicity. Biometals 8. https://doi.org/10.1007/BF00156151.

Hei, T.K., Filipic, M., 2004. Role of oxidative damage in the genotoxicity of arsenic. Free Radic. Biol. Med. 37, 574–581. https://doi.org/10.1016/j.freeradbiomed.2004.02.003.

Hengstler, J.G., 2003. Occupational exposure to heavy metals: DNA damage induction and DNA repair inhibition prove co-exposures to cadmium, cobalt and lead as more dangerous than hitherto expected. Carcinogenesis 24, 63–73. https://doi.org/10.1093/carcin/24.1.63.

Honda, A., Tsuji, K., Matsuda, Y., Hayashi, T., Fukushima, W., Sawahara, T., et al., 2015. Effects of air pollution-related heavy metals on the viability and inflammatory responses of human airway epithelial cells. Int. J. Toxicol. 34, 195–203. https://doi.org/10.1177/1091581815575757.

Hou, L., Zhang, X., Wang, D., Baccarelli, A., 2012. Environmental chemical exposures and human epigenetics. Int. J. Epidemiol. 41, 79–105. https://doi.org/10.1093/ije/dyr154.

Hsu, C.-C., Wu, Y., 2022. Recent advances in nanotechnology-enabled biosensors for detection of exosomes as new cancer liquid biopsy. Exp. Biol. Med. 247, 2152–2172. https://doi.org/10.1177/15353702221110813.

Hu, G., Li, P., Cui, X., Li, Y., Zhang, J., Zhai, X., et al., 2018. Cr(VI)-induced methylation and down-regulation of DNA repair genes and its association with markers of genetic damage in workers and 16HBE cells. Environ. Pollut. 238, 833–843. https://doi.org/10.1016/j.envpol.2018.03.046.

Hui, F., Pang, Z., Viau, C., Balcke, G.U., Fobil, J.N., Basu, N., Xia, J., 2025. Integrative modeling of urinary metabolomics and metal exposure reveals systemic impacts of electronic waste in exposed populations. Metabolites 15, 456. https://doi.org/10.3390/metabo15070456.

Hussain, S., Saadi, S., Hassan, W., Akitsu, T., 2025. Therapeutic potential of oxaloacetic acid against toxic metals and sodium nitroprusside induced phospholipids peroxidation. Next Research 2, 100333. https://doi.org/10.1016/j.nexres.2025.100333.

Iqbal, B., Ahmad, N., Li, G., Jalal, A., Khan, A.R., Zheng, X., et al., 2024. Unlocking plant resilience: Advanced epigenetic strategies against heavy metal and metalloid stress. Plant Sci. 349, 112265. https://doi.org/10.1016/j.plantsci.2024.112265.

Islam, S., Mansha, A., Asim, S., 2024. Effects of metal ions and substituents on HOMO–LUMO gap evident from UV–visible and fluorescence spectra of anthracene derivatives. J. Fluoresc. 34, 2495–2512. https://doi.org/10.1007/s10895-023-03482-y.

Järup, L., 2003. Hazards of heavy metal contamination. Br. Med. Bull. 68, 167–182. https://doi.org/10.1093/bmb/ldg032.

Jomova, K., Alomar, S.Y., Nepovimova, E., Kuca, K., Valko, M., 2025. Heavy metals: toxicity and human health effects. Arch. Toxicol. 99, 153–209. https://doi.org/10.1007/s00204-024-03903-2.

Kar, S., Leszczynski, J., 2019. Exploration of computational approaches to predict the toxicity of chemical mixtures. Toxics 7, 15. https://doi.org/10.3390/toxics7010015.

Kaur, I., Behl, T., Aleya, L., Rahman, M.H., Kumar, A., Arora, S., et al., 2021. Role of metallic pollutants in neurodegeneration: effects of aluminum, lead, mercury, and arsenic in mediating brain impairment events and autism spectrum disorder. Environ. Sci. Pollut. Res. 28, 8989–9001. https://doi.org/10.1007/s11356-020-12255-0.

Kaur, R., Garkal, A., Sarode, L., Bangar, P., Mehta, T., Singh, D.P., et al., 2024. Understanding arsenic toxicity: implications for environmental exposure and human health. J. Hazard. Mater. Lett. 5, 100090. https://doi.org/10.1016/j.hazl.2023.100090.

Khalid, F., Azmat, H., 2025. Acute arsenic exposure induces cyto-genotoxicity and histological alterations in Labeo rohita. J. Trace Elem. Med. Biol. 88, 127600. https://doi.org/10.1016/j.jtemb.2025.127600.

Kim, N., Filipovic, D., Bhattacharya, S., Cuddapah, S., 2024. Epigenetic toxicity of heavy metals – implications for embryonic stem cells. Environ. Int. 193, 109084. https://doi.org/10.1016/j.envint.2024.109084.

Kong, X., Li, C., Pan, Y., 2025. Association between heavy metals mixtures and life's essential 8 score in general US adults. Cardiovasc. Toxicol. 25, 592–603. https://doi.org/10.1007/s12012-025-09969-3.

Kotnala, S., Tiwari, S., Nayak, A., Bhushan, B., Chandra, S., Medeiros, C.R., et al., 2025. Impact of heavy metal toxicity on the human health and environment. Sci. Total Environ. 987, 179785. https://doi.org/10.1016/j.scitotenv.2025.179785.

Koyama, H., Kamogashira, T., Yamasoba, T., 2024. Heavy metal exposure: molecular pathways, clinical implications, and protective strategies. Antioxidants 13, 76. https://doi.org/10.3390/antiox13010076.

Kozak, J., 2025. The role of heavy metals in the biology of female cancers. Int. J. Mol. Sci. 26, 5155. https://doi.org/10.3390/ijms26115155.

Kucab, J.E., Zou, X., Morganella, S., Joel, M., Nanda, A.S., Nagy, E., et al., 2019. A compendium of mutational signatures of environmental agents. Cell 177, 821–836. https://doi.org/10.1016/j.cell.2019.03.001.

Kumar, A., Roy, J., Ojha, P.K., 2025. The first report on chronic toxicity assessment of metals towards Ceriodaphnia dubia using QSTR technique: a step towards healthier and safer human health and eco-system. Comput. Toxicol. 34, 100357. https://doi.org/10.1016/j.comtox.2025.100357.

Kurita, H., Ohuchi, K., Inden, M., 2025. Effects of environmental non-essential toxic heavy metals on epigenetics during development. Toxics 13, 167. https://doi.org/10.3390/toxics13030167.

Kwon, J.-Y., Lee, S., Surenbaatar, U., Lim, H.-J., Kim, B.-G., Eom, S.-Y., et al., 2023. Association between levels of exposure to heavy metals and renal function indicators of residents in environmentally vulnerable areas. Sci. Rep. 13, 2856. https://doi.org/10.1038/s41598-022-27292-7.

La-Ane, R., Amqam, H., Susilawaty, A., Natsir, M., Yusbud, M., Adhariana, H., Aris, A., 2025. Heavy metals exposure and risk factors for autism spectrum disorder (ASD) among school children in Makassar City. Indonesia. Toxicol. Rep. 14, 102032. https://doi.org/10.1016/j.toxrep.2025.102032.

Li, T., Zhuang, Y., Yang, W., Xie, Y., Shang, W., Su, S., et al., 2021. Silencing of METTL3 attenuates cardiac fibrosis induced by myocardial infarction via inhibiting the activation of cardiac fibroblasts. FASEB J. 35. https://doi.org/10.1096/fj.201903169R.

Li, Y., Ye, Z., Yu, Y., Li, Ye, Jiang, J., Wang, L., et al., 2023. A combined method for human health risk area identification of heavy metals in urban environments. J. Hazard. Mater. 449, 131067. https://doi.org/10.1016/j.jhazmat.2023.131067.

Liu, J.-X., Liu, F.-H., Qin, X., Bao, Q., Zheng, W.-R., Xing, W.-Y., et al., 2024. Urinary heavy metals and overall survival of advanced high-grade serous ovarian cancer: a nested case-control study in China. Ecotoxicol. Environ. Saf. 286, 117176. https://doi.org/10.1016/j.ecoenv.2024.117176.

Liu, R., Guo, X., Yang, G., Lu, S., Chen, F., Jia, W., et al., 2025. Formation of metal-microplastic complexes in lung adenocarcinoma is associated with increased risk of cancer progression. J. Hazard. Mater. 494, 138517. https://doi.org/10.1016/j.jhazmat.2025.138517.

Liu, S.-X., Davidson, M.M., Tang, X., Walker, W.F., Athar, M., Ivanov, V., et al., 2005. Mitochondrial damage mediates genotoxicity of arsenic in mammalian cells. Cancer Res. 65, 3236–3242. https://doi.org/10.1158/0008-5472.CAN-05-0424.

Lu, H., Li, H., Liu, T., Fan, Y., Yuan, Y., Xie, M., et al., 2019. Simulating heavy metal concentrations in an aquatic environment using artificial intelligence models and physicochemical indexes. Sci. Total Environ. 694, 133591. https://doi.org/10.1016/j.scitotenv.2019.133591.

Ma, L., Li, Jun, Zhan, Z., Chen, L., Li, D., Bai, Q., et al., 2016. Specific histone modification responds to arsenic-induced oxidative stress. Toxicol. Appl. Pharmacol. 302, 52–61. https://doi.org/10.1016/j.taap.2016.03.015.

Maji, C., Biswas, S., Sarkar, P.K., Patra, P.H., Bandyopadhyay, S., Sarkar, S., et al., 2020. Evaluation of ameliorative effect of two selected plant drugs on experimentally induced arsenic toxicity in sheep. Environ. Sci. Pollut. Res. 27, 36744–36753. https://doi.org/10.1007/s11356-020-09569-4.

Maliyam, P., Laphookhieo, S., Koedrith, P., Puttarak, P., 2024. Antioxidative and anti-cytogenotoxic potential of Lysiphyllum strychnifolium (Craib) A. Schmitz extracts against cadmium-induced toxicity in human embryonic kidney (HEK293) and dermal fibroblast (HDF) cells. Heliyon 10, e34480. https://doi.org/10.1016/j.heliyon.2024.e34480.

Maraschi, A.C., Adorno, H.A., Gonçalves, Y.C., Souza, I.C., Monferrán, M.V., Wunderlin, D.A., et al., 2024. Effects of metallic dust on Nile tilapia: exploring the relationship between metal bioaccumulation, metallothionein levels, and oxidative stress responses. Sci. Total Environ. 956, 177423. https://doi.org/10.1016/j.scitotenv.2024.177423.

Martinez-Zamudio, R., Ha, H.C., 2011. Environmental epigenetics in metal exposure. Epigenetics 6, 820–827. https://doi.org/10.4161/epi.6.7.16250.

Meng, Y., Wang, K., Wang, T., Tu, Y., Gong, S., Zhang, Y., et al., 2021. Early occupational exposure to lead on neutrophil-to-lymphocyte ratio and genotoxicity. Environ. Int. 151, 106448. https://doi.org/10.1016/j.envint.2021.106448.

Mesnage, R., 2025. Environmental health is overlooked in longevity research. Antioxidants 14, 421. https://doi.org/10.3390/antiox14040421.

Mezencev, R., Gibbons, C., 2023. Interactions between chromium species and DNA *in vitro* and their potential role in the toxicity of hexavalent chromium. Metallomics 15, mfad045. https://doi.org/10.1093/mtomcs/mfad045.

Mo, Y., Zhang, Y., Zhang, Q., 2024. The pulmonary effects of nickel-containing nanoparticles: cytotoxicity, genotoxicity, carcinogenicity, and their underlying mechanisms. Environ. Sci. Nano 11, 1817–1846. https://doi.org/10.1039/D3EN00929G.

Mukherjee, A., Vankamamidi, S.E., Ks, M., 2025. Probing the microRNA landscape in cadmium chloride induced renal toxicity through an in silico approach. Sci. Rep. 15, 25251. https://doi.org/10.1038/s41598-025-11473-1.

Nava-Rivera, L.E., Betancourt-Martínez, N.D., Lozoya-Martínez, R., Carranza-Rosales, P., Guzmán-Delgado, N.E., Carranza-Torres, I.E., et al., 2021. Transgenerational effects in DNA methylation, genotoxicity and reproductive phenotype by chronic arsenic exposure. Sci. Rep. 11, 8276. https://doi.org/10.1038/s41598-021-87677-y.

Nemati-Mansour, S., Mosaferi, M., Babaie, J., Mohammadpoorasl, A., Dehghanzadeh, R., Nikniaz, L., et al., 2024. Interplay of arsenic exposure and cigarette smoking on oxidative DNA damage in healthy males. Environ. Sci. Eur. 36, 86. https://doi.org/10.1186/s12302-024-00915-8.

Njale, E., Mahugija, J.A.M., Shilla, D.J., Dharsee, N., Mbare, N.S., 2024. Association between exposure to heavy metals and the risk of breast cancer: a case-control study in Tanzania. Environ. Toxicol. Pharmacol. 110, 104498. https://doi.org/10.1016/j.etap.2024.104498.

O'Hara, K.A., Nemec, A.A., Alam, J., Klei, L.R., Mossman, B.T., Barchowsky, A., 2006. Chromium (VI) inhibits heme oxygenase-1 expression in vivo and in arsenic-exposed human airway epithelial cells. J. Cell. Physiol. 209, 113–121. https://doi.org/10.1002/jcp.20710.

Otsuka, A., Minami, K., Higashi, K., Kawaguchi, A., Tamura, S., Ide, S., Hendzel, M.J., Kurokawa, K., Maeshima, K., 2024. Chromatin organization and behavior in HRAS-transformed mouse fibroblasts. Chromosoma 133, 135–148. https://doi.org/10.1007/s00412-024-00817-x.

Ozturk, M., Metin, M., Altay, V., Bhat, R.A., Ejaz, M., Gul, A., et al., 2022. Arsenic and human health: genotoxicity, epigenomic effects, and cancer signaling. Biol. Trace Elem. Res. 200, 988–1001. https://doi.org/10.1007/s12011-021-02719-w.

Pal, S., Firdous, S.M., 2024. Unraveling the role of heavy metals xenobiotics in cancer: a critical review. Discov. Oncol. 15, 615. https://doi.org/10.1007/s12672-024-01417-y.

Parida, L., Patel, T.N., 2023. Systemic impact of heavy metals and their role in cancer development: a review. Environ. Monit. Assess. 195, 766. https://doi.org/10.1007/s10661-023-11399-z.

Queiroz, M.I.C., Lazaro, C.M., Dos Santos, L.M.B., Rentz, T., Virgilio-da-Silva, J.V., Moraes-Vieira, P.M.M., et al., 2024. In vivo chronic exposure to inorganic mercury worsens hypercholesterolemia, oxidative stress and atherosclerosis in the LDL receptor knockout mice. Ecotoxicol. Environ. Saf. 275, 116254. https://doi.org/10.1016/j.ecoenv.2024.116254.

Rajpoot, A., Sharma, V., 2025. Protective effects of Allium sativum essential oil against lead nitrate-induced cardiotoxicity: modulation of lipid metabolism, nitric oxide dynamics, inflammatory mediators, and histological profiles in Swiss albino mice. Toxicol. Rep. 14, 101950. https://doi.org/10.1016/j.toxrep.2025.101950.

Ramos-Pan, L., Touzani, A., Fernández-Bertólez, N., Fraga, S., Laffon, B., Valdiglesias, V., 2024. Impact of gold nanoparticle exposure on genetic material. Mutat. Res. - Genet. Toxicol. Environ. Mutagen. 900, 503827. https://doi.org/10.1016/j.mrgentox.2024.503827.

Raza, A., Salehi, H., Bashir, S., Tabassum, J., Jamla, M., Charagh, S., et al., 2024. Transcriptomics, proteomics, and metabolomics interventions prompt crop improvement against metal(loid) toxicity. Plant Cell Rep. 43, 80. https://doi.org/10.1007/s00299-024-03153-7.

Rehman, M.Y.A., Van Herwijnen, M., Malik, R.N., Briedé, J.J., Krauskopf, J., 2025. The miRNA-based liquid biopsies for environmental health risk assessment: insights into arsenic-induced hepatotoxicity and systemic health impacts. Sci. Total Environ. 988, 179825. https://doi.org/10.1016/j.scitotenv.2025.179825.

Ren, X., McHale, C.M., Skibola, C.F., Smith, A.H., Smith, M.T., Zhang, L., 2011. An emerging role for epigenetic dysregulation in arsenic toxicity and carcinogenesis. Environ. Health Perspect. 119, 11–19. https://doi.org/10.1289/ehp.1002114.

Ryoo, S.-W., Choi, B.-Y., Son, S.-Y., Lee, J.-H., Min, J.-Y., Min, K.-B., 2025. Lead and cadmium exposure was associated with faster epigenetic aging in a representative sample of adults aged 50 and older in the United States. Chemosphere 374, 144194. https://doi.org/10.1016/j.chemosphere.2025.144194.

Salnikow, K., Zhitkovich, A., 2008. Genetic and epigenetic mechanisms in metal carcinogenesis and cocarcinogenesis: nickel, arsenic, and chromium. Chem. Res. Toxicol. 21, 28–44. https://doi.org/10.1021/tx700198a.

Senthilnathan, S., Nallusamy, G., Varadaraj, P., 2024. An interesting case of acquired renal fanconi syndrome. Cureus 16 (7), e65208. https://doi.org/10.7759/cureus.65208.

Sha, J., Hao, L., Fang, X., Zhu, N., Yang, Y., Peng, K., et al., 2025. pH stability improved electrochemical biosensor based on dopamine/aptamer imprinted polymer modified electrodes for simultaneous detection of four toxic metal ions. Food Control 171, 111099. https://doi.org/10.1016/j.foodcont.2024.111099.

Shaji, E., Santosh, M., Sarath, K.V., Prakash, P., Deepchand, V., Divya, B.V., 2021. Arsenic contamination of groundwater: a global synopsis with focus on the Indian Peninsula. Geosci. Front. 12, 101079. https://doi.org/10.1016/j.gsf.2020.08.015.

Sharma, A., Khan, S., Pooja, Ansari, L., Asija, D., Nain, N., Singh, A., 2025. Mechanistic insight into genotoxicity mediated DNA damage: fate of DNA. J. Biochem. Mol. Toxicol. 39, e70429. https://doi.org/10.1002/jbt.70429.

Shetty, B.R., Jagadeesha, P.B., Salmataj, S.A., 2025. Heavy metal contamination and its impact on the food chain: exposure, bioaccumulation, and risk assessment. CyTA - J. Food 23, 2438726. https://doi.org/10.1080/19476337.2024.2438726.

Shi, C., Cheng, L., Yu, Y., Chen, S., Dai, Y., Yang, J., Zhang, H., Chen, J., Geng, N., 2024. Multi-omics integration analysis: tools and applications in environmental toxicology. Environ. Pollut. 360, 124675. https://doi.org/10.1016/j.envpol.2024.124675.

Shilnikova, N., Momoli, F., Karyakina, N., Krewski, D., 2025. Review of non–invasive biomarkers as a tool for exposure characterization in human health risk assessments. J. Toxicol. Environ. Health Part B 28, 122–150. https://doi.org/10.1080/10937404.2024.2428206.

Sobral, A.F., Cunha, A., Costa, I., Silva-Carvalho, M., Silva, R., Barbosa, D.J., 2025. Environmental xenobiotics and epigenetic modifications: implications for human health and disease. J. Xenobiotics 15, 118. https://doi.org/10.3390/jox15040118.

Stepanyan, A., Arakelyan, A., Schug, J., 2025. Transcriptome alterations in long-term mining region residents: insights into immune response and molecular pathways. Environ. Int. 197, 109344. https://doi.org/10.1016/j.envint.2025.109344.

Taboureau, O., El M'Selmi, W., Audouze, K., 2020. Integrative systems toxicology to predict human biological systems affected by exposure to environmental chemicals. Toxicol. Appl. Pharmacol. 405, 115210. https://doi.org/10.1016/j.taap.2020.115210.

Taghizadeh, S.F., Goumenou, M., Rezaee, R., Alegakis, T., Kokaraki, V., Anesti, O., et al., 2019. Cumulative risk assessment of pesticide residues in different Iranian pistachio cultivars: applying the source specific HQS and adversity specific HIA approaches in real life risk simulations (RLRS). Toxicol. Lett. 313, 91–100. https://doi.org/10.1016/j.toxlet.2019.05.019.

Thévenod, F., Lee, W.-K., 2013. Cadmium and cellular signaling cascades: interactions between cell death and survival pathways. Arch. Toxicol. 87, 1743–1786. https://doi.org/10.1007/s00204-013-1110-9.

Valko, M., Jomova, K., Rhodes, C.J., Kuča, K., Musílek, K., 2016. Redox- and non-redox-metal-induced formation of free radicals and their role in human disease. Arch. Toxicol. 90, 1–37. https://doi.org/10.1007/s00204-015-1579-5.

VonHandorf, A., Zablon, H.A., Puga, A., 2021. Hexavalent chromium disrupts chromatin architecture. Semin. Cancer Biol. 76, 54–60. https://doi.org/10.1016/j.semcancer.2021.07.009.

Waisberg, M., Joseph, P., Hale, B., Beyersmann, D., 2003. Molecular and cellular mechanisms of cadmium carcinogenesis. Toxicology 192, 95–117. https://doi.org/10.1016/S0300-483X(03)00305-6.

Wang, D., Liang, Q., Tai, D., Wang, Y., Hao, H., Liu, Z., et al., 2023. Association of urinary arsenic with the oxidative DNA damage marker 8-hydroxy-2 deoxyguanosine: a meta-analysis. Sci. Total Environ. 904, 166600. https://doi.org/10.1016/j.scitotenv.2023.166600.

Wang, F., Yang, P., Xu, L., Han, X., Zhang, M., 2025a. Effects of cadmium on female drosophila melanogaster and its transgenerational inheritance effects. J. Environ. Manage. 374, 124076. https://doi.org/10.1016/j.jenvman.2025.124076.

Wang, H., Gan, X., Tang, Y., 2025b. Mechanisms of heavy metal cadmium (Cd)-induced malignancy. Biol. Trace Elem. Res. 203, 608–623. https://doi.org/10.1007/s12011-024-04189-2.

Wang, X., Lai, X., Pan, C., Yin, T., Zhu, W., Zhu, X., et al., 2025c. The association between metal exposure in follicular fluid and PCOS risk in reproductive-aged women: mediating roles of LH/FSH ratio and androgens. Ecotoxicol. Environ. Saf. 300, 118473. https://doi.org/10.1016/j.ecoenv.2025.118473.

Wangmo, S., Talukdar, A., Dutta, N., Nabumpenje, A., Bastola, J.K., Bhattacharya, S., 2025. Biosensors in heavy metal and metalloid detection: genetic, cellular, and nanomaterial applications. In: Development in Waste Water Treatment Research and Processes. Elsevier, pp. 721–736.

Wedrychowski, A., Schmidt, W.N., Hnilica, L.S., 1986. The in vivo cross-linking of proteins and DNA by heavy metals. J. Biol. Chem. 261, 3370–3376. https://doi.org/10.1016/S0021-9258(17)35792-7.

Wei, S., Tao, H.-Y., Duan, Z., Wang, Y., 2025. Environmental exposure, epitranscriptomic perturbations, and human diseases. Environ. Sci. Technol. 59, 6387–6399. https://doi.org/10.1021/acs.est.5c00907.

Winder, C., Bonin, T., 1993. The genotoxicity of lead. Mutat. Res. Mol. Mech. Mutagen. 285, 117–124. https://doi.org/10.1016/0027-5107(93)90059-O.

Wise, J.P., Wise, S.S., Little, J.E., 2002. The cytotoxicity and genotoxicity of particulate and soluble hexavalent chromium in human lung cells. Mutat. Res. Toxicol. Environ. Mutagen. 517, 221–229. https://doi.org/10.1016/S1383-5718(02)00071-2.

Wu, K., Chen, Y., Huang, W., 2025. Combined molecular toxicity mechanism of heavy metals mixtures. In: Pei, D., Liu, Y. (Eds.), Toxicological Assessment of Combined Chemicals in the Environment. Wiley, pp. 125–172.

Wu, R., Veličković, M., Burnum-Johnson, K.E., 2024. From single cell to spatial multi-omics: unveiling molecular mechanisms in dynamic and heterogeneous systems. Curr. Opin. Biotechnol. 89, 103174. https://doi.org/10.1016/j.copbio.2024.103174.

Yang, Y., Hassan, M.F., Ali, W., Zou, H., Liu, Z., Ma, Y., 2025. Effects of cadmium pollution on human health: a narrative review. Atmosphere 16, 225. https://doi.org/10.3390/atmos16020225.

Yıldız, B., Demirel, R., Havadar, H.B., Yıldız, G., Öziç, C., Kamiloğlu, N.N., et al., 2024. Blocking SIG1R along with low cadmium exposure display anti-cancer qualities in both MCF7 and MDA-MB-231 cells. Biol. Trace Elem. Res. 202, 3588–3600. https://doi.org/10.1007/s12011-023-03947-y.

Yu, M., Xun, J., Ge, Y., Li, X., Chen, X., Cui, L., et al., 2025. Relationship between internal metal exposure and thyroid cancer incidence: a case-control study simultaneously validated by BKMR and WQS models. Food Chem. Toxicol. 201, 115443. https://doi.org/10.1016/j.fct.2025.115443.

Yuan, J., Mo, Y., Zhang, Y., Zhang, Y., Zhang, Q., 2022. Nickel nanoparticles induce epithelial-mesenchymal transition in human bronchial epithelial cells via the HIF-1α/HDAC3 pathway. Nanotoxicology 16, 695–712. https://doi.org/10.1080/17435390.2022.2142169.

Zhang, L., Li, N., Zhang, X., Wu, H., Yu, S., 2023a. Hexavalent chromium caused DNA damage repair and apoptosis via the PI3K/AKT/FOXO1 pathway triggered by oxidative stress in the lung of rat. Ecotoxicol. Environ. Saf. 267, 115622. https://doi.org/10.1016/j.ecoenv.2023.115622.

Zhang, P., Yang, M., Lan, J., Huang, Y., Zhang, J., Huang, S., et al., 2023b. Water quality degradation due to heavy metal contamination: health impacts and eco-friendly approaches for heavy metal remediation. Toxics 11, 828. https://doi.org/10.3390/toxics11100828.

Zhang, Q., Feng, H., Hu, G., Zheng, P., Su, Z., Zhang, Y., et al., 2023c. Hexavalent chromium induces γH2AX and RAD51 involved in DNA damage repair in BEAS-2B cells by modulating LNC-DHFR-4:1. Environ. Int. 174, 107895. https://doi.org/10.1016/j.envint.2023.107895.

Zhang, Q., Yang, Y., Liu, J., Wu, Y., Liu, Y., Zhang, J., 2024. Testicular dysfunction and "its recovery effect" after cadmium exposure. Food Chem. Toxicol. 188, 114656. https://doi.org/10.1016/j.fct.2024.114656.

Zhao, X., Zhao, L., Lin, H., Wang, X., 2025. Histone modification changes upon exposure of human lung adenocarcinoma cells to nanoparticles. Sci. Rep. 15, 20724. https://doi.org/10.1038/s41598-025-07206-z.

Zhou, C., Liu, M., Mei, X., Li, Q., Zhang, W., Deng, P., et al., 2021. Histone hypoacetylation contributes to neurotoxicity induced by chronic nickel exposure in vivo and in vitro. Sci. Total Environ. 783, 147014. https://doi.org/10.1016/j.scitotenv.2021.147014.

Zhu, Y., Costa, M., 2020. Metals and molecular carcinogenesis. Carcinogenesis 41, 1161–1172. https://doi.org/10.1093/carcin/bgaa076.

Zou, Z., Yoshimura, Y., Yamanishi, Y., Oki, S., 2023. Elucidating disease-associated mechanisms triggered by pollutants via the epigenetic landscape using large-scale ChIP-Seq data. Epigenetics Chromatin 16, 34. https://doi.org/10.1186/s13072-023-00510-w.

CHAPTER

17

Apoptosis: Cell death and damage

Nimisha Joly Chennakadan[a], *Yadhu Krishna Menon K*[a], *Chandrabose Selvaraj*[b], *Yuvaraj Sivamani*[c] *and Sumitha Elayaperumal*[a]

[a]Department of Biotechnology and Bioinformatics, JSS Academy of Higher Education and Research, Mysore, Karnataka, India [b]Department of Marine Biotechnology, CSRDD Lab, Bioinformatics Division, AMET University (Deemed to be University), Kanathur, Tamil Nadu, India [c]Crescent School of Pharmacy, B.S. Abdur Rahman Crescent Institute of Science & Technology, Chennai, Tamil Nadu, India

17.1 Introduction

Apoptosis, also called programmed cell death, is a fundamental process important for the development and maintenance of homeostasis in multicellular organisms. Compared to necrosis, which is the form of cell death that can be spontaneous and inflammatory due to some form of external damage done to the cell, apoptosis is systematic and uses energy to remove damaged, dysfunctional, or excess cells without causing an immune system reaction. Apoptosis is performed by intrinsic (mitochondrial) and extrinsic (death receptor) pathways, resulting in effector caspases that split and deactivate proteins, the DNA body fragmenting, cell membranes bulging, and the dying cells being engulfed by the surrounding cells (Fleisher, 1997). Apoptosis is crucial for normal development, immunosuppression, and for malignancies and autoantibodies diseases. The loss of regulation of the apoptotic process is often observed in biosystemic defects, giving reason as to why apoptosis is a major topic of interest in medically focused studies.

Apoptosis is one biological process that has undergone a remarkable change in definition over the past 100 years. Observations of programmed cell death can be traced back to the mid-19th century. However, it was not until Kerr, Wyllie, and Currie published their work in 1972 that the term "apoptosis" was first used along with the detailed arguments that explain its usage. In their landmark paper, the discontinuation of blood supply was illustrated by identifying the "key" apoptotic elements; among them are "the condensation of chromatin" as well as "the fragmentation of cells into apoptotic bodies." Further research in the 1990s was crucial in discovering important molecular features associated with apoptosis, like the Bcl-2 family of

Genome Damage: Repair and Adaptation Across Life.
DOI: https://doi.org/10.1016/B978-0-443-42857-9.00026-8

proteins, caspases, death receptors, and others. Identifying caspase-3 as an "executioner" of apoptosis and understanding the contribution of mitochondria in releasing cytochrome c were critical steps forward toward highlighting apoptotic elements (Gibellini and Moro, 2021). Other research increased understanding of the role of apoptosis in the development of disease; nearly all advanced cancers stem in some form from apoptosis dysregulation, wherein tumor cells actively avoid death instead of being destroyed as is intended. These discoveries could be the beginning of creating a new breed of cancer and neurodegenerative disease treatment based on targeted manipulation of apoptosis.

The objective of this assessment is to evaluate in detail apoptosis, its molecular mechanisms, control elements, and consequences concerning health issues. The study will describe intrinsic and extrinsic apoptotic pathways as well as ER-mediated apoptosis, paying particular attention to the main proteins and signaling pathways involved in cell death. Apoptotic activity in the advancement of cellular homeostasis and the disruption of certain pathological states, including malignant neoplasms, other degenerative brain disorders, heart diseases, and autoimmune diseases, will be appraised. In addition, novel strategies and drugs aimed at apoptosis modulation, like small-molecule inhibitors, gene editing, and immunomodulation, will be presented. This review highlights current methods of apoptosis identification and outlines research objectives for the following years. This review integrates previous scoped literature with recent literature published with the hope of advancing the understanding of apoptosis as a marker of cellular death and disease progression.

17.1.1 Physiological significance of apoptosis in cellular homeostasis and organismal development

Apoptosis serves the unique function of a quality control mechanism by destroying damaged or dangerous cells, making it essential not just to the health, but the very existence of multicellular organisms. For example, in embryogenesis, sculpting of the tissue, such as at the fingers and toes, is done through apoptosis of interdigital cells. Apoptosis is also important in the immune system for the removal of self-reactive T and B lymphocytes, which, if uncontrolled, could lead to autoimmune disorders. In addition, apoptosis controls the rate of cell replacement in the skin and the intestinal epithelium, which helps to maintain homeostasis (Gibellini and Moro, 2021). The balance of cell death and division helps in retaining tissue structure and function. Apoptosis acts as a protective mechanism against malignancies by destroying infected and genotypically unstable cells. Apart from these, the lack or excess of apoptosis can result in several types of disease, illustrating the importance of its regulation.

Cell death may take place in several ways, with apoptosis, necrosis, and autophagy being the main categories. Apoptosis is a programmed, noninflammatory, and self-cleaning process that entails a meticulously orchestrated dismantling of cells. On the other hand, necrosis occurs because of acute cellular damage, which causes the cell membranes to rupture, and inflammation and tissue damage follow such events. Necrosis is virtually always accompanied by some pathological processes, such as ischemia and infection, whereas apoptosis occurs as an ordinary and natural physiological event. However, autophagy is a form of survival mechanism that entails breaking down and reusing parts of the cell under challenging conditions, like loss of nutrients. Different from apoptosis, autophagy does not lead to cell death but could eventually result in triggering apoptosis after extensive cellular damage (Sorice, 2022). These

TABLE 17.1 Comparison between intrinsic, extrinsic, and ER-mediated apoptosis.

Feature	Intrinsic pathway	Extrinsic pathway	ER-mediated pathway
Trigger	DNA damage and stress	Ligand-receptor interaction	ER stress and calcium depletion
Key players	Mitochondria and Bcl-2 family	Death receptors (Fas and TNFR)	CHOP and caspase-12
Initiator caspase	Caspase-9	Caspase-8	Caspase-12
Mitochondrial involvement	Yes	No	Sometimes

processes finely interact with one another; for instance, autophagy in some cases acts as a pro-survival mechanism, while in others, it may lead to cell death by apoptosis or necrosis. Making sense of these distinctions is particularly important in biologically understanding how to devise therapeutic approaches for harnessing cell death in various disease processes.

17.2 Molecular mechanisms of apoptosis

Apoptosis refers to an orderly biological process where individual cells undergo self-destruction, serve an imperative role in the maintenance of tissue homeostasis, and aid in the removal of damaged or infected cells. Diverse molecular mechanisms of apoptosis are controlled predominantly by two broad types of pathways: the intrinsic (mitochondrial) and extrinsic (death receptor) pathways (Table 17.1) (Kutuk, 2010). The intrinsic pathway is induced by stress signals inside the cells, which disrupts the mitochondrial outer membrane. This leads to the release of cytochrome C and subsequent activation of the apoptosome-mediated caspase-dependent cell disassembly. An apoptosis-inducing ligand bound to a death receptor, like Fas or TNFR1, sets off the extrinsic pathway, or coupling of the ligand death receptors with Fas and TNFR1 activates apoptosis via caspase 8 cleavage. Furthermore, ER and calcium signaling-associated apoptosis occurs from the burden of misfolded proteins and involves the UPR. (Hajnóczky et al., 2003) In the final execution phase, cellular material is degraded by caspases 3, 6, and 7, resulting in orderly cell dismantling. The knowledge gained from these mechanistic processes is fundamental in the understanding of apoptosis-related diseases like cancer and neurodegenerative diseases, along with aiding in the development of treatment strategies.

17.2.1 Intrinsic (mitochondrial) pathway

There are many apoptosis, which are a form of programmed cell death, uses highly regulated mechanisms to ensure efficiency in homeostasis and stop the growth of unhealthy cells. The intrinsic, or mitochondrial, pathway of apoptosis is primarily triggered by intracellular stress signals, including DNA damage, oxidative stress, and nutrient deprivation. This specific pathway is different in comparison to the other pathways because it relies on mitochondria for its regulation instead of the external signaling of death receptors. The most demonstrative event in this pathway is the permeabilization of the outer membrane of the mitochondria, which enables the release of apoptogenic factors into the cytoplasm. This process triggers multiple biochemical processes, which result in the reactivation of caspases and the breakdown of the cell (Kutuk, 2010). The permeabilization of the outer membrane of the mitochondria is controlled mostly by the Bcl-2 protein family, which consists of both pro and antiapoptotic proteins that decide the outcome of the cell. It is the ratio of these conflicting proteins that determines the cell's fate with regard to avoiding apoptosis or undergoing the process. As soon as MOMP starts, the release of cytochrome c into the cytosol triggers apoptosome formation, which acts as a platform for caspase activation (Garrido et al., 2006). Within this complex, initiator caspase-9 is activated, which further activates the executioner caspases, procaspase-3, procaspase-6, and procaspase-7. They destroy cellular structures in ways that guarantee cell death without inflammation and in an orderly fashion.

Mitochondrial outer membrane permeabilization (MOMP) serves as a significant regulatory event in the intrinsic apoptotic pathway. It is often considered an irreversible step in the process of apoptosis since, following permeabilization, the cell is destined to die. The step is completed by the opening of mitochondrial pores due to the action of pro-apoptotic proteins, Bax and Bak (Gillies and Kuwana, 2014). These proteins self-associate in the mitochondrial membrane to form oligomeric pores that are large enough to permit the efflux of cytochrome c, Smac/DIABLO, and apoptosis-inducing factor (AIF). Smac/DIABLO facilitates apoptosis by degrading the inactive forms of the apoptosis proteins IAPs, which are responsible for inhibiting the activation of apoptotic caspases. AIF translocates to the nucleus, where it can facilitate chromatin condensation and large-scale DNA fragmentation without the action of caspases. MOMP is heavily controlled, as too much mitochondrial permeabilization can lead to cell death, while not enough can allow defective cells to exist and lead to disease, including cancer (Jourdain and Martinou, 2009). The family of Bcl-2 proteins is critical in the regulation of MOMP. Their roles can be split into three functional categories: Pro-apoptotic effectors, antiapoptotic regulators, and BH3-only sensitizer/activators. The pro-apoptotic proteins Bax and Bak are the chief executors of mitochondrial permeabilization. They are able to oligomerize in the mitochondrial membrane during apoptosis. Antiapoptotic proteins Bcl-2, Bcl-xL, and Mcl-1 block apoptosis by binding to Bax and Bak, preventing permeabilization of the mitochondria (Garrido et al., 2006). The third group consists of BH3-only proteins such as Bid, Bim, Puma, and Noxa. These proteins act as "apoptotic sensors" because they can respond to stress signals, which shift the balance toward the neutralization of antiapoptotic proteins or directly activating Bax and Bak. The interactions among these proteins can determine the threshold of apoptosis to be executed. In a healthy state, the cell has greater levels of antiapoptotic proteins, which enable survival. However, following apoptotic signals, the level of BH3-only proteins increases. These proteins promote MOMP through the activation of pro-apoptotic proteins Bax and Bak.

The release of cytochrome c after MOMP has occurred is a critical step of the intrinsic pathway. In the cytosol, cytochrome c attaches to Apaf-1 in the presence of ATP, forming a large multisubunit structure known as the apoptosome. This complex can incorporate and activate apoptotic initiator caspase-9, which is critical for this step. Just like any other important cellular process, concern is elevated when considering the safety of activating caspase-9, which induces the collapse of the cell. After the formation of the apoptosome, several proteins, including the X-linked inhibitor of apoptosis protein XIAP, prevent executioner caspases from cleaving. Death of the cell is delayed by such means. For receptors that capture the execution death procaspases, Smac/DIABLO and Omi/HtrA2 act against these inhibitors. Together with apoptosis, this allows orderly death of the cell (Jourdain and Martinou, 2009).

The process of apoptosis is executed by three executioner caspases; apoptosis is characterized by nuclear fragmentation, chromatin condensation, and disintegration of the cytoskeleton. Caspases are enzymes that cut proteins known as key cellular substrates. For example, these processes are accomplished by caspase-3, which in my case is responsible for severing structural proteins like Lamin and actin, which undergird both the Zeiss and cytoplasm. With decimation of ICAD through cleavage, caspase-activated DNase (CAD) is released to pulverize chromosomal DNA into oligonucleosomal fragments, which create the signature "ladder of DNA." This phenomenon is CAD's default state, where it is bound to its inhibitory protein, blanketed in two forms of adhesion bonds before being solved by caspase-3. Following on from this, CAD activates andaves (Zhang et al., 2021). The active oligo-nucleosome stimulating endonuclease (CADeq) is a multienzyme complex structure that resides in the nucleoplasm and oversees two other structures responsible for severing chromosomal DNA into oligonucleosomal fragments. After caspase-3's cleavage, ICAD undergoes adhesion bond splitting, enabling CAD to render chromosomal DNA into oligonucleosomal fragments. The process generates a sequence of nucleosomes artificially packaged on deoxyribonucleic acid repeatedly, as depicted by dissection. The phenomenon of rupturing DNA chromosomal units along the stalk of DNA nucleosomes is defined in the dissection as apoptosis.

Mitochondria have an additional function in supporting the apoptotic signals through calcium homeostasis. The release of calcium from the mitochondria increases the sensitivity to apoptosis by affecting calcium signaling pathways. For example, higher levels of calcium in the cytosol can trigger calpains, which are a family of calcium-dependent proteases, thus promoting apoptosis. In addition, calcium overload can disrupt mitochondrial activity, which worsens the release of cytochrome c and increases caspase activation. This relationship between calcium and the mitochondria is one aspect of the intricate system that controls apoptosis (Orth et al., 2012).

The pathological interruption of the intrinsic pathway poses dangerous consequences. In the context of cancer, disruptions in the signaling of apoptosis allow tumorigenic cells to escape programmed cell death, resulting in their uncontrolled proliferation and destruction of tissues. Loss-of-function mutations in pro-apoptotic genes like Bax or overexpression of the antiapoptotic Bcl-2 protein are well-known in many types of cancer, making these proteins important therapeutic targets. Treatment strategies using specific nuclide BH3-mimic drugs (like the venetoclax drug) have been designed to block Bcl-2 to induce sensitivity to apoptosis in tumoral cells. On the other hand, too much stimulation of the intrinsic pathway is observed with neurodegenerative disorders like Alzheimer's and Parkinson's diseases, where unregulated apoptosis results in neuronal death. In these disorders, therapeutics are designed to decrease apoptosis using antiapoptosis proteins or by inhibiting the activity of executioner apoptotic proteases known as caspases.

17.2.2 Extrinsic (death receptor) pathway

The extrinsic apoptotic pathway represents another approach that cells employ to carry out programmed cell death upon receiving instructions from outside the cell. This phenomenon is mainly driven by the existence of specific transmembrane death receptors within the family of tumor necrosis factor (TNF) receptors. Death receptors act like sentinel molecules which detect and bind certain ligands and, as a result, start a chain of signaling processes within the cell, which will ultimately end up with apoptosis. The primary distinction between this pathway and the intrinsic pathway, which is controlled by mitochondria, is that stimuli for the extrinsic pathway come from the outside. It is now clear that this form of programmed cell death is critically important for immune surveillance, embryogenesis, and destroying infected and dysfunctional cells (Prahalad and Bernards, 2016). The pathological conditions associated with this imbalance range from cancer and autoimmune disease to chronic inflammation. Components of this pathway include death receptors, death-inducing signaling complex (DISC), initiator caspases, and some elements of the intrinsic pathway.

Death receptors are defined by the existence of an intracellular death domain (DD), which is vital for the reception of apoptotic signals. Some of the most studied death receptors are Fas (CD95/APO-1), TNF receptor 1 (TNFR1), and Tumor Necrosis Factor–Related Apoptosis-Inducing Ligand (TRAIL), receptors (DR4 and DR5), which are bound by their respective Fas ligand (FasL), TNF-α and TNF-ß, and the apoptosis-inducing ligand (TRAIL). Death receptors are tri-subject to trimerization, which aids in the binding of adaptor proteins and triggers the creation of a signaling hub referred to as DISC that enables the initiation of caspase signal transduction. The formation of DISC marks an essential transition in extrinsic apoptosis, where it acts as a scaffold to anchor and permit the activation of initiator caspases, caspase 8, and caspase 10. One of the most researched death receptor pathways is the Fas-mediated apoptotic pathway. The Fas receptor is known to be present in many tissues and is known to trigger apoptosis of activated lymphocytes, thus playing an important role in immune regulation. FasL binding to Fas causes receptor oligomerization and the recruitment of Fas-associated DD (FADD) adaptor protein. FADD serves as a bridge between Fas and caspase-8. FADD has both a DD, which binds to Fas, and a death effector domain (DED), which binds to and recruits caspase-8 (Gibellini and Moro, 2021). The dimerization and auto-processing of caspase-8 are facilitated by its cleavage and activation upon forming a homodimeric structure. After cleavage and activation of the initiator procaspase 8, Fas signaling is transmitted through downstream effector caspases, like caspase-3, caspase-6, and caspase-7. They execute apoptosis by cleaving cellular substrates and components such as structural proteins, cytoskeletal proteins, and DNA.

A mechanism akin to this is found in TNFR1-induced apoptosis. Recruitment of tumor necrosis factor receptor-associated DD (TRADD) and receptor-interacting protein kinase 1 (RIPK1) follows after TNF-α binds to TNFR1. This leads to the formation of DISC as well as caspase-8 activation. YET, TNFR1 signaling is even more intricate than Fas signaling because it includes apoptosis and necrosis in addition to cell death and has dichotomous outcomes depending on whether additional proteins are recruited or not. Active caspase-8 leads to apoptosis, whereas activated NF-ЬB signaling leads to inflammation and cell survival. The dichotomy of signaling in TNF-a receptors is more pronounced in cancer, where unregulated NF- Б signals will subvert the effects of apoptosis and enhance tumor development. The TRAIL-related research area of interest for inducing apoptosis selectively in cancer cells without affecting normal cells has captured the most attention. TRAIL interacts with its receptors DR4 (TRAIL-R1) and DR5 (TRAIL-R2), which bind to FADD and result in the downstream activation of caspase-8. Due to its selective apoptotic effects on tumors, TRAIL has been studied in the context of cancer treatment. Preclinical trials have shown some success with recombinant TRAIL as well as with agonistic antibodies against receptors of TRAIL; however, resistance mechanisms such as the overexpression of antiapoptotic proteins pose a challenge for their effective application in clinical settings.

Regulation of the apoptosis protein executioner level, or "caspase-8", is one of the key control points for the extrinsic apoptotic pathway. Cellular FLICE-like inhibitory protein (c-FLIP) is a caspase-8 homologue that is devoid of proteolytic activity and thus, c-FLIP inhibits the effector caspase-8 by complexing with it. Elevated amounts of c-FLIP cause dramatic competitions at the

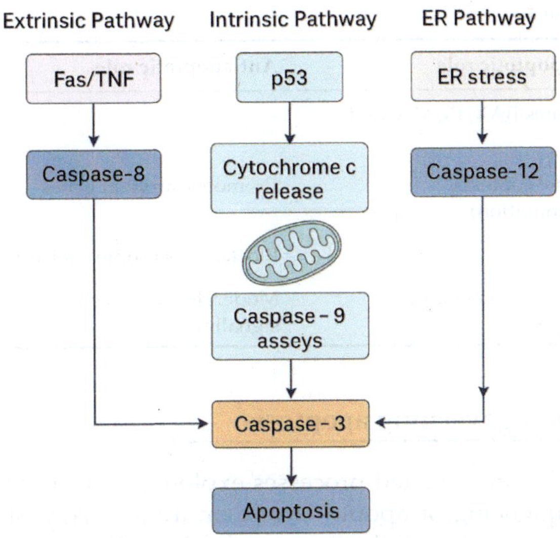

FIGURE 17.1 Integrated diagram of apoptotic pathways.

DISC levels, which results in the blockade of apoptosis (Fleisher, 1997). Most tumors manipulate this system to escape extrinsic apoptosis by increasing the expression of c-FLIP. Reduced c-FLIP expression, on the other hand, augments sensitivity toward intrinsic apoptotic signaling, emphasizing the notion that c-FLIP profoundly determines cell fate. The activation of Bid, a BH3-only protein, serves as a connection between extrinsic and intrinsic pathways of apoptosis. Bid is cleaved by caspase-8 to form tBid, which can migrate to the mitochondria where it activates MOMP. This switches on the intrinsic apoptotic pathway by promoting the release of cytochrome c and the formation of the apoptosome, thus integrating both apoptotic pathways. Bid's ability to enhance the apoptotic signal guarantees that the process of apoptosis is effective, especially in cases where the extrinsic signal is inadequate to induce cell death.

In disease development, the loss of control of the extrinsic apoptotic pathway is of great concern. In cancer, a primary method for tumor development and evasion of treatment is resistance to extrinsic apoptosis. To avoid apoptosis, many tumors underexpress death receptors, c-FLIP is overexpressed, and/or survival pathways such as NF-κB are turned on. Attempts to treat by restoring extrinsic apoptosis have included TRAIL therapies, death receptor agonists, and caspase-8 activators. Still, clinical effectiveness is limited because of TRAIL resistance and tumor heterogeneity. In autoimmune conditions, uncontrolled extrinsic apoptosis can result in too few immune cells and the loss of tolerance to self-antigens. For instance, the Fas or FasL mutations that cause autoimmune lymphoproliferative syndrome (ALPS) are defined by inadequate lymphocyte apoptosis and immune system dysregulation. In inflammatory diseases, the continuous stimulation of TNFR1 signaling can lead to damage and progression of the disease. For example, rheumatoid arthritis, inflammatory bowel disease, and sepsis are some conditions where TNF-α is over-signaled and results in chronic inflammation along with immune cells which are resistant to apoptosis (Jain et al., 2010). The development of monoclonal antiTNF therapies, like infliximab and adalimumab, has greatly reduced the inflammation and improved outcomes in patients. On the other hand, the long-term suppression of TNF-α signaling may render the patients more prone to infections, which emphasizes the necessity for a carefully devised strategy (Fig. 17.1).

TABLE 17.2 Pro- and antiapoptotic roles of key regulators.

Molecule	Type	Pro-apoptotic role	Antiapoptotic role
p53	Tumor suppressor	Activates BAX, PUMA, and Noxa	-
NF-κB	Transcription factor	In some contexts (inflammation)	Promotes survival genes
Bcl-2	Mitochondrial protein	-	Blocks cytochrome c release
ROS	Reactive species	Induces mitochondrial damage	Moderate ROS = cell signaling

17.3 Regulatory networks governing apoptosis

Immunological, developmental, and cellular balance-related processes exploit apoptosis as a highly managed step. The triggering and sequencing of apoptosis rely on a vast array of signaling molecules implementing intracellular and extracellular communications. Key regulators such as p53, NF-κB, autophagy, and oxidative stress are critical in determining the cell fate decisions (Table 17.2). These molecular checkpoints have an important role to play in ensuring that apoptosis is set in motion when there is DNA damage, infection, oxidative stress, or metabolic dysregulation (Schlatter et al., 2009). Such conditions foster the emergence of a wide range of pathologies, including cancer, neurodegenerative disorders, and autoimmune diseases, due to the dysregulation of these pathways. These networks reveal the ins and outs of apoptosis and why they matter in disease pathology and therapy, which makes them important to comprehend (Mai and Liu, 2009).

17.3.1 p53 and its role in apoptosis induction

The tumor suppressor p53 is one of the most highly characterized apoptotic regulators. In a cellular setting, p53 functions as a transcription factor and is responsible for the coordination of certain cellular processes, such as stress responses, to maintain genomic integrity. In the case of DNA lesions, oncogenic stimuli, and oxidative stress, p53 coordinates the response. In the absence of stress, p53 levels are kept low due to MDM2, which is an E3 ubiquitin ligase that directs p53 to be degraded in the proteasome. In contrast, when cells are stressed, particular posttranslational changes like acetylation and phosphorylation increase the half-life of p53 by preventing it from being degraded, allowing it to accumulate and become transcriptionally active (Sheikh and Fornace, 2000). Transcriptional control of pro-apoptotic genes is one of the primary methods by which p53 triggers apoptosis. Through several transcriptional mechanisms, activated p53 promotes expression of some members of the Bcl-2 protein family, such as Bax, Puma, and Noxa, which results in the outer membrane of mitochondria being permeabilized (MOMP) and the release of cytochrome c, which results in the execution of caspase-dependent apoptosis (Bellamy, 1997). In addition, p53 also binds to Fas (CD95) and TRAIL receptors (DR4 and DR5), p53 is therefore linked to the extrinsic type of apoptotic pathways. In addition to transcriptional regulation, p53 can also promote pro-apoptotic behavior by migrating to the mitochondria and binding to Bcl-2 family proteins, enabling the liberated cytochrome c to leave the mitochondria. Loss of p53 function is common to a variety of cancers, aiding their survival and enabling resistance to apoptosis-inducing therapies. Disruptive mutations in TP53, the gene that encodes p53, hinder the activation of apoptotic pathways, causing an escape from programmed cell death. Consequently, p53 has become a prominent focal point of anticancer therapies directed toward restoring its apoptotic activity. Compounds such as nutlin-3, which blocks p53-MDM2 binding, and PRIMA-1, which reactivates the function of mutant p53, are gaining recognition for their novel ability to reactivate p53-dependent apoptosis in cancerous cells.

17.3.2 Autophagy-apoptosis interplay

The integration of autophagy and apoptosis represents a dual cellular process that manages and decides the cell's action on being challenged by stress. Autophagy is a form of survival since it preserves the cellular equilibrium by removing defective organelles and proteins. There is a possibility of excessive, uncontrolled autophagy promoting cell death through apoptosis. Critical molecular components with dual roles, such as key players in Beclin-1, Bcl-2, and p62, control the interplay between autophagy and apoptosis, determining life or death for the cell (Sorice, 2022). Beclin-1, as a fundamental autophagy initiator, works with Bcl-2 to promote suppression of apoptosis. In the case when cells have plenty of nutrients, Bcl-2 will bind with Beclin-1, blocking autophagy and triggering the survival of the cell. However, after a given period of stress, Beclin-1 can be cleaved by caspases, and its link with Bcl-2 is lost, which leads the cell toward apoptosis. Also, modulating apoptosis is the autophagy receptor p62, which can degrade apoptotic proteins and promotes apoptosis. On the other hand, we have hyperactive autophagy that leads to depletion of important things to survive, in turn activating apoptosis or the caspase-dependent cell death.

In the context of cancer, impaired autophagy may foster tumor advancement by permitting the buildup of dysfunctional mitochondria and genomic instability. On the other hand, excessive autophagy can facilitate apoptosis of a cancer cell, which makes modulation of autophagy a viable strategy for therapy (Fleisher, 1997). Certain autophagy inhibitors, like chloroquine and hydroxychloroquine, have been investigated to increase the rate of apoptosis of cancer cells, while some autophagy activators may shield normal cells from undergoing apoptosis due to stress in neurodegenerative disorders.

17.3.3 Oxidative stress and reactive oxygen species in apoptosis

Due to the damage caused by reactive oxygen species (ROS) and the lack of necessary antioxidant responses, oxidative stress serves as a powerful stimulus of apoptosis. Mitochondria are mostly responsible for the generation of ROS during oxidative phosphorylation. Under normal physiological conditions, ROS act as secondary messengers during intercellular communication; nevertheless, when overproduced, ROS can cause severe oxidative harm to proteins, lipids, and DNA, eventually resulting in apoptosis. Apoptosis induced by ROS occurs due to mPTP pore opening, cytochrome c release, or pro-apoptotic kinases activation, typically marked by the presence of Apoptosis Signal-Regulating Kinase 1 (ASK1) or c-Jun N-terminal kinase (JNK). By the action of redox-sensitive transcription factors, such as p53 and NF-κB, modulatory control of apoptosis is possible thanks to ROS. NF-κB, in the context of low ROS levels, triggers protective signaling, but high levels of ROS fix the p53 and fuel responses related to DNA damage alongside p53-induced apoptosis (Fleisher, 1997). The Engagement of excess oxidative stress to apoptosis pertains to important implications for several diseases. In Alzheimer's and Parkinson's disease, oxidative stress-mediated apoptosis results in loss of neurons. On the other hand, a great many cancer cells resist apoptosis induced by ROS due to the overexpression of antioxidant enzymes, superoxide dismutase (SOD), and glutathione peroxidase (GPX). Splitting oxidative stress with pro-oxidant therapeutic agents, like arsenic trioxide and curcumin, has demonstrated success in selectively causing apoptotic death in certain cancer cells.

17.4 Apoptosis in developmental biology and tissue homeostasis

A biological phenomenon called apoptosis plays a key role in the sculpting of an embryo, tissue homeostasis, and immune system functioning. For instance, during the early skeletal and organ systems development, apoptosis is necessary for morphogenesis to enable functioning tissues and organs to form structures. In the immune system, apoptosis removes self-reactive lymphocytes, allowing for immune tolerance while simultaneously preventing autoimmunity. In adult tissues, apoptosis helps eliminate old or damaged senescent cells during epithelial turnover and organ regeneration. The balance of apoptosis needs to be managed with care, as unregulated apoptosis can lead to cell death, contributing to developmental malformations, degenerative conditions, or cancer (Robert A Weinberg, 2007).

17.4.1 Apoptotic mechanisms in embryogenesis and morphogenesis

Apoptosis is crucial during embryogenesis and is required for processes such as tissue sculpting, organogenesis, and removing excess cells. The developing embryo transforms into a more complex structure through morphogenetic processes that depend on apoptosis to get rid of the cells that either serve no purpose or hamper the proper tissue organization. There are several classical examples of limb development, closure of the neural tube, and heart development that show the importance of apoptosis in sculpting embryonic structure (Gibellini and Moro, 2021). One of the best studied examples of apoptosis during morphogenesis is limb development. During the early stages of bud formation, mesenchymal cells within the limb buds undergo rapid proliferation while apoptosis occurs at later stages to remove interdigital cells to enable the formation of separate digits. Bone morphogenetic proteins (BMPs) are primarily responsible for the tight control of apoptosis in this case, as they initiate apoptosis in the interdigital regions (Huppertz and Herrler, 2005). The opposite can also happen, with overactive apoptosis resulting in limb malformations and other congenital defects, including syndactyly, which is the fusion of digits.

Another important role of apoptosis during development is noted during neural tube closure, which is a critical event in central nervous system development. The neural tube is the embryonic structure that develops into the brain and spinal cord, and apoptosis is necessary in this regard for the elimination of superfluous neural crest cells and blocking inadequate closure of the neural tube. If apoptosis does not occur in this area, it can result in serious developmental defects like spina bifida or anencephaly. In addition, apoptosis is also important in the formation of the heart, particularly in the septation of the cardiac chambers and the remodeling of primitive blood vessels. Apoptotic processes during heart development have been associated with noncompletions of heart development, such as atrial and ventricular septal defects.

17.4.2 Tissue homeostasis in epithelial turnover and organ regeneration

Apart from development, apoptosis is essential in adults for preserving tissue homeostasis by controlling epithelial turnover and organ regeneration. Histological renewal processes in many tissues require the loss of aged and/or damaged tissues by apoptosis so that proper structural integration is supported. It is most obvious in epithelial tissues of the skin, intestine, and lungs, which all have high functional cell turnover rates. In the intestinal epithelium, epithelial cells are generated in the Lieberkühn crypts and migrate toward the villi, where they are lost by apoptosis into the intestinal lumen. This turnover allows the intestinal epithelium to absorb nutrients optimally while at the same time preventing excessive proliferation, which may result in tumors. Disruptions in apoptosis regulation of the intestinal epithelium have been associated with inflammatory bowel disease (IBD) and colorectal carcinoma (Robert A Weinberg, 2007).

An analogous phenomenon takes place in the epidermis, where keratinocytes are gradually subject to apoptotic cell death as they migrate toward the surface of the skin. This cell death is essential in carving out the skin's protective outer layer, which can be a defensive shield against external threats. Failure to execute keratinocyte apoptosis leads to the development of psoriasis, a chronic inflammatory skin disease marked by unregulated cell proliferation coupled with reduced apoptosis. Apoptosis is active during the regeneration of organs in the elimination of unwanted or damaged cells to aid in tissue repair. In the liver, for example, apoptosis is involved in the hairline imbalance of hepatocytes during their regenerative phase after an injury. As with lung diseases, liver disorders such as hepatitis and cirrhosis tend not to exhibit apoptosis and therefore are associated with necrotic tissue and fibrosis; other liver tissue tends to get excess tissue damage or scar material being placed around them due to fibrosis. In the same fashion, the lungs undergo apoptosis for the purpose of controlling the turnover of alveolar cells as well as disposing of damaged cells after the period of infection or toxic exposure.

17.4.3 Senescence-associated apoptosis in aging tissues

The role of apoptosis in aging is two-fold, as it both eliminates cells that need to be removed while also making the issue worse. Aging itself is accompanied by the gradual deterioration of the body's cells, and its ability to regenerate also declines. Apoptosis helps facilitate this by getting rid of any damaging or aged cells that would only serve to compound the tissue, along with serving as a preventative measure. But with that said, there can be overzealous depletion of necessary cell populations that succumb to age-related degenerative diseases as a result of apoptosis (Baar et al., 2017). Beyond the descriptive association with aging, cumulative genotoxic insults progressively overwhelm DNA *adenine excision of* repair pathways, leading to persistent double-strand breaks and activation of p53-dependent checkpoints. Defects or decline in key repair systems such as nucleotide excision repair, nonhomologous end joining, and homologous recombination lower the threshold for apoptosis or senescence in repeatedly stressed cells. This shift in damage sensing and repair capacity promotes loss of stem and progenitor cells, clonal expansion of surviving but genomically unstable cells, and tissue dysfunction, thereby functionally linking DNA damage accumulation to organismal aging phenotypes. Experimental models in which DNA repair factors are mutated or exhausted show premature aging, supporting this mechanistic connection between genomic instability, altered apoptotic responses, and age-associated decline.

Factors like the shortening of telomeres, oxidative stress, and DNA damage can trigger a state known as irreversible growth arrest, which is termed cellular senescence. In the early stages,

TABLE 17.3 Apoptosis dysregulation in major diseases.

Disease type	Apoptosis status	Consequences
Cancer	Inhibited	Tumor progression and drug resistance
Neurodegeneration	Excessive	Neuronal loss and cognitive decline
Autoimmune disorders	Faulty clearance	Autoantigen persistence and inflammation
Cardiovascular disease	Overactivation	Cardiac cell loss and heart failure

senescent cells survive the onslaught of apoptosis but eventually succumb to it with prolonged stress. The feature of senescence-associated apoptosis is vital in the elderly and the aged patients in domains of the muscles, brain, and the cardiovascular system, which, as the name suggests, is at its core the frame that sustains our body. Abnormal activation of apoptotic pathways in the body can result in recurrent loss of neurons, which leads to various neurodegenerative disorders like Parkinson's or Alzheimer's (Baar et al., 2017). While excessive frustration within the nervous system can also compel undesirable neuronal apoptosis. On the other hand, cancer can be developed by poor triggering of these pathways, as it can permit the accumulation of senescent cells that acquire cancerous characteristics.

In older age, the apoptosis of endothelial and smooth muscle cells can lead to heart failure and atherosclerosis. Increased apoptosis of vascular endothelial cells results in insufficient repair of blood vessels, which can lead to hypertension and ischemic diseases. Apoptosis also plays a role in sarcopenia, the age-related degeneration of skeletal muscle. The age-related loss of muscle mass and strength can be attributed to increased apoptotic activity of muscle fibers. Some researchers are studying antiapoptotic therapies alongside senolytic drugs to target age-related issues with tissue degeneration and severe frailty.

17.5 Apoptosis and disease pathogenesis

Apoptosis is a highly controlled form of programmed cell death that is critical for cellular homeostasis and healthy living. Nonetheless, its dysregulation is involved in many illnesses such as cancer, neurodegenerative disorders, cardiovascular diseases, autoimmune disorders, and even infectious diseases. Problems with apoptotic signaling can result in insurmountable cell growth, overwhelming cell death, or faulty immune function, all of which perpetuate disease development. Grasping the processes of how apoptosis is changed in different pathologies was essential in devising certain therapeutic approaches that aim to restore balance in apoptosis (Table 17.3).

17.5.1 Apoptosis dysregulation in cancer

To put it simply, uncontrolled cell reproduction defines cancer. Among the features of cancer, the bypassing of apoptosis by the tumor cells stands out as a prominent characteristic. This bypass enables tumor cells to survive and reproduce in endless quantities, even in the presence of cellular stress or chemotherapeutic factors working against them. Resisting apoptosis involves multiple execution mechanisms, such as the overexpression of antiapoptotic proteins, mutations in tumor-suppressive genes, and the dysregulation of the apoptosis signaling cascade. Apoptosis is the preferred method for normal cells to deal with damaged or potentially cancerous cells. In contrast, cancer cells master the art of subverting these controls. One of the main ways through which cancer cells manage to neutralize the control is through the disruption of intrinsic (mitochondrial) apoptosis pathways. Tumor cells, for example, can repress pro-apoptotic factors Bax and Bak while overexpressing antiapoptotic factors like Bcl-2 and Bcl-xL. Preventing Outer Mitochondrial Membrane Permeabilization (MOMP), followed by cytochrome c release, halts the caspase activation sequence from proceeding (Robert A Weinberg, 2007).

An example of this is the downregulation of death receptors, Fas (CD95), or decoy receptors by tumor cells. This means that they bypass the extrinsic (death receptor) pathway. In this manner, the ability of the immune system to kill tumor cells using cytotoxic T cells or natural killer (NK) cells is greatly impaired.

17.5.2 Neurodegenerative disorders

Neurodegenerative diseases such as Alzheimer's Disease (AD), Parkinson's Disease (PD), and Huntington's Disease (HD) all have one thing in common; the neurodegenerative processes involve progressive neuron loss, with apoptosis prominently featuring the neurodegenerative processes in these processes. Neurons as a tissue have a very limited capacity to regenerate, rendering apoptosis a severely harmful process in the case of neurodegenerative disorders. Mitochondrial impairment, oxidative stress, excitotoxicity, and protein aggregation are several leading causes of apoptosis in neuronal tissues (Zhao et al., 2022).

Within the framework of AD, cognitive impairment is accompanied by overall synaptic loss because of caspase-dependent apoptosis as well as mitochondrial dysfunction alongside the progressive impact of β-amyloid plaque and tau tangles. In the case of Parkinson's disease, however, oxidative stress paired with damage to mitochondria alongside excessive release of pro-apoptotic proteins (Bax and Bad) directs the gradual obliteration of dopaminergic neurons in the substantia nigra. Some strategies focus on neuroprotection and reducing the apoptotic processes in question with Bcl-2 mimetics, caspase inhibitors, and antioxidants portraying hypothesis for therapeutic development who have focus on oxidative stress in neurodegenerative disorders.

17.5.3 Autoimmune and inflammatory disorders

Apoptosis assists in the homeostasis of immune cells and the control of inflammation, making it crucial for the functioning of autoimmune and inflammatory disorders. A disruption in the apoptotic processes results in the immune inefficacy, uncontrolled inflammation, and tissue damage characteristic of Systemic Lupus Erythematosus (SLEs) and Rheumatoid Arthritis (RAs). The incompletely cleared apoptotic cells result in an overload of the cellular debris in which nuclear antigens are exposed, causing the production of autoantibodies, which further lead to complex formation of the immune and inflammatory cytokines and organ damage. Moreover, inflammation is aggravated by impaired phagocyte action, mutations in the genes responsible for apoptotic clearance, and over-stimulation of Toll-like receptors (TLRs). The modulation of therapeutic approaches is directed to B-cell depletion (Rituximab), DNase I therapy, and Type I Interferon inhibitors, which seek to control apoptotic dysregulation. On the other hand, the hallmark of RA is excessive apoptosis of the synovial fibroblasts, which leads to destruction and inflammation of the joints. The increased turnover of apoptotic cells results in the inflammatory responses that cause synovial hyperplasia and the degradation of cartilage. Employing some apoptotic pathways in the treatment of RA by switching the signaling of TNF-α, or the activation of caspase, brings hope for potential treatments.

17.5.4 Infectious diseases and apoptosis modulation

Apoptosis offers a double-edged sword in infectious diseases as it can provide host defense by ridding the body of infected cells, but can also conveniently be harnessed by pathogens for their survival. Several viruses, for instance HIV, Influenza, and SARS-CoV-2, appear to have some evolved capacity to suppress or induce apoptosis for their replication advantage. HIV enhances the apoptosis of uninfected bystander CD4+ T cells via FasL and TNF-α signaling, which causes immune depletion while simultaneously enforcing apoptosis's inhibition in infected cells by activating antiapoptotic protein (Bcl-2 and Bcl-xL) expressions to enable prolonged viral replication. Apoptosis in infected cells is also provoked by the Influenza virus with the PB1-F2 protein, which causes the release of mitochondrial cytochrome c, consequently activating caspases, which allows for the dispersal of the virus. SARS-CoV-2, known as the agent of COVID-19, uses apoptosis for immune evasion just like its predecessors. It is known to induce apoptosis of lung epithelium and immune cells, which leads to marked inflammation and the destruction of tissues through viral proteins, while at the same time subduing apoptosis in infected cells in order to further viral replication (Fleisher, 1997).

Even bacterial pathogens try to manipulate apoptosis to their benefit. While Mycobacterium tuberculosis inhibits macrophage apoptosis to remain in the host, Salmonella and Listeria regards immune clearance by manipulating choreographic apoptotic processes. This form of modulation during infectious diseases has possible therapeutic prospects by manipulating apoptotic control mechanisms to either strengthen immune responses or worsen pathogen containment. There is ongoing research to achieve better results in dealing with illness by controlling infection-associated apoptosis with caspase inhibitors, p53 activators, and apoptosis-stimulating medications. It is crucial for novel antiviral, antibacterial, and immunomodulatory treatments to understand how pathogens manipulate apoptosis.

17.6 Methodologies for apoptosis detection and quantification

Quantifying apoptosis accurately holds significant value within both clinical and preclinical research to unravel the intricacies surrounding disease, its progression, and the corresponding treatment response. Numerous structural, biochemical, and molecular techniques have been established to recognize and classify the various stages of apoptosis, such as membrane and mitochondrial changes, DNA fragmentation, and mid/late-stage apoptotic signs. These methods enhance the understanding of the mechanistic features of apoptosis and are essential in pharmacology, oncology, and immunology (Save et al., 2004). With regard to traditional morphological evaluation, light microscopy and electron microscopy permit visualization of several apoptotic cell features such as chromatin condensation, nuclear fragmentation, and budding of the membrane (apoptotic blebs). Biochemical methods include DNA cleavage through fragmentation

TABLE 17.4 Modern techniques for apoptosis detection.

Method	Marker detected	Application	Limitations
TUNEL assay	DNA fragmentation	Tissue sections	May detect necrosis
Annexin V staining	Phosphatidylserine flip	Flow cytometry of early apoptosis	Nonspecific
Caspase-3 activity	Active caspase-3	Biochemical assays	Needs fresh sample
Western blot	Bcl-2, PARP, caspases	Pathway profiling	Semiquantitative

assays using transferase dUTP nick-end labeling (TUNEL), also known as DNA laddering. Flow cytometry coupled with an Annexin V/Propidium Iodide (PI) staining is a high-throughput technique that allows quantification of a population of cells that have undergone apoptosis. Other molecular methods, such as western blotting and qPCR quantification of activity, use signaling proteins for apoptosis, like caspases and proteins of the Bcl-2 family. More recently, imaging modalities capable of capturing live cells have made it feasible to observe apoptosis in real-time using fluorescent markers and biosensors (White and Cinti, 2004; Kyrylkova et al., 2012).

All methods have some benefits and some drawbacks. Even though microscopy offers high-resolution visualization, it is slow and often needs specialized skills. Although flow cytometry and molecular techniques are quantitative and capable of high-throughput analyses, they usually have complicated sample preparation steps. Advanced imaging platforms, which capture images of live cells, are essential to real-time cellular imaging and capturing dynamically changing cellular appearance during apoptosis, which has profoundly improved the study of apoptosis. In both basic and clinical research, the study of apoptosis is always done with a combination of many methods to make sure that the results are valid and dependable (Table 17.4).

17.6.1 Morphological assessment: Light and electron microscopy techniques

Morphological assessment is among the oldest and most popular methods for identifying apoptosis. It enables the researcher to notice the specific cellular alterations linked to cell death at both the structural and ultrastructural levels. These changes comprise the following: fragmentation of the nucleus, condensation of the chromatin, blebbing of the cell membrane, and creation of apoptotic bodies. The primary two tools employed to analyze the apoptotic cells are light microscopy and electron microscopy, with each one offering certain favorable aspects in accordance with the cell, e.g., resolution, visualization (White & Cinti, 2004; Lee and Kang, 2019). Due to its simplicity and ability to provide cellular changes, light microscopy is often used in apoptosis studies. With haematoxylin and eosin (H&E) staining, apoptotic cells can be visually identified by their condensed and fragmented nuclei, which stain dark and appear as small dark stains within the cytoplasm. Use of DNA-binding dyes like DAPI (4′, 6-diamidine 2-phenylindole) and Hoechst 33342, which stain condensed chromatin, further enhances the visualization of apoptosis through fluorescence microscopy because they make it easier to differentiate between healthy and apoptotic nuclei. The strength of fluorescence microscopy is the ability to see the apoptotic process in fixed and living cells, therefore providing a real-time analysis of the apoptotic cell death process over time. Another technique is phase-contrast microscopy, which is useful for live-cell imaging because it makes it possible to see the apoptotic relief changes without the need for fluorescent stains and fixation steps. Although light microscopy is useful for identifying structural features that apoptotic cells have, it is not possible to examine their details, such as organelle changes (Save et al., 2004; Lazar et al., 2007).

Apoptotic cells and their associated subcellular alterations can be examined much more precisely with the use of electron microscopy, especially transmission electron microscopy (TEM). Apoptotic progression includes a series of events: first, condensation of chromatin along the nuclear membrane, then fragmentation of the nucleus, and finally, shrinkage of the cytoplasm. Swollen mitochondria in cell apoptosis, with their cristae structures disrupted, are a common sight and indicate participation in the intrinsic pathway of apoptosis. Apoptotic bodies—membrane vesicles encapsulating apoptotic cellular debris, which are subject to phagocytosis by some other cell, or immune cells, are another feature of apoptotic cells are detected through TEM. Examination of apoptotic cells through scanning electron microscopy (SEM) adds to the knowledge gleaned from other electron microscopy techniques by showing the three-dimensional surface structure; membrane blebbing, which is a prominent feature of apoptosis and is not found in necrosis, can also be seen (Kari et al., 2022). One of the greatest challenges to electron microscopy remains, its unparalleled resolution aside. The processes preceding the actual imaging steps for TEM or SEM are often the most arduous, consisting of fixation, dehydration, sectioning, and staining, which may cause artifacts if handled incorrectly. Moreover, the expenses associated with conducting electron microscopy and the high level of skill and troubleshooting knowledge needed further complicate its use for routine apoptosis exams. This is why many researchers combine light microscopy and high electron microscopy with biochemical and molecular assays in the attempt to bridge the gap in their understanding of apoptosis (Riccardi and Nicoletti, 2006).

17.6.2 Biochemical methods: DNA fragmentation assays (TUNEL, DNA laddering)

The fragmentation of DNA is a fundamental marker of apoptosis; thereby, it is an essential parameter for the detection and measurement of cell death. The phenomenon of apoptosis involves the systematic activation of endonucleases, which results in the cleavage of the chromosomal DNA into oligonucleosomal fragments, usually in multiples of 180–200 base pairs. This fragmentation occurs as a result of CAD being activated, which is the process of nuclear degradation attendant with DNA. When analyzed through electrophoresis, this degradation produces a classic "ladder" pattern. Two of the most common biochemical approaches to detect DNA fragmentation in apoptotic cells include the terminal deoxynucleotidyl TUNEL assay and the DNA laddering assay. Both methods yield important information on the level of apoptosis present in a heterogeneous cell population (Kyrylkova et al., 2012).

The TUNEL assay is among the most widely adopted approaches for the in-situ detection of DNA fragmentation. It uses the enzyme terminal deoxynucleotidyl transferase (TdT), which adds a labeled deoxyuridine triphosphate (dUTP) to the free 3′ OH end of the fragmented DNA. TUNEL-labeled nucleotides are detectable either with fluorescence techniques or colorimetric detection methods, which enable the detection of dying cells with great precision. The primary benefit of the TUNEL assay is its capability to detect apoptosis within single cells, which is advantageous with tissue sections, cultured cells, and even paraffin wax-embedded sections. TUNEL fluorescence microscope visualization is possible with nucleotides conjugated to fluorescein or rhodamine and colorimetric detection with peroxidase or alkaline phosphatase-based methods, or for a quantitative readout via spectrophotometry. However sensitive the TUNEL assay, it may also yield false positive results due to necrotic tissues and tissues in DNA repair cycles having TdT labeled strand breaks. TUNEL is frequently combined with other apoptotic markers, such as activated caspases or externalized phosphatidylserine, for a broader characterization of apoptosis (Denton and Kumar, 2015).

Unlike the TUNEL assay which reveals the cell's fragmentation of DNA, the DNA laddering is an apoptosis-associated analytic biochemistry technique that detects the presence of apoptotic DNA fragmentation on a population level. This method consists of the isolation of genomic DNA from both apoptotic and nonapoptotic control cells, followed by separation of the DNA fragments on an agarose gel using electrophoresis. Apoptotic cells that have been stained with either ethidium bromide or SYBR Green, have a well-defined ladder pattern, with the bands representing oligonucleosomal DNA fragments of approximately 180–200 base pairs, and spaced out at regular intervals. This fragmentation pattern is characteristic of apoptosis and is one of the features that differentiate it from necrosis where DNA fragmentation is random and results in a diffuse smear instead of distinct bands. The most important benefits of the DNA laddering assay are identification and verification of apoptosis that occurs within a given cell population. Like any other technique, there are certain limitations, such as insensitivity compared to TUNEL, since a certain threshold of apoptotic cells needs to be achieved in order to visualize the laddering pattern. Furthermore, DNA laddering is mostly a qualitative approach and therefore not ideal for exact measurement of the degree of apoptosis unless combined with some qualitative densitometric or other apoptosis markers (Kari et al., 2022).

Regardless of the drawbacks involved, both TUNEL and DNA laddering continue to play pivotal roles in the study of apoptosis. The TUNEL assay is particularly helpful when examining apoptosis in tissue or single-cell samples since it gives spatial context to apoptotic processes, while DNA laddering serves to verify the existence of apoptotic DNA fragmentation in homogenized cell samples. In a number of instances, both techniques are used alongside other apoptosis assays, including measurements of caspase activity or Annexin V staining, to achieve multifaceted and thus more reliable detection of apoptosis. In addition, more recent developments in molecular biology have resulted in the creation of more sophisticated DNA fragmentation assays, analysis of a cell's DNA breaks, more commonly referred to as the Comet assay, as well as real-time PCR methods that more accurately measure the amount of apoptotic DNA cleavage (Lee and Kang, 2019). Incorporating biochemical approaches to identify DNA fragmentation is vital in advancing the study of apoptosis. On one hand, the TUNEL assay has sensitivity and single-cell resolution ideal for tissue-based studies, whereas the DNA laddering assay visually confirms apoptotic DNA fragmentation in cell populations. The combination of these techniques with the rest of the apoptosis detection methods increases their dependability, resulting in more insight into the absorption of cell death processes in different normative and pathological situations (Kari et al., 2022).

17.6.3 Flow cytometry-based approaches: Annexin v/propidium iodide staining

Offering the ability to analyze thousands of cells in seconds, Flow cytometry is one of the most powerful detection methods in apoptosis studies and offers exceptional sensitivity.

Among several flow cytometry-related apoptosis assays, one of the most recognized techniques is Annexin V/PI staining, which enables distinguishing apoptotic, necrotic, and viable cells. This technique emphasizes how cells undergoing apoptosis suffer distinct changes in phospholipid membrane, that is, the activity of phosphatidylserine (PS) translocase that externalizes PS to the outer layer of the cell membrane. For additional distinction of early and late apoptotic, necrotic, and viable cells, PI, a DNA-binding fluorescent dye, is utilized along with Annexin V, which allows for distinguishing the complex dynamics of cell death (Riccardi and Nicoletti, 2006). One of the first changes that can be noticed during apoptosis is the movement of PS from the inner side of the plasma membrane to its exterior. Healthy cells retain PS on the cytoplasmic face of the membrane because of amino phospholipid translocases. However, in the case of apoptosis, the activation of caspases results in these translocases being turned off and scramblases turned on, which leads to the externalization of PS. This is considered one of the apoptotic hallmark changes since the binding of Annexin V, a phospholipid-binding protein that has an affinity for PS, marks it. Annexin V can be bound to a fluorophore such as FITC or PE so that it can be visualized in flow cytometry and thus quantitatively determine the number of cells undergoing apoptosis in a population.

An example of PI usage is alongside Annexin V for differentiating between early apoptosis, late apoptosis, and necrosis. As a nucleic acid stain, PI does not permeate cell membranes, indicating death when fluorescent binding to DNA occurs. Only viable, and early apoptotic cells can exclude PI binding with the membrane and subsequently exhibit no fluorescence. Compromised late apoptotic and necrotic cells exhibit membrane integrity, which results in PI binding and stimulated fluorescence. During a typical Annexin V/PI assay, cells are divided into four populations: alive (Annexin V-negative and PI-negative), early apoptotic (positive for Annexin V and negative for PI), late apoptosis (positive for Annexin V and PI), and necrosis (negative for Annexin V and positive for PI). This enables further analysis and quantification of apoptosis and alleviates the difficulties of distinguishing it from necrotic cell death, which is always useful when studying the cell death mechanisms in response to drug treatments or pathological cases.

The simplicity, speed, and ability to assess large populations of cells make the Annexin V/PI staining assay extremely common. Nevertheless, some caveats in results interpretation should be considered. The primary limitation is that Annexin V staining cannot, by itself, differentiate between apoptosis and other forms of cell death, such as necroptosis, which also involves PS externalization. Furthermore, during late stages of apoptosis, some cells may not have membrane integrity and therefore, based on PI uptake alone, are indistinguishable from necrotic cells. To address these issues, other methods such as caspase activation assays or mitochondrial membrane potential assays or even DNA fragmentation assays are used along with Annexin V/PI staining for a more integrated evaluation of apoptosis (Lugli et al., 2005; Rieger et al., 2010)

To increase its specificity and sensitivity, the Annexin V/PI assay is subject to various modifications. For example, applying Annexin V linked with different fluorophores like A-PC or Alexa Fluor dyes enables multicolor flow cytometry analysis alongside other markers of apoptosis and cell cycle phases. Moreover, Annexin V is applicable in live-cell imaging where the process of apoptosis can be visualized in real-time through time-lapse fluorescence microscopy. This is particularly useful for studies looking at the kinetics of apoptosis in relation to various drug-induced, immune cell interactions, or other rapidly changing cellular events. Another approach uses Annexin V staining with the viability dye 7-aminoactinomycin D (7-AAD) instead of PI. Just like PI, 7-AAD is a DNA-binding dye; however, it is more advanced because it has better spectral qualities, which makes it more ideal for multicolur flow cytometry experiments. Using 7-AAD instead of PI minimizes spectral overlap with probes that are frequently used, so apoptosis can be monitored simultaneously with other cellular parameters like mitochondrial function or oxidative stress (Wilkins et al., 2002).

In practical terms, the Annexin V/PI staining technique is widely applied in both fundamental and clinical research activities. In cancer research, for instance, this assay is used to determine the ability of chemotherapeutic drugs to induce apoptosis in tumor cells in cancer patients. In immunology, Annexin V/PI staining is used to study the apoptotic death of immune cells after infection, inflammation, or immune surveillance is over. There is also the use of the technique to quantify the extent of apoptosis in neurons in models of AD, PD, and stroke in neurodegenerative disease research. The ability of the staining technique to provide quantitative results immediately and the flexibility in application make Annexin V/PI apoptosis detection invaluable in research of this nature in various biological and medical disciplines (Wilkins et al., 2002).

17.6.4 Molecular assays: Western blot, qPCR for apoptotic markers (caspases, Bcl-2 family proteins)

Detection and quantification of apoptotic markers at the gene and protein levels requires molecular assays. Western blotting and quantitative polymerase chain reaction (qPCR) are among the commonly utilized methods. They accurately measure the apoptotic regulatory proteins such as Bcl-2, caspases, and other members of the Bcl-2 family. These techniques aid in understanding the apoptotic signaling pathways, evaluating the effectiveness of drugs, and determining possible therapeutic targets in pathological conditions associated with dysregulated cell death (Vaux and Korsmeyer, 1999; Alshiraihi and Kato, 2023).

Western blots have recently been employed as a protein assessment method to check the presence of apoptosis markers at different stages of the cell death process. This technique aids in resolving cleaved caspases, Bcl-2 family pro-apoptotic as well as antiapoptotic proteins, and many more apoptosis regulatory proteins using the specific monoclonal antibodies to their corresponding proteins. The method starts with cell or tissue protein harvesting and then proceeds to the protein's separation by sodium dodecyl sulphate-polyacrylamide gel electrophoresis (SDS-PAGE). Upon the electrophoresis step, the proteins are deposited on a membrane which is commonly made of nitrocellulose or polyvinylidene difluoride (PVDF). They are then probed with primary antibodies to the specific primary apoptotic markers. The primary antibodies are then exposed with secondary antibodies to fluorescent or chemiluminescent dyes, which enable detection of the proteins of interest. Images captured by the imaging systems reveal the signal, which quantifies the levels of proteomic expression of interest (Alshiraihi and Kato, 2023).

Caspases' activation, which characterizes apoptosis, is easily detectable using western blotting technique. As inactive pro-enzymes, proteolytic cleavage of caspases is associated with their activation. For instance, executioner caspase-3 is produced in the form of a pro-caspase (~32 kDa) and cleaves to the active form (~17 and 12 kDa) during apoptosis. The presence of cleaved antibodies to specific caspase-3 confirms the executioner phase of apoptosis. Detection by western blotting of caspase-9 and caspase-8 that participate in intrinsic and extrinsic pathways of apoptosis, respectively, is also possible. Cleavage of caspase-9 or caspase-8 shows the passing of signals for apoptosis from upper levels (Pessoa, 2021). Another aspect where western blotting is applied in apoptosis research is the analysis of Bcl-2 family proteins that control mitochondrial apoptosis. The Bcl-2 family consists of both pro-apoptotic (Bax, Bak, Bid, Bim, Puma, and Noxa) and antiapoptotic (Bcl-2, Bcl-xL, and Mcl-1) proteins. How the balance of these proteins shifts is often associated with the development of cancer as well as neurodegenerative diseases. For example, Bcl-2 overexpression in cancer facilitates cell survival by preventing MOMP, which leads to cytochrome c release and apoptosome formation. Western blot quantification of Bcl-2 and Bax is indicative of cell apoptotic thresholds and gives an idea of therapeutic efficacy in cancer treatment (Wilkins et al., 2002; Alshiraihi and Kato, 2023).

Besides protein-level apoptosis detection, it is also studied at the level of gene expression using quantitative polymerase chain reaction (qPCR). Apoptotic gene expression can be quantified by mRNA levels, which are the endpoint of the highly sensitive technique of qPCR. It makes possible the study of transcriptional alterations that occur when apoptosis is induced. The technique begins with total RNA isolation from cells or tissues, followed by reverse transcription to produce cDNA. In a real-time PCR system, the cDNA is then amplified with gene-specific primers and fluorescent probes. The intensity of the fluorescence measured is proportional to the amount of target mRNA, thus giving a quantification measurement of gene expression. qPCR is known to be one of the methods that can analyze the expression of pro- and antiapoptotic genes. For example, the ratio of expression of Bax and Bcl-2 genes is often measured to determine sensitivity to apoptosis. A high Bax/Bcl-2 ratio is indicative of cells undergoing apoptosis, and a low ratio is representative of apoptosis-resistant cells, as seen among some chemoresistant cancer cells. Apoptotic effector caspases transcriptional members also include caspase-3, caspase-8, and caspase-9, and their transcriptional levels figure to some extent what stage the apoptotic pathways are in. Increased apoptosis-inducing caspase mRNA levels point to the execution of apoptosis, while lower expression may lend itself toward suggestive phenomena where cancer cells avoid or suppress apoptosis (Vaux and Korsmeyer, 1999).

It is well-known that quantitative PCR is capable of analyzing mark pro and antiapoptotic gene expression. For instance, the apoptotic sensitivity is frequently assessed using the expression ratio of Bax to Bcl-2 genes. A high Bax/Bcl-2 ratio is considered to reflect cells that actively undergo apoptosis, while low ratios signal toward apoptosis-resistant/avoiding cells, as is the case with some chemotherapy-resistant cancer patients. Other transcriptional caspases effector of apoptotic marc caspases has mRNA, such as caspase-3, as well as caspases 8 and 9, and to some degree, their rate of transcription correlates with the stage of execution of the apoptotic pathways. Increased mRNA expression of apoptosis-inducing caspase heralds the completion of apoptosis, while reduced expression may give rise to suggestive phenomena where cancer cells survive or mask apoptotic pathways. Even though advantageous, both western blotting and qPCR lack certain characteristics. In western blotting, protein extraction and antibody specificity, along with detection conditions, need to be optimized for reliable results. Moreover, sample loading variation can interfere with quantification, requiring the use of controls such as β -actin or Glyceraldehyde-3-Phosphate Dehydrogenase (GAPDH). Conversely, the integrity of RNA, the efficiency of primers, and normalization methods can affect the results of qPCR. To reduce variability, housekeeping genes like GAPDH, β-actin, and 18S rRNA are frequently utilized for normalization (Gelles and Edward Chipuk, 2016). In order to enhance the detection of apoptosis, many researchers combine western blotting and qPCR with other approaches. A good example is flow cytometry-based real-time assessment of apoptosis using Annexin V/PI staining and TUNEL assays, which detect DNA fragmentation. Previous all, with the expansion

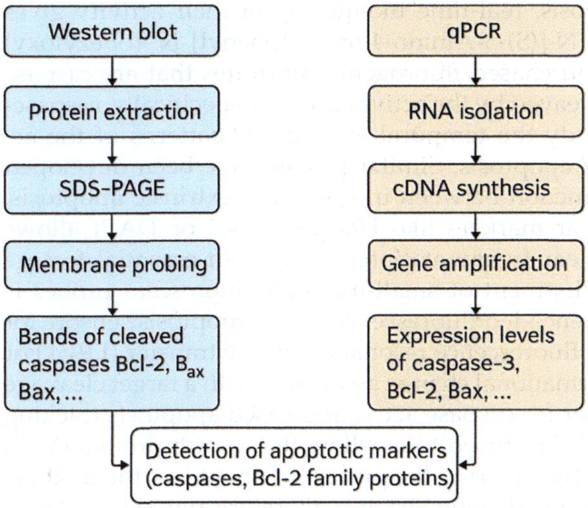

FIGURE 17.2 Schematic representation of protein- and transcript-based detection of apoptotic markers. Western blotting (*left*) involves protein extraction, SDS-PAGE, and membrane probing to visualize bands of cleaved caspases and Bcl-2 family members (Bcl-2, Bax, etc.), whereas qPCR (*right*) includes RNA isolation, cDNA synthesis, and gene amplification to quantify mRNA expression levels of caspase-3, Bcl-2, Bax, and related apoptotic genes, together providing complementary assessment of apoptosis-associated markers.

of novel apoptotic regulators and biomarkers, high-throughput transcriptomic and proteomic methods, such as RNA sequencing and mass spectrometry have furthered the apoptosis research (Fig. 17.2).

17.6.5 Live-cell imaging techniques: Real-time fluorescence microscopy for apoptosis tracking

Live-cell imaging techniques have advanced the research into the apoptosis process by allowing the real-time visualization of cellular activities. Unlike endpoint assays that merely capture a cell's death, live-cell imaging enables the observer to examine the full process of the apoptotic cascade. This technique offers essential information on the time course of apoptosis, the spatial distribution of apoptotic markers, and the influence of drugs in contextually relevant ways. Real-time fluorescence microscopy is one of the most commonly used imaging methods to track the process of apoptosis because of its specificity, sensitivity, and quantification of the apoptotic process at the single-cell level (Vaux and Korsmeyer, 1999). Tracking apoptosis using fluorescence microscopy relies on applying fluorescent probes, dyes, and genetically encoded fluorescent sensors (GEFS) that tag pivotal markers of apoptosis. One of the most important dyes used in early apoptosis detection is the Annexin V conjugated with fluorophores like fluorescein isothiocyanate (FITC) or phycoerythrin (PE). V-substituted anhydrides, adduct Annexin V dyes are now readily available. This reagent is capable of binding to PS, which is a translocated phospholipid located at the inner surface of the bilayer and is transported to the outer surface of the cell's plasma membrane in early apoptosis. This enables the detection of apoptotic cells in real-time as compared to necrotic or viable cells, when used along with PI staining that detects necrotic cells devoid of the cell membrane. The staining assay of Annexin V with PI is common in both static fluorescence microscopy and "slow" movies, tracking changes in the cells to assess the rate of apoptosis occurring in the cells over time (Bastian et al., 2013).

Another method frequently applied in the live-cell imaging of apoptosis is the use of highly specific fluorescent probes that are sensitive to caspases. Given that the activation of caspases marks one of the later key events in apoptosis, real-time monitoring of their activity gives direct proof for the induction of apoptosis. N-[(S)-3-Amino-4-methylphenyl] N′-(obezyloxy) carboxylic acid is one of the more popular purchased fluorogenic substrates that are caspase reporters. It becomes fluorescent after being cleaved by the active caspases, specifically, caspase-3 or 7. With this, researchers are able to study the temporal and spatial patterns of the active caspases during the execution phase of apoptosis. Similar probes have been developed for caspases 8 and 9, which permit the distinction between intrinsic and extrinsic apoptosis. Coupling caspase-sensitive dyes with nuclear markers like Hoechst 33342 or DAPI allows researchers to visualize simultaneously nuclear fragmentation, which is an essential feature of apoptosis (Vermes et al., 1995). The development of small-molecule fluorescent probes is paralleled by the development of genetically encoded fluorescence-based apoptosis sensors for imaging live apoptosis. These sensors utilize fluorescence resonance energy transfer (FRET) or split fluorescent proteins, which have a conformational change associated with a target cleavage by a caspase. One of the examples is the sensor for caspase activation-based apoptosis tracking (SCAT3) that has a cyan fluorescent protein (CFP) fused to a yellow fluorescent protein (YFP) with a caspase-3 cleavage site in between the two. During the process of apoptosis, the linker is cleaved by caspase-3, which results in a FRET signal being lost. It is, therefore, possible to detect apoptosis in real-time. There are also similar FRET-based biosensors for caspase-8 and -9, which are very helpful for studying apoptotic processes in live cells.

One development of interest in live-cell apoptosis imaging is the application of mitochondrial membrane potential ($\Delta\psi m$) dyes like Tetramethyl rhodamine methyl ester (TMRM) and JC-1. Depolarization of the mitochondria, which is a characteristic of intrinsic apoptosis due to the disruption of mitochondria and cytochrome c release, accompanies. The advantage of JC-1 dye is that it emits red fluorescence (healthy mitochondria) when mitochondria are functional and green fluorescence (depolarized mitochondria) when the $\Delta\psi m$ is lost. Thus, it can serve as a marker for real-time induction of apoptosis. This strategy is common in the study of neurodegenerative and cancer diseases for the evaluation of the apoptosis implicating mitochondria (Hunter et al., 2005). The use of confocal and super-resolution microscopy, which offers enhanced spatial resolution and reduced phototoxicity when compared to traditional wide-field fluorescence microscopy, can further improve imaging of live cells undergoing apoptosis. Confocal microscopes' ability to optically section across the z-axis of the cell facilitates the localization of apoptotic markers within subcellular structures. In addition, super-resolution imaging techniques like STED microscopy, along with SIM, have advanced imaging of apoptotic structures, capturing detail with nanometer resolution, such as highly detailed imaging of chromatin condensation, apoptotic body formation, and disassembly of the cytoskeleton. An equally powerful form of real-time fluorescence microscopy in apoptosis research is the utilization of time-lapse imaging. Image capture over prolonged periods allows researchers to view aspects of apoptosis in action, including the sequence of caspase activation, nuclear disintegration, and cell volumetric reduction (shrinkage). Time-lapse imaging enables observation of in vitro drug screening assays where the quantitative comparison of a cell's apoptotic response to various compounds is analyzed. Apoptosis quantification through cell morphological alteration and fluorescence intensity measurement, along with other features such as annotations, is automated using ImageJ, Cell Profiler, or Imaris software (Alshiraihi and Kato, 2023).

Although beneficial, live-cell fluorescence microscopy used for apoptosis tracking comes with certain limitations. Overexposure to fluorescent light can lead to interference with cellular behavior and influence the kinetics of apoptosis through photobleaching and phototoxicity. To alleviate these problems, some researchers turn to low-light imaging modalities such as spinning disk confocal microscopy or two-photon microscopy that minimize photodamage at high imaging resolution. Moreover, the selected fluorescent probes must undergo optimization to reduce nonspecific binding and background fluorescence. Another difficulty in live-cell apoptosis imaging is differentiating apoptosis from cell death through necroptosis or ferroptosis, which may have some overlapping components. In order to solve this problem, most researchers incorporate an apoptosis-specific marker along with either a necroptosis inhibitor known as Necrostatin-1 or a ferroptosis inhibitor referred to as Ferrostatin-1 in order to capture apoptosis-specific phenomena. Additionally, multicolor imaging using multiple fluorescent probes in one imaging session can enable a more holistic view of apoptosis by monitoring caspase activity, mitochondrial movement, and nuclear fragmentation at the same time (Vaux and Korsmeyer, 1999; Hunter et al., 2005). In recent years have observed the implementation of artificial intelligence (AI) and deep learning in various fields, which has further improved apoptosis imaging through automated detection and classification of apoptotic cells in extensive datasets. AI-enabled algorithms perform image analysis by distinguishing apoptotic traits in captured images with high accuracy due to predefined machine learning algorithms, which enhances the exactitude and repetitiveness of the quantification activity of apoptosis. This is especially beneficial in high-throughput drug screening, where big data sets of apoptosis imaging have to be imaged and analyzed within a specified timeframe.

17.6.7 Crosstalk of apoptosis with ferroptosis, necroptosis, and immunogenic cell death

Apoptosis does not just operate through the conventional cellular death pathways. It is, in fact, a regulated cell death (RCD) modality that goes beyond the classical intrinsic and extrinsic pathways and consists of ferroptosis and necroptosis, which share the same upstream signals and molecular switches. One important player in the network is caspase 8. By the fact that its action not only leads to the progression of apoptosis but also to the blocking of necroptosis by cutting kinases RIPK1 and RIPK3 into pieces that perform the opposite, caspase 8 is regarded as a central point in the network. When caspase 8 is genetically or through drugs inhibited, then the signaling of death receptors can transition from apoptosis to necroptosis, indicating how slight changes in apoptotic signaling can redirect cells to different death fates.

Ferroptosis is a newly discovered kind of RCD that is iron-dependent and characterized by the uncontrolled destruction of lipids through peroxidation. It is biochemically different from apoptosis, which is caspase-dependent but it often occurs in stressed tumor cells with apoptosis. Mitochondrial dysfunction, accumulation of ROS, and depletion of glutathione/GPX4 can change the prevailing from apoptotic caspase activation to ferroptotic lipid damage and more and more organelle centered pathways (mitochondria, ER, and lysosomes) are being acknowledged as common decision points that determine either apoptotic or ferroptotic outcomes. This feature is being taken advantage of in therapy as the combination of cytotoxic that induce apoptosis with the ones that promote ferroptosis could make resistant apoptosis refractory tumors susceptible, and also work together with immune checkpoint blockade.

The nonapoptotic types of RCD also have a functional connection with the immunogenic cell death (ICD), which is a way of death of cancer cells that not only kills the cells but also triggers the adaptive antitumor immunity through the same process of dying—release of ATP and HMGB1 and the cell surface being marked by the calreticulin. After a long time, apoptosis was thought to be an immunologically "silent" event; however, certain specific apoptotic and necroptotic situations are now clearly seen as ICD, and even at the early stage of cell death by ferroptosis, cancer cells can be very immunogenic. This, in turn, widens the range of death modalities that can activate the dendritic cells and cytotoxic T lymphocytes. The co-regulation of this crosstalk—by controlling if the cells undergo apoptosis, necroptosis, or ferroptosis, and by enhancing the ICD hallmarks to the maximum—is becoming a main strategy for the immunologically "cold" tumors to be turned into the "hot" ones and for raising immunity to immunotherapy.

17.6.8 Apoptopsis-based strategies

On a clinical level, apoptosis-based therapies are represented by BH3 mimetics such as venetoclax (ABT-199), which have received approval for chronic lymphocytic leukemia and specific acute myeloid leukemia regimens, respectively, thereby establishing the concept that the pharmacological manipulation of apoptotic thresholds is clinically possible, while earlier and broader BCL-2 family inhibitors such as navitoclax and ABT-737 exhibited very strong pro-apoptotic effects but were restricted by on-target side effects like thrombocytopenia. Apart from the BCL-2 pathway, small-molecule Smac mimetics [inhibitor of apoptosis proteins (IAP) antagonists] such as Debio 1143/xevinapant are being tried in human patients, mainly in combination with chemo- and radiotherapy. These agents degrade the cIAP1/2 and XIAP, which not only leads to caspase activation but may even induce necroptosis in apoptosis-resistant cells, thus making tumors susceptible again to cytotoxic agents. Nevertheless, the long-term durability of apoptosis-based therapies is limited due to inherent and developed resistance which is mainly caused by the upregulation of alternative antiapoptotic proteins (for instance MCL-1 or BCL-XL), mutations or downregulation of drug targets, activation of parallel survival pathways (PI3K–AKT and MAPK), tumor diversity, and modified drug transport via ATP-binding cassette pumps, which results in relapse after the initial response. Thus, the present studies are being directed toward the use of rational combination regimens (for instance, BH3 mimetics with hypomethylating agents, or kinase inhibitors, or additional BH3/Smac mimetics), the selection of patients based on markers, the application of less toxic doses, and the taking advantage of communication with other RCD forms such as necroptosis, ferroptosis, and ICD in order to increase both the long-lasting and the safe application of apoptosis-targeted therapies in cancer and perhaps in aging-related contexts.

Despite these promising approaches, targeting DNA damage and apoptotic pathways in aging tissues faces important limitations. Many interventions that enhance DNA repair, activate p53, or clear senescent cells show efficacy mainly in short-lived animal models, and their benefits in humans remain uncertain. Chronic manipulation of DNA damage responses carries safety concerns, including impaired tumor surveillance, hematopoietic toxicity, and off-target genomic alterations with CRISPR-based strategies. Senolytic and pro-apoptotic therapies may also disrupt tissue homeostasis by depleting stem and progenitor pools or exacerbating inflammation in aged organs. These translational barriers highlight the need for precise dosing, tissue-specific delivery, and long-term follow-up before such antiaging strategies can be widely applied in the clinic.

17.7 Conclusion

Apoptosis is a highly controlled cellular process which is often referred to as programmed cell death and is significantly important in disease prevention, maintenance of overall cellular fitness, and development of an organism. A failure to regulate apoptosis appropriately leads to excessive cell death or inadequate apoptotic signaling, which is characteristic of many pathological processes, including cancer, neurodegenerative disorders, autoimmune diseases, and cardiovascular issues. Understanding of apoptosis opened our eyes to the molecular basis of these pathologies, which allowed the design of therapies to circumvent the many issues related to the uncontrolled apoptosis. While the research on apoptosis has expanded the understanding of the biology of the cell profoundly, it has also raised great interest in other fields like regenerative medicine, immunotherapy, and orthopedic oncology. In the context of cancer, the failure to undergo apoptosis is used as a mechanism employed by tumor cells to enhance their chances of survival and subsequently proliferate uncontrollably while developing resistance toward conventional therapies. The discovery of important apoptotic regulators like Bcl-2 family proteins, IAPs, and p53 has led to the design of new anticancer interventions. Apoptosis-inducing agents like BH3 mimetics (e.g., Venetoclax) and death receptor agonists have been designed to selectively kill cancerous tissues without harming normal tissues. In addition, chemotherapy and radiotherapy mainly work through the induction of apoptosis in tumor cells, further emphasizing the fundamental role of apoptosis modulation in the treatment of cancer. Still, a major issue in the treatment of cancer is therapy resistance, because cancer cells, or the tumors themselves, use mutations that cause them to be unresponsive to the apoptotic signals. Further research strives to create new treatment combinations with better abilities to induce apoptosis in aggressive malignancies. On the other hand, increased apoptosis results in tissue damage and a decline in organ function in neurodegenerative conditions, which include Alzheimer's, Parkinson's, and Huntington's diseases. As postmitotic cells, neurons are especially sensitive to apoptotic stimuli from mitochondrial damage, oxidative stress, and protein aggregates. Preclinical studies have shown promise using caspase inhibitors and neuroprotective compounds aimed at preserving mitochondrial function. Although translation to the clinic has proven difficult because of the multifaceted nature of the pathology of neurodegenerative diseases, these approaches are still very relevant in neurobiology research. Parallel to these, in cardiovascular diseases, there is abundant evidence that excessive apoptosis of cardiomyocytes leads to heart failure, myocardial infarction, and atherosclerosis. Interventions directed at apoptosis mediators, particularly Bcl-2 proteins and caspases, may provide novel approaches toward the prevention and treatment of heart disease and thus improve survival and quality of life.

Research in stem cell therapy and regenerative medicine is poised to benefit from apoptosis investigation far beyond disease pathogenesis. The selective removal of unwanted or damaged cells through controlled induction of apoptosis greatly facilitates functional integration in tissue engineering, where transplanted tissues are provided. Moreover, the removal of apoptotic cells by phagocytes is essential for immune tolerance and inflammation resolution, which may be useful in treating autoimmune diseases and inflammatory disorders. Researchers are proposing novel methods of tissue regeneration, immune response alteration, and precision medicine with the use of mechanistic approaches of apoptosis for tissue-directed purposes. It can be expected that there would be significant changes in apoptosis-based clinical usages. The rise of personalized medicine through genomic and proteomic profiling allows for the focus on particular apoptotic pathways in patients, which can reduce side effects while enhancing treatment outcomes. Manipulation of apoptotic mediators in the context of a selective therapeutic intervention, such as precision therapies, enables direct gene alterations through the development of editing technologies like CRISPR-Cas9. In addition, the application of AI technologies to the dissection of apoptosis is facilitating the identification of new drugs through high-throughput screening of compounds that modulate apoptosis processes. Future studies will further investigate the complex interactions between the processes of apoptosis and necrosis and necroptosis and ferroptosis for the purpose of crafting more sophisticated treatment techniques. The implementation of new imaging technologies, bioinformatics, and synthetic biology integrated with stronger insights into apoptotic signals will most certainly improve our capacity to control apoptosis for positive clinical outcomes. While new therapies that target apoptosis make their way through clinical trials, the implementation of research surrounding apoptosis into real-world treatments for cancers, neurodegenerative disorders, cardiovascular diseases, and immune-related diseases are still very possible and hopeful in biomedical science.

References

Alshiraihi, I., Kato, T.A., 2023. Apoptosis detection assays. Methods Mol. Biol. 2519, 53–63. https://doi.org/10.1007/978-1-0716-2433-3_6.

Baar, M.P., Brandt, R.M.C., Putavet, D.A., Klein, J.D.D., Derks, K.W.J., Bourgeois, B.R.M., et al., 2017. Targeted apoptosis of senescent cells restores tissue homeostasis in response to chemotoxicity and aging. Cell 169 (1), 132–147. https://doi.org/10.1016/j.cell.2017.02.031.

Bastian, A.M., Yogesh, T.L., Kumaraswamy, K.L., 2013. Various methods available for detection of apoptotic cells—a review. Indian J. Cancer 50 (3), 274–283. https://doi.org/10.4103/0019-509X.118720.

Bellamy, C.O.C., 1997. p53 and apoptosis. Br. Med. Bull. 53 (3), 522–538. https://academic.oup.com/bmb/article/53/3/522/285145.

Denton, D., Kumar, S., 2015. Terminal deoxynucleotidyl transferase (TdT)-mediated dutp nick-end labeling (TUNEL) for detection of apoptotic cells in drosophila. Cold Spring Harb. Protoc. 2015 (6), 568–571. https://doi.org/10.1101/pdb.prot086199.

Fleisher, T.A., 1997. Apoptosis. In: Annals of Allergy, Asthma and Immunology, 78. American College of Allergy, Asthma and Immunology, pp. 245–250.

Garrido, C., Galluzzi, L., Brunet, M., Puig, P.E., Didelot, C., Kroemer, G., 2006. Mechanisms of cytochrome c release from mitochondria. Cell Death Differ. 13 (9), 1423–1433. https://doi.org/10.1038/sj.cdd.4401950.

Gelles, J.D., Edward Chipuk, J., 2016. Robust high-throughput kinetic analysis of apoptosis with real-time high-content live-cell imaging. Cell Death Dis. 7 (12). https://doi.org/10.1038/cddis.2016.332.

Gibellini, L., Moro, L., 2021. Programmed cell death in health and disease. Cells 10 (7). MDPI https://doi.org/10.3390/cells10071765 .

Gillies, L.A., Kuwana, T., 2014. Apoptosis regulation at the mitochondrial outer membrane. J. Cell. Biochem. 115 (4), 632–640. https://doi.org/10.1002/jcb.24709.

Hajnóczky, G., Davies, E., Madesh, M., 2003. Calcium signaling and apoptosis. Biochem. Biophys. Res. Commun. 304 (3), 445–454. https://doi.org/10.1016/S0006-291X(03)00616-8.

Hunter, A.L., Choy, J.C., Granville, D.J., 2005. Detection of apoptosis in cardiovascular diseases. Methods Mol. Med. 112, 277–289. https://doi.org/10.1385/1-59259-879-x:277.

Huppertz, B., Herrler, A., 2005. Regulation of proliferation and apoptosis during development of the preimplantation embryo and the placenta. Birth Defects Res C Embryo Today 75 (4), 249–261. https://doi.org/10.1002/bdrc.20056.

Jain, A., Lamark, T., Sjøttem, E., Larsen, K.B., Awuh, J.A., Øvervatn, A., McMahon, M., Hayes, J.D., Johansen, T., 2010. p62/SQSTM1 is a target gene for transcription factor NRF2 and creates a positive feedback loop by inducing antioxidant response element-driven gene transcription. J. Biol. Chem. 285 (29), 22576–22591. https://doi.org/10.1074/jbc.M110.118976.

Jourdain, A., Martinou, J.C., 2009. Mitochondrial outer-membrane permeabilization and remodelling in apoptosis. Int. J. Biochem. Cell Biol. 41 (10), 1884–1889. https://doi.org/10.1016/j.biocel.2009.05.001.

Kari, S., Subramanian, K., Altomonte, I.A., Murugesan, A., Yli-Harja, O., Kandhavelu, M., 2022. Programmed cell death detection methods: a systematic review and a categorical comparison. Apoptosis 27 (7–8), 482–508. https://doi.org/10.1007/S10495-022-01735-Y.

Kutuk, O., Basaga, H., 2010. Apoptotic pathways in mitochondria. Apoptosis. Springer. https://doi.org/10.1201/9781439845431-3.

Kyrylkova, K., Kyryachenko, S., Leid, M., Kioussi, C., 2012. Detection of apoptosis by TUNEL assay. Methods Mol. Biol. 887, 41–47. https://doi.org/10.1007/978-1-61779-860-3_5.

Lazar, N., Manara, C., Navarro, S., Bleau, A.-M., Llombart-Bosch, A., Scotlandi, K., et al., 2007. Domain-specific CCN3 antibodies as unique tools for structural and functional studies. J. Cell Commun. Signal. 1 (2), 91–102. https://doi.org/10.1007/s12079-007-0009-8.

Lee, K.H., Kang, T.B., 2019. The molecular links between cell death and inflammasome. Cells 8 (9). https://doi.org/10.3390/cells8091057.

Lugli, E., Troiano, L., Ferraresi, R., Roat, E., Prada, N., Nasi, M., et al., 2005. Characterization of cells with different mitochondrial membrane potential during apoptosis. Cytometry Part A 68 (1), 28–35. https://doi.org/10.1002/cyto.a.20188.

Mai, Z., Liu, H., 2009. Boolean network-based analysis of the apoptosis network: Irreversible apoptosis and stable surviving. J. Theor. Biol. 259 (4), 760–769. https://doi.org/10.1016/j.jtbi.2009.04.024.

Orth, J.D., Loewer, A., Lahav, G., Mitchison, T.J., 2012. Prolonged mitotic arrest triggers partial activation of apoptosis, resulting in DNA damage and p53 induction. Mol. Biol. Cell 23 (4), 567–576. https://doi.org/10.1091/mbc.E11-09-0781.

Pessoa, J., 2021. Live-cell visualization of cytochrome c: a tool to explore apoptosis. Biochem. Soc. Trans. 49 (6), 2903–2915. https://doi.org/10.1042/BST20211028.

Prahallad, A., Bernards, R., 2016. Opportunities and challenges provided by crosstalk between signalling pathways in cancer. In: Oncogene, 35. Nature Publishing Group, pp. 1073–1079.

Riccardi, C., Nicoletti, I., 2006. Analysis of apoptosis by propidium iodide staining and flow cytometry. Nat. Protoc. 1 (3), 1458–1461. https://doi.org/10.1038/NPROT.2006.238.

Rieger, A.M., Hall, B.E., Luong, L.T., Schang, L.M., Barreda, D.R., 2010. Conventional apoptosis assays using propidium iodide generate a significant number of false positives that prevent accurate assessment of cell death. J. Immunol. Methods 358 (1–2), 81–92. https://doi.org/10.1016/J.JIM.2010.03.019.

Robert, A., Weinberg, R.A.W., 2007. The Biology of Cancer, first ed. W.W. Norton & Company. https://doi.org/10.1201/9780203852569.

Save, V., Hall, P.A., Coates, P.J., 2004. Detecting and quantifying apoptosis in tissue sections. Methods Mol. Biol. 282, 67–84. https://doi.org/10.1385/1-59259-812-9:067.

Schlatter, R., Schmich, K., Vizcarra, I.A., Scheurich, P., Sauter, T., Borner, C., et al., 2009. ON/OFF and beyond—a boolean model of apoptosis. PLoS Comput. Biol. 5 (12). https://doi.org/10.1371/journal.pcbi.1000595.

Sheikh, M.S., Fornace, A.J., 2000. Role of p53 family members in apoptosis. J. Cell. Physiol. 182 (2), 171–181. https://doi.org/10.1002/(SICI)1097-4652(200002)182:2⟨171::AID-JCP5⟩3.0.CO;2-3.

Sorice, M., 2022. Crosstalk of autophagy and apoptosis. Cells 11 (9). MDPI https://doi.org/10.3390/cells11091479 .

Vaux, D.L., Korsmeyer, S.J., 1999. Cell death in development. Cell 96 (2), 245–254. https://doi.org/10.1016/S0092-8674(00)80564-4.

Vermes, I., Haanen, C., Steffens-Nakken, H., Reutellingsperger, C., 1995. A novel assay for apoptosis flow cytometric detection of phosphatidylserine expression on early apoptotic cells using fluorescein labelled Annexin V. J. Immunol. Methods 184 (1), 39–51. https://doi.org/10.1016/0022-1759(95)00072-I.

White, M.K., Cinti, C., 2004. A morphologic approach to detect apoptosis based on electron microscopy. Methods Mol. Biol. 285, 105–111. https://doi.org/10.1385/1-59259-822-6:105.

Wilkins, R.C., Kutzner, B.C., Truong, M., Sanchez-Dardon, J., McLean, J.R.N., 2002. Analysis of radiation-induced apoptosis in human lymphocytes: flow cytometry using Annexin V and propidium iodide versus the neutral comet assay. Cytometry 48 (1), 14–19. https://doi.org/10.1002/CYTO.10098.

Zhang, F., Cheng, N., Du, J., Zhang, H., Zhang, C., 2021. MicroRNA-200b-3p promotes endothelial cell apoptosis by targeting HDAC4 in atherosclerosis. BMC Cardiovasc. Disord. 21 (1). https://doi.org/10.1186/s12872-021-01980-0.

Zhao, Q., Feng, H., Yang, Z., Liang, J., Jin, Z., Chen, L., et al., 2022. The central role of a two-way positive feedback pathway in molecular targeted therapies-mediated pyroptosis in anaplastic thyroid cancer. Clin. Transl. Med. 12 (2). https://doi.org/10.1002/ctm2.727.

CHAPTER

18

Phytochemical modulation of DNA repair: Mechanisms, epigenetics, and therapeutic potential

Prasad Andhare[a], *Khushbu Verma*[a], *Bhargavi Sonavane*[b], *Bhavtosh Kikani*[b], *Niraj Kumar Singh*[c,d] *and Datta Madamwar*[b]

[a]Department of Life Sciences, School of Science, GSFC University, Vadodara, Gujarat, India [b]Department of Biological Sciences, P. D. Patel Institute of Applied Sciences, Charotar University of Science and Technology, Anand, Gujarat, India [c]Department of Science and Technology, Gujarat Biotechnology Research Centre (GBRC), Government of Gujarat, Gandhinagar, Gujarat, India [d]School of Forensic Sciences, National Forensic Sciences University (NFSU), Nagpur, Maharashtra, India

18.1 Introduction: Importance of genome stability

Genome integrity must be properly maintained to ensure that cellular systems function correctly and that genetic information is transmitted. It extends to the preservation of the sequence of DNA, chromosomal architecture, and the epigenome during the cell cycle and over the duration of the life of an organism (Soto-Palma et al., 2022). The human genome is subjected continuously to a variety of destabilizing agents such as replication errors, oxidative stress, and genotoxic environmental agents. Unless these disruptions are corrected, they lead to mutations, chromosomal aberrations, or epigenetic deregulation that are drivers for tumorigenesis, degenerative diseases, and premature aging. Therefore genome health is far more than just a molecular matter; indeed, genome integrity plays a crucial role around which physiological soundness and evolutionary fitness revolve (Gauthier and Comaills, 2021). In view of the pervasive nature of genomic insults, cellular organisms have evolved complex systems for monitoring and

Genome Damage: Repair and Adaptation Across Life.
DOI: https://doi.org/10.1016/B978-0-443-42857-9.00023-2

repairing damaged DNA. These include DNA-repair pathways, cell-cycle checkpoints, telomere maintenance, and chromatin-remodeling systems that collectively comprise an interrelated regulatory network for detecting and repairing DNA damage (Sheikh et al., 2025). In higher eukaryotes, genome integrity is additionally regulated by transcriptional accuracy, noncoding RNAs, and metabolic control. The inactivation of these protective mechanisms frequently results in genomic instability syndromes, diseases characterized by an increased rate of mutations and susceptibility to cancer or neurodegenerative diseases. Thus we describe genome stability as a dynamic and multidimensional process that lies at the intersection of molecular biology, cell physiology, and disease pathogenesis (Zaher and Mosammaparast, 2024).

18.1.1 Overview of DNA damage sources: Endogenous (reactive oxygen species, replication errors) and exogenous (ultraviolet, ionizing radiation, and chemicals)

Genomic injury originates from a wide variety of endogenous and exogenous sources, continuously challenging genomic stability. The most abundant endogenous sources are reactive oxygen species (ROS) that accumulate during normal cellular respiration and metabolic activities (Nath and Roy, 2021). These reactive intermediates generate oxidative adducts, e.g., 8-oxo-guanine, single-strand breaks, and abasic sites. Furthermore, spontaneous hydrolytic reactions, such as the deamination of cytosine to uracil and depurination, generate miscoding and noncoding lesions (Li et al., 2022). Mistakes in DNA replication are commonly eliminated by proofreading and mismatch (MMR) repair enzymes, but can bypass such corrections, causing insertion, deletion, or mispairing of bases that reduce genome fidelity (Lózsa et al., 2023).

The exogenous factors add to the weight of DNA damage by exposure to ambient ultraviolet (UV) radiation, ionizing radiation (IR), and genotoxic chemicals. The UV-radiation creates a bulky cyclobutane pyrimidine dimer and 6-4 photoproducts, which distort the DNA helix and ultimately block replication and transcription (Wang et al., 2022). IR produces double-strand breaks and base modifications by radiolysis of water. In contrast, alkylation or intercalating lesions are introduced by industrial pollutants such as polycyclic aromatic hydrocarbons and chemotherapeutic agents by modulation of base pairs and interference with chromatin architecture. These insults will inevitably stress the cells by the combined effect and thereby require strong cellular surveillance and repair mechanisms to avoid damaging effects (Wadgaonkar, 2024). Schematic representation of major sources of DNA damage (endogenous and exogenous) and the corresponding repair pathways (BER, NER, MMR, HR, NHEJ) is shown in Fig. 18.1.

18.1.2 Role of DNA repair in preventing disease

DNA-repair pathways are central to protecting the integrity of the genome by counteracting mutagenic lesions and preventing disease. These pathways encompass base excision repair (BER), nucleotide excision repair (NER), MMR, homologous recombination (HR), and nonhomologous end joining (NHEJ). Each pathway is designed to repair some type of DNA damage, and their concerted and coordinated action is the driving force that ensures DNA replication is more than 99% accurate in maintaining genomic information (Oh and Myung, 2022). For example, BER deals with oxidized and small base lesions, and HR and NHEJ handle double-strand breaks produced by replication stress or radiation. This functional specialization underscores the variety present in DNA damage response (DDR) networks, as well as their conserved nature at the origin of their essential functions in cellular homeostasis

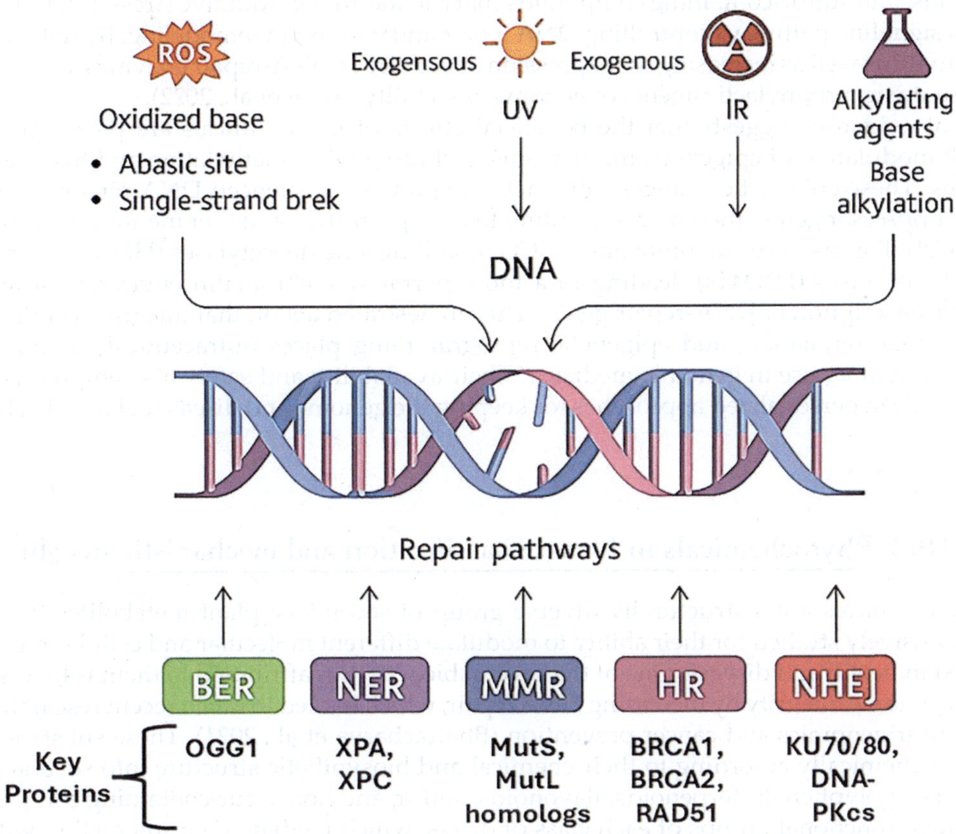

FIGURE 18.1 Overview of DNA damage and repair pathways.

(Wang et al., 2023; Prime et al., 2024). When these repair pathways are impaired by genetic mutations or epigenetic downregulation, the effects can be observed as human diseases ranging in severity. Mutated genes such as those encoding BRCA1/2, MLH1, or ataxia-telangiectasia mutated (ATM) result in hereditary cancer predisposition in breast, ovarian, and colorectal cancers (Catalano et al., 2022). In addition, the compromised DNA repair is involved in premature aging, immune deficiency, and neurodegenerative disorders in the context of xeroderma pigmentosum, ataxia-telangiectasia (A-T), and Cockayne syndrome. Therefore the fidelity and efficiency of DNA repair are not only at the heart of genome maintenance, but they are also closely associated with pathogenesis and development of most human diseases (Schulz et al., 2023).

18.1.3 Emergence of nutraceuticals in genomic maintenance

Advances in molecular nutrition and biomedical studies have drawn greater attention to the positive effects of dietary phytochemicals, called nutraceuticals, on genome stability. Nutraceuticals are bioactive compounds that are derived from a wide range of natural food products such as fruits, vegetables, herbs, and spices (AlAli et al., 2021). These include polyphenols, flavonoids, terpenoids, and sulfur-containing compounds that can modulate oxidative stress, inflammation, and cell signaling pathways controlling DNA repair and chromatin remodeling. By reducing the ROS activity as well as increasing the expression or activity of DNA-repair enzymes, these agents act as nontoxic, prophylactic agents of genomic instability (Yuan et al., 2022).

Recent evidence suggests that the beneficial effects of nutraceuticals are partly mediated through modulation of epigenetic mechanisms, including DNA methylation and histone modifications. These epigenetic changes can reactivate previously silenced DNA-repair genes and tumor-suppressor genes, thereby contributing to their protective and therapeutic effects (Aslam et al., 2024). For instance curcumin and EGCG inhibit histone deacetylases (HDACs) and DNA methyltransferases (DNMTs), leading to a more permissive chromatin environment and increased transcription of DNA-repair genes. This orchestrated action that integrates antioxidant defense, gene regulation, and epigenetic reprogramming places nutraceuticals as bona fide agents for adjunct use in genomic medicine. Their availability and safety also emphasize their possible role in personalized approaches for keeping the genome and disease at bay (Hsieh et al., 2025).

18.2 Phytochemicals in focus: Classification and mechanistic insight

Phytochemicals are a structurally diverse group of secondary plant metabolites that have been extensively studied for their ability to modulate different molecular and cellular pathways involved in health and disease. One of the several bioactivities attributed to them is their ability to affect genome integrity by increasing DNA repair, which has led to much recent research in the field of nutrigenomics and cancer prevention (Bhattacharya et al., 2021). These substances are classified chemically according to their chemical and biosynthetic structure into several major groups, i.e., polyphenols, terpenoids, flavonoids, sulfur, and nonsulfur-containing compounds. The unique functional groups of each class of drugs, which mediate their interaction with cell signaling pathways, transcriptional apparatus, or factors regulating epigenetic control of DNA-repair mechanisms, are not mutually exclusive (El Omari et al., 2021).

Mechanistically, phytochemicals interact in genome maintenance through modulation of ROS, enhance the expression of DNA-repair genes, stimulate checkpoint kinases, and remodel chromatin structure. These effects can also be direct, through binding to DNA-repair proteins, or indirect by the modulation of redox-sensitive transcription factors, including NF-κB, Nrf2, and p53 (Dong et al., 2025). Moreover, phytochemicals tend to possess pleiotropic behavior by modulating diverse stress-response pathways and mediating cellular responses to genotoxic insults. These versatile activities have been classified as prospective coadjuvants in therapy as well as dietary factors with potential disease preventive roles. The subsequent subchapters are devoted to the representative members of each phytochemical class, indicating DNA-repair modulation (Chatterjee et al., 2025).

18.2.1 Polyphenols: Curcumin, rosmarinic acid, and p-coumaric acid

Polyphenolic compounds are one of the most widely studied classes of phytochemicals, due to their strong antioxidant, anti-inflammatory, and anticancer activities. With several hydroxylated phenyl rings as their structural feature, polyphenols such as curcumin, rosmarinic acid, and p-coumaric acid provide protection to the genome by regulating oxidative stress, DDR signaling, and epigenetic pathways (Rudrapal et al., 2022). Curcumin, a diarylheptanoid from *Curcuma longa*, serves as a ROS scavenger and modulates the expression of molecules of the BER and HR pathways. It also upregulates tumor-suppressor genes such as p53 and BRCA1 and downregulates prosurvival signaling by inhibiting NF-κB and accordingly offers both protection against alkylation damage and sensitization of cancer cells to DNA-damaging agents (Svolacchia et al., 2023).

Rosmarinic acid, a phytochemical present in herbs such as rosemary and basil, imparts its genome-protective effects mainly because of the modulation of oxidative and inflammatory signals. It upregulates antioxidant enzymes such as superoxide dismutase and glutathione peroxidase to neutralize ROS-induced DNA injuries (Czerwińska and Radziejewska, 2024). Additionally, rosmarinic acid is shown to inhibit DNA fragmentation and chromosomal breaks in several in vitro and in vivo bioassays. Likewise, p-coumaric acid, a hydroxycinnamic acid derivative in cereals and fruits, exerts radioprotective effects by retarding lipid peroxidation and DNA strand breaks induced by IR stress. Together, these polyphenols are capable of targeting both upstream oxidative insults and downstream DNA-repair pathways, demonstrating their pleiotropic genomic stability modulating potential (Ceylan et al., 2025).

18.2.2 Terpenoids: Limonene

Isoprene-unit-based terpenoids are one of the largest groups of phytochemicals and are ubiquitous among plant species. Limonene, a monocyclic monoterpene present mainly in the peel of citrus fruits, has also become a promising candidate for chemopreventive and DNA-protective interventions (Gbadebo et al., 2025). It functions through modulation of phase I and phase II detoxifying enzymes, inhibition of DNA adduct formation, and activation of the ATM/ATR checkpoint pathway after exposure of DNA under stress conditions (Jurkovicova et al., 2022). Thereby, limonene increases the expression of repair enzymes OGG1 and XRCC1, which are involved in BER and reduce the level of mutagenic lesions. It also affects mitochondrial ROS generation and cytochrome c release and protects the cells by regulating cellular redox homeostasis, which is important for preventing oxidative DNA damage (Di Guilmi et al., 2023).

Besides antioxidant activity, limonene has been shown to regulate epigenetic gene regulation, predominantly via histone acetylation and DNA methylation. Limonene has been demonstrated to modulate chromatin structure by increasing acetylation of histones H3 and H4, thereby enhancing the expression of genes related to DNA repair and cell-cycle control (Sharma et al., 2024). Moreover, limonene's effect on transcriptional factors, such as AP-1 and Nrf2, also contributes to the mechanism by which it positively regulates the expression of genes associated with cellular defenses. These multifactorial effects emphasize the limonene's role as a dietary phytochemical with genome protection activity, particularly in the case of oxidative stress-associated genotoxicity (Lee et al., 2021).

18.2.3 Flavonoids: Hesperidin and quercetin

Flavonoids, as a subgroup of polyphenols with a flavone backbone, are rich in fruits, vegetables, and tea. Hesperidin of the flavanone glycoside group of phytonutrients in citrus fruits is reported to regulate oxidative stress and inflammation, which are two of the pivotal factors contributing to DNA damage and chromosomal instability (Ahmad et al., 2023). Apart from that, it also possesses radio-protective and anti-mutagenic effects through reducing intracellular ROS generation and increasing the activity of antioxidant enzymes. Hesperidin also modulates the expression of DNA-repair genes and has been shown to prevent γ-H2AX formation (a marker of DNA double-strand breaks), thereby indicating a potential role in avoiding or mitigating genomic damage (Jiang et al., 2021; Rivera-Yanez et al., 2021).

Quercetin, one of the most common dietary flavonols, has demonstrated multitargeted bioactivities in cancer, inflammation, and oxidative damage. Mechanistically, it targets DDR central nodes such as ATM/ATR signaling, p53 stabilization, and PARP1 activation (Ye et al., 2023). Furthermore, it has been reported that quercetin normalizes defective NER in cells irradiated with UV and sensitizes cancer cells to chemotherapy by blocking the repair of DNA strand breaks occurring after treatment with these drugs. It also disrupts epigenetic factors such as HDACs, favoring repair gene transcription. Collectively, Hesperidin and quercetin provide powerful dietary control of DNA repair that has relevance for either disease prevention or even therapeutic synergy (Zhu et al., 2025).

18.2.4 Sulfur-containing compounds: Allicin

Sulfur-containing phytochemicals, commonly those that are obtained from *Allium* plants, have distinctive chemical reactivity that allows them to modulate both cellular redox status and gene expression (Gorrepati et al., 2024). Allicin, a thiosulfinate compound derived from alliin in crushed garlic, possesses powerful antimicrobial, anti-inflammatory, and anticancer activities. It plays a role in maintaining the stability of the genome by controlling the oxidative equilibrium, activating phase II detoxification enzymes, and inducing DNA-repair genes (Pandey et al., 2023). Allicin can increase glutathione (GSH) and scavenges ROS, resulting in the inhibition of oxidative modifications of bases and fragmentation of the DNA strands. In addition, its electrophilic character may allow it to react with the thiol groups of DNA-repair enzymes, thereby changing their catalytic efficiency and the stability of repair complexes (Sarvizadeh et al., 2021).

Recent studies show that allicin also modulates gene expression by epigenetic mechanisms, such as miRNA alteration, the suppression of proinflammatory gene transcription, and putative inhibition of DNMTs (Shoaib et al., 2023). This unique bipotent nature, which combines redox manipulation with epigenetic inflection, makes allicin a very special group of phytochemicals with systemic action on genome defense activities. In animal models, dietary allicin decreased DNA fragmentation in the liver of carcinogen-treated animals and enhanced the repair of alkylated bases in colonic epithelial cells. These data underscore the functional significance of sulfur-containing compounds in both prevention and regulation of the DNA-repair network (Zhou et al., 2022).

18.3 Mechanisms of DNA-repair modulation by phytochemicals

The influential role of phytochemicals on DNA repair is modulated based on their ability to interfere with ROS generation, cellular redox, signaling pathways, and genomic regulators. These bioactive phytochemicals mediate pleiotropic effects, including those involved in scavenging of ROS, modulation of DDR pathways, altered gene transcription, and communication

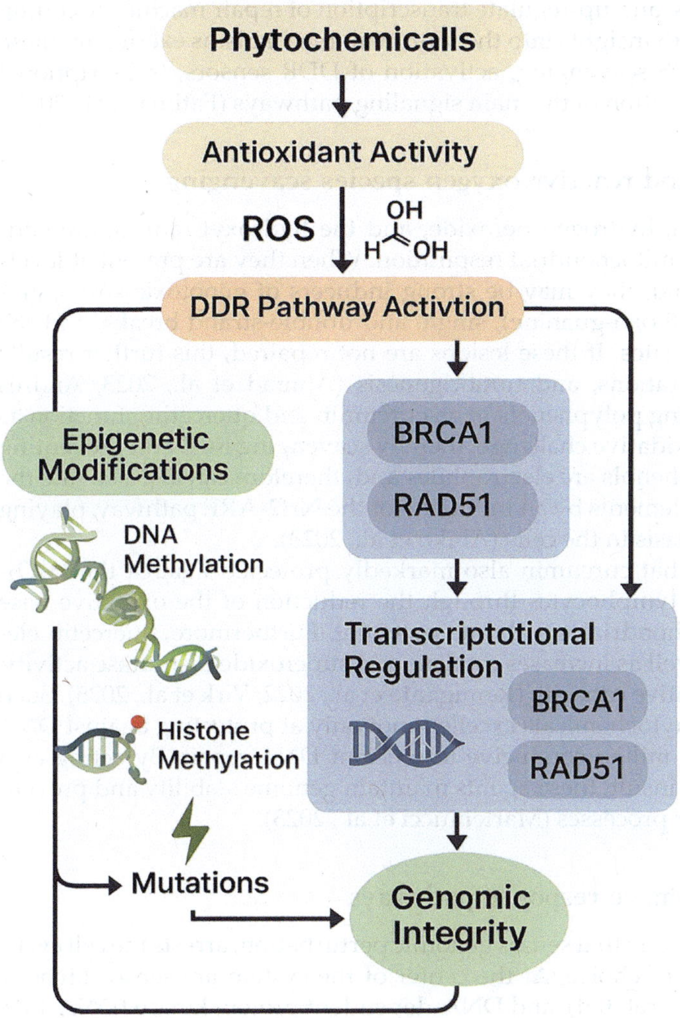

FIGURE 18.2 Mechanisms of DNA-repair modulation by phytochemicals.

with intracellular signaling pathways (Murai and Matsuda, 2023). Integrated schematic of phytochemical actions on genome maintenance is represented in Fig. 18.2. These interventions not only suppress DNA lesions, but also enhance the efficient activation of repair machineries, BER, NER, HR, and NHEJ. Phytochemicals also indeed impact these bioprocesses, which have potential from the point of view of chemoprevention, radioprotection, and therapeutic sensitization (Rall-Scharpf et al., 2025).

Mechanisms of action of phytochemicals are not limited to their direct actions; instead, they cross-link with the endogenous antioxidant enzymes, epigenetic regulators, and DNA-binding proteins. Their modulatory function is concentration, chemical structure, and cellular redox status dependent (Khan et al., 2024). A few phytochemicals, including curcumin, quercetin, allicin, and limonene, have been found to mediate the recruitment of repair proteins to the damage site, up-regulate chromatin openness, and up-regulate transcription of repair machinery coding genes. The subheadings offer in-depth insights into the four main mechanisms exerted by these phytochemicals on DNA repair: ROS scavenging, activation of DDR sensors, transcriptional regulation of repair genes, and modulation of the main signaling pathways (Fatima et al., 2021).

18.3.1 Antioxidative defense and reactive oxygen species scavenging

ROS, including superoxide anion, hydrogen peroxide, and the hydroxyl radical, are end-products of aerobic metabolism and mitochondrial respiration. When they are present at levels exceeding the physiological threshold, they may be strong inducers of genotoxic stress, giving rise to base modifications (e.g., 8-oxo-guanine), single and double-strand breaks, and the formation of apurinic/apyrimidinic sites. If these lesions are not repaired, this further results in mutagenesis, chromosomal aberrations, and tumorigenesis (Ahmad et al., 2023; Andrés et al., 2023). Phytochemicals, including polyphenols (e.g., curcumin and quercetin), function as electron donors in response to this oxidative challenge, thereby scavenging ROS and preventing interaction with nucleic acids. Polyphenols are electrophiles and, therefore can also activate the expression of antioxidant response elements by an induction of the Nrf2-ARE pathway, playing a role in maintaining redox homeostasis in the cells (Abiko et al., 2023).

Experimental evidence showed that curcumin also markedly protected against the H_2O_2-induced DNA strand breaks in the lymphocytes through the reduction of the oxidative base lesions and the restoration of mitochondrial membrane potential. Furthermore, quercetin elevates cellular glutathione levels as well as increases catalase and superoxide dismutase activity, strengthening intracellular antioxidative capacity (Remigante et al., 2022; Virk et al., 2025). Such properties make ROS-scavenging phytochemicals excellent not only at protecting against DNA damage but also creating a cellular milieu conducive to efficient DNA repair. By acting as a scavenger to eliminate the oxidative insult, these agents maintain genome stability and prevent the mobilization of mutagenic repair processes (Martemucci et al., 2023).

18.3.2 Modulation of DNA damage response pathways

DDR is an organized signaling cascade that senses genomic perturbation, arrests the cell cycle, and prompts the repair mechanism of choice. At the center of the system are sensor kinases, including ATM, ATR (ATM and Rad3-related), and DNA-dependent protein kinase (DNA-PK), which, upon detection of DNA strand breaks, activate kinase signaling cascades (Molinaro et al., 2021). Phytochemicals modify the DDR by induction of these kinase activities, promotion of phosphorylation of their downstream targets, such as CHK1/CHK2, and increasing the local recruitment of DNA-repair proteins. Terpenoids such as limonene and flavonoids such as kaempferol and apigenin have been demonstrated to sensitize cells to genotoxic stress through regulation of these pathways (Manna et al., 2020).

For example, limonene increases ATM phosphorylation upon IR treatment, allowing HR repair to work effectively. These responses are associated with the elevation of repair proteins, such as RAD51 and BRCA2, after foci formation of repair proteins, suggesting that the repair machinery for strand breaks is further facilitated (Ghosh and Ghosh, 2021). Also, flavonoids may help in maintaining the MRN complex (MRE11-RAD50-NBS1) at double-strand break sites to facilitate ATM timely activation, leading to the engagement of the checkpoint signaling pathway. These data indicate that phytochemicals do not simply protect DNA from damage but coordinately enhance cellular repair processes, enhancing repair fidelity and ensuring prevention of erroneous repair events (McCarthy-Leo et al., 2022).

18.3.3 Regulation of DNA-repair gene expression

Phytochemicals not only act on the initial repair signaling, but they also modulate gene expression patterns that direct long-term genomic maintenance. This is regulated by and depends on transcription factors, promoter elements, and the presence of chromatin modifiers. Some of the phytochemicals also stimulate the expression of key DNA-repair genes, including BRCA1, BRCA2, RAD51, p53, and PARP1, through direct effects via transcriptional activation or indirectly by epigenetic modification (Wang et al., 2022; Fratantonio et al., 2021). For example, polyphenolic compounds, such as curcumin and epigallocatechin gallate (EGCG), have shown the ability to associate with nuclear receptors and activate transcriptional coactivators, thereby increasing promoter accessibility and gene transcription in the presence of genotoxic stress (Rajendran et al., 2022).

Curcumin, for example, upregulates the tumor-suppressor p53, leading to the induction of DNA-repair genes for BER as well as HR. Furthermore, quercetin may induce RAD51 mRNA expression, in addition to an increase in oxidative stress conditions, suggesting an enhancement of HR repair (Huminiecki, 2022). Transcriptional consequences may be further enhanced by simultaneous repression of HDACs, which can result in a hypomethylated condition and hence a relaxed architectural chromatin state that is more conducive to initiation of transcription. This coordinated control of gene expression reveals a major advantage of phytochemicals. Furthermore, they not only regulate repair at the enzymatic or signaling level but also maintain prolonged genomic robustness through epigenetically programmed responses (Babar et al., 2022).

18.3.4 Signaling pathways involved

18.3.4.1 NF-κB pathway

Nuclear factor kappa-light-chain-enhancer of activated B cells (NF-κB) is a redox-sensitive transcription factor that regulates inflammation and cell survival. Its activation paradoxically has tumor-promoting effects during genotoxic stress, where it upregulates anti-apoptotic and prosurvival genes (Uyeda, 2022). Curcumin, resveratrol, and EGCG are also phytochemicals that block NF-κB translocation to the nucleus by inhibiting upstream kinases such as IκB kinase (IKK) and subsequently repress the transcription of genes that promote disruption of DNA-repair fidelity.

They thereby restore pro-apoptotic signaling in damaged cells, shifting the balance toward either repair or apoptosis depending on the injury context (Bhutadiya and Mistry, 2021; Chimento et al., 2023).

18.3.4.2 Mitogen-activated protein kinase pathway

Mitogen-activated protein kinases (MAPKs) are key components between the sensors of DNA damage and response elements that modulate transcription. This pathway involves kinases such as ERK, JNK, and p38, which act to varying degrees on cell-cycle arrest, apoptosis, or repair gene expression (Bahar et al., 2023). Flavonoids, such as quercetin and apigenin, modulate MAPK signaling by suppressing JNK/p38 and activating ERK in certain circumstances. Such effects modulate the phosphorylation status of transcription factors, including c-Jun and ATF2, which in turn affect the expression of DNA repair and stress-response genes. The context dependence of the pathway thus permits alteration of repair decisions by phytochemicals, fine-tuning an adjustable response to DNA damage (Engin and Engin, 2021; Morgan et al., 2022; Pua et al., 2022).

18.3.4.3 Phosphoinositide 3-kinase/Akt pathway

Phosphoinositide 3-kinase (PI3K)/Akt signaling is a major signaling pathway that is involved in cell survival, metabolic actions, and DNA repair following oxidative and genotoxic stress. Activation of PI3K/Akt may increase DNA-repair through induction of NHEJ and HR factors, although sustained activation may also maintain tumorigenic survival (Abd El-Hafeez et al., 2023). Allicin and curcumin have been demonstrated to inhibit aberrant PI3K/Akt signaling in cancer cells, thus reducing the known repair inhibitors MDM2, as well as promoting p53 stabilization. This modulation of DNA-repair capability is also useful in sensitizing the malignant cells for therapeutic DNA damage, which gives phytochemicals special importance as an adjuvant in combination therapy (Bhuker et al., 2024).

18.4 Epigenetic regulation by dietary phytochemicals

Heritable and reversible changes in gene expression that do not involve changes to the DNA sequence are known as epigenetic modifications (Fig. 18.3). These changes consist of DNA methylation and histone acetylation/methylation and noncoding RNA-driven regulatory mechanisms, all of which play an essential role in the context of chromatin dynamics and transcriptional availability (Castro-Muñoz et al., 2023). Considering the DNA damage repairing system, the activity and expression of repair genes can be significantly regulated epigenetically compared to a normal state of the cell, which either enhances the relationship between the mechanism and genomic stability or disrupts their interaction, ultimately causing genomic instability (Laxmi et al., 2023). Plant phytochemicals have been highlighted as epigenetic modifiers that can reset abnormal epigenetic signatures, prominently in premalignant and malignant cells. The ability to regulate histone remodeling and DNA methylation allows them to stand at the crossroads between the field of nutrition and the realm of therapies targeting the epigenome (Dorna et al., 2025).

Through their pleiotropic effects, phytochemicals not only serve as antioxidants and signaling molecules but also as epigenetic regulators. Compounds including curcumin, resveratrol, and EGCG have an epigenetic effect by targeting critical enzymes, such as HDACs and DNMTs, to counteract the transcriptional down-regulation of DNA repair and tumor-suppressor genes (Prasanth et al., 2024). Such regulatory command permits dietary constituents to reset endogenous cancer-gene expression toward a homeostatic state in cells that are overwhelmed with epigenetic silencing, which is frequently seen in cancers and chronic inflammation (Vezzani et al., 2022).

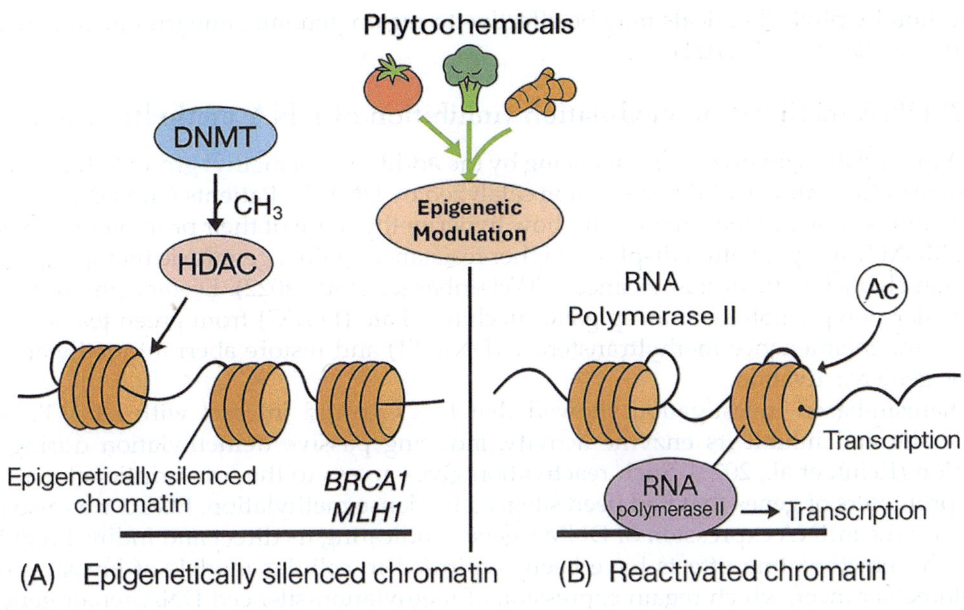

FIGURE 18.3 Mechanisms of DNA-repair modulation by phytochemicals: (A) Epigenetically silenced chromatin showing DNMT- and HDAC-mediated gene repression; (B) Reactivated chromatin showing transcriptional activation via histone acetylation.

18.4.1 Histone modification (inhibition of histone deacetylases)

Histone acetylation is a key post-translational modification that modulates chromatin architecture and accessibility of genes. Lysine acetylation on the histone tails by histone acetyltransferases (HATs) can relax the chromatin and thus promote transcriptional activation. This process is reversed by the action of HDACs that deacetylate the histones, leading to chromatin compaction and gene repression (Fang et al., 2021). In cancer and other diseases, hyperfunction of HDACs is responsible for general DNA hypoacetylation and results in silencing of tumor-suppressor genes and DNA repair processes. Previously reported studies indicated that some phytochemicals, including curcumin and resveratrol, could suppress HDAC activity and induce hyperacetylation of histones H3 and H4. This modification loosens the chromatin structure and induces re-expression of genes implicated in DNA-repair, apoptosis, and cell-cycle arrest (Roy and Datta, 2022).

Curcumin, an extract obtained from *C. longa*, exhibits a broad-spectrum HDAC inhibitory activity, which has been shown mainly against class I and II HDACs. Its modulating effect has been ascribed to the rise of acetylation levels at promoter regions of genes such as p21, GADD45, and BRCA1, restoring their repressed expression in cancer cells (Bouyahya et al., 2022; Jalli et al., 2023). A polyphenolic stilbene in grapes and berries, resveratrol has the same effects via inhibition of HDACs and acetylation-dependent recruitment of transcription factors to repair gene promoters. In both instances, the HDAC inhibitory effect is associated with increased DNA damage recognition and repair efficiency, thereby verifying that the strategy of histone modification by phytochemicals may be effective to regain genomic integrity in epigenetically stressed cells (Saleh et al., 2021).

18.4.2 DNA methylation modulation (inhibition of DNA methyltransferases)

DNA methylation involves gene silencing by the addition of a methyl group to the 5-position of cytosine in CpG dinucleotides, a reaction catalyzed by DNMTs. Patients with DNA repair and tumor-suppressor genes that frequently show hypermethylation of their promoter regions (e.g., MLH1, MGMT, and p16) often display a higher resistance against genotoxic therapies, which is a common characteristic of many cancers (Weisenberger et al., 2022). Dietary phytochemicals, in particular polyphenols such as epigallocatechin gallate (EGCG) from green tea, are known to inhibit the maintenance methyltransferase (DNMT1) and restore aberrant methylation and genomic gene reactivation.

Mechanism-based investigation showed that EGCG could interact with DNMT1 on the catalytic site and inhibit its enzyme activity, inducing passive demethylation during DNA replication (Kciuk et al., 2023). Such reactivation gives access to the transcriptional machinery for the promoters of genes that had been silenced by dense methylation. EGCG was also found to decrease the mRNA expression of DNMT genes, indicating its direct and indirect regulatory role. Such demethylating effects have been confirmed in cellular models of breast, prostate, and colorectal cancer, which regain expression of methylation-silenced DNA-repair genes and increased cellular sensitivity to chemotherapeutic agents following EGCG treatment. The ability to alter the methylation of DNA without introducing global genomic instability highlights what can be called phytochemical drug design (Mireles-Elizalde et al., 2023; Qadir Nanakali et al., 2023).

18.4.3 Reactivation of tumor-suppressor genes

Reactivation of silenced tumor-suppressor genes (TSGs) through an epigenetic approach is a major goal in cancer epigenetics. Other TSGs, such as BRCA1, MLH1, and MGMT, are frequently silenced in cancer cells by promoter hypermethylation and histone deacetylation. By dual-targeting DNMTs and HDACs, the dietary phytochemicals would provide a double-modal strategy for reactivating this epigenetic repression (Zhang et al., 2022). Curcumin, EGCG, and sulforaphane were able to synergistically reactivate TSGs by remodeling of chromatin and demethylation of DNA. These agents, thus, remediate DNA-repair proficiency, strengthen genomic surveillance, and sensitize tumor cells to DNA damage-based therapies (Dachani et al., 2024).

BRCA1, a central player in HR repair, is often repressed in breast and ovarian cancers by hypermethylation and histone deacetylation. Phytochemical drugs have been able to restore their expression by reprogramming their epigenetic landscape, thus leading to such genetic repairs of double-strand breaks and inhibiting the genomic instability (Zhang, 2024). Likewise, MGMT, essential for removing alkyl adducts of guanine, has been reactivated in glioblastomas using phytochemical-mediated epigenetic reprogramming. This reactivation enhances not only the DNA-repair ability but also suppresses the mutation rate and restores the chemosensitivity. The capacity of dietary components to epigenetically reactivate latent TSGs provides an exciting opportunity for nontoxic treatment of cancer as well as other diseases characterized by a defective DNA-repair machinery (Tessmer and Margison, 2023).

18.5 Clinical and preclinical evidence

A spectrum of in vitro cell culture studies, in vivo animal models, and disease-specific investigations, particularly in oncology, tends to support the therapeutic potential of dietary phytonutrients in enhancing DNA repair. These studies together showed that dietary phytochemicals such as curcumin, quercetin, EGCG, and sulforaphane not only confer antioxidant production but can directly influence DNA-repair machines and improve how cells respond to genotoxic stress (Sichetti et al., 2025). Preclinical evidence has led to their roles in modulating gene expression, restoring epigenetic balance, and enhancing the fidelity of DNA-repair pathways. Importantly, these effects have been validated across a variety of cellular systems, including epithelial, neuronal, and immune cells, thereby reinforcing the broad biological applicability of these compounds (Patra et al., 2024).

At the translational stage, three key medical aspects are evaluated to strengthen existing therapeutic methods. They are reactivating "silenced" tumor-suppressor gene, decreasing mutational burden in HeLa-like tumor cells and making resistant cells more sensitive to agents that damage DNA (Pramodh et al., 2022). Insights into the mechanisms underlying endothelial cell differentiation derived from preclinical research are becoming a significant part of clinical study design, particularly in the context of combined therapies. In addition, the safety and cost of phytochemicals, as well as the ease with which they can be incorporated into diets, make it feasible to use them outside experimental models for clinical purposes (Senrung et al., 2024). A compilation of experimental studies reporting enhanced DNA-repair activity in cell lines and animal models following treatment with dietary phytochemicals is given in Table 18.1.

18.5.1 In vitro studies on DNA-repair enhancement

In vitro studies establish the underlying mechanisms by which phytochemicals modulate DNA repair. Human cell lines become more efficient at repairing DNA damage, such as that caused by hydrogen peroxide, UV irradiation, and alkylating agents, when they have been pretreated or cotreated together with the polyphenolic compounds curcumin and quercetin (Hasan et al., 2022). These phytochemicals repair genomic integrity through induction of repair proteins (e.g., BRCA1, RAD51, PARP1), enhancement of strand break repair kinetics, and by modulating the function of DNA damage sensors. Phytochemical-exposed cells always show decreased DNA fragmentation in comet assays and faster disappearance of DNA damage foci in γ-H2AX foci quantification (Saitoh and Oda, 2021).

Mechanistically, these enhancements have been associated with several cellular events, including ROS scavenging and epigenetic reprogramming, as well as the activation of checkpoint kinases. For example, EGCG has been shown to increase XRCC1 and OGG1 in oxidative stress models to assist in BER (Piscone et al., 2025). In addition, flavonoids such as apigenin and kaempferol activate the ATM-CHK2-p53 axis to induce cell-cycle arrest, which in turn enhances damage recognition and repair more effectively. These in vitro observations provide evidence of a direct cause-and-effect relationship between a phytochemical intervention and enhanced DNA repair and will likely give impetus to their study in preclinical and clinical cancer therapy (Yasmin et al., 2023).

TABLE 18.1 In vitro and in vivo evidence of DNA-repair enhancement by phytochemicals.

Phytochemical	Model system	Type	DNA damage agent	Repair marker	Mechanism of action	References
Curcumin	Human lymphocytes	In vitro	Topoisomerase inhibitors, Platinum drugs, and base analogs	γ-H2AX	Increases the proteins of the base excision repair and NHEJ	Prabhu et al. (2024)
Resveratrol	Human breast cancer cell line (MCF-7)	In vivo	Ionizing radiation	p53, γ-H2AX	Upregulation of Tumor-suppressor genes-p53, p21, and p16	Merlin et al. (2021)
Allicin	Liver cancer cells (HepG2 cells)	In vitro	Cisplatin	γ-H2AX	Decreased cytoplasmic p53, PI3K/mTOR pathway, and Bcl-2 expression, while increasing AMPK/TSC2 and Beclin-1 expression in Hep G2 cells	Savairam et al. (2023), Talib et al. (2024)
Quercetin	Human cervical cancer cell line (HeLa)	In vivo	Ionizing radiation	XRCC1, OGG1	Block ATM activation and increase antioxidant response by modulating NRF2 signaling	Piscone et al. (2025)

(*continued on next page*)

Sulforaphane	Human fibroblasts	In vitro	UV	γ-H2AX, 53BP1	Increased transcription of phase 2 detoxifying enzymes	Qin et al. (2021)
Rosmarinic acid	MCF-7 cells	In vitro	Doxorubicin	ATM, BRCA1	Inhibition of DNA methyltransferases, as well as increased levels of apoptotic indicators such as p53	Ijaz et al. (2023), Helvacioglu et al. (2024)
Hesperidin	Zebrafish	In vivo	Benzo(a)pyrene	ZHE1	NRF2 activation, increasing GSH level, removal of H_2O_2 before it can form $^{\bullet}OH$ or HO_X	Fu et al. (2022), Bukowska et al. (2022)
Limonene	NIH-3T3 fibroblast cells	In vitro	Genotoxic compounds	γ-H2AX	Enhanced antioxidant response by increased production of antioxidant enzymes	Warman et al. (2023)
p-Coumaric acid	Mice	In vivo	UVB	CPDs	Photoprotection observed	Biswas et al. (2021)

18.5.2 In vivo studies in animal models

Animal models in vivo have additionally supported the positive effects of dietary phytochemicals in the repair of DNA, providing information on pharmacokinetics, tissue targets, and systemic effects (Dinda and Chakraborty, 2022). Phytochemical prevention with dietary phytochemicals, such as those found in fruits and vegetables, has been shown to lower the incidence of tumors that develop in rodents after treatment with chemicals or with radiation by decreasing the occurrence of tumors and the level of genetic damage in the treated animals (De et al., 2023). For example, when mice were fed with curcumin or sulforaphane, oxidative DNA lesions were decreased both in the liver and colon, concomitant with the induction of repair enzymes and the modification in the levels of inflammatory cytokines. These effects are typically associated with improved antioxidant defense, increased glutathione availability, and redox balance, which is favorable for genome stability (Laurindo et al., 2023).

Similarly, phytochemicals have been reported to influence DNA-repair pathways in a tissue-specific manner (i.e., in the brain, liver, digestive tract, and reproductive organs). Phytochemical treatment restores the DNA-repair activity in well-known DNA-repair-deficient transgenic mice and moderately moderates the mutational effective load (Huppert et al., 2022). Furthermore, histological investigations showed a decrease in chromosomal abnormalities, micronuclei formation, and apoptotic cell death in phytochemical-treated animals in comparison to the control. These in vivo results confirm the in vitro findings and thus reveal the likelihood of dietary phytochemicals as systemic genomic protectants, particularly in individuals with high risk for the pathologies related to DNA damage (Seo et al., 2022).

18.5.3 Effect on chemotherapy resistance

18.5.3.1 Breast cancer

Phytochemicals are reported to be potent in overcoming chemotherapy resistance in breast cancer, a malignancy with a compromised DNA-repair system or epigenetic modulation. The drug behavior is generally related to the drug resistance, and the resistance to drugs (such as doxorubicin or cisplatin) is often implicated with the enhanced expression of efflux transporters and the downregulated HR repair, due to the downregulation of BRCA1 (Mecca et al., 2024). Curcumin and EGCG can resensitize resistant breast cancer cells by reactivating silenced BRCA1 and recovering HR ability. Such reactivation enhances the repair of DNA double-strand breaks and allows apoptosis upon chemotherapeutic challenge. In addition, these compounds downregulate the drug transport system and suppress NF-κB signaling, i.e., enhancing the intracellular drug accumulation and cytotoxic activity (Farghadani et al., 2021).

Preclinical investigations have also indicated that combinations of curcumin or quercetin and DNA-damaging agents exhibit synergistic antiproliferative activity and induction of apoptosis in both triple-negative and hormone-dependent breast cancer cells (Malabadi et al., 2024). At the mechanistic level, such effects are explained by epigenetic reactivation of tumor suppressors, as well as regulation of PI3K/Akt signaling and the induction of mitochondrial insufficiency. These results would contribute to the encouraging evidence for using phytochemicals in clinical practice as a chemosensitizer, suggesting a potential ability to prevent the acquired drug resistance and to lower the therapeutic dose to mitigate side effects (Ganguly et al., 2021).

18.5.3.2 Lung cancer

In lung cancer, especially in nonsmall cell lung carcinomas, resistance to platinum-based chemotherapy presents one of the greatest clinical problems. This resistance is frequently associated with dysfunctional MMR and changes in detoxifying enzyme expression (Saqib et al., 2024). Some phytochemicals, such as resveratrol and allicin, have potential for their abilities to resensitize through repair gene expression modification and attenuation of antiapoptotic signaling. The allicin specifically increases the oxidative stress in cancer cells and upregulates repair proteins such as XRCC1 and p53 so that the balance favors apoptosis in the presence of chemotherapy drugs (Shiau et al., 2022).

In xenograft models of lung cancer in mice, cisplatin combined with phytochemicals has shown synergistic effect leading to remarkable inhibition of tumor growth, inhibition of angiogenesis, and increased expression of DNA-repair enzymes (Boța et al., 2024). Furthermore, they block critical prosurvival signaling pathways such as NF-κB and PI3K/Akt, thereby rendering tumors more susceptible to DNA damage-induced apoptosis. Considering their low systemic toxicity and high mechanistic specificity, phytochemicals could constitute a promising coadjuvant therapy for the management of lung cancer, particularly for the resistant/recurrent ones. These data support a strong preclinical rationale for clinical translation and the necessity of carefully designed human trials (Mustafa et al., 2024).

18.6 Anti-inflammatory and genotoxic protection

The close relationship of chronic inflammation and genomic instability in the context of cancer and degenerative diseases has been well established. Inflammatory processes produce a prooxidative and promutagenic microenvironment due to the constitutive liberation of cytokines, chemokines, ROS, and nitrogen intermediates, able to directly damage DNA and to interfere with repair accuracy (Ashrafi et al., 2024). Chronic inflammation also epigenetically silences important tumor suppressors and DNA-repair proteins. Phytochemicals, owing to their anti-inflammatory and antioxidant nature, can offer a dual protection against genotoxic stress. These molecules not only quench free radicals but also dampen inflammatory signaling pathways, such as reducing DNA damage at the site of origin (Roy et al., 2025).

Inhibition of those signaling pathways that maintain the inflammatory state represents a major mechanism by which phytochemicals exert their anti-inflammatory effects. Of these, nuclear factor-kappa B (NF-κB) serves as a major transcription factor that is involved in cross talk between inflammation and genomic instability (Prasad et al., 2023). Natural plant compounds, including curcumin, resveratrol, and EGCG, inhibit both NF-κB activation and its downstream targets, thereby down-regulating expression of inflammatory cytokines, cyclooxygenase, and adhesion molecules. The section discusses the mechanisms for disruption of inflammatory-genotoxic signaling by phytochemicals with implications in the maintenance of sustainable genomic stability (Ghobadi and Asoodeh, 2024).

18.6.1 Suppression of NF-κB signaling

NF-κB is a major transcription factor that dictates the expression of an entire constellation of genes, including those responsible for inflammation, immune responses, cell survival, and proliferation. When a cell is at rest, NF-κB is sequestered in the cytoplasm by its inhibitor IκBα. On induction by proinflammatory cytokines, ROS, or microorganisms' products, the IκBα protein is phosphorylated and degraded, thereby allowing its translocation of NF-κB into the nucleus and initiating gene transcription (Almowallad et al., 2022). Chronic activation of NF-κB leads to a chronic inflammatory environment, exacerbates oxidative stress, and causes DNA damage. Compounds such as curcumin or resveratrol act in this process through their ability to block IKK (IκB kinase), a molecule with enzymic activity. By stabilizing IκBα, the inhibitor protein from which NF-κB is held inside cells, these compounds prevent its entry into the nucleus. This inhibits the generation of multiple proinflammatory mediators, including inducible nitric oxide synthase (iNOS) and tumor necrosis factor-α (TNF-α), ultimately serving to reduce the number of free radicals being released (Lyubitelev and Studitsky, 2023; Kannan et al., 2025).

In cellular models of inflammation-induced genotoxicity, curcumin has displayed a capability to abrogate most NF-κB-mediated transcription of genes encoding inducible nitric oxide synthase (iNOS), tumor necrosis factor- α(TNF-α), and interleukin-6 (IL-6), all of which figure in DNA damage propagation (Mirzaei et al., 2023). More importantly, flavonoids such as apigenin can lower NF-κB DNA-binding activity as well as decrease inflammation. Phytochemicals shut off the NF-κB signal, which in turn can restore cellular redox balance and mitigate the formation of new DNA lesions. These anti-inflammatory measures facilitate the maintenance of genomic stability directly, suggesting a therapeutic role for phytochemicals in managing an inflammation-driven mutagenesis (Hemarangan et al., 2025).

18.6.2 Downregulation of proinflammatory cytokines

Proinflammatory and pro-oxidative cytokines, such as TNF-α, IL-1β, and IL-6, are major mediators of inflammation, upregulating and promoting high levels of ROS production, mitochondrial dysfunction, and DNA damage. These cytokines not only help maintain the proinflammatory microenvironment but also drive signaling pathways that inhibit the DNA-repair process and enhance the mutation rate (Albano et al., 2022). Dietary phytochemicals act on several fronts to inhibit cytokine-induced genotoxicity, such as transcriptional down-regulatory effect, signal transduction block, and receptor antagonism. For example, curcumin decreases TNF-α and IL-6 expression at both the mRNA and protein level and suppresses STAT3 phosphorylation -a key transcription factor associated with cytokine gene regulation (Xu et al., 2021).

Like curcumin, other polyphenols, including quercetin, EGCG, and baicalin, have shown suppression of cytokine expression in vitro and in vivo. These agents reduce the amounts of cytokines released by activated macrophages, dendritic cells, and epithelial cells and attenuate the ensuing activation of genotoxic enzymes, such as NADPH oxidase and myeloperoxidase (Liu et al., 2023). Phytochemicals interfere with the cycle of cytokine release and induction of oxidative DNA damage, protecting cellular homeostasis and DNA-repair mechanisms. This cytokine-directed anti-inflammatory effect provides additional support for the genome-protective activity of phytochemicals in chronic inflammatory-related disease states (Paudel et al., 2023).

18.6.3 Role in reducing inflammation-driven mutagenesis

Inflammation-induced mutagenesis is defined as the acquisition of DNA damage because of persistent inflammatory stimulation. This is promoted by chronic exposure to ROS and reactive nitrogen species (RNS), causing nucleotide oxidation, deamination, and strand breaks (Hung et al., 2024). Meanwhile, inflammatory signals commonly inhibit DNA-repair pathways by activating epigenetic and post-translational repression of repair proteins. Phytochemicals oppose this pro-mutagenic cascade by both reducing ROS/RNS production and simultaneously enhancing the repair ability. Such as activation of antioxidant enzymes expression and reactivation of silenced DNA-repair genes by histone acetylation and promoter demethylation by sulforaphane and resveratrol, respectively (Thiruvengadam et al., 2021).

Furthermore, phytochemicals suppress the activity of inflammation-mediated enzymes, such as COX-2 and iNOS, implicated in DNA alkylation and nitrosation. In models of colitis-associated cancer, the addition of phytochemicals reduced the frequency of mutations in oncogenic and TSGs, decreased microsatellite instability, and increased DDR signaling (Mintz et al., 2021). Together, these effects indicate that phytochemicals not only block the induction of mutagenic events but also stimulate the repair of inflammation-induced lesions. Therefore the targeted administration of anti-inflammatory phytochemicals through diet or therapy has the potential to attenuate the genomic impact of chronic inflammation (Ramos et al., 2023).

18.7 Therapeutic and nutrigenomic applications

The use of phytochemicals in genome stability, in nutrigenomics, and in precision nutrition is becoming increasingly important. Nutrigenomics is the science of how diet-derived compounds impact and modify gene expression and health outcomes at the genomic level. Phytochemicals, in this perspective, offer an exceptional class of dietary bioactives that contribute to cellular homeostasis, DNA-repair pathways, epigenetic activity, and oxidative balance (Agrawal et al., 2024). Their incorporation into personalized nutrition may offer new opportunities, not only for disease prevention, but also for use as an adjunctive therapeutic option in genotoxic stress-associated disease states such as cancer, neurodegeneration, and metabolic syndromes. The ability of these compounds to regulate gene networks associated with genome maintenance suggests their utility in the realm of therapeutic nutrition (Choudhary et al., 2025).

The clinical potential of phytochemicals, in fact, goes beyond therapy and involves chemoprevention - interventions where agents interfere before the occurrence of disease through acting on initial molecular events induced by DNA damage and mutation accumulation. For example, the polyphenolic compounds such as quercetin and curcumin, if used regularly in food or dietary supplements, can prestimulate endogenous free radical scavenging systems, as well as DNA-repair enzymes (Biswas et al., 2022). These molecular changes, which are commonly regulated by transcription factors such as Nrf2 and p53, lead to the activation of a more robust cellular defense system, limiting the risk of mutagenesis and carcinogenesis. Hence, incorporation of these bioactive compounds into functional foods and nutraceuticals is a potential frontier of genomic medicine and public health nutrition (Jazvinšćak Jembrek et al., 2021).

18.7.1 Nutritional modulation for genome integrity

Nutritional modulation refers to the purposeful manipulation of ingredients in the diet that can influence certain biological events that are important for the maintenance of genome integrity. On the other hand, consumption of fruits, vegetables, spices, and herbs, which are natural sources of phytochemicals, has been consistently correlated with lowered burden of

DNA adducts and aging-related degenerative disorders (Choi and Friso, 2023). This protective effect has been associated with the ability of dietary phytochemicals to modulate redox-sensitive transcription factors, DNA-repair mechanisms, and inflammation. Sulforaphane, EGCG, and apigenin were shown to trigger the expression of detoxification and repair genes, thus reinforcing the ability of cells to address genotoxic insults arising from internal metabolism as well as from the external environment (Yadav et al., 2021).

Studies in nutrigenomics have demonstrated beyond a reasonable doubt that variations from one individual to another in genetic background affect their responsiveness to phytochemicals. Single-nucleotide polymorphisms in genes encoding for enzymes, such as GSTs and SODs, and repair factors such as XRCC1 and OGG1 will modulate the efficacy of these compounds (Galvez et al., 2021). Cellular interventions must take account of an individual's genotypic profile to optimize phytochemical efficacy contributing to genome maintenance. This individualized treatment strategy would be especially appropriate for populations with high genetic risk of genome instability, including those with familial cancer syndromes or DNA-repair defects. Thus nutritional genomics is leading the evolution of preventive health with a new era in the use of phytochemicals to support DNA-repair fidelity (Dall'Asta et al., 2022).

18.7.2 Dietary recommendations and safety considerations

Although the health-promoting effects of phytochemicals are well known, application of these chemicals as a clinical or dietary guideline will require well-thought-out safety, bioavailability, and dosage. Phytochemicals frequently produce biphasic dose-response curves (hormesis), in which low to moderate doses are protective and high doses can have cytotoxic or antagonist effects (Pande et al., 2022). For instance curcumin has been reported to have strong therapeutic potential in preclinical models but has poor systemic bioavailability because of both poor solubility and rapid metabolism. Approaches to enhance the delivery of berberine using nano formulations, liposomal carriers, and adjuvant conjugation with piperine have been promising to circumvent these pharmacokinetic constraints (Rajput et al., 2022).

Moreover, there is still a lack of long-term safety information for the high-dose supplementation of phytochemicals. Most of the compounds, however, when isolated and concentrated, interact with drug-metabolizing enzymes or exert some prooxidant effects under specific situations. Dietary guidance should therefore focus on whole-food sources and modest consumption reminiscent of traditional dietary patterns that have low rates of disease, such as the Mediterranean or Asian diet (Spanakis et al., 2022). Performance of toxicological testing, standardization of the content of phytochemicals, cumulative dietary exposure, and consideration of clinical translation are essential. Including these safety considerations in dietary policies and public health programs will guarantee the optimal therapeutic value of the phytochemicals and their long-term genomic health benefits (Ali et al., 2025).

18.8 Challenges and future directions

Despite the significant advances in our understanding of the translational status of the phytochemical-induced genomic and epigenomic effects, there remain several translational issues that are obstructing the progress of phytochemicals into the clinic or as targets for the treatment of cancer. Central to this, is the problem of bioavailability—many phytochemicals are poorly absorbed, rapidly metabolized, and are not well systemically dispersed after oral administration despite being potent in vitro (Açar et al., 2023). Molecules such as curcumin and resveratrol are subjected to extensive first-pass metabolism and have poor plasma levels, preventing them from exerting their full biological effect in vivo. Nanoparticle encapsulation, employing liposomal delivery systems, and coadministration with an absorption enhancer (e.g., piperine) are some of the strategies that have shown promising, albeit inconsistent, results in clinical studies developed to improve bioavailability. These bioformulations also add multiple dimensions in terms of pharmacokinetic profiling, dose optimization, and long-term safety assessment, necessitating strict standardization and regulatory control (Robertson et al., 2022).

Another important challenge is to standardize the phytochemical composition among different sources and preparations. Natural variability in plant secondary metabolism due to cultivar, climate, harvest time, and extraction technique leads to inconsistent concentrations of active constituents, which impairs reproducibility across studies. Moreover, a lack of standardization in dosing further adds to the difficulties in clinical trial design and incorporation into standardized therapy protocols based on evidence (Raudone and Savickiene, 2024). Validated biomarkers for expanded DNA repair and for genome editing will be critical to evaluate clinical activity. Dealing with the mentioned challenges will necessitate multidisciplinary efforts integrating plant biochemistry, pharmacology, clinical nutrition, and regulatory sciences for the generation of potent, scalable and safe delivery systems of phytochemicals that are genome-protective functions (Fenech et al., 2023).

18.8.1 Bioavailability and metabolism of phytochemicals/standardization and dosing

The pharmacokinetic behavior of phytochemicals presents a major challenge for their therapeutic application. Many phytochemicals are unstable in the gastrointestinal tract, undergo rapid metabolic inactivation (conjugation) in the liver, and are subsequently excreted, resulting in plasma concentrations that are often below therapeutic levels (Zuccari and Alfei, 2023). Consequently, experimental studies frequently employ supra-physiological doses to elicit measurable biological effects, doses that may not be safe or ethically feasible in human populations.

To overcome these limitations, researchers have explored advanced drug-delivery systems such as solid lipid nanoparticles, micelles, phytosomes, and hydrogel-based carriers. These approaches aim to enhance solubility, protect phytochemicals from premature metabolism, and allow controlled or sustained release. However, significant challenges remain, including issues related to scalability, cost-effectiveness, and long-term toxicity, which currently limit their clinical translation (Farasati et al., 2023).

In addition to delivery challenges, phytochemical preparations require rigorous standardization across different sources and formulations. Because phytochemicals are extracted from complex botanical matrices, their concentration and biological activity can vary widely depending on plant species, cultivation conditions, extraction methods, and formulation processes. Establishing pharmacopoeial benchmarks for key genome-modulating phytochemicals, along with evidence-based, haplotype- or gene-specific therapeutic thresholds, will therefore be critical (Dhami et al., 2015).

Furthermore, personalized dosing strategies based on pharmacogenomic profiling may enhance therapeutic responses while reducing toxicity compared with empirical dosing approaches. Such precision-based strategies must be supported by the identification and validation of robust biomarkers and surrogate endpoints that reliably reflect biological effectiveness. These measures will be essential for evaluating genome-targeted phytochemical interventions and for meeting regulatory requirements for clinical approval (Jauhari et al., 2024).

18.8.2 Gaps in current research and prospective studies

Despite a growing body of preclinical evidence demonstrating the role of phytochemicals in modulating DNA repair and epigenetic regulation, clinical validation of these effects remains limited. Human studies conducted to date are few in number and are often constrained by small sample sizes, heterogeneous study designs, and reliance on surrogate rather than direct genomic endpoints (Ghasemi et al., 2021; Mukherjee et al., 2022). There is a clear need for well-designed longitudinal studies that systematically assess DNA damage biomarkers, DNA-repair capacity, and mutation burden in response to dietary phytochemical interventions.

Another critical gap lies in the underexplored relationship between phytochemicals and the gut microbiota, which is increasingly recognized as an important regulator of xenobiotic metabolism and epigenetic signaling. Microbiota-mediated biotransformation of phytochemicals may substantially influence their bioavailability and biological activity in the host; however, this microbiome–phytochemical axis is rarely incorporated into current nutrigenomic studies (Rahman et al., 2022).

In addition, the mechanistic understanding of how phytochemicals interact at the molecular level with DNA-repair complexes, transcriptional regulators, and epigenetic enzymes remains incomplete. While high-throughput omics approaches—including transcriptomics, epigenomics, and proteomics—have provided initial insights, these datasets are often not integrated with functional validation studies (Mukherjee et al., 2021). Emerging methodologies such as CRISPR-based reporter systems, single-cell sequencing, and high-throughput screening platforms offer promising tools to dissect these mechanisms with greater precision.

Future research should therefore emphasize systematic and interdisciplinary approaches that integrate mechanistic insights with translational relevance. Such efforts will be essential to advance phytochemicals from promising bioactive compounds to clinically validated agents for genome health management and disease prevention (Salame et al., 2022).

18.9 Conclusion

Phytochemicals represent a potential model for modulating DNA-repair pathways by coordinating a multidimensional response including antioxidant protection, DDR activation, transcriptional gene regulation, and epigenetic modification. These natural bioactives, by modulation of critical genetic pathways and repair gene expression, augment the integrity of the genome and suppress mutagenic processes, which are the primary causes of cancer and degenerative diseases. Their potential to reactivate epigenetically silenced TSGs as well as DNA-repair enzymes underlies their therapeutic value not only in chemoprevention but also in reversing therapy resistance. The chapter will cover the molecular basis, experimental evidence, and translational implications of the action of phytochemicals in genome maintenance, connecting mechanisms of action with nutrigenomic utilities. Despite being hindered by bioavailability, standardization, and clinical validation challenges, recent developments in delivery vehicles, biomarker identification, and personalized treatments are promising and foretell that phytochemicals could be the molecules of choice as lead agents in precision nutrition and genomic medicine.

References

Abd El-Hafeez, A.A., Sun, N., Chakraborty, A., Ear, J., Roy, S., Chamarthi, P., et al., 2023. Regulation of DNA damage response by trimeric G-proteins. iScience 26 (2), 105973.

Abiko, Y., Toriba, A., Kumagai, Y., 2023. Phytochemicals to regulate oxidative and electrophilic stress through Nrf2 activation. Redox Experimental Medicine 2023 (1). doi:10.1530/REM-22-0021.

Açar, Y., Akbulut, G., 2023. Nutritional epigenetics and phytochemicals in Cancer formation. J. Am. Nutrit. Assoc. 42 (7), 700–705.

Agrawal, P., Kaur, J., Singh, J., Rasane, P., Sharma, K., Bhadariya, V., et al., 2024. Genetics, nutrition, and health: a new frontier in disease prevention. J. Am. Nutrit. Assoc. 43 (4), 326–338.

Ahmad, A., Afzaal, M., Saeed, F., Ali, S.W., Imran, A., Zaidi, S.Y.R., et al., 2023. A comprehensive review of the therapeutic potential of citrus bioflavonoid hesperidin against lifestyle-related disorders. Cogent Food Agri. 9 (1), 2226427.

AlAli, M., Alqubaisy, M., Aljaafari, M.N., AlAli, A.O., Baqais, L., Molouki, A., et al., 2021. Nutraceuticals: transformation of conventional foods into health promoters/disease preventers and safety considerations. Molecules 26 (9), 2540.

Albano, G.D., Gagliardo, R.P., Montalbano, A.M., Profita, M., 2022. Overview of the mechanisms of oxidative stress: impact in inflammation of the airway diseases. Antioxidants 11 (11), 2237.

Ali, D.S., Elfalleh, W., Smaoui, S., 2025. Nutraceutical and functional foods: developmental of toxicity testing and safety assessment. Euro-Mediterr. J. Environ. Integr. 10, 1–18.

Almowallad, S., Alqahtani, L.S., Mobashir, M., 2022. NF-kB in signaling patterns and its temporal dynamics encode/decode human diseases. Life 12 (12), 2012.

Andrés, C.M.C., Lastra, J.M.P.D.L., Juan, C.A., Plou, F.J., Pérez-Lebeña, E., 2023. Chemical insights into oxidative and nitrative modifications of DNA. Int. J. Mol. Sci. 24 (20), 15240.

Ashrafi, S., Zahmatkesh, N., Tehrani, M.K., Hadizadeh, N., Rahimzadeh, S., Mahdavi, M., et al., 2024. How Oxidative Stress Modifies the Immune Response Landscape in Cancer. Biomed. Res. Bull. 2 (3), 140–161.

Aslam, S., Iqbal, R., Saeed, R.F., Akram, N., Ijaz, F., Liaqat, I., et al., 2024. Nutritional genomics and cancer prevention. Nutr. Diet. Interv. Cancer 191, 217–244 Springer, Cham.

Babar, Q., Saeed, A., Tabish, T.A., Pricl, S., Townley, H., Thorat, N., 2022. Novel epigenetic therapeutic strategies and targets in cancer. Biochim. Biophys. Acta - Mol. Basis Dis. 1868 (12), 166552.

Bahar, M.E., Kim, H.J., Kim, D.R., 2023. Targeting the RAS/RAF/MAPK pathway for cancer therapy: from mechanism to clinical studies. Signal Transduct. Targeted Therapy 8 (1), 455.

Bhattacharya, T., Dutta, S., Akter, R., Rahman, M.H., Karthika, C., Nagaswarupa, H.P., et al., 2021. Role of phytonutrients in nutrigenetics and nutrigenomics perspective in curing breast cancer. Biomolecules 11 (8), 1176.

Bhuker, S., Kaur, A., Rajauria, K., Tuli, H.S., Saini, A.K., Saini, R.V., et al., 2024. Allicin: a promising modulator of apoptosis and survival signaling in cancer. Med. Oncol. 41 (9), 210.

Bhutadiya, V.L., Mistry, K.N., 2021. A review on bioactive phytochemicals and it's mechanism on cancer treatment and prevention by targeting multiple cellular signaling pathways. Int. J. Pharm. Pharm. Sci. 13 (12). doi:10.22159/ijpps.2021v13i12.42798, 15-9.

Biswas, P., Dey, D., Biswas, P.K., Rahaman, T.I., Saha, S., Parvez, A., et al., 2022. A comprehensive analysis and anticancer activities of quercetin in ROS-mediated cancer and cancer stem cells. Int. J. Mol. Sci. 23 (19), 11746.

Biswas, S., Mukherjee, P.K., Kar, A., Bannerjee, S., Jana, S.N., Haldar, P.K., et al., 2021. Enhanced permeability and photoprotective potential of optimized p-Coumaric acid-phospholipid complex loaded gel against UVA mediated oxidative stress. J. Photochem. Photobiol. B 221, 112246.

Boţa, M., Vlaia, L., Jîjie, A.R., Marcovici, I., Crişan, F., Oancea, C., et al., 2024. Exploring synergistic interactions between natural compounds and conventional chemotherapeutic drugs in preclinical models of lung cancer. Pharmaceuticals 17 (5), 598.

Bouyahya, A., El Hachlafi, N., Aanniz, T., Bourais, I., Mechchate, H., Benali, T., et al., 2022. Natural bioactive compounds targeting histone deacetylases in human cancers: recent updates. Molecules 27 (8), 2568.

Bukowska, B., Duchnowicz, P., 2022. Molecular mechanisms of action of selected substances involved in the reduction of benzo [a] pyrene-induced oxidative stress. Molecules 27 (4), 1379.

Castro-Muñoz, L.J., Vázquez Ulloa, E., Sahlgren, C., Lizano, M., De La Cruz-Hernández, E., Contreras-Paredes, A., 2023. Modulating epigenetic modifications for cancer therapy. Oncol. Rep. 49 (3), 59.

Catalano, F., Borea, R., Puglisi, S., Boutros, A., Gandini, A., Cremante, M., et al., 2022. Targeting the DNA damage response pathway as a novel therapeutic strategy in colorectal cancer. Cancers 14 (6), 1388.

Ceylan, F.D., Günal-Köroğlu, D., Saricaoglu, B., Ozkan, G., Capanoglu, E., Calina, D., et al., 2025. Anticancer potential of hydroxycinnamic acids: mechanisms, bioavailability, and therapeutic applications. Naunyn-Schmiedeberg's Arch. Pharmacol. 398 (1), 469–495.

Chatterjee, P., Gupta, S., Banerjee, S., 2023. Understanding the role of the natural warriors: phytochemicals in breast cancer chemoprevention. In: Pati, S., Sarkar, Y., Lahiri, D. (Eds.), Recent Frontiers of Phytochemicals. Elsevier, Netherlands, pp. 261–293.

Chimento, A., D'Amico, M., De Luca, A., Conforti, F.L., Pezzi, V., De Amicis, F., 2023. Resveratrol, epigallocatechin gallate and curcumin for cancer therapy: challenges from their pro-apoptotic properties. Life 13 (2), 261.

Choi, S.W., Friso, S., 2023. Modulation of DNA methylation by one-carbon metabolism: a milestone for healthy aging. Nutrit. Res. Pract. 17 (4), 597–615.

Choudhary, M.K., Pancholi, B., Kumar, M., Babu, R., Garabadu, D., 2025. A review on endoplasmic reticulum-dependent anti-breast cancer activity of herbal drugs: possible challenges and opportunities. J. Drug Targeting 33 (2), 206–231.

Czerwińska, K., Radziejewska, I., 2024. Rosmarinic acid: a potential therapeutic agent in gastrointestinal cancer management—a review. Int. J. Mol. Sci. 25 (21), 11704.

Dachani, S.R., Kaleem, M., Mujtaba, M.A., Mahajan, N., Ali, S.A., Almutairy, A.F., et al., 2024. A comprehensive review of various therapeutic strategies for the management of skin cancer. ACS Omega 9 (9), 10030–10048.

Dall'Asta, M., Barbato, M., Rocchetti, G., Rossi, F., Lucini, L., Marsan, P.A., et al., 2022. Nutrigenomics: an underestimated contribution to the functional role of polyphenols. Current Opin. Food Sci. 47, 100880.

De, S., Paul, S., Manna, A., Majumder, C., Pal, K., Casarcia, N., et al., 2023. Phenolic phytochemicals for prevention and treatment of colorectal cancer: a critical evaluation of in vivo studies. Cancers 15 (3), 993.

Dhami, N., Mishra, A.D., 2015. Phytochemical variation: how to resolve the quality controversies of herbal medicinal products? J. Herb. Med. 5 (2), 118–127.

Di Guilmi, A.M., Fonknechten, N., Campalans, A., 2023. OGG1 at the crossroads between repair and transcriptional regulation. In: Sugimoto, N. (Ed.), Handbook of Chemical Biology of Nucleic Acids. Springer, Singapore, pp. 1–34.

Dinda, B., Chakraborty, A., 2022. Pharmacokinetics and metabolism of phytochemicals having anti-obesity and antidiabetic activity. In: Pathak, S., Banerjee, A. (Eds.), Natural Products in Obesity and Diabetes: Therapeutic Potential and Role in Prevention and Treatment. Springer International Publishing, Cham, pp. 469–505.

Dong, R., Wang, J., Guan, R., Sun, J., Jin, P., Shen, J., 2025. Role of oxidative stress in the occurrence, development, and treatment of breast cancer. Antioxidants 14 (1), 104.

Dorna, D., Grabowska, A., Paluszczak, J., 2025. Natural products modulating epigenetic mechanisms by affecting histone methylation/demethylation: targeting cancer cells. Br. J. Pharmacol. 182 (10), 2137–2158.

El Omari, N., Bakrim, S., Bakha, M., Lorenzo, J.M., Rebezov, M., Shariati, M.A., et al., 2021. Natural bioactive compounds targeting epigenetic pathways in cancer: a review on alkaloids, terpenoids, quinones, and isothiocyanates. Nutrients 13 (11), 3714.

Engin, A.B., Engin, A., 2021. Protein Kinase-Mediated Decision Between the Life and Death. Springer, Cham, pp. 1–33.

Fang, Z., Wang, X., Sun, X., Hu, W., Miao, Q.R., 2021. The role of histone protein acetylation in regulating endothelial function. Front. Cell Dev. Biol. 9, 672447.

Farasati Far, B., Naimi-Jamal, M.R., Sedaghat, M., Hoseini, A., Mohammadi, N., Bodaghi, M., 2023. Combinational system of lipid-based nanocarriers and biodegradable polymers for wound healing: an updated review. J. Funct. Biomater. 14 (2), 115.

Farghadani, R., Naidu, R., 2021. Curcumin: modulator of key molecular signaling pathways in hormone-independent breast cancer. Cancers 13 (14), 3427.

Fatima, N., Baqri, S.S.R., Alsulimani, A., Fagoonee, S., Slama, P., Kesari, K.K., et al., 2021. Phytochemicals from Indian ethnomedicines: promising prospects for the management of oxidative stress and cancer. Antioxidants 10 (10), 1606.

Fenech, M.F., Bull, C.F., Van Klinken, B.J.W., 2023. Protective effects of micronutrient supplements, phytochemicals and phytochemical-rich beverages and foods against DNA damage in humans: a systematic review of randomized controlled trials and prospective studies. Adv. Nutr. 14 (6), 1337–1358.

Fratantonio, D., Virgili, F., Benassi, B., 2021. Diet and epigenetics: dietary effects on DNA methylation, histone remodeling and mRNA stability. Compreh. Foodom. 1, 364–379.

Fu, C., Li, Y., Xi, H., Niu, Z., Chen, N., Wang, R., et al., 2022. Benzo (a) pyrene and cardiovascular diseases: an overview of pre-clinical studies focused on the underlying molecular mechanism. Front. Nutr. 9, 978475.

Galvez, H.R.C., Flores, J.S., Sanchez, E.D.T., Bravo, D.R., Villela, M.Z.R., Uribe, E.R., 2021. Genetic profile for the detection of susceptibility to poisoning by exposure to pesticides. Ann. Agric. Environ. Med. 28 (2), 208–213.

Ganguly, S., Arora, I., Tollefsbol, T.O., 2021. Impact of stilbenes as epigenetic modulators of breast cancer risk and associated biomarkers. Int. J. Mol. Sci. 22 (18), 10033.

Gauthier, B.R., Comaills, V., 2021. Nuclear envelope integrity in health and disease: consequences on genome instability and inflammation. Int. J. Mol. Sci. 22 (14), 7281.

Gbadebo, O.S., Oke, E.D., Ajibuwa, F.A., 2025. Anticancer properties of beta-caryophyllene and d-limonene terpenes: a review. Asian Pacific J. Trop. Biomed. 15 (4), 129–140.

Ghasemi, S., Xu, S., Nabavi, S.M., Amirkhani, M.A., Sureda, A., Tejada, S., et al., 2021. Epigenetic targeting of cancer stem cells by polyphenols (cancer stem cells targeting). Phytother. Res. 35 (7), 3649–3664.

Ghobadi, N., Asoodeh, A., 2024. Curcumin and epigallocatechin-3-gallate: targeting receptor tyrosine kinases and their signaling pathways in cancer therapy. Current Pharmacol. Reports 10 (6), 297–311.

Ghosh, S., Ghosh, A., 2021. Activation of DNA damage response signaling in mammalian cells by ionizing radiation. Free Radical Res. 55 (8), 814–827.

Gorrepati, K., Krishna, R., Singh, S., Shirsat, D.V., Soumia, P.S., Mahajan, V., 2024. Harnessing the nutraceutical and therapeutic potential of Allium spp.: current insights and future directions. Front. Nutr. 11, 1497953.

Hasan, A.A., Tatarskiy, V., Kalinina, E., 2022. Synthetic pathways and the therapeutic potential of quercetin and curcumin. Int. J. Mol. Sci. 23 (22), 14413.

Helvacioglu, S., Charehsaz, M., Bankoglu, E.E., Stopper, H., Aydin, A., 2024. The ameliorative effect of rosmarinic acid and epigallocatechin gallate against doxorubicin-induced genotoxicity. Drug Chem. Toxicol. 47 (6), 1087–1099.

Hemarangan, J., Yashoda, K., Karthik, S.S., Ganesh, H.N., Kanth, T.V., Murali, R.V., et al., 2025. Mechanism of phytochemicals in modulating signalling pathways involved in inflammation and cancer. In: Plant Derived Bioactive Compounds in Human Health and Disease. CRC Press, Boca Raton, FL, pp. 49–65.

Hsieh, H.H., Kuo, M.Z., Chen, I.A., Lin, C.J., Hsu, V., HuangFu, W.C., et al., 2025. Epigenetic modifications as novel therapeutic strategies of cancer chemoprevention by phytochemicals. Pharm. Res. 42 (1), 69–78.

Huminiecki, L., 2022. Evidence for multilevel chemopreventive activities of natural phenols from functional genomic studies of curcumin, resveratrol, genistein, quercetin, and luteolin. Int. J. Mol. Sci. 23 (23), 14957.

Hung, J.H., Teng, C.F., Hung, H.C., Chen, Y.L., Chen, P.J., Ho, C.L., et al., 2024. Genomic instabilities in hepatocellular carcinoma: biomarkers and application in immunotherapies. Ann. Hepatol. 29, 101546.

Huppert, L.A., Green, M.D., Kim, L., Chow, C., Leyfman, Y., Daud, A.I., et al., 2022. Tissue-specific Tregs in cancer metastasis: opportunities for precision immunotherapy. Cellul. Mol. Immunol. 19 (1), 33–45.

Ijaz, S., Iqbal, J., Abbasi, B.A., Ullah, Z., Yaseen, T., Kanwal, S., et al., 2023. Rosmarinic acid and its derivatives: Current insights on anticancer potential and other biomedical applications. Biomed. Pharmacother. 162, 114687.

Jalli, D.R., Alone, D.P., 2023. Nontoxic Natural Products as Regulators of Tumor Suppressor Gene Function. In: Biomedical Applications and Toxicity of Nanomaterials. Springer Nature, Singapore, pp. 229–256.

Jauhari, S., Jauhari, R., Rupanagunta, G.P., Nandave, M., Upadhyay, J., Joshi, R., 2024. Insights on the Integration of Ethnopharmacology and Omics in Medicinal Plant Research. In: Ethnopharmacology and OMICS Advances in Medicinal Plants Volume 2: Revealing the Secrets of Medicinal Plants. Springer Nature, Singapore, pp. 501–514.

Jazvinšćak Jembrek, M., Oršolić, N., Mandić, L., Sadžak, A., Šegota, S., 2021. Anti-oxidative, anti-inflammatory and anti-apoptotic effects of flavonols: targeting Nrf2, NF-κB and p53 pathways in neurodegeneration. Antioxidants 10 (10), 1628.

Jiang, W., Xia, T., Liu, C., Li, J., Zhang, W., Sun, C., 2021. Remodeling the epigenetic landscape of cancer—application potential of flavonoids in the prevention and treatment of cancer. Front. Oncol. 11, 705903.

Jurkovicova, D., Neophytou, C.M., Gašparović, A.Č., Gonçalves, A.C., 2022. DNA damage response in cancer therapy and resistance: challenges and opportunities. Int. J. Mol. Sci. 23 (23), 14672.

Kannan, G., Paul, B.M., Thangaraj, P., 2025. Stimulation, regulation, and inflammaging interventions of natural compounds on nuclear factor kappa B (NF-kB) pathway: a comprehensive review. Inflammopharmacology 33, 1–18.

Kciuk, M., Alam, M., Ali, N., Rashid, S., Głowacka, P., Sundaraj, R., et al., 2023. Epigallocatechin-3-gallate therapeutic potential in cancer: mechanism of action and clinical implications. Molecules 28 (13), 5246.

Khan, A., Khan, A., Khan, M.A., Malik, Z., Massey, S., Parveen, R., et al., 2024. Phytocompounds targeting epigenetic modulations: an assessment in cancer. Front. Pharmacol. 14, 1273993.

Laurindo, L.F., de Maio, M.C., Minniti, G., de Góes Corrêa, N., Barbalho, S.M., Quesada, K., et al., 2023. Effects of medicinal plants and phytochemicals in Nrf2 pathways during inflammatory bowel diseases and related colorectal cancer: a comprehensive review. Metabolites 13 (2), 243.

Laxmi, A., Garg, S.S., Singh, A., Prabhakar, P.K., Gupta, J., 2023. Histone modifying potential of dietary phytochemicals: implications in treating breast cancer. Current Pharmacol. Reports 9 (6), 489–510.

Lee, D.Y., Song, M.Y., Kim, E.H., 2021. Role of oxidative stress and Nrf2/KEAP1 signaling in colorectal cancer: mechanisms and therapeutic perspectives with phytochemicals. Antioxidants 10 (5), 743.

Li, X., Cao, G., Liu, X., Tang, T.S., Guo, C., Liu, H., 2022. Polymerases and DNA repair in neurons: implications in neuronal survival and neurodegenerative diseases. Front. Cellul. Neurosci. 16, 852002.

Liu, H.M., Cheng, M.Y., Xun, M.H., Zhao, Z.W., Zhang, Y., Tang, W., et al., 2023. Possible mechanisms of oxidative stress-induced skin cellular senescence, inflammation, and cancer and the therapeutic potential of plant polyphenols. Int. J. Mol. Sci. 24 (4), 3755.

Lózsa, R., Németh, E., Gervai, J.Z., Márkus, B.G., Kollarics, S., Gyüre, Z., et al., 2023. DNA mismatch repair protects the genome from oxygen-induced replicative mutagenesis. Nucleic. Acids. Res. 51 (20), 11040–11055.

Lyubitelev, A., Studitsky, V., 2023. Inhibition of cancer development by natural plant polyphenols: molecular mechanisms. Int. J. Mol. Sci. 24 (13), 10663.

Malabadi, R.B., Sadiya, M.R., Kolkar, K.P., Mammadova, S.S., Chalannavar, R.K., Baijnath, H., et al., 2024. Triple negative breast cancer (TNBC): signalling pathways-role of plant-based inhibitors. Open Access Res. J. Biol. Pharm. 10 (2), 028–071.

Manna, I., Das, D., Mondal, S., Bandyopadhyay, M., 2020. Potential pharmacotherapeutic phytochemicals from Zingiberaceae for cancer prevention. In: Kumar, M., Sharma, A., Kumar, P. (Eds.), Pharmacotherapeutic Botanicals for Cancer Chemoprevention. Springer Nature, Singapore, pp. 221–281.

Martemucci, G., Portincasa, P., Centonze, V., Mariano, M., Khalil, M., D'Alessandro, A.G., 2023. Prevention of oxidative stress and diseases by antioxidant supplementation. Med. Chem. 19 (6), 509–537.

McCarthy-Leo, C., Darwiche, F., Tainsky, M.A., 2022. DNA repair mechanisms, protein interactions and therapeutic targeting of the MRN complex. Cancers 14 (21), 5278.

Mecca, M., Sichetti, M., Giuseffi, M., Giglio, E., Sabato, C., Sanseverino, F., et al., 2024. Synergic role of dietary bioactive compounds in breast cancer chemoprevention and combination therapies. Nutrients 16 (12), 1883.

Merlin, J.J., Rupasinghe, H.V., Dellaire, G., Murphy, K., 2021. Role of dietary antioxidants in p53-mediated cancer chemoprevention and tumor suppression. Oxid. Med. Cell. Long. 2021 (1), 9924328.

Mintz, J., Vedenko, A., Rosete, O., Shah, K., Goldstein, G., Hare, J.M., et al., 2021. Current advances of nitric oxide in cancer and anticancer therapeutics. Vaccines 9 (2), 94.

Mireles-Elizalde, P., Martínez-Flores, R., Zugasti-Cruz, A., Salinas-Santander, M., Cepeda-Nieto, A.C., Vera-Reyes, I., et al., 2023. Polyphenol-mediated epigenetic regulation of human cancer breast. Stud. Nat. Prod. Chem. 76, 149–172.

Mirzaei, S., Gholami, M.H., Zabolian, A., Saleki, H., Bagherian, M., Torabi, S.M., et al., 2023. Resveratrol augments doxorubicin and cisplatin chemotherapy: a novel therapeutic strategy. Current Mol. Pharmacol. 16 (3), 280–306.

Molinaro, C., Martoriati, A., Cailliau, K., 2021. Proteins from the DNA damage response: regulation, dysfunction, and anticancer strategies. Cancers 13 (15), 3819.

Morgan, D., Berggren, K.L., Spiess, C.D., Smith, H.M., Tejwani, A., Weir, S.J., et al., 2022. Mitogen-activated protein kinase-activated protein kinase-2 (MK2) and its role in cell survival, inflammatory signaling, and migration in promoting cancer. Mol. Carcinog. 61 (2), 173–199.

Mukherjee, P.K., Banerjee, S., Kar, A., 2021. Molecular combination networks in medicinal plants: understanding synergy by network pharmacology in Indian traditional medicine. Phytochem. Rev. 20 (4), 693–703.

Mukherjee, P.K., Banerjee, S., Gupta, B.D., Kar, A., 2022. Evidence-based validation of herbal medicine: translational approach. In: Mukherjee, P.K. (Ed.), Evidence-based Validation of Herbal Medicine. Elsevier, Netherlands, pp. 1–41.

Murai, T., Matsuda, S., 2023. Pleiotropic signaling by reactive oxygen species concerted with dietary phytochemicals and microbial-derived metabolites as potent therapeutic regulators of the tumor microenvironment. Antioxidants 12 (5), 1056.

Mustafa, M., Ahmad, R., Tantry, I.Q., Ahmad, W., Siddiqui, S., Alam, M., et al., 2024. Apoptosis: a comprehensive overview of signaling pathways, morphological changes, and physiological significance and therapeutic implications. Cells 13 (22), 1838.

Nath, S., Roy, S., 2021. Genomic instability in carcinogenesis: the role of oxidative stress. In: *Handbook of Oxidative Stress in Cancer*: Mechanistic Aspects. Springer Nature, Singapore, pp. 1–19.

Oh, J.M., Myung, K., 2022. Crosstalk between different DNA repair pathways for DNA double strand break repairs. Mutation Res./Genetic Toxicol. Environ. Mutagen. 873, 503438.

Pande, 2 S., Raisuddin, S., 2022. The underexplored dimensions of nutritional hormesis. Current Nutrit. Reports 11 (3), 386–394.

Pandey, P., Khan, F., Alshammari, N., Saeed, A., Aqil, F., Saeed, M., 2023. Updates on the anticancer potential of garlic organosulfur compounds and their nanoformulations: plant therapeutics in cancer management. Front. Pharmacol. 14, 1154034.

Patra, A., Choudhuri, I., Paria, P., Samanta, A., Khanra, K., Chakraborty, A., 2024. Phytochemicals as potential DNA polymerase β inhibitors for targeted ovarian cancer therapy: an in-silico approach. Biosci. Biotechnol. Res. Asia 21 (2), 617–631.

Paudel, S., Mishra, N., Agarwal, R., 2023. Phytochemicals as immunomodulatory molecules in cancer therapeutics. Pharmaceuticals 16 (12), 1652.

Piscone, A., Gorini, F., Ambrosio, S., Noviello, A., Scala, G., Majello, B., et al., 2025. Targeting the 8-oxodG base excision repair pathway for cancer therapy. Cells 14 (2), 112.

Prabhu, K.S., Kuttikrishnan, S., Ahmad, N., Habeeba, U., Mariyam, Z., Suleman, M., et al., 2024. H2AX: a key player in DNA damage response and a promising target for cancer therapy. Biomed. Pharmacother. 175, 116663.

Pramodh, S., Raina, R., Hussain, A., Bagabir, S.A., Haque, S., Raza, S.T., et al., 2022. Luteolin causes 5′ CpG demethylation of the promoters of TSGs and modulates the aberrant histone modifications, restoring the expression of TSGs in human cancer cells. Int. J. Mol. Sci. 23 (7), 4067.

Prasad, S., Kumar, V., Singh, C., Singh, A., 2023. Crosstalk between phytochemicals and inflammatory signaling pathways. Inflammopharmacology 31 (3), 1117–1147.

Prasanth, M.I., Sivamaruthi, B.S., Cheong, C.S.Y., Verma, K., Tencomnao, T., Brimson, J.M., et al., 2024. Role of epigenetic modulation in neurodegenerative diseases: implications of phytochemical interventions. Antioxidants 13 (5), 606.

Prime, S.S., Darski, P., Hunter, K.D., Cirillo, N., Parkinson, E.K., 2024. A review of the repair of DNA double strand breaks in the development of oral cancer. Int. J. Mol. Sci. 25 (7), 4092.

Pua, L.J.W., Mai, C.W., Chung, F.F.L., Khoo, A.S.B., Leong, C.O., Lim, W.M., et al., 2022. Functional roles of JNK and p38 MAPK signaling in nasopharyngeal carcinoma. Int. J. Mol. Sci. 23 (3), 1108.

Qadir Nanakali, N.M., Maleki Dana, P., Sadoughi, F., Asemi, Z., Sharifi, M., Asemi, R., et al., 2023. The role of dietary polyphenols in alternating DNA methylation in cancer. Crit. Rev. Food Sci. Nutr. 63 (33), 12256–12269.

Qin, S., He, X., Lin, H., Schulte, B.A., Zhao, M., Tew, K.D., et al., 2021. Nrf2 inhibition sensitizes breast cancer stem cells to ionizing radiation via suppressing DNA repair. Free Radical Biol. Med. 169, 238–247.

Rahman, M.M., Islam, M.R., Shohag, S., Ahasan, M.T., Sarkar, N., Khan, H., et al., 2022. Microbiome in cancer: role in carcinogenesis and impact in therapeutic strategies. Biomed. Pharmacother. 149, 112898.

Rajendran, P., Abdelsalam, S.A., Renu, K., Veeraraghavan, V., Ben Ammar, R., Ahmed, E.A., 2022. Polyphenols as potent epigenetics agents for cancer. Int. J. Mol. Sci. 23 (19), 11712.

Rajput, A., Mandlik, S., Dawre, S., Mandlik, D., Pingale, P., Butani, S., 2022. Herbal bioenhancers in nanoparticulate drug delivery system. In: Pingale, P.L. (Ed.), Drug Delivery Technology. De Gruyter, Boston, MA, pp. 45–86.

Rall-Scharpf, M., Schlotter, D., Koch, P., Szafranski, K., Groth, M., Sahm, A., et al., 2025. Replication stress responses in human lymphocytes change sex-specifically during aging. Nucleic. Acids. Res. 53 (11), gkaf498.

Ramos, L., Palacios, J., Barrientos, R., Gómez, J., Castagnini, J., Barba, F., Simirgiotis, M., 2023. UHPLC-MS phenolic fingerprinting, aorta endothelium relaxation effect, antioxidant, and enzyme inhibition activities of azara dentata ruiz & pav berries. Foods 12 (3), 643. doi:10.3390/foods12030643.

Raudone, L., Savickiene, N., 2024. Phytochemical profiles of plant materials: from extracts to added-value ingredients. Plants 13 (7), 964. doi:10.3390/plants13070964.

Remigante, A., Spinelli, S., Straface, E., Gambardella, L., Caruso, D., Falliti, G., et al., 2022. Antioxidant activity of quercetin in a H2O2-induced oxidative stress model in red blood cells: functional role of band 3 protein. Int. J. Mol. Sci. 23 (19), 10991.

Rivera-Yanez, C.R., Ruiz-Hurtado, P.A., Mendoza-Ramos, M.I., Reyes-Reali, J., Garcia-Romo, G.S., Pozo-Molina, G., et al., 2021. Flavonoids present in propolis in the battle against photoaging and psoriasis. Antioxidants 10 (12), 2014.

Robertson, I., Hau, T.W., Sami, F., Ali, M.S., Badgujar, V., Murtuja, S., et al., 2022. The science of resveratrol, formulation, pharmacokinetic barriers and its chemotherapeutic potential. Int. J. Pharm. 618, 121605.

Roy, M., Datta, A., 2022. Phytochemicals in ROS-mediated epigenetic modulation of cancer. In: Chakraborti, S., Ray, B.K., Roychowdhury, S. (Eds.), Handbook of Oxidative Stress in Cancer: Mechanistic Aspects. Springer Nature, Singapore, pp. 1583–1600.

Roy, S., Deka, D., Kondaveeti, S.B., Ayyadurai, P., Siripragada, S., Philip, N., et al., 2025. An overview of potential of natural compounds to regulate epigenetic modifications in colorectal cancer: a recent update. Epigenetics 20 (1), 2491316.

Rudrapal, M., Khairnar, S.J., Khan, J., Dukhyil, A.B., Ansari, M.A., Alomary, M.N., et al., 2022. Dietary polyphenols and their role in oxidative stress-induced human diseases: Insights into protective effects, antioxidant potentials and mechanism (s) of action. Front. Pharmacol. 13, 806470.

Saitoh, T., Oda, T., 2021. DNA damage response in multiple myeloma: the role of the tumor microenvironment. Cancers 13 (3), 504.

Salame, N., Fooks, K., El-Hachem, N., Bikorimana, J.P., Mercier, F.E., Rafei, M., 2022. Recent advances in cancer drug discovery through the use of phenotypic reporter systems, connectivity mapping, and pooled CRISPR screening. Front. Pharmacol. 13, 852143.

Saleh, H.A., Yousef, M.H., Abdelnaser, A., 2021. The anti-inflammatory properties of phytochemicals and their effects on epigenetic mechanisms involved in TLR4/NF-κB-mediated inflammation. Front. Immunol. 12, 606069.

Saqib, M., Din, Z.S., Zafar, S., Munawar, N., Nawaz, R., Ahmed, S., et al., 2024. Lung cancer, platinum analog-based frontline treatment and pharmacogenetic limitations. Personal. Med. 21 (6), 385–400.

Sarvizadeh, M., Hasanpour, O., Naderi Ghale-Noie, Z., Mollazadeh, S., Rezaei, M., Pourghadamyari, H., et al., 2021. Allicin and digestive system cancers: from chemical structure to its therapeutic opportunities. Front. Oncol. 11, 650256.

Savairam, V.D., Patil, N.A., Borate, S.R., Ghaisas, M.M., Shete, R.V., 2023. Allicin: a review of its important pharmacological activities. Pharmacol. Res.-Modern Chin. Med. 8, 100283.

Schulz, W.A., 2023. DNA damage and DNA repair. In: Schulz, W.A. (Ed.), Molecular Biology of Human Cancers. Springer, Cham, pp. 51–73.

Senrung, A., Tripathi, T., Aggarwal, N., Janjua, D., Yadav, J., Chaudhary, A., et al., 2024. Phytochemicals showing antiangiogenic effect in pre-clinical models and their potential as an alternative to existing therapeutics. Curr. Top. Med. Chem. 24 (4), 259–300.

Seo, C.S., Jung, M.S., Shin, H.K., Lee, M.Y., 2022. In vitro and in vivo genotoxicity assessments and phytochemical analysis of the traditional herbal prescription Siryung-Tang. Molecules 27 (13), 4066.

Sharma, E., Tewari, M., Sati, P., Sharma, I., Attri, D.C., Rana, S., et al., 2024. Serving up health: how phytochemicals transform food into medicine in the battle against cancer. Food Front. 5 (5), 1866–1908.

Sheikh, A.N., Bhat, G.A., Kumar, S., Mudassar, S., 2025. Role of signaling pathways regulating genomic stability in cancer. In: Rehman, M.U., Khan, M.S. (Eds.), Cell Signaling Pathways and Their Therapeutic Implication in Cancers. Springer Nature, Singapore, pp. 395–435.

Shiau, J.P., Chuang, Y.T., Tang, J.Y., Yang, K.H., Chang, F.R., Hou, M.F., et al., 2022. The impact of oxidative stress and AKT pathway on cancer cell functions and its application to natural products. Antioxidants 11 (9), 1845.

Shoaib, S., Ansari, M.A., Ghazwani, M., Hani, U., Jamous, Y.F., Alali, Z., et al., 2023. Prospective epigenetic actions of organo-sulfur compounds against cancer: perspectives and molecular mechanisms. Cancers 15 (3), 697.

Sichetti, M., Giuseffi, M., Giglio, E., Marino, G., Mecca, M., 2025. Effect of natural polyphenols on breast cancer chemoprevention and treatment. Mol. Nutr. Food Res. 69 (16), e70055.

Soto-Palma, C., Niedernhofer, L.J., Faulk, C.D., Dong, X., 2022. Epigenetics, DNA damage, and aging. J. Clin. Invest. 132 (16), e158446.

Spanakis, M., Patelarou, E., Patelarou, A., 2022. Drug-food interactions with a focus on Mediterranean diet. Appl. Sci. 12 (20), 10207.

Svolacchia, F., Brongo, S., Catalano, A., Ceccarini, A., Svolacchia, L., Santarsiere, A., et al., 2023. Natural products for the prevention, treatment and progression of breast cancer. Cancers 15 (11), 2981.

Talib, W.H., Baban, M.M., Azzam, A.O., Issa, J.J., Ali, A.Y., AlSuwais, A.K., et al., 2024. Allicin and cancer hallmarks. Molecules 29 (6), 1320.

Tessmer, I., Margison, G.P., 2023. The DNA alkyltransferase family of DNA repair proteins: common mechanisms, diverse functions. Int. J. Mol. Sci. 25 (1), 463.

Thiruvengadam, M., Venkidasamy, B., Subramanian, U., Samynathan, R., Ali Shariati, M., Rebezov, M., et al., 2021. Bioactive compounds in oxidative stress-mediated diseases: targeting the NRF2/ARE signaling pathway and epigenetic regulation. Antioxidants 10 (12), 1859.

Uyeda, M., 2022. Oxidative stress and its biological significance. In: Maurya, P.K., Qamarn, I. (Eds.), Novel Therapeutic Approaches Targeting Oxidative Stress. Academic Press, Cambridge, MA, pp. 27–76.

Vezzani, B., Carinci, M., Previati, M., Giacovazzi, S., Della Sala, M., Gafà, R., et al., 2022. Epigenetic regulation: a link between inflammation and carcinogenesis. Cancers 14 (5), 1221.

Virk, T.L., Liu, Q., Yuan, Y., Xu, X., Chen, F., 2025. Curcumin as therapeutic modulator of impaired antioxidant defense system: implications for oxidative stress-associated reproductive dysfunction. Biology 14 (7), 750.

Wadgaonkar, P., 2024. Environmental causes of cancer. In: Kesharwani, P., Thakur, C. (Eds.), Cancer Epigenetics and Nanomedicine. Academic Press, Cambridge, MA,, pp. 69–92.

Wang, C., Gao, P., Xu, J., Liu, S., Tian, W., Liu, J., et al., 2022. Natural phytochemicals prevent side effects in BRCA-mutated ovarian cancer and PARP inhibitor treatment. Front. Pharmacol. 13, 1078303.

Wang, R., Sun, Y., Li, C., Xue, Y., Ba, X., 2023. Targeting the DNA damage response for cancer therapy. Int. J. Mol. Sci. 24 (21), 15907.

Wang, Y., Deng, X., Zhou, M., 2022. DNA damage mediated by UV radiation and relative repair mechanisms in mammals. Genome Instab. Dis. 3 (6), 331–337.

Warman, D.J., Jia, H., Kato, H., 2023. Effects of thyme (Thymus vulgaris L.) essential oil on aging-induced brain inflammation and blood telomere attrition in chronologically aged C57BL/6J mice. Antioxidants 12 (6), 1178.

Weisenberger, D.J., Lakshminarasimhan, R., Liang, G., 2022. The role of DNA methylation and DNA methyltransferases in cancer. In: Jeltsch, A., Jurkowska, R.Z. (Eds.), DNA Methyltransferases-Role and Function, Vol. 1389. Springer International Publishing, Cham, pp. 317–348.

Xu, J., Lin, H., Wu, G., Zhu, M., Li, M., 2021. IL-6/STAT3 is a promising therapeutic target for hepatocellular carcinoma. Front. Oncol. 11, 760971.

Yadav, E., Yadav, P., Kamal, M.A., Verma, A., 2021. Polyphenols as modulators of oxidative stress in cancer disease. In: Tabrez, S., Khan, M.I. (Eds.), Polyphenols-based Nanotherapeutics for Cancer Management. Springer, Singapore, pp. 143–188.

Yasmin, R., Gogoi, S., Bora, J., Chakraborty, A., Dey, S., Ghaziri, G., et al., 2023. Novel insight into the cellular and molecular signalling pathways on cancer preventing effects of Hibiscus sabdariffa: a review. J. Cancer Prevent. 28 (3), 77.

Ye, Y., Ma, Y., Kong, M., Wang, Z., Sun, K., Li, F., 2023. Effects of dietary phytochemicals on DNA damage in cancer cells. Nutr. Cancer 75 (3), 761–775.

Yuan, M., Zhang, G., Bai, W., Han, X., Li, C., Bian, S., 2022. The role of bioactive compounds in natural products extracted from plants in cancer treatment and their mechanisms related to anticancer effects. Oxid. Med. Cell. Long. 2022 (1), 1429869.

Zaher, H.S., Mosammaparast, N., 2024. RNA damage responses in cellular homeostasis, genome stability, and disease. Annu. Rev. Pathol.: Mech. Dis. 20, 433–457.

Zhang, X., 2024. Epigenetic pathways in breast cancer: diagnostic markers, transgenerational impacts, and emerging therapies. Int. J. High School Res. 6 (11), 1584087.

Zhang, Z., Wang, G., Li, Y., Lei, D., Xiang, J., Ouyang, L., et al., 2022. Recent progress in DNA methyltransferase inhibitors as anticancer agents. Front. Pharmacol. 13, 1072651.

Zhou, Y., Li, X., Luo, W., Zhu, J., Zhao, J., Wang, M., et al., 2022. Allicin in digestive system cancer: from biological effects to clinical treatment. Front. Pharmacol. 13, 903259.
Zhu, J., Cheng, X., Naumovski, N., Hu, L., Wang, K., 2025. Epigenetic regulation by quercetin: a comprehensive review focused on its biological mechanisms. Crit. Rev. Food Sci. Nutr. 65 (4), 627–646.
Zuccari, G., Alfei, S., 2023. Development of phytochemical delivery systems by nano-suspension and nano-emulsion techniques. Int. J. Mol. Sci. 24 (12), 9824.

CHAPTER

19

Significance of algal green chemicals in repair mechanisms

Ram Lal, Megha Jaiswal, Sukriti Chauhan and Vinod K. Kannaujiya

Department of Botany, Mahila Mahavidyalaya, Banaras Hindu University, Varanasi, Uttar Pradesh, India

19.1 Introduction

The term algae refers to a diverse collection of microorganisms that can vary in size from tiny blue-green algae to massive, intricate seaweeds that span several meters. Based on physical dimensions, algae are distinguished as macroalgae and microalgae. Macroalgae are large multicellular organisms that can be seen with the naked eye. Microalgae, on the other hand, are composed of minuscule individual cells that may have either prokaryotic (cyanobacteria) or eukaryotic, such as green algae (Chlorophyta) (Sarker and Kaparaju, 2024). These photosynthetic organisms (macro and microalgae) thrive in diverse ecological niches, ranging from terrestrial to aquatic ecosystems. They can adapt to habitats such as reservoirs or deserts and show resilience in a variety of environmental circumstances, including salinity, temperature, pH, and illumination levels (Sarker and Salam, 2020). Algal diversity adapts to different environmental stresses and flourishes by synthesizing various types of green chemicals. The algal green chemicals refer to a diverse group of bioactive compounds synthesized by both macroalgae and microalgae, which exhibit significant ecological and physiological functions (Cadar et al., 2025; Lal and Kannaujiya, 2025). There are various kinds of algal-derived green chemicals, such as pigments (chlorophyll, carotenoids, phycobiliproteins [PBPs]) (Patel et al., 2022; Chen et al., 2023), sterols/polyphenols, polysaccharides, lipid derivatives (fatty acids) (Matin et al., 2024; Pandey and Kannaujiya, 2024), and mycosporine-like amino acids (MAAs) (Fuentes-Tristan et al., 2019; Amin et al., 2025). Collectively, these eco-friendly green chemicals play a significant role in protecting and repairing cells, which in turn enhances environmental resilience and holds considerable potential for the development of sustainable biotechnological applications.

Genome Damage: Repair and Adaptation Across Life.
DOI: https://doi.org/10.1016/B978-0-443-42857-9.00007-4

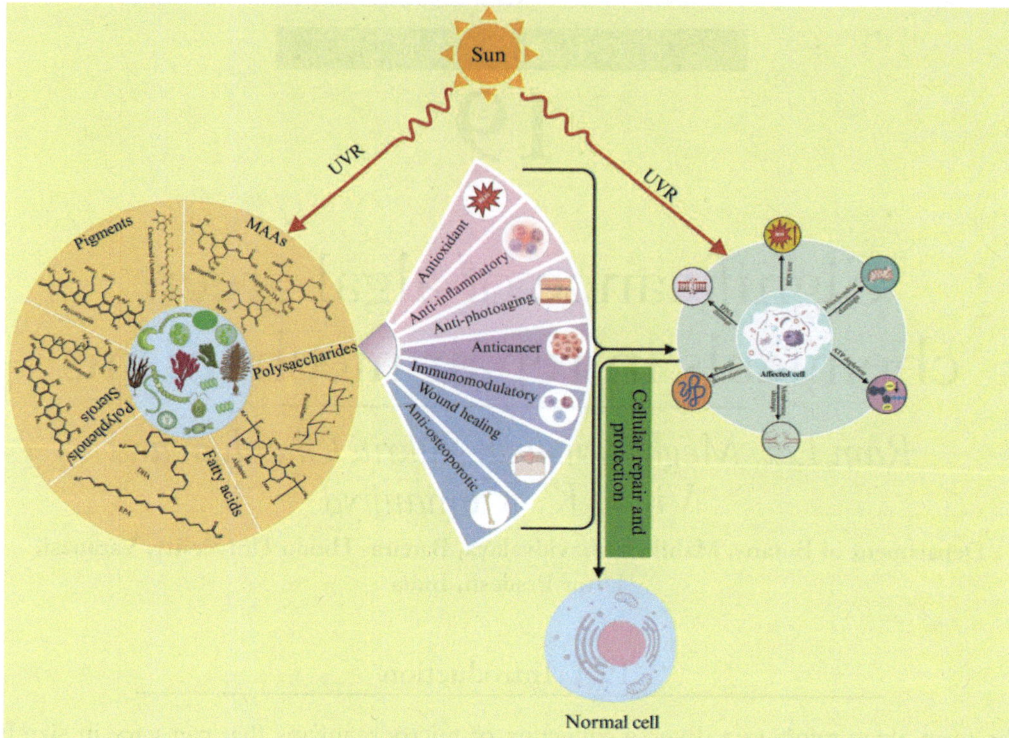

FIGURE 19.1 Algal-derived green chemicals and their biochemical properties and mechanistic role in the repair system. Figure created in https://biorender.com/.

19.2 Algal green chemicals: An overview

Green chemicals, such as pigments, MAAs, polysaccharides, fatty acids, sterols, and polyphenols, are synthesized by various kinds of macroalgae and cyanobacteria (Table 19.1). These algal-derived chemicals have various significant biological functions such as antioxidative, anti-inflammatory, antimicrobial, antiphotoaging, anticancerous, immunomodulatory, wound healing, antiosteoporotic, and regenerative effects to promote repair, regeneration, and stress mitigation (Fig. 19.1; Liu et al., 2024; Visuddho et al., 2024; Aher et al., 2025).

19.2.1 Sensitization of cancer cells to chemotherapy by certain algal compounds

Algal-derived bioactive compounds have been shown to potentiate the effects of chemotherapeutic agents by sensitizing cancer cells. Through specific interactions with cellular pathways, these metabolites enhance the susceptibility of cancer cells to chemotherapy (Visuddho et al., 2024).

For instance phycocyanin derived from *Spirulina* has been reported to induce apoptosis in multiple cancer cell types. C-phycocyanin, recognized as a TAM-targeted photosensitizer

TABLE 19.1 Major algal green chemicals and their repair roles.

Compounds/class	Representative molecules	Algal sources	Role and mechanism	References
Pigments	Phycobiliproteins (PBPs) and carotenoids	Micro as well as macroalgae	Antioxidant, photoprotection, antiaging, skin care	Kannaujiya et al. (2020, 2023), Jaiswal et al. (2026)
MAAs	Shinorine, porphyra-334	Cyanobacteria, brown, and red algae	Absorb and dissipate UVR, DNA protection, sunscreens, wound healing, anticancerous, antiphotoaging, help in cell proliferation, stimulate skin renewal, antiadipogenic	Sun et al. (2020), Punchakara et al. (2023), Pandey et al. (2024), Lal and Kannaujiya (2025), Lal et al. (2026)
Polysaccharides	Alginate, fucoidan, carrageenan, laminarins, ulvans	Brown and red algae	Wound healing, antioxidant, anti-inflammatory, antibacterial, and antifungal, anticancerous, angiogenesis, dressings, hydrogels, drug delivery	Kim et al. (2017), Anupama et al. (2023), Al-Fakeh et al. (2024), Pandey and Kannaujiya (2024), Chinjoo and Golzary (2025)
Fatty acids	Eicosapentaenoic acids (EPA) and docosahexaenoic acids (DHA)	Diatoms and green algae	Membrane stabilization, anti-inflammatory, nutraceuticals, neuroprotection, bone and cartilage repair	Thurn et al. (2022), Jesionowska et al. (2023)
Sterols/polyphenols	Fucosterol, phlorotannins, flavonoids, and nonflavonoids	Brown and red algae	Antioxidant, tissue repair, anti-inflammatory, antiphotoaging, anticancerous, antiviral, antimicrobial, antidiabetic, immunomodulatory, neuroprotective activity, hepatoprotective, antiobesity, and therapeutic supplements	Shrestha et al. (2020), Meinita et al. (2021), Jayawardhana et al. (2023), Edo et al. (2025)

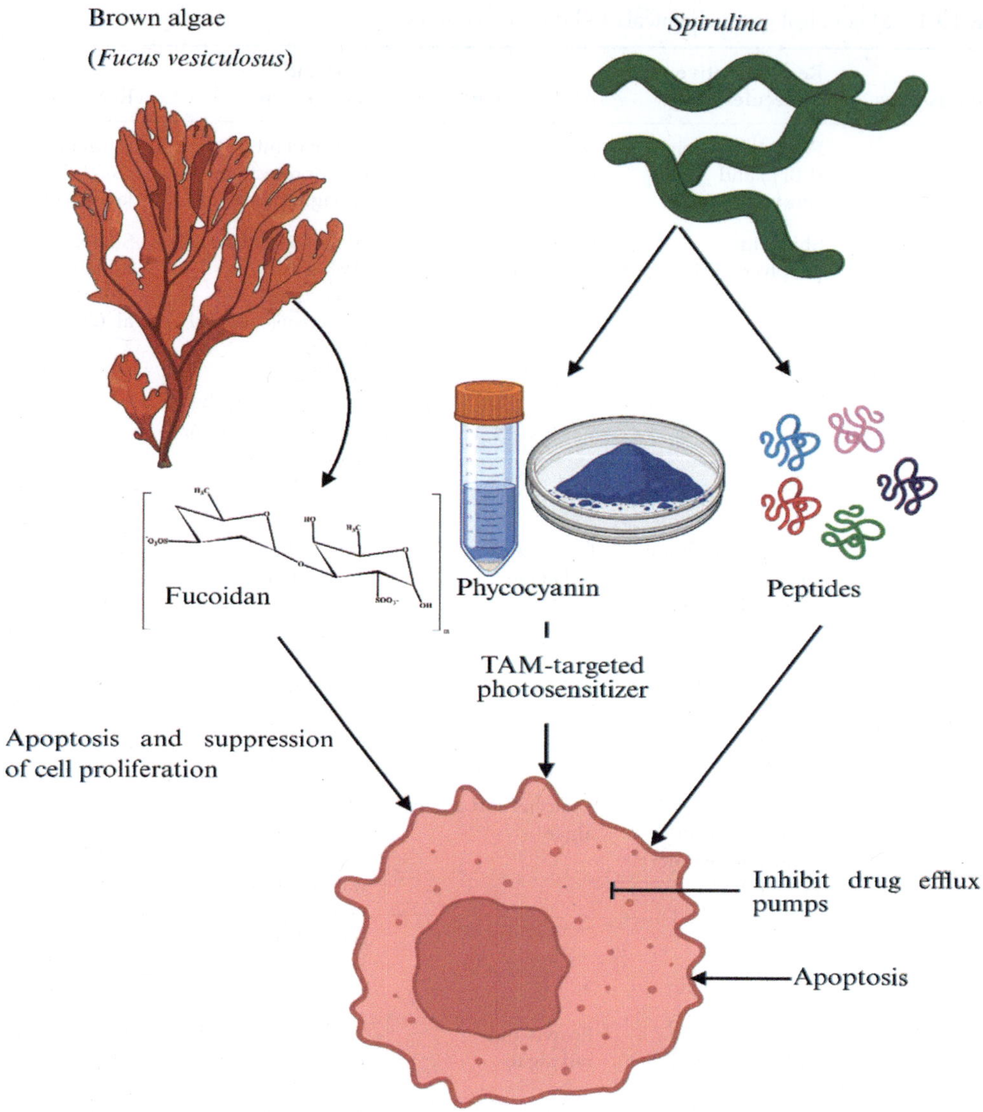

FIGURE 19.2 Schematic illustration of algal bioactive compounds sensitize cancer cells to chemotherapy.

(Fig. 19.2), exhibits strong *in vitro* photodynamic efficacy and selectively gathers in tumor tissues owing to its attraction to tumor-associated macrophages (TAMs), thereby offering a promising strategy for enhancing cancer therapeutic efficacy (Wan et al., 2017; Patel et al., 2022). In addition, peptides present in *Spirulina* have shown the ability to sensitize cancer cells for chemotherapy by regulating cell cycle arrest, apoptosis, and inhibition of drug efflux pumps (Jiang et al., 2017). Similarly, seaweeds are rich in bioactive constituents capable of triggering cancer cell death, thereby improving susceptibility to chemotherapeutic agents

(Ghorbani et al., 2018). Notably, fucoidan, a sulfated polysaccharide derived from brown algae such as *Fucus vesiculosus* (Fig. 19.2), has been demonstrated to enhance the responsiveness of cancer cells to drugs such as cisplatin and doxorubicin through the induction of apoptosis and suppression of cell proliferation (Reyes et al., 2020; Jin et al., 2021).

19.2.2 Membrane disruption

Marine anticancer peptides exert their cytotoxic effects primarily by depolarizing cancer cell membranes through membrane disruption, which results in loss of osmotic balance and leakage of cytoplasmic contents. As a consequence, cell lysis or cell death occurs through necrotic pathways. Furthermore, these peptides, having low reactive oxygen species (ROS) activity, are suggested to contribute to cancer prevention (Huang et al., 2011; Gaspar et al., 2013; Harris et al., 2013; Raucher and Ryu, 2015; Teerasak et al., 2016; Xie et al., 2020; Shaik and Sarbon, 2022).

19.2.3 Bioactive metabolites and modulation of in vitro antioxidant activity

ROSs are highly reactive oxygen-derived ions and molecules that cause significant threats to cellular components such as lipids, proteins, and nucleic acids. They possess the ability to oxidize these macromolecules by disturbing their functional integrity. These species include oxygen-containing radicals such as peroxyl ($ROO^{\bullet}$), alkoxyl ($RO^{\bullet}$), sulfonyl ($ROS^{\bullet}$), hydroxyl ($OH^{\bullet}$), hydroperoxyl ($HO_2^{\bullet}$), superoxide ($O_2^{\bullet}$), thiyl peroxyl ($RSOO^{\bullet}$), and nitric oxide ($NO^{\bullet}$), along with nonradical oxidants like singlet oxygen ($^1O_2$), hydrogen peroxide (H_2O_2), hypochlorous acid (HOCl), and organic hydroperoxides (ROOH) (Halliwell, 2003; Bayir, 2005; Storz, 2005; Pradhan et al., 2020). An increase in ROS above the physiological level results in the onset of various chronic disorders, such as cancer, aging, neurodegenerative disorders, and inflammatory diseases (Bayir, 2005). Although cells are equipped with antioxidant defense systems to counteract ROS and preserve cellular redox balance via enzymatic and nonenzymatic pathways (Amin et al., 2024; Amin and Kannaujiya, 2025). Antioxidants are therefore essential for protecting biomolecules from oxidative damage, thus maintaining balanced cellular metabolic function (Gechev et al., 2006). However, the widespread use of synthetic antioxidants (butylated hydroxyanisole, butylated hydroxytoluene, tertiary-butylhydroquinone, and propyl gallate) has raised safety concerns due to their potential side effects (Ribeiro et al., 2019). This has allowed a shift in research interest toward natural alternatives. Marine algae are prospective sources of antioxidants and have a wide range of bioactive chemicals, such as phycobilins, phycocyanin, carotenoids, astaxanthin, polyphenols, and vitamins (Mourelle et al., 2017; Coulombier et al., 2021). These bioactive chemicals have the ability to neutralize ROS and confer protective effects against cancer, diabetes, inflammation, aging, and immune dysfunction (Takamatsu et al., 2003). The antioxidant capacity of micro and macroalgal-derived chemicals has been extensively documented, which suggests that they might be useful in therapeutic and nutraceutical applications (Pradhan et al., 2020; Choudhary et al., 2021). The antioxidant activity of algal compounds has been evaluated using multiple assays, including DPPH (2,2-diphenyl-1-picrylhydrazyl), FRAP (ferric reducing antioxidant power), ABTS (2,2′-azino-bis (3-ethylbenzothiazoline-6-sulfonic acid), SRSA (superoxide radical scavenging activity), and HRSA (hydroxyl radical scavenging assays), lipid peroxide inhibition, and nitric oxide (NO) scavenging activity (Coulombier et al., 2021; Tziveleka et al., 2021). For instance methanolic extracts of blue-green algae have shown considerable DPPH radical scavenging activity. In addition, phycocyanin isolated from *Spirulina platensis* exhibited significant H_2O_2 scavenging potential (Estrada et al., 2001). Likewise, the antioxidant characteristics of green algae such as *Ulva fasciata* and *Ulva reticulata* are attributed to flavonoid content, which imparts free-radical-scavenging activities (Rao et al., 2004; Meenakshi et al., 2009; Chakraborty and Paulraj, 2010). Brown algal phlorotannins, including *Eisenia bicyclis, Ecklonia cava,* and *Ecklonia kurome,* exhibited significant antioxidant properties (Shibata et al., 2007; Li et al., 2011). Similarly, ethanolic extracts derived from red algae, such as *Gracilaria tenuistipitata* and *Callophyllis japonica,* has also exhibited significant antioxidant potentials (Kang et al., 2005; Yang et al., 2012).

19.2.3.1 Phenolic compounds

Phenols are a class of chemical compounds characterized by a hydroxyl substituent that is directly bonded to an aromatic ring. The amount of phenolic units is the primary factor that determines the classification of these compounds, such as simple phenols and polyphenols (Coulombier et al., 2021; Jimenez-Lopez et al., 2021). Among marine-derived phenolics, bromophenols (BPs) represent a notable group of secondary metabolites composed of one or more phenolic rings substituted with bromine atoms. Numerous BPs have been isolated from diverse marine macroalgae (red, brown, and green), as well as ascidians and sponges (Liu et al., 2011). One other significant category of marine polyphenols is known as phlorotannins. Phlorotannins are a group of secondary metabolites that include marine algae and are characterized by their structural diversity (Li et al., 2011; Agarwal et al., 2023).

Recent studies have identified BPs as promising candidates for the prevention of diseases linked to free radical-induced damage (Liu et al., 2011). More than 60 different BPs primarily isolated from marine red algae exhibit considerable in vitro antioxidant activities. The DPPH radical scavenging test is the most frequent method used to evaluate antioxidant activity (Barciela et al., 2024). Previous investigations found that nonbromophenolic compounds, including 5-(hydroxymethyl)-2-methoxybenzene-1,3-diol and 3,4-dihydroxybenzoic acid (DBA), exhibit notable antioxidant properties (Lee et al., 2007; Cha et al., 2014). Human HaCaT keratinocyte cells exposed to UVB or H_2O_2, lead to the generation of ROS. It has been shown that DBA may reduce the amount of ROS produced inside. This provides protection against the oxidative stress that is caused by UVB (Cha et al., 2014).

Ryu et al. (2019) reported that 3-bromo-4,5-dihydroxybenzaldehyde protects human keratinocytes from oxidative stress by activating the protein kinase B (Akt) and extracellular signal-regulated kinase (ERK) pathways, leading to the enhancement of nuclear factor erythroid 2-related factor 2 (Nrf2)-mediated transcription of the antioxidant enzyme heme oxygenase-1 (HO-1).

Human umbilical vein endothelial cells (HUVECs), which are very vulnerable to oxidative damage, have also shown that dieckol and diphlorethohydroxycarmalol protect against high-glucose-induced oxidative stress (Heo et al., 2010; Lee et al., 2010). Also, treatment with 6,6′-bieckol significantly reduced high-glucose-induced cell death in HUVECs, along with decreased intracellular ROS and NO levels, and thiobarbituric acid reactive substances (TBARS) when given in a dose-dependent manner. Moreover, high glucose levels enhanced the production of nitric oxide synthase (iNOS), cyclooxygenase-2 (COX-2), and nuclear factor-kappa B (NF-κB), but the administration of 6,6′-bieckol significantly inhibited their overexpression (Park et al., 2014; Tziveleka et al., 2021).

19.2.3.2 Meroterpenoids

Algal meroterpenoids are terpenoid-derived secondary metabolites, which have gained attention over the last two decades for their high antioxidative activities (Fuloria et al., 2022). De los Reyes et al. (2013, 2016) reported the isolation of meroterpenoids with notable radical scavenging properties from the brown alga *Cystoseira usneoides*. Among these, cystodiones A, B, G, and H, cystomexicone B, amentadione, amentadione 1′-methyl ether, 6-cis-amentadione 1′-methyl ether, and 11-hydroxyamentadione demonstrated strong antioxidant activity in the ABTS assay, exhibiting 77%–115% efficacy relative to the standard antioxidant Trolox.

Fisch et al. (2003) reported that the isolation of several triprenyltoluquinol derivatives from *Cystoseira crinita*, a brown alga, has shown to exhibit strong radical scavenging activity at a concentration of 230 μM (92.5%–96.7%), comparable to that of α-tocopherol (95.2%) (Fisch et al., 2003). In contrast, the co-occurring quinones displayed considerably lower DPPH radical scavenging activity than both α-tocopherol and the hydroquinones, with values comparable to butylated hydroxytoluene (BHT; 35.6% at 230 μM). The differences observed in the DPPH assay were attributed to minor impurities in the samples (autoxidation) and challenges associated with handling small amounts of material, rather than inherent structural variations. The TBARS assay showed that all hydroquinones exhibited a significant inhibition of linolenic acid methyl ester peroxidation. The inhibition ranged from 66.5% to 74.9% at a concentration of 164 μM, which was equivalent to the inhibition of α-tocopherol (72.7%) and BHT (69.3%) (Giner et al., 2022). Furthermore, these compounds exhibited activities corresponding to 13%–59% of α-tocopherol in the Trolox equivalent antioxidant capacity (TEAC) assay and 40%–112% of α-tocopherol in the photochemiluminescence (PCL) assay.

19.2.3.3 Nitrogenous compounds

Peptides and alkaloids isolated from *Porphyra yezoensis, Porphyra dioica, and Dictyota coriacea* are documented to exhibit moderate to significant antioxidant activity (Tamura et al., 1998; Cermeño et al., 2019; Yan et al., 2019). Histidine-related dipeptides such as carnosine and anserine have been shown to exhibit antioxidant activities comparable to those of α-tocopherol and BHT, as demonstrated by ferric thiocyanate and TBARS assays (Tamura et al., 1998). Similarly, Cermeño et al. (2019) reported the isolation of bioactive peptides from *Porphyra dioica* that displayed strong antioxidant activity, as estimated via oxygen radical absorbance capacity (ORAC) assay (Cermeño et al., 2019). Notably, peptides containing tyrosine residues were found to possess particularly high levels of antioxidant activity.

Similarly, MAAs demonstrated relatively weak free-radical-scavenging capacity compared with standard antioxidants such as Trolox and ascorbic acid (Gacesa et al., 2018). MAAs act as UV-absorbing photoprotective compounds that reduce ROS production by dissipating high energy via heat (Lal et al., 2025). Interestingly, heat stress of porphyra-334 at temperatures exceeding 100°C led to its dehydrated derivative, which exhibited more than a 100-fold increase in DPPH radical scavenging activity (Yoshiki et al., 2009).

19.2.3.4 Carbohydrates and polysaccharides

Mannitol is the simplest sugar alcohol found abundantly in macroalgae, accounting for up to 59%, 47%, and 9% of the dry weight in species belonging to Ochrophyta, Rhodophyta, and Chlorophyta, respectively (Andrade et al., 2013). The results of antioxidant activity assays that involved both enzymatic targets (α-glucosidase, acetylcholinesterase [AChE], butyrylcholinesterase [BuChE]) and free radicals (DPPH, NO, OH, O_2^-) demonstrated that an increased mannitol content is strongly associated with the inhibition of cholinesterase and the scavenging of DPPH radicals. Additionally, increased mannitol content contributed, albeit to a lesser extent, to the inhibition of α-glucosidase and the scavenging of OH, O_2^-, and NO radicals (Agarwal et al., 2023; Pandey and Kannaujiya, 2025). Two simple glucosides, D-isofloridoside and floridoside, were isolated from the red alga *Laurencia undulata*, and their antioxidant properties were extensively characterized (Li et al., 2009). Evaluations included suppression of intracellular ROS levels, free-radical-scavenging activity, inhibition of membrane protein oxidation, and myeloperoxidase (MPO) activity. Additionally, they modulate the gene expression for GSH and SOD, and protein expression of MMP2 and MMP9. Both compounds showed significant antioxidant activity, which harnessed potent inhibition of MMP2 and MMP9, underscoring their therapeutic relevance (Li et al., 2009; Yang et al., 2020).

Marine macroalgae represent the primary nonanimal source of sulfated polysaccharides (SPs), which are classified mainly as ulvans from green algae, fucoidans from brown algae, and carrageenans and porphyrans from red algae. These SPs exhibit notable in vitro antioxidant activity, encompassing both radical scavenging and metal chelating properties (Rupérez et al., 2002; Wang et al., 2007; Hu et al., 2009). Their antioxidant potential is strongly influenced by structural characteristics, including degree of sulfation, molecular weight (MW), monosaccharide composition, and glycosidic branching patterns (Zhang et al., 2003; Qi et al., 2005; Wang et al., 2007). For instance low-MW SPs have been shown to display greater antioxidant activity than their high-MW counterparts (Sun et al., 2014), likely due to their enhanced cellular uptake and superior proton-donating capacity (Tziveleka et al., 2021).

19.2.4 Tissue and organismal repair

Tissue damage triggers a series of intricate cellular and molecular reactions. These damaged tissues are essential for restoring function and maintaining structural integrity. Various traditional approaches have been employed for tissue repairing, but frequently exhibit some drawbacks, such as a slow healing rate, prone to infection, and the formation of scars. Recently, algal-derived chemicals such as pigments, MAAs, polysaccharides, fatty acids, and polyphenols (Table 19.2) have become popular as biocompatible and efficient alternatives that influence tissue microenvironment during the healing process (Favier and Nikovics, 2023; Tavares et al., 2023). These green chemicals play a significant role in biological repair mechanisms, including skin (wound healing), bone, and internal organs (Liu et al., 2024; Aher et al., 2025; Lal and Kannaujiya, 2025). Their effects are extensive because they exhibit antioxidant, anti-inflammatory, antibacterial, and regenerative properties, which lead to repair, regeneration, and homeostasis (Kumari et al., 2023).

19.2.4.1 Algal chemicals in repair and wound healing

In the process of healing a wound, there are a number of biological phases that take place. These steps include clotting, inflammation, proliferation of tissues, maturation, and reformation. Conventional methods of wound management often include the use of biosynthetic dressings and topical medicines, both of which may have limited efficacy and may also have adverse consequences. Polysaccharides, phenolics, and fatty acids are some of the beneficial compounds that

TABLE 19.2 Comparative roles of algal and marine-derived bioactive compounds in wound healing.

Bioactive compound	Source organism	Antioxidant activity	Anti-inflammatory action	Cell migration and proliferation	ECM synthesis (collagen/elastin)	Angiogenesis	Antimicrobial properties	Key references
Phycocyanin	*Spirulina platensis* (Cyanobacteria)	Scavenges ROS during inflammation	Reduces oxidative stress and cytokine levels	Stimulates keratinocytes and fibroblasts	Enhances collagen synthesis	Not reported	Inhibits bacterial/fungal growth	Wu et al. (2016), Liu et al. (2024), Pandian et al. (2024)
Mycosporine-like amino acids (MAAs)	*Porphyra spp., Chondrus crispus, Gracilaria spp.* (Red algae)	Scavenges ROS; protects the ECM	Suppresses MMPs and cytokines	Activates FAK/ERK/JNK pathways	Preserves collagen/elastin structure	Not reported	Not the primary mechanism	Choi et al. (2015), Singh et al. (2021), Punchakara et al. (2023)
Omega-3 PUFAs (EPA, DHA)	*Schizochytrium* spp., *Nannochloropsis* spp. (Microalgae); Fish oils	Regulates ROS balance; enhances antioxidant defenses	Resolvin-mediated cytokine suppression	Enhances membrane signaling for migration	Stimulates fibroblast proliferation and collagen synthesis	Upregulates VEGF	Indirect via inflammation control	Holzinger and Karsten (2013), Pandian et al. (2024), Chinjoo and Golzary (2025)

(continued on next page)

TABLE 19.2 Comparative roles of algal and marine-derived bioactive compounds in wound healing—cont'd

Bioactive compound	Source organism	Antioxidant activity	Anti-inflammatory action	Cell migration and proliferation	ECM synthesis (collagen/elastin)	Angiogenesis	Antimicrobial properties	Key references
Fucoidan	*Fucus vesiculosus, Undaria pinnatifida* (Brown algae)	Reduces oxidative stress	Inhibits NF-κB and MAPK pathways	Promotes keratinocyte/fibroblast migration	Stimulates collagen/elastin synthesis	Induces VEGF and TGF-β	Not the primary mechanism	O'Leary et al. (2004), Apostolova et al. (2020), Lee et al. (2023), Hou et al. (2025), Naghipour et al. (2025)
Alginates	*Laminaria spp., Macrocystis pyrifera* (Brown algae)	Maintains a moist ROS-buffering environment	Binds TNF-α and MMPs	Supports migration via a moist gel matrix	Indirect via fibroblast support	Supports endothelial proliferation	Barrier, antimicrobial agent incorporation	Lee and Mooney (2011), Percival et al. (2015), Aderibigbe and Buyana (2018), Mazurek et al. (2025)
Carrageenan	*Kappaphycus alvarezii, Gigartina spp.* (Red algae)	Moisture retention reduces oxidative damage	Physical anti-inflammatory effect	Promotes migration via hydrated matrix	Supports ECM organization	Not reported	Not the primary mechanism	Neamtu et al. (2022), Amruth et al. (2024)

FAK, focal adhesion kinase; *MAPK*, mitogen-activated protein kinase; PUFA, polyunsaturated fatty acid; *VEGF*, vascular endothelial growth factor; JNK, c-Jun N-terminal kinases; MMPs, matrix metalloproteinases.

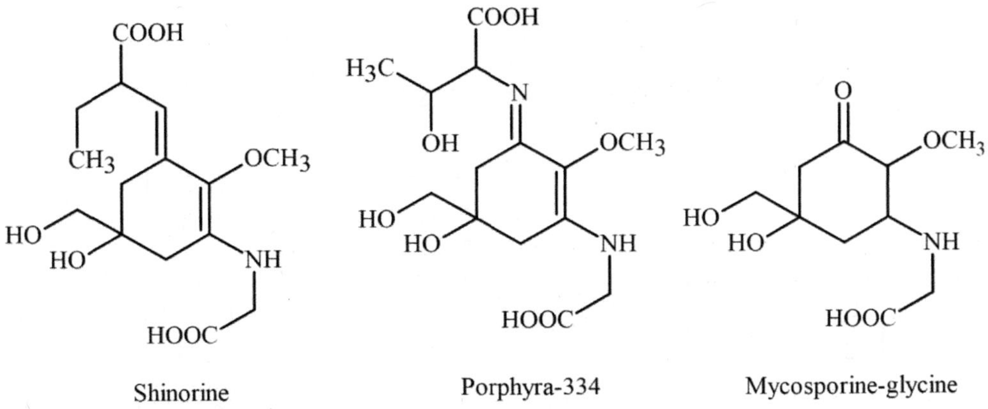

FIGURE 19.3 Molecular structure of shinorine, porphyra-334, and mycosporine-glycine.

may be found in algae. These molecules have antibacterial, anti-inflammatory, and antioxidant properties, and they can help in the healing process of wounds (Aher and Thitame, 2025).

Various types of algal green chemicals have been described in Table 19.2, which play a multifunctional role, including the wound-healing process. The blueish color of PBPs pigment, classified as phycocyanin (PC), is most prevalent in cyanobacteria like *Spirulina platensis*. It attracts much interest due to the multifaceted function responsible for the processes of wound healing (Wu et al., 2016). It's widely recognized that antioxidant benefits improve tissue healing. The inflammatory phase of wound healing produces an abundance of ROS, including hydroxyl radicals and superoxide anions. Phycocyanin is excellent in scavenging these ROS. When these ROS remain unregulated, they hinder the healing process by damaging cellular lipids, proteins, and DNA via oxidative stress. PC counteracts ROS, so shielding cells from the damaging effects of oxidative stress and fostering succeeding stages of proliferative and remodeling processes that are essential for wound healing (Liu et al., 2024).

Additionally, PC activates important cellular components in tissue regeneration, such as keratinocytes and fibroblasts. It improves keratinocyte proliferation and migration essential in re-epithelialization and restoration of the skin barrier function. Concurrently, PC augments fibroblast proliferation and extracellular matrix synthesis, notably collagen, thereby contributing to the strength and architectural integrity of the regenerating tissue (Pandian et al., 2024). In addition, PC exhibits significant antimicrobial activity that inhibits the growth of various bacterial and fungal pathogens, to be commonly associated with wound infections. This summarizes the effect of antimicrobial, antioxidant, and cell-regulatory activities that mitigate the infection risks by preventing chronic wounds and biofilm formation (Chinjoo and Golzary, 2025).

MAAs support wound healing through stimulating cell motility, reducing inflammation, and acting as antioxidants. Through the activation of certain intracellular signaling cascades, resulting directly promote wound closure and tissue regeneration. These cascades refer to focal adhesion kinase (FAK) and mitogen-activated protein kinases (MAPKs), such as ERK and c-Jun N-terminal kinase (JNK) (Choi et al., 2015). MAAs (shinorine, porphyra-334, and mycosporine-glycine) (Fig. 19.3) stimulate phosphorylation of FAK at Tyr397 in skin cells (keratinocytes and fibroblasts), initiating downstream activation of ERK and JNK MAP kinases. This signaling cascade accelerates cell migration (key for re-epithelialization) and proliferation, both of which

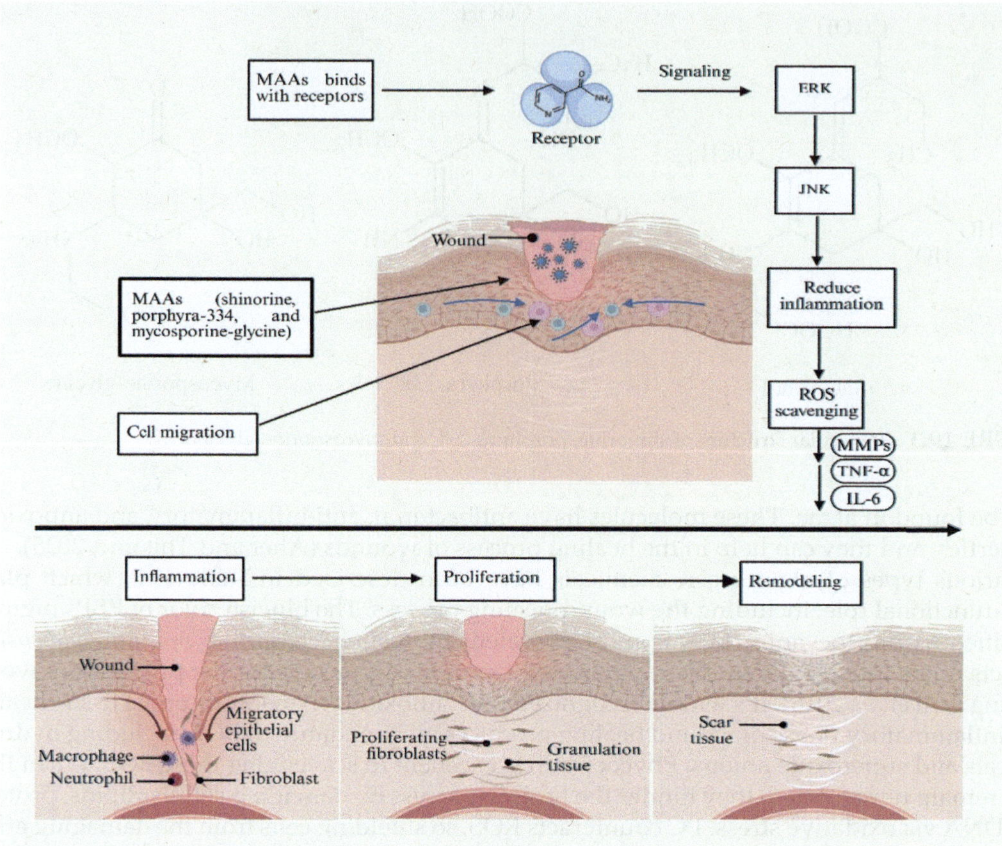

FIGURE 19.4 Mechanistic role of mycosporine-like amino acids (MAAs) in wound healing. *ERK*, Extracellular signal-regulated kinase; *IL-6*, interleukin-6; *JNK-c*, Jun N-terminal kinases; *MMPs*, matrix metalloproteinases; *TNF-α*, tumor necrosis factor-alpha.

are critical for wound closure (Fig. 19.4) (Punchakara et al., 2023). The process of wound healing is accelerated using MAAs because they stimulate the proliferation of keratinocytes and fibroblasts, resulting in a change in the wound bed. In vitro scratch wound assays have shown dose-dependent acceleration of wound closure with various MAAs. MAAs efficiently eliminate ROS produced during the inflammatory phase of wound healing, hence reducing further cellular and matrix damage. They suppress the expression of matrix metalloproteinases (MMPs) and pro-inflammatory cytokines, preserving collagen and elastin structure, and reducing scar formation (Singh et al., 2021). By inhibiting degradation and promoting synthesis of extracellular matrix (ECM) proteins,such as collagen and elastin, MAAs contribute to the proliferation and remodeling phases of wound healing, aiding in the restoration of tissue strength and elasticity.

Omega-3 polyunsaturated fatty acids (PUFAs), including eicosapentaenoic acid (EPA) and docosahexaenoic acid (DHA), help in wound healing via various interrelated pathways. Omega-3 PUFAs modulate inflammation by serving precursors to specialized pro-resolving mediators (SPMs), including resolvins and protectins. Through the reduction of neutrophil infiltration, they inhibit the pro-inflammatory cytokines,such as tumor necrosis factor-alpha (TNF-α) and interleukin-1 beta (IL-1β), and promote macrophage-mediated clearance of cellular debris; these mediators resolve inflammation. This supports the rapid progression from inflammation to tissue repair (Pandian et al., 2024; Chinjoo and Golzary, 2025). The incorporation of Omega-3 PUFAs into cellular membranes results in enhanced membrane fluidity. This, in turn, has an effect on the function of receptors and the intracellular signaling pathways involved in the processes of cell proliferation, migration, and differentiation, which are necessary for tissue regeneration. In addition, omega-3 fatty acids promote angiogenesis through upregulation of vascular endothelial growth factor (VEGF). This ensures a sufficient oxygen and nutrient supply in the healing tissues. Omega-3 PUFAs also mitigate oxidative stress through ROS regulation and enhance the cellular antioxidant defenses (Holzinger and Karsten, 2013). Furthermore, the presence of these fatty acids facilitates the proliferation of fibroblasts and the production of collagen, thus strengthening the extracellular matrix and restoring the strength and integrity of the tissue (Matin et al., 2024).

Fucoidan is a sulfated polysaccharide mostly found in brown seaweed. Fucoidan exhibits significant activity in wound healing. It modulates inflammatory responses by inhibiting key pro-inflammatory cytokines, including tumor necrosis factor-alpha (TNF-α), interleukin-6 (IL-6), and interleukin-1β (IL-1β). This modulation occurs via suppressing the nuclear factor-kappa B (NF-κB) and mitogen activated protein kinase (MAPK) signaling pathway. This reduces inflammation in the wound site and inhibits chronic wounds (Apostolova et al., 2020; Hou et al., 2025). Furthermore, fucoidan enhances the proliferation as well as migration of fibroblasts and keratinocytes, which are essential for re-epithelialization and granulation tissue formation. It also stimulates the expression of growth factors such as vascular endothelial growth factor (VEGF) and transforming growth factor-beta (TGF-β), thereby promoting tissue regeneration (O'Leary et al., 2004; Lee et al., 2023).

Fucoidan also helps synthesize and organize the extracellular matrix parts like collagen and elastin. These collagen and elastin make tissues stronger and more flexible (Naghipour et al., 2025). During the process of wound healing, the antioxidant activity of the substance helps to protect cells and extracellular structures from damage that is caused by ROS. This is accomplished by scavenging free radicals and lowering oxidative stress simultaneously.

Alginates are naturally synthesized anionic polysaccharides mostly found in brown seaweed. They have a well-known ability to help with wound care because of their unique physical, chemical, and biological features. Upon application to a wound, alginate dressings interact with wound exudate, resulting in the formation of a hydrophilic gel that maintains a moist wound environment, which is critical for optimal healing (Aderibigbe and Buyana, 2018; Mazurek et al., 2025). The gel form helps in autolytic healing by eliminating dead tissue and keeping living cells from drying out. Also, the moist environment helps keratinocytes and fibroblasts move around, which is necessary for re-epithelialization and the formation of granulation tissue (Mazurek et al., 2025). Additionally, calcium ions released from alginate dressings participate in hemostasis through interactions with the coagulation cascade, thereby aiding in bleeding control (Lee and Mooney, 2011). Beyond these physical benefits, alginates exhibit significant bioactive effects, including modulation of inflammation. They bind and reduce the levels of excessive pro-inflammatory cytokines such as tumor necrosis factor-alpha (TNF-α) and matrix metalloproteinases (MMPs), mitigating chronic inflammation and tissue degradation common in nonhealing wounds. Alginate matrices also function as a barrier to microbial infiltration and can be incorporated with antimicrobial agents to further reduce infection risk (Percival et al., 2015).

Furthermore, alginate dressings promote angiogenesis by maintaining a moist and oxygen-permeable environment that supports endothelial cell proliferation and vascularization, crucial for nutrient and oxygen delivery during tissue repair. Their excellent biocompatibility, biodegradability, and ease of painless removal render alginates among the most versatile biomaterials in modern wound care (Mazurek et al., 2025). Clinical trials and meta-analyses consistently report that alginate dressings reduce healing time, pain, and the frequency of dressing changes compared to conventional materials, underscoring their therapeutic superiority (Percival et al., 2015).

Carrageenan, a sulfated polysaccharide extracted from red seaweeds, is widely acknowledged for its beneficial role in wound healing due to its unique biochemical and physical properties. Classified mainly into kappa (κ), iota (ι), and lambda (λ) types based on sulfate group content, carrageenans exhibit excellent water solubility, biocompatibility, and gel-forming capabilities that make them ideal for wound dressings (Neamtu et al., 2022). Upon application to wounds, carrageenan retains moisture effectively and absorbs wound exudate to maintain a hydrated environment, which is essential for efficient cell migration and proliferation necessary for tissue regeneration (Amruth et al., 2024). The moist interface prevents dressing adhesion to the wound bed, thereby minimizing pain upon dressing changes and further tissue trauma (Neamtu et al., 2022).

Carrageenans exert anti-inflammatory effects by modulating local cytokine production, reducing tumor necrosis factor-alpha (TNF-α), interleukin-6 (IL-6), and other mediators, thus preventing prolonged inflammation that can delay healing (Vuković et al., 2025). They also possess antioxidant and antimicrobial activities, reducing oxidative stress and hindering microbial colonization within the wound, which supports faster wound closure and minimizes infection risks (Neamtu et al., 2022; Postolović et al., 2022). Furthermore, carrageenans stimulate fibroblast and keratinocyte proliferation and migration, key cellular processes for re-epithelialization and new tissue formation (Biranje et al., 2022). Growth factor signaling, including upregulation of VEGF, enhances angiogenesis at the wound site, ensuring adequate nutrient and oxygen supply for robust healing (Vuković et al., 2025). Carrageenan-based composite materials, such as films loaded with antimicrobial agents or incorporated into hydrogels, have demonstrated improved hemostasis, accelerated wound contraction, and enhanced collagen deposition in preclinical studies, highlighting their potential as advanced wound care biomaterials (Chanpuria et al., 2023).

19.3 Summary and prospects

Fucoidan and other algal metabolites-pigments, MAAs, polysaccharides, fatty acids, sterols, polyphenols-offer diverse bioactivities. They fight cancer and inflammation, kill bacteria and viruses, and aid wound healing while tackling oxidative stress. Algal green chemicals include pigments, MAAs, polysaccharides, and more. They exhibit anticancer, anti-inflammatory, antibacterial, and wound-healing effects by reducing oxidative stress, boosting cell growth, and promoting new blood vessels.

Progress in understanding the biochemistry and mechanisms of algal compounds has been substantial, but translational challenges persist. Key issues include biosafety, bioavailability, standardization, and scalable, sustainable production, all of which require further investigation. In addition, exploring synergistic interactions among algal metabolites and optimizing delivery platforms, such as nanoparticle-based systems and biocompatible dressings, may enhance therapeutic performance. Future studies should prioritize identifying precise molecular targets and signaling pathways, strengthening in vivo validation, and conducting rigorously designed clinical trials.

The emerging biotechnological approaches such as genetic engineering and modification in metabolic pathways, offer promising strategies in improving the yield and consistency of high-value bioactive compounds. Integrative frameworks that combine systems biology with multi-omics analyses can clarify the complex interactions of these molecules within biological systems. Overall, algal green chemicals represent promising, eco-friendly, sustainable candidates for use as therapeutics and functional ingredients in pharmaceutical, nutraceutical, and cosmeceutical applications. Progress in interdisciplinary research and technology will be crucial to fully harness their potential to improve health, accelerate tissue repair, and combat chronic diseases in ways that also support environmental sustainability.

Acknowledgements

Ram Lal is grateful to the Council of Scientific and Industrial Research (CSIR), Ministry of Science and Technology, Government of India for the award of a Junior Research Fellowship (JRF) (File No: 09/ 0013(21769)/2025-EMR-I). Megha Jaiswal is thankful to Joint CSIR-UGC for the award of JRF (NTA Ref. No: 231620098896). Sukriti Chauhan is thankful to Joint CSIR-UGC for the award of JRF (NTA Ref. No: 241610157292). Vinod K. Kannaujiya is gratefully acknowledged for financial support from ANRF (ANRF/IRG/2024/000209/LS), New Delhi, India.

References

Aderibigbe, B., Buyana, B., 2018. Alginate in wound dressings. Pharmaceutics 10, 42.

Agarwal, S., Singh, V., Chauhan, K., 2023. Antidiabetic potential of seaweed and their bioactive compounds: a review of developments in last decade. Crit. Rev. Food Sci. Nutr. 63, 5739–5770.

Aher, A.A., Thitame, S.N., 2025. Biological applications of algae in osteogenic therapies: a review of recent advances. J. Pharm. Bioallied Sci. 17, S16–S19.

Aher, A.A., Thitame, S.N., Viddyasagar, M., Wabale, A.S., 2025. Bioactive algae in wound healing: natural solutions for modern medicine. J. Pharm. Bioallied Sci. 17, S20–S23.

Al-Fakeh, M.S., Al-Subaie, N.S., El-Ghoul, Y., Hamden, Z., 2024. Preparation and properties of a novel alginate/carrageenan crosslinked coordination polymer and evaluation of the antibacterial, antioxidant and anticancer potential of its Co(II), and Cr(III) polymeric complexes. RSC Adv. 14, 38934–38943.

Amin, N., Jaiswal, M., Kannaujiya, V.K., 2024. Effects of temperature on morphology, physiology, and metabolic profile of diazotrophic cyanobacteria inhabiting diverse habitats. Plant Physiol. Biochem. 216, 109186.

Amin, N., Kannaujiya, V.K., 2025. Effect of ultraviolet radiation on morphology, physiology and antioxidant defense system of the rice-field cyanobacterium *Anabaena* sp. strain VKB01. Physiol. Plant. 177, e70544.

Amin, N., Singh, V.K., Kannaujiya, V.K., 2025. Mycosporine-like amino acids as a potential inhibitor of tyrosinase-related protein 1: computational screening, pharmacokinetics, and molecular dynamics simulation. Mol. Biotechnol. 67, 1674–1694.

Amruth, P., Akshay, P., Rosemol Jacob, M., 2024. Developmental prospects of carrageenan-based wound dressing films: unveiling techno-functional properties and freeze-drying technology for the development of absorbent films—a review. Int. J. Biol. Macromol. 276, 33668.

Andrade, P.B., Barbosa, M., Matos, R.P., Lopes, G., Vinholes, J., Mouga, T., et al., 2013. Valuable compounds in macroalgae extracts. Food Chem. 138, 1819–1828.

Anupama, D., Udduttula, A., Jaiswal, A.K., 2023. Unveiling the secrets of marine-derived fucoidan for bone tissue engineering-A review. Front. Bioeng. Biotechnol. 10, 1100164.

Apostolova, E., Lukova, P., Baldzhieva, A., Katsarov, P., Nikolova, M., Iliev, I., et al., 2020. Immunomodulatory and anti-inflammatory effects of fucoidan: a review. Polymers 12, 2338.

Balaji Raghavendra Rao, H., Sathivel, A., Devaki, T., 2004. Antihepatotoxic nature of *Ulva reticulata* (Chlorophyceae) on acetaminophen-induced hepatoxicity in experimental rats. J. Med. Food 7, 495–497.

Barciela, P., Carpena, M., Perez-Vazquez, A., Silva, A., Jorge, A.O.S., Prieto, M.A., 2024. Bromophenols in red algae: exploring the chemistry and uncovering biological benefits of these unknown compounds. Biol. Life Sci. Forum 35, 11.

Bayir, H., 2005. Reactive oxygen species. Crit. Care Med. 33, S498–S501.

Biranje, S.S., Sun, J., Cheng, L., Cheng, Y., Shi, Y., Yu, S., et al., 2022. Development of cellulose Nanofibril/Casein-based 3D Composite hemostasis scaffold for potential wound-healing application. ACS Appl. Mater. Interfaces 14, 3792–3808.

Cadar, E., Popescu, A., Dragan, A.-M.-L., Pesterau, A.-M., Pascale, C., Anuta, V., et al., 2025. Bioactive compounds of marine algae and their potential health and nutraceutical applications: a review. Mar. Drugs 23, 152.

Cermeño, M., Stack, J., Tobin, P.R., O'Keeffe, M.B., Harnedy, P.A., Stengel, D.B., et al., 2019. Peptide identification from a *Porphyra dioicaprotein* hydrolysate with antioxidant, angiotensin converting enzyme and dipeptidyl peptidase IV inhibitory activities. Food Funct 10, 3421–3429.

Cha, J.W., Piao, M.J., Kim, K.C., Zheng, J., Yao, C.W., Hyun, C.L., et al., 2014. Protective Effect of 3,4-dihydroxybenzoic acid isolated from *Cladophora wrightiana* Harvey against ultraviolet B radiation-induced cell damage in human HaCaT keratinocytes. Appl. Biochem. Biotechnol. 172, 2582–2592.

Chakraborty, K., Paulraj, R., 2010. Sesquiterpenoids with free-radical-scavenging properties from marine macroalga *Ulva fasciata* Delile. Food Chem. 122, 31–41.

Chanpuria, A., Bajpai, N.S.K., Bajpai, M., Dubey, S., 2023. Kappa-carrageenan/ciprofloxacin loaded hydrogel patch for wound healing application. Curr. J. Appl. Sci. Technol. 42, 148–164.

Chen, Z., Wu, W., Wen, Y., Zhang, L., Wu, Y., Farid, M.S., et al., 2023. Recent advances of natural pigments from algae. Food Prod. Process. Nutr. 5, 39.

Chinjoo, N., Golzary, A., 2025. Microalgae: revolutionizing skin repair and enhancement. Biotechnol. Rep. 47, e00911.

Choi, Y.-H., Yang, D., Kulkarni, A., Moh, S., Kim, K., 2015. Mycosporine-like amino acids promote wound healing through focal adhesion kinase (FAK) and mitogen-activated protein kinases (MAP Kinases) signaling pathway in keratinocytes. Mar. Drugs 13, 7055–7066.

Choudhary, B., Chauhan, O.P., Mishra, A., 2021. Edible seaweeds: a potential novel source of bioactive metabolites and nutraceuticals with human health benefits. Front. Mar. Sci. 8, 740054.

Coulombier, N., Jauffrais, T., Lebouvier, N., 2021. Antioxidant compounds from microalgae: a review. Mar. Drugs 19, 549.

De Los Reyes, C., Zbakh, H., Motilva, V., Zubía, E., 2013. Antioxidant and anti-inflammatory meroterpenoids from the brown alga *Cystoseira usneoides*. J. Nat. Prod. 76, 621–629.

De Los Reyes, C., Ortega, M.J., Zbakh, H., Motilva, V., Zubía, E., 2016. *Cystoseira usneoides*: a brown alga rich in antioxidant and anti-inflammatory merediterpenoids. J. Nat. Prod. 79, 395–405.

Edo, G.I., Jikah, A.N., Elooghene, E.M., Oberhiri, S.O., Yousif, E., Gaaz, T.S., et al., 2025. Exploring carrageenan: biological activities and applications in drug delivery and tissue engineering. Curr. Pharmacol. Rep. 11, 34.

Estrada, J.P., Bescós, P.B., Del Fresno, A.V., 2001. Antioxidant activity of different fractions of *Spirulina platensis* protean extract. Il Farmaco 56, 497–500.

Favier, A.L., Nikovics, K., 2023. Molecular and cellular mechanisms of inflammation and tissue regeneration. Biomedicines 11, 1416.

Fisch, K.M., Böhm, V., Wright, A.D., König, G.M., 2003. Antioxidative meroterpenoids from the brown alga *Cystoseira crinita*. J. Nat. Prod. 66, 968–975.

Fuentes-Tristan, S., Parra-Saldivar, R., Iqbal, H.M.N., Carrillo-Nieves, D., 2019. Bioinspired biomolecules: mycosporine-like amino acids and scytonemin from *Lyngbya* sp. with UV-protection potentialities. J. Photochem. Photobiol. B 201, 111684.

Fuloria, N.K., Raheja, R.K., Shah, K.H., Oza, M.J., Kulkarni, Y.A., Subramaniyan, V., Sekar, M., Fuloria, S., et al., 2022. Biological activities of meroterpenoids isolated from different sources. Front. Pharmacol. 13, 830103.

Gacesa, R., Lawrence, K.P., Georgakopoulos, N.D., Yabe, K., Dunlap, W.C., Barlow, D.J., et al., 2018. The mycosporine-like amino acids porphyra-334 and shinorine are antioxidants and direct antagonists of Keap1-Nrf2 binding. Biochimie 154, 35–44.

Gaspar, D., Veiga, A.S., Castanho, M.A.R.B., 2013. From antimicrobial to anticancer peptides. A review. Front. Microbiol. 4, 294.

Gechev, T.S., Van Breusegem, F., Stone, J.M., Denev, I., Laloi, C., 2006. Reactive oxygen species as signals that modulate plant stress responses and programmed cell death. Bioessays 28, 1091–1101.

Ghorbani, J., Rahban, D., Aghamiri, S., Teymouri, A., Bahador, A., 2018. Photosensitizers in antibacterial photodynamic therapy: an overview. Laser Therapy 27, 293–302.

Giner, R.M., Ríos, J.L., Máñez, S., 2022. Antioxidant activity of natural hydroquinones. Antioxidants 11, 343.

Halliwell, B., 2003. Free radicals and other reactive species in disease. Encyclopedia of Life Sciences. Wiley, New York doi:10.1038/npg.els.0002269.

Harris, F., Dennison, S.R., Singh, J., Phoenix, D.A., 2013. On the selectivity and efficacy of defense peptides with respect to cancer cells. Med. Res. Rev. 33, 190–234.

Heo, S.J., Hwang, J.Y., Choi, J.I., Lee, S.H., Park, P.J., Kang, D.H., et al., 2010. Protective effect of diphlorethohydroxycarmalol isolated from *Ishige okamurae* against high glucose-induced-oxidative stress in human umbilical vein endothelial cells. Food Chem. Toxicol. 48, 1448–1454.

Holzinger, A., Karsten, U., 2013. Desiccation stress and tolerance in green algae: consequences for ultrastructure, physiological and molecular mechanisms. Front. Plant Sci. 4, 327.

Hou, T., Liu, X., Zhang, S., Tang, K., Liu, S., Liu, J., et al., 2025. Anti-inflammatory effects of the fucoidan from sea cucumber *Apostichopus japonicus*. Mar. Biotechnol. 27, 32.

Hu, T., Liu, D., Chen, Y., Wu, J., Wang, S., 2009. Antioxidant activity of sulfated polysaccharide fractions extracted from *Undaria pinnitafida* in vitro. Int. J. Biol. Macromol. 46, 193–198.

Huang, Y.-B., Wang, X.-F., Wang, H.-Y., Liu, Y., Chen, Y., 2011. Studies on mechanism of action of anticancer peptides by modulation of hydrophobicity within a defined structural framework. Mol. Cancer Ther. 10, 416–426.

Jaiswal, M., Lal, R., Pandey, S., Amin, N., Pandey, A., Kannaujiya, V.K., 2026. Phycobiliproteins-algae-derived pigments: advances in research on the therapeutic potential. In: Ramamoorthy, S., Efferth, T. (Eds.), Bioprospecting of Natural Pigments in Human Health. Academic Press, Cambridge, MA, pp. 217–260.

Jayawardhana, H.H.A.C.K., Jayawardena, T.U., Sanjeewa, K.K.A., Liyanage, N.M., Nagahawatta, D.P., Lee, H.-G., et al., 2023. Marine algal polyphenols as skin protective agents: current status and future prospectives. Mar. Drugs 21, 285.

Jesionowska, M., Ovadia, J., Hockemeyer, K., Clews, A.C., Xu, Y., 2023. EPA and DHA in microalgae: health benefits, biosynthesis, and metabolic engineering advances. J. Am. Oil Chem. Soc. 100, 831–842.

Jiang, L., Wang, Y., Yin, Q., Liu, G., Liu, H., Huang, Y., et al., 2017. Phycocyanin: a potential drug for cancer treatment. J. Cancer 8, 3416–3429.

Jimenez-Lopez, C., Pereira, A.G., Lourenço-Lopes, C., Garcia-Oliveira, P., Cassani, L., Fraga-Corral, M., et al., 2021. Main bioactive phenolic compounds in marine algae and their mechanisms of action supporting potential health benefits. Food Chem. 341, 128262.

Jin, J.O., Chauhan, P.S., Arukha, A.P., Chavda, V., Dubey, A., Yadav, D., 2021. The therapeutic potential of the anticancer activity of fucoidan: current advances and hurdles. Mar. Drugs 19, 265.

Kang, K.A., Bu, H.D., Park, D.S., Go, G.M., Jee, Y., Shin, T., et al., 2005. Antioxidant activity of ethanol extract of *Callophyllis japonica*. Phytother. Res. 19, 506–510.

Kannaujiya, V.K., Singh, P.R., Kumar, D., Sinha, R.P., 2020. Phycobiliproteins in microalgae: occurrence, distribution, and biosynthesis. In: Jacob-Lopes, E., Queiroz, M., Zepka, L. (Eds.), Pigments from Microalgae Handbook. Springer, Germany, pp. 43–68.

Kannaujiya, V.K., Amin, N., Pandey, S., Jaiswal, M., Sinha, R.P., 2023. Phycocyanin from *Arthrospira*: production, applications, and advances. In: Jacob-Lopes, E., Queiroz, M.I., Maroneze, M.M., Queiroz Zepka, L. (Eds.), Handbook of food and feed from microalgae. Academic Press, Cambridge, MA, pp. 237–252.

Kim, B., Yang, S., You, H., Shin, H., Lee, J., 2017. Fucoidan-induced osteogenic differentiation promotes angiogenesis by inducing vascular endothelial growth factor secretion and accelerates bone repair. J. Tissue Eng. Regen. Med. 12, e1311–e1324.

Kumari, A., Garima, N., Bharadvaja, N., 2023. A comprehensive review on algal nutraceuticals as prospective therapeutic agent for different diseases. 3 Biotech 13, 44.

Lal, R., Jaiswal, M., Amin, N., Kannaujiya, V.K., 2025. Ecological screening and functional analysis of pigments, antioxidants and potent sunscreen compounds in cyanobacteria inhabiting diverse habitats. AIMS Mol. Sci. 12, 272–291.

Lal, R., Amin, N., Shrivastava, U.P., Sinha, R.P., Kannaujiya, V.K., 2026. Characterization and in silico anti-melanogenic potentials of mycosporine-like amino acids of epiphytic lichen isolated from central Nepal. In: Vegetos, 39, pp. 644–655.

Lal, R., Kannaujiya, V.K., 2025. Green photoprotective chemicals: diversity, recent developments, challenges and therapeutic significance. Total Environ. Microbiol. 1, 100047.

Lee, J.H., Lee, T.-K., Kang, R.-S., Shin, H.J., Lee, H.S., 2007. The in vitro antioxidant activities of the bromophenols from the red alga *Tichocarpus crinitus* and phenolic derivatives. J. Kor. Magn. Reson. 11, 56–63.

Lee, K.Y., Mooney, D.J., 2011. Alginate: properties and biomedical applications. Prog. Polym. Sci. 37, 106–126.

Lee, S., Lee, E.J., Lee, G.M., Yun, J.-H., Yoo, W., 2023. Inhibitory effect of fucoidan on TNF-α-induced inflammation in human retinal pigment epithelium cells. Front. Nutr. 10, 1162934.

Lee, S.H., Han, J.S., Heo, S.J., Hwang, J.Y., Jeon, Y.J., 2010. Protective effects of dieckol isolated from *Ecklonia cava* against high glucose-induced oxidative stress in human umbilical vein endothelial cells. Toxicol. In Vitro 24, 375–381.

Li, Y.X., Li, Y., Lee, S.-H., Qian, Z.J., Kim, S.K., 2009. Inhibitors of oxidation and matrix metalloproteinases, floridoside, and d-isofloridoside from marine red alga *Laurencia undulata*. J. Agric. Food Chem. 58, 578–586.

Li, Y.X., Wijesekara, I., Li, Y., Kim, S.K., 2011. Phlorotannins as bioactive agents from brown algae. Process Biochem. 46, 2219–2224.

Liu, M., Hansen, P.E., Lin, X., 2011. Bromophenols in marine algae and their bioactivities. Mar. Drugs 9, 1273–1292.

Liu, S., Shi, L., Luo, H., Chen, K., Song, M., Wu, Y., et al., 2024. Processed microalgae: green gold for tissue regeneration and repair. Theranostics 14, 5235–5261.

Matin, M., Koszarska, M., Atanasov, A.G., Król-Szmajda, K., Jóźwik, A., Stelmasiak, A., et al., 2024. Bioactive potential of algae and algae-derived compounds: focus on anti-inflammatory, antimicrobial, and antioxidant effects. Molecules 29, 4695.

Mazurek, Ł., Kuś, M., Jurak, J., Rybka, M., Kuczeriszka, M., Stradczuk-Mazurek, M., et al., 2025. Biomedical potential of alginate wound dressings-from preclinical studies to clinical applications: a review. Int. J. Biol. Macromol. 309, 142908.

Meenakshi, S., Gnanambigai, D.M., Mozhi, S.T., Arumugam, M., Balasubramanian, T., 2009. Total flavanoid and in vitro antioxidant activity of two seaweeds of Rameshwaram coast. Glob. J. Pharm. 3, 59–62.

Meinita, M.D.N., Harwanto, D., Tirtawijaya, G., Negara, B.F.S.P., Sohn, J.-H., Kim, J.-S., et al., 2021. Fucosterol of marine macroalgae: bioactivity, safety and toxicity on organism. Mar. Drugs 19, 545.

Mourelle, M.L., Gómez, C.P., Legido, J.L., 2017. The potential use of marine microalgae and cyanobacteria in cosmetics and thalassotherapy. Cosmetics 4, 46.

Naghipour, A., Zare, P., Langarizadeh, M.A., Mottaghi, A., Ghaedi, B., Tavakoli, M.R., et al., 2025. Dermo-cosmetic potential of fucoidans. J. Appl. Phycol. 37, 1431–1446.

Neamtu, B., Barbu, A., Negrea, M.O., Berghea-Neamțu, C.Ș., Popescu, D., Zăhan, M., et al., 2022. Carrageenan-based compounds as wound healing materials. Int. J. Biol. Macromol. 23, 9117.

O'Leary, R., Rerek, M., Wood, E.J., 2004. Fucoidan modulates the effect of transforming growth factor (TGF)-BETA.1 on fibroblast proliferation and wound repopulation in in vitro models of dermal wound repair. Biol. Pharm. Bull. 27, 266–270.

Pandey, A., Amin, N., Kannaujiya, V.K., Sinha, R.P., 2024. Extraction, characterization and antioxidative potentials of UV-screening compound, mycosporine-like amino acids from epilithic cyanobacterium *Lyngbya* sp. HKAR-15. World J. Microbiol. Biotechnol. 40, 378.

Pandey, S., Kannaujiya, V.K., 2024. Bacterial extracellular biopolymers: eco-diversification, biosynthesis, technological development and commercial applications. Int. J. Biol. Macromol. 279, 135261.

Pandey, S., Kannaujiya, V.K., 2025. Biochemical, thermal, and functional characterization of released polymeric substances of *Neowestiellopsis* sp. strain VKB03. Algal Res. 90, 104114.

Pandian, N., Chaurasia, R., Chatterjee, S., Biswas, B., Patra, P., Tiwari, A., et al., 2024. Roadmap of algal autotrophic tissue engineering in the avenue of regenerative wound therapy. Mater. Adv. 5, 7516–7533.

Park, M.-H., Heo, S.-J., Park, P.-J., Moon, S.-H., Sung, S.-H., Jeon, B.-T., et al., 2014. 6,6′-bieckol isolated from *Ecklonia cava* protects oxidative stress through inhibiting expression of ROS and proinflammatory enzymes in high-glucose-induced human umbilical vein endothelial cells. Appl. Biochem. Biotechnol. 174, 632–643.

Patel, A.K., Albarico, F.P.J.B., Perumal, P.K., Vadrale, A.P., Nian, C.T., Chau, H.T.B., et al., 2022. Algae as an emerging source of bioactive pigments. Bioresour. Technol. 351, 126910.

Percival, S.L., McCarty, S.M., Lipsky, B., 2015. Biofilms and wounds: an overview of the evidence. Adv. Wound Care 4, 373–381.

Postolović, K., Ljujić, B., Kovačević, M.M., Đorđević, S., Nikolić, S., Živanović, S., et al., 2022. Optimization, characterization, and evaluation of carrageenan/alginate/poloxamer/curcumin hydrogel film as a functional wound dressing material. Mater. Today Commun. 31, 103528.

Pradhan, B., Nayak, R., Patra, S., Jit, B.P., Ragusa, A., Jena, M., 2020. Bioactive metabolites from marine algae as potent pharmacophores against oxidative stress-associated human diseases: a comprehensive review. Molecules 26, 37.

Punchakara, A., Prajapat, G., Bairwa, H.K., Jain, S., Agrawal, A., 2023. Applications of mycosporine-like amino acids beyond photoprotection. Appl. Environ. Microbiol. 89, e00740-23.

Qi, H., Zhao, T., Zhang, Q., Li, Z., Zhao, Z., Xing, R., 2005. Antioxidant activity of different molecular weight sulfated polysaccharides from *Ulva pertusa* Kjellm (Chlorophyta). J. Appl. Phycol. 17, 527–534.

Raucher, D., Ryu, J.S., 2015. Cell-penetrating peptides: strategies for anticancer treatment. Trends Mol. Med. 21, 560–570.

Reyes, M.E., Riquelme, I., Salvo, T., Zanella, L., Letelier, P., Brebi, P., 2020. Brown seaweed fucoidan in cancer: implications in metastasis and drug resistance. Mar. Drugs 18, 232.

Ribeiro, J.S., Santos, M.J.M.C., Silva, L.K.R., Pereira, L.C.L., Santos, I.A., da Silva, M.V., 2019. Natural antioxidants used in meat products: a brief review. Meat Sci. 148, 181–188.

Rupérez, P., Ahrazem, O., Leal, J.A., 2002. Potential antioxidant capacity of sulfated polysaccharides from the edible marine brown seaweed *Fucus vesiculosus*. J. Agric. Food Chem. 50, 840–845.

Ryu, Y.S., Fernando, P.D.S.M., Kang, K.A., Piao, M.J., Zhen, A.X., Kang, H.K., et al., 2019. Marine compound 3-Bromo-4,5-dihydroxybenzaldehyde protects skin cells against oxidative damage via the Nrf2/HO-1 pathway. Mar. Drugs 17, 234.

Sarker, N.K., Kaparaju, P., 2024. Microalgal bioeconomy: a green economy approach towards achieving sustainable development goals. Sustainability 16, 11218.

Sarker, N.K., Salam, P.A., 2020. Design of batch algal cultivation systems and ranking of the design parameters. Energy Ecol. Environ. 5, 196–210.

Shaik, M.I., Sarbon, N.M., 2022. A review on purification and characterization of anti-proliferative peptides derived from fish protein hydrolysate. Food Rev. Int. 38, 1389–1409.

Shibata, T., Ishimaru, K., Kawaguchi, S., Yoshikawa, H., Hama, Y., 2007. Antioxidant activities of phlorotannins isolated from Japanese Laminariaceae. J. Appl. Phycol. 20, 705–711.

Shrestha, S., Zhang, W., Smid, S.D., 2020. Phlorotannins: a review on biosynthesis, chemistry and bioactivity. Food Biosci. 39, 100832.

Singh, A., Čížková, M., Bišová, K., Vítová, M., 2021. Exploring mycosporine-like amino acids (MAAs) as safe and natural protective agents against UV-induced skin damage. Antioxidants 10, 683.

Storz, P., 2005. Reactive oxygen species in tumor progression. Front. Biosci. 10, 1881–1896.

Sun, L., Wang, L., Li, J., Liu, H., 2014. Characterization and antioxidant activities of degraded polysaccharides from two marine chrysophyta. Food Chem 160, 1–7.

Sun, Y., Zhang, N., Zhou, J., Dong, S., Zhang, X., Guo, L., et al., 2020. Distribution, contents, and types of mycosporine-like amino acids (MAAs) in marine macroalgae and a database for MAAS based on these characteristics. Mar. Drugs 18, 43.

Takamatsu, S., Hodges, T.W., Rajbhandari, I., Gerwick, W.H., Hamann, M.T., Nagle, D.G., 2003. Marine natural products as novel antioxidant prototypes. J. Nat. Prod. 66, 605–608.

Tamura, Y., Takenaka, S., Sugiyama, S., Nakayama, R., 1998. Occurrence of anserine as an antioxidative dipeptide in a red alga, *Porphyra yezoensis*. Biosci. Biotechnol. Biochem. 62, 561–563.

Tavares, J.O., Cotas, J., Valado, A., Pereira, L., 2023. Algae food products as a healthcare solution. Mar. Drugs 21, 578.

Teerasak, E., Thongararm, P., Roytrakul, S., Meesuk, L., Chumnanpuen, P., 2016. Prediction of anticancer peptides against MCF-7 breast cancer cells from the peptidomes of *Achatina Fulica* mucus fractions. Comput. Struct. Biotechnol. J. 14, 49–57.

Thurn, A.L., Stock, A., Gerwald, S., Weuster-Botz, D., 2022. Simultaneous photoautotrophic production of DHA and EPA by *Tisochrysis lutea* and *Microchloropsis salina* in co-culture. Bioresour. Bioprocess. 9, 130.

Tziveleka, L.-A., Tammam, M.A., Tzakou, O., Roussis, V., Ioannou, E., 2021. Metabolites with antioxidant activity from marine macroalgae. Antioxidants 10, 1431.

Visuddho, V., Halim, P., Helen, H., Muhar, A.M., Iqhrammullah, M., Mayulu, N., et al., 2024. Modulation of apoptotic, cell cycle, DNA repair, and senescence pathways by marine algae peptides in cancer therapy. Mar. Drugs 22, 338.

Vuković, J.S., Perišić, S., Nikolić, A., Milošević, I., Mirilović, M., Prokić, B.B., et al., 2025. Toward natural wound healing therapy: honey and *Calendula officinalis* loaded κ-carrageenan films with promising hemostatic potential. Pharmaceutics 17, 578.

Wan, D.H., Zheng, B.Y., Ke, M.R., Duan, J.Y., Zheng, Y.Q., Yeh, C.K., et al., 2024. C-phycocyanin as a tumour-associated macrophage-targeted photosensitiser and a vehicle of phthalocyanine for enhanced photodynamic therapy. Chem. Commun. 53, 4112–4115.

Wang, P., Jiang, X., Jiang, Y., Hu, X., Mou, H., Li, M., et al., 2007. In vitro antioxidative activities of three marine oligosaccharides. Nat. Prod. Res. 21, 646–654.

Wu, Q., Liu, L., Miron, A., Klímová, B., Wan, D., Kuča, K., 2016. The antioxidant, immunomodulatory, and anti-inflammatory activities of *Spirulina*: an overview. Arch. Toxicol. 90, 1817–1840.

Xie, M., Liu, D., Yang, Y., 2020. Anti-cancer peptides: classification, mechanism of action, reconstruction and modification. Open Biol. 10, 200004.

Yan, P., Li, G., Wang, C., Wu, J., Sun, Z., Martin, G.E., et al., 2019. Characterization by empirical and computational methods of dictyospiromide, an intriguing antioxidant alkaloid from the marine alga *Dictyota coriacea*. Org. Lett. 21, 7577–7581.

Yang, J.-I., Yeh, C.-C., Lee, J.-C., Yi, S.-C., Huang, H.-W., Tseng, C.-N., et al., 2012. Aqueous extracts of the edible *Gracilaria tenuistipitata* are protective against H_2O_2-induced DNA damage, growth inhibition, and cell cycle arrest. Molecules 17, 7241–7254.

Yang, S., Chen, M.F., Ryu, B., Chen, J., Xiao, Z., Hong, P., et al., 2020. The Protective effect of the polysaccharide precursor, d-isofloridoside, from *Laurencia undulata* on alcohol-induced hepatotoxicity in HepG2 cells. Molecules 25, 1024.

Yoshiki, M., Tsuge, K., Tsuruta, Y., Yoshimura, T., Koganemaru, K., Sumi, T., et al., 2009. Production of new antioxidant compound from mycosporine-like amino acid, porphyra-334 by heat treatment. Food Chem. 113, 1127–1132.

Zhang, Q., Yu, P., Li, Z., Zhang, H., Xu, Z., Li, P., 2003. Antioxidant activities of sulfated polysaccharide fractions from *Porphyra haitanesis*. J. Appl. Phycol. 15, 305–310.

CHAPTER

20

Phyco-chemicals in DNA repair: Algal and cyanobacterial antioxidants and epigenetic modulators in cellular defense

Bhargavi Sonavane[a], *Stuti Nareshkumar Patel*[b], *Datta Madamwar*[a] and *Niraj Kumar Singh*[c,d]

[a]**Department of Biological Sciences, P. D. Patel Institute of Applied Sciences, Charotar University of Science and Technology, Anand, Gujarat, India** [b]**Małopolska Centre of Biotechnology, Jagiellonian University, Krakow, Poland** [c]**Department of Science and Technology, Gujarat Biotechnology Research Centre (GBRC), Government of Gujarat, Gandhinagar, Gujarat, India** [d]**School of Forensic Sciences, National Forensic Sciences University (NFSU), Nagpur, Maharashtra, India**

20.1 Introduction

Genomic instability, a hallmark of cancer, involves mutations, chromosomal aberrations, and DNA lesions arising from disrupted replication, repair, and cell-cycle regulation (Collins, 2001; Astley et al., 2002). Oxidative stress from excess reactive oxygen species (ROS) is a major driver, promoting oncogenic transformation, heterogeneity, and therapy resistance (Santocono et al., 2006; Lorenzo et al., 2009; Azqueta and Collins, 2012). ROS include radical species such as superoxide and hydroxyl radicals, and nonradicals like hydrogen peroxide and singlet oxygen. While essential at physiological levels, excess ROS causes oxidative stress, inducing base modifications, strand breaks, and DNA–protein crosslinks (Picardo and Dell'Anna, 2010; Sies, 2019). Lesions such as 8-oxo-guanine lead to transversion mutations, while ROS-oxidized dNTPs impair replication fidelity and trigger double-strand breaks (Collins et al., 1998). ROS also activate oncogenes and pathways such as NF-κB, MAPK, and HIF-1α, driving proliferation and apoptosis evasion (Perillo et al., 2020) (Fig. 20 1).

Genome Damage: Repair and Adaptation Across Life.
DOI: https://doi.org/10.1016/B978-0-443-42857-9.00009-8

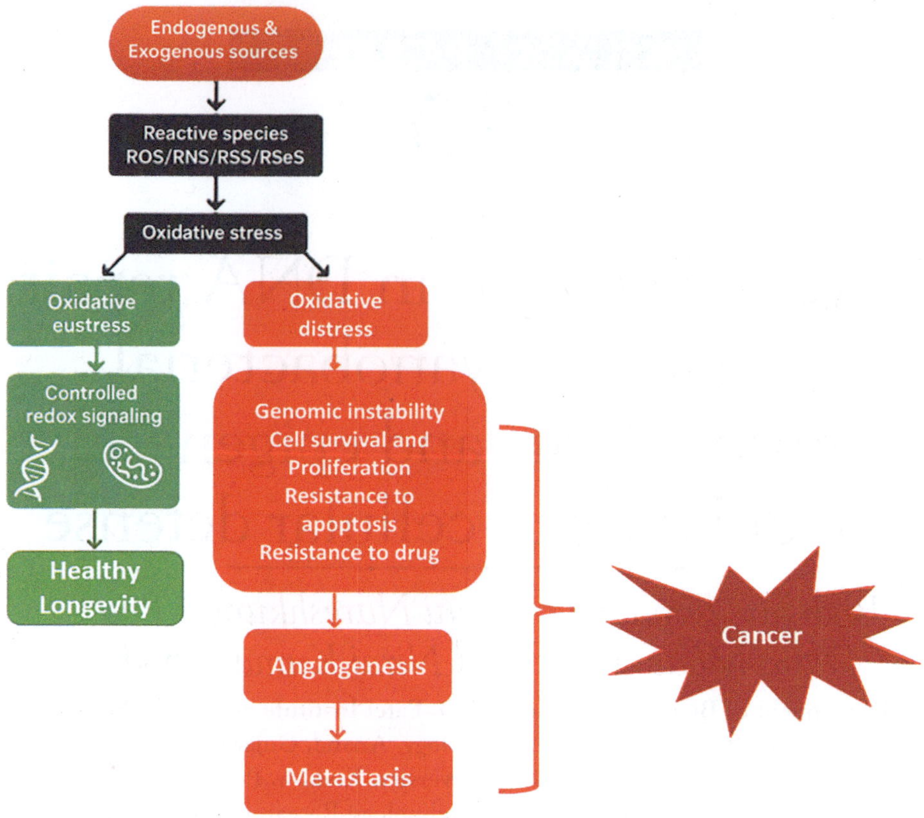

FIGURE 20.1 Oxidative stress and its relation to genomic instability and cancer.

DNA repair pathways, including base excision repair (BER), NER, MMR, and HR, safeguard genome stability by correcting lesions through coordinated action of glycosylases, polymerases, ligases, and endonucleases (Di Meo et al., 2016). Defects such as BRCA1/2 mutations or p53 loss heighten cancer risk, while antioxidants like carotenoids and polyphenols both scavenge ROS and enhance repair enzyme activity (Raza et al., 2017).

Microalgae and cyanobacteria provide bioactive compounds with antioxidant, antiinflammatory, and anticancer properties. Species such as *Chlorella vulgaris, Spirulina platensis*, and *Tetraselmis sp.* yield carotenoids, phycobiliproteins, polyphenols, vitamins, and polysaccharides that scavenge ROS, reduce DNA damage, and modulate redox pathways (Ferdous and Yusof, 2021; Severo et al., 2022; Smriti and Singh, 2024). These compounds promote apoptosis, inhibit angiogenesis and metastasis, and enhance chemotherapy efficacy while protecting normal cells. Marine macroalgae also supply fucoidans, phlorotannins, and laminarin with strong immunomodulatory effects (Yaacob et al., 2022). Although concerns remain regarding antioxidants in cancer therapy, algal compounds show selective redox modulation, supporting their potential as chemopreventive and therapeutic agents (De Jesus Raposo et al., 2013; Niki, 2016; Basily et al., 2018; Khavari et al., 2021; Huang et al., 2024; Zhang et al., 2024).

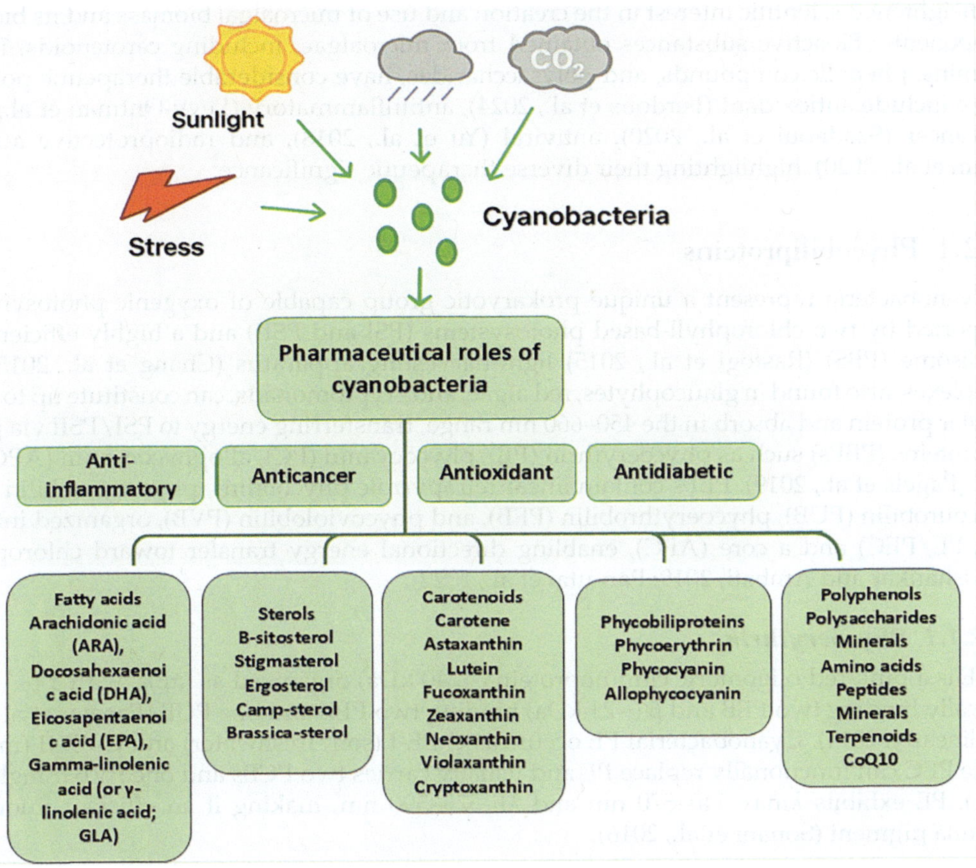

FIGURE 20.2 Pharmaceutical roles of cyanobacterial and algal products.

20.2 Algal and cyanobacterial metabolites: Biochemical classification

Microalgae are unicellular photosynthetic entities, often ranging from a few micrometres to several hundred micrometres in size. Owing to their exceptional ecological flexibility and widespread dispersion, they inhabit practically all environmental settings (Sumi, 2009; Shalaby, 2011). Their industrial applicability encompasses various sectors, including food, cosmetics, pharmaceuticals, nutraceuticals, and biomedicine, due to their abundant array of bioactive compounds, such as proteins, essential vitamins, polyunsaturated fatty acids (PUFAs), and various secondary metabolites (Fig. 20 2). Moreover, some microalgal species have beneficial biological characteristics such as effective photosynthesis, a reactive cellular surface, and superior biocompatibility (Sathasivam et al., 2019; Zhang et al., 2019).

Despite significant disruptions to the global economy caused by the COVID-19 pandemic in 2020, the microalgae industry exhibited robust and steady development. Industry projections indicate that the worldwide microalgae market may attain an estimated value of USD 4.6 billion by 2027, underscoring its increasing economic and application potential (Ziyaei et al., 2020). This has heightened scientific interest in the creation and use of microalgal biomass and its bioactive components. Bioactive substances obtained from microalgae, including carotenoids, PUFAs, vitamins, phenolic compounds, and polysaccharides, have considerable therapeutic potential. These include antioxidant (Ferdous et al., 2024), antiinflammatory (Levy-Ontman et al., 2017), anticancer (Saadaoui et al., 2020), antiviral (Yu et al., 2018), and radioprotective activities (Alam et al., 2020), highlighting their diverse therapeutic significance.

20.2.1 Phycobiliproteins

Cyanobacteria represent a unique prokaryotic group capable of oxygenic photosynthesis, supported by two chlorophyll-based photosystems (PSI and PSII) and a highly efficient phycobilisome (PBS) (Rastogi et al., 2015) light-harvesting apparatus (Chang et al., 2015). PBS complexes, also found in glaucophytes, red algae, and cryptomonads, can constitute up to 60% of cellular protein and absorb in the 450–660 nm range, transferring energy to PSI/PSII via phycobiliproteins (PBPs) such as phycoerythrin (PE), phycocyanin (PC), allophycocyanin (APC), and PEC (Pagels et al., 2019). PBPs contain linear tetrapyrrole phycobilins, phycocyanobilin (PCB), phycourobilin (PUB), phycoerythrobilin (PEB), and phycoviolobilin (PVB), organized into rods (PC, PE/PEC) and a core (APC), enabling directional energy transfer toward chlorophyll a (Ravishankar and Ambati, 2019; Pan-utai et al., 2022).

20.2.1.1 *Phycoerythrin*

PE is a pink-red oligomeric chromoprotein (~240 kDa) organized as $(\alpha\beta)_6$, with α (~19 kDa) typically binding two PEB and β (~21 kDa) binding two PEB and one PUB (Parmar et al., 2013; Mullineaux, 2014). Cyanobacterial PE occurs as C-PE-I (soil/freshwater) and C-PE-II (marine), while PEC can functionally replace PE and usually carries two PCBs and one PVB (Singh et al., 2022). PE exhibits λmax 540–570 nm and λF 575–590 nm, making it an efficient fluorescent antenna pigment (Sonani et al., 2016).

20.2.1.2 *Phycocyanin*

PC is the major blue PBP in cyanobacteria, mainly forming $(\alpha\beta)_3$ trimers, with α and β subunits of ~10–19 kDa and 14–21 kDa, each binding one to four PCBs (Jiang et al., 2017). It shows λmax 610–620 nm and λF 640–653 nm, with strain-specific variations such as λF 635 nm in Nostoc sp. R76DM (Patel et al., 2019).

20.2.1.3 *Allophycocyanin*

APC forms the PBS core, accepting energy from PC and transferring it to chlorophyll a in the reaction center (Ma et al., 2020b). Typically trimeric, with α ~15.5 kDa and β ~17.0 kDa, APC binds PCBs and exhibits λmax 650–655 nm and λF 657–660 nm, with spectral differences from PC attributed to distinct protein environments (Sonani, 2015).

20.2.2 Carotenoids

Carotenoids are C_{30}–C_{50} isoprenoid pigments, predominantly C_{40} tetraterpenes, built from eight C_5 isoprene units with extended conjugated double bonds absorbing at 400–500 nm (Ankita Jain and V.L. Sirisha, 2020; Nascimento et al., 2022). They include hydrocarbon carotenes (e.g., β-carotene) and oxygenated xanthophylls (e.g., lutein and zeaxanthin), with additional structural diversity (allenic/acetylenic bonds) in microalgal carotenoids such as fucoxanthin and neoxanthin (Dembitsky and Maoka, 2007; Takaichi, 2011). Their structures determine solubility, aggregation, and strong antioxidant/photoprotective properties (Guedes et al., 2011; Bulut et al., 2019).

20.2.2.1 *Xanthophylls and lutein*

Xanthophylls are oxygenated carotenoids (e.g., hydroxyl, epoxide, and keto groups) classified as primary (essential for photosynthesis) or secondary (stress-induced), and in organisms like Haematococcus pluvialis, secondary astaxanthin accumulates massively in lipid vesicles (Orejuela-Escobar et al., 2021). In cyanobacteria and microalgae, xanthophylls associate with chlorophyll–protein complexes, enhancing light harvesting and protecting against photooxidative stress by quenching singlet oxygen and free radicals while stabilizing pigment–protein assemblies (Rodriguez-Amaya, 2015). Lutein, abundant in *Scenedesmus and Chlorella sp.*, is a free, nonesterified carotenoid in microalgae, whose conjugated polyene backbone underlies its strong antioxidant and singlet oxygen–quenching capacity, often enhanced under nitrogen deprivation or heterotrophic conditions (Jain et al., 2020).

20.2.2.2 *β-carotene*

β-carotene is a red-orange, nonpolar C_{40} carotenoid with β-ionone rings, synthesized from geranylgeranyl pyrophosphate and acting as a key provitamin A source. *Dunaliella salina* is especially rich in β-carotene and co-produces α-carotene, zeaxanthin, cryptoxanthin, lutein, and lycopene. Beyond its role as a natural colorant, β-carotene functions as a potent antioxidant, mitigating oxidative stress linked to cancer, UV damage, inflammatory disorders, and immune decline (Nascimento et al., 2022).

20.2.2.3 *Astaxanthin*

Astaxanthin ($C_{40}H_{52}O_4$) is a red-orange keto-xanthophyll with hydroxyl and ketone groups, derived from IPP and DMAPP, and structurally related to zeaxanthin and canthaxanthin. In Haematococcus pluvialis, it can accumulate up to 2–20 g/kg dry weight and exhibits antioxidant capacity exceeding vitamins C and E and far surpassing β-carotene, lutein, and zeaxanthin, supporting applications in UV protection, anticancer strategies, neuroprotection, and antiinflammatory interventions (Rao et al., 2013; Zhang and Wang, 2015).

20.2.2.4 *Fucoxanthin*

Abundance in brown seaweeds and microalgae such as *Phaeodactylum tricornutum* is distinguished by an allenic bond, nine conjugated double bonds, a 5,6-monoepoxide, and multiple oxygenated groups. These structural features underpin its roles in marine light harvesting and photoprotection, as well as its antioxidant, antiinflammatory, hepatoprotective, antitumor, antidiabetic, and antiobesity activities (Galasso et al., 2017; Jain and Sirisha, 2020).

20.2.2.5 *Zeaxanthin*

Zeaxanthin ($C_{40}H_{56}O_2$; 568.88 Da) is a yellow carotenoid with a fully conjugated double-bond system, widely distributed in corn, egg yolk, marigold, peppers, and leafy greens. Although nonprovitamin A, it is crucial for ocular health, particularly in reducing age-related macular degeneration risk, and is extensively used as a natural pigment and nutraceutical, with additional microalgal sources including *Rhodophyta* and *Spirulina sp* (De Jesus Raposo et al., 2015).

20.2.3 Polyunsaturated fatty acids

PUFAs are long-chain unsaturated fatty acids with two or more double bonds, classified into ω-3, ω-6, and ω-9 series based on the first double-bond position. They are synthesized from acetyl-CoA through elongation and desaturation pathways, with ω-3 and ω-6 routes being metabolically distinct (Scanferlato et al., 2019). EPA and DHA derive from ALA (ω-3), while GLA and ARA arise from LA (ω-6). Humans lack Δ12 and Δ15 desaturases and must obtain LA and ALA from the diet. Competition for shared enzymes necessitates balanced ω-3 and ω-6 intake. ALA supplementation increases EPA/DHA levels (Ansari et al., 2019). Essential PUFAs contribute to membrane structure, eicosanoid synthesis, immunity, tissue regeneration, and dermatological health. DHA and EPA lower cardiovascular risk, hypertension, and triglycerides while increasing HDL, and DHA is vital for nervous system development (Djuricic and Calder, 2021). Microalgae such as *Arthrospira platensis, Isochrysis galbana, Porphyridium cruentum, Phaeodactylum, Pavlova lutheri,* and various diatoms accumulate EPA, DHA, ARA, and GLA at high levels (Ryckebosch et al., 2012; Sun et al., 2016). EPA, DHA, and ALA show anticancer benefits across multiple cancer types, improve survival when combined with CAF chemotherapy, benefit colorectal cancer patients with KRAS wild-type or MMR deficiency, and reduce inflammatory markers with vitamin D co-supplementation (Haidari et al., 2020). ω-3 and ω-6 PUFAs also alleviate cancer-related fatigue (Darwito et al., 2019; Peppone et al., 2019).

20.2.4 Sulfated polysaccharides

Microalgae and cyanobacteria produce polysaccharides with antioxidant, antiinflammatory, and anticancer activities. Extracts from *C. stigmatophora, P. tricornutum, Tetraselmis spp., Pavlova viridis, Sarcinochrysis marina,* and *Porphyridium sp.* show strong antioxidant or antiinflammatory properties (Amna Kashif et al., 2018). Isochrysis galbana and Nannochloropsis oculata polysaccharides exhibit antioxidant and antiproliferative effects against HeLa cells (Ben Hafsa et al., 2017). Nostoglycan from *Nostoc sphaeroides* provides oxidative stress protection and induces apoptosis in lung cancer cells (Li et al., 2018), and *P. viridis* fractions display in vivo antitumor activity (Sun et al., 2016). About 45 of 166 screened strains produce EPS with notable bioactivity (Gaignard et al., 2019). EPS from *Graesiella sp., Chlorella pyrenoidosa, Chlorococcum* sp., and *Scenedesmus* sp. show antioxidant and colon cancer-cytotoxic properties (Trabelsi et al., 2016). Sulfated polysaccharides (spS) from *Navicula* sp., *Tribonema* sp., and *Porphyridium cruentum* exhibit antioxidant, antiproliferative, and apoptosis-inducing effects, including hepatic carcinoma inhibition (Fimbres-Olivarria et al., 2018).

20.2.5 Phenolic compounds

Phenols are typically classified by their structural complexity, and this framework is strongly predictive of their biological and ecological functions. Simple monohydric phenols (e.g., phenol and cresols) commonly exhibit antiseptic and membrane-disruptive antimicrobial effects, whereas dihydric (catechol, resorcinol, and hydroquinone) and trihydric phenols (pyrogallol and phloroglucinol) show enhanced redox reactivity, frequently cycling to quinone forms and thus acting as effective antioxidants or, in specific microenvironments, pro-oxidants that generate reactive intermediates capable of inhibiting microbial enzymes (Hamed et al., 2015; Zhang et al., 2019). More complex polyphenol molecules bearing multiple phenolic hydroxyl groups across one or more aromatic rings include phenolic acids, flavonoids, stilbenes, lignans, and tannins; their diversity arises from differences in backbone architecture and degree of polymerization. Phenolic acids represent relatively simple polyphenols and are commonly grouped into hydroxybenzoic and hydroxycinnamic acids, while flavonoids comprise the largest family, unified by a C6–C3–C6 scaffold and subdivided into flavones, flavanols, flavanones, anthocyanins, and isoflavones. Stilbenes possess a characteristic C6–C2–C6 framework, lignans are formed by coupling two phenylpropanoid units, and tannins are high-molecular-weight polyphenols categorized as hydrolysable or condensed based on their bonding chemistry (Khotimchenko et al., 2010). Functionally, polyphenols act as chemical defenses, UV-protectants, and signaling/metabolic regulators, mediating metal chelation, radical scavenging, microbiome shaping, and pathogen suppression via protein precipitation, enzyme inhibition, and interference with cell envelopes; conjugated forms such as phenolic glycosides serve as stable storage/transport derivatives that can be hydrolyzed to release active aglycones, while polymeric phenolics like lignin strengthen cell walls and increase resistance to biodegradation (Roohinejad et al., 2017; Usman et al., 2017). In marine brown algae (Kadam et al., 2015), phlorotannin's polymers of phloroglucinol represent a distinctive polyphenol group that contributes notably to antioxidant capacity and antiherbivory protection, emphasizing how hydroxylation pattern, substitution, and polymerization collectively govern phenolic bioactivity across terrestrial and marine systems (Ngo and Kim, 2013; Sun et al., 2016; Wells et al., 2017; Sanjeewa et al., 2018).

20.2.6 Algal sterols

Marine microalgae produce sterols of nutritional and pharmaceutical significance. *Pavlova lutheri* produces brassicasterol, campesterol, stigmasterol, and sitosterol (Randhir et al., 2020; Abdelkarim et al., 2025). Microalgae such as *Chlorella, Chlamydomonas, Scenedesmus, Ankistrodesmus, Nannochloropsis, Stephanodiscus*, and *Cryptomonas* are important sterol sources (Martin-Creuzburg and Merkel, 2016). Sterol-rich fractions from *Nannochloropsis oculata* and *N. salina* show cytotoxicity against multiple cancer cell types, and fatty acids from *S. maxima* also demonstrate anticancer effects (Sayegh et al., 2020). Mechanistically, sterols inhibit tumor growth, angiogenesis, and metastasis and promote apoptosis via caspase-3 activation and Bax/Bcl-2 modulation (Sanjeewa et al., 2016). Dietary β-sitosterol suppresses colon, lung, liver, prostate, and breast cancer cell proliferation (Sayegh et al., 2016).

20.3 Antioxidative and geno-protective mechanisms

Antioxidants synthesized within the body as part of routine metabolism are classified as endogenous antioxidants, whereas those supplied from outside sources, primarily through diet and nutraceutical supplementation, are termed exogenous antioxidants. In addition, a distinct

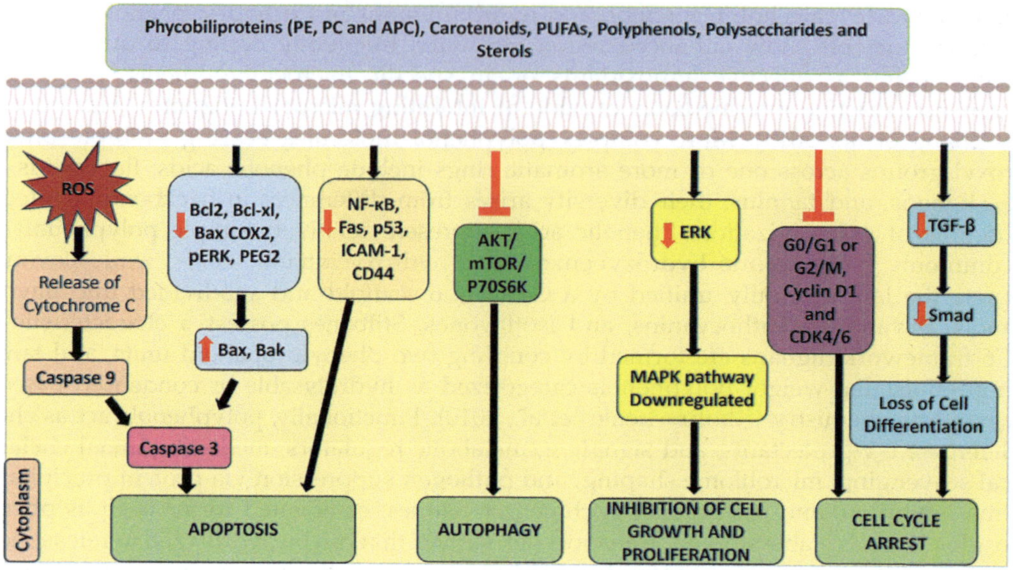

FIGURE 20.3 Schematic overview of the anticancer/antiproliferative mechanisms attributed to cyanobacterial/algal bioactives phycobiliproteins (PE, PC, and APC), carotenoids, PUFAs, polyphenols, polysaccharides, and sterols, highlighting their coordinated modulation of redox status and major signaling networks. *PUFAs*, Polyunsaturated fatty acids.

class of synthetic antioxidants is industrially produced and widely applied in food processing to retard oxidation. Antioxidants protect cells through three coordinated tiers of defense. The first line limits the generation of new free radicals by controlling reactive intermediates and metal-catalyzed reactions; this includes enzymatic systems such as SOD, CAT, GPx, and reduced glutathione, metal-sequestering proteins like ferritin and ceruloplasmin, and supportive antioxidant minerals including selenium, copper, and zinc. The second line neutralizes radicals that have already formed, thereby curbing oxidative stress; contributors here include endogenous and dietary molecules such as glutathione, albumin, CoQ, flavonoids, carotenoids, uric acid, and vitamins A, C, and E. The third line focuses on restoring oxidative damage by repairing or removing compromised biomolecules, particularly DNA, proteins, and lipids, primarily via enzymes such as DNA repair enzymes, proteases/peptidases, lipases, and transferases. Microalgal antioxidants are primarily composed of carotenoids, phenolics, flavonoids, PUFAs, SPs, sterols, and phycobiliproteins (Fig. 20.3) (Mut-Salud et al., 2016; Calvani et al., 2020).

Phycocyanin (PC) from *S. platensis* and its bilin chromophore PCB exhibit strong antioxidative and geno-protective effects that also translate into hepatoprotection. Their chromophores scavenge key ROS/RNS-including hydroxyl radicals, singlet oxygen, peroxynitrite ($ONOO^-$), and HOCl-with PCB reported to be particularly effective against $ONOO^-$, thereby limiting oxidative DNA strand breaks (Bhat and Madyastha, 2001). In liver injury models, PC reduces lipid peroxidation and malondialdehyde (MDA) while enhancing MnSOD, catalase, and GST and activating the Nrf2/HO-1 pathway, changes associated with reduced γ-H2AX signaling and chromosomal damage during CCl_4-induced hepatotoxicity (Ou et al., 2010). Beyond redox control, C-phycocyanin (C-PC) modulates apoptosis and cell-cycle progression in cancer cells by downregulating Bcl-2, upregulating Bax/death signaling, promoting cytochrome c release, caspase activation, and PARP-1 cleavage, and inducing G1/S arrest via cyclin D1/CDK4 suppression and p21 induction (Roy et al., 2007). Phycobiliproteins more broadly (PC, APC, PE, and PEC) show antioxidant, antiinflammatory, and anticancer activities (Pagels et al., 2019), including radical scavenging (Pratita et al., 2024) by PC from *A. platensis* (Pan-utai et al., 2022) and growth inhibition of melanoma and laryngeal cancer cells by PC from *Porphyra yezoensis* (Zhang et al., 2011). Proposed mechanisms include caspase-3/9 activation, NF-κB suppression, and autophagy induction via Akt/mTOR inhibition, with reports of chemo-sensitization; importantly, C-PC can protect plasmid DNA from oxidative damage more effectively than trolox, reinforcing its geno-protective potential (Kaur et al., 2020; Patel et al., 2022).

Algae-derived polysaccharides, especially SPs and exopolysaccharides, are widely categorized as bioactive macromolecules whose antioxidant capacity is shaped by sulfate content, molecular weight, and polymer conformation. Functionally, they mitigate oxidative stress through direct free-radical scavenging (e.g., DPPH/ABTS/hydroxyl radicals) and metal-ion chelation (e.g., Fe^{2+}), thereby limiting metal-catalyzed ROS generation and downstream oxidative injury to lipids and proteins (Roohinejad et al., 2017). Their sulfate ester groups and sugar backbones enhance electron donation and radical stabilization, which collectively contribute to broad protection against oxidative degradation of cellular macromolecules. Importantly, these antioxidative actions translate into geno-protection by lowering ROS burden and lipid peroxidation (often reflected by reduced malondialdehyde) and by strengthening endogenous antioxidant defenses (increased activities of enzymes such as SOD, catalase, and related systems) in experimental models. In parallel, several microalgal polysaccharides also show antitumor-relevant effects ranging from immune stimulation [enhanced macrophage/natural killer (NK) activity] to antiinvasive actions and direct suppression of tumor cell growth, suggesting that oxidative stress control and immunomodulation may act together in their broader protective profile (Muñoz-Almagro et al., 2020).

Microalgae synthesize a wide range of phenolic compounds, flavonoids (flavones, flavanols, and isoflavones), and phenolic acids that contribute to strong antioxidant and anticancer activities. Apigenin and genistein have been reported in *Phaeodactylum tricornutum, Diacronema lutheri, Porphyridium purpureum, Haematococcus pluvialis, Tetraselmis suecica*, and *C. vulgaris* (Goiris et al., 2014). Quercetin from *Scenedesmus* sp. is noted as a key antioxidant (Bulut et al., 2019), and *P. tricornutum* produces protocatechuic acid with high radical scavenging activity (Haoujar et al., 2019). Phenolic fractions from *Euglena cantabrica* have been described as more effective than conventional antioxidants, while phenolic acids from *Spirulina maxima* reduce lipid peroxidation. Antioxidant potential is also documented across *Nannochloropsis, Spirulina, Dunaliella, Chlorella, Tetraselmis*, and *Porphyridium* (Leong et al., 2024; Buonvino et al., 2025; Vasilakis et al., 2025).

20.4 Epigenetic modulation by algal products

Epigenetics refers to heritable changes in gene expression that occur without alterations in the underlying DNA sequence. These regulatory mechanisms include DNA methylation, histone modifications (acetylation and methylation), and noncoding RNA-mediated regulation. Collectively, they govern chromatin structure, transcriptional activity, and cellular responses to stress. Disruption of epigenetic regulation is a hallmark of many pathological conditions,

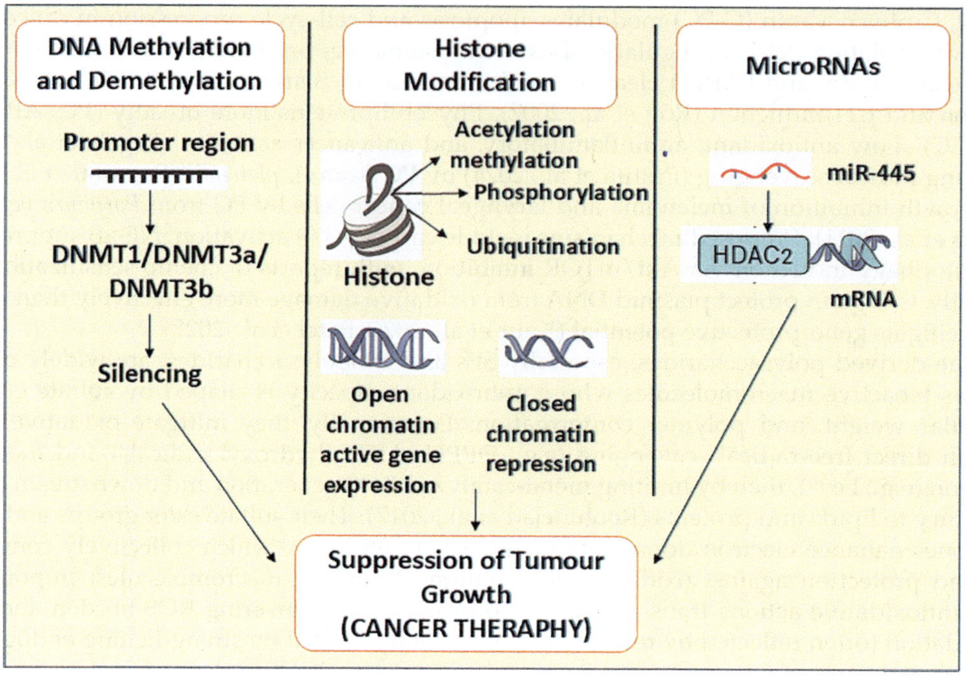

FIGURE 20.4 Epigenetic modulation by algal products.

including cancer, neurodegenerative disorders, and metabolic diseases (Kichouh-Aiadi et al., 2024).

A major contributor to epigenetic dysregulation and genome instability is DNA damage, which arises through processes such as DNA oxidation, hydrolysis, and aberrant methylation. ROS, generated predominantly under aerobic conditions, accelerate spontaneous mutagenesis compared to anaerobic states. Guanine is particularly vulnerable, forming the mutagenic lesion 8-oxo-7,8-dihydroguanine (8-oxo-G). This lesion mispairs with adenine, producing G→T transversions during replication. The accumulation of 8-oxo-G is oncogenic, strongly linked to mitochondrial dysfunction, telomeric instability, and tumorigenesis. Moreover, ROS interferes with the activity of epigenetic enzymes (Singh et al., 2024) such as DNA methyltransferases (DNMTs) and histone deacetylases (HDACs), causing abnormal chromatin remodeling and altered transcriptional programs. Such oxidative DNA lesions can simultaneously trigger local hypomethylation and hypermethylation of tumor suppressor gene promoters, further promoting cancer progression (Fig. 20.4) (Lorenzo et al., 2009; Singh et al., 2024).

Algal metabolites exert protective roles by modulating epigenetic pathways

Histone modification (inhibition of HDACs): Several bioactive compounds act as natural HDAC inhibitors, leading to hyperacetylation of histones such as H3 and H4. This acetylation neutralizes the positive charge on histones, weakens DNA–histone interactions, and produces a relaxed chromatin state. The open structure enhances transcription factor binding at promoter regions, thereby promoting the expression of protective genes (Li et al., 2017).

DNA methylation modulation (inhibition of DNMTs): Polyphenols, flavonoids, and phenolic metabolites also inhibit DNMTs (DNMT1, DNMT3A/3B), either through direct enzymatic blockade or by altering methyl donor availability. This results in hypomethylation of hypermethylated CpG islands in the promoters of regulatory genes(Ceccarelli et al., 2020). The loss of repressive methylation marks drives a transcriptionally active chromatin state, counteracting gene silencing observed in cancer and degenerative diseases.

Reactivation of tumor suppressor genes: By simultaneously inhibiting HDACs and DNMTs, phytochemicals restore the expression of silenced tumor suppressor genes such as p21, p53, and BRCA1/2. This reactivation reinstates DNA repair pathways, controls aberrant cell proliferation, and induces apoptosis in malignant cells. Ultimately, these epigenetic modifications contribute to genomic stability and cancer-preventive effects (Benton et al., 2017).

At the nuclear level, ROS additionally influences histone modifications. For example, H3K9 demethylation and H3S10 phosphorylation serve as regulatory checkpoints that coordinate transcriptional activity with DNA repair. When this balance is disrupted, excessive DNA oxidation and chromatin remodeling occur, leading to programmed cell death. Perturbation in the histone deacetylation and miRNA elements of the epigenetic system is one of the key disruptive events in Aβ neurotoxicity (Formisano et al., 2015). We observed that HDAC3 and miRNA-335 function differentially: in the Aβ-intoxicated mice, which showed that HDAC3 expression is downmodulated, 76% in the miRNA-335 expression was observed, and brain-derived neurotrophic factor (BDNF) expression is up-modulated in the CPC group against Aβ toxicity (Li et al., 2020).

PC, one of the phycobiliproteins isolated from *Porphyra yezoensis*, demonstrates significant anticancer activity against human melanoma and laryngeal cancer cells in a dose-dependent manner (Zhang et al., 2011). Its mechanism involves cell-cycle arrest at G0/G1 or G2/M phases and induction of apoptosis through activation of caspase-3 and caspase-9. These effects are accompanied by reduced Bcl-2/Bax ratio, downregulation of COX-2, p-ERK, PEG2, cyclin D1, and CDK4, DNA fragmentation, cytochrome c release, increased ROS generation, NF-κB suppression, and modulation of apoptotic mediators including Fas, p53, ICAM-1, and CD44, culminating in chromatin condensation and programmed cell death (Patel et al., 2022).

Similarly, higher dietary intake of marine ω-3 polyunsaturated fatty acids (MO3PUFAs) has been linked with improved survival in stage III colon cancer patients carrying wild-type *KRAS* and deficient DNA mismatch repair, molecular profiles often associated with tumor progression and poor prognosis (Song et al., 2019).

Carotenoids such as β-carotene, astaxanthin, lycopene, lutein, and β-cryptoxanthin contribute to genomic stability by mitigating oxidative stress and reducing 8-hydroxy-2′-deoxyguanosine (8-oxodG) and related DNA lesions (Lyons and O'Brien, 2002). This promotes BER efficiency, preserves telomeric integrity, and decreases chromosomal instability. Beyond DNA protection, carotenoids modulate gene expression relevant to cellular defense: β-carotene induces haem oxygenase-1, while lycopene and α-carotene upregulate connexin-43, enhancing intercellular communication and tumor suppression. Carotenoids exert antiinflammatory effects mainly by suppressing redox-sensitive inflammatory signaling and thereby reducing cytokine production. By limiting oxidative stress, they decrease activation of key transcriptional hubs such as NF-κB, which otherwise drives expression of major pro-inflammatory mediators. In experimental LPS-induced inflammation models, the carotenoid fucoxanthin inhibited upstream TLR4/MyD88-linked signaling (including phosphorylation of IKK and IκBα), reduced NF-κB activation and its nuclear translocation, and consequently lowered the secretion of core cytokines such as

TNF-α, IL-1β, and IL-6. (Astley et al., 2002; Santocono et al., 2006; Azqueta and Collins, 2012; Su et al., 2019; Kim et al., 2019; Jain and Sirisha, 2020)

The sulphated polysaccharides (SPs) from marine algae also reduces intracellular ROS levels, thereby preventing oxidative base modifications and DNA strand breaks. For example, polysaccharides from *Codium* species and related algae were shown to protect nucleic acids from oxidative stress, suggesting a geno-protective role. Although direct DNA repair assays are limited, these findings indicate that SPs helps preserve genomic integrity by maintaining redox balance and controlling inflammation, particularly under stress conditions associated with aging and environmental exposure. Although genome stability is not always addressed as a primary endpoint, accumulating evidence suggests that the antioxidant and antiinflammatory actions of algal SPs indirectly support geno-protective functions. By attenuating oxidative stress and chronic inflammation, two major drivers of DNA damage, SPs contribute to the preservation of genomic integrity. For example, fucoidan-mediated reduction in ROS and inflammatory cytokines limits oxidative DNA strand breaks and minimizes replication stress. Moreover, SP-induced modulation of immune responses reduces cytokine-storm–associated genotoxicity, particularly relevant in viral infections such as COVID-19, where excessive inflammation is linked to DNA damage and tissue injury. The production of short-chain fatty acids (SCFAs) following microbial fermentation of algal SPs further reinforces genome stability by maintaining intestinal barrier integrity and suppressing pro-inflammatory signaling pathways (Singh et al., 2024). Collectively, these mechanisms position algae-derived SPs as promising geno-protective agents with relevance to cancer prevention, aging-associated genomic instability, and protection against inflammation-driven DNA damage (Ziyaei et al., 2020).

The antioxidant and anticancer activities demonstrated in the study imply an important role for *Chaetomorpha* sp. in maintaining genome stability. The potent antioxidant activity of *Chaetomorpha* extracts, particularly the ethanolic fraction, suggests a protective role against ROS-induced DNA damage. Moreover, the observed growth inhibition of MDA-MB-231 breast cancer cells, with an IC_{50} of 225.18 ± 0.61 μg mL^{-1}, indicates selective cytotoxicity toward malignant cells, likely mediated through oxidative stress modulation and apoptosis-inducing compounds such as dichloroacetic acid (DCA). By reducing oxidative pressure and suppressing cancer cell proliferation, *Chaetomorpha* sp. contributes indirectly to the preservation of genomic integrity, highlighting its geno-protective potential in the context of cancer prevention and cellular homeostasis (Haq et al., 2019).

Collectively, these findings highlight that algal chemicals, together with natural compounds such as phycobiliproteins, carotenoids, polyphenols, and algal lipids, act as dual modulators, exerting both antioxidant and epigenetic regulation. By stabilizing telomeres, reducing mutagenic lesions, reactivating silenced tumor suppressor genes, and restoring redox homeostasis, they provide promising therapeutic avenues for cancer prevention, genomic stability, and mitochondrial health.

20.5 Immunomodulation and inflammation control

Microalgae represent a rich and sustainable source of immunomodulatory and antiinflammatory bioactive compounds, including carotenoids, phycobiliproteins, polysaccharides, peptides, and PUFAs, whose efficacy has been demonstrated across cellular, animal, and limited clinical studies (Calvani et al., 2020). Diverse algal metabolites modulate both innate and adaptive immune responses, contributing to immune homeostasis and inflammation control. For example, diatoxanthin derived from diatoms has been reported to suppress SARS-CoV-2-induced cytokine storms, while *Dunaliella salina* enhances NK cell activity and promotes T- and B-cell proliferation in leukemic mouse models (Khavari et al., 2021). In human studies, *C. vulgaris* supplementation significantly increased NK cell activity and elevated cytokines such as IFN-γ, IL-1β, and IL-12, indicating enhanced immune competence(Barbalace et al., 2019). Similarly, extracts from *S. platensis*, sulfated polysaccharides from *Tribonema sp.*, and β-glucans from *Euglena gracilis* further amplify immune responses by regulating key inflammatory mediators, including IL-6, IL-10, and TNF-α (Barbalace et al., 2019; Checa and Aran, 2020; Smriti and Singh, 2024).

Among algal-derived immunomodulators such as phycobiliproteins, PC from *S. platensis* exhibits particularly strong antiinflammatory activity through both direct and indirect mechanisms. Directly, PC modulates Toll-like receptor (TLR) dependent signaling by suppressing TLR2/4-mediated NF-κB activation, thereby reducing the expression of pro-inflammatory cytokines such as IL-6, TNF-α, and IL-1β. This TLR-dependent mechanism has been validated in murine models of bleomycin-induced lung inflammation, where PC-mediated protection was absent in TLR-deficient mice. PC also inhibits NF-κB nuclear translocation in macrophages and disease models, resulting in broad suppression of inflammatory gene transcription. Indirectly, PC exerts antiinflammatory effects via activation of the Keap1–Nrf2–ARE pathway, leading to upregulation of cytoprotective enzymes including HO-1, NQO1, GSH-Px, and NADPH-generating systems (Liu et al., 2020). Through these redox-sensitive mechanisms, PC mitigates oxidative stress and inflammation in diverse pathological contexts, including X-ray–induced hepatic injury, UV-induced skin apoptosis, methylglyoxal-induced mitochondrial dysfunction, retinal degeneration, and neuroinflammation (Chang et al., 2011; Minić, 2021). PC further acts as a natural cyclooxygenase-2 (COX-2) inhibitor, selectively suppressing pro-inflammatory mediators such as prostaglandin E_2, leukotriene B_4, nitric oxide (NO), and histamine, with inhibitory potency (IC_{50} = 80–180 nM) exceeding that of some conventional NSAIDs (Saadaoui et al., 2020). In vivo models, including carrageenan-induced paw edema and arachidonic acid–induced inflammation, demonstrated significant reductions in edema, myeloperoxidase activity, and cytokine expression following PC administration. In parallel, PC and its chromophore PCB modulate immune balance by downregulating pro-inflammatory cytokines (IL-6 and IL-8) while enhancing antiinflammatory mediators such as IL-10 and TGF-β1, increasing regulatory T cell populations, and promoting secretory IgA production, thereby supporting both systemic and mucosal immunity (Roy et al., 2007; Kannaujiya et al., 2017; Minić, 2021).

Algal polysaccharides constitute a major class of bioactive macromolecules with well-documented roles in immune regulation and inflammation control, primarily through modulation of innate and adaptive immune responses. Polysaccharides isolated from microalgae such as *Coccomyxa gloeobotrydiformis* markedly suppress lipopolysaccharide-induced inflammatory responses in RAW 264.7 macrophages by balancing pro- and antiinflammatory signaling pathways (Zhang et al., 2019). Similarly, exopolysaccharides from *Crypthecodinium cohnii* attenuate inflammatory signaling via coordinated regulation of TLR4, MAPK, and NF-κB pathways, thereby limiting excessive immune activation. Polysaccharides from *Chlorella pyrenoidosa* enhance immune surveillance by stimulating phagocytic activity in vivo, inducing IL-1β secretion, and upregulating antigen-presenting and co-stimulatory molecules in immune cells, with their bioactivity closely dependent on molecular weight and structural features. High-molecular-weight arabinogalactans and polysaccharides from *Chlorella ellipsoidea* exhibit particularly strong immunomodulatory effects through activation of NF-κB and MAPK signaling in macrophages, underscoring the importance of structure–activity relationships (Usman et al., 2017; Severo et al., 2022). In parallel, spS from marine macroalgae such as *Ulva rigida, Codium fragile*, and *Caulerpa lentillifera* fine-tune immune signaling by stimulating macrophage activation, NO production, and cytokine secretion (IL-6, IL-10, TNF-α, and IL-1β) without provoking chronic inflammation (Ou et al., 2010). Fucoidan from *Undaria pinnatifida* and *Fucus vesiculosus* further regulates TLR-2/4–mediated NF-κB and MAPK pathways, contributing to controlled inflammatory responses and enhanced antiviral immunity, while κ- and β-carrageenans from red algae such as *Tichocarpus crinitus* promote Th1-biased immunity by elevating IFN-γ and IL-12 levels (Khotimchenko et al., 2010). Green algal sulfated ulva also stimulates proliferation of T and B lymphocytes, enhances IFN-γ production via antigen-presenting cell activation, and modulates NK cell–associated cytokines. Notably, low-molecular-weight ulvan fractions from *Ulva ohnoi, U. intestinalis*, and *U. armoricana* consistently demonstrate superior immunostimulatory efficacy, significantly upregulating cytokines such as CCL20, IL-8, and TNF-α, particularly in intestinal epithelial models (Ciancia et al., 2020). Additional anti-inflammatory validation comes from polysaccharides such as GNP, which reduce intracellular ROS and NO levels while suppressing TNF-α and IL-6 expression through downregulation of iNOS and COX-2 via inhibition of MAPK and NF-κB signaling (Chi et al., 2020; Tang et al., 2021). Collectively, these findings establish algal polysaccharides, especially sulfated and structurally optimized forms, as potent natural immunomodulators capable of balancing immune activation and inflammation, supporting their development as functional nutraceuticals and immune-supportive agents (Amna Kashif et al., 2018; Yu et al., 2018; Chen et al., 2021).

Carotenoids derived from microalgae and cyanobacteria further contribute to inflammation control and immune regulation (Yaacob et al., 2022). Fucoxanthin, isolated from diatoms such as *Phaeodactylum tricornutum* and *Thalassiosira weissflogii*, markedly suppresses lipopolysaccharide-induced inflammation in macrophages and microglial cells by inhibiting COX-2 and inducible nitric oxide synthase (iNOS), reducing prostaglandin E_2, NO, and pro-inflammatory cytokines, and attenuating NF-κB and MAPK signaling (Kaulmann and Bohn, 2014). These effects translate in vivo to reduced systemic inflammation in sepsis and colitis models, protection against UV-induced tissue damage, and mitigation of inflammation-associated behavioral deficits. Comparable antiinflammatory activities have been reported for violaxanthin from *Chlorella ellipsoidea* and fucoxanthin from *Nitzschia laevis*, reinforcing the role of algal carotenoids as effective natural modulators of chronic inflammation (Galasso et al., 2017; Jain and Sirisha, 2020).

Phenolic compounds like Phlorotannins from *Ecklonia stolonifera* and *Ecklonia cava*, such as eckol and dieckol, significantly suppress the expression of pro-inflammatory enzymes, including inducible iNOS and cyclooxygenase-2 (COX-2), along with reducing NO production in activated macrophages. These compounds also inhibit NF-κB activation, a central transcription factor governing inflammatory cytokine expression. In immune modulation, algal polyphenols influence cytokine secretion and macrophage activation, thereby shaping both innate and adaptive immune responses. For instance, polyphenol-rich extracts from *Ascophyllum nodosum* reduce TNF-α and IL-6 production in lipopolysaccharide-stimulated macrophages, demonstrating their capacity to attenuate excessive inflammatory responses while maintaining immune balance (Punampalam et al., 2018; Lorente-Cebrián et al., 2025).

Beyond inflammation control, algal immunomodulators intersect with cancer therapy and tissue repair. PC promotes apoptosis in cancer cells and selectively accumulates in tumor-associated macrophages, functioning as a TAM-targeted photosensitizer for photodynamic

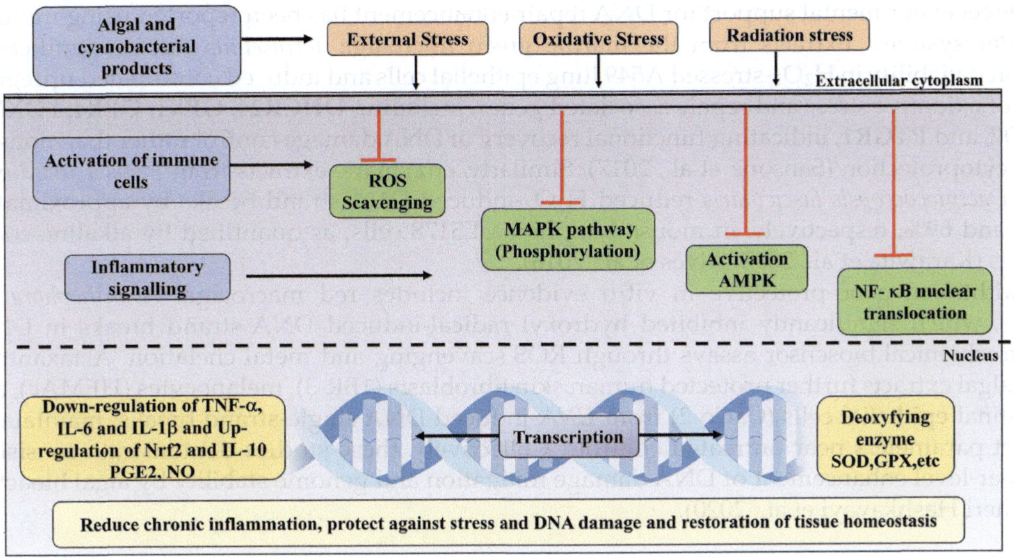

FIGURE 20.5 Schematic representation of algal products modulation in inflammation.

therapy. Fucoidan from brown algae such as *Fucus vesiculosus* enhances chemosensitivity to cisplatin and doxorubicin by inducing apoptosis and inhibiting proliferation. In dermatological and wound-healing contexts, microalgal metabolites, including carotenoids, polysaccharides, and PUFAs, accelerate re-epithelialization, angiogenesis, and oxidative stress resolution, with innovative systems such as living *Chlorella pyrenoidosa-based* hydrogels improving hypoxic wound environments and reducing inflammation (Pérez et al., 2016).

Collectively, these findings establish microalgae as multifunctional immunomodulatory platforms capable of simultaneously regulating inflammation, enhancing immune competence, and mitigating disease-associated immune dysregulation (Fig. 20.5). The broad-spectrum efficacy, low systemic toxicity, and sustainability of algal bioactives underscore their strong translational potential for applications in immunotherapy, inflammatory and autoimmune diseases, neurodegeneration, dermatology, and oncology.

20.6 Preclinical and translational evidence

20.6.1 In vitro studies: Cellular models showing DNA repair enhancement

In vitro studies provide strong evidence that marine algal peptides, pigments, and extracts enhance cellular DNA damage control by modulating oxidative stress, cell-cycle regulation, and genome surveillance mechanisms. Peptides derived from *C. vulgaris*, *S. platensis*, and cyanobacteria such as *Lyngbya majuscula* consistently activate the p53–p21 signaling pathway, resulting in post-G1 cell-cycle arrest and controlled apoptosis in cancer cell models. This checkpoint activation facilitates DNA damage sensing and repair or selectively eliminates genomically unstable cells, thereby promoting genome integrity (Saadaoui et al., 2020; Visuddho et al., 2024).

Direct experimental support for DNA repair enhancement has been reported using multiple cellular systems. Extracts from the marine green microalga *Tetraselmis suecica* significantly restored viability in H_2O_2-stressed A549 lung epithelial cells and induced coordinated upregulation of oxidative stress and repair-associated genes, including **DHCR24, GPX1, GPX4, FOXM1, SOD2,** and **PTGR1**, indicating functional recovery of DNA damage control rather than nonspecific cytoprotection (Sansone et al., 2017). Similarly, enzymatic extracts from *Pediastrum duplex* and *Dactylococcopsis fascicularis* reduced H_2O_2-induced DNA strand breaks by approximately 80% and 69%, respectively, in mouse lymphoma L5178 cells, as quantified by alkaline comet assays (Karawita et al., 2007; Alves et al., 2016).

Additional geno-protective in vitro evidence includes red macroalgal *Acanthophora* extracts, which significantly inhibited hydroxyl radical-induced DNA strand breaks in DNA-electrochemical biosensor assays through ROS scavenging and metal chelation. Astaxanthin-rich algal extracts further protected human skin fibroblasts (1BR-3), melanocytes (HEMAc), and intestinal epithelial cells (CaCo-2) from UVA-induced DNA single-strand breaks, maintaining comet parameters near untreated controls. Collectively, these studies demonstrate consistent cellular-level enhancement of DNA damage mitigation and genome stability by algal bioactive (Bagheri Hashkavayi et al., 2020).

20.6.2 In vivo models: Animal studies validating geno-protective functions

Animal studies provide supportive, though often indirect, validation of geno-protective effects under physiologically relevant conditions. In rodent cancer models, administration of algal peptide formulations resulted in reduced DNA fragmentation, attenuation of oxidative stress markers, and protection of normal tissues from chemotherapy-induced genotoxicity, reflected by improved survival and decreased systemic toxicity. While direct measurements of DNA repair kinetics in vivo remain limited, these functional outcomes strongly suggest protection against genome instability (Hamed et al., 2015).

Carotenoid-based interventions further support in vivo geno-protection. Fucoxanthin supplementation suppressed colorectal tumor formation in azoxymethane/dextran sulfate sodium–induced mouse models and reduced inflammation in LPS-induced sepsis models through inhibition of NF-κB and COX-2 signaling. In metabolic disease models, fucoxanthin derived from *Phaeodactylum tricornutum* and *Melosira nummuloides* significantly reduced adiposity, inflammatory cytokines, and oxidative stress in high-fat-diet-fed mice. β-Carotene from *Dunaliella bardawil* improved cholesterol efflux and reduced atherogenic inflammation in LDL receptor–deficient mice, while zeaxanthin derivatives attenuated cardiac inflammation and oxidative damage in aging rat models (Lyons and O'Brien, 2002; Rao et al., 2013; Zhang and Wang, 2015).

Neuroprotective in vivo validation has also been reported for phycobiliproteins. C-PC from *S. platensis* ameliorated cognitive deficits in amyloid-β–induced Alzheimer's disease mouse models by restoring hippocampal BDNF expression, suppressing neuroinflammation, and rebalancing apoptotic signaling. These findings collectively demonstrate organism-level protection against DNA damage–associated pathological processes (Li et al., 2020).

20.6.3 Clinical potential: Translational research in cancer, aging, and toxicology

Taken together, preclinical evidence highlights significant translational potential for marine algal bioactives in cancer prevention, aging-related genome instability, neurodegenerative disorders, and toxicological protection. Algal peptides, carotenoids, and phycobiliproteins exhibit antioxidant, antiinflammatory, and selective anticancer activities that are directly relevant to the modulation of DNA damage responses (DDR) (Leong et al., 2024) (Table 20.1).

Human studies, although still limited, are encouraging. Fucoxanthin has demonstrated a high safety margin in animal toxicology studies and favorable tolerability in human supplementation trials. Similarly, *Arthrospira* (Spirulina) has shown clinically measurable benefits in dyslipidemia and type 2 diabetes patients, supporting its regulatory acceptance and feasibility for therapeutic development. However, many promising cyanobacterial and microalgal compounds remain untested in human trials, underscoring a critical translational gap that standardized and bioengineered algal products may help bridge (Rastogi et al., 2024).

20.6.4 Formulation and bioavailability: Challenges in delivery and therapeutic dosage optimization

Despite promising bioactivity, clinical translation of algal-derived compounds is constrained by poor bioaccessibility, stability, and systemic bioavailability. Carotenoid absorption is limited by rigid algal cell walls, resulting in low intestinal uptake, while peptides are prone to enzymatic degradation and rapid clearance (do Nascimento et al., 2021).

PC represents a notable exception, as its chromophore PCB survives gastrointestinal digestion, is absorbed as a PCB–peptide complex, and is metabolized to phycocyanorubin (PCR). PCR activates the Keap1–Nrf2–HO-1 pathway, leading to systemic antioxidant, antiinflammatory, and immunomodulatory effects across the gut–liver, gut–lung, and vascular axes. This mechanistic insight explains how orally administered PC exerts effects beyond the gastrointestinal tract (Liu et al., 2022). In addition to it, C57BL/6 mice CPC (200mg/kg for 2 weeks, intraperitoneal injection of saline), and oligomeric Aβ1–42 + CPC. However, CPC treatment significantly ($P < 0.05$) attenuated the spatial memory deficits in the mice injected with (Romay et al., 2003; Li et al., 2020).

Recent advances in encapsulation technologies, nanocarrier-based delivery systems, recombinant expression platforms, and formulation optimization have improved stability and efficacy in preclinical models. These strategies emphasize the importance of precise dosage optimization and delivery design to enable effective therapeutic translation. Future research should focus on enhancing organ-specific bioavailability, defining therapeutic dosing windows, and integrating algal bioactives into multimodal treatment regimens to fully realize their clinical potential (Ashaolu, 2024).

20.7 Future directions and biotechnological applications

The long-term and excessive consumption of nutraceutical supplements raises important safety concerns, as high daily intake may lead to adverse or toxic effects, highlighting the necessity to clearly distinguish between prophylactic and therapeutic dosing regimens. While prophylactic doses may confer broad cellular protection to both healthy and malignant cells, therapeutic doses must be precisely optimized to selectively inhibit cancer cell proliferation. Although a substantial body of in vitro and in vivo evidence suggests that appropriately dosed antioxidant nutraceuticals can reduce cancer risk and improve therapeutic outcomes, results from randomized clinical trials remain inconsistent. These variable outcomes are likely driven by multiple interacting factors, including dosage, compound synergism, bioavailability,

TABLE 20.1 In vitro and in vivo evidence of DNA repair enhancement by algal products.

Sr. no.	Algal source/peptide	Model system	Key outcome	References
1.	*Chlorella vulgaris* peptides (VECYGPNRPQF) *Spirulina platensis* (C-phycocyanin) *Lyngbya majuscula* peptides (Isomalyngamide A) Nanocarrier-delivered algal peptides	Human gastric cancer AGS cells Multiple cancer cell lines Breast cancer cell lines Rodent tumor models Murine cancer models	PostG1 arrest via p53/p21, reduced DNA fragmentation ROS reduction, protection from oxidative DNA damage Genome instability suppression via apoptosis induction Reduced DNA fragmentation and chemotherapy-induced genotoxicity	Calvani et al. (2020)
2.	*Tetraselmis suecica* extracts		Activation of DHCR24, GPX1/4, FOXM1, SOD2; reversal of H_2O_2-induced oxidative DNA stress Restoration of cell viability and suppression of DNA damage-linked inflammatory mediators	Sansone et al. (2017)
3.	*Acanthophora* extract	Human cancer cell lines (HT-29, MCF-7, SKOV3, SKMES-1)	Near-complete inhibition of Fenton-induced DNA strand damage at 0.20 mg mL^{-1} ROS scavenging and metal chelation preventing oxidative DNA breakage Controlled cytotoxicity with DNA protection in nontarget systems	Bagheri Hashkavayi et al. (2020)

(continued on next page)

TABLE 20.1 In vitro and in vivo evidence of DNA repair enhancement by algal products—cont'd

Sr. no.	Algal source/peptide	Model system	Key outcome	References
4.	Astaxanthin-rich algal extract	Human fibroblasts (1BR-3) Human melanocytes (HEMAc) Human intestinal CaCo-2 cells	Significant reduction of UVA-induced DNA strand breaks Suppression of DNA migration and oxidative DNA damage Protection against oxidative DNA damage and GSH depletion	Jain et al. (2020)
5.	*Spirulina platensis*-Phycocyanin	C57BL/6 mice	Downregulation of histone deacetylase-3 (3), restores miRNA-335 levels, and upregulates brain-derived neurotrophic factor (BDNF) expression in the hippocampus, thereby promoting synaptic plasticity and memory function	Li et al. (2020)

patient health status, cancer heterogeneity, lifestyle factors, supplementation behavior, and study duration. Accordingly, rigorously designed and well-controlled clinical trials employing advanced and standardized methodologies are urgently required to establish safe and effective antioxidant supplementation strategies in oncology.

Recombinant phycobiliproteins (PBPs) produced through heterologous expression in *Escherichia coli* represent a high-purity, cost-effective, and scalable alternative to conventional extraction from algal biomass, a process that is labor-intensive and often yields proteins with reduced bioactivity. For example, recombinant phycoerythrin (RPE) expressed in *E. coli* BL21 has been shown to markedly suppress tumor growth in H22 tumor-bearing mice, while simultaneously modulating gut microbiota by enriching beneficial genera such as *Bacteroides*, *Alloprevotella*, *Alistipes*, and *Ruminococcus*, and reducing potentially harmful taxa including *Prevotella* and *Barnesiella*. These findings highlight the dual antitumor and immune-regulatory potential of recombinant PBPs (Qi et al., 2019).

Similarly, recombinant PCB, generated via co-expression of the *ho1* and *pcyA* genes from *Synechocystis*, exhibited strong antioxidative and antiinflammatory activities, achieving approximately 90% DPPH radical scavenging at 819.2 μg/mL, thereby demonstrating therapeutic relevance in oxidative stress–associated and metabolic disorders (Ma et al., 2020a). Advanced engineering of holo-α-phycocyanin (rHHPC) using *cpcA, cpcE, cpcF, ho1*, and *pcyA* further enhanced its functional properties, resulting in superior hydroxyl radical scavenging activity ($IC_{50} = 277.5 \pm 25.8$ μg/mL), improved stability, and performance exceeding that of native PC, supporting its application in photodynamic therapy and imaging (Guan et al., 2009; Alvey et al., 2011; Ge et al., 2013).

In the context of cancer metastasis, the recombinant β-subunit of C-phycocyanin (C-PC/β), expressed using the pET43.1a + vector, effectively inhibited cancer cell proliferation and induced apoptosis through cytoskeletal depolymerization and caspase activation, underscoring its antimetastatic potential. Recombinant PBPs have also shown considerable promise in diagnostics; notably, recombinant holo-ApcA conjugated with PEB exhibited enhanced fluorescence and high analytical sensitivity, enabling alpha-fetoprotein (AFP) detection at concentrations as low as 0.11 ng/mL in immunoassays (Liu et al., 2009; Shoja et al., 2015).

To overcome technical challenges such as plasmid instability, innovations including single-plasmid expression systems and alternative eukaryotic hosts such as *Pichia pastoris* and *Saccharomyces cerevisiae* are being explored. Additional improvements involve fusion strategies, such as Strep-tag II, which facilitate affinity purification under native conditions, achieving over 75% binding efficiency in *Anabaena*-expressed PBPs. Furthermore, streptavidin-fused PBPs—including SLA–PCB, SLA–PEB, SA–PCA–PCB, and SA–PCA–PEB have demonstrated strong biotin-binding capacity and high fluorescence, enabling ultrasensitive detection of AFP, carcinoembryonic antigen (CEA), and miRNA at picomolar levels. Collectively, these advances position recombinant PBPs as versatile and multifunctional biomolecules with broad applications in cancer therapy, microbiota modulation, diagnostics, and bioimaging (Ge et al., 2018).

20.8 Conclusion

This chapter synthesizes emerging evidence demonstrating that algal and cyanobacterial metabolites constitute a multifunctional defense system against genomic instability by integrating antioxidant activity, modulation of DDR pathways, epigenetic reprogramming, and immune-inflammatory control. Across diverse classes of bioactives-including phycobiliproteins, carotenoids, PUFAs, phenolic compounds, spS, and sterols—these metabolites consistently attenuate ROS–driven DNA lesions, preserve telomeric integrity, and support the fidelity of core repair mechanisms such as BER, checkpoint signaling, and apoptosis–cell-cycle coupling. Importantly, their actions extend beyond passive radical scavenging; algal compounds actively influence key regulatory axes involving p53, ATM/ATR, PARP, NF-κB, and Nrf2, thereby coordinating redox balance with genome surveillance and cellular fate decisions. Equally significant is the epigenetic dimension highlighted in this chapter. Algal-derived polyphenols, carotenoids, and phycobiliproteins modulate DNA methylation and histone acetylation landscapes through inhibition of DNMTs and HDACs, enabling reactivation of silenced tumor suppressor genes and restoration of DNA repair gene expression. This dual antioxidant–epigenetic mode of action provides a mechanistic basis for the observed chemopreventive, radioprotective, and neuroprotective effects reported in preclinical models. In parallel, immunomodulatory functions, particularly those mediated by PC and spS, reduce inflammation-driven genotoxic stress, reinforcing genome stability at the tissue and organismal levels. The convergence of in vitro, in vivo, and early translational data positions algal bioactives as promising candidates for integration into preventive nutrition, functional foods, and adjunctive therapeutic strategies targeting cancer, aging-associated genomic instability, and environmental or radiation-induced DNA damage. Nevertheless, critical challenges remain, including variability in bioavailability, dose-dependent duality of antioxidant effects, and limited human clinical validation. Advances in recombinant production, nanoformulation, and delivery systems, coupled with rigorously designed clinical trials, will be essential to translate these molecules from experimental promise to clinical reality. Overall, algal and cyanobacterial phyco-chemicals emerge not merely as supportive antioxidants but as systems-level modulators of genome integrity, bridging redox biology, epigenetics, and immune regulation. Their sustainable origin and molecular versatility underscore their potential role in next-generation interventions aimed at safeguarding genomic stability and cellular resilience in the face of increasing environmental and metabolic stressors.

References

Abdelkarim, O.H., Wijffels, R.H., Barbosa, M.J., 2025. Exploiting microalgal diversity for sterol production. Front. Plant Sci. 16, 1–23. https://doi.org/10.3389/fpls.2025.1616863.

Alam, M.A., Xu, J.L., Wang, Z., 2020. Microalgae biotechnology for food, health and high value products, microalgae biotechnology for food, health and high value products. Springer. https://doi.org/10.1007/978-981-15-0169-2.

Alves, C., Pinteus, S., Horta, A., Pedrosa, R., 2016. High cytotoxicity and anti-proliferative activity of algae extracts on an in vitro model of human hepatocellular carcinoma. SpringerPlus 5. https://doi.org/10.1186/s40064-016-2938-2.

Alvey, R.M., Biswas, A., Schluchter, W.M., Bryant, D.A., 2011. Attachment of noncognate chromophores to CpcA of synechocystis sp. PCC 6803 and synechococcus sp. PCC 7002 by heterologous expression in *Escherichia coli*. Biochemistry 50, 4890–4902. https://doi.org/10.1021/bi200307s.

Amna Kashif, S., Hwang, Y.J., Park, J.K., 2018. Potent biomedical applications of isolated polysaccharides from marine microalgae Tetraselmis species. Bioprocess. Biosyst. Eng. 41, 1611–1620. https://doi.org/10.1007/s00449-018-1987-z.

Ansari, H., Nouranian, M., Hajigholami, A., Mahmudian, A., 2019. Effect of omega-3 fatty acid supplements on cancer-related fatigue in an outpatient setting: a randomized controlled trial. Middle East J. Cancer 10, 333–340. https://doi.org/10.30476/mejc.2019.78605.

Ashaolu, T.J., 2024. The powerful phycobiliproteins-phycocyanin and phycoerythrin: Pleiotropic applications and biofunctional uses. Algal. Res. 82, 103636. https://doi.org/10.1016/j.algal.2024.103636.

Astley, S.B., Elliott, R.M., Archer, D.B., Southon, S., 2002. Increased cellular carotenoid levels reduce the persistence of DNA single-strand breaks after oxidative challenge. Nutr. Cancer 43, 202–213. https://doi.org/10.1207/S15327914NC432_11.

Azqueta, A., Collins, A.R., 2012. Carotenoids and DNA damage. Mutat. Res. Fundam. Mol. Mech. Mutagen. 733, 4–13. https://doi.org/10.1016/j.mrfmmm.2012.03.005.

Bagheri Hashkavayi, A., Hashemnia, S., Osfouri, S., 2020. Investigations of antioxidant potential and protective effect of acanthophora algae on DNA damage: an electrochemical approach. Microchem. J. 159. https://doi.org/10.1016/j.microc.2020.105455.

Barbalace, M.C., Malaguti, M., Giusti, L., Lucacchini, A., Hrelia, S., Angeloni, C., 2019. Anti-inflammatory activities of marine algae in neurodegenerative diseases. Int. J. Mol. Sci. 20. https://doi.org/10.3390/ijms20123061.

Basily, H., Nassar, M., Diwani, G.E., Abo El-Enin, S., 2018. Exploration of using the algal bioactive compounds for cosmeceuticals and pharmaceutical applications. Egypt. Pharm. J. 17, 109. https://doi.org/10.4103/epj.epj_6_18.

Ben Hafsa, M., Ben Ismail, M., Garrab, M., Aly, R., Gagnon, J., and Naghmouchi, K., 2017. Antimicrobial, antioxidant and cytotoxic properties of water-soluble polysaccharides extracted from microalgae Isochrysis galbana and Nannochloropsis oculata. J. Serb. Chem. Soc. 82 (5), 509–522. https://doi.org/10.2298/JSC161016036B.

Benton, C.B., Fiskus, W., Bhalla, K.N., 2017. Targeting histone acetylation: readers and writers in leukemia and cancer. Cancer J. 23, 286–291. https://doi.org/10.1097/PPO.0000000000000284.

Bhat, V.B., Madyastha, K.M., 2001. Scavenging of peroxynitrite by phycocyanin and phycocyanobilin from Spirulina platensis: protection against oxidative damage to DNA. Biochem. Biophys. Res. Commun. 285, 262–266. https://doi.org/10.1006/bbrc.2001.5195.

Bulut, O., Akın, D., Sönmez, Ç., Öktem, A., Yücel, M., Öktem, H.A., 2019. Phenolic compounds, carotenoids, and antioxidant capacities of a thermo-tolerant Scenedesmus sp. (Chlorophyta) extracted with different solvents. J. Appl. Phycol. 31, 1675–1683. https://doi.org/10.1007/s10811-018-1726-5.

Buonvino, S., Trinca, C., Leu, S., Licoccia, S., Melino, S., 2025. A new algal-friendly extract from euglena cantabrica with potential applications in biomedical field. Mar. Drugs 23, 1–21. https://doi.org/10.3390/md23120453.

Calvani, M., Pasha, A., Favre, C., 2020. Nutraceutical boom in cancer: inside the labyrinth of reactive oxygen species. Int. J. Mol. Sci. 21. https://doi.org/10.3390/ijms21061936.

Ceccarelli, V., Ronchetti, S., Marchetti, M.C., Calvitti, M., Riccardi, C., Grignani, F., et al., 2020. Molecular mechanisms underlying eicosapentaenoic acid inhibition of HDAC1 and DNMT expression and activity in carcinoma cells. Biochim Biophys. Acta. Gene. Regul. Mech. 1863, 194481. https://doi.org/10.1016/j.bbagrm.2020.194481.

Chang, C.J., Yang, Y.H., Liang, Y.C., Chiu, C.J., Chu, K.H., Chou, H.N., et al., 2011. A novel phycobiliprotein alleviates allergic airway inflammation by modulating immune responses. Am. J. Respir. Crit. Care Med. 183, 15–25. https://doi.org/10.1164/rccm.201001-0009OC.

Chang, L., Liu, X., Li, Y., Liu, C., Yang, F., Zhao, J., et al., 2015. Structural organization of an intact phycobilisome and its association with photosystem II. Nat. Publ. Group 1–12. https://doi.org/10.1038/cr.2015.59.

Checa, J., Aran, J.M., 2020. Reactive oxygen species: drivers of physiological and pathological processes. J. Inflamm. Res. 13, 1057–1073. https://doi.org/10.2147/JIR.S275595.

Chen, J., Zeng, W., Gan, J., Li, Y., Pan, Y., Li, J., et al., 2021. Physicochemical properties and anti-oxidation activities of ulvan from Ulva pertusa Kjellm. Algal. Res. 55, 102269. https://doi.org/10.1016/j.algal.2021.102269.

Chi, Y., Li, H., Wang, Pei, Du, C., Ye, H., Zuo, S., et al., 2020. Structural characterization of ulvan extracted from Ulva clathrata assisted by an ulvan lyase. Carbohydr. Polym. 229, 115497. https://doi.org/10.1016/j.carbpol.2019.115497.

Ciancia, M., Fernández, P.V., Leliaert, F., 2020. Diversity of sulfated polysaccharides from cell walls of coenocytic green algae and their structural relationships in view of green algal evolution. Front. Plant Sci. 11, 1–15. https://doi.org/10.3389/fpls.2020.554585.

Collins, A.R., 2001. Carotenoids and genomic stability. Mutat. Res. Fundam. Mol. Mech. Mutagen. 475, 21–28. https://doi.org/10.1016/S0027-5107(01)00071-9.

Collins, A.R., Olmedilla, B., Southon, S., Granado, F., Duthie, S.J., 1998. Serum carotenoids and oxidative DNA damage in human lymphocytes. Carcinogenesis 19, 2159–2162. https://doi.org/10.1093/carcin/19.12.2159.

Darwito, D., Dharmana, E., Riwanto, I., Budijitno, S., Suwardjo, S., Purnomo, J., et al., 2019. Effects of Omega-3 supplementation on Ki-67 and VEGF expression levels and clinical outcomes of locally advanced Breast Cancer patients treated with Neoadjuvant CAF chemotherapy: a randomized controlled trial report. Asian Pac. J. Cancer Prev. 20, 911–916. https://doi.org/10.31557/APJCP.2019.20.3.911.

De Jesus Raposo, M.F., De Morais, A.M.M.B., De Morais, R.M.S.C., 2015. Carotenoids from marine microalgae: a valuable natural source for the prevention of chronic diseases. Mar. Drugs 13, 5128–5155. https://doi.org/10.3390/md13085128.

De Jesus Raposo, M.F., De Morais, R.M.S.C., De Morais, A.M.M.B., 2013. Health applications of bioactive compounds from marine microalgae. Life Sci. 93, 479–486. https://doi.org/10.1016/j.lfs.2013.08.002.

Dembitsky, V.M., Maoka, T., 2007. Allenic and cumulenic lipids. Prog. Lipid Res. 46, 328–375. https://doi.org/10.1016/j.plipres.2007.07.001.

Di Meo, S., Reed, T.T., Venditti, P., Victor, V.M., 2016. Role of ROS and RNS sources in physiological and pathological conditions. Oxid Med. Cell Longev. 2016. https://doi.org/10.1155/2016/1245049.

Djuricic, I., Calder, P.C., 2021. Beneficial outcomes of omega-6 and omega-3 polyunsaturated fatty acids on human health: an update for 2021. Nutrients 13. https://doi.org/10.3390/nu13072421.

do Nascimento, T.C., Pinheiro, P.N., Fernandes, A.S., Murador, D.C., Neves, B.V., de Menezes, C.R., et al., 2021. Bioaccessibility and intestinal uptake of carotenoids from microalgae Scenedesmus obliquus. Lwt 140, 110780. https://doi.org/10.1016/j.lwt.2020.110780.

Ferdous, U.T., Yusof, Z.N.B., 2021. Medicinal prospects of antioxidants from Algal Sources in Cancer Therapy. Front. Pharmacol. 12, 1–22. https://doi.org/10.3389/fphar.2021.593116.

Ferdous, U.T., Yusof, Z.N.B., Smriti, A.R., Singh, G., Khavari, F., Saidijam, M., et al., 2024. Medicinal prospects of antioxidants from algal sources in cancer therapy. Biochim. Biophys. Acta. Gene. Regul. Mech. 93, 109. https://doi.org/10.1016/j.lfs.2013.08.002.

Fimbres-Olivarria, D., Carvajal-Millan, E., Lopez-Elias, J.A., Martinez-Robinson, K.G., Miranda-Baeza, A., Martinez-Cordova, L.R., et al., 2018. Chemical characterization and antioxidant activity of sulfated polysaccharides from Navicula sp. Food Hydrocoll. 75, 229–236. https://doi.org/10.1016/j.foodhyd.2017.08.002.

Formisano, L., Guida, N., Laudati, G., Mascolo, L., Di Renzo, G., Canzoniero, L.M.T., 2015. MS-275 inhibits aroclor 1254-induced SH-SY5Y neuronal cell toxicity by preventing the formation of the HDAC3/REST complex on the synapsin-1 promoter. J. Pharmacol. Exp. Ther. 352, 236–243. https://doi.org/10.1124/jpet.114.219345.

Gaignard, C., Laroche, C., Pierre, G., Dubessay, P., Delattre, C., Gardarin, C., et al., 2019. Screening of marine microalgae: Investigation of new exopolysaccharide producers. Algal. Res. 44, 1–62. https://doi.org/10.1016/j.algal.2019.101711.

Galasso, C., Corinaldesi, C., Sansone, C., 2017. Carotenoids from marine organisms: biological functions and industrial applications. Antioxidants 6. https://doi.org/10.3390/antiox6040096.

Ge, B., Chen, Y., Yu, Q., Lin, X., Li, J., Qin, S., 2018. Regulation of the heme biosynthetic pathway for combinational biosynthesis of phycocyanobilin in Escherichia coli. Process Biochem. 71, 23–30. https://doi.org/10.1016/j.procbio.2018.05.011.

Ge, B., Li, Y., Sun, H., Zhang, S., Hu, P., Qin, S., Huang, F., 2013. Combinational biosynthesis of phycocyanobilin using genetically-engineered Escherichia coli. Biotechnol. Lett 35, 689–693. https://doi.org/10.1007/s10529-012-1132-z.

Goiris, K., Muylaert, K., Voorspoels, S., Noten, B., De Paepe, D., E Baart, G.J., De Cooman, L., 2014. Detection of flavonoids in microalgae from different evolutionary lineages. J. Phycol. 50, 483–492. https://doi.org/10.1111/jpy.12180.

Guan, X.Y., Zhang, W.J., Zhang, X.W., Li, Y.X., Wang, J.F., Lin, H.Z., et al., 2009. A potent anti-oxidant property: fluorescent recombinant α-phycocyanin of Spirulina. J. Appl. Microbiol. 106, 1093–1100. https://doi.org/10.1111/j.1365-2672.2008.04051.x.

Guedes, A.C., Amaro, H.M., Malcata, F.X., 2011. Microalgae as sources of carotenoids. Mar. Drugs 9, 625–644. https://doi.org/10.3390/md9040625.

Haidari, F., Abiri, B., Iravani, M., Ahmadi-Angali, K., Vafa, M., 2020. Effects of vitamin D and omega-3 fatty acids co-supplementation on inflammatory factors and tumor marker cea in colorectal cancer patients undergoing chemotherapy: a randomized, double-blind, placebo-controlled clinical trial. Nutr. Cancer 72, 948–958. https://doi.org/10.1080/01635581.2019.1659380.

Hamed, I., Özogul, F., Özogul, Y., Regenstein, J.M., 2015. Marine bioactive compounds and their health benefits: a review. Compr. Rev. Food Sci. Food Saf. 14, 446–465. https://doi.org/10.1111/1541-4337.12136.

Haoujar, I., Cacciola, F., Abrini, J., Mangraviti, D., Giuffrida, D., Oulad El Majdoub, Y., Kounnoun, A., Miceli, N., Fernanda Taviano, M., Mondello, L., Rigano, F., 2019. The contribution of carotenoids, phenolic compounds, and flavonoids to the antioxidative properties of marine microalgae isolated from Mediterranean Morocco. Molecules 24 (22), 4037. https://doi.org/10.3390/molecules24224037.

Haq, S.H., Al-Ruwaished, G., Al-Mutlaq, M.A., Naji, S.A., Al-Mogren, M., Al-Rashed, S., et al., 2019. Antioxidant, anticancer activity and phytochemical analysis of green algae, chaetomorpha collected from the Arabian Gulf. Sci. Rep. 9, 1–7. https://doi.org/10.1038/s41598-019-55309-1.

Huang, H., Lang, Y., Zhou, M., 2024. A comprehensive review on medical applications of microalgae. Algal. Res. 80, 103504. https://doi.org/10.1016/j.algal.2024.103504.

Jain, A., Sirisha, V.L., 2020. Algal carotenoids: understanding their structure, distribution and potential applications in human health. Encycl. Mar. Biotechnol. 33–64.

Jain, A., Sirisha, V.L., Rathod, N.B., Hamed, I., Cagalj, M., Nascimento, T.C., et al., 2020. Influence of astaxanthin, zeaxanthin and lutein on DNA damage and repair in UVA-irradiated cells. Carcinogenesis 85, 4–13. https://doi.org/10.1207/S15327914NC432_11.

Jiang, L., Wang, Y., Yin, Q., Liu, G., Liu, H., Huang, Y., et al., 2017. Phycocyanin: a potential drug for cancer treatment. J. Cancer 8, 3416–3429. https://doi.org/10.7150/jca.21058.

Kadam, S.U., Tiwari, B.K., O'Donnell, C.P., 2015. Extraction, structure and biofunctional activities of laminarin from brown algae. Int. J. Food Sci. Technol. 50, 24–31. https://doi.org/10.1111/ijfs.12692.

Kannaujiya, V.K., Sundaram, S., Sinha, R.P., 2017. Phycobiliproteins: recent developments and future applications, phycobiliproteins: recent developments and future applications. Springer, Singapore. https://doi.org/10.1007/978-981-10-6460-9.

Karawita, R., Senevirathne, M., Athukorala, Y., Affan, A., Lee, Y.J., Kim, S.K., et al., 2007. Protective effect of enzymatic extracts from microalgae against DNA damage induced by H_2O_2. Mar. Biotechnol. 9, 479–490. https://doi.org/10.1007/s10126-007-9007-3.

Kaulmann, A., Bohn, T., 2014. Carotenoids, inflammation, and oxidative stress-implications of cellular signaling pathways and relation to chronic disease prevention. Nutr. Res. 34, 907–929. https://doi.org/10.1016/j.nutres.2014.07.010.

Kaur, P., Dhandayuthapani, S., Venkatesan, T., Gantor, M., Rathinavelu, A., 2020. Molecular mechanism of C-phycocyanin induced apoptosis in LNCaP cells. Bioorg. Med. Chem. 28, 115272. https://doi.org/10.1016/j.bmc.2019.115272.

Khavari, F., Saidijam, M., Taheri, M., Nouri, F., 2021. Microalgae: therapeutic potentials and applications. Mol. Biol. Rep. 48, 4757–4765. https://doi.org/10.1007/s11033-021-06422-w.

Khotimchenko, Y.S., Khozhaenko, E.V., Khotimchenko, M.Y., Kolenchenko, E.A., Kovalev, V.V., 2010. Carrageenans as a new source of drugs with metal binding properties. Mar. Drugs 8, 1106–1121. https://doi.org/10.3390/md8041106.

Kichouh-Aiadi, S., Gallardo-Rodríguez, J.J., Cerón-García, M.C., López-Rosales, L., García-Camacho, F., Sánchez-Mirón, A., 2024. Exploring the potential of epigenetic chemicals to increase metabolite production in the dinoflagellate microalga Amphidinium carterae. J. Appl. Phycol. 36, 1169–1179. https://doi.org/10.1007/s10811-024-03189-y.

Kim, D., Kim, Yerin, Kim, Yuri, 2019. Effects of β-carotene on expression of selected microRNAs, histone acetylation, and DNA methylation in colon cancer stem cells. J. Cancer Prev. 24, 224–232. https://doi.org/10.15430/jcp.2019.24.4.224.

Leong, H.J.Y., Teoh, M.L., Beardall, J., Convey, P., 2024. Green beauty unveiled: Exploring the potential of microalgae for skin whitening, photoprotection and anti-aging applications in cosmetics. J. Appl. Phycol. 36, 3315–3328. https://doi.org/10.1007/s10811-024-03345-4.

Levy-Ontman, O., Huleihel, M., Hamias, R., Wolak, T., Paran, E., 2017. An anti-inflammatory effect of red microalga polysaccharides in coronary artery endothelial cells. Atherosclerosis 264, 11–18. https://doi.org/10.1016/j.atherosclerosis.2017.07.017.

Li, H., Su, L., Chen, S., Zhao, L., Wang, H., Ding, F., Chen, H., Shi, R., Wang, Y., Huang, Z., 2018. Physicochemical characterization and functional analysis of the polysaccharide from the edible microalga Nostoc sphaeroides. Molecules 23 (2), 508. https://doi.org/10.3390/molecules23020508.

Li, F., Zhu, Y., Wan, Y., Xie, X., Ke, R., Zhai, C., et al., 2017. Activation of PPARγ inhibits HDAC1-mediated pulmonary arterial smooth muscle cell proliferation and its potential mechanisms. Eur. J. Pharmacol. https://doi.org/10.1016/j.ejphar.2017.08.045.

Li, Z., Gan, L., Yan, S., Yan, Y., Huang, W., 2020. E ff ect of C - phycocyanin on HDAC3 and miRNA - 335 in Alzheimer's disease. Transl. Neurosci. 11 (1), 161–172.

Liu, Q., Li, W., Qin, S., 2020. Therapeutic effect of phycocyanin on acute liver oxidative damage caused by X-ray. Biomed. Pharmacother. 130, 110553. https://doi.org/10.1016/j.biopha.2020.110553.

Liu, R., Qin, S., Li, W., 2022. Phycocyanin: anti-inflammatory effect and mechanism. Biomed. Pharmacother. 153, 113362. https://doi.org/10.1016/j.biopha.2022.113362.

Liu, S., Chen, H., Qin, S., Zhang, W., Guan, X., Lu, Y., 2009. Highly soluble and stable recombinant holo-phycocyanin alpha subunit expressed in Escherichia coli. Biochem. Eng. J. 48, 58–64. https://doi.org/10.1016/j.bej.2009.08.006.

Lorente-Cebrián, S., Costa, A.G.V., Castillo-Rivas, J.A., Castro, M., Arbonés-Mainar, J.M., Goñi, S., et al., 2025. Phenolic compounds and epigenetic mechanisms regulating gene expression: effects on human health. J. Physiol. Biochem. https://doi.org/10.1007/s13105-025-01105-7.

Lorenzo, Y., Azqueta, A., Luna, L., Bonilla, F., Domínguez, G., Collins, A.R., 2009. The carotenoid β-cryptoxanthin stimulates the repair of DNA oxidation damage in addition to acting as an antioxidant in human cells. Carcinogenesis 30, 308–314. https://doi.org/10.1093/carcin/bgn270.

Lyons, N.M., O'Brien, N.M., 2002. Modulatory effects of an algal extract containing astaxanthin on UVA-irradiated cells in culture. J. Dermatol. Sci. 30, 73–84. https://doi.org/10.1016/S0923-1811(02)00063-4.

Ma, C., Li, W., Ge, B., Lin, J., Qin, S., 2020a. Biosynthesis of phycocyanobilin in recombinant Escherichia coli. J. Oceanol. Limnol. 38, 529–538. https://doi.org/10.1007/s00343-019-9060-6.

Ma, J., You, X., Sun, S., Wang, X., Qin, S., Sui, S., 2020b. Structural basis of energy transfer in porphyridium purpureum phycobilisome. Nature. https://doi.org/10.1038/s41586-020-2020-7.

Martin-Creuzburg, D., Merkel, P., 2016. Sterols of freshwater microalgae: potential implications for zooplankton nutrition. J. Plankton Res. 38, 865–877. https://doi.org/10.1093/plankt/fbw034.

Minić, S., 2021. The bioactive properties of Spirulina -derived phycobiliproteins and phycobilins. Biologia. Serbica 43, 32–42. https://doi.org/10.5281/zenodo.5512528.

Mullineaux, C.W., 2014. Electron transport and light-harvesting switches in cyanobacteria. Front. Plant Sci. 5, 1–6. https://doi.org/10.3389/fpls.2014.00007.

Muñoz-Almagro, N., Gilbert-López, B., Pozuelo-Rollón, M.C., García-Fernandez, Y., Almeida, C., Villamiel, M., et al., 2020. Exploring the microalga euglena cantabrica by pressurized liquid extraction to obtain bioactive compounds. Mar. Drugs 18, 1–17. https://doi.org/10.3390/md18060308.

Mut-Salud, N., Álvarez, P.J., Garrido, J.M., Carrasco, E., Aránega, A., Rodríguez-Serrano, F., 2016. Antioxidant intake and antitumor therapy: toward nutritional recommendations for optimal results. Oxid. Med. Cell. Longev. 2016. https://doi.org/10.1155/2016/6719534.

Nascimento, T.C., Nass, P.P., Fernandes, A.S., Nörnberg, M.L., Zepka, Q.Z., Jacob-Lopes, E., 2022. Microalgae carotenoids: an overview of biomedical applications. Algal Biotechnol. Integr. Algal Eng. Bioenergy Bioremediat. Biomed. Appl. 409–425. https://doi.org/10.1016/B978-0-323-90476-6.00013-3.

Ngo, D.H., Kim, S.K., 2013. Sulfated polysaccharides as bioactive agents from marine algae. Int. J. Biol. Macromol. 62, 70–75. https://doi.org/10.1016/J.IJBIOMAC.2013.08.036.

Niki, E., 2016. Oxidative stress and antioxidants: distress or eustress? Arch. Biochem. Biophys. 595, 19–24. https://doi.org/10.1016/J.ABB.2015.11.017.

Orejuela-Escobar, L., Gualle, A., Ochoa-Herrera, V., Philippidis, G.P., 2021. Prospects of microalgae for biomaterial production and environmental applications at biorefineries. Sustainability 13, 1–19. https://doi.org/10.3390/su13063063.

Ou, Y., Zheng, S., Lin, L., Jiang, Q., Yang, X., 2010. Protective effect of C-phycocyanin against carbon tetrachloride-induced hepatocyte damage in vitro and in vivo. Chem. Biol. Interact. 185, 94–100. https://doi.org/10.1016/j.cbi.2010.03.013.

Pagels, F., Guedes, A.C., Amaro, H.M., Kijjoa, A., Vasconcelos, V., 2019. Phycobiliproteins from cyanobacteria: chemistry and biotechnological applications. Biotechnol. Adv. 37, 422–443. https://doi.org/10.1016/J.BIOTECHADV.2019.02.010.

Pan-utai, W., Iamtham, S., Boonbumrung, S., Mookdasanit, J., 2022. Improvement in the sequential extraction of phycobiliproteins from arthrospira platensis using green technologies. Life 12, 1–19. https://doi.org/10.3390/life12111896.

Parmar, P., Kumari, N., Sharma, V., 2013. Structural and functional alterations in photosynthetic apparatus of plants under cadmium stress. Bot. Stud. 54, 1–6. https://doi.org/10.1186/1999-3110-54-45.

Patel, H.M., Rastogi, R.P., Trivedi, U., Madamwar, D., 2019. Cyanobacterial diversity in mat sample obtained from the hypersaline desert, Rann of Kachchh. 3 Biotech 9 (8), 304.

Patel, S.N., Sonani, R.R., Roy, D., Kumar, N., Subudhi, S., Pabbi, S., 2022. Exploring the structural aspects and therapeutic perspectives of cyanobacterial phycobiliproteins. 3 Biotech 1–15. https://doi.org/10.1007/s13205-022-03284-2.

Peppone, L.J., Inglis, J.E., Mustian, K.M., Heckler, C.E., Padula, G.D.A., Mohile, S.G., et al., 2019. Multicenter randomized controlled trial of omega-3 fatty acids versus omega-6 fatty acids for the control of cancer-related fatigue among breast cancer survivors. JNCI Cancer Spectr. 3, 1–9. https://doi.org/10.1093/JNCICS/PKZ005.

Pérez, M.J., Falqué, E., Domínguez, H., 2016. Antimicrobial action of compounds from marine seaweed. Mar. Drugs 14, 1–38. https://doi.org/10.3390/md14030052.

Perillo, B., Di Donato, M., Pezone, A., Di Zazzo, E., Giovannelli, P., Galasso, G., et al., 2020. ROS in cancer therapy: the bright side of the moon. Exp. Mol. Med. 52, 192–203. https://doi.org/10.1038/s12276-020-0384-2.

Picardo, M., Dell'Anna, M.L., 2010. Oxidative stress. Vitiligo 231–237. https://doi.org/10.1007/978-3-540-69361-1_27.

Pratita, A.T.K., Djahrah, S.M., Fathurohman, M., 2024. Isolation and antiradical activity of phycobiliprotein compounds from Chlorella pyrinoidosa. Food Res. 8, 85–95. https://doi.org/10.26656/fr.2017.8(6).524.

Punampalam, R., Soo, K., Weng, N., 2018. Evaluation of antioxidant properties of phycobiliproteins and phenolic compounds extracted from Bangia atropurpurea. Malays. J. Fundam. Appl. Sci. 14, 289–297.

Qi, H., Liu, Y., Qi, X., Liang, H., Chen, H., Jiang, P., Wang, D., 2019. Dietary recombinant phycoerythrin modulates the gut microbiota of H22 tumor-bearing mice. Mar. Drugs 17, 1–17. https://doi.org/10.3390/md17120665.

Randhir, A., Laird, D.W., Maker, G., Trengove, R., Moheimani, N.R., 2020. Microalgae: a potential sustainable commercial source of sterols. Algal Research. Elsevier. https://doi.org/10.1016/j.algal.2019.101772.

Rao, A.R., Sindhuja, H.N., Dharmesh, S.M., Sankar, K.U., Sarada, R., Ravishankar, G.A., 2013. Effective inhibition of skin cancer, tyrosinase, and antioxidative properties by astaxanthin and astaxanthin esters from the green alga haematococcus pluvialis. J. Agric. Food Chem. 61, 3842–3851. https://doi.org/10.1021/jf304609j.

Rastogi, R.P., Sonani, R.R., Madamwar, D., 2015. Effects of PAR and UV radiation on the structural and functional integrity of phycocyanin, phycoerythrin and allophycocyanin isolated from the marine cyanobacterium lyngbya sp. A09DM. In: Photochem. Photobiol., 91, pp. 837–844.

Rastogi, R.P., Sonani, R.R., Madamwar, D., Singh, V., Shirbhate, E., Kore, R., et al., 2024. Medicinal prospects of antioxidants from algal sources in cancer therapy. Algal. Res. 93, 109. https://doi.org/10.1016/j.lfs.2013.08.002.

Ravishankar, G.A., Ambati, R.R., 2019. Handbook of algal technologies and phytochemicals, handbook of algal technologies and phytochemicals: volume II phycoremediation, biofuels and global biomass production. CRC Press. https://doi.org/10.1201/9780429057892.

Raza, M.H., Siraj, S., Arshad, A., Waheed, U., Aldakheel, F., Alduraywish, S., Arshad, M., 2017. ROS-modulated therapeutic approaches in cancer treatment. J. Cancer Res. Clin. Oncol. 143, 1789–1809. https://doi.org/10.1007/s00432-017-2464-9.

Rodriguez-Amaya, D.B., 2015. Carotenes and xanthophylls as antioxidants. Handb. Antioxid. Food Preserv. 17–50. https://doi.org/10.1016/B978-1-78242-089-7.00002-6.

Romay, C., González, R., Ledón, N., Remirez, D., Rimbau, V., 2003. C-phycocyanin : a biliprotein with antioxidant, anti-inflammatory and neuroprotective effects. Curr. Protein Pept. Sci. 4 (3), 207–216.

Roohinejad, S., Koubaa, M., Barba, F.J., Saljoughian, S., Amid, M., Greiner, R., 2017. Application of seaweeds to develop new food products with enhanced shelf-life, quality and health-related beneficial properties. Food Res. Int. 99, 1066–1083. https://doi.org/10.1016/j.foodres.2016.08.016.

Roy, K.R., Arunasree, K.M., Reddy, N.P., Dheeraj, B., Reddy, G.V., Reddanna, P., 2007. Alteration of mitochondrial membrane potential by Spirulina platensis C-phycocyanin induces apoptosis in the doxorubicinresistant human hepatocellular-carcinoma cell line HepG2. Biotechnol. Appl. Biochem. 47, 159–167. https://doi.org/10.1042/ba20060206.

Ryckebosch, E., Bruneel, C., Muylaert, K., Foubert, I., 2012. Microalgae as an alternative source of omega-3 long chain polyunsaturated fatty acids. Lipid Technol. 24, 128–130. https://doi.org/10.1002/lite.201200197.

Saadaoui, I., Rasheed, R., Abdulrahman, N., Bounnit, T., Cherif, M., Al Jabri, H., Mraiche, F., 2020. Algae-derived bioactive compounds with anti-lung cancer potential. Mar. Drugs 18. https://doi.org/10.3390/md18040197.

Sanjeewa, K.K.A., Fernando, I.P.S., Samarakoon, K.W., Lakmal, H.H.C., Kim, E.A., Kwon, O.N., et al., 2016. Anti-inflammatory and anti-cancer activities of sterol rich fraction of cultured marine microalga nannochloropsis oculata. Algae 31, 277–287. https://doi.org/10.4490/algae.2016.31.6.29.

Sanjeewa, K.K.A., Kang, N., Ahn, G., Jee, Y., Kim, Y.T., Jeon, Y.J., 2018. Bioactive potentials of sulfated polysaccharides isolated from brown seaweed Sargassum spp in related to human health applications: a review. Food Hydrocoll. 81, 200–208. https://doi.org/10.1016/J.FOODHYD.2018.02.040.

Sansone, C., Galasso, C., Orefice, I., Nuzzo, G., Luongo, E., Cutignano, A., et al., 2017. The green microalga Tetraselmis suecica reduces oxidative stress and induces repairing mechanisms in human cells. Sci. Rep. 7, 1–12. https://doi.org/10.1038/srep41215.

Santocono, M., Zurria, M., Berrettini, M., Fedeli, D., Falcioni, G., 2006. Influence of astaxanthin, zeaxanthin and lutein on DNA damage and repair in UVA-irradiated cells. J. Photochem. Photobiol. B 85, 205–215. https://doi.org/10.1016/j.jphotobiol.2006.07.009.

Sathasivam, R., Radhakrishnan, R., Hashem, A., Abd_Allah, E.F., 2019. Microalgae metabolites: a rich source for food and medicine. Saudi J Biol. Sci. 26, 709–722. https://doi.org/10.1016/j.sjbs.2017.11.003.

Sayegh, F., Elazzazy, A., Bellou, S., Moustogianni, A., Elkady, A.I., Baeshen, M.N., et al., 2016. Production of polyunsaturated single cell oils possessing antimicrobial and anticancer properties. Ann. Microbiol. 66, 937–948. https://doi.org/10.1007/s13213-015-1176-0.

Sayegh, F., Elazzazy, A., Bellou, S., Moustogianni, A., Elkady, A.I., Baeshen, M.N., et al., 2020. Microalgae: a potential sustainable commercial source of sterols. Algal. Res. 31, 937–948. https://doi.org/10.4490/algae.2016.31.6.29.

Scanferlato, R., Bortolotti, M., Sansone, A., Chatgilialoglu, C., Polito, L., de Spirito, M., et al., 2019. Hexadecenoic fatty acid positional isomers and De Novo PUFA synthesis in colon cancer cells. Int. J. Mol. Sci. 20, 1–20. https://doi.org/10.3390/ijms20040832.

Severo, I.A., Dias, R.R., do Nascimento, T.C., Deprá, M.C., Maroneze, M.M., Zepka, L.Q., et al., 2022. Microalgae-derived polysaccharides: potential building blocks for biomedical applications. World J. Microbiol. Biotechnol. 38, 1–17. https://doi.org/10.1007/s11274-022-03342-0.

Shalaby, E.A., 2011. Algae as promising organisms for environment and health. Plant. Signal. Behav. 6, 1338–1350. https://doi.org/10.4161/psb.6.9.16779.

Shoja, Z., Memari, H.R., Ardakani, M.R., 2015. Cloning and expression of beta subunit gene of phycocyanin from spirulina platensis in Escherichia coli. Jundishapur J. Microbiol. 8, 10–15. https://doi.org/10.5812/jjm.17809.

Sies, H., 2019. Oxidative stress: eustress and distress in redox homeostasis. Stress Physiol. Biochem. Pathol. Handb. Stress Ser. 3, 153–163. https://doi.org/10.1016/B978-0-12-813146-6.00013-8.

Singh, K.B., Kaushalendra, Rajan, J.P., 2022. Therapeutical and nutraceutical roles of cyanobacterial tetrapyrrole chromophore: recent advances and future implications. Front. Microbiol. 13, 1–5. https://doi.org/10.3389/fmicb.2022.932459.

Singh, V., Shirbhate, E., Kore, R., Vishwakarma, S., Parveen, S., Veerasamy, R., et al., 2024. Microbial metabolites-induced epigenetic modifications for inhibition of colorectal cancer: current status and future perspectives. Mini Rev. Med. Chem. 24. https://doi.org/10.2174/0113895575320344240625080555.

Smriti, A.R., Singh, G., 2024. Exploring the potential of microalgae as food supplements : a comprehensive review. J. Sci. Ind. Res. 83, 688–702. https://doi.org/10.56042/jsir.v83i6.9786.

Sonani, R.R., 2015. Antioxidant potential of phycobiliproteins: role in anti-aging research. Biochem. Anal. Biochem. 4, 4–11. https://doi.org/10.4172/2161-1009.1000172.

Sonani, R.R., Rastogi, R.P., Patel, R., Madamwar, D., 2016. Recent advances in production, purification and applications of phycobiliproteins. World J. Biol. Chem. 2016 (7), 100–110. https://doi.org/10.4331/wjbc.v7.i1.100.

Song, M., Ou, F.S., Zemla, T.J., Hull, M.A., Shi, Q., Limburg, P.J., Alberts, S.R., Sinicrope, F.A., Giovannucci, E.L., Van Blarigan, E.L., Meyerhardt, J.A., 2019. Marine omega-3 fatty acid intake and survival of stage III colon cancer according to tumor molecular markers in NCCTG Phase III trial N0147 (Alliance). Int. J. Cancer 145 (2), 380–389. https://doi.org/10.1002/ijc.32113.

Su, J., Guo, K., Huang, M., Liu, Y., Zhang, J., Sun, L., et al., 2019. Fucoxanthin, a marine xanthophyll isolated from Conticribra weissflogii ND-8: preventive anti-inflammatory effect in a mouse model of sepsis. Front. Pharmacol. 10, 1–17. https://doi.org/10.3389/fphar.2019.00906.

Sumi, Y., 2009. Microalgae pioneering the future — application and utilization — NISTEP science & technology foresight center. quarterly review No.34, 9–21.

Sun, L., Chu, J., Sun, Z., Chen, L., 2016. Physicochemical properties, immunomodulation and antitumor activities of polysaccharide from Pavlova viridis. Life Sci. 144, 156–161. https://doi.org/10.1016/J.LFS.2015.11.013.

Takaichi, S., 2011. Carotenoids in algae: distributions, biosyntheses and functions. Mar. Drugs 9, 1101–1118. https://doi.org/10.3390/md9061101.

Tang, T., Effiong, K., Hu, J., Li, C., Xiao, X., 2021. Chemical prevention and control of the green tide and fouling organism ulva: key chemicals, mechanisms, and applications. Front. Mar. Sci. 8. https://doi.org/10.3389/fmars.2021.618950.

Trabelsi, L., Chaieb, O., Mnari, A., Abid-Essafi, S., Aleya, L., 2016. Partial characterization and antioxidant and antiproliferative activities of the aqueous extracellular polysaccharides from the thermophilic microalgae Graesiella sp. BMC Complement. Altern. Med. 16, 1–10. https://doi.org/10.1186/s12906-016-1198-6.

Usman, A., Khalid, S., Usman, Atif, Hussain, Z., Wang, Y., 2017. Algal polysaccharides, novel application, and outlook. algae based polymers, blends, and composites: chemistry. Biotechnol. Mater. Sci. 115–153. https://doi.org/10.1016/B978-0-12-812360-7.00005-7.

Vasilakis, G., Marka, S., Ntzouvaras, A., Zografaki, M.E., Kyriakopoulou, E., Kalliampakou, K.I., et al., 2025. A new algal friendly extract from euglena cantabrica with potential applications in biomedical field. Mar. Drugs 23, 1–17. https://doi.org/10.3390/md23020077.

Visuddho, V., Halim, P., Helen, H., Muhar, A.M., Iqhrammullah, M., Mayulu, N., et al., 2024. Modulation of apoptotic, cell cycle, DNA repair, and senescence pathways by marine algae peptides in cancer therapy. Mar. Drugs 22 (8), 338.

Wells, M.L., Potin, P., Craigie, J.S., Raven, J.A., Merchant, S.S., Helliwell, K.E., et al., 2017. Algae as nutritional and functional food sources: revisiting our understanding. J. Appl. Phycol. 29, 949–982. https://doi.org/10.1007/s10811-016-0974-5.

Yaacob, N.S., Abdullah, H., Ahmad, M.F., Maniyam, M.N., Sjahrir, F., 2022. Microalgae biotechnology: emerging biomedical applications. Algal biotechnology: integrated algal engineering for bioenergy, bioremediation, and biomedical applications. Elsevier Inc. https://doi.org/10.1016/B978-0-323-90476-6.00017-0.

Yu, Y., Shen, M., Song, Q., Xie, J., 2018. Biological activities and pharmaceutical applications of polysaccharide from natural resources: a review. Carbohydr. Polym. 183, 91–101. https://doi.org/10.1016/J.CARBPOL.2017.12.009.

Zhang, J., Liu, L., Ren, Y., Chen, F., 2019. Characterization of exopolysaccharides produced by microalgae with antitumor activity on human colon cancer cells. Int. J. Biol. Macromol. 128, 761–767. https://doi.org/10.1016/J.IJBIOMAC.2019.02.009.

Zhang, L., Wang, H., 2015. Multiple mechanisms of anti-cancer effects exerted by astaxanthin. Mar. Drugs 13, 4310–4330. https://doi.org/10.3390/md13074310.

Zhang, L.X., Cai, C.E., Guo, T.T., Gu, J.W., Xu, H.L., Zhou, Y., et al., 2011. Anti-cancer effects of polysaccharide and phycocyanin from porphyra yezoensis. J. Mar. Sci. Technol. 19, 377–382. https://doi.org/10.51400/2709-6998.2178.

Zhang, T., Liu, D., Zhang, Y., Chen, L., Zhang, W., Sun, T., 2024. Biomedical engineering utilizing living photosynthetic cyanobacteria and microalgae: Current status and future prospects. Mater. Today Bio. 27, 101154. https://doi.org/10.1016/j.mtbio.2024.101154.

Ziyaei, K., Ataie, Z., Mokhtari, M., Adrah, K., (2020). Since January 2020 Elsevier has created a COVID-19 resource centre with free information in English and Mandarin on the novel coronavirus COVID-19. The COVID-19 resource centre is hosted on Elsevier Connect, the company ' s public news and information.

CHAPTER

21

DNA damage, aging concept: Feasibility and significance

Nimisha Joly Chennakadan[a], Yadhu Krishna Menon K[a], Chandrabose Selvaraj[b], Yuvaraj Sivamani[c], Kishan Nanga Srihari Kishan Reddy[a] and Sumitha Elayaperumal[a]

[a]Department of Biotechnology and Bioinformatics, JSS Academy of Higher Education and Research, Mysore, Karnataka, India [b]Department of Marine Biotechnology, CSRDD Lab, Bioinformatics Division, AMET University (Deemed to be University), Kanathur, Tamil Nadu, India [c]Crescent School of Pharmacy, B.S. Abdur Rahman Crescent Institute of Science & Technology, Chennai, Tamil Nadu, India

21.1 Introduction

Aging is a dynamic and multifaceted biological phenomenon that involves an irreversible decline in physiological functions as well as a diminished regenerative capacity. Furthermore, a susceptibility to a range of chronic diseases is associated with the condition. From the molecular and cellular levels to tissues and organs, the biological changes are all-encompassing. Aging is the result of a blend of intrinsic factors, like programmed genetics, and extrinsic factors, like environmental exposure, causing a slow decline in the body's homeostatic systems. Out of the many cellular and molecular processes involved in aging, the accrual of DNA damage and the decline in maintenance systems are particularly crucial regarding genome efficiency.

Over the course of a lifetime, DNA faces constant threats from a wide variety of damaging factors, both internal and external in origin. Internally, the factors include spontaneous DNA replication errors, base deamination, hydrolytic reactions, and reactive oxygen species (ROS). These factors add to a chronic burden of genomic harm. The forces of lipid peroxidation, advanced glycation end-products (AGEs), and atypical one-carbon metabolism also contribute to these internal stresses (Yemini et al., 2021). Externally, lifestyle choices and the environment also play a role: UV radiation, ionizing radiation, industrial and environmental pollutants, tobacco, carcinogenic foods, and even heavy metals pose a threat to the DNA. These internal

Genome Damage: Repair and Adaptation Across Life.
DOI: https://doi.org/10.1016/B978-0-443-42857-9.00006-2

and external factors are highly likely to cause damage to DNA structure and cause single-strand breaks (SSBs) and double-strand breaks (DSBs). If not fixed, these errors and cellular changes will eventually lead to cellular processes being disrupted, causing transcription, replication, and chromosome segregation, and in turn, triggering a cell to either undergo programmed cell death, enter permanent cell cycle arrest, or turn cancerous—undergo a transformation to a cancer cell.

The DNA damage theory of aging argues that a basic mechanism of aging and age-related diseases is the accumulation of DNA lesions over a given period of time, coupled with an effectiveness decline in DNA repair pathways (Chen et al., 2022). With advancing age, progressively accumulated DNA lesions combined with an age-related decline in genome maintenance pathways, such as base excision and nucleotide excision repair, drive persistent genomic instability that manifests as impaired tissue homeostasis, stem cell exhaustion, and functional decline at the organismal level. This theory is validated with a combination of human genetic disorders, transgenic animal studies, and aging studies. Inherited syndromes like Werner, Hutchinson-Gilford Progeria, xeroderma pigmentosum, and Cockayne syndrome display aggressive aging with early hair graying, osteoporosis, neurodegeneration, and a heightened risk for cancer. Likewise, animal models lacking vital DNA repair proteins like XPA, ERCC1, or ataxia-telangiectasia mutated (ATM) suffer from early onset of frailty and an array of age-related ailments. Repair of damaged DNA or increased NAD + metabolism also enhances lifespan and improves age-related pathologies and has been shown to delay aging phenotypes. This all accentuates the importance of genomic stability in the maintenance of tissue function and the progressive decline due to aging.

Beyond its direct mutagenic and cytotoxic actions, genomic instability impacts other hallmarks of aging. Persistent DNA damage breaks apart and remodels chromatin structure, causing epigenetic drift associated with abnormal histone changes, reconfigured DNA methylation patterns, and diminished transcription accuracy. This form of epigenetic deregulation helps erode a cell's distinct identity and enables the acquisition of transcriptional noise. Also, cytosolic DNA damage sensors like cGAS-stimulator of interferon genes (STING), cGAS-STING, and others triggered DNA damage, which in turn activates the innate immune system and perpetuates "inflammaging," low-grade chronic inflammation. These inflammatory processes then exacerbate oxidative and genotoxic stress (Esposito et al., 2022). Mitochondrial DNA (mtDNA) is particularly prone to oxidative lesions because it is unprotected by histones and is situated near the electron transport chain. The increased mutations to mtDNA weaken its ability to defend against oxidative damage, which in turn increases the generation of ROS, diminishes the integrity of the nuclear and mitochondrial genomes, and the cycle continues. This cycle of instability of the genomic structure and the dysfunction of mitochondria is a fundamental enhancer of age-related decline.

The goal of this article is to present an integrated approach regarding the molecular and cellular aspects of an association between DNA damage and aging. In the other sections of this paper, we focus on the different types and sources of DNA damage with emphasis on internal and external origins. We describe the elaborate systems of DNA repair, which include base excision repair (BER), nucleotide excision repair (NER), mismatch repair (MMR), homologous recombination (HR), and nonhomologous end joining (NHEJ), devised to counter these lesions and genomic instability (Sahin & DePinho, 2010). In addition, we look at the way the different pathways of DNA repair contribute to the different characteristics of aging, which include stem cell decline, dysregulated cell-to-cell signaling, altered cellular homeostasis, and imbalance of protein turnover. There is a focus on the systems of DNA damage, mitochondrial failure, and epigenetics as critical elements of aging biology.

Focusing on insights derived from progeroid mouse strains, DNA repair knockout lines, and other models exposed to genotoxic stressors. These models not only replicate numerous facets of human aging but are also useful as testing grounds for therapeutic approaches. In the final parts of the work, we present new approaches for reducing DNA damage and increasing repair capacity, such as pharmacological sirtuin and PARP activators, NAD + boosters NMN and nicotinamide riboside (NR), senolytic and antioxidant therapies, as well as some more advanced gene editing techniques. We also discuss the increasing prospects of biomarkers of DNA damage as indicators of biological age and monitoring the effectiveness of treatment (Table 21.1) (Geense et al., 2017).

Thus, cumulative DNA damage together with reduced repair capacity forms a central mechanistic axis linking molecular lesions to cellular dysfunction and clinically observable features of aging.

21.2 Genotoxic stress and the molecular anatomy of DNA lesions

The integrity of DNA can be viewed as being at risk continuously, and from a plethora of both internal and external sources (Fig. 21.1). Different types of DNA lesions, each with its own distinct structural implications and biological consequences, arise from DNA-damaging challenges. The ability of a cell to mitigate and repair the damage is a function of the effectiveness and accuracy of the repair processes (López-Otín et al., 2016). The effectiveness of the repair processes, their accuracy, and the plethora of sources and types of damage one can be subjected to reconstruct the cumulative impact of damage on DNA and inform one on the impact of genomic stability and aging.

Due to regular metabolic processes, DNA is consistently damaged internally. This is chiefly aggravated by ROS, which result from oxidative phosphorylation in the mitochondria. These types of ROS can lead to oxidation of DNA bases, causing changes like 8-oxo-2′-deoxyguanosine (8-oxo-dG), single and DSBs, or creation of abasic sites (Burhans and Weinberger, 2007). Another major endogenous factor stemming from DNA damage is replication stress, which is particularly common in proliferative cells. DNA polymerase, replication fork stalling, secondary and repetitive structures, as well as replication and transcription machinery collisions, contribute to strand breakage and the development of mutations. Additionally, spontaneous hydrolytic reactions, such as hydrolytic deamination of cytosine to uracil or depurination resulting in the loss of guanine and adenine bases from DNA, can occur and generate disrupted base pairs. Byproducts of lipid peroxidation, for example, malondialdehyde and 4-hydroxy-2-nonenal, can also give rise to the formation of bulky DNA adducts, which have the potential to be mutagenic (Yemini et al., 2021).

Exogenous DNA damage is attributed to environmental and lifestyle factors. Ultraviolet (UV) radiation from sunlight, specifically UVB, causes adjacent pyrimidines on the same DNA strand to crosslink covalently. These lesions distort the DNA helix and impede replication and transcription, which requires urgent repair (Burhans and Weinberger, 2007). Ionizing radiation (IR), such as X-rays and gamma rays, penetrates the tissues and induces a variety of DNA lesions, including DSBs and complex clusters of damage that are difficult to

TABLE 21.1 The table compares key aging theories—Wear-and-tear, free radical, programmed aging, and genomic instability—based on causes, mechanisms, and implications. It highlights their unique and overlapping contributions to the aging process and disease.

Feature	Wear-and-tear theory	Free radical theory	Programmed aging theory	Molecular/genomic instability theory
Core concept	Aging results from accumulated physical damage due to daily use	Aging is caused by the accumulation of damage from ROS and free radicals	Aging is a genetically regulated and time-programmed biological process	Aging results from accumulating DNA damage and loss of genomic integrity
Primary cause of aging	Mechanical and biochemical strain on tissues and organs	Oxidative stress from metabolic byproducts	Genetic programming and hormonal signals	DNA damage, telomere attrition, chromosomal abnormalities
Key processes involved	Tissue degradation, joint wear, organ fatigue	Mitochondrial dysfunction, ROS accumulation	Epigenetic regulation, telomere shortening, apoptosis	Mutations, impaired DNA repair, chromatin changes
Biological focus	Whole organs, systems	Mitochondria, oxidative pathways	Genes, endocrine regulation	Nuclear genome, repair systems, DNA–protein integrity
Evidence supporting	Observed physical wear with age (e.g., cartilage erosion)	Correlation of ROS and aging, lifespan extension with antioxidants in some models	Progeroid syndromes, conserved aging genes, and reproductive timing	Increased DNA damage with age, progeroid diseases, and mutation load in aged tissues
Strengths	Intuitive, simple, aligns with physical decline	Links metabolism to aging, a mechanistic insight	Explains species-specific lifespan patterns	Strong molecular basis, links aging with age-related diseases
Limitations	Does not explain variable aging rates across species	Antioxidant therapies not consistently effective	Lacks a unified mechanism, can appear teleological	Complexity of DNA damage interactions, hard to separate cause from consequence
Relation to disease	Degenerative diseases (arthritis, cataracts)	Neurodegeneration, cardiovascular disease	Age-related hormone imbalances, neuroendocrine decline	Cancer, neurodegeneration, immunosenescence
Reversibility potential	Limited; focuses on damage limitation	Possibly reversible with antioxidants or ROS modulators	Theoretically modifiable via genetic/epigenetic interventions	Potential with gene therapy, DNA repair enhancement

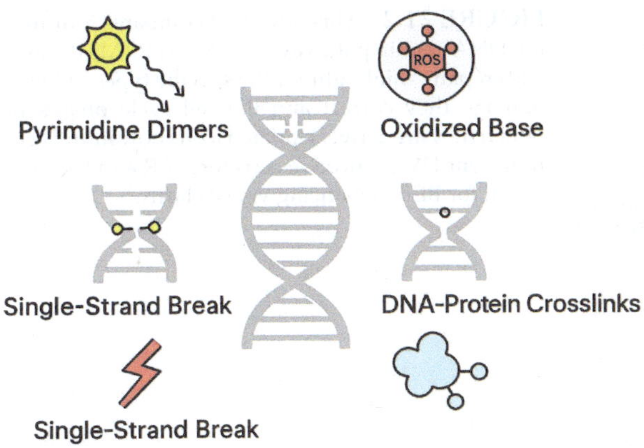

FIGURE 21.1 The figure illustrates major DNA lesion types—Pyrimidine dimers, oxidized bases, strand breaks, and DNA–protein crosslinks—Each linked to sources like UV light, ROS, and chemicals. It highlights how diverse insults compromise genome integrity and contribute to aging and disease.

repair. Chemical mutagens such as polycyclic aromatic hydrocarbons (benzo[a]pyrene), alkylating agents (methyl methanesulfonate), and crosslinking agents (cisplatin) act on DNA to induce adducts, interstrand crosslinks, and strand breaks. Heavy metals (arsenic, cadmium, and lead) are environmental pollutants that interfere with the enzymes involved in cellular repair and indirectly increase genomic instability. Lifestyle choices, such as smoking and diets high in fats, exacerbate this damage by increasing ROS levels and exposure to dietary carcinogens.

The fact that DNA damages are structurally diverse implies that several types of DNA repair systems must exist in nature. SSBs are among the most common, and BER is primarily responsible for fixing them (Sahin and DePinho, 2010). DSBs are less common but more perilous because they can result in chromosomal rearrangements and deletions due to erroneous repair, potentially leading to oncogenic chromosome translocations. Base modifications such as oxidation, alkylation, and deamination alter the hydrogen bonding potential and cause mispairing during replication. Abasic or apurinic/apyrimidinic (AP) sites are generated through the spontaneous hydrolysis of the N-glycosidic bond or as an intermediate product during BER. Bulky adducts and DNA crosslinks (predominantly exogenous) prevent strand separation or block replication forks. Their removal necessitates the concerted actions of the NER, translesion synthesis, or HR systems (Royer et al., 2012).

Together with exogenous factors, they shape the mutational landscape of cells and determine the rate of functional decline with age (Geense et al., 2017). The cellular machinery for dealing with DNA damage, which consists of several overlapping and specific DNA repair pathways (discussed in detail in the following section), is referred to as a DNA repair system.

21.3 DNA repair circuitry: Guardians of genomic stability

The integrity of the genome is ensured by several levels of DNA repair systems, which are tightly coordinated and interlinked to detect and mend various types of DNA damage. The lesions result from intrinsic cellular processes and exposure to external factors such as UV light, environmental toxins, and IR. DNA repair pathways operate both at baseline and in response to injury signals to uphold genomic integrity by facilitating DNA replication and transcription

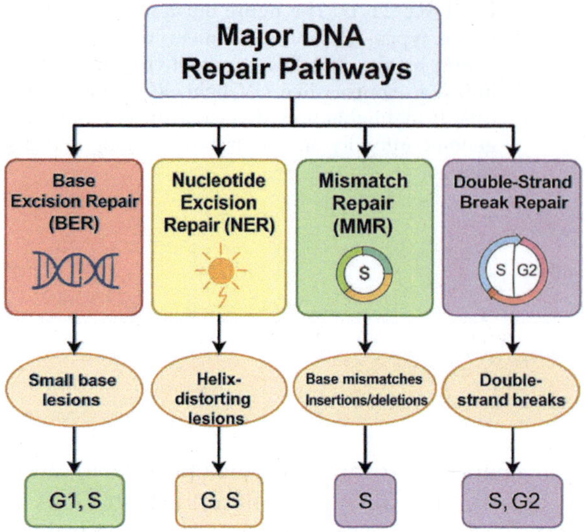

FIGURE 21.2 This flowchart compares four major DNA repair pathways—BER, NER, MMR, and DSB repair—highlighting the specific types of DNA damage they correct and the cell cycle phases in which they are active. Updated icons now intuitively represent UV-induced lesions for NER and base excision for BER, enhancing visual clarity.

throughout an organism's lifespan (Fig. 21.2). The prevention of mutations that threaten cellular vital functions, differentiation, or existence is called genomic maintenance (Chen et al., 2022). As organisms age, the efficiency, coordination, and fidelity of these pathways progressively decline. This loss is critically linked to five well-documented aging hallmarks: Genomic instability, cellular senescence, stem cell exhaustion, chronic inflammation, and cancer.

BER is one of the most basic and evolutionarily conserved DNA repair pathways. It targets small, nonhelix-distorting base lesions resulting from oxidation, deamination, and alkylation—damage types that occur frequently due to normal cellular metabolism. The BER pathway is initiated when a DNA glycosylase removes a damaged base, resulting in the formation of an abasic site known as an apurinic/apyrimidinic (AP) site. The site is then cleaved by AP endonuclease 1 (APE1) to generate a single-strand break. DNA polymerase β then fills in the gap with the correct nucleotide, and DNA ligase III, frequently in conjunction with XRCC1, seals the remaining nick (Okereke et al., 2013). BER is especially important in postmitotic tissues such as the brain, which are highly susceptible to oxidative stress due to their intense metabolic activity.

Studies have shown that the expression and activity of the key BER enzymes OGG1, APE1, and DNA polymerase β are significantly reduced with aging. This decreased repair capacity leads to the increased accumulation of oxidative DNA lesions such as 8-oxoguanine, which causes mutagenesis, mitochondrial dysfunction, and altered cellular signaling. Experimental data from aged rodent models and human fibroblasts confirm this tendency (Vijg and Suh, 2013). Moreover, boosting BER capacity by overexpressing repair-facilitating proteins such as SIRT6 restores the repair system's efficiency, delays tissue aging, and prolongs lifespan in mice.

NER is a DNA repair pathway that removes bulky helix-distorting DNA lesions caused by UV radiation or chemical carcinogens. Unlike BER, which identifies specific lesions on DNA bases, NER targets the structural distortions of the DNA helix. The NER pathway is further categorized into global genome NER (GG-NER), which scans the entire genome for DNA damage, and transcription-coupled NER (TC-NER), which specifically targets lesions that impede the transcription process (Okereke et al., 2013). In GG-NER, the DNA distortion is recognized by a protein complex called XPC-RAD23B, while in TC-NER, stalled transcription complexes are identified by proteins known as CSA and CSB. In both cases, DNA duplex unwinding is performed by the XPB and XPD helicases, which are part of the TFIIH complex. The damaged DNA strand is then cut on both sides of the lesion by XPG and XPF-ERCC1 endonucleases, leading to the removal of a 24–32 nucleotide fragment. The gap is subsequently filled by DNA polymerases δ or ε and sealed by DNA ligase I (Vijg & Suh, 2013). Defects in NER proteins lead to several progeroid syndromes known as xeroderma pigmentosum (XP), cockayne syndrome (CS), and trichothiodystrophy (TTD). The diseases are characterized by features of premature aging that include neurodegeneration, photosensitivity, developmental delay, and early onset of age-related pathology. Importantly, these syndromes constitute a compelling demonstration in humans that deficient DNA repair can cause premature aging. Moreover, during normal aging, NER activity declines in various tissues, including the skin, liver, and brain.

Mismatch repair (MMR) functions as a postreplicative DNA repair system to rectify base–base mismatches and insertion–deletion loops that escape proofreading during DNA replication and recombination (Burhans and Weinberger, 2007). The MutS and MutL homologs bind DNA as heterodimeric complexes, known as MutSα (MSH2-MSH6) for single base mismatches and MutSβ (MSH2-MSH3) for loop recognition. Upon detection, the MutLα complex (MLH1-PMS2 heterodimer) is recruited, and exonuclease-mediated excision of the error-containing DNA strand is initiated. DNA polymerase δ fills the gap, and DNA ligase I seals the strand. Defective MMR leads to a cellular phenotype referred to as microsatellite instability (MSI), which is observed in several cancer types, such as colorectal, endometrial, and gastric. While the role of MMR in aging is not well established, defective MMR indirectly promotes age-related accumulation of somatic mutations (Vijg and Suh, 2013). Moreover, new research indicates that MMR efficiency might decline with age because of decreased expression of MMR proteins or epigenetic repression of their genes, which further contributes to the accumulation of mutations in aging tissues and increases cancer risk.

DSBs are among the most dangerous forms of DNA damage, and improper repair can lead to chromosomal translocations, deletions, and cell death. Two main pathways are responsible for DSB repair: Nonhomologous end joining (NHEJ) and HR. NHEJ is the primary DSB repair pathway in nondividing cells and functions throughout the cell cycle. It acts by directly ligating broken DNA ends independently or with minimal homology (Hoeijmakers, 2009). The Ku70/Ku80 heterodimer initiates the process by binding to DNA ends and recruiting DNA-PKcs. This is followed by the end processing (usually mediated by Artemis) and final ligation (DNA ligase IV/XRCC4). While NHEJ is fast, it is error-prone and can lead to small insertions or deletions at the target site. In contrast, HR is an error-free repair pathway needing a homologous DNA template (typically the sister chromatid) and therefore functions only in S and G2 phases of the cell cycle.

HR involves the following steps: end resection, strand invasion (Rad51-mediated), DNA synthesis, and holliday junction resolution. As an organism ages, both the efficiency and accuracy of DSB repair diminish. NHEJ, a repair pathway which is already somewhat error-prone, becomes even less efficient (Young et al., 2013). Additionally, the HR pathway is downregulated in senescent cells due to lower expression of critical proteins such as RAD51. The loss of DSB repair fidelity is especially profound in stem cells and results in their functional exhaustion and decreased regenerative potential. Ku80 or DNA ligase IV-deficient mice display accelerated aging, supporting the link between DSB repair defects and aging biology.

The DNA damage response (DDR) is a broad signaling network that detects DNA lesions and coordinates the resulting cellular actions, such as cell cycle arrest, repair, senescence, or apoptosis. The DDR is primarily governed by the concerted action of two phosphatidylinositol 3-kinase-related kinases (PIKKs), namely, ATM and ATM- and Rad3-related (ATR). ATM is mostly responsible for DSBs, while ATR is responsible for replication stress and single-stranded DNA (Gorbunova and Seluanov, 2016). Once activated, they phosphorylate a plethora of downstream effectors, such as tumor suppressor p53, checkpoint kinases CHK1 and CHK2, H2AX histone variant (γH2AX), and BRCA1. These alterations aid in chromatin remodeling, the recruitment of repair complexes, and gene regulation (Obrenovac and Seluanov, 2016). Persistent DDR activation in senescent cells causes the secretion of pro-inflammatory cytokines, growth factors, and proteases known as the senescence-associated secretory phenotype (SASP). SASP has negative effects on the body because it causes chronic inflammation and tissue dysfunction. Importantly, DDR signaling can transmit DNA damage signals to neighbouring cells by paracrine pathways, augmenting the aging effects.

21.4 Chronological accrual of DNA lesions and aging hallmarks

Aging is a time-dependent process, and one of the hallmarks of aging at the molecular level is the accumulation of DNA damage throughout the genome. Importantly, it is not merely a byproduct of aging but a direct cause of age-related functional deterioration. As DNA damage accrues due to replication cycles, metabolic byproducts, and environmental stressors, the capacity of cells to uphold genome integrity steadily wanes. The imbalance between DNA damage and repair mechanisms causes genomic instability, which exacerbates other aging hallmarks, such as epigenetic changes, telomere attrition, cellular senescence, and mitochondrial dysfunction (Nature, 2021).

One of the earliest and most well-explored outcomes of DNA damage accumulation is the induction of cellular senescence. When the DNA damage is extensive or irreparable, cells stop dividing due to the activation of p53 and p16^INK4a^ pathways, as well as changes in chromatin structure and gene expression (Gorbunova and Seluanov, 2016). Senescent cells acquire a pro-inflammatory secretory profile termed SASP, which promotes tissue dysfunction by driving systemic chronic inflammation in aged organisms. Persistent DNA damage signals originating from DSBs or impaired telomeres maintain the SASP and drive aging both at cellular and systemic levels.

Additionally, the build-up of DNA damage hinders epigenetic stability. DNA damage triggers chromatin remodeling and histone modifications necessary for repair, but these changes may result in permanent epigenetic alterations over time. For instance, the loss of heterochromatin and histone marks (H3K9me3 and H4K20me) as well as disruption of DNA methylation patterns caused by damage-induced redistribution are the factors that contribute to age-related epigenetic drift. These changes affect gene transcription, transposable element repression, and cell identity, which are all the hallmarks of aged tissues. Additionally, DNA damage can alter the function of epigenetic enzymes, such as SIRT1 and SIRT6.

Another major consequence of persistent DNA damage is mitochondrial dysfunction. Mitochondrial DNA (mtDNA) is especially vulnerable because mitochondria lack histones and are close to the electron transport chain (Brown and Jackson, 2015). Accumulated mutations

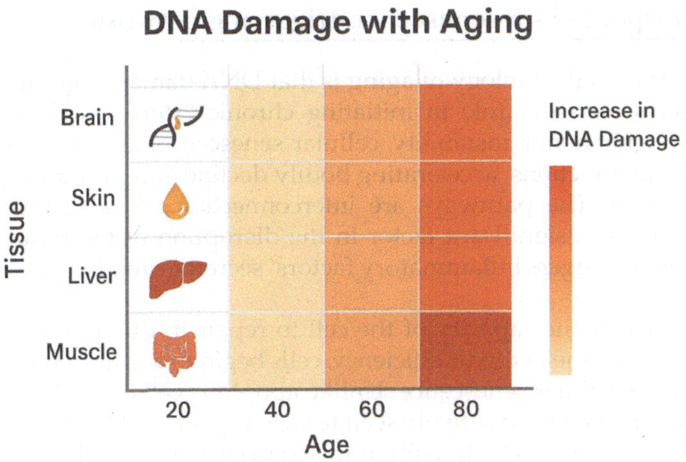

FIGURE 21.3 This heatmap illustrates the progressive increase in DNA damage across different tissues—brain, skin, liver, and muscle—with advancing age from 20 to 80 years. Darker shades indicate higher damage levels, highlighting that tissues like the brain and muscle show significant accumulation of genomic instability in older age.

impair mitochondrial function, resulting in higher ROS generation and subsequent DNA damage, which closes the "vicious" cycle. A loss in mitochondrial function lowers ATP synthesis, interferes with metabolic signaling, and promotes age-related organ dysfunction and cellular senescence. In aged cells, impaired mitophagy fails to remove dysfunctional mitochondria, which causes energetic and redox imbalance (Fig. 21.3).

Telomere attrition is also a cause and consequence of DNA damage. Telomeres progressively shorten with every cell division since the termini of human chromosomes undergo incomplete replication and oxidative damage (Nature, 2021). Critically short telomeres are identified as DSBs, triggering DDR and senescence. If telomeres are not properly capped, chromosomal ends can fuse, leading to genomic rearrangements. While telomerase can counteract telomere attrition to a limited extent in germ cells and some types of stem cells, it is inactive in the majority of somatic tissues, which explains why accumulating telomeric DNA damage drives replicative aging.

Consequently, loss of tissue homeostasis and the emergence of age-related pathologies result from the diminished regenerative ability. For instance, aged hematopoietic stem cells (HSCs) demonstrate lower reconstitution capability because of accumulated DNA damage and altered differentiation propensity (Wilson, 2021). Likewise, the aging brain with neurodegeneration is more prone to DNA breaks and diminished repair responses in postmitotic neurons, accompanied by glial inflammatory activation. Age-related declines in stem cell and progenitor activity, cardiovascular function, and muscle strength have also been connected to DNA damage, which causes loss of metabolic control.

Importantly, the connections are not merely origin-associations. Experimental induction of DNA damage in young animals, through irradiation or genetic manipulation, leads to premature aging, while enhanced repair capacity delays the age-related functional decline (McKinnon, 2009). Therefore, the accumulation of DNA damage acts as both an upstream and downstream regulator of several aging hallmarks, placing it at the center stage of the aging process.

21.5 The DNA damage response as a senescence-triggering network

One of the groundbreaking discoveries in the biology of aging is that DNA damage, besides compromising genomic integrity, plays a critical role in initiating chronic "sterile" inflammation. The interconnectedness between genomic instability, cellular senescence, and chronic inflammation significantly impacts tissue functions, accelerating bodily decline and promoting age-related diseases (Sharon et al., 2016). The pathways are interconnected into a vicious cycle. DDR activation due to senescence is a significant factor in the disruption of the tissue microenvironment homeostasis because it triggers inflammatory factors' secretion and hampers immune surveillance.

When the level of DNA damage exceeds the capacity of the cell to repair it, due to either excessive genotoxic stress or age-related decline in repair efficiency, cells begin to accumulate in a state of permanent growth arrest called cellular senescence. Unlike apoptotic cells, which die and are eliminated, senescent cells remain viable and actively secrete various factors that change the tissue microenvironment (Hasty et al., 2003). The transition to senescence is controlled by the continuous activation of DDR kinases, such as ATM and ATM and Rad3-related (ATR). These enzymes phosphorylate several downstream targets, including the tumor suppressor p53, checkpoint kinases 1 and 2 (CHK1 and CHK2), and the histone variant H2AX. This results in the stabilization of p53 and induction of the cyclin-dependent kinase inhibitor p21, which enforces irreversible cell cycle arrest. Senescent cells also exhibit widespread chromatin remodeling and epigenetic reprogramming, which is associated with the formation of senescence-associated heterochromatic foci (SAHF), global DNA hypomethylation, and redistribution of histone modifications.

An important hallmark of senescent cells is the SASP: a pro-inflammatory tissue remodeling secretome. The SASP encompasses a variety of molecules such as cytokines (e.g., IL-6, IL-1β, and IL-8), chemokines (e.g., MCP-1), growth factors (e.g., VEGF and HGF), and matrix-degrading enzymes (e.g., MMPs). The expression of these factors is determined by the action of transcription factors (NF-kB and C/EBPβ), which are activated in response to the DDR and oxidative stress (Niedernhofer et al., 2006). Notably, SASP not only stabilizes the senescent state (autocrine activity) but also affects other cells and tissues (paracrine activity). The SASP can also induce secondary senescence in adjacent healthy cells, recruit immune cells such as macrophages and neutrophils, alter tissue structure, and promote chronic inflammation. Thus, even a small number of senescent cells can have significant effects on tissue function and the body as a whole.

Importantly, DNA damage-induced cellular senescence is not limited to telomere erosion but can also be triggered by various factors such as oxidative or oncogenic stress, radiation, and mitochondrial or chromatin dysfunction. Senescent cells accumulate in various tissues and organs of mammals (including humans, mice, and nonhuman primates) with age. They are usually present in stem cell niches, endothelial layers, and stromal compartments. Moreover, senescent cells evade apoptosis, and they increase with age when the immune system's ability to identify and eliminate them becomes impaired.

One of the particularly intriguing mechanisms linking DNA damage to inflammation is based on the sensing of cytosolic DNA fragments. Genomic stress causes nuclear DNA to leak into the cytoplasm via micronuclei or nuclear envelope rupture. Additionally, impaired mitochondria may release mitochondrial DNA (mtDNA) into the cytosol. These abnormal DNA fragments are detected by the cyclic GMP–AMP synthase (cGAS), which, upon binding to the

DNA, catalyzes the synthesis of cyclic GMP–AMP (cGAMP). cGAMP activates the adaptor protein STING, leading to a signaling cascade that culminates in IRF3 and NF-κB activation (Niedernhofer et al., 2006). This pathway triggers the production of type I interferons and pro-inflammatory cytokines, thus establishing a direct link between DNA damage and innate immune activation at the molecular level. Chronic activation of the cGAS-STING pathway has been attributed to impaired immune homeostasis and systemic inflammation in both aged tissues and progeroid mouse models. This is termed "inflammaging" and is regarded as a key hallmark of aging and an important contributor to disease susceptibility in the elderly.

The persistent DNA damage and DDR signaling likewise have reciprocal effects on inflammation. Inflammatory cytokines secreted during SASP can induce the formation of reactive oxygen and nitrogen species in nearby cells, causing DNA damage and reinforcing DDR signaling. Furthermore, the telomeric DNA and other parts of the genome are oxidized by ROS during inflammation, which leads to genomic instability and cellular senescence. This creates a feed-forward cycle, which eventually leads to widespread tissue loss and organ dysfunction. Many age-related diseases, like atherosclerosis, osteoarthritis, type 2 diabetes, Alzheimer's disease, and Parkinson's disease, are linked to this vicious cycle. In organs like the brain, senescent glial cells produce neurotoxic substances that impair neuronal function and connectivity, negatively affecting health and cognition (Young et al., 2013).

The inability of the immune system to clear the accumulating damage and senescent cells also contributes to this problem. Immune cells like natural killer (NK) cells, CD8+ cytotoxic T lymphocytes, and macrophages utilize mechanisms such as perforin-granzyme and immune checkpoints to recognize and eliminate senescent cells. However, the aging process is linked to immunosenescence, which entails T-cell exhaustion, diminished NK cell cytotoxicity, impaired antigen presentation, and chronic low-grade inflammation (Burhans and Weinberger, 2007). These factors impede the immune system's ability to clear senescent cells, allowing them to accumulate. These impairments impede the proper elimination of senescent cells, leading to their excessive accumulation and consequently aggravating age-related tissue damage and dysfunction.

Because of the profound contribution of senescent cells and SASP to aging and age-related disease pathogenesis, there is a significant interest in the development of therapeutic approaches targeting these processes. Senolytics, which are agents that selectively eliminate senescent cells, have been shown to improve tissue function and healthspan in preclinical studies. Examples are dasatinib and quercetin, which target the pathways that senescent cells employ to prevent apoptosis. Senomorphics aim to mitigate the harmful pro-inflammatory effects of SASP. These include rapamycin, metformin, and NF-κB inhibitors. Furthermore, due to the tendency of pro-inflammatory signaling activated by cytosolic DNA sensors to cause systemic damage, suppressors of the STING pathway are also being investigated (Chen, 2021). These novel approaches target the mechanisms that link DNA damage to inflammation and cellular senescence, offering the potential to significantly delay or even reverse certain aspects of aging.

21.6 Mitochondrial genome vulnerability and the reactive oxygen species feedback loop

Mitochondria, known as the powerhouses of the cell, are major sources of aging-related damage because they generate both energy and ROS. Unlike nuclear DNA (nDNA), mtDNA

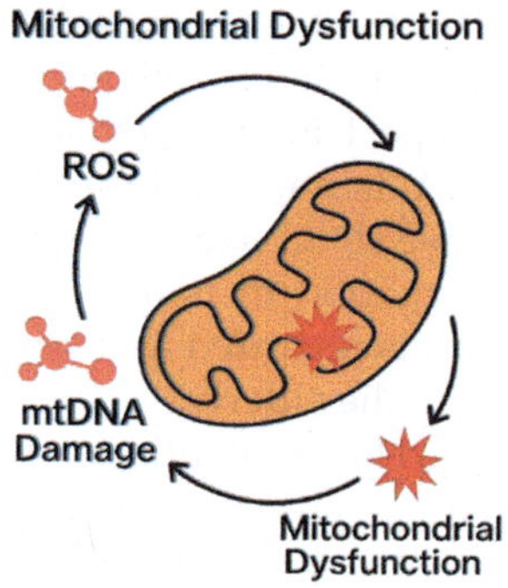

FIGURE 21.4 This diagram illustrates a vicious cycle where mitochondrial dysfunction leads to excessive production of ROS, which in turn damages mitochondrial DNA (mtDNA). The resulting mtDNA damage further impairs mitochondrial function, perpetuating the cycle of ROS generation and cellular decline.

is prone to damage because of its close proximity to the electron transport chain (ETC) and lack of protective histones and a limited capacity for DNA repair. Consequently, mtDNA mutations gradually accumulate over time, impairing mitochondrial function and initiating a vicious cycle of increased ROS production (Fig. 21.4). mtDNA is a small circular genome that codes for key proteins of the oxidative phosphorylation system. Due to its close proximity to the ROS-generating ETC complexes I and III, mtDNA is constantly exposed to oxidative damage (Chekulaeva and Landthaler, 2016). Lesions like 8-oxo-dG, strand breaks, and deletions are prevalent in aged tissues. While mitochondria contain some BER enzymes, they do not have the full array of DNA repair pathways present in the nucleus, such as NER and mismatch repair (MMR). As a result, oxidative lesions and replication errors are more likely to remain unrepaired or be misrepaired in mtDNA.

The repeated mtDNA damage and the resultant defective assembly of ETC complexes cause impaired electron flow, which further increases ROS levels. The subsequent oxidative stress harms lipids, proteins, and nuclear DNA in addition to accelerating mtDNA mutation (Yemini et al., 2021). The cross-talk between the two is also a major driving force of cellular aging (McKinnon, 2009). Specifically, high ROS levels trigger nuclear DDR pathways, p53 signaling, and the inflammation cascade, which promote senescence and loss of tissue homeostasis.

Tissues with high energy demands, such as skeletal muscle, heart, and brain, are particularly susceptible to the age-related consequences of mtDNA damage. In aged muscle, increased mtDNA deletions are highly correlated with the atrophy of muscle fibers and loss of their regenerative potential (Hasty et al., 2003). In the brain, mtDNA mutations hinder neuronal bioenergetics and synaptic effectiveness, contributing to neurodegenerative diseases like Parkinson's and Alzheimer's. Mitochondrial dysfunction also compromises the stability of stem cell niches, which impairs the ability of stem cells to self-renew and differentiate.

The mitochondrial-nuclear genome axis is essential for cellular homeostasis and stress responses. Under normal circumstances, mitochondrial retrograde signaling regulates the expression of nuclear genes in response to energy depletion and redox imbalance (Sedelnikova et al., 2004). Chronic mitochondrial stress caused by mtDNA damage continuously activates the pathway, forcing cells to rely on anaerobic glycolysis (the Warburg effect) and lowering intracellular NAD+/NADH ratios needed for DNA repair (Maynard et al., 2009). Sirtuins, which are a family of NAD+-dependent deacetylases involved in genome maintenance and mitochondrial biogenesis, are particularly sensitive to nutrient availability. As NAD+ levels decrease with age, sirtuin activity diminishes, which accelerates the loss of stability in both genomes. Animal models with impaired mtDNA polymerase γ proofreading (PolgAmut/mut mice) accumulate mtDNA mutations and display premature aging phenotypes, such as kyphosis and anemia, as well as a strongly reduced lifespan. These results demonstrate that compromised mtDNA integrity causes aging. On the contrary, measures that improve mitochondrial quality control (such as caloric restriction, NAD+ replenishment, and exercise) enhance mitochondrial turnover and reduce mtDNA damage (Hoeijmakers, 2009).

21.7 Epigenomic instability as a collateral of DNA damage

Epigenome refers to a network of heritable and reversible modifications on DNA or chromatin that regulates structure and gene expression. The relationship between DNA damage and epigenetic changes is an important yet often overlooked aspect of basic aging research (Liu et al., 2007). Due to the genotoxic stress, cells need to mobilize the enzymatic machinery and ensure its physical access to DNA. While such remodeling is necessary for efficient DNA repair, it also poses a threat to the long-term epigenomic stability, especially upon frequent or chronic damage.

One major consequence of DNA damage is the loss or alteration of histones, which are proteins that serve as spools for DNA threads. DSBs trigger the recruitment of chromatin modifiers, including ATM, γH2AX, and TIP60, which facilitate chromatin relaxation and histone H1 eviction (Lombard et al., 2005). During repair, posttranslational modifications of histones (e.g., acetylation, methylation, phosphorylation, and ubiquitination) and incorporation of histone variants promote chromatin remodeling. However, the original labels and marks are not always fully restored. Over time, as cycles of damage and repair continue, they can lead to changes in the patterns of histone modifications, which may eventually silence previously active genomic regions.

DNA methylation, another crucial epigenetic mark, can also be disrupted by genotoxic stress. Oxidative base damage affects DNA methyltransferase binding, and the BER enzymes TET and TDG regulate methylation turnover during demethylation. In aged tissues, global hypomethylation and site-specific promoter hypermethylation of several genes, such as tumor suppressors and key developmental regulators, are commonly observed (Schumacher et al., 2008). These changes lead to aberrant gene expression, reduced plasticity, and increased disease susceptibility. The accelerated loss of this youthful methylation pattern can be measured by so-called "epigenetic clocks" that correlate with chronological age and predict remaining lifespan.

Persistent DNA damage also promotes the formation of SAHF, which are compact chromatin domains that silence proliferation-driving genes. The redistribution of HP1, H3K9me3, and macroH2A proteins is involved in SAHF formation. While SAHFs initially support the cell cycle arrest by reinforcing it, they also cause a loss of cellular identity and decrease transcriptional plasticity. This is especially true for aging stem cells, whose epigenetic profiles are altered in ways that prevent them from adequately self-renewing and differentiating (Mansoor et al., 2016).

TEs are known to reactivate in response to DNA damage and chromatin relaxation. Their mobilization causes inflammation (due to cytoplasmic DNA sensing pathways) and additional genomic instability (due to insertional mutagenesis). Increased expression of endogenous retroelements, such as LINE-1, in aging tissues contributes to innate immune activation and the SASP.

Collectively, these examples show how DNA damage can drive epigenetic dysfunction and stochastic gene expression, impair lineage fidelity, and increase susceptibility to transformation.

While some epigenetic changes are reversible, others become permanent and heritable, promoting dysfunction in perpetuity. The overlap of DNA damage and epigenomic instability in cellular aging also affects the pathogenesis of age-associated diseases, such as cancer, neurodegeneration, and metabolic disorders (Xu et al., 2021).

The interplay between genome stability and epigenetic regulation, considering its bidirectional nature, can open new avenues for intervention. Thus, approaches aimed at enhancing chromatin stability and promoting the maintenance of youthful epigenetic states during aging hold promise for delaying or reversing age-related functional decline.

21.8 Transgenic and knockout models decoding DNA repair–aging axis

Experimental models have been pivotal in demonstrating that DNA damage is a causal factor in aging (Patz et al., 2000). Transgenic and knockout mouse models have demonstrated that impaired DNA repair causes premature aging, whereas improved repair delays aging. In addition, these models suggest that there is a causal link between the ability to maintain genomic stability and organismal longevity (Table 21.2).

Progeroid syndromes are a group of rare inherited disorders characterized by features mimicking physiological aging. Many of these syndromes are caused by mutations in DNA repair genes. For example, mice lacking the NER components XPA, XPC, or CSA exhibit phenotypes analogous to XP and CS patients, such as growth retardation and neurodegeneration. CSB-deficient mice, which are more vulnerable to oxidative damage and have defective transcription-coupled repair, exhibit progressive cerebellar atrophy. These models show that defects in even one repair system result in damage accumulation and premature aging (Bloom et al., 2023).

Additionally, mice with mutations in DNA polymerase γ (Polg^mut/mut^) exhibit an early onset aging phenotype because of accelerated accumulation of mtDNA mutations. The mice show hair graying, sarcopenia, osteoporosis, anemia, and reduced lifespan. Importantly, the premature aging in Polg mutant mice occurs without overt DNA strand breaks, indicating that mitochondrial genome instability is sufficient to induce systemic aging via metabolic and inflammatory pathways.

Another elegant model involves Ku80 and DNA ligase IV knockout mice, which are devoid of critical components of the nonhomologous end joining (NHEJ) pathway. These mice show an accumulation of DNA DSBs, severe growth retardation, stem cell depletion, and immune system deficiencies, and die prematurely (Liochev, 2015). When such mice are crossed with p53-deficient animals, the resulting progeny develop tumors rather than exhibit accelerated aging, thereby illustrating that p53 can direct cells with damaged DNA either toward senescence and apoptosis or toward transformation.

ATM-deficient mice, modeling ataxia-telangiectasia, exhibit cerebellar degeneration, thymic atrophy, insulin resistance, and cancer predisposition. These phenotypes support the notion that the ability to detect DNA damage and activate a checkpoint response is critical for preventing aging-related diseases (Hoeijmakers, 2009). Likewise, BRCA1- and BRCA2-deficient models highlight the consequences of HR deficiency, such as chromosomal instability, infertility, and increased cancer risk.

Conversely, genetic models that boost DNA repair efficiency strongly support a beneficial role of genome maintenance. SIRT6 is a histone deacetylase that regulates the BER pathway and promotes DSB repair. Its overexpression reportedly extends lifespan and improves glucose

TABLE 21.2 This table compares key mouse models used in aging research, highlighting their genetic mutations, associated phenotypes, effects on lifespan, and relevance to human aging mechanisms. These models replicate various aspects of DNA repair deficiency, mitochondrial dysfunction, telomere attrition, and nuclear architecture defects seen in premature and natural aging.

Mouse model	Genetic mutation/target	Phenotype	Lifespan effect	Relevance to aging research
Ercc1−/−	Deficient in **ERCC1-XPF** endonuclease	Growth retardation, neurodegeneration, kyphosis, sterility, liver pathology	Severely shortened (4–6 months)	Mimics accelerated aging; DNA repair defect model (NER, ICL repair); progeroid phenotype
XPG−/−	Deletion of **XPG** (NER endonuclease)	Progressive neurodegeneration, weight loss, cachexia	~6 months	Models Cockayne syndrome and NER dysfunction; used for studying DNA repair and aging
Ku80−/−	Loss of **Ku80** (NHEJ DSB repair)	Osteoporosis, skin atrophy, reduced organ mass, growth delay	Shortened (~25–30% less than WT)	Models age-related tissue degeneration; NHEJ deficiency mimics stem cell aging
LmnaG609G/G609G	Progerin-producing mutation in **Lmna**	Features of Hutchinson-Gilford Progeria: growth delay, vascular calcification	~6 months	Gold-standard model for HGPS; mimics lamin A mutation seen in human premature aging
SIRT6−/−	Knockout of **SIRT6** (chromatin modifier)	Hypoglycemia, genomic instability, wasting, premature death	Shortened (~1 month)	Shows the link between chromatin stability, DNA repair, and aging
p53+/− or p53−/−	Partial or complete loss of **p53**	Cancer-prone, reduced apoptosis, increased cell proliferation	Variable; often reduced due to tumors	Studies link tumor suppression, DNA damage response, and lifespan

(*continued on next page*)

TABLE 21.2 This table compares key mouse models used in aging research, highlighting their genetic mutations, associated phenotypes, effects on lifespan, and relevance to human aging mechanisms. These models replicate various aspects of DNA repair deficiency, mitochondrial dysfunction, telomere attrition, and nuclear architecture defects seen in premature and natural aging—cont'd

Mouse model	Genetic mutation/target	Phenotype	Lifespan effect	Relevance to aging research
Terc−/− (G4-G6)	Telomerase RNA component knockout	Telomere shortening, infertility, and stem cell depletion in late generations	Reduced (only after 4–6 generations)	Telomere biology and replicative senescence model; mimics human telomeropathies
PolgD257A/D257A	Proofreading-deficient mitochondrial DNA polymerase	Mitochondrial dysfunction, premature aging, kyphosis, and anemia	Shortened (~12–14 months)	Mitochondrial theory of aging; accumulation of mtDNA mutations
Zmpste24−/−	Deficiency in **Zmpste24** (lamin A processing)	Progeroid phenotype, low body mass, abnormal bone development	~4–6 months	Mimics nuclear lamina defects; related to HGPS; chromatin and nuclear architecture studies

metabolism in mice. Mice that overexpress DNA glycosylases or APE1 display reduced mutation burden and improved cognitive performance with aging. Caloric restriction, a powerful lifespan-extending intervention, also affects other nutrient-sensitive pathways and induces the expression of DNA repair enzymes (Bekaert et al., 2005). Importantly, they help to delineate tissue-specific consequences of impaired DNA repair. For example, conditional deletion of repair genes in HSCs and neural progenitors causes anemia and neurodegeneration, respectively. These findings suggest that the effect of DNA damage is cell-specific and depends on factors such as proliferation rate, metabolic activity, and microenvironment.

Thus, transgenic and knockout models collectively highlight the significance of DNA damage as an aging mechanism (Freitas and de Magalhães, 2011). These models have been instrumental in elucidating the consequences of loss-of-function tissue repair on the organismal level, such as inflammation and lifespan reduction. They are also indispensable for testing the interventions based on enhanced DNA repair, modulation of DDR signaling, and clearance of senescent cells. Future models incorporating inducible, tissue-specific, and age-dependent manipulations will sharpen our understanding of the role genome maintenance mechanisms play in shaping organismal aging trajectories (Jinesh et al., 2025).

21.9 Feasibility of DNA lesion profiling as an aging biomarker

There is an increasing interest in identifying molecular markers that accurately reflect biological age and predict susceptibility to age-related diseases (Table 21.3). Unlike chronological age, biological aging reflects the accumulated insults and the ensuing functional losses that organisms experience as they grow older. DNA damage and repair markers play a key role in systemic aging due to their involvement in genome maintenance.

One of the most extensively used markers for DNA damage is γH2AX, a phosphorylated histone variant that forms foci at the sites of DSBs. The number of γH2AX foci and their persistence demonstrate a strong correlation with age across different tissues (Stepanova et al., 2024). Additionally, they are elevated in pathological conditions, including cancer, neurodegeneration, and cardiovascular dysfunction. Immunostaining of γH2AX in PBMCs or skin biopsies is a minimally invasive method for genomic stress assessment. Furthermore, the comet assay (single-cell gel electrophoresis), which detects DNA single- and DSBs in individual cells, has been used to measure age- or environment-related accumulation of DNA damage.

Oxidative DNA lesions, specifically 8-oxo-2′-deoxyguanosine (8-oxo-dG), are a widely accepted biomarker for endogenous damage. It is possible to measure them in urine, blood, or tissue extracts by high-performance liquid chromatography or mass spectrometry (Blouin and Saini, 2024). High 8-oxo-dG levels are always found in aged people and patients with diabetes, as well as in subjects exposed to pollutants and cigarette smoke. Urinary 8-oxo-dG levels increase with age, and this biomarker is a suitable candidate for noninvasive monitoring of systemic oxidative stress and biological aging.

Telomere length (usually assessed by qPCR or FISH) is another widely used parameter. Critically short telomeres induce DNA damage responses and cellular senescence. Shorter telomeres are associated with increased risk of cardiovascular disease, poor immune function, and decreased lifespan (Nakamura and Nakamura, 2020). While telomere shortening is genetically and environmentally determined, it remains one of the most studied hallmarks of aging and has been used to develop several aging clocks. Other markers may include DNA repair

TABLE 21.3 This table summarizes key laboratory methods used to detect DNA damage in human samples, detailing the type of damage each technique identifies, required sample types, detection principles, and practical considerations. It highlights how different assays, from comet tests to oxidative base quantification, provide complementary insights into genome integrity and damage sources.

Method	Damage detected	Sample type	Principle	Advantages/ limitations
Comet assay (single-cell gel electrophoresis)	Single-strand breaks, alkali-labile sites, DSBs	Peripheral blood mononuclear cells, tissue biopsies	Electrophoresis of lysed cells causes fragmented DNA to migrate, forming a "comet" tail	Sensitive, semiquantitative; requires careful handling and interpretation
γH2AX immunostaining	DNA double-strand breaks	Cells and tissue sections	Phosphorylated H2AX is visualized via fluorescent antibodies at DSB sites	Specific, allows quantification; needs microscopy and controls
TUNEL assay	DNA fragmentation (late-stage damage/apoptosis)	Tissue sections and cultured cells	Label-free 3′-OH DNA ends using terminal deoxynucleotidyl transferase (TdT)	Detects apoptosis; may not distinguish between repairable and irreversible damage
8-oxo-dG ELISA or HPLC	Oxidative DNA base modifications (e.g., 8-oxoguanine)	Serum, urine, and DNA extract	Detects 8-oxo-dG using antibodies or chromatographic separation	Specific for oxidative stress; may require DNA purification
qPCR-based damage quantification	Lesions in specific genes or mtDNA	Genomic or mitochondrial DNA	Reduced amplification efficiency due to DNA lesions is measured quantitatively	Gene- or region-specific; indirect but sensitive
FPG-modified comet assay	Oxidized purines (e.g., 8-oxoG)	Blood and cell lines	Uses FPG enzyme to convert oxidized bases into strand breaks	Oxidative damage-specific; adds resolution to the standard comet assay
Alkaline gel electrophoresis	Strand breaks, alkali-labile sites	Extracted DNA	Separation of damaged DNA fragments under alkaline conditions	Simple; lower resolution than cell-based methods
DNA Adduct Detection (e.g., ^32P-postlabeling)	Bulky chemical adducts	Extracted DNA	Labeling and chromatographic separation of modified nucleotides	Very sensitive to environmental exposures; labor-intensive

genes (XRCC1, OGG1, and PARP1) whose expression levels decrease with age and predict repair capacity. Additionally, age-related methylation changes in the promoters of DNA repair genes may serve as epigenetic biomarkers of genome maintenance. Inflammatory markers (e.g., IL-6, CRP, and interferon-stimulated genes) associated with DNA damage act as complements to the other measures.

Recent efforts to construct integrated "aging clocks" combine DNA damage metrics with other hallmark features like mitochondrial dysfunction, impairment of proteostasis, and senescence load (Oka et al., 2024). These indices aim to yield more precise and individualized biological age estimations. For example, the DNA methylation GrimAge and DunedinPACE clocks use the damage-associated CpG sites to predict lifespan, healthspan, and morbidity. However, more work needs to be done to validate these markers in different populations and ensure their reproducibility across laboratories (Stepanova et al., 2025). Factors such as tissue type, timing of sample collection, and method sensitivity can affect the outcomes. Longitudinal studies are needed to determine whether DNA damage markers predict future decline or if they merely reflect the cumulative insults. Additionally, interventional studies (e.g., on antioxidants, exercise, or senolytics) would facilitate testing the causal links between biomarker expression and health outcomes.

21.10 Therapeutic modulation of DNA damage in age-associated disorders

Because DNA damage is a central driver of biological aging and age-related diseases, approaches to reduce genomic damage or enhance its repair are being actively investigated. The approaches can be broadly divided into the following categories: reducing the rate of DNA damage, enhancing repair efficiency, modulating the response to damage, removing senescent cells resulting from accumulated damage, and bolstering systemic capacity to withstand genotoxic stress. Together, they address multiple aspects of health and longevity. One of the easiest approaches is to avoid excessive exposure to factors that cause DNA damage, as well as minimize endogenous stress (Rossner et al., 2011). Caloric restriction (CR) is the most studied nongenetic intervention for lifespan extension. CR decreases oxidative stress and enhances mitochondrial efficiency while upregulating DNA repair genes, including SIRT1, OGG1, and APE1. CR mimetics, such as resveratrol, metformin, and rapamycin, partially mimic the effects of caloric restriction by activating AMPK/sirtuin signaling pathways. These agents increase NAD+ availability, enhance mitochondrial quality control, and lower inflammation levels.

Another strategy is to directly enhance the cell's DNA repair capacity. NAD+ precursors like NR and nicotinamide mononucleotide (NMN) replenish NAD + pools, which are vital for PARP enzymes and sirtuins in DNA BER. Supplementation improves genomic stability in aged animals, cognitive function, and muscle regeneration. Gene therapy strategies to overexpress key DNA repair enzymes like SIRT6, APE1, or MGMT were shown to have protective effects in models of neurodegeneration, cancer, and premature aging (Schumacher et al., 2008).

Another approach is to modulate the DDR. Although DDR is essential for genome integrity, sustained activation due to continuous cellular damage promotes cell senescence and tissue deterioration. Drugs targeting specific DDR components (such as ATM or CHK1 inhibitors) are under investigation to fine-tune the balance between repair and survival. However, these techniques need to be carefully dosed in order not to impair normal repair functions or favor tumorigenesis. Senolytic therapies, which target the selective removal of senescent cells, are a potent tool for breaking the cycle of DNA damage–inflammation–senescence (Chen, 2021). Compounds like dasatinib, quercetin, and navitoclax interfere with antiapoptotic systems protecting senescent cells. In preclinical models, senolytics decreased SASP and inflammation related to senescence and improved tissue function. By eliminating cells with persistent DNA damage signaling, these approaches partially restore the local tissue environment and reduce the spread of secondary damage.

The potential of antioxidants to mitigate oxidative DNA damage has also been investigated. While dietary antioxidants such as vitamins C and E have not been successful in large-scale trials, the targeted mitochondrial antioxidants MitoQ and SkQ1 reduce mtDNA damage and improve metabolic health in aging models. Enhancing endogenous antioxidant defenses through NRF2 activation to upregulate antioxidant enzymes. Approaches aimed at broad epigenome rejuvenation and the reversal of damage-associated transcriptional drift. Epigenetic reprogramming via transient expression of Yamanaka factors (OSKM) is capable of erasing some age-related epigenetic marks and improving DNA repair in cells and animals without inducing pluripotency. While still in its infancy, the potential application of transient reprogramming to restore genomic and epigenomic stability needs to be explored in the future.

Lifestyle factors, such as exercise and sleep (adequate rest), also affect DNA repair. Regular exercise also increases mitochondrial turnover, upregulates BER enzymes, and lowers inflammation. Quality sleep is required for the nocturnal DNA repair cycles and the circadian regulation of the genes involved in genome maintenance (Niedernhofer et al., 2006; Sharon et al., 2016). These behavioral techniques stand alongside molecular methodologies as a cornerstone of the antiaging strategy. Ultimately, strategies to combine efforts to minimize DNA damage, boost its repair, and alleviate the consequences hold the most significant promise for fostering healthy aging. Personalized interventions based on DNA damage biomarker status and genetic profiling will probably be important for maximizing the safety and efficacy of therapeutic strategies (Table 21.4).

21.11 Neurodegeneration and genomic instability: A causal continuum

As the brain ages, it becomes increasingly susceptible to genomic lesions such as DNA double-strand (DSBs) and SSBs, oxidative base modifications, and DNA–protein crosslinks. Neurons are especially vulnerable because of their high metabolic activity, extensive oxidative phosphorylation, and postmitotic nature. Unlike proliferative cells, neurons cannot dilute DNA damage through cell division (Gorbunova and Seluanov, 2016). Thus, even modest levels of genomic lesions can accumulate over time and interfere with cellular functions. This is further exacerbated by the fact that the antioxidant defense and DNA repair enzyme activity/expression decline with age, and that the aging brain has an elevated background of inflammation. Consequently, DNA damage accumulates in neurons, leading to impaired synaptic plasticity and loss of dendritic spines, axonal degeneration, and eventually neuronal cell death.

A growing body of evidence indicates that impaired DNA damage repair contributes to the progression of major neurodegenerative diseases such as Alzheimer's disease (AD), Parkinson's disease (PD), and amyotrophic lateral sclerosis (ALS). For example, neurons in the brains of AD patients display increased levels of phosphorylated H2AX (γH2AX) and 53BP1 foci, which are markers of persistent DSBs and the activated DDR. These sites overlap with regions of tau hyperphosphorylation and amyloid beta deposition (Royer et al., 2012; Geense et al., 2017).

TABLE 21.4 This table outlines how therapeutic agents target senescent cells and DNA repair mechanisms to combat aging and age-related diseases. While senolytics eliminate damaged cells, senomorphics and DNA repair enhancers work to suppress harmful inflammation or restore genomic stability, ultimately aiming to preserve tissue function and delay degeneration.

Drug class/type	Target	Mechanism of action	Therapeutic goal
Senolytics	Antiapoptotic pathways in senescent cells (e.g., BCL-2, BCL-xL)	Induce apoptosis in senescent cells by inhibiting pro-survival pathways (e.g., via dasatinib, navitoclax)	Clear senescent cells to reduce SASP and chronic inflammation
Senomorphics	SASP transcription factors (e.g., NF-κB, mTOR, JAK/STAT)	Suppress the senescence-associated secretory phenotype without killing the cell	Alleviate inflammatory signaling and tissue damage
DNA repair enhancers	PARP1, SIRT1/6, APE1, Ku70/80	Boost endogenous DNA repair by enhancing chromatin remodeling, BER, or DSB repair	Reduce genomic instability, delay aging-associated mutations
Telomerase activators	Telomerase reverse transcriptase (TERT)	Stimulate telomerase to elongate telomeres and prevent replicative senescence	Improve replicative lifespan of stem/progenitor cells
p53 modulators	p53 signaling pathway	Restore or enhance p53 function to balance DNA repair and cell fate decisions	Promote accurate repair or removal of heavily damaged cells
STING inhibitors	cGAS-STING innate immune axis	Block the inflammatory response triggered by cytosolic DNA from damaged nuclei/mtDNA	Suppress inflammation-driven aging and neurodegeneration

Similarly, dopaminergic neurons of the substantia nigra, which have a very high oxidative burden and are highly vulnerable in PD brains, exhibit elevated DNA DSBs and increased levels of DNA damage markers such as γH2AX and PARP1. In ALS, spinal motor neurons exhibit nuclear and mitochondrial DNA damage, oxidative lesions, and DDR activation, aligning with the idea of a common mechanism of DNA damage build-up in different neurodegenerative diseases. The convergence of these observations suggests that DNA damage is not merely a consequence of neuronal degeneration, but rather an active driver of neuronal vulnerability and dysfunction.

Beyond the direct cytotoxic effects, DNA damage also alters the neuronal epigenetic landscape, which is central for transcriptional programs, and thus leads to their massive dysregulation. Thus, the loss of heterochromatin and repressive histone marks (e.g., H3K9me3 and H3K27me3) together with enhanced histone acetylation can induce the aberrant expression of genes regulating cell cycle re-entry, inflammation, and apoptosis. Such alterations are especially harmful to neurons, where accurate gene expression regulation is crucial for synaptic plasticity, neurotransmitter signaling, and neuronal identity (Yemini et al., 2021; Chen et al., 2022). Age-related DNA methylation drift, which includes global hypomethylation and site-specific hypermethylation, changes the genomic regulatory landscape and affects neuronal gene expression. Notably, the TC-NER pathway, which targets transcription-blocking lesions in active gene bodies for elimination, is critical for brain health. CS and XP, caused by inherited mutations in NER components, exhibit severe neurodevelopmental and neurodegenerative abnormalities. These clinical findings demonstrate the importance of active DNA repair for neuronal survival and genomic stability throughout life.

Mitochondrial dysfunction is another prominent feature of neurodegeneration that closely interacts with genomic instability. Neurons are highly dependent on mitochondrial function since the bulk of their energy needs is covered by ATP generated in the oxidative phosphorylation pathway. However, the mitochondrial genome (mtDNA) is particularly vulnerable to oxidative damage because it lacks protective histones and is located in close proximity to the ETC. Age- and disease-related impairments of mtDNA replication and repair cause mtDNA deletions, point mutations, and loss of the expression of genes critical for energy production (López-Otín et al., 2016). The ROS-induced mtDNA damage then leads to increased ROS production and the spread of ROS into the nucleus, where they cause further DNA lesions. Furthermore, damaged mtDNA can enter the cytosol, especially from dysfunctional or ruptured mitochondria. The extranuclear DNA is recognized as a danger-associated molecular pattern (DAMP) by the innate immune system in the cytoplasm. The primary mechanism is the cyclic GMP–AMP synthase–stimulator of interferon genes (cGAS-STING) pathway, which activates type I interferons and NF-κB signaling, leading to a robust inflammatory response. Chronic activation of this pathway in glial and neuronal cells enhances the inflammatory response, accelerates senescence-like phenotype development, and further impairs neuronal functions.

The inability to regenerate is another factor that worsens neuron susceptibility to DNA damage. Unlike tissues such as the skin or the intestine, the adult brain has a very limited capacity to replace neurons lost to injury, disease, or aging. This means that the damage accumulated in a small subset of long-lived neurons would lead to loss of neuronal function, which would progressively worsen with age (Schumacher et al., 2008; Chen et al., 2022). In this regard, the glial cell responses become critically important. DNA damage and protein aggregates in neurons activate astrocytes and microglia, which trigger the neuroinflammatory response mediated by pro-inflammatory cytokines (e.g., IL-1β, TNF-α), chemokines, and ROS. However, the chronic activation of glial cells promotes an inflammatory environment, which is toxic to neurons. Senescent glial cells (microglia) also display SASP, which amplifies inflammatory signaling and causes damage to adjacent neurons and oligodendrocytes.

Given the complexity and multifactorial nature of neuronal genomic instability, several therapeutic avenues are being investigated. Enhancing neuronal DNA repair is one of the targets. Small molecule activators of the NAD+-dependent enzymes (PARP1 and SIRT1) target to enhance the efficacy of DNA damage repair and to decrease its accumulation. Similarly, efforts are underway to develop gene therapy approaches targeting DNA repair proteins such as APE1, OGG1, or XRCC1 in animal models of neurodegeneration. Antioxidant therapies aiming to reduce oxidative DNA damage through mitochondrial-targeted antioxidants (e.g., MitoQ, SkQ1) are also in active development (Hoeijmakers, 2009). Moreover, senolytic drugs that target and eliminate senescent glial cells (e.g., dasatinib, navitoclax, and quercetin) could potentially ameliorate the inflammatory environment of the aging brain and promote a more homeostatic state. Emerging data also support the use of "senomorphics," which suppress the SASP and dampen neuroinflammation without inducing cell death. These strategies target different points along the DNA damage–neurodegeneration axis and have considerable potential to reduce, delay, or even reverse age-related cognitive and functional decline.

In conclusion, DNA damage is a primary pathological factor in neurodegenerative diseases, closely interconnected with mitochondrial dysfunction, chronic inflammation, and impaired proteostasis. Because neurons are both long-lived and nondividing, they are especially vulnerable to the effects of genomic instability, and the consequences of persistent DNA damage in these cells are profound (Mansoor et al., 2016; Nakamura and Nakamura, 2020). Targeting DNA damage in the nervous system is a promising therapeutic approach to alleviate neurodegeneration. The interventions targeting DNA repair, mitochondrial function, and inflammatory signaling could potentially enhance quality of life by safeguarding neuronal integrity and health.

21.12 Future perspectives in DNA damage–aging interface: From bench to biogerontology

Aging, which has long been considered an inevitable consequence of the passage of time, is now understood as an active biological process that results from the gradual accumulation of molecular damage. Research over the past several decades has shown that DNA lesions caused by replication stress, oxidative threats, and environmental exposure accumulate over time and drive cellular senescence, apoptosis, and inflammation. These responses become maladaptive when chronically activated, leading to tissue dysfunction, degenerative disease, and loss of organismal resilience.

DNA repair pathways are the first line of defense against this constant molecular entropy. Nevertheless, their effectiveness fades with age primarily due to transcriptional inhibition, epigenetic alterations, and metabolic changes. The loss of genome integrity increases mutagenesis and cancer risk, and this is known to affect other aging hallmarks, such as mitochondrial dysfunction, telomere shortening, and stem cell depletion. In this light, the accumulation of DNA damage is one of the main mechanisms of aging and a potential target for therapeutic intervention (Xu et al., 2021).

Animal models lacking DNA repair genes show accelerated aging, while those with enhanced repair tend to age more slowly or have an extended lifespan. This dichotomy underscores the potential to target DNA damage for the amelioration of age-related loss-of-function in humans. Already, several biomarkers such as γH2AX (phosphorylated H2AX), 8-oxo-dG (8-oxo-deoxyguanosine), and telomere length are being used to assess biological age and DNA repair efficiency in a noninvasive way. These tools can aid in implementing early intervention methods and assessing the effectiveness of antiaging treatments.

Despite these promising approaches, therapeutic modulation of DNA damage in aging faces important limitations, including incomplete target specificity, off-target genomic effects, and variability in responses among different tissues and individuals. Safety challenges such as long-term carcinogenic risk, unintended impacts on stem cell function, and interference with normal stress-response pathways remain major concerns that require rigorous preclinical and clinical evaluation. Furthermore, significant translational barriers persist, including difficulties in optimal dosing and timing across the lifespan, restricted delivery of agents to specific organs or cell types, and the complexity of designing clinical trials that capture meaningful aging-related outcomes. Addressing these limitations will be essential for moving DNA damage–targeted interventions from conceptual promise to safe and effective strategies for promoting healthy aging in humans.

Future research needs to focus on therapeutics that can enhance the body's natural repair mechanisms without decreasing genomic stability. Advances in gene editing, RNA therapeutics, and epigenetic reprogramming hold promise to precisely enhance the repair pathways and rejuvenate aged cells. Furthermore, senolytics, DDR modulators, and NAD + enhancers are moving toward clinical use as preclinical data continue to show promise. However, the long-term safety and tissue specificity of such approaches need to be carefully assessed (Lombard et al., 2005).

Moreover, systems-level approaches like multiomics and machine learning will facilitate the integration of DNA damage biology with other aging pillars, hence improving our understanding of how genome instability interlinks with other aging hallmarks. This will pave the way for personalized aging interventions, identify at-risk populations, and formulate public health policies to enhance healthspan alongside lifespan.

In conclusion, DNA damage is a biomarker and a driver of aging. Addressing the factors that impair it, enhancing its repair, and mitigating its downstream effects will be key to delivering on the promise of healthy aging. Efforts are needed to bring the understanding of biological aging targets from the laboratory to the real world in order to slow down, stop, or reverse aging (Sedelnikova et al., 2004).

References

Bekaert, S., Van Pottelbergh, I., De Meyer, T., Zmierczak, H., Kaufman, J.M., Van Oostveldt, P., et al., 2005. Telomere length versus hormonal and bone mineral status in healthy elderly men. Mech. Ageing Dev. 126 (11), 1115–1122. https://doi.org/10.1016/j.mad.2005.04.007.

Bloom, S.I., Tucker, J.R., Machin, D.R., Abdeahad, H., Adeyemo, A.O., Thomas, T.G., et al., 2023. Reduction of double-strand DNA break repair exacerbates vascular aging. Aging 15, 9913–9947. https://doi.org/10.18632/aging.205066.

Blouin, T., Saini, N., 2024. Aldehyde-induced DNA-protein crosslinks: DNA damage, repair and mutagenesis. Front. Oncol. 14, 1478373. https://doi.org/10.3389/fonc.2024.1478373.

Brown, J.S., Jackson, S.P., 2015. Ubiquitylation, neddylation and the DNA damage response. Open Biology 5, 150018. https://doi.org/10.1098/rsob.150018.

Burhans, W.C., Weinberger, M., 2007. DNA replication stress, genome instability and aging. Nucleic Acids Res. 35 (22), 7545–7556. https://doi.org/10.1093/nar/gkm1059.

Chekulaeva, M., Landthaler, M., 2016. Eyes on translation. Mol. Cell 63 (6), 918–925. https://doi.org/10.1016/j.molcel.2016.08.031.

Chen, Y.H., 2021. Background and thoughts on "The optimal use of glucocorticoids for patients with chronic obstructive pulmonary disease: a Chinese expert consensus (2021)" [in Chinese]. Zhonghua Jie He He Hu Xi Za Zhi 44, 1034–1036. https://doi.org/10.3760/cma.j.cn112147-20210929-00686.

Chen, J., Ying, J., Xiao, Y., Dong, Y., Ozoemena, K.I., Lenaerts, S., Yang, X., 2022. Stoichiometry design in hierarchical CoNiFe phosphide for highly efficient water oxidation. Sci. China Mater. 65 (10), 2685–2693. https://doi.org/10.1007/s40843-022-2061-x.

Esposito, G., Mauri, P.A., Cipriani, S., Franchi, M., Corrao, G., Parazzini, F., 2022. Correction to: the role of maternal age on the risk of preterm birth among singletons and multiples: a retrospective cohort study in Lombardy, Northern Italy. BMC Pregnancy and Childbirth 22 (366). https://doi.org/10.1186/s12884-022-04695-y.

Freitas, A.A., de Magalhães, J.P., 2011. A review and appraisal of the DNA damage theory of ageing. Mutat. Res. Rev. Mutat. Res. 728 (1–2), 12–22. https://doi.org/10.1016/j.mrrev.2011.05.001.

Geense, W., Zegers, M., Vermeulen, H., van den Boogaard, M., van der Hoeven, J., 2017. MONITOR-IC study, a mixed methods prospective multicentre controlled cohort study assessing 5-year outcomes of ICU survivors and related healthcare costs: a study protocol. BMJ Open 7, e018006. https://doi.org/10.1136/bmjopen-2017-018006.

Gorbunova, V., Seluanov, A., 2016. DNA double strand break repair, aging and the chromatin connection. Mutat. Res. 788, 2–6. https://doi.org/10.1016/j.mrfmmm.2016.02.004.

Hasty, P., Campisi, J., Hoeijmakers, J., van Steeg, H., Vijg, J., 2003. Aging and genome maintenance: lessons from the mouse? Science 299 (5611), 1355–1359. https://doi.org/10.1126/science.1079161.

Hoeijmakers, J.H.J., 2009. DNA damage, aging, and cancer. N. Engl. J. Med. 361 (15), 1475–1485. https://doi.org/10.1056/NEJMra0804615.

Jinesh, S., Özüpek, B., Aditi, P., 2025. Premature aging and metabolic diseases: the impact of telomere attrition. Front. Aging 6, 1541127. https://doi.org/10.3389/fragi.2025.1541127.

Liochev, S.I., 2015. Which is the most significant cause of aging? Antioxidants 4 (4), 793–810. https://doi.org/10.3390/antiox4040793.

Liu, X., Takahashi, H., Harada, Y., Ogawara, T., Ogimura, Y., Mizushina, Y., et al., 2007. 3′-Azido-2′,3′-dideoxynucleoside 5′-triphosphates inhibit telomerase activity in vitro, and the corresponding nucleosides cause telomere shortening in human HL60 cells. Nucleic Acids Res. 35 (21), 7140–7149. https://doi.org/10.1093/nar/gkm859.

Lombard, D.B., Chua, K.F., Mostoslavsky, R., Franco, S., Gostissa, M., Alt, F.W., 2005. DNA repair, genome stability, and aging. Cell 120 (4), 497–512. https://doi.org/10.1016/j.cell.2005.01.028.

López-Otín, C., Galluzzi, L., Freije, J.M.P., Madeo, F., Kroemer, G., 2016. Metabolic control of longevity. Cell 166 (4), 802–821. https://doi.org/10.1016/j.cell.2016.07.031.

Mansoor, S.E., Lü, W., Oosterheert, W., Shekhar, M., Tajkhorshid, E., Gouaux, E., 2016. X-ray structures define human P2 × 3 receptor gating cycle and antagonist action. Nature 538 (7623), 66–71. https://doi.org/10.1038/nature19367.

Maynard, S., Schurman, S.H., Harboe, C., de Souza-Pinto, N.C., Bohr, V.A., 2009. Base excision repair of oxidative DNA damage and association with cancer and aging. Carcinogenesis 30 (1), 2–10. https://doi.org/10.1093/carcin/bgn250.

McKinnon, P.J., 2009. DNA repair deficiency and neurological disease. Nat. Rev. Neurosci. 10 (2), 100–112. https://doi.org/10.1038/nrn2559.

Nakamura, J., Nakamura, M., 2020. DNA-protein crosslink formation by endogenous aldehydes and AP sites. DNA Repair (Amst.) 88, 102806. https://doi.org/10.1016/j.dnarep.2020.102806.

Nature, 2021. Africa's vaccines revolution must have research at its core. Nature 592 (7855), 487–488. https://doi.org/10.1038/d41586-021-01038-3.

Niedernhofer, L.J., Garinis, G.A., Raams, A., Lalai, A.S., Robinson, A.R., Appeldoorn, E., et al., 2006. A new progeroid syndrome reveals that genotoxic stress suppresses the somatotroph axis. Nature 444 (7122), 1038–1043. https://doi.org/10.1038/nature05456.

Oka, Y., Nakazawa, Y., Shimada, M., Ogi, T., 2024. Endogenous aldehyde-induced DNA-protein crosslinks are resolved by transcription-coupled repair. Nat. Cell Biol. 26 (6), 784–796. https://doi.org/10.1038/s41556-024-01401-2.

Okereke, E., Aradeon, S., Akerele, A., Tanko, M., Yisa, I., Obonyo, B., 2013. Knowledge of safe motherhood among women in rural communities in northern Nigeria: implications for maternal mortality reduction. Reprod. Health 10, 57. https://doi.org/10.1186/1742-4755-10-57.

Patz, J.A., Graczyk, T.K., Geller, N., Vittor, A.Y., 2000. Effects of environmental change on emerging parasitic diseases. Int. J. Parasitol. 30 (12–13), 1395–1405. https://doi.org/10.1016/S0020-7519(00)00141-7.

Rossner, P., Uhlirova, K., Beskid, O., Rossnerova, A., Svecova, V., Sram, R.J., 2011. Expression of XRCC5 in peripheral blood lymphocytes is upregulated in subjects from a heavily polluted region in the Czech Republic. Mutat. Res. Fundam. Mol. Mech. Mutagen. 713 (1–2), 76–82. https://doi.org/10.1016/j.mrfmmm.2011.06.001.

Royer, D.L., Pagani, M., Beerling, D.J., 2012. Geobiological constraints on Earth system sensitivity to CO_2 during the Cretaceous and Cenozoic. Geobiology 10 (4), 298–310. https://doi.org/10.1111/j.1472-4669.2012.00320.x.

Sahin, E., DePinho, R.A., 2010. Linking functional decline of telomeres, mitochondria and stem cells during ageing. Nature 464 (7288), 520–528. https://doi.org/10.1038/nature08982.

Schumacher, B, Garinis, G.A., Hoeijmakers, J.H.J., 2008. Age to survive: DNA damage and aging. Trends Genet. 24 (2), 77–85. https://doi.org/10.1016/j.tig.2007.11.004.

Sedelnikova, O.A., Horikawa, I., Zimonjic, D.B., Popescu, N.C., Bonner, W.M., Barrett, J.C., 2004. Senescing human cells and ageing mice accumulate DNA lesions with unrepairable double-strand breaks. Nat. Cell Biol. 6 (2), 168–170. https://doi.org/10.1038/ncb1095.

Sharon, G., Sampson, T.R., Geschwind, D.H., Mazmanian, S.K., 2016. The central nervous system and the gut microbiome. Cell 167 (4), 915–932. https://doi.org/10.1016/j.cell.2016.10.027.

Stepanova, D., Brunet Guasch, M., Byrne, H.M., Alarcón, T., 2025. Understanding how chromatin folding and enzyme competition affect rugged epigenetic landscapes. Bull. Math. Biol. 87, 59. https://doi.org/10.1007/s11538-025-01434-0.

Stepanova, D., Byrne, H. M., & Alarcón, T. (2025). Stochastic modelling reveals that chromatin folding buffers epigenetic landscapes against sirtuin depletion during DNA damage (preprint). arXiv. https://arxiv.org/abs/2508.16548

Vijg, J., Suh, Y., 2013. Genome instability and aging. Annu. Rev. Physiol. 75, 645–668. https://doi.org/10.1146/annurev-physiol-030212-183715.

Wilson, L.A., 2021. Students from Garcia Summer Program present research at MRS virtual meeting. MRS Bull. 46 (5), 451–453. https://doi.org/10.1557/s43577-021-00106-3.

Xu, H., Li, S., Liu, Y.-S., 2021. Roles and mechanisms of DNA methylation in vascular aging and related diseases. Front. Cell Dev. Biol. 9, 699374. https://doi.org/10.3389/fcell.2021.699374.

Yemini, E., Lin, A., Nejatbakhsh, A., Varol, E., Sun, R., Mena, G.E., Samuel, A.D.T., Paninski, L., Venkatachalam, V., Hobert, O., 2021. NeuroPAL: a multicolor atlas for whole-brain neuronal identification in *C. elegans*. Cell 184 (1), 272–288. https://doi.org/10.1016/j.cell.2020.12.012.

Young, S., Rhodes, P., Touyz, S., Hay, P., 2013. The relationship between obsessive-compulsive personality disorder traits, obsessive-compulsive disorder and excessive exercise in patients with anorexia nervosa: a systematic review. J. Eat. Disord. 1, 16. https://doi.org/10.1186/2050-2974-1-16.

CHAPTER

22

Structural and dynamic signatures of DNA damage in *Leishmania* : Mechanisms, implications, and therapeutic perspectives

Aviral Kaushik[a,b], *Naureenbanu Makarani*[a,b], *Krupanshi Bharadava*[a,b], *Sanjay Kumar*[c] and *Radhey Shyam Kaushal*[a,b]

[a] **Biophysics & Structural Biology, Research & Development Cell, Parul University, Vadodara, Gujarat, India** [b] **Department of Life Sciences, Parul Institute of Applied Science Research & Development Cell, Parul University, Vadodara, Gujarat, India** [c] **Structural and Molecular Biology Centre for Discovery and Innovation Building, The City College of New York, NY, United States**

22.1 Introduction

22.1.1 Overview of DNA damage and its biological significance

Deoxyribonucleic acid (DNA) is the basic molecule that contains the genetic information essential for all living things to develop, operate, and reproduce (Hoeijmakers, 2009). Numerous internal and environmental factors can induce DNA damage, which poses a major risk to the integrity of cells and organisms. Genome stability is constantly threatened by endogenous sources such as metabolic byproducts, reactive oxygen species (ROS), and replication mistakes. This burden is further increased by exogenous factors such as chemical mutagens, ionizing radiation, and UV radiation. Cells use a precisely calibrated DNA damage response (DDR) system that consists of lesion recognition, signal transduction, and repair processes to fend off this genomic attack. (such as nonhomologous end joining, homologous recombination, and base excision repair) and apoptosis if the damage cannot be repaired. The DDR preserves the equilibrium between genetic flexibility and stability, which is essential for cancer prevention,

Genome Damage: Repair and Adaptation Across Life.
DOI: https://doi.org/10.1016/B978-0-443-42857-9.00017-7

stress tolerance, ageing, and development. Understanding the DNA damage process is very important for infections like *Leishmania* since their ability to survive, adapt, and cause disease largely depends on how they handle and respond to genomic insults (Silva et al., 2021).

21.1.2 Unique genomic and molecular features of *Leishmania* influencing DNA damage responses

Leishmania exhibits unique features that distinguish it from classic eukaryotic models. Unlike most eukaryotes, *Leishmania's* genes are organized as large polycistronic transcription units, and the gene expression is mostly regulated at the posttranscriptional level. As a result, *Leishmania* lacks the conventional promoter-driven transcriptional regulation characteristics of other organisms (Florêncio et al., 2025). Cells within a clonal population display distinct chromosomal copy numbers, a phenomenon termed mosaic aneuploidy. Chromosome 31 is frequently noted for elevated copy numbers among several species of *Leishmania*. *Leishmania* exploits specific forms of genomic plasticity, including segmental amplifications and copy number variations (CNVs), under pharmacological and environmental stress (Bussotti et al., 2018; Florêncio et al., 2025).

Leishmania employs recombinant-dependent DNA replication techniques. In the presence of genotoxic stress, essential homologous recombination proteins, such as RAD51 and BRCA1-related factors, are required for genome replication and repair (Gorodetska et al., 2019; Black et al., 2023). The protein kinase ataxia-telangiectasia and rad3-related (ATR kinase), a key participant in the DDR, has been shown to modulate DNA repair and replication timing in *Leishmania*, facilitating a more accurate response to damage (Da Silva et al., 2025). Moreover, kinases like casein kinase 1 isoform 2 (LmCK1.2), while largely related to parasite vitality and intracellular survival, have growing links with DDR pathways. These various features indicate how *Leishmania* combines genomic integrity with the evolutionary benefits of genome variability (Rachidi et al., 2014).

22.1.3 Relevance of studying DNA damage in *Leishmania*: Parasite survival, drug resistance, and virulence

Understanding DNA damage and subsequent repair in *Leishmania* is crucial for understanding the parasite's ability to survive in highly oxidative environments, including host macrophages (Kaushal et al., 2025). The parasite is exposed to nitric oxide and ROS after being ingested by host immune cells, both of which are capable of gravely harming DNA. A strong and adaptable DDR is needed for survival in such circumstances. For instance, effective homologous recombination repair paired with ATR-kinase-mediated signaling ensures that replication proceeds without precipitating a catastrophic genomic collapse. These steps are critical for effective differentiation and establishment inside host tissues in addition to survival (Bussotti et al., 2018; Da Silva et al., 2025; Florêncio et al., 2025; Sadr et al., 2025).

Drug resistance in *Leishmania* is closely linked to its ability to undergo quick genetic modifications. Mosaic aneuploidy and genomic amplifications enable the parasite to change the expression of genes involved in drug metabolism and efflux. For instance, downregulation of aquaglyceroporin 1 (AQP1), which limits drug uptake, and overexpression of ABC transporters and thiol-dependent detoxification pathways have been connected to resistance to antimonials, a first-line treatment for leishmaniasis (Santi and Murta, 2022; Bhusal et al., 2025). Recent research has shown that *Leishmania* drug resistance can modify their lipid metabolic pathways

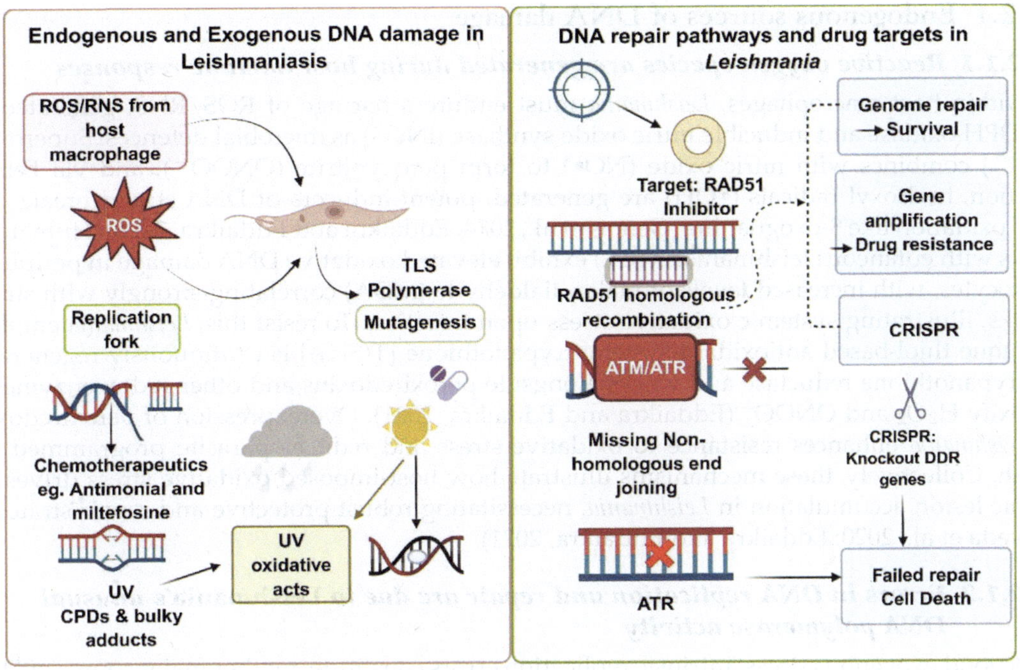

FIGURE 22.1 DNA damage and repair in *Leishmania* (schematics prepared by using BioRender tool).(Perkel, 2020)

and translational machinery even before drug exposure, demonstrating a form of anticipatory resistance strategy. These results underscore the essential role that transitional and genomic plasticity play in developing resistance to chemotherapeutic therapy (Moncada-Diaz et al., 2024).

The genetic dynamics of the parasite affect virulence. Specific deletions, single-nucleotide polymorphisms (SNPs), and gene amplification in pathways involved with redox balance, nucleotide metabolism, and signaling have been revealed by comparative genomic study between virulent and avirulent strains (Zheng et al., 2020; Moncada-Diaz et al., 2024). These modifications might affect the expression of virulence factors such as GP63 metalloprotease, lipophosphoglycan (LPG), and proteophosphoglycans (PPG), which modulate host immune responses and facilitate intracellular persistence. The stability and expression of these virulence genes can be altered by variations in DNA repair ability and DDR pathway efficiency, further binding genome maintenance mechanisms to negative impacts (Elmahallawy and Alkhaldi, 2021; Wijnant et al., 2022; Da Silva et al., 2025).

22.2 Mechanisms of DNA damage in *Leishmania*

Intrinsic mechanisms that cause DNA lesions continuously compromise Leishmania's genomic integrity. The stability and plasticity of the *Leishmania* genome are substantially regulated by two primary endogenous processes: ROS produced during host-parasite interactions and replication/repair errors resulting from the parasite's unique DNA polymerases (Fig. 22.1) (Bharadava et al., 2024).

22.2.1 Endogenous sources of DNA damage

22.2.1.1 *Reactive oxygen species are generated during host immune responses*

Within host macrophages, *Leishmania* must endure a barrage of ROS/RNS, generated by NADPH oxidase and inducible nitric oxide synthase (iNOS) as microbial defences. Superoxide ($O_2^{-\bullet}$) combines with nitric oxide (NO•) to form peroxynitrite ($ONOO^-$), and via Fenton reaction, hydroxyl radicals (•OH) are generated, potent inducers of DNA strand breaks and base oxidation like 8-oxoguanine (Genois et al., 2014; Eddaikra and Eddaikra, 2021). Human patients with cutaneous leishmaniasis (CL) exhibit elevated oxidative DNA damage in peripheral leukocytes, with increased levels of malondialdehyde (MDA) correlating strongly with strand breaks, illustrating systemic oxidative stress upon infection. To resist this, *Leishmania* employs a unique thiol-based antioxidant system: trypanothione [$T(SH)_2$] is continuously regenerated by trypanothione reductase and works alongside peroxiredoxins and other redox enzymes to detoxify H_2O_2 and $ONOO^-$(Eddaikra and Eddaikra, 2021). Overexpression of peroxiredoxins in *Leishmania* enhances resistance to oxidative stress and reduces parasite programmed cell death. Collectively, these mechanisms illustrate how host-imposed oxidative stress drives genomic lesion accumulation in *Leishmania*, necessitating robust protective and repair strategies (Poveda et al., 2020; Eddaikra and Eddaikra, 2021).

22.2.1.2 *Errors in DNA replication and repair are due to* Leishmania's *unusual DNA polymerase activity*

Beyond exogenous stress, intrinsic replication errors constitute another major source of DNA damage in *Leishmania*, attributable to the parasite's expanded and atypical DNA polymerase and repertoire. *Leishmania* also encodes multiple translesion synthesis (TLS) polymerases, such as Pol κ, Rev1, Pol η, and Pol ζ, with *L. major* harboring three Pol κ copies, enabling bypass of DNA lesions during replication (Passos-Silva et al., 2010; Poveda et al., 2020). While TLS polymerase aids the parasite to tolerate replication-blocking lesions like oxidative base modifications, they are inherently error-prone, introducing mutations that increase genetic variability and aid adaptation (Passos-Silva et al., 2010).

Also, the nuclear DNA polymerase β (Pol β) in *Leishmania infantum* and *Leishmania donovani*, a family of X polymerase, appears tailored for both base excision repair (BER) and replication-related tasks. Structural and expression analysis confirm a hand-shaped structure with regulated levels peaking during the infective stage, hinting at a replication-coupled repair function (Taladriz, 2001; Khan et al., 2019). Functional studies show that the overexpression of Pol β enhances survival against oxidative stress and the drug amphotericin B, as well as increased infectivity in macrophages (Mishra et al., 2018; Khan et al., 2019). Oxidative stress and pharmacological pressure similarly elevate *L. donovani* uracil DNA glycosylase (UNG), a BER enzyme, thereby augmenting infectivity and conferring resistance to DNA-damaging drugs (Taladriz, 2001; Mishra et al., 2018; Khan et al., 2019).

22.2.2 Exogenous sources of DNA damage

Leishmania parasites encounter various external DNA-damaging chemicals during their complicated life cycle, particularly during the transfer between the sandfly vector and the mammalian host. These external insults represent considerable challenges to the parasite's genomic integrity and are primary contributors to DNA lesions that require robust repair mechanisms (Taladriz, 2001; Kaushik et al., 2025). Environmental elements like UV light and exposure to external oxidative stress in natural reservoirs or when parasites are manipulated in vitro are among the most detrimental external stressors, as are the oxidative and nitrosative species produced by host macrophages during the infection and chemotherapeutic agents like pentavalent antimonials, miltefosine, and amphotericin B (Farrell et al., 2011; Mishra et al., 2018).

22.2.2.1 *Oxidative stress induced by host macrophages*

Upon being phagocytosed by host macrophages, *Leishmania* parasites are resistant to extreme oxidative and nitrosative stressors. When iNOS and NADPH oxidase are activated, superoxide and nitric oxide are produced. These then oxidize peroxynitrite, hydroxyl radicals, and other reactive species (ROS/RNS), resulting in DNA damage that is characterized by abasic sites, single and double-strand breaks, and base oxidation (e.g., 8-oxoG) (Mishra et al., 2018; Kaushal et al., 2025). Experimental evidence shows that oxidative stress causes the upregulation of BER enzymes like UNG and Pol β in *L. donovani*. Overexpression of UNG increases survival during oxidative and drug-induced stress, while upregulation of Polβ shields parasites from ROS and lesions caused by amphotericin B, enhancing infectivity (Mishra et al., 2018; Khan et al., 2019).

22.2.2.2 *Chemotherapeutic agents*

22.2.2.2.1 Pentavalent antimonials (e.g., meglumine antimonate)

Antimonials remain frontline antileishmanial drugs, but their genotoxic side effects extend to both parasite and host. Meglumine antimonate (20 mg/kg for 20 days) caused oxidative base damage in host leukocytes in *L. infantum*-infected mouse models, as demonstrated by comet tests in conjunction with Fpg/EndoIII digestion. This was accompanied by increased micronucleated cells, elevated oxidative stress markers, and depletion of glutathione-dependent antioxidant defences (SOD-GPx axis) (Moreira et al., 2017). These results suggest that antimonials increase oxidative stress, which leads to ROS-mediated genomic lesions and may contribute to mutagenesis in the parasite and the host (Sudhandiran and Shaha, 2003; Moreira et al., 2017).

22.2.2.2.2 Miltefosine

By disrupting mitochondrial activity and parasite membrane integrity, the amphiphilic oral alkylphosphocholine miltefosine causes promastigotes to die apoptotically. Its cytotoxic effects are associated with metabolic interference and membrane disruption, as well as nuclear DNA fragmentation and oligonucleotide cleavage, an indication of programmed cell death (Paris et al., 2004). Even though the exact mechanisms of direct DNA damage are still unclear, miltefosine may indirectly cause DNA lesions like fragmentation and strand breaks by encouraging apoptosis-related processes (Paris et al., 2004; Verma et al., 2007).

22.2.2.2.3 Amphotericin B

Though anticancer in mammals, similar protective upregulation of Pol β in *L. donovani* also confers resistance against amphotericin B, indicating the drug's capability to inflict oxidative/genotoxic stress (Khan et al., 2019). While the precise molecular damage caused by amphotericin B is less defined, its association with DDR underscores its role as an exogenous genotoxic agent (Khan et al., 2019; Eddaikra and Eddaikra, 2021).

22.2.2.3 Environmental factors

Although *Leishmania* is primarily transmitted by sandflies and largely resides in shielded environments, UV exposure in the environment, especially in CL, can damage parasite DNA during promastigote stages on host skin or sandfly surfaces. UVB and UVA induce cyclobutene pyrimidine dimers (CPDs), 6-4 photoproducts, and 8-oxo-guanine lesions (Sarkar and Gaddameedhi, 2020). Typically, UV-damaged DNA is repaired via nucleotide excision repair (NER). Although NER has not yet been deeply characterized in *Leishmania*, the universal occurrence of UV lesions and potential for functional homologs suggest similar repair requirements (Mullenders, 2018). Through secondary genotoxic pathways, UVA-driven ROS production intensifies oxidative DNA damage and adds to genomic instability (Ramasamy et al., 2017).

22.3 Structural signatures of DNA damage in *Leishmania*

The term "DNA damage" describes structural changes to the DNA molecule that affect its stability, transcription, or replication. Both endogenous metabolic processes and foreign stresses, such as host immunological responses or chemotherapeutic treatments, can cause DNA damage in *Leishmania*. Structural indicators of this type of damage include a number of lesions that physically break the DNA strands, change base pairing, or deform the DNA helix. Despite having a relatively basic genome, *Leishmania* demonstrates remarkable resistance to DNA damage, which it uses to support environmental adaptation, drug resistance, and genome plasticity (Chatterjee and Walker, 2017; Bussotti et al., 2023).

22.3.1 Types of DNA damage

22.3.1.1 Single-strand breaks and double-strand breaks

One DNA strand cleaving, which commonly occurs during the BER process or as a result of oxidative damage, results in single-strand breaks (SSBs). Cells in many organisms come into contact with around 10,000 SSBs each day. *Leishmania* probably goes through similar processes, especially considering the high level of oxidative stress inside macrophages (Nickoloff et al., 2020; Da Silva, 2021). SSBs, while more catastrophic than DSBs, can stall replication and, if left unrepaired or processed through overlapping sites, escalate into double-strand breaks (DSBs) (Da Silva, 2021).

DSBs, where both strands are severed, are among the most dangerous lesions, potentially resulting in genomic rearrangements or cell death. Idiosyncratically, *Leishmania* utilizes DSBs not only for DNA repair but also to induce gene CNVs via homologous recombination, supporting gene amplification events vital for adaptation and drug resistance (Da Silva, 2021). DSBs arise spontaneously through replication stress or are actively generated alongside CNVs, both as byproducts of oxidative lesions and by cellular strategy. Radiation exposure in *L. major* leads to G1 cell cycle arrest with high mortality due to inefficient DSB repair, reinforcing the importance of DSB processing in parasite survival (Garcia et al., 2016; Herrmann May et al., 2025).

22.3.1.2 Base modifications

Oxidative base damage is among the most prevalent and deleterious forms of DNA lesions. 8-oxoguanine (8-oxoG), resulting from guanine oxidation, mispairs with adenine during replication, promoting G:C→T:A transversions. In general, in cellular systems, baseline levels of 8-oxoG approach 10^3–10^4 lesions per cell per day (Furtado et al., 2012). In *Leishmania*, elevated

ROS during host infection likely increases these lesions, necessitating BER via UNG and Pol β to preserve genome fidelity and support postreplication survival under redox stress. Other base modifications, such as deamination events, abasic sites, and oxidized pyrimidines, pose significant threats, although they are inferentially similar to oxidative stress-described lesions (Furtado et al., 2012; Poveda et al., 2020).

22.3.1.3 *Crosslinking and bulky adducts*

Cross-links, both intra-strand and inter-strand, are covalent linkages between nucleotides or between DNA strands, profoundly obstructing replication and transcription. Exogenous chemicals, such as platinum-based agents, and endogenous byproducts like lipid peroxidation products (e.g., malondialdehyde), induce such damage (Zhu et al., 2009; Yousefzadeh et al., 2021). Though understudied in *Leishmania*, similar lesions likely arise through oxidative metabolism and chemical exposures. Intriguingly, bulky adducts such as pyrimidine dimers following UV radiation, despite being less studied in *Leishmania*, are universally problematic, necessitating NER or homologous recombination to remove them.

Platinum-based crosslinking adducts in other organisms extensively distort DNA's helical structure, unwinding around the lesion site and widening grooves, particularly with complex trifunctional cross-links (Zhu et al., 2009). If similar lesions occurred in *Leishmania*, they would greatly impede replication and transcription, compelling the parasite to rely heavily on recombination and repair pathways. Furthermore, proteolytic and recombinational activities typically break down complex adducts called DNA-protein cross-links (DCPs), which include covalently trapping transcription or replication enzymes (Garcia et al., 2016; Kojima and Machida, 2020).

22.3.2 Chromatin remodeling and DNA damage

22.3.2.1 *Chromatin compaction versus relaxation*

DNA damage detection and repair are greatly impacted by the dynamic alternation between accessible euchromatin and confined heterochromatin in *Leishmania's* chromatin. While relaxed chromatin improves DNA accessibility and repair factor recruitment, highly compact nucleosomal arrays limit access to DNA lesions. Increased sensitivity to micrococcal nuclease digestion in cells overexpressing HAT2 indicates that the histone acetyltransferase HAT2 in *L. donovani* preferentially acetylates histone H4 at lysine 4 (H4K4ac), a change that broadly enhances chromatin recondensation (Jha et al., 2017; Dacher et al., 2019). Similarly, transcription and DNA repair are facilitated by HAT4-mediated H4K14 acetylation, which preserves a euchromatic state (Afrin et al., 2019). These alterations cause nucleosomes to relax, allowing DNA repair tools to penetrate lesions induced by ROS, chemotherapy, or replication stress. However, histone deacetylases exacerbate chromatin compaction during amastigote stages, potentially making the parasite more resistant to genotoxic damage in macrophages (Bhaskara, 2015; Da Silva et al., 2025).

22.3.2.2 *Role of histone modifications and histone variants*

Following DNA damage, chromatin structure and integrity are largely controlled by histone posttranslational modifications (PTMs) and the incorporation of histone variants. Histone variants like H2A.Z, H2B.V, and H3.V are expressed by *Leishmania* and are enriched close to divergent strand-switch regions and transcriptionally active polycistronic units (Afrin et al., 2019). *Leishmania*'s DNA damage is identified by phosphorylation of H2A at Thr128, similar to H2AX phosphorylation (γH2X in higher eukaryotes), which facilitates the recruitment of RAD51 and other repair factors to double-strand break sites. This illustrates a damage-sensing mechanism that is conserved and specific to kinetoplastid chromatin (Saha, 2020).

Additional PTMs implicated in DNA repair include acetylation of histone H3 and H4, which correlate with chromatin relaxation at transcription start sites and possibly at lesion sites following DNA damage (McDonald et al., 2022). Methylation marks, including H3K4me3 and H4K2me2/3, are associated with DNA repair and chromatin compaction states across eukaryotes. While *Leishmania*-specific mapping has not been completed, conserved residues suggest analogous roles in break recognition and chromatin modulation (Afrin et al., 2019).

In other eukaryotes, histone variants like H2A are dynamically evicted or inserted at DNA lesion sites to fine-tune chromatin openness. Though not yet directly demonstrated in *Leishmania*, the presence of H2A.Z and its parallels in kinetoplastids infer that variant remodeling likely enhances repair accessibility. Conversely, macroH2A variants, known to promote chromatin compaction around damage in mammals, may have functional counterparts that modulate parasite genome accessibility (Hauer and Gasser, 2017; Ferrand et al., 2020).

22.3.3 Telomere and kinetoplast DNA stability

22.3.3.1 *Kinetoplast DNA damage signatures*

Leishmania, like other trypanosomatids, harbors a unique mitochondrial DNA structure called kDNA, comprized of a dense network of thousands of minicircles(~0.5–2kb) and dozens of maxicircles (~20–40kb) interlocked into a disk-like structure (Camacho et al., 2019). Damage to these circles, such as strand breaks, base oxidation, or deletions, impairs mitochondrial function and guide RNA (gRNA) editing essential for respiratory complex assembly and energy production (Amodeo et al., 2023). Recent long-read sequencing of *L. martiniquensis* has revealed 214 distinct minicircle classes. Perturbation of their copy number, as seen with certain drugs, causes "transkinetoplastidy" and defective RNA editing, linking structural instability to functional disruption. Drugs like diminazene aceturate have been shown to cause irreversible crosslinking and degradation of kDNA, further destabilizing the network (Da Silva Oliveira and De Freitas, 2015; Anuntasomboon et al., 2022). These damage-specific signatures, loss of specific minicircle classes, compromised maxicircle-encoded mRNAs, and impaired network integrity represent key markers of kDNA compromise and are directly associated with decreased parasite survival and drug sensitivity.

22.3.3.2 *Telomere damage and parasite viability*

Leishmania chromosomes terminate with telomeric repeat tracts (5′-TTAGGG-3′), similar to vertebrates (Klebanov-Akopyan et al., 2018). Telomere integrity is crucial for chromosomal stability, preventing end-to-end fusions and degradation. The parasite encodes telomere-associated factors, including replication protein A (RPA-1), which binds to G-rich telomeric strands, and the Ku70/80 heterodimer, traditionally associated with nonhomologous end joining (NHEJ) but implicated here primarily in telomere maintenance (Aksenova and Mirkin, 2019). In *L. amazonesis*, RPA-1 protects telomeres even without the RPA1N domain, indicating its specialized adaptation for telomere maintenance amid DDR. In *L. maxicana*, Ku80 disruption caused unexpected telomere lengthening, likely due to an alternative telomere elongation mechanism, without compromising overall genomic stability over prolonged culture (Da Silveira et al., 2013; Poláková et al., 2021). However, telomere dysfunction in other models triggers DNA damage signaling, cell cycle arrest, and reduced viability. Thus, while *Leishmania* appears capable of compensating for telomeric stress, disruption in telomere maintenance proteins ultimately influences cell proliferation, genome integrity, and parasite survival.

22.4 Dynamic responses to DNA damage in *Leishmania*

22.4.1 DNA damage response pathways

The DDR in *Leishmania* orchestrates how the parasite detects DNA lesions, activates checkpoints, recruits repair proteins, and restores genome integrity. Although these processes are broadly conserved across eukaryotes, *Leishmania* exhibits significant divergence in DDR structure and regulation, reflecting adaptations to its life cycle and genomic architecture (Fig. 22.2) (Garcia et al., 2016). The key differences between *Leishmania* and higher eukaryotes are shown in Table 22.1.

22.4.1.1 *Key sensor kinases: Ataxia telangiectasia and rad3 related and ataxia telangiectasia mutated*

Leishmania harbors homologs of the primary eukaryotic DDR kinases ataxia telangiectasia and rad3 related (ATR) and ataxia telangiectasia mutated (ATM). In particular, recent studies in *L. donovani* highlight ATR's pivotal role; it regulates replication timing under unperturbed conditions and upon genotoxic stress, thereby limiting genome instability (Da Silva et al., 2025). Importantly, *Leishmania* ATR differs structurally from its mammalian counterpart: it lacks the PIKK regulatory domain (PRD) necessary for TOPBP1-mediated activation in higher eukaryotes. Functional studies in *T. brucei*, a related kinetoplastid, illuminate ATR's conserved role: Upon depletion, parasites accumulate γH2A (analogous to γH2AX) from RAD51 and RPA foci, and exhibit S-phase arrest and heightened sensitivity to DNA damage (Marin et al., 2020; Silva et al., 2021). These findings strongly suggest that ATR acts as the apical DDR kinase in *Leishmania*, initiating checkpoint signaling and repair factor recruitment.

ATM homologs in kinetoplastid are less well-characterized; however, inference from *T. brucei* and related organisms supports their role in responding chiefly to DSBs and oxidative stress, analogous to mammalian ATM, phosphorylating downstream effectors like histone H2A (Marin et al., 2020; Da Silva et al., 2025).

22.4.1.2 *Homologous recombination and RAD51*

Downstream of ATR/ATM activation, homologous recombination is the principal pathway for repairing DSBs in *Leishmania*. Central to this pathway is RAD51, a recombinase that mediates strand invasion and repair templating. In *T. brucei*, ATR depletion impairs RAD51 foci formation and upregulation following DSBs, indicating ATR-dependent recruitment of homologous recombination (HR) machinery. *Leishmania*, which shares HR reliance, uses RAD51 not only for genome integrity but also for generating gene amplifications and CNVs, mechanisms tied to drug resistance (Marin et al., 2020; Da Silva, 2021; Da Silva et al., 2025).

22.4.1.3 *Replication protein A and single-stranded DNA sensing*

RPA, the single-stranded DNA-binding complex, binds ssDNA at stalled replication forks or processed DSBs, acting as a platform for ATR recruitment via ATRIP. In kinetoplastids, the

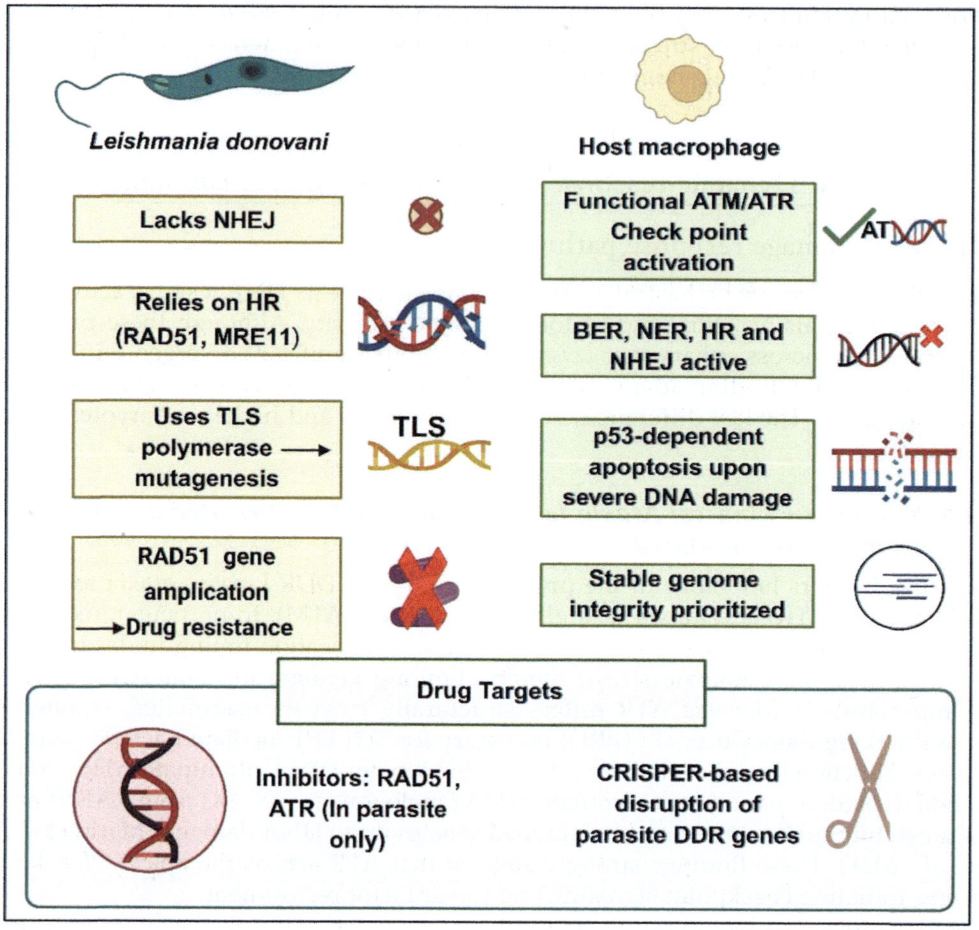

FIGURE 22.2 Contrasting DNA damage responses—*Leishmania* versus host macrophage(schematics prepared by using BioRender tool)(Perkel, 2020)

large RPA-1 lacks its conventional N-terminal domain, yet remains functional in DNA damage-sensing and telomere protection. RPA phosphorylation, mediated by ATR/ATM kinases, alters its interaction with ssDNA and repair proteins, though detailed studies in *Leishmania* remain pending; analogous mechanisms are likely at play (Da Silveira et al., 2013).

22.4.1.4 *Checkpoints and cell cycle arrest*

ATR activation halts cell cycle progression to allow DNA repair. In *T. brucei*, ATR loss disrupts both intra-S-phase checkpoints and the G1/S checkpoint, leading to fork collapse and impaired recruitment of DNA repair factors (Marin et al., 2020). In *Leishmania*, genome instability and copy number dynamics are tightly linked to replication timing and damage responses, suggesting similar checkpoint mechanisms preserve genomic integrity.

TABLE 22.1 Differences between Leishmania and higher eukaryotes.

Feature	*Leishmania/kinetoplastids*	Higher eukaryotes	References
ATR structure	Lacks PIKK regulatory PRD; likely alternative activation	ATR contains the PRD domain and requires TOPBP1 for activation	Burrows and Elledge (2008); Mordes and Cortez (2008)
ATM Kinase	Poorly characterized; presumed to respond to DSBs and oxidative stress, similar to other kinetoplastids	Well-characterized; activated by DSBs, phosphorylates downstream targets like CHK2, p53 and γH2AX	Da Silva et al. (2025); Marin et al. (2020)
RAD51 and HR pathway	Central to DSB repair; also mediates gene amplification and drug resistance. Upregulated under genotoxic stress	Primarily used for precise DSB repair in S/G2 phase; does not promote gene amplification	Orhan et al. (2021); Wassing and Esashi (2021)
Replication protein A (RPA)	RPA1 lacks a canonical N-terminal domain but retains function in ssDNA sensing and telomere binding	RPA is a conserved ssDNA binding complex, with N-terminal phosphorylation sites for checkpoint coordination	Da Silva et al. (2025); Da Silveira et al. (2013)
Checkpoint pathways	ATR-dependent checkpoint activation observed; S-phase and G1/S arrest reported in related trypanosomatids	Both ATM and ATR initiate checkpoints via CHK1/CHK2 and control transitions across G1, S, G2 and M phases	Da Silva et al. (2025); Marin et al. (2020)
Use of HR for genome plasticity	RAD51-mediated homologous recombination contributes to extrachromosomal amplification and drug resistance	HR maintains genome integrity, but excessive recombination is typically suppressed	Chatterjee and Walker (2017); Wassing and Esashi (2021)

22.4.2 DNA repair mechanisms in *Leishmania*

22.4.2.1 Base excision repair and nucleotide excision repair

BER is essential in *Leishmania*, serving to correct small base lesions such as oxidized, deaminated, and alkylated bases (Fig. 22.3). Key enzymes include uracil DNA glycosylase (UNG) and Polβ. Studies in *L. infantum* demonstrate that Polβ exhibits intrinsic 5′-deoxyribose-5-phosphate lyase activity, essential for nucleotide incision and gap-filling during BER. Overexpression of Polβ increases oxidative stress tolerance and enhances infectivity in macrophages (López-Camarillo et al., 2009; Roshanara et al., 2024). UNG is also induced under oxidative conditions, likely facilitating the removal of uracil and abasic sites generated by ROS.

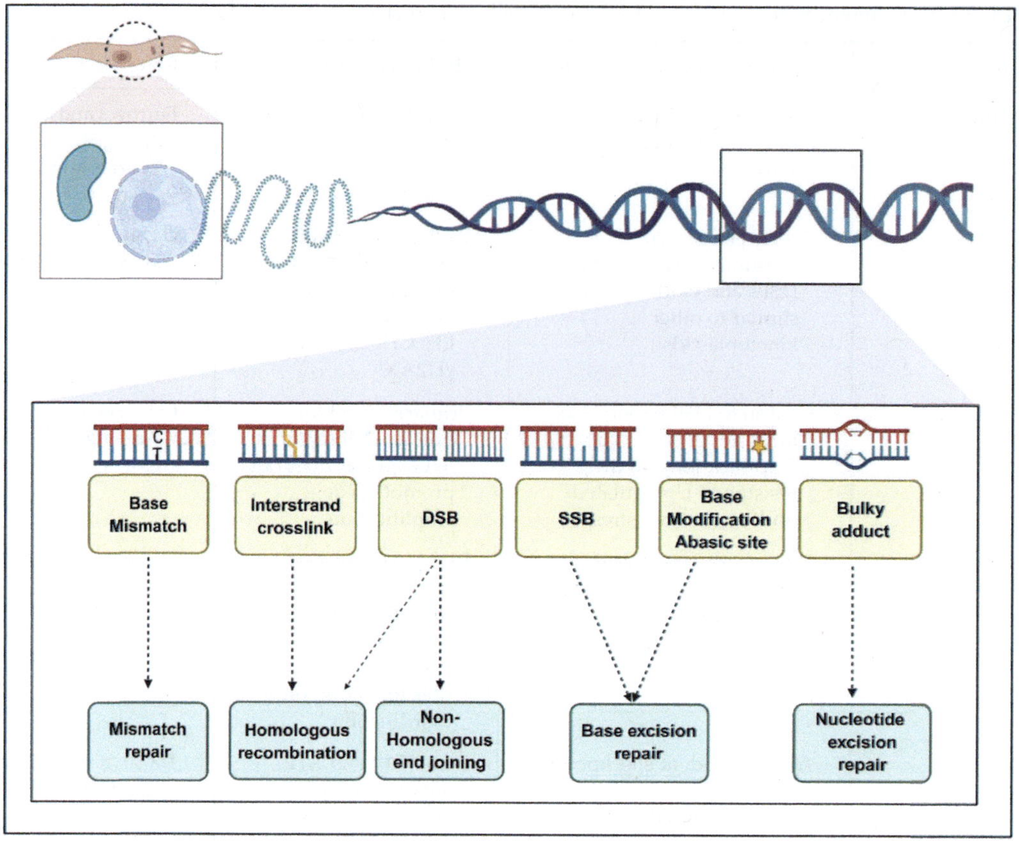

FIGURE 22.3 Overview of DNA damage types and corresponding repair pathways in eukaryotic cells (schematics prepared by using BioRender tool).(Perkel, 2020)

NER (Fig. 22.3) addresses bulky helix-distorting lesions such as UV-induced thymine dimers. Recent proteomic work in *L. donovani* identifies a centrin-associated DNA damage recognition complex (LdCen1-LdDRP), homologous to human XPC-centrin2, that recognizes UV damage and facilitates NER. The activation of global genomic NER pathways is supported by the recruitment of γH2A and the upregulation of these proteins during the G1/S transition when exposed to UV-C (Roshanara et al., 2024).

22.4.2.2 *Homologous recombination versus nonhomologous end joining*

Since *Leishmania* lacks a functional classical NHEJ system, HR is the primary mechanism for repairing DSBs in the parasite. Resection, strand invasion mediated by RAD51, and ssDNA binding by RPA are all necessary for this HR pathway. HR is also necessary for *Leishmania's* genome to be plastic: Drug-resistant gene amplification events and CNVs are caused by RAD51-dependent recombination (Duncan et al., 2017; Zhang and Matlashewski, 2019). Furthermore, Cas9-induced DSBs are commonly repaired by single-strand annealing, a variant HR technique that typically results in deletions (Zhang and Matlashewski, 2019). In *Leishmania*, nonhomologous end joining, or NHEJ, is essentially nonexistent. Rather, the parasite uses single-strand annealing and microhomology-mediated end joining (MMEJ), both of which necessitate brief homologous sequences at break ends. Ku70 and Ku80 are still present in *Leishmania*, but they do not facilitate traditional NHEJ. Instead of using conventional DSB repair, Ku might help with telomere maintenance or alternative end joining methods (Zhang and Matlashewski, 2019; Roshanara et al., 2024).

22.4.2.3 *Mechanisms of lesion bypass and tolerance*

Translesion synthesis (TLS) polymerases are used by *Leishmania* to replicate lesions that block the replication fork. A characterized polymerase theta (LiPolΘ) in *L. infantum* exhibits lesion bypass capabilities, aiding the parasite in surviving oxidative stress. Error-prone TLS polymerases increase the risk of mutagenesis, which could encourage genomic variety and adaptation, but they also enable replication to proceed even in the presence of DNA damage (Fernández-Orgiler et al., 2016; Maiorano et al., 2021).

22.4.3 Regulation of DNA damage response in developmental stages

22.4.3.1 *Stage-specific DNA repair efficiencies: Promastigotes versus amastigotes*

Leishmania parasites transform between promastigotes, which proliferate in the vector, and amastigotes, the transformed form for survival in the mammalian macrophages. These stages display variable DDR efficiencies, shaped by differential expression and activity of repair enzymes (Garcia et al., 2016; Makarani et al., 2025).

In promastigotes, DDR pathways such as BER, NER, and HR are highly active. Using the ATR and ATM inhibitors in *L. major* promastigote form increases sensitivity to oxidative damage, indicating that efficient DDR is essential for promastigote persistency (Da Silva et al., 2018). In the amastigote form, studies suggest a shift in DDR regulation. Amastigotes exhibit reduced global transcription and translation activity compared to promastigotes. This downregulation likely impacts DDRs, wherein certain repair components, such as those involved in BER, remain functionally active. However, DNA repair pathways closely tied to replication, such as HR, may be less prominent in amastigotes due to their slower rate of cell division. While stage-specific expression of DDR genes in *L. infantum* amastigotes is not yet thoroughly profiled, analogous systems in *Trypanosoma* propose that reduced replication correlates with reduced HR activity during intracellular adaptation (Silva et al., 2021; Santi and Murta, 2022).

22.4.3.2 *Implications for parasite survival in the host*

The divergence in DDR activity between stages shows functional adaptation to environmental pressures. Promastigotes are relatively free from host immune assault, but still require lesion bypass mechanisms, such as PolΘ-mediated TLS, to survive during transmission and maturation. These mechanisms likely support genetic plasticity, facilitating adaptation during stage transitions. Upon entering the mammalian host, amastigotes face a totally different environment dominated by ROS/RNS. Their increased dependence on BER and NER systems is crucial to neutralize oxidative DNA lesions and bulky adducts. For instance, amastigotes overexpress UNG and Polβ, providing effective removal of oxidized bases and SSBs, a life-or-death necessity within macrophages (Passos-Silva et al., 2010; Hauer and Gasser, 2017; ; Poveda et al., 2020).

22.5 DNA damage and antileishmanial drug resistance

The persistence and virulence of *Leishmania* in hostile host environments are closely tied to its unique capacity to undergo genotoxic stress. Many frontline antileishmanial therapies, such as antimonials, amphotericin B, and miltefosine, evoke cytotoxicity primarily by inducing oxidative and DNA damage (Kaushik et al., 2025; Makarani et al., 2025). However, *Leishmania* has developed strong and flexible DDR mechanisms that help the parasite avoid drug-induced stress by not only repairing lesions but also promoting genomic plasticity. These mechanisms have a direct impact on the emergence and persistence of drug resistance (Afrin et al., 2019; Khan et al., 2019).

22.5.1 Modified DNA repair pathways contributing to drug resistance

Leishmania uses its DNA repair machinery, especially HR, to adapt to drug-induced RAD51, its paralogs, and the MRN (meiotic recombination) complex. According to research, RAD51 expression is triggered by DSBs produced by antileishmanial medications, which promotes gene amplification and the development of extrachromosomal circles—processes that are intimately linked to drug resistance. RAD51 disruption dramatically reduces the synthesis of linear or circular amplicons that promote resistance in *L. infantumn* (Kamran et al., 2023). These HR elements are therefore essential for both repairing and modifying the CNVs needed for drug-induced survival.

Due of the lack of traditional NHEJ, *Leishmania* simultaneously depends on MMEJ and single-strand annealing (SSA).These mechanisms may help produce linear amplicons that increase resistance (Genois et al., 2014; Kamran et al., 2023).

22.5.2 Role of mutations in DNA damage response genes in resistance development

Mutational events with DDR genes themselves can confer drug resistance. A CRISPR Cas9 genome-wide screen in *L. infantum* exposed to miltefosine and amphotericin B highlighted the involvement of genes including DDR elements specifically, a RING variant protein associated with ubiquitination pathways (Queffeulou et al., 2024). Moreover, resistance to topoisomerase I inhibitors has been traced to specific point mutation in the TOP1B gene, especially F187Y, G191A and W232R, which reduce drug binding and thereby confer resistance (Rosa-Teijeiro et al., 2021). Mutations in transporter genes also intersect with DDR pathways. Mutations in the MDR1 efflux pump genes are linked to antimonial resistance, likely by reducing intracellular accumulation of ROS-inducing drug metabolites and secondary DNA damage (Moncada-Diaz et al., 2024).

22.5.3 Emerging strategies to overcome drug resistance by targeting DNA damage response

Targeting DDR elements offers a promising route to counteract resistance, some of which include inhibition of RAD51 and MRE11, TOP1B targeted drugs, combination therapies with DDR inhibitors and CRISPR screening to identify resistance determinants.

Disabling HR machinery (e.g., RAD51, BRCA2, and MRE11) reduces extrachromosomal gene amplification and drug resistance. Small molecules, RAD51 inhibitors, are effective in trypanosomatids, and the MRE11 inhibitor Mirin is a potential lead (Genois et al., 2014). Designing next-generation inhibitors that overcome TOP1B mutations (e.g. F187Y and G191A) can sustain efficacy where current inhibitors fail (Rosa-Teijeiro et al., 2021). Co-administration of the antileishmanial drugs, such as ATR/ATM kinase inhibitors or TLS polymerase blockers, could prevent adaptive gene amplification and resistance. Preliminary studies in related parasites showed that ROS-induced drugs combined with DNA damage sensitizers dramatically reduce survival (Barman et al., 2023). Genome-wide CRISPR-Cas9 libraries have uncovered novel resistance-associated genes such as RING domain proteins and sterol methyltransferases, which may interface with DDR pathways, providing a new therapeutic target (Queffeulou et al., 2024).

22.6 DNA damage and parasite evolution

DNA damage in *Leishmania* drives genetic diversity, adaptive evolution, and phenotypic plasticity in addition to endangering genomic integrity. *Leishmania* can quickly adapt to adverse circumstances, including host immunity and drug exposure, because it tolerates and even takes advantage of DNA lesions and incomplete repair, in contrast to higher eukaryotes that emphasize genomic integrity. This is achieved through mechanisms including oxidative mutagenesis, recombination-mediated genome rearrangements, and regulated aneuploidy (López-Camarillo et al., 2009; Genois et al., 2014; Talimi et al., 2024).

22.6.1 Role of DNA damage in generating genetic diversity

The primary cause of *Leishmania's* genetic diversity is DNA damage, particularly oxidative lesions and replication errors. Sandfly infection reduces genetic variability by causing population congestion for the parasites. Research indicates that ROS-induced DNA damage at this stage increases the number of SNPs and encourages gain of heterozygosity (GOH), a crucial mechanism for restoring genetic variation through crowding (Bussotti et al., 2023). The parasite uses this alternating pattern of gain (GOH) and loss (LOH) in heterozygosity as part of its evolutionary strategy to balance genomic integrity and flexibility.The predominance of GOH events in transition mutations linked to oxidative damage (A↔G and T↔C) suggests regulated mutational mechanisms as opposed to random drift. Such repeated mutagenesis serves as a potent catalyst for genetic innovation and is most likely caused by oxidative stress in the host or vector (Poetsch et al., 2018).

22.6.2 Genome plasticity impact on adaptability and virulence

DNA damage and repair-mediated recombination is critical for *Leishmania's* genome plasticity, which includes extrachromosomal DNA, gene CNVs, and mosaic aneuploidy. In response to environmental or drug-induced stress, HR-mediated CNVs rapidly form from direct repeats and homologous regions to enhance the expression of genes associated with fitness, such as virulence factors or drug transporters. When they are no longer useful, they are reversibly removed (Laffitte et al., 2016). These CNVs are not random: ploidy and amplification patterns recur consistently under stress, suggesting that genome rearrangements follow deterministic adaptive paths. Moreover, experimental evolution studies in *L. donovani*show that CNV-based adaptation keeps parasites genetically diverse. This is achieved through epistatic interactions among multiple gene amplifications, with some amplifications reinforcing others functionally, and by posttranscriptional buffering of deleterious dosage effects, enabling simultaneous adaptability and genomic robustness. Dynamic amplification of snoRNA genes can modulate ribosomal function, enhance translation efficiency and contribute to parasite fitness under culture or host pressure (Black et al., 2023; Bussotti et al., 2021).

These mechanisms also support virulence. For example, macrophage survival and infectivity are increased by amplification of the GP63 or amastin gene families. Cross-species hybridization during sandfly co-infection results in offspring with mixed ploidy and novel phenotypic features, such as altered tissue tropism, further illustrating how sexual recombination and the DDR can cause phenotypic variety and virulence shifts (Bussotti et al., 2023; Kamran et al., 2023).

22.7 Therapeutic implications

A unique way to treat infectious illnesses, especially those caused by parasites like *Leishmania*, is the therapeutic exploitation of DNA damage and repair mechanisms. These parasites have a specific DNA repair pathway because of the evolutionary desire to live in adverse conditions, especially the genotoxic stress within host macrophages. *Leishmania's* DDR machinery is a desirable target for next-generation antileishmanial therapy because of this biological peculiarity (Genois et al., 2014; Silva et al., 2021).

22.7.1 Targeting DNA repair pathways for drug development

Due to its absence of typical NHEJ, *Leishmania* is predominantly dependent on HR, BER, and MMEJ for genome maintenance. In particular, the survival of parasites under genotoxic environments depends on proteins such as DNA polymerase β, MRE11, RPA, and RAD50. Studies have demonstrated that RAD51 in *L. donovani* can be deleted by CRISPR-Cas9 to induce lethal phenotypes, indicating that it may be a promising therapeutic target (Poveda et al., 2020; Zhang et al., 2022). Similarly, it has been observed that inhibiting ATR kinase, which controls DNA replication stress and genome integrity, may render *Leishmania* more sensitive to oxidative stress. These proteins offer parasite-selective therapeutic windows, minimizing off-target toxicity in human hosts because they usually display little similarity with their mammalian counterparts (Da Silva et al., 2025; Zou, 2017).

22.7.2 Use of DNA-damaging agents combined with DNA damage response inhibitors

Combination medicines that combine DDR inhibitors with DNA-damaging compounds, like antimonials, amphotericin B, and miltefosine, constitute a potential therapeutic approach. The effectiveness of these traditional treatments can be increased by limiting the parasite's capacity to repair the damage caused by ROS or alkylation-induced lesions (Biswas et al., 2023; Mavroeidi et al., 2024). By co-administering DNA-damaging drugs with inhibitors of repair enzymes like RAD51, DNA polymerase β, or ATR, synthetic lethality in *Leishmania* can be generated, drawing inspiration from cancer models. The translational potential of this method in leishmaniasis is evidenced by the success of such tactics in trypanosomatids and apicomplexan parasites. Crucially, because they prevent the parasite from evolving through gene amplification and chromosomal rearrangements, these combination approaches also have the added benefit of lowering drug resistance (Marin et al., 2020; Poveda et al., 2020).

22.7.3 CRISPR-based gene editing prospects to study and disrupt DNA damage response pathways

The development of CRISPR-Cas9 technology in *Leishmania* has provided functional validation of DDR genes at unprecedented detail. Tools like LeishBASEedit, which promotes high-efficiency mutagenesis utilizing cytosine base editors (CBEs), allow researchers to explore critical genes that are tricky to knock out using conventional approaches (Engstler and Beneke, 2023; Queffeulou et al., 2024).

Conditional inhibition of MRE11 and BRCA2 or point mutation of DDR proteins, such as RAD51, has been revealed to have a substantial influence in DNA repair and parasite viability through the application of CRISPR. These genetic techniques verify DDR proteins as therapeutic targets and enable screening of chemical libraries for inhibitors that precisely impair repair pathways. Furthermore, genome-wide CRISPR screens carried out under drug pressure (e.g., amphotericin B or miltefosine) have uncovered novel DDR-related resistance pathways, indicating their role in drug tolerance and virulence (Genois et al., 2014; Marin et al., 2020; Queffeulou et al., 2024).

22.8 Future directions and challenges

22.8.1 Integrating multiomics approaches to study DNA damage response in *Leishmania*

The issue of DDR research in *Leishmania* has arrived in the era of systems biology, where multi-omics integration, including genomes, transcriptomics, proteomics, metabolomics, and epigenomics, enables a thorough understanding of how these parasites respond to genotoxic stress. Proteomic experiments have identified posttranslational changes, including phosphorylation and ubiquitination in the DDR pathway, whereas high-throughput genomics and transcriptomics have already mapped several essential DDR genes (such as RAD51, MRE11, ATR, and RPA) (Kaur and Rajput, 2014; Poveda et al., 2020; Zhang et al., 2022).

In addition, new metabolomics findings suggest a strong link between DDR and the parasite's metabolic adaptability and redox balance. For example, the way *Leishmania* reacts to DNA damage was altered by particular alterations in trypanothione and glutathione metabolism brought on by oxidative stress from ROS produced by macrophages. Through the use of integrative omics platforms such as MixOmics and MOFA+, dimensionality reduction, grouping, and correlation analysis across datasets are now feasible, enabling the identification of DDR-associated molecular signatures that transcend single omics layers (Rohart et al., 2017; Argelaguet et al., 2020). When the parasite progresses from the promastigote to the amastigote phases, its tolerance for DNA damage changes, making it tough to create time-resolved, stage-specific DDR models. The building of multiomics atlases suited to drug resistance phenotypes and parasite lifecycle stages will be required for future therapeutic targeting.

22.8.2 Understanding host-parasite interactions influencing DNA damage response

Leishmania, which is present in macrophages, is subjected to nitrosative and oxidative bursts that induce a variety of DNA damage. By modifying metabolic flux and stress response pathways, host immunological signaling—specifically involving kinases like mTOR and MAPK—can affect the parasite's DDR, according to current transcriptome and phosphoproteomic research (Nandan et al., 2024). Furthermore, chromatin remodeling and parasite epigenetics may be altered by host-pathogen interaction, which may have an effect on DNA accessibility and repair fidelity. Co-infection models and dual RNA-Seq profiling approaches are currently being employed to better understand bidirectional interactions, which reveal that host microenvironments have a direct impact on the parasite's expression of the DDR gene. IFN-γ-activated macrophages, for example, can lower the expression of parasite RAD, increasing their sensitivity to DNA damage and modifying their mode of life (Rezaei et al., 2017; Engstler and Beneke, 2023; Sadr et al., 2025). Despite these improvements, there are still substantial gaps in separating parasite-intrinsic DDR pathways from host-indirectly controlled ones. It may be feasible to precisely identify these entangled events in both space and time by creating single-cell dual-omics approaches (such as scRNA-seq + ATAC-seq) (Karagiannis et al., 2023).

22.8.3 AI and machine learning applications to predict DNA damage response targets

The increased availability of high-throughput omics enables the application of AI and machine learning (ML) to decipher DDR networks and predict treatment targets in *Leishmania*. ML models have already been used to predict antileishmanial activity based on chemical descriptors of known compounds with good accuracy (AUC > 0.85), utilizing support vector machines (SVM), random forests, and deep learning architectures. More recently, omics layers have integrated data from AI-assisted genome-wide association studies (GWAS) and CRISPR screening to identify synthetic lethal connections involving DDR pathways. For example, important genes like RAD51, RPA1, and TOP1B have been identified as high-value targets by predictive algorithms that consider transcript expression, gene essentiality, and protein-protein interaction networks (Akpunarlieva et al., 2017; Sundar and Singh, 2018; Bussotti et al., 2021; Harigua-Souiai et al., 2022).

To more accurately predict DDR vulnerabilities, future AI tools will need to include spatial data (like subcellular localization), temporal modeling (like DDR dynamics during stage transitions), and multimodal input (like sequence, structure, expression, and metabolite flux). Among the challenges are cross-validation across species, integrating sparse or noisy *Leishmania* omics datasets, and guaranteeing model interpretability (Sundar and Singh, 2018; Harigua-Souiai et al., 2022; Wei et al., 2024).

The advancement of DDR research in *Leishmania* depends on pan-omics integration, host-parasite interface studies, and AI-driven target prediction. By combining ML with high-quality multilayered data, the field can identify key DDR pathways, model resilience mechanisms, and prioritize new therapeutic targets. This collaborative, data-intensive strategy is intended to usher in a new era of precision antiparasitic techniques that are more intelligent, targeted, and resistant.

22.9 Conclusion

This broad examination of the structural and dynamic indicators of DNA damage in *Leishmania* sheds light on the role of genomic stability and plasticity in parasite survival, treatment resistance, and virulence. *Leishmania* is distinct from classical eukaryotic systems in that it has developed a highly specific and adaptive DDR apparatus as a result of the unique pressure of its digenetic life cycle, frequent exposure to host-derived oxidative and nitrosative stress, and the requirement to withstand chemotherapeutic insults. Constitutive mosaic aneuploidy, an unusually tolerant replication machinery, and an over-reliance on HR, BER, and MMEJ in the absence of NHEJ are some examples of these adaptations.

Exogenous pressures include host-derived ROS, UV radiation, and chemotherapy, as well as endogenous variables such as oxidative metabolism and replication faults, which result in a constant burden of DNA damage. The parasite's ability to endure and adjust to these hurdles is further boosted by particular features such as lesion bypass polymerases, histone variant dynamics, and repair activities in specialized genomic regions, including telomeres and kinetoplastid DNA.

DDR mechanisms have emerged as major contributors to virulence and drug resistance, particularly RAD51-mediated gene amplification and chromosomal rearrangements. Recent breakthroughs in CRISPR-Cas9 editing, multiomics integration, and AI-based modeling have made it possible to control and therapeutically target these repair systems employing powerful techniques.

In summary, the DNA damage induced by *Leishmania* provides a problem as well as an opportunity, encouraging flexibility and detecting flaws. By focusing on parasite-specific DDR pathways, there is a considerable likelihood that more effective and robust treatments will be generated.

References

Afrin, F., Khan, I., Hemeg, H.A., 2019. Leishmania-host interactions—an epigenetic paradigm. Front. Immunol. 10, 492. https://doi.org/10.3389/fimmu.2019.00492.

Akpunarlieva, S., Weidt, S., Lamasudin, D., Naula, C., Henderson, D., Barrett, M., Burgess, K., Burchmore, R., 2017. Integration of proteomics and metabolomics to elucidate metabolic adaptation in Leishmania. J. Proteomics 155, 85–98. https://doi.org/10.1016/j.jprot.2016.12.009.

Aksenova, A.Y., Mirkin, S.M., 2019. At the beginning of the end and in the middle of the beginning: structure and maintenance of telomeric DNA repeats and interstitial telomeric sequences. Genes 10, 118. https://doi.org/10.3390/genes10020118.

Amodeo, S., Bregy, I., Ochsenreiter, T., 2023. Mitochondrial genome maintenance—the kinetoplast story. FEMS Microbiol. Rev. 47, fuac047. https://doi.org/10.1093/femsre/fuac047.

Anuntasomboon, P., Siripattanapipong, S., Unajak, S., Choowongkomon, K., Burchmore, R., Leelayoova, S., et al., 2022. Identification of a conserved maxicircle and unique minicircles as part of the mitochondrial genome of Leishmania martiniquensis strain PCM3 in Thailand. Parasit. Vectors 15, 459. https://doi.org/10.1186/s13071-022-05592-1.

Argelaguet, R., Arnol, D., Bredikhin, D., Deloro, Y., Velten, B., Marioni, J.C., Stegle, O., 2020. MOFA+: a statistical framework for comprehensive integration of multi-modal single-cell data. Genome Biol 21, 111. https://doi.org/10.1186/s13059-020-02015-1.

Barman, M., Dandasena, D., Suresh, A., Bhandari, V., Kamble, S., Singh, S., Subudhi, M., Sharma, P., 2023. Artemisinin derivatives induce oxidative stress leading to DNA damage and caspase-mediated apoptosis in Theileria annulata-transformed cells. Cell Commun. Signal. 21, 78. https://doi.org/10.1186/s12964-023-01067-7.

Bharadava, K., Upadhyay, T.K., Kaushal, R.S., Ahmad, I., Alraey, Y., Siddiqui, S., et al., 2024. Genomic insight of leishmania parasite: in-depth review of drug resistance mechanisms and genetic mutations. ACS Omega acsomega 3, c09400. https://doi.org/10.1021/acsomega.3c09400.

Bhaskara, S., 2015. Histone deacetylases 1 and 2 regulate DNA replication and DNA repair: potential targets for genome stability-mechanism-based therapeutics for a subset of cancers. Cell Cycle 14, 1779–1785. https://doi.org/10.1080/15384101.2015.1042634.

Bhusal, C.K., Sinha, S., Kaur, D., Sehgal, R., 2025. Unravelling drug resistance in leishmaniasis: genomic adaptations and emerging therapies. Front. Mol. Biosci. 12, 1573618. https://doi.org/10.3389/fmolb.2025.1573618.

Biswas, H., Makinwa, Y., Zou, Y., 2023. Novel cellular functions of ATR for therapeutic targeting: embryogenesis to tumorigenesis. Int. J. Mol. Sci. 24, 11684. https://doi.org/10.3390/ijms241411684.

Black, J.A., Reis-Cunha, J.L., Cruz, Angela.K., Tosi, Luiz.R.O., 2023. Life in plastic, it's fantastic! how leishmania exploit genome instability to shape gene expression. Front. Cell. Infect. Microbiol. 13, 1102462. https://doi.org/10.3389/fcimb.2023.1102462.

Burrows, A.E., Elledge, S.J., 2008. How ATR turns on: TopBP1 goes on ATRIP with ATR: Figure 1. Genes Dev 22, 1416–1421. https://doi.org/10.1101/gad.1685108.

Bussotti, G., Gouzelou, E., Côrtes Boité, M., Kherachi, I., Harrat, Z., Eddaikra, N., et al., 2018. Leishmania genome dynamics during environmental adaptation reveal strain-specific differences in gene copy number variation, karyotype instability, and telomeric Amplification. mBio 9, e01399-18. https://doi.org/10.1128/mBio.01399-18.

Bussotti, G., Li, B., Pescher, P., Vojtkova, B., Louradour, I., Pruzinova, K., et al., 2023. Leishmania allelic selection during experimental sand fly infection correlates with mutational signatures of oxidative DNA damage. Proc. Natl. Acad. Sci 120, e2220828120. https://doi.org/10.1073/pnas.2220828120.

Bussotti, G., Piel, L., Pescher, P., Domagalska, M.A., Rajan, K.S., Cohen-Chalamish, S., et al., 2021. Genome instability drives epistatic adaptation in the human pathogen Leishmania. Proc. Natl. Acad. Sci. 118, e2113744118. https://doi.org/10.1073/pnas.2113744118.

Camacho, E., Rastrojo, A., Sanchiz, Á., González-de La Fuente, S., Aguado, B., Requena, J.M., 2019. Leishmania mitochondrial genomes: maxicircle structure and heterogeneity of minicircles. Genes 10, 758. https://doi.org/10.3390/genes10100758.

Chatterjee, N., Walker, G.C., 2017. Mechanisms of DNA damage, repair, and mutagenesis. Environ. Mol. Mutagen. 58, 235–263. https://doi.org/10.1002/em.22087.

Da Silva, G.L.A., Damasceno, J.D., Black, J.A., McCulloch, R., Tosi, L.R.O., 2025. ATR, a DNA damage kinase, modulates DNA replication timing in Leishmania major. https://doi.org/10.1101/2025.01.13.632695

Da Silva, M.S., 2021. DNA double-strand breaks: a double-edged sword for trypanosomatids. Front. Cell Dev. Biol. 9, 669041. https://doi.org/10.3389/fcell.2021.669041.

Da Silva Oliveira, G.L., De Freitas, R.M., 2015. Diminazene aceturate—an antiparasitic drug of antiquity: advances in pharmacology & therapeutics. Pharmacol. Res. 102, 138–157. https://doi.org/10.1016/j.phrs.2015.10.005.

Da Silva, R.B., Machado, C.R., Rodrigues, A.R.A., Pedrosa, A.L., 2018. Selective human inhibitors of ATR and ATM render Leishmania major promastigotes sensitive to oxidative damage. PLoS One 13, e0205033. https://doi.org/10.1371/journal.pone.0205033.

Da Silveira, R.D.C.V., Da Silva, M.S., Nunes, V.S., Perez, A.M., Cano, M.I.N., 2013. The natural absence of RPA1N domain did not impair Leishmania amazonensis RPA-1 participation in DNA damage response and telomere protection. Parasitology 140, 547–559. https://doi.org/10.1017/S0031182012002028.

Dacher, M., Tachiwana, H., Horikoshi, N., Kujirai, T., Taguchi, H., Kimura, H., Kurumizaka, H., 2019. Incorporation and influence of Leishmania histone H3 in chromatin. Nucleic Acids Res 47, 11637–11648. https://doi.org/10.1093/nar/gkz1040.

Duncan, S.M., Jones, N.G., Mottram, J.C., 2017. Recent advances in leishmania reverse genetics: manipulating a manipulative parasite. Mol. Biochem. Parasitol. 216, 30–38. https://doi.org/10.1016/j.molbiopara.2017.06.005.

Eddaikra, A., Eddaikra, N., 2021. Endogenous enzymatic antioxidant defense and pathologies. In: Waisundara, V. (Ed.), Antioxidants - benefits, sources, mechanisms of action. IntechOpen. https://doi.org/10.5772/intechopen.95504.

Elmahallawy, E.K., Alkhaldi, A.A.M., 2021. Insights into leishmania molecules and their potential contribution to the virulence of the parasite. Vet. Sci. 8, 33. https://doi.org/10.3390/vetsci8020033.

Engstler, M., Beneke, T., 2023. Gene editing and scalable functional genomic screening in Leishmania species using the CRISPR/Cas9 cytosine base editor toolbox LeishBASEedit. eLife 12, e85605. https://doi.org/10.7554/eLife.85605.

Farrell, A.W., Halliday, G.M., Lyons, J.G., 2011. Chromatin structure following UV-induced DNA damage—repair or death? Int. J. Mol. Sci. 12, 8063–8085. https://doi.org/10.3390/ijms12118063.

Fernández-Orgiler, A., Martínez-Jiménez, M.I., Alonso, A., Alcolea, P.J., Requena, J.M., Thomas, M.C., et al., 2016. A putative Leishmania DNA polymerase theta protects the parasite against oxidative damage. Nucleic Acids Res 44, 4855–4870. https://doi.org/10.1093/nar/gkw346.

Ferrand, J., Rondinelli, B., Polo, S.E., 2020. Histone Variants: Guardians of Genome Integrity. Cells 9, 2424. https://doi.org/10.3390/cells9112424.

Florêncio, M., Cupolillo, E., Boité, M.C., 2025. Understanding the genetic complexity of Leishmania infantum in the Americas: a focus on 3′NT/NU gene deletion. Mem. Inst. Oswaldo Cruz 120, e240160. https://doi.org/10.1590/0074-02760240160.

Furtado, C., Kunrath-Lima, M., Rajão, M.A., Mendes, I.C., De Moura, M.B., Campos, P.C., et al., 2012. Functional Characterization of 8-Oxoguanine DNA Glycosylase of Trypanosoma cruzi. PLoS One 7, e42484. https://doi.org/10.1371/journal.pone.0042484.

Garcia, J.B.F., Rocha, J.P.V.D., Costa-Silva, H.M., Alves, C.L., Machado, C.R., Cruz, A.K., 2016. Leishmania major and Trypanosoma cruzi present distinct DNA damage responses. Mol. Biochem. Parasitol. 207, 23–32. https://doi.org/10.1016/j.molbiopara.2016.05.004.

Genois, M.-M., Paquet, E.R., Laffitte, M.-C.N., Maity, R., Rodrigue, A., Ouellette, M., et al., 2014. DNA Repair Pathways in Trypanosomatids: from DNA Repair to Drug Resistance. Microbiol. Mol. Biol. Rev. 78, 40–73. https://doi.org/10.1128/MMBR.00045-13.

Gorodetska, I., Kozeretska, I., Dubrovska, A., 2019. BRCA Genes: The Role in Genome Stability, Cancer Stemness and Therapy Resistance. J. Cancer 10, 2109–2127. https://doi.org/10.7150/jca.30410.

Harigua-Souiai, E., Oualha, R., Souiai, O., Abdeljaoued-Tej, I., Guizani, I., 2022. Applied machine learning toward drug discovery enhancement: leishmaniases as a case study. Bioinforma. Biol. Insights 16, 11779322221090349. https://doi.org/10.1177/11779322221090349.

Hauer, M.H., Gasser, S.M., 2017. Chromatin and nucleosome dynamics in DNA damage and repair. Genes Dev 31, 2204–2221. https://doi.org/10.1101/gad.307702.117.

Herrmann May, N., Cao, A., Schmid, A., Link, F., Arias-del-Angel, J., Meiser, E., Beneke, T., 2025. Improved base editing and functional screening in Leishmania via co-expression of the AsCas12a ultra variant, a T7 RNA polymerase, and a cytosine base editor. eLife 13, RP97437. https://doi.org/10.7554/eLife.97437.

Hoeijmakers, J.H.J., 2009. DNA damage, aging, and cancer. N. Engl. J. Med. 361, 1475–1485. https://doi.org/10.1056/NEJMra0804615.

Jha, P.K., Khan, Mohd.I., Mishra, A., Das, P., Sinha, K.K., 2017. HAT2 mediates histone H4K4 acetylation and affects micrococcal nuclease sensitivity of chromatin in Leishmania donovani. PLoS One 12, e0177372. https://doi.org/10.1371/journal.pone.0177372.

Kamran, Mohd., Bhattacharjee, R., Das, S., Mukherjee, S., Ali, N., 2023. The paradigm of intracellular parasite survival and drug resistance in leishmanial parasite through genome plasticity and epigenetics: perception and future perspective. Front. Cell. Infect. Microbiol. 13, 1001973. https://doi.org/10.3389/fcimb.2023.1001973.

Karagiannis, K., Gannavaram, S., Verma, C., Pacheco-Fernandez, T., Bhattacharya, P., Nakhasi, H.L., et al., 2023. Dual-scRNA-seq analysis reveals rare and uncommon parasitized cell populations in chronic L. donovani infection. Cell Rep 42, 113097. https://doi.org/10.1016/j.celrep.2023.113097.

Kaur, G., Rajput, B., 2014. Comparative analysis of the omics technologies used to study antimonial, amphotericin b, and pentamidine resistance in leishmania. J. Parasitol. Res. 2014, 1–11. https://doi.org/10.1155/2014/726328.

Kaushal, R.S., Kaushik, A., Prabha, C.R., 2025. Intricate role of reactive oxygen species and reactive nitrogen species in leishmaniasis: mechanisms, implications, and therapeutic potential. In: Prabhakar, P.K. (Ed.), Advances in medical diagnosis, treatment, and care. IGI Global, pp. 269–310.

Kaushik, A., Makarani, N., Bharadava, K., Gehlot, J., Naik, B.V., Singh, A., et al., 2025. Antiprotozoal agents – Integration of drug discovery, medicinal chemistry, and advanced computational approaches: an in-depth review. The Microbe 7, 100395. https://doi.org/10.1016/j.microb.2025.100395.

Khan, M.I., Mishra, A., Jha, P.K., Abhishek, K., Chaba, R., Das, P., et al., 2019. DNA polymerase β of Leishmania donovani is important for infectivity and it protects the parasite against oxidative damage. Int. J. Biol. Macromol. 124, 291–303. https://doi.org/10.1016/j.ijbiomac.2018.11.159.

Klebanov-Akopyan, O., Mishra, A., Glousker, G., Tzfati, Y., Shlomai, J., 2018. Trypanosoma brucei UMSBP2 is a single-stranded telomeric DNA binding protein essential for chromosome end protection. Nucleic Acids Res 46, 7757–7771. https://doi.org/10.1093/nar/gky597.

Kojima, Y., Machida, Y.J., 2020. DNA–protein crosslinks from environmental exposure: Mechanisms of formation and repair. Environ. Mol. Mutagen. 61, 716–729. https://doi.org/10.1002/em.22381.

Laffitte, M.-C.N., Leprohon, P., Papadopoulou, B., Ouellette, M., 2016. Plasticity of the Leishmania genome leading to gene copy number variations and drug resistance. F1000Research 5, 2350. https://doi.org/10.12688/f1000research.9218.1.

López-Camarillo, C., Lopez-Casamichana, M., Weber, C., Guillen, N., Orozco, E., Marchat, L.A., 2009. DNA repair mechanisms in eukaryotes: Special focus in Entamoeba histolytica and related protozoan parasites. Infect. Genet. Evol. 9, 1051–1056. https://doi.org/10.1016/j.meegid.2009.06.024.

Maiorano, D., El Etri, J., Franchet, C., Hoffmann, J.-S., 2021. Translesion synthesis or repair by specialized DNA polymerases limits excessive genomic instability upon replication stress. Int. J. Mol. Sci. 22, 3924. https://doi.org/10.3390/ijms22083924.

Makarani, N., Bharadava, K., Kaushik, A., Dave, A., Gangawane, A.K., Kaushal, R.S., 2025. Leishmaniasis: a multifaceted approach to diagnosis, maladies, drug repurposing and way forward. The Microbe, 100239. https://doi.org/10.1016/j.microb.2025.100239.

Marin, P.A., Obonaga, R., Pavani, R.S., Da Silva, M.S., De Araujo, C.B., Lima, A.A., et al., 2020. ATR kinase is a crucial player mediating the DNA damage response in Trypanosoma brucei. Front. Cell Dev. Biol. 8, 602956. https://doi.org/10.3389/fcell.2020.602956.

Mavroeidi, D., Georganta, A., Panagiotou, E., Syrigos, K., Souliotis, V.L., 2024. Targeting ATR pathway in solid tumors: evidence of improving therapeutic outcomes. Int. J. Mol. Sci. 25, 2767. https://doi.org/10.3390/ijms25052767.

McDonald, J.R., Jensen, B.C., Sur, A., Wong, I.L.K., Beverley, S.M., Myler, P.J., 2022. Localization of epigenetic markers in Leishmania chromatin. Pathogens 11, 930. https://doi.org/10.3390/pathogens11080930.

Mishra, A., Khan, Mohd.I., Jha, P.K., Kumar, A., Das, S., Das, Prolay, Das, Pradeep, Sinha, K.K., 2018. Oxidative stress-mediated overexpression of uracil DNA glycosylase in leishmania donovani confers tolerance against antileishmanial drugs. Oxid. Med. Cell. Longev. 2018, 4074357. https://doi.org/10.1155/2018/4074357.

Moncada-Diaz, M.J., Rodríguez-Almonacid, C.C., Quiceno-Giraldo, E., Khuong, F.T.H., Muskus, C., Karamysheva, Z.N., 2024. Molecular mechanisms of drug resistance in Leishmania spp. Pathogens 13, 835. https://doi.org/10.3390/pathogens13100835.

Mordes, D.A., Cortez, D., 2008. Activation of ATR and related PIKKs. Cell Cycle 7, 2809–2812. https://doi.org/10.4161/cc.7.18.6689.

Moreira, V.R., De Jesus, L.C.L., Soares, R.-E.P., Silva, L.D.M., Pinto, B.A.S., Melo, M.N., et al., 2017. Meglumine antimoniate (glucantime) causes oxidative stress-derived DNA damage in BALB/c mice infected by Leishmania (Leishmania) infantum. Antimicrob. Agents Chemother. 61, e02360-16. https://doi.org/10.1128/AAC.02360-16.

Mullenders, L.H.F., 2018. Solar UV damage to cellular DNA: from mechanisms to biological effects. Photochem. Photobiol. Sci. 17, 1842–1852. https://doi.org/10.1039/c8pp00182k.

Nandan, D., Brar, H.K., Reiner, N., 2024. Manipulation of macrophages: emerging mechanisms of Leishmaniasis. Front. Biosci.-Landmark 29, 292. https://doi.org/10.31083/j.fbl2908292.

Nickoloff, J.A., Sharma, N., Taylor, L., 2020. Clustered DNA double-strand breaks: biological effects and relevance to cancer radiotherapy. Genes 11, 99. https://doi.org/10.3390/genes11010099.

Orhan, E., Velazquez, C., Tabet, I., Sardet, C., Theillet, C., 2021. Regulation of RAD51 at the transcriptional and functional levels: what prospects for cancer therapy? Cancers 13, 2930. https://doi.org/10.3390/cancers13122930.

Paris, C., Loiseau, P.M., Bories, C., Bréard, J., 2004. Miltefosine induces apoptosis-like death in Leishmania donovani Promastigotes. Antimicrob. Agents Chemother. 48, 852–859. https://doi.org/10.1128/AAC.48.3.852-859.2004.

Passos-Silva, D.G., Rajão, M.A., Nascimento De Aguiar, P.H., Vieira-da-Rocha, J.P., Machado, C.R., Furtado, C., 2010. Overview of DNA repair in Trypanosoma cruzi, Trypanosoma brucei, and Leishmania major. J. Nucleic Acids 2010, 840768. https://doi.org/10.4061/2010/840768.

Perkel, J.M., 2020. The software that powers scientific illustration. Nature 582 (7810), 137–138. https://doi.org/10.1038/d41586-020-01404-7.

Poetsch, A.R., Boulton, S.J., Luscombe, N.M., 2018. Genomic landscape of oxidative DNA damage and repair reveals regioselective protection from mutagenesis. Genome Biol 19, 215. https://doi.org/10.1186/s13059-018-1582-2.

Poláková, E., Albanaz, A.T.S., Zakharova, A., Novozhilova, T.S., Gerasimov, E.S., Yurchenko, V., 2021. Ku80 is involved in telomere maintenance but dispensable for genomic stability in Leishmania mexicana. PLoS Negl. Trop. Dis. 15, e0010041. https://doi.org/10.1371/journal.pntd.0010041.

Poveda, A., Méndez, M.Á., Armijos-Jaramillo, V., 2020. Analysis of DNA polymerases reveals specific gene expansion in Leishmania and Trypanosoma spp. Front. Cell. Infect. Microbiol. 10, 570493. https://doi.org/10.3389/fcimb.2020.570493.

Queffeulou, M., Leprohon, P., Fernandez-Prada, C., Ouellette, M., Mejía-Jaramillo, A.M., 2024. CRISPR-Cas9 high-throughput screening to study drug resistance in Leishmania infantum. mBio 15, e00477-24. https://doi.org/10.1128/mbio.00477-24.

Rachidi, N., Taly, J.F., Durieu, E., Leclercq, O., Aulner, N., Prina, E., et al., 2014. Pharmacological assessment defines Leishmania donovani casein kinase 1 as a drug target and reveals important functions in parasite viability and intracellular infection. Antimicrob. Agents Chemother. 58, 1501–1515. https://doi.org/10.1128/AAC.02022-13.

Ramasamy, K., Shanmugam, M., Balupillai, A., Govindhasamy, K., Gunaseelan, S., Muthusamy, G., et al., 2017. Ultraviolet radiation-induced carcinogenesis: Mechanisms and experimental models. J. Radiat. Cancer Res. 8, 4. https://doi.org/10.4103/0973-0168.199301.

Rezaei, R., Hazrati Tappeh, K., Seyyedi, S., Mikaili, P., 2017. The Anti-leishmanial Efficacy of Artemisia dracunculus Ethanolic Extract in Vitro and Its Effects on IFN-γ and IL-4 Response. Iran. J. Parasitol. 12, 398–407.

Rohart, F., Gautier, B., Singh, A., Lê Cao, K.-A., 2017. mixOmics: An R package for 'omics feature selection and multiple data integration. PLOS Comput. Biol. 13, e1005752. https://doi.org/10.1371/journal.pcbi.1005752.

Rosa-Teijeiro, C., Wagner, V., Corbeil, A., d'Annessa, I., Leprohon, P., Do Monte-Neto, R.L., et al., 2021. Three different mutations in the DNA topoisomerase 1B in Leishmania infantum contribute to resistance to the antitumor drug topotecan. Parasit. Vectors 14, 438. https://doi.org/10.1186/s13071-021-04947-4.

Roshanara, R., Tandon, R., Puri, N., Selvapandiyan, A., 2024. Interaction between LdCen1 and LdDRP in Leishmania donovani is crucial for survival in host by facilitating the parasite's UV-induced DNA damage repair. https://doi.org/10.21203/rs.3.rs-5118904/v1

Sadr, S., Sharifi, I., Morovati, S., Sepahvand, H., Nazemian, S., Bamorovat, M., Rezaeian, Z., Akhtardanesh, B., 2025. An inclusive assessment of apoptosis mechanisms in Leishmania species: a narrative literature review. Curr. Res. Parasitol. Vector-Borne Dis. 7, 100260. https://doi.org/10.1016/j.crpvbd.2025.100260.

Saha, S., 2020. Histone modifications and other facets of epigenetic regulation in trypanosomatids: leaving their mark. mBio 11, e01079-20. https://doi.org/10.1128/mBio.01079-20.

Santi, A.M.M., Murta, S.M.F., 2022. Impact of genetic diversity and genome plasticity of leishmania spp. in treatment and the search for novel chemotherapeutic targets. Front. Cell. Infect. Microbiol. 12, 826287. https://doi.org/10.3389/fcimb.2022.826287.

Sarkar, S., Gaddameedhi, S., 2020. Solar ultraviolet-induced DNA damage response: melanocytes story in transformation to environmental melanomagenesis. Environ. Mol. Mutagen. 61, 736–751. https://doi.org/10.1002/em.22370.

Silva, G.L.A., Tosi, L.R.O., McCulloch, R., Black, J.A., 2021. Unpicking the roles of DNA damage protein kinases in trypanosomatids. Front. Cell Dev. Biol. 9, 636615. https://doi.org/10.3389/fcell.2021.636615.

Sudhandiran, G., Shaha, C., 2003. Antimonial-induced increase in intracellular Ca2+ through non-selective cation channels in the host and the parasite is responsible for apoptosis of intracellular Leishmania donovani amastigotes. J. Biol. Chem. 278, 25120–25132. https://doi.org/10.1074/jbc.M301975200.

Sundar, S., Singh, B., 2018. Understanding Leishmania parasites through proteomics and implications for the clinic. Expert Rev. Proteomics 15, 371–390. https://doi.org/10.1080/14789450.2018.1468754.

Taladriz, S., 2001. Nuclear DNA polymerase beta from Leishmania infantum. cloning, molecular analysis and developmental regulation. Nucleic Acids Res 29, 3822–3834. https://doi.org/10.1093/nar/29.18.3822.

Talimi, H., Daoui, O., Bussotti, G., Mhaidi, I., Boland, A., Deleuze, J.-F., Fissoune, R., Lemrani, M.F., Späth, G., 2024. A comparative genomics approach reveals a local genetic signature of Leishmania tropica in Morocco. Microb. Genomics 10. https://doi.org/10.1099/mgen.0.001230.

Verma, N.K., Singh, G., Dey, C.S., 2007. Miltefosine induces apoptosis in arsenite-resistant Leishmania donovani promastigotes through mitochondrial dysfunction. Exp. Parasitol. 116, 1–13. https://doi.org/10.1016/j.exppara.2006.10.007.

Wassing, I.E., Esashi, F., 2021. RAD51: Beyond the break. Semin. Cell Dev. Biol. 113, 38–46. https://doi.org/10.1016/j.semcdb.2020.08.010.

Wei, L., Barrie, U., Aloisio, G.M., Khuong, F.T.H., Arang, N., Datta, A., Kaushansky, A., Wetzel, D.M., 2024. Using machine learning to dissect host kinases required for Leishmania internalization and development. Mol. Biochem. Parasitol. 260, 111651. https://doi.org/10.1016/j.molbiopara.2024.111651.

Wijnant, G.-J., Dumetz, F., Dirkx, L., Bulté, D., Cuypers, B., Van Bocxlaer, K., Hendrickx, S., 2022. Tackling drug resistance and other causes of treatment failure in Leishmaniasis. Front. Trop. Dis 3, 837460. https://doi.org/10.3389/fitd.2022.837460.

Yousefzadeh, M., Henpita, C., Vyas, R., Soto-Palma, C., Robbins, P., Niedernhofer, L., 2021. DNA damage—how and why we age? eLife 10, e62852. https://doi.org/10.7554/eLife.62852.

Zhang, W.-W., Matlashewski, G., 2019. Single-strand annealing plays a major role in double-strand DNA break repair following CRISPR-Cas9 cleavage in Leishmania. mSphere 4, e00408–e00419. https://doi.org/10.1128/mSphere.00408-19.

Zhang, W.-W., Wright, D.G., Harrison, L., Matlashewski, G., 2022. Reconstitution of mycobacterium marinum non-homologous DNA end joining pathway in Leishmania. mSphere 7, e00156-22. https://doi.org/10.1128/msphere.00156-22.

Zheng, Z., Chen, J., Ma, G., Satoskar, A.R., Li, J., 2020. Integrative genomic, proteomic and phenotypic studies of Leishmania donovani strains revealed genetic features associated with virulence and antimony-resistance. Parasit. Vectors 13, 510. https://doi.org/10.1186/s13071-020-04397-4.

Zhu, Y., Wang, Y., Chen, G., 2009. Differences in conformational dynamics of [Pt3(HPTAB)]6+-DNA adducts with various cross-linking modes. Nucleic Acids Res 37, 5930–5942. https://doi.org/10.1093/nar/gkp618.

Zou, L., 2017. DNA replication checkpoint: new atr activator identified. Curr. Biol 27, R33–R35. https://doi.org/10.1016/j.cub.2016.11.025.

CHAPTER

23

Impacts of nanoparticles on DNA damage and repair

Amol Gholap[a], *Pankaj Khuspe*[b], *Dipali Mane*[c] *and Md Jasim Uddin*[d]

[a]Department of Pharmaceutics, St. John Institute of Pharmacy and Research, Palghar, Maharashtra, India [b]Department of Pharmaceutics, Shriram Shikshan Sanstha's College of Pharmacy, Paniv, Solapur, Maharashtra, India [c]Department of Pharmacology, Shriram Shikshan Sanstha's College of Pharmacy, Paniv, Solapur, Maharashtra, India [d]Department of Pharmaceutical Technology, Faculty of Pharmacy, Universiti Malaya, Kuala Lumpur, Malaysia

23.1 Introduction

23.1.1 Nanoparticles background

Nanoparticles (NPs) are particulate structures having one or more dimensions that fall within the nanometer range, usually between 1 and 100 nm. At this magnitude, the matter has special physicochemical features and special characteristics, namely the high surface-to-volume ratios, changed electronic configuration, and quantum size effect. Depending on composition and structure, NPs may be broadly categorized in several groups: Metallic NPs (e.g., gold, silver, and platinum), metal oxide NP (e.g., TiO_2, ZnO, and Fe_3O_4), polymeric NPs (e.g., polylactic-co-glycolic acid and polyethylene glycol-based systems), carbon-based nanomaterials (e.g., fullerenes, carbon nanotubes, and graphene), and quantum dots (semiconductor nanocrystals such as CdSe and PbS) (Wang et al., 2024; Santonastaso et al., 2025).

The unique qualities on which they are based support a growing diversity of uses. NPs are being exploited in medicine as a controlled delivery system (NPs as drug formulation), diagnostic imaging (NPs as contrast agent), and theranostic (NPs as an agent) platform. They increase the ultraviolet (UV) blocking properties and skin penetration and formulation stability in cosmetics, as presented in Fig. 23.1. They are incorporated in the automotive industry, coatings, catalysts, and electronics. Potential environmental uses are in the treatment of water, destruction of pollutants, and agricultural output improvement (Joseph et al., 2023). These technological advantages are, nevertheless, contrasted with growing interest in their safety.

Genome Damage: Repair and Adaptation Across Life.
DOI: https://doi.org/10.1016/B978-0-443-42857-9.00019-0

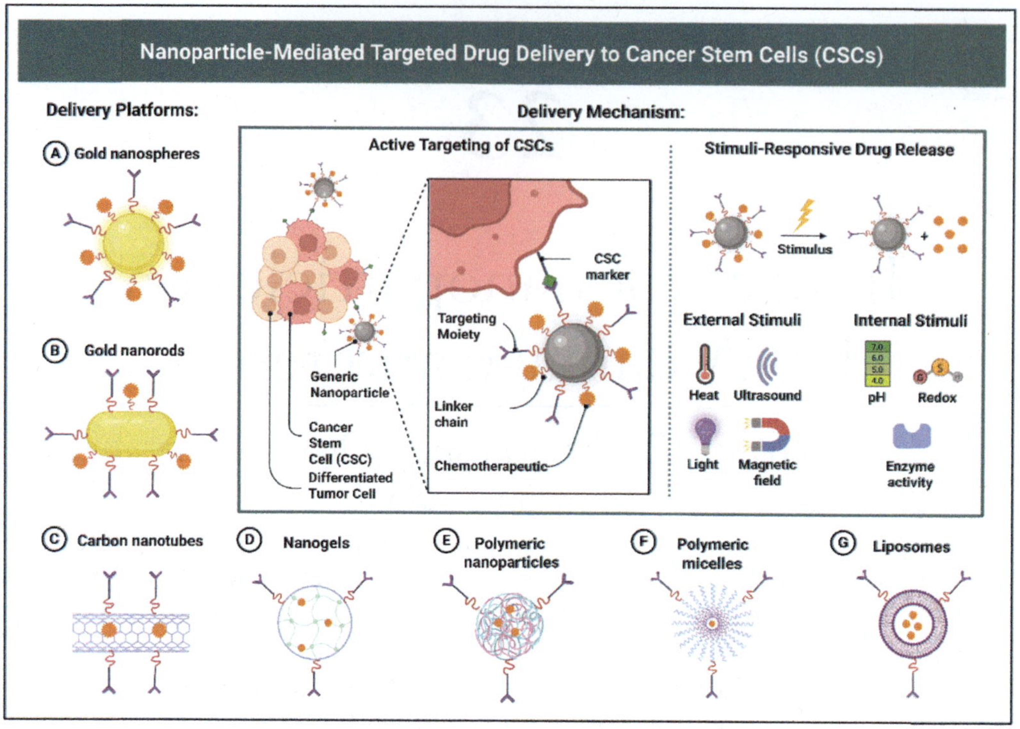

FIGURE 23.1 Figure shows a schematic representation of nanoparticle-mediated targeted drug delivery to cancer stem cells (CSCs). Various nanocarrier platforms, including (A) gold nanospheres, (B) gold nanorods, (C) carbon nanotubes, (D) nanogels, (E) polymeric nanoparticles, (F) polymeric micelles, and (G) liposomes, are employed to enhance drug delivery. The key panel depicts direct CSCs targeting with nanoparticles attached with targeting molecules that identify CSC markers, and conjugated with chemotherapeutics to deliver the nanoparticles selectively. On the right, stimuli-responsive drug release is depicted, triggered by external (heat, ultrasound, light, magnetic field) or internal (pH, redox, and enzyme activity) stimuli, ensuring spatiotemporal control and minimizing off-target effects in cancer therapy. (Created by using BioRender.com).

Small size/high reactivity, the biological stability of some of the NPs allows them to reach physiological barriers, or enter an interaction with subcellular structures, and arguably disrupt biomolecular activities. Such is the interplay between material science and biology that they require a multidisciplinary methodology to examine their usefulness as well as its implications on human and ecological wellness (Raheem et al., 2023).

23.1.2 Genotoxicity concerns

The possible genotoxic effect of NPs is also connected with their specific physicochemical properties that determine the manner of interaction of NPs with the cellular and molecular systems. Parameters like the size of the particle, surface charge, morphology, and the chemical composition, as well as solubility of the particle and surface functionalization, all play a crucial role in cellular uptake, cellular distribution, and reactivity. As an example, smaller NPs could enter the nuclear envelope, though higher aspect ratio objects such as carbon nanotubes may mechanically interfere with the structure of cellular membranes through to chromatin (Landsiedel et al., 2022).

After internalization, NPS may cause oxidative stress through the reactive oxygen species (ROS); catalize reactive redox reactions, directly interact with nucleic acids, and destabilize the genome. Certain oxides NPs, when kept in UV rays, are photocatalytically active and also augment the DNA damaging capacility (Khursheed et al., 2022; Xuan et al., 2023).

In addition, genotoxic potential may be modulated by surface coatings and functional groups, which themselves may either mitigate or enhance damage induced according to their chemical reactivity. These are especially meaningful interactions regarding public health. Exposure of NPs in human parameters can be multifactorial, i.e., occupational, environmental, and use of consumer products, which makes genotoxicity the decisive indicator of long-term human health consequences. In addition to interfering with the health of human beings, DNA modification of other environmental organisms caused by NP may interfere with the ecological balance and biodiversity. Hence, it is crucial that we learn more about NP-DNA interactions so that we can develop recommendations and guidelines that safeguard people (Siivola et al., 2022; Cao et al., 2023).

23.1.3 Importance of nanoparticle effects on DNA damage and repair

Analysis of DNA integrity is a key biological indicator of nitric oxide safety, since genomic integrity forms the basis of cell health, tissue, and organismic harmony. Disruption of the DNA damage-repair balance is capable of triggering mutagenesis, genomic instability, and changes in gene expression patterns leading to carcinogenesis and premature aging as well as hereditary diseases (Płuciennik et al., 2024). The resultant NP-inflicted DNA damage can take both direct and indirect forms, with the former being able to facilitate this by straightforward affinity with DNA strands or interfering with the process of replication, and the latter being the ability to cause damage through oxidative lesions produced by ROS, mitochondrial impairment, and epigenetic modifications. Cellular repair networks identified and modulated to repair such damage include base excision repair (BER), nucleotide excision repair (NER), mismatch repair (MMR), and pathways such as homologous recombinations (HR) and nonhomologous end joining (NHEJ) of broken covalent bonds in double strands. Nevertheless, such damage can be made worse by NP-mediated interference with the repair systems: this further increases mutational burden (Liu et al., 2024).

A study on the NP effect on DNA damage and repair has wide implications. In the topic of oncology, the understanding of NP interactions with the genome will aid in the process of nanotherapeutic construction, as well as in the mechanism of the NP carcinogenic risk. It helps with the advancement of the understanding of how chronic NP exposure can act to contribute to age-related genomic degeneration in gerontology. In addition, the response to DNA damage serves as a biomarker of risk assessment and toxicity, which can be employed as measurable endpoints in regulatory assessment of nanomaterials and nanotechnology consumer products, environmental exposure, and clinical diagnosis and applications. These insights facilitate the creation of less hazardous nanomaterials, in vitro and in vivo predictive-testing approaches, and evidence-based regulatory decisions to make technological advances consistent with the necessities of biosafety (Xuan et al., 2023).

23.2 Physicochemical properties of nanoparticles relevant to DNA interactions

23.2.1 Size and surface area effects

When it comes to the biological interactions, that is, the possibility of DNA damage and interference of DNA repair, which is of critical importance in NPs, the nanoscale dimension plays a vital role. The motion of particles in the 1000 nm to 100 nm range will have an increased rate of cellular entry through a clathrin-mediated mechanism, endocytosis, caveolae-mediated endocytosis, or piercing the membrane. The smaller size enables endosomal escape and localization into the nucleus, which is connected to an increased potential of direct genomic interaction (Peng et al., 2025). The smaller NPs would possess a considerable contribution of the surface to volume ratio, which considerably heightens the reactivity of smaller sizes and facilitates the production of the ROS and the catalytic effects with the nucleic acids. In the case of quantum dots (<10 nm) and ultrasmall metal oxides have been discovered capable of reaching the nuclear compartment more effectively, thus increasing the probability of breaking the DNA strands and base alterations. In addition, the small NPs effect goes further to DNA repair regulation- the low amount of NPs can sterically block the binding of repair proteins or sequester cofactors necessary to repair processes, including BER or NER. On the other hand, when NPs are larger, >50 nm, they can be compartmentalized to the interior of the cell, resulting in indirect genotoxicity due to oxidative stress instead of a direct binding to the DNA. On balance it has been seen that the interaction of NP with DNA does not only depend on the virulence of the encounter in a given NP but also the mechanism of the reaction is solely defined by the interaction between the NP surface area and the NP size itself and hence optimization of size of the NPs is an essential variable in nanotoxicological risk assessment (Ewii et al., 2025).

23.2.2 Shape and morphology

There are highly elevated preferences for the NPs geometry that affect their interaction pattern with cells and nuclear structures. NPs appear with isotropic uptake and a relatively equal distribution inside the cells in most cases, whereas rod-shaped particles have an increased stickiness to the membranes but are slow in internalization as the entry is also orientation-dependent. Mechanical stress caused by the high-aspect-ratio of nanorods or nanowires may also attract cell membranes and even chromatin in anisotropic binding domains (Harun-Ur-Rashid et al., 2023). The large interaction surfaces of two-dimensional nanostructures, such as a graphene oxide sheet, allow binding with DNA backbones, which potentially can affect transcription or replication fidelity, through multivalent binding. Endosomal escape efficiencies and intracellular trafficking are also determined by morphological features, and the proximity of nuclear DNA is affected. An example here is the case of the tubular carbon nanotubes that can translocate or move toward the cytoskeleton and to the perinuclear region, raising the risk of direct contact with the genome. This means that morphology has a significant effect on size and surface chemistry in determining the outcome of NP-DNA interaction (Kazantseva et al., 2021).

23.2.3 Surface charge and functionalization

The surface charge has a significant influence on the NP-cell interactions, internalization kinetics, and DNA binding affinity. Nearly all NPs bearing a positive surface charge (amine-functionalized gold or silica NPs) are highly electrostatically attracted to both the cell membranes (anionic) and the phosphate backbone of the DNA (anionic). This means that positively charged NPs can be absorbed readily as well as directly embedded in the genome (Singh et al., 2024). The cationic charge, however, may elicit membrane damage and mitochondrial depolarization, causing the worsening of oxidative DNA damage. Neutral or negatively charged NPs show low cellular uptake in general, but might have genotoxic effects due to the production of ROS in the intracellular environment. Targeting ligands can be used to functionalize the NPs to achieve superior nuclear localization (e.g., NLS and peptides) and increase the chances of NP-direct contact with DNA. Also, the surface coating, including polyethylene glycol (PEG), can improve systemic circulation with community immunogenicity and aggregation behavior, which influence DNA exposure indirectly. Customization of surface charge and functionalities, thus becomes a defining factor in the therapeutic opportunity that balances with the genotoxic risk (Huang et al., 2021).

23.2.4 Composition and core chemistry

The core chemistry and elemental composition of the NPs determine physicochemical behaviors and interactivity with biological systems and their genotoxicity. Commonly used metallic NPs include silver (AgNPs) and gold (AuNPs), which have high electron density, which is associated with a unique plasmonic behavior that can lead to localized heating or electron transfer at the NP surface and damage DNA. AgNPs trigger silver ions with high antimicrobial and oxidative effects that could be transformed into oxidative DNA damage, such as lesions of 8-oxoguanine. AuNPs are normally believed to represent nongenotoxic controls but become genotoxic upon functionalization with reactive moieties or upon photoactivation (Dhaka et al., 2023). Transitional metal oxides like TiO_2, especially in the anatase phase, do serve as photocatalysts, forming ROS in the presence of UV light, which results in the formation of strand breaks and oxidation of bases. Nanomaterials constructed of carbon, comprised of carbon nanotubes or carbon derivative versions of graphene, interact with the DNA through a process known as π-π stacking and can interfere with the replication and repair enzymes' access. Additionally, reactivity can also be tuned via doping or alloying the cores, e.g., TiO_2-Fe has higher ROS production. Dissolution potential and catalytic activity of the NP core are intrinsic properties inherent to the chemical reactivity and determine, finally, whether genotoxicity can occur either through a direct chemical interaction with DNA or through secondary pathways of oxidative stress and repair pathway perturbation (Nain et al., 2025).

23.2.5 Solubility, aggregation, and dissolution

The aggregation and solubility of NP is also key factors affecting bioavailability and dynamic interactions of DNA. Some metal salts form highly soluble NPs that are capable of snappy ions liberation, which triggers a genotoxic issue for acute oxidative and alkylative pathways. Slowly soluble particles can languish in cellular compartments to produce chronic, low-level DNA stress. The process of aggregation minimizes the active surface area and cellular uptake, as well as enhancing localized toxicity when sedimentation occurs. Sustained release of ions and long-term genotoxic exposure depend on the dissolution kinetics of NPs, in the case of metal-based NPs. Aggregation and dispersibility control are therefore vital in reducing random DNA damage that causes variable biological response profiles (Altammar, 2023).

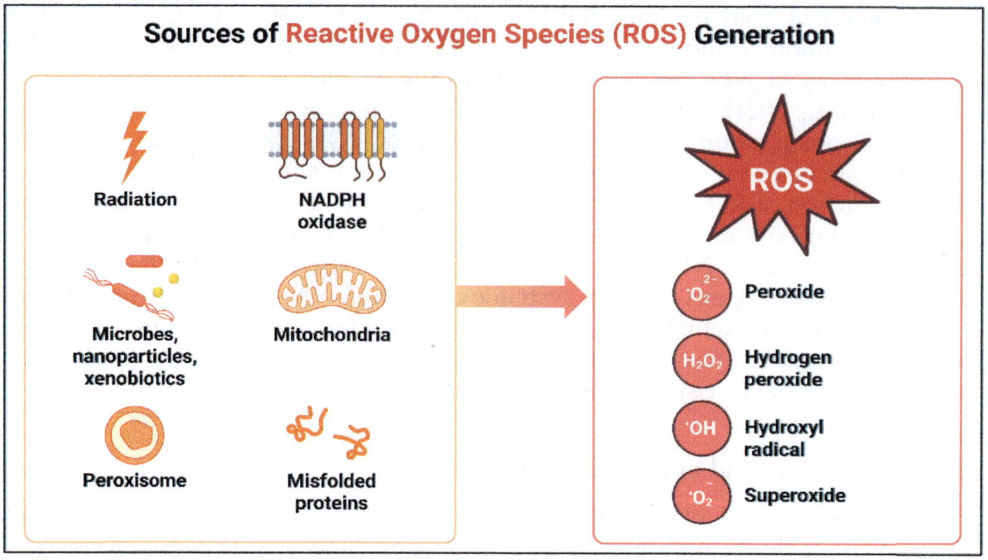

FIGURE 23.2 Figure illustrates the major sources and types of reactive oxygen species (ROS) generation. The endogenous and exogenous factors that can cause ROS are radiation, microbes, nanoparticles, xenobiotics, peroxisomes, NADPH oxidase, mitochondria, and misfolded protein accumulations. Such sources will lead to the synthesis of various forms of ROS, including peroxide (O^2), hydrogen peroxide (H_2O_2), hydroxyl radical (•OH), and superoxide (O_2•$^-$). Although ROS are vital in cellular signaling and in host responses, their unregulated overload causes oxidative stress, biomolecular damage, and pathophysiological effects, especially in cancer, neurodegeneration, and inflammatory disorders. This number brings to light the mechanistic genesis and diversity of ROS. (Created by using BioRender.com).

23.3 Mechanisms of nanoparticle-induced DNA damage

23.3.1 Oxidative stress and reactive oxygen species generation

The excessive generation of ROS is one of the major pathways of NPs induced genotoxicity. The redox cycling reaction is possible because many metallic and metal oxide NP with catalytic reactivity and a large surface area to volume ratio are used, including TiO_2, ZnO, and Fe_3O_4 NP. Fenton and Fenton-like reactions, especially those of transition metals (e.g., Fe^{2+} and Cu^+), produce highly reactive hydroxyl radicals (•OH), which cause indiscriminate oxidative damage to DNA bases and sugar-phosphate backbone (Mishra and Panda, 2021).

Another mechanism of ROS production is electron transfer processes at NP surfaces where the valence band holes and conduction band electrons take part in redox reactions with nearby biomolecules, as presented in Fig. 23.2. Also, NPs may inhibit mitochondrial electron transport chains (ETC), and especially at complexes I and III, they create leakage of electrons and production of superoxide anion (O•). Superoxide dismutase and an increase in the amount of H_2O_2 generate amplification of oxidative stress in the nuclear and mitochondrial compartments because it forms •OH. What oxidative stress causes is the base changes, notably the 8-oxoguanine (8-oxoG), thymine glycol lesions, which facilitate these processes. Single-strand breaks (SSBs) through oxidizing sugar and double-strand breaks (DSBs) upon lesion to opposite strands in close proximity are also caused by ROS. The build-up of such DNA lesions exceeds the capacity of BER and other DNA repair mechanisms and may result in an elevated risk of association with chromosome aberration, genome instability, and cancer (Balkrishna et al., 2021).

23.3.2 Direct physical interaction with DNA

Other than oxidative reactions, some NPs also bind DNA with physical association. Electrostatic binding Nonmetallic and carbon-based nanomaterials NPs of metals and nanomaterials based on carbon can conjugate by means of the electrostatic interaction of functional groups on their surface with the negatively charged phosphate backbone. Also, planar or aromatic structures of the NPs can intercalate between base pairs, disrupting 2-center 2-electron interactions in the stacking of the purine and pyrimidine rings and causing distortions in the double helix structural geometry (Hua et al., 2023). Such interactions may affect conformational changes, including conversion of B-DNA to alternative helical forms (A- or Z-DNA), local deformation and unwinding, or distorted supercoiling that disrupts replication and transcription events. Sustained binding may also disrupt the process of recruiting repair proteins to damaged lesions, stalling lesion recognition and repair. Structural perturbation is dependent on the size, surface charge, hydrophobicity, and functionalization of the NPs, where smaller and cationic NPs elicit greater nuclear penetration and DNA affinity (Zhang et al., 2023).

23.3.3 Inflammatory pathways and secondary genotoxicity

Pro-inflammatory signaling may be induced by the exposure to NP, specifically two toll-like receptors (TLRs) and inflammasome signaling in immune and epithelial cells. This culminates in the secretion of cytokines such as TNF-α, IL-1β, and IL-6, which subsequently stimulate the transcription factors (e.g., NF-kB and STAT3) (Kawai et al., 2024). Inflammation can persist as a chronic process that leads to persistently high rates of ROS and reactive nitrogen species (RNS) production by infiltrating macrophages and neutrophils, causing secondary oxidative damage to DNA in target (and bystander) cells. This indirect genotoxicity is important because it may take place even when a great proportion of NPs remain extracellularly blocked, confirming the systemic character of the NP-driven genomic stress. Inflammatory microenvironments experienced repeatedly can also modify the signaling of DNA repair, thus augmenting damage accumulation further (Jin et al., 2023).

23.3.4 Metal ion release from NP cores

A large number of metallic NPs are dissolved in physiological conditions, liberating toxic metallic ions Ag^+, Zn^{2+}, or Cu^{2+}. These ions have many mechanisms to interfere with genomic integrity. There are, firstly, their involvement in redox cycling reactions, which promotes the generation of ROS and worsen oxidative damage to DNA (Peng et al., 2025). Secondly, some of these ions directly destabilize nucleic acids by saturating available nucleobases or phosphate groups and interfering with the formation and maintenance of hydrogen bonding and base-pairing fidelity. Metal ion release is very sensitive to composition, crystallinity, surface coating, and the pH or ionic composition of the surrounding fluid. In acidic microenvironments (including lysosomes), there is increased proton-mediated degradation of NPs, which contributes to long-term genotoxic pressures across the cellular environment (Zhang et al., 2022).

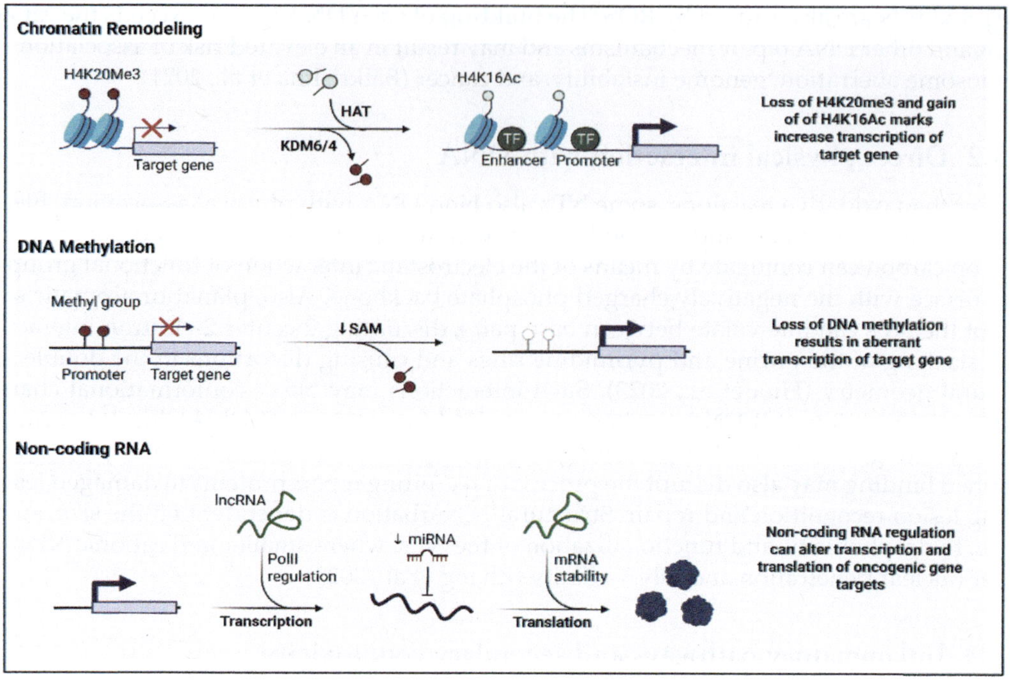

FIGURE 23.3 Figure illustrates epigenetic regulation mechanisms that play a role in the expression of oncogenes. (*Top*) Chromatin remodeling: Loss of H4K20me3 repressive and gain of H4K16Ac active marks via histone demethylase (KDM6/4) and histone acetyltransferase (HAT) activity facilitate transcription factor binding in enhancers and promoters and enhance target gene expression. (*Middle*) DNA methylation: Loss of DNA methylation by reduced CpG site S-adenosylmethionine (SAM) availability, or DNA demethylation, at promoters results in aberrant oncogenic target activation. (*Bottom*) Noncoding RNAs: RNA polymerase II activity, mRNA stability, and translation have been dysregulated by long noncoding RNAs (lncRNAs) and microRNAs (miRNAs), which alter transcription and translation of oncogenic gene targets. (Created by using BioRender.com)

23.3.5 Epigenetic alterations

The effect of NP exposure is the possibility of disturbing the epigenetic landscape, creating a lasting consequence in terms of epigenomic regulation in DNA damage response and repair genes. Among the most significant processes is the changes in DNA methylation, in which NPs regulate the activity or expression of DNA methyltransferases (DNMTs) and ten-eleven translocation (TET) enzymes, as presented in Fig. 23.3 (Käver et al., 2024).

Dysregulation of repetitive elements through hypomethylation contributes to genomic instability, and DNA repair through hypermethylation of CpG islands in promoters of DNA repair genes, such as BRCA1, MLH1, and OGG1, disables repair gene expression without altering gene sequence. Further, the histone modification patterns, including acetylation, methylation, and phosphorylation, can be changed by NPs by means of impacting histone-modifying enzymes, i.e., histone acetyltransferases (HATs) and histone deacetylases (HDACs). The chromatin remodeling processes affect the access of DNA repair machinery to damaged locations and may prolong the recognition and repair of the lesions. This type of epigenetic dysregulation not only contributes to chronic DNA damage but also oncogenic transformation, in that it allows mutagenic events to accumulate. Crucially, such effects might become inheritable with cell divisions, suggesting that epigenomic remodeling induced by the NP can harbor long-lasting biological ramifications, following NP clearance (Li et al., 2024).

23.4 Nanoparticles induced DNA damage

23.4.1 Single-strand breaks

The SSBs are among the most common DNA lesions caused by NP exposure. They happen as a result of the cleavage of the phosphodiester backbone of one of the strands, frequently because of the ROS formed at the NP surfaces or through a secondary oxidative stress due to the response (Ameixa and Bald, 2024). Despite being usually repaired through the BER system, the high load of SSBs that overwhelm repair machines may stall the replication fork and convert SSBs into more harmful types of damage, namely, DNA DSBs. Some metallic NPs (silver and titanium dioxide) induce Fenton-like effects that increase the amount of hydroxyl radical formation and the extent of SSB formation. The repair kinetics of SSB depend on NP size, surface charge, and coating that regulates cellular uptake and distribution in the subnucleus. These ongoing SSBs lead to genomic instability, mutagenesis, and changed transcription patterns, particularly in dividing cells, stressing the value of sensitive assays like alkaline comet assay to measure insertion/deletion strand discontinuities caused by NPs (Kaphle et al., 2025).

23.4.2 Double-strand breaks

One of the most cytotoxic DNA lesions caused by exposure to NPs is the DSB, which implies the cleavage of both strands of DNA in proximity to one another. Such end breaks can occur as a direct result of high-energy NP-mediated interactions, or indirectly as a result of replication fork collapse to non-repaired SSBs. DNA damage induces the DNA damage response (DDR), which first involves the ATM-H2AX-MDC1 signaling pathway to repair via NHEJ or HR (May et al., 2018). Yet, NP exposure is capable of disrupting DDR pathways by disrupting protein recruitment, altering chromatin architecture, or affecting protein posttranslational modifications. As an example, zinc oxide NP and cerium oxide NP were found to diminish the efficiency of HR and enhance error-prone NHEJ and accelerate mutational load. Failure of repair during the interference of NP to the repair fidelity leads to cell cycle arrest, apoptosis, or oncogenic transformation of the DSBs. Quantitative assessment of the damaged γ-H2AX foci is currently considered as a gold-standard method applied to evaluate the effect of NPs in terms of induced DSBs and dynamics of their repair (Vítor et al., 2020).

23.4.3 Oxidative base modifications (e.g., 8-oxoG)

Oxidative base alterations and specifically the generation of 8-oxo-7,8-dihydroguanine (8-oxoG) are one of the characteristic lesions involved in NP-induced genotoxicity. ROS formed by ROS, including singlet oxygen and the hydroxyl radical, form 8-oxoG, which mispairs with adenine during replication and results in G:C→T:A transversions. NPs based on metals such as iron oxide and copper oxide that are highly catalytically active can also significantly cause such lesions because of their pro-oxidant nature. Recognition and excision of 8-oxoG is mediated principally through the BER pathway, in part by the enzyme 8-oxoguanine DNA glycosylase 1 (OGG1), but exposure to NPs can prevent this underlying repair by down-regulating OGG1 expression or activity. Chronic oxidative base damage is related to transcriptional mutagenesis, mitochondrial dysfunction, and persistent inflammation. Large detection methods like high-performance liquid chromatography with electrochemical detection (HPLC-ECD) or mass spectrometry have played central roles in the quantification of NP-induced levels of 8-oxoG and in their mechanistic contribution to genomic instability (Wang et al., 2025a).

23.4.4 DNA protein crosslinks and adducts

The DNA protein crosslinks (DPCs) and covalent DNA adducts are more complicated lesions in which DNA is covalently attached to structural or enzymatic proteins, and as such, the DNA adduct jeopardizes replication and transcription. Oxidative formation of reactive intermediates by NPs, as well as direct chemical reactions between NP-generated radicals and the nucleophilic residues in proteins, are possible. Gold and platinum-based NPs have, by way of illustration, been shown to bind to thiol/amine functional amino acids present in proteins bound to DNA, trapping them on the DNA helix. DPC repair is conducted first by proteolytic destruction of the protein part by either SPRTN metalloprotease or the ubiquitin-proteasome system and then by NER or via HR (Carnie et al., 2025). NPs are likely to block these pathways either through an effect on the proteasome or by exhausting repair factors. The expression of DPCs is associated with replication distress, mitotic errors, and induction of apoptosis, which makes DPCs central lesions in the flux of NP-associated cytotoxicity and risk evaluation of carcinogenesis (Blouin and Saini, 2024).

23.4.5 Chromosomal aberrations and aneuploidy

Deletions, translocations, formation of dicentric chromosomes, and other chromosomal aberrations are also observed when this damage induced by NPs in the DNA is misrepaired or cannot be repaired at all. Such instability is centered on DSB misrepair and oxidative stress-initiated telomere erosion (Scalabrin et al., 2024). Some NPs also interfere with the mechanism of the spindle apparatus in the mechanism of aneuploidy through mis-segregation of chromosomes. Such effects are most notable under high-aspect-ratio NPs, including carbon nanotubes, which can actually contact mitotic machines. In detecting these aberrations, cytogenetic assays, such as micronucleus and metaphase spread analysis, are very crucial. The long-lasting chromosome instability even after being exposed to NP has implications for long-term effects on tumorigenesis (Montilla-Rojo et al., 2025).

23.5 Effects of nanoparticles on DNA repair

23.5.1 Base excision repair

BER is the dominant mechanism that has the capacity to repair minor base alterations, comprising oxidation, alkylation, and deamination. The exposure of NPs, especially the metal-based NPs including TiO_2, Ag, and ZnO, was found to affect BER by inhibiting the activity of major enzymes involved in its process. Glycosylases, such as OGG1 and MUTYH, are the first step in recognizing lesions and removing them; NPs may change glycosylase catalytic efficiency by binding (directly) or oxidation or by destabilizing the protein. Likewise, apurinic/apyrimidinic endonuclease 1 (APE1), which processes the DNA backbone cleavage in abasic locations, may fall victim to the release of metal ions or overgeneration of ROS prompted by NPs, with the result showing decreased cleavage functionality (Lai et al., 2024). Gap-filling synthesis requires DNA polymerase β (Pol β) that is also blocked by sequestration of important cofactors (e.g., Mg^{2+}) or sterical occlusion in the active site by the NPs. These impairments have an additive effect in delaying lesion repair and a higher mutagenesis risk and damage to genome stability. Permanent inhibition of BER through NP can interact with a physiological oxidative stress, causing a worsening genomic instability of exposed cells (Kustra et al., 2025).

23.5.2 Nucleotide excision repair

This comes in handy during the removal of the bulky lesions in the DNA like UV-induced cyclobutane pyrimidine dimers (CPDs) and chemical adducts by NER Crosslighting. This can trigger and also hinder NER by NPs, in particular, containing high ROS-generating capabilities (e.g., quantum dots, TiO_2 with UV). Experimental findings indicate that NPs compromise lesion recognition as destabilization of XPC-RAD23B and DDB2 complexes decreases the lesion-recognizing capacity of these proteins when destabilizing helix-distorting lesions (An et al., 2025). Moreover, DNA helicase (transcription factor IIH) activity is interfered with, impairing the unwinding of DNA during the verification stage. In the downstream, exposure to NP was found to result in the slowed excision progression because of reduced expression of XPA and ERCC1XPF required to form dual incision events. This suppression could be following oxidative repair proteins remodeling or impaired nuclear import. Taken together, it means there will be a decreased efficiency of removal of bulky adducts, which extends the lesions that block the progress of transcription, increasing the likelihood of mutagenic bypass upon replication. In NP exposure, NER suppression is of special concern to the skin, lung, and ocular tissues, which face nanoparticulate pollutants in the environment (D'Souza et al., 2024).

23.5.3 Mismatch repair

The consistency of replication is preserved by MMR through editing of base base mismatch and insertion deletion loops. Several mechanisms of action between MMR and NPs are possible, with the mutation of MMR possible via direct protein-NP interactions and subsequent transcriptional downregulations, and by increasing oxidation mediated by ROS produced following NP exposure. Reduced nuclear localization/functional activity of MutSα (MSH2–MSH6) and MutLα (MLH1–PMS2) after exposure to NP reduces mismatch recognition and excision (An et al., 2025). In other instances, NPs can cause posttranslational alterations that interfere with the formation of protein-protein interactions necessary to form a repair complex, including carbonylation or nitrosylation. Inhibition of MMR not only hastens the accumulation of mutations, but it also impairs the DDR needed to trigger cell cycle arrest and apoptosis following replication misincorporations, in theory predisposing to genomic instability in proliferating tissue (D'Souza et al., 2024).

23.5.4 Homologous recombination

HR is a high-efficiency repair pathway of a DNA DSBs using a sister chromatid as the template. Both repression of transcription as well as posttranslational interference of key mediators lead to NP interference with HR. The main recombinase in the process, RAD51, has been indicated as having decreased expression and strand invasion capacity following NP exposure; presumably through oxidative stress, competitive binding of metal ions, or disruption of cytoskeleton involvement in nuclear transport. Also, BRCA1 and BRCA2, which help to locate and stabilize RAD51, can be downregulated either by epigenetic changes that occur through promoter hypermethylation or proteasomal destruction caused by NP (Toh and Ngeow, 2021). UPstream sensor impairment, like MRN complex (MRE11RAD50NBS1) impairment and ATM kinase impairment, can also impede HR initiation. The resultant effect is maintenance of a high dependence on the error-prone repair pathways, resulting in chromosomal aberrations, loss of heterozygote, or an augmented carcinogenic vulnerability. Of note, the HR suppression of NPs in the separation of stem and progenitor cells has long-term clonal implications, which may either move toward malignant transformation or toward the loss of tissue regulatory potentials due to a lack of cell recovery (Yamamoto and Hirasawa, 2021).

23.5.5 Nonhomologos end joining

NHEJ fills in DSBs during the cell cycle without a template, and therefore, NHEJ is essential to the maintenance of genome stability during stress. Critical NHEJ proteins, Ku70 / Ku80 heterodimer and the DNA-dependent protein kinase catalytic subunit (DNA-PKcs) are susceptible to the NP-mediated inhibition. Direct binding of metal and metal oxide NPs to Ku proteins will compromise DNA end-binding specificity, whereas end processing and synapsis are disrupted by altering the DNA-PKcs phosphorylation potential induced by ROS-mediated oxidation. The XRCC4-Ligase IV downstream ligation may also be damaged because the NP depletes the cofactors adenosine triphosphate (ATP) and Mg^{2+}. NHEJ inhibition is especially harmful when combined with a deficiency in HR, since cells will have either unusually numerous and irreverting DSBs or erroneously repaired ends that may again amplify the rate of translocations. Besides, known involvement of NHEJ in V(D)J recombination, chronic exposure to NP would hypothetically compromise adaptive immune-system diversification, which to date has not been thoroughly examined (Vogt and He, 2023).

23.6 Factors influencing nanoparticles genotoxicity

23.6.1 Dosage and exposure duration

Genotoxicity caused by NP depends highly on the dosage-time of administration and exposure levels. The critically sublethal DNA damages and temporary activation of repair routes can be caused by low levels of exposure, whereas frequent or immense doses can exceed repair functioning, causing lasting mutations or chromosomal aberrations. A long-term low-dose exposure in particular is nasty stuff, potentially conferring both cumulative oxidative damage and replication stress, neither of which is immediately cytotoxic. Besides, dose-response curves of NPs are not always linear and depend on the size, surface reactivity, and dissolution kinetics. The accumulation of oxidative DNA damage can be enhanced by time-dependent accumulation in subcellular compartments, especially the nucleus and the mitochondria, and lead to decreases in the fidelity of repair. Acute DSBs can be caused by short-term high-dose exposure, but more commonly by chronic exposure at low doses, there is subtle and progressive genomic instability. Thus, toxicological testing has to combine the dose level as well as its time variables to be predictive of DNA insult and repair under conditions of real-life exposures(Teixeira et al., 2024; Pathak et al., 2025).

23.6.2 Cell type and tissue specificity

The NP interactions with the genome are highly heterogeneous across cellular and tissue landscapes. The conductive proliferating cells, like those found in bone marrow, intestinal crypt, or tumor microenvironments, are usually more vulnerable because of the increased vulnerability of the replication and activation of DNA repair channels. On the contrary, terminally differentiated cells are likely to develop lesions that cannot be repaired with time, compromising their functions (Suba, 2024). NP uptake and distribution in the body, as well as their subsequent genotoxicity, are also influenced by NP-specific microenvironments formed by vascularization, metabolic activity, and oxidative conditions. An example is that in hepatocytes, because of their dominance in xenobiotic disposition, they may have pronounced DNA damage, whereas neural tissues are exquisitely sensitive to an oxidative insult because they do not have significant regenerative capacity. Also, the repair efficiency of tissue-specific DNA repair genes is affected due to their tissue-specific expression. These context-specific susceptibilities are important to understand to allow effective risk assessment and targeted nanomedicine application design that has a reduced genotoxic risk (Youhanna et al., 2022).

23.6.3 Aggregation state in biological media

The ensemble of NPs in the biological fluid plays an important role in their bioavailability and cellular uptake, as well as their genotoxic potential. NPs may become surrounded by a corona of biomolecules in protein-rich fluids, such as plasma, and in so doing can change the hydrodynamic diameter and surface reactivity. Each of the aggregated forms can have low rates of cellular internalization and instead can settle locally, thus causing local disproportionate exposure and localized DNA damage (Chandoliya et al., 2024). On the other hand, particles dispersed optimally reach the cellular and nuclear compartments more effectively, and this may pose extensive genomic lesions. NP surface chemistry, ionic strength, and pH of the surrounding medium determine the level of aggregation, and it is therefore essential to understand the appropriate ways to determine genotoxicity of NP through in situ characterization before it can be predicted with precision (Mo et al., 2024).

23.6.4 Co-exposure with other environmental stressors

The DNA damage mediated by NP may either be enhanced/regulated when co-exposure to other genotoxic environmental agents occurs, including UV radiation, polycyclic aromatic hydrocarbons, heavy metals, or ionizing radiation. Synergism can occur, either through interaction of the NP with cellular uptake or bioactivation of the co-pollutants, or when there is combined oxidative stress overcoming the repair systems (Downie et al., 2024). As an illustration, some cases of metal oxide NPs can be photoactivated by UV and increase the generation of ROS and breakages of DNA strands. Also, NP-associated oxidative stress may be enhanced by co-exposure to inflammatory mediators, thus leading to a microenvironment in which genomic instability can develop. The nature of such interactions proves the need to employ multistressor models in the realistic nanotoxicology evaluation (Wang et al., 2025b).

23.6.5 Genetic predisposition and individual variability

Genetic polymorphism on gene repair, antioxidant defense system, xenobiotic metabolism pathways, and inter-individual variability in NP-induced genotoxicity can be facilitated in the genes. Mutations in the BER or NER genes, e.g., can lower the efficiency in removing lesions, leading to enhanced susceptibility to mutagenesis. Patients with defective p53 signaling or genetic cancer disorders can be at risk of augmented genomic injury when subjected to NP. Biological responses are also further regulated by epigenetic factors, age, and preexisting diseases. This variability will require personalized risk assessment methods to be integrated, which may include genomic profiling, allowing the possibility to recognize high-risk populations within both an occupational and environmental NP exposure contexts (Schmidt et al., 2022).

23.7 Advances in Nanoparticle -induced DNA damage detection and analysis

23.7.1 Comet assay and variants

Single-cell gel electrophoresis (SCGE), also known as the comet assay, is a sensitive test that detects DNA strand breaks at the individual cell level. In an alkaline state, it measures SSBs and alkali-labile sites with high sensitivity, and its neutral form is typically used to determine DSBs. Cells are placed in an agarose gel and noncompartmentalized, and cells are lysed to discard membranes and proteins and put through electrophoresis; fragmented DNA will migrate, leaving a comet-like tail, the intensity of which is proportional to damage (Collins et al., 2023). Modifications to the original comet assay, including the enzyme comet assay [enabling detection of oxidative base lesions via the use of either formamidopyrimidine DNA glycosylase (FPG) or endonuclease III (EndoIII)], aid in the ability to detect oxidative lesions to the base. The flexibility, low cell consumption, and affordability of the assay make it ideal to screen genotoxicity caused by NPs. Nevertheless, electrophoretic conditions, scoring, DNA repair procedures, and environmental conditions lead to significant variability; hence, strict standardization must be observed when assessing sublethal exposure to NP (Ladeira et al., 2024).

23.7.2 γ-H2AX foci assay

Much more specific markers are the γ-H2AX foci assay, which relates to the fast phosphorylation of the histone H2AX at serine 139 (2-H2AX) after DNA damage. When NP is exposed, DSBs induce the activation of the ATM/ATR kinases and stimulate phosphorylation of H2AX and the appearance of specifically located nuclear foci of breakpoints. Measuring γ-H2AX foci is quantified by immunofluorescence microscopy or flow cytometry to provide single-cell resolution. This is an important method, specifically in detecting the biomarkers of early genotoxic events due to the quick induction of γ-H2AX within minutes of an insult. Variants with 53BP1 co-staining are more accurate by ensuring that true DSBs can be confirmed as opposed to signaling due to replication stress. Disadvantages involve the possibility of confounding because of apoptotic DNA fragmentation, and transient foci resolution as a part of the active repair, and hence, the analysis should be timed appropriately. When used to study NP, γ-H2AX assays may be useful in elucidating the mechanistic nature of NP genome interactions, including dose-dependent induction of DSB, the repair kinetics, and chromatin remodeling of NP (Yachi et al., 2023).

23.7.3 High-throughput sequencing for mutational signatures

High-throughput sequencing (HTS) opens up the possibility of complete profiling of mutational signatures of genomic lesions induced by NP. Whole-genome sequencing (WGS) and its subset, whole-exome sequencing (WES), call single-nucleotide variants (SNVs), small insertions/deletions (indels), and structural variants at the base level (Maslov et al., 2022). Computational deconvolution of mutation spectra may permit attribution of given patterns to oxidative stress, the generation of bulky adducts, or replication errors brought on by NPs. There is also the possibility of inference of damage origin in terms of mechanism using integration to reference mutational databases on damage signatures, e.g., COSMIC. Panels in targeted sequencing offer greater sensitivity to hotspot mutations, and duplex sequencing can reduce to a minimum possible sequencing errors, which is necessary to identify low-frequency variants in nonhomogeneous cell populations. Although the analytical strength is high, issues are the need to discriminate NP-specific mutational patterns against background noise, and direct versus indirect oxidation or inflammatory insult. HTS can therefore form a relevant instrument to associate NP exposure to lasting changes at the genomic level and the eventual carcinogenic effects (Wan et al., 2022).

23.7.4 Single-cell genomics and imaging

The capabilities to resolve NP-induced alterations in the genome have now been transformed by single-cell genomics and imaging technologies. Single-cell whole-genome or transcriptome sequencing allows cell-to-cell variation in mutational burdens, copy-number alterations, and transcriptional responses to stress to be captured, and heterogeneous susceptibility to exposures to be identified across exposed populations. DNA damage and repair complex formation can be visualized in individual cells by either super-resolution microscopy or live cell microscopy modalities in the presence and absence of repair enzymes to dissect the spatiotemporal dynamics of the repair complex and marker dynamics (Zhuang, 2021). Reporters Multiplexed fluorescence in situ hybridization (FISH) and CRISPR-based live reporters can additionally be used to visualize the rearrangements that occur in the chromosomes in real-time, or to monitor the resolution of the activation of repair pathways. In the toxicology of NP, these methods determine subpopulations that have persistent DNA damage or malfunctioning repairs that could be lost to bulk analyses. Current issues mean there is still work to do in integrating data, throughput, and differentiating between primary and secondary NP-mediated effects and the stress response, but continual new developments hold the promise of many gains in the prediction of genotoxicity on a single-cell scale (Watson et al., 2022).

23.8 Risk assessment and regulatory aspects

23.8.1 Current safety guidelines

The globally harmonized test frameworks used to assess the safety of nanomaterials that may be genotoxic include OECD and the International Organization for Standardization (ISO). OECD has tailored its test guidelines (TGs), like TG 471 (ames test), TG 487 (in vitro micronucleus assay), and TG 489 (in vivo comet assay), to take account of nano aspects, e.g., agglomeration, dispersion stability, and dosimetry. Their physical properties can be characterized according to ISO standards. ISO / TR 16197 and ISO 19007 offer methodology to characterize NPs, dispersion procedures, and cytotoxicity/genotoxicity test rules (Doak et al., 2023). The features pointed out in these typologies forward the comprehensive physicochemical characterization, previous to biological testing, which includes size distribution, surface charge, and dissolution kinetics. Both OECD and ISO emphasize the need for tiered approaches to testing that embrace the use of both in vitro and in vivo models, substantiated by quality control and reproducibility requirements to guarantee rational hazard determination and regulatory decision-making about nanomaterial safety in the context of biomedical, industrial, and environmental environments (Andreoli et al., 2025).

23.8.2 In vitro versus in vivo genotoxicity models

The in vitro genotoxicity models, namely human lymphocyte micronucleus assays, comet assays in primary cells, and high-content imaging systems, are easy to use, cost-effective, and enable NP screening under controlled exposure conditions. Such systems make mechanistic study, oxidative stress induction, DNA strand breaks, and repair pathway modulation possible and lessen ethical considerations. They are, however, usually simplistic in comparison to whole-organism pharmacokinetics, biodistribution, and secondary tissue effects (Vignard et al., 2023).

In vivo model, involving the micronucleus and transgenic mutation assays of rodents, combines exposures with physiological barriers, immune reactions, and body elimination, and thus provides more realistic exposure-related conditions. Nevertheless, they are costly, prone to interspecies variability, and perhaps need to be adapted to nanospecific endpoints, like organotropism and persistence. The best overall genotoxicity testing strategy may incorporate a weighted-evidence selection regime, with the in vitro data helping to both select an exposure dose and to understand dose-response and mode of action, and the in vivo data determining potential to cause a hazard under bio-relevant exposure pathways (Di Ianni et al., 2021).

23.8.3 Challenges in translating lab to real-world

To translate laboratory results into contexts of real exposure to NPs is one of the challenges that are faced since there are differences that are seen in environments of exposure, the dosimetry, and biological complexity. All too often, in vitro assays require higher, bolus doses in order to generate measurable effects and are not necessarily representative of low-level, chronic environmental or occupational exposure. Moreover, dynamic changes of NPs could occur-protein corona build-up, agglomeration, and dissolution under physiological fluid conditions, which change bioavailability and genotoxicity relative to pristine laboratory suspensions (Balkrishna et al., 2021; Mishra and Panda, 2021; Hua et al., 2023).

The method of in vivo is more integrative but limited since one will not have the direct extrapolation across species due to the differences in NP uptake, metabolism, and effects on DNA repair. Moreover, co-exposures (e.g., pollutants, UV radiation, and chemicals) that occur in the real-world and genetic susceptibilities, which are variable, are seldom simulated in controlled studies. The second most important gap is the poor knowledge of long-term genotoxic effects, particularly those of long-acting NPs resistant to dissolution in the tissues. Regulatory mechanisms are beginning to promote an integrated testing strategy (ITS) and adverse outcome pathway (AOP)-based models in an attempt to translate experimental results into human-relevant maladies. Improving predictive capability through further development of physiologically based pharmacokinetic (PBPK) modeling and next-generation organ-on-chip systems potentially could increase predictive accuracy, although durable translation will be hampered until consistent TGs, long-term follow-up, and epidemiological data collections can be established at the population level (Dhaka et al., 2023; Xuan et al., 2023; Liu et al., 2024; Ewii et al., 2025).

23.9 Mitigation, safety, and design strategies

23.9.1 Surface coating and functionalization

Chemical functionalization of NPs and surface coating belong to the strategic approaches to reduce ROS overproduction and, therefore, genotoxicity (Chaudhari et al., 2025; Khuspe et al., 2025b, 2025a). NPs of Ag, TiO_2, and ZnO are considered inorganic materials that are highly reactive on their surfaces and can catalyze a redox reaction to generate ROS. The inert and biocompatible coating of these particles with materials, like PEG, silica, or phospholipids, is successful in blocking the reactive surface and minimizing the electron transfer and the generation of ROS (Khan et al., 2025; Uddin et al., 2025). In addition, antioxidant molecules (e.g., catechins and ascorbate derivatives) can be used to functionalize cells or tissues to work at a local level; they can have direct antagonistic effects on oxidative stress. Improving hydrophilicity, surface charge, and the protein corona composition may further optimize these targets through ligand engineering as a means of alleviating nonspecific cellular uptake and oxidative burst (Hatvate et al., 2025; Vardikar et al., 2025). Improved happenings in bioinspired coatings, such as cell membrane cloaking and zwitterionic polymers, have shown improved ROS blockage with maintenance of NP capacity (Faiyazuddin et al., 2024; Kumar et al., 2024; Pardeshi et al., 2024a, 2024d). Notably, the selection of coating has to support a trade-off between ROS scavenging and preservation of targeted therapeutic or diagnostic effect and often requires accurate engineering of the material biological interface(Pardeshi et al., 2024b; Patil et al., 2024; Sawant et al., 2024a, 2024b).

23.9.2 Biodegradable and clearance-optimized nanoparticles

The main focus on NP-caused DNA damage is biodegradability and good clearance. Chronic NP macromolecules may build-up in the tissues, sustaining a long-chain oxidative cellular stress and disrupting DNA repair mechanisms. The long-term exposure to ROS is minimized by debasing biodegradable platforms like poly[lactic-co-glycolic acid (PLGA), polycaprolactone, chitosan, and lipid-based carriers] using enzymatic or hydrolytic means into nontoxic metabolites (Mesut et al., 2024; Nimmy et al., 2024; Rojekar et al., 2024b; Shaikh et al., 2024; Vardikar et al., 2025). Further prevention is made based on toxin dispersal to other sensitive tissues like the liver, spleen, and the bone marrow by means of engineering NPs with size, shape, and surface properties that allow clearance by the kidney or hepatobiliary tract. Degradation structures that are responsive to stimuli (e.g., pH-sensitive bonds and redox-cleavable bonds) allow for the control disassembly following target binding, which reduces the persistence in the system (Bhowmik et al., 2023; Guptaet al., 2023; Saidet al., 2023; Hatvate et al., 2024; Khuspe et al., 2024; Rojekar et al., 2024; Uddin et al., 2024). The more recent approaches promise an understanding of the preparation of hybrid nanostructures comprised of biodegradable core-shell NPs, in which the degradation rate can be facilely altered by modifying the shell. Optimization of clearance is important in achieving the balance between therapeutic efficacy and biosafety, especially in chronic/repeated occasions of exposure, and this decreases the genotoxic risk profiles (Hatvate et al., 2023b; Pardeshi et al., 2023; Pardeshi and Giram, 2023; Sawant et al., 2023; Parit et al., 2024).

23.9.3 Dose minimization and targeted delivery

Reducing the dose of NPs without impairing the therapeutic or diagnostic effects can be considered one of the main strategies to curb their ROS-related DNA damage (Magar Tushar V., 2011a; Rajurkar, 2012a). Targeted delivery vehicles increase NP deposition at the therapeutic target and minimize off-target exposure, resulting in decreased systemic ROS burden (Borude and Mahajan, 2011b; Gupta et al., 2024; Kapare et al., 2024; Pardeshi et al., 2024c; Rojekar et al., 2024a). Targeting focuses on ligand targeting via, e.g., antibodies, aptamers, and peptides, magnetic steerage of superparamagnetic NPs, and external stimuli initiation (light, ultrasound, or temperature). This precise delivery translates into the local therapeutic index, resulting in the ability to reduce the dose with the same or improved performance of functional outcomes (Magar Tushar V., 2011b; Khedkar Ajay, 2012a, 2012b; Rajurkar Vikas G., 2012). Stimuli-responsive release profiles may also be incorporated into nanocarrier design, meaning that the payload/active surface of the NP is only exposed in the microenvironment of interest (e.g., hypoxic tumors or inflammation sites, where these pH and redox gradients are unique) (Borude and Mahajan, 2011a; Bhagat, 2012; Bothiraja et al., 2014; Hatvate et al., 2023a; Sayyad et al., 2023).

This is achieved using pharmacokinetic optimization of the half-life of circulation and the prevention of early clearance by means of PEGylation or protein corona modulation, a facilitation of distribution in the target with little systemic effect. Studies of the effect of the drug dose are further detailed with computational modeling and microdosing regimens to optimize efficacy dose exposure (Anand, 2011; Borude, 2011; Dattatray, 2012a; Gholap, 2012; Rajurkar, 2012b; Rajurkar Vikas G., 2012a, 2012b). By implementing dose minimization and high-specificity delivery together, there is potential to significantly minimize the generation of NP-induced ROS and the risks of DNA damage, without compromising or possibly even improving overall therapeutic efficacy (Dattatray, 2012b; Dighe, 2012b, 2012a; Rajurkar, 2012c; Singh et al., 2016; Dattatray, 2012; Puri, 2018).

23.10 Future opportunities

Owing to requirements to identify the genotoxic potential of NPs, artificial intelligence (AI) is becoming an emerging, critical tool in planning to combine quantities of physicochemical and toxicological information with molecular interaction dynamics into complex solutions using AI models. Computational methods such as machine learning (ML) algorithms trained on large databases covering the NP size, shape, surface charge, composition and functionalization can be used to quantify complex and non-linear relationships between NP properties with DNA damage outputs such as oxidative DNA lesions, strand breaks and chromosomal aberrations (Chaudhari et al., 2025; Gholap et al., 2025; Jadhav et al., 2025; Vardikar et al., 2025; Khuspe et al., 2026a, 2026b; Khuspe and Mane, 2026a, 2026b; Pardeshi et al., 2026). The in silico models based on AI algorithms augment both in vitro and in vivo tests, allowing to screen and prioritize high-risks nanomaterials much faster and, thus, decreasing the amount of experimentation required and its ethical issues. Further, high-throughput omics data are subjected to deep learning methods, also making mechanistic findings on NP binding patterns and disruption of repair pathways possible. Use of AI predictions in conjunction with systems biology models can further the field into predictive toxicology and result in an improved basis on which to make regulatory decisions toward safe use of nanomaterials in precision nanomedicine.

Based on the integration of nanotoxicology and personalized medicine, there are prospects to develop NP-based diagnostics and therapeutics in customizing them to specific genetic and epigenetic predispositions to DNA damage. Polymorphism in DNA repair genes, oxidative stress regulators, and xenobiotic metabolic pathways, which are modulating NP bioreactivity and genotoxic susceptibility, can be discovered using genomic profiling. The use of AI combined with patient stratification allows selection of NP formulations with the highest therapeutic activity and smallest genotoxic liability with defined patient subgroups with precision. It is especially applicable in the field of oncology, where tumor-specific DNA repair deficiencies may be targeted to deliver NPs to tumors with chemotherapeutic or radiosensitizing agents. Moreover, the real-time monitoring of the biomarker of DNA damage that is integrated with an AI-based predictive analytics algorithm enables adapting dosing and safety optimization. This synergy between personalized medicine and nanotechnology, therefore, promises to ensure that maximum therapeutic outcomes and minimal genotoxic side effects are realized, opening the door to the clinical use of individualized nanomedicine protocols.

23.11 Challenges

Interactions between NP and biomolecules are also multifaceted by several factors because NP size, shape, surface charge, functionalization, and aggregation state, among other factors, are imminent factors, in

addition to the dynamic physicochemical conditions in the biological milieu. When subjected to a protein, lipid, and nucleic acid environment, NPs bind these biomolecules rapidly to produce a unique proteinaceous, lipid, and nucleic acid coat, termed a biomolecular corona, which contributes to the modulation of cellular recognition, uptake, and intracellular trafficking (Ewii et al., 2025; Peng et al., 2025). This corona has the capability to modify the NP reactivity, redox potential, as well as the genesis of ROS and can hence affect the outcomes of DNA damage. In addition, interferences with nuclear proteins, histones, and DNA repair enzymes may directly or indirectly damage genomic stability, the consequences generally reliant on small conformational and kinetic variables (Abdelsalam et al., 2022).

Although the effects of acute NP toxicity are well described, chronic and low-dose long-term effects of exposure are not well characterized. Persistent oxidative stress, lasting inflammation, and accumulating DNA damage in the form of SSBs, DSBs, and oxidative base lesions can be brought about by the doses of NP that are less than the cytotoxic levels. It is still possible that the effects can be aggravated by continual interruption of the pathways of repairing DNA, in the regulation of epigenetics, or even the maintenance of telomeres, causing genomic impairment in the long-run. It may also be possible that low-dose NP exposure induces adaptive cellular responses, such as changes in the antioxidant defenses and DNA damage tolerance mechanisms, that may obscure genotoxicity in short-term assays, but more likely predispose to mutagenesis in vivo. Tissue storage in the liver, spleen, and brain also increases the risk of carcinogenesis, neurodegeneration, or reproductive interference with a delayed onset and justifies the necessity of a longitudinal study incorporating realistic exposures, including molecular endpoints, in a multifaceted study (Di Ianni et al., 2022; Ansari et al., 2024).

23.12 Conclusion

NPs are a mixed blessing; on the one hand, they are a potentially very useful technology, used (or pointed to) in biomedical, environmental, and industrial contexts, but on the other hand, their genotoxic threats should be looked into with a fine-tooth comb. This is expressed in their small scale at the nanoscale, large surface area to volume ratio, and interaction with biomolecular structures that can help it enter cellular and nuclear compartments, which can directly or indirectly impair the integrity of DNA. The mechanisms of occurrence of DNA damage by oxidative stress include overproduction of ROS, attack of the oxidant on the DNA itself, disruption of repair or replication proteins, or infliction of aberrations of the chromosomal structure. DNA repair is an elaborate process subject to regulation, and experimental evidence showed that any NP can interfere with the efficiency of DNA repair pathways that include BER, NER, and HR, which may result in persistent cells and mutagenesis. Even when some engineered NPs are engineered to be biocompatible and, in fact, used in nanomedicine to enhance targeted repair of DNA at genotoxic loci, off-target genotoxicity is a risk, especially in chronic cases. New technologies in high-throughput genotoxicity testing, omics-based profiling, and advanced imaging offer insights into NP genome interactions, but lingering translation missing links remain between in vitro and in vivo risk assessment in people. Predictive toxicology is made more strenuous by the diversity of NP physicochemical characteristics and the intricacy of biological systems. The next-generation studies are needed to incorporate multidisciplinary methodology-which incorporation of nanomaterial characterization, mechanistic toxicology and modeling, building predictive frameworks about the safety of NPs. The development of regulatory rules will center on the inclusion of nanospecific parameters as a basis of evaluating risk. The bottom line here is that safe and sustainable nanotechnology demands a balance between innovation and genomic safety, where realization of the potential of NPs does not jeopardize genetic stability in human and environmental systems.

References

Abdelsalam, N.R., Abdel-Megeed, A., Ghareeb, R.Y., Ali, H.M., Salem, M.Z.M., Akrami, M., et al., 2022. Genotoxicity assessment of amino zinc nanoparticles in wheat (Triticum aestivum L.) as cytogenetical perspective. Saudi J. Biol. Sci. 29, 2306–2313.

Altammar, K.A., 2023. A review on nanoparticles: characteristics, synthesis, applications, and challenges. Front. Microbiol. 14, 1155622.

Ameixa, J., Bald, I., 2024. Unraveling the complexity of DNA radiation damage using DNA nanotechnology. Acc. Chem. Res. 57, 1608–1619.

An, S., Kusakabe, M., Kim, H.-S., Kozono, H., Cheon, N.Y., Kim, J., et al., 2025. XPC-RAD23B enhances UV-DDB binding to DNA to facilitate lesion search in nucleotide excision repair. Nucleic Acids Res 53, gkaf463.

Anand, M., 2011. Smart crystals technology: a review. Pharmacologyonline 1, 238–243.

Andreoli, C., Dusinska, M., Bossa, C., Battistelli, C.L., Silva, M.J., Louro, H., 2025. Regulatory practices on the genotoxicity testing of nanomaterials and outlook for the future. Regul. Toxicol. Pharmacol. 162, 105881.

Ansari, A.A., Parchur, A.K., Li, Y., Jia, T., Lv, R., Wang, Y., et al., 2024. Cytotoxicity and genotoxicity evaluation of chemically synthesized and functionalized upconversion nanoparticles. Coord. Chem. Rev. 504, 215672.

Balkrishna, A., Kumar, A., Arya, V., Rohela, A., Verma, R., Nepovimova, E., et al., 2021. Phytoantioxidant functionalized nanoparticles: a green approach to combat nanoparticle-induced oxidative stress. Oxid. Med. Cell. Longev. 2021 (1), 3155962.

Bhagat, V., 2012. Gold nanorods: a promising drug delivery vehicle. Inven. Rapid NDDS 1, 1–4.

Bhowmik, D.D., Deshmukh, A.Y., Hatvate, N.T., 2023. Quintessential impact of dendrimer bioconjugates in targeted drug delivery. In: Verma, S.K., Das, A.K. (Eds.), Bioconjugated materials part 2 applications in drug delivery, vaccine formulations and important conjugates for cancer therapy, comprehensive analytical chemistry. Elsevier, Amsterdam, Netherlands, pp. 257–302.

Blouin, T., Saini, N., 2024. Aldehyde-induced DNA-protein crosslinks- DNA damage, repair and mutagenesis. Front. Oncol. 14, 1478373.

Borude, S., 2011. Aquasomes: a potential drug delivery carrier. Pharmacologyonline 1, 230–237.

Borude, S.S., Mahajan, A.M., 2011a. Recent updates on q fever: diagnosis and treatment. Pharmacologyonline 1, 254–264.

Borude, S.S., Mahajan, A.M., 2011b. POEMS syndrome: a review. Pharmacologyonline 1, 265–274.

Bothiraja, C., Gholap, K.S., Pawar, A.P., 2014. Investigation of ethyl cellulose microsponge gel for topical delivery of eberconazole nitrate for fungal therapy. Ther. Deliv. 5, 781–794.

Cao, Y., Chen, J., Bian, Q., Ning, J., Yong, L., Ou, T., et al., 2023. Genotoxicity evaluation of titanium dioxide nanoparticles in vivo and in vitro: a meta-analysis. Toxics 11, 882.

Carnie, C.J., Jackson, S.P., Stingele, J., 2025. Transcription-coupled repair of DNA–protein crosslinks. Trends Cell Biol 35, 316–329.

Chandoliya, R., Sharma, S., Sharma, V., Joshi, R., Sivanesan, I., 2024. Titanium dioxide nanoparticle: a comprehensive review on synthesis, applications and toxicity. Plants 13, 2964.

Chaudhari, Y.S., Chaudhari, M.Y., Alam, M.I., Khalid, M., Webster, T.J., Sundaram, G.S., et al., 2025. Surface engineering of nano magnesium alloys for orthopedic implants: a systematic review of strategies to mitigate corrosion and promote bone regeneration. Front. Bioeng. Biotechnol. 13, 1617585.

Collins, A., Møller, P., Gajski, G., Vodenková, S., Abdulwahed, A., Anderson, D., et al., 2023. Measuring DNA modifications with the comet assay: a compendium of protocols. Nat. Protoc. 18, 929–989.

Dattatray, T., 2012a. Potential applications of botanical extract. Inven. Rapid Ethnopharmacol 1, 1–4.

Dattatray, T., 2012b. Artificial viral envelopes: emerging vehicle for gene delivery. PHARMATUTOR-ART-1439 1, 1–3.

Dattatray, Tupe, 2012. The threat of photoaging: clinical conditions, evaluation and treatment. Inven. Rapid Cosmeceuticals 1, 1–3.

Dhaka, A., Chand Mali, S., Sharma, S., Trivedi, R., 2023. A review on biological synthesis of silver nanoparticles and their potential applications. Results Chem. 6, 101108.

Dighe, S.B., 2012a. Dandruff: pathogenesis and treatment. Inven. Rapid Cosmeceuticals 1, 1–4.

Dighe, S.B., 2012b. Role of sebum in human health & acne. Inven. Rapid Cosmeceuticals 1, 1–3.

Di Ianni, E., Erdem, J.S., Møller, P., Sahlgren, N.M., Poulsen, S.S., Knudsen, K.B., et al., 2021. In vitro-in vivo correlations of pulmonary inflammogenicity and genotoxicity of MWCNT. Part. Fibre Toxicol. 18, 25.

Di Ianni, E., Møller, P., Cholakova, T., Wolff, H., Jacobsen, N.R., Vogel, U., 2022. Assessment of primary and inflammation-driven genotoxicity of carbon black nanoparticles in vitro and in vivo. Nanotoxicology 16, 526–546.

Doak, S.H., Andreoli, C., Burgum, M.J., Chaudhry, Q., Bleeker, E.A.J., Bossa, C., et al., 2023. Current status and future challenges of genotoxicity OECD Test Guidelines for nanomaterials: a workshop report. Mutagenesis 38, 183–191.

Downie, A.T., Cramp, R.L., Franklin, C.E., 2024. The interactive impacts of a constant reef stressor, ultraviolet radiation, with environmental stressors on coral physiology. Sci. Total Environ. 907, 168066.

D'Souza, A., Kim, M., Chazin, W.J., Schärer, O.D., 2024. Protein-protein interactions in the core nucleotide excision repair pathway. DNA Repair (Amst) 141, 103728.

Ewii, U.E., Attama, A.A., Olorunsola, E.O., Onugwu, A.L., Nwakpa, F.U., Anyiam, C., et al., 2025. Nanoparticles for drug delivery: insight into in vitro and in vivo drug release from nanomedicines. Nano TransMed 4, 100083.

Faiyazuddin, M., Maqsood, S., Abbas, S., Chaudhari, Y.S., Sharmila, S., Webster, T.J., 2024. Prospects and ethics of ChatGPT in biomedical research and medical practice: an emphasis on nanotechnology. J. Biomed. Nanotechnol. 20, 1661–1678.

Gholap, A.D., 2012. Translational nanomedicine: status assessment and opportunities. Int. J. Res. Pharm. Biomed. Sci. 4, 1454–1463.

Gholap, A.D., Khuspe, P.R., Mane, D.V., Galave, V., Mali, A., Mohite, P., et al., 2025. Challenges and future directions biomaterial-based additive manufacturing in tissue engineering and regeneration. In: Singh, S. (Ed.), Biomaterial-based additive manufacturing in tissue engineering and regeneration. biomaterials, bioengineering and sustainability. Springer Nature Singapore, Singapore, pp. 491–529.

Gupta, J., Kamandar, P., Bhowmik, D.D., Rojekar, S., Faiyazuddin, M., Hatvate, N.T., et al., 2024. Harnessing nanovaccines for effective immunization—a special concern on covid-19: facts, fidelity, and future prospective. ACS Biomater. Sci. Eng. 10, 271–297.

Gupta, J.S., Kamandar, P.A., Banchhod, G.V., Hatvate, N.T., 2023. Antibody-drug conjugates for cancer therapy: an up-to-date review on the chemistry and pharmacology. In: Verma, S.K., Das, A.K. (Eds.), Bioconjugated materials Part 2 applications in drug delivery, vaccine formulations and important conjugates for cancer therapy, comprehensive analytical chemistry. Elsevier, Amsterdam, Netherlands, pp. 105–190.

Harun-Ur-Rashid, M., Jahan, I., Foyez, T., Imran, A.B., 2023. Bio-inspired nanomaterials for micro/nanodevices: a new era in biomedical applications. Micromachines 14, 1786.

Hatvate, N.T., Bhowmik, D., Sawant, S.D., Londhe, K.M., Rojekar, S., Faiyazuddin, M., 2025. Engineered artificial MicroRNA in plant: an overview. In: Datta, R. (Ed.), MicroRNA advances and application in plant biology. Elsevier, USA, pp. 311–341.

Hatvate, N.T., Khuspe, P.R., Mandhare, T.A., Kashid, P., Gaikwad, V.D., 2023a. Proteomics in oncology: retrospect and prospects. In: Kulkarni, S., Haghi, A.K., Manwatkar, S. (Eds.), Novel technologies in biosystems, biomedical & drug delivery. Springer Nature Singapore, Singapore, pp. 243–269 .

Hatvate, N.T., Khuspe, P.R., Mandhare, T.A., Kashid, P., Vora, L.K., 2024. Decoding of nanoneutraceuticals and their delivery. In: Mohanta, Y.K. (Ed.), Sustainable green nanotechnology. CRC Press, Florida, New York, pp. 284–309.

Hatvate, N.T., Sawant, S.D., Khemka, P.N., Vora, L.K., 2023b. Unlocking the potential of nanobiohybrids to combat environmental pollution. In: Lens, P. (Ed.), Nanobiohybrids for advanced wastewater treatment and energy recovery. IWA Publishing, UK., pp. 169–190.

Hua, X., Feng, X., Hua, Y., Wang, D., 2023. Paeoniflorin attenuates polystyrene nanoparticle-induced reduction in reproductive capacity and increase in germline apoptosis through suppressing DNA damage checkpoints in Caenorhabditis elegans. Sci. Total Environ. 871, 162189.

Huang, R., Luther, D.C., Zhang, X., Gupta, A., Tufts, S.A., Rotello, V.M., 2021. Engineering the interface between inorganic nanoparticles and biological systems through ligand design. Nanomaterials 11, 1001.

Jadhav, K., Shevalkar, G., Sugandhi, V.V., Pai, R., Parikh, K., Prajapati, M.K., et al., 2025. Exploring protein aggregation in biological products: from mechanistic understanding to practical solutions. AAPS PharmSciTech 26, 189.

Jin, M., Fang, J., Wang, J., Shao, X., Xu, S., Liu, P., et al., 2023. Regulation of toll-like receptor (TLR) signaling pathways in atherosclerosis: from mechanisms to targeted therapeutics. Acta Pharmacol. Sin. 44, 2358–2375.

Joseph, T., Kar Mahapatra, D., Esmaeili, A., Piszczyk, Ł., Hasanin, M., Kattali, M., et al., 2023. Nanoparticles: taking a unique position in medicine. Nanomaterials 13, 574.

Kapare, H.S., Pagar, S., Kamandar, P., Bhowmik, D., Vishwakarma, N., Raikwar, S., et al., 2024. Exploring modified chitosan-based gene delivery technologies for therapeutic advancements. Int. J. Biol. Macromol. 260, 129581.

Kaphle, A., Jayarathna, S., Krishnan, S., Cho, S.H., 2025. Monte carlo study of gold nanoparticle-mediated radiosensitization effects using nanoscale cell model combined with fractal-based DNA model. Med. Phys. 52, 4605–4620.

Käver, L., Hinney, A., Rajcsanyi, L.S., Maier, H.B., Frieling, H., Steiger, H., et al., 2024. Epigenetic alterations in patients with anorexia nervosa—a systematic review. Mol. Psychiatry 29, 3900–3914.

Kawai, T., Ikegawa, M., Ori, D., Akira, S., 2024. Decoding toll-like receptors: recent insights and perspectives in innate immunity. Immunity 57, 649–673.

Kazantseva, N.E., Smolkova, I.S., Babayan, V., Vilčáková, J., Smolka, P., Saha, P., 2021. Magnetic nanomaterials for arterial embolization and hyperthermia of parenchymal organs tumors: a review. Nanomaterials 11, 3402.

Khan, A.R., Grewal, N.S., Jun, Z., Khalid, M., Zhang, H.-J., 2025. Advances in smart hybrid scaffolds: a strategic approach for regenerative clinical applications. Eng. Regen. 6, 85–110.

Khedkar Ajay, T.D., 2012a. Pharmacosomes: an emerging vesicular system. Inven. Rapid NDDS 1, 1–3.

Khedkar Ajay, T.D., 2012b. Cubosomes: emerging drug delivery system. Inven. Rapid NDDS 1, 1–3.

Khursheed, R., Dua, K., Vishwas, S., Gulati, M., Jha, N.K., Aldhafeeri, G.M., et al., 2022. Biomedical applications of metallic nanoparticles in cancer: current status and future perspectives. Biomed. Pharmacother. 150, 112951.

Khuspe, P.R., Faiyazuddin, M., Uddin, M.J., Bhowmik, D.D., Said, R.P., Sonawane, K.S., et al., 2025a. Lipid-based vaccines against viruses including COVID-19. In: Lipids in pulmonary drug delivery. Elsevier, USA, pp. 409–433.

Khuspe, P.R., Faiyazuddin, M., Uddin, M.J., Gupta, J., Kamandar, P., Bhowmik, D.D., et al., 2025b. Current scenario and challenges of lipid-based nanomedicines for pulmonary drug delivery. In: Faiyazuddin Hasan Ali, M.D., Akbar Babar Iqbal, M.D. (Eds.), Lipids in pulmonary drug delivery. Elsevier, USA, pp. 21–47.

Khuspe, P.R., Mane, D.V., 2026a. Emerging machine learning techniques for the development and analysis of anticancer drugs from plants. In: Mahendr, R., Shandesh, B. (Eds.), Emerging trends in phytotherapy of cancer. Elsevier, USA, pp. 461–493.

Khuspe, P.R., Mane, D.V., 2026b. Artificial intelligence driven discovery of anticancer phytochemicals: from plants to lead compounds. In: Mahendr, R., Shandesh, B. (Eds.), Emerging trends in phytotherapy of cancer. Elsevier, USA, pp. 441–460.

Khuspe, P.R., Pardeshi, S.R., Uddin, M.J., Das, U., Hatvate, N.T., Rojekar, S., et al., 2024. Achieving optimal health with host-directed therapies (HDTs) in infectious diseases—a new horizon. Adv. Ther. 8, 2400169.

Khuspe, P.R., Rebello, N.L., Jain, M.N., Desai, H.R., Giram, P., 2026a. Photon-avalanching nanoparticles for biomedical applications. In: Kalim, D., Kevin, D.B., Chaudhery, M.H. (Eds.), Upconversion nanoparticles for biomedical applications. Elsevier, USA, pp. 185–225.

Khuspe, P.R., Rebello, N.L., Jain, M.N., Desai, H.R., Giram, P., 2026b. Upconversion nanoparticles for drug and gene delivery. In: Kalim, D., Kevin, D.B., Chaudhery, M.H. (Eds.), Upconversion nanoparticles for biomedical applications. Elsevier, USA, pp. 517–575.

Kumar, D., Kumar, S., Kumari, R., Tanwar, R., Kumar, V., Khalid, M., et al., 2024. A comprehensive review on the biomass-mediated synthesis of silver nanoparticles: opportunities and challenges. J. Environ. Chem. Eng 13, 115133.

Kustra, A., Zając, M., Bednarczyk, P., Maliszewska-Olejniczak, K., 2025. Exposure to polystyrene nanoparticles leads to dysfunction in DNA repair mechanisms in Caco-2 cells. Biol. Res. 58, 49.

Ladeira, C., Møller, P., Giovannelli, L., Gajski, G., Haveric, A., Bankoglu, E.E., et al., 2024. The comet assay as a tool in human biomonitoring studies of environmental and occupational exposure to chemicals—a systematic scoping review. Toxics 12, 270.

Lai, R., Pan, X., Qin, Y., Liang, J., Wu, L., Dong, M., et al., 2024. Highly integrated, self-powered and activatable bipedal DNA nanowalker for imaging of base excision repair in living cells. J. Nanobiotechnology 22, 636.

Landsiedel, R., Honarvar, N., Seiffert, S.B., Oesch, B., Oesch, F., 2022. Genotoxicity testing of nanomaterials. WIREs Nanomedicine and Nanobiotechnology 14 (6), e1833.

Li, C.M.-C., Cordes, A., Oliphant, M.U.J., Quinn, S.A., Thomas, M., Selfors, L.M., et al., 2024. Brca1 haploinsufficiency promotes early tumor onset and epigenetic alterations in a mouse model of hereditary breast cancer. Nat. Genet. 56, 2763–2775.

Liu, Z., Hua, X., Zhao, Y., Bian, Q., Wang, D., 2024. Polyethylene nanoplastics cause reproductive toxicity associated with activation of both estrogenic hormone receptor NHR-14 and DNA damage checkpoints in C. elegans. Sci. Total Environ. 906, 167471.

Magar Tushar V., M.D.S., 2011a. Advances in the field of nanooncology: an overview. Inven. Impact NDDS 4, 1–3.

Magar Tushar V., M.D.S., 2011b. Elastic vesicles as drugs carriers through the skin. Inven. Rapid NDDS 1, 1–2.

Maslov, A.Y., Makhortov, S., Sun, S., Heid, J., Dong, X., Lee, M., et al., 2022. Single-molecule, quantitative detection of low-abundance somatic mutations by high-throughput sequencing. Sci. Adv. 8, eabm3259.

May, S., Hirsch, C., Rippl, A., Bohmer, N., Kaiser, J.-P., Diener, L., et al., 2018. Transient DNA damage following exposure to gold nanoparticles. Nanoscale 10, 15723–15735.

Mesut, B., Al-Mohaya, M., Yeşilkaya, E., Das, U., Akhtar, M.S., Sah, R., et al., 2024. Demystifying the potential of lipid-based nanocarriers in targeting brain malignancies. Naunyn. Schmiedebergs. Arch. Pharmacol. 397, 9243–9279.

Mishra, M., Panda, M., 2021. Reactive oxygen species: the root cause of nanoparticle-induced toxicity in Drosophila melanogaster. Free Radic. Res. 55, 919–935.

Mo, Y., Zhang, Y., Zhang, Q., 2024. The pulmonary effects of nickel-containing nanoparticles: cytotoxicity, genotoxicity, carcinogenicity, and their underlying mechanisms. Environ. Sci. Nano 11, 1817–1846.

Montilla-Rojo, J., Hillenius, S., Eleveld, T.F., Salvatori, D.C.F., Looijenga, L.H.J., 2025. Chromosome-specific aberrations, rather than general aneuploidy, may drive mouse embryonic stem cell-derived teratoma metastasis. Stem Cell Res. Ther. 16, 181.

Nain, R., Patel, H., Chahar, M., Kumar, S., Rohilla, D., Pal, M., 2025. Biosynthesized metallic nanoparticles for sustainable environmental remediation: mechanisms, applications, and future perspectives. Discov. Chem. 2, 124.

Nimmy, S., Salam, A., Ibrahim, M.H.A.N., 2024. Manipulating the rhizosphere microbiome for plant health. In: Nimmy, S., Salam, A., Ibrahim (Eds.), Microbiome engineering. CRC Press, Florida, p. 19.

Pardeshi, S., Giram, P., 2023. Preparation, antibacterial and antiviral activity measurements and detection methods. In: Deshmukh, K. (Ed.), Antibacterial and antiviral functional materials. ACS Publications, USA, pp. 33–64.

Pardeshi, S., Kapare, H., More, A., Rebello, N., Giram, P., 2024a. Coating approaches of functionalized magnetic nanoparticles for theranostic applications. In: Mayank, P., Kalim, D., Chaudhery, M.H. (Eds.), Functionalized magnetic nanoparticles for theranostic applications. Wiley, pp. 131–171. doi:10.1002/9781394172917.ch5.

Pardeshi, S., More, M., Togre, N., Rebello, N., Giram, P., 2023. Dendrimers based antibacterial and antiviral materials. In: Kalim, D., Chaudhery, M.H. (Eds.), Antibacterial and antiviral functional materials. ACS Publications, USA, pp. 139–169.

Pardeshi, S., Pillai, D., Giram, P., Pardeshi, C., Naik, J., 2024b. Nanocarriers for ocular drug delivery. In: Prashant, L.P., Sanjula, B., Javed, A. (Eds.), Nanocarrier drug delivery systems. De Gruyter, Germany, pp. 469–500.

Pardeshi, S.R., Hatvate, N.T., Dhorkule, N., Sayyad, S.F., Faiyazuddin, M., Khalid, M., 2024c. Environmental implications and nanotechnological advances in octocrylene-enriched sunscreen formulations: a comprehensive review. Chemosphere 358, 142235.

Pardeshi, S.R., Hatvate, N.T., Gharat, K.D., Naik, J.B., Omri, A., 2024d. Advances in dorzolamide hydrochloride delivery: harnessing nanotechnology for enhanced ocular drug delivery in glaucoma management. Discov. Nano 19, 199.

Pardeshi, S.R., Sen, N.K., Khandagale, P.D., Khadka, I.M., Pardeshi, P.R., Datta, D., et al., 2026. Prospects and applications of nanoemulsions in cosmeceuticals. In: Lai, W.C. (Ed.), Nanotechnology in cosmeceuticals. Elsevier, USA, pp. 369–386.

Parit, S., Manchare, A., Mundhe, P., Hatvate, N., Rojekar, S., Patravale, V., 2024. Antibody-drug conjugates: a promising breakthrough in cancer therapy. Int. J. Pharm. 659, 124211.

Pathak, G., Mangla, S., Gupta, G.K., Bhan, V., Kapoor, R.K., 2025. Toxicological assessment and risk management of nanoparticles mediated composite materials-critical review: state of the art. Discov. Polym. 2, 12.

Patil, M., Dhumal, D., Rojekar, S., Jain, H., Pingale, P., Kumar Prajapati, M., 2024. Nanocarriers for antimicrobial therapy. In: Nanocarrier Drug Delivery Systems. De Gruyter, Germany, pp. 561–602.

Peng, Y., Yang, Z., Sun, H., Li, J., Lan, X., Liu, S., 2025. Nanomaterials in medicine: understanding cellular uptake, localization, and retention for enhanced disease diagnosis and therapy. Aging Dis 16, 168.

Płuciennik, K., Sicińska, P., Misztal, W., Bukowska, B., 2024. Important factors affecting induction of cell death, oxidative stress and dna damage by nano- and microplastic particles in vitro. Cells 13, 768.

Puri, A.V., 2018. Comparative evaluation of papaya and banana starch as disintegrant in diclofenac sodium tablet formulation. Int. J. Pharm. Biol. Sci. 8, 512–522.

Raheem, M.A., Rahim, M.A., Gul, I., Zhong, X., Xiao, C., Zhang, H., et al., 2023. Advances in nanoparticles-based approaches in cancer theranostics. OpenNano 12, 100152.

Rajurkar, V.G., 2012a. Recent technologies in nanomedicine for treatment of cancer. Inven. Rapid Pharm Tech 4, 1–3.

Rajurkar, V.G., 2012b. Potential applications of capillary electromigration methods for the improvement of separation efficiency. Inven. Rapid Pharm Anal. Qual. Assur. 1, 1–3.

Rajurkar, V.G., 2012c. Potential applications of superparmagnetic nanoparticles technology. Inven. Rapid NDDS 1, 1–3.

Rajurkar Vikas, G., 2012a. New threat of clinical toxicities evoked from nanocarrier systems. Inven. Rapid NDDS 1, 1–4.

Rajurkar Vikas, G., 2012b. Lipid nanocapsules: a novel promising tool as drug delivery system. Inven. Rapid NDDS 1, 1–4.

Rajurkar Vikas G., D.S.B., 2012. Phonophoresis: ultrasound enhanced transdermal drug delivery. Inven. Rapid NDDS 1, 1–3.

Rojekar, S., Kapare, H.S., Vishwakarma, N., Raikwar, S., Garkal, A., Mehta, T.A., et al., 2024a. Chitosan scaffolds: Expanding horizons in biomedical applications. Carbohydr. Polym. 323, 121394.

Rojekar, S., Parit, S., Manchare, A., Nangare, S.N., Hatvate, N., Sugandhi, V.V., et al., 2024b. Revolutionizing eye care: exploring the potential of microneedle drug delivery. Pharmaceutics 16, 1398.

Rojekar, S., Togre, N., Bhoj, P., Haeck, C., Hatvate, N., Singh, N., et al., 2024. Current status of mannose receptor-targeted drug delivery for improved anti-HIV therapy. J. Control. Release 372, 494–521.

Said, R.P., Pawar, R.D., Ambore, G.S., Hatvate, N.T., 2023. Importance of carbohydrate-drug conjugates in vaccine development: a detailed review. In: Verma, S.K., Das, A.K. (Eds.), Bioconjugated materials part 2 applications in drug delivery, vaccine formulations and important conjugates for cancer therapy, comprehensive analytical chemistry. Elsevier, Amsterdam, Netherlands, pp. 191–256.

Santonastaso, M., Mottola, F., Iovine, C., Genualdo, V., Montano, L., Piscopo, M., et al., 2025. Protective effects of anthocyanin and α-tocopherol against titanium dioxide nanoparticle-induced DNA damage in human sperm cells. Expo. Heal. 17, 523–535.

Sawant, S.D., Pardeshi, S.R., Khuspe, P.R., Sayyad, S.F., Chavan, M., Hatvate, N.T., et al., 2024a. Omics tools for determining microbial composition: a systematic review. Microbiome Eng. 1, 35–50.

Sawant, S.D., Pardeshi, S.R., Khuspe, P.R., Sayyad, S.F., Chavan, M., Hatvate, N.T., et al., 2024b. The role of inulin in altering the animal microbiome: a comprehensive review. Microbiome Eng. 1, 79–102.

Sawant, S.D., Sayyad, S.F., Hatvate, N.T., Chavan, M., Rojekar, S., Faiyazuddin, M., 2023. Food biopackaging for human benefits: status and perspectives. In: Ahmed, S. (Ed.), Biobased packaging materials. Springer Nature Singapore, Singapore, pp. 1–35.

Sayyad, S.F., Hatvate, N.T., Dhumal, V.V., Pardeshi, S.R., Chavda, V.P., Vora, L.K., 2023. Drug delivery strategies for avobenzone: a case study of photostabilization. Pharmaceutics 15, 1008.

Scalabrin, S., Magris, G., Liva, M., Vitulo, N., Vidotto, M., Scaglione, D., Del Terra, L., et al., 2024. A chromosome-scale assembly reveals chromosomal aberrations and exchanges generating genetic diversity in coffea arabica germplasm. Nat. Commun. 15, 463.

Schmidt, A., Groh, A.M., Frick, J.S., Vehreschild, M.J.G.T., Ludwig, K.U., 2022. Genetic predisposition and the variable course of infectious diseases. Dtsch. Arztebl. Int. 119 (8), 117–123.

Shaikh, M.S., Faiyazuddin, M., Khan, M.S., Pathan, S.K., Syed, I.J., Akhtar, M.S., et al., 2024. Chikungunya virus vaccine: a decade of progress solving epidemiological dilemma, emerging concepts, and immunological interventions. Front. Microbiol. 15, 1413250.

Siivola, K.M., Burgum, M.J., Suárez-Merino, B., Clift, M.J.D., Doak, S.H., Catalán, J., 2022. A systematic quality evaluation and review of nanomaterial genotoxicity studies: a regulatory perspective. Part. Fibre Toxicol. 19, 59.

Singh, R., Srinivas, S.P., Kumawat, M., Daima, H.K., 2024. Ligand-based surface engineering of nanomaterials: trends, challenges, and biomedical perspectives. OpenNano 15, 100194.

Singh, T., Kumar, V.R., Sanap, G., Anjaiah, B., Roja, K., 2016. In vitro antipyrial activity of psidium guajava leaf extraCT. Eur. J. Pharm. Med. Res. 3, 312–315.

Suba, Z., 2024. DNA damage responses in tumors are not proliferative stimuli, but rather they are DNA repair actions requiring supportive medical care. Cancers (Basel) 16, 1573.

Teixeira, M.I., Lopes, C.M., Amaral, M.H., Costa, P.C., 2024. Navigating neurotoxicity and safety assessment of nanocarriers for brain delivery: strategies and insights. Acta Biomater 189, 25–56.

Toh, M., Ngeow, J., 2021. Homologous recombination deficiency: cancer predispositions and treatment implications. Oncologist 26, e1526–e1537.

Uddin, M.J., Faiyazuddin, M., Omri, A., Gowri, S., Khalid, M., 2024. Advances in artificial intelligence for drug delivery and development: a comprehensive review. Comput. Biol. Med. 178, 108702.

Uddin, M.J., Faiyazuddin, M., Saeed, S.M., Sadiq, S.B., Koly, I.A.C., 2025. Lipid-based drug delivery for systemic effect through pulmonary route of administration. In: MD Faiyazuddin (Ed.), Lipids in pulmonary drug delivery. Elsevier, USA, pp. 255–299.

Vardikar, A., Das, U., Mandal, S., Mandlik, S., Bari, D., Pardeshi, C., et al., 2025. Surface engineered multimodal magnetic nanoparticles for neurodegenerative diseases. In: Targeted therapy for the central nervous system. Elsevier, USA, pp. 121–153.

Vignard, J., Pettes-Duler, A., Gaultier, E., Cartier, C., Weingarten, L., Biesemeier, A., et al., 2023. Food-grade titanium dioxide translocates across the buccal mucosa in pigs and induces genotoxicity in an in vitro model of human oral epithelium. Nanotoxicology 17, 289–309.

Vítor, A.C., Huertas, P., Legube, G., de Almeida, S.F., 2020. Studying DNA double-strand break repair: an ever-growing toolbox. Front. Mol. Biosci. 7, 24.

Vogt, A., He, Y., 2023. Structure and mechanism in non-homologous end joining. DNA Repair (Amst) 130, 103547.

Wan, J.C.M., Stephens, D., Luo, L., White, J.R., Stewart, C.M., Rousseau, B., et al., 2022. Genome-wide mutational signatures in low-coverage whole genome sequencing of cell-free DNA. Nat. Commun. 13, 4953.

Wang, J., Li, C., Han, J., Xue, Y., Zheng, X., Wang, R., et al., 2025a. Reassessing the roles of oxidative DNA base lesion 8-oxoGua and repair enzyme OGG1 in tumorigenesis. J. Biomed. Sci. 32, 1.

Wang, W., Yang, F., Zhang, L., Wang, M., Yin, L., Dong, X., et al., 2024. Targeting DNA damage and repair machinery via delivering WEE1 inhibitor and platinum (IV) prodrugs to stimulate STING pathway for maximizing chemo-immunotherapy in bladder cancer. Adv. Mater. 36 (1), 2308762.

Wang, Y., Yang, C., Shi, Q., Zhang, L., Liu, H., You, J., et al., 2025b. Co-exposure to enrofloxacin and atrazine enhances the hepatotoxicity in Larimichthys crocea by targeting the hypothalamic–pituitary–thyroid and gut–liver axes. J. Hazard. Mater. 489, 137548.

Watson, E.R., Taherian Fard, A., Mar, J.C., 2022. Computational methods for single-cell imaging and omics data integration. Front. Mol. Biosci. 8, 768106.

Xuan, L., Ju, Z., Skonieczna, M., Zhou, P., Huang, R., 2023. Nanoparticles-induced potential toxicity on human health: applications, toxicity mechanisms, and evaluation models. MedComm 4 (4), e327.

Yachi, Y., Matsuya, Y., Yoshii, Y., Fukunaga, H., Date, H., Kai, T., 2023. An analytical method for quantifying the yields of DNA double-strand breaks coupled with strand breaks by γ-H2AX focus formation assay based on track-structure simulation. Int. J. Mol. Sci. 24, 1386.

Yamamoto, H., Hirasawa, A., 2021. Homologous recombination deficiencies and hereditary tumors. Int. J. Mol. Sci. 23, 348.

Youhanna, S., Kemas, A.M., Preiss, L., Zhou, Y., Shen, J.X., Cakal, S.D., et al., 2022. Organotypic and microphysiological human tissue models for drug discovery and development—current state-of-the-art and future perspectives. Pharmacol. Rev. 74, 141–206.

Zhang, H., Zhou, H., Zhang, N., Jia, Y., Qiu, M., Yao, S., et al., 2023. Circ_0089282 inhibits carbon black nanoparticle-induced DNA damage by promoting DNA repair protein in the lung. Toxicol. Sci. 192, 71–82.

Zhang, N., Xiong, G., Liu, Z., 2022. Toxicity of metal-based nanoparticles: challenges in the nano era. Front. Bioeng. Biotechnol. 10, 1001572.

Zhuang, X., 2021. Spatially resolved single-cell genomics and transcriptomics by imaging. Nat. Methods 18, 18–22.

CHAPTER

24

DNA mutagenesis and genomic instability induced by heavy metals: Biological and environmental perspectives

Naureenbanu Makarani[a,b], *Krupanshi Bharadava*[a,b], *Aviral Kaushik*[a,b], *Sanjay Kumar*[c] and *Radhey Shyam Kaushal*[a,b]

[a]**Biophysics & Structural Biology, Research & Development Cell, Parul University, Vadodara, Gujarat, India** [b]**Department of Life Sciences, Parul Institute of Applied Science Research & Development Cell, Parul University, Vadodara, Gujarat, India** [c]**Structural and Molecular Biology, Center for Discovery and Innovation Building, The City College of New York, NY, United States**

24.1 Introduction

Human health is negatively impacted by heavy metals (HMs), and modern industrialization and anthropogenic and industrial activities have increased exposure to these metals. Toxic metal contamination of air and water is a global environmental issue that affects hundreds of millions of people. Heavy metal pollution of food is another issue that affects both human and animal health. In this context, the concentration of heavy metals in food, air, and water resources is evaluated (Ghorani-Azam et al., 2016; Luo et al., 2020). HMs are naturally occurring metals with a comparatively high density. These metals that contain metalloids, such as arsenic, are harmful and can have negative health effects. Exposure to these dangerous substances has significantly increased as a result of the growing use of these metals in various fields, such as industries and agriculture; this has become a global health problem (Tchounwou et al., 2012). Even though there are trace levels of these metals in the environment, the exposure to them rises as mining and smelting activities expand. Additionally, there is a growth in the usage of these metals

Genome Damage: Repair and Adaptation Across Life.
DOI: https://doi.org/10.1016/B978-0-443-42857-9.00001-3

in a variety of industries, such as pharmaceuticals and agriculture, which exposes people to them. The severity of HM poisoning cases in the US is depicted. These metals may also become contaminated in the environment as a result of air deposition, corrosion, HM leaching, and soil erosion (He et al., 2005).

These nonessential metals, which include arsenic, lead (Pb), cadmium (Cd), and mercury (Hg), interact with various cell components such as proteins, mitochondria, cell membranes, and DNA to cause transcriptional changes, reactive oxygen species (ROS)-related damage, cell DNA damage that results in numerous diseases, and conditions like cancer and multiorgan damage (Dasharathy et al., 2022). There may be cumulative effects from concurrent exposure to two or more metals (Costa, 2019; Gazwi et al., 2020). Excessive exposure to heavy metals, especially Pb and Hg, can cause serious side effects such as kidney failure, bloody diarrhea, and abdominal colic. However, until it is done frequently, low-dose exposure poses a subtle and concealed threat that can be identified by its consequences, such as neuropsychiatric illnesses that include anxiety, exhaustion, and negative effects on children's IQ and intellectual function. Another significant element of long-term exposure is the discovery that a number of metals are carcinogenic to humans.

The toxicity mechanism varies depending on the HM. Therefore, it is critical to comprehend the exposure and toxicity of various metals in order to treat illnesses and lessen the pollution that HMs create. It is impossible to go over and discuss every HM and all of its dangerous effects. Thus, we chose to focus on Cd, chromium (Cr), arsenic, Pb, and Hg. In addition to discussing the present treatments for HM toxicity and alternative medical systems for treating it, this study focuses on the metal stress and its mutagenic, carcinogenic, and teratogenic effects of HM toxicity as well as its mode of action and issues related to certain organs.

24.2 Heavy metal toxicity

24.2.1 Chromium

In industrial operations, Cr is a naturally occurring heavy metal that can be found in saltwater and the Earth's crust (Tchounwou et al., 2012). Among the several oxidation states of Cr, which range from -2 to + 6, the most prevalent and stable forms are trivalent and hexavalent (Shekhawat et al., 2015). While trace levels of Cr (III) are necessary for natural lipid and protein metabolism as well as a cofactor for insulin action, Cr (VI) is linked to some illnesses and pathologies (Achmad and Auerkari, 2017; Vincent, 2017; Vincent, 2019). Hexavalent Cr has been recognized as a category I occupational carcinogen by the International Agency for Research on Cancer (IARC) in their 2018 report (Loomis et al., 2018). Ingestion of Cr-containing food and water or skin contact with Cr-containing goods are the main ways that nonoccupational human populations are exposed. In addition, a significant quantity of Cr is released into the soil, groundwater, and air by the metallurgical, refractory, and chemical sectors, which harms human, animal, and marine health (Fang et al., 2014). Because Cr bioaccumulates in the human body, it can lead to several illnesses. This includes lung, laryngeal, bladder, kidney, testicular, bone, and thyroid malignancies, as well as neurological, GI, and dermatological disorders.

According to a recent meta-analysis, exposure to Cr (VI) may raise the incidence and mortality of certain malignancies in humans, such as those of the lung, throat, bladder, kidney, testicles, bone, and thyroid (Deng et al., 2019). Furthermore, an ecological study conducted

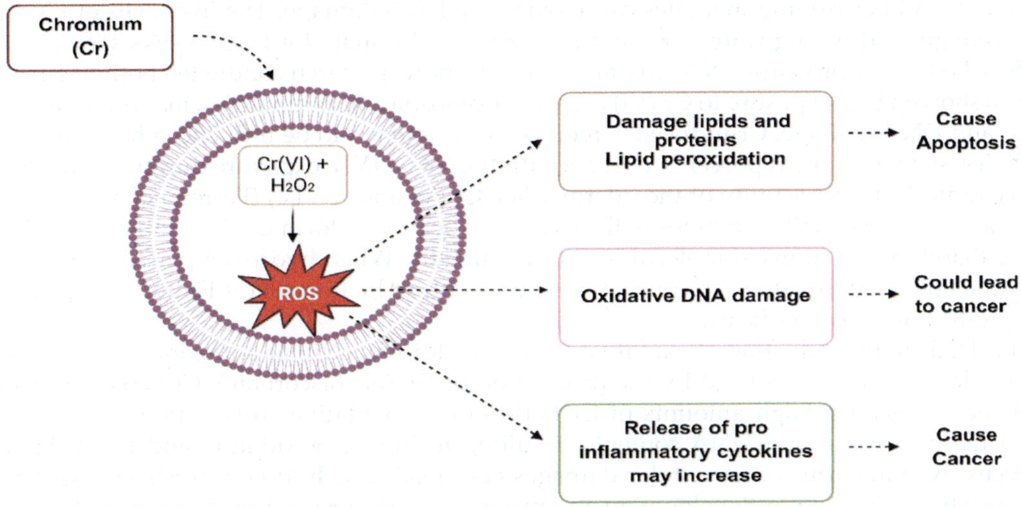

FIGURE 24.1 Hexavalent chromium induces DNA damage through oxidative mechanisms.

in Greece in 2011 discovered that drinking water had high levels of Cr (IV), ranging from 41 to 156 μg/L, and that women were more likely to develop liver ($p < 0.001$), lung (p 0.047), and genitourinary malignancies (p 0.025) (Linos et al., 2011). Another study conducted in India in 2012 revealed that people exposed to groundwater contaminated with Cr (VI) had a higher frequency of gastrointestinal and dermatological issues (Sharma et al., 2012). It is acknowledged that the mechanisms of Cr toxicity and carcinogenicity include ROS production, genomic instability, and DNA damage. ROS can be produced by both Cr (VI) and Cr (III) (Pavesi and Moreira, 2020). Following Cr carcinogenicity, transcription regulation disruption results in DNA damage. Moreover, disruptions in chromatin architecture led to changes in chromatin accessibility as well as changes in nucleosomal positions both locally and throughout the genome. Single- and double-strand breaks (SSBs and DSBs, respectively), DNA–Cr–protein crosslinks, DNA inter/intrastrand crosslinks, and p53 point mutations are all examples of Cr-induced DNA damage (Urbano et al., 2012; Thompson et al., 2013). The ability of Cr (III) to bind nucleic acids has been shown in in vitro studies (Urbano et al., 2012). Binary and ternary DNA adducts are formed by Cr (III). In mammalian cells, ternary adducts are more common and pertinent to toxicology. There are four different kinds of ternary adducts: Cr (III) glutathione, Cr (III) histidine, Cr (III) cysteine, and Cr (III) ascorbate DNA adducts. In human lung cancer cells (A549), Cr (III) creates DNA–protein crosslinks. Since the collapse of the replication fork during DNA replication causes cell cycle arrest and apoptosis via p53 activation, SSBs and DSBs have been identified in rapidly dividing cells, including cancer cells (Wilhelm et al., 2020). In one study, Ueno et al. found that a single injection of Cr(VI) caused SSBs to be found in the liver and kidney of male albino mice, but no discernible amounts of DNA strand breaks were found in the brain, lung, or spleen (Urbano et al., 2012). Fig. 24.1 illustrate the hexavalent Cr causes DNA damage.

Strand breaks and hydrogen abstraction from the ribose moiety might result from hydroxyl radicals produced when GSH reduces Cr (VI). Generally speaking, ROS activity or the oxidation states of Cr(V) peroxo intermediates cause oxidative DNA damage. The liver cells of F344 rats that were given drinking water containing potassium chromate for three weeks showed signs of SSBs, DSBs, and protein-rDNA crosslinks. Furthermore, research conducted both in vitro and in vivo showed that exposure to Cr (VI) caused chromosome abnormalities, microsatellite instability, and DNA damage. Cr-induced carcinogenicity appears to be influenced by a number of variables, such as tissue type, cell type, concentration of Cr (VI), exposure duration, production of free radicals, and reactivity of the intermediate Cr (V) and Cr (IV) (Ferreira et al., 2019). The hexavalent cation Cr (VI) combines with cellular reductants to form Cr (V), Cr (IV), and Cr (III) rather than binding to macromolecules or DNA directly. When hydrogen peroxide is present, these metabolic intermediates produced during Cr (VI) reduction set off Fenton-type reactions that produce hydroxyl radicals.

This leads to the induction of oxidative stress. In addition, cellular antioxidants are depleted and oxidative stress is induced by the production of Cr-Asc (ascorbate), Cr-GSH, and Cr-cys (cysteine) crosslinks. High amounts of Cr (VI)-induced oxidative stress and ROS generation target the cells' lipid and DNA contents, resulting in lipid peroxidation and DNA damage, respectively. Numerous other cellular damages and deaths, including necrosis and apoptosis, may also transpire. Tumor development and progression can result from medium-to-low levels of ROS generation in cells. According to one study, the ratio of reduced-to-oxidized glutathione (GSH/GSSG) depends on both dose and duration (Thompson et al., 2011).

24.2.2 Lead

Pb's physicochemical characteristics, including its softness, high malleability, ductility, low melting point, poor conductibility, and corrosion resistance, make it useful in a variety of industrial, agricultural, and household settings. Being nonbiodegradable, it is among the oldest recognized environmental and occupational toxins. Pb metal is classified as a suspected human carcinogen under group 2B by the IARC, while inorganic Pb compounds are classified as likely human carcinogens under group 2A (Brewer and Prasad, 2020). Over 10 years, Pb builds up in teeth and bones after exposure. In many places of the world, Pb has been phased out of use, but it is still needed for battery manufacture and smelting. It is one of the most harmful metals that can harm the body's main organs (Acharya et al., 2013). Elemental pollution has resulted from several homosexual activities such as mining, production, and industrialization. Pb has been used to produce Pb-acid batteries, ammunition, painted glass, metal oxides, pigments, chemicals, and X-ray shielding material. Even though the leading industries (battery manufacturers) are managing lead exposures, there are still infrequent occurrences of exposure in some of the industries (demolition industry). Natural Earth deposits contain lead, which is then discharged in the environment due to industrialization (Papanikolaou et al., 2005). The two main ways that lead poisoning can happen are through ingestion and inhalation. Pb can be inhaled by fumes or dust from activities in the automotive, smelting, storage battery, and Pb-glazed ceramics industries. Pb is also released by smoking cigarettes. Pb enters the food chain by ingestion through the use of tainted food or drinking water. Pb solder and water pipes could contaminate drinking water. Compared to adults, children are more susceptible to Pb poisoning. Toys, soiled paint, dust, or exposed contaminated soil are some of the ways they are exposed. Pb components released by cars are absorbed by the plants. Additionally, as a result of increased urbanization, Pb is now more prevalent in soil and sewage sludge, which can enter the food chain (Sharma and Dubey, 2005). Because absorbed Pb can be passed on to the fetus, exposure to Pb during pregnancy may be harmful (Tchounwou et al., 2012). The animals most impacted in cities are dogs and cats (Levin et al., 2021).

Numerous investigations have documented the various ways in which Pb causes toxicity, including oxidative stress. Oxidative stress, or an imbalance in the production of ROS and antioxidant defense to counteract it, is brought on by Pb exposure (Table 24.1) (Timbrell, 2008). The initiation of oxidative stress is represented by two simultaneous pathways: (1) Depletion and removal of antioxidant reserves, and (2) ROS production. Under normal circumstances, glutathione, a significant intracellular antioxidant, can exist in both reduced (GSH, 90%) and oxidized (GSSG, glutathione disulfide, 10%) forms. The sulfhydryl groups of antioxidant enzymes such as glutathione, glutathione reductase (GR), and glutathione peroxidase (GPx), which are the targets for inactivation, create a covalent connection with Pb by electron sharing (Sharma and Dubey, 2005). By increasing free radical levels, triggering lipid peroxidation, and decreasing many antioxidants (MDA, SOD, etc.), Pb puts hepatocytes under oxidative stress. Pb disrupts the transcription of several genes, including CYP2E1 and NOX2. Pb causes changes to DNA, DNA methylation, mitochondrial stress, the electron transport chain, the IRE1JNK pathway, RER damage in the ER, nuclear pyknosis, inflammation, and overexpression of NF-κB. It also changes the expression of the CYP7A1 gene and HMGR, which changes the metabolism of cholesterol and results in hepatocytic necrosis (Renu et al., 2021). The second potential way that Pb can harm the hemopoietic system is by way of δ-aminolaevulinic acid (ALA). A chemical called ALA contributes to the creation of porphyrins, which in turn leads to the synthesis of heme. It is created in the mitochondria when the ALA synthase enzyme condenses glycine and succinyl-CoA. Porphobilinogen is formed by the cytosolic enzyme δ-Aminolevulinic acid dehydratase (ALAD) (Piomelli, 2002). Pb may inhibit ALAD, which leads to the buildup of ALA and the production of free radicals, such as hydrogen peroxide (H_2O_2) and superoxide (O2 −). Hydroxyl (•OH) radicals are produced when the produced free radicals interact with oxyhemoglobin. Hemolysis follows hemoglobin oxidation due to increased lipid peroxidation. An indicator of oxidative stress is lipid peroxidation. Phosphatidylcholine in the cellular membrane can change its characteristics when Pb binds to it (Flora et al., 2012). Fig. 24.2 shows the process by which Pb toxicity promotes DNA damage.

The degree of exposure determines the compound's harmful effects. While chronic exposure has negative effects on the neurological system, hematological, renal, cardiovascular, and reproductive systems, acute exposure, however rare or through occupational exposure, can harm the gastrointestinal tract, kidney, and brain (Kalia and Flora, 2005). Because of its sensitivity, the nervous system—which is essential to the body's regular operation—is vulnerable to the effects of Pb poisoning. Pb primarily affects the peripheral nervous system (PNS) in adults and the central nervous system (CNS) in children. Muscle tremor, dullness, anger, poor concentration, memory loss, and hallucinations are some of the main signs of acute encephalopathy, a disorder linked to Pb exposure to the CNS (Hsiang, 2011). Lower levels of Pb toxicity can produce hyperactivity and irritable behavior in fetuses and early children due to the Pb's ability to cross the blood-brain barrier (BBB), while greater levels can result in lifelong brain damage and death (Ara and Usmani, 2015). Pb levels as low as 10μg/dl in the blood can have an impact on the hematopoietic system. Pb inhibits the production of hemoglobin, alters the heme biosynthesis pathway, and shortens the life of erythrocytes by making them brittle, which results in anemia. Pb toxicity results in hemolytic anemia (acute high-level Pb exposure) and frank anemia (blood Pb level elevated for an extended period of time). Three essential enzymes in the heme biosynthesis pathway are impacted by Pb: δ-aminolevulinic acid ferrochelatase,

TABLE 24.1 Mutagenic mechanisms and genomic instability induced by metals

Heavy metal	Mechanisms of DNA damage	Pathways affected	Genomic/mutagenic effects	Associated disorders
Lead (Pb)	Oxidative stress, enzyme inactivation (δ-ALAD, ferrochelatase), impaired heme biosynthesis, and methylation interference	Heme synthesis, mitochondrial DNA, antioxidant enzymes	DNA strand breaks, methylation changes, and chromosomal damage	Anemia, neurotoxicity, and immune dysfunction
Arsenic (As)	ROS generation, DNA methylation alteration, inhibition of DNA repair (PARP-1), formation of methylated arsenicals (MMA III)	DNA repair (NER and BER), epigenetic regulation	Chromosomal aberrations, gene silencing, 8-OHdG formation	Skin/lung/liver cancers, neurotoxicity
Mercury (Hg)	Binds to thiol/selenol groups, disrupts mitochondrial function, depletes GSH, and interferes with transcription	Mitochondrial DNA, antioxidant defense, and transcription machinery	Hypomethylation, DNA damage, and impaired gene regulation	CNS damage, nephrotoxicity, and immunosuppression
Cadmium (Cd)	ROS generation, miRNA dysregulation, apoptosis resistance, disruption of Ca/Zn/Fe homeostasis	DNA repair (NER and BER), β-catenin pathway, ER stress response	DNA strand breaks, methylation changes, chromosomal instability	Kidney, lung, prostate cancers; osteoporosis; nephrotoxicity
Chromium (Cr)	Cr(VI) is reduced to Cr(III) via ROS-producing intermediates, crosslinks with DNA and proteins, and strand breaks	DNA replication/repair, chromatin structure, and redox homeostasis	DNA–protein crosslinks, SSBs, DSBs, point mutations, and altered chromatin	Lung, liver, bladder, kidney cancers; skin disorders

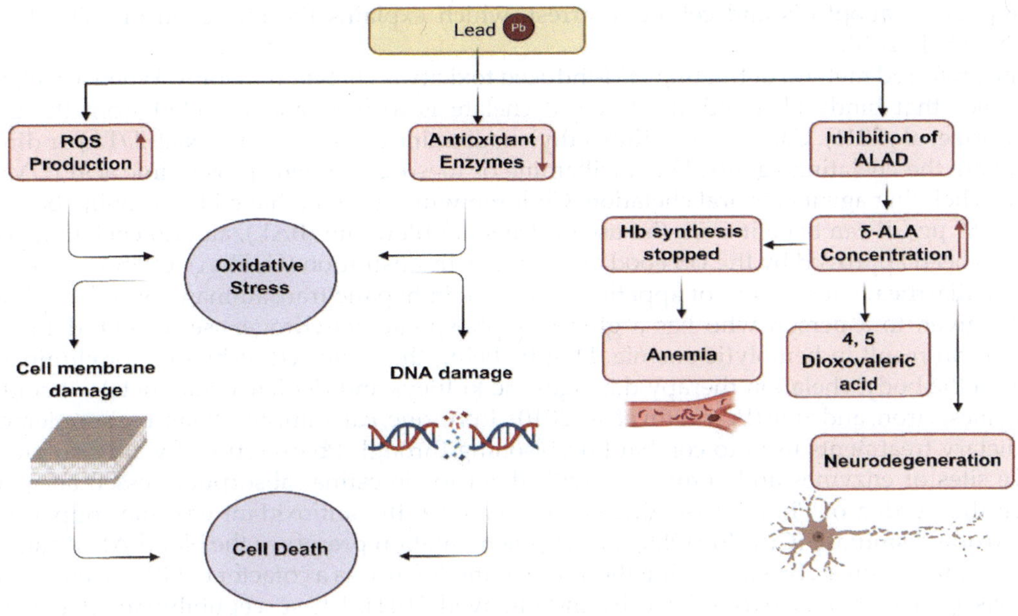

FIGURE 24.2 Lead toxicity promotes DNA damage through ROS generation and antioxidant enzyme suppression. ROS, reactive oxygen species.

aminolevulinic acid synthetase (ALAS) (Mitra et al., 2017). Pb exposure affects the renal system as low as 10 μg/dl and results in renal dysfunction when the concentration is higher than 60 μg/dl (Ekong et al., 2006). Although the exact role that Pb plays in the cardiovascular system is still unknown, blood Pb levels have been found to positively correlate with hypertension (Vaziri and Gonick, 2008). According to studies, exposure to Pb alters the synthesis of cytokines, raises the generation of ROS, and reduces the bioavailability of nitric oxide, all of which promote atherogenesis and the development of plaque due to elevated cholesterol levels (Sevim et al., 2020). Pb can have negative effects on both men's and women's reproductive systems. Exposure to Pb can increase a woman's risk of miscarriage, infertility, preterm birth, preeclampsia, and low birth weight. Increased toxicity can result from Pb in the blood entering the fetus through the placenta and breast milk. Men who are exposed to low levels of Pb have fewer sperm, and those who are exposed to high levels may have fewer sperm and motile sperm. According to studies, Pb can delay puberty and change the beginning of menopause by acting as an endocrine disruptor molecule (Mazumdar et al., 2017). It is well-known that Pb poisoning affects both the innate and adaptive immune systems. Pb exposure at work has a detrimental effect on the innate immune system because it induces neutrophils to produce ROS and alters mature macrophages' eicosanoid metabolism (Theron et al., 2012). Pb disrupts the vital T-helper 1 (*1)/T-helper 2 (*2) lymphocyte balance necessary for resistance against infectious illnesses, which impacts the adaptive immune system through T-lymphocytes. It enhances the expression of 2,3-dioxygenase, which is necessary for the creation of N-formyl-kynurenine, while decreasing CD4+T-helper cells. The tryptophan metabolic metabolite N-formyl-kynurenine causes

T-lymphocyte apoptosis and cell cycle arrest, which explains the reduction of CD4+T cells (Mishra et al., 2010).

The preferred method of treating HM-induced toxicity is chelation therapy. When a chelating substance that binds Pb is administered, a chelate is formed and excreted from the body (Gunturu et al., 2011). Calcium disodium ethylenediamine tetraacetate (CaNa2EDTA) or dimercaprol are the chelating agents; D-penicillamine or meso-2,3 dimercaptosuccinic acid (DMSA) are the chelating agents for oral chelation. Children with BLL who have Pb poisoning between 45 and 69 μg/dl can benefit from the use of British antilewisite (BAL), an oral chelation agent that has been approved by the US Food and Drug Administration (USFDA). DMSA side effects include diarrhea, nausea, loss of appetite, and a rise in hepatic transaminase. Succimer should not be given to a person who has a glucose-6-phosphate-dehydrogenase (G6PD) deficiency since it can result in hemolytic anemia. Despite being the preferred technique for eliminating Pb from the body, chelation therapy damages the kidneys and depletes vital metals, including manganese, iron, and zinc (Prasanthi et al., 2010). Iron, zinc, calcium, and thiamine supplements are dietary treatments used to combat Pb poisoning through Pb excretion. By adhering to the active sites of enzymes and competing with them for intestinal absorption, essential metals lessen the burden of Pb. Because zinc is a cofactor for the antioxidant enzyme copper zinc-superoxide dismutase (Cu/Zn SOD), zinc supplementation preserves the blood ALAD and reduces oxidative stress. By supporting the antioxidant defense as a cofactor of GPx, selenium also protects against Pb toxicity (El-Neweshy and El-Sayed, 2011). The susceptibility to Pb toxicity is increased by thiamine, pyridoxine, and ascorbic acid deficiencies. Thiamine's pyrimidine ring interacts with Pb to increase its excretion from the body and lessen its toxicity. Pb is chelated by the nitrogen atom of the pyridoxine ring before absorption, which lessens its buildup in tissues and lowers the inhibition of ALAD action. Since vitamin E is a naturally occurring nonenzymatic antioxidant that can reduce oxidative damage, it has a beneficial effect on lowering BLL (Farag et al., 2010). Commercial strains of probiotics include Lactobacillus, Bifidobacterium, Saccharomyces, and Bacillus, respectively. Bifidobacterium longum, Lactobacillus rhamnosus, and Lactobacillus plantarum are a few lactic acid bacteria (LAB) that bind HMs in vitro. It has been reported that Lactobacillus plantarum CCFM8661 has a protective effect against Pb toxicity, lowering BLL and preventing oxidative stress brought on by the metal.

24.2.3 Cadmium

Despite being uncommon, Cd can be found naturally in water, soil, and minerals like hydroxide, carbonate, chloride, and sulfide salts. High concentrations of Cd in soil, water, and air can happen after industrial operations, which may expose people to significant amounts of Cd. Furthermore, consuming tainted food will result in significant Cd exposure. Smoking can also result in Cd exposure, as it can raise Cd levels in the blood and urine. The body's essential processes may be disrupted by the presence of Cd in tainted water, potentially leading to either short-term or long-term illnesses (Jiang et al., 2015; Richter et al., 2017; Cao et al., 2018).

According to Kim et al. (2020), the IARC has categorized Cd as carcinogenic to humans (Group 1). The manufacturing of alloys, batteries, glass, and electroplating can all expose workers to Cd. In certain nations, the level of Cd in the air is regularly measured due to the significance of the topic. Although it has been discovered that Cd pollutes rice, cereals, and seafood (Chunhabundit, 2016), a tiny amount of Cd is absorbed after oral ingestion.

Unfortunately, widespread Cd contamination of food and water supplies was the cause of the Itai-itai illness outbreak in Japan. The patients had GI and lung disorders, kidney failure, and excruciating degenerative bone disease (Nishijo et al., 2017). Industrial dust is a more effective way for Cd to be removed from the lungs than low GI absorption. Chronic or acute Cd exposure in industrial settings may cause lung damage and renal tubular failure. Smokers' blood levels of CD are nearly twice as high as nonsmokers'. In the adult population, nonsmokers had Cd blood levels of 0.4 versus 1.3 μg/L compared to smokers, according to a study. This appears to be connected to tobacco plants' propensity to collect comparatively large levels of Cd in their tissues, particularly in their leaves (Proshad et al., 2020).

It is known that Cd binds to the metallothionein (MT) protein to generate dispositional tolerance. By restricting the distribution of Cd to delicate cellular macromolecules, such as MT, induction shields the liver. However, extremely high Cd levels cause unbound and free Cd, which is why the harmful effects happen. Cd levels in human tissues were reported by Lech and Sadlik (2017) in a 10-year study of a population that was not poisoned. The brain, stomach, small intestine, liver, kidneys, heart, and lungs were all autopsied. The brain has the lowest concentration of Cd (0.02μg/g digested material). Since Cd is mostly transported into the kidneys and liver, high quantities of the metal were found in the kidneys (16.0 μg/g) and liver (1.5 μg/g) (Lech and Sadlik, 2017). The kidneys easily absorb Cd bound to MT. The complex contributes mechanistically to the chronic nephrotoxicity of Cd and is extremely harmful to the kidney. Cd has the ability to behave and operate similarly to critical metals. For example, Cd binds to albumin in plasma similarly to zinc. As a result, iron, zinc, and calcium homeostasis are dysregulated (Schaefer et al., 2020).

Exposure to Cd has been linked to serious bone health issues, including osteoporosis and an increased risk of fractures. A study by Chen et al. (2019) compared people living in two Cd-contaminated areas with those from a noncontaminated region. The results highlighted a clear connection between high Cd exposure and osteoporosis. While both men and women were affected, the impact was especially significant among women. One possible explanation for Cd's harmful effects on bones is its ability to lower serum parathyroid hormone (PTH) levels when present in high amounts in the body. This hormonal disruption can lead to the release of calcium from bones, weakening them over time. Supporting this, Lv et al. (2017) reported that higher Cd levels in the body were associated with lower bone density, reinforcing the link between Cd exposure and osteoporosis. Beyond its impact on bones, Cd also contributes to cardiovascular diseases related to smoking, such as peripheral artery disease and coronary heart disease (Table 24.1).

Using mediation analysis, Li et al. (2019) discovered a correlation between Cd exposure and the cardiovascular disease pathway in a group of 4304 middle-aged men and women. They explained that Cd has a role in the development of cardiovascular illnesses brought on by smoking (Li et al., 2019). Cd may cause malignancies of the kidney, lungs, pancreas, breast, prostate, and gastrointestinal tract. According to a suggested mechanism, a series of events, including the production of ROS, an increase in TNF-α, the overexpression of Nrf2, and, in the end, abnormal gene expression, apoptosis resistance, and dysregulation of cell proliferation (Wang Y et al., 2018). After extensive exposure in the chemical industry, Cd may become nephrogenic. According to a theory in this area, renal carcinogenesis, particularly Cd-induced nephrocarcinogenesis, is influenced by the β-catenin signaling system. This signaling leads to the accumulation of the transcriptional factor

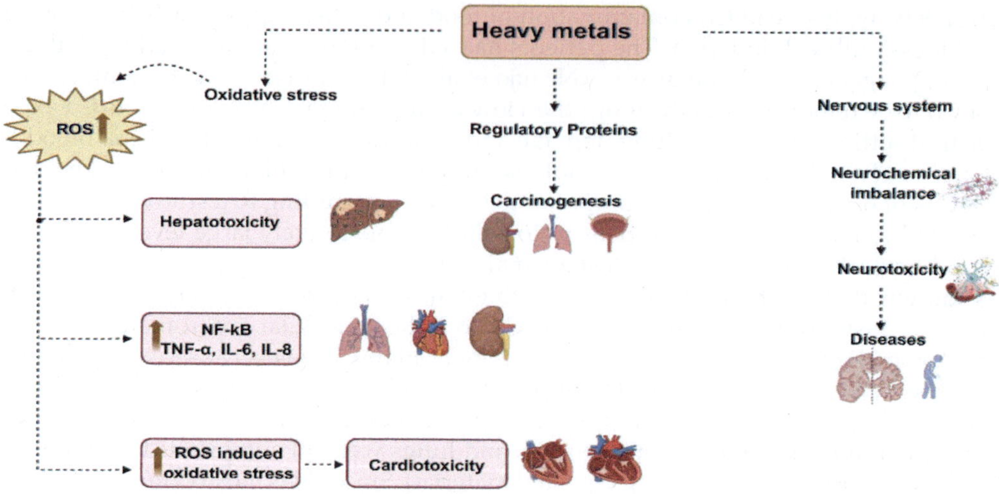

FIGURE 24.3 Mechanistic overview of heavy metal-induced toxicity via oxidative stress and regulatory pathways.

β-catenin protein in the cytoplasm. The avoidance of ROS-dependent/β-catenin signaling caused by overexpression of β-catenin contributes to kidney carcinogenesis (Fig. 24.3) (Son et al., 2012). A widely used approach for evaluating Cd exposure and its associated effects is urinary Cd. According to certain research, urinary Cd levels and the prevalence of renal failure or lung function deficiencies are related. In this context, there may be a dose-response relationship (Moitra et al., 2013). Urine containing beta 2-microglobulin may be a helpful indicator of tubular proteinuria, especially when evaluating toxicity from Cd. Given that urine contains as little as 360μg/L of beta 2 microglobulin, it plays a potential role (Argyropoulos et al., 2017).

24.2.4 Arsenic

Throughout history, arsenic has been used for a wide range of things, including poisons and medications. Before antibiotics were created in the early 1900s, arsphenamine (sold under the brand name Salvarsan) was used to treat amoebic dysentery, trypanosomiasis, and syphilis. Medicinal solutions such as potassium arsenite, arsenic iodide, and arsphenamine were used to treat rheumatism, arthritis, asthma, malaria, trypanosome infection, tuberculosis, and diabetes (Dasharathy et al., 2022). Humans can become contaminated by arsenic through natural mineral deposition in soil, air, food, and drinking water, industrial effluent that contains arsenic, or agricultural pesticides. Paddy, green vegetables, and underground veggies are shown to acquire more arsenic. Other foods that contain arsenic include fish, shellfish, pork, chicken, dairy products, and cereals; however, exposure from these sources is frequently much lower than that from tainted groundwater (Kumar, 2021). The biggest threat to public health originates from tainted groundwater. Higher amounts of arsenic in the form of inorganic ions are naturally present in the groundwater of about 108 nations, including Bangladesh, China, Mexico, Argentina, the United States of America, India, and Chile (Shaji et al., 2021).

The digestive tract swiftly absorbs arsenic into the bloodstream after it enters the body through food. The blood's capacity to deliver oxygen to vital organs is restricted by arsenic, which binds to oxyhemoglobin and causes severe red blood cell death. It also interferes with the kidney and liver's normal functions (Chandrasekaran et al., 2010). The sulphydryl group of cells is the specific target of arsenic, a protoplastic poison that disrupts cellular respiration and enzyme function (Jan et al., 2015). After long-term exposure to arsenic, humans and organisms (bacteria, algae, and fungi) enzymatically methylate poisonous inorganic arsenic molecules to produce less harmful monomethylarsonic acid (MMA) and dimethylarsinic acid (DMA), which are then eliminated by urine. Because monomethylarsonic acid (MMA III) has a strong affinity for sulfhydryl groups, it is highly hazardous. Its buildup is connected to arsenic-induced carcinogenesis, and it cannot be eliminated in the urine (Jaishankar et al., 2014).

Anemia, gastrointestinal symptoms, skin lesions, neuropathy, hyperpigmentation, vascular lesions, liver damage, and renal damage are among the known side effects of arsenic exposure. It has been demonstrated that long-term exposure to arsenic poisoning damages the neurological, respiratory, reproductive, endocrine, and cardiovascular systems. Maternal exposure to arsenic impairs T cell and B cell activities (Tung et al., 2014). The initial manifestations of acute arsenic poisoning include diarrhea, vomiting, and abdominal pain. In extreme situations, they are followed by muscle cramps, mortality, and tingling and numbness in the limbs (Dasharathy et al., 2022).

In addition to skin cancer, prolonged exposure to arsenic can cause lung and bladder cancer. Among the additional detrimental health effects associated with long-term ingestion of inorganic arsenic are developmental effects, diabetes, lung illness, and cardiovascular disease. One of the leading causes of death is arsenic-induced myocardial infarction. The serious condition known as "Blackfoot disease," which damages blood vessels and causes gangrene, has been linked to arsenic exposure (Hong et al., 2014). Due to their similar chemical makeup, arsenic and phosphorus can be substituted in important molecules or reactions like DNA methylation disruption, DNA repair inhibition, oxidative stress disruption in ATP generation, and more. Arsenic and phosphorus are members of the same periodic table group (Shen et al., 2013). Since the beginning of time, plants and plant-based products have been used to treat illnesses; the main benefits of plant-based medicine seem to be perceived efficacy, low frequencies of serious side effects, and low cost. The first line of treatment for arsenic poisoning is based on the symptoms that have appeared; the patient must be removed from the source of exposure. Chelating agents like succimer and dimercaprol are essential for patients with significant burdens, and hemodialysis should be considered for patients with kidney failure (Flora et al., 2005). According to the current literature review, natural products like rutin, β-carotene, α-tocopherol, curcumin, resveratrol, and quercetin, as well as medicinal plants like *Withania somnifera, Ipomea aquatica, Phyllanthus emblica, Triticum aestivum, Tephrosia purpurea, Camellia sinensis, Vitis vinifera*, and *Moringa oleifera* (Kumar et al., 2015; Dua et al., 2015; Sayed et al., 2015) and natural products like these substances also showed a good intrinsic antioxidant effect by suppressing arsenic-induced oxidative stress through multimodal augmentation of endogenous defense mechanisms, which alleviated arsenic toxicity (Sárközi et al., 2015; Ruma et al., 2015). The natural substances, also known as phytochemicals, are well-known antioxidants and nutraceuticals. This supports the idea of antioxidant therapy in people and highlights the necessity of antioxidant supplements. The biological benefits of these substances must be verified in humans who have been exposed to arsenic on a long-term basis, though. Following the administration of Arsenicum album-30, a potentized homeopathic remedy, to a group of arsenic patients, the levels of arsenic in their blood and urine were routinely assessed. The results suggest that the medication might be effective in treating human arsenic poisoning.

24.2.5 Mercury

According to the periodic table, Hg is a block element. At room temperature, it is the only common metal that is liquid. There are two oxidation states in which it can exist: I (mercurous) and II (mercuric). Hg's toxic profile varies slightly among its inorganic, organic, and elemental forms; for example, organomercuric compounds are thought to be the most dangerous due to their capacity to cross the BBB (Maqbool et al., 2017). Because of its great volatility, gaseous Hg—which is used in argon lamps, electron tubes, and other devices—is the most challenging to manage. Commonly found in sediments and water, dissolved elemental Hg, Hg (0), can be broken down by microbes into organic Hg compounds like methyl Hg, which can then accumulate in the food chain (Yang et al., 2020). Although the majority of Hg emissions are caused by human activities like burning coal, burning fuels containing Hg (Bourtsalas and Themelis, 2019), burning waste, mining, and the production of iron and steel, there are also natural sources like volcanic activity, weathering of rocks, oceans, soil, and biosphere accumulations. Approximately 65% of global Hg emissions in 2000 came from burning fossil fuels, 11% from the manufacture of gold, and only 7% and 6%, respectively, from the production of nonferrous metals and cement. The WHO states that eating tainted seafood and dental amalgams are two main ways that people might be exposed to Hg (Siblerud et al., 2019). Compared to fish and animals that eat herbivorous foods, carnivorous animals and fish are said to be more susceptible to Hg toxicity (de Oliveira et al., 2021). Higher species (such as mammals) can be poisoned by HM, much like humans. When HM interacts with vertebrates, it frequently causes genetic variation for tolerance and changes life-history features. Invertebrates that have adapted to HM live shorter lives and reproduce more laboriously (Rahman et al., 2019). Hg's solubility and its capacity to form bonds with sulfhydryl groups are the two factors that influence its toxicity. First and foremost, organic Hg is nearly entirely absorbed into the bloodstream, while Hg that is inorganic is absorbed into the body at a rate of about 2%–38%. It can enter the CNS and the BBB, as indicated before. Additionally, the toxin can easily infiltrate the cell membrane due to its fat-friendly characteristics. Hg often interacts with nucleophile components like sulfhydryl (-SH) or selenohydryl (-SeH) groups and forms stable complexes because it is electrophilic. Oxidative stress is caused by the chemical reactions that result in the production of ROS. By destroying lipids, nucleic acids, and enzymes, ROS are known to result in cell death (Wu et al., 2016). MeHg is also known to target and interfere with the mitochondrial electron transport chain, which lowers the body's glutathione levels by building a GS-HgCH3 complex with the thiol group in GSH. Hg inhibits the physiological and metabolic processes of the cell, causing oxidative stress and harming calcium homeostasis since the free thiol group is especially crucial for proper membrane functioning (Ynalvez et al., 2016). One of the main causes of neurotoxicity and neuroinflammation is Hg's inhibition of free thiol groups. In addition, Hg compounds also target the tubulin protein of microtubules (Syversen and Kaur 2012). The effects of lipid peroxidation and the antioxidant system are weakened by Hg, which also raises superoxide levels. According to reports, there are changes in the Nrf2-Keap1 molecules' signaling. One of the genetic changes is DNA damage that results from DNA breaks. Notable ultrastructural changes include RER dilatation, excessive H_2O_2 generation, mitochondrial edema, and changes in JNK phosphorylation and GRP78 protein activation.

It is well-known that secondary alterations in the structure of DNA and RNA, as well as structural alterations in ribosomal proteins, occur when Hg forms a bond with SH groups (Goodrich et al., 2013). Numerous neurotransmitters, including acetylcholine, serotonin, and norepinephrine, are impacted by Hg (Richard et al., 2010). It has been demonstrated to impact the immune system by lowering the absorption of vitamin B12 and plasma ascorbic acid (Zeneli et al., 2016; De Oliveira et al., 2021), as well as by triggering cytokine expression and IL-1 release (De Oliveira et al., 2021). By altering cytokine production, Hg has been demonstrated to cause immunosuppression, type-4 hypersensitivity, and autoimmunity (Aaseth et al., 2021). Children who are exposed to organic Hg during pregnancy may have mental retardation, deafness, blindness, and dysarthria (Boucher et al., 2014). On the other hand, it has been demonstrated that exposure to inorganic Hg is linked to neuropsychiatric problems, excessive sweating and hypersalivation, erythema, pruritus, sleeplessness, and renal tubular dysfunction (Ynalvez et al., 2016). Hg buildup in the kidney, liver, brain, and muscles has been documented. Additionally, Hg has a preference for sulfhydryl groups and binding sites on the surface of T cells, which affects T cell activity. Neurotoxicity, immunotoxicity, cardiovascular problems, carcinogenicity, and reproductive problems are among the health impacts of Hg (Zheng et al., 2017; Ma et al., 2018; Sanders et al., 2019).

BAL, DMPS, DMSA, penicillamine, calcium EDTA, and others are some of the most often utilized chelating agents for Hg toxicity. The chelating agents DMPS and DMSA combine with a variety of heavy metals, including Hg, to create complexes. Compared to chelators like BAL, a lot of data indicate that treatment with DMPS and DMSA increased urinary Hg production (Bernhoft et al., 2012; Cao Yang et al., 2015; Kim et al., 2019). Deferasirox and deferiprone together are an effective treatment for Hg toxicity in vivo by Iranmanesh (Iranmanesh et al.,2013). Oral administration of DMSA and DMPS is known to efficiently eliminate Hg through the kidney (Iranmanesh et al.,2013). Furthermore, methyl Hg is said to be more effectively removed by DMSA, even from the brain region (Bernhoft et al., 2012; Kosnett, 2013; Rafati-Rahimzadeh et al., 2014). Selenium is involved in the body's metabolism of Hg. Selenium has been demonstrated to lessen the toxicity of Hg, while the precise mechanisms are yet unknown (Bjørklund et al., 2015). DMSA is the favored chelating agent since it is significantly less toxic than DMPS and CaEDTA (Bjørklund et al., 2017). These chelating agents have many disadvantages, even though they are less harmful. There is a need to investigate alternative natural sources of treatment because adverse effects like fever, muscle soreness, and in severe cases, breathing difficulties, heart failure, and kidney damage, are known to occur during the course of treatment (Cao Yang et al., 2015; Bjørklund et al., 2017). Due to chelation's adverse effects, many alternative treatments have become more popular when used in conjunction with chelation. Plasma exchange treatment is an effective substitute in cases of extreme toxicity. For inorganic Hg toxicity, it works best. Remarkably, hemodialysis was shown to be useless, while induced sweating was proven to be a successful treatment (Kim et al., 2019). In addition, Ayurvedic medicine is becoming more and more popular for the same purpose (Breeher et al., 2015). It has been demonstrated that the extracts of *Connarus favosus* and *Plathymenia* reticulata decrease methylHg-induced lipid peroxidation and acetylcholinesterase activity, hence lowering Hg poisoning in zebrafish (Gombeau et al., 2019). The preventive effects of zinc (Zn) and N-acetylcysteine (NAC) against Hg toxicity in nursing rats were assessed in a recent study (Oliveira et al., 2020). It is well-established that a variety of natural antioxidants, including carotenoids, uric acid, vitamin E, and vitamin C, can lessen the ROS generated as a result of Hg exposure (Ghosh et al., 2013; Unsal, 2018; Subramaniam et al., 2019).

24.3 Mechanisms of heavy metal toxicity and associated health effects

24.3.1 Heavy metals binding to thiol group (binding to thiol groups and enzymatic inhibition)

One of the fundamental mechanisms by which heavy metals exert toxicity is their strong affinity for thiol (-SH) groups of cystine residues in proteins and of low molecular weight antioxidants such as glutathione (GSH). Heavy metals such as arsenic (As^{3+}), mercury (Hg^{2+}/MeHg), cadmium (Cd^{2+}), and lead (Pb^{2+}) form stable metal-thiol complexes that disrupt the native structures and function of enzymes. For instance, arsenite binds with lipoic acid in the pyruvate dehydrogenase complex, resulting in disruption of energy metabolism and apoptosis. Methylmercury is highly reactive with thiol groups and selenium in selenoenzymes such as glutathione peroxidase and thioredoxin reductase, leading to redox imbalance and oxidative neuronal damage (Koyama et al., 2024; Jomova et al., 2025). Cd and Pb are also known to displace zinc from critical zinc-finger structures in transcription factors and DNA repair proteins, disrupting gene expression and genomic stability. This thiol-directed mechanism not only inhibits enzymatic activity but also contributes further to cellular stress by depleting intracellular antioxidants such as GSH, thereby increasing susceptibility to oxidative damage (Kennedy et al., 2008; Witkiewicz-Kucharczyk et al., 2020; Makarani et al., 2025). Table 24.2 shows a Comparative mechanistic pathway across metal toxicity.

24.3.2 Generation of reactive oxygen species

Many heavy metals induce oxidative stress directly through redox cycling (as with iron or Cr) or indirectly by interfering with mitochondrial functions and antioxidant defenses. Even nonredox active metals such as Cd and arsenic can provoke substantial ROS production by depleting glutathione, inhibiting antioxidant enzymes (such as superoxide dismutase and catalase), and inducing NADPH oxidase activity (Samet et al., 2020; Zhu and Costa, 2020; Mansoor et al., 2023). Arsenic has been shown to increase the mitochondrial superoxide production and inhibit complex I and III of the electron transport chain (Hu et al., 2020). Cd interferes with the Fenton reaction through iron and copper release from intracellular stores, resulting in the indirect generation of hydrogen peroxide and hydroxyl radicals. Hg disrupts thiol homeostasis and mitochondrial function, exacerbating oxidative damage in high-energy-demanding tissues such as the brain (Briffa et al., 2020; Kaushik et al., 2025). The elevated ROS causes extensive biomolecular damage, including DNA strand breaks, lipid peroxidation, and protein carbonylation. Persistent oxidative stress also activates pro-inflammatory pathways (e.g., NF-κB), contributing to chronic inflammation and degenerative diseases (Rauf et al., 2024; Kaushal et al., 2025).

24.3.3 The carcinogenicity of heavy metals

Heavy metals such as arsenic, Cd, Cr [especially Cr (VI)], and nickel are well-established human carcinogens. These metals promote carcinogenesis not by directly causing point mutations like classical genotoxins but by inducing oxidative DNA damage, disrupting DNA repair pathways, altering cell signaling, and modifying the epigenome. Arsenic does not form DNA adducts but induces chromosomal mutation and inhibits base excision and nucleotide

TABLE 24.2 Comparative mechanistic pathways across metal toxicity.

Pathway	Mercury (Hg^{2+} / MeHg)	Arsenic (As^{3+} / As^{5+})	Cadmium (Cd^{2+})	Chromium (VI)	Lead (Pb^{2+})	References
Thiol-binding	Strong; MeHg forms stable complexes with cysteine, GSH, thioredoxin, lipoic acid; inhibits selenoenzymes and mitochondrial enzymes	High affinity for protein-bound—SH; binds lipoic acid in PDH, inhibits redox enzymes	Binds protein thiols; displaces Zn from zinc fing ers; inhibits SOD, catalase	Consumes GSH during intracellular reduction to Cr(III); generates DNA-binding intermediates	Weak direct thiol affinity; indirectly impairs proteins with—SH groups and chelates Zn/Ca	Witkiewicz-Kucharczyk et al. (2020)
ROS generation	Indirect disrupts mitochondrial function, induces lipid peroxidation and H_2O_2, and enhances NOX activity	Direct via NADPH oxidase stimulation, mitochondrial ROS depletes antioxidants like GSH	Indirect via mitochondrial damage, Fe release, and redox enzyme inhibition	Strong ROS from redox cycling ($Cr^{6+} \rightarrow Cr^{3+}$); hydroxyl radical generation via Fenton chemistry	Indirect via δ-ALA accumulation, mitochondrial dysfunction, and oxidative stress	Hu et al. (2020); Mittal et al. (2014); Mansoor et al. (2023)
DNA repair inhibition	Inhibits PARP-1, DNA polymerases; crosslinks DNA and histones; hinders NER, BER	Disrupts NER and BER by Zn displacement from DNA repair proteins	Inhibits XRCC1, ERCC1, OGG1 expression; blocks BER	Blocks ATM, DNA-PK; prevents repair foci formation; causes DNA–protein crosslinks	Binds DNA polymerase β; interferes with ligase activity; causes chromosomal aberrations	Demin et al. (2021); Wang et al. (2023); Lebedeva et al. (2024);

(*continued on next page*)

TABLE 24.2 Comparative mechanistic pathways across metal toxicity—cont'd

Pathway	Mercury (Hg^{2+} / MeHg)	Arsenic (As^{3+} / As^{5+})	Cadmium (Cd^{2+})	Chromium (VI)	Lead (Pb^{2+})	References
Cancer links	Kidney, brain, and immune cancers (less documented in humans); strong evidence in animal models	Skin, bladder, liver, prostate, lung; Group 1 carcinogen	Lung, kidney, prostate; Group 1 carcinogen	Respiratory cancers (lung, nasal) in occupational settings; DNA damage-driven carcinogenesis	Brain, kidney, lung; classified as a probable human carcinogen (Group 2A/2B)	Salvatore Gallicchio and Harper (2021)
Epigenetic alterations	Changes in DNA methylation (hypo/hyper), miRNA dysregulation (e.g., miR-92a), histone deacetylation, and chromatin remodeling	Genome-wide hypomethylation, promoter hypermethylation (p16, p53), altered miRNA (e.g., miR-222)	Promoter hypermethylation (p16), miRNA disruption (let-7, miR-101), histone PTM changes	Promoter hyper-methylation, H3K9me2 upregulation, global acetylation suppression	Alters DNA methylation and chromatin condensation; impairs histone acetylation	Arita and Costa (2009; Zhu and Costa (2020)

excision repair systems. It also modulates oncogenic pathways such as p53, c-Fos/c-Jun and NF-κB, enhancing proliferation and survival (Zhu and Costa, 2020). Cd inhibits DNA repair enzymes (e.g., XRCC1, OGG1) and displaces zinc from zinc-finger proteins, compromising genome maintenance (Alnajjar and Sweasy, 2019). Cr (VI) enters cells through sulfate channels and is reduced intracellularly to Cr (III), generating intermediates that produce ROS and DNA adducts. Cr (VI) inter- and intrastrand crosslinks and suppresses p53 (Sawicka et al., 2020; Mezencev and Gibbons, 2023). Nickel compounds silence tumor suppressor genes by promoting histone deacetylation and DNA hypermethylation. These metals are associated with organ-specific cancers; arsenic with skin, bladder, and lung cancer; Cd with lung and prostate cancers, Cr with respiratory tract tumors, and nickel with nasal and renal carcinomas (Arita and Costa, 2009).

24.3.4 Epigenetic mechanisms of heavy metals

Beyond classical genotoxicity, heavy metals alter gene expression through epigenetic modifications, including DNA methylation, histone posttranslational modifications (PTMs), and noncoding RNA expression changes. Arsenic is particularly well documented in this regard, showing both global hypomethylation (which may activate photo oncogenesis) and promoter-specific hypermethylation of tumor suppressors such as p53 and p16INK4a. It also modifies histone marks (e.g., increasing H3K9me2 and H3K4me3), affecting chromatin accessibility and transcriptional regulation (Romagosa et al., 2011; Zhu Costa, 2020). Cd has been shown to increase DNA methyltransferase expression and hypermethylated genes involved in DNA repair and apoptosis. Cr (VI) induces histone H3 lysine 9 demethylations (H3K9me2), associated with gene silencing, and disrupts histone acetylation. Nickel alters chromatin architecture by promoting H3K9 methylation and histone deacetylation, resulting in long-term transcriptional repression. These modifications are heritable during cell division, explaining how brief exposure can result in persistent gene expression changes, potentially contributing to diseases such as cancer, neurodegeneration, and immune dysfunction (Chen et al., 2019; Xiao et al., 2024).

24.3.5 Comparison of the mechanistic action

When comparing the mechanisms by which heavy metals exert toxicity, several converging themes emerge. Thiol-binding is a shared feature of arsenic, Cd, and Hg, leading to enzyme inactivation and depletion of antioxidants (Balali-Mood et al., 2021). Oxidative stress is another common outcome, although initiated by different pathways, direct redox cycling in the case of Cr and indirect mitochondrial disruption in the case of Cd and arsenic. DNA repair inhibition is also prevalent, with arsenic and Cd disrupting both BER and NER pathways through protein structural alteration or gene repression (Koedrith and Seo, 2011). The carcinogenicity of these metals is largely nonmutagenic but is instead driven by chronic inflammation, suppression of tumor suppressors, and epigenetic dysregulation. Arsenic, Cr, and nickel share the ability to silence gene expression via promoter hypermethylation and histone modifications. Despite these similarities, the degree of effect and the affected tissues can vary. Arsenic primarily targets the skin and bladder; Cd affects the lungs and kidneys; Cr is especially harmful to respiratory tissues; and nickel impacts the upper respiratory tract and kidneys. Understanding these shared and unique pathways is critical for designing biomarkers of exposure, therapeutic interventions, and regulatory policies (Xiao et al., 2015; Zhu and Costa, 2020).

24.4 Omics approaches in heavy metal toxicity

The integration of omics technologies, including proteomics and metabolomics, has provided novel insight into the complex biological disruptions caused by heavy metal exposure. These high-throughput approaches allow researchers to identify biomarkers of exposure, elucidate molecular mechanisms, and understand adaptive and pathological responses at a systems biology level (Jamla et al., 2021). Omics-driven toxicology fills the gap between environmental exposure and disease onset by revealing pathway-level perturbations that may remain hidden in traditional assays (Jamla et al., 2021; Ochoa-Leite et al., 2024).

24.4.1 Proteomic studies of heavy metal toxicity

Mohanraj et al. (2022) studied and performed quantitative proteomics on synaptosomal fractions from rats exposed to Pb via maternal lactation. It revealed significant dysregulation of 289 proteins linked to synaptic function and memory consolidation. Among these, 45 were identified as SUMOylated proteins, particularly GRM3, GRIA2/3, and FLOT1, implicating impaired posttranslational SUMO modification in Pb-induced cognitive deficits (Mohanraj et al., 2022).

A rat model study of 2023 combined exposure to Pb, Cd, and Hg resulted in 363 differentially expressed proteins of the brain. Key proteins include LPCAT3, SLC7A11, ASCL4, and KYAT1, linked to ferroptosis and kynurenine pathway dysregulation, suggesting these mechanisms contribute to neurotoxicity (Xie et al., 2023). Algahmadi et al., 2024 studied *Penicillium chrysogenum* and identified the differential expression of metal-binding and detoxification proteins when exposed to Pb. These studies reveal fungal proteomic adaptations and potential enzymatic pathways for Pb biosorption (Algahmadi et al., 2024).

24.4.2 Metabolomic studies of heavy metal toxicity

A study of 2023 on rats using nontargeted metabolomics showed exposure to Pb/Hg/Cd altered 206 serum metabolites. Affected pathways included unsaturated fatty acid biosynthesis, linolic/phenylalanine/tryptophan metabolism, and a disrupted kynurenic acid to quinolinic acid ratio, showing neuroinflammatory responses (Xie et al., 2023). Another 2023 study examined liver, kidney, and plasma metabolite profiles in rats exposed to multiheavy-metal mixtures. It reported dysregulation in purine metabolism and aminoacyl-tRNA biosynthesis. Chemotherapeutic treatments (EGCG, citrate, and glutathione) partially reversed modified purine metabolites like adenine, guanine, AMP, and hypoxanthine, indicating therapeutic potential (Xie et al., 2023; Zhou et al., 2024).

24.5 Conclusion

In today's world, exposure to heavy metals has quietly become part of our daily lives—through the air we breathe, the water we drink, and the food we eat. Over time, these metals can build up in our bodies, silently affecting our health. They do not just harm one part of us—they can damage our cells, interfere with how our bodies grow and heal, and even cause changes in our DNA that may Pb to serious diseases. Acute and chronic symptoms of metal toxicity are also possible. Growth, proliferation, differentiation, damage-repairing mechanisms, and apoptosis are among the biological processes that heavy metals interfere with. Additionally, toxic metals may encourage epigenetic changes that affect how genes are expressed. A comparison of the mechanisms of action shows that these metals can cause toxicity through similar routes, such as oxidative stress, enzyme inactivation, ROS production, and weakened antioxidant defense. Research initiatives have raised awareness of the harm caused by HM. This review also emphasizes the value and efficacy of alternative medicine. While it is impossible to totally prevent, HM exposure has become a normal part of our daily lives and may be greatly reduced. The negative consequences of these HMs have been clarified by this communication. Modern medications change the natural metabolic process and can sometimes make the illness much worse, even when they are effective in treating HM-induced toxicities, which sometimes have serious adverse effects. However, the main focus of study in the upcoming days will be on alternative medical systems that have negligible or no impact to treat HM-induced toxicity. The greatest resources for managing metabolic diseases, such as HM toxicity, are plant-based products.

References

Aaseth, J., Alexander, J., Alehagen, U., Tinkov, A., Skalny, A., Larsson, A., et al., 2021. The aging kidney-as influenced by heavy metal exposure and selenium supplementation. Biomolecules 11 (8), 1078. https://doi.org/10.3390/biom11081078.

Acharya, S., 2013. Lead between the lines. Nat. Chem. 5 (10), 894. https://doi.org/10.1038/nchem.1761. 894.

Achmad, R.T., Auerkari, E.I., 2017. Effects of chromium on human body. Annu. Res. Rev. Biol. 13 (2), ARRB-33462. https://doi.org/10.9734/ARRB/2017/33462.

Algahmadi, A., Mohammed, A.E., Alfadda, A.A., Alanazi, I.O., Alwehaibi, M.A., Scaria Joy, S., et al., 2024. Proteomics of Penicillium chrysogenum for a deeper understanding of lead (Pb) metal bioremediation. ACS Omega 9, 26245–26256. https://doi.org/10.1021/acsomega.4c02006.

Alnajjar, K.S., Sweasy, J.B., 2019. A new perspective on oxidation of DNA repair proteins and cancer. DNA Repair (Amst.) 76, 60–69. https://doi.org/10.1016/j.dnarep.2019.02.006.

Ara, A., Usmani, J.A., 2015. Lead toxicity: a review. Interdiscip. Toxicol. 8 (2), 55. https://doi.org/10.1515/intox-2015-0009.

Argyropoulos, C.P., Chen, S.S., Ng, Y.H., Roumelioti, M.E., Shaffi, K., Singh, P.P., et al., 2017. Rediscovering beta-2 microglobulin as a biomarker across the spectrum of kidney diseases. Front. Med. 4, 73. https://doi.org/10.3389/fmed.2017.00073.

Arita, A., Costa, M., 2009. Epigenetics in metal carcinogenesis: nickel, arsenic, chromium and cadmium. Metallomics 1, 222. https://doi.org/10.1039/b903049b.

Balali-Mood, M., Naseri, K., Tahergorabi, Z., Khazdair, M.R., Sadeghi, M., 2021. Toxic mechanisms of five heavy metals: mercury, lead, chromium, cadmium, and arsenic. Front. Pharmacol. 12, 643972. https://doi.org/10.3389/fphar.2021.643972.

Bernhoft, R.A., 2012. Mercury toxicity and treatment: a review of the literature. J. Environ. Public Health 2012 (1), 460508. https://doi.org/10.1155/2012/460508.

Bjørklund, G., 2015. Selenium as an antidote in the treatment of mercury intoxication. Biometals 28 (4), 605–614. https://doi.org/10.1007/s10534-015-9857-5.

Bjørklund, G., Mutter, J., Aaseth, J., 2017. Metal chelators and neurotoxicity: lead, mercury, and arsenic. Arch. Toxicol. 91 (12), 3787–3797. https://doi.org/10.1007/s00204-017-2100-0.

Boucher, O., Muckle, G., Jacobson, J.L., Carter, R.C., Kaplan-Estrin, M., Ayotte, P., et al., 2014. Domain-specific effects of prenatal exposure to PCBs, mercury, and lead on infant cognition: results from the environmental contaminants and child development study in Nunavik. Environ. Health Perspect. 122 (3), 310–316. https://doi.org/10.1289/ehp.1206323.

Bourtsalas, A.T., Themelis, N.J., 2019. Major sources of mercury emissions to the atmosphere: the US case. Waste Manage. (Oxford) 85, 90–94. https://doi.org/10.1016/j.wasman.2018.12.008.

Breeher, L., Mikulski, M.A., Czeczok, T., Leinenkugel, K., Fuortes, L.J., 2015. A cluster of lead poisoning among consumers of Ayurvedic medicine. Int. J. Occup. Environ. Health 21 (4), 303–307. https://doi.org/10.1179/2049396715Y.0000000009.

Brewer, G.J., Prasad, A.S (Eds.), 2020. Essential and toxic trace elements and vitamins in human health. Academic Press.

Briffa, J., Sinagra, E., Blundell, R., 2020. Heavy metal pollution in the environment and their toxicological effects on humans. Heliyon 6, e04691. https://doi.org/10.1016/j.heliyon.2020.e04691.

Bui Huy, T., Tuyet-hanh, T.T., Johnston, R., Nguyen-Viet, H., 2014. Assessing health risk due to exposure to arsenic in drinking water in Hanam Province, Vietnam. In: Int. J. Environ. Res. Public Health, 11, pp. 7575–7591.

Cao, Y., Skaug, M.A., Andersen, O., Aaseth, J., 2015. Chelation therapy in intoxications with mercury, lead, and copper. J. Trace Elem. Med. Biol. 31, 188–192. https://doi.org/10.1016/j.jtemb.2014.04.010.

Cao, Z.R., Cui, S.M., Lu, X.X., Chen, X.M., Yang, X., Cui, J.P., Zhang, G.H., 2018. Effects of occupational cadmium exposure on workers' cardiovascular system. Chin. J. Ind. Hyg. Occup. Dis. 36 (6), 474–477. https://doi.org/10.3760/cma.j.issn.1001-9391.2018.06.025.

Chandrasekaran, V.R.M., Muthaiyan, I., Huang, P.C., Liu, M.Y., 2010. Using iron precipitants to remove arsenic from water: is it safe? Water Res. 44 (19), 5823–5827. https://doi.org/10.1016/j.watres.2010.06.063.

Chen, Q.Y., DesMarais, T., Costa, M., 2019. Metals and mechanisms of carcinogenesis. Annu. Rev. Pharmacol. Toxicol. 59, 537–554. https://doi.org/10.1146/annurev-pharmtox-010818-021031.

Chunhabundit, R., 2016. Cadmium exposure and potential health risk from foods in contaminated area, Thailand. Toxicol. Res. 32 (1), 65–72. https://doi.org/10.5487/TR.2016.32.1.065.

Costa, M., 2019. Review of arsenic toxicity, speciation and polyadenylation of canonical histones. Toxicol. Appl. Pharmacol. 375, 1–4. https://doi.org/10.1016/j.taap.2019.05.006.

Das, R., Das, A., Roy, A., Kumari, U., Bhattacharya, S., Haldar, P.K., 2015. β-Carotene ameliorates arsenic-induced toxicity in albino mice. Biol. Trace Elem. Res. 164 (2), 226–233. https://doi.org/10.1007/s12011-014-0212-4.

Dasharathy, S., Arjunan, S., Maliyur Basavaraju, A., Murugasen, V., Ramachandran, S., Keshav, R., et al., 2022. Mutagenic, carcinogenic, and teratogenic effect of heavy metals. Evid.-Based Complement. Altern. Med. 2022 (1), 8011953. https://doi.org/10.1155/2022/8011953.

Demin, A.A., Hirota, K., Tsuda, M., Adamowicz, M., Hailstone, R., Brazina, J., et al., 2021. XRCC1 prevents toxic PARP1 trapping during DNA base excision repair. Mol. Cell 81, 3018–3030. https://doi.org/10.1016/j.molcel.2021.05.009.

Deng, Y., Wang, M., Tian, T., Lin, S., Xu, P., Zhou, L., et al., 2019. The effect of hexavalent chromium on the incidence and mortality of human cancers: a meta-analysis based on published epidemiological cohort studies. Front. Oncol. 9, 24. https://doi.org/10.3389/fonc.2019.00024.

de Oliveira, D.F., de Castro, B.S., do Nascimento Recktenvald, M.C.N., da Costa Júnior, W.A., da Silva, F.X., et al., 2021. Mercury in wild animals and fish and health risk for indigenous Amazonians. Food Addit. Contam. Part B 14 (3), 161–169. https://doi.org/10.1080/19393210.2020.1849410.

Dua, T.K., Dewanjee, S., Gangopadhyay, M., Khanra, R., Zia-Ul-Haq, M., De Feo, V., 2015. Ameliorative effect of water spinach, Ipomea aquatica (Convolvulaceae), against experimentally induced arsenic toxicity. J. Transl. Med. 13, 1–17. https://doi.org/10.1186/s12967-015-0430-3.

Ekong, E.B., Jaar, B.G., Weaver, V.M., 2006. Lead-related nephrotoxicity: a review of the epidemiologic evidence. Kidney Int. 70 (12), 2074–2084. https://doi.org/10.1038/sj.ki.5001809.

El-Neweshy, M.S., El-Sayed, Y.S., 2011. Influence of vitamin C supplementation on lead-induced histopathological alterations in male rats. Exp. Toxicol. Pathol. 63 (3), 221–227. https://doi.org/10.1016/j.etp.2009.12.003.

Fang, H., Jing, T., Liu, Z., Zhang, L., Jin, Z., Pei, Y., 2014. Hydrogen sulfide interacts with calcium signaling to enhance the chromium tolerance in Setaria italica. Cell Calcium 56 (6), 472–481. https://doi.org/10.1016/j.ceca.2014.10.004.

Farag, A.G., Elhalwagy, M.E., Farid, H.E., 2010. Effect of ginger supplementation on developmental toxicity induced by fenitrothion insecticide and/or lead in albino rats. Pestic. Biochem. Physiol. 97 (3), 267–274. https://doi.org/10.1016/j.pestbp.2010.03.007.

Ferreira, L.M., Cunha-Oliveira, T., Sobral, M.C., Abreu, P.L., Alpoim, M.C., Urbano, A.M., 2019. Impact of carcinogenic chromium on the cellular response to proteotoxic stress. Int. J. Mol. Sci. 20 (19), 4901. https://doi.org/10.3390/ijms20194901.

Flora, G., Gupta, D., Tiwari, A., 2012. Toxicity of lead: a review with recent updates. Interdiscip. Toxicol. 5 (2), 47. http://doi.org/10.2478/v10102-012-0009-2.

Flora, S.J., Bhadauria, S., Pant, S.C., Dhaked, R.K., 2005. Arsenic induced blood and brain oxidative stress and its response to some thiol chelators in rats. Life Sci. 77 (18), 2324–2337. https://doi.org/10.1016/j.lfs.2005.04.016.

Gazwi, H.S., Yassien, E.E., Hassan, H.M., 2020. Mitigation of lead neurotoxicity by the ethanolic extract of Laurus leaf in rats. Ecotoxicol. Environ. Saf. 192, 110297. https://doi.org/10.1016/j.ecoenv.2020.110297.

Ghorani-Azam, A., Riahi-Zanjani, B., Balali-Mood, M., 2016. Effects of air pollution on human health and practical measures for prevention in Iran. J. Res. Med. Sci. 21 (1), 65. https://doi.org/10.4103/1735-1995.189646.

Ghosh, U., Nagalakshmi, M., Sadhukhan, R., Girigoswami, A., 2013. Biocapped nanoparticles are more bioactive to HeLa cells than its chemical counterpart. Adv. Sci. Eng. Med. 5 (8), 783–788. https://doi.org/10.1166/asem.2013.1359.

Gombeau, K., De Oliveira, R.B., Sarrazin, S.L.F., Mourão, R.H.V., Bourdineaud, J.P., 2019. Protective effects of plathymenia reticulata and Connarus favosus aqueous extracts against cadmium-and mercury-induced toxicities. Toxicol. Res. 35, 25–35. https://doi.org/10.5487/TR.2019.35.1.025.

Goodrich, J.M., Basu, N., Franzblau, A., Dolinoy, D.C., 2013. Mercury biomarkers and DNA methylation among Michigan dental professionals. Environ. Mol. Mutagen. 54 (3), 195–203. https://doi.org/10.1002/em.21763.

Gunturu, K.S., Nagarajan, P., McPhedran, P., Goodman, T.R., Hodsdon, M.E., Strout, M.P., 2011. Ayurvedic herbal medicine and lead poisoning. J. Hematol. Oncol. 4, 1–6. https://doi.org/10.1186/1756-8722-4-51.

He, Z.L., Yang, X.E., Stoffella, P.J., 2005. Trace elements in agroecosystems and impacts on the environment. J. Trace Elem. Med. Biol. 19 (2-3), 125–140. https://doi.org/10.1016/j.jtemb.2005.02.010.

Hong, Y.S., Song, K.H., Chung, J.Y., 2014. Health effects of chronic arsenic exposure. J. Prev. Med. Public Health 47 (5), 245. https://doi.org/10.3961/jpmph.14.035.

Hsiang, J., Díaz, E., 2011. Lead and developmental neurotoxicity of the central nervous system. Curr. Neurobiol. 2 (1), 35–42.

Hu, Y., Li, J., Lou, B., Wu, R., Wang, G., Lu, C., et al., 2020. The role of reactive oxygen species in arsenic toxicity. Biomolecules 10, 240. https://doi.org/10.3390/biom10020240.

Iranmanesh, M., Fatemi, S.J.A., Golbafan, M.R., Dahooee Balooch, F., 2013. Treatment of mercury vapor toxicity by combining deferasirox and deferiprone in rats. Biometals 26, 783–788. https://doi.org/10.1007/s10534-013-9656-9.

Jaishankar, M., Tseten, T., Anbalagan, N., Mathew, B.B., Beeregowda, K.N., 2014. Toxicity, mechanism and health effects of some heavy metals. Interdiscip. Toxicol. 7 (2), 60. 10.2478/intox-2014-0009.

Jamla, M., Khare, T., Joshi, S., Patil, S., Penna, S., Kumar, V., 2021. Omics approaches for understanding heavy metal responses and tolerance in plants. Curr. Plant Biol. 27, 100213. https://doi.org/10.1016/j.cpb.2021.100213.

Jan, A.T., Azam, M., Siddiqui, K., Ali, A., Choi, I., Haq, Q.M.R., 2015. Heavy metals and human health: mechanistic insight into toxicity and counter defense system of antioxidants. Int. J. Mol. Sci. 16 (12), 29592–29630. https://doi.org/10.3390/ijms161226183.

Jiang, J.H., Ge, G., Gao, K., Pang, Y., Chai, R.C., Jia, X.H., et al., 2015. Calcium signaling involvement in cadmium-induced astrocyte cytotoxicity and cell death through activation of MAPK and PI3K/Akt signaling pathways. Neurochem. Res. 40, 1929–1944. https://doi.org/10.1007/s11064-015-1686-y.

Jomova, K., Alomar, S.Y., Nepovimova, E., Kuca, K., Valko, M., 2025. Heavy metals: toxicity and human health effects. Arch. Toxicol. 99, 153–209. https://doi.org/10.1007/s00204-024-03903-2.

Kalia, K., Flora, S.J., 2005. Strategies for safe and effective therapeutic measures for chronic arsenic and lead poisoning. J. Occup. Health 47 (1), 1–21. https://doi.org/10.1539/joh.47.1.

Kaushal, R.S., Kaushik, A., Prabha, C.R., 2025. Intricate role of reactive oxygen species and reactive nitrogen species in leishmaniasis: mechanisms, implications, and therapeutic potential. In: Prabhakar, P.K. (Ed.), Advances in medical diagnosis, treatment, and care. IGI Global, pp. 269–310.

Kaushik, A., Peshrana, A., Barapatre, R., Pansheriya, S., Kaushal, R.S., 2025. Unveiling the properties of ascorbic acid against M. tb through in silico approach: A comparative drug-based study. J. Mol. Model. 31, 94. https://doi.org/10.1007/s00894-025-06322-x.

Kennedy, P.J., Vashisht, A.A., Hoe, K.-L., Kim, D.-U., Park, H.-O., Hayles, J., et al., 2008. A genome-wide screen of genes involved in cadmium tolerance in schizosaccharomyces pombe. Toxicol. Sci. 106, 124–139. https://doi.org/10.1093/toxsci/kfn153.

Kim, J.J., Kim, Y.S., Kumar, V., 2019. Heavy metal toxicity: an update of chelating therapeutic strategies. J. Trace Elem. Med. Biol. 54, 226–231. https://doi.org/10.1016/j.jtemb.2019.05.003.

Kim, T.H., Kim, J.H., Le Kim, M.D., Suh, W.D., Kim, J.E., Yeon, H.J., et al., 2020. Exposure assessment and safe intake guidelines for heavy metals in consumed fishery products in the Republic of Korea. Environ. Sci. Pollut. Res. 27 (26), 33042–33051. https://doi.org/10.1007/s11356-020-09624-0.

Koedrith, P., Seo, Y.R., 2011. Advances in carcinogenic metal toxicity and potential molecular markers. Int. J. Mol. Sci. 12, 9576–9595. https://doi.org/10.3390/ijms12129576.

Kosnett, M.J., 2013. The role of chelation in the treatment of arsenic and mercury poisoning. J. Med. Toxicol. 9, 347–354. https://doi.org/10.1007/s13181-013-0344-5.

Koyama, H., Kamogashira, T., Yamasoba, T., 2024. Heavy metal exposure: molecular pathways, clinical implications, and protective strategies. Antioxidants 13, 76. https://doi.org/10.3390/antiox13010076.

Kumar, A., Kumar, R., Rahman, M.S., Iqubal, M.A., Anand, G., Niraj, P.K., et al., 2015. Antidote effects of plants of Himalayan sub-origin against arsenic induced toxicity. J Bio Chem Res 2, 99–109.

Kumar, N. (Ed.), 2021. Arsenic toxicity: challenges and solutions. Springer, Singapore.

Lebedeva, N.A., Anarbaev, R.O., Maltseva, E.A., Sukhanova, M.V., Rechkunova, N.I., Lavrik, O.I., 2024. DNA repair protein XRCC1 stimulates activity of DNA polymerase λ under conditions of microphase separation. Int. J. Mol. Sci. 25, 6927. https://doi.org/10.3390/ijms25136927.

Lech, T., Sadlik, J.K., 2017. Cadmium concentration in human autopsy tissues. Biol. Trace Elem. Res. 179, 172–177. https://doi.org/10.1007/s12011-017-0959-5.

Levin, R., Vieira, C.L.Z., Rosenbaum, M.H., Bischoff, K., Mordarski, D.C., Brown, M.J., 2021. The urban lead (Pb) burden in humans, animals and the natural environment. Environ. Res. 193, 110377. https://doi.org/10.1016/j.envres.2020.110377.

Li, H., Fagerberg, B., Sallsten, G., Borné, Y., Hedblad, B., Engström, G., et al., 2019. Smoking-induced risk of future cardiovascular disease is partly mediated by cadmium in tobacco: malmö diet and Cancer Cohort Study. Environ. Health 18, 1–10. https://doi.org/10.1186/s12940-019-0495-1.

Linos, A., Petralias, A., Christophi, C.A., Christoforidou, E., Kouroutou, P., Stoltidis, M., et al., 2011. Oral ingestion of hexavalent chromium through drinking water and cancer mortality in an industrial area of Greece-an ecological study. Environ. Health 10, 1–8. https://doi.org/10.1186/1476-069X-10-50.

Loomis, D., Guha, N., Hall, A.L., Straif, K., 2018. Identifying occupational carcinogens: an update from the IARC Monographs. Occup. Environ. Med. 75 (8), 593–603. 10.1136/oemed-2018-105189.

Luo, S., Calderón-Urrea, A., Yu, J., Liao, W., Xie, J., Lv, J., ... Tang, Z., 2020. The role of hydrogen sulfide in plant alleviates heavy metal stress. Plant and Soil 449 (1), 1–10. https://doi.org/10.1007/s11104-020-04471-x.

Lv, Y., Wang, P., Huang, R., Liang, X., Wang, P., Tan, J., et al., 2017. Cadmium exposure and osteoporosis: a population-based study and benchmark dose estimation in southern China. J. Bone Miner. Res. 32 (10), 1990–2000. https://doi.org/10.1002/jbmr.3151.

Ma, Y., Zheng, Y.X., Dong, X.Y., Zou, X.T., 2018. Effect of mercury chloride on oxidative stress and nuclear factor erythroid 2-related factor 2 signalling molecule in liver and kidney of laying hens. J. Anim. Physiol. Anim. Nutr. (Berl.) 102 (5), 1199–1209. https://doi.org/10.1111/jpn.12920.

Makarani, N., Bharadava, K., Kaushik, A., Dave, A., Gangawane, A.K., Kaushal, R.S., 2025. Leishmaniasis: a multifaceted approach to diagnosis, maladies, drug repurposing and way forward. The Microbe, 100239. https://doi.org/10.1016/j.microb.2025.100239.

Mansoor, S., Ali, A., Kour, N., Bornhorst, J., AlHarbi, K., Rinklebe, J., et al., 2023. Heavy metal induced oxidative stress mitigation and ROS scavenging in plants. Plants 12, 3003. https://doi.org/10.3390/plants12163003.

Maqbool, F., Niaz, K., Hassan, F.I., Khan, F., Abdollahi, M., 2017. Immunotoxicity of mercury: pathological and toxicological effects. J. Environ. Sci. Health, Part C 35 (1), 29–46. https://doi.org/10.1080/10590501.2016.1278299.

Mazumdar, I., Goswami, K., Ali, M.S., 2017. Status of serum calcium, vitamin D and parathyroid hormone and hematological indices among lead exposed jewelry workers in Dhaka, Bangladesh. Indian J. Clin. Biochem. 32, 110–116. https://doi.org/10.1007/s12291-016-0582-9.

Mezencev, R., Gibbons, C., 2023. Interactions between chromium species and DNA in vitro and their potential role in the toxicity of hexavalent chromium. Metallomics 15, mfad045. https://doi.org/10.1093/mtomcs/mfad045.

Mishra, K., Rani, R., Yadav, V.S., Naik, S., 2010. Effect of lead exposure on lymphocyte subsets and activation markers. Immunopharmacol. Immunotoxicol. 32 (3), 446–449. https://doi.org/10.3109/08923970903503668.

Mitra, P., Sharma, S., Purohit, P., Sharma, P., 2017. Clinical and molecular aspects of lead toxicity: an update. Crit. Rev. Clin. Lab. Sci. 54 (7-8), 506–528. https://doi.org/10.1080/10408363.2017.1408562.

Mittal, M., Siddiqui, M.R., Tran, K., Reddy, S.P., Malik, A.B., 2014. Reactive oxygen species in inflammation and tissue injury. Antioxid. Redox Signal. 20, 1126–1167. https://doi.org/10.1089/ars.2012.5149.

Mohanraj, N., Joshi, N.S., Poulose, R., Patil, R.R., Santhoshkumar, R., Kumar, A., et al., 2022. A proteomic study to unveil lead toxicity-induced memory impairments invoked by synaptic dysregulation. Toxicol. Rep. 9, 1501–1513. https://doi.org/10.1016/j.toxrep.2022.07.002.

Moitra, S., Blanc, P.D., Sahu, S., 2013. Adverse respiratory effects associated with cadmium exposure in small-scale jewellery workshops in India. Thorax 68 (6), 565–570.

Nishijo, M., Nakagawa, H., Suwazono, Y., Nogawa, K., Kido, T., 2017. Causes of death in patients with Itai-itai disease suffering from severe chronic cadmium poisoning: a nested case–control analysis of a follow-up study in Japan. BMJ Open 7 (7), e015694.

Ochoa-Leite, C., Rodrigues, S., Ramos, A.S., Ribeiro, F., Barbosa, J., Jerónimo, C., et al., 2024. Metabolomics and proteomics in occupational medicine: a comprehensive systematic review. J. Occup. Med. Toxicol. 19, 38. https://doi.org/10.1186/s12995-024-00436-3.

Oliveira, V.A., de Souza da Costa, N., Mesquita, M., Pedroso, T.F., da Luz Fiuza, T., Peixoto, N.C., et al., 2020. Mercury toxicity in pregnant and lactating rats: zinc and N-acetylcysteine as alternative of prevention. Environ. Sci. Pollut. Res. 27, 40563–40572. https://doi.org/10.1007/s11356-020-09836-4.

Papanikolaou, N.C., Hatzidaki, E.G., Belivanis, S., Tzanakakis, G.N., Tsatsakis, A.M., 2005. Lead toxicity update. a brief review. Med. Sci. Monit. 11 (10), RA329.

Pavesi, T., Moreira, J.C., 2020. Mechanisms and individuality in chromium toxicity in humans. J. Appl. Toxicol. 40 (9), 1183–1197.

Piomelli, S., 2002. Childhood lead poisoning. Pediatric Clinics 49 (6), 1285–1304.

Prasanthi, R.J., Devi, C.B., Basha, D.C., Reddy, N.S., Reddy, G.R., 2010. Calcium and zinc supplementation protects lead (Pb)-induced perturbations in antioxidant enzymes and lipid peroxidation in developing mouse brain. Int. J. Dev. Neurosci. 28 (2), 161–167. https://doi.org/10.1016/j.ijdevneu.2009.12.002.

Proshad, R., Zhang, D., Uddin, M., Wu, Y., 2020. Presence of cadmium and lead in tobacco and soil with ecological and human health risks in Sichuan province, China. Environ. Sci. Pollut. Res. 27 (15), 18355–18370. https://doi.org/10.1007/s11356-020-08160-1.

Rafati-Rahimzadeh, M., Rafati-Rahimzadeh, M., Kazemi, S., Moghadamnia, A.A., 2014. Current approaches of the management of mercury poisoning: need of the hour. DARU J. Pharm. Sci. 22, 1–10. https://doi.org/10.1186/2008-2231-22-46.

Rahman, Z., Singh, V.P., 2019. The relative impact of toxic heavy metals (THMs)(arsenic (As), cadmium (Cd), chromium (Cr)(VI), mercury (Hg), and lead (Pb)) on the total environment: an overview. Environ. Monit. Assess. 191, 1–21. https://doi.org/10.1007/s10661-019-7528-7.

Rauf, A., Khalil, A.A., Awadallah, S., Khan, S.A., Abu-Izneid, T., Kamran, M., et al., 2024. Reactive oxygen species in biological systems: Pathways, associated diseases, and potential inhibitors—a review. Food Sci. Nutr. 12, 675–693. https://doi.org/10.1002/fsn3.3784.

Renu, K., Chakraborty, R., Myakala, H., Koti, R., Famurewa, A.C., Madhyastha, H., et al., 2021. Molecular mechanism of heavy metals (lead, chromium, arsenic, mercury, nickel and cadmium)-induced hepatotoxicity–a review. Chemosphere 271, 129735. https://doi.org/10.1016/j.chemosphere.2021.129735.

Richard Pilsner, J., Lazarus, A.L., NAM, D.H., Letcher, R.J., Sonne, C., Dietz, R., et al., 2010. Mercury-associated DNA hypomethylation in polar bear brains via the luminometric methylation assay: a sensitive method to study epigenetics in wildlife. Mol. Ecol. 19 (2), 307–314. https://doi.org/10.1111/j.1365-294X.2009.04452.x.

Richter, P., Faroon, O., Pappas, R.S., 2017. Cadmium and cadmium/zinc ratios and tobacco-related morbidities. Int. J. Environ. Res. Public Health 14 (10), 1154. https://www.mdpi.com/1660-4601/14/10/1154#.

Romagosa, C., Simonetti, S., López-Vicente, L., Mazo, A., Lleonart, M.E., Castellvi, J., et al., 2011. p16Ink4a overexpression in cancer: a tumor suppressor gene associated with senescence and high-grade tumors. Oncogene 30, 2087–2097. https://doi.org/10.1038/onc.2010.614.

Salvatore Gallicchio, V., Harper, J., 2021. Role of heavy metals in the incidence of human cancers. In: Nazal, M.K., Zhao, H. (Eds.), Heavy metals - their environmental impacts and mitigation. IntechOpen. https://doi.org/10.5772/intechopen.98259.

Samet, J.M., Chen, H., Pennington, E.R., Bromberg, P.A., 2020. Non-redox cycling mechanisms of oxidative stress induced by PM metals. Free Radic. Biol. Med. 151, 26–37. https://doi.org/10.1016/j.freeradbiomed.2019.12.027.

Sanders, A.P., Mazzella, M.J., Malin, A.J., Hair, G.M., Busgang, S.A., Saland, J.M., et al., 2019. Combined exposure to lead, cadmium, mercury, and arsenic and kidney health in adolescents age 12–19 in NHANES 2009–2014. Environ. Int. 131, 104993. https://doi.org/10.1016/j.envint.2019.104993.

Sárközi, K., Papp, A., Máté, Z., Horváth, E., Paulik, E., Szabó, A., 2015. Rutin, a flavonoid phytochemical, ameliorates certain behavioral and electrophysiological alterations and general toxicity of oral arsenic in rats. Acta Biol. Hung 66, 14–26. https://doi.org/10.1556/ABiol.66.2015.1.2.

Sawicka, E., Jurkowska, K., Piwowar, A., 2020. Chromium (III) and chromium (VI) as important players in the induction of genotoxicity – current view. Ann. Agric. Environ. Med. 28 (1), 1–10. https://doi.org/10.26444/aaem/118228.

Sayed, S., Ahsan, N., Kato, M., Ohgami, N., Rashid, A., Akhand, A.A., 2015. Protective effects of Phyllanthus emblica leaf extract on sodium arsenite-mediated adverse effects in mice. Nagoya J. Med. Sci. 77 (1-2), 145.

Schaefer, H.R., Dennis, S., Fitzpatrick, S., 2020. Cadmium: Mitigation strategies to reduce dietary exposure. J. Food Sci. 85 (2), 260–267. https://doi.org/10.1111/1750-3841.14997.

Sevim, Ç., Doğan, E., Comakli, S., 2020. Cardiovascular disease and toxic metals. Curr. Opin. Toxicol. 19, 88–92. https://doi.org/10.1016/j.cotox.2020.01.004.

Shaji, E., Santosh, M., Sarath, K.V., Prakash, P., Deepchand, V., Divya, B.V., 2021. Arsenic contamination of groundwater: a global synopsis with focus on the Indian Peninsula. Geosci. Front. 12 (3), 101079. https://doi.org/10.1016/j.gsf.2020.08.015.

Sharma, P., Bihari, V., Agarwal, S.K., Verma, V., Kesavachandran, C.N., Pangtey, B.S., et al., 2012. Groundwater contaminated with hexavalent chromium [Cr (VI)]: a health survey and clinical examination of community inhabitants (Kanpur, India). PLoS One 7 (10), e47877. https://doi.org/10.1371/journal.pone.0047877.

Sharma, P., Dubey, R.S., 2005. Lead toxicity in plants. Braz. J. Plant Physiol. 17, 35–52. https://doi.org/10.1590/S1677-04202005000100004.

Shekhawat, K., Chatterjee, S., Joshi, B., 2015. Chromium toxicity and its health hazards. Int. J. Adv. Res. 3 (7), 167–172.

Shen, S., Li, X.F., Cullen, W.R., Weinfeld, M., Le, X.C., 2013. Arsenic binding to proteins. Chem. Rev. 113 (10), 7769–7792. https://doi.org/10.1021/cr300015c.

Siblerud, R., Mutter, J., Moore, E., Naumann, J., Walach, H., 2019. A hypothesis and evidence that mercury may be an etiological factor in Alzheimer's disease. Int. J. Environ. Res. Public Health 16 (24), 5152. https://doi.org/10.3390/ijerph16245152.

Son, Y.O., Wang, L., Poyil, P., Budhraja, A., Hitron, J.A., Zhang, Z., et al., 2012. Cadmium induces carcinogenesis in BEAS-2B cells through ROS-dependent activation of PI3K/AKT/GSK-3β/β-catenin signaling. Toxicol. Appl. Pharmacol. 264 (2), 153–160. https://doi.org/10.1016/j.taap.2012.07.028.

Subramaniam, V.D., Prasad, S.V., Banerjee, A., Gopinath, M., Murugesan, R., Marotta, F., et al., 2019. Health hazards of nanoparticles: understanding the toxicity mechanism of nanosized ZnO in cosmetic products. Drug Chem. Toxicol. 42 (1), 84–93. https://doi.org/10.1080/01480545.2018.1491987.

Syversen, T., Kaur, P., 2012. The toxicology of mercury and its compounds. J. Trace Elem. Med. Biol. 26 (4), 215–226. https://doi.org/10.1016/j.jtemb.2012.02.004.

Tchounwou, P.B., Yedjou, C.G., Patlolla, A.K., Sutton, D.J., 2012. Heavy metal toxicity and the environment. In: Molecular, clinical and environmental toxicology, pp. 133–164.

Theron, A.J., Tintinger, G.R., Anderson, R., 2012. Harmful interactions of non-essential heavy metals with cells of the innate immune system. J. Clin. Toxicol S 3 (005), 0495–2161. http://dx.doi.org/10.4172/2161-0495.S3-005.

Thompson, C.M., Proctor, D.M., Haws, L.C., Hébert, C.D., Grimes, S.D., Shertzer, H.G., et al., 2011. Investigation of the mode of action underlying the tumorigenic response induced in B6C3F1 mice exposed orally to hexavalent chromium. Toxicol. Sci. 123 (1), 58–70. https://doi.org/10.1093/toxsci/kfr164.

Thompson, C.M., Proctor, D.M., Suh, M., Haws, L.C., Kirman, C.R., Harris, M.A., 2013. Assessment of the mode of action underlying development of rodent small intestinal tumors following oral exposure to hexavalent chromium and relevance to humans. Crit. Rev. Toxicol. 43 (3), 244–274. https://doi.org/10.3109/10408444.2013.768596.

Timbrell, J.A., 2008. Principles of biochemical toxicology. CRC Press. https://doi.org/10.3109/9781420007084.

Unsal, V., 2018. Natural phytotherapeutic antioxidants in the treatment of mercury intoxication-a review. Adv. Pharm. Bull. 8 (3), 365. https://doi.org/10.15171/apb.2018.043.

Urbano, A.M., Ferreira, L.M., Alpoim, M.C., 2012. Molecular and cellular mechanisms of hexavalent chromium-induced lung cancer: an updated perspective. Curr. Drug Metab. 13 (3), 284–305.

Vaziri, N.D., Gonick, H.C., 2008. Cardiovascular effects of lead exposure. Indian J. Med. Res. 128 (4), 426–435.

Vincent, J.B., 2017. New evidence against chromium as an essential trace element. J. Nutr. 147 (12), 2212–2219. https://doi.org/10.3945/jn.117.255901.

Vincent, J.B., 2019. Effects of chromium supplementation on body composition, human and animal health, and insulin and glucose metabolism. Curr. Opin. Clin. Nutr. Metab. Care 22 (6), 483–489. 10.1097/MCO.0000000000000604.

Wang, R., Sun, Y., Li, C., Xue, Y., Ba, X., 2023. Targeting the DNA damage response for cancer therapy. Int. J. Mol. Sci. 24, 15907. https://doi.org/10.3390/ijms242115907.

Wang, Y., Mandal, A.K., Son, Y.O., Pratheeshkumar, P., Wise, J.T., Wang, L., et al., 2018. Roles of ROS, Nrf2, and autophagy in cadmium-carcinogenesis and its prevention by sulforaphane. Toxicol. Appl. Pharmacol. 353, 23–30. https://doi.org/10.1016/j.taap.2018.06.003.

Wilhelm, T., Said, M., Naim, V., 2020. DNA replication stress and chromosomal instability: dangerous liaisons. Genes 11 (6), 642. https://doi.org/10.3390/genes11060642.

Witkiewicz-Kucharczyk, A., Goch, W., Olędzki, J., Hartwig, A., Bal, W., 2020. The Reactions of H_2O_2 and GSNO with the zinc finger motif of XPA. not a regulatory mechanism, but no synergy with cadmium toxicity. Molecules 25, 4177. https://doi.org/10.3390/molecules25184177.

Wu, X., Cobbina, S.J., Mao, G., Xu, H., Zhang, Z., Yang, L., 2016. A review of toxicity and mechanisms of individual and mixtures of heavy metals in the environment. Environ. Sci. Pollut. Res. 23, 8244–8259. https://doi.org/10.1007/s11356-016-6333-x.

Xiao, C., Liu, Y., Xie, C., Tu, W., Xia, Y., Costa, M., et al., 2015. Cadmium induces histone H3 lysine methylation by inhibiting histone demethylase activity. Toxicol. Sci. 145, 80–89. https://doi.org/10.1093/toxsci/kfv019.

Xiao, X., Fay, A., Molina, P.S., Kovach, A., Glickman, M.S., Li, H., 2024. Structure of the M. tuberculosis DnaK–GrpE complex reveals how key DnaK roles are controlled. Nat. Commun. 15, 660. https://doi.org/10.1038/s41467-024-44933-9.

Xie, J., Zhou, F., Ouyang, L., Li, Q., Rao, S., Su, R., et al., 2023. Insight into the effect of a heavy metal mixture on neurological damage in rats through combined serum metabolomic and brain proteomic analyses. Sci. Total Environ. 895, 165009. https://doi.org/10.1016/j.scitotenv.2023.165009.

Yang, L., Zhang, Y., Wang, F., Luo, Z., Guo, S., Strähle, U., 2020. Toxicity of mercury: molecular evidence. Chemosphere 245, 125586. https://doi.org/10.1016/j.chemosphere.2019.125586.

Ynalvez, R., Gutierrez, J., Gonzalez-Cantu, H., 2016. Mini-review: toxicity of mercury as a consequence of enzyme alteration. Biometals 29, 781–788. https://doi.org/10.1007/s10534-016-9967-8.

Zeneli, L., Sekovanić, A., Ajvazi, M., Kurti, L., Daci, N., 2016. Alterations in antioxidant defense system of workers chronically exposed to arsenic, cadmium and mercury from coal flying ash. Environ. Geochem. Health 38, 65–72. https://doi.org/10.1007/s10653-015-9683-2.

Zheng, L.Y., Sanders, A.P., Saland, J.M., Wright, R.O., Arora, M., 2017. Environmental exposures and pediatric kidney function and disease: A systematic review. Environ. Res. 158, 625–648. https://doi.org/10.1016/j.envres.2017.06.029.

Zhou, J., Hong, H., Zhao, J., Fang, R., Chen, S., Tang, C., 2024. Metabolome analysis to investigate the effect of heavy metal exposure and chemoprevention agents on toxic injury caused by a multi-heavy metal mixture in rats. Sci. Total Environ. 906, 167513. https://doi.org/10.1016/j.scitotenv.2023.167513.

Zhu, Y., Costa, M., 2020. Metals and molecular carcinogenesis. Carcinogenesis 41, 1161–1172. https://doi.org/10.1093/carcin/bgaa076.

CHAPTER

25

Significance of plant nutraceuticals in DNA repair

Rinkal Mulani[a], *Chandani Kaushik*[b] *and Jitendriya Panigrahi*[b]

[a]Department of Biochemistry, Shri Alpesh N. Patel Postgraduate Institute of Science and Research, Anand, Gujarat, India [b]Department of Biotechnology, Shri Alpesh N. Patel Postgraduate Institute of Science and Research, Anand, Gujarat, India

25.1 Introduction

DNA integrity is paramount for living organisms to realize their survival and proper functioning. The plants and humans equally undergo continuous exposure to an array of stressors that impinge upon genomic stability. Thus the ability to detect, repair, and somehow even adapt to DNA damage is key to assuring health, longevity, and the ability to withstand disturbances. Throughout the decades, plant-based nutraceuticals have been increasingly considered to positively influence DNA repair systems and thus offer advantages to human medicine and agricultural sustainability. This chapter, therefore, discusses the concept, evidence to back it, as well as some practical applications of plant nutraceuticals in genome stability.

25.1.1 DNA damage and genomic stability

25.1.1.1 Definition and role of DNA repair in maintaining genome integrity

DNA is continuously at risk from chemical and physical insults that may alter its structure and sequence. Organisms have developed a highly efficient and sophisticated DNA repair network comprising mismatch repair (MMR), nucleotide excision repair (NER), base excision repair (BER), and double-strand break repair (homologous recombination [HR], nonhomologous end joining [NHEJ]) (Allkanjari and Baldock, 2021; Aykan et al., 2022; Dosouky, 2023). These mechanisms protect genomic stability by assuring the correct replication and the authentic transmission of genetic information from one generation to another (Errico and Costanzo, 2012).

Genome Damage: Repair and Adaptation Across Life.
DOI: https://doi.org/10.1016/B978-0-443-42857-9.00005-0

25.1.1.2 *Consequences of unrepaired DNA: Mutations, cancer, aging, and crop deterioration*

The failure to repair DNA damage permits mutation accumulation that in humans leads to carcinogenesis, premature ageing, and neurodegeneration (Madabhushi et al., 2014; Maynard et al., 2015). In plants, DNA repair deficiencies translate into diminished yield, lesser tolerance to stress, and pathogen susceptibility (Balestrazzi et al., 2011). Therefore DNA repair is not just a molecular event but the determination of organismal health, productivity, and survival.

25.1.1.3 *Shared relevance in plant and human systems*

Whether for humans or plants that are utterly dissimilar in physiology, both are highly conserved DNA repair mechanisms. The shared importance given to maintaining genomic stability across kingdoms offers a unique opportunity to study plant-derived bioactive compounds not only as protectors of plant genomes but also as potential agents in human DNA repair and, accordingly, forge an agriculture-to-medicine axis in the area of genomic maintenance.

25.1.2 Sources of DNA damage

25.1.2.1 *Endogenous: Reactive oxygen/nitrogen species and metabolic by-products*

Endogenous DNA damage comes into existence as by-products of normal cellular metabolism. During mitochondrial respiration, reactive oxygen species (ROS) form and target nucleobases to cause lesions such as 8-hydroxy-2′-deoxyguanosine (8-OHdG) (Urbaniak et al., 2020). Reactive nitrogen species (RNS), lipid peroxidation by-products, and DNA replication errors also contribute to the constant background DNA damage.

25.1.2.2 *Exogenous: Ultraviolet radiation, pollutants, xenobiotics, and pathogens*

Exogenous stressors add further insult. UV radiation brings about thymine dimers, while environmental pollutants, heavy metals, and xenobiotics bring about bulky adducts and strand breaks. The oxidative burst by pathogens within plants and chemical exposures in humans seriously threaten genome integrity (Dutta et al., 2018). Such a two-pronged offence from endogenous and exogenous causes substantiates the existence of numerous protection mechanisms, supplemented by dietary and environmental factors.

25.1.3 Introduction to plant nutraceuticals

25.1.3.1 *Definition and differentiation from pharmaceuticals and supplements*

Nutraceuticals are the name given to bioactive compounds that come from food or plants that confer some health benefit apart from basic nutritional value (Riar and Panesar, 2024). Such are the opposite of pharmaceuticals, which are usually synthetic and intended for curing or treating a specific disease. These are natural, prevent disease before it even arrives, and play multiple roles. That separates them from general dietary supplements that are taken without concrete measures of physiological or therapeutic effects, commonly through modulation of oxidative stress, inflammation, and DNA repair pathways.

25.1.3.2 Importance in preventive medicine and sustainable agriculture

Polyphenols, carotenoids, and glucosinolates are among the compounds in nutraceuticals that, in humans, have been correlated with the decreased occurrence of cancer, cardiovascular disease, and neurodegeneration (Kamal et al., 2022). From an agricultural point of view, plant-based compounds defend crops against UV stress, pathogens, and pollutants so that biocontrol farming practices may be fostered (Aremu et al., 2024). Hence, nutraceuticals sit right where preventive medicine and agricultural biotechnology meet to treat human health as well as food security.

25.1.4 Scope of the chapter

The chapter looks at the significance of plant nutraceuticals in DNA repair, which are broadly classified into the following areas:

- Mechanisms of DNA repair modulation: How plant compounds protect DNA, boost repair enzymes, modulate stress pathways, and alter epigenetic landscapes.
- Experimental and clinical evidence: Summarizing in vitro, animal, and human studies, focusing on compounds such as curcumin, resveratrol, sulforaphane, and green tea polyphenols.
- Applying health and agriculture: Assessing the different roles in cancer chemoprevention, radioprotection, ageing, neurodegeneration, and crop improvement.
- Challenges and future directions: Highlighting limitations such as bioavailability, standardization, and regulation, alongside emerging trends like personalized nutrition, omics, and novel delivery systems.
- From molecular insight, experimental evidence to translational application, this chapter holds plant nutraceuticals as important agents for genomic stability in human welfare and sustainable agriculture

25.2 DNA damage and repair: Molecular framework

The whole set of genomic DNAs of living cells from endogenous sources, such as metabolism by-products and exogenous environmental agents, faces its damaging insults. Endogenous threats include free radicals or ROS, alkylating agents, and spontaneous deamination of bases, while UV light, ionizing radiation, and agents for DNA cross-linking are on the extrinsic side (Liu et al., 2024). Lesions may change the base chemical composition, deform the helical structure, or act against the smoothness of the DNA sugar-phosphate backbone (Yang, 2008). All of these pose threats to genomic stability and the viability of a living organism. Cells, therefore, have devised a multitiered molecular defense system to protect the genetic information, in which the specific, significant type of lesions influences the choice of pathways employed for repair (Barzilai et al., 2008). Within this organization, there is even patterned communication between the repair process and the DNA damage response (DDR) signaling network. The hierarchical architecture encompassing DNA damage type at the top, repair pathways at the center, and DDR regulatory nodes downstream is illustrated (Fig. 25.1).

25.2.1 Types of DNA damage and repair

The DNA lesions are among the most digitized, characterized by discontinuities in one strand of the duplex, often accompanied by ends at the terminus that cannot be ligated, like 3′-phosphate or 5′-hydroxyl (Datta et al., 2011). They arise from endogenous sources such as oxidative damage by ROS, spontaneous hydrolysis, and abortive topoisomerase I activity or from exogenous agents like ionizing radiation and alkylating chemicals (Liu et al., 2024). The repair takes place by way of the Single-Strand Break Repair (SSBR) pathway, which starts with recognition of the lesion by PARP1 and poly (ADP-ribose) synthesis to induce the scaffold protein XRCC1 (Caldecott, 2022). In some cases, processing enzymes such as PNKP or APE1 remove and restore the proper 5′-phosphates and 3′-hydroxyl groups, carrying out single-strand break repair. DNA polymerase β fills the gap, and, more often than not, DNA ligase IIIα, working in complex with XRCC1, seals the nick (Caldecott, 2022), visualized (Fig. 25.1). Efficient SSBR is required to eliminate replication-associated double-strand breaks in the first line, thereby maintaining genomic stability; defects in the pathway are associated with neurodegenerative disorders such as SCAN1 and MCSZ (Winczura and Reynolds, 2018; Li et al., 2022).

DNA double-strand breaks are the most cytotoxic form of damage to DNA, in which the enzyme acts simultaneously on both strands with no complementary strand left to serve as a repair template (Khan and Ali, 2017). They are induced by ionizing radiation, radiomimetic drugs (such as bleomycin), replication fork collapse, and programmed processes, for instance, meiotic and V(D)J recombination (Cannan and Pederson, 2016; Nickoloff et al., 2023). There are two key mechanisms of repair: HR, an error-free process active in the S/G2 phase, starts by MRN complex recognition, followed by the formation of a transient break with CtIP, long-range resection by Exo1 or DNA2-BLM to form 3′ ssDNA. RAD51 filaments, assisted by BRCA1/2-PALB2, replace RPA-coated ssDNA for strand invasion and template-directed synthesis (Moynahan and Jasin, 2010). The Cell cycle is fully associated with NHEJ and commences with Ku70-Ku80 binding, followed by the recruitment of DNA-PKcs. This is then processed by Artemis and other enzymes, and ligated by the XRCC4-Ligase IV-XLF complex (Chang et al., 2017). The further stages of the repair process (Fig. 25.1). Pathway choice influences genome stability and, with defects in HR or NHEJ, creates conditions for syndromes like ataxia-telangiectasia and Fanconi anemia, which in turn are therapeutically exploited through PARP inhibitors for BRCA-mutated cancers. Base modifications such as 8-oxoguanine (oxidation), O 6-methylguanine (alkylation), and uracil (deamination) alter base-pairing and result in mutations if not repaired

FIGURE 25.1 Detailed breakdown of DNA single-strand break (SSB) and double-strand break (DSB) repair processes. (A) Single-strand break repair (SSBR): The repair of single-strand breaks begins with the detection of SSBs by PARP1, which then synthesizes poly (ADP-ribose) or PAR to recruit additional repair factors to the site of damage. XRCC1 then functions as a central coordinator by assembling the repair complex and guiding various end-processing enzymes, including PNKP, APE1, TDP1, and APTX, to the damage site to process the DNA ends. DNA polymerase β, or DNA polymerases δ/ε along with FEN1 and in the case of long-patch repair, fill the gap. This is then sealed by XRCC1-Ligase IIIα or Ligase I. Lastly, PARG/ARH3 cleaves PAR chains from proteins to restore chromatin structure. (B) Double-strand break repair (DSBR): (a) At one end, MRN complexes detect DSBs and activate ATM, which leads to γH2AX signaling and the activation of a cell cycle checkpoint. The method of repair depends on the stage of the cell cycle: (b) During the G1 phase, NHEJ, involving Ku70–Ku80, DNA-PKcs, Artemis, and Ligase IV, is dominant, and although it is fast, it is more prone to errors. In contrast, (c) HR, which involves CtIP, RPA, BRCA1/2, and RAD51, is dominant in the S/G2 phases, and it utilizes strand invasion and DNA synthesis to perform accurate repair. *ATM, Ataxia telangiectasia* mutated.

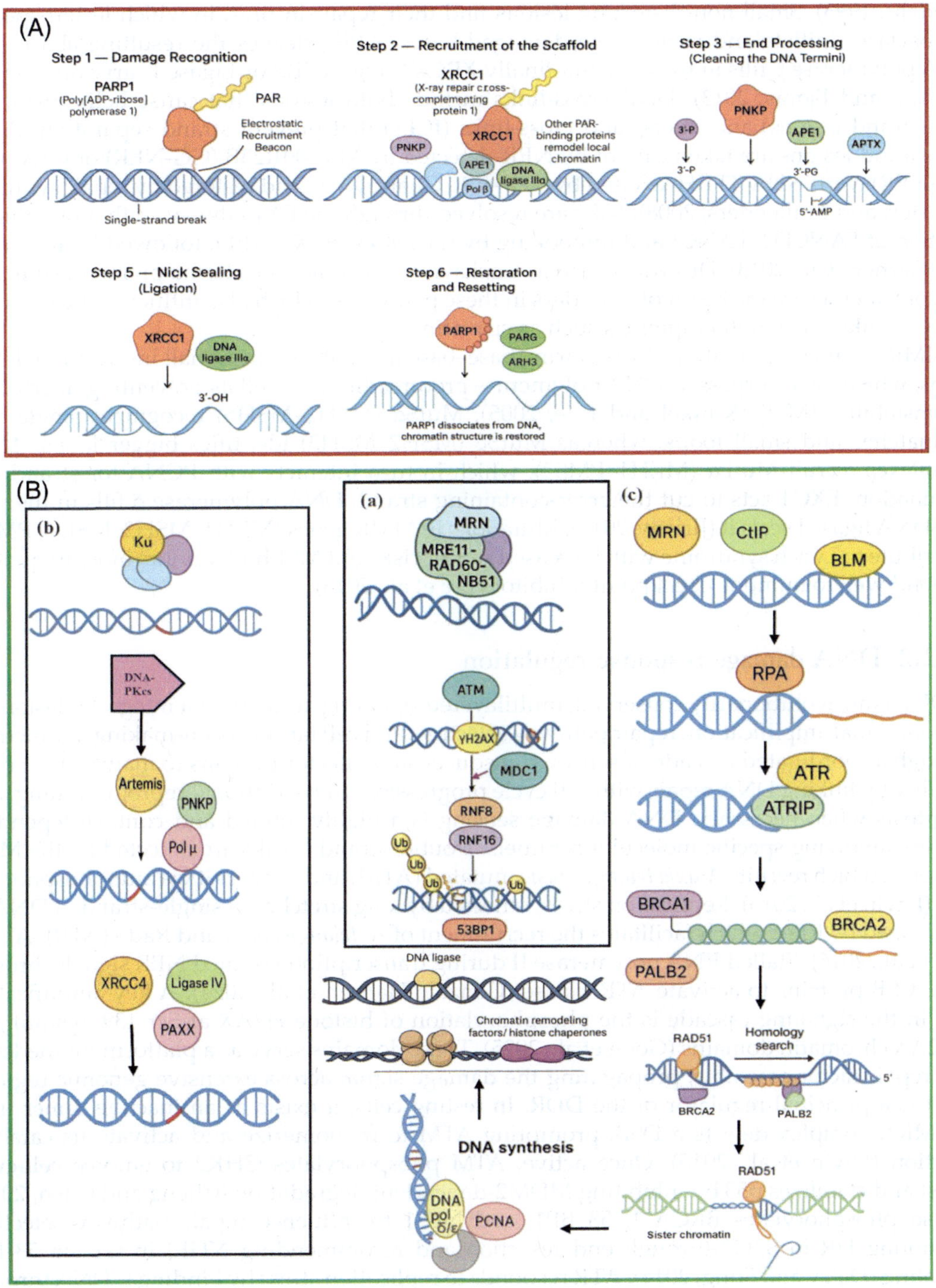
(A)
Step 1 — Damage Recognition
PARP1
(Poly[ADP-ribose] polymerase 1)
PAR
Electrostatic Recruitment Beacon
Single-strand break
Step 2 — Recruitment of the Scaffold
XRCC1
(X-ray repair cross-complementing protein 1)
PNKP
XRCC1
APE1
Pol β
DNA ligase IIIα
Other PAR-binding proteins remodel local chromatin
Step 3 — End Processing
(Cleaning the DNA Termini)
3'-P
PNKP
APE1
APTX
3'-P
3'-PG
5'-AMP
Step 5 — Nick Sealing
(Ligation)
XRCC1
DNA ligase IIIα
3'-OH
Step 6 — Restoration and Resetting
PARP1
PARG
ARH3
PARP1 dissociates from DNA, chromatin structure is restored
(B)
(b)
Ku
DNA-PKcs
Artemis
PNKP
Pol μ
XRCC4
Ligase IV
PAXX
(a)
MRN
MRE11-RAD60-NB51
ATM
γH2AX
MDC1
RNF8
RNF16
Ub
53BP1
DNA ligase
Chromatin remodeling factors/ histone chaperones
DNA synthesis
DNA pol δ/ε
PCNA
(c)
MRN
CtIP
BLM
RPA
ATR
ATRIP
BRCA1
BRCA2
PALB2
RAD51
Homology search
5'
BRCA2
PALB2
RAD51
Sister chromatin

(Lindahl, 1993). Small non-distorting lesions find their repair in BER, in which lesion-specific glycosylases initiate and remove the damaged bases, APE1 cleaves the resulting abasic site, DNA polymerase β fills in the gap, and finally XRCC1–Ligase IIIα or Ligase I carry out ligation (Krokan and Bjørås, 2013). DNA cross-links include both lesions that cause helix distortion (intrastrand lesions) and interstrand cross-links (ICLs) that prevent strand separation. Bulky intrastrand lesions are taken care of by NER-detected by XPC–HR23B (GG-NER) or CSA/CSB (TC-NER), verified by TFIIH, excised by XPF–ERCC1 and XPG, and restored by repair synthesis (Fousteri and Mullenders, 2008). ICLs are resolved through the FA pathway, with monoubiquitination of FANCD2/FANCI and unhooking by nucleases (SLX4, XPF), followed by repair via HR (Garner et al., 2013). Disorders such as Cockayne syndrome, xeroderma pigmentosum, and Fanconi anemia are the result of disorders in these pathways, which also influence the response to cross-linking chemotherapeutics such as cisplatin.

MMR eliminates mutations by repairing base–base mismatches and small insertion/deletion loops, which have surpassed DNA polymerase proofreading, as well as preventing microsatellite instability (MSI) (Kunkel and Erie, 2005). MutSα (MSH2–MSH6) recognizes single-base mismatches and small loops, whereas MutSβ (MSH2–MSH3) identifies bigger loops. These complexes recruit MutLα (MLH1–PMS2), which in turn interacts with PCNA for strand discrimination. EXO1 acts to cut the error-containing strand, DNA polymerase δ fills in the gap, and DNA ligase I seals it (Jiricny, 2013). Mutations in MMR genes (MLH1, MSH2, MSH6, PMS2) contribute to Lynch syndrome with increased cancer risk, and MSI-high tumors more frequently respond well to immune checkpoint inhibitors (Le et al., 2015).

25.2.2 DNA damage response regulation

DDR is an evolutionarily conserved, multilayered signaling network that integrates lesion detection, signal amplification, repair pathway selection, and cell fate decision-making. It functions through a coordinated cascade of sensor, transducer, and effector proteins to maintain genomic stability by linking DNA repair with cell cycle progression, transcriptional reprogramming, and apoptosis when necessary. DNA damage sensing is a highly refined and context-dependent process involving specific molecular sentinels. Double-strand breaks are detected by the MRN complex, which recruits *Ataxia telangiectasia* mutated (ATM) and triggers it via autophosphorylation (Lavin et al., 2015). Replication stress, reflected by long stretches of single-stranded DNA, is recognized by RPA, which facilitates the recruitment of *A. telangiectasia* and Rad3 (ATR)–ATRIP (Lee et al., 2016). Stalled RNA polymerase II during transcription-coupled NER signals through CSA/CSB proteins to activate ATR kinase (Gonzalo-Hansen et al., 2024). A key amplification step in the signaling cascade is the phosphorylation of histone H2AX at Ser 139, resulting in γ-H2AX chromatin domains (Gong et al., 2025). These domains serve as a platform for mediator and repair factor assembly, propagating the damage signal across extensive genomic regions. ATM is a principal regulator of the DDR. In resting cells, it exists as an inactive dimer until the MRN complex detects a DSB, prompting ATM to monomerize and activate its catalytic function (Lavin et al., 2015). Once active, ATM phosphorylates CHK2 to enforce cell-cycle arrest and stabilizes p53 by inhibiting MDM2-dependent degradation (Cheng and Chen, 2010). It also phosphorylates BRCA 1, 53 BP1, and NBS1 to influence repair pathway selection, promoting HR in S/G2 through end resection and recommending NHEJ in G1 via 53 BP1 signaling (Daley and Sung, 2014). ATR responds to replication stress by binding ssDNA through the ATRIP–RPA complex and is activated by cofactors such as Topbp1 or ETAA1. Activated

ATR phosphorylates CHK1, which enforces intra-S and G 2/m checkpoints by suppressing origin firing, stabilizing forks, and inhibiting CDK1 via Wee1 (Lee et al., 2016). In addition to checkpoint control, ATR facilitates fork restart through HR, and safeguards reversed forks from MRE 11-mediated degradation, thereby acting as the main protector of replication integrity (Leung et al., 2023). p53, positioned at pathway crossroads, is the ultimate decision-maker in DDR (Rizzotto et al., 2021). Stabilized downstream of ATM/ATR-CHK 1/CHK 2 signaling, it promotes DNA repair by inducing repair genes such as XPC, DDB2, POLβ, and causes cell cycle arrest through inhibition of CDKs via p21 unless it triggers apoptosis (BAX, PUMA, NOXA) or senescence to prevent the proliferation and survival of genetically unstable cells (Fousteri and Mullenders, 2008).

Cross-talk occurred between DDR and the cell cycle to ensure that repair occurred in the proper phase. HR was restricted in S/G2 when sister chromatids were present; NHEJ takes precedence in G1 (Daley and Sung, 2014). The very checkpoints that enact this control: G1/S to prevent damaged DNA from being replicated; intra-S to slow down fork progression; G2/M to halt mitosis until repair is complete. Another dimension that influences pathway choice is the battle of signaling: ATM–53BP1 for NHEJ, ATR–BRCA1 for HR. Once a successful repair happens, there shall be the silencing of checkpoint signaling. If lesions persist, apoptosis or senescence may be activated. DDR functions through several intersecting pathways rather than a simple cascade. Lesions commonly overlap in signaling: SSBs detected by PARP1 can evolve into DSBs at stalled forks, leading to the activation of ATM and ATR; DSBs activate ATM while ATR is recruited to the sites of resection; BER and NER intermediates also trigger checkpoints when forks stall; and MMR comes into play with persistent ssDNA. These diverse pathways link damage recognition with repair and cellular responses and together provide a dynamic, highly integrated mechanism.

25.3 Definition and characteristics

Plant nutraceuticals are bioactive compounds derived from botanical sources, typically considered secondary metabolites, that promote health and prevent chronic diseases, genetic damage, and cancer. These phytochemicals are not typical nutrients like carbohydrates or proteins but possess therapeutic potential, particularly in DNA repair, reducing oxidative stress, and gene regulation (Biesalski et al., 2009). A common characteristic of plant nutraceuticals is that they are a highly diverse group of bioactive compounds mainly originating from the secondary metabolism of plants. Unlike primary metabolites such as carbohydrates, lipids, or proteins, secondary metabolites are not directly essential for a plant's growth or survival; however, they play important ecological roles in defense, pigmentation, and signaling (Böttner et al., 2021). These secondary metabolites, inter alia flavonoids, polyphenols, alkaloids, and terpenoids, are extensively exploited for their therapeutic benefits in humans, especially in disease prevention and mechanisms that involve restoring DNA integrity and facilitating repair. Primarily derived from plant secondary metabolism, plant nutraceuticals are prominent for their antioxidant properties. These molecules safeguard against ROS and free radicals that bring about DNA damage through oxidation, strand breaks, and base modifications. Quercetin, a flavonoid, and resveratrol, a polyphenol, can scavenge ROS and also boost antioxidant activities by stimulating the expression levels of enzymes such as catalase, superoxide dismutase (SOD), and glutathione peroxidase (Baur and Sinclair, 2006; Boots et al., 2008).

Mutating antioxidants are essentially plant-based nutraceuticals with anti-inflammatory or anti-mutagenic effects. They work by modulating pro-inflammatory cytokines (TNF-α and IL-6), inhibiting NF-κB activation, and suppressing inducible nitric oxide synthase (iNOS) (Aggarwal et al., 2007). For example, curcumin has been shown to reduce chronic inflammation and prevent inflammation-induced DNA damage, a step towards cancer development (Goel et al., 2008). They also affect gene expression and enzyme activity involved in DNA repair and maintenance pathways. For instance resveratrol activates sirtuins (SIRT1) and p53, which play significant parts in maintaining genomic stability and responding to DNA damage (Lagouge et al., 2006), while sulforaphane and glucosinolates induce phase II detoxifying enzymes (such as NQO1 and GST) to enhance the cell's ability to process and eliminate genotoxic agents (Myzak et al., 2006). These regulatory roles highlight their potential in nutrigenomics, where dietary bioactives influence gene activity associated with DNA repair, apoptosis, and cell cycle control. They are found at varying levels across herbs, fruits, vegetables, spices, and seeds (Li and Schluesener, 2017). Their distribution among different edible and medicinal plants is another intriguing feature. Phytochemicals typically vary in concentration among different fruits (viz. berries and grapes), vegetables (like broccoli and spinach), legumes, seeds, and spices (including turmeric and ginger). These variations can be influenced significantly by plant part, cultivar, growing conditions, and processing methods (Li and Schluesener, 2017). Traditional diets rich in these agents, such as Mediterranean and Asian diets, have long been associated with a lessened occurrence of chronic diseases like cancers and neurodegenerative disorders, possibly due to their nutraceutical content (Scalbert et al., 2005).

25.3.1 Major classes of plant nutraceuticals

Plant nutraceuticals consist of various bioactive compounds, with the main classes influencing DNA repair and genomic stability being flavonoids, polyphenols, glucosinolates, carotenoids, alkaloids, as well as essential vitamins and minerals. Flavonoids such as quercetin, found in onions, apples, and capers, are potent antioxidants. They protect DNA from oxidative damage by inducing repair enzymes like PARP1 (Boots et al., 2008). Kaempferol, present in kale, tea, and broccoli, promotes DNA strand break repair and helps maintain the redox balance within cells (Calderón-Montaño et al., 2011). Among polyphenols, resveratrol—a stilbene found in grapes and berries—supports genome stability by activating sirtuins, primarily SIRT1 (Baur and Sinclair, 2006); while curcumin from turmeric aids NER and reduces oxidative DNA lesions (Goel et al., 2008). Part of the glucosinolates group, these sulfur-containing compounds are mainly observed in cruciferous vegetables. Examples include sulforaphane in broccoli sprouts, which promotes phase II enzyme activity for detoxification, such as NQO1 and GST, and activates certain genes involved in DNA repair (Myzak et al., 2006). Carotenoids like β-carotene and lycopene provide provitamin A benefits along with antioxidant effects. β-Carotene, present in carrots and sweet potatoes, shields DNA from UV-induced damage (Stahl and Sies, 2005), while lycopene, present in tomatoes, fights oxidative stress and prevents DNA strand breakage (Agarwal and Rao, 2000). Berberine, derived from Berberis species and classified as an alkaloid, enhances DNA repair by regulating mRNA expression of tumor suppressor genes such as p53 (Liu et al., 2022); meanwhile, caffeine in tea and coffee disrupts DNA damage checkpoints and influences repair processes (Sarkaria et al., 1999). Finally, essential vitamins and minerals are vital for DNA metabolism. Vitamin C acts as an antioxidant, preventing oxidative DNA damage and regenerating vitamin E (Carr and Frei, 1999).

25.3.2 Sources and traditional uses

Plants possess nutraceuticals as they participate in providing nutrition alongside disease-preventing aspects. The Table depicts the various nutraceutical classes with their respective sources and uses. Plant nutraceuticals—including those of Ayurveda, TCM, and folk applications in Africa, Europe, and the Americas—have been traditionally considered a part of every healing system. across the globe. The bioactive compounds were used previously for nutrition as well as for health prevention and cure. Table 25.1 represents the major classes of nutraceuticals along with their uses.

25.4 Mechanisms of DNA repair modulation by plant nutraceuticals

Plant-based nutraceuticals help to maintain genomic stability, working through complementary mechanisms. They serve as antioxidants but also as modulators of DNA repair enzymes, regulators of stress and inflammatory signaling pathways, and epigenetic modifiers of gene expression. These mechanisms synergistically work to reduce mutagenesis, preserve DNA integrity, and impart long-term benefits upon health.

25.4.1 Antioxidant and free radical scavenging

25.4.1.1 Protection against oxidative DNA lesions

Oxidative stress is a major cause of DNA damage, through lesions such as 8-hydroxy-2′-deoxyguanosine (8-oxo-dG), wherein the erroneous base carries mispairing during replication, thus enhancing carcinogenesis. These nutraceuticals from plants, amongst which are the polyphenols, prove to be great scavengers of free radicals that directly dispose of ROS, together with hydroxyl radicals and superoxide anions. For instance quercetin and green tea catechins scavenge hydroxyl radicals to prevent DNA strand breaks from ROS (Pham-Huy et al., 2008). Resveratrol, a stilbene of grape origin, has also been shown to reduce oxidative DNA damage markers in vitro and in vivo, thereby protecting against mutagenesis (Win et al., 2002).

25.4.1.2 Maintenance of redox homeostasis

Beyond direct scavenging, nutraceuticals preserve genomic stability by supporting the cell's endogenous antioxidant defense systems. The higher the levels of these compounds, viz. curcumin of turmeric or green tea polyphenol epigallocatechin gallate (EGCG), the greater the enhancement of the antioxidant enzymes produced by a cell (SOD, catalase, or glutathione peroxidase). This could maintain the redox balance and reduce chronic accumulation of oxidative DNA lesions (Hosseini et al., 2020). Sulforaphane from cruciferous vegetables proceeds to induce the phase II detoxification enzymes that reduce the oxidative load within the cells and indirectly reduce DNA damage. In establishing this kind of balance, the nutraceuticals offer a cellular environment that is unfriendly to mutagenic intrusion.

TABLE 25.1 Major classes of nutraceuticals, the most important compounds, food sources, and ethnobotanical uses.

Nutraceutical class	Bioactive compounds	Primary sources	Traditional/ethnobotanical uses	Reference(s)
Flavonoids	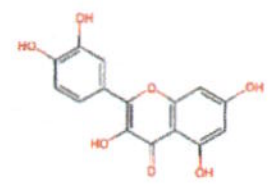 Quercetin 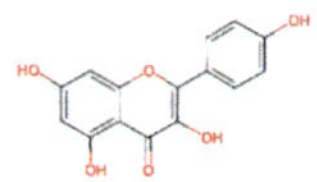 Kaempferol Quercetin Kaempferol	Onion, kale, tea	Anti-inflammatory	Lombard et al. (2005), Wang et al. (2012), Lučić et al. (2023)
Polyphenols	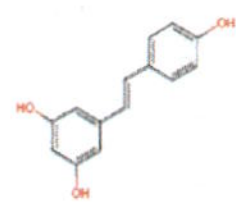 Resveratrol 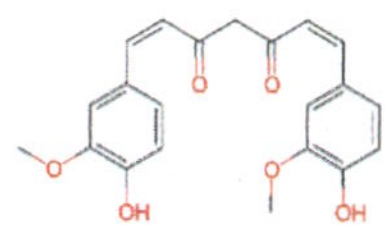 Curcumin Resveratrol Curcumin	Grapes, turmeric	Resveratrol in traditional Japanese remedies, turmeric in Indian medicine for wound healing and immunity	Hasan and Bae (2017), Kocaadam and Şanlier (2017)
Glucosinolates	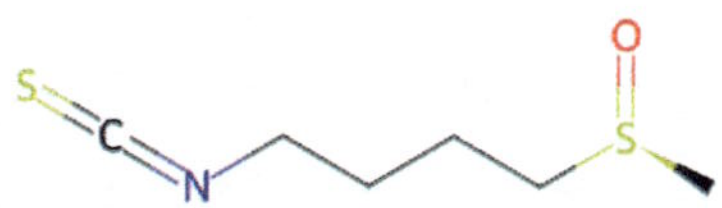	Broccoli, cabbage	Used traditionally in Europe for detoxification and digestion	Tabart et al. (2018)

(continued on next page)

Carotenoids	β-carotene β-carotene 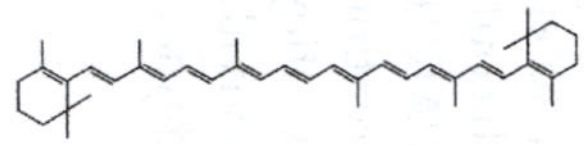 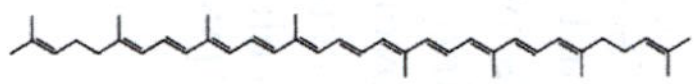 lycopene lycopene	Carrot, tomato	Carrot juice for eye health and immune support in folk medicine	Haftel et al. (2015)
Alkaloids	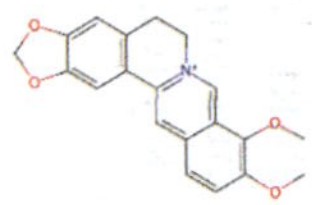 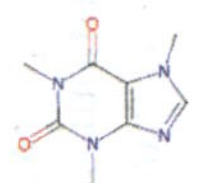 Berberine Caffeine Berberine Caffeine	Berberis root, tea leaves	Berberine is used in TCM for infections; caffeine for alertness	Waheeb et al. (2025)
Vitamins/minerals	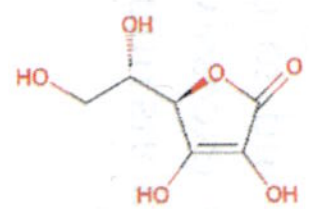 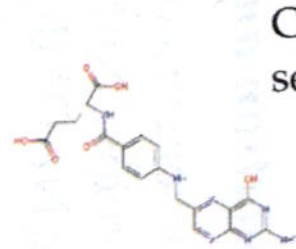 Vitamin C Folate Folate	Citrus, legumes, seeds	Used in ethnomedicine for healing wounds and fatigue prevention	Degerud et al. (2015)

Structures are drawn with ChemDraw Professional 16.0 version 2016.

25.4.2 Activation of DNA repair enzymes

25.4.2.1 Enhancement of base excision repair enzymes (e.g., OGG1, APE1)

The BER pathway is a crucial repair mechanism that addresses oxidative base lesions. Plants and their nutraceuticals enhance and regulate BER efficiency by either upregulating or activating specific enzymes such as OGG1 and Apn1 (APE1). Resveratrol, for example, activates OGG1, the enzyme responsible for removing oxidized guanine residues (Gerszon et al., 2014). Sulforaphane activates Nrf2, which further induces genomic expression for BER, promoting the repair of oxidative damage. Experimental animal models have shown that diets rich in nutraceuticals can increase BER enzyme expression, thereby reducing DNA strand break accumulation in the presence of oxidative stress (Cooke et al., 2003). Nutraceuticals like resveratrol and sulforaphane support DNA maintenance through a set of well-coordinated molecular actions. While sulforaphane partly relies on diffusion and specific transport proteins, resveratrol's cellular uptake is mainly passive. Both compounds, however, reach the nuclei of cells to assist with redox regulation and signaling preparation. Resveratrol positively influences the activity of SIRT1 and AMPK, helping in the loosening of chromatin and the optimization of BER machinery, while sulforaphane enhances Nrf2 stability by modifying Keap1 (Fig. 25.2). This allows Nrf2 to enter the nucleus and activate the transcription of DNA and antioxidant repair genes (OGG1, APE1, XRCC1, Pol β). The functionality of repair proteins is adjusted and optimized through post-translational modifications, which strengthen their catalytic functions and protein–protein interactions. In BER, the oxidative lesion 8-oxoguanine is recognized and removed by OGG1, APE1 cuts the DNA backbone, and the resulting gap is filled by Pol β or Pol δ/ε and sealed by XRCC1–Ligase complexes. PARP1/2 detect DNA breaks within seconds, remodel the chromatin with PARylation, and, while scaffolding repair proteins, reset themselves by autorelease and PAR hydrolysis. In summary, nutraceuticals bring together transcriptional, posttranslational, and chromatin adjustments to improve the BER process (Fig. 25.2).

25.4.2.2 PARP modulation and DNA damage sensing

PARP proteins act as sensors of DNA damage, detecting single-strand breaks and recruiting appropriate repair proteins. Nutraceuticals modulate PARP activity to ensure effective DNA repair while preventing NAD^+ depletion beyond cellular needs, which could disrupt cell energy metabolism. For instance resveratrol has been reported to bind to PARP1, promoting its role in DNA strand break sensing and repair without causing overactivation (Virág and Szabó, 2002). Flavonoids are said to exhibit mild PARP inhibitory activity; they may help prevent hyperactivation under excessive damage, thus supporting cell survival. Consequently, this precise regulation positions nutraceuticals as key regulators of DNA damage sensing and repair orchestration.

FIGURE 25.2 A stepwise narrative of the DNA repair activation via nutraceuticals: sulforaphane and resveratrol enter cells. They alter the redox state and trigger specific signaling pathways, including SIRT1, AMPK, and Nrf2. Nrf2-driven transcription amplifies the expression of antioxidant and base excision repair (BER) genes, while post-translational modifications regulate the activity of repair enzymes. BER follows recognition of a lesion by OGG1, incision of the strand by APE1, gap filling by Pol β or Pol δ/ε with XRCC1, ligases, and FEN1, and ending with nick sealing. PARP1/2 also detect DNA breaks, participate in chromatin remodeling through PARylation, and serve as scaffolds for repair complexes. Overall, the effect is improved BER efficiency and a lower DNA damage load.

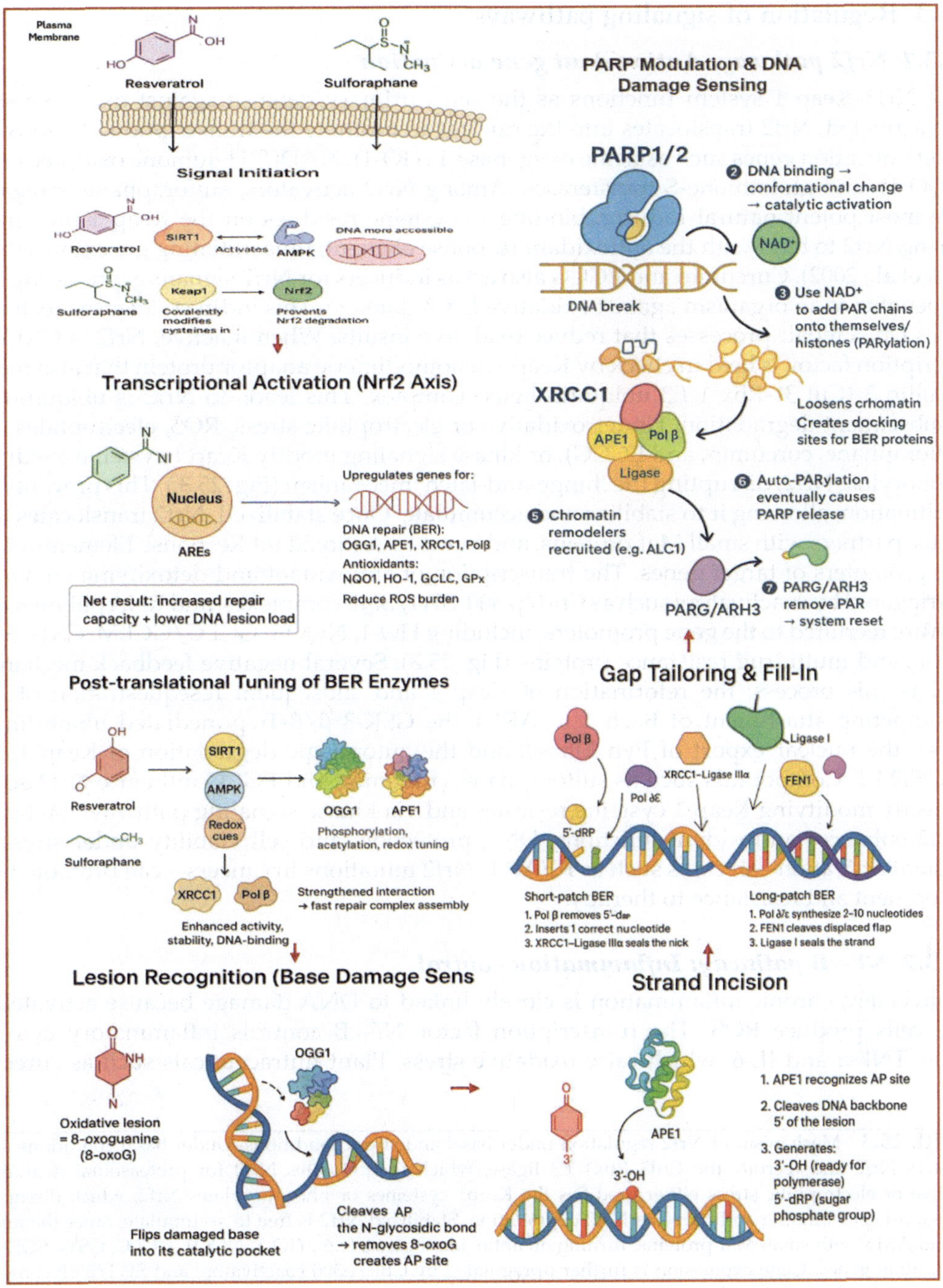

Plasma Membrane
Resveratrol
Sulforaphane
Signal Initiation
Resveratrol
SIRT1
Activates
AMPK
DNA more accessible
Sulforaphane
Keap1
Covalently modifies cysteines in
Nrf2
Prevents Nrf2 degradation
Transcriptional Activation (Nrf2 Axis)
Nucleus
AREs
Upregulates genes for:
DNA repair (BER): OGG1, APE1, XRCC1, Polβ
Antioxidants: NQO1, HO-1, GCLC, GPx
Reduce ROS burden
Net result: increased repair capacity + lower DNA lesion load
Post-translational Tuning of BER Enzymes
Resveratrol
Sulforaphane
SIRT1
AMPK
Redox cues
OGG1
APE1
Phosphorylation, acetylation, redox tuning
XRCC1
Pol β
Strengthened interaction → fast repair complex assembly
Enhanced activity, stability, DNA-binding
Lesion Recognition (Base Damage Sens
Oxidative lesion = 8-oxoguanine (8-oxoG)
OGG1
Flips damaged base into its catalytic pocket
Cleaves AP the N-glycosidic bond → removes 8-oxoG creates AP site
PARP Modulation & DNA Damage Sensing
PARP1/2
2 DNA binding → conformational change → catalytic activation
NAD+
DNA break
3 Use NAD+ to add PAR chains onto themselves/ histones (PARylation)
XRCC1
APE1
Pol β
Ligase
Loosens crhomatin
Creates docking sites for BER proteins
6 Auto-PARylation eventually causes PARP release
5 Chromatin remodelers recruited (e.g. ALC1)
7 PARG/ARH3 remove PAR → system reset
PARG/ARH3
Gap Tailoring & Fill-In
Pol β
Ligase I
XRCC1-Ligase IIIα
FEN1
Pol δ/ε
5'-dRP
Short-patch BER
1. Pol β removes 5'-dRP
2. Inserts 1 correct nucleotide
3. XRCC1–Ligase IIIα seals the nick
Long-patch BER
1. Pol δ/ε synthesize 2–10 nucleotides
2. FEN1 cleaves displaced flap
3. Ligase I seals the strand
Strand Incision
APE1
3'-OH
1. APE1 recognizes AP site
2. Cleaves DNA backbone 5' of the lesion
3. Generates: 3'-OH (ready for polymerase) 5'-dRP (sugar phosphate group)

25.4.3 Regulation of signaling pathways

25.4.3.1 Nrf2 pathway: Antioxidant gene activation

The Nrf2–Keap 1 system functions as the cell's primary defense against oxidative stress. When activated, Nrf2 translocates into the nucleus to promote the transcription of antioxidant and detoxification genes such as heme oxygenase-1 (HO-1), NAD(P)H quinone oxidoreductase-1 (NQO 1), and glutathione-S-transferases. Among Nrf2 activators, sulforaphane is regarded as the most potent natural inducer, binding to cysteine residues on the Keap 1 protein and releasing Nrf2 to bind with the antioxidant response element, thus initiating gene transcription (Kwak et al., 2002). Curcumin and EGCG also act as inducers for Nrf2 signaling, providing long-term benefits to an organism against oxidative DNA damage. This indirect defense mechanism enables DNA repair processes that reduce oxidative insults. When inactive, Nrf2, a CNC-bzip transcription factor, is held in check by Keap 1, a homodimeric adaptor protein that also recruits the Cullin 3 (Cul 3)–Rbx 1 E3 ubiquitin ligase complex. This leads to Nrf2' s ubiquitination and subsequent degradation. Under oxidative or electrophilic stress, ROS, electrophiles (such as sulforaphane, curcumin, and EGCG), or kinase signaling modify Keap 1 cysteine residues or phosphorylate Nrf2, disrupting the 'hinge-and-latch' mechanism (Fig. 25.3). This prevents Nrf2 ubiquitination, allowing it to stabilize and accumulate. Once stabilized, Nrf2 translocates to the nucleus, partners with small Maf proteins, and binds to Antioxidant Response Elements (AREs) in the promoters of target genes. The transcription of antioxidant and detoxifying enzymes is then triggered by coactivators such as Cbp/p 300, SWI/SNF complexes, and RNA polymerase II, which are recruited to the gene promoters, including HO-1, NQO 1, GCLC/GCLM, GSTs, SODs, catalase, and multidrug resistance proteins (Fig. 25.3). Several negative feedback mechanisms complete this process: the reformation of Keap 1 and subsequent resequestration of Nrf2, the competing attachment of Bach 1 to AREs, the GSK-3 β/β-Trcp-mediated ubiquitination process, the nuclear export of Fyn kinase, and the autophagic degradation of Keap 1 via p 62/SQSTM 1. Compounds such as sulforaphane, curcumin, and EGCG influence Nrf2 activity by directly modifying Keap 1 cysteine residues and via kinase signaling pathways. Activation of Nrf2 enhances antioxidant functions, DNA protection, and cell viability under stress, but persistent activation—in cases such as KEAP 1/Nrf2 mutations in cancers—can promote tumor development and resistance to therapy.

25.4.3.2 NF-κB pathway: Inflammation control

Conversely, chronic inflammation is closely linked to DNA damage because activated immune cells produce ROS. The transcription factor NF-κB controls inflammatory cytokines such as TNF-α and IL-6, which raise oxidative stress. Plant nutraceuticals such as curcumin,

FIGURE 25.3 Mechanism of Nrf2 regulation under basal and stress conditions. Under basal conditions, Keap1 sequesters Nrf2 and recruits the Cul3–Rbx1 E3 ligase, which ubiquitinates Nrf2 for proteasomal degradation. Oxidative or electrophilic stress either modifies the Keap1 cysteines or phosphorylates Nrf2, which disrupts the "hinge-and-latch" binding and prevents Nrf2 degradation. Stabilized Nrf2 is free to accumulate, enter the nucleus, and bind AREs with small Maf proteins, turning on antioxidant genes (e.g., HO-1, NQO1, GCLC, GSTs, SODs) and detoxification genes. Gene expression is further upregulated by CBP/p300 coactivators and SWI/SNF complexes. Negative feedback is mediated by Keap1 resynthesis, Bach1 competition, GSK-3β/β-TrCP ubiquitination, and p62-mediated autophagic degradation. Cellular defense is enhanced by the action of small molecules such as sulforaphane, curcumin, and EGCG in this pathway, but they can also contribute to tumorigenesis if constantly activated.

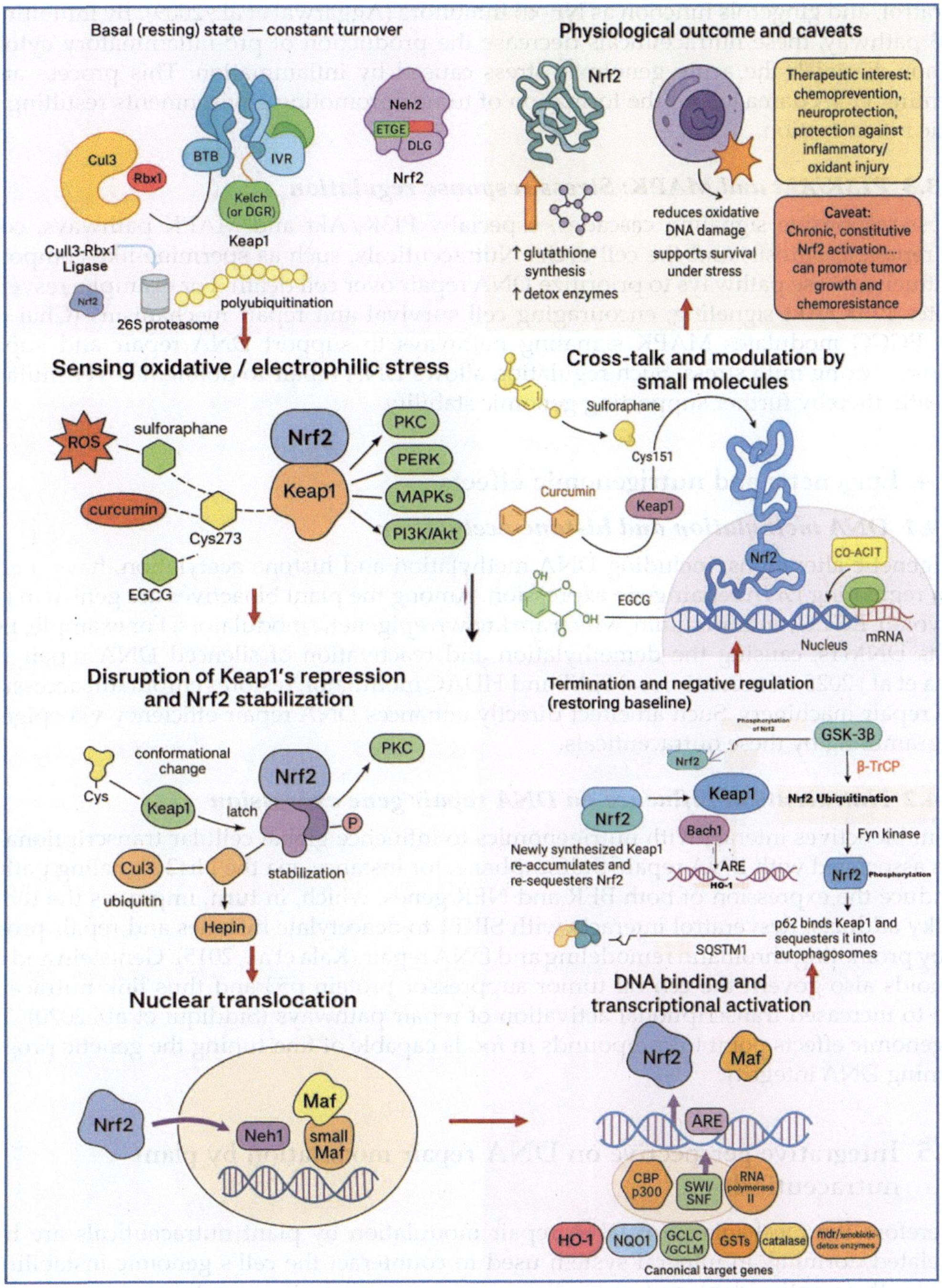
Basal (resting) state — constant turnover
Cul3
Rbx1
BTB
IVR
Kelch (or DGR)
Keap1
Neh2
ETGE
DLG
Nrf2
Cull3-Rbx1 Ligase
Nrf2
26S proteasome
polyubiquitination
Physiological outcome and caveats
Nrf2
↑ glutathione synthesis
↑ detox enzymes
reduces oxidative DNA damage
supports survival under stress
Therapeutic interest: chemoprevention, neuroprotection, protection against inflammatory/ oxidant injury
Caveat: Chronic, constitutive Nrf2 activation... can promote tumor growth and chemoresistance
Sensing oxidative / electrophilic stress
ROS
sulforaphane
curcumin
Cys273
EGCG
Nrf2
Keap1
PKC
PERK
MAPKs
PI3K/Akt
Cross-talk and modulation by small molecules
Sulforaphane
Cys151
Curcumin
Keap1
Nrf2
CO-ACIT
EGCG
mRNA
Nucleus
Disruption of Keap1's repression and Nrf2 stabilization
conformational change
Cys
Keap1
latch
Nrf2
PKC
P
Cul3
ubiquitin
stabilization
Hepin
Termination and negative regulation (restoring baseline)
GSK-3β
Nrf2
β-TrCP
Keap1
Nrf2 Ubiquitination
Nrf2
Bach1
Newly synthesized Keap1 re-accumulates and re-sequesters Nrf2
ARE
HO-1
Fyn kinase
Nrf2
p62 binds Keap1 and sequesters it into autophagosomes
SQSTM1
Nuclear translocation
Nrf2
Neh1
Maf
small Maf
DNA binding and transcriptional activation
Nrf2
Maf
ARE
CBP p300
SWI/ SNF
RNA polymerase II
HO-1
NQO1
GCLC /GCLM
GSTs
catalase
mdr/ xenobiotic detox enzymes
Canonical target genes

resveratrol, and gingerols function as NF-κB inhibitors (Aggarwal et al., 2009). By inhibiting the NF-κB pathway, these nutraceuticals decrease the production of pro-inflammatory cytokines and thus diminish the acute genotoxic stress caused by inflammation. This process aids in preventing DNA damage and the formation of tumor-promoting environments resulting from chronic inflammation.

25.4.3.3 PI3K/Akt and MAPK: Stress response regulation

Stress-responsive signaling cascades, especially PI3K/Akt and MAPK pathways, control DNA repair, apoptosis, and the cell cycle. Nutraceuticals, such as spermine-like compounds, can influence these pathways to prioritize DNA repair over cell death. For example, resveratrol activates PI3K/Akt signaling, encouraging cell survival and repair mechanisms (Chai et al., 2017). EGCG modulates MAPK signaling pathways to support DNA repair and suppress apoptosis during mild stress. Such regulation allows DNA repair to dominate over mutagenic cell death, thereby further supporting genomic stability.

25.4.4 Epigenetic and nutrigenomic effects

25.4.4.1 DNA methylation and histone acetylation

Epigenetic alterations, including DNA methylation and histone acetylation, have a crucial role in regulating DNA repair gene expression. Among the plant bioactives are genistein (a soy isoflavone), EGCG, and curcumin, which are known epigenetic modulators. For example, EGCG inhibits DNMTs, causing the demethylation and reactivation of silenced DNA repair genes (Dorna et al., 2025). Curcumin, as a HAT and HDAC modulator, restores chromatin accessibility to the repair machinery. Such an effect directly enhances DNA repair efficiency via epigenetic reprogramming by these nutraceuticals.

25.4.4.2 Nutraceutical influence on DNA repair gene expression

Plant bioactives interact with nutrigenomics to influence global cellular transcriptional networks associated with DNA repair. Sulforaphane, for instance, via the Nrf2 signaling pathway, can induce the expression of both BER and NER genes, which, in turn, improves the removal of bulky adducts. Resveratrol interacts with SIRT1 to deacetylate histones and repair proteins, thereby promoting chromatin remodeling and DNA repair (Kala et al., 2015). Genistein and other flavonoids also govern the central tumor suppressor protein p53 and thus link nutraceutical intake to increased transcriptional activation of repair pathways (Siddiqui et al., 2020). These nutrigenomic effects point to compounds in foods capable of fine-tuning the genetic programs governing DNA integrity.

25.4.5 Integrative perspective on DNA repair modulation by plant nutraceuticals

Therefore the mechanisms of DNA repair modulation by plant nutraceuticals are highly interrelated, forming an integral system used to counteract the cell's genomic instability. On one side, their antioxidant and redox-balancing roles act further upstream in reducing the initial overload of DNA lesions. By contrast, on the downstream end, plant nutraceuticals directly activate the DNA repair enzymes that carry out the removal of already formed damage. Simultaneous modulation of the intracellular stress response by nutraceuticals via signaling pathways, including Nrf2, NF-κB, and PI3K/Akt, keeps the balance between pro-survival and repair mechanisms tipped toward apoptosis in case of extensive and irreversible damage. In addition, based on epigenetic and nutrigenomic effects, they provide possibilities for long-term adaptation by reprogramming the expression of DNA repair genes and sustaining chromatin conformations feasible for repair. Such a multifactorial synergy emphasizes the importance of dietary phytochemicals in regulating genome instability and suggests that they can be used as adjuvants in preventing and treating diseases.

25.5 Evidence from experimental and clinical studies

A multitude of experimental and clinical evidence has arisen to demonstrate the protective feature of DNA repair-associated plant nutraceuticals. The studies, ranging from in vitro cellular systems to animal models, and then human clinical trials, as well as population-based investigations, provide examples for both mechanistic insight and translational potential. Collectively, deposited knowledge constitutes an encompassing picture of plant nutraceuticals and genome maintenance.

25.5.1 In vitro and animal studies

25.5.1.1 *Reduction in DNA damage biomarkers (e.g., 8-OHdG)*

The plant nutraceuticals have been demonstrated in cell culture and animal studies to reduce DNA damage biomarkers, particularly 8-hydroxy-2′-deoxyguanosine (8-OHdG), a well-known marker of oxidative DNA damage. In cultured human lymphocytes treated with hydrogen peroxide, green tea catechins (EGCG) decreased levels of 8-OHdG (Riso et al., 2002). Moreover, following exposure to a carcinogen, rodents supplemented with resveratrol showed a reduction in oxidative DNA adducts and an enhancement in DNA repair activity (Athar et al., 2007). Such results strongly imply that nutraceuticals act against the formation of DNA lesions under oxidative-stress conditions.

25.5.1.2 *Mechanistic insights from cell line and mouse models*

Experimentally, some models can even aid in mechanistic considerations of nutraceutical action on DNA repair pathways. For example, in mouse models, sulforaphane enhanced Nrf2-dependent transcription of detoxification and DNA repair genes that ultimately led to the efficient repair of DNA adducts induced by chemical carcinogens (Gwon et al., 2020). Curcumin was also shown in cancer cell lines to restore MMR and BER activity through upregulation of repair proteins such as OGG1 and MLH1 (Hosseini et al., 2020). Together, in vitro and in vivo systems provide mechanistic evidence that nutraceuticals reduce DNA damage as well as stabilize DNA repair.

25.5.2 Human trials and epidemiological insights

25.5.2.1 *Clinical studies on green tea polyphenols and cruciferous vegetables*

Several encouraging findings from clinical research have been reported with the nutraceutical-rich diet in DNA protection. Intervention studies with green tea polyphenols have indicated reduced 8-OHdG in urinary excretion by healthy volunteers and smokers, indicating lower systemic DNA oxidation (Luo et al., 2006). Similarly, the consumption of cruciferous vegetables such as broccoli, Brussels sprouts, etc., has been associated with greater activity of detoxifying enzymes and a reduced incidence of DNA strand breaks in peripheral blood lymphocytes (Zhu and Loft, 2001). These findings add to the translational relevance of plant bioactives in human genome protection.

25.5.2.2 *Population studies linking plant-based diets to lower DNA damage*

Large-scale epidemiological studies further support the protective link between plant-based diets and DNA integrity. Populations that consume higher amounts of fruits and vegetables and foods rich in polyphenols consistently report lower levels of DNA strand breaks and oxidative lesions relative to those consuming Western diets laden with processed foods (Azqueta and Collins, 2016). For example, cohorts characterized by Mediterranean or Asian dietary patterns show fewer biomarkers of DNA damage that fit well with the increased intakes of resveratrol, isoflavones, and flavonoids. These data at the population level lend credence to the claim that nutraceuticals, as part of whole diets, carry genomic benefits over the long haul.

25.5.3 Case studies of specific compounds

25.5.3.1 *Curcumin*

Curcumin is one of the potential polyphenol compounds that may be able to enhance DNA repair highly. Preclinical evidence suggests that curcumin increases the BER and NER pathways while blocking NF-κB, bringing about anti-inflammatory effects (Hosseini et al., 2020). On the other hand, human pilot studies report that DNA damage levels of peripheral lymphocytes decrease after supplementation with curcumin formulations; however, issues with bioavailability appear to persist.

25.5.3.2 *Resveratrol*

Extensively studied for the activation of SIRT1, resveratrol enhances DNA repair via chromatin remodeling along with p53 regulation. Animal trials following oxidative stress show resveratrol treatment improved DNA strand break repair, while human studies demonstrate lower DNA oxidation markers correlated with moderate consumption of wine rich in resveratrol; however, there are still only very few trials on actual supplementation (Oommen et al., 2016).

25.5.3.3 *Sulforaphane*

Sulforaphane obtained from broccoli sprouts is an excellent Nrf2 activator and has been shown in both animal and human studies to increase detoxification enzyme expression and DNA repair capacity. Two clinical studies from China, conducted on populations exposed to air pollutants, have shown that consuming broccoli sprout beverages enhances excretion of carcinogen-derived metabolites and diminishes biomarkers of oxidative DNA damage (Kendler et al., 2013). This therefore provides very strong translational evidence demonstrating how sulforaphane can enhance DNA protection under real-world environmental stress conditions.

25.5.4 Integrative perspective on experimental and clinical evidence

Evidence drawn from in vitro, animal, and human studies, as well as from epidemiological observations, makes a case for plant nutraceuticals in DNA repair and protection from genomic instability. While cellular and animal studies provide mechanistic insights into how specific compounds activate repair enzymes and signaling pathways, human intervention studies support the reductions in DNA damage biomarkers following nutraceutical-rich diets. Other studies at the population level have linked long-term intake of fruits, vegetables, and phytochemical-rich foods with lowered DNA damage and chronic disease risk. Case studies of agents such as curcumin, resveratrol, and sulforaphane back up and give real-world translational possibilities to the mechanistic plausibility. When considered together, such findings point to nutraceuticals as dietary agents for genome maintenance, potentially acting synergistically when consumed as part of a broad, plant-based diet.

25.6 Applications in human health and agriculture

The modulation of DNA repair by plant nutraceuticals goes beyond simple cellular protection, providing implications in human health and agriculture. Leveraging the genomic stabilizing property of nutraceuticals can provide avenues for preventing diseases, assisting therapeutically, ensuring healthy aging, and improving crops.

25.6.1 Cancer chemoprevention and radioprotection

25.6.1.1 DNA repair enhancement as a therapeutic strategy

Cancer, on the other hand, is associated with genomic instability and impaired DNA repair. Plant nutraceuticals capable of stimulating repair pathways, therefore, represent good options for chemoprevention of cancer. Curcumin and resveratrol stimulate enzyme activity in BER, including OGG1, which quickly removes oxidative lesions before mutagenesis (Hosseini et al., 2020). Furthermore, sulforaphane stimulates Nrf2 signaling in the induction of detoxification and DNA repair genes, resulting in reduced amounts of DNA adducts induced by carcinogens (Kendler et al., 2013). This makes nutraceuticals potential natural agents in preventing cancer initiation by preserving genomic integrity.

25.6.1.2 Adjunct roles in chemotherapy and radiotherapy

Radiation and chemotherapeutic agents, while effectively controlling tumors, also induce collateral DNA damage in normal tissues. Nutraceuticals of plant origin, therefore, would serve as radioprotectors and chemosensitizers, with fewer side effects for the patient, better therapeutic effect. For instance resveratrol and EGCG protect normal cells against radiation-induced DNA strand breaks while sensitizing tumor cells to DNA damage therapies (Carsten et al., 2008). Curcumin has also been studied as an adjunct to cancer therapy, where it may lessen toxicity through its anti-inflammatory and DNA repair-modulating effects. Thus nutraceuticals can potentially be utilized in integrative oncology, assisting DNA repair of normal cells while working in conjunction with conventional therapy.

25.6.2 Neurodegeneration and aging

25.6.2.1 Delaying cellular senescence via genomic maintenance

DNA damage accumulation and defective repair capacity have quite strong links with aging and neurodegenerative diseases, especially Alzheimer's and Parkinson's. Plant nutraceuticals delay the onset of senescence by reducing the number of lesions and keeping repair efficiency high. For instance polyphenols such as resveratrol promote the activation of SIRT1, an important regulator of DNA repair and longevity, leading to increased replicative lifespan of cells (Baur and Sinclair, 2006). In the same way, curcumin blocks p53 activation after DNA damage, allowing cells to escape premature senescence. Thus nutraceuticals will help maintain DNA integrity and healthy ageing pathways.

25.6.2.2 Protection of neural tissue from oxidative stress

Neurons are highly susceptible to oxidative DNA damage since they are post-mitotic cells with a high metabolic requirement. Nutraceuticals protect the neurons by reducing oxidative lesions and promoting their repair. EGCG crosses the blood-brain barrier to reach brain neuronal cells, reducing oxidative DNA breaks while enhancing synaptic plasticity and, consequently, improving memory in animal models (Kwak et al., 2002). Sulforaphane also activates Nrf2 in neural tissues, promoting antioxidant defenses and DNA repair in the presence of neurotoxic stress. These results suggest the possibility of applying nutraceuticals for the prevention and/or delay of neurodegenerative diseases through the maintenance of genomic integrity.

25.6.3 Crop improvement and agricultural benefits

25.6.3.1 Enhancement of plant DNA repair to increase stress tolerance

In the case of plants, just as in humans, DNA repair is essential for them to survive stress situations such as UV radiations or, drought, and pathogen attacks. Plant-based nutraceuticals and phytochemicals can modulate DNA repair enzymes inside the plant system. Polyphenols act as antioxidants for sun protection against oxidative UV damage, while isoflavones and other related chemicals affect DNA repair pathways in plants (Azqueta and Collins, 2016). Increasing DNA repair potential in plants makes them able to withstand abiotic and biotic stresses.

25.6.3.2 Role in improving yield, longevity, and resilience

Strengthening DNA repair will restore the plants' ability to adjust and, consequently, enhance the stability of yield and longevity. More DNA stability means less mutation, further enhanced by environmental stress; therefore crops can maintain their productivity. Thus application of extracts rich in nutraceuticals or biostimulants of a natural origin could potentially boost crop resilience. As an example, it has been reported that supplementation of flavonoids in plants would enhance photosynthetic activity and DNA protection when plants are exposed to high light (Ferreyra et al., 2021).

25.6.3.3 Potential for developing stress-resilient cultivars

These advances in plant biotechnology offer possibilities to incorporate nutraceutical pathways into crop breeding. Overexpression of genes for phytochemical biosynthesis or DNA repair enzymes affected by nutraceuticals can lead to the development of cultivars resilient to stress and able to survive climate change. Genetically enhancing the Nrf2-like regulatory pathways of plants or flavonoid synthesis might give rise to cultivars that can better endure oxidative and genotoxic stress. It marks the promising agricultural application of nutraceuticals, connecting food security to genomic stability.

25.6.4 Integrative perspective on human health and agricultural applications

The scope of plant nutraceuticals in DNA repair stretches across human health and agriculture, indicating their vast purview of sustainability and disease prevention areas. In cancer, these substances act as chemopreventives and radioprotectors, outlining how enhancement of DNA repair can be channeled into therapeutic means and support treatment. In a neurodegenerative and ageing context, nutraceuticals suppress oxidative DNA damage, delay cellular senescence, and protect neurons for a healthier lifespan. In addition to health-related applications, their uses in agriculture are just as important. Support for plant DNA repair increases stress tolerance and secondary yield stability, along with crop resilience, among others. These findings thus characterize nutraceuticals as unique bioactive substances bridging biomedical and agricultural sciences, with a dual income for human welfare and global food security.

25.7 Challenges and future directions

Despite promising evidence in favor of plant nutraceuticals in DNA repair and genomic maintenance, the transformation of these into medicine and agriculture still faces many hurdles that need to be addressed to enhance their medicinal and agricultural potential.

25.7.1 Bioavailability and metabolic stability

25.7.1.1 Poor absorption of compounds like curcumin

The most formidable hurdles in nutraceutical research are those related to their low bioavailability. Numerous phytochemicals, including curcumin, resveratrol, and EGCG, suffer low intestinal absorption and are rapidly metabolized and eliminated systemically; this limits their therapeutic concentrations in the tissues involved (Cione et al., 2019). In the case of curcumin, a great deal of it is metabolized into glucuronide and sulfate conjugates, which exhibit very little biological activity, leading to a mostly in vitro effect of curcumin that does not really translate in vivo. Resveratrol is yet another phytochemical that is familiar with a rapid hepatic metabolism that limits its systemic effects, notwithstanding potent cellular activity.

25.7.1.2 Strategies: Nanoformulations, liposomes, adjuvants

To explore novel drug delivery mechanisms, these drawbacks are being addressed. Nanoformulations, liposomes, and polymeric nanoparticles may serve to improve the solubility of nutraceuticals, increase their permeability across membranes, and extend their circulation time. Piperine—an alkaloid from black pepper—has been used as an adjuvant to inhibit glucuronidation while considerably improving the bioavailability of curcumin (Gorgani et al., 2017). Encapsulation mechanisms like micelles loaded with curcumin or liposomes containing resveratrol can employ delivery targeting tissues such as the brain or tumor sites. These approaches have the promise to shorten the distance between experimentally obtained efficacy and clinical relevance.

25.7.2 Standardization and dosing

25.7.2.1 Variability in plant composition

Another major hurdle lies in the variation in the concentration of bioactive compounds in plants. Factors such as geographic origin, growing conditions, harvesting time, and extraction methods can significantly alter phytochemical content (Palit and Mandal, 2021). For instance the amount of precursor for sulforaphane in broccoli is reported to show more than a tenfold variation among cultivars, thus posing a problem in the reproducibility of clinical results.

25.7.2.2 Need for robust extraction, purification, and dosage frameworks

This variation requires the development of standards, guaranteeing equal efficacy and safety. Validated methods for the extraction, purification, and analytical measurement of nutraceuticals bear significance and should be developed for clinical and commercial formulations alike. Further, ideal dosage regimens for a given nutraceutical remain unclear, as most studies report arbitrary dosages not achievable through diet alone. Controlled dose–response studies are needed to formulate therapeutic windows for therapeutic application, thus reducing hazard and rejection from clinical practice.

25.7.3 Regulatory, ethical, and research gaps

25.7.3.1 Lack of global regulatory consensus

Unlike pharmaceuticals, nutraceuticals are in a grey area of regulation; across regions, classification remains inconsistent. In some regions, they are treated as dietary supplements. In others, they are under the guise of functional foods or botanical drugs (Thakkar et al., 2020). The lack of compromise on a global basis hinders research, patenting, and commercialization, thus adding to the already existing limitation on collaborative research on an international scale.

25.7.3.2 Ethical considerations in human trials

Human nutraceutical trials also pose ethical challenges. Since such compounds are mostly obtained from food items that are consumed daily by people, it becomes very difficult to define placebo controls and set up blinding strategies. Long-term interventions raise serious ethical concerns, especially when patients with grave conditions such as cancer or aged persons are involved. Furthermore, insufficient funding and industry-favoring research leave gaps in independent corroboration of health claims.

25.7.4 Emerging trends

25.7.4.1 Personalized nutrition based on genomics

One major direction of future research is personalized nutrition, wherein nutraceutical interventions are matched with an individual's genetic and epigenetic profiles. For example, polymorphisms of DNA repair genes such as OGG1 or even XRCC1 variants may modulate the responses to antioxidants such as resveratrol or EGCG. Nutrigenomic approaches should aid in the best-fit selection and dosing of nutraceuticals so as to increase efficacy and decrease side effects (Bauman et al., 2011).

25.7.4.2 *Integration of omics technologies: Proteomics, metabolomics, and transcriptomics*

Integrating omics approaches such as transcriptomics, proteomics, and metabolomics offers us an opportunity not only to resolve the issue of how nutraceuticals act at and above system levels but also to do so in ways never achieved before. DNA repair modulation biomarkers can be identified through high-throughput methods to assess nutraceutical efficacy; thus elucidating these biomarkers provides mechanistic insights that can lead to developing models predicting the effectiveness of nutraceuticals. The combination of omics techniques and computational biology could also pave the way for discovering synergistic interactions among dietary phytochemicals within complex biological networks. Advancements in delivery systems for targeted therapy may also mark a new era in clinically focused nutraceutical applications. Targeting delivery sites such as tumors or specific brain regions with smart nanocarriers, or stimuli-responsive polymers and nanoparticles conjugated with ligands, is likely to soon enter clinical practice for nutraceuticals. This approach will enhance absorption of plant actives and further reduce side effects on non-target tissues, aligning these nutraceuticals more closely with the principles of precision medicine. As these methodologies mature, plant bioactives may evolve from dietary supplements into bona fide therapeutics with explicit DNA repair-modulating functions.

25.7.5 Integrative outlook

Although attempting to address issues related to plant-based nutraceuticals in DNA repair modulation highlights the complexity of deriving tangible health and agricultural benefits from promising laboratory findings, these pose obstacles to the clinical and commercial use of DNA repair-modulating plant nutraceuticals—such as limited bioavailability, high variability in plant composition, and regulatory challenges. However, rapid progress in delivery technologies, standardization protocols, and omics-based personalization is opening new possibilities for the reliable and targeted use of these compounds. Nonetheless, a multidisciplinary approach blending molecular biology, pharmacology, food science, systems biology, and regulatory science will be necessary to unlock the full potential of these compounds. In this context, nutraceuticals could move beyond being merely classified as dietary supplements to become precision tools for genome stability, bridging the gap between preventative healthcare and sustainable agriculture. Given current technologies and global health challenges, the next decade may witness a disruptive leap in the application of nutraceuticals, integrating traditional dietary knowledge with advanced genomic medicine.

25.8 Conclusion

Based on the evidence presented in earlier chapters, it is clear that plant nutraceuticals play a diverse role in DNA repair. Such compounds act at different levels—scavenging reactive species, enhancing repair enzymes, activating key signaling pathways, and reconfiguring epigenetic landscapes—collectively providing a safeguard for genomic stability. Their use in human health, especially for cancer chemoprevention, radioprotection, and neuroprotection, extends into agriculture, where they bolster plant stress tolerance and resilience, promoting crop productivity. This dual significance positions plant nutraceuticals as a bridge between preventive medicine and sustainable agriculture. By reducing DNA damage and supporting genomic maintenance, plant nutraceuticals address some of the core issues linked to ageing, chronic diseases, and environmental stressors in both humans and crops. Moving forward, plant nutraceuticals may find their greatest value in personalized nutrition, functional foods, and biotechnological innovations. Concurrent advancements in nanotechnology, omics, and precision agri-food will undoubtedly supply the necessary tools to improve their bioavailability, efficacy, and application. These transformative processes could elevate nutraceuticals from being merely dietary supplements to strategic tools for genome stabilization and disease prevention. Achieving this requires interdisciplinary and translational research, uniting molecular biologists, nutritionists, clinicians, plant scientists, and policymakers. Supporting the potential of nutraceuticals through such collaboration will contribute to human health and longevity, as well as to food security and the sustainability of global agriculture.

References

Agarwal, S., Rao, A.V., 2000. Tomato lycopene and its role in human health and chronic diseases. CMAJ 163 (6), 739–744.

Aggarwal, B.B., Sundaram, C., Malani, N., Ichikawa, H., 2007. Curcumin: the Indian solid gold. Mol. Targets Ther. Uses Curcumin Health Dis. 1–75.

Aggarwal, B.B., Van Kuiken, M.E., Iyer, L.H., Harikumar, K.B., Sung, B., 2009. Molecular targets of nutraceuticals derived from dietary spices: potential role in suppression of inflammation and tumorigenesis. Exp. Biol. Med. 234 (8), 825–849.

Allkanjari, K., Baldock, R.A., 2021. Beyond base excision repair: an evolving picture of mitochondrial DNA repair. Biosci. Rep. 41 (10), BSR20211320.

Aremu, A.O., Omogbene, T.O., Fadiji, T., Lawal, I.O., Opara, U.L., Fawole, O.A., 2024. Plants as an alternative to the use of chemicals for crop protection against biotic threats: trends and future perspectives. Eur. J Plant Pathol. 170 (4), 711–766.

Athar, M., Back, J.H., Tang, X., Kim, K.H., Kopelovich, L., Bickers, D.R., et al., 2007. Resveratrol: a review of preclinical studies for human cancer prevention. Toxicol. Appl. Pharmacol. 224 (3), 274–283.

Aykan, K., Altuntaş, İ., Erbaş, O., 2022. DNA repair mechanisms: DNA repair defects and related diseases. Demiroglu Sci. Univ. Florence Nightingale J Med. 8 (3), 130–140.

Azqueta, A., Collins, A., 2016. Polyphenols and DNA damage: a mixed blessing. Nutrients 8 (12), 785.

Balestrazzi, A., Confalonieri, M., Macovei, A., Donà, M., Carbonera, D., 2011. Genotoxic stress and DNA repair in plants: emerging functions and tools for improving crop productivity. Plant Cell Rep. 30 (3), 287–295.

Barzilai, A., Biton, S., Shiloh, Y., 2008. The role of the DNA damage response in neuronal development, organization, and maintenance. DNA Repair (Amst.) 7 (7), 1010–1027.

Bauman, D.E., Harvatine, K.J., Lock, A.L., 2011. Nutrigenomics, rumen-derived bioactive fatty acids, and the regulation of milk fat synthesis. Annu. Rev. Nutr. 31 (1), 299–319.

Baur, J.A., Sinclair, D.A., 2006. Therapeutic potential of resveratrol: the in vivo evidence. Nat. Rev. Drug Discov. 5 (6), 493–506.

Biesalski, H.K., Dragsted, L.O., Elmadfa, I., Grossklaus, R., Müller, M., Schrenk, D., et al., 2009. Bioactive compounds: definition and assessment of activity. Nutrition 25 (11-12), 1202–1205.

Boots, A.W., Haenen, G.R., Bast, A., 2008. Health effects of quercetin: from antioxidant to nutraceutical. Eur. J. Pharmacol. 585 (2-3), 325–337.

Böttner, L., Grabe, V., Gablenz, S., Böhme, N., Appenroth, K.J., Gershenzon, J., et al., 2021. Differential localization of flavonoid glucosides in an aquatic plant implicates different functions under abiotic stress. Plant Cell Environ. 44 (3), 900–914.

Caldecott, K.W., 2022. DNA single-strand break repair and human genetic disease. Trends Cell Biol. 32 (9), 733–745.

Calderon-Montano, J.M., Burgos-Moron, E., Pérez-Guerrero, C., Lopez-Lazaro, M., 2011. A review on the dietary flavonoid kaempferol. Mini Rev. Med. Chem. 11 (4), 298–344.

Cannan, W.J., Pederson, D.S., 2016. Mechanisms and consequences of double-strand DNA break formation in chromatin. J. Cell. Physiol. 231 (1), 3–14.

Carr, A.C., Frei, B., 1999. Toward a new recommended dietary allowance for vitamin C based on antioxidant and health effects in humans. Am. J. Clin. Nutr. 69 (6), 1086–1107.

Carsten, R. E., Bachand, A. M., Bailey, S. M., Ullrich, R. L., 2008. Resveratrol reduces radiation-induced chromosome aberration frequencies in mouse bone marrow cells. Radiat Res 169 (6), 633–638.

Chai, R., Fu, H., Zheng, Z., Liu, T., Ji, S., Li, G., 2017. Resveratrol inhibits proliferation and migration through SIRT1 mediated post translational modification of PI3K/AKT signaling in hepatocellular carcinoma cells. Mol. Med. Rep 16 (6), 8037–8044.

Chang, H.H., Pannunzio, N.R., Adachi, N., Lieber, M.R., 2017. Non-homologous DNA end joining and alternative pathways to double-strand break repair. Nat. Rev. Mol. Cell Biol. 18 (8), 495–506.

Cheng, Q., Chen, J., 2010. Mechanism of p53 stabilization by ATM after DNA damage. Cell Cycle 9 (3), 472–478.

Cione, E., La Torre, C., Cannataro, R., Caroleo, M.C., Plastina, P., Gallelli, L., 2019. Quercetin, epigallocatechin gallate, curcumin, and resveratrol: from dietary sources to human microRNA modulation. Molecules 25 (1), 63.

Cooke, M.S., Evans, M.D., Dizdaroglu, M., Lunec, J., 2003. Oxidative DNA damage: mechanisms, mutation, and disease. FASEB J. 17 (10), 1195–1214.

Daley, J.M., Sung, P., 2014. 53BP1, BRCA1, and the choice between recombination and end joining at DNA double-strand breaks. Mol. Cell. Biol. 34 (8), 1380–1388.

Datta, K., Purkayastha, S., Neumann, R.D., Pastwa, E., Winters, T.A., 2011. Base damage immediately upstream from double-strand break ends is a more severe impediment to nonhomologous end joining than blocked 3′-termini. Radiat. Res. 175 (1), 97–112.

Degerud, E.M., Manger, M.S., Strand, T.A., Dierkes, J., 2015. Bioavailability of iron, vitamin A, zinc, and folic acid when added to condiments and seasonings. Ann. N Y Acad. Sci 1357 (1), 29–42.

Dorna, D., Grabowska, A., Paluszczak, J., 2025. Natural products modulating epigenetic mechanisms by affecting histone methylation/demethylation: targeting cancer cells. Br. J. Pharmacol. 182 (10), 2137–2158.

Dosouky, M.T., 2023. Overview of DNA damage and various genetic repair mechanisms. Recent Res. Genet. Genomics. 5 (1), 1–30.

Dutta, S., Mitra, M., Agarwal, P., Mahapatra, K., De, S., Sett, U., et al., 2018. Oxidative and genotoxic damages in plants in response to heavy metal stress and maintenance of genome stability. Plant. Signal. Behav. 13 (8), e1460048.

Errico, A., Costanzo, V., 2012. Mechanisms of replication fork protection: a safeguard for genome stability. Crit. Rev. Biochem. Mol. Biol. 47 (3), 222–235.

Ferreyra, M.L.F., Serra, P., Casati, P., 2021. Recent advances on the roles of flavonoids as plant protective molecules after UV and high light exposure. Physiol. Plant. 173 (3), 736–749.

Fousteri, M., Mullenders, L.H., 2008. Transcription-coupled nucleotide excision repair in mammalian cells: molecular mechanisms and biological effects. Cell Res. 18 (1), 73–84.

Garner, E., Kim, Y., Lach, F.P., Kottemann, M.C., Smogorzewska, A., 2013. Human GEN1 and the SLX4-associated nucleases MUS81 and SLX1 are essential for the resolution of replication-induced Holliday junctions. Cell Rep. 5 (1), 207–215.

Gerszon, J., Rodacka, A., Puchała, M., 2014. Antioxidant properties of resveratrol and its protective effects in neurodegenerative diseases. Adv. Cell Biol. 4 (2), 97–117.

Goel, A., Kunnumakkara, A.B., Aggarwal, B.B., 2008. Curcumin as "Curecumin": from kitchen to clinic. Biochem. Pharmacol. 75 (4), 787–809.

Gong, P., Guo, Z., Wang, S., Gao, S., Cao, Q., 2025. Histone phosphorylation in DNA damage response. Int. J. Mol. Sci. 26 (6), 2405.

Gonzalo-Hansen, C., Steurer, B., Janssens, R.C., Zhou, D., Van Sluis, M., Lans, H., et al., 2024. Differential processing of RNA polymerase II at DNA damage correlates with transcription-coupled repair syndrome severity. Nucleic. Acids. Res. 52 (16), 9596–9612.

Gorgani, L., Mohammadi, M., Najafpour, G.D., Nikzad, M., 2017. Piperine—the bioactive compound of black pepper: from isolation to medicinal formulations. Compr. Rev. Food Sci. Food Saf. 16 (1), 124–140.

Gwon, Y., Oh, J., Kim, J.S., 2020. Sulforaphane induces colorectal cancer cell proliferation through Nrf2 activation in a p53-dependent manner. Appl. Biol. Chem 63 (1), 86.

Haftel, L., Berkovich, Z., Reifen, R., 2015. Elevated milk β-carotene and lycopene after carrot and tomato paste supplementation. Nutrition 31 (3), 443–445.

Hasan, M.M., Bae, H., 2017. An overview of stress-induced resveratrol synthesis in grapes: Perspectives for resveratrol-enriched grape products. Molecules 22 (2), 294.

Hosseini, S.M., Mohammadiasl, J., Talaiezadeh, A., Alidadi, R., Bijanzadeh, M., 2020. Influence of two DNA repair pathway polymorphisms in colorectal cancer risk in southwest Iran. Asian Pac. J. Cancer Prev. 21 (7), 1919.

Jiricny, J., 2013. Postreplicative mismatch repair. Cold Spring Harb. Perspect. Biol. 5 (4), a012633.

Kala, R., Shah, H.N., Martin, S.L., Tollefsbol, T.O., 2015. Epigenetic-based combinatorial resveratrol and pterostilbene alters DNA damage response by affecting SIRT1 and DNMT enzyme expression, including SIRT1-dependent γ-H2AX and telomerase regulation in triple-negative breast cancer. BMC Cancer 15 (1), 672.

Kamal, R.M., Abdull Razis, A.F., Mohd Sukri, N.S., Perimal, E.K., Ahmad, H., Patrick, R., et al., 2022. Beneficial health effects of glucosinolates-derived isothiocyanates on cardiovascular and neurodegenerative diseases. Molecules 27 (3), 624.

Kendler, K.S., Aggen, S.H. and Neale, M.C., 2013. Evidence for multiple genetic factors underlying DSM-IV criteria for major depression. JAMA psychiatry, 70 (6), 599–607.

Khan, F.A., Ali, S.O., 2017. Physiological roles of DNA double-strand breaks. J. Nucleic Acids 2017 (1), 6439169.

Kocaadam, B., Şanlier, N., 2017. Curcumin, an active component of turmeric (*Curcuma longa*), and its effects on health. Crit. Rev. Food Sci. Nutr. 57 (13), 2889–2895.

Krokan, H.E., Bjørås, M., 2013. Base excision repair. Cold Spring Harb. Perspect. Biol. 5 (4), a012583.

Kunkel, T.A., Erie, D.A., 2005. DNA mismatch repair. Annu. Rev. Biochem. 74 (1), 681–710.

Kwak, M.K., Itoh, K., Yamamoto, M., Kensler, T.W., 2002. Enhanced expression of the transcription factor Nrf2 by cancer chemopreventive agents: role of antioxidant response element-like sequences in the Nrf2 promoter. Mol. Cell. Biol. 22 (9), 2883–2892.

Lagouge, M., Argmann, C., Gerhart-Hines, Z., Meziane, H., Lerin, C., Daussin, F., et al., 2006. Resveratrol improves mitochondrial function and protects against metabolic disease by activating SIRT1 and PGC-1α. Cell 127 (6), 1109–1122.

Lavin, M.F., Kozlov, S., Gatei, M., Kijas, A.W., 2015. ATM-dependent phosphorylation of all three members of the MRN complex: from sensor to adaptor. Biomolecules 5 (4), 2877–2902.

Le, D.T., Uram, J.N., Wang, H., Bartlett, B.R., Kemberling, H., Eyring, A.D., et al., 2015. PD-1 blockade in tumors with mismatch-repair deficiency. N. Engl. J. Med. 372 (26), 2509–2520.

Lee, Y.C., Zhou, Q., Chen, J., Yuan, J., 2016. RPA-binding protein ETAA1 is an ATR activator involved in DNA replication stress response. Curr. Biol. 26 (24), 3257–3268.

Leung, W., Simoneau, A., Saxena, S., Jackson, J., Patel, P.S., Limbu, M., et al., 2023. ATR protects ongoing and newly assembled DNA replication forks through distinct mechanisms. Cell Rep. 42 (7), 112792.

Li, C., Schluesener, H., 2017. Health-promoting effects of the citrus flavanone hesperidin. Crit. Rev. Food Sci. Nutr. 57 (3), 613–631.

Li, X., Cao, G., Liu, X., Tang, T.S., Guo, C., Liu, H., 2022. Polymerases and DNA repair in neurons: implications in neuronal survival and neurodegenerative diseases. Front. Cell Neurosci 16, 852002.

Lindahl, T., 1993. Instability and decay of the primary structure of DNA. Nature 362 (6422), 709–715.

Liu, N., Du, J., Ge, J., Liu, S.B., 2024. DNA damage-inducing endogenous and exogenous factors and research progress. Nucleosides Nucleotides Nucleic Acids 44, 1–33.

Liu, Y., Nwafor, E.O., Li, Z., Wang, J., Feng, X., Li, H., et al., 2022. Effects of berberine on liver cancer. Nat. Prod. Commun. 17 (5), 1934578.

Lombard, K., Peffley, E., Geoffriau, E., Thompson, L., Herring, A., 2005. Quercetin in onion (Allium cepa L.) after heat-treatment simulating home preparation. J. Food Compos. Anal. 18 (6), 571–581.

Lučić, D., Pavlović, I., Brkljačić, L., Bogdanović, S., Farkaš, V., Cedilak, A., et al., 2023. Antioxidant and antiproliferative activities of kale (*Brassica oleracea* L. var. acephala DC.) and wild cabbage (*Brassica incana* Ten.) polyphenolic extracts. Molecules 28 (4), 1840.

Luo, H., Tang, L., Tang, M., Billam, M., Huang, T., Yu, J., et al., 2006. Phase IIa chemoprevention trial of green tea polyphenols in high-risk individuals of liver cancer: modulation of urinary excretion of green tea polyphenols and 8-hydroxydeoxyguanosine. Carcinogenesis 27 (2), 262–268.

Madabhushi, R., Pan, L., Tsai, L.H., 2014. DNA damage and its links to neurodegeneration. Neuron 83 (2), 266–282.

Maynard, S., Fang, E.F., Scheibye-Knudsen, M., Croteau, D.L., Bohr, V.A., 2015. DNA damage, DNA repair, aging, and neurodegeneration. Cold Spring Harb Perspect. Med 5 (10), a025130.

Moynahan, M.E., Jasin, M., 2010. Mitotic homologous recombination maintains genomic stability and suppresses tumorigenesis. Nat. Rev. Mol. Cell Biol. 11 (3), 196–207.

Myzak, M.C., Dashwood, W.M., Orner, G.A., Ho, E., Dashwood, R.H., 2006. Sulforaphane inhibits histone deacetylase in vivo and suppresses tumorigenesis in Apcmin mice. FASEB J. 20 (3), 506.

Nickoloff, J.A., Sharma, N., Allen, C.P., Taylor, L., Allen, S.J., Jaiswal, A.S., et al., 2023. Roles of homologous recombination in response to ionizing radiation-induced DNA damage. Int. J. Radiat. Biol. 99 (6), 903–914.

Oommen, D., Dodd, N. J., Yiannakis, D., Moyeed, R., Jha, A. N., 2016. Linking genotoxicity and cytotoxicity with membrane fluidity: A comparative study in ovarian cancer cell lines following exposure to auranofin. Mutat Res 809, 43–49.

Palit, P., Mandal, S.C., 2021. Climate change, geographical location, and other allied triggering factors modulate the standardization and characterization of traditional medicinal plants: a challenge and prospect for phyto-drug development. In: Evidence-Based Validation of Traditional Medicines: A Comprehensive Approach. Springer, Singapore, pp. 359–369.

Pham-Huy, L.A., He, H., Pham-Huy, C., 2008. Free radicals, antioxidants in disease and health. Int. J. Biomed. Sci. 4 (2), 89.

Riar, C.S., Panesar, P.S., 2024. Bioactive compounds and nutraceuticals: Classification, potential sources, and application status. In: Bioactive Compounds and Nutraceuticals from Dairy, Marine, and Nonconventional Sources. Apple Academic Press, pp. 3–60.

Riso, P., Erba, D., Criscuoli, F., Testolin, G., 2002. Effect of green tea extract on DNA repair and oxidative damage due to H2O2 in Jurkat T cells. Nutr. Res. 22 (10), 1143–1150.

Rizzotto, D., Englmaier, L., Villunger, A., 2021. At a crossroads to cancer: how p53-induced cell fate decisions secure genome integrity. Int. J. Mol. Sci. 22 (19), 10883.

Sarkaria, J.N., Busby, E.C., Tibbetts, R.S., Roos, P., Taya, Y., Karnitz, L.M., et al., 1999. Inhibition of ATM and ATR kinase activities by the radiosensitising agent, caffeine. Cancer Res. 59 (17), 4375–4382.

Scalbert, A., Johnson, I.T., Saltmarsh, M., 2005. Polyphenols: antioxidants and beyond. Am. J. Clin. Nutr. 81 (1), 215S–217S.

Siddiqui, S.S., Rahman, S., Rupasinghe, H.V., Vazhappilly, C.G., 2020. Dietary flavonoids in p53-mediated immune dysfunctions linking to cancer prevention. Biomedicines 8 (8), 286.

Stahl, W., Sies, H., 2005. Bioactivity and protective effects of natural carotenoids. Biochim. Biophys. Acta Mol Basis Dis. 1740 (2), 101–107.

Tabart, J., Pincemail, J., Kevers, C., Defraigne, J.O., Dommes, J., 2018. Processing effects on antioxidant, glucosinolate, and sulforaphane contents in broccoli and red cabbage. Eur. Food Res. Technol. 244 (12), 2085–2094.

Thakkar, S., Anklam, E., Xu, A., Ulberth, F., Li, J., Li, B., et al., 2020. Regulatory landscape of dietary supplements and herbal medicines from a global perspective. Regul. Toxicol. Pharmacol. 114, 104647.

Urbaniak, S.K., Boguszewska, K., Szewczuk, M., Kaźmierczak-Barańska, J., Karwowski, B.T., 2020. 8-Oxo-7,8-dihydro-2′-deoxyguanosine (8-oxodG) and 8-hydroxy-2′-deoxyguanosine (8-OHdG) as a potential biomarker for gestational diabetes mellitus (GDM) development. Molecules 25 (1), 202.

Virág, L., Szabó, É., Bakondi, E., Bai, P., Gergely, P., Hunyadi, J., Szabó, C., 2002. Nitric oxide–peroxynitrite–poly (ADP-ribose) polymerase pathway in the skin. Exp. Dermatol. 11 (3), 189–202.

Waheeb, T.S., Abdulkader, M.A., Ghareeb, D.A., Moustafa, M.E., 2025. Neuroprotective efficacy of berberine and caffeine against rotenone-induced neuroinflammatory and oxidative disturbances associated with Parkinson's disease via inhibiting α-synuclein aggregation and boosting dopamine release. Inflammopharmacology 33, 1–22.

Wang, P., Heber, D., Henning, S.M., 2012. Quercetin increased bioavailability and decreased methylation of green tea polyphenols in vitro and in vivo. Food Funct 3 (6), 635–642.

Win, W., Cao, Z., Peng, X., Trush, M.A., Li, Y., 2002. Different effects of genistein and resveratrol on oxidative DNA damage in vitro. Mutat. Res. Genet. Toxicol Environ Mutagen. 513 (1-2), 113–120.

Winczura, A., Reynolds, J.J., 2018. The repair of DNA single-strand breaks and DNA adducts: mechanisms and links to human disease. In: DNA Repair and Replication. Garland Science, New York, NY, pp. 63–92.

Yang, W., 2008. Structure and mechanism for DNA lesion recognition. Cell Res. 18 (1), 184–197.

Zhu, C.Y., Loft, S., 2001. Effects of Brussels sprouts extracts on hydrogen peroxide-induced DNA strand breaks in human lymphocytes. Food Chem. Toxicol. 39 (12), 1191–1197.

CHAPTER

26

DNA damage and repair: Future perspectives

Yogendra Singh Rajpurohit[a,b,*], *Aanchal Mishra*[a,b], *Dhirendra Kumar Sharma*[a] *and Ishu Soni*[a,b]

[a]Molecular Biology Division, Bhabha Atomic Research Centre, Mumbai, Maharashtra, India [b]Life Sciences, Homi Bhabha National Institute (Department of Atomic Energy, Deemed to be University), Mumbai, Maharashtra, India

***Corresponding author: ysraj@barc.gov.in**

26.1 A brief revisit of a long journey of DNA damage, repair, and cancer

The pursuit to understand how DNA damage, repair mechanisms, and cancer development intertwine has been a long and winding road. This journey has seen major discoveries and shifts in how we think about these connections. A lot of what we know comes from studying patients and epidemiological research. Percival Pott (1775), the first epidemiologist, demonstrated a clear link between occupational exposure and cancer by reporting a high incidence of scrotal carcinoma among chimney sweeps, which he attributed to prolonged exposure to soot (Doll, 1975). This observation was among the earliest to recognize that external factors could contribute to the development of cancer. Later, in 1914 Theodor Boveri suggested that cancer is linked with abnormal chromosomal distributions in neoplastic cells, thus laying a foundation for a genetic paradigm in cancer research (Hardy and Zacharias, 2005; Hansford and Huntsman, 2014). The field of experimental biology opens up in the late 1920s when Hermann Müller reported that *Drosophila melanogaster* exposed to ionizing radiation from X-rays develops gene mutations and chromosomal abnormalities (Muller, 1927, 1928). This work was the first to introduce the idea of radiation as a mutagen for the deliberate induction of gene mutations. Moreover, Oswald Avery and his colleagues Colin MacLeod and Maclyn McCarty provided strong evidence in 1944, confirming that DNA is the hereditary material (Avery et al., 2000; Ghose, 2004; Portugal, 2010; Portin, 2014; Rijkers et al., 2024) and was probably the most important discovery about the storage and transmission of genetic information. This critical discovery created the ground for another ground-breaking moment in the history of heredity, which occurred in 1953 when James Watson and Francis Crick deciphered the double helix structure of DNA, which opened

Genome Damage: Repair and Adaptation Across Life.
DOI: https://doi.org/10.1016/B978-0-443-42857-9.00002-5

a window to understand the structural basis of heredity (Watson and Crick, 1953). It explained how DNA can replicate and store genetic information, thus forming the basis for understanding DNA damage and repair. The formal basis of "DNA repair" began in 1964 when the "Dark Repair" and photoreactivating "repair-replication" were identified in UV light-exposed *Escherichia coli* (Boyce and Howard-Flanders, 1964; Howard-Flanders et al., 1964; Pettijohn and Hanawalt, 1964; Setlow and Carrier, 1964; Harm, 1968). These discoveries collectively allowed researchers to begin understanding how cells actively counteract DNA damage. Studies on DNA damage and repair expanded rapidly into various fields of biology, including bacteriology, radiation, and cancer biology. The outcome of contemporary research suggested that DNA repair was crucial to restore genomic integrity (Hakem, 2008; O'Connor, 2015; Tian et al., 2015; Chatterjee and Walker, 2017; Huang and Zhou, 2021) and showed that both bacteria and eukaryotes could repair X-ray-induced damage to DNA (Dean et al., 1966; Brunk and Hanawalt, 1967; Samoilenko, 1983). Between 1970 and till date, many important research milestones have been recorded, for example, defining apoptosis as a mode of programmed death in 1972 (Kopeina and Zhivotovsky, 2022), oncogenes in 1981 (Bishop, 1981), tumor suppressor genes in 1984 (Murphree and Benedict, 1984), proposing checkpoints in the cell cycle in 1989, and identifying mutations in p53 in different cancers in 1990 (Chen et al., 1990; Casey et al., 1991; Dasika et al., 1999). In 1997 the roles of key DNA repair genes such as BRCA1/2 and RAD51 were suggested (Kinzler and Vogelstein, 1997; Milner et al., 1997; Sharan et al., 1997; Roth and Gellert, 2000). By 2002 research noted that oncogene expression could induce the production of reactive oxygen species (ROS) and damaged DNA due to deregulated metabolism (Vafa et al., 2002). The DNA damage response (DDR) was conceptualized as an anticancer barricade in early tumorigenic processes, studied in 2005. DDR-related genes function either altered or absent in late-stage tumors (Bartkova et al., 2005; Rai et al., 2007). The discovery of a new mechanism for the repair of double-strand breaks (DSBs) in DNA called "extended synthesis-dependent strand annealing" (ESDSA) was made in the radioresistant bacterium *Deinococcus radiodurans* in 2006 (Misra et al., 2006; Zahradka et al., 2006; Slade et al., 2009; Bentchikou et al., 2010; Sharma et al., 2024; Sharma et al., 2025). Furthermore, in 2012 Jennifer Doudna and Emmanuelle Charpentier established a template for using CRISPR to gene-edit (Jinek et al., 2013). A further understanding of the process called nuclear metamorphosis is currently being studied—the repair of double-strand DNA breaks at the nuclear envelope (Shokrollahi et al., 2024). New protein complex identification involved in DNA damage recognition and the role of autophagy (termed nucleophagy) in removing harmful DNA-protein lesions from the nucleus will potentially provide avenues within new cancer therapies to target specific DNA repair pathways. In 2019 another completely unknown phenomenon of adaptive mutability in cancer was found. Inhibition of EGFR/BRAF induces DNA damage, results in increased mutability, and triggers microsatellite instability (Russo et al., 2019). Thus, these historical perspectives depict the gradual evolution of understanding DNA damage, cellular responses, and repair, within the context of cancer, from the earliest epidemiological observation to a thorough molecular understanding of DDR in the development of targeted cancer therapies.

26.2 Cracking the code: DNA repair, cancer resistance, and new developments

Both prokaryotic and eukaryotic cells have evolved numerous mechanisms that recognize and repair various types of DNA damage arising either spontaneously or through various external and internal DNA damaging factors. These mechanisms include direct reversal of damage, base excision repair (BER), nucleotide excision repair (NER), mismatch repair (MMR), single-strand break repair, and DSB repair. These various DNA repair pathways correct a spectrum of DNA lesions and maintain genomic integrity [Recent reviews may be consulted for a detailed understanding of different DNA repair pathways and their implications] (Li and Heyer, 2008; Krokan and Bjoras, 2013; Scharer, 2013; Chatterjee and Walker, 2017; Ait Saada et al., 2018; Wright et al., 2018; Li et al., 2020; Duan et al., 2021; Gutierrez and O'Connor, 2021; Spies et al., 2021; Chakraborty et al., 2023; Gohil et al., 2023).

Cancer cells' genomes have an inherent instability tendency due to changes in the genetic circuits and loss of regulatory control over the cell cycle. The genomic instability appears due to the hyperactivation of proto-oncogenes or the inactivation of tumor suppressor gene functions (Ferguson et al., 2015). This causes aberrations in the cell cycle that could afford an accumulation of genetic mutations, permitting tumor growth and progression (Chen et al., 1990; Kontomanolis et al., 2020; Dakal et al., 2024). Our understanding of how DNA damage and cancer are interrelated has paved the way for the development of therapeutic strategies that induce DNA damage, including chemotherapy and radiotherapy (Aktas et al., 2019; Alhmoud et al., 2020; Reuvers et al., 2020; Melia and Parsons, 2023). Nevertheless, cancer cells eventually also become resistant to these approaches, giving rise to the concept of targeting DNA repair pathways along with chemo- or radiotherapy for clinical advantage (Biau et al., 2019; Pan et al., 2024). The therapeutic theory behind synthetic lethality drives the use of poly(ADP-ribose) polymerase(PARP) inhibitors in BRCA 1/2 mutant tumors or homologous recombination (HR) deficient ones, where the selective targeting of one gene is tolerated, while simultaneous targeting of two genes results in cell death (Faraoni and Graziani, 2018; Rose et al., 2020; Lau et al., 2022). Several inhibitors of PARP have been tested in various clinical trials, including the irreversible inhibition of PARP1 and PARP2 (Olaparib, Niraparib, Rucaparib, Talazoparib, Veliparib) that are aimed at trapping of PARP on single-strand breaks (SSBs) with subsequent potential to conversion into DSBs through inhibition of replication fork movement in HR-deficient tumors, including those with BRCA1/2 mutations (Zhu et al., 2009; Kaufman et al., 2015; Oza et al., 2015; Munroe and Kolesar, 2016; Li et al., 2020; Loizzi et al., 2020; Lee, 2021; Principe, 2022; Fizazi et al., 2025; Oza et al., 2025). Olaparib/Rucaparib is FDA-approved for advanced ovarian cancer and metastatic castration-resistant prostate cancer with HR gene mutations (Kim et al., 2015; Balasubramaniam et al., 2017). PARP inhibitors combined with ATR inhibitors (Kim et al., 2015; Balasubramaniam et al., 2017), CHK1 inhibitors (Qiu et al., 2018; Huang, Li, et al., 2024), and WEE1 inhibitors (Yano and Shiotani, 2023) offer an effective synthetic lethal combination to treat resistant tumors (Cleary et al., 2020). Additionally, inhibitors that aim at DNA damage signaling responsible for modulating the DDR are under exploration as monotherapy or in combination with other therapies against various cancers under evaluation for their therapeutic benefit. This include checkpoint kinase inhibitors (Kamali et al., 2023); APE1 inhibitors that act against the BER pathway by blocking APE1 cleavage at AP sites, potentiating the cytotoxicity of alkylating agents (Fishel and Kelley, 2007; Fishel et al., 2007; Montaldi and Sakamoto-Hojo, 2013; Fishel et al., 2019); DNA-PK inhibitors targeting nonhomologous end joining (NHEJ), promoting sensitivity of cancer cells to ionizing radiation and drugs (Dong et al., 2017; Matsumoto, 2022); ATM inhibitors targeting HR, therefore sensitizing tumor cells to DNA damaging therapies (Zimmermann et al., 2022; Priya et al., 2023).

Two newly emerging hypotheses, "environmental gear selection" and "DNA damage baseline drift," suggested new ways of interpreting DNA repair dynamics in cancerous cells

(O'Connor, 2015; Huang and Zhou, 2021). The "environmental gear selection" hypothesis proposes that differing environmental stressors activate different DNA repair pathways. For example, alkylating agents may primarily activate the NER and BER pathways; radiation-induced damage may activate BER, HR, canonical nonhomologous end joining (c-NHEJ), or alternative end joining (alt-EJ) pathways. Thus this mechanism highlights how cells change their response with respect to the DNA damage type encountered. The "DNA damage baseline drift" hypothesis suggests that cancer cells may display a shift in their baseline DNA repair capacity compared to that of the normal populations, contributing to tumor progression and resistance to treatment. This implies that normal cells maintain baseline DNA repair, which ensures genomic stability. However, if repair mechanisms become deregulated, DNA damage may persist and promote tumor growth rather than prevent it. Furthermore, the theory specifies that repair baseline reset caused by accumulated DNA damage sites might also contribute to therapy-induced resistance by allowing cancer cells to evade the treatment-induced DNA damage. Together, these theories illustrate the complex and dynamic nature of DNA damage and repair in cancer cells. Current and future research is employing cutting-edge technologies, including cancer genome profiling, deep sequencing, and single-cell sequencing, to elucidate mutations and misregulation in DDR-related genes (Caravagna et al., 2020; Jiang et al., 2020; Zhong et al., 2020; Tanner et al., 2021; Yu et al., 2021; Angaroni et al., 2022). The ultimate ambition is to develop therapies that are really well tailored to target particular DNA repair defects and eventually lay the groundwork for personalized strategies.

26.3 Replication gap suppression: A key hurdle in effective cancer therapy

Historically, cancer therapy resistance was associated with DNA DSBs repair and the HR proficiency of cancer cells (Liu and Tewari, 2016; Faraoni and Graziani, 2018; Trenner and Sartori, 2019). However, the newest findings suggest that it is actually the ssDNA gaps that qualify as the most important toxic lesions created by genotoxic chemotherapeutic agents (Nagaraju and Scully, 2007; Ramirez-Otero and Costanzo, 2024). Therefore, the replication gap suppression (RGS) phenomenon is important in determining whether a cancer responds to treatment (Fig. 26.1; Cantor, 2021). Tumors with a BRCA1 or BRCA2 mutation background display increased sensitivity to genotoxic chemotherapy (Chalasani and Livingston, 2013). The BRCA-RAD51 pathway is required for repair of ssDNA discontinuities through template switch while DNA replication is ongoing, and its deficiency has been connected with a loss of replication restraint and increased ssDNA gap formation (Kolinjivadi et al., 2017; Jackson and Moldovan, 2022). In addition to their well-characterized defect in DSB repair via HR, BRCA1/2-deficient cells are also deficient in RGS, resulting in excess ssDNA gap accumulation and increased sensitivity to PARP inhibitors (PARPi) (Fig. 26.1, step A; Dibitetto et al., 2024; Li and Zou, 2024). Traditionally, restoration of DSB repair and/or fork protection (FP) contributed to the fate of BRCA1/2-mutated cancers with respect to chemotherapy resistance. However, it has been seen that cancer cells can restore RGS directly related to acquired chemotherapy resistance, making gap suppression superior over DSB repair (Cong et al., 2021; Jackson and Moldovan, 2022; Wu et al., 2024). Furthermore, emerging evidence suggests two points: on the one hand, HR and FP proficiency do not always guarantee chemoresistance; on the other, RGS is a consistent predictor of drug response, even when HR and FP mechanisms remain intact (Canman, 2021; Cantor, 2021; Cong et al., 2021; Panzarino et al., 2021). Additionally, the ssDNA

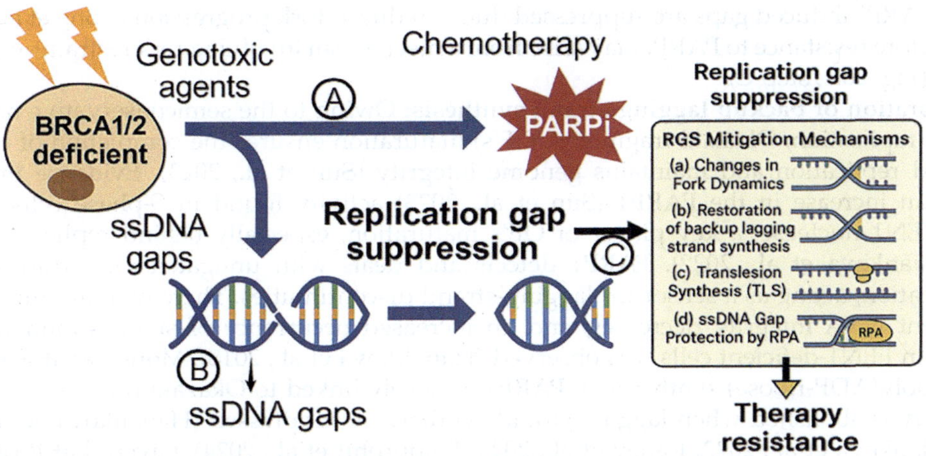

FIGURE 26.1 Replication gap suppression (RGS) as a critical determinant of therapy resistance in BRCA1/2-deficient cancer cells. PARP inhibitors (PARPi) exploit single-stranded DNA (ssDNA) gaps vulnerability (A), whereas genotoxic agents and chemotherapy induce ssDNA gaps during replication, especially in cells lacking BRCA1/2 function (B). Cancer cells evade treatment by activating replication gap suppression pathways, reducing ssDNA gap accumulation. The right panel highlights four key RGS mitigation mechanisms that contribute to resistance (C): (a) changes in replication fork dynamics, (b) restoration of backup lagging-strand synthesis, (c) translesion synthesis (TLS), and (d) ssDNA gap protection by replication protein A (RPA). These adaptive responses prevent gap persistence and promote survival under chemotherapy or PARPi treatment, ultimately driving therapeutic resistance.

gap model provides a better rationale for the effectiveness of PARPi in tumors that do not express classical BRCAness (a genetic mimic condition similar to tumors arising from mutations in the BRCA1 or BRCA2 genes) (Fig. 26.1, step B). Therefore, RGS mechanisms are key to developing therapy resistance in cancer cells. The following sections will discuss the key RGS mechanisms by which cells can adjust the therapy response (Fig. 26.1, step C).

1. **Changes in Fork Dynamics:** The maintenance of genomic integrity is closely integrated with the speed at which replication forks progress (Maya-Mendoza et al., 2018). Many factors, including DNA lesions, the availability of nucleotide pools, or regeneration dynamics of replisome components, may slow forks. If such slowing occurs, fork stability can be diminished as nucleases like MRE11 and EXO1 may get access to slowed or stalled replication fork (Saxena and Zou, 2022; Nelligan and Dungrawala, 2024). Conversely, excessive speed in replication can exhaust nucleotide pools and other factors, leading to an incomplete replication within those regions and display characteristic under-replicated foci where mitotic DNA synthesis occurs (Bournaka et al., 2024). Thus, a requirement for balanced replication fork speed to avoid over-accumulation of ssDNA gaps and ensure the faithful duplication of DNA is needed. ssDNA gaps may be contained by restoration of fork responses and slowing down replication in situations of stress (Mocanu and Chan, 2021; Schreuder et al., 2024). Dynamic changes in fork speed, however, have an importance in RGS and cancer therapy resistance. For instance, in BRCA-deficient cells, ssDNA gaps accumulate due to unhindered progression of the replication fork in response to stress, which otherwise may be slowed by HR machinery (Panzarino et al., 2021). On the contrary, hyper-restraint of elongation of forks could prevent ssDNA gaps, as in the case of the FANCJ-deficient cell type, where it showed

that PARPi-induced gaps are suppressed due to reduced fork progression (Cong et al., 2024). Therefore resistance to PARPi may arise from a mechanism involving fork restraint-mediated RGS (Fig. 26.1, step C).

2. **Restoration of backup lagging-strand synthesis:** Owing to the semiconservative nature of DNA replication, Okazaki fragments (OF's) maturation ensures the completion of lagging-strand replication and maintains genome integrity (Sun et al., 2023). Evidence indicated that an increase in the PARP1 (Sun et al., 2023) activity found in S-phase cells lacking the FEN1 nuclease, a key player of OF's maturation, especially behind replication forks (Vaitsiankova et al., 2022). PARP1 detects and deals with unligated OF's during DNA replication, acting as a sensor for lagging-strand discontinuities. Thereby inhibiting PARP1, nascent DNA integrity decreases, and an increased occurrence of single-strand nicks or gaps in FEN1-deficient cells was observed (Hanzlikova et al., 2018). Moreover, it was noted that poly(ADP-ribose) synthesis by PARP1 is closely linked to Okazaki processing since its activity is abolished when lagging-strand synthesis is suppressed (Hanzlikova et al., 2018; Vaitsiankova et al., 2022; Kanev et al., 2024; Rajpurohit et al., 2024). Given that PARP1 is a major player in the maturation of Okazaki fragments, the inhibition of this enzyme leads to aberrant Okazaki fragment maturation (OFM) and synergies with PARPi therapy, primarily in BRCA1-deficient cells. Nevertheless, backup lagging-strand synthesis, an important RGS mechanism, has been shown to contribute to PARPi resistance, in particular when cells are deficient in BRCA1 (Cong et al., 2021; Huang, Chen, et al., 2024). Cells deficient in BRCA1 showed defect in OFM,elevated PARP1, and diminished XRCC1 (a BER factor) activity (Vaitsiankova et al., 2022; Sun et al., 2023). This is perhaps due to excess accumulation of 53BP1, which obstructs the XRCC1 recruitment and ability to process the lagging strand (Cong et al., 2021; Li and Zou, 2024) and therefore, deletion of 53BP1 restores XRCC1 levels in tandem with LIG3, reviving backup OFP and supporting vigorous replication (Cantor, 2021; Cong et al., 2021). Additionally, 53BP1 is suggested to sequester or block Polα (polymerase alpha); therefore in BRCA-RAD51-deficient cells, this would indicate that loss of 53BP1 may reinstate Polα-LIG3-dependent OFP (Mirman et al., 2022). Therefore, lagging-strand synthesis moves toward a backup XRCC1-LIG3 mechanism in cells devoid of both BRCA1 and 53BP1, making those cells very sensitive to PARPi when they lose LIG3(Paes Dias et al., 2021).

   Additionally, loss of canonical primary OFM factors like FEN1 or LIG1 increases PAR in control cells, similar to increased PAR when noncanonical LIG3 is depleted (Paes Dias et al., 2021). Hence, together, canonical and alternative OFM pathways are key to lagging-strand synthesis, thereby forming an important part of RGS and PARPi therapy resistance (Fig. 26.1, step C; Cong and Cantor, 2022; Li and Zou, 2024).

3. **Translesion Synthesis (TLS):** TLS plays a critical role in the context of RGS, as it allows DNA replication to proceed despite the presence of DNA lesions or other obstacles (Venkadakrishnan et al., 2023). Thus, TLS can significantly impact the effectiveness of cancer therapies by enabling DNA replication to continue even when DNA damage persists, which otherwise in the absence of TLS, would typically lead to replication fork stalling or collapse and activation of apoptosis (Nayak et al., 2021; Patel et al., 2021). TLS involves specialized DNA polymerases, mainly from the Y-family of polymerases (Pol Kappa [κ], Eta [η], Iota [ι], and Rev1), B-family member (Zeta [ζ]), and minor roles performing Pol theta (Polθ) from A-family and Pol mu (Pol μ) and Pol lambda (Pol λ) of X-family (Rajpurohit et al., 2025). These TLS polymerases can bypass DNA damage and promote damage tolerance, ensuring that DNA replication continues (Tonzi and Huang, 2019; Inomata et al., 2021). TLS polymerases

are specifically recruited to areas of DNA damage and are capable of synthesizing past lesions when replication fork preservation pathways fail to protect against fork collapse (Quinet et al., 2018; Yang and Gao, 2018). The family of polymerases Y has a conserved N-terminal polymerase domain, which shows structural similarity to replicative polymerases Delta (δ) and Epsilon (ε), with important differences, lacking a 3′-5′ exonuclease domain and possessing a wide solvent-exposed large active site (Pata, 2010; Yang and Gao, 2018). These characteristics enable them to hold a large distorted helix to bypass several DNA lesions with lesser fidelity (Boudsocq et al., 2002; Xu et al., 2009; Sherrer et al., 2012; Maxwell and Suo, 2014; Vyas et al., 2015). The C-terminal region of Y-family TLS polymerases contains PCNA-interacting peptide (PIP) boxes that mediate weak binding to PCNA (proliferating cell nuclear antigen) and ubiquitin-binding domains that bind to monoubiquitinated PCNA on lysine 164 (PCNA-Ub) in a regulated manner (Kannouche and Lehmann, 2004; Hishiki et al., 2009; Masuda et al., 2015). Rev1 Interacting Regions (RIRs), also present in the C-terminal regions of TLS polymerases, help establish a link with Rev1, particularly during recruitment of Pol η/Pol κ at the fork (Edmunds et al., 2008; Qin et al., 2013; Yang et al., 2013; Pustovalova et al., 2016; Niu et al., 2019). Together, the combination of these motifs, along with other minor interacting regions, allows fine regulation of TLS polymerase activity at the replication fork. Additionally, replication fork remodeling factors such as HLTF and SHPRH may also affect the TLS recruitment in response to UV light or methyl methane sulfonate (Lin et al., 2011).

The active site of TLS polymerases is amenable to accommodate noncanonical DNA bases that favor their replication beyond damaged bases, but with a possible increase in mutagenesis (Ling et al., 2001). Therefore TLS polymerases are subjected to careful regulation by cells, and they are invoked only as and when required during their activities (Paniagua and Jacobs, 2023). For example, Pol η is upregulated for stress, particularly UV-induced lesions and bypasses Cyclobutane pyrimidine dimer lesions accurately (Stary et al., 2003; Menck et al., 2024). Pol κ sustains replication fork progression under both conditions of endogenous or exogenous DNA replication stress, promoting both FP and rescue following HU treatment (Jones et al., 2012; Tonzi et al., 2018). REV1-POL-zeta (ζ) complex is connected to stalled FP. REV1 uses deoxycitidine opposite a DNA lesion, no matter what the base is, resulting in error-prone TLS via POLζ (Rizzo and Korzhnev, 2019; Stolyarenko et al., 2024). Additionally, REV1-POLζ repairs ssDNA gaps created by replication restart by PRIMPOL primase, which accumulate in cancer cells lacking HR repair (Taglialatela et al., 2021; Tirman et al., 2021) and for effective replication of template DNA containing non-B DNA structures such as hairpins and G-quadruplexes (Eddy et al., 2014; Northam et al., 2014; Ketkar et al., 2023). Interestingly, Pol ι localizes to replication foci in undamaged cells, suggesting its important contribution to genomic integrity maintenance even without exogenous stress-induced conditions (Kannouche et al., 2003; Mansilla et al., 2023). Thus, altogether TLS polymerases constitute a critical aspect of RGS, whereby their biochemical activity helps cells to overcome replication stress due to ssDNA gaps and allows cancer cells to develop therapy resistance by mitigating ssDNA gaps and allowing unhindered replication (Fig. 26.1, step C; Quinet et al., 2014; Sertic et al., 2018; Anand et al., 2023).

4. **ssDNA Gap Protection by RPA:** RPA (replication protein A) is considered critical for RGS because it protects ssDNA gaps formed during DNA replication. RPA binds with ssDNA regions, stabilizes and protects them from nucleases, and promotes repair or bypass mechanisms (Marechal and Zou, 2015; Dueva and Iliakis, 2020). The protective role of RPA becomes even more critical in BRCA-deficient cells, where gap formation propensity has increased

due to the absence of HR mediated ssDNA gap removal mechanism, and concurrent increase in FANCJ helicase activity may encourage ssDNA formation (Vugic et al., 2023; Cong et al., 2024). On this note, the gap formation versus RPA protection balance determines cell survival and therapy response, excess gaps, sensitizing cells to genotoxic stress, and PARPi, while good RPA protection allows for replication stress tolerance and makes possible therapy resistance (Cong et al., 2021). Poorer clinical outcomes correlate with higher RPA levels in the tumor (Byrne and Oakley, 2019; Zhang and Yu, 2020). This fact suggests that these cancer cells may be able to develop genotoxic resistance by using RPA to stabilize gaps and activate mitigation mechanisms later. Targeting RPA would increase synthetic lethality between PARPi and BRCA1 deficiency, as RPA inhibitors render BRCA1-deficient cells more sensitive to PARPi (Previtali et al., 2024). In BRCA-deficient cells, there are more replication gaps, making RPA essential for survival. This emphasizes the importance of RPA in RGS and RPA as a very relevant therapeutic target for overcoming resistance in cancer (Fig. 26.1, step C).

26.4 Current and future approaches for cancer treatment based on DNA damage and repair

26.4.1 Harnessing the DNA damage defect and synthetic lethality for cancer treatment

Synthetic lethality (SL) is an anticancer strategy of interest, wherein the simultaneous inactivation of two or more genes leads to cell death while individually targeting these genes has minor or no effect on cell viability (Previtali et al., 2024; Karami Fath et al., 2025). The strategy enables selective killing of cancer cells with mutated or inactivated driver genes such as BRCA1 or 2, potentially limiting therapy side effects (Fig. 26.2). The effectiveness of SL in drug discovery is dependent on the identification of new SL pairs in cancer cells, with strategies such as phenotypic profiling, experimental approaches, and computational methods (O'Neil et al., 2017; Yar et al., 2020). The recent developments in big data strategies allow scientists to systematically screen each gene in many cancer cell lines, identifying cancer-specific vulnerabilities and synthetic lethal relationships (Luo et al., 2024). The definition of SL has recently expanded, leading to the introduction of various subtypes. "Collateral SL" refers to targeting a nontumor suppressor gene (nTSG) alongside a tumor suppressor gene. "Metabolic SL" involves the simultaneous disruption of two metabolic genes (Ju et al., 2022). Meanwhile, "contextual" or "conditional" SL, where the simultaneous inactivation of genetic and nongenetic abnormal conditions to obtain an enhanced decisive anticancer action (Aly and Ganesan, 2011). For example, hypoxic aggressive solid tumors showed HR-deficiency due to the hypoxic tumor microenvironment and were found to be sensitive to PARPi and ATRi (Chan et al., 2010; Pires et al., 2012).

One of the glaring examples of SL in cancer treatment is the application of PARPi in a BRCA1/2 mutant genetic background (Cerrato et al., 2016; Voutsadakis and Stravodimou, 2023). The number of studies exploring PARPi with BRCA1/2-deficient tumors has increased manifold since the initial proposal of this hypothesis (Fig. 26.2, steps A, B). The class of PARPi has proven to be useful, with numerous PARPi being evaluated in the clinic or already approved to treat BRCA1/2- and/or somatic HR-deficient cancers, including ovarian, pancreatic, breast, and prostate cancer, as a single agent or combination (Rose et al., 2020; Ragupathi et al., 2023; Dibitetto et al., 2024). Various PARPi have received European Union and/or United

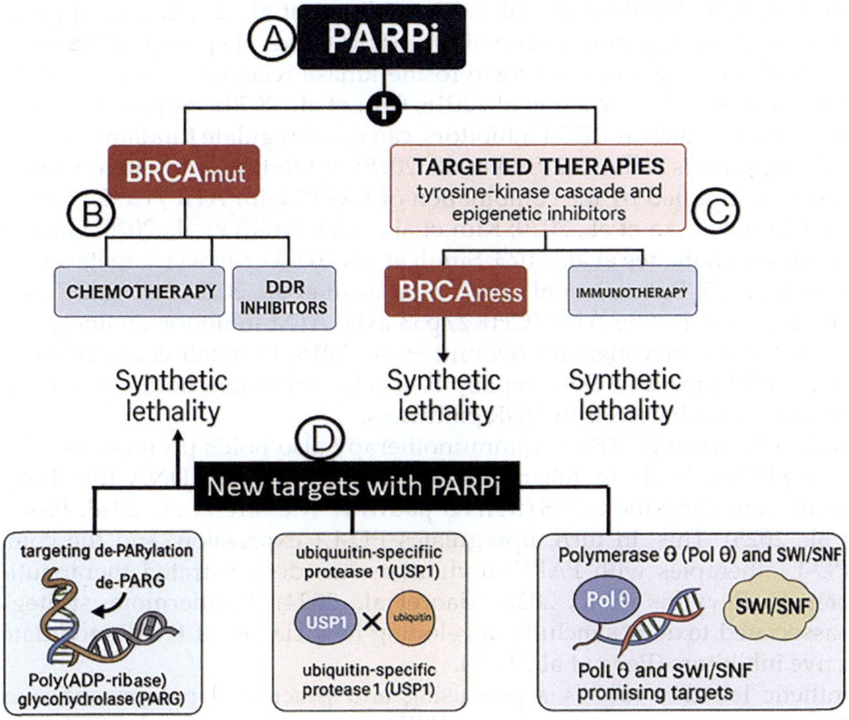

FIGURE 26.2 The evolving landscape of poly(ADP-ribose) polymerase inhibitor (PARPi)-based therapies, emphasizing synthetic lethality as a central concept. PARPi (A) exploit vulnerabilities in DNA repair-deficient tumors, particularly those with BRCA1/2 mutations (BRCAmut), where co-treatment with chemotherapy or DDR inhibitors leads to synthetic lethality (B). Additionally, targeted therapies induce BRCAness (C), a phenotype mimicking BRCA deficiency, enabling synthetic lethality with PARPi when combined with immunotherapy or other agents. Expanding on this, new therapeutic strategies (D) include inhibition of poly(ADP-ribose) glycohydrolase (PARG) to block de-PARylation, targeting the USP1-UAF1 deubiquitinase complex to disrupt protein degradation regulation, and inhibition of polymerase theta (Polθ) or components of the SWI/SNF chromatin remodeling complex, which are particularly relevant in homologous recombination-deficient cancers. These novel targets enhance the efficacy of PARPi and broaden their clinical application in overcoming therapy resistance.

States approvals, including rucaparib (Rubraca), niraparib (Zejula), and talazoparib (Talzenna), while fuzuloparib (AiRuiYi) and pamiparib (Partruvix) have received approval in China (Previtali et al., 2024). Emerging work has focused on the combination with PARPi to exploit synergistic mechanisms, enhance efficacy, and sensitize PARPi-resistant tumors. The current clinic effort toward PARPi combination with DNA damaging chemotherapy, such as Topo-I inhibitor (Topotecan) (Pratz et al., 2017; Chowdhuri and Das, 2021), Gemcitabine (a pyrimidine nucleoside analog) (Jiang et al., 2019), or inhibitors targeting ribonucleotide reductase (RNR) and ATR / WEE1 (Fordham et al., 2018; Sturm et al., 2023; Ziener et al., 2025) are examples of "contextual SL." However, hematologic toxicities (myelosuppression) have been seen in some of these combinations, highlighting the need for new, rational, and more targeted drugs that offer promising efficacy while minimizing normal tissue toxicity. Combining PARPi with other therapies may represent a promising strategy to enhance efficacy and overcome resistance and normal tissue toxicity (Abbotts et al., 2022; Bhamidipati et al., 2023). One approach involves inducing a homologous recombination-deficient (HRD)-like phenotype (BRCAness) by combining PARPi with inhibitors of receptor tyrosine kinase receptors (e.g., EGFR, VEGFR, and FLT3-ITD) (Fig. 26.2, step C; Cerrato et al., 2016; Chu et al., 2020). Additionally, modulators of epigenetic mechanisms, such as BRD4 inhibitors, can downregulate fundamental DDR proteins, further sensitizing tumors to PARPi (Sun et al., 2018). Inhibiting alternative DDR pathways is another avenue, exemplified by the combination of PARPi with ATR /CHK1/ WEE1 inhibitor (Kim et al., 2017; Gralewska et al., 2020; Kim et al., 2020; Smith et al., 2020; Gupta et al., 2022; Li et al., 2022; Wikiniyadhanee et al., 2023; Smith et al., 2024) or newer targets such as Polθ and USP1 (Belan et al., 2022; Bubenik et al., 2022; Simoneau et al., 2023; Sullivan-Reed et al., 2023; Cadzow et al., 2024).Under the ATM/CHK2/p53 axis, ATM inhibitors, namely AZD1390 and M4076 are under active investigation (Durant et al., 2018; Previtali et al., 2024). Considering the key roles of ATM kinase in DNA repair, its specific inhibitors hold considerable promise, especially for cancers with defects in DDR pathways.

The synergism between PARPi and immunotherapy also holds promise, as PARPi-induced DNA repair inhibition leads to the release of double-stranded DNA (dsDNA) fragments into the cytosol, activating the cGAS/STING pathway (Cesaire et al., 2018; Post et al., 2020; Phillipps et al., 2023). This, in turn, upregulates PD-L1 expression, and the combination of anti-PD-1/PD-L1 therapies with PARP modulators has demonstrated therapeutic relevance (Fig. 26.2, step C; Revythis et al., 2022; Xiao et al., 2024). Furthermore, strategies to mitigate PARPi-associated toxicities include developing new classes of PARP modulators, such as PARP1-selective inhibitors (Peng et al., 2022).

Thus, synthetic lethality (SL) is a promising and practically proven anticancer strategy enhancing the precision of cancer therapies. PARPi exemplifies successful SL-based treatment, especially in BRCA1/2-deficient tumors, with several approved PARPi are now in clinical use. Combining PARPi with chemotherapy, DDR inhibitors, or agents that induce a HRD phenotype improves therapeutic efficacy but may pose toxicity challenges. Emerging approaches aim to overcome resistance and toxicity by using PARP1-selective inhibitors, integrating immunotherapies, or modulating epigenetic and alternative DNA repair pathways to broaden and refine the application of SL in cancer treatment.

26.4.2 Targeting de-PARylation, offering a therapeutic intervention within DNA damage response

Poly(ADP-ribose) glycohydrolase (PARG) is an enzyme involved in genome maintenance against replication stress through de-PARylation catalysis and offering a therapeutic target in DDR (Feng and Koh, 2013; Slade, 2020). PARG is generally overexpressed in cancer and directly associated with PARPi resistance (Tanuma et al., 2019). PARG inhibitors (PARGi) were originally anticipated to act against BRCA1/2-deficient tumor cells, providing a potential SL parallelism with PARPi, but recent studies indicate divergent pharmacology and genetic dependencies as a function of HRD status upon PARG or PARP inhibition (Fig. 26.2, step D; Gravells et al., 2018). AstraZeneca discovered a cell-permeable, potent PARGi, PDD00017273, from a high-throughput biochemical screen, followed by iterative chemical optimization and modification of its chemotype. However, PDD00017273 was found to be not suitable for in vivo studies due to poor metabolic stability. IDE161, an orally bioavailable PARGi, likely a 4-substituted indole and indazole sulfonamide derivative, with potent PAR chain hydrolysis inhibition and anti-proliferative activity in breast and ovarian cancer cells, including PARPi-resistant cell lines (Abed et al., 2023). IDE161 has been found to exhibit antitumor activity in PARPi-resistant xenograft models, allowing it to progress into clinical trials as the first PARGi.

26.4.3 Ubiquitin-specific protease 1 has emerged as a potential target in cancer therapy

Ubiquitination and deubiquitination participate in cellular homeostasis and DDR, and ubiquitin-specific protease 1 (USP1) was found to be overexpressed in various human cancers, such as NSCLC, melanoma, and recurrent sarcoma, justifying that USP1 could be a target for cancer treatment (Chen et al., 2011; Garcia-Santisteban et al., 2013; Huang et al., 2023). USP1-UAF1 deubiquitinase complex inhibitor (ML323) was optimized from quantitative high-throughput screening (qHTS) and medicinal chemistry, making it one of the most effective USP1 inhibitors (Liang et al., 2014). It has been recently demonstrated that USP1 inhibition is effective in a subset of BRCA1/2 wild-type lung and ovarian tumors, depending on the USP1 dependency on PCNA ubiquitination and subsequent protein degradation (Simoneau et al., 2023). In BRCA1 mutant cells, USP1 inhibition by another potent inhibitor (I-138, a structurally related small molecule to ML323) results in the accumulation of mono- and polyubiquitinated PCNA, with decreased DNA synthesis and DNA damage (Simoneau et al., 2023). The results indicate that I-138 binds to USP1-UAF1 at an allosteric site and inhibition of USP1 preferentially decreased the viability of BRCA1 mutant breast cancer cell line MDA-MB-436 and BRCA1 mutant ovarian cancer cell line UWB1.289 to constitute an SL partner with PARPi (Simoneau et al., 2023). Similarly, highly selective USP1, KSQ-4279, exhibited dose-dependent antitumor activity in ovarian- and TNBC-derived tumor xenograft models, and combination with olaparib produced greater and sustained tumor regression than monotherapy in partially sensitive tumors to PARPi (Cadzow et al., 2024). These findings resulted in KSQ-4279 being added to a phase 1 clinical trial in tumor patients with BRCA mutation-harboring or other HRD deficiency, either alone or in combination with PARPi (Trial ID-NCT05240898). Altogether, USP1-specific inhibitors like ML323, I-138, and KSQ-4279 disrupt USP1-UAF1 function, impairing DNA repair and showing SL in BRCA1-mutant cancers (Fig. 26.2, step D).

26.4.4 Polθ and SWI/SNF chromatin remodeling complex emerged as a promising target for synthetic lethality in HRD cancers

Polymerase θ (Polθ) is an A-family DNA polymerase DNA repair enzyme that has been a potential target for SL in cancer therapy (Pismataro et al., 2023). Polθ is with a unique structure that includes an N-terminal helicase-like domain and a polymerase domain, and Polθ-mediated end joining (TMEJ) activity is crucial for the unwinding of replicative obstacles, particularly in the absence of NHEJ and HR repair (Campbell and Li, 2017; Mateos-Gomez et al., 2017; Ozdemir et al., 2018; Schrempf et al., 2021). Polθ is overexpressed in various human cancers, such as colorectal, breast, papillary thyroid, ovarian, lung, and gastric cancer, whereas in the majority of normal cells, it is low-level or absent, making it an excellent target for oncology (Higgins et al., 2010; Lemee et al., 2010; Pillaire et al., 2010; Allera-Moreau et al., 2012; Frye et al., 2024). Some research has emerged on the development of Polθ inhibitors based on either inhibition of helicase domain (Novobiocin, GSK101) or Polθ-Pol domain. Polθ helicase domain inhibitor candidate GSK101 (IDE705) has shown promising results with PARPi (niraparib) (Fig. 26.2, step D).

SWI/SNF chromatin remodeling complexes regulate gene transcription and are mutated in 20%–25% of human cancers, often driving oncogenesis. These mutations create vulnerabilities to disruptions in other SWI/SNF components or related pathways. Exploiting such synthetic lethal interactions offers a promising targeted cancer therapy approach (Fig. 26.2, step D; Wanior et al., 2021; Li et al., 2023).

26.5 DNA damage and its role in neuronal aging and neurodegeneration

Among other factors, DNA damage is one of the principal determinants of loss of neuronal function and the development of neurodegenerative disease with age progression (Delint-Ramirez and Madabhushi, 2025). While the mechanisms involved remain the focus of active research, there is mounting evidence to suggest that genomic instability led by DNA damage and chromatin rearrangement is an underlying cause that allows age-related neuronal degeneration to progress (Welch and Tsai, 2022; Sharma et al., 2025). Mechanistically, dysfunctions of the DDR have the potential to produce stable DNA lesions, mutations, and trigger of stress responses like apoptosis, cellular senescence, and chronic inflammation, all of which have negative impacts on neuronal survival and function (Fig. 26.3, step A; Schumacher et al., 2021).

ROS produce oxidative DNA lesions like 8-oxoguanine (8-oxoG) that are mainly processed by the BER pathway. In spite of the energy requirement of the brain, ROS are stringently regulated and even serve as signaling molecules in synaptic transmission. Aging, however, is accompanied by reduced BER capacity, resulting in enhanced accumulation of 8-oxoG in neurons, particularly in Alzheimer's disease (AD), Parkinson's disease (PD), Huntington's disease (HD), and amyotrophic lateral sclerosis (ALS) (Fig. 26.3, step B; Hegde et al., 2012). Genetic analysis has identified polymorphism in BER genes (OGG1, APE1, XRCC1) as corresponding to increased risk for disease (Li et al., 2011; Li et al., 2013), and animal models with deficiency of BER components show cognitive dysfunction and neuropathological features (Tayebati, 2006; Akbari et al., 2015). Enhancing OGG1 has been found to suppress AD cognitive dysfunction in an animal model (Mizuno et al., 2022). Active DNA demethylation is another critical process in the brain implicated in dynamic gene regulation and neuronal plasticity (Wang et al., 2022). This process produces transient SSBs as intermediates, and their repair is critical to ensure genomic stability. Neuronal activity stringently regulates this demethylation process, and dysregulation of SSB repair proteins like XRCC1 leads to neurodegeneration. For instance, XRCC1 mutations cause spinocerebellar ataxia autosomal recessive 26 (SCAR26), a neurodegenerative disorder with seizures and motor dysfunction, in part due to hyperactivation of poly(ADP-ribose) polymerase 1 (PARP1) at unrepaired SSBs (Hoch et al., 2017; Komulainen et al., 2021). In XRCC1-deficient conditions, PARP1 is overactivated at repair intermediates, causing NAD+ depletion, cellular energy collapse, and neuron death (Fig. 26.3, step D; Murata et al., 2019; Komulainen et al., 2021). This process has been implicated in a variety of neurodegenerative diseases, including AD and PD. NAD+ supplementation or inhibition of PARP1 has been therapeutic in models (Zhou et al., 2015).

DSBs, stochastic and programmed, are increasingly recognized as early events in neurodegenerative disease. Elevated levels of DSBs are present in old brains and in the gray and white matter of nonhuman primates (Tse and Herrup, 2017). Programmed DSBs are generated during

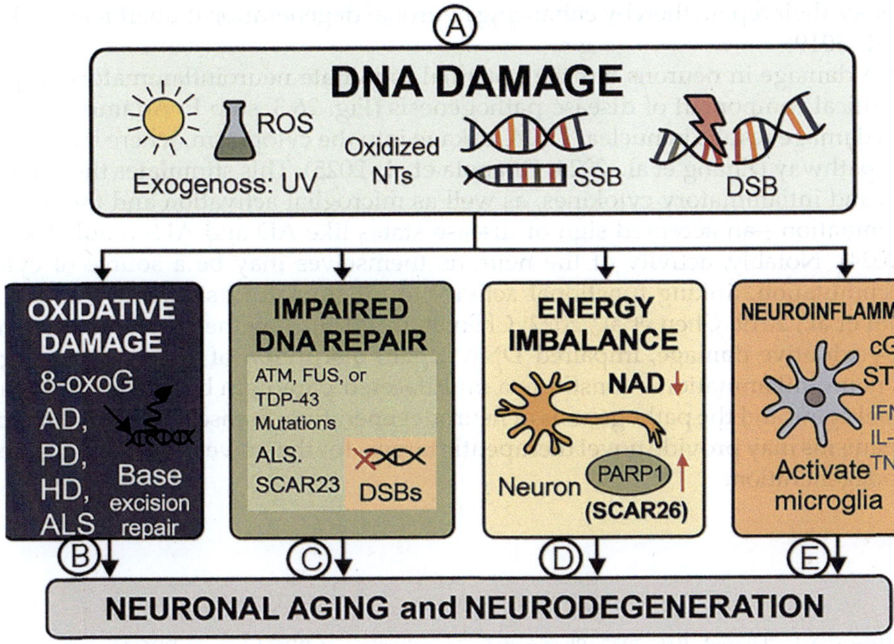

FIGURE 26.3 DNA damage driver roles in neuronal aging and neurodegeneration. DNA damage drives neuronal aging and neurodegeneration through four interconnected mechanisms. Exogenous factors like UV radiation and endogenous sources such as reactive oxygen species (ROS) lead to the formation of oxidized nucleotides, single-strand breaks (SSBs), and double-strand breaks (DSBs) (A). Oxidative damage, exemplified by 8-oxoguanine (8-oxoG), accumulates due to impaired base excision repair (BER), contributing to disorders like Alzheimer's disease (AD), Parkinson's disease (PD), Huntington's disease (HD), and amyotrophic lateral sclerosis (ALS) (B). Impaired DNA repair, resulting from mutations in genes such as ATM, FUS, and TDP-43, causes unresolved DSBs associated with ALS and SCAR23 (C). Energy imbalance occurs when hyperactivation of PARP1 at unrepaired DNA lesions depletes NAD+, leading to neuronal dysfunction, as seen in SCAR26 (D). Additionally, chronic DNA damage activates the cGAS-STING pathway, triggering neuroinflammatory responses marked by increased cytokine production (e.g., IFNs, IL-1, IL-6, TNF-α) and microglial activation (E). These molecular and cellular responses converge to promote progressive neuronal dysfunction, aging, and degeneration.

neuronal activity and are part of gene regulation; however, when DSB repair mechanisms are compromised—such as by ATM, FUS, or TDP-43 mutations—these breaks are persistent and cause disease (Fig. 26.3, step C; Provasek et al., 2022). In neurodegenerative disorders like ALS and spinocerebellar ataxia autosomal recessive 23 (SCAR23), defects in the mechanism of DSBs DNA damage repair pathways are central to disease pathogenesis. In ALS, TDP-43 mislocalization and aggregation block the repair pathway of DSBs, which in turn disrupts genomic integrity (Da Cruz and Cleveland, 2011; Guerrero et al., 2016). In addition, mutations in FUS, another RNA-binding protein, block the resolution of R-loop-induced DSBs and destabilize DNA repair pathways (Perego et al., 2019; Beghè et al., 2025). In SCAR23, biallelic TDP2 mutations (a DSB repair enzyme required for TOP2B-generated DSBs) lead to severe neurodevelopmental delay and severe neurodegeneration (Ciaccio et al., 2019; Erratum: Confirming TDP2 mutation in spinocerebellar ataxia autosomal recessive 23 (SCAR23), 2018). In addition, synaptic dysfunction, a common feature of the disorders, may enhance the build-up of DSBs caused by neuronal activity and block their repair, thereby enhancing neuronal degeneration (Suberbielle et al., 2013; Shanbhag et al., 2019).

Lastly, DNA damage in neurons has the potential to initiate neuroinflammatory responses, which are a critical component of disease pathogenesis (Fig. 26.3, step E; Adamu et al., 2024). Chronic DNA damage results in nuclear DNA leakage into the cytoplasm, where it activates the cGAS-STING pathway (Zhang et al., 2024; Dhapola et al., 2025). This stimulates the production of interferons and inflammatory cytokines, as well as microglial activation and the promotion of neuroinflammation—an accepted sign of disease states like AD and ALS (Paul et al., 2021; Guo et al., 2024). Notably, activity of the neurons themselves may be a source of cytoplasmic DNA accumulation, linking functional activity to inflammatory signaling through DNA damage (Khan et al., 2019; Chen et al., 2022; Chen et al., 2025). Together, these interdependent mechanisms—oxidative damage, impaired DNA repair, disruption of chromatin, energy imbalance, and neuroinflammation—constitute a multifaceted paradigm by which DNA damage drives neuronal aging and the pathogenesis of neurodegenerative disease (Fig. 26.3). Elucidation of these mechanisms may provide novel therapeutic targets for the prevention or slowing of age-related neurodegeneration.

References

Abbotts, R., Dellomo, A.J., Rassool, F.V., 2022. Pharmacologic induction of BRCAness in BRCA-proficient cancers: expanding PARP inhibitor use. Cancers (Basel) 14 2640.

Abed, M., Muñoz, D., Seshadri, V., Rao, A.A., Maskey, R., Federowicz, S., et al., 2023. IDE161, a potential first-in-class clinical candidate PARG inhibitor, selectively targets homologous-recombination-deficient and PARP inhibitor-resistant breast and ovarian tumors. Cancer Res. 83: 6093.

Adamu, A., Li, S., Gao, F., Xue, G., 2024. The role of neuroinflammation in neurodegenerative diseases: current understanding and future therapeutic targets. Front. Aging Neurosci. 16, 1347987.

Ait Saada, A., Lambert, S.A.E., Carr, A.M., 2018. Preserving replication fork integrity and competence via the homologous recombination pathway. DNA Repair (Amst.) 71, 135–147.

Akbari, M., Morevati, M., Croteau, D., Bohr, V.A., 2015. The role of DNA base excision repair in brain homeostasis and disease. DNA Repair (Amst.) 32, 172–179.

Aktas, B.Y., Guner, G., Guven, D.C., Arslan, C., Dizdar, O., 2019. Exploiting DNA repair defects in breast cancer: from chemotherapy to immunotherapy. Expert Rev. Anticancer Ther. 19, 589–601.

Alhmoud, J.F., Woolley, J.F., Al Moustafa, A.E., Malki, M.I., 2020. DNA damage/repair management in cancers. Cancers (Basel) 12 1050.

Allera-Moreau, C., Rouquette, I., Lepage, B., Oumouhou, N., Walschaerts, M., Leconte, E., et al., 2012. DNA replication stress response involving PLK1, CDC6, POLQ, RAD51 and CLASPIN upregulation prognoses the outcome of early/mid-stage non-small cell lung cancer patients. Oncogenesis 1, e30.

Aly, A., Ganesan, S., 2011. BRCA1, PARP, and 53BP1: conditional synthetic lethality and synthetic viability. J. Mol. Cell Biol. 3, 66–74.

Anand, J., Chiou, L., Sciandra, C., Zhang, X., Hong, J., Wu, D., et al., 2023. Roles of trans-lesion synthesis (TLS) DNA polymerases in tumorigenesis and cancer therapy. NAR Cancer 5, zcad005.

Angaroni, F., Guidi, A., Ascolani, G., d'Onofrio, A., Antoniotti, M., Graudenzi, A., 2022. J-SPACE: a Julia package for the simulation of spatial models of cancer evolution and of sequencing experiments. BMC Bioinf. 23, 269.

Avery, O.T., Macleod, C.M., McCarty, M., 2000. Studies on the chemical nature of the substance inducing transformation of pneumococcal types: Induction of transformation by a desoxyribonucleic acid fraction isolated from Pneumococcus type III. Rockefeller University Press, New York, NY, pp. S3–S8.

Balasubramaniam, S., Beaver, J.A., Horton, S., Fernandes, L.L., Tang, S., Horne, H.N., et al., 2017. FDA Approval summary: rucaparib for the treatment of patients with deleterious BRCA mutation-associated advanced ovarian cancer. Clin. Cancer Res. 23, 7165–7170.

Bartkova, J., Horejsi, Z., Koed, K., Kramer, A., Tort, F., Zieger, K., et al., 2005. DNA damage response as a candidate anti-cancer barrier in early human tumorigenesis. Nature 434, 864–870.

Beghè, C., Harpham, H., Barberic, Y., Gromak, N., 2025. R-loops in neurodegeneration. Curr. Opin. Genet. Dev. 92, 102345.

Belan, O., Sebald, M., Adamowicz, M., Anand, R., Vancevska, A., Neves, J., et al., 2022. POLQ seals post-replicative ssDNA gaps to maintain genome stability in BRCA-deficient cancer cells. Mol. Cell 82, 4664–4680e9.

Bentchikou, E., Servant, P., Coste, G., Sommer, S., 2010. A major role of the RecFOR pathway in DNA double-strand-break repair through ESDSA in Deinococcus radiodurans. PLoS Genet. 6, e1000774.

Bhamidipati, D., Haro-Silerio, J.I., Yap, T.A., Ngoi, N., 2023. PARP inhibitors: enhancing efficacy through rational combinations. Br. J. Cancer 129, 904–916.

Biau, J., Chautard, E., Verrelle, P., Dutreix, M., 2019. Altering DNA repair to improve radiation therapy: specific and multiple pathway targeting. Front. Oncol. 9, 1009.

Bishop, J.M., 1981. Enemies within: the genesis of retrovirus oncogenes. Cell 23, 5–6.

Boudsocq, F., Ling, H., Yang, W., Woodgate, R., 2002. Structure-based interpretation of missense mutations in Y-family DNA polymerases and their implications for polymerase function and lesion bypass. DNA Repair (Amst.) 1, 343–358.

Bournaka, S., Badra-Fajardo, N., Arbi, M., Taraviras, S., Lygerou, Z., 2024. The cell cycle revisited: DNA replication past S phase preserves genome integrity. Semin. Cancer Biol. 99, 45–55.

Boyce, R.P., Howard-Flanders, P., 1964. Release of ultraviolet light-induced thymine dimers from DNA in E. coli K-12. Proc. Natl. Acad. Sci. USA 51, 293–300.

Brunk, C.F., Hanawalt, P.C., 1967. Repair of damaged DNA in a eucaryotic cell: tetrahymena pyriformis. Science 158, 663–664.

Bubenik, M., Mader, P., Mochirian, P., Vallee, F., Clark, J., Truchon, J.F., et al., 2022. Identification of RP-6685, an orally bioavailable compound that inhibits the DNA polymerase activity of poltheta. J. Med. Chem. 65, 13198–13215.

Byrne, B.M., Oakley, G.G., 2019. Replication protein A, the laxative that keeps DNA regular: the importance of RPA phosphorylation in maintaining genome stability. Semin. Cell Dev. Biol. 86, 112–120.

Cadzow, L., Brenneman, J., Tobin, E., Sullivan, P., Nayak, S., Ali, J.A., et al., 2024. The USP1 inhibitor KSQ-4279 overcomes PARP Inhibitor resistance in homologous recombination-deficient tumors. Cancer Res. 84, 3419–3434.

Campbell, J.L., Li, H., 2017. Poltheta helicase: drive or reverse. Nat. Struct. Mol. Biol. 24, 1007–1008.

Canman, C.E., 2021. Which holds the key to BRCAness: inability to repair the break, protect the fork, or prevent the gap? Cancer Res. 81, 1214–1215.

Cantor, S.B., 2021. Revisiting the BRCA-pathway through the lens of replication gap suppression: "Gaps determine therapy response in BRCA mutant cancer. DNA Repair (Amst.) 107, 103209.

Caravagna, G., Heide, T., Williams, M.J., Zapata, L., Nichol, D., Chkhaidze, K., et al., 2020. Subclonal reconstruction of tumors by using machine learning and population genetics. Nat. Genet. 52, 898–907.

Casey, G., Lo-Hsueh, M., Lopez, M.E., Vogelstein, B., Stanbridge, E.J., 1991. Growth suppression of human breast cancer cells by the introduction of a wild-type p53 gene. Oncogene 6, 1791–1797.

Cerrato, A., Morra, F., Celetti, A., 2016. Use of poly ADP-ribose polymerase [PARP] inhibitors in cancer cells bearing DDR defects: the rationale for their inclusion in the clinic. J. Exp. Clin. Cancer Res. 35, 179.

Cesaire, M., Thariat, J., Candeias, S.M., Stefan, D., Saintigny, Y., Chevalier, F., 2018. Combining PARP inhibition, radiation, and immunotherapy: a possible strategy to improve the treatment of cancer? Int. J. Mol. Sci. 19 3793.

Chakraborty, S., Schirmeisen, K., Lambert, S.A., 2023. The multifaceted functions of homologous recombination in dealing with replication-associated DNA damages. DNA Repair (Amst.) 129, 103548.

Chalasani, P., Livingston, R., 2013. Differential chemotherapeutic sensitivity for breast tumors with "BRCAness": a review. Oncologist 18, 909–916.

Chan, N., Pires, I.M., Bencokova, Z., Coackley, C., Luoto, K.R., Bhogal, N., et al., 2010. Contextual synthetic lethality of cancer cell kill based on the tumor microenvironment. Cancer Res. 70, 8045–8054.

Chatterjee, N., Walker, G.C., 2017. Mechanisms of DNA damage, repair, and mutagenesis. Environ. Mol. Mutagen. 58, 235–263.

Chen, J., Dexheimer, T.S., Ai, Y., Liang, Q., Villamil, M.A., Inglese, J., et al., 2011. Selective and cell-active inhibitors of the USP1/UAF1 deubiquitinase complex reverse cisplatin resistance in non-small cell lung cancer cells. Chem. Biol. 18, 1390–1400.

Chen, K., Lai, C., Su, Y., Bao, W.D., Yang, L.N., Xu, P.P., et al., 2022. cGAS-STING-mediated IFN-I response in host defense and neuroinflammatory diseases. Curr. Neuropharmacol. 20, 362–371.

Chen, P.L., Chen, Y.M., Bookstein, R., Lee, W.H., 1990. Genetic mechanisms of tumor suppression by the human p53 gene. Science 250, 1576–1580.

Chen, Z., Behrendt, R., Wild, L., Schlee, M., Bode, C., 2025. Cytosolic nucleic acid sensing as driver of critical illness: mechanisms and advances in therapy. Signal Transduct. Target Ther. 10, 90.

Chowdhuri, S.P., Das, B.B., 2021. Top1-PARP1 association and beyond: from DNA topology to break repair. NAR Cancer 3, zcab003.

Chu, Y.Y., Yam, C., Chen, M.K., Chan, L.C., Xiao, M., Wei, Y.K., et al., 2020. Blocking c-Met and EGFR reverses acquired resistance of PARP inhibitors in triple-negative breast cancer. Am. J. Cancer Res. 10, 648–661.

Ciaccio, C., Castello, R., Esposito, S., Pinelli, M., Nigro, V., Casari, G., et al., 2019. Consolidating the role of TDP2 mutations in recessive Spinocerebellar ataxia associated with pediatric onset drug resistant epilepsy and intellectual disability (SCAR23). Cerebellum 18, 972–975.

Cleary, J.M., Aguirre, A.J., Shapiro, G.I., D'Andrea, A.D., 2020. Biomarker-Guided Development of DNA Repair Inhibitors. Mol. Cell 78, 1070–1085.

Cong, K., Cantor, S.B., 2022. Exploiting replication gaps for cancer therapy. Mol. Cell 82, 2363–2369.

Cong, K., MacGilvary, N., Lee, S., MacLeod, S.G., Calvo, J., Peng, M., et al., 2024. FANCJ promotes PARP1 activity during DNA replication that is essential in BRCA1 deficient cells. Nat. Commun. 15, 2599.

Cong, K., Peng, M., Kousholt, A.N., Lee, W.T.C., Lee, S., Nayak, S., et al., 2021. Replication gaps are a key determinant of PARP inhibitor synthetic lethality with BRCA deficiency. Mol. Cell 81, 3128–3144e7.

Da Cruz, S., Cleveland, D.W., 2011. Understanding the role of TDP-43 and FUS/TLS in ALS and beyond. Curr. Opin. Neurobiol. 21, 904–919.

Dakal, T.C., Dhabhai, B., Pant, A., Moar, K., Chaudhary, K., Yadav, V., et al., 2024. Oncogenes and tumor suppressor genes: functions and roles in cancers. MedComm (2020) 5, e582.

Dasika, G.K., Lin, S.C., Zhao, S., Sung, P., Tomkinson, A., Lee, E.Y., 1999. DNA damage-induced cell cycle checkpoints and DNA strand break repair in development and tumorigenesis. Oncogene 18, 7883–7899.

Dean, C.J., Feldschreiber, P., Lett, J.T., 1966. Repair of x-ray damage to the deoxyribonucleic acid in Micrococcus radiodurans. Nature 209, 49–52.

Delint-Ramirez, I., Madabhushi, R., 2025. DNA damage and its links to neuronal aging and degeneration. Neuron 113, 7–28.

Dhapola, R., Paidlewar, M., Kumari, S., Sharma, P., Vellingiri, B., Medhi, B., et al., 2025. cGAS-STING and neurodegenerative diseases: a molecular crosstalk and therapeutic perspective. Int. Immunopharmacol. 159, 114902.

Dibitetto, D., Widmer, C.A., Rottenberg, S., 2024. PARPi, BRCA, and gaps: controversies and future research. Trends Cancer 10, 857–869.

Doll, R., 1975. Pott and the path to prevention. Arch. Geschwulstforsch. 45, 521–531.

Dong, J., Zhang, T., Ren, Y., Wang, Z., Ling, C.C., He, F., et al., 2017. Inhibiting DNA-PKcs in a non-homologous end-joining pathway in response to DNA double-strand breaks. Oncotarget 8, 22662–22673.

Duan, M., Speer, R.M., Ulibarri, J., Liu, K.J., Mao, P., 2021. Transcription-coupled nucleotide excision repair: new insights revealed by genomic approaches. DNA Repair (Amst.) 103, 103126.

Dueva, R., Iliakis, G., 2020. Replication protein A: a multifunctional protein with roles in DNA replication, repair and beyond. NAR Cancer 2, zcaa022.

Durant, S.T., Zheng, L., Wang, Y., Chen, K., Zhang, L., Zhang, T., et al., 2018. The brain-penetrant clinical ATM inhibitor AZD1390 radiosensitizes and improves survival of preclinical brain tumor models. Sci. Adv. 4, eaat1719.

Eddy, S., Ketkar, A., Zafar, M.K., Maddukuri, L., Choi, J.Y., Eoff, R.L., 2014. Human Rev1 polymerase disrupts G-quadruplex DNA. Nucleic. Acids. Res. 42, 3272–3285.

Edmunds, C.E., Simpson, L.J., Sale, J.E., 2008. PCNA ubiquitination and REV1 define temporally distinct mechanisms for controlling translesion synthesis in the avian cell line DT40. Mol. Cell 30, 519–529.

Erratum, 2018. Confirming TDP2 mutation in spinocerebellar ataxia autosomal recessive 23 (SCAR23). Neurol. Genet. 4, e277.

Faraoni, I., Graziani, G., 2018. Role of BRCA mutations in cancer treatment with poly(ADP-ribose) polymerase (PARP) inhibitors. Cancers (Basel) 10 487.

Feng, X., Koh, D.W., 2013. Roles of poly(ADP-ribose) glycohydrolase in DNA damage and apoptosis. Int. Rev. Cell. Mol. Biol. 304, 227–281.

Ferguson, L.R., Chen, H., Collins, A.R., Connell, M., Damia, G., Dasgupta, S., et al., 2015. Genomic instability in human cancer: molecular insights and opportunities for therapeutic attack and prevention through diet and nutrition. Semin. Cancer Biol. 35, S5–S24.

Fishel, M.L., He, Y., Smith, M.L., Kelley, M.R., 2007. Manipulation of base excision repair to sensitize ovarian cancer cells to alkylating agent temozolomide. Clin. Cancer Res. 13, 260–267.

Fishel, M.L., Kelley, M.R., 2007. The DNA base excision repair protein Ape1/Ref-1 as a therapeutic and chemopreventive target. Mol. Aspects Med. 28, 375–395.

Fishel, M.L., Xia, H., McGeown, J., McIlwain, D.W., Elbanna, M., Craft, A.A., et al., 2019. Antitumor activity and mechanistic characterization of APE1/Ref-1 inhibitors in bladder cancer. Mol. Cancer Ther. 18, 1947–1960.

Fizazi, K., Azad, A.A., Matsubara, N., Carles, J., Fay, A.P., De Giorgi, U., et al., 2025. Talazoparib plus enzalutamide versus placebo plus enzalutamide for patients with advanced prostate cancer and changes in specific DNA repair genes: a plain language summary of the results from the TALAPRO-2 study. Future Oncol. 21, 637–652.

Fordham, S.E., Blair, H.J., Elstob, C.J., Plummer, R., Drew, Y., Curtin, N.J., et al., 2018. Inhibition of ATR acutely sensitizes acute myeloid leukemia cells to nucleoside analogs that target ribonucleotide reductase. Blood Adv. 2, 1157–1169.

Frye, C.C., Tennant, L., Yeager, A., Azimzadeh, P., Bhardwaj, P., Xu, Y., et al., 2024. Overexpression of human DNA polymerase theta is a biomarker of aggressive and DNA repair-deficient papillary thyroid cancers. Surgery 176, 1380–1387.

Garcia-Santisteban, I., Peters, G.J., Giovannetti, E., Rodriguez, J.A., 2013. USP1 deubiquitinase: cellular functions, regulatory mechanisms and emerging potential as target in cancer therapy. Mol. Cancer 12, 91.

Ghose, T., 2004. Oswald Avery: the professor, DNA, and the Nobel Prize that eluded him. Can. Bull. Med. Hist. 21, 135–144.

Gohil, D., Sarker, A.H., Roy, R., 2023. Base excision repair: mechanisms and impact in biology, disease, and medicine. Int. J. Mol. Sci. 24 14186.

Gralewska, P., Gajek, A., Marczak, A., Mikula, M., Ostrowski, J., Sliwinska, A., et al., 2020. PARP inhibition increases the reliance on ATR/CHK1 checkpoint signaling leading to synthetic lethality-an alternative treatment strategy for epithelial ovarian cancer cells independent from HR effectiveness. Int. J. Mol. Sci. 21 9715.

Gravells, P., Neale, J., Grant, E., Nathubhai, A., Smith, K.M., James, D.I., et al., 2018. Radiosensitization with an inhibitor of poly(ADP-ribose) glycohydrolase: a comparison with the PARP1/2/3 inhibitor olaparib. DNA Repair (Amst.) 61, 25–36.

Guerrero, E.N., Wang, H., Mitra, J., Hegde, P.M., Stowell, S.E., Liachko, N.F., et al., 2016. TDP-43/FUS in motor neuron disease: complexity and challenges. Prog. Neurobiol. 145-146, 78–97.

Guo, X., Yang, L., Wang, J., Wu, Y., Li, Y., Du, L., et al., 2024. The cytosolic DNA-sensing cGAS-STING pathway in neurodegenerative diseases. CNS Neurosci. Ther. 30, e14671.

Gupta, N., Huang, T.T., Horibata, S., Lee, J.M., 2022. Cell cycle checkpoints and beyond: exploiting the ATR/CHK1/WEE1 pathway for the treatment of PARP inhibitor-resistant cancer. Pharmacol. Res. 178, 106162.

Gutierrez, R., O'Connor, T.R., 2021. DNA direct reversal repair and alkylating agent drug resistance. Cancer Drug Resist. 4, 414–423.

Hakem, R., 2008. DNA-damage repair; the good, the bad, and the ugly. EMBO J. 27, 589–605.

Hansford, S., Huntsman, D.G., 2014. Boveri at 100: theodor boveri and genetic predisposition to cancer. J. Pathol. 234, 142–145.

Hanzlikova, H., Kalasova, I., demin, a.a., pennicott, l.e., cihlarova, z., caldecott, k.w., 2018. the importance of poly(adp-ribose) polymerase as a sensor of unligated Okazaki fragments during DNA replication. Mol. Cell 71, 319–331e3.

Hardy, P.A., Zacharias, H., 2005. Reappraisal of the Hansemann-Boveri hypothesis on the origin of tumors. Cell Biol. Int. 29, 983–992.

Harm, W., 1968. Dark repair of photorepairable UV lesions in Escherichia coli. Mutat. Res. 6, 25–35.

Hegde, M.L., Mantha, A.K., Hazra, T.K., Bhakat, K.K., Mitra, S., Szczesny, B., 2012. Oxidative genome damage and its repair: implications in aging and neurodegenerative diseases. Mech. Ageing Dev. 133, 157–168.

Higgins, G.S., Harris, A.L., Prevo, R., Helleday, T., McKenna, W.G., Buffa, F.M., 2010. Overexpression of POLQ confers a poor prognosis in early breast cancer patients. Oncotarget 1, 175–184.

Hishiki, A., Hashimoto, H., Hanafusa, T., Kamei, K., Ohashi, E., Shimizu, T., et al., 2009. Structural basis for novel interactions between human translesion synthesis polymerases and proliferating cell nuclear antigen. J. Biol. Chem. 284, 10552–10560.

Hoch, N.C., Hanzlikova, H., Rulten, S.L., Tetreault, M., Komulainen, E., Ju, L., et al., 2017. XRCC1 mutation is associated with PARP1 hyperactivation and cerebellar ataxia. Nature 541, 87–91.

Howard-Flanders, P., Simson, E., Theriot, L., 1964. The excision of thymine dimers from DNA, filament formation and sensitivity to ultraviolet light in Escherichia coli K-12. Mutat. Res. 106, 219–226.

Huang, P., Wang, Y., Zhang, P., Li, Q., 2023. Ubiquitin-specific peptidase 1: assessing its role in cancer therapy. Clin. Exp. Med. 23, 2953–2966.

Huang, R., Zhou, P.K., 2021. DNA damage repair: historical perspectives, mechanistic pathways and clinical translation for targeted cancer therapy. Signal Transduct. Target Ther. 6, 254.

Huang, Y., Chen, S., Yao, N., Lin, S., Zhang, J., Xu, C., et al., 2024. Molecular mechanism of PARP inhibitor resistance. Oncoscience 11, 69–91.

Huang, Z., Li, W., Wu, Y., Cheng, B., Huang, S., 2024. Inhibition of Chk1 with prexasertib enhances the anticancer activity of ciclopirox in non-small cell lung cancer cells. Cells 13 1752.

Inomata, Y., Abe, T., Tsuda, M., Takeda, S., Hirota, K., 2021. Division of labor of Y-family polymerases in translesion-DNA synthesis for distinct types of DNA damage. PLoS One 16, e0252587.

Jackson, L.M., Moldovan, G.L., 2022. Mechanisms of PARP1 inhibitor resistance and their implications for cancer treatment. NAR Cancer 4, zcac042.

Jiang, H., Chen, H., Chen, N., 2020. Construction and validation of a seven-gene signature for predicting overall survival in patients with kidney renal clear cell carcinoma via an integrated bioinformatics analysis. Anim Cells Syst (Seoul) 24, 160–170.

Jiang, Y., Dai, H., Li, Y., Yin, J., Guo, S., Lin, S.Y., et al., 2019. PARP inhibitors synergize with gemcitabine by potentiating DNA damage in non-small-cell lung cancer. Int. J. Cancer 144, 1092–1103.

Jinek, M., East, A., Cheng, A., Lin, S., Ma, E., Doudna, J., 2013. RNA-programmed genome editing in human cells. eLife 2, e00471.

Jones, M.J., Colnaghi, L., Huang, T.T., 2012. Dysregulation of DNA polymerase kappa recruitment to replication forks results in genomic instability. EMBO J. 31, 908–918.

Ju, S.H., Lee, S.E., Kang, Y.E., Shong, M., 2022. Development of metabolic synthetic lethality and its implications for thyroid cancer. Endocrinol. Metab. (Seoul) 37, 53–61.

Kamali, A.N., Bautista, J.M., Eisenhut, M., Hamedifar, H., 2023. Immune checkpoints and cancer immunotherapies: insights into newly potential receptors and ligands. Ther. Adv. Vaccines Immunother. 11, 25151355231192043.

Kanev, P.B., Atemin, A., Stoynov, S., Aleksandrov, R., 2024. PARP1 roles in DNA repair and DNA replication: the basi(c)s of PARP inhibitor efficacy and resistance. Semin. Oncol. 51, 2–18.

Kannouche, P., Fernandez de Henestrosa, A.R., Coull, B., Vidal, A.E., Gray, C., Zicha, D., et al., 2003. Localization of DNA polymerases eta and iota to the replication machinery is tightly co-ordinated in human cells. EMBO J. 22, 1223–1233.

Kannouche, P.L., Lehmann, A.R., 2004. Ubiquitination of PCNA and the polymerase switch in human cells. Cell Cycle 3, 1011–1013.

Karami Fath, M., Najafiyan, B., Morovatshoar, R., Khorsandi, M., Dashtizadeh, A., Kiani, A., et al., 2025. Potential promising of synthetic lethality in cancer research and treatment. Naunyn Schmiedebergs Arch. Pharmacol. 398, 1403–1431.

Kaufman, B., Shapira-Frommer, R., Schmutzler, R.K., Audeh, M.W., Friedlander, M., Balmana, J., et al., 2015. Olaparib monotherapy in patients with advanced cancer and a germline BRCA1/2 mutation. J. Clin. Oncol. 33, 244–250.

Ketkar, A., Sewilam, R.S., McCrury, M.J., Hall, J.S., Bell, A., Paxton, B.C., et al., 2023. Conservation of the insert-2 motif confers Rev1 from different species with an ability to disrupt G-quadruplexes and stimulate translesion DNA synthesis. RSC Chem. Biol. 4, 466–485.

Khan, S., Godfrey, V., Zaki, M.H., 2019. Cytosolic nucleic acid sensors in inflammatory and autoimmune disorders. Int. Rev. Cell Mol. Biol. 344, 215–253.

Kim, G., Ison, G., McKee, A.E., Zhang, H., Tang, S., Gwise, T., et al., 2015. FDA approval summary: olaparib monotherapy in patients with deleterious germline BRCA-mutated advanced ovarian cancer treated with three or more lines of chemotherapy. Clin. Cancer Res. 21, 4257–4261.

Kim, H., George, E., Ragland, R., Rafail, S., Zhang, R., Krepler, C., et al., 2017. Targeting the ATR/CHK1 axis with PARP inhibition results in tumor regression in BRCA-mutant ovarian cancer models. Clin. Cancer Res. 23, 3097–3108.

Kim, H., Xu, H., George, E., Hallberg, D., Kumar, S., Jagannathan, V., et al., 2020. Combining PARP with ATR inhibition overcomes PARP inhibitor and platinum resistance in ovarian cancer models. Nat. Commun. 11, 3726.

Kinzler, K.W., Vogelstein, B., 1997. Cancer-susceptibility genes. Gatekeepers and caretakers. Nature 386 (761), 63.

Kolinjivadi, A.M., Sannino, V., de Antoni, A., Techer, H., Baldi, G., Costanzo, V., 2017. Moonlighting at replication forks - a new life for homologous recombination proteins BRCA1, BRCA2 and RAD51. FEBS Lett. 591, 1083–1100.

Komulainen, E., Badman, J., Rey, S., Rulten, S., Ju, L., Fennell, K., et al., 2021. Parp1 hyperactivity couples DNA breaks to aberrant neuronal calcium signalling and lethal seizures. EMBO Rep. 22, e51851.

Kontomanolis, E.N., Koutras, A., Syllaios, A., Schizas, D., Mastoraki, A., Garmpis, N., et al., 2020. Role of oncogenes and tumor-suppressor genes in carcinogenesis: a review. Anticancer Res. 40, 6009–6015.

Kopeina, G.S., Zhivotovsky, B., 2022. Programmed cell death: past, present and future. Biochem. Biophys. Res. Commun. 633, 55–58.

Krokan, H.E., Bjoras, M., 2013. Base excision repair. Cold Spring Harb. Perspect. Biol. 5, a012583.

Lau, C.H., Seow, K.M., Chen, K.H., 2022. The molecular mechanisms of actions, effects, and clinical implications of PARP inhibitors in epithelial ovarian cancers: a systematic review. Int. J. Mol. Sci. 23 8125.

Lee, A., 2021. Niraparib: a review in first-line maintenance therapy in advanced ovarian cancer. Target Oncol. 16, 839–845.

Lemee, F., Bergoglio, V., Fernandez-Vidal, A., Machado-Silva, A., Pillaire, M.J., Bieth, A., et al., 2010. DNA polymerase theta up-regulation is associated with poor survival in breast cancer, perturbs DNA replication, and promotes genetic instability. Proc. Natl. Acad. Sci. USA 107, 13390–13395.

Li, L.Y., Guan, Y.D., Chen, X.S., Yang, J.M., Cheng, Y., 2020. DNA repair pathways in cancer therapy and resistance. Front. Pharmacol. 11, 629266.

Li, Q., Wang, J.M., Peng, Y., Zhang, S.H., Ren, T., Luo, H., et al., 2013. Association of DNA base-excision repair XRCC1, OGG1 and APE1 gene polymorphisms with nasopharyngeal carcinoma susceptibility in a Chinese population. Asian Pac. J. Cancer Prev. 14, 5145–5151.

Li, S., Wang, L., Wang, Y., Zhang, C., Hong, Z., Han, Z., 2022. The synthetic lethality of targeting cell cycle checkpoints and PARPs in cancer treatment. J. Hematol. Oncol. 15, 147.

Li, X., Heyer, W.D., 2008. Homologous recombination in DNA repair and DNA damage tolerance. Cell Res. 18, 99–113.

Li, X., Zou, L., 2024. BRCAness, DNA gaps, and gain and loss of PARP inhibitor-induced synthetic lethality. J. Clin. Invest. 134.

Li, Z., Guan, W., Li, M.X., Zhong, Z.Y., Qian, C.Y., Yang, X.Q., et al., 2011. Genetic polymorphism of DNA base-excision repair genes (APE1, OGG1 and XRCC1) and their correlation with risk of lung cancer in a Chinese population. Arch. Med. Res. 42, 226–234.

Li, Z., Zhao, J., Tang, Y., 2023. Advances in the role of SWI/SNF complexes in tumours. J. Cell. Mol. Med. 27, 1023–1031.

Liang, Q., Dexheimer, T.S., Zhang, P., Rosenthal, A.S., Villamil, M.A., You, C., et al., 2014. A selective USP1-UAF1 inhibitor links deubiquitination to DNA damage responses. Nat. Chem. Biol. 10, 298–304.

Lin, J.R., Zeman, M.K., Chen, J.Y., Yee, M.C., Cimprich, K.A., 2011. SHPRH and HLTF act in a damage-specific manner to coordinate different forms of postreplication repair and prevent mutagenesis. Mol. Cell 42, 237–249.

Ling, H., Boudsocq, F., Woodgate, R., Yang, W., 2001. Crystal structure of a Y-family DNA polymerase in action: a mechanism for error-prone and lesion-bypass replication. Cell 107, 91–102.

Liu, F.W., Tewari, K.S., 2016. New targeted agents in gynecologic cancers: synthetic lethality, homologous recombination deficiency, and PARP inhibitors. Curr. Treat. Options Oncol. 17, 12.

Loizzi, V., Ranieri, G., Laforgia, M., Gadaleta, C.D., Gargano, G., Kardhashi, A., et al., 2020. PARP inhibitors and epithelial ovarian cancer: molecular mechanisms, clinical development and future prospective. Oncol. Lett. 20, 90.

Luo, J., Li, Y., Zhang, Y., Wu, D., Ren, Y., Liu, J., et al., 2024. An update on small molecule compounds targeting synthetic lethality for cancer therapy. Eur. J. Med. Chem. 278, 116804.

Mansilla, S.F., Bertolin, A.P., Venerus Arbilla, S., Castano, B.A., Jahjah, T., Singh, J.K., et al., 2023. Polymerase iota (Pol iota) prevents PrimPol-mediated nascent DNA synthesis and chromosome instability. Sci. Adv. 9, eade7997.

Marechal, A., Zou, L., 2015. RPA-coated single-stranded DNA as a platform for post-translational modifications in the DNA damage response. Cell Res. 25, 9–23.

Masuda, Y., Kanao, R., Kaji, K., Ohmori, H., Hanaoka, F., Masutani, C., 2015. Different types of interaction between PCNA and PIP boxes contribute to distinct cellular functions of Y-family DNA polymerases. Nucleic. Acids. Res. 43, 7898–7910.

Mateos-Gomez, P.A., Kent, T., Deng, S.K., McDevitt, S., Kashkina, E., Hoang, T.M., et al., 2017. The helicase domain of Poltheta counteracts RPA to promote alt-NHEJ. Nat. Struct. Mol. Biol. 24, 1116–1123.

Matsumoto, Y., 2022. Development and evolution of DNA-dependent protein kinase inhibitors toward cancer therapy. Int. J. Mol. Sci. 23 4264.

Maxwell, B.A., Suo, Z., 2014. Recent insight into the kinetic mechanisms and conformational dynamics of Y-family DNA polymerases. Biochemistry 53, 2804–2814.

Maya-Mendoza, A., Moudry, P., Merchut-Maya, J.M., Lee, M., Strauss, R., Bartek, J., 2018. High speed of fork progression induces DNA replication stress and genomic instability. Nature 559, 279–284.

Melia, E., Parsons, J.L., 2023. DNA damage and repair dependencies of ionising radiation modalities. Biosci. Rep. 43 BSR20222586.

Menck, C.F.M., Galhardo, R.S., Quinet, A., 2024. The accurate bypass of pyrimidine dimers by DNA polymerase eta contributes to ultraviolet-induced mutagenesis. Mutat. Res. 828, 111840.

Milner, J., Ponder, B., Hughes-Davies, L., Seltmann, M., Kouzarides, T., 1997. Transcriptional activation functions in BRCA2. Nature 386, 772–773.

Mirman, Z., Sasi, N.K., King, A., Chapman, J.R., de Lange, T., 2022. 53BP1-shieldin-dependent DSB processing in BRCA1-deficient cells requires CST-Polalpha-primase fill-in synthesis. Nat. Cell Biol. 24, 51–61.

Misra, H.S., Khairnar, N.P., Kota, S., Shrivastava, S., Joshi, V.P., Apte, S.K., 2006. An exonuclease I-sensitive DNA repair pathway in Deinococcus radiodurans: a major determinant of radiation resistance. Mol. Microbiol. 59, 1308–1316.

Mizuno, Y., Abolhassani, N., Mazzei, G., Saito, T., Saido, T.C., Yamasaki, R., et al., 2022. Deficiency of MTH1 and/or OGG1 increases the accumulation of 8-oxoguanine in the brain of the App(NL-G-F/NL-G-F) knock-in mouse model of Alzheimer's disease, accompanied by accelerated microgliosis and reduced anxiety-like behavior. Neurosci. Res. 177, 118–134.

Mocanu, C., Chan, K.L., 2021. Mind the replication gap. R. Soc. Open Sci. 8, 201932.

Montaldi, A.P., Sakamoto-Hojo, E.T., 2013. Methoxyamine sensitizes the resistant glioblastoma T98G cell line to the alkylating agent temozolomide. Clin. Exp. Med. 13, 279–288.

Muller, H.J., 1927. Artificial transmutation of the gene. Science 66, 84–87.

Muller, H.J., 1928. The production of mutations by x-rays. Proc. Natl. Acad. Sci. USA 14, 714–726.

Munroe, M., Kolesar, J., 2016. Olaparib for the treatment of BRCA-mutated advanced ovarian cancer. Am. J. Health Syst. Pharm. 73, 1037–1041.

Murata, M.M., Kong, X., Moncada, E., Chen, Y., Imamura, H., Wang, P., et al., 2019. NAD+ consumption by PARP1 in response to DNA damage triggers metabolic shift critical for damaged cell survival. Mol. Biol. Cell 30, 2584–2597.

Murphree, A.L., Benedict, W.F., 1984. Retinoblastoma: clues to human oncogenesis. Science 223, 1028–1033.

Nagaraju, G., Scully, R., 2007. Minding the gap: the underground functions of BRCA1 and BRCA2 at stalled replication forks. DNA Repair (Amst.) 6, 1018–1031.

Nayak, S., Calvo, J.A., Cantor, S.B., 2021. Targeting translesion synthesis (TLS) to expose replication gaps, a unique cancer vulnerability. Expert Opin. Ther. Targets 25, 27–36.

Nelligan, A., Dungrawala, H., 2024. SNF2L suppresses nascent DNA gap formation to promote DNA synthesis. Nucleic. Acids. Res. 52, 13003–13018.

Niu, X., Chen, W., Bi, T., Lu, M., Qin, Z., Xiao, W., 2019. Rev1 plays central roles in mammalian DNA-damage tolerance in response to UV irradiation. FEBS J. 286, 2711–2725.

Northam, M.R., Moore, E.A., Mertz, T.M., Binz, S.K., Stith, C.M., Stepchenkova, E.I., et al., 2014. DNA polymerases zeta and Rev1 mediate error-prone bypass of non-B DNA structures. Nucleic. Acids. Res. 42, 290–306.

O'Connor, M.J., 2015. Targeting the DNA damage response in cancer. Mol. Cell 60, 547–560.

O'Neil, N.J., Bailey, M.L., Hieter, P., 2017. Synthetic lethality and cancer. Nat. Rev. Genet. 18, 613–623.

Oza, A.M., Cibula, D., Benzaquen, A.O., Poole, C., Mathijssen, R.H., Sonke, G.S., et al., 2015. Olaparib combined with chemotherapy for recurrent platinum-sensitive ovarian cancer: a randomised phase 2 trial. Lancet Oncol. 16, 87–97.

Oza, A.M., Lisyanskaya, A., Fedenko, A., de Melo, A.C., Shparyk, Y., Rakhmatullina, I., et al., 2025. Rucaparib versus chemotherapy for treatment of relapsed ovarian cancer with deleterious BRCA1 or BRCA2 mutation (ARIEL4): final results of an international, open-label, randomised, phase 3 trial. Lancet Oncol. 26, 249–264.

Ozdemir, A.Y., Rusanov, T., Kent, T., Siddique, L.A., Pomerantz, R.T., 2018. Polymerase theta-helicase efficiently unwinds DNA and RNA-DNA hybrids. J. Biol. Chem. 293, 5259–5269.

Paes Dias, M., Tripathi, V., van der Heijden, I., Cong, K., Manolika, E.M., Bhin, J., et al., 2021. Loss of nuclear DNA ligase III reverts PARP inhibitor resistance in BRCA1/53BP1 double-deficient cells by exposing ssDNA gaps. Mol. Cell 81, 4692–4708e9.

Pan, D., Wang, Q., Shen, A., Qi, Z., Zheng, C., Hu, B., 2024. When DNA damage responses meet tumor immunity: from mechanism to therapeutic opportunity. Int. J. Cancer 155, 384–399.

Paniagua, I., Jacobs, J.J.L., 2023. Freedom to err: the expanding cellular functions of translesion DNA polymerases. Mol. Cell 83, 3608–3621.

Panzarino, N.J., Krais, J.J., Cong, K., Peng, M., Mosqueda, M., Nayak, S.U., et al., 2021. Replication gaps underlie BRCA deficiency and therapy response. Cancer Res. 81, 1388–1397.

Pata, J.D., 2010. Structural diversity of the Y-family DNA polymerases. Biochim. Biophys. Acta 1804, 1124–1135.

Patel, S.M., Dash, R.C., Hadden, M.K., 2021. Translesion synthesis inhibitors as a new class of cancer chemotherapeutics. Expert Opin. Investig. Drugs 30, 13–24.

Paul, B.D., Snyder, S.H., Bohr, V.A., 2021. Signaling by cGAS-STING in neurodegeneration, neuroinflammation, and aging. Trends Neurosci. 44, 83–96.

Peng, X., Pan, W., Jiang, F., Chen, W., Qi, Z., Peng, W., et al., 2022. Selective PARP1 inhibitors, PARP1-based dual-target inhibitors, PROTAC PARP1 degraders, and prodrugs of PARP1 inhibitors for cancer therapy. Pharmacol. Res. 186, 106529.

Perego, M.G.L., Taiana, M., Bresolin, N., Comi, G.P., Corti, S., 2019. R-loops in motor neuron diseases. Mol. Neurobiol. 56, 2579–2589.

Pettijohn, D., Hanawalt, P., 1964. Evidence for repair-replication of ultraviolet damaged DNA in bacteria. J. Mol. Biol. 9, 395–410.

Phillipps, J., Zhou, A.Y., Butt, O.H., Ansstas, G., 2023. PARP inhibition and immunotherapy: a promising duo in fighting cancer. Transl. Cancer Res. 12, 2433–2437.

Pillaire, M.J., Selves, J., Gordien, K., Gourraud, P.A., Gentil, C., Danjoux, M., et al., 2010. A 'DNA replication' signature of progression and negative outcome in colorectal cancer. Oncogene 29, 876–887.

Pires, I.M., Olcina, M.M., Anbalagan, S., Pollard, J.R., Reaper, P.M., Charlton, P.A., et al., 2012. Targeting radiation-resistant hypoxic tumour cells through ATR inhibition. Br. J. Cancer 107, 291–299.

Pismataro, M.C., Astolfi, A., Barreca, M.L., Pacetti, M., Schenone, S., Bandiera, T., et al., 2023. Small molecules targeting DNA polymerase theta (POLtheta) as promising synthetic lethal agents for precision cancer therapy. J. Med. Chem. 66, 6498–6522.

Portin, P., 2014. The birth and development of the DNA theory of inheritance: sixty years since the discovery of the structure of DNA. J. Genet. 93, 293–302.

Portugal, F., 2010. Oswald T. Avery: nobel lureate or noble luminary? Perspect. Biol. Med. 53, 558–570.

Post, C.C.B., Westermann, A.M., Bosse, T., Creutzberg, C.L., Kroep, J.R., 2020. PARP and PD-1/PD-L1 checkpoint inhibition in recurrent or metastatic endometrial cancer. Crit. Rev. Oncol. Hematol. 152, 102973.

Pratz, K.W., Rudek, M.A., Gojo, I., Litzow, M.R., McDevitt, M.A., Ji, J., et al., 2017. A phase I study of topotecan, carboplatin and the PARP Inhibitor veliparib in acute leukemias, aggressive myeloproliferative neoplasms, and chronic myelomonocytic leukemia. Clin. Cancer Res. 23, 899–907.

Previtali, V., Bagnolini, G., Ciamarone, A., Ferrandi, G., Rinaldi, F., Myers, S.H., et al., 2024. New horizons of synthetic lethality in cancer: current development and future perspectives. J. Med. Chem. 67, 11488–11521.

Principe, D.R., 2022. Precision medicine for BRCA/PALB2-mutated pancreatic cancer and emerging strategies to improve therapeutic responses to PARP inhibition. Cancers (Basel) 14 897.

Priya, B., Ravi, S., Kirubakaran, S., 2023. Targeting ATM and ATR for cancer therapeutics: inhibitors in clinic. Drug Discov. Today 28, 103662.

Provasek, V.E., Mitra, J., Malojirao, V.H., Hegde, M.L., 2022. DNA double-strand breaks as pathogenic lesions in neurological disorders. Int. J. Mol. Sci. 23 4653.

Pustovalova, Y., Magalhaes, M.T., D'Souza, S., Rizzo, A.A., Korza, G., Walker, G.C., et al., 2016. Interaction between the Rev1 C-terminal domain and the PolD3 subunit of polzeta suggests a mechanism of polymerase exchange upon Rev1/Polzeta-dependent translesion synthesis. Biochemistry 55, 2043–2053.

Qin, Z., Lu, M., Xu, X., Hanna, M., Shiomi, N., Xiao, W., 2013. DNA-damage tolerance mediated by PCNA*Ub fusions in human cells is dependent on Rev1 but not Poleta. Nucleic. Acids. Res. 41, 7356–7369.

Qiu, Z., Oleinick, N.L., Zhang, J., 2018. ATR/CHK1 inhibitors and cancer therapy. Radiother. Oncol. 126, 450–464.

Quinet, A., Lerner, L.K., Martins, D.J., Menck, C.F.M., 2018. Filling gaps in translesion DNA synthesis in human cells. Mutat. Res. Genet. Toxicol. Environ. Mutagen. 836, 127–142.

Quinet, A., Vessoni, A.T., Rocha, C.R., Gottifredi, V., Biard, D., Sarasin, A., et al., 2014. Gap-filling and bypass at the replication fork are both active mechanisms for tolerance of low-dose ultraviolet-induced DNA damage in the human genome. DNA Repair (Amst.) 14, 27–38.

Ragupathi, A., Singh, M., Perez, A.M., Zhang, D., 2023. Targeting the BRCA1/2 deficient cancer with PARP inhibitors: clinical outcomes and mechanistic insights. Front. Cell Dev. Biol. 11, 1133472.

Rai, R., Peng, G., Li, K., Lin, S.Y., 2007. DNA damage response: the players, the network and the role in tumor suppression. Cancer Genom. Proteom. 4, 99–106.

Rajpurohit, Y.S., Lal, M., Sharma, D.K., Soni, I., 2025. Human TLS DNA polymerase: saviors or threats under replication stress? Mol. Cell. Biochem. 480, 4991–5008.

Rajpurohit, Y.S., Sharma, D.K., Lal, M., Soni, I., 2024. A perspective on tumor radiation resistance following high-LET radiation treatment. J. Cancer Res. Clin. Oncol. 150, 226.

Ramirez-Otero, M.A., Costanzo, V., 2024. Bridging the DNA divide": understanding the interplay between replication-gaps and homologous recombination proteins RAD51 and BRCA1/2. DNA Repair (Amst.) 141, 103738.

Reuvers, T.G.A., Kanaar, R., Nonnekens, J., 2020. DNA damage-inducing anticancer therapies: from global to precision damage. Cancers (Basel) 12 2098.

Revythis, A., Limbu, A., Mikropoulos, C., Ghose, A., Sanchez, E., Sheriff, M., et al., 2022. Recent insights into PARP and immuno-checkpoint inhibitors in epithelial ovarian cancer. Int. J. Environ. Res. Public Health 19 8577.

Rijkers, G., Dekker, K., Berbers, G., 2024. Oswald Avery: pioneer of bacterial vaccines and the first to discover the function of DNA. Cureus 16, e71465.

Rizzo, A.A., Korzhnev, D.M., 2019. The Rev1-Polzeta translesion synthesis mutasome: structure, interactions and inhibition. Enzymes 45, 139–181.

Rose, M., Burgess, J.T., O'Byrne, K., Richard, D.J., Bolderson, E., 2020. PARP inhibitors: clinical relevance, mechanisms of action and tumor resistance. Front. Cell Dev. Biol. 8, 564601.

Roth, D.B., Gellert, M., 2000. New guardians of the genome. Nature 404, 823–825.

Russo, M., Crisafulli, G., Sogari, A., Reilly, N.M., Arena, S., Lamba, S., et al., 2019. Adaptive mutability of colorectal cancers in response to targeted therapies. Science 366, 1473–1480.

Samoilenko, I.I., 1983. [Radioresistance mechanisms of Micrococcus radiodurans]. Nauchnye Doki. Vyss. Shkoly Biol. Nauki 4–17.

Saxena, S., Zou, L., 2022. Hallmarks of DNA replication stress. Mol. Cell 82, 2298–2314.

Scharer, O.D., 2013. Nucleotide excision repair in eukaryotes. Cold Spring Harb. Perspect. Biol. 5, a012609.

Schrempf, A., Slyskova, J., Loizou, J.I., 2021. Targeting the DNA repair enzyme polymerase theta in cancer therapy. Trends Cancer 7, 98–111.

Schreuder, A., Wendel, T.J., Dorresteijn, C.G.V., Noordermeer, S.M., 2024. (Single-stranded DNA) gaps in understanding BRCAness. Trends Genet. 40, 757–771.

Schumacher, B., Pothof, J., Vijg, J., Hoeijmakers, J.H.J., 2021. The central role of DNA damage in the ageing process. Nature 592, 695–703.

Sertic, S., Mollica, A., Campus, I., Roma, S., Tumini, E., Aguilera, A., et al., 2018. Coordinated activity of Y family TLS polymerases and EXO1 protects non-S phase cells from UV-induced cytotoxic lesions. Mol. Cell 70, 34–47e4.

Setlow, R.B., Carrier, W.L., 1964. The disappearance of thymine dimers from DNA: an error-correcting mechanism. Proc. Natl. Acad. Sci. USA 51, 226–231.

Shanbhag, N.M., Evans, M.D., Mao, W., Nana, A.L., Seeley, W.W., Adame, A., et al., 2019. Early neuronal accumulation of DNA double strand breaks in Alzheimer's disease. Acta Neuropathol. Commun. 7, 77.

Sharan, S.K., Morimatsu, M., Albrecht, U., Lim, D.S., Regel, E., Dinh, C., et al., 1997. Embryonic lethality and radiation hypersensitivity mediated by Rad51 in mice lacking Brca2. Nature 386, 804–810.

Sharma, D.K., Soni, I., Misra, H.S., Rajpurohit, Y.S., 2024. Natural transformation-specific DprA coordinate DNA double-strand break repair pathways in heavily irradiated D. radiodurans. Appl. Environ. Microb. 90, e0194823.

Sharma, D.K., Soni, I., Rajpurohit, Y.S., 2025. Surviving the storm: exploring the role of natural transformation in nutrition and DNA repair of stressed Deinococcus radiodurans. Appl. Environ. Microb. 91, e0137124.

Sharma, H., Koirala, S., Chew, Y.L., Konopka, A., 2025. DNA damage and chromatin rearrangement work together to promote neurodegeneration. Mol. Neurobiol. 62, 1282–1290.

Sherrer, S.M., Sanman, L.E., Xia, C.X., Bolin, E.R., Malik, C.K., Efthimiopoulos, G., et al., 2012. Kinetic analysis of the bypass of a bulky DNA lesion catalyzed by human Y-family DNA polymerases. Chem. Res. Toxicol. 25, 730–740.

Shokrollahi, M., Stanic, M., Hundal, A., Chan, J.N.Y., Urman, D., Jordan, C.A., et al., 2024. DNA double-strand break-capturing nuclear envelope tubules drive DNA repair. Nat. Struct. Mol. Biol. 31, 1319–1330.

Simoneau, A., Engel, J.L., Bandi, M., Lazarides, K., Liu, S., Meier, S.R., et al., 2023. Ubiquitinated PCNA drives USP1 synthetic lethality in cancer. Mol. Cancer Ther. 22, 215–226.

Slade, D., 2020. PARP and PARG inhibitors in cancer treatment. Genes Dev. 34, 360–394.

Slade, D., Lindner, A.B., Paul, G., Radman, M., 2009. Recombination and replication in DNA repair of heavily irradiated deinococcus radiodurans. Cell 136, 1044–1055.

Smith, H.L., Prendergast, L., Curtin, N.J., 2020. Exploring the synergy between PARP and CHK1 inhibition in matched BRCA2 mutant and corrected cells. Cancers (Basel) 12 878.

Smith, H.L., Willmore, E., Prendergast, L., Curtin, N.J., 2024. ATR, CHK1 and WEE1 inhibitors cause homologous recombination repair deficiency to induce synthetic lethality with PARP inhibitors. Br. J. Cancer 131, 905–917.

Spies, J., Polasek-Sedlackova, H., Lukas, J., Somyajit, K., 2021. Homologous recombination as a fundamental genome surveillance mechanism during DNA replication. Genes (Basel) 12 1960.

Stary, A., Kannouche, P., Lehmann, A.R., Sarasin, A., 2003. Role of DNA polymerase eta in the UV mutation spectrum in human cells. J. Biol. Chem. 278, 18767–18775.

Stolyarenko, A.D., Novikova, A.A., Shilkin, E.S., Poltorachenko, V.A., Makarova, A.V., 2024. The catalytic activity of human REV1 on undamaged and damaged DNA. Int. J. Mol. Sci. 25 4107.

Sturm, M.J., Henao-Restrepo, J.A., Becker, S., Proquitte, H., Beck, J.F., Sonnemann, J., 2023. Synergistic anticancer activity of combined ATR and ribonucleotide reductase inhibition in Ewing's sarcoma cells. J. Cancer Res. Clin. Oncol. 149, 8605–8617.

Suberbielle, E., Sanchez, P.E., Kravitz, A.V., Wang, X., Ho, K., Eilertson, K., et al., 2013. Physiologic brain activity causes DNA double-strand breaks in neurons, with exacerbation by amyloid-beta. Nat. Neurosci. 16, 613–621.

Sullivan-Reed, K., Toma, M.M., Drzewiecka, M., Nieborowska-Skorska, M., Nejati, R., Karami, A., et al., 2023. Simultaneous targeting of DNA polymerase theta and PARP1 or RAD52 triggers dual synthetic lethality in homologous recombination-deficient leukemia cells. Mol. Cancer Res. 21, 1017–1022.

Sun, C., Yin, J., Fang, Y., Chen, J., Jeong, K.J., Chen, X., et al., 2018. BRD4 inhibition is synthetic lethal with PARP inhibitors through the induction of homologous recombination deficiency. Cancer Cell 33, 401–416e8.

Sun, H., Ma, L., Tsai, Y.F., Abeywardana, T., Shen, B., Zheng, L., 2023. Okazaki fragment maturation: DNA flap dynamics for cell proliferation and survival. Trends Cell Biol. 33, 221–234.

Taglialatela, A., Leuzzi, G., Sannino, V., Cuella-Martin, R., Huang, J.W., Wu-Baer, F., et al., 2021. REV1-Polzeta maintains the viability of homologous recombination-deficient cancer cells through mutagenic repair of PRIMPOL-dependent ssDNA gaps. Mol. Cell 81, 4008–4025e7.

Tanner, G., Westhead, D.R., Droop, A., Stead, L.F., 2021. Benchmarking pipelines for subclonal deconvolution of bulk tumour sequencing data. Nat. Commun. 12, 6396.

Tanuma, S.I., Shibui, Y., Oyama, T., Uchiumi, F., Abe, H., 2019. Targeting poly(ADP-ribose) glycohydrolase to draw apoptosis codes in cancer. Biochem. Pharmacol. 167, 163–172.

Tayebati, S.K., 2006. Animal models of cognitive dysfunction. Mech. Ageing Dev. 127, 100–108.

Tian, H., Gao, Z., Li, H., Zhang, B., Wang, G., Zhang, Q., et al., 2015. DNA damage response–a double-edged sword in cancer prevention and cancer therapy. Cancer Lett. 358, 8–16.

Tirman, S., Quinet, A., Wood, M., Meroni, A., Cybulla, E., Jackson, J., et al., 2021. Temporally distinct post-replicative repair mechanisms fill PRIMPOL-dependent ssDNA gaps in human cells. Mol. Cell 81, 4026–4040e8.

Tonzi, P., Huang, T.T., 2019. Role of Y-family translesion DNA polymerases in replication stress: implications for new cancer therapeutic targets. DNA Repair (Amst.) 78, 20–26.

Tonzi, P., Yin, Y., Lee, C.W.T., Rothenberg, E., Huang, T.T., 2018. Translesion polymerase kappa-dependent DNA synthesis underlies replication fork recovery. eLife 7 e41426.

Trenner, A., Sartori, A.A., 2019. Harnessing DNA double-strand break repair for cancer treatment. Front. Oncol. 9, 1388.

Tse, K.H., Herrup, K., 2017. DNA damage in the oligodendrocyte lineage and its role in brain aging. Mech. Ageing Dev. 161, 37–50.

Vafa, O., Wade, M., Kern, S., Beeche, M., Pandita, T.K., Hampton, G.M., et al., 2002. c-Myc can induce DNA damage, increase reactive oxygen species, and mitigate p53 function: a mechanism for oncogene-induced genetic instability. Mol. Cell 9, 1031–1044.

Vaitsiankova, A., Burdova, K., Sobol, M., Gautam, A., Benada, O., Hanzlikova, H., et al., 2022. PARP inhibition impedes the maturation of nascent DNA strands during DNA replication. Nat. Struct. Mol. Biol. 29, 329–338.

Venkadakrishnan, J., Lahane, G., Dhar, A., Xiao, W., Bhat, K.M., Pandita, T.K., et al., 2023. Implications of translesion DNA synthesis polymerases on genomic stability and human health. Mol. Cell. Biol. 43, 401–425.

Voutsadakis, I.A., Stravodimou, A., 2023. Homologous recombination defects and mutations in DNA damage response (DDR) genes besides BRCA1 and BRCA2 as breast cancer biomarkers for PARP inhibitors and other DDR targeting therapies. Anticancer Res. 43, 967–981.

Vugic, D., Dumoulin, I., Martin, C., Minello, A., Alvaro-Aranda, L., Gomez-Escudero, J., et al., 2023. Replication gap suppression depends on the double-strand DNA binding activity of BRCA2. Nat. Commun. 14, 446.

Vyas, R., Efthimiopoulos, G., Tokarsky, E.J., Malik, C.K., Basu, A.K., Suo, Z., 2015. Mechanistic basis for the bypass of a bulky DNA adduct catalyzed by a Y-family DNA polymerase. J. Am. Chem. Soc. 137, 12131–12142.

Wang, D., Wu, W., Callen, E., Pavani, R., Zolnerowich, N., Kodali, S., et al., 2022. Active DNA demethylation promotes cell fate specification and the DNA damage response. Science 378, 983–989.

Wanior, M., Kramer, A., Knapp, S., Joerger, A.C., 2021. Exploiting vulnerabilities of SWI/SNF chromatin remodelling complexes for cancer therapy. Oncogene 40, 3637–3654.

Watson, J.D., Crick, F.H., 1953. Molecular structure of nucleic acids; a structure for deoxyribose nucleic acid. Nature 171, 737–738.

Welch, G., Tsai, L.H., 2022. Mechanisms of DNA damage-mediated neurotoxicity in neurodegenerative disease. EMBO Rep. 23, e54217.

Wikiniyadhanee, R., Lerksuthirat, T., Stitchantrakul, W., Chitphuk, S., Takeda, S., Dejsuphong, D., 2023. ATR inhibitor synergizes PARP inhibitor cytotoxicity in homologous recombination repair deficiency TK6 cell lines. Biomed. Res. Int. 2023, 7891753.

Wright, W.D., Shah, S.S., Heyer, W.D., 2018. Homologous recombination and the repair of DNA double-strand breaks. J. Biol. Chem. 293, 10524–10535.

Wu, S., Yao, X., Sun, W., Jiang, K., Hao, J., 2024. Exploration of poly (ADP-ribose) polymerase inhibitor resistance in the treatment of BRCA1/2-mutated cancer. Genes Chromosomes Cancer 63, e23243.

Xiao, F., Wang, Z., Qiao, L., Zhang, X., Wu, N., Wang, J., et al., 2024. Application of PARP inhibitors combined with immune checkpoint inhibitors in ovarian cancer. J. Transl. Med. 22, 778.

Xu, C., Maxwell, B.A., Brown, J.A., Zhang, L., Suo, Z., 2009. Global conformational dynamics of a Y-family DNA polymerase during catalysis. PLoS Biol. 7, e1000225.

Yang, K., Weinacht, C.P., Zhuang, Z., 2013. Regulatory role of ubiquitin in eukaryotic DNA translesion synthesis. Biochemistry 52, 3217–3228.

Yang, W., Gao, Y., 2018. Translesion and Repair DNA Polymerases: diverse structure and mechanism. Annu. Rev. Biochem. 87, 239–261.

Yano, K., Shiotani, B., 2023. Emerging strategies for cancer therapy by ATR inhibitors. Cancer Sci. 114, 2709–2721.

Yar, M.S., Haider, K., Gohel, V., Siddiqui, N.A., Kamal, A., 2020. Synthetic lethality on drug discovery: an update on cancer therapy. Expert Opin Drug Discov 15, 823–832.

Yu, J., Mao, W., Xu, B., Chen, M., 2021. Construction and validation of an autophagy-related long noncoding RNA signature for prognosis prediction in kidney renal clear cell carcinoma patients. Cancer Med. 10, 2359–2369.

Zahradka, K., Slade, D., Bailone, A., Sommer, S., Averbeck, D., Petranovic, M., et al., 2006. Reassembly of shattered chromosomes in deinococcus radiodurans. Nature 443, 569–573.

Zhang, Y., Yu, C., 2020. Distinct expression and prognostic values of the replication protein a family in gastric cancer. Oncol. Lett. 19, 1831–1841.

Zhang, Y., Zou, M., Wu, H., Zhu, J., Jin, T., 2024. The cGAS-STING pathway drives neuroinflammation and neurodegeneration via cellular and molecular mechanisms in neurodegenerative diseases. Neurobiol. Dis. 202, 106710.

Zhong, W., Huang, C., Lin, J., Zhu, M., Zhong, H., Chiang, M.H., et al., 2020. Development and validation of nine-RNA binding protein signature predicting overall survival for kidney renal clear cell carcinoma. Front. Genet. 11, 568192.

Zhou, M., Ottenberg, G., Sferrazza, G.F., Hubbs, C., Fallahi, M., Rumbaugh, G., et al., 2015. Neuronal death induced by misfolded prion protein is due to NAD+ depletion and can be relieved in vitro and in vivo by NAD+ replenishment. Brain 138, 992–1008.

Zhu, Y., Hu, J., Hu, Y., Liu, W., 2009. Targeting DNA repair pathways: a novel approach to reduce cancer therapeutic resistance. Cancer Treat. Rev. 35, 590–596.

Ziener, J., Henao-Restrepo, J.A., Leonhardi, J., Sturm, M.J., Becker, S., Morales-Prieto, D.M., et al., 2025. Combined inhibition of ribonucleotide reductase and WEE1 induces synergistic anticancer activity in Ewing's sarcoma cells. BMC Cancer 25, 277.

Zimmermann, A., Zenke, F.T., Chiu, L.Y., Dahmen, H., Pehl, U., Fuchss, T., et al., 2022. A new class of selective ATM inhibitors as combination partners of DNA double-strand break inducing cancer therapies. Mol. Cancer Ther. 21, 859–870.

Index

Page numbers followed by "*f*" and "*t*" indicate, figures and tables respectively.

L

M

N

9780443428579